Lecture Notes in Computer Science 4208

Commenced Publication in 1973
Founding and Former Series Editors:
Gerhard Goos, Juris Hartmanis, and Jan van Leeuwen

Michael Gerndt Dieter Kranzlmüller (Eds.)

High Performance Computing and Communications

Second International Conference, HPCC 2006
Munich, Germany, September 13-15, 2006
Proceedings

Springer

Volume Editors

Michael Gerndt
Technische Universität München
Institut für Informatik
Boltzmannstr. 3, 85748 Garching, Germany
E-mail: gerndt@in.tum.de

Dieter Kranzlmüller
Johannes Kepler Universität Linz
Institut für Graphische und Parallele Datenverarbeitung (GUP)
Altenbergerstr. 69, 4040 Linz, Austria
E-mail: kranzlmueller@gup.jku.at

Library of Congress Control Number: 2006932039

CR Subject Classification (1998): D, F.1-2, C.2, G.1-2, H.4-5

LNCS Sublibrary: SL 1 – Theoretical Computer Science and General Issues

ISSN 0302-9743
ISBN-10 3-540-39368-4 Springer Berlin Heidelberg New York
ISBN-13 978-3-540-39368-9 Springer Berlin Heidelberg New York

Springer is a part of Springer Science+Business Media

springer.com

Printed in Germany

Typesetting: Camera-ready by author, data conversion by Scientific Publishing Services, Chennai, India
Printed on acid-free paper SPIN: 11847366 06/3142 5 4 3 2 1 0

Preface

Welcome to the proceedings of the 2006 International Conference on High-Performance Computing and Communications (HPCC 2006), which was held in Munich, Germany, September 13–15, 2006. This year's conference marks the second edition of the HPCC conference series, and we are honored to serve as the Chairmen of this event with the guidance of the HPCC Steering Chairs, Beniamino Di Martino and Laurence T. Yang.

With the rapid growth in computing and communication technology, the past decade has witnessed a proliferation of powerful parallel and distributed systems and an ever-increasing demand for the practice of high-performance computing and communication (HPCC). HPCC has moved into the mainstream of computing and has become a key technology in future research and development activities in many academic and industrial branches, especially when the solution of large and complex problems must cope with very tight time constraints. The HPCC 2006 conference provides a forum for engineers and scientists in academia, industry, and government to address all resulting profound challenges and to present and discuss their new ideas, research results, applications, and experience on all aspects of HPCC.

There was a very large number of paper submissions (328), not only from Europe, but also from Asia and the Pacific, and North and South America. This number of submissions represents a substantial increase of contributions compared to the first year of HPCC, which clearly underlines the importance of this domain.

All submissions were reviewed by at least three Program Committee members or external reviewers. It was extremely difficult to select the presentations for the conference because there were so many excellent and interesting submissions. In order to allocate as many papers as possible and keep the high quality of the conference, we finally decided to accept 95 regular papers for oral technical presentations. We believe that all of these papers and topics not only provide novel ideas, new results, work in progress, and state-of-the-art techniques in this field, but also stimulate the future research activities in the area of HPCC.

The exciting program for this conference was the result of the hard and excellent work of many others, such as the Program Vice-Chairs, the Program Committee members, and the external reviewers. We would like to express our sincere appreciation to all authors for their valuable contributions and to all Program Committee members and external reviewers for their cooperation in completing the program under a very tight schedule. We are also grateful to the members of the Organizing Committee for supporting us in handling the many organizational tasks and to the keynote speakers for accepting to come to the conference with enthusiasm.

Last but not least, we hope that the attendees enjoyed the conference program, and the attractions of the city of Munich, together with the social activities of the conference.

September 2006

Michael Gerndt
Dieter Kranzlmüller

Organization

HPCC 2006 was organized by the Technical University of Munich, Germany, in collaboration with the Johannes Kepler University Linz, Austria.

Executive Committee

General Chair:	Michael Gerndt (Technical University Munich, Germany)
Program Chair:	Dieter Kranzlmüller (Joh. Kepler University Linz, Austria)
Local Chair:	Karl Fürlinger (Technical University Munich, Germany)
Steering Chairs:	Beniamino Di Martino (Second University of Naples, Italy)
	Laurence T. Yang (St. Francis Xavier University, Canada)
Organizing Committee:	Roland Hopferwieser (Joh. Kepler University Linz, Austria)
	Christian Glasner (Joh. Kepler University Linz, Austria)

Program Vice-Chairs

Parallel and Distributed Architectures
Roland Wismüller (University of Siegen, Germany)

Embedded Systems
Francois Bodin (University of Rennes, France)

Networking Protocols, Routing, Algorithms
Michel Diaz (LAAS-CNRS, Toulouse, France)

Reliability and Fault-Tolerance
Erik Maehle (Medical University of Luebeck, Germany)

Security and Authentication
Antonino Mazzeo (University of Naples, Italy)

Wireless and Mobile Computing
Marios Dikaiakos (University of Cyprus, Cyprus)

Pervasive Computing and Communications
Frank Stajano (University of Cambridge, UK)

Web Services and Internet Computing
Laurent Lefevre (INRIA, France)

Grid and Cluster Computing
Marian Bubak (AGH University of Science and Technology, Cracow, Poland)

Peer-to-Peer Computing
Yunhao Liu (Hong Kong University of Science and Technology, Hong Kong)

Tools and Environments for Software Development
Martin Schulz (Lawrence Livermore National Laboratory, USA)

Performance Evaluation and Measurement
Allen Malony (University of Oregon, USA)

Programming Interfaces for Parallel Systems
Rudolf Eigenmann (Purdue University, USA)

Languages and Compilers for HPC
Hans Zima (California Institute of Technology, USA and University of Vienna, Austria)

Parallel and Distributed Algorithms
Jack Dongarra (University of Tennessee, USA)

Applications in High-Performance Scientific and Engineering Computing
Alan Sussman (University of Maryland, USA)

Database Applications and Data Mining
Peter Brezany (University of Vienna, Austria)

Biological and Molecular Computing
Harald Meier (Technische Universität München, Germany)

Collaborative and Cooperative Environments
Vassil Alexandrov (University of Reading, UK)

Special Session Chairs

High-Performance Simulation of Reactive Flows
Salvatore Filippone (University of Rome Tor Vergata, Italy)

Service Level Agreements
Frances Brazier (Vrije Universiteit, Amsterdam, The Netherlands)
Benno Overeinder (Vrije Universiteit, Amsterdam, The Netherlands)
Omer F. Rana (Cardiff University and Welsh eScience Center, UK)
Jianhua Shao (Cardiff University and Welsh eScience Center, UK)

Pervasive Computing Application and Security Service
Byoung-Soo Koh (Digicaps Co., Ltd., Seoul, South Korea)
Ilsun You (Department of Information Science, Korean Bible University, South Korea)

Automatic Performance Analysis of Parallel/Distributed Applications
Emilio Luque (Universitat Autonoma de Barcelona, Spain)
Tomas Margalef (Universitat Autonoma de Barcelona, Spain)

Program Committee

David Abramson (Monash University, Australia)
Georg Acher (Technische Universität München, Germany)
Hyo beom Ahn (Kongju National University, Korea)
Rashid Al-Ali (QCERT, Qatar)
Raad S. Al-Qassas (University of Glasgow, UK)
Ilkay Altintas (University of California, USA)
Henrique Andrade (IBM T.J. Watson Research Center, USA)
Cosimo Anglano (Univ. Piemontale, Italy)
Rocco Aversa (Seconda Università di Napoli, Italy)
Irfan Awan (University of Bradford, UK)
Eduard Ayguade (Univ. Politecnica de Catalunya, Barcelona, Spain)
Rosa M. Badia (Universitat Politecnica Catalunya, Spain)
Piotr Bala (N. Copernicus University, Torun, Poland)
Frank Ball (Bournemouth University, UK)
Richard Barrett (Oak Ridge National Laboratory, USA)
Dominique Bayart (Alcatel, France)
Alessio Bechini (University of Pisa, Italy)
Micah Beck (LOCI Labs, USA)
Siegfried Benkner (University of Vienna, Austria)
Alastair Beresford (University of Cambridge, UK)
Erik Berglund (Linkping University, Sweden)
Massimo Bernaschi (IAC-CNR, Italy)
Thomas Bonald (France Telecom, France)
Olivier Bonaventure (Université Catholique de Louvain, Belgium)
Luciano Bononi (University of Bologna, Italy)
Cristian Borcea (New Jersey Institute of Technology, USA)
Thomas Brandes (SCAI, Fraunhofer Gesellschaft, Germany)
Olivier Brun (LAAS, France)
Holger Brunst (Technical University Dresden, Germany)
Wojciech Burakowski (Warsaw University, Poland)
Helmar Burkhart (University of Basel, Switzerland)
Fabian Bustamante (Northwestern University, Illinois, USA)
Rajkumar Buyya (University of Melbourne, Australia)
David Callahan (Microsoft, USA)
Calin Cascaval (IBM Research, Yorktown Heights, USA)
Valentina Casola (University of Naples, Italy)
Bradford Chamberlain (Cray Research, USA)
Barbara Chapman (University of Houston, USA)
Lei Chen (Hong Kong University of Science and Technology, Hong Kong)

YongRak Choi (University of Daejeon, Korea)
I-Hsin Chung (IBM T.J. Watson Research Center, USA)
Yeh-Ching Chung (National Tsing Hua University, Taiwan)
Michele Colajanni (University of Modena, Italy)
Toni Cortes (UPC Barcelona, Spain)
Domenico Cotroneo (University of Naples, Italy)
Valentin Cristea (Polytechnic University of Bucharest, Romania)
Marilia Curado (University of Coimbra, Portugal)
Jan-Thomas Czornack (Technische Universität München, Germany)
Pasqua D'Ambra (ICAR-CNR, Italy)
Alessandro De Maio (NUMIDIA s.r.l., Italy)
Giuseppe De Pietro (ICAR CNR, Italy)
Geert Deconinck (University of Leuven, Belgium)
Ewa Deelman (University of Southern California, USA)
Isabelle Demeure (ENST, France)
Luiz DeRose (CRAY Inc., USA)
Frdric Desprez (ENS Lyon, France)
Daniela di Serafino (Second University of Naples, Italy)
Roxana Diaconescu (California Institute of Technology, USA)
Ivan Dimov (University of Reading, UK)
Chen Ding (University of Rochester, New York, USA)
Karim Djemame (University of Leeds, UK)
Dirk Duellmann (CERN, Geneva, Switzerland)
Olivier Dugeon (France Telecom R&D, France)
Marc Duranton (Philips Research, Eindhoven, The Netherlands)
Ernesto Exposito (LAAS-CNRS, France)
Thomas Fahringer (University of Innsbruck, Austria)
Wu-chun Feng (Virginia Tech, USA)
Xinwen Fu (Dakota State University, USA)
Karl Fürlinger (Technische Universität München, Germany)
Fabrizio Gagliardi (Microsoft, Switzerland)
Luis Javier Garcia Villalba (Complutense University of Madrid, Spain)
Maria Garzaran (University of Illinois at Urbana-Champaign, USA)
Jean-Patrick Gelas (University of Lyon 1, France)
Alexander Gelbukh (National Polytechnic Institute, Mexico)
Vladimir Getov (University of Westminster, UK)
Olivier Gluck (ENS Lyon, France)
Frank-Oliver Glöckner (MPI für Marine Mikrobiologie, Bremen, Germany)
Ananth Grama (Purdue University, USA)
Karl-Erwin Grosspietsch (Fraunhofer AIS, Germany)
Amitava Gupta (Jadavpur University, Kolkata, India)
Vesna Hassler (European Patent Office, Austria)
Zdenek Havlice (Technical University of Kosice, Slovakia)
Hermann Hellwagner (Universität Klagenfurt, Austria)
Volker Heun (Ludwig-Maximilians-Universität München, Germany)

Ladislav Hluchy (Institute of Informatics, Slovak Academy of Sciences, Slovakia)
Ralf Hofestaedt (Bielefeld University, Germany)
Jeff Hollingsworth (University of Maryland, USA)
Chunming Hu (Beihang University, China)
Tai-Yi Huang (National Tsing Hua University, Taiwan)
Zhiyi Huang (University of Otago, New Zealand)
Marty Humphrey (University of Virginia, USA)
HoJae Hyun (Korea Information Security Agency, Korea)
Valerie Issarny (INRIA, Rocquencourt, France)
Zhen Jiang (West Chester University of Pennsylvania, USA)
Josep Jorba (Universitat Oberta de Catalunya, Spain)
Guy Juanole (LAAS, France)
YunHee Kang (Cheonan University, Korea)
Helen Karatza (Aristotle University of Thessaloniki, Greece)
Karen L. Karavanic (Portland State University, USA)
Wolfgang Karl (Universität Karlsruhe, Germany)
Constantine Katsinis (Drexel University, USA)
Yeong-Deok Kim (Woosong University, Korea)
Jacek Kitowski (AGH University of Science and Technology, Cracow, Poland)
Paris Kitsos (Hellenic Open University, Greece)
Peter Knijnenburg (Leiden University, The Netherlands)
Harald Kornmayer (Forschungszentrum Karlsruhe, Germany)
Stefan Kramer (Technische Universität München, Germany)
Jerome Lacan (ENSICA, France)
Nicolas Larrieu (LAAS, France)
Craig Lee (AeroSpace Org., USA)
Deok-Gyu Lee (Soonchunyang University, Korea)
Jenq-Kuen Lee (National Tsing Hua University, Taiwan)
Johun Lee (Dong-Ah Broadcasting College, Korea)
Yiming Li (National Chiao Tung University, Taiwan)
Jie Lian (University of Waterloo, Canada)
Xiaofei Liao (Huazhong University of Science and Technology, China)
Wei Lin (Australian Taxation Office, Sydney, Australia)
Antonio Liotta (University of Essex, UK)
Bin Lu (West Chester University of Pennsylvania, USA)
Simone Ludwig (Concordia University, Canada)
Thomas Ludwig (University of Heidelberg, Germany)
Bob Mann (University of Edinburgh, UK)
Jesus Marco (CSIC IFCA Santander, Spain)
Eva Marin (UPC Barcelona, Spain)
Muneer Masadah (University of Glasgow, UK)
Xavier Masip (UPC Barcelona, Spain)
Laurent Mathy (Lancaster University, UK)
John May (Lawrence Livermore National Laboratory, USA)
Piyush Mehrotra (NASA Ames Research Center, USA)

Harald Meier (Technische Universität München, Germany)
Xiaoqiao Meng (University of California, Los Angeles, USA)
Barton Miller (University of Wisconsin Madison, USA)
Bernd Mohr (Research Centre Juelich, ZAM, Germany)
Nikolay Moldovyan (Specialized Center of Program Systems, Russia)
Edmundo Monteiro (University of Coimbra, Portugal)
Anna Morajko (Universitat Autonoma de Barcelona, Spain)
Jose Moreira (IBM T.J. Watson Research Center, USA)
Frank Mueller (North Carolina State University, USA)
Henk Muller (University of Bristol, UK)
Dong Myung Shin (Korea Information Security Agency, Korea)
Wolfgang Mühlbauer (Technische Universität München, Germany)
Jarek Nabrzyski (PSNC, Poznan, Poland)
Tatsuo Nakajima (Waseda University, Japan)
Laura Nett Carrington (San Diego Supercomputing Center, USA)
Nobuhiko Nishio (Ritsumei University, Japan)
Teresa Oh (Cheongju University, Korea)
Yong-Chul Oh (Korea Polytechnic University, Korea)
Vicent Oria (New Jersey Institute of CLIPS-IMAG, USA)
Jose Orlando Pereira (University of Minho, Portugal)
Salvatore Orlando (University of Venice "Ca′ Foscari," Italy)
Michael Ott (Technische Universität München, Germany)
Mohamed Ould-Khaoua (University of Glasgow, UK)
Julian Padget (Bath University, UK)
Stylianos Papanastasiou (University of Glasgow, UK)
Myung-Chan Park (International Graduate University for Peace, Korea)
SuJin Park (University of Daejeon, Korea)
Rubem Pereira (Liverpool John Moores University, UK)
Ron Perrott (Queen's University, Belfast, UK)
Antonio Picariello (University of Naples, Italy)
Jean-Marc Pierson (INSA, France)
Evaggelia Pitoura (University of Ioannina, Greece)
Thomas Plagemann (University of Oslo, Norway)
Gilles Pokam (University of California, San Diego, USA)
Balakrishna Prabhu (VTT TRC, Finland)
Isabelle Puaut (University of Rennes, France)
Aaron Quigley (University College Dublin, Ireland)
Khaled Ragab (University of Tokyo, Japan)
Massimiliano Rak (Seconda Università degli Studi di Napoli, Italy)
Omer Rana (Cardiff University, UK)
Thomas Rattei (Technische Universität München, Germany)
Andrew Rau-Chaplin (Dalhousie University, Canada)
Thomas Rauber (University of Bayreuth, Germany)
Lawrence Rauchwerger (Texas A&M University, USA)
Kasim Rehman (Cambridge Systems Associates, UK)

Vincent Roca (INRIA Rhone-Alpes, France)
Daniel Rodriguez Garcia (University of Reading, UK)
Paul Roe (Queensland Univ. of Technology, Australia)
Mathilde Romberg (University of Ulster, UK)
Philip Roth (Oak Ridge National Laboratory, USA)
Stefano Russo (University of Naples, Italy)
Rizos Sakellariou (University of Manchester, UK)
Kave Salamatian (Laboratoire d'Informatique de Paris 6, France)
Yahya Sanadidi (UCLA, USA)
Miguel Santana (Central R&D, ST Microelectronics, France)
Vivek Sarkar (IBM T.J. Watson Research Center, USA)
Mitsuhisa Sato (University of Tsukuba, Japan)
Ichiro Satoh (National Institute of Informatics, Japan)
Olaf Schenk (University of Basel, Switzerland)
Erich Schikuta (University of Vienna, Austria)
David Schmidt (University of Massachussetts at Amherst, USA)
Assaf Schuster (TECHNION, Israel Institute of Technology, Haifa, Israel)
James Scott (Intel Research, UK)
Stephen Scott (ORNL, USA)
Sameer Shende (University of Oregon, USA)
Qi Shi (Liverpool John Moores University, UK)
Nicolas Sklavos (University of Patras, Greece)
Peter M.A. Sloot (University of Amsterdam, The Netherlands)
Peter Sobe (Universität Lübeck, Germany)
Matthew Sottile (Los Alamos National Laboratory, USA)
Alexandros Stamatakis (Foundation for Research and Technology-Hellas, Greece)
Thomas Sterling (Caltech and Louisiana Sate University, USA)
Kurt Stockinger (Lawrence Berkeley National Laboratory, USA)
Daniel Stodden (Technische Universität München, Germany)
Vaidy Sunderam (EMORY University, USA)
Mee Young Sung (University of Incheon, Korea)
Martin Swany (University of Delaware, USA)
Domenico Talia (Università della Calabria, Rende, Italy)
Kun Tan (Microsoft Research, China)
David Taniar (Monash University, Melbourne, Australia)
Jie Tao (Universität Karlsruhe, Germany)
Renata Teixeira (Laboratoire d'Informatique de Paris 6, France)
Dan Terpstra (University of Tennessee, Knoxville, USA)
Nigel A. Thomas (University of Newcastle, UK)
Carsten Trinitis (Technische Universität München, Germany)
Ziga Turk (University of Ljubljana, Slovenia)
Hanjo Täubig (Technische Universität München, Germany)
Stefano Ubertini (University of Rome, Italy)
Andreas Uhl (University of Salzburg, Austria)
Theo Ungerer (University of Augsburg, Germany)

Shmuel Ur (IBM Research Haifa, Israel)
Marian Vajtersic (University of Salzburg, Austria)
Sudharshan Vazhkudai (Oak Ridge National Laboratory, USA)
Daniel Veit (University of Karlsruhe, Germany)
Iakovos Venieris (National Technical University of Athens, Greece)
Salvatore Venticinque (Seconda Università di Napoli, Italy)
Pascale Vicat-Blanc (ENS Lyon, France)
Pablo Vidales (Deutsche Telekom Labs, Germany)
Umberto Villano (Università del Sannio, Italy)
Valeria Vittorini (University of Naples, Italy)
Max Walter (Technische Universität München, Germany)
Xiaofang Wang (New Jersey Institute of Technology, USA)
Xin-Gang Wang (University of Bradford, UK)
Greg Watson (Los Alamos National Laboratory, USA)
Josef Weidendorfer (Technische Universität München, Germany)
Andrew L. Wendelborn (University of Adelaide, Australia)
Marianne Winslett (University of Illinois, USA)
Felix Wolf (Forschungszentrum Juelich, Germany)
Dan Wu (University of Windsor, Canada)
Brian Wylie (Research Centre Juelich, ZAM, Germany)
Tao Xie (New Mexico Institute of Mining and Technology, USA)
Baijian Yang (Ball State University, USA)
Kun Yang (Univ. of Essex, UK)
Marcelo Yannuzzi Sanchez (UPC Barcelona, Spain)
Wai Gen Yee (Illinois Institute of Technology, USA)
Hao Yin (Tsinghua University, China)
Martin Zacharias (International University Bremen, Germany)
Ning Zhang (University of Manchester, UK)
Hongbo Zhou (Slippery Rock University, USA)
Hai Zhuge (Chinese Academy of Science, China)
Wolfgang Ziegler (Fraunhofer Institute, Germany)
Anna Zygmunt (AGH University of Science and Technology, Cracow, Poland)

Sponsoring Institutions

(as of July 10, 2006)

HP
Intel
Megware
ParTec
Transtec

Table of Contents

Introducing Combustion-Turbulence Interaction in Parallel Simulation of Diesel Engines

Paola Belardini[1], Claudio Bertoli[1], Stefania Corsaro[2], and Pasqua D'Ambra[3]

[1] Istituto Motori (IM)-CNR
Via Marconi, 8 I-80125 Naples, Italy
{p.belardini, c.bertoli}@im.cnr.it

[2] Department of Statistics and Mathematics for Economic Research
University of Naples "Parthenope"
Via Medina 40, I-80133 Naples, Italy
stefania.corsaro@uniparthenope.it

[3] Institute for High Performance Computing and Networking (ICAR)-CNR
Via Pietro Castellino 111, I-80131 Naples, Italy
pasqua.dambra@na.icar.cnr.it

Abstract. In this work we focus on parallel combustion simulation in modern Common Rail Diesel engines when the interaction between complex chemical kinetics and turbulence is taken into account. We introduce a turbulence term in a detailed chemical reaction model and analyze the impact on the reliability of pollutant emission predictions and on the efficiency and scalability of our combustion software. The parallel combustion software we developed adaptively combines numerical schemes based either on Backward Differentiation Formulas or semi-implicit Runge-Kutta methods for the solution of ODE systems arising from the chemical reaction model. It is based on CHEMKIN-II package for managing detailed chemistry and on two general-purpose solvers for adaptive solution of the resulting ODE systems. Furthermore, it is interfaced with KIVA3V-II code in order to simulate the entire engine cycle.

1 Introduction

In recent years stringent limits on engines pollutant emissions have been imposed by Government laws, strongly conditioning motor industry. For this reason, the design of modern engines relies on even more sophisticated, complex technologies. In particular, in latest-generation Diesel engines the Common Rail technology is usually employed and an high level of Exhaust Gas Recirculation (EGR) rate is also established; moreover, multiple fuel injection mechanism is typically adopted in order to reduce soot emissions and combustion noise. The impact on engine modeling is the need of accurately simulating highly complex, different physical-chemical phenomena occurring in each engine cycle. Mathematical models for the description of the overall problem typically involve unsteady Navier-Stokes equations for turbulent multi-component mixtures of ideal gases, coupled with suitable equations for fuel spray and combustion modeling. The solution

M. Gerndt and D. Kranzlmüller (Eds.): HPCC 2006, LNCS 4208, pp. 1–10, 2006.

of the complex overall model usually relies on a time-splitting approximation technique, where different physical phenomena are conceptually decoupled and, consequently, different sub-models are solved separately one from each other on a suitable 3d computational grid representing engine cylinder and piston bowl.

In recent years much attention has been addressed to combustion: great effort has been devoted to the design of detailed chemical reaction models suitable to predict emissions in sophisticated, last-generation Common Rail Diesel engines. Therefore, the numerical solution of chemistry has become one of the most computational demanding parts in simulations, thus leading to the need of efficient parallel combustion solvers [1,15]. The typical main computational kernel in this framework is the solution of systems of non-linear Ordinary Differential Equations, characterized by a very high stiffness degree. The reaction model does not introduce any coupling among grid cells, in other words cells are supposed to be isolated when chemical reactions occur. This assumption actually neglects turbulence effects in combustion process, a quite severe approximation in the high temperature combustion phase.

In this work we discuss the impact of introducing, in multidimensional modeling of Diesel engines, based on a decoupled solution of transport and reaction phenomena, a model describing the combustion-turbulence interaction. We model the interaction between complex chemistry and turbulence following the approach discussed in [13,14], that is, splitting the characteristic times of the species involved in combustion into the sum of a laminar term and a properly smoothed turbulent term. This model does not introduce any coupling among computational grid cells, preserving locality in the solution of the turbulent combustion process. The simulations are obtained by a parallel combustion software [4,5,6], based on CHEMKIN-II package for managing detailed chemistry and on a multi-method ODE solver [7] for the solution of the ODE systems arising from the turbulent chemical reaction model. The software is interfaced with the KIVA3V-II code for the simulation of the entire engine cycle. In section 2 we briefly outline the turbulent combustion model, in section 3 we describe the solution procedure and the main features of the parallel software, in section 4 we discuss some results of numerical simulations on realistic test cases.

2 Turbulent Combustion Model

Diesel engine combustion is characterized by liquid fuel injection in a turbulent environment. Two main phases can be distinguished in the overall phenomenon. Fuel injection gives raise to chemical reactions that, under suitable temperature and pressure conditions, lead to fuel ignition. The period from the starting of fuel injection and fuel ignition is named ignition delay: in this phase chemical reactions occur without giving strong energy contributions, but high stiffness is the main feature, due to very different reaction rates among the reactant species. Kinetics occurring before ignition is usually referred to as low temperature combustion or cold phase, while combustion phase, or high temperature combustion, is the chain of reactions subsequent to ignition.

In a time-splitting solution procedure for simulation of Diesel engines, computational grid cells are supposed to be isolated during the solution of the chemical reaction equations driving combustion, thus, neglecting their interaction, turbulence is not properly considered. On the other hand, combustion is strongly influenced by turbulence, since it has significant effects on the transport properties and on the mixing of reactants. Neglecting turbulence can seriously affect results concerning the numerical simulation of combustion phase, when turbulence effects are more relevant: indeed, chemical species conversion rates estimation does not take into account mixture inhomogeneities, thus leading to overestimated combustion rates. As a consequence, the stiffness degree of the arising ODE systems increases in this case, that is, neglecting turbulence has considerable effects from the mathematical point of view as well.

We consider a turbulent combustion model in order to accurately predict the effects of both chemical kinetics and turbulent mixing. The model is based on a recent detailed kinetic scheme, which considers N-dodecane as primary fuel. It involves 62 chemical species and 285 reactions. The kinetic scheme considers the H abstraction and the oxidation of the primary fuel, with production of alchil-peroxy-radicals, followed by the ketoydroperoxide branching. In the model the fuel pirolysis determines the chetons and olefins formation. Moreover, a scheme of soot formation and oxidation is provided, together with a classical scheme of NOx formation. The reaction system is expressed by the following system of non-linear Ordinary Differential Equations:

$$\dot{\rho}_m = W_m \sum_{r=1}^{R} (b_{mr} - a_{mr}) \dot{\omega}_r(\rho_1, \ldots, \rho_m, T), \qquad m = 1, \ldots, M, \tag{1}$$

where R is the number of chemical reactions involved in the system, M is the number of species, $\dot{\rho}_m$ is the production rate of species m, W_m is its molecular weight, a_{mr} and b_{mr} are integral stoichiometric coefficients for reaction r and $\dot{\omega}_r$ is the kinetic reaction rate.

Production rate terms can be separated into creation rates and destruction rates[12]:

$$\dot{\rho}_m = \dot{C}_m - \dot{D}_m, \qquad m = 1, \ldots M, \tag{2}$$

where $\dot{C}_m$, $\dot{D}_m$ are the creation and the destruction rate of species m respectively. The latter can be expressed as

$$\dot{D}_m = \frac{X_m}{\tau_m}, \qquad m = 1, \ldots M, \tag{3}$$

where X_m, τ_m are respectively the molar concentration and the characteristic time for destruction rate of species m. Expression (3) shows that the eigenvalues of the Jacobian matrix of the right-hand side of system (1) are related to the characteristic times for destruction rates of species involved in the combustion model. Detailed reaction models involve a great number of intermediate species and no equilibrium assumption is made. Thus, the overall reaction systems include species varying on very different timescales one from each other;

this motivates the high stiffness degree that typically characterizes ODE systems arising in this framework. Moreover, relation (3) shows that if combustion rates are overestimated, that is, characteristic times are underestimated, then it results in a higher stiffness degree: this explains the impact of neglecting turbulence from the computational point of view.

We model interaction between complex kinetics and turbulence following the approach discussed in [13,14]. The model relies on the assumption that the characteristic time τ_m^c of each species involved in the combustion model depends both on a *kinetic timescale* and a *turbulent timescale*. The former is defined as the time needed by a species to reach equilibrium state under perfectly homogeneous conditions, the latter is the eddy breakup time. More precisely, we suppose that it holds

$$\tau_m^c = \tau_m^k + f\tau_m^t, \qquad m = 1, ...M \tag{4}$$

where τ_m^k, τ_m^t are the kinetic and the turbulent timescales of species m, respectively. The turbulent timescale is considered proportional to the eddy turnover time as estimated by the standard $k - \epsilon$ turbulence model employed in the KIVA3V-II code. The factor f serves as a delay coefficient that slows down reactions according to turbulence effects. It is assumed to be

$$\frac{1 - e^r}{0.632}, \tag{5}$$

where r is the ratio between combustion products and total reactant concentrations. It indicates the stage of combustion within specific regions: the value $r = 1$ corresponds to complete consumption of fuel and oxygen. Note that a reliable estimate of r is a key issue when detailed chemical kinetic models are used, since in that case combustion products have to be well established. The delay coefficient f changes accordingly to r, depending on the local conditions.

From relation (4) it follows that the densities ρ_m satisfy the equation

$$\frac{\partial \rho_m}{\partial t} = \frac{\rho_m^* - \rho_m}{\tau_m^k + f\tau_m^t}, \qquad m = 1, ...M \tag{6}$$

where ρ_m^* is the equilibrium concentration. Therefore, the main computational kernel in the turbulent combustion model is the solution, in each grid cell and at each splitting time step, of system (1) where the right-hand side is properly scaled, according to (6). Note that, for sake of efficiency, the kinetic timescale for all the species is assumed to be equal to that of the slowest species involved in the oxidation scheme.

In Figure 1 two graphics reporting stiffness degree estimations during the ignition delay period of a typical engine simulation are shown. One refers to a simulation where the detailed combustion model, without turbulence term, was employed. The other one shows the results of a simulation involving interaction between complex kinetics and turbulence. Figure reveals that, as expected, since the introduction of turbulence term in the model slows down reaction rates, it has, from the computational point of view, a smoothing effect on ODE systems.

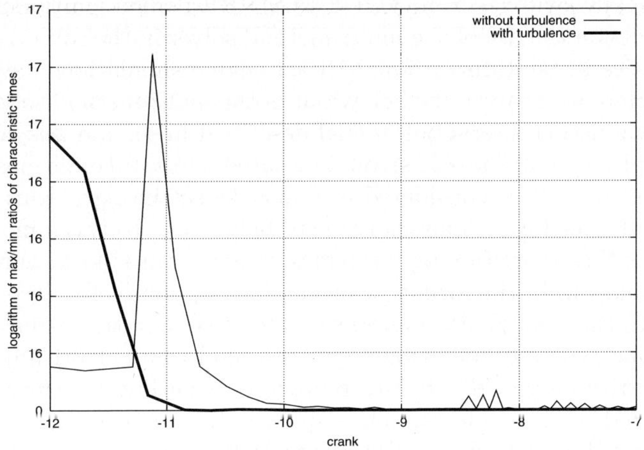

Fig. 1. Stiffness estimate

3 Parallel Solution of Turbulent Combustion

In this section we describe the main features of the parallel package we developed for the numerical simulation of turbulent combustion in Diesel engines. As also proposed in [1,15], parallelism is based on a domain decomposition technique where the computational grid is partitioned among the parallel processors; the main computational kernel arising from the turbulent combustion model is the solution of a system of non-linear Ordinary Differential Equations per each grid cell, at each time step of the splitting solution procedure, which can be solved concurrently, since there is no coupling among the cells. On the other hand, the approach followed for accounting interaction between complex kinetics and turbulence does not affect inherent parallelism in the solution process: indeed, the scaling procedure described in section 2 preserves the locality with respect to grid cells.

In our software, main contribution is related to the local stiff ODE solver. Indeed, we proposed a multi-method solver, based on an adaptive combination of a 5-stages Singly Diagonally Implicit Runge-Kutta (SDIRK) method [11] and variable coefficient Backward Differentiation Formulas (BDF) [8].

We tested SDIRK4 and VODE packages when no interaction between turbulence and chemical reactions was considered; main features of both solvers and some results are described in [6]. Those results showed that the VODE package is more accurate than SDIRK4 in the cold phase. From the mathematical point of view, low temperature combustion corresponds to the transient phase; Runge-Kutta based methods are well-known to loose accuracy in the very stiff phase [3], indeed we observed that SDIRK4 could overestimate ignition delay [6] and, consequently, underestimate pressure rise. On the other hand, SDIRK4 was, in all of our numerical simulations, more efficient than VODE in the high temperature combustion phase, therefore, we proposed and developed a multi-method solver that automatically switches from VODE to SDIRK4 when ignition is approaching. First results on the use of the multi-method solver in the solution of detailed chemical kinetics in multidimensional Diesel engine simulations have been presented at [7]. Here we analyze the behaviour of the multi-method solver when the turbulence combustion interaction model described in section 2 is considered.

Note that physical stiffness is strongly related to local conditions, therefore, when adaptive solvers are considered in a parallel setting, grid partitioning becomes a critical issue for computational load balancing. To this aim, our parallel solver supports three partitioning strategies, namely, pure block, pure cyclic and random distribution. In the former, a block of contiguous cells, according to cell numbering into the grid, is distributed to each process, in the latter, cells are distributed among processes following a typical round-robin algorithm. In the case of random partitioning grid cells are reordered according to a permutation of indexes, deduced by a pseudo-random sequence, before grid distribution. Experiments on our test cases, when VODE or SDIRK4 were used as solver, revealed that random partitioning produces the best parallel performance results. Details on the performance analysis of different grid partinioning strategies on our test cases can be found in [5].

The parallel combustion software is written in Fortran and uses standard MPI API for message passing. Furthermore, it is based on CHEMKIN-II [12], a software package for managing large models of chemical reactions in the context of simulation software. It provides a database and a software library for computing model parameters involved in system (1).

Our parallel software has been interfaced with the KIVA3V-II code [2], in order to properly test it within real simulations. Details on the integration between the parallel combustion software and KIVA3V-II can be found in [4,6].

4 Numerical Results

In this section we show some results concerning simulations performed on a prototype, single cylinder Diesel engine, with IV valves, having characteristics similar to the 16 valves EURO IV Fiat Multijet. Simulations have been carried out at 1500 rpm. In order to fit a wide range of operating conditions both in the experimental measurements and in the numerical simulations, the production engine has been modified for having a swirl variable head. The engine is equipped with an external supercharging system to simulate intake conditions deriving from turbo-charging application. In addition, the exhaust pipe is provided with a motored valve to simulate the backpressure due to the turbocharger operation. In the experiments three test cases, corresponding to different operating conditions, reported in Table 1, have been considered. The position of the piston into the cylinder is measured by means of crank angle values, therefore, injection timing and injection period are expressed with respect to them. The limit positions of the piston, that is, the lowest point from which it can leave and the highest point it can reach, correspond to -180° and 0° crank angle values respectively. Numerical experiments have been carried out on a Beowulf-class Linux cluster,

Table 1. Engine operating conditions

	Rail Pressure (bar)	EGR	Injection timing (crank angle)	Injection period (crank angle)	Injected fuel (mg)
Test case 1	500	40%	-12.7	7.3	8.0
Test case 2	900	0%	-2.3	5.7	8.7
Test case 3	500	0%	-2.1	7.3	8.5

made of 16 PCs connected via a Fast Ethernet switch, available at IM-CNR. Eight PCs are equipped with a 2.8GHz Pentium IV processor, while the others have a 3.8GHz Pentium IV processor. All the processors have a RAM of 1 GB and an L2 cache of 256 KB. We used the GNU Fortran compiler (version 3.2.2) and the LAM implementation (version 7.0) of MPI.

ODE systems have been solved by means of the multi-method solver we developed. In the stopping criteria, both relative and absolute error control tolerances were considered; at this purpose, we defined two vectors, *rtol* and *atol*, respectively. In all the experiments here analyzed *atol* values were fixed in dependence of the particular chemical species. The reason motivating this choice relies on the very different concentrations characterizing chemical species involved in detailed reaction models. All the components of *rtol* were set to 10^{-3}, in order to satisfy the application accuracy request.

In Fig. 2 the in-cylinder combustion pressure graph is shown. Experimental pressure values are compared to predicted ones for testing results reliability. In order to analyze the impact of accounting for kinetics-turbulence interaction, two graphics are shown: on the left, results concerning simulations performed without turbulence term in kinetics model are reported. On the right, the represented pressure curves refer to numerical simulations involving the turbulent combustion model described in section 2. We observe that, for all the considered test cases, the introduction of chemistry-turbulence interaction term in the combustion model provides a better agreement between experimental and simulated pressure curves, confirming that the complex physical phenomena occurring during combustion are more accurately described in this case.

In Figure 3 we show some performance results on the Test case 1. In the experiments we are discussing, computational grid has been distributed according to random grid partitioning strategy. We show results of numerical simulations involving the detailed combustion model without the kinetics-turbulence interaction term and compare them with the ones obtained via the numerical solution of the turbulent combustion model in order to investigate the impact of accounting for kinetics-turbulence interaction on performance as well. In Fig. 3, on the top-left, we reported the total number of performed function evaluations for each process configuration, that is a measure of the total computational complexity of the combustion solver. More precisely, for a fixed number of processes varying from one to sixteen, we added the number of function evaluations performed by each process. On the top-right, simulation time, expressed in hours, versus number of processes is represented. We note that considering the turbulence term in

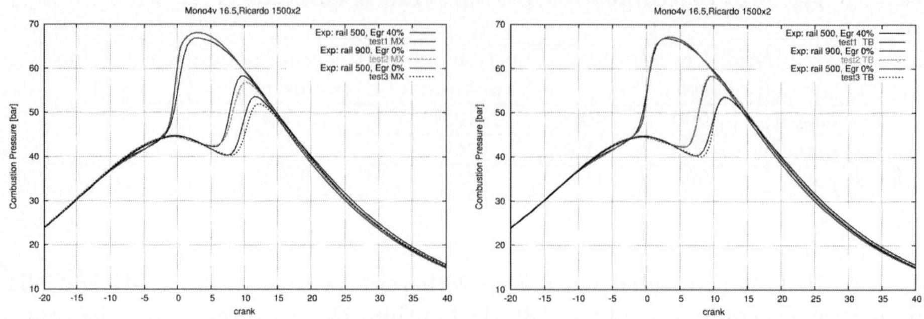

Fig. 2. Comparison between experimental and simulated in-cylinder pressure. Left: combustion simulation without turbulence term. Right: turbulent combustion model simulation results.

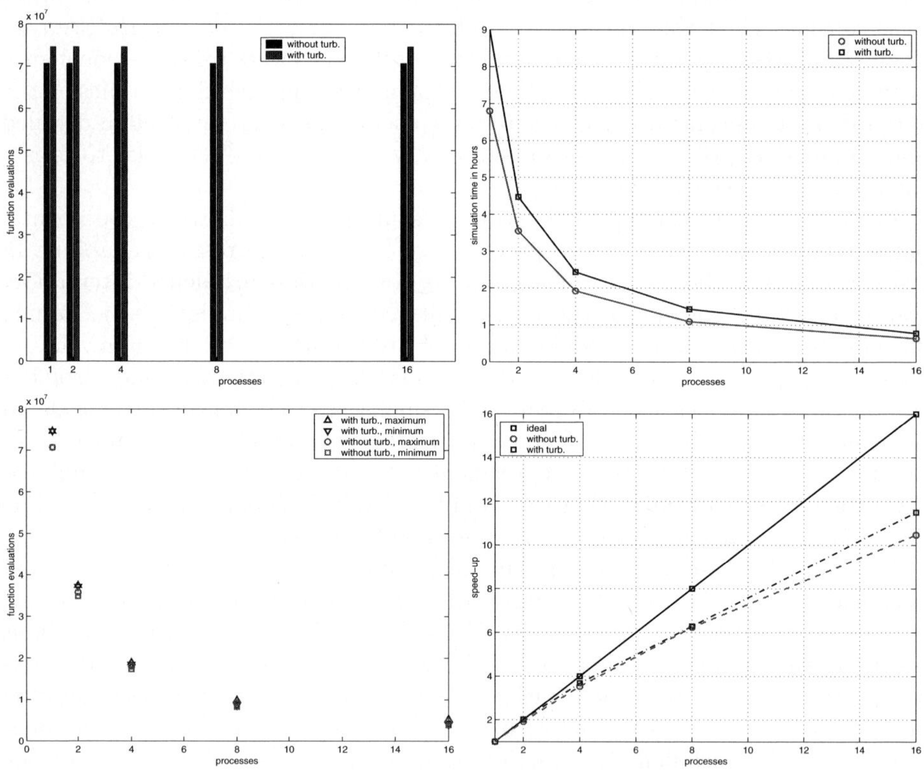

Fig. 3. Top-left: total number of performed function evaluations versus number of processes. Top-right: simulation time expressed in hours. Bottom-left: maximum and minimum number of performed function evaluations versus number of processes. Bottom-right: speed-up.

the combustion model results in an higher overall simulation time. Accounting for turbulence-chemistry interaction turns in a local scaling of the right-hand side of systems (1), which affects the properties of the involved ODE systems, leading to an increase of the computational load per processor. On the other hand, communication costs and serial computation overheads are not affected. Furthermore, accounting for turbulence-chemistry interaction seems to produce better load balancing: on the bottom-left of the figure we analyze the load balancing among the processes for each configuration. We reported the minimum and the maximum number of performed function evaluations, for each process configuration. More precisely, for a fixed number of processes, we computed the total number of function evaluations performed by each process and considered the maximum and the minimum among such values. We note that, when the turbulent term is introduced in the combustion model, the gap between those values is reduced. The whole previous analysis is in agreement with the speed-up lines, represented on the bottom-right of the figure, where we can observe that higher speed-up values are obtained when turbulence is considered.

5 Some Conclusions

In this work we have shown first results related to the effort of improving reliability of parallel simulations of combustion in Diesel engines. A chemistry-turbulence interaction model has been introduced in a decoupled solution of the chemical reaction and of the Navier-Stokes equations for the reactive fluid flow. Even though the chemistry-turbulence interaction term does not affect the inherent parallelism in the combustion solver, it seems to produce larger local computational complexity and better load balancing. Numerical results on realistic test cases show that the use of the model provides a better agreement between experimental and simulated results. Therefore, the impact of the chemistry-turbulence interaction term on the ODE systems arising in detailed combustion models has to be deeply investigated.

References

1. A. Ali, G. Cazzoli, S. Kong, R. Reitz, C. J. Montgomery, Improvement in Computational Efficiency for HCCI Engine Modeling by Using Reduced Mechanisms and Parallel Computing, *13th International Multidimensional Engine Modeling User's Group Meeting*, (Detroit 2003).
2. A. A. Amsden, KIVA-3V: A Block-Structured KIVA Program for Engines with Vertical or Canted Valves, *Los Alamos National Laboratory Report No. LA-13313-MS*, (1997).
3. U. M. Ascher, L. R. Petzold, Computer Methods for Ordinary Differential Equations and Differential-Algebraic Equations, *SIAM*, (1998).
4. P. Belardini, C. Bertoli, S. Corsaro, P. D'Ambra, Parallel Simulation of Combustion in Common Rail Diesel Engines by Advanced Numerical Solution of Detailed Chemistry, in *Applied and Industrial Mathematics in Italy*, Proc. of the 7th Conference of SIMAI, M. Primicerio, R. Spigler, V. Valente eds., World Scientific Pub., (2005).

5. P. Belardini, C. Bertoli, S. Corsaro, P. D'Ambra, Multidimensional Modeling of Advanced Diesel Combustion System by Parallel Chemistry, *Society for Automotive Engineers (SAE) Paper*, 2005-01-0201, (2005).
6. P. Belardini, C. Bertoli, S. Corsaro, P. D'Ambra, The Impact of Different Stiff ODE Solvers in Parallel Simulation of Diesel Combustion, in Proc. of HPCC'05, L. Yang, O. Rana, B. Di Martino, J. Dongarra eds., *Lecture Notes in Computer Science*, Springer Pub., vol. 3726, pp. 958-968, 2005.
7. P. Belardini, C. Bertoli, S. Corsaro, P. D'Ambra, A Multi-Method ODE Software Component for Parallel Simulation of Diesel Engine Combustion, SIAM Conference on *Parallel Processing for Scientific Computing*, San Francisco, February 2006.
8. Brown,P.N., Byrne,G.D., Hindmarsh, A.C., VODE: A Variable Coefficient ODE Solver, *SIAM J. Sci. Stat. Comput.*, **10**, (1989).
9. C. W. Gear, Numerical Initial Value Problems in Ordinary Differential Equations, *Prentice-Hall*, Englewood Cliffs, NY, (1973).
10. Gustavsson, J., Golovitchev, V.I., Spray Combustion Simulation Based on Detailed Chemistry Approach for Diesel Fuel Surrogate Model, *Society for Automotive Engineers (SAE) Paper*, 2003-0137, (2003).
11. E. Hairer, G. Wanner, Solving Ordinary Differential Equations II. Stiff and Differential-Algebraic Problems, second edition, *Springer Series in Comput. Mathematics*, Vol. 14, Springer-Verlag, (1996).
12. R. J. Kee, F. M. Rupley, J. A. Miller, Chemkin-II: A Fortran chemical kinetics package for the analysis of gas-phase chemical kinetics, *SAND89-8009, Sandia National Laboratories*, (1989).
13. S. C. Kong, Z. Han and R. D. Reitz, The Development and Application of a Diesel Ignition and Combustion Model for Multidimensional Engine Simulation, *SAE 950278*, (1995).
14. S. C. Kong, C. D., Marriott, R. D. Reitz and M. Christensen, Modeling and Experiments of HCCI Engine Combustion Using Detailed Chemical Kinetics with Multidimensional CFD, *SAE 2001-01-1026*, (2001).
15. P. K. Senecal, E. Pomraning, K. J. Richards, T. E. Briggs, C. Y. Choi, R. M. McDavid, M. A. Patterson, Multi-dimensional Modeling of Direct-Injection Diesel Spray Liquid Lenght and Flame Lift-off Lenght using CFD and Parallel Detailed Chemistry, *SAE 2003-01-1043*, (2003).

An Enhanced Parallel Version of Kiva–3V, Coupled with a 1D CFD Code, and Its Use in General Purpose Engine Applications

Gino Bella[1], Fabio Bozza[2], Alessandro De Maio[3], Francesco Del Citto[1], and Salvatore Filippone[1,⋆]

[1] DIM, University of Rome "Tor Vergata"
salvatore.filippone@uniroma2.it
[2] N.U.M.I.D.I.A s.r.l.
[3] DIME, University of Naples "Federico II"

Abstract. Numerical simulations of reactive flows are among the most computational demanding applications in the scientific computing world.

KIVA-3V, a widely used computer program for CFD, specifically tailored to engine applications, had been deeply modified in order to improve accuracy and stability, while reducing computational time.

The original methods included in KIVA to solve equations of fluid dynamics had been fully replaced by new solvers, with the aim of both improving performance and writing a fully parallel code. Almost every feature of original KIVA-3V has been partially or entirely rewritten, a full 1D code has been included and a strategy to link directly 3D zones with zero dimensional models has been developed.

The result is a reliable program, noticeably faster than the original KIVA-3V in serial mode and obviously even more in parallel, capable of treating more complex cases and bigger grids, with the desired level of details where required.

1 Computational Models

1.1 3D Code

The three dimensional CFD code used in this work is a highly modified version of the well known Kiva 3V code; this is the result of a long development process, which we will briefly outline in Sec. 2. Most of the code features have been reimplemented, starting from the inner solver kernels, to the treatment of mesh movement and boundary conditions; moreover the code is now fully parallelized.

1.2 1D Code

The one-dimensional model (1Dime code) has been developed at the DIME of University of Naples "Federico II" during several years [5, 6, 7, 8, 9, 10]. It can handle all main engines configurations, thanks to a very versatile and modular structure.

⋆ Corresponding author.

M. Gerndt and D. Kranzlmüller (Eds.): HPCC 2006, LNCS 4208, pp. 11–20, 2006.

The numerical procedure includes a reliable 1D flow model [14, 16] for the characterization of wave propagation phenomena in the external pipes, described by the equations:

$$\mathbf{U}_t + [\mathbf{F}(\mathbf{U})]_x = \mathbf{S}. \tag{1}$$

The system of eqs. (1) is numerically solved in a finite volume Total Variation Diminishing (TVD) scheme, reaching a second and a fourth order accuracy in space and time respectively [14, 16].

The one-dimensional model also includes a multi–zone zero–dimensional treatment of the cylinders. In particular, burning rate, in the case of an SI engine, is computed by a turbulent combustion model, based on a fractal description of the flame front area [6, 7, 8, 9, 10, 18]. According to this approach, an initially smooth flame surface of spherical — the laminar flame A_L — is wrinkled by the presence of turbulent eddies of different length scales. The interactions between the turbulent flow field and the flame determine the development of a turbulent flame, A_T, which propagates at the laminar flame speed S_L. The combustion model has been recently extended to the analysis of twin-spark engines too [19]. The interested reader will find model and application details in the references.

2 Parallelization of Kiva–3V

The parallelization of KIVA has been carried out according to the strategies laid out in [11] and later in [3], based on the parallel linear algebra toolkit of [12]. The basic idea is to apply domain decomposition techniques to partition and balance the computations among processors in the parallel machine, according to the "owner computes" paradigm. The toolkit employed provides the facilities to build and maintain a representation of the discretization mesh that can be used not only to support the linear system solvers, but also to handle other parts of the computations, such as the explicit flux phase and the spray dynamics, thanks to the fact that the adjacency matrix of the discretization graph is isomorphic to the linear system coefficient matrix.

While modifying the main strcture of the work we also included more modern solvers for the innermost lineaer systems. The original KIVA code employs the Conjugate Residual method, one member of the Krylov subspace projection family of methods [15,20,13,21]; the field of iterative solvers has witnessed many advances in recent years, with many new methods for non symmetric matrices, and more effective preconditioners, i.e. transformations of the systems aimed at improving the convergence speed. By employing the PSBLAS software we were able to include better solvers; moreover, the solver and preconditioner choices are now parametrized, so that the choices made may be revised as further software developments become availbale.

For further details please refer to [3].

2.1 Moving Meshes

Changes in Local Node Numbering. Kiva–3V originally stored all the hidden nodes of the full grid before *ifirst*, which coincided with the first active cell. We now have to take into account the fact that each process has its own local numbering, mapping one section of the global address space into a local contiguous space.

General Improvements. In the original version of Kiva, the piston was considered topologically flat, while a bowl below the piston surface was marked and treated separately. In the current code the the piston surface may have any required topological profile; hence, the valves can now move through any fluid cell in the squish zone, even below the original flat piston.

In the original code on only the head of the valve was moved; this could lead to deforming the shape of the valve. We removed this limitation, and now treat the entire valve, including the stem, as a rigid object.

Graph Partitioning. When running in parallel we do apply a domain partitioning technique based on the usage of the Metis graph partitioning tool. To illustrate this let us consider Fig. 1 where we show the wireframe model of a computational mesh with the various partitions color-coded according to the legend shown.

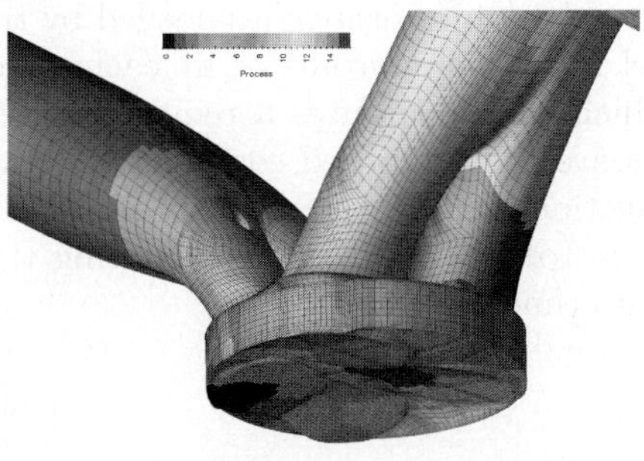

Fig. 1. Domain partitioning on 16 processes

2.2 Data Management Facilities

A major problem in using a parallel application such as ours in a production environment, is chosing the correct strategy to store, manage and retrieve all the data output generated by the simulations.

To avoid serial bottlenecks each process should write on local disk(s) its own part of data, delegating to a post–processor the task of collecting and merging them, when needed. Data from each job has to be uniquely identified, to allow the usage of a database to store simulation data for future reference.

We thus implemented an application environment with the possibility of monitoring and altering the status of a running job retrieving information about a terminated job, managing output data and generating mpeg or XviD video; as a by-product of this activity we also introduced enhanced dump and restart capabilities, taking into account the features of parallel data distribution.

3 Code Coupling

3.1 3D–0D

The boundary condition between 3D and 0D volumes, namely the valve cross section, is implemented as a special kind of velocity boundary, where the velocity distribution imposed at each time step is deduced from a full 3D simulation, while the mean normal velocity is computed in order to respect the mass flow imposed by the 0D.

Hence, a three dimensional simulation, with the whole cylinder and the piston, is initially performed, in order to save the velocities of grid nodes defining the valve cross section, for each valve. These velocity vectors are represented in a cylindrical coordinates system with the vertical axis coincident with the displacement axis of each valve and the origin of angles is chosen as the bottom – left topological corner of each valve.

For each 3D – 0D boundary used, KIVA needs a file produced by the full 3D simulation, a phase shift between the homologue valve lifts in 3D and 0D runs and a 0D model. The files are loaded at the start of each simulation, the velocities are transformed into the local system of coordinates of the corresponding valve and normalized with respect to the average radial component.

Mean values of thermodynamics properties needed by the 0D model are computed, using a parallel recursive subroutine, at each time step and passed to the 0D subroutine, which then computes a required mass flow. The velocities of the boundary nodes are then imposed so that the mean normal velocity on the whole valve cross section satisfies the requested mass flow, while the components of each velocity vector are deduced interpolating the values saved in the reference file, both along time and space.

3.2 3D–1D

At the current development stage, it is possible to link an arbitrary number of 1D elements to an equal number of boundaries of the three dimensional grid, without wondering of the direction of the flow. Obviously, connection between 1D and 3D elements should be set where the flow is supposed to be uniform enough along the section, or the mono dimensional approximation fails.

The coupling strategy is explicit and performed by a special boundary condition for both 3D and 1D code. While passing any thermodynamic property from the three–dimensional part to the mono–dimensional one, its value is integrated along the two internal layers of cells close to the boundary. Both values obtained (one for each layer) is used by the 1D solver as the boundary condition for next time step, reaching a quite accurate description even of rapidly changing conditions within the flow, as shown later on. In the other direction, the value of any property computated on the internal node of the 1D description closest to the link with the 3D section is applied as boundary condition for the next time step of the 3D solver.

The resulting computer program is an embedded application in which the 3D and the 1D parts are driven at the same time and exchange information via the boundary conditions. The integration of the two models, mono and three dimensional, is complete and robust, using the same inputs as the two original programs, having only added the handling of these new kind of links and boundaries.

The result is one single application that, together with the 3D – 0D coupling described above, permits to handle complex and complete problems, permitting of a different level of approximations for different parts, and taking advantage of parallel computing, concerning the time consumptive 3D simulation, in order to considerably reduce the computing time, or to increase the dimension of the three dimensional part of the problem.

4 Results

4.1 Performance of Modified Version of Kiva–3V

In analyzing performance of the implicit solvers in KIVA we had to search for the best compromise between preconditioning and solution methods in the application context. We settled on the Bi-CGSTAB method for all of the linear systems; the critical solver is that for the pressure correction equation, where we employed a block ILU preconditioner, i.e. an incomplete factorization based on the local part of A. The BiCGSTAB method always converged, usually in less than 10 iterations, and practically never in more than 30 iterations, whereas the original solver quite often would not converge at all within the specified maximum number of iterations. Moreover, the termination of the SIMPLE loop is based on the amount of the pressure correction; thus a better solution for the pressure equation reduces the number of SIMPLE iterations needed.

4.2 3D–1D Coupling

1D–3D Pipe with Variable Boundary Conditions. The first test for the 1D–3D coupling method described above is a flow through a pipe, composed by three segments (figure 2). The diameter doesn't vary along the whole pipe and the connection with the volumes at the ends of the pipe are supposed to be lossless. Let besides the duct be adiabatic and without friction.

The pipe can be represented both with all 1D segments and with one 3D segment, in the middle or at one end.

For this test conditions we measured both total and static pressure and temperature.

Let chose the rightmost boundary condition imposing a variable total pressure given by the relation $p_0 = 1.15 + 0.3 \cdot sin\left(\frac{1}{T} \cdot t\right)$.

The results obtained are shown in figures 4 – 5, in terms of pressure, temperature and velocity in the first two sections. It's important to notice that, during the simulation, the velocity vector changes direction, hence the mass flow through the boundaries both with the external volumes and between 1D and 3D sections changes.

Fig. 2. Test case 1D-3D-1D

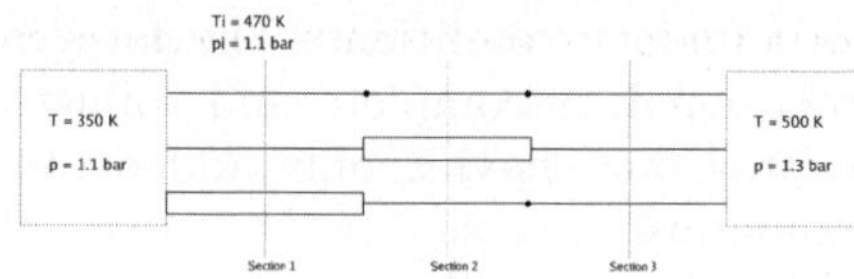

Fig. 3. Test case 1D-3D-1D with boundary conditions

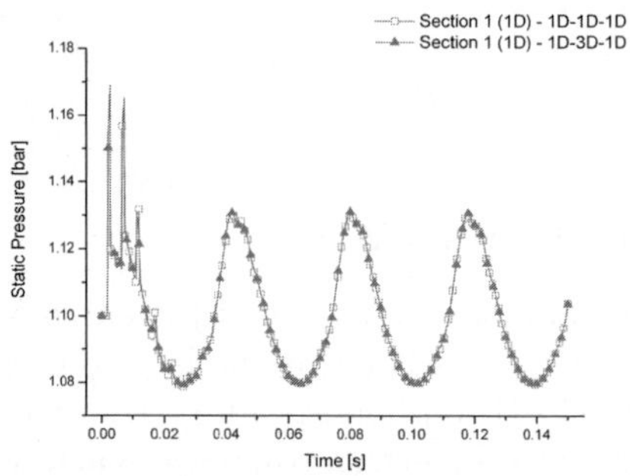

Fig. 4. Static pressure in section 1

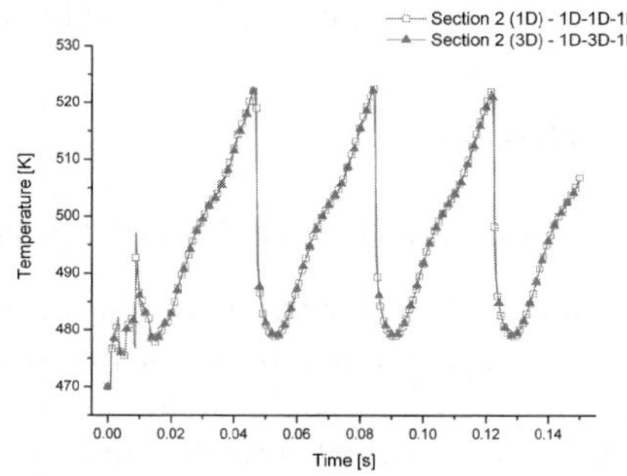

Fig. 5. Temperature in section 2

4.3 3D–0D Coupling

In order to test the new boundary created for coupling Kiva with a 0D model, a flow trough an inlet duct is examined, both with and without a cylinder. The comparison is made paying attention not to have reverse flow from the cylinder to the intake duct during the full 3D simulation, in order to avoid the description of the thermodynamic properties of the 0D cylinder connected to the duct.

As explained above, a full 3D simulation is required to obtain a map of velocities through the valve cross section. The reference run is driven during the end of the intake process of the four-valves engine represented in figure 4.2, namely

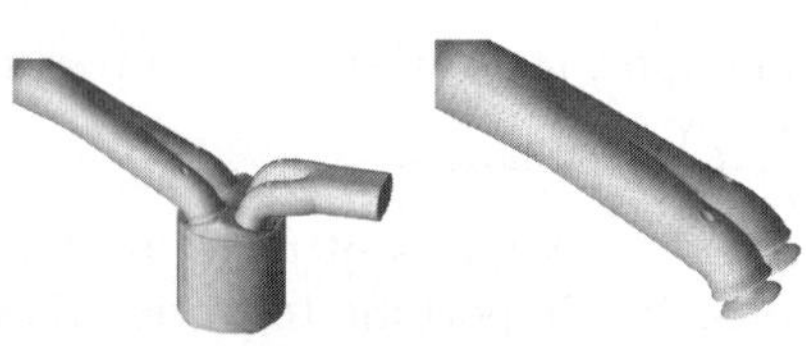

Fig. 6. Full 3D model and 3D intake duct with special velocity boundaries

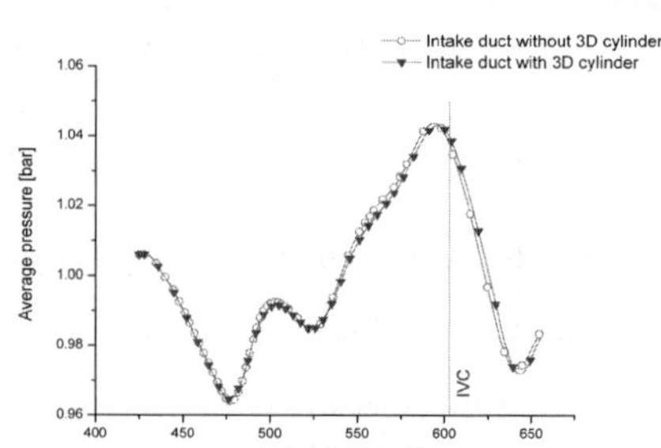

Fig. 7. Average pressure in intake duct

around the bottom-center crank position, till 50 degrees after the inlet valve close crank angle.

The same intake duct, without cylinder and exhaust duct, is then taken into consideration, as shown in figure 6. The boundary conditions imposed are the same as for the full 3D simulation at the inlet of the duct, but the new special boundary of velocity at the valves cross section, where the vector field is obtained from the first run.

A comparison between the two simulations, in term of average pressure in the intake duct during the simulation is shown in Figure 7. Notice that for this simulation the time of the combined 3D-1D simulation was 24 min, whereas the full 3D simulation took 189 min

4.4 Full Engine Test

1D Engine with 3D Airbox. In order to validate the code against a more exhaustive test case, we have decided to simulate the whole automotive spark–ignition engine represented in figure 8.

This is a 4 cylinders, 16 valve engine, with EGR valve in the exhaust line. We have compared the results obtained with a pure 1D–0D modeling with those obtained by substituting the 0D airbox (*P4* in the scheme) with a 3D model, as represented in figure 9. The airbox has six ports, each one connected to a 1D pipe. Notice that there is no practical limit to the number of ports of the 3D mesh, connected with 1D elements or having their own boundary conditions, and that we can connect different 3D elements by one or more 1D lines. While in a full 1D–0D simulation the cylinders have exactly the same behaviour, in a 1D simulation with a 3D airbox we expect to have different values of variables in the four cylinders, depending on the airbox geometry.

As shown in figure 10, the amount of trapped mass in the four cylinders is indeed varying, while remaining close to the value returned by the 1D model.

Parallel Performance. Our parallel performance and scalability test were conducted on the Linux cluster available at the computing center of CASPUR

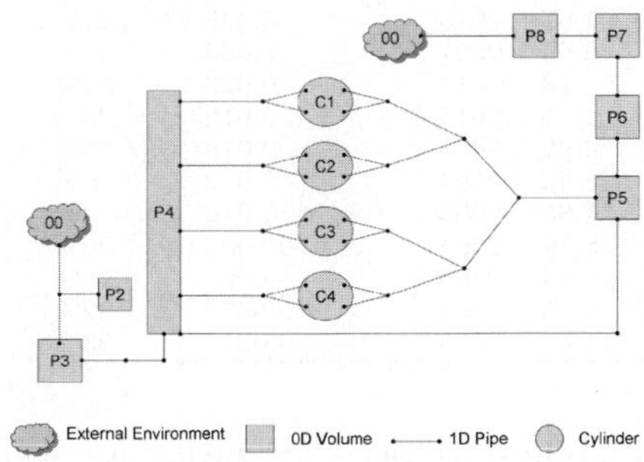

Fig. 8. 1D Scheme of test engine

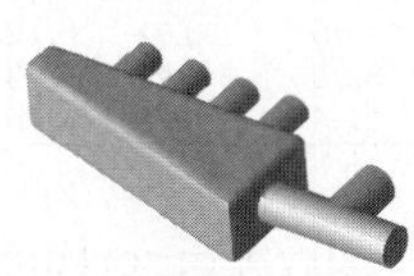

Fig. 9. 3D airbox

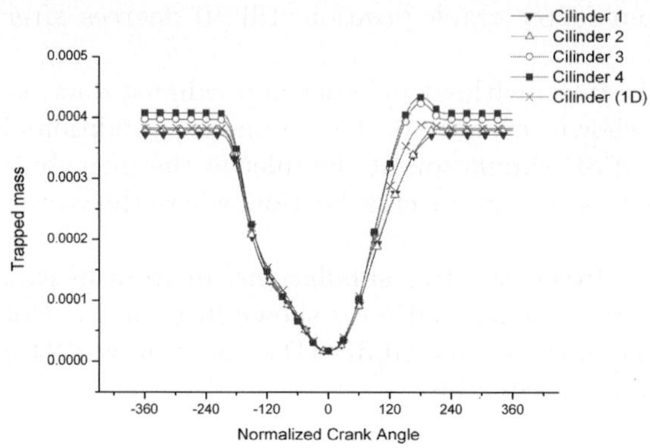

Fig. 10. Trapped mass at 3000 rpm

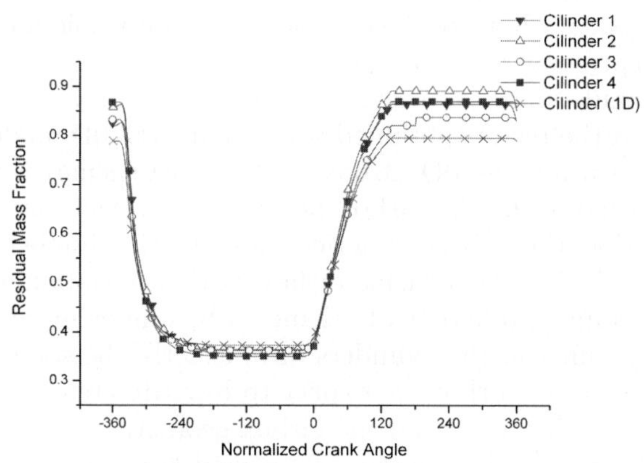

Fig. 11. Burned mass fraction at 3000 rpm

Table 1. Airbox, ncells=281863

NP	Total time	Time 1D	Time 3D	Speedup	Speedup 3D	Cycles	Time/cycle (m)	Efficiency
1	126.76	0.73	126.03	1	1	2016	0.063	100%
2	77.26	0.72	76.54	1.64	1.65	2016	0.038	82%
4	76.46	1.1	75.36	1.66	1.67	2015	0.038	41%
6	61.16	0.82	60.34	2.07	2.09	2007	0.030	35%
8	46.15	0.81	45.34	2.75	2.78	2018	0.023	34%
12	36.31	0.84	35.47	3.49	3.55	2015	0.018	29%
16	26.44	0.85	25.59	4.79	4.92	2006	0.013	30%
20	24.02	0.77	23.25	5.28	5.42	2004	0.012	26%
24	19.24	0.83	18.41	6.59	6.85	2003	0.010	27%
30	16.36	0.87	15.49	7.75	8.14	2004	0.0082	26%
36	13.46	0.78	12.68	9.42	9.94	2003	0.0067	26%
42	13.96	0.77	13.19	9.08	9.55	2004	0.0070	22%
48	11.4	0.79	10.61	11.12	11.88	2005	0.0057	23%

(Inter-University Consortium for the Application of Super-Computing for Universities and Research) in Rome, comprising dual-processor nodes based on the AMD Opteron 250 with a 4 GB RAM and a 1024 KB cache memory. The nodes are connected via InfiniBand Silverstorm InfiniHost III Ex HCA; this network interface has a user level latency of approx. 5 μsec and a measured sustained bandwidth of 960 MB/sec.

In a pure 3D setting with combustion our code is capable of a sustained parallel efficiency well over 50 %. When coupled with the 1D code and without combustion, due to the characteristics and smaller size of the computational mesh, the efficiency is lower, but still around 25 % with more than 40 processors nodes, as shown in Table 1.

5 Conclusions

We have demonstrated an effective coupling of a general-purpose 3D fluid dynamics code with specialized 0D and 1D codes, providing an effective strategy to attack difficult simulations, giving good performance on parallel machine architectures.

References

1. Amsden, A.A., *KIVA 3: A KIVA with Block Structured Mesh for Complex Geometry,* 1992, Los Alamos National Laboratory report LA 12503-Ms.
2. Amsden, A.A., *KIVA 3-V: A Block Structured KIVA Program for Engine with Vertical or Canted Valves,* 1997, Los Alamos National Laboratory report LA 13313-Ms.
3. G. Bella, A. Buttari, A. De Maio, F. Del Citto, S. Filippone, F. Gasperini *FAST-EVP: an Engine Simulation Tool,* Proceedings of HPCC '05.
4. Courant R., Friederichs K.O., Lewy H., *Uber die partiellendifferenz–leichungen der mathematischen phisik,* Mathematische Annalen 1928, vol. 100, pagg. 32–74. English Translation in *IBM journal* 1967, pagg. 215–234.
5. Bozza F., Cameretti M.C., Tuccillo R., *Numerical Simulation of In–Cylinder Processes and Duct Flow in a Light–Duty Diesel Engine,* Fourth Int. Symp. on Small Diesel Engines, Varsavia, 1996, in "Journal of POLISH CIMAC", Vol. 2, n. 1, pp. 51–66, 1996.
6. Bozza F., Tuccillo R., de Falco D., *A Two–Stroke Engine Model Based on Advanced Simulation of Fundamental Processes,* SAE paper 952139, also in "Design and Emissions of Small Two– and Four–Stroke Engines", SAE SP-1112, pp. 87–98, 1995.
7. Bozza F., Gimelli A., Senatore A., Caraceni A., *A Theoretical Comparison of Various VVA Systems for Performance and Emission Improvements of SI Engines,* SAE Paper 2001-01-0671, 2001, also in "Variable Valve Actuation 2001", SAE SP-1599, ISBN 0-7680-0746-1.
8. Bozza F., Cardone M., Gimelli A., Senatore A., Tuccillo R., *A Methodology for the Definition of Optimal Control Strategies of a VVT–Equipped SI Engine* proc. of. 5th Int. Conf. ICE2001 Internal Combustion Engines: Experiments and Modeling, September 2001.
9. Bozza F., Gimelli A., Tuccillo R., *The Control of a VVA Equipped SI–Engine Operation by Means of 1D Simulation and Mathematical Optimization,* SAE Paper 2002-01-1107, 2002 SAE World Congress, Detroit, March 2002, also in "Variable Valve Actuation 2002", SAE SP-1692, ISBN 0-7680-0960-X. Published also on SAE 2002 Transactions - Journal of Engines.

10. Bozza F., Torella E., *The Employment of a 1D Simulation Model for A/F Ratio Control in a VVT Engine*, SAE Paper 2003-01-0027, 2003 SAE World Congress, Detroit, March 2003, also in “Variable Valve Actuation 2003”.
11. S. Filippone, P. D’Ambra, M. Colajanni: *Using a Parallel Library of Sparse Linear Algebra in a Fluid Dynamics Applications Code on Linux Clusters.* Parallel Computing - Advances & Current Issues, G. Joubert, A. Murli, F. Peters, M. Vanneschi eds., Imperial College Press Pub. (2002), pp. 441–448.
12. S. Filippone and M. Colajanni. *PSBLAS: A library for parallel linear algebra computation on sparse matrices.* ACM Trans. Math. Softw., 26(4):527–550, December 2000.
13. A. Greenbaum: Iterative Methods for Solving Linear Systems, SIAM, 1997.
14. Harten A., *On a Class of High Resolution Total Variation Stable Finite Difference Schemes*, Jrl of Computational Physics, Vol. 21, 21–23, 1984.
15. C. T. Kelley: Iterative Methods for Linear and Nonlinear Equations, SIAM, 1995.
16. Manna M., *A Three Dimensional High Resolution Compressible Flow Solver*, PhD Thesis, Catholic Univ. of Louvain - Von Karman Institute for Fluid Dynamics, 1992, also in TN 180, VKI, 1992.
17. Miller D.S., *Internal Flow Systems*, Second Edition, BHR Group Limited, 1990.
18. Matthews R.D., Chin Y.W., *A Fractal–Based SI Engine Model: Comparisons of Predictions with Experimental Data*, SAE Paper 910075, 1991.
19. Bozza F., Gimelli A., Siano D., Torella E. Mastrangelo G., *A Quasi–Dimensional Three–Zone Model for Performance and Combustion Noise Evaluation of a Twin–Spark High–EGR Engine* SAE Paper 2004-01-0619, SAE World Congress, Detroit, March 2004, also in “Modeling of Spark–Ignition Engines”, SAE SP-1830, ISBN 0-7680-1366-6, pp. 63–73.
20. Y. Saad, Iterative Methods for Sparse Linear Systems, PWS Pub., Boston, 1996.
21. R. Barrett, M. Berry, T. Chan, J. Demmel, J. Donat, J. Dongarra, V. Eijkhout, R. Pozo, C. Romine, and H. van der Vorst. *Templates for the solution of linear systems.* SIAM, 1993.

A Distributed, Parallel System for Large-Scale Structure Recognition in Gene Expression Data

Jens Ernst

Lehrstuhl für Effiziente Algorithmen,
Institut für Informatik,
Technische Universität München
ernstj@in.tum.de

Abstract. Due to the development of very high-throughput lab technology, known as *DNA microarrays*, it has become feasible for scientists to monitor the transcriptional activity of all known genes in many living organisms. Such assays are typically conducted repeatedly, along a timecourse or across a series of predefined experimental conditions, yielding a set of *expression profiles*. Arranging these into subsets, based on their pair-wise similarity, is known as *clustering*. Clusters of genes exhibiting similar expression behavior are often related in a biologically meaningful way, which is at the center of interest to research in functional genomics.

We present a distributed, parallel system based on spectral graph theory and numerical linear algebra that can solve this problem for datasets generated by the latest generation of microarrays, and at high levels of experimental noise. It allows us to process hundreds of thousands of expression profiles, thereby vastly increasing the current size limit for unsupervized clustering with full similarity information.

Keywords: computational biology, structure recognition, gene expression analysis, unsupervized clustering, spectral graph theory.

1 Introduction

Computational Biology has provided a great wealth of complex and challenging algorithmic problems for researchers in the field of combinatorial optimization to work on. Moreover, the development of high-throuput lab technology has brought about a massive increase in the rate at which experimental data is generated and has to be processed by suitable algorithms. This not only calls for a great deal of effort in optimizing these algorithms for efficiency, but also presents a natural motivation for exploiting their parallelism and for distributing work across a network of computers. In this paper we consider *unsupervized clustering* of gene expression profiles as a particular case of a structure recognition problem where input data is generated at an industrial, ever increasing rate and at great cost, while efficient processing is indispensible to handling an otherwise unmanageable amount of information.

M. Gerndt and D. Kranzlmüller (Eds.): HPCC 2006, LNCS 4208, pp. 21–30, 2006.

The DNA of all organisms contains a number of relatively short regions which code for protein, the main chemical ingredient to life. Genes are DNA sequences composed of such regions. The process by which the information encoded in a gene is transcribed and protein molecules are produced by cellular machinery is known as *gene expression.* This process is highly dynamic and very complex in that there exist networks of genes that mutually promote or suppress each other's expression, particularly in response to environmental changes affecting the cell. While it is possible to sequence entire genomes and to predict the location of genes at a reasonably high level of reliability, many questions on gene function and interaction are still unanswered. A great deal of research is directed towards understanding gene expression on a gobal level, taking into account many – perhaps all – genes of a given organism in parallel. This gives rise to the term *functional genomics.* The technology of *DNA microarrays* [5] has made it possible to observe the levels of expression activity of a very large number of genes in a single experiment and has become a standard research tool. In simplified terms, short but sufficiently representative nucleic acid subsequences (*probes*) of genes are physically spotted or photolithographically synthesized on a glass substrate. Then, in a hybridization reaction, these probes attract and bind to fluorescently labeled counterparts of their respective sequences from, say, a tissue sample. Later, fluorescence scanning of the array allows quantitative inferences on the activity of each gene on the microarray in terms of its level of transcription, and hence its expression. Often, but not always, probes correspond to genes.

Typically, a series of multiple experiments using identical arrays is performed under different experimental conditions or along a timecourse, yielding a vector of expression scores for each probe. Vectors of this type are known as *expression profiles.* Let n be the number of probes and let m be the number of array experiments. Then the profiles can be written in the form of an $n \times m$ matrix of expression values. Typical values for n range up to $4 \cdot 10^6$, whereas $m \leq 50$ is usually a consequence of array cost and other practical limitations. The goal of clustering in this context is to partition the set of probes, represented by their corresponding expression profiles, into subsets (or *clusters*) in such a way that members of the same cluster exhibit highly similar expression behavior (or *coexpression*), while similarity between clusters should be low. To formally capture this, we assume to have a similarity measure $s : \mathbf{R}^m \times \mathbf{R}^m \longrightarrow [0, 1]$ which maps each pair of expression profiles to a similarity score. A score of 1 indicates perfect similarity. This function is symmetric but not necessarily metric. In practice, Pearson correlation and measures based on L_2 distances are commonly applied.

The input for the clustering problem is hence a symmetric $n \times n$ *similarity matrix* A containing values between 0 and 1 and can be thought of as the edge weight matrix of an undirected graph. The diagonal elements all have a value of 1. The problem hence consists in partitioning the index set $\{1, 2, \ldots, n\}$ into subsets representing the clusters. But the partitioning problem is complicated by the fact that the number of clusters is not known a priori and that microarray technology is extremely prone to experimental error, leading to incorrect expression values, which translates into false positive and false negative similarity scores in A.

Extensive work has been done on clustering algorithms for gene expression data since microarray technology was introduced. For a comprehensive survey of the most popular methods, the reader is referred to [3,6,8]. The main limitation shared by all existing algorithms is on the maximum feasible value of n. The author is not aware of any current commercial or academic software system that can handle more than some tens of thousands of probes. Our system is designed to improve this limit on n by an order of magnitude. We shall restate the clustering problem more formally in the following Section. In Section 3 we briefly derive the sequential *SR-algorithm* ("Spectral Reduction") which is well suited for this clustering problem. We then propose a set of techniques for parallelizing and distributing this algorithm in Section 4. The final Section presents experimental results to demonstrate the quality and performance of the method.

2 The Clustering Problem

The following graph theoretical framework shall henceforth be used for modeling the input data to be processed by clustering algorithms.

Definition 1. *Given a set $S = \{X_1, \ldots, X_n\}$ of expression profiles $X_i \in \mathbf{R}^m$ and a simliarity measure $s : \mathbf{R}^m \times \mathbf{R}^m \to \mathbf{R}$, the associated* similarity graph *for S is a complete, undirected graph $G = (V, E, w)$ with self-loops, where $|V| = n$ and $E = \big\{\{v, v'\} : v, v' \in V\big\}$. For $1 \le i, j \le n$, each vertex v_i is bijectively associated with one expression profile X_i and each edge $\{v_i, v_j\}$ is weighted with the score of the similarity $w(v_i, v_j) := w(v_j, v_i) := s(X_i, X_j)$ between X_i and X_j.*

Within this formal framework we now define a data model for our clustering problem. We assume that an unknown cluster structure exists in the input data and is perturbed by some random noise, modeling experimental data error.

Definition 2. *A* cluster graph *is an edge-weighted, complete, undirected graph $G_0 = (V, E, w_0)$ with self-loops whose vertex set V can be partitioned into K clusters C_k such that $w_0(v_i, v_j) = 1$ for all edges $\{v_i, v_j\}$ within the same cluster C_k, $1 \le k \le K$, and $w_0(v_i, v_j) = 0$ for edges $\{v_i, v_j\}$ connecting different clusters. Let p^{int} and p^{ext} be two different distributions on the interval $[0, 1]$ with respective expectations μ^{int} and μ^{ext}. A (p^{int}, p^{ext})–*perturbation, *applied to G_0, yields a randomly weighted* perturbed cluster graph *$G = (V, E, w)$ where each edge $\{v_i, v_j\}$ is independently assigned a weight $w(v_i, v_j) = w(v_j, v_i)$ with respect to distribution p^{int}, if $w_0(v_i, v_j) = 1$ and with respect to p^{ext}, if $w_0(v_i, v_j) = 0$.*

This allows us to formally state the clustering problem: Let $G = (V, E, w)$ be the result of some (p^{int}, p^{ext})-perturbation with $\mu^{int} \neq \mu^{ext}$, applied to a fixed but unknown cluster graph $G_0 = (V, E, w_0)$ whose vertex set consists of the clusters $C_1, \ldots, C_K$. Let A_0 be the unknown edge weight matrix of G_0. Given G or its edge weight matrix A, our task is to reconstruct G_0 by partitioning V into the K original clusters C_k. The value of K is not necessarily known. It is assumed that each cluster C_k is of size $c_k n$ for some constant $c_k > 0$. The vector $(c_1, \ldots, c_K)$ is called *cluster structure* of G. Also, we assume that $\{c_1, \ldots, c_K\} = \{c_1, \ldots, c_r\}$ where $c_1, \ldots, c_r$ are each unique and m_i is the multiplicity of c_i for $1 \le i \le r$.

3 Spectral Properties of Perturbed Cluster Graphs

In this Section we survey some fundamental mathematical properties of the input similarity matrix A, which will help us in deriving an algorithm for the above clustering problem. To this end, we first consider the unperturbed matrix A_0 and then examine the effect of random noise.

Lemma 1. *The spectrum of A_0 consists of 0 and the set $\{c_1 n, c_2 n, \ldots, c_r n\}$. Each eigenvalue $c_i n$ has multiplicity m_i, and its associated eigenspace is spanned orthogonally by characteristic vectors of the m_i clusters of size $c_i n$.*

By multiplying A_0 with the characteristic vectors of the individual clusters, one can immediately verify that the characteristic vectors belonging to the K individual clusters are indeed eigenvectors, with the respective clusters sizes $c_i n$ as associated eigenvalues. Orthonormality is also obvious. As Rank $A_0 = K$, it follows that no additional non-zero eigenvalues exist.

Note that this immediately suggests an algorithm for identifying clusters. A simple rotation of any set of dominant eigenvectors of A_0 yields a set of characteristic vectors indicating cluster membership. A different way of thinking of the situation is to consider in a column-wise fashion a matrix $(K \times n)$-matrix Z_0 whose rows contain a set of mutually orthonormal vectors spanning the dominant eigenspaces. Each column, representing a vertex, corresponds to a point in $\mathbf{R}^K$, and two vertices belong to the same cluster if and only if they correspond to the same point. So instead of solving the trivial clustering problem for G_0 by looking for connected components of edges with weight 1, one can obtain the desired partition by examining the dominant eigenvectors of A_0.

Next, we can ask about the effects of (p^{int}, p^{ext})-perturbation which transforms A_0 into A. It is a well-known fact of matrix theory that the spectrum of a symmetric matrix is relatively insensitive to component-wise perturbation – in fact, eigenvalues are said to be *perfectly conditioned.* We will show that they are stable enough to still reflect the number of clusters, K. The eigenspaces, on the other hand, will be shown to be sufficiently stable to allow the reconstruction of the unperturbed cluster structure with high probability. The following theorem summarizes these results. Due to page restrictions, we omit the proof here and refer the reader to [1] where we provide this proof in great detail.

Theorem 1. *With probability $1 - o(1)$, matrix A has the following properties: The spectrum of A consists of K dominant eigenvalues of magnitude $\Theta(n)$ and $n-K$ eigenvalues of magnitude $O(n^{\frac{1}{2}})$, counting multiplicity. The eigenspaces associated with the dominant eigenvalues are spanned by vectors which are constant within the index sets associated with the individual clusters, up to a component-wise variation of $o(n^{-\frac{1}{2}})$. Let Z be a $(K \times n)$-matrix whose rows contain a set of mutually orthonormal vectors spanning the dominant eigenspaces. Then the columns z_t of Z $(1 \le t \le n)$ constitute n points in $\mathbf{R}^K$, and two columns t, t' satisfy $\|z_t - z_{t'}\|_\infty = o(n^{-\frac{1}{2}})$ if vertices v_t and $v_{t'}$ belong to the same cluster. Otherwise, it holds that $\|z_t - z_{t'}\|_\infty = \Omega(n^{-\frac{1}{2}})$.*

Algorithm 1. SR-algorithm

Input: (p^{int}, p^{ext})*-perturbed cluster graph* $G = (V, E, w)$
Output: *Adjacency matrix of* G_0

(*i*) $A :=$ edge weight matrix of G;
(*ii*) $\gamma := |V|^{2/3}$;
(*iii*) K:=number of eigenvalues $\lambda \geq \gamma$;
(*iv*) $\{x_1, x_2, \ldots, x_K\}$:=orthonormal set of eigenvectors associated with eigenvalues $> \gamma$;
(*v*) $Z := [x_1, x_2, \ldots, x_K]^{\mathrm{T}}$;
(*vi*) Identify the K-clusters defined by the column vectors of Z;
(*vii*) Construct adjacency matrix A_0 based on the K-clusters;

As a consequence, even in the presence of (p^{int}, p^{ext})-perturbation, the number of clusters can be obtained by counting the number of dominant eigenvalues. Any threshold γ satisfying $\gamma = o(n)$ and $\gamma = \omega(n^{\frac{1}{2}})$, can be used to distinguish them from the others. The columns z_t of Z define n points in $\mathbf{R}^K$ which form K groups named K*-clusters.* For sufficiently large n and with probability $1 - o(1)$, the K-clusters are disjoint and grow arbitrarily tight as n grows. The task of assigning each vertex to its cluster can be accomplished by identifying the K-clusters in $\mathbf{R}^K$. This constitutes a Euclidean clustering problem and can be solved by applying the K-means algorithm, a linear-time farthest-first method or thresholding. This leads us to the *Spectral Reduction* (*SR*) Algorithm 1.

A great advantage of the SR-algorithm over direct Euclidean clustering on the expression vectors is that it does not rely on a particular choice of the similarity measure. Regardless of how s is defined, the structure recognition problem is *reduced* to a simple Euclidean clustering problem, and the correct number of clusters is obtained as a by-product. For instance, one could envision designing a custom similarity measure such as the absolute value of Pearson Correlation (in order to define both, exact correlation and exact anti-correlation as maximal relatedness between expression profiles). Moreover, similarity matrix A need not even stem from a gene expression data set. Even if the biological entities whose similarity is described by the matrix do not have a vector representation at all, the clustering problem is transformed into a Euclidean one, and according vector representations in K-space are generated for these entities.

The only remaining issue to be addressed is how to compute an orthonormal basis of the dominant eigenspaces. It is not necessary to compute the entire eigen decomposition of matrix A as only a few dominant eigenvectors are needed. Instead we resort to an iterative numerical method based on Krylov subspaces and the well-known Rayleigh-Ritz procedure. Algorithm 2 gives a sketch of the method. Here we apply it as a black box and refer the reader to [2] for a detailed treatment of the theory behind it and [4] for implementation hints.

Algorithm 2. Compute eigenpairs

Input: $A \in \mathbf{R}^{n \times n}$
Output: *Approximate dominant eigenpairs* $(\widetilde{\lambda}_i, \widetilde{x}_i)$

(*i*) Choose $v_1 \in \mathbf{R}^n$ with $\|v_1\| = 1$;
(*ii*) $\beta_1 := 0$; $v_0 := \mathbf{0}$;
(*iii*) **for** $j = 1, 2, \ldots$ **do**
$w_j := Av_j - \beta_j v_{j-1}$;
$\alpha_j := \langle w_j, v_j \rangle$;
$w_j := w_j - \alpha_j v_j$;
$\beta_{j+1} := \|w_j\|$;
$v_{j+1} := \frac{1}{\beta_{j+1}} w_j$;
Compute $(\widetilde{\lambda}_i^{(j)}, \widetilde{x}_i^{(j)})$, $1 \le i \le j$ by the Rayleigh-Ritz procedure
if $|\widetilde{\lambda}_j^{(j)}| < n^{\frac{2}{3}}$ **then** stop **fi**;
od

4 The Parallel and Distributed SR-Algorithm

To identify the parts of the SR-Algorithm that can benefit the most from parallelization, let us first examine its complexity, assuming that similarity matrix A is given as input. Identifying the K tightly concentrated clusters of the columns of Z in Euclidean K-space requires no more than a simple linear-time computation, and storing matrix Z requires only linear space. While this could be parallelized in a rather straightforward way, the gain would be negligible. Computing a set of dominant eigenvalues of A, however, costs $\Omega(n^2)$ time and space. This severely limits the maximum feasible value of n. In the scenario of a typical uniprocessor with 4GB of RAM and without use secondary storage, we find this limit to be around $n \approx 40,000$. The processing time in this case is still within the range of minutes. As an aside, a limitation of n to roughly $40,000$ is also seen in all other academic and commercial clustering systems that the author is aware of. This leads us to focus our efforts on parallelizing Algorithm 2.

Notice that all quadratic-time operations within this procedure are applications of matrix A as an operator to given column vectors. Therefore, the best speedup can be expected from parallelizing all matrix-vector multiplications. Notice further that A is not involved in any other operations. This allows us to distribute A among multiple machines (and their RAM), thereby increasing the maximum permissible n. As a third essential observation, note that for all practical purposes, the number of iterations of Algorithm 2, and hence the overall number of matrix-vector multiplications, is a very small number $<< n$.[1]

[1] In [1] we show that the number of iterations is bounded by $O(\log n)$, and we give perturbation theoretical arguments suggesting that it is even a constant independent of n, although a formal proof of this conjecture is not yet completed.

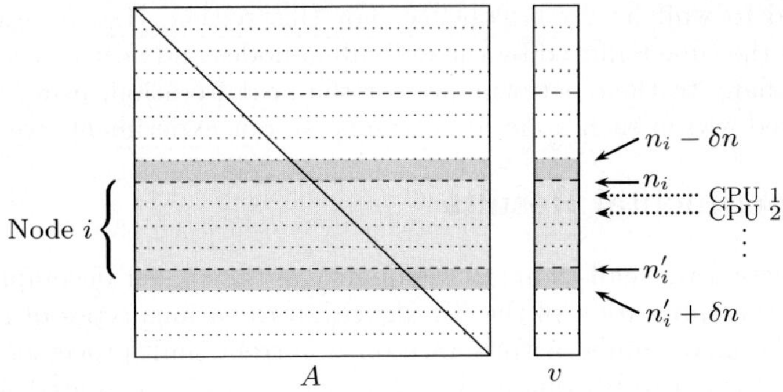

Fig. 1. Decomposition of matrix A and vector v for parallel multiplication

This last observation means that storing A and multiplying A with a series of vectors can be performed on a distributed-memory architecture consisting of machines that only communicate via a network protocol: The resulting amount of communication is of magnitude $O(n)$ and therefore not a serious limitation.

We choose our target architecture to be a general network of multiprocessor machines (*nodes*). The individual machines in the network communicate by generic message passing, for instance using the standard TCP/IP protocol. The processors within one node communicate via shared memory. This covers most of the architectures commonly used for scientific computation today, including the special cases of networks of uniprocessors and single multiprocessors.

One node is designated to play the part of a *coordinator* (or master). The gene expression profiles are distributed by the coordinator among all other machines which compute horizontal bands of the similarity matrix A to be held in local storage. The number of rows (indices block $[n_i, \ldots, n'_i]$) that each node i should store is determined prior to the distribution process by a static load balancing step in which the RAM capacity and the node's overall performance (number of inner products of length-n-vectors per millisecond) are measured. The index range associated with each machine is chosen proportional to the measured performance. In addition to the core index block, δn extra rows preceding it and δn rows following it are also stored by each node for subsequent dynamic load balancing (see Figure 1). All work to be performed by an individual node is evenly distributed among all the processors of this node. This requires no additional synchronization because computations on different rows of A are independent.

After all matrix bands have been computed, the coordinator generates the first vector v to which A should be applied (see Algorithm 2) and distributes it to all other nodes. The nodes perform the inner products required for generating their associated index block of Av. Again, this work is shared within each node by all the processors that communicate via shared memory. Then the coordinator collects the result blocks from each node, assembles and uses Av to finish the current iteration of Algorithm 2. Note that this presents a synchronization point for all nodes. Should the load on an individual node change significantly during this process (due to use in a multi user environment), the other nodes may be caused to wait as a consequence. For this reason, the coordinator process measures the time required by the individual nodes and adjusts the index ranges proportionally to their performance for the next iteration, using the $2\delta n$ extra rows stored within each node, if necessary. In our experiments we use $\delta = 0.05$.

5 Experimental Results

In this Section we shall examine the quality of the cluster decomposition resulting from the application of the SR-algorithm to various types of input data, as well as the algorithm's performance on a current multiprocessor architecture. We assess the clustering quality based on synthetically generated similarity data consisting of some known, planted structure which was obscured by adding significant amounts of noise. This data complies with the data model introduced in Section 2. A cluster structure $(c_1, c_2, \ldots, c_K)$ is imposed, resulting in clusters of respective sizes $c_1 n, c_2 n, \ldots, c_K n$. Then, perturbation is modeled by the two distributions p^{int} and p^{ext}. We choose binary similarity scores and Bernoulli distributions, causing false positive similarity with some probability α and false negative scores with probability β. This is referred to as *(α, β)-perturbation.* After the clustering process, we quantify the quality of the resulting partition, using the *Matching coefficient* and the *Minkowski measure* defined as follows.

Definition 3. *Given the two matrices $A_0, A_1 \in \{0,1\}^{n \times n}$ (where A_0 represents the true solution and A_1 represents the solution to be evaluated), we define*

$$n_{k,l} := \left|\{(i,j) : 1 \le i,j \le n, [A_0]_{i,j} = k, [A_1]_{i,j} = l\}\right|$$

for $k, l \in \{0,1\}$. Based on these quantities, we define the following measures:

(i) Minkowski Measure $M_1 := \sqrt{\frac{n_{0,1}+n_{1,0}}{n_{1,1}+n_{1,0}}}$

(ii) Matching Coefficient $M_2 := \frac{n_{0,0}+n_{1,1}}{n_{0,0}+n_{0,1}+n_{1,0}+n_{1,1}}$

Note that $n_{1,0}$ is the number of *false negatives* in the solution, $n_{0,1}$ is the number of *false positives*, whereas $n_{0,0}$ and $n_{1,1}$ are the true negatives and positives, respectively. In the ideal case where $A_0 = A_1$, the measures given in the above definition have the respective values of 0 and 1. Only the Minkowski Measure can potentially assume values greater than 1.

The results of our experiments can be seen in Figures 2 and 3. For growing values of n, the situation rapidly improves due to tighter concentrations in the Chernoff-type random variables which are the key to the correctness of Theorem 1. In this sense the choice of $n = 1{,}000$ presents a worst-case scenario.

To measure the performance of our implementation, we used a machine consisting of 32 *nodes*, each containing four AMD Opteron 850 processors at 2.4GHz with 8GB of RAM running Linux. The nodes are interconnected by Gigabit Ethernet. For this analysis we rely on a data set generated for the analysis of cigarette smoking-induced changes in bronchial epithelia, and reversibility of effects when smoking is discontinued. This data set may provide insight to molecular events leading to chronic obstructive pulmonary disease (COPD) and lung cancer [7].

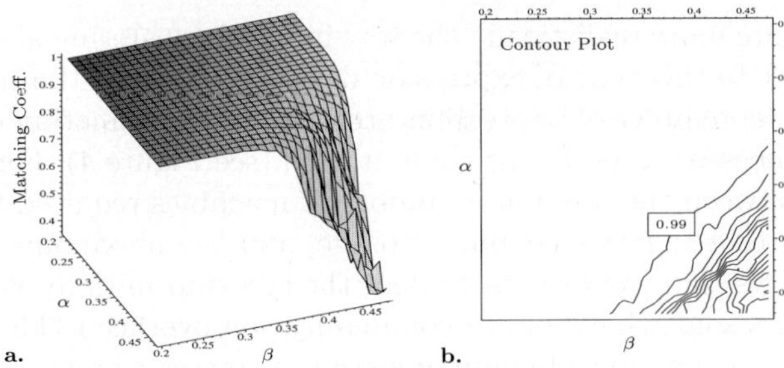

Fig. 2. a. The Matching Coefficient as a function of α and β for an (α, β)-perturbed cluster graph with relative cluster sizes $(0.1, 0.2, 0.3, 0.4)$ and $n = 1000$. **b.** Contour plot of the Matching Coefficient, showing parameter pairs with equal Matching scores.

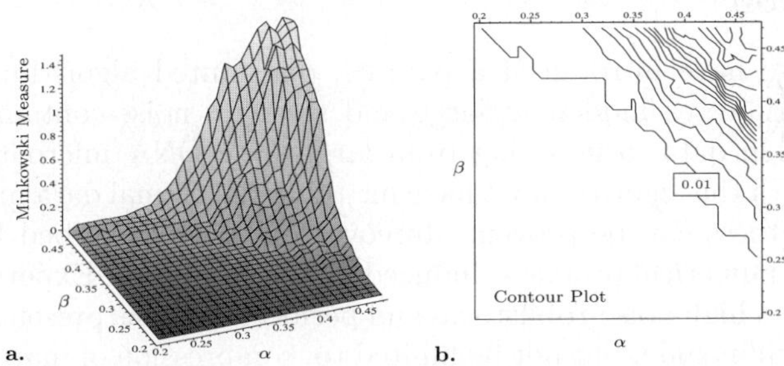

Fig. 3. a. Minkowski Measure as a function of α and β for an (α, β)-perturbed cluster graph with relative cluster sizes $(0.1, 0.2, 0.3, 0.4)$ and $n = 1000$. **b.** Contour plot of **a.**

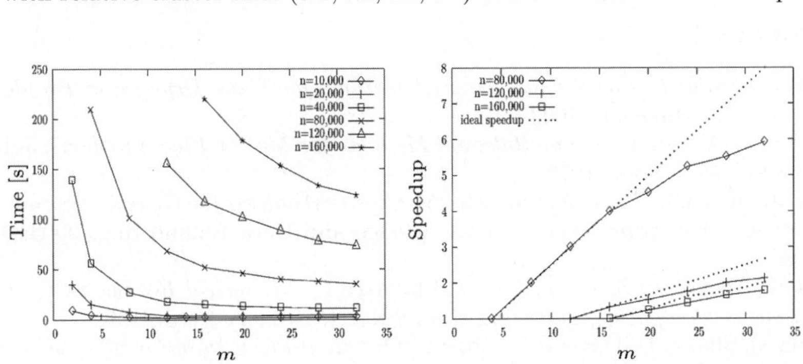

Fig. 4. a. Running times as a function of the number n of gene expression profiles and as a function of the number m of nodes. **b.** Speedup diagram for multiple input sizes

It was created using an Affymetrix GeneChip® Human Genome U133 microarray set HG-U133A. We conduct our cluster analysis in single-probe resolution, meaning that probes belonging to the same transcript of the same gene are considered separately rather than as components of a sum signal. In this experiment we generate 1,000 clusters, bypassing the automatic detection of the number of clusters, to simulate more realistically the way in which clustering algorithms are applied in practice to this type of expression data. We measure the running time as a function of the number of expression profiles and as a function of the number m of multiprocessors used in the computation (see Figure 4). For each value of n, m ranges between the minimum number of machines required to complete the run without resorting to secondary storage, and the maximum value of 32. This analysis shows that, as m is increased, the speedup initially is near-linear and then decreases somewhat, due to communication overhead. This should be expected because the amount of communication increases as a function of m.

Its ability to handle values of n on the order of $k \times 10^5$ with very moderate running time (e.g. $n = 160,000$ in less than 3 minutes) allows the parallel SR-Algorithm to process microarray data of the latest generation, e.g. Affymetrix Exon and Tiling Arrays®, without requiring a priori data filtering.

6 Conclusion

In this paper we have introduced a parallel, distributed algorithm for recovering cluster structures hidden in large and strongly noise-contaminated sets of gene expression data, originating from large-scale DNA microarray experiments. Along with the algorithm, we have introduced a formal data model within which its correctness can be proven. Moreover, we have described how to efficiently solve the numerical problems induced by the algorithm. Experimental evidence for both, its high noise-robustness and performance was presented. Further developments will include, but not be limited to, compression of matrix A to further increase the limit on parameter n. The software will soon be made accessible at `http://wwwmayr.informatik.tu-muenchen.de/personen/ernstj/`.

References

1. Ernst J. *Similarity-Based Clustering Algorithms for Gene Expression Profiles*, Dissertation, TU München, 2003
2. Gourlay A, Watson G. *Computational Methods for Matrix Eigenproblems*, John Wiley & Sons, New York, 1973
3. Daxin Jiang, Chun Tang, Aidong Zhang. *Cluster Analysis for Gene Expression Data: A Survey*, IEEE Transactions on Knowledge and Data Engineering, 2004; 16(11), 1370-86
4. Press W, Teukolsky S, Vetterling W, Flannery B. *Numerical Recipes in C: The Art of Scientific Computing*, 2nd edn., Cambridge University Press, 1992
5. Schena M, Shalon D, Davis RW, Brown PO. *Quantitative monitoring of gene expression patterns with a complementary DNA microarray*, Science, 1995; 270:467-470
6. Shamir R, Sharan R. *Algorithmic approaches to clustering gene expression data*, Current Topics in Computational Biology, 2002; 269-300
7. Spira A, Beane J, Shah V, Liu G, Schembri F, Yang X, Palma J, Brody JS. *Effects of cigarette smoke on the human airway epithelial cell transcriptome*, Proc Natl Acad Sci US, 2004; 101(27):10143-8
8. Valafar F. *Pattern Recognition Techniques in Microarray Data:A Survey*, Special Issue of Annals of New York, Techniques in Bioinformatics and Medical Informatics, 2002; 980:41-64

Cluster Design in the Earth Sciences Tethys

Jens Oeser, Hans-Peter Bunge, and Marcus Mohr

Department of Earth and Environmental Sciences, Geophysics Section, Ludwigs-Maximilians-University, Munich, Germany
jens.oeser@geophysik.uni-muenchen.de

Abstract. Computational modeling is a powerful tool in the Earth Sciences. In the solid Earth important simulation areas include seismic wave propagation, rupture and fault dynamics in the lithosphere, creep in the mantle, and magneto-hydrodynamic flow linked to magnetic field generation in the core. These problems rank among the most demanding calculations computational physicists can perform today. They exceed the limitations of the largest high-performance computing systems by a factor of ten to one hundred measured both in terms of the demands on capacity and capability of systems. Off-the-shelf high-performance Linux clusters are useful to ease the limitations in capacity computing by exploiting price advantages in mass produced PC hardware. Here we review our experience of building a 128 processor AMD Opteron Gigabit Ethernet Linux cluster. The machine is operated at the scientific department level, targeted directly at large-scale geophysical and tectonic modeling and is funded by the German Ministry of Education and Science and the Free State of Bavaria. We observe an aggregate system performance of 140 GFLOPs out of a theoretical 624 GFLOPs peak.

1 Introduction

The Earth's interior is complex, consisting of three distinct regions nested one inside the other. Starting from the outside, there is first the brittle crust, then the solid mantle, and finally the (mostly) liquid core. As a result of convective and other forcings, all three regions are in motion, albeit on different time scales. On the longest time scale solid state convection (creep) overturns the solid mantle once in about every 100-200 million years [1]. This overturn is the primary means by which our planet rids itself of primordial and radioactive heat. It gives rise to large-scale geologic activity such as plate tectonics and continental drift. Reflecting this importance geophysicists have performed computer-based studies of mantle convection for decades, initially with simple 2D approaches [2], and recently in fully 3D spherical models [3,4,5,6].

On a shorter time scale of perhaps 1 to 1000 years convection of the liquid iron core generates Earth's magnetic field. The field is sustained by a complicated dynamo process that probably operated throughout much of Earth's history. Only recently have geophysicists been able to study dynamo action in sophisticated

M. Gerndt and D. Kranzlmüller (Eds.): HPCC 2006, LNCS 4208, pp. 31–40, 2006.

magneto-hydrodynamic models of the core [7,8]. On an even shorter time scale of hours to seconds both the core and the mantle are traversed by seismic sound waves, and seismologists are now turning to computer models to study seismic wave propagation through our planet [9,10].

Geophysical modeling has benefitted greatly from the advent of large-scale parallel computers. Focusing on the Earth's mantle, computing resources are now sufficient to simulate convection at full vigor in 3D spherical geometry. One current frontier lies in applying a range of data-assimilation techniques to study the geologic history of deep Earth flow [11,12]. Data-assimilation is, of course, a familiar tool in numerical weather prediction. A powerful approach to data-assimilation exists through variational methods, see e.g. [13,14]. In mantle convection this technique requires one to solve a numerical adjoint code and the forward model iteratively to find solutions to a non-linear inverse problem [15]. Unfortunately adjoint modeling comes at a heavy computational price. Weeks to months of dedicated integration time are needed to solve these problems even on some of the most powerful parallel machines currently in use at national compute centers. Such resources are often out of reach for individual research groups and academic departments.

While parallel computing has moved into the mainstream, it is not economically feasible on a department budget to invest in dedicated commercial high-performance parallel machines sufficient to handle, for example, forward and adjoint mantle circulation models with earth-like Rayleigh numbers on the order of 10^9. The spatial resolution in such a simulation needs to be at least 20 km throughout the mantle, involving about 100,000,000 numerical grid points. However, it appears practical to address such problems at the academic department level by building cost-efficient *departmental* supercomputers.

In this paper we report on our approach on building such a system for capacity computing in the Earth sciences. In order to give an impression of the requirements and challenges of cluster computing in this field we consider as an example application the above mentioned problem of mantle convection. We start in Chap. 2 with a brief overview of the governing equations before describing our solution algorithm and its parallelization in Chap. 3. We then report how these requirements influenced the design of our Tethys cluster in Chap. 4 and provide some first benchmarks.

2 A Model for Mantle Convection

Generally speaking mantle convection is a flow process driven by temperature variations. As such it can be described by the time-dependent compressible Navier-Stokes equations that constitute a system of partial differential equations (PDEs) describing the conservation of mass and momentum in combination with an equation for energy conservation, see e.g. [16]. However, several simplifications to this model are possible. The first one is to assume a quasi-static flow field, i.e. the time-derivative is dropped from the momentum equations. One then exploits the small magnitude of the flow velocities to also drop here the non-linear convection terms. Finally one employs the so called Boussinesq approximation, see again e.g. [16], whose main assumption is that density variations may be neglected except for the buoyancy terms. Thus, one arrives at a generalized Stokes problem coupled to a time-dependent energy equation. Using standard notation in which the divergence of a matrix field A is understood as a vector whose k-th entry is the (scalar) divergence of the k-th column of A and $(\operatorname{grad} \boldsymbol{u})_{ij} = \partial u_j / \partial x_i$ this system can be written as:

$$\operatorname{div} \boldsymbol{u} = 0 \tag{1}$$

$$\operatorname{div} \left[\nu \left(\operatorname{grad} \boldsymbol{u} + (\operatorname{grad} \boldsymbol{u})^T\right)\right] - \operatorname{grad} p + \varrho\alpha\,(T - T_0)\,\boldsymbol{g} = 0 \tag{2}$$

$$\varrho\, c_p \left(\frac{\partial T}{\partial t} + \boldsymbol{u} \cdot \operatorname{grad} T\right) - \operatorname{div}\,(\kappa \operatorname{grad} T) - \varrho H = 0 \tag{3}$$

Here the dependent variables are velocity $\boldsymbol{u}$, pressure p and temperature T, α is the coefficient of thermal expansion, c_p the specific heat at constant pressure, H the rate of internal heat production per unit volume, $\boldsymbol{g}$ the gravitational acceleration, κ the thermal diffusivity, ϱ the density and ν the kinematic viscosity of the fluid. Note that we have also dropped the inertial and coriolis term from the momentum equation, because of their small relative amplitude. Their absence distinguishes the slow creeping flow of the mantle (where viscous forces dominate) from other perhaps more familiar geophysical flows such as ocean circulation or the magneto-hydrodynamic flow problem of the core.

3 TERRA: Algorithm and Parallel Issues

In order to solve the coupled non-linear PDE system of mantle convection given in Sect. 2 we use a Finite Element technique. Our numerical modeling code TERRA is a parallel version of the vectorized model introduced in [17]. The code is widely used for mantle convection studies. Supported by NASA's High Performance Computing and Communication *(HPCC)* initiative TERRA demonstrated sustained parallel performance better than 100 GFLOPs (10^{11} floating point operations per second) in 1998 on a 1024 Processor Cray T3E-1200.

In TERRA the mantle is discretized using a computational grid adopted from the atmospheric community. The mesh is based on the regular icosahedron [18] and displays the remarkable property of providing an almost uniform triangulation of the spherical surface. The icosahedral grid allows one to avoid the *pole-problem* of conventional latitude-longitude grids, where narrow wedge-shaped computational cells near the pole impose a severe Courant limitation on the time step. Starting from the icosahedron a Finite Element surface mesh is built recursively by splitting nodal distances in half and inserting new nodes at the midpoints, see e.g. [19] for a detailed description. A graphical representation of the first three steps is given in Fig. 1. Note that each time we repeat this process lateral resolution in the domain is doubled. We can thus achieve arbitrary degrees of mesh refinement. The surface grid is then expanded into 3D by radial

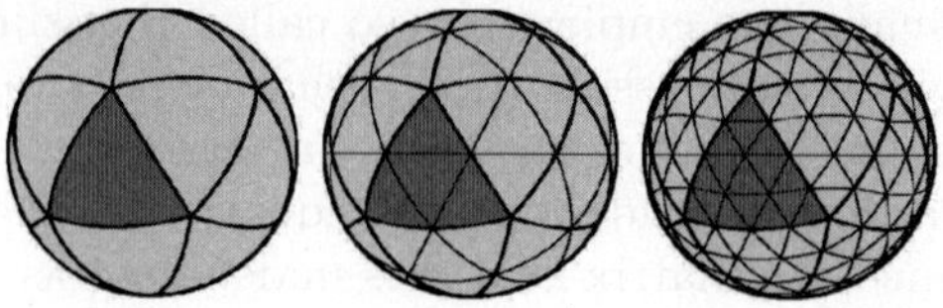

Fig. 1. Three successive mesh-refinements of the icosahedral grid

replication down to the inner surface of the mantle. In this fashion one obtains prismatic elements with spherical triangles on top and bottom. The icosahedral discretization also yields a convenient data structure for our code. Combining pairs of the original twenty icosahedral triangles to form ten diamonds on the sphere, we obtain ten logically rectangular blocks. Topologically the spherical mesh thus appears as one single logically rectangular problem domain.

After converting the problem into its weak formulation and choosing piecewise linear Finite Element ansatz functions the resulting system of equations is solved using the following approach. The temperature T is integrated in time with an explicit second order approach, i.e. the modified Euler method. Thus, for each time step the non-differential part of (3) must be evaluated twice. This requires knowledge of $\boldsymbol{u}$ at the respective simulation time. The latter is determined by solving the Stokes problem (1), (2) for the velocity field $\boldsymbol{u}$ and the pressure p via a Schur complement approach, see e.g. [20]. This is accomplished by an inner-outer iteration pair, where the outer conjugate gradient (CG) iteration is used to compute the pressure and in each CG step a multigrid method is employed to determine the new search direction. The velocity is computed together with p in the CG method. For details of the algorithm see e.g. [21,22].

We, thus, see that at the core of the mantle convection code TERRA lies the problem of efficiently solving a discrete linear system stemming from an elliptic boundary value problem. This task arises not only in mantle convection simulations, but also e.g. in geo-potential problems and (quasi-)static viscoelastic analysis. In fact it can be found at the heart of numerous applications ranging from computational fluid dynamics over chip layout to bioelectric field simulations.

It is well-known that multigrid methods, see e.g. [23], are among the most efficient methods for solving large elliptic systems. Their key advantage is the possibility to reach an optimal linear scaling of computational expense relative to the number of unknowns of the problem. This makes these methods at least competitive in cost with the fast transform schemes (FFT) available for spectral codes. We note that high computational efficiency is essential for global mantle flow problems that involve millions of grid points. The nested structure of the icosahedral grid lends itself naturally to multigrid, as each grid is a subset of the next finer grid level.

Contrary to standard FFT approaches for PDEs both the CG and MG method can be parallelized in a straightforward manner using a domain decomposition (or more precisely a grid partitioning approach) in our case, see e.g. [23,24]. This is because the FEM approach leads only to a spatially restricted coupling of unknowns at nodes, which is similar to the local stencils of a Finite Difference setting. Splitting the grid into sub-domains and associating each sub-domain with one parallel process the main effort of parallelization lies in the co-ordination of information exchange along the sub-domain boundaries. In a distributed memory setting this can be achieved by explicit message passing based on the MPI standard[1].

In TERRA grid partitioning focusses on the ten diamonds that compose the icosahedral grid. Each associated 3D block is partitioned by powers of four. If four processes are used they work on the blocks associated with the upper, left, right and lower quarter of each diamond. Hence each process works on one quarter of the global problem domain. This approach can be repeated further, separately for the upper and the lower hemisphere, to achieve domain decompositions for any power of two.

Due to the above mentioned locality most of the work within each sub-domain can be performed independently of the others, with communication limited mostly to the exchange of boundary data among nearest neighbors. Depending on the volume-to-surface ratio of the sub-domains this local communication property assures high parallel efficiency. However, without going into too much detail, we must mention two aspects here. The first one is that due to the grid hierarchy employed in the MG method the volume-to-surface ratio naturally degrades on coarser grid levels. TERRA employs the standard approach of coarse grid agglomeration to counter this effect, we again point to [24] for this and alternative solutions. The second remark concerns the fact that TERRA uses a 2D partitioning of the grid. From a pure volume-to-surface point of view this is of course less favorable than a 3D splitting, where the grid is also partitioned in the radial direction. However, 2D splitting also has two distinct advantages. First, TERRA uses line-smoothing in the radial direction in order to ensure good MG convergence rates in the presence of strong viscosity variations. The choice of 2D decomposition avoids cutting these lines by sub-domain interfaces, which is a considerable performance and implementation advantage for the smoother. Second, 2D splitting results in a significant reduction of the number of messages that must be passed between processes compared to the 3D case. Using e.g. a partitioning with 64 sub-domains we need approximately 2,560 (2D) versus 3.520 (3D) messages for one boundary update after a single smoothing step. This reduction of about 27% may be considerable in a system with (relatively) high latency interconnects, cf. the design of Tethys in Sect. 4 in this respect.

4 Munich Earth Modeling Cluster Tethys

4.1 Design of Tethys

Global mantle flow studies rely on modern parallel computers. Taking mantle convection simulations with 100 million finite element grid points (to approximate the dynamic regime of vigorous mantle convection) as representative, about

[1] A PVM implementation of TERRA is also still available.

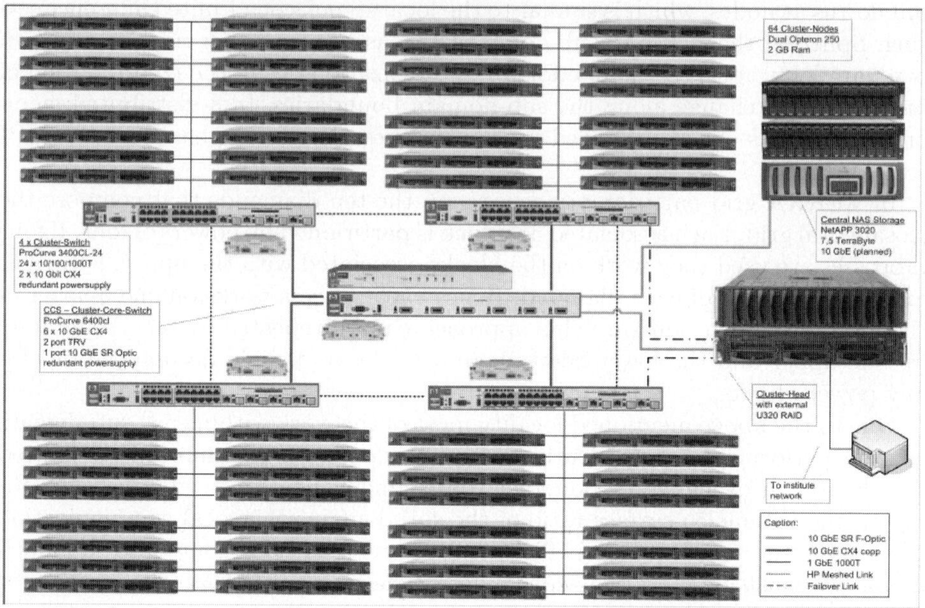

Fig. 2. Schematic representation of the setup of Tethys and its interconnect topology

70 Gigabytes (GB) of main memory and a sustained system performance of about 100 GFLOPs are needed to complete at least one convective overturn per hour. These requirements are now well in reach of dedicated PC-clusters. Such systems are called *Beowulf*, after the initial *Beowulf* PC-cluster project at the Goddard Space Flight Center [`http://www.beowulf.org`].

Our current cluster, named *Tethys* for Tectonic High Performance Simulator, includes 64 AMD Opteron 250 (64 bit, single core) 2.4 GHz dual-processors, each equipped with 2 GB RAM and 160 GB of disc-space. A dual-processor configuration improves the price performance ratio of our system and allows for some flexibility for those codes that require very large per-processor memory. Table 1 summarizes the Tethys configuration, Figure 2 gives an overview of the setup of the cluster and its network topology. The 64 compute nodes of Tethys

Table 1. Hardware specification of the 64 compute nodes

no. of processors	two per node
type of processors	AMD Opteron 250 (64 bit, single core)
clock speed	2.4 GHz
L1 cache	64/64KB (data/instruction)
L2 cache	1MB (data + instruction)
local memory	2 GB RAM (DDR1)
local storage	160 GB
network interface	1000T Ethernet (2 ports)

are arranged in four bricks consisting of 16 nodes each. The nodes within one such brick communicate via a single 1000T cluster node switch. For inter-brick communication the four cluster node switches are attached to a central cluster core switch.

4.2 Some Benchmarks

We start our exposition by considering the communication performance of the Tethys cluster. In order to evaluate the latter we employ the Intel MPI benchmarks (IMB) suite[2]. For the sake of brevity we restrict ourselves to the *Exchange* and *Allreduce* benchmarks, since they represent exemplary communication patterns in domain decomposition approaches for parallel multigrid. In the *Exchange* test processes are assumed to form a periodic process chain. Each process exchanges data with both of its neighbors in the chain. This resembles the update of internal boundary values e.g. after one MG smoothing step. The *Allreduce* test, on the other hand, represents a collective communication involving all processes. The corresponding MPI_ALLREDUCE method gathers data from all processes performing a global reduction operation, e.g. a summation, on the fly and returns the result to all processes. This situation typically occurs in the computation of the residual norm in parallel iterative methods. Further details of the benchmarks can be found in the user's guide.

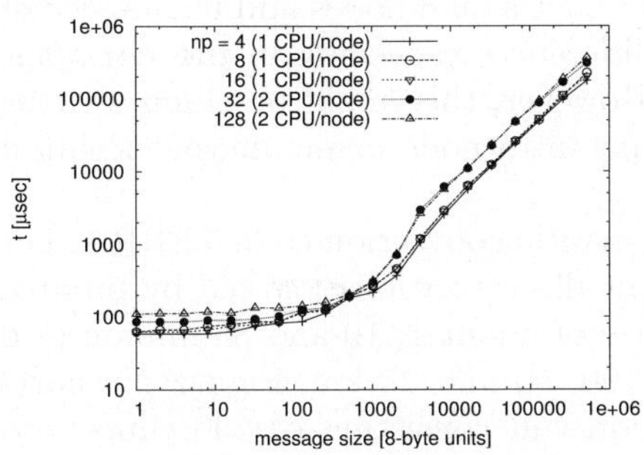

Fig. 3. Performance for *Exchange* benchmark

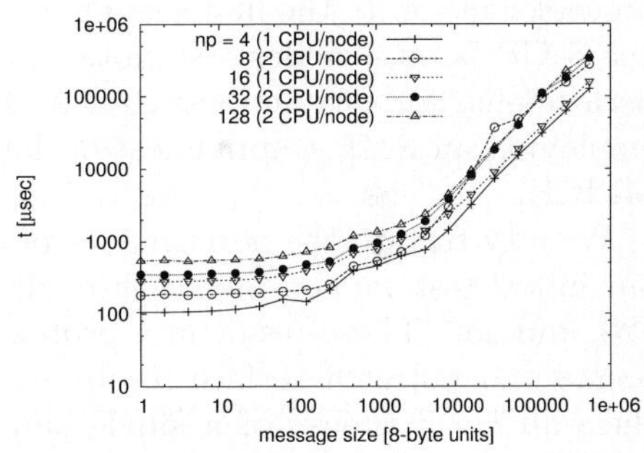

Fig. 4. Performance for *Allreduce* benchmark

Figures 3 and 4 give timing results for both tests for different message sizes and increasing number of processes (np). Performance is in line with expectations for TCP/IP based communication over GBit-Ethernet and there is a linear scaling with increasing message size. Note that the first four cases (np = 4, 8, 16, 32) involve only nodes belonging to a single brick. Thus, communication involves a single cluster node switch only. The case having 128 CPUs (np = 128),

[2] We employ IMB version 2.3 in combination with MPICH 1.2.5.3 an the Intel Compiler suite version 9.0 for these tests.

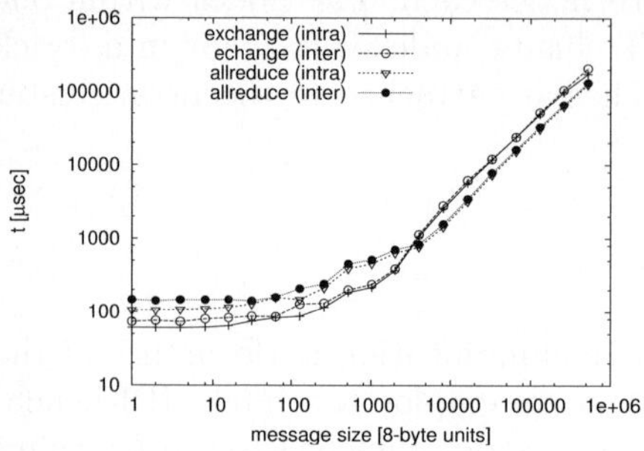

Fig. 5. Comparison between intra- and inter-brick communication

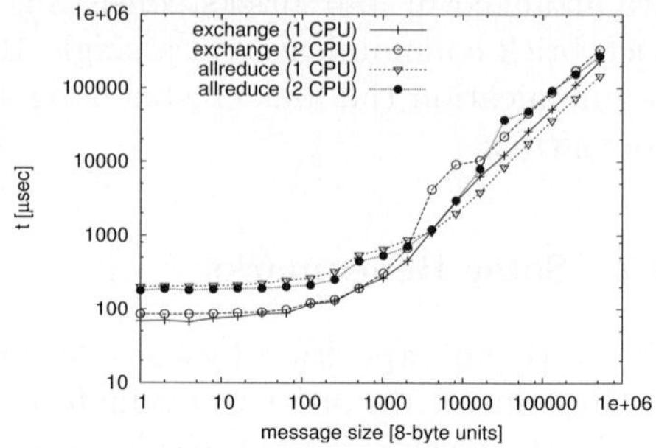

Fig. 6. Influence of double-CPU nodes on communication performance

instead, involves all four bricks and hence the entire cluster network including the cluster core switch, see Fig. 2. We also show the direct comparison of a setting with 4 processes running either on 4 nodes within one brick or running on four nodes each in a separate brick in Fig. 5. We verify from the similar scaling of the two configurations that the hierarchical system of our network topology works quite well, and does not present any serious bottle-neck to global communication.

Finally we measure in Fig. 6 the effect of using our cluster nodes in dual-processor mode. In the first setting we use 8 CPUs on 8 nodes and in the second one 8 CPUs on 4 nodes. We note small differences especially in the *Exchange* performance for certain message lengths. However, this might be eliminated by employing an SMP communication layer for intra-node communication within MPICH.

We now turn to the performance of our mantle convection code TERRA. For our initial test we considered three different discretizations marked by mt=64, 128, and 256. These result in a problem size of about 1, 10 and 85 million grid points and reflect a surface resolution of 100, 50 and 25 km. Figure 7 reports the wall clock times[3] of a single simulation run consisting of 500 time-steps for different numbers of processes np. We observe a nice, approximately linear decay with increasing np. This is directly reflected in Fig. 8 where the speedup for mt=64 and 128 is plotted. From the Earth scientist's point of view the scaleup of the application is of course more interesting than the mere speedup. In this respect we point out that we obtain a run-time of 2002 s (for the mt=128 case on 16 processes) and 2564 s (for the mt=256 case on 128 processes), which both lead to the same workload per process, see again Fig. 7. Based on an approximate requirement of 8000 operations per node and time-step the latter example results in an aggregate system performance of 140 GFLOPs out of a theoretical 624 GFLOPs peak. These are initial results and an improved scale-up might be obtained by tuning the code to the Opteron architecture.

[3] The value for mt=128 on a single process is extrapolated (due to memory limitations).

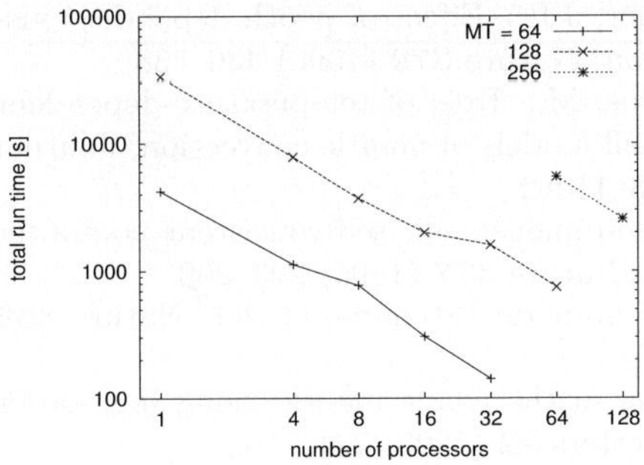

Fig. 7. Run times for different problem sizes and increasing number of processes

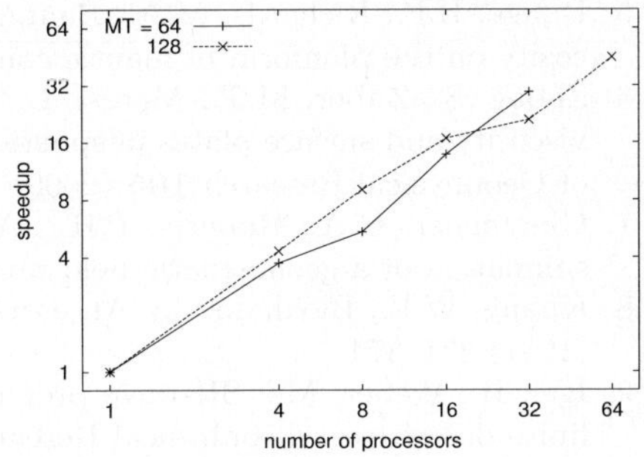

Fig. 8. Speedup of the TERRA code on the Tethys cluster

5 Conclusions and Outlook

We have built a large-scale geophysical modeling cluster at Munich University's (LMU) Geosciences department. The machine serves as a departmental supercomputer to perform a range of geosciences simulations, including compute intensive variational data-assimilation calculations for global mantle convection studies. We observe parallel efficiencies of better than 80% and an aggregate system performance of 200 GFLOPs. We conclude that cost-efficient Beowulf clusters should take an increasing role in performing large-scale capacity-oriented geosciences simulations.

Acknowledgment

Tethys was funded by the German Ministry of Education and Research (BMBF) and the Free State of Bavaria by means of the HBFG program. We would also like to thank Microstaxx GmbH and the High-Performance Group of Fujitsu-Siemens Computers for their support.

References

1. Bunge, H.P., Richards, M.A., Lithgow-Bertelloni, C., Baumgardner, J.R., Grand, S., Romanowicz, B.: Time scales and heterogeneous structure in geodynamic earth models. Science **280** (1998) 91–95
2. Jarvis, G.T., McKenzie, D.P.: Convection in a compressible fluid with infinite prandtl number. Journal of Fluid Mechanics **96** (1980) 515–583
3. Glatzmaier, G.A.: Numerical simulations of mantle convection: Time-dependent, three-dimensional, compressible, spherical shell. Geophysical and Astrophysical Fluid Dynamics **43** (1988) 223–264
4. Tackley, P.J., Stevenson, D.J., Glatzmaier, G.A., Schubert, G.: Effects of an endothermic phase transition at 670 km depth on a spherical model of convection in Earth's mantle. Nature **361** (1993) 699–704

5. Bunge, H.P., Richards, M.A., Baumgardner, J.R.: Effect of depth dependent viscosity on the planform of mantle convection. Nature **379** (1996) 436–438
6. Zhong, S., Zuber, M.T., Moresi, L., Gurnis, M.: Role of temperature-dependent viscosity and surface plates in spherical shell models of mantle convection. Journal of Geophysical Research **105** (2000) 11063–11082
7. Glatzmaier, G.A., Roberts, P.H.: A three-dimensional, self-consistent computer simulation of a geomagnetic field reversal. Nature **377** (1995) 203–209
8. Kuang, W.L., Bloxham, J.: An earth-like numerical dynamo model. Nature **389** (1997) 371–374
9. Igel, H., Weber, M.: SH-wave propagation in the whole mantle using high-order finite differences. Geophysical Research Letters **22** (1995) 731–734
10. Komatitsch, D., Tromp, J.: Introduction to the spectral element method for three-dimensional seismic wave propagation. Geophysical Journal International **139** (1999) 806–822
11. Bunge, H.P., Richards, M.A., Baumgardner, J.R.: Mantle circulation models with sequential data-assimilation: Inferring present-day mantle structure from plate motion histories. Philosophical Transactions of the Royal Society of London: Series A **360** (2002) 2545–2567
12. McNamara, A.K., Zhong, S.: Thermochemical structures beneath Africa and the Pacific Ocean. Nature **437** (2005) 1136
13. Courtier, P., Talagrand, O.: Variational assimilation of meterological observations with the adjoint vorticity equation I: Numerical results. Quarterly Journal of the Royal Meteorological Society **113** (1987) 1329–1347
14. Wunsch, C.: The Ocean Circulation Inverse Problem. Cambridge University Press (1996)
15. Bunge, H.P., Hagelberg, C.R., Travis, B.J.: Mantle circulation models with variational data-assimilation: Inferring past mantle flow and structure from plate motion histories and seismic tomography. Geophysical Journal International **152** (2003) 280–301
16. Landau, L., Lifschitz, E.: Fluid mechanics. Pergamon Press (1987)
17. Baumgardner, J.R.: Three-Dimensional Treatment of Convective Flow in the Earth's Mantle. Journal of Statistical Physics **39** (1985) 501–511
18. Williamson, D.: Integration of the barotropic vorticity equations on a spherical geodesic grid. Tellus **20** (1968) 642–653
19. Baumgardner, J.R., Frederickson, P.O.: Icosahedral discretization of the two-sphere. SIAM Journal on Numerical Analysis **22** (1985) 1107–1115
20. Benzi, M., Golub, G., Liesen, J.: Numerical solution of saddle point problems. Acta Numerica **14** (2005) 1–137
21. Verfürth, R.: A Combined Conjugate Gradient-Multigrid Algorithm for the Numerical Solution of the Stokes Problem. IMA Journal of Numerical Analysis **4** (1984) 441–455
22. Yang, W.S., Baumgardner, J.R.: A matrix-dependent transfer multigrid method for strongly variable viscosity infinite Prandtl number thermal convection. Geophysical and Astrophysical Fluid Dynamics **92** (2000) 151–195
23. Trottenberg, U., Oosterlee, C., Schüller, A.: Multigrid. Academic Press (2001) ISBN: 0-12-701070-X.
24. Hülsemann, F., Kowarschik, M., Mohr, M., Rüde, U.: Parallel Geometric Multigrid. In Bruaset, A.M., Tveito, A., eds.: Numerical Solution of Partial Differential Equations on Parallel Computers. Number 51 in Lecture Notes in Computational Science and Engineering. Springer (2005) ISBN: 3-540-29076-1.

A Streaming Implementation of Transform and Quantization in H.264

Haiyan Li, Chunyuan Zhang, Li Li, and Ming Pang

School of Computer Science, National University of Defense Technology,
Changsha, Hunan, P. R. China, 410073
hy_lee@163.com

Abstract. The H.264 video coding standard uses a 4*4 multiply-free integer transform, minimizing computational complexity. The emerging programmable stream architecture provides a powerful mechanism to achieve high performance in media processing and signal processing. This paper analyzes the algorithm characteristics of transform and quantization in H.264 and presents a streaming implementation of transform and quantization on Imagine stream processor. We evaluate our implementation on a cycle-accurate simulator of Imagine and demonstrate stream processing efficiency by comparing its performance against other implementations. Experimental results show that our streaming implementation deals with transform and quantization of a 4*4 block in 6.875ns. The coding efficiency can satisfy the real-time requirement of current video applications.

1 Introduction

H.264 [1], proposed by Joint Video Team (JVT), is a new digital video coding standard. Some highlighted features are applied in H.264 for improved coding efficiency. Small block-size integer transform is one of the enhanced techniques to avoid inverse transform mismatch problem. It uses a 4*4 transform block size to somewhat reduce the block artifacts. All operations in transform process can be carried out in integer arithmetic only requiring additions and shifts. While a scaling multiplication is integrated into the following quantizer to decrease the total number of multiplications. By short tables, a set of new scalar quantization formulas use multiplications but avoid divisions [2].

Transform and quantization is a computationally-intensive component. H.264 adopts block-based motion prediction, so residual difference between current frame and predicted frame is organized in block. Each block is independent of others, exposing a great deal of data parallelism. The issue of optimizing transform and quantization in H.264 has been addressed in various research domains. For example, enhanced SIMD technologies such as MMX and SSE2 are used to improve coding rate of H.264 [3,4]. DSP acceleration is always the favor of product researchers [5,6]. In addition, special-purpose hardware implementations for transform and quantization in H.264 have emerged in succession [7,8,9,10].However, a new exploration on stream processing for H.264 has still remained.

M. Gerndt and D. Kranzlmüller (Eds.): HPCC 2006, LNCS 4208, pp. 41 – 50, 2006.

In this paper we develop a streaming implementation of transform and quantization in H.264 on the programmable Imagine stream processor [11]. Two computational kernels, corresponding to transform and quantization respectively, are constructed to operate on large homogeneous stream elements. In the end, we evaluate its performance by comparing it against other implementations. Experimental results show that the streaming transform and quantization implementation deals with the transform and quantization of a 4*4 block in 6.875ns, namely processing 145.5 millions of inputs per second. The coding efficiency can satisfy the real-time require-ment of current video applications.

Implementing on stream processor requires us to modify algorithms on the basis of stream processing and to arrange data in a streaming sequence in order to efficiently utilize the SIMD manner of stream architecture. Exploiting data parallelism and locality are encouraging for high performance applications.

The remainder of this paper is organized as follows. In Section 2, the principle and the algorithm of transform and quantization in H.264 are given. Section 3 describes the details of Imagine stream architecture. In Section 4, we discuss a streaming implementation of transform and quantization. Our experimental results and discussions are presented in Section 5. Finally, the paper is summarized in Section 6.

2 Transform and Quantization of H.264

The 4*4 transform adopted in the H.264 standard is an integer orthogonal computation [12]. This allows for bit-exact implementation for all encoders and decoders. Another important feature in the new standard is the removal of computationally-expensive multiplication that appears in the conventional standards.

The 4*4 integer transform of an input array X is shown in Equation (1).

$$W = C_f X C_f^T \tag{1}$$

where the matrix C_f is given by Equation (2).

$$C_f = \begin{bmatrix} 1 & 1 & 1 & 1 \\ 2 & 1 & -1 & -2 \\ 1 & -1 & -1 & 1 \\ 1 & -2 & 2 & -1 \end{bmatrix} \tag{2}$$

In Equation (2), the absolute values of all the coefficients of the C_f matrix are either 1 or 2. Thus, the transform operation presented by Equation (1) can be computed using signed additions and left-shifts only to avoid expensive multiplications. While a scaling multiplication is integrated into the following quantizer to minimize the total number of multiplications.

H.264 uses a scalar quantizer. The post-scaling and quantization formulas are shown in Equation (3) and (4).

$$qbits = 15 + (QP \quad DIV \quad 6) \tag{3}$$

$$Z_{ij} = round\left(W_{ij} \frac{MF}{2^{qbits}} \right) \tag{4}$$

where QP is a quantization parameter. It can take any integer value from 0 up to 51. The wide range of 52 quantizer step sizes makes it possible for an encoder to accurately and flexibly control the trade-off between bit rate and image quality. Z_{ij} is a quantized coefficient. MF is a multiplication factor that depends on QP and the position (i, j) of the element in the matrix, referred to [12].

Fast algorithms for traditional DCT are also suited for integer transform, such as butterfly transform [2]. It converts matrix multiply into matrix-vector multiply, ensuring that there is no dependency between different vector columns in a matrix. And its algorithm structure is relatively simple (see Fig.1), regarded as a good algorithm for hardware implementation.

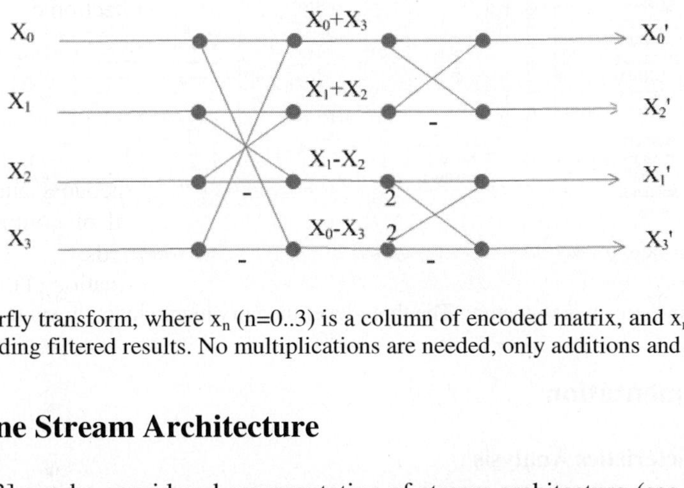

Fig. 1. Butterfly transform, where x_n (n=0..3) is a column of encoded matrix, and x_n' (n=0..3) is its corresponding filtered results. No multiplications are needed, only additions and shifts.

3 Imagine Stream Architecture

Imagine [13] can be considered representative of stream architecture (see Fig.2). We have done thorough research on Imagine [14], and proposed a new MASA [15] supporting multiple execution morphs. Imagine is programmable and flexible since it directly maps applications into streams and kernels. We have mapped many applications, such as fluid computation [15], Reed-solomon decoder [16], and motion estimation of H.264 [17] and so on. Imagine can provide high performance in so many domains including media processing and signal processing. For example, Imagine is able to sustain performance of 15.35 giga operations per second (GOPS) in MPEG-2 encoding application, corresponding to 287 frames per second (fps) on a 320*288-pixel, 24-bit color image [18].

In fact, the implementation of stream application is casted by a collection of *streams* passing through a series of computational *kernels*. A kernel is a small program executed in arithmetic clusters that is repeated for each successive element of its input streams to produce output stream for the next kernel in the application. Streams are ordered finite-length sequences of data records. Each record in a stream is a set of related data elements of a single arbitrary data type. The semantics of applying a kernel to a stream are completely parallel, so the computation of a kernel can be performed on different independent elements in the input stream(s) in parallel. Kernel reads its inputs from *stream register file* (SRF). During computation, all temporary data are stored in the *local register file* (LRF) of each cluster. And the output stream of a kernel are sent back to SRF. Only the initial and final data streams need to be transferred through streaming memory to the off-chip SDRAM. This three level memory hierarchy is able to meet the large instruction and data bandwidth demands of computationally intensive applications well.

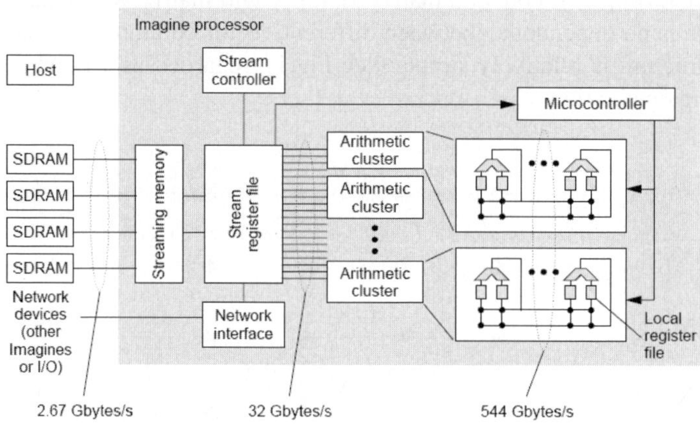

Fig. 2. Imagine stream architecture. The three-level memory hierarchy is shown in the figure.

4 Implementation

4.1 Characteristics Analysis

Stream programming model can match the requirement of media processing very well. Before mapping transform and quantization of H.264 video encoder on Imagine stream processor, it is necessary to analyze the inherent characteristics especially in computation intensity, parallelism and locality.

Computation Intensity
Integer transform is a computationally intensive module like block searching in motion estimation except for extra decision-making. If using butterfly algorithm mentioned in Section 2, a transform needs 64 additions and 16 shifts. As a result, 12.16 million additions and 3.04 million shifts are executed in one second for a CIF image of 352*288 pixels at 30fps.

Parallelism

Large data parallelism exists in transform process. Based on 8-cluster structure of Imagine, different columns of encoded matrix can be performed in parallel. Efficient data organization may help magnify the advantage of data-level parallelism. Besides, transform has obvious instruction-level parallelism. The pure additions and shifts can be packed into VLIW compactly. An additional level of task parallelism can be discerned from the pipelining of kernels.

Locality

Video coding is processed orderly frame by frame and block by block, like a stream of data flowing through every sequential processing module. Kernels encapsulate short-term kernel locality, and allow efficient use of the LRFs. For example, the intermediate results of butterfly transform can be stored in the inner LRFs. At the same time, stream capture long-term producer-consumer locality in the transfer of data form one kernel (e.g. *transform* kernel) to another kernel (e.g. *quantization* kernel) through the SRF without requiring costly memory operations, as well as spatial locality by the nature of stream being a series of data records.

4.2 Implementation

Programming model of Imagine includes two levels: stream level (in StreamC) and kernel level (in KernelC) [19]. According to the relationship of input and output streams, the kernel diagram of transform and quantization is described as Fig.3.

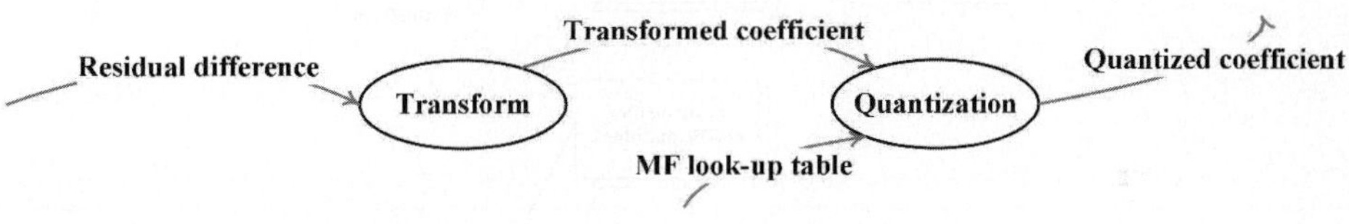

Fig. 3. Kernel diagram. An ellipse represents a computational kernel. And the connection line represents the input or output streams of the kernel.

These two kernels are chained together, where the output stream from the *transform* kernel is fed into the next *quantization* kernel as an input stream. Producer-consumer locality is exploited by consuming the result of one kernel as soon as it is produced. MF look-up table is organized as a constant stream and loaded with transformed coefficient stream into the *quantization* kernel. Note that a kernel can take more than one streams as its input, and the output may be one or more streams for different kernels.

Integer transform $Y=C_fXC_f^T$ performs matrix multiplications twice. Assume that $B=C_fX$, then $Y= C_fXC_f^T= BC_f^T= (C_fB^T)^T$. Based on transpose we keep the block that is to be transformed as right matrix while the left matrix is C_f. The matrix X can be divided into four vectors by column. Inter-column independence makes the butterfly algorithm suited for stream processing. The main loop in the *transform* kernel is illustrated in Fig.4.

```
For x in input residual difference stream:
    Call butterfly transform computation
    Transpose (communication operations)
    Recall butterfly transform computation
    Output transformed coefficient stream
```

Fig. 4. Pseudocode of *transform* kernel

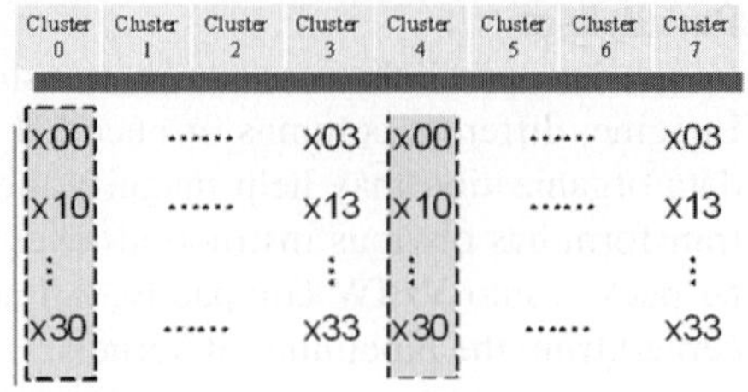

Fig. 5. Stream distribution in eight clusters

The input stream of *transform* kernel consists of 4*4 matrix blocks. Fig.5 illustrates the distribution of each stream record in clusters, where x_{ij} represent one stream record with the relative position in its affiliated block. We choose a simple solution-replicating a 4*4 matrix twice. It brings half waste of cluster resources because the computation of four clusters is redundant. The better data records in the input stream are organized, the better performance stream architecture will get [20]. The way of organizing the data records is explicit for stream programmers. So it requires programmers to understand the algorithm characteristics in order to map it efficiently.

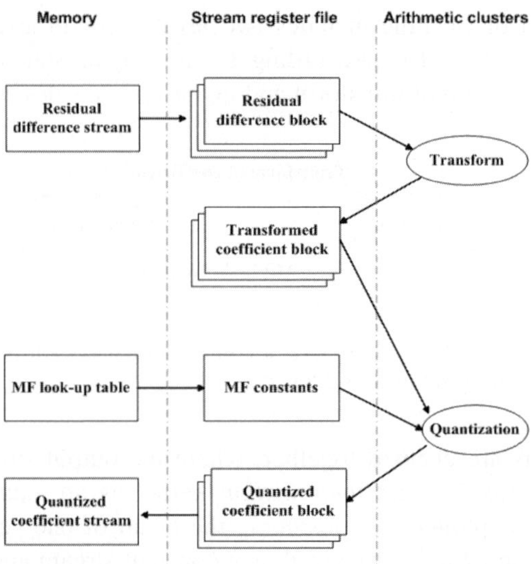

Fig. 6. Three-level memory hierarchy diagram

The *transform* kernel reads each data element from memory (In the whole encoder, the input of *transform* kernel is also fed by its prior kernel. So, it is reasonable to suppose the input stream exists in SRF not memory.). After kernel execution, the transformed coefficient stream produced by *transform* is sent to the following *quantization* kernel directly. MF is loaded by a series of communication operations and multiplies Y by index. Finally, the output quantized coefficient stream can be written back to memory or stored in SRF as an intermediate stream for other kernels. The three-level bandwidth hierarchy corresponds to the three columns of Fig.6.

5 Simulations and Results

5.1 Experimental Results

We run our streaming implementation of transform and quantization on ISim, a cycle-accurate simulator ISim (500MHz), which is provided by the Imagine Project of Stanford University [21]. The simulator can accurately model all aspects of stream processing and stream memory system. And the execution cycles obtained by ISim is convictive enough to evaluate the performance of a streaming implementation. The correctness of our implemented transform and quantization is also checked. This is done by passing different input sequences to our stream program and comparing the experimental result and the mathematical value.

Simulated results show that $3.485*10^5$ cycles is needed for a CIF image. Thus, dealing with the transform and quantization process of a 4*4 block requires 6.875ns, thereinto 5.79ns for integer transform. It means that our streaming implementation is able to process 145.5 millions of inputs per second. The performance may be optimized by some advanced techniques such as loop unrolling and software pipelining [20]. The processing rate is higher than that defined to HDTV video sequence (HDTV must process 124.5 millions of inputs per second [9]).

We compare our streaming implementation with other different improvements, mentioned in Section 1. Take the time of 4*4 integer transform as the criterion, shown in Fig.7. Obviously, Imagine obtains comparative performance with special-purpose hardware designed for transform and quantization application, much better than MMX and DSP improvements. However, Imagine is more flexible than special-purpose hardware. Thus, it has good adaptability and scalability.

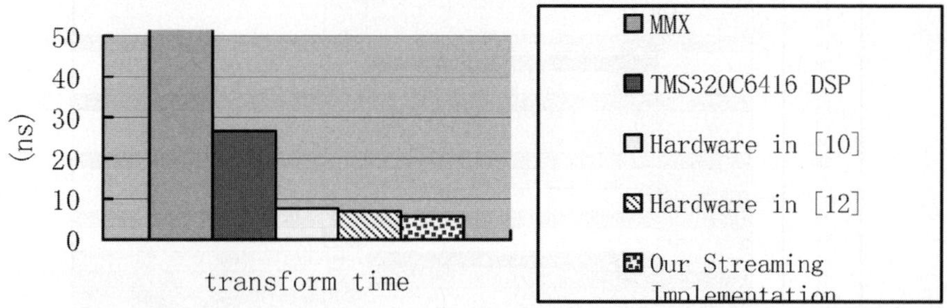

Fig. 7. Comparison among different implementations, where the data of processing time refer to [4,6,7,9] resprectively

Our streaming implementation has the shortest processing time of five cases in Fig.7. Imagine can achieve high performance for three reasons. First, stream computation is efficient when operated on homogeneous data elements. Stream processing mechanism ensures to overlap between kernel computation and memory access, hiding the latency of memory operations. Second, Imagine performs in the SIMD manner. Large data parallelism and little global data reuse may explore the powerful computing capability of Imagine. Third, kernel locality and producer-consumer locality are captured in LRF and SRF of Imagine. The three-level memory hierarchy can afford the bandwidth requirement well.

5.2 Discussions

For an actual image, the residual difference input stream is a very large data set. Processing each element in a single computation is impractical because the size of data set may greatly exceed the size of on-chip storage. Instead, most Imagine applications use the common technique of *stripmining* [18]. In our implementation, residual difference pixel-blocks are divided into input batches, stream operations are applied to an entire input batch at a time. The size of a batch in our implementation is almost 14000 pixels. Fig.8 gives the utilization of SRF in the execution of transform and quantization.

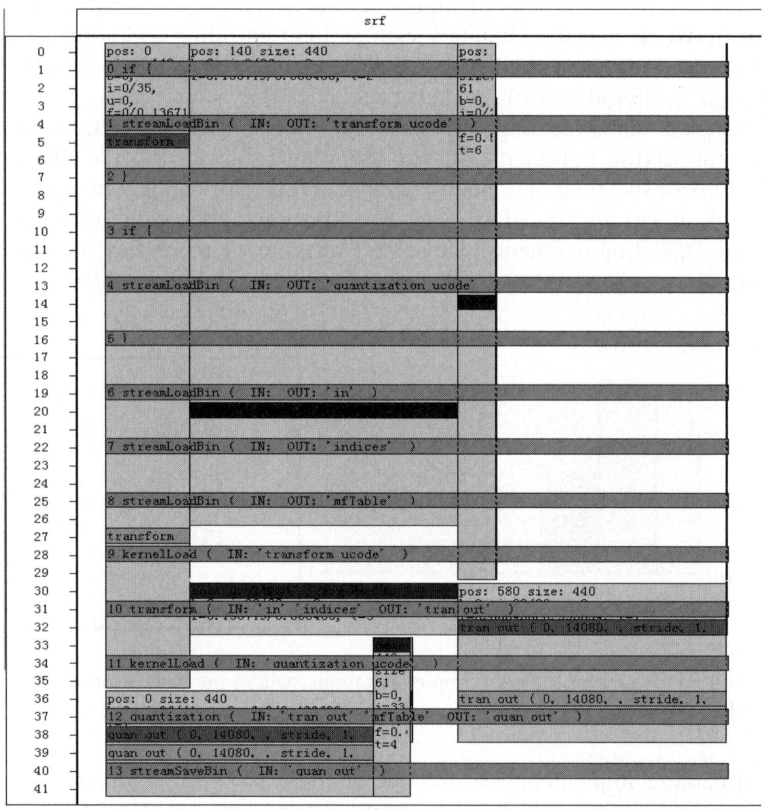

Fig. 8. SRF utilization, where the horizontal axis represents SRF size and colorful bars represent usage of SRF. Vertical axis represents stream execution time. The inputs in a batch are 14080 residual difference pixels in this figure. But blue bar indicates a read or write that requires a memory access when SRF is spilled over.

The utilization of functional units is given in Table 1. The result is matched with the algorithm characteristics: large amount of additions and shifts make great use of adders only expect for some stalls before initial operands are prepared. While multipliers are only used for quantization, so the utilization ratio is not very high.

By taking advantage of unique three-level memory hierarchy and large numbers of functional units, Imagine can achieve so high performance. Combined with the accelerated implementation of motion estimation on Imagine [17], we can infer that the whole H.264 encoder will get better performance and meet real-time requirement of current video applications.

Table 1. Arithmetic unit utilization

	Adders	Multipliers
Utilization ratio	83%	26%

6 Conclusion

In this paper we have presented a streaming implementation of transform and quantization in H.264. Experimental results show that processing transform and quantization for a 4*4 block needs 6.875ns. As a result, our implementation is able to process 145.5 millions of inputs per second. It can satisfy the real-time requirement of video applications. And it proves that the programming model including memory hierarchy of stream architecture is helpful for large numbers of data to repeat the same or similar operations. But some issues are still needed to be researched deeply, such as slice partition granularity and stream algorithm optimization. We will pay more attention to its further improvement.

Acknowledgements. We thank Imagine project group of Stanford University for providing the Imagine simulator. We also thank the reviewers for their insightful comments. This work was sponsored by National Natural Science Foundation of China under Grant 60573103.

References

1. JVT, Draft ITU-T Recommendation and Final Draft International Standard of Joint Video Specification (ITU-T Rec. H.264| ISO/IEC 14496-10 AVC). May 2003.
2. Henrique S. Malvar, Antti Hallapuro, Marta Karczewicz, Louis Kerofsky: Low-Complexity Transform and Quantization in H.264/AVC. IEEE Transactions on Circuits and Systems for Video Technology, Vol.13, No.7, July 2003.
3. Cui Yansong, Duan Dagao, Deng Zhongliang: The Analysis of Transform and Quantization in H.264. Modern Cable Transmission, 2004.5, pp 71-74
4. Wei Fang, Li Xueming: SIMD Optimization of Transform and Quantization in H.264. Computer Engineering and Applications, 2004.17, pp 24-27
5. Liu Baolan, Liu Guizhong, Su Rui: Implementation and Optimization of Pixel-Compression Module in H.264 Based on DSP System. Microelectronics, Vol. 22, 2005, No.6, pp200-205

6. Shen Haitao, Fan Yangyu, Wang Fengqin, Hao Chongyang: An Implementation of Transform Encoding on DSP in H.264. 2004
7. Liu Ling-zhi, Qiu Lin, Rong Meng-tian, Jiang Li: A 2-D Forward/Inverse Integer Transform Processor of H.264 Based on Highly-parallel Architecture. In Proceedings of the 4th IEEE International Workshop on System-on-chip for Real-Time Applications, 2004
8. Ihab Amer, Wael Badawy, and Graham Jullien: Hardware Prototyping for the H.264 4*4 Transformation. ICASSP 2004.
9. Roger Endrigo Carvalho Porto, Marcelo Schiavon Porto, Thaisa Leal da Silva, Leandro Zanetti Paiva da Rosa, Jose Luis Almada Guntzel, Luciano Volcan Agostini: An Integer 2-D DCT Architecture for H.264/AVC Video Coding Standard. XX SIM-South Symposium on Microeletronics.
10. Young-hun Lim, Yong-jin Jeong: Hardware Implementation of Integer Transform and Quantization for H.264. December 2003.
11. Ujval J. Kapasi, William J. Dally, Scott Rixner, John D. Owens, Brucek Khailany: The Imagine Stream Processor. Appears in the Proceedings of the 2002 International Conference on Computer Design, September 2002.
12. "H.264/MPEG-4 Part 10: Transform&Quantization"www.vcodex.com
13. Brucek Khailany, William J. Dally, Ujval J. Kapasi, Peter Mattson, Jinyung Namkoong, John D. Owens, Brian Towles, Andrew Chang: Imagine: Media Processing with Streams. IEEE Micro, March-April 2001.
14. Mei Wen, Nan Wu, Haiyan Li, Chunyuan Zhang: Research and Evaluation of Imagine Stream Architecture. Advances on Computer Architecture, ACA'04
15. Mei Wen, Nan Wu, Haiyan Li, Li Li, Chunyuan Zhang: Multiple-morghs Adaptive Stream Architecture. Journal of Computer Science and Technologgy, 2005
16. Mei Wen, Nan wu, Haiyan Li, Li Li, Chunyuan Zhang: A Parallel Reed-solomon Decoder on the Imagine Stream Processor. Second International symposium on Parallel and Distributed Processing and Applications, Hongkong, 2004.12
17. Haiyan Li, Mei Wen, Chunyuan Zhang, Nan Wu, Li Li, Changqing Xun: Accelerated Motion Estimation of H.264 on Imagine Stream Processor. International Conference on Image Analysis and Recognition 2005
18. John D. Owens, Scott Rixner, Ujval J. Kapasi, Peter Mattson, Brian Towles, Ben Serebrin, William J. Dally: Media Processing Applications on the Imagine Stream Processor. In the Proceedings of the 2002 International Conference on Computer Design, 2002
19. Abhishek Das, Peter Mattson, Ujval Kapasi, John Owens, Scott Rixner, Nuwan Jayasena: Imagine Programming System User's Guide 2.0, June 2004
20. Haiyan Li, Chunyuan Zhang, Li Li, Ming Pang: Stream Algorithm of 4*4 Integer Transform. Conference on Virtual Reality and Vision 2006.
21. The Imagine Project, Stanford University, http://cva.stanford.edu/imagine/

A Parallel Transferable Uniform Multi-Round Algorithm in Heterogeneous Distributed Computing Environment

Hiroshi Yamamoto*, Masato Tsuru, and Yuji Oie

Department of Computer Science and Electronics,
Kyushu Institute of Technology, Kawazu 680-4, Iizuka, 820-8502 Japan
yamamoto@infonet.cse.kyutech.ac.jp, {tsuru, oie}@cse.kyutech.ac.jp

Abstract. The performance of parallel computing systems using the master/worker model for distributed grid computing tends to be degraded when large data sets have to be dealt with, due to the impact of data transmission time. In our previous study, we proposed a parallel transferable uniform multi-round algorithm (PTUMR), which efficiently mitigated this impact by allowing chunks to be transmitted in parallel to workers in environments that were homogeneous in terms of workers' computation and communication capacities. The proposed algorithm outperformed the uniform multi-round algorithm (UMR) in terms of application turnaround time, but it could not be directly adapted to heterogeneous environments. In this paper, therefore, we propose an extended version of PTUMR suitable for heterogeneous environments. This algorithm divides workers into appropriate groups based on both computation and communication capacities of individual workers, and then treats each group of workers as one virtual worker. The new PTUMR algorithm is shown through performance evaluations to significantly mitigate the adverse effects of data transmission time between master and workers compared with UMR, achieving turnaround times close to the theoretical lower limits even in heterogeneous environments.

Keywords: Grid Computing, Master/Worker Model, Divisible Workload, Multi-Round Scheduling, UMR.

1 Introduction

Grid computing has recently increased in popularity for distributed applications [1,2]. The master/worker model is suited to grid computing environments involving a large number of computers that differ in resource capacities. In this model, a master with application tasks dispatches subtasks to several workers, which process the data allocated by the master. A typical instance of applications based on the master/worker model is a divisible workload application

* The author is now working in FUJITSU LABORATORIES LTD. The contact e-mail address is hiro-yamamoto@jp.fujitsu.com

M. Gerndt and D. Kranzlmüller (Eds.): HPCC 2006, LNCS 4208, pp. 51–60, 2006.

[3,4,5], where the master divides the application data into an arbitrary number of chunks and then dispatches them to multiple workers. For a given application, it is assumed that computation and transmission times for a chunk are roughly proportional to the size of the chunk.

The existing uniform multi-round algorithm (UMR), can handle an application having a large amount of data in a 'multiple-round' manner to overlap the time required for communication with that required for computation [3,6,7,8]. However, it utilizes sequential transmission model, i.e. the master transmits data to one worker at a time [9,10]. In actual networks where the master and workers are connected via a heterogeneous network, it is unable to minimize the adverse effects of data transmission on the application turnaround time.

Therefore, in our previous study, we proposed a new scheduling algorithm, parallel transferable uniform multi-round (PTUMR) adapted to the heterogeneous network [11]. The proposed algorithm fully utilizes the high-speed data transmission capacity of the heterogeneous network by allowing the master to transmit application data to multiple workers simultaneously. However it is not adaptive to a heterogeneous environment containing workers with varying resource capacities.

The contributions of this paper are two-fold. Firstly, we extend the PTUMR so that it can be applied in a heterogeneous environment. The master divides workers into appropriate groups based on both computation and communication capacities of individual workers, and then treats the set of workers in a group as one virtual worker. After that, the master optimally transmits chunks to the virtual workers sequentially as in UMR. Secondly, we evaluate the efficiency of the new PTUMR in various environments. The proposed algorithm reduces the adverse effects of data transmission time on application turnaround time to a greater extent than the conventional UMR by handling heterogeneity in terms of workers' capacities and the network model, allowing turnaround times close to the theoretical lower limit to be achieved.

This paper is organized as follows. In Section 2, the conceptual basis for multiple-round scheduling and the conventional UMR algorithm are introduced. The proposed PTUMR algorithm is presented in Section 3, and its performance is investigated in Section 4. The conclusion follows in Section 5.

2 The Conventional UMR Scheduling Algorithm

Recently, a number of scheduling methods have been proposed in which the master dispatches data to workers in a multiple-round manner in order to overlap communication with computation and thereby reduce the application turnaround time. Figure 1 shows a simple example of this scenario where the master dispatches a workload of the application to a worker. In this figure, a black rectangle represents the fixed-length overhead for one round of computation and a gray rectangle represents the fixed-length overhead in one round of data transmission. In multiple-round scheduling the entire application data set W [units] is divided into multiple chunks of arbitrary size and processed in M rounds, which can reduce

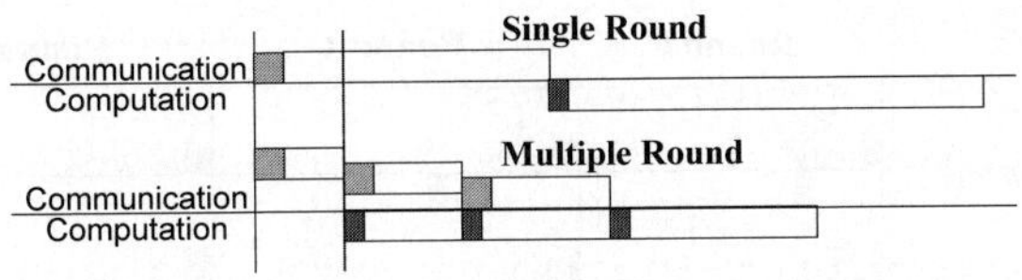

Fig. 1. Multiple-round scheduling

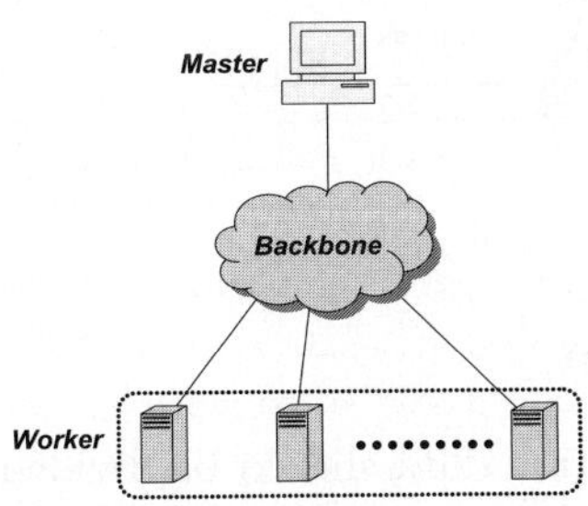

Fig. 2. Distributed computing model

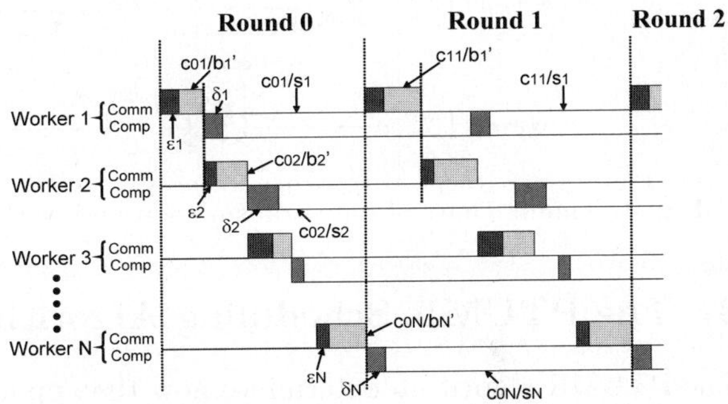

Fig. 3. Timing chart of data transmission and worker processing under UMR

the adverse effects of data transmission time on the application turnaround time. However, the use of a large number of rounds results in an increase in the total overhead. Thus, optimizing the number of rounds so as to minimize the application turnaround time is a key issue in multiple-round scheduling.

UMR is an example of a multiple-round scheduling algorithm [6,7]. The distributed computing model for UMR is shown in Fig. 2. The master and N workers are connected to a high-speed network that is free of bottlenecks. This model has heterogeneity in terms of the computation and communication capacities of workers: each worker i has associated with its computation speed s_i [units/s], data transmission capacity b_i [units/s] of the link attached to the worker, and overheads δ_i [s], ϵ_i [s] added to the computation time and data transmission time, respectively. Furthermore, the data transmission capacity of the link attached to the master is denoted by b_0 [units/s].

UMR adopts the sequential transmission model whereby the master transmits a chunk to one worker at a time. Therefore, the actual data transmission rate b'_i between the master and worker i has to be bounded above by $\min\{b_i, b_0\}$. Figure 3 illustrates how the data is transmitted to workers and then processed under UMR, where the size of the chunk allocated to the worker i in Round j is denoted by c_{ji} [units]. The master determines the amount of chunks allocated to each worker in such a way that the computation time becomes identical for all workers during a round. To reduce data transmission time in the first round, relatively small chunks are transmitted to workers in this round, and the size of chunks then grows exponentially in subsequent rounds.

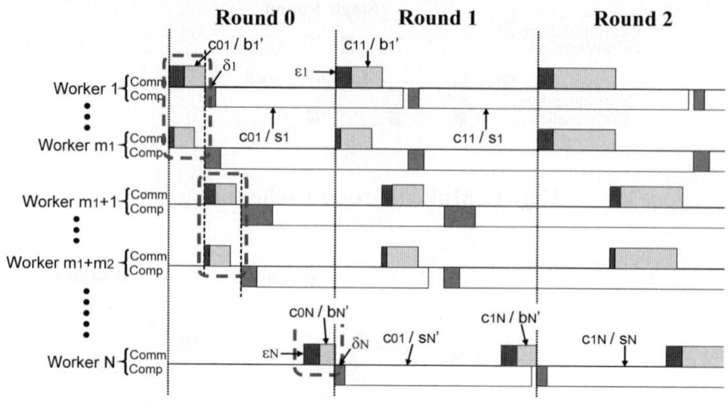

Fig. 4. Timing chart of data transmission and worker processing under PTUMR

3 The PTUMR Scheduling Algorithm

The PTUMR algorithm determines how the application data should be divided and when the data should be transmitted to workers in a network environment that allows the master to transmit data to multiple workers in parallel (Fig. 4), assuming that ϵ_i can be overlapped among concurrent transmissions. More precisely, the PTUMR divides workers into appropriate groups, and treats the set of workers in each group as a single virtual worker. Then the master transmits chunks to virtual workers sequentially, as in UMR.

After appropriately grouping the workers, the PTUMR algorithm analytically determines the appropriate number M^+ of rounds so that the application turnaround time for the total amount W of application data is minimized.

We assume that the values b_0 and $b_i (i = 1, 2, \ldots, N)$ are known to the master, and also that it can control the rate of data transmission b'_i to worker i to be within the range $[0, \min(b_0, b_i)]$. Note that such control can be achieved under the TCP by constraining the sending socket buffer size.

3.1 Computation and Data Transmission Capacities of a Virtual Worker

This subsection shows how to derive the resource capacity of a virtual worker in terms of the resource capacities of its members. The set of workers composing the virtual worker k is denoted by L_k and the number of workers in L_k by m_k.

In order to minimize the computation time of the virtual worker, the size c_{ji} of chunk allocated to worker i in Round $j (= 0, \ldots, M-1)$ should be proportional to its computation speed s_i [3]. In addition, we take the overhead Δ_k of virtual worker k to be the largest overhead δ_i among all workers in L_k. Then, the computation time of the virtual worker k in Round j is given by

$$T_{comp_{jk}} = \frac{C_{jk}}{\sum_{i \in L_k} s_i} + \max_{i \in L_k}\{\delta_i\} = \frac{C_{jk}}{S_k} + \Delta_k. \quad \left(C_{jk} = \sum_{i \in L_k} c_{ji} \right) \tag{1}$$

where C_{jk} is defined as the total size of chunks allocated to all workers in L_k in Round j, and S_k denotes the computation speed of virtual worker k.

Next, we assume that the data transmission time in each round is identical for all workers in L_k by limiting the data transmission rate b'_i ($\leq b_i$) to each worker i. In addition, we define that the overhead E_k of virtual worker k is the largest overhead ϵ_i among all workers in L_k. Then, the data transmission time of the virtual worker k in Round j is given by

$$T_{comm_{jk}} = \frac{C_{jk}}{\sum_{i \in L_k} b'_i} + \max_{i \in L_k}\{\epsilon_i\} = \frac{C_{jk}}{B_k} + E_k. \quad (2)$$

3.2 The Grouping Method

Since the resource capacity of the virtual worker depends on those of its members, the grouping method strongly affects the performance of PTUMR. The grouping method of PTUMR consists of two steps.

1. All workers are sorted and given serial numbers in ascending order of $r_i = s_i / \min\{b_0, b_i\}$. The workers are then divided into several groups according to the following equation.

$$m_k = \max\left\{ m \;\middle|\; \sum_{l=l_k}^{l_k+m-1} b_l \leq b_0 \right\} + x, \quad l_{k+1} = l_k + m_k. \quad (3)$$

   where l_k indicates the serial number of the first worker composing the virtual worker k and x indicates the number of additional workers added to the group after the number of worker has been chosen in a way to make full use of the master-network bandwidth. Note that the optimal value of x cannot be derived analytically. However, in our extensive performance evaluation, we found ten or more additional workers improved the performance to a nearly optimal point in various conditions.

1'. The master groups workers whose resource capacities are close to each other according to the following equation.

$$m_k = \min\{m_k \text{ in } (3),\ m'_k\},\ m'_k = \max\left\{ m \;\middle|\; r_{l_k+m-1} \leq \frac{1.5 \times \sum_{l=l_k}^{l_k+m-2} r_l}{m-1} \right\}. \quad (4)$$

   If r_i of the next worker is 1.5 times larger than the average r_i of all workers already selected for the group, this next worker will not be included.

2. The virtual workers are sorted in ascending order of $R_k = S_k/B_k$, and the number N_v of virtual workers utilized for application processing is chosen according to the following equation.

$$N_v = \max\left\{ n \;\middle|\; \sum_{k=1}^{n} R_k < 1 \right\}. \quad (5)$$

Step 1 presents a basic grouping method which attempts to preferentially select workers with larger r_i, as happens in UMR [7], and to fully utilize the bandwidth of the master-network link. Furthermore, the additional number x of workers aims at reducing the number of steps required to transmit data to all workers, which allows overlapping of the overhead for more workers.

However, in more heterogeneous environments, a virtual worker determined by Step 1 may include some workers with much lower r_i than others, which leads to critical degradation of the resource capacity of the virtual worker. Therefore, when heterogeneity is high, the basic grouping method does not result in efficient execution of the application (shown later in Section 4.1). Step 1' proposes a modified PTUMR, PTUMR with Grouping Threshold (GT), which is restricted to make a group of workers with similar resource capacities.

Step 2 then preferentially selects virtual workers with larger R_k, and limits the number of virtual workers so as to prevent the allocation of application tasks to a virtual worker having low capacity.

3.3 Derivation of Parameters That Result in Almost Minimal Turnaround Time

The new scheduling algorithm, PTUMR, determines the number M^+ of rounds that is nearly optimal in terms of minimizing application turnaround time. The application turnaround time T_{real} is determined by given parameters (the number M of rounds and the size C_{jk} of chunk allocated to virtual worker k in Round j). However, since T_{real} is difficult to express analytically, we instead derive the ideal application turnaround time T_{ideal} under the (ideal) assumption that no virtual worker ever enters the idle computation state once it has received its first chunk of data. In addition, we also assume that the time required to compute chunks received in each round is identical for all virtual workers.

We denote by $w_j(=\sum_{k=1}^{N_v} C_{jk})$ the total amount of chunk to be allocated to virtual workers in Round j, and from Eq. (1), the relation between w_j and the size C_{jk} of chunk allocated to the virtual worker k is given by

$$C_{jk} = \alpha_k \times w_j + \beta_k, \tag{6}$$
$$\left(\alpha_k = \frac{S_k}{\sum_{k=1}^{N_v} S_k}, \quad \beta_k = \frac{S_k \times \sum_{k=1}^{N_v}\{S_k \times \Delta_k\}}{\sum_{k=1}^{N_v} S_k} - S_k \times \Delta_k . \right)$$

In addition, from Eqs. (1) and (2), the total amount of chunk w_j for Round j can be determined by the total chunk size w_0 in the first round as follows.

$$w_j = \theta^j (w_0 - \gamma) + \gamma, \quad \left(\theta = \frac{1/\sum_{k=1}^{N_v} S_k}{\sum_{k=1}^{N_v}\{\alpha_k/B_k\}}, \right. \tag{7}$$
$$\left. \gamma = \frac{\sum_{k=1}^{N_v}\{S_k \times \Delta_k\}/\sum_{k=1}^{N_v} S_k - \sum_{k=1}^{N_v}\{\beta_k/B_k\} - \sum_{k=1}^{N_v} E_k}{\sum_{k=1}^{N_v}\{\alpha_k/B_k\} - 1/\sum_{k=1}^{N_v} S_k} . \right)$$

The application turnaround time T_{ideal} under the ideal assumption can be derived as a function of the number M of rounds, as follows.

$$T_{ideal} = \frac{1}{\sum_{k=1}^{N_v} S_k} \left\{ W + M \times \sum_{k=1}^{N_v} (S_k \times \Delta_k) + \sum_{k=1}^{N_v} \left[S_k \times \sum_{t=1}^{k} \left(\frac{\alpha_t \times \left(\frac{1-\theta}{1-\theta^M} \times (W - M\gamma) + \gamma \right) + \beta_t}{B_t} + E_t \right) \right] \right\}. \quad (8)$$

Due to space limitation, derivation of the application turnaround time in detail is omitted.

Let M^* denote the real value minimizing T_{ideal} in Eq. (8), which can be obtained by solving $\frac{\partial T_{ideal}}{\partial M} = 0$. Then, it is necessary to determine an appropriate number M^+ of rounds as an integer expected to nearly minimize T_{real} if M^* is not an integer. There are four possible integers to consider: $\lfloor M^* \rfloor - 1$, $\lfloor M^* \rfloor$, $\lceil M^* \rceil$ and $\lceil M^* \rceil + 1$. We can choose one among them in such a way as to minimize T_{real}.

4 Performance Evaluation

In this study, we assume, as was the case in the study proposing UMR [7], that the computation speed s_i, the worker-network link capacity b_i, and the latency parameters ϵ_i and δ_i corresponding to the related overheads of workers, are distributed uniformly within the following range.

$$\left((1 - \sqrt{3} \times het) \times mean, (1 + \sqrt{3} \times het) \times mean \right). \quad (9)$$

where *het* represents the heterogeneity of each resource capacity in the environment. We employ a coefficient of variation of each resource capacity as *het*. In addition, *mean* can be set to the average capacity over all workers, namely $\bar{s}$, $\bar{b}$, $\bar{\delta}$, and $\bar{\epsilon}$ listed in Tab 1.

The effectiveness of PTUMR is evaluated by comparing the achievable turnaround time $T_{real}(M^+)$ with the lower bound T_{bound}, which corresponds to the best possible turnaround time in an environment where the network resources are sufficient to ensure negligible data transmission times and any latency corresponding to related overheads is ignored. From Eq. (8), T_{bound} is obtained as follows.

$$T_{bound} = \frac{W}{\sum_{i=1}^{N} s_i}. \quad (10)$$

For a given parameter set (heterogeneity *het* and average resource capacities $\bar{s}$, $\bar{b}$, $\bar{\delta}$ and $\bar{\epsilon}$), 100 experiments were conducted. The average and maximum application turnaround time T_{real} in the 100 experiments was then used as a measure of performance of the scheduling algorithms.

Table 1. Model parameters and their values examined in performance evaluation

W	$100, 500, 1000, 5000, 10000$ [units]
$\bar{s}$	1 [units/s]
b_0	$200, 400, \cdots, 2000$ [units/s]
$\bar{b}$	200 [units/s]
$\bar{\epsilon}$	$0.001, 0.01$ [s]
δ	0.1 [s]

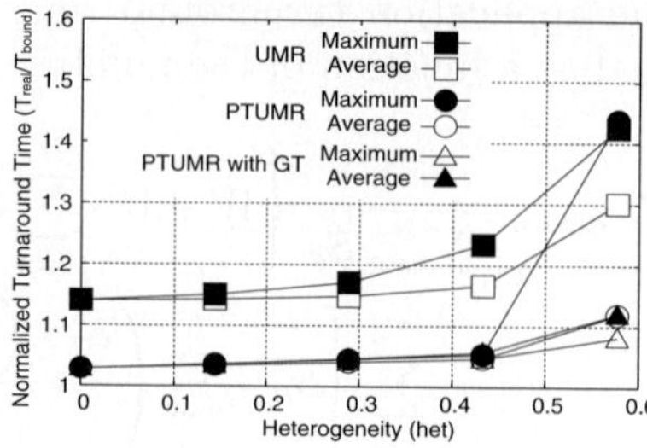

Fig. 5. Impact of heterogeneity

4.1 The Impact of Heterogeneous Resource Capacities

First, we investigated the effect of the heterogeneity *het* on the performance of our scheduling algorithms, UMR and PTUMR. In our evaluation model, we assumed a total amount W of application data of 1000, a master-network transmission capacity b_0 of 1000, an average overhead $\bar{\epsilon}$ at the start of the data transmission of 0.01, and 100 workers (N). When we evaluated the impact of the heterogeneity *het* of each resource capacity, all resource capacities s_i, b_i, δ_i and ϵ_i of each worker were randomly chosen according to Eq. (9).

Figure 5 shows the average and maximum normalized turnaround times T_{real}/T_{bound} for 100 experiments as a function of the heterogeneity *het*, where the number x of additional workers under PTUMR was set to 10. As shown in Fig. 5, regardless of *het*, normal PTUMR is superior to conventional UMR in terms of average normalized turnaround time. However, when the heterogeneity is high, the application turnaround time of normal PTUMR in the worst case becomes larger than that of UMR. In contrast, PTUMR with GT can achieve an application turnaround time close to the lower bound even in the worst case. It is apparent from these results that PTUMR with GT can achieve an excellent turnaround time by grouping workers in an appropriate way.

In the following section, we will consider only PTUMR with GT with the number x of addition workers of 10. In addition we will investigate the performance of the scheduling algorithms in highly heterogeneous environment, namely $het = \frac{\sqrt{3}}{4}$.

4.2 The Impact of Network Resources

The impact of network resources is examined here by assuming a total workload W of 1000 and 100 workers (N). Figure 6 shows the average normalized turnaround time T_{real}/T_{bound}, as a function of the master-network link capacity b_0. Even if b_0 increases, the UMR algorithm cannot effectively utilize the additional network capacity. By contrast, the application turnaround time under PTUMR decreases with increasing b_0 because the algorithm can utilize the full capacity of b_0 by transmitting chunks to multiple workers in parallel. Furthermore, PTUMR achieves T_{real} close to its lower bound T_{bound} across a wide range

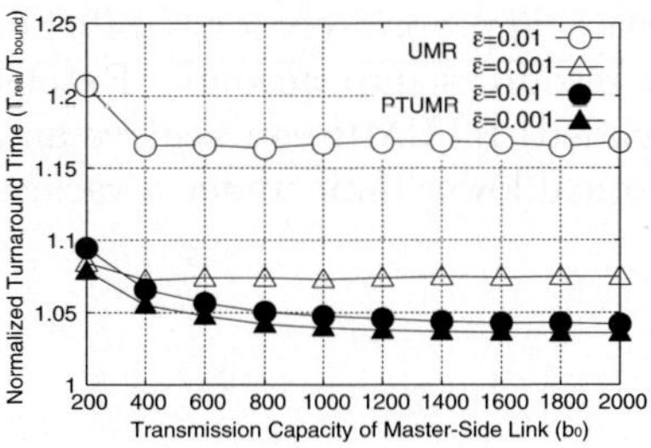

Fig. 6. Impact of network resources

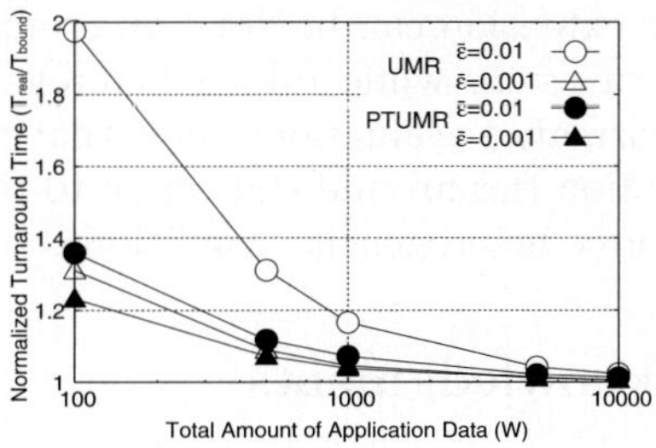

Fig. 7. Impact of total workload

of overhead $\bar{\epsilon}$. This is because PTUMR can reduce the impact of the overhead by aggressively overlapping the overhead ϵ_i for multiple workers.

This evaluation demonstrates that the PTUMR algorithm can achieve application turnaround time quite close to the lower bound through effective utilization of the transmission capacity of the master-network link and the overlapping of overheads for multiple workers.

4.3 The Impact of Total Workload

Finally, the effect of the total amount W of application data is evaluated assuming a master-network transmission capacity b_0 of 1000 and 100 workers (N). Figure 7 shows the average normalized turnaround time T_{real}/T_{bound}, as a function of the application data size W. PTUMR provides excellent performance quite close to the lower bound for any W and any $\bar{\epsilon}$, that is, the PTUMR algorithm effectively eliminates the performance degradation associated with these factors. Under UMR, the normalized turnaround time becomes quite poor as the total data size W decreases, although good performance is achieved for large W. The degradation of performance for low values of W under UMR can be attributed to the increase of the overhead ratio which comes about as a result of decreasing the chunk size. This increase in the overhead ratio can be neutralized by PTUMR. These results therefore show that the PTUMR algorithm can effectively schedule applications of any size by minimizing the adverse effect of overheads on the application turnaround time.

5 Conclusion

We have proposed a novel scheduling algorithm called PTUMR for divisible workload-type applications based on a master-worker model in grid computing environments. The PTUMR allows the master to optimally transmit data to workers in parallel in a multi-round manner, which can considerably reduce application turnaround time compared with conventional multi-round scheduling algorithms such as UMR. The PTUMR presented in this paper is greatly extended from that in our previous study [11] in terms of adaptability to heterogeneous environments with respect to computation and communication resources.

This extension can be done by grouping workers appropriately for parallel data transmission, while taking heterogeneity in resources into account. Extensive performance evaluations show that the (extended) PTUMR can achieve an application turnaround time close to the theoretical lower limit under a variety of resource heterogeneity conditions.

Acknowledgments

This work was supported in part by the Ministry of Education, Culture, Sports, Science and Technology, Japan, under the Project National Research GRID Initiative (NAREGI) and in part by the Ministry of Internal Affairs and Communications, Japan.

References

1. I. Foster and C. Kesselman, *The GRID Blueprint for a New Computing Infrastructure,* Morgan Kaufmann Publishers, 1998.
2. I. Foster, C. Kesselman, and S. Tuecke, "The Anatomy of the Grid," *International Journal of Supercomputer Applications*, Vol. 15, No. 3, pp. 200–222, 2001.
3. V. Bharadwaj, D. Ghose, V. Mani, and T. G. Robertazzi, "Scheduling Divisible Loads in Parallel and Distributed Systems," *IEEE Computer Society Press,* 1996.
4. T. G. Robertazzi, "Ten Reasons to Use Divisible Load Theory," *Jounal of IEEE Computer,* Vol. 36, No. 5, pp. 63–68, May 2003.
5. D. Gerogiannis and S. C. Orphanoudakis, "Load Balancing Requirements in Parallel Implementations of Image Feature Extraction Tasks," *IEEE Trans. Parallel and Distributed Systems,* Vol. 4, No. 9, pp. 994–1013, 1993.
6. Y. Yang and H. Casanova, "UMR: A Multi-Round Algorithm for Scheduling Divisible Workloads," *Proc. of International Parallel and Distributed Processing Symposium (IPDPS'03),* Nice, France, April 2003.
7. Y. Yang and H. Casanova, "A Multi-Round Algorithm for Scheduling Divisible Workload Applications: Analysis and Experimental Evaluation," *Technical Report of Dept. of Computer Science and Engineering, University of California CS20020721,* 2002.
8. O. Beaumont, A. Legrand, and Y. Robert, "Optimal Algorithms for Scheduling Divisible Workloads on Heterogeneous Systems," *Proc. of International Parallel and Distributed Processing Symposium (IPDPS'03),* Nice, France, April 2003.
9. C. Cyril, O. Beaumont, A. Legrand, and Y. Robert, "Scheduling Strategies for Master-Slave Tasking on Heterogeneous Processor Grids," *Technical Report 2002-12,* LIP, March 2002.
10. A. L. Rosenberg, "Sharing Partitionable Workloads in Heterogeneous NOWs: Greedier Is Not Better," *Proc. of the 3rd IEEE International Conference on Cluster Computing (Cluster 2001),* pp. 124–131, California, USA, October 2001.
11. H. Yamamoto, M. Tsuru, and Y. Oie, "Parallel Transferable Uniform Multi-Round Algorithm for Achieving Minimum Application Turnaround Times for Divisible Workload," *Proc. of the 2005 International Conference on High Performance Computing and Communication (HPCC-05), LNCS 3726,* pp. 817–828, Capri-Sorrento Penisular, Italy, September 2005.

Clustering Multicast on Hypercube Network*

Lu Song, Fan BaoHua, Dou Yong, and Yang XiaoDong

College of Computer Science, National University of Defense Technology,
Changsha, Hunan 410073, People's Republic of China
lusong@nudt.edu.cn

Abstract. Multicast communication is one of the general patterns of collective communication in multiprocessors. On hypercube network, the optimal multicast tree problem is NP-hard and all existing multicast algorithms are heuristic. And we find that the existing works are far away from optimal. So this paper aims to design an more efficient algorithm to reduce the communication traffic of multicast in hypercube network. We propose a clustering model and an efficient clustering multicast algorithm. Compared with the existing related works by simulation experiments, our heuristic algorithm reduces the communication traffic significantly.

1 Introduction

A multicast communication in networks means that a source node sends a message to some destination nodes. The multicast algorithms are to determine the paths routing the message to the destination nodes. One goal of the researches on multicast is to reduce the communication traffic.

Depending on the different underlying switching, Lin and Ni[1] formulated the multicast communication problem in multicomputers as three different graph problems: the Steiner tree (ST) problem, the multicast tree (MT) problem, and the multicast path (MP) problem. The optimization of all these three multicast problems on hypercube networks has been proved to be NP-hard [1,2,9]. Lan, Esfahanian and Ni proposed a famous algorithm (LEN's MT algorithm)[2] for multicast tree problem on hypercube networks. Another heuristic algorithm for multicast tree problem was proposed by Sheu[3].

However, the existing algorithms are optimal or approximately optimal only under some special conditions. Moveover, routing following Sheu's algorithm may lead to a cycle in some case, which means the correctness of the algorithm can not be always guaranteed. In this paper, we propose a clustering model for hypercube and design a heuristic multicast algorithm. The idea is putting the neighboring nodes together to form a cluster. Then choose for the routing path based on the clusters.

In section 2, the definition of multicast tree and optimal multicast tree [7] are presented. We also propose two properties of multicast set, global-info and

* Supported by PCSIRT (Program for Changjiang Scholars and Innovative Research Team in University) of China under Grant No. IRT0446.

M. Gerndt and D. Kranzlmüller (Eds.): HPCC 2006, LNCS 4208, pp. 61–70, 2006.

locality, which are used for message routing. Then we outline the LEN's MT algorithm and Sheu's algorithm. In section 3, we propose a clustering model for hypercube, the relevant clustering algorithm and the clustering multicast algorithm with an illustration. Sections 4 presents the performance analysis and the result of simulation experiments. The final section is about the conclusions.

2 Related Works

A multicast communication can be supported by many one-to-one communications, which is called unicast-based multicast. In this method, system resources are wasted due to the unnecessary blocking caused by nondeterminism and asynchrony [5,6,7]. Even without blocking, multicast may reduce communication traffic and latency considerably. In this section, we state the multicast tree problem and outline LEN's and Sheu's algorithms.

2.1 Multicast Tree

Multicast tree problem is formulated as a graph problem by Lin and Ni[1]. And the optimal multicast problem is originally defined by [4].

Definition 1 (Multicast Tree). *Given a graph $G = (V, E)$, a source node u_0, and a multicast set $M \subseteq V$, a multicast tree is a subtree $G_T = (V_T, E_T)$ of G such that $M \subseteq V_T$ and for each $u \in M$, the path length from u_0 to u is equal to length on G.*

Definition 2 (Optimal Multicast Tree). *Given a graph $G = (V, E)$, the optimal multicast tree (OMT) $G_{OMT} = (V_{OMT}, E_{OMT})$ from source vertex u_0 to destination set $M = \{u_1, u_2, \cdots, u_k\}$ is a multicast tree of G such that $|E_{OMT}|$ is as small as possible.*

Here, we propose two properties of multicast set, global-info and locality. The global-info describes the distribution of the destination nodes. The locality describes the extent of neighboring. They are both used for message routing.

Definition 3 (Global-info). *The global-info consists of (1) the number of destination vertices $\|M\|$, (2) the counter of relative address (see definition 9) of all destination vertices on each dimension, $t \triangleq t_1 t_2 \cdots t_n$, $t = \sum_{i=1}^{k}(bitxor(u_i, u_0))$.*

Definition 4 (Locality). *The locality is the extent of nearness between vertices inside destination set, which can be measured by an array of distance D–array. There is no locality between deferent destination subset.*

$$D\text{–}array = \begin{bmatrix} H(u_1,u_1) & H(u_1,u_2) & \cdots & H(u_1,u_k) \\ H(u_2,u_1) & H(u_2,u_2) & \cdots & H(u_2,u_k) \\ \cdots & \cdots & \cdots & \cdots \\ H(u_n,u_1) & H(u_n,u_2) & \cdots & H(u_n,u_n) \end{bmatrix}.$$

2.2 LEN's MT Algorithm

In LEN's algorithm [2], when an intermediate node w receives the message and the destination set M, it has to check if it is a destination node itself. If so, it accepts the message locally and deletes itself from M. Then, it has to compute the relative address of all the destination nodes. For a destination u, the ith bit of u's relative address is 1 if u is different from w on dimension i. Hence, for each dimension $i(0 \leqslant i \leqslant n-1)$, LEN's algorithm counts how many destination nodes whose ith bit of the relative address is 1. After that, it always chooses a particular dimension j with the maximum count value. All destination nodes whose jth bit of the relative address is 1 are sent to the neighbor of w on jth dimension. Then, these nodes are deleted from destination set M. This procedure is repeated until the multicast set M becomes empty.

LEN's MT algorithm votes for the paths to route the message depending on the global-info, without considering the locality between the destination nodes. Hence, the result from LEN's algorithm is far from optimal. Let us give an example. Consider Q_5, suppose the source node is 00000(0) and the multicast destination set M = {01010(10), 11101(29), 10001(17), 11111(31), 11100(28), 01011(11), 00010(2), 10110(22), 11110(30), 11011(27)}. The result from LEN's algorithm is shown in figure 1(a), where the gray nodes are destination nodes and the nodes with dashed line are intermediate nodes. A better result from our algorithm is shown figure 1(b).

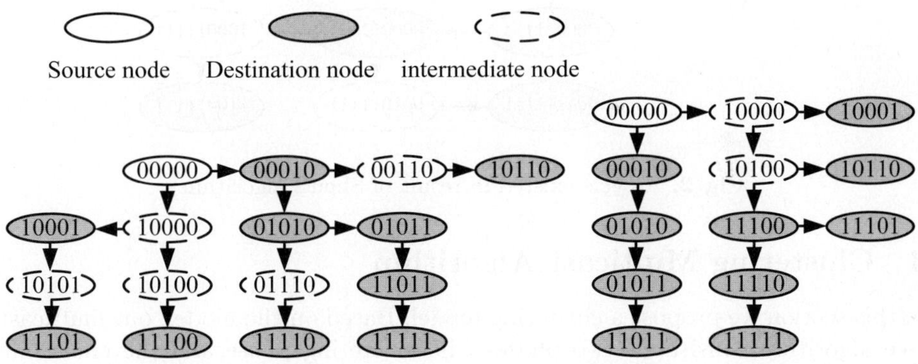

Fig. 1. The results from LEN's MT algorithm and our algorithm

2.3 Sheu's Algorithm

Sheu's algorithm [3] consists of two phases, i.e. the neighbors linking phase and the message routing phase. The neighbors linking phase is executed only on the source node. It links the destination nodes in the multicast set M which are adjacent. After this phase, the multicast set becomes a neighboring forest in which each element is a root of tree. The routing phase is executed on each intermediate nodes in the multicast tree. Similar to LEN's MT algorithm, this phase votes for the paths to routing the message.

Sheu' algorithm votes for the paths depending on the global-info and the locality between the nodes. However, the neighbor linking phase make the global-info injured without reserving weight of the trees. Under particular condition, the result from Sheu's algorithm has a cycle. Here is an example. Consider Q_9, suppose the source node is 000000000(0) and the multicast destination set $M = \{000010000(16), 000011000(24), 000011100(28), 000001111(15), 000011111(31), 110011111(415), 001111111(127)\}$. There is a cycle in the result from Sheu's algorithm shown by figure 2, where two branches arrive node 000011111(31). In the neighbor linking phase, the global-info losses the node 000011111(31) which should be used to vote for the paths to node 110011111(415) and 001111111(127). It means that Sheu's MT algorithm is not always right. In the rest of this paper, we leave Sheu's algorithm out of account.

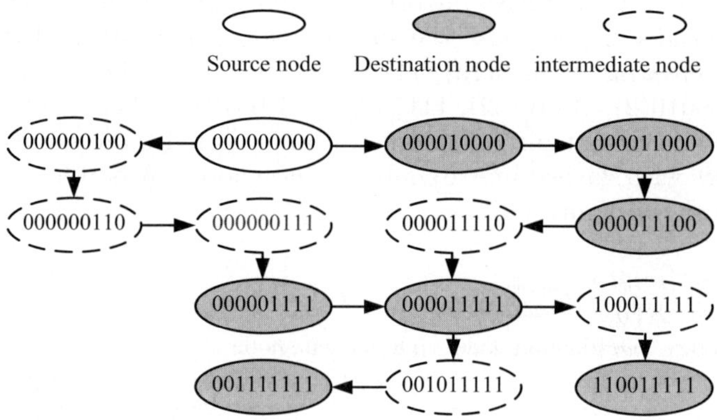

Fig. 2. A cycle existed in result of Sheu's algorithm

3 Clustering Multicast Algorithm

In this section we propose a clustering model. Based on the model, our multicast tree algorithm consists of two phases, the clustering phase and the clustering multicast phase. The Algorithm 1 describes the clustering phase executed only on the source node. The clustering multicast phase is executed on each intermediate node in the multicast tree. This procedure is shown by algorithm 2.

3.1 Clustering Model

Since one of the goals in the MT problem is to minimize the traffic, a multicast tree is better if it has fewer nodes in the multicast tree. Actually, there are at least $k + 1$ nodes to form a tree for a 1-to-k multicast communication. Suppose the multicast set contains only two adjacent node u and v. The multicast tree formed by putting two nodes on the same path will never be worse than that formed by putting them on different paths, because the former case never leads to additional traffic. It is also suitable for the case that node u and node v are neighboring, $(H(u, u_0) \gg H(u, v), H(v, u_0) \gg H(u, v))$. The nearness between nodes is called locality (Definition 4). Based on the idea of locality, we propose a clustering model which can split the multicast set into clusters.

Based on the definition of hypercube [8,10], we propose the definition of subcube, expansion subcube, distance and relative address.

Definition 5 (Hypercube). *A hypercube is defined as a graph $Q_n = (V, E)$. The vertex set V of Q_n consists of all binary sequence of length n on the set $\{0, 1\}$, $V = \{x_1x_2\cdots x_n | x_i \in \{0, 1\}, i = 1, 2, \cdots, n\}$. Two vertices $u = u_1u_2\cdots u_n$ and $v = v_1v_2\cdots v_n$ are linked by an edge if and only if u and v differ exactly in one coordinate, $E = \{(u, v)| \sum_{i=1}^{n} |u_i - v_i| = 1\}$.*

Definition 6 (Subcube of Q_n). *A subcube H_k of Q_n is a binary sequence $b_1b_2\cdots b_{n-k}$ of length $(n - k)$ which presents a subgraph containing 2^k vertices and $k \cdot 2^{k-1}$ edges. H_k can be denoted as $b_1b_2\cdots b_{n-k}\star\star\star$ inside of which the vertex has a form like $b_1b_2\cdots b_{n-k}x_{n-k+1}\cdots x_n (x_j \in \{0, 1\}, n-k+1 \leqslant j \leqslant n)$.*

Definition 7 (Expansion subcube). *A expansion subcube of destination set M denoted as $expan(M)$ is the minimal subcube containing M. $expan(M) = \{H_k | k \leqslant i (\forall i, M \subseteq H_i)\}$.*

Definition 8 (Distance). *The distance between vertices is defined as the Hamming distance, $H(u, v) = \sum_{i=1}^{n} |u_i - v_i| (u = u_1u_2\cdots u_n, v = v_1v_2\cdots v_n)$. The distance between vertex sets is defined as $H(U, V) = \min_{u_i \in U, v_j \in V}(H(u_i, v_j))$.*

Definition 9 (Relative address). *The relative address of vertex u to vertex v is a binary sequence $R(u, v) \triangleq r_1r_2\cdots r_n$, where $r_i = u_i \oplus v_i (u = u_1u_2\cdots u_n, v = v_1v_2\cdots v_n)$.*

Using definition 5,6,7,8,9, we present the definition of cluster.

Definition 10 (Cluster). *Cluster is a set of destination nodes that are near each other. There are two metrics of a cluster c, weight and degree. (1)Weight. Weight is defined by the number of nodes in the cluster. $W(c) = \|c\|$.(2)Degree. Degree is defined as the dimension of the expansion subcube of cluster c. $D(c) = k(H_k = expan(c))$.*

Actually, considering each node as a cluster, LEN's algorithm is a kind of generalized clustering multicast algorithm where the degree of each cluster is zero. It means that LEN's algorithm doesn't use the locality between destination nodes. As mentioned above, the result from LEN's algorithm is far away from optimal.

In next two subsections, we propose a clustering algorithm to split the multicast set into clusters and a clustering multicast algorithm based on clusters with their priorities.

3.2 Clustering Algorithm

In this subsection, we propose a clustering algorithm (algorithm 1) which is used to split the multicast set into clusters. In algorithm 1, all the destination nodes are classified by their distance to the source node. Then we find adjacent nodes that are directly linked by an edge in the graph of hypercube, and put these adjacent nodes together to form a tree. After step 2, we get a set of trees. At step 3, we combine the trees with the one whose expansion subcube contains the other's. The combination forms a cluster. Thus, the cluster set is generated by our clustering algorithm.

Algorithm 1. Clustering algorithm on Q_n

```
Input  : source node u_0, multicast set M = {u_1, u_2, ..., u_k}
Output: cluster set C

step 1: M_0 <- {u_0}, M_1 <- M_2 <- ... <- M_n <- ∅
        for i ∈ [1, n] do
        |  M_i = {u_j | H(u_j, u_0) = i}
        end
step 2: ∀i ∈ [1, n];  for u ∈ M_i do
        |  if ∃v ∈ M_{i-1} && H(u, v) == 1 then
        |  |  M_{i-1} = M_{i-1} ∪ {u}
        |  |  M_i = M_i - {u}
        |  end
        end
        C = M_1 ∪ M_2 ∪ ... ∪ M_n
step 3: ∀i, j ∈ [1, n]; if expan(c_i) ⊆ expan(c_j) then
        |  c_j <- c_j ∪ c_i
        |  c_i <- ∅
        end
```

3.3 Clustering Multicast

In this subsection, we present the clustering multicast algorithm (algorithm 2). At step 1, if local node is contained in the multicast set, it deletes itself from the multicast set. At step 2, for each cluster, we find out the head that is nearest to local node in the cluster. Then compute the relative address of each head to local node. Now we can use the relative addresses as the message routing direction and the weight of clusters as the priority. At step 3, for each cluster and dimension i, we counts the product of wight and ith bit of relative address. Similar to LEN's algorithm, we choose a particular dimension j with the maximum count value. All clusters whose relative address of the head has 1 on jth bit are sent to the neighbor of local node on jth dimension. This procedure is repeated until the cluster set C becomes empty.

The following theorem shows the correctness of clustering multicast Algorithm.

Algorithm 2. Message routing phase

```
Input   : Local node address w, cluster set C
Output: None
step 1: if ∃c ∈ C, w ∈ c then
            send the message to local processor
            delete w from c
        end
step 2: foreach c_i ∈ C do
            u_i ← min(H(u_j, w)); (u_j ∈ c_i)
            u_i ← bitxor(u_i, w)
            w_i ← ||c_i||
        end
step 3: while C ≠ ∅ do
            t ← u_i * w_i; (t ≜ t_1 t_2 ··· t_n)
            t_j ← max_{1≤i≤n}(t_i)
            C' ← {c_i | c_i ∈ C, u_i and w diff on dimension j }
            if ∃c ∈ C, H({w}, c) ⩾ H(c, C') then
                C' ← C' ∪ {c}
                C ← C − c
            end
            Transmit C' and the message to node (w ⊕ 2^j)
            C ← C − C'
        end
```

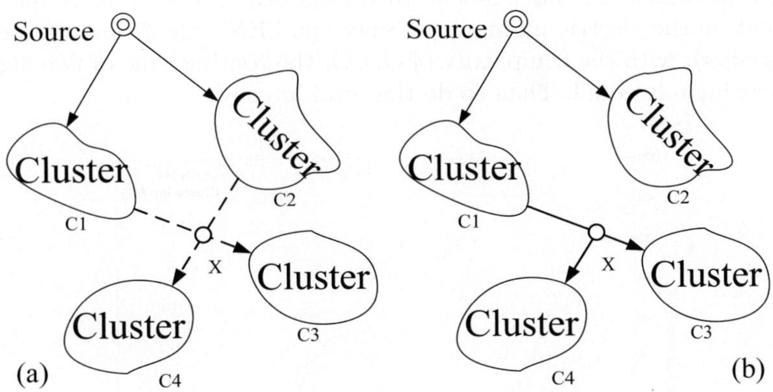

Fig. 3. Routing between clusters

Theorem 1. *Given a cluster set C in Q_n, the edges selected by clustering multicast algorithm form a multicast tree.*

Proof. (Proof by Contradiction.) Assume to the contrary that there is a cycle in the result from clustering multicast algorithm. It means that paths between clusters intersect at a node. We may suppose that the path from cluster c_1 to cluster c_3 and the path from cluster c_2 to cluster c_4 intersect at node x (see figure 3(a)). The distance between each cluster and node x is denoted as d_1,d_2,d_3 and d_4. There are two cases to consider. (1) $d_1 == d_2$. Because the algorithm 2 executed in sequence, cluster c_3 and c_4 can only be both appeared in the branch from cluster c_1(or c_2), contradicting our assumption. (2) $d_1 \neq d_2$. We may assume $d_1 < d_2$. Then we get $d(c_1, c_4) < (c_2, c_4)$. It means that cluster c_4 is closer to c_1 than c_2. According to clustering multicast algorithm, cluster c_3 and c_4 are both appeared in the branch from cluster c_1 (shown in figure 3(b)), again a contradiction. This completes the proof. □

4 Performance Analysis

4.1 Complexity Compare

Consider the 1-to-k multicast on Q_n. The complexities of step 1 and step 2 in clustering algorithm (algorithm 1) are both $O(nk)$. Since the elements of C are far less than M, the complexity of step 3 is $O(step3) < O(nk)$. Hence the complexity of clustering algorithm is $O(nk)$. Similarly in message routing algorithm, $O(step1) = O(k)$, $O(stpe2) = O(step3) < O(nk)$, the complexity of clustering multicast is $O(nk)$. So the complexity to finish multicast is $O(nk)$. And the complexity of LEN's algorithm is also $O(nk)$[1,2].

And the transfer latency of LEN's algorithm and our algorithm are approximately equal. The latency of multicast was composed of transferring time and routing time which are independent with each other. The transferring time is dependent on the electric mechanism. Since the LEN's algorithm and our algorithm are both with the complexity of $O(nk)$, the routing time of two algorithm are approximately equal. Then so do the total latency.

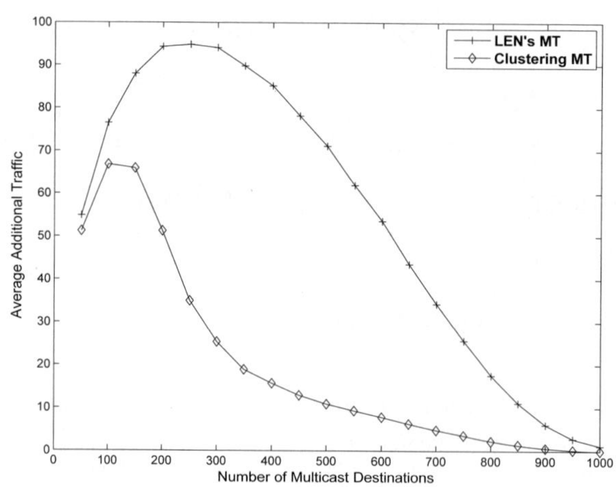

Fig. 4. Average additional traffic generated by LEN's MT algorithm and clustering multicast

4.2 Simulation

In this subsection, we analysis the performance of our multicast algorithms on Q_{10} by simulation experiments. In general, the distribution of destination nodes has a great effect on the total traffic of multicast communication. But, similar to the performance studies in existing works [1,2,3], we assume that the routing distribution is *uniform*. For a 1-to-k multicast, it requires at least k *units of traffic*, where a *unit of traffic* is measured as one message transmitted over one link. As in the existing studies [1,2,3], we use *average additional traffic* to evaluate the performance of a multicast communication, where the *average additional traffic* is defined as the average amount of total traffic minus k. The number of destination nodes k is ranged from 50 to 1000 by the step of 50. For each k, we perform the simulation 500 times and the amount of traffic generated for a given k is averaged over the 500 runs. Figure 4 shows the comparison of average additional traffic generated by LEN's MT algorithm and our multicast algorithm.

5 Conclusions

In this paper, we propose a clustering model and a clustering algorithm to achieve a better balance between the global-info and locality which are two essential properties of the multicast set. Based on the clustering model, an efficient heuristic multicast algorithm is presented. By simulation experiments, our clustering multicast algorithm has significant improvements compared to the existing algorithms. And the generation of resulting multicast tree is fully distributed.

References

1. X. Lin and L. M. Ni: Multicast communication in multicomputer networks. IEEE Trans. Parallel Distrib. Systems **4** (1993) 1105-1117
2. Y. Lan, A. H. Esfahanian, and L. M. Ni: Multicast in hypercube multiprocessors. J. Parallel Distrib. Comput. **8** (1990) 30-41
3. Shih-Hsien Sheu and Chang-Biau Yang: Multicast Algorithms for Hypercube Multiprocessors. Journal of Parallel and Distributed Computing **61** (2001) 137-149
4. Y. Lan, A. H. Esfahanian, and L. M. Ni: Distributed multi-destination routing in hypercube multiprocessors. Proceedings of the Third Conference on Hypercube Concurrent Computers and Applications 1988 631-639
5. Choi Y., Esfahanian A. H., and Ni L. M.: One-to-k communication in distributed-memory multiprocessors. Proc. 25th Annual Allerton Conference on Communication, Control, and Computing 1987 268-270
6. William James Dally and Brian Patrick Towles: Principles and Practices of Interconnection Networks. Morgan Kaufmann Publishers 2003
7. J. Duato, S. Yalamanchili, and L. M. Ni: Interconnection Networks: An Engineering Approach. Morgan Kaufmann Publishers 2002

8. Junming Xu: Topological structure and Analysis of Interconnection Networks. Kluwer Academic Publishers 2001
9. R. L. Graham and L. R. Foulds: Unlikelihood that minimal phylogenies for realistic biological study can be constructed in reasonable computational time. Math. Biosci. **60** (1982) 133-142
10. Jianer Chen, Guojun Wang and Songqiao Chen: Locally subcube-connected hypercube networks: theoretical analysis and experimental results. IEEE Transactions on Computers **5** (2002) 530-540

Checkpointing and Communication Pattern-Neutral Algorithm for Removing Messages Logged by Senders

JinHo Ahn

Dept. of Computer Science, College of Science, Kyonggi University
San 94-6 Iuidong, Yeongtonggu, Suwonsi Kyonggido 443-760, Republic of Korea
jhahn@kyonggi.ac.kr

Abstract. The traditional sender-based message logging protocols use a garbage collection algorithm to result in a large number of additional messages and forced checkpoints. So, in our previous work, an algorithm was introduced to allow each process to autonomously remove useless log information in its volatile storage by piggybacking only some additional information without requiring any extra message and forced checkpoint. However, even after a process has executed the algorithm, its storage buffer may still be overloaded in some communication and checkpointing patterns. This paper proposes a new garbage collection algorithm CCPNA for sender-based message logging to address all the problems mentioned above. The algorithm considerably reduces the number of processes to participate in the garbage collection by using the size of the log information of each process. Thus, CCPNA incurs more additional messages and forced checkpoints than our previous algorithm. However, it can avoid the risk of overloading the storage buffers regardless of the specific checkpointing and communication patterns. Also, CCPNA reduces the number of additional messages and forced checkpoints compared with the traditional algorithm.

Keyword: message-passing system, fault-tolerance, message logging, checkpointing, garbage collection.

1 Introduction

With the remarkable advance of processor and network technologies, message-passing distributed systems composed of heterogenous networked computers are becoming a cost-effective solution for high performance parallel computing compared with expensive special-purpose supercomputers. However, one of the big challenges the distributed systems should address is providing fault-tolerance. In other words, even if the failure of a single process in a distributed application occurs, it may lead to restarting the application from its initial state, which is critical to long-running scientific and engineering applications. Rollback-recovery techniques such as checkpointing-based recovery and log-based recovery are very attractive for supporting transparent fault-tolerance to the applications because

M. Gerndt and D. Kranzlmüller (Eds.): HPCC 2006, LNCS 4208, pp. 71–80, 2006.

the techniques require fewer special resources compared to process replication techniques [5]. In checkpointing-based recovery, when some processes crash, the processes affected by the failures roll back to their last checkpoints such that the recovered system state is consistent. But, this technique may not restore the maximum recoverable state because it relies only on checkpoints of processes saved on the stable storage. Log-based recovery performing careful recording of messages received by each process with its checkpoints enables a system to be recovered beyond the most recent consistent set of checkpoints. This feature is desirable for the applications that frequently interact with the outside world consisting of input and output components that cannot roll back [5]. In this technique, messages can be logged either by their senders or by their receivers. First, receiver-based message logging (RBML) approach [8,14] logs the recovery information of every received message to the stable storage before the message is delivered to the receiving process. Thus, the approach simplifies the recovery procedure of failed processes. However, its main drawback is the high failure-free overhead caused by synchronous logging. Sender-based message logging (SBML) approach [2,4,9,11,13] enables each message to be logged in the volatile memory of its corresponding sender for avoiding logging messages to stable storage. Therefore, it reduces the failure-free overhead compared with the RBML approach. However, the SBML approach forces each process to maintain in its limited volatile storage the log information of its sent messages required for recovering receivers of the messages when they crash. Thus, as enough empty buffer space for logging messages sent in future should be ensured in this approach, it requires an efficient algorithm to garbage collect log information of each process [1]. Traditional SBML protocols [2,4,9,11,13] use one between two message log management procedures to ensure system consistency despite future failures according to each cost. The first procedure just flushes the message log to the stable storage. It is very simple, but may result in a large number of stable storage accesses during failure-free operation and recovery. The second procedure forces messages in the log to be useless for future failures and then removes them. In other wards, the procedure checks whether receivers of the messages has indeed received the corresponding messages and then taken no checkpoint since. If so, it forces the receivers to take their checkpoints. Thus, this behavior may lead to high communication and checkpointing overheads as inter-process communication rate increases. To address their problems, in our previous work, a low-cost algorithm called PGCA [1] was presented to have the volatile memory of each process for message logging become full as late as possible with no extra message and forced checkpoint. The algorithm allows each process to locally and independently remove useless log information from its volatile storage by piggybacking only some additional information. However, the limitation of the algorithm is that after a process has performed the algorithm, the storage buffer of the process may still be overloaded in some communication and checkpointing patterns. In this paper, we propose an active garbage collection algorithm called CCPNA to lift the limitation. For this, the algorithm CCPNA uses an array recording the size of the log information for each process. When the free buffer space in the volatile storage is needed, the algorithm selects a small number of processes based on the array that take part in having the messages previously logged for them be useless despite their future failures. Thus, CCPNA results in low communication and checkpointing overheads compared with the traditional ones while avoiding the disadvantage of the algorithm PGCA.

2 System Model

A distributed computation consists of a set P of $n(n > 0)$ sequential processes executed on hosts in the system and there is a stable storage that every process can always access that persists beyond processor failures [5]. Processes have no global memory and global clock. The system is asynchronous: each process is executed at its own speed and communicates with each other only through messages at finite but arbitrary transmission delays. We assume that the communication network is immune to partitioning, there is a stable storage that every process can always access and hosts fail according to the fail stop model [10]. Events of processes occurring in a failure-free execution are ordered using Lamport's *happened before* relation [6]. The execution of each process is *piecewise deterministic* [12]: at any point during the execution, a state interval of the process is determined by a *non-deterministic* event, which is delivering a received message to the appropriate application. The k-th state interval of process p, denoted by ${si_p}^k(k > 0)$, is started by the delivery event of the k-th message m of p, denoted by ${dev_p}^k(m)$. Let p's state, ${s_p}^i =< {si_p}^0, {si_p}^1, \cdots, {si_p}^i >$, represent the sequence of all state intervals up to ${si_p}^i$. Therefore, given p's initial state, ${s_p}^0$, and the non-deterministic events, $[{dev_p}^1, {dev_p}^2, \cdots, {dev_p}^i]$, its corresponding state ${s_p}^i$ is uniquely determined. ${s_p}^i$ and ${s_q}^j (p \neq q)$ are *mutually consistent* if all messages from q that p has delivered to the application in ${s_p}^i$ were sent to p by q in ${s_q}^j$, and vice versa. A set of states, which consists of only one from each process in the system, is *a globally consistent state* if any pair of the states is mutually consistent [3].

The log information of each message kept by its sender consists of four fields, its receiving process' identifier(rid), send sequence number(ssn), receive sequence number(rsn) and data($data$). In this paper, the log information of message m and the message log in process p's volatile memory are denoted by $e(m)$ and log_p.

3 The Proposed Algorithm

The sender-based message logging needs an algorithm to allow each process to remove the log information in its volatile storage while ensuring system consistency in case of failures. This algorithm should force the log information to become useless for future recovery to satisfy the goal. In the traditional sender-based message logging protocols, to garbage collect every $e(m)$ in log_p, p requests that the receiver of m ($m.rid$) takes a checkpoint if it has indeed received m and taken no checkpoint since. Also, processes occasionally exchange the state interval indexes of their most recent checkpoints for garbage collecting the log information in their volatile storages. However, this algorithm may result in a large number of additional messages and forced checkpoints needed by the forced garbage collection. To illustrate how to remove the log information in the algorithm, consider the example shown in figure 1. Suppose p_3 intends to remove the log information in log_{p3} at the marked point. In this case, the algorithm forces p_3 to send checkpoint requests to p_1, p_2 and p_4. When receiving the request, p_1, p_2 and p_4 take their checkpoints, respectively. Then, the three processes send each a checkpoint reply to p_3. After receiving all the replies, p_3 can remove $(e(m_1), e(m_2), e(m_3), e(m_4), e(m_5), e(m_6), e(m_7), e(m_8))$ from log_{p3}. Also, in this checkpointing and communication pattern, the algorithm proposed in [1] cannot allow p_3 to autonomously decide whether log information of each sent message is useless for recovery of the receiver of the message by using some piggybacking information. Thus, even after executing the algorithm, p_3 should maintain all the log information of the eight messages in log_{p3}.

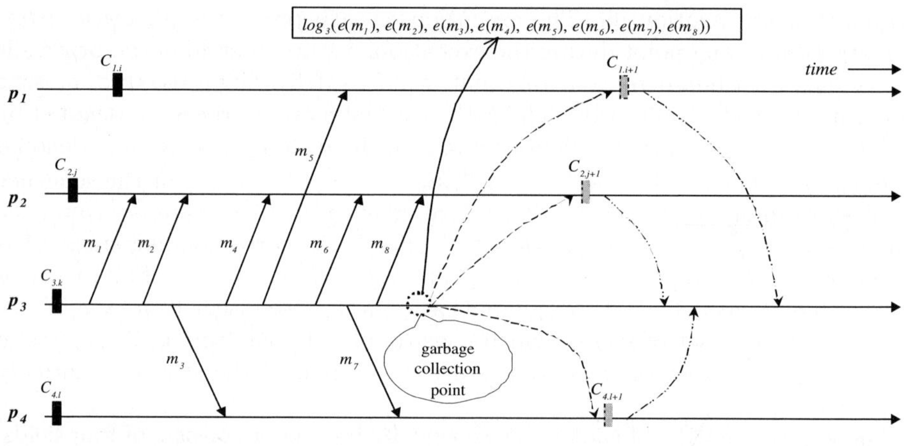

Fig. 1. An example showing the problem of the traditional sender-based message logging protocols

To solve the problem, we present an algorithm CCPNA based on the following observation: if the requested empty space (=E) is less than or equal to the sum (=Y) of sizes of $e(m_1)$, $e(m_2)$, $e(m_4)$, $e(m_6)$ and $e(m_8)$, p_3 has only to force p_2 to take a checkpoint. This observation implies that the number of extra messages and forced checkpoints may be reduced if p_3 knows sizes of the respective log information for p_1, p_2 and p_4 in its volatile storage. CCPNA obtains such information by maintaining an array, $LogSize_p$, to save the size of the log information in the volatile storage by process. Thus, CCPNA can reduce the number of additional messages and forced checkpoints by using the vector compared with the traditional algorithm.

- log_p: It is a set saving $e(rid, ssn, rsn, data)$ of each message sent by p. It is initialized to $\emptyset$.
- $Lssn_p$: It is the send sequence number of the latest message sent by p. It is initialized to 0.
- $Lrsn_p$: It is the receive sequence number of the latest message delivered to p. It is initialized to 0.
- $LssnVec_p$: It is a vector where $LssnVec_p[q]$ records the send sequence number of the latest message received by p that q sent. Each element of the vector is initialized to 0.
- $LogSize_p$: It is a vector where $LogSize_p[q]$ is the sum of sizes of all $e(m)$s in log_p such that p sent m to q. $LogSize_p[q]$ is initialized to 0.
- $LrsnInLchkpt_p$: It is the rsn of the latest message delivered to p before p's having taken its last checkpoint. It is initialized to 0.
- $ENsend_p$: It is a set of rsns that aren't yet recorded at the senders of their messages. It is initialized to an empty set Φ. It is used for indicating whether p can send messages to other processes(when $ENsend_p = \Phi$) or not.

Fig. 2. Data Structures for every process p in CCPNA

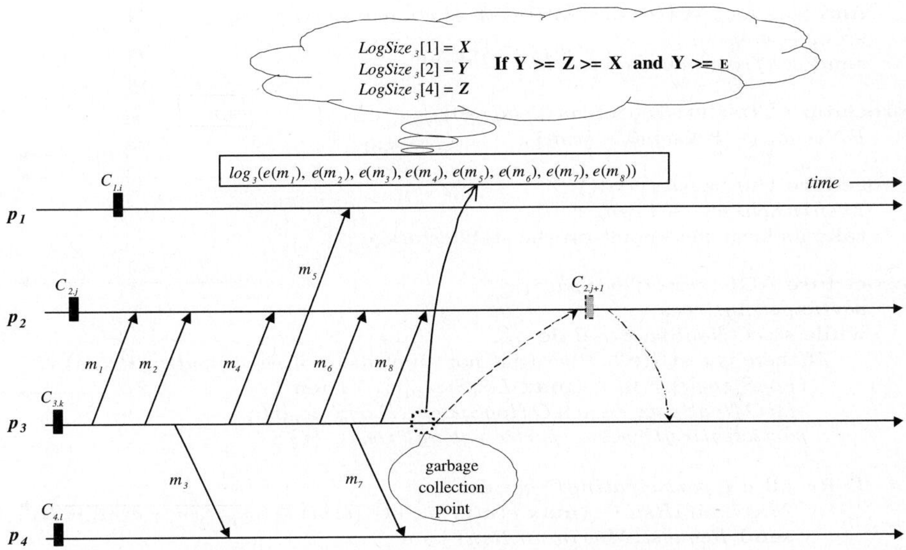

Fig. 3. An example of executing our algorithm CCPNA

In CCPNA, each process p should maintain the data structures shown in figure 2. First, $LogSize_p$ is a vector where $LogSize_p[q]$ is the sum of sizes of all $e(m)$s in log_p, such that p sent message m to q. Whenever p sends m to q, it increments $LogSize_p$ by the size of $e(m)$. When p needs more empty buffer space, it executes CCPNA. It first chooses a set of processes, denoted by $participatingProcs$, which will participate in the forced garbage collection. It selects the largest, $LogSize_p[q]$, among the remaining elements of $LogSize_p$, and then appends q to $participatingProcs$ until the required buffer size is satisfied. Then p sends a request message with the rsn of the last message, sent from p to q, to all $q \in participatingProc$ such that the receiver of m is q for $\exists e(m) \in log_p$. When

```
procedure MSend(data, q)
    wait until(ENsend_p = Φ) ;
    Lssn_p ← Lssn_p + 1 ;
    send m(Lssn_p, data) to q ;
    log_p ← log_p ∪ {(q, Lssn_p, -1, data)} ;
    LogSize_p[q] ← LogSize_p[q] + size of (q, Lssn_p, -1, data) ;

procedure MRecv(m(ssn, data), sid)
    if(LssnVec_p[sid] < m.ssn) then {
      Lrsn_p ← Lrsn_p + 1 ;
      LssnVec_p[sid] = m.ssn ;
      send ack(m.ssn, Lrsn_p) to sid ;
      ENsend_p ← ENsend_p ∪ {Lrsn_p} ;
      deliver m.data to the application ;
    }
    else discard m ;

procedure Ack-Recv(ack(ssn, rsn), rid)
    find ∃e ∈ log_p st ((e.rid = rid) ∧ (e.ssn = ack.ssn)) ;
    e.rsn ← ack.rsn ;
    send confirm(ack.rsn) to rid ;

procedure Confirm-Recv(confirm(rsn))
    ENsend_p ← ENsend_p - {rsn} ;

procedure Checkpointing()
    LrsnInLchkpt_p ← Lrsn_p ;
    take its local checkpoint on the stable storage ;

procedure AGC(sizeOflogSpace)
    participatingProcs ← ∅ ;
    while sizeOflogSpace > 0 do
        if(there is r st ((r ∈ P) ∧ (r is not an element of participatingProcs) ∧
          (LogSize_p[r] ≠ 0) ∧ (max LogSize_p[r]))) then {
          sizeOflogSpace ← sizeOflogSpace − LogSize_p[r] ;
          participatingProcs ← participatingProcs ∪ {r} ;
        }
    T: for all u ∈ participatingProcs do {
            MaximumRsn ← (max e(m).rsn) st ((e(m) ∈ log_p)∧(u = e(m).rid)) ;
            send Request(MaximumRsn) to u ;
       }
       while participatingProcs ≠ ∅ do {
            receive Reply() from u st (u ∈ participatingProcs) ;
            for all et(m) ∈ log_p st (u = e(m).rid) do
                remove e(m) from log_p ;
            LogSize_p[u] ← 0 ;
            participatingProcs ← participatingProcs − {u} ;
       }

procedure CheckLrsnInLchkpt(Request(MaximumRsn), q)
    if(LrsnInLchkpt_p < MaximumRsn) then
      Checkpointing() ;
    send Reply() to q ;
```

Fig. 4. Procedures for every process p in CCPNA

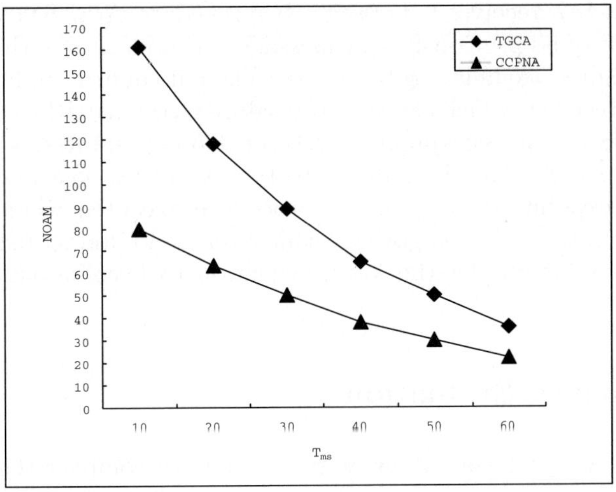

Fig. 5. *NOAM* vs. T_{ms}

q receives the request message with the rsn from p, it checks whether the rsn is greater than $LrsnInLchkpt_p$. If so, it should take a checkpoint and then send p a reply message. Otherwise, it has only to send p a reply message. When p receives the reply message from q, it removes all $e(m)$s from log_p such that the receiver of m is q.

For example, in figure 3, when p_3 attempts to execute CCPNA at the marked point after it has sent m_8 to p_2, it should create $participatingProcs$. In this figure, we can see that $LogSize_{p3}[p_2](= Y)$ is the largest $(Y \geq Z \geq X)$ among all the elements of $LogSize_{p3}$ due to $e(m_1)$, $e(m_2)$, $e(m_4)$, $e(m_6)$ and $e(m_8)$ in log_{p3}. Thus, it first selects and appends p_2 to $participatingProcs$. Suppose that the requested empty space E is less than or equal to Y. In this case, it needs to select any process like p_1 and p_4 no longer. Therefore, p_3 sends a checkpoint request message with $m_8.rsn$ to only p_2 in $participatingProcs$. When p_2 receives the request message, it should take a forced checkpoint like in this figure because the rsn included in the message is greater than $LrsnInLchkpt_{p2}$. Then it sends p_3 a reply. When p_3 receives a reply message from p_2, it can remove $e(m_1)$, $e(m_2)$, $e(m_4)$, $e(m_6)$ and $e(m_8)$ from log_{p3}. From this example, we can see that CCPNA chooses a small number of processes to participate in the garbage collection based on $LogSize_{p3}$ compared with the traditional algorithm. Thus, CCPNA may reduce the number of additional messages and forced checkpoints.

3.1 Algorithmic Description

The procedures for process p in our algorithm are formally described in figure 4. MSend() is the procedure executed when each process p sends a message m and logs the message to its volatile memory. In this case, p adds the size of $e(m)$ to $LogSize_p$[q] after transmitting the message. Procedure MRecv()

is executed when p receives a message. In procedure ACK-RECV(), process p receives the rsn of its previously sent message and updates the third field of the element for the message in its log to the rsn. Then, it confirms fully logging of the message to its receiver, which executes procedure CONFIRM-RECV(). If process p attempts to take a local checkpoint, it calls procedure CHECKPOINTING(). In this procedure, $LrsnInLchkpt_p$ is updated to the rsn of the last message received before the checkpoint. AGC() is the procedure executed when each process attempts to initiate the forced garbage collection, and CHECKLRSNINLCHKPT() is the procedure for forcing the log information to become useless for future recovery.

4 Performance Evaluation

In this section, we perform extensive simulations to compare the proposed algorithm CCPNA with the traditional algorithm TGCA using simjava discrete-event simulation language [7]. Two performance indexes are used for comparison; the average number of additional messages ($NOAM$) and the average number of forced checkpoints ($NOFC$) required for garbage collection per process. In the literature, these two indexes dominate the overhead caused by garbage collection during failure-free operation [5]. A system with 20 nodes connected through a general network was simulated. Each node has one process executing on it and, for simplicity, the processes are assumed to be initiated and completed together. The message transmission capacity of a link in the network is 100Mbps. For the simulation, 20 processes have been executed for 72 hours per simulation run. Every process has a 10MB buffer space for storing its log_p. The message size ranges from 50KB to 200KB. Normal checkpointing is initiated at each process with an interval following an exponential distribution with a mean T_{ckpt}=360 seconds. The simulation parameter is the mean message sending interval, T_{ms}, following an exponential distribution.

Figure 5 shows $NOAM$ for the various T_{ms} values. In this figure, we can see that $NOAM$s of the two algorithms increase as T_{ms} decreases. The reason is that forced garbage collection should frequently be performed because the high inter-process communication rate causes the storage buffer of each process to be overloaded quickly. However, $NOAM$ of CCPNA is much lower than that of TGCA. CCPNA reduces about 38% - 50% of $NOAM$ compared with TGCA.

Figure 6 illustrates $NOFC$ for the various T_{ms} values. In this figure, we can also see that $NOFC$s of the two algorithms increase as T_{ms} decreases. The reason is that as the inter-process communication rate increases, a process may take a forced checkpoint when it performs forced garbage collection. In the figure, $NOFC$ of CCPNA is lower than that of TGCA. CCPNA reduces about 25% - 51% of $NOFC$ compared with TGCA.

Therefore, we can conclude from the simulation results that regardless of the specific checkpointing and communication patterns, CCPNA enables the garbage collection overhead occurring during failure-free operation to be significantly reduced compared with TGCA.

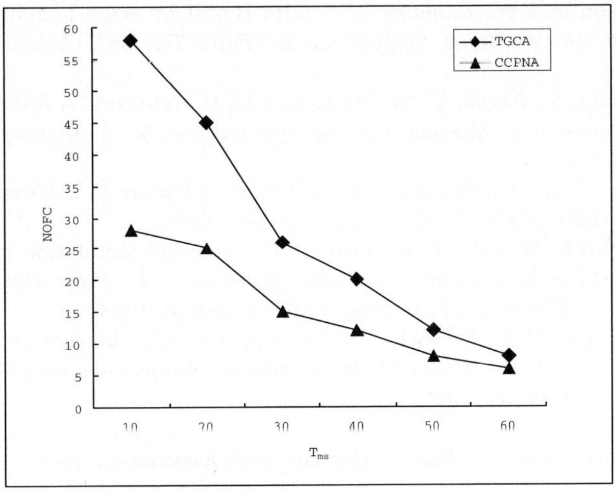

Fig. 6. *NOFC* vs. T_{ms}

5 Conclusion

In this paper, we presented a garbage collection algorithm CCPNA for efficiently removing log information of each process in sender-based message logging. CCPNA allows each process to keep an array to save the size of the log information for every process in its storage by process. It chooses a minimum number of processes to participate in the forced garbage collection based on the array. Thus, it incurs more additional messages and forced checkpoints than our previous algorithm. However, it can avoid the risk of overloading the storage buffers unlike the latter. Moreover, CCPNA reduces the number of additional messages and forced checkpoints needed by the garbage collection compared with the traditional algorithm TGCA. From our simulation experiments, we can see that CCPNA significantly reduces about 38% - 50% of NOAM and 25% - 51% of NOFC regardless of the communication patterns compared with TGCA.

References

1. JinHo Ahn. An Efficient Algorithm for Removing Useless Logged Messages in SBML Protocols. *Lecture Notes In Computer Science*, Vol. 3816, pp. 166-171, Dec. 2005.
2. A. Bouteiller, F. Cappello, T. Hérault, G. Krawezik, P. Lemarinier and F. Magniette. MPICH-V2: a Fault Tolerant MPI for Volatile Nodes based on Pessimistic Sender Based Message Logging. *In Proc. of the 15th International Conference on High Performance Networking and Computing(SC2003)*, November 2003.
3. K. M. Chandy, and L. Lamport. Distributed Snapshots: Determining Global States of Distributed Systems. *ACM Transactions on Computer Systems*, 3(1): 63-75, 1985.

4. D. B. Johnson and W. Zwaenpoel. Sender-Based Message Logging. *In Digest of Papers: 17th International Symposium on Fault-Tolerant Computing*, pp. 14-19, 1987.
5. E. N. Elnozahy, L. Alvisi, Y. M. Wang and D. B. Johnson. A Survey of Rollback-Recovery Protocols in Message-Passing Systems. *ACM Computing Surveys*, 34(3), pp. 375-408, 2002.
6. L. Lamport. Time, Clocks, and the Ordering of Events in a Distributed System. *Communications of the ACM*, 21, pp. 558-565, 1978.
7. R. McNab and F. W. Howell. simjava: a discrete event simulation package for Java with applications in computer systems modelling. *In Proc. First International Conference on Web-based Modelling and Simulation*, 1998.
8. M. L. Powell and D. L. Presotto. Publishing: A reliable broadcast communication mechanism. *In Proc. of the 9th International Symposium on Operating System Principles*, pp. 100-109, 1983.
9. P. Sens and B. Folliot. The STAR Fault Tolerant manager for Distributed Operating Environments. *Software Practice and Experience*, 28(10), pp. 1079-1099, 1998
10. R. D. Schlichting and F. B. Schneider. Fail-stop processors: an approach to designing fault-tolerant distributed computing systems. *ACM Transactions on Computer Systems*, 1, pp. 222-238, 1985.
11. R.E. Strom, D.F. Bacon and S.A. Yemeni. Volatile Logging in n-Fault-Tolerant Distributed Systems. *In Digest of Papers: the 18th International Symposium on Fault-Tolerant Computing*, pp. 44-49, 1988.
12. R.E. Strom and S.A. Yemeni. Optimistic recovery in distributed systems. *ACM Transactions on Computer Systems*, 3, pp. 204-226, 1985.
13. J. Xu, R.B. Netzer and M. Mackey. Sender-based message logging for reducing rollback propagation. *In Proc. of the 7th International Symposium on Parallel and Distributed Processing*, pp. 602-609, 1995.
14. B. Yao, K. -F. Ssu and W. K. Fuchs. Message Logging in Mobile Computing. *In Proc. of the 29th International Symposium on Fault-Tolerant Computing*, pp. 14-19, 1999.

The Design of a Dynamic Efficient Load Balancing Algorithm on Distributed Networks*

Yeojin Lee[1], Okbin Lee[1], Wankyoo Choi[1], Chunkyun Youn[2], and Ilyong Chung[1,**]

[1] Dept. of Computer Science and BK Team, Chosun University, Gwangju, Korea
iyc@chosun.ac.kr

[2] Department of Internet Software, Honam University, Gwangju, Korea
chqyoun@itc.honam.ac.kr

Abstract. In order to maintain load balancing in a distributed network, each node should obtain workload information from all the nodes in the network. To accomplish this, this processing requires $O(v^2)$ communication complexity, where v is the number of nodes. First, we present a new synchronous dynamic distributed load balancing algorithm on a $(v, k+1, 1)$-configured network applying a symmetric balanced incomplete block design, where $v = k^2 + k + 1$. Our algorithm designs a special adjacency matrix and then transforms it to $(v, k+1, 1)$-configured network for an efficient communication. It requires only $O(v\sqrt{v})$ communication complexity and each node receives workload information from all the nodes without redundancy since each link has the same amount of traffic for transferring workload information. Later, this algorithm is reconstructed on distributed networks, where v is an arbitrary number of nodes and is analyzed in terms of efficiency of load balancing.

1 Introduction

In a distributed network it is likely that some nodes are heavily loaded while others are lightly loaded or idle. It is desirable that workload be balanced between these nodes so that utilization of nodes can be increased and response time can be reduced. A load balancing scheme[1]-[3] determines whether a task should be executed locally or by a remote node. This decision can be made in a centralized or distributed manner. In a distributed network, distributed manner is recommended. In order to make this decision, each node can be informed about the workload information of other nodes. Also this information should be the latest because outdated information may cause an inconsistent view of the system state. So disseminating load information may incur a high link cost or a significant communication traffic overhead. For example, the ARPANET[4] routing algorithm is a distributed adaptive algorithm using estimated delay as

* This research was supported by the Program for the Training of Graduate Students in Regional Innovation which was conducted by the Ministry of Commerce Industry and Energy of the Korean Government.

** Corresponding author.

M. Gerndt and D. Kranzlmüller (Eds.): HPCC 2006, LNCS 4208, pp. 81–90, 2006.

the performance criterion and a version of the backward-search algorithm[5]. For this algorithm, each node maintains a delay vector and a successor node vector. Periodically, each node exchanges its delay vector with all of its neighbors. On the basis of all incoming delay vectors, a node updates both of its vectors.

In order to decrease communication overhead for obtaining workload information from all the nodes in the network, messages should be exchanged between adjacent nodes and then load balancing process be performed periodically by using these local messages. So each processor balances the workload with its neighbors so that the whole system will be balanced after a number of iterations. CWA(Cube Walking Algorithm)[6] is employed for load balancing on hypercube network. It requires $O(v^2)$ communication complexity and a communication path is $O(log_2 v)$. To reduce communication cost, flooding scheme is applied. However, the overlap of workload information occurs[7]-[8]. Based on SBN(Symmetric Broadcast Networks), communication patterns between nodes are constructed. It also needs $O(v^2)$ communication complexity for collecting workload information from all the nodes and a communication path is $O(log_2 v)$[9]-[10].

In this paper we design the network topology consisting of v nodes and $v \times k$ links and each node of which is linked to $2k$ nodes, where $v = k^2 + k + 1$. On this network, each node receives information from k adjacent nodes and then sends these information to other k adjacent nodes periodically. So, each node receives workload information for $k^2 + k$ nodes with two-round message interchange. Our algorithm needs only $O(v\sqrt{v})$ communication complexity. Later, this algorithm is revised for distributed networks and is analyzed in terms of efficiency of load balancing.

2 About (v, k, λ)-Configuration

Let $V = \{0, 1, ..., v-1\}$ be a set of v elements. Let $B = \{B_0, B_1, ..., B_{b-1}\}$ be a set of b blocks, where B_i is a subset of V and $|B_i| = k$. For a finite incidence structure $\sigma = \{V, B\}$, if σ satisfies following conditions, then it is a balanced incomplete block design(BIBD)[11], which is called a (b, v, r, k, λ)-configuration.

1. B is a collection of b k-subsets of V and these k-subsets are called the blocks.
2. Each element of V appears exactly r of the b blocks.
3. Every two elements of V appears simultaneously in exactly λ of the b blocks.
4. $k < v$.

For a (b, v, r, k, λ)-configuration, if it satisfies $k = r$ and $b = v$, then it is a symmetric balanced incomplete block design (SBIBD)[12] and it is called a (v, k, λ)-configuration. There are some relations among parameters b, v, r, k, λ that are necessory conditions for existence of this configuration, $bk = vr$ and $r(k-1) = \lambda(v-1)$.

3 Generation of a $(v, k+1, 1)$-Configuration

We now present an algorithm to generate an incidence structure $\sigma = \{V, B\}$ satisfying the condition for a $(v, k+1, 1)$-configuration in the case that k is a prime number. This $(v, k+1, 1)$-configuration is employed for constructing network topology below.

3.1 Design of an Algorithm to Construct $(v, k+1, 1)$-Configuration

In order to construct $(v, k+1, 1)$-configuration, $(k+1)$ sectors are designed. The first sector is composed of $(k+1)$ blocks and the others k blocks. Each block in the first sector contains element 0. The remaining elements of the first block, the second block,..., the kth block of the first sector are $(1, 2, ..., k)$,$(k+1, k+2, ..., 2k)$,...,$(k^2+1, k^2+2, ..., k^2+k)$,respectively. Each block in the ith sector contains element $(i-1)$ and the remaining k elements are chosen from $(k+1, k+2, ..., k^2+k)$. According to Algorithm 1, incidence structure X generates the first sector and incidence structure Y the remaining sectors.

Algorithm 1 for Generating an incidence structure

Incidence structure $T = \{V, B\}$, where $V = \{0, 1,.., v-1\}$, $B = \{B_0, B_1,.., B_{b-1}\}$, $|B_i| = k+1$. $B_{i,j}$ is the jth element of B_i

1. Select a prime number k and compute $v = k^2 + k + 1$.
2. Construct two incidence structures $X = \{V, C\}$ and $Y = \{V, D\}$.
   (a) $C_{i,j} = \begin{bmatrix} 0 & if\ j = 0 \\ & \\ t, t = i \times k + j & if\ j \geq 1 \end{bmatrix}$
   $0 \leq i, j \leq k$.
   (b) $D_{i,j} = \begin{bmatrix} C_{0,t}\ , t = \lfloor i/k \rfloor + 1 & if\ j = 0 \\ & \\ C_{j,t}\ , t = 1 + (i + (j-1) \times \lfloor i/k \rfloor)\ mod\ k & if\ j \geq 1 \end{bmatrix}$
   $0 \leq i \leq (k^2 - 1),\ 0 \leq j \leq k$.
3. Generate $Z = \{V, B\}$ from X and Y.
   $B_i \longleftarrow C_i$
   $B_{i+k+1} \longleftarrow D_i$

The table below illustrates how to create $Z = \{V, B\}$, $V = \{0, 1, ..., 12\}$. We now prove that this structure satisfies the conditions of a $(v, k+1, 1)$-configuration.

Definition 1. On incidence structure Y, Sector S_i is the ith family of k blocks, $D_j \in S_i$, $i = \lfloor j/k \rfloor$.

For example, If k equals 3 , then $\lfloor 0/k \rfloor = \lfloor 1/k \rfloor = \lfloor 2/k \rfloor = 0$. So, $S_0 = \{D_0, D_1, D_2\}$. There are k sectors in Y.

Lemma 1. For two elements $D_{i1,j1}$ and $D_{i2,j2}$, $D_{i1,j1} \neq D_{i2,j2}$, $if\ j1 \neq j2$.
Proof: From Algorithm 1-2-(a), if $0 < j \leq k$, $0 \leq i \leq k$ then $C_{i,j} = i \times k + j$.

Table 1. A set of blocks on Z generated from Algorithm 1

X	Y		Z
			$B_0 = \{0, 1, 2, 3\}$
			$B_1 = \{0, 4, 5, 6\}$
	$D_0 = \{1, 4, 7, 10\}$		$B_2 = \{0, 7, 8, 9\}$
	$D_1 = \{1, 5, 8, 11\}$		$B_3 = \{0, 10, 11, 12\}$
	$D_2 = \{1, 6, 9, 12\}$		$B_4 = \{1, 4, 7, 10\}$
$C_0 = \{0, 1, 2, 3\}$	$D_3 = \{2, 4, 8, 12\}$		$B_5 = \{1, 5, 8, 11\}$
$C_1 = \{0, 4, 5, 6\}$	$D_4 = \{2, 5, 9, 10\}$	$\Longrightarrow$	$B_6 = \{1, 6, 9, 12\}$
$C_2 = \{0, 7, 8, 9\}$	$D_5 = \{2, 6, 7, 11\}$		$B_7 = \{2, 4, 8, 12\}$
$C_3 = \{0, 10, 11, 12\}$	$D_6 = \{3, 4, 9, 11\}$		$B_8 = \{2, 5, 9, 10\}$
	$D_7 = \{3, 5, 7, 12\}$		$B_9 = \{2, 6, 7, 11\}$
	$D_8 = \{3, 6, 8, 10\}$		$B_{10} = \{3, 4, 9, 11\}$
			$B_{11} = \{3, 5, 7, 12\}$
			$B_{12} = \{3, 6, 8, 10\}$

This means if $j > 0$ then all the elements are distinct. And as shown in Algorithm 1-2-(b), an element of C_j is placed on the jth element of a certain block of Y if $D_{i,j} = C_{j,t}, t \neq 0$.

Lemma 2. For a sector consisting of k blocks, the first element of each block has the same value and the other k^2 elements are equal to $V - C_0$.
Proof: In the case that $D_{i,0} = C_{0,\lfloor i/k \rfloor+1}$, the first element of k blocks on a sector has the same value. According to Algorithm 1-2-(b), $D_{i,j} = C_{j,t}, t = 1 + (i + (j-1)\lfloor i/k \rfloor) \ mod \ k$. Since k is a prime number, each element except the first element of each block is distinct and these distinct k^2 elements are equal to $V - C_0$.

Lemma 3. For incidence structure Y, $D_{a,j} = D_{b,j}, j \geq 1$, if $b = ((a - c(j-1)) \ mod \ k \ + k(\lfloor a/k \rfloor + c)) \ mod \ k^2$.
Proof: From Algorithm 1-2-(b), $D_{a,j} = C_{j,t}$. We now prove that $D_{b,j} = C_{j,t}$. t can be calculated from parameters b, j below. Then t obtained on this lemma is equal to that from Algorithm 1-2-(b). Therefore, $D_{a,j} = D_{b,j}$.

$$
\begin{aligned}
t &= 1 + (b + (j-1) \times \lfloor b/k \rfloor) \ mod \ k \\
&= 1 + (((a - c(j-1)) \ mod \ k + k(\lfloor a/k \rfloor + c)) + (j-1)\lfloor ((a - c(j-1)) \ mod \ k + k(\lfloor a/k \rfloor + c))/k \rfloor) \ mod \ k \\
&= 1 + (a - c(j-1)) + (j-1) \times (\lfloor a/k \rfloor + c) \ mod \ k \\
&= 1 + (a + (j-1)\lfloor a/k \rfloor) \ mod \ k
\end{aligned}
$$

Here, if $D_{a,j}$ is in sector S_s then $D_{b,j}$ is in $S_{(s+c) \ mod \ k}$. In case of $c \equiv 0 \ (mod \ k)$, then $a = b$.

Lemma 4. Each element of V appears in exactly $k+1$ times in Z.
Proof: According to Algorithm 1-2-(a), $C_{i,0} = 0$. Since $0 \leq i \leq k$, 0 appears $k+1$ times. The other $v-1$ elements, $V - \{0\}$, appear exactly once on X. From Lemma 3, each element of $C_{0,j}, 1 \leq j \leq k$, appears k times in a sector of Y and the rest k^2 elements appear once in every sector of Y. Therefore, each element appears $k+1$ times in Z.

Lemma 5. Any pair of elements of V appears in exactly only once in Z.
Proof: The first element of V makes a pair with all the other elements and this pair appears once by designing rule of incidence structure(see Algorithm 1-2-(a)). Each element of $C_{0,j}, 1 \leq j \leq k$ makes a pair with $V - C_0$ elements and it also appears once proven by Lemma 3. The rest k^2 elements are now considered. For an arbitrary pair $D_{a,j1} = D_{a,j2}$, $j1, j2 \geq 1$, in order to make the same pair on other block D_b, the two elements should be on the same block. According to Lemma 4, if $j1 = j2$, then they are located on D_b. However, this case does not occur since $j1 \neq j2$. Therefore, any pair of elements of V appears in exactly once in Z.

Theorem 1. Z designed by Algorithm 1 satisfies the conditions of a $(v, k+1, 1)$-configuration.
Proof: Z satisfies the conditions of the SBIBD by Lemma 4 and Lemma 5.

3.2 Design of Network Configuration

In order to construct a network topology which has minimum link cost and traffic overhead, we imported $(v, k+1, 1)$-configuration. An incidence structure $Z = \{V, B\}$ satifies the conditions for a $(v, k+1, 1)$-configuration and M is a binary incidence matrix of Z . Then this matrix M can be transformed to an adjacency matrix of a graph $G = \{V, E\}$. Based on this idea, network topology can be designed as follows.

Algorithm 2 for Design of Network Configuration

1. Create an incidence structure $Z = \{V, B\}$ by Algorithm 1.
2. Generate $L = \{V, E\}$ from Z by exchanging blocks so that each block E_i includes element i.

```
E_0 ⟵ B_0
for ( i = 1 ; i ≤ k ; i = i + 1 )
    E_(i+1)k ⟵ B_i
for ( i = k + 1 ; i < v ; i = i + 1 ) {
    if ( (B_{i,⌈i/k⌉−1} mod k) = 0) then
        E_{⌈i/k⌉−1} ⟵ B_i
    else {
        if ( (i mod k) = 0) then   t = i − k
      else  t = ⌊i/k⌋ ∗ k
        E_{t+(i mod k)} ⟵ B_i }
}
```

3. Create an adjacency matrix $A = (a_{i,j})$ for graph G from L , where G is a network topology containing v nodes.

$$a_{i,j} = \begin{bmatrix} 1 \ if \ (i \neq j) \ and \ (j \in E_i) \\ 0 \ otherwise \end{bmatrix}$$

Table 2. Blocks of L generated from Z of Table 1

L
E_0= { 0, 1, 2, 3 }
E_1= { 1, 6, 9, 12 }
E_2= { 2, 5, 9, 10 }
E_3= { 3, 5, 7, 12 }
E_4= { 1, 4, 7, 10 }
E_5= { 1, 5, 8, 11 }
E_6= { 0, 4, 5, 6 }
E_7= { 2, 6, 7, 11 }
E_8= { 2, 4, 8, 12 }
E_9= { 0, 7, 8, 9 }
E_{10}= { 3, 6, 8, 10 }
E_{11}= { 3, 4, 9, 11 }
E_{12}= { 0, 10, 11, 12 }

Lemma 6. The ith row of the adjacency matrix obtained from Algorithm 2 contains element i.

Proof: In Fig. 1, $(k+1)$ sectors are described - one sector on X and k sectors on Y. Each sector on X and Y is composed of $(k+1)$ and k blocks, respectively. New sectors on Z' would be generated. Since the ith block on $X(i > 0)$ includes element $(i+1)k$, it is relocated on S_i on Z' - $E_0, E_{2k}, E_{3k}, ..., E_{(k+1)k}$. We now look into S_i on Z, $1 \leq i \leq k$. $(k-1)$ blocks of it can be placed on S_i on Z' and the remainder on the ith block of S_0 on Z' because B_{ik+j}, $1 \leq j \leq k$, the jth block of S_i on Z, includes one of $\{ik+1, ik+2, ..., ik+(k-1), i\}$. If a certain block contains $(ik+2)$, then it is placed on E_{ik+2}. Therefore, the ith block on Z' can include element i.

G has v nodes since G is created from $(v, k+1, 1)$-configuration. Each block E_i is composed of $k+1$ elements including i. Each node obtains $2k$ links from Step 3 of Algorithm 2. So, G becomes a 2k-regular graph. Therefore there are $(2k \times v)/2 = vk$ links in G. Given $Z = \{V, B\}$ obtained from Algorithm 1, performance of Algorithm 2 is shown on Table 2.

4 Design of an Efficient Load Balancing Algorithm on Distributed Networks

An efficient load balancing algorithm is now constructed on $(v, k+1, 1)$-configured networks generated by Algorithm 2.

Definition 2. Construct two sets P_i and R_i consisting of adjacent k nodes, where P_i is a set of nodes to send workload information to node i at time T_{2t}, and R_i is a set of nodes to send workload information to node i at time T_{2t+1}.

$P_i = \{j \mid j \in E_i - \{i\}\}$

$R_i = \{j \mid i \in E_j, (0 \leq j \leq n-1) \ and \ (i \neq j)\}$

The following example will provide a better understanding of the concept given by Definition 2. Nodes $3, 6, 9$ and node 12 send their workload information to node 1 at time T_{2t}, and P_1 is $\{3,6,9,12\}$, while nodes $0, 4$ and node 5 send the information to node 1 at time T_{2t+1}, and R_1 is $\{0,4,5\}$.

Definition 3. Generate two sets SF_i and RF_i, where $SF_i(j)$ is a set of workload information transmitted from node j to node i at time T_{2t} and $RF_j(i)$ is workload information transmitted from node j to node i at time T_{2t+1}.
$SF_i = \{SF_i(j) \mid j \in P_i, SF_i(j) = j\}$.
$RF_i = \{RF_j(i) \mid j \in R_i, RF_j(i) = \{E_j - \{i\}\}$.

Algorithm 3 for Construction of an Efficient Load Balancing Algorithm on (v,k+1,1)-configured networks

1. Node i receives a set of workload information $SF_i(j)$ from node $j \in P_i$ at T_{2t} and renews a table of workload information.
2. Node i receives a set of workload information $RF_j(i)$ from node $j \in R_i$ at T_{2t+1} and renews a table of workload information.
3. Repeat the first step.

The following table indicates that node i receives workload information $SF_i(j)$ and $RF_j(i)$ from node j at times T_{2t} and T_{2t+1}, respectively. For example, node 1 receives information on $\{6\}$, $\{9\}$ and $\{12\}$ from $SF_1(6)$, $SF_1(9)$ and $SF_1(12)$ at T_{2t}, and receives information on $\{0,2,3\}$, $\{4,7,10\}$ and $\{5,8,11\}$ from $RF_0(1)$, $RF_4(1)$ and $RF_5(1)$ at T_{2t+1}. Therefore, node 1 can obtain workload information for all the nodes at T_{2t+2} and this fact is proven in Theorem 2.

Theorem 2. According to Algorithm 3, every node obtains workload information for all the nodes at T_{2t+2}.
Proof: At T_{2t}, node j sends workload information $SF_i(j)$ to node i and then node i receives k workload information. At T_{2t+1}, node i receives workload information from node j by Algorithm 3-2. On an arbitrary pair $(RF_{i1}(j), RF_{i2}(j))$, $i1 \neq i2$, intersection of these sets is empty since on $(v, k+1, 1)$-configuration, every pair of two objects appears simultaneously in exactly one of v blocks and node j is an element of R_{i1} and R_{i2}, respectively. So node i obtains workload information for k^2 nodes at T_{2t+1}. Therefore, node i receives workload information for $k^2 + k$ nodes at T_{2t+2}.

In case that $v1 \neq k^2 + k + 1$, $(v - v1)$ nodes are deleted starting from node v node to node $v1$. The load balancing algorithm with $v1$ nodes is the same as Algorithm 3 except node i receives information from node j, where $i, j < v1$.

5 Analysis of an Efficient Load Balancing Algorithm

Algorithm 3 can be performed well when $v = k^2 + k + 1$. Each node receives workload information from all the nodes in $O(v\sqrt{v})$ time. However, the number of nodes may not be v in real-world application. Then we can not gurantee a desired result for an efficient load balancing. The algorithm now is analyzed in terms of how much a node receives information.

Definition 3. w_0 is the number of information on missing nodes for $S_{\lceil n/k \rceil}$ and w_1 from $S_{\lceil n/k \rceil +1}$ to the S_{k+1}.

Lemma 7. If $(n \ mod \ k) = 0$ then $w_0 = {}_kP_2$. Otherwise, $w_0 = k2 *_{k1} P_2 +_{k-k2} P_2$, where $k1 = n/k$ and $k2 = (k*(k1+1)-n)$.
Proof: In case that $n = (i+1)k$, $E_n = (0, ik+1, ik+2, ..., (i+1)k)$. Suppose that node n is deleted. According to Algorithm 3, node 0 does not receive workload information on (k-1) nodes - node $(ik+1)$, node $(ik+2)$,..., node $((i+1)k-1)$. Neither the other $(k-1)$ nodes in E_n do. So w_0, the number of information on missing nodes, is $k*(k-1)$. In case that $n \neq (i+1)k$ and node n is deleted. $(k2+1)$ nodes are deleted from $E_{\lceil n/k \rceil * k}$. $(k-k2)$ nodes do not receive information. Then the number of missing information is ${}_{k-k2}P_2$. Because of deleting node n from E_n, $k1$ nodes do not receive information and the number of missing information is ${}_{k1}P_2$. $(k2-1)$ blocks - $E_{n+1}, E_{n+2}, ..., E_{\lceil n/k \rceil * k-1}$ are the same as E_n and the number of missing information for these blocks is $(k2-1) *_{k1} P_2$. Therefore, $w_0 = k2 *_{k1} P_2 +_{k-k2} P_2$.

Lemma 8. If $(n \ mod \ k) = 0$ then $w_1 = (k-k1+1)*((k-2)*_{k1} P_2 +_{k1-1} P_2)$. Otherwise, $w_1 = (k-k1)*((k-k3-2)*_{k1+1} P_2 + (k3+1)*_{k1} P_2)$, where if $(n \leq E_{(k1+1),k1})$ then $k3 = k2-1$ else $k3 = k2$.
Proof: In case that $(n \ mod \ k) = 0$ and node n is deleted. The number of n is $(k-k1+1)$ in $(k-k1+1)$ sectors - from S_{k1+1} to S_{k+1}. In other word, a sector consisting of $(k-1)$ blocks has one n. For a block deleting n, $(k1-1)$ nodes do not receive information and the number of missing information is ${}_{k1-1}P_2$ and for the remaining $(k-2)$ blocks, $k1$ nodes do not receive information and the number of missing information is ${}_{k1}P_2$. Since w_1 for a sector is $(k-2) *_{k1} P_2 +_{k1-1} P_2$, w_1 for $(k-k1+1)$ sectors is $(k-k1+1)*((k-2)*_{k1} P_2 +_{k1-1} P_2)$. In case that $n \neq (i+1)k$ and node n is deleted. Similarly, the number of n is $(k-k1)$ in $(k-k1)$ sectors - from the S_{k1} to the S_{k+1} sector. For a block deleting $m, n \leq m \leq k*(n/k+1)$, $k1$ nodes do not receive information and the number of missing information is ${}_{k1}P_2$. Because the number of blocks deleting m in a sector is $(k2+1)$, the number of missing information is $(k2+1) *_{k1} P_2$. For the $(k-k2-2)$ remaining blocks, $(k1+1)$ nodes do not receive information and that is $(k-k2-2) *_{k1+1} P_2$. One important thing must be considered. According to Algorithm 2, a certain block in Sector $i, 1 \leq i \leq k$ moves to the E_i - the ith block in S_0. $E_{i,i-1}$ does not appear in S_i. Where $i = k1+1$, if $(n \leq E_{(k1+1),k1})$ then $k3 = k2-1$ else $k3 = k2$. Therefore, w_1 is $(k-k1)*((k-k3-2)*_{k1+1} P_2 + (k3+1)*_{k1} P_2)$.

Algorithm 4 for Computing the number of information on missing nodes given n nodes

$n \neq k^2 + k + 1$
$k1 = n/k$
$k2 = (k * (k1 + 1) - n)$
if ($(n \ mod \ k) = 0$) then {
  $w_0 \ = \ _kP_2$
  $w_1 \ = (k - k1 + 1) * ((k - 2) *_{k1} P_2 +_{k1-1} P_2)$
  }
  else {
if $(n \leq E_{(k1+1),k1})$ then
  $k3 = k2 - 1$
  else $k3 = k2$
  $w_0 \ = \ k2 *_{k1} P_2 +_{k-k2} P_2$
  $w_1 \ = \ (k - k1) * ((k - k3 - 2) *_{k1+1} P_2 + (k3 + 1) *_{k1} P_2)$
  }
$w \ = \ w_0 + w_1$

Given the number of nodes - from 43 nodes to 32 nodes, the following table illustrates the result of executing Algorithm 4.

Table 3. The number of Information on missing nodes required for load balancing

the number of nodes	w_0	w_1
43	$6 *_6 P_2 +_1 P_2$	$6 *_6 P_2$
42	$_7P_2$	$2 * (5 *_6 P_2 + 1 *_5 P_2)$
41	$1 *_5 P_2 +_6 P_2$	$2 * (4 *_6 P_2 + 2 *_5 P_2)$
40	$2 *_5 P_2 +_5 P_2$	$2 * (3 *_6 P_2 + 3 *_5 P_2)$
39	$3 *_5 P_2 +_4 P_2$	$2 * (2 *_6 P_2 + 4 *_5 P_2)$
38	$4 *_5 P_2 +_3 P_2$	$2 * (1 *_6 P_2 + 5 *_5 P_2)$
37	$5 *_5 P_2 +_2 P_2$	$2 * (1 *_6 P_2 + 5 *_5 P_2)$
36	$6 *_5 P_2 +_1 P_2$	$2 * (0 *_6 P_2 + 6 *_5 P_2)$
35	$_7P_2$	$3 * (5 *_5 P_2 + 1 *_4 P_2)$
34	$1 *_4 P_2 +_6 P_2$	$3 * (4 *_5 P_2 + 2 *_4 P_2)$
33	$2 *_4 P_2 +_5 P_2$	$3 * (3 *_5 P_2 + 3 *_4 P_2)$
32	$3 *_4 P_2 +_4 P_2$	$3 * (2 *_5 P_2 + 4 *_4 P_2)$

6 Conclusion

Researches have shown that distributed load balancing schemes can reduce task turnaround time substantially by performing tasks at lightly loaded nodes in the network. However, maintaining workload information on all the nodes is necessary for such scheme, while it may incur large communication overhead. In this paper, in order for the system to increase utilization and to reduce response time, we present an efficient load balancing algorithm on $(v, k + 1, 1)$-configured networks consisting of v nodes and vk links by employing the symmetric balanced incomplete block design. To accomplish this, each node should receive workload information from all the nodes without redundancy and each link should have the same amount of traffic for transferring workload information. Our algorithm requires two rounds of message interchange and $O(v\sqrt{v})$ communication complexity, as opposed to $O(v^2)$ communication complexity needed by protocols which use only one round of message interchange. Later, this algorithm is reconstructed on distributed networks, where v is an arbitrary number of nodes and is analyzed in terms of efficiency of load balancing.

References

1. C.Hui, S.Chanson, Hydrodynamic Load Balancing, IEEE Transactions on Parallel and Distributed System, vol. 10, no. 11, pp. 1118-1137, 1999.
2. V. Bharadwaj, D. Ghose and T. Robertazzi, "A New Paradigm for Load Scheduling in Distributed Systems," Cluster Computing, vol. 6, no. 1, pp. 7-18, 2003.
3. A. Legrand, et al., "Mapping and Load-Balancing Iterative Computing," IEEE Transactions on Parallel and Distributed System, vol. 15, no. 6, pp. 546-558, 2004.
4. M. Padlipsky, A perspective on the ARPANET Reference Model, Proc. of INFOCOM, IEEE, 1983.
5. L. Ford, D. Fulkerson, Flow in Network, Princeton University Press, 1962.
6. M. Wu, On Runtime Parallel Scheduling for processor Load balancing, IEEE Transactions on Parallel and Distributed System, vol. 8, no. 2, pp. 173-185, 1997.
7. K. Nam, J. Seo, Synchronous Load balancing in Hypercube Multicomputers with Faulty Nodes, Journal of Parallel and Distributed Computing, vol. 58, pp. 26-43, 1999.
8. H. Rim, J. Jang, S. Kim, Method for Maximal Utilization of Idle links for Fast Load Balancing, Journal of Korea Information Science Society, vol. 28, no. 12, pp. 632-641, 2001.
9. S. Das, D. Harvey, and R. Biswas, Adaptive Load-Balancing Algorithms Using Symmetric Broadcast Networks, Journal of parallel and Distributed Computing, vol. 62, no. 6, pp. 1042-1068, 2002.
10. S. Das, D. Harvey, and R. Biswas, Parallel Processing of Adaptive Meshes with Load Balancing, IEEE Transactions on Parallel and Distributed Systems, vol. 12, no. 12, pp. 1269-1280, 2001.
11. C.L.Liu, Introduction to Combinatorial Mathematics, pp. 359-383, McGraw-Hill, 1968.
12. I. Chung, W. Choi, Y. Kim, M. Lee, The Design of conference key distribution system employing a symmetric balanced incomplete block design, Information Processing Letters, vol. 81, no. 6, pp. 313-318, 2002.3.

Distributed Resource Allocation for Stream Data Processing

Ao Tang[1], Zhen Liu[2], Cathy Xia[2], and Li Zhang[2]

[1] California Institute of Technology, Department of Electrical Engineering, Pasadena, CA 91125

[2] IBM T.J. Watson Research Center, 19 Skyline Drive, Hawthorne, NY 10532

Abstract. Data streaming applications are becoming more and more common due to the rapid development in the areas such as sensor networks, multimedia streaming, and on-line data mining, etc. These applications are often running in a decentralized, distributed environment. The requirements for processing large volumes of streaming data at real time have posed many great design challenges. It is critical to optimize the ongoing resource consumption of multiple, distributed, cooperating, processing units. In this paper, we consider a generic model for the general stream data processing systems. We address the resource allocation problem for a collection of processing units so as to maximize the weighted sum of the throughput of different streams. Each processing unit may require multiple input data streams *simultaneously* and produce one or many valuable output streams. Data streams flow through such a system after processing at multiple processing units. Based on this framework, we develop distributed algorithms for finding the best resource allocation schemes in such data stream processing networks. Performance analysis on the optimality and complexity of these algorithms are also provided.

Keywords: Stream Processing, Distributed Algorithm, Resource Allocation.

1 Introduction

The rapid development of the network technologies has triggered the emergence of many new applications. Stream data processing is one of the most interesting and challenging applications that are under extensive study by the research community. In such applications, continuous data streams arriving to the system need to be processed by multiple processing units in real-time to generate streams of desirable results. One example of this type of application is network monitoring and management. Continuous streams of network usage information are collected from various monitoring points in the network. These information need to be analyzed and correlated on the fly to determine whether the network is in a normal running mode, or is under intrusive attacks. Many financial applications such as the stock quote and trading systems also exhibit this type of characteristics. Continuous quote and trade data streams need to

M. Gerndt and D. Kranzlmüller (Eds.): HPCC 2006, LNCS 4208, pp. 91–100, 2006.

be processed in real-time. Sensor networks is another area where many stream data processing applications arise. The nature of these continuous processing applications to process a large volume of data has lead to many new design challenges.

Stream data processing systems typically require a large amount of processing power with many different computers in order to achieve satisfactory performance levels. Multiple processing units often share a pool of computing resource. One important problem is to find the best resource allocation scheme for the multiple processing units to efficiently utilize the available resources. As in most real time systems, applications are often running in a decentralized environment. The resource allocation scheme also has to be decentralized in nature.

This paper addresses some of the fundamental resource allocation problems raised above. We formulate a generic stream data processing model with data streams passing through multiple processing units to generate the result streams. Sub-optimal allocation of the resources may lead to the under-utilization of certain processing units and over-utilization of some others. Our goal is to obtain a distributed mechanism that maximizes the weighted sum of the throughput of different output streams. In our model, each processing unit may require multiple input data streams *simultaneously* and produce one or many valuable output streams. Such kind of simultaneous flow consumption is related to the fork-join mechanism in queueing applications and supply chain management [1,2,4,10,8]. It is an important feature in many streaming processing applications. For example, the network usage information from multiple routers need to be correlated to derive the overall user flow information. Another distinct characteristic in our model is the introduction of the shrink/expansion factor for the flows at each processing units. The volume of the output data stream can be different from the volume of the input data stream at each processing unit. Such a phenomenon naturally occurs in the join, filter and selection mechanisms in streaming query like applications [12].

In this paper, we present an analytical approach to solve the generic stream data processing problem. We first develop the optimal solution for several special cases, including the case with a single output and the case with a tree topology. For the single output case, we propose a backward algorithm which produces an optimal solution in linear time. For the tree case, we provide a backward shrink algorithm which also yields an optimal solution in linear time. Based on the algorithm for trees, we propose two distributed algorithms to find the best, or close to optimal solutions in a general network with multiple streams. The algorithms are based on an aggregation heuristic that aggregates local subgraphs into equivalent super nodes, where the super nodes can play the role as a cluster head or local manager. We present experimental results to demonstrate the quality of our distributed solutions.

The paper is organized as follows. Section 2 presents the general model. We then investigate the structural properties of the optimal solutions for special cases (a single output stream case and the tree case) in Sections 3. In Section 4, we propose two distributed solutions for the general resource allocation problem based on the optimal algorithms derived previously. Experimental evaluations of the effectiveness of these solutions are also presented. Concluding remarks are provided in Section 5.

2 Model

In a stream data processing system, incoming data flow continuously from several sources. These data needs to go through several levels of processing, such as selection, filtering, or combining, to generate the expected output. We use a directed acyclic graph, referred to as *stream processing graph*, to describe the producer-consumer relationship among processing units associated with the streams. There are source nodes, sink nodes and processing nodes in the graph, where directed edges represent the information flow between various nodes. The source nodes correspond to the source of the input data streams. These nodes only have edges going out, and do not have any edges between them. The sink nodes correspond to the receivers of the eventual processed information. These nodes only have edges going to them, and do not have any edges in between. Processing nodes stand for processing units. A processing unit may require inputs from multiple data streams simultaneously and produce one or many valuable output streams. Such a graph can be plotted in a way such that all the directed edges are pointing downward. We can now view the system as information coming from the top and passing through the processing units in the middle and eventually leading to the output streams at the bottom, see Figure 1.

Denote $\mathcal{I}, \mathcal{P}, \mathcal{O}$ respectively the set of source, processing, and sink nodes in the graph, as illustrated in Figure 1. Let $\mathcal{E}$ denote the set of all the directed edges. Each node in $\mathcal{I}$ is a source node. Each node in $\mathcal{O}$ is a sink node. For convenience, we will refer to the underlying graph, $G = (\mathcal{N}, \mathcal{E})$, where $\mathcal{N} = \mathcal{I} \cup \mathcal{P} \cup \mathcal{O}$, and graph G is assumed to be connected. For each $j \in \mathcal{N}$, let $\mathcal{I}^j$ denote the set of immediate predecessors, i.e. all nodes i such that the directed edge (i, j) is in $\mathcal{E}$.

Let $\mathcal{O}^j$ denote the set of immediate successors, i.e. all the nodes k such that the directed edge (j, k) is in $\mathcal{E}$. Without loss of generality, we assume that each source node produces a single stream as the input to the processing nodes, and there is exactly one output stream leading to each sink node. Therefore, $|\mathcal{O}^i| = 1$ for all source nodes $i \in \mathcal{I}$, and $|\mathcal{I}^k| = 1$ for all sink nodes $k \in \mathcal{O}$.

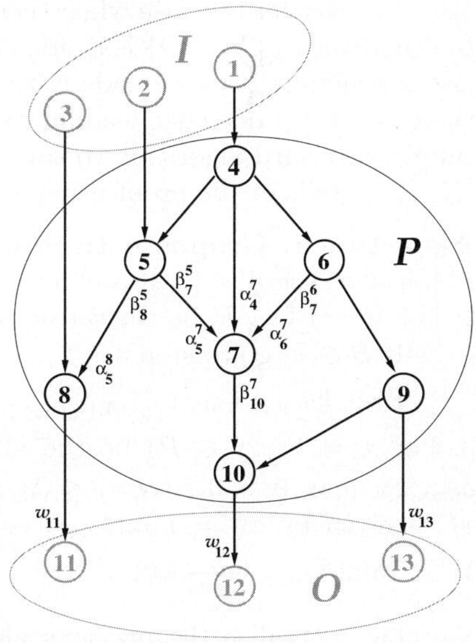

Fig. 1. A graph representation of the problem

We now describe the quantitative relationship between the input, output and resource consumption. In our model, each processing unit processes data flows from its upstream nodes *simultaneously* at a given proportion and generate output flows to its downstream nodes at a possibly different proportion. Each processing unit $j \in \mathcal{P}$, with a unit of CPU resource, will process α_i^j amount of flow from node i for all $i \in \mathcal{I}^j$, and generate β_k^j amount of flow to node k for all $k \in \mathcal{O}^j$. Here, the superscript j always represents the current node under consideration. For all the source nodes $j \in \mathcal{I}$, let λ_j be their flow input rates, where $0 < \lambda_j \leq \infty$. Each unit of output flow to node $k \in \mathcal{O}$ has value w_k.

We assume all the parameters λ, α, β and w are positive, as is the case in most real applications. In general, quantities α_i^j and β_k^j, although measurable, are not deterministic. They typically depend on the input data. Throughout this paper, unless specifically stated, we shall assume that this dependence is stationary. The quantities α_i^j and β_k^j are defined as the average consumption and production rates, respectively. The case of changing consumption and production rates will be discussed in Section 4.

Assume we have a total of R units of CPU resource available. Our goal is to find optimal or approximate solutions of allocating the resource among all the processing units to maximize the weighted sum (e.g. based on the importance) of the throughput of the output streams. We look for distributed solutions capable of adapting to local changes in the consumption and production rates.

3 The Single Output Case and Trees

We first consider the case when there is only one final output stream of interest. In other words, $\mathcal{O} = \{O\}$ is a singleton, where O is the only sink node. Without loss of generality, denote node N to be the *last processing node* reaching O (since there is exactly one edge leading to each sink node). In this case, we can have a simple backward algorithm to solve the problem in time $O(|\mathcal{E}|)$. Please refer to [11] for details of the proof using a backward tracing argument.

Algorithm 1. Graphs with Single Output
1. Initialize set $A = \{N\}$, and let $x_N = 1$.
2. Let $B := \bigcup_{i \in A} \mathcal{I}^i$ be the set of all predecessors of nodes in A.
   - If $B \subset \mathcal{I}$, go to step 3;
   - Else, let $x_i = \max_{\{j \in A:(i,j) \in \mathcal{E}\}} \frac{\alpha_i^j x_j}{\beta_j^i}, \forall i \in B$; set $A = B$; go back to step 2.
3. Let $\mathbf{x} = (x_j, j \in \mathcal{P})$ be the allocation produced by steps 1 & 2. Denote $\delta_{\max} := \max\{\delta > 0 : \delta \alpha_i^j x_j \leq \lambda_i, i \in \mathcal{I}, (i,j) \in \mathcal{E}\}$. Then the final allocation $\mathbf{x}^*$ is given by $x_i^* = \min(\delta_{\max}, \frac{R}{\sum_i x_i}) \cdot x_i, \quad i \in \mathcal{P}$, and the total return is $V^* = \min(\delta_{\max}, \frac{R}{\sum_i x_i}) \beta_O^N$.

We now generalize the previous algorithm to address the cases with multiple output nodes, i.e., $|\mathcal{O}| > 1$. In this setting, there is a decision between generating output for one stream versus generating output for another stream, or both. This kind of trade-off is not easy to evaluate due to the simultaneous flow consumption and output. We will first derive the algorithms to treat certain simpler cases. And then extend the solution to address the general cases.

Example 2: A Binary Tree
We now define a linear program as follows,

$$\begin{aligned} \max\ & w_4\beta_4^2x_2 + w_5\beta_5^3x_3 \\ \text{s.t. } & x_1 + x_2 + x_3 \le R, \quad \alpha_0^1x_1 \le \lambda_0, \\ & \alpha_1^2x_2, \le \beta_2^1x_1, \qquad \alpha_1^3x_3 \le \beta_3^1x_1 \qquad (1) \\ & x1 \ge 0, \quad x2 \ge 0, \quad x_3 \ge 0. \end{aligned}$$

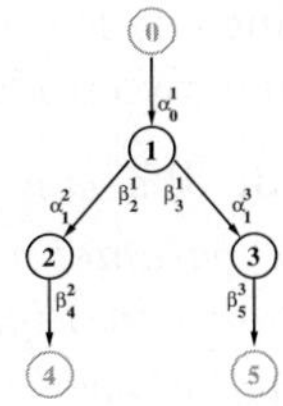

Fig. 2. A 3-node binary tree

In the event this maximization problem has more than one optimal solution, we let $(x_1^\#, x_2^\#, x_3^\#)$ be the optimal solution that minimizes the total allocated capacity $x_1+x_2+x_3$. We use this convention for the optimal solutions in all the optimization problems considered in this paper.

Theorem 1. *The solution to the above problem is*
i) If $\frac{w_4\beta_4^2\beta_2^1}{\alpha_1^2+\beta_2^1} > w_5\beta_5^3$, *then,* $x_2^\# = \frac{\beta_2^1}{\alpha_1^2+\beta_2^1}r$, $x_3^\# = 0$, $x_1^\# = \frac{\alpha_1^2}{\alpha_1^2+\beta_2^1}r$, *where* $r = \min(R, \frac{\alpha_1^2+\beta_2^1}{\alpha_0^1\alpha_1^2}\lambda_0)$.
ii) If $\frac{w_5\beta_5^3\beta_3^1}{\alpha_1^3+\beta_3^1} > w_4\beta_4^2$, *then,* $x_2^\# = 0$, $x_3^\# = \frac{\beta_3^1}{\alpha_1^3+\beta_3^1}r$, $x_1^\# = \frac{\alpha_1^3}{\alpha_1^3+\beta_3^1}r$, *where* $r = \min(R, \frac{\alpha_1^3+\beta_3^1}{\alpha_0^1\alpha_1^3}\lambda_0)$.
iii) Else, $x_2^\# = \frac{\alpha_1^3\beta_2^1}{\alpha_1^2\alpha_1^3+\alpha_1^3\beta_2^1+\alpha_1^2\beta_3^1}r$, $x_3^\# = \frac{\alpha_1^2\beta_3^1}{\alpha_1^2\alpha_1^3+\alpha_1^3\beta_2^1+\alpha_1^2\beta_3^1}r$, $x_1^\# = \frac{\alpha_1^2\alpha_1^3}{\alpha_1^2\alpha_1^3+\alpha_1^3\beta_2^1+\alpha_1^2\beta_3^1}r$, *where* $r = \min(R, \frac{\alpha_1^2\alpha_1^3+\alpha_1^3\beta_2^1+\alpha_1^2\beta_3^1}{\alpha_0^1\alpha_1^2\alpha_1^3}\lambda_0)$.

Proof. This result can be proved case by case with linear algebra using contradiction techniques. Please refer to [11] for details. □

Theorem 2. *The problem in Figure 2 is equivalent to the simpler model in Figure 3. The equivalent parameters α, β and w are given as follows:*

i) If $\frac{w_4\beta_4^2\beta_2^1}{\alpha_1^2+\beta_2^1} > w_5\beta_5^3$, *then* $\hat\alpha_0^1 = \alpha_0^1, \hat\beta_2^1 = \beta_2^1$, $\hat\alpha_1^2 = \alpha_1^2, \hat\beta_3^2 = \beta_4^2$, $\hat w_3 = w_4$.
ii) If $\frac{w_5\beta_5^3\beta_3^1}{\alpha_1^3+\beta_3^1} > w_4\beta_4^2$, *then,* $\hat\alpha_0^1 = \alpha_0^1, \hat\beta_2^1 = \beta_3^1$, $\hat\alpha_1^2 = \alpha_1^3, \hat\beta_4^2 = \beta_5^3$, $\hat w_3 = w_5$.
iii) Else, $\hat\alpha_0^1 = \alpha_0^1, \hat\beta_2^1 = \beta_2^1 + \beta_3^1, \hat\alpha_1^2 = \frac{\frac{\alpha_1^2}{\beta_2^1}\alpha_1^3+\frac{\alpha_1^3}{\beta_3^1}\alpha_1^2}{\frac{\alpha_1^2}{\beta_2^1}+\frac{\alpha_1^3}{\beta_3^1}}, \hat\beta_3^2 = \frac{\frac{\beta_2^1}{\alpha_1^2}\beta_4^2+\frac{\beta_3^1}{\alpha_1^3}\beta_5^3}{\frac{\beta_2^1}{\alpha_1^2}+\frac{\beta_3^1}{\alpha_1^3}}, \hat w_3 = \frac{\frac{\beta_2^1}{\alpha_1^2}\beta_4^2w_4+\frac{\beta_3^1}{\alpha_1^3}\beta_5^3w_5}{\frac{\beta_2^1}{\alpha_1^2}\beta_4^2+\frac{\beta_3^1}{\alpha_1^3}\beta_5^3}$.

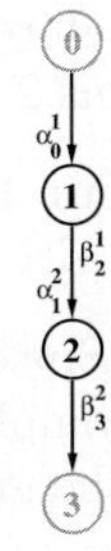

Fig. 3. 2-node Representation

Notice that these parameter mappings are independent of the parameters λ_0 and R. This is a key property for the later algorithms.

Proof. The proof is straightforward by checking feasibility conditions both ways. □

After merging the leaf nodes into a single leaf, we also have another basic reduction to reduce two node in tandem into a single node.

Theorem 3. *We can further aggregate the model in Figure 3 with parameters* $\hat{\alpha}_0^1, \hat{\alpha}_1^2, \hat{\beta}_2^1, \hat{\beta}_3^2, \hat{w}_3$, *into a simpler model as shown in Figure 4 with the equivalent parameters* $\tilde{\alpha}_0^1, \tilde{\beta}_2^1, \tilde{w}_2$ *as follows:*
$\tilde{\alpha}_0^1 = \frac{\hat{\alpha}_0^1\hat{\alpha}_1^2}{\hat{\alpha}_1^2+\hat{\beta}_2^1}$, $\tilde{\beta}_2^1 = \frac{\hat{\beta}_3^2\hat{\beta}_2^1}{\hat{\alpha}_1^2+\hat{\beta}_2^1}$, $\tilde{w}_2 = \hat{w}_3$, $\tilde{x}_1^* = \frac{\hat{\alpha}_1^2+\hat{\beta}_2^1}{\hat{\alpha}_1^2}\hat{x}_1^*$, $\tilde{x}_1^* = \frac{\hat{\alpha}_1^2+\hat{\beta}_2^1}{\hat{\beta}_2^1}\hat{x}_2^*$.

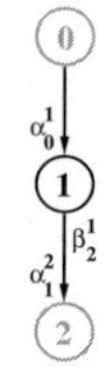

Fig. 4. 1-node Representation

Proof. The proof can be easily carried out by showing that the solution obtained from the optimal solution for one problem is feasible for the other problem, and the two solutions have the same objective value. The details can be found in [11]. □

Besides binary trees, Theorem 2 can also be applied repeatedly to handle general fork trees with arbitrary out-degree(≥ 2). It is straight forward to check formula that result of the merging process in Theorem 2 does not depend on the order of the merging process. It is also straight forward to prove a similar theorem as Theorem 1. The idea of dealing with general tree is to apply Theorem 1 to unit two layer subtree and then replace them with a new node. The following algorithm states the whole process.

Algorithm 2. Backward Shrink Algorithm for Trees
1. If there are 2 leaves with a common predecessor, apply Theorem 2 to these 3 nodes (2 leaves and their predecessor) to find the equivalent 2 node structure. Otherwise, Use Theorem 3 to aggregate the 2 nodes(a leave and its predecessor)to be a single node structure.
2. Repeat from step 1 until there is only one node left.
3. Set all resource to that node, and map resource allocation back according to Theorems 2 and 3.

Theorem 4. *Algorithm 2 terminates and yields the optimal solution. It runs in time* $O(|\mathcal{E}|)$.

Proof. Since each round of execution of step 1 decreases the number of links by 1, the complexity is $O(|\mathcal{E}|)$. The optimality can be proved by induction on the size of the graph. The details are omitted due to limited space. □

4 Distributed Solutions

In this section, we present distributed solutions for the problem. Simulation experiments demonstrate that they perform well even for general network topologies that do not have a tree structure.

4.1 Distributed Algorithms

We develop two heuristics to solve the general problem. These heuristics are based on the the optimal solutions for the tree case and for the single-output case. Experimental results are provided to illustrate their effectiveness. As we will see in the next section, these heuristics can be implemented easily in a distributed way.

The first heuristic is based on the optimal solution for trees. As assumed earlier, all the nodes have been labeled from 1 to N such that all the edges (i, j) satisfy $i < j$. This algorithm will start from the bottom of the graph and move up to the top. At each step, the algorithm examines each node, generate aggregated information based on information from its children, and pass this information up to its parents.

Heuristic A

- initialize graph G to be the whole graph;
- for *node* $= N$ to 1 (compute bottom up for the aggregated solution)
  - if *node* is a leaf in G then pass its parameters α, β, w to its parents;
  - else (all the children of *node* must be leaves in G;)
    - apply Theorem 2 repeatedly to remove one leaf at a time from G;
    - apply Theorem 3 to obtain the updated parameters α, β, w for *node*;
    - pass the updated parameters to all its parents;
      (*node* has no children left in G;)
- G has one node left, with aggregated parameters;
- solve this single node problem;
- for *node* $= 1$ to N (compute a solution for original problem from top down)
  - apply Theorem 1 and 3 to compute solution for *node* and the flow amount to all its children;

If the original graph is a tree, it can be shown that the above algorithm obtains the optimal solution. For the general graph case, we will present experimental results to demonstrate the quality of this distributed algorithm.

Another heuristic for the general problem with multiple output streams is developed based on the single output algorithm combined with the general gradient decent algorithm. Assume there are multiple output streams, $O_1, \ldots, O_k$. We define a function $f(u_1, \ldots, u_k)$ to be the best objective value if the solutions are generating flows for the output streams according to the relative proportion given by $(u_1, \ldots, u_k)$. Finding $f(u_1, \ldots, u_k)$ is the same as solving a modified problem with a new final sink node O_{k+1}, and making all the original output flows to flow into this final sink node. The β parameters for all the flows from $O_1, \ldots, O_k$ to O_{k+1} are all set to be 1. The α proportions at O_{k+1} are given by $(u_1, \ldots, u_k)$ for flows from $O_1, \ldots, O_k$. The β parameter at O_{k+1} is $w_1 u_1 + \ldots + w_k u_k$. The weight factor w at O_{k+1} is 1. The equivalence of these two problems can be easily checked. Since we can apply the backtrack algorithm in the earlier sections to find the optimal solution for the single output problem, we can find the value of $f(u_1, \ldots, u_k)$ for any given $(u_1, \ldots, u_k)$. We now apply the gradient decent algorithm to find the maximum value for function $f(u_1, \ldots, u_k)$.

Heuristic B

0) initialize $(u_1, \ldots, u_k)$ to be $(w_1, \ldots, w_k)$;
1) call Algorithm 1 for the single output problem with $(u_1, \ldots, u_k)$;
2) estimate the gradient for $f(u_1, \ldots, u_k)$;
3) move point $(u_1, \ldots, u_k)$ along the gradient direction;
4) repeat from step 1) until relative difference between consecutive solutions is smaller than a given threshold.

Note that in Heuristic B, the gradient method can be replaced by other search techniques such as simulated annealing, Tabu search, genetic algorithms, smart hill-climbing [13], etc.

Heuristic A has the advantage that it can quickly generate high quality solutions for simple graph topologies. However, when the graph is complex, the quality may degrade. Heuristic B is expected to be able to handle more effectively complex graph structures.

4.2 Experimental Results

We present below experimental results to compare the performance of these two heuristics and the optimal solution. The setting of the experiment is as follows. First, directed acyclic graphs with N nodes are generated randomly using the following 4 steps:

1) Randomly generate N points (x_i, y_i) in the unit square $[0, 1] \times [0, 1]$;
2) For $i = 1, \ldots, N$, generate its successor set $S_i := \{j : x_j \geq x_i, y_j \geq y_i\}$;
3) For $i = 1, \ldots, N$, generate its immediate successor set $s_i := S_i - \cup_{k \in S_i} S_k$;
4) For $i = 1, \ldots, N$, create a link from i to j if $j \in s_i$.

This algorithm is inspired by a scheme to generate random partial orders among N elements. Once the graph is generated, the parameters α, β, w are then generated from independent uniform random samples.

We randomly generate graphs with 20, 50, and 100 nodes. For each fixed number of nodes, we generate 1000 instances of the problem with random topology and random parameter values. We apply the two heuristics to obtain the corresponding objective values. We also obtain the optimal solution through a static linear program formulation. We have collected the characteristics of the random graphs, as well as the quality of the two heuristics. Because the problem is a maximization problem, the quality of the heuristics is reflected by the achieved percentage of the optimal solution. The results from Heuristic A is presented in Table 1. We can see that Heuristic A generates reasonably good solutions for small size graphs. However, the quality of the solutions degrades as the size of the graph grows. This behavior is consistent with our earlier intuitions.

Table 2 presents the results for Heuristic B. We used 10% relative difference as the stopping criterion for the gradient algorithm. We observe that Heuristic B is consistently better than Heuristic A. It is also important that the average number of iterations is small. This means Heuristic B does not require too much additional time to compute compared with Heuristic A. It is very promising

Table 1. Results for Heuristic A

# of Nodes	20	50	100
Avg # of edges	74.2	507.4	2063.7
Avg # of source nodes	4.6	5.8	7.0
Avg # of sink nodes	3.1	3.7	4.9
% optimality (avg)	74.4	57.6	54.3
% optimality (std)	26.6	33.2	34.3
% cases > 90% optimal	42.1	29.5	25.1

Table 2. Results for Heuristic B

# of Nodes	20	50	100
Avg # of edges	79.1	520.1	1912.7
Avg # of source nodes	4.4	4.9	7.6
Avg # of sink nodes	3.4	3.6	5.0
% optimality (avg)	82.4	68.9	59.0
% optimality (std)	25.4	36.4	39.2
% cases > 90% optimal	61	41.1	32.2
Avg # of iterations	5.2	10.1	10.0

to find out that Heuristic B consistently generates quality solutions, and more importantly, its effectiveness can be improved through the use of more sophisticated search methods. Keeping in mind that we are interested in the distributed nature and the efficiency of the algorithm. Heuristic B seems to be a preferable solution.

5 Concluding Remarks

This paper solves the CPU resource allocation problem in stream processing systems with the objective of maximizing the total return of multiple output streams. We explore structural properties of the optimal solution for the the problem under different network topologies, and develop efficient, yet simple to implement algorithms to solve them. Detailed performance analysis on optimality and complexity of those algorithms are also provided.

We further present two distributed solutions to the general problem and give the corresponding measurement-based distributed implementation. Our experimental results show that the algorithms are highly robust and capable of quickly adapting to real-time fluctuations in the consumption and production rates and changes in resource consumption requirements, while achieving high quality solutions even in non-stationary systems.

References

1. F. Baccelli, Z. Liu. On the Execution of Parallel Programs on Multiprocessor Systems-A Queuing Theory Approach. Journal of the ACM, Vol.37, No.2, April 1990, pp.373-417
2. F. Baccelli, Z. Liu. On the Stability Condition of a Precedence-based Queueing Discipline. *Adv. Appl. Prob.*, Vol 21, 1989, pp. 883-887.
3. F. Baccelli, F. Makowski, and D. Towsley. Acyclic Fork-Join Queueing Networks. *J. ACM*, Vol. 36, 3, 1989, pp. 615-642.
4. C. Baldwin, K.B. Clark, J. Magretta, J.H. Dyer, M. Fisher, D.V. Fites *Harvard Business Review on Managing the Value Chain*, Harvard Business School Press, 2000.
5. A. Brandstdt, V. Bang Le, and J.P. Spinrad. Graph Classes: A Survey. *SIAM*, 1999.

6. E. Coffman and G. Lueker. *Probabilistic Analysis of Packing and Partitioning Algorithms.* Wiley, 1991.
7. T.H. Cormen, C.E. Leiserson, R.L. Rivest, and C. Stein, *Introduction to Algorithms*, Second Edition, MIT Press and McGraw-Hill, 2001.
8. A. Mas-Collel,M. Whinston and J. Green. Microeconomic Theory. Oxford University Press, 1995.
9. J. A. Sharp, *Data Flow Computing,* (Ed.), Ablex Publication Corp., 1991.
10. D. Simchi-Levi, P. Kaminsky, E. Simchi-Levi. *Designing and Managing the Supply Chain*, 2 edition, McGraw-Hill/Irwin, 2002.
11. K. Tang, Z. Liu, C. Xia and L. Zhang. Distributed Resource Allocation for Stream Processing Systems. *IBM Research Report.* 2006.
12. S. Viglas and J. Naughton. Rate-Based Query Optimization for Streaming Information Sources. *ACM SIGMOD*, 2002.
13. B. Xi, Z. Liu, M. Raghavachari, C. Xia, and L. Zhang. A Smart Hill-Climbing Algorithm for Application Server Configuration. Proceedings of the 13th International Conference on World Wide Web. 2004, pp. 287-296.

Network Probabilistic Connectivity: Expectation of a Number of Disconnected Pairs of Nodes

Alexey S. Rodionov* and Olga K. Rodionova

Institute of Computational Mathematics and Mathematical Geophysics
Siberian Division of the Russian Academy of Science
Novosibirsk, Russia
Tel.: +383-3326949
alrod@rav.sscc.ru

Abstract. The task of calculating the expectation of a number of disconnected pairs of nodes (EDP) in unreliable network is discussed. The task is NP-hard that is it requires complete enumeration of subgraphs. The techniques for decreasing a number of enumerated subgraphs by using the branching (factoring) method and taking advantage from possible structural features are discussed. Usage of chains, bridges, cutnodes and dangling nodes is considered.

1 Introduction

Random graphs is an acknowledged model for networks of different kinds. For the analysis of network's reliability the probability of connectivity is used mostly ([1,2,3,4,5,6,7], for example) while the expectation of a number of disconnected pairs of nodes (EDP) is sometimes more valuable and informative index. For example all trees are equal from the point of the probability of connectivity while they are not from the point of EDP. Complementary index to EDP is a number of *connected* pairs of nodes (ECP), their sum is equal to the whole number of pairs of nodes in a network. Finding EDP requires complete enumeration of network destructions for its calculation. This is one of the reasons why few of researches deal with it. We can refer to the paper [8] where this index is mentioned among other valuable indexes of a network reliability. We investigate how to decrease the enumeration by using the branching (factoring) method and taking advantage from possible structural features.

As in [1] we mostly gain from considering simple chains and dangling nodes. Existence of bridges or cutnodes can help also.

The rest of the paper is organized as follows: in Section 2 we give main definitions and notations. In Section 3 we discuss reduction of the task dimension by considering bridges, cutnodes and dangling nodes. In Section 4 we give the exact equations for EDP for some kinds of graphs and in Section 5 we present equations for branching by chain. Section 6 is the brief conclusion.

* Corresponding author.

M. Gerndt and D. Kranzlmüller (Eds.): HPCC 2006, LNCS 4208, pp. 101–109, 2006.

2 Definitions and Notations

We consider random graphs with reliable nodes and unreliable edges. Let us denote:

$G(n, m) = (V, U, P, WT)$ – non-oriented graph with a set of nodes V, set of edges U, matrix of edges reliabilities P and vector of nodes weights WT.
$n = |V|$, $m = |U|$ – number of nodes and edges, respectively.
$w_i = w(v_i)$ – weight of a node v_i, $WT = w_1, \ldots, w_n$.
$W(G)$ – total weight of all nodes of G.
p_{ij} – probability of an edge e_{ij} being existent (being in a working state, edge's *reliability*), $P = ||p_{ij}||$; $q_{ij} = 1 - p_{ij}$.
$M(G, P)$ **and** $N(G, P)$ – ECP and EDP of a random graph G.

We use simply G for a graph if its n and m are clear from the context. If needed we use $G(P)$, $G(WT)$ or $G(P, WT)$ also. If we need refer to the weight of i-th node in some graph G, then we use $WT_i(G)$. If we consider some special edges then we usually assign them personal numbers that is use notation e_k (k-th edge) instead of e_{ij} (an edge that connects v_i and v_j). Notation p_k is used for corresponding edge's reliability.

For simplifying some equations we assume that $\prod_{s=i+1}^{i} p_s \equiv 1$.

A weight w_i equal to a number of nodes that were contracted to form a special node v_i (initial weight of each node is 1) is needed for keeping the number of disconnected pairs in case when this node is separated from some other nodes in a graph.

It is obvious that

$$N(G) = \sum_{i=1}^{n-1} \sum_{j=i+1}^{n} a_{ij} w_i w_j, \tag{1}$$

where a_{ij} is a probability of v_i and v_j be disconnected in G.

The branching (factoring) method for calculating an expected value of any function of a random graph G is based on the equation of composite probability by two alternative hypothesis: existence or non-existence of some edge e_{ij}. Thus

$$N(G) = p_{ij} N(G^*_{ij}) + (1 - p_{ij}) N(G \backslash e_{ij}), \tag{2}$$

where G^*_{ij} is a graph obtained from G by contracting v_i and v_j by an edge e_{ij}, $G \backslash e_{ij}$ – graph obtained from G by deleting the edge e_{ij}. Recursions go on until deriving a graph for which a $N(G)$ is easily obtained.

We say that a random graph is connected (disconnected) when its structure is connected, not realization. In last case we say that "realization of a random graph is connected (disconnected)."

3 EDP for Graphs of a Small Dimension ($n = 2, 3, 4$)

Case of $n = 2$ is obvious:

$$N(G) = (1 - p_{12}) w_1 \cdot w_2. \tag{3}$$

For $n = 3$ we use (1):

$$N(G) = w_1 w_2 (1 - p_{12})(1 - p_{23} p_{13}) + \\ w_1 w_3 (1 - p_{13})(1 - p_{12} p_{23}) + w_2 w_3 (1 - p_{23})(1 - p_{12} p_{13}). \tag{4}$$

For $n = 4$ we use the equation of composite probability considering all possible ways of a graph destruction as hypotheses. After collecting terms we have:

$$\begin{aligned} N(G) = & \\ & w_1 w_2 q_{12}[(1-p_{13}p_{23})(1-p_{14}p_{24}) - p_{13}q_{14}q_{23}p_{24}p_{34} - q_{13}p_{14}p_{23}q_{24}p_{34}] + \\ & w_2 w_3 q_{23}[(1-p_{24}p_{34})(1-p_{12}p_{13}) - q_{12}p_{13}p_{14}p_{24}q_{34} - p_{12}q_{13}p_{14}q_{24}p_{34}] + \\ & w_3 w_4 q_{34}[(1-p_{13}p_{14})(1-p_{23}p_{24}) - p_{12}p_{13}q_{14}q_{23}p_{24} - p_{12}q_{13}p_{14}p_{23}q_{24}] + \\ & w_1 w_4 q_{14}[(1-p_{12}p_{24})(1-p_{13}p_{34}) - q_{12}p_{13}p_{23}p_{24}q_{34} - p_{12}q_{13}p_{23}q_{24}p_{34}] + \\ & w_1 w_3 q_{13}[(1-p_{12}p_{23})(1-p_{14}p_{34}) - p_{12}q_{14}q_{23}p_{24}p_{34} - q_{12}p_{14}p_{23}p_{24}q_{34}] + \\ & w_2 w_4 q_{24}[(1-p_{23}p_{34})(1-p_{12}p_{14}) - q_{12}p_{13}p_{14}p_{23}q_{34} - p_{12}p_{13}q_{14}q_{23}p_{34}]. \end{aligned} \tag{5}$$

4 Using Structural Peculiarities

As in the case of calculation of a probabilistic connectivity we can take advantage from some peculiarities of a graph under consideration.

First let us make the following derivation. During factoring process by equation (2) we may obtain several graphs with the same structure and matrix P but with different weights of nodes. In this case we can gain from the following useful lemma.

Lemma 1. *If during a graph G factoring process some subgraphs $G_1, \ldots, G_k$ with same structure and matrix P are obtained in which only one special node v_s has different weight w_{si} in G_i, $i = 1, \ldots, k$, then the total contribution of these subgraphs into $N(G)$ is equal to*

$$\left(\sum_{i=1}^{k} p_i\right) \cdot N(G^o), \tag{6}$$

where p_i is a probability of G_i's realization, and graph G^o has the same structure and P as G_i and

$$WT_s(G^o) = \sum_{i=1}^{k} p_i w_{si} / \sum_{i=1}^{k} p_i. \tag{7}$$

Proof. From (1) we have that for any selected index s

$$N(G) = \Big(\sum_{i \in \{1 \ldots n\} \setminus s} w_i \Big) w_s + \sum_{i \in \{1 \ldots n\} \setminus s} \; \sum_{(j>i) \& (j \neq s)} w_i w_j = A w_s + B. \tag{8}$$

So, if we change a weight w_s of v_s, then coefficients A and B remain unchanged. Thus the total contribution D of all G_i into $N(G)$ is

$$D = \sum_{i=1}^{k} p_i N(G_i) = \sum_{i=1}^{k} p_i (A w_{si} + B) \tag{9}$$
$$= A\Big(\sum_{i=1}^{k} p_i w_{si}\Big) + \sum_{i=1}^{k} p_i B = \sum_{i=1}^{k} p_i \Big(A \sum_{i=1}^{k} p_i w_{si} / \sum_{i=1}^{k} p_i + B\Big),$$

from which we have what was to be proved. □

4.1 *k*-Component Graphs

If a graph G consist of k mutually disconnected subgraphs $G_1, \ldots, G_k$ then, obviously

$$N(G) = \sum_{i=1}^{k} N(G_i) + \sum_{i=1}^{k-1} \sum_{j=i+1}^{k} W(G_i) W(G_j). \tag{10}$$

4.2 Deleting Dangling Nodes

Let a connected graph $G(n, m)$ have a dangling node v_i adjacent to some node v_j. Then deletion of the edge e_{ij} leads to occurrence of $w_i \cdot W(G \backslash \{e_{ij}\}) = w_i \cdot \sum_{k \neq i} w_k$ pairs of disconnected nodes. Hence,

$$N(G) = p_{ij} N(G_{ij}^*) + (1 - p_{ij})(w_i W(G \backslash \{e_{ij}\}) + N(G \backslash \{e_{ij}\})). \tag{11}$$

Note that in this case a graph G_{ij}^*, obtained by contracting nodes v_i and v_j by e_{ij}, is by its structure *the same as* $G \backslash \{e_{ij}\}$, and $WT_j(G^*) = w_i + w_j$. Thus from lemma 1 we obtain:

$$N(G) = N(G^o) + (1 - p_{ij}) w_i W(G \backslash e_{ij}), \tag{12}$$

where G^o — a graph which has the same structure and a weight of v_j

$$WT_j(G^o) = p_{ij}(w_j + w_i) + (1 - p_{ij}) w_j = w_j + p_{ij} w_i. \tag{13}$$

4.3 Using Cutnodes

Let G consists from two subgraphs (blocks) G_1 and G_2 that are jointed through a node v_s (cutnode). Then obviously for any pair of nodes v_i and v_j a path between them or lies in one of the blocks if these nodes are in the same block, or goes through v_s otherwise. Thus from (1) we have

$$N(G) = N(G_1) + N(G_2) + \sum_{i \in X(G_1)} \sum_{j \in X(G_2)} (a_{is} + a_{sj} - a_{is} a_{sj}) w_i w_j, \tag{14}$$

where a_{is} and a_{sj} are probabilities of v_i and v_j being disconnected with v_s in G_1 and G_2 correspondingly.

4.4 Bridge Removing

Let a connected graph $G(n, m)$ have an edge c_{st} such that its deletion leads to dividing the graph into two separated components $G_1(k, f)$ and $G_2(n-k, m-f-1)$ (they are obviously connected graphs). Such edges are known as bridges. Then

$$N(G) = p_{st}N(G_{st}^{*}) + (1-p_{st})\big[W(G_1)W(G_2) + N(G_1) + N(G_2)\big]. \tag{15}$$

Using (14) we obtain

$$\begin{aligned} N(G) = (1-p_{st})\big[W(G_1)W(G_2)\big] + N(G_1) + N(G_2) + \\ p_{st}\sum_{i\in X(G_1)}\sum_{j\in X(G_2)} (a_{is} + a_{sj} - a_{is}a_{tj})w_i w_j, \end{aligned} \tag{16}$$

where a_{is} and a_{tj} are probabilities of v_i and v_j being disconnected with v_s in G_1 and with v_t in G_2 correspondingly.

4.5 Case of a Chain

Let us discuss the case of a chain with k nodes and $k-1$ edges (Ch_k). Let nodes be enumerated in increasing order thus that nodes v_1 and v_k have a degree 1 while all other nodes have degree 2. For simplicity we denote $e_{i,i+1}$ as e_i here.

From (1) we have

$$N(Ch_k) = \sum_{i=1}^{k-1}\sum_{j=i+1}^{k} w_i w_j\Big(1 - \prod_{s=i}^{j-1} p_s\Big). \tag{17}$$

4.6 Case of a Cycle

Now we consider a cycle with k nodes and k edges (C_k). Let nodes be enumerated in order thus that node v_{i+1} follows v_i, $i = 1, \ldots, k-1$ and v_1 follows v_k. For simplicity we denote $e_{i,i+1}$, $i = 1, \ldots, k-1$ as e_i and $e_{k,1}$ as e_k and denote reliability of e_i as p_i. Now we can use equation (1) directly. Between each pair of nodes v_i and v_j there are two pathes, clockwise and counterclockwise. Thus

$$N(C_k) = \sum_{i=1}^{k-1}\sum_{j=i+1}^{k} w_i w_j\Big(1 - \prod_{s=i}^{j-1} p_s\Big)\Big(1 - \prod_{s=j}^{k} p_s \prod_{s=1}^{i-1} p_s\Big). \tag{18}$$

4.7 Case of a Tree

The EDP for a n-nodes tree T_n we obtain from (1):

$$N(T_n) = \sum_{i=1}^{n-1}\sum_{j=i+1}^{n} w_i w_j\Big(1 - \prod_{e_{st}\in Pt_{ij}} p_{st}\Big), \tag{19}$$

where Pt_{ij} is a path from v_i to v_j.

For partial cases we can obtain simpler expressions. The case of a chain that is the special case of a tree has been discussed earlier. Now let us have a star-like tree S_k in which k nodes are adjacent to a single node (root). Thus we have some node v_0 with weight w_0 that is adjacent to dangling nodes v_i, $i = 1, \ldots, k$ with weights w_i. For simplicity let us denote edges (v_0, v_i) as e_i and their reliabilities as p_i, correspondingly. Using (19) we obtain:

$$N(Ch_k) = w_0 \sum_{i=1}^{k} p_i w_i + \sum_{i=1}^{k-1} \sum_{j=i+1}^{k} w_i w_j (1 - p_i p_j). \tag{20}$$

5 Branching by Chain

As in [1] where we consider the probabilistic connectivity, we can gain from considering simple chains. Because of limited area of this paper we consider only the case of 2-edges chains here.

For further consideration we need the following lemma.

Lemma 2. *If during a graph G factoring process some subgraphs $G_1, \ldots, G_k$ are obtained with probabilities $p_1, \ldots, p_k$, such that they have the same structure and matrix P, and two special nodes v_s and v_t have weights w_{si} and w_{ti} in G_i, $i = 1, \ldots, k$, then the total contribution of these subgraphs into $N(G)$ is equal to*

$$D = \sum_{i=1}^{k} p_i N(G^o) + a_{st} \frac{\sum\limits_{i=1}^{k} p_i \sum\limits_{i=1}^{k} p_i w_{si} w_{ti} - \sum\limits_{i=1}^{k} p_i w_{si} \cdot \sum\limits_{i=1}^{k} p_i w_{ti}}{\left(\sum\limits_{i=1}^{k} p_i\right)^2}. \tag{21}$$

where graph G^o has the same structure and P as G_i and

$$WT_s(G^o) = \sum_{i=1}^{k} p_i w_{si} / \sum_{i=1}^{k} p_i, \tag{22}$$

$$WT_t(G^o) = \sum_{i=1}^{k} p_i w_{ti} / \sum_{i=1}^{k} p_i.$$

Here a_{st} is a probability of v_s and v_t being disconnected in G^o.

Proof. Proof is similar to that of lemma 1 but more complex because of existence of production $w_s w_t$ in (1). □

Now we continue to the main theorem.

Theorem 1. *If a connected random graph has a simple chain $C = e_{sx}, e_{xt}$ connecting nodes s and t through a node v_x with degree 2 then the following equation is true.*

$$N(G) = \left[p_{st}(1 - p_{sx}p_{xt}) + p_{sx}p_{xt}\right] N(G^*) + \tag{23}$$
$$(1 - p_{st} - p_{sx}p_{xt}) \left\{ N(G^o) + a_{st} \frac{(1 - p_{st})(1 - p_{xt}) p_{sx} p_{xt}}{(1 - p_{st} - p_{sx}p_{xt})^2} w_x^2 \right\}.$$

where a_{st} is a probability of v_s and v_t being disconnected in G.

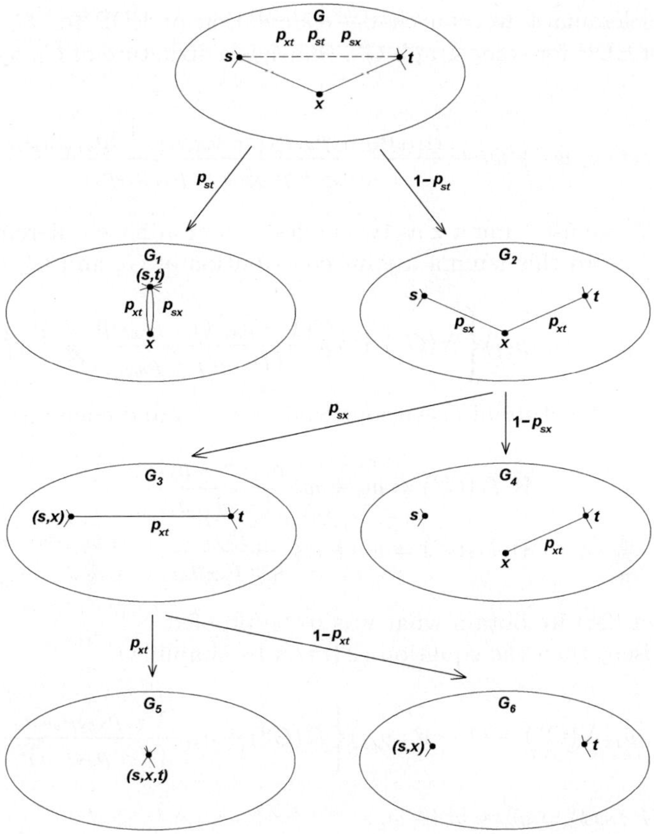

Fig. 1. Branching in a graph with a 2-edge chain

Proof. If e_{st} exists then we first choose it for factoring. Then we consequently make factoring by e_{sx} and e_{xt} (see Fig. 1). Two terminal graphs G_1 and G_4 we can easy replace by graphs with a structure of G_5 and G_6, correspondingly, using (12) and (13) (note that in G_1 we first replace multi-edge by a single one with equivalent reliability):

$$N(G_1) = N(G_5^0),\ WT(G_5^0, st) = w_s + w_t + (p_{sx} + p_{xt} - p_{sx}p_{xt})w_x, \quad (24)$$
$$N(G_4) = N(G_6^0),\ WT(G_6^0, t) = w_t + p_{xt}w_x.$$

For graphs G_5 and G_6 themselves we have:

$$WT(G_5, xst) = w_s + w_t + w_x, \quad (25)$$
$$WT(G_6, s) = w_s + w_x,$$
$$WT(G_6, t) = w_t.$$

It is clear that graphs G_1, G_4, G_5 and G_6 are obtained with probabilities p_{st}, $(1 - p_{st})(1 - p_{sx})$, $(1 - p_{st})p_{sx}p_{xt}$ and $(1 - p_{st})(1 - p_{xt})p_{sx}$, correspondingly.

According to lemma 1 we can change calculation of EDP for G_5 and G_5^0 to calculation of EDP for some graph G^*, that has a structure of G_5 with a weight of joint node

$$WT(G^*, sxt) = w_s + w_t + \frac{(p_{sx}p_{st} + p_{sx}p_{xt} + p_{st}p_{xt} - 2p_{sx}p_{st}p_{xt})}{p_{st} + p_{sx}p_{st} - p_{sx}p_{st}p_{xt}} w_x. \quad (26)$$

For G_6 and G_6^0 we use lemma 2 as two nodes v_s and v_t have different weights in them. According to this lemma a joint contribution of G_6 and G_6^0 into EDP of G is

$$(1 - p_{st} - p_{sx}p_{xt}) \left\{ N(G^o) + a_{st} \frac{(1 - p_{st})(1 - p_{xt})p_{sx}p_{xt}}{(1 - p_{st} - p_{sx}p_{xt})^2} w_x^2 \right\}. \quad (27)$$

where G^o has a structure of G without chain e_{sx}, e_{xt} and edge e_{st} and

$$WT_s(G^o) = w_s + w_x \frac{p_{sx}(1 - p_{xt})}{1 - p_{sx}p_{xt}}; \quad (28)$$
$$WT_t(G^o) = w_t + w_x \frac{p_{xt}}{1 - p_{sx}p_{xt}}.$$

From this and (26) we obtain what was to be proofed. □

If e_{st} is absent then the equation (23) can be simplified:

$$N(G) = p_{sx}p_{xt}N(G^*) + (1 - p_{sx}p_{xt}) \left\{ N(G^o) + a_{st} \frac{(1 - p_{xt})p_{sx}p_{xt}}{(1 - p_{sx}p_{xt})^2} w_x^2 \right\}, \quad (29)$$

where $WT(G^*, sxt) = w_s + w_t + w_x$.

Now let us discuss obtaining of a_{st} that is a probability of v_s and v_t being disconnected in G. Obviously

$$a_{st} = (1 - p_{st})(1 - p_{sx}p_{xt})P(v_s \text{ and } v_t \text{ are disconnected in } G_6). \quad (30)$$

There are well-known algorithms for finding 2-terminal probabilistic connectivity (see [6], for example). Note that a complexity of this task is obviously less then complexity of calculation of EDP for G_1 or G_4 which makes use of equation (23) effective.

6 Conclusion

Thus we have presented some useful equations that can help in calculating the EDP of a random graph. Most advantage for speeding up is gained from branching by chains. Note that chains are inevitable during the factoring process as a result of edge deletion and, sometimes, as a result of contracting by edge (refer to [1]). Experiments shows that calculation of EDP for 30 random $G(10, 15)$ is in average more than 20 times faster with use of our equations then by using equation (1).

References

1. Rodionova, O.K., Rodionov, A.S., and Choo, H.: Network Probabilistic Connectivity: Exact Calculation with Use of Chains. ICCSA-2004, Springer LNCS. **3046** (2004) 315–324
2. Krivoulets, V.G., Polesskii, V.P., "What is the Theory of Bounds for Network Reliability?," *Information Processes, ISSN: 1819-5822, Vol. 1, n. 2,* pp. 199-203, 2001.
3. Cancela, H., Petingi L., "Diameter constrained network reliability: exact evaluation by factorization and bounds," *Proc. of the Int. Conf. on Industrial Logistics, Japan, 2001,* pp. 359-366, 2001.
4. Lucet,C., Manouvrier, J.-F., "Exact Methods to compute Network Reliability, *Proc. of 1st International Conf. on Mathematical Methods in Reliability, Bucharest, Roumanie, Sep. 1997,* (paper available via http:// ramp.ucsd.edu/resources.html)
5. Satyanarayana, A. and Chang, M.K. "Network reliability and the factoring theorem," *Networks, Vol. 13,* pp.107120, 1983.
6. Shooman, A.M.: Algorithms for Network Reliability and Connection Availability Analysis. Electro/95 Int. Professional Program Proc. (1995) 309–333
7. Moore, E.F., Shannon, C.E., "Reliable Circuits Using Less Reliable Relays," *J. Franclin Inst., 262, n. 4b,* pp. 191-208, 1956.
8. Palmer C.R., Siganos G., Faloutsos M., Faloutsos C., and Gibbons P.: The connectivity and faulttolerance of the Internet topology. Workshop on Network-Related Data Management (NRDM-2001), http://www.research. att.com/ divesh/papers/cjfgs98-ir.ps.
9. Rodionov, A.S., Choo H.: On Generating Random Network Structures: Connected Graphs. International Conference on Information Networking ICOIN 2004, Revised selected papers. Springer LNCS. **3090** (2004) 483–491
10. Shooman, A.M., Kershenbaum, A.: Exact Graph-Reduction Algorithms for Network Reliability Analysis. Proc. GLOBECOM' 91. **2** (1991) 1412–1420

Parallel LU Factorization of Band Matrices on SMP Systems*

Alfredo Remón[1], Enrique S. Quintana-Ortí[1], and Gregorio Quintana-Ortí[1]

Depto. de Ingeniería y Ciencia de Computadores
Universidad Jaume I
12.071–Castellón, Spain
{remon, quintana, gquintan}@icc.uji.es

Abstract. In this paper we present two routines for the LU factorization of band matrices that target (parallel) SMP architectures and expose coarser-grain parallelism than their LAPACK counterpart. Preliminary experimental results on two different parallel platforms show some initial benefits for the new approach.

Keywords: Band matrix, LU factorization, LAPACK, multithreaded BLAS, Symmetric Multiprocessor (SMP).

1 Introduction

Linear systems with band coefficient matrix have to be solved, among others, in static and dynamic analyses and linear equations in structural engineering, finite element analysis in structural mechanics, and domain decomposition methods for partial differential equations in civil engineering. Exploiting the structure of the coefficient matrix for band linear systems yields huge savings, both in number of computations and storage space. This is recognized in LAPACK, which includes unblocked and blocked routines for the LU factorization of band matrices [1,2]. While the first routine performs most of the operations in terms of BLAS-2, the second routine reorganizes the computations so as to carry out a major part in terms of BLAS-3.

In this paper we analyze the parallel efficacy of the (double-precision) routine DGBTRF in LAPACK. As the parallelism in LAPACK is extracted by using multithreaded implementations of BLAS, the fragmentation (partitioning) of the operations in the LAPACK routine may potentially reduce its parallel performance. Our approach consists in *merging* operations so that parallelism with coarser-grain is exposed. In order to do so, our first routine requires a simple modification of the storage scheme for band matrices so that a few rows of zeros are added to the bottom of the matrix. By doing some additional copies and manipulation of the matrix blocks, the second routine does not require this workspace.

* This research was supported by the CICYT project TIN2005-09037-C02-02 and FEDER, and project No. P1B-2004-6 of the *Fundación Caixa-Castelló/Bancaixa and UJI.*

M. Gerndt and D. Kranzlmüller (Eds.): HPCC 2006, LNCS 4208, pp. 110–118, 2006.

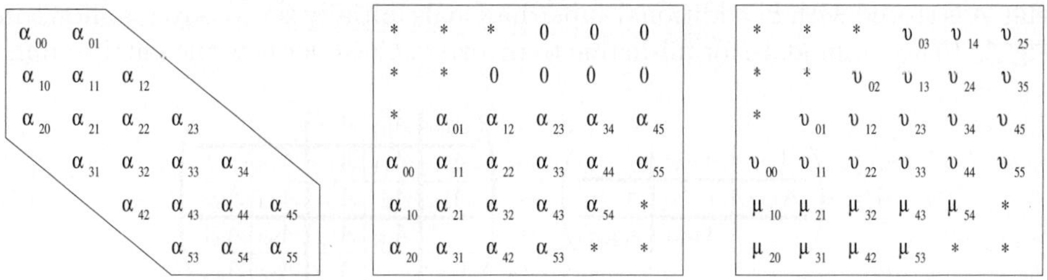

Fig. 1. 6×6 band matrix with upper and lower bandwidths $k_l = 2$ and $k_u = 1$, respectively (left); packed storage scheme used in LAPACK (center); result of the LU factorization where $\upsilon_{i,j}$ and $\mu_{i,j}$ stand, respectively, for the entries of the upper triangular factor U and the multipliers of the Gauss transformations

The paper is structured as follows. The blocked factorization xGBTRF routine in LAPACK is reviewed in Section 2. Our new routines are then presented in Section 3. The preliminary experiments on two Intel SMP architectures, based on Xeon™ and Itanium2™ processors, show some benefits of this approach in Section 4. Finally, some concluding remarks are given in Section 5.

2 LAPACK Blocked Routine for the Band LU Factorization

Given a matrix $A \in \mathbb{R}^{n\times n}$, with lower and upper bandwidth k_l and k_u respectively, routine xGBTRF computes the *LINPACK-style LU factorization with partial pivoting*

$$L_{n-2}^{-1} \cdot P_{n-2} \cdots L_1^{-1} \cdot P_1 \cdot L_0^{-1} \cdot P_0 \cdot A = U \tag{1}$$

where $P_0,\ P_1, \ldots, P_{n-2} \in \mathbb{R}^{n\times n}$ are permutation matrices, $L_0,\ L_1, \ldots, L_{n-2} \in \mathbb{R}^{n\times n}$ are Gauss transformations, and $U \in \mathbb{R}^{n\times n}$ is upper triangular with upper bandwidth $k_l + k_u$. In order to reduce the storage needs, the lower triangular factor of the (traditional) *LAPACK-style LU factorization with partial pivoting*

$$PA = LU \tag{2}$$

is not explicitly constructed. Instead, the multipliers corresponding to the Gauss transformations are stored overwriting the subdiagonal entries of A. This corresponds to the permutations matrices not being applied to the Gauss transformations, as in (1). Figure 1 illustrates the packed storage scheme used for band matrices in LAPACK and how this scheme accommodates for the result of the LU factorization with pivoting.

In order to describe routine xGBTRF, for simplicity, we will assume that the algorithmic block size, n_b, is an exact multiple of both k_l and k_u; we also consider that A is stored with k_l additional superdiagonals initially set to zero (as shown in Fig. 1) to accommodate for fill-in due to pivoting. Consider now the partitionings

$$A = \left(\begin{array}{c|c|c} A_{TL} & A_{TM} & \\ \hline A_{ML} & A_{MM} & A_{MR} \\ \hline & A_{BM} & A_{BR} \end{array}\right) \rightarrow \left(\begin{array}{c|c|c|c|c} A_{00} & A_{01} & A_{02} & & \\ \hline A_{10} & A_{11} & A_{12} & A_{13} & \\ \hline A_{20} & A_{21} & A_{22} & A_{23} & A_{24} \\ \hline & A_{31} & A_{32} & A_{33} & A_{34} \\ \hline & & A_{42} & A_{43} & A_{44} \end{array}\right), \tag{3}$$

where $A_{TL}, A_{00} \in \mathbb{R}^{k \times k}$, $A_{11}, A_{33} \in \mathbb{R}^{n_b \times n_b}$, and $A_{22} \in \mathbb{R}^{l \times u}$, with $l = k_l - n_b$ and $u = \bar{k}_u - n_b = (k_u + k_l) - n_b$. With this partitioning, A_{02}, A_{13}, and A_{24} are lower triangular.

Routine xGBTRF corresponds to what is usually known as a right-looking algorithm; that is, an algorithm where, at a certain iteration, A_{TL} has been already factorized, A_{ML} and A_{TM} have been overwritten, respectively, by the multipliers and the corresponding block of U, and A_{MM} has been updated correspondingly. In order to move forward in the computation by n_b rows/columns, during the current iteration of the routine the following operations are performed (the annotations to the right of some of these operations correspond to the name of the BLAS routine that is used):

1. Obtain $W_{31} :=$ TRIU(A_{31}), a copy of the upper triangular part of A_{31}; compute the LU factorization with partial pivoting

$$P_1 \begin{pmatrix} A_{11} \\ A_{21} \\ W_{31} \end{pmatrix} = \begin{pmatrix} L_{11} \\ L_{21} \\ L_{31} \end{pmatrix} U_{11}. \tag{4}$$

   The blocks of L and U overwrite the corresponding blocks of A and W_{31}. (In the actual implementation, the copy W_{31} is obtained as this factorization is being computed.)

2. Apply the permutations in P_1 to the remaining columns of the matrix:

$$\begin{pmatrix} A_{12} \\ A_{22} \\ A_{32} \end{pmatrix} := P_1 \begin{pmatrix} A_{12} \\ A_{22} \\ A_{32} \end{pmatrix} \quad \text{and} \qquad (\text{xLASWP}) \tag{5}$$

$$\begin{pmatrix} A_{13} \\ A_{23} \\ A_{33} \end{pmatrix} := P_1 \begin{pmatrix} A_{13} \\ A_{23} \\ A_{33} \end{pmatrix}. \tag{6}$$

   A careful application of permutations is needed in the second expression as only the lower triangular structure of A_{13} is physically stored. As a result of the application of permutations, A_{13}, which initially equals zero, may become lower triangular. No fill-in occurs in the strictly upper part of this block.

3. Compute the updates:

$$A_{12}(= U_{12}) := L_{11}^{-1} A_{12}, \qquad (\text{xTRSM}) \qquad (7)$$
$$A_{22} := A_{22} - L_{21} U_{12}, \qquad (\text{xGEMM}) \qquad (8)$$
$$A_{32} := A_{32} - L_{31} U_{12}. \qquad (\text{xGEMM}) \qquad (9)$$

4. Obtain the copy of the lower triangular part of A_{13}, $W_{13} := \text{TRIL}(A_{13})$; compute the updates

$$W_{13}(= U_{13}) := L_{11}^{-1} W_{13}, \qquad (\text{xTRSM}) \qquad (10)$$
$$A_{23} := A_{23} - L_{21} W_{13}, \qquad (\text{xGEMM}) \qquad (11)$$
$$A_{33} := A_{33} - L_{31} W_{13}; \qquad (\text{xGEMM}) \qquad (12)$$

and copy back $A_{13} := \text{TRIL}(W_{13})$.
5. Undo the permutations on $\left[L_{11}^T, L_{21}^T, W_{31}^T\right]^T$ so that these blocks store the multipliers used in the LU factorization in (6) and W_{31} is upper triangular; copy back $A_{31} := \text{TRIU}(W_{31})$.

In our notation, after these operations are carried out, A_{TL} (the part that has been already factorized) grows in n_b rows/columns so that

$$A = \left(\begin{array}{c|c|c} A_{TL} & A_{TM} & \\ \hline A_{ML} & A_{MM} & A_{MR} \\ \hline & A_{BM} & A_{BR} \end{array}\right) \leftarrow \left(\begin{array}{cc|cc|c} A_{00} & A_{01} & A_{02} & & \\ A_{10} & A_{11} & A_{12} & A_{13} & \\ \hline A_{20} & A_{21} & A_{22} & A_{23} & A_{24} \\ & A_{31} & A_{32} & A_{33} & A_{34} \\ \hline & & A_{42} & A_{43} & A_{44} \end{array}\right) \qquad (13)$$

in preparation for the next iteration.

From this particular implementation we observe that:

– Provided $n_b \ll k_u, k_l$, a major part of the floating-point arithmetic operations (flops) are performed in terms of the BLAS-3 matrix product in (8). Thus, by calling a tuned implementation of xGEMM, high performance is to be expected from xGBTRF.
– No attempt is made to exploit the lower triangular structure of A_{13}/U_{13} in the computations corresponding to (10)–(12) as there is no appropriate BLAS kernel for these types of operations. Moreover, the additional space W_{13} and the extra copies in Step 4 are only required in order to use BLAS-3 to compute these updates.
– In Step 1, the LAPACK-style LU factorization of $\left[A_{11}^T, A_{21}^T, W_{31}^T\right]^T$ is computed. This allows the use of BLAS-3 in the application of the triangular factors to the blocks to the right. At the end, in Step 4, the permutations are undone on $\left[L_{11}^T, L_{21}^T, L_{31}^T\right]^T$ so that these blocks store the multipliers from the LINPACK-style LU factorization. In this way, L_{31} recovers the upper triangular form, and no additional space is needed to store it overwriting the corresponding block of A (Step 5).

3 New Algorithms for the Band LU Factorization

The exposition of the algorithm underlying routine xGBTRF in the previous section shows that, in order to advance the computation by n_b rows/columns, two invocations of routine xTRSM and four invocations of xGEMM are necessary for the updates in Steps 3 and 4. This is needed because of the special storage pattern used for band matrices in LAPACK. As all parallelism is extracted in LAPACK from calling multithreaded implementations of BLAS routines, the fragmentation of the computations that need to be performed in a single iteration of the algorithm into several small operations (and a large one) is potentially harmful for the efficacy of the codes. This is specially so on SMP platforms, where the threshold between what is considered "small" and "large" is considerably high; see [3] for details.

In this section we describe how to merge the operations corresponding to Steps 2–4 so that higher performance is likely to be obtained on SMP architectures. Our first algorithm requires additional storage space in the data structure containing A so that n_b rows of zeros are present at the bottom of the matrix. By doing some extra copies and manipulation of the matrix blocks, the second algorithm does not require this workspace.

3.1 Routine xGBTRF+M

Consider the data structure containing A (see Fig. 1-center) is padded with n_b rows at the bottom with all the entries in this additional space initially set to zero. (Interestingly, that corresponds, e.g., to the structure that would be necessary in the LAPACK storage scheme to keep a band matrix with square blocks of order n_b in the lower band. Block band matrices are frequently encountered in practice as well, but no specific support is provided for them in the current version of LAPACK). Then, Steps 1–4 in xGBTRF are transformed as follows:

1. In the first step, the LU factorization with partial pivoting
$$P_1 \begin{pmatrix} A_{11} \\ A_{21} \\ A_{31} \end{pmatrix} = \begin{pmatrix} L_{11} \\ L_{21} \\ L_{31} \end{pmatrix} U_{11} \tag{14}$$
is computed and the blocks of L and U overwrite the corresponding blocks of A. There is no longer need for workspace W_{31} nor copies to/from it as the additional rows at the bottom accommodate for the elements in the strictly lower triangle of L_{31}.
2. Apply the permutations in P_1 to the remaining columns of the matrix:
$$\begin{pmatrix} A_{12} & A_{13} \\ A_{22} & A_{23} \\ A_{32} & A_{33} \end{pmatrix} := P_1 \begin{pmatrix} A_{12} & A_{13} \\ A_{22} & A_{23} \\ A_{32} & A_{33} \end{pmatrix} \qquad \text{(xLASWP)}. \tag{15}$$
A single call to xLASWP suffices now as the zeros at the bottom of the data structure and the additional k_l superdiagonal set to zero in the structure

ensure that fill-in may only occur in the elements in the lower triangular part of A_{13}.

3. Compute the updates:

$$(A_{12},\ A_{13})\,(=(U_{12},\ U_{13})) := L_{11}^{-1}\,(A_{12},\ A_{13}) \qquad (\text{xTRSM}), \qquad (16)$$

$$\begin{pmatrix} A_{22}\ A_{23} \\ A_{32}\ A_{33} \end{pmatrix} := \begin{pmatrix} A_{22}\ A_{23} \\ A_{32}\ A_{33} \end{pmatrix} - \begin{pmatrix} L_{21} \\ L_{31} \end{pmatrix} (U_{12},\ U_{13}) \qquad (\text{xGEMM}). \qquad (17)$$

The lower triangular system in (16) returns a lower triangular block in A_{13}.

4. Undo the permutations on $\left[L_{11}^T, L_{21}^T, L_{31}^T\right]^T$ so that these blocks store the multipliers used in the LU factorization in (6) and L_{31} is upper triangular.

Indeed, the procedure just described would allow the block lower triangular matrix L to be explicitly constructed using the additional space provided so that Step 4 could be eliminated. In case multiple linear systems (right-hand sides) were to be solved with the same coefficient matrix, this in turn would facilitate the use BLAS-3 during the forward substitution stage (with L).

3.2 Routine xGBTRF+C

The previous approach, though efficient in the sense of grouping as many computations as possible in coarse-grain blocks, requires a non-negligible workspace. Although a padding of k_l rows is already necessary at the top of the matrix to accommodate for fill-in in U, we recognize this is not completely satisfactory. Therefore we propose the following algorithm which also aims at clustering blocks but does not require storage for n_b rows:

1. Obtain W_{31} := STRIL(A_{31}), a copy of the physical space that would be occupied by the strictly lower triangular part of A_{31} and set STRIL(A_{31}) := 0; compute the LU factorization with partial pivoting

$$P_1 \begin{pmatrix} A_{11} \\ A_{21} \\ A_{31} \end{pmatrix} = \begin{pmatrix} L_{11} \\ L_{21} \\ L_{31} \end{pmatrix} U_{11}. \qquad (18)$$

The blocks of L and U overwrite the corresponding blocks of A. A copy of the elements in the physical storage overwritten in this factorization is kept at W_{31}.

2. Obtain W_{13} := STRIU(A_{13}), a copy of the physical space that would be occupied by the strictly upper triangular part of A_{13} and set STRIL(A_{13}) := 0; apply the permutations in P_1 to the remaining columns of the matrix:

$$\begin{pmatrix} A_{12}\ A_{13} \\ A_{22}\ A_{23} \\ A_{32}\ A_{33} \end{pmatrix} := P_1 \begin{pmatrix} A_{12}\ A_{13} \\ A_{22}\ A_{23} \\ A_{32}\ A_{33} \end{pmatrix} \qquad (\text{xLASWP}). \qquad (19)$$

3. Compute the updates:

$$(A_{12},\ A_{13})\,(=(U_{12},\ U_{13})) := L_{11}^{-1}\,(A_{12},\ A_{13}) \qquad (\text{xTRSM}), \tag{20}$$

$$\begin{pmatrix} A_{22} & A_{23} \\ A_{32} & A_{33} \end{pmatrix} := \begin{pmatrix} A_{22} & A_{23} \\ A_{32} & A_{33} \end{pmatrix} - \begin{pmatrix} L_{21} \\ L_{31} \end{pmatrix} (U_{12},\ U_{13}) \qquad (\text{xGEMM}). \tag{21}$$

4. Undo the permutations on $\left[L_{11}^T, L_{21}^T, L_{31}^T\right]^T$ so that these blocks store the multipliers used in the LU factorization in (6) and L_{31} is upper triangular; restore STRIL(A_{31}) and STRIU(A_{13}) from W_{31} and W_{13}, respectively.

4 Experimental Results

All experiments in this section were performed using IEEE double-precision (real) arithmetic and band matrices of order $n = 10000$ with $k_l = k_u$. In the evaluation, for each bandwidth dimension, we employed values from 1 to 200 to determine the optimal block size, n_b^{opt}; only those results corresponding to n_b^{opt} are shown.

The usual measure to report performance of linear algebra codes is the MFLOP (millions of flops per second) rate. However, the actual number of flops done in the band LU factorization depends on the pivoting sequence. Therefore, in order to reduce the number of results, we will report the speed-up the new routines attain over DGBTRF. Thus, e.g., a speed-up factor of 1.2 means the execution time (MFLOP rate) of the new routine is 20% lower (higher) than that of DGBTRF.

We report the performance of the routines on two different SMP architectures, with 2 and 4 processors; see Table 1. Two threads were employed on XEON and 4 on ITANIUM. As the efficacy of the kernels in BLAS is crucial, for each platform we use the implementations listed in Table 2.

Table 1. SMP architectures employed in the evaluation

Platform	Architecture	#Proc.	Frequency (GHz)	L2 cache (KBytes)	L3 cache (MBytes)	RAM (GBytes)
XEON	Intel Xeon	2	2.4	512	–	1
ITANIUM	Intel Itanium2	4	1.5	256	4	4

Figure 2 illustrates the speed-up achieved by routines DGBTRF+M and DGBTRF+C for matrices with bandwidth ranging from 50 to 500. The lower limit was selected because that was experimentally determined to be (approximately) the threshold from which the blocked codes outperformed the unblocked routine DGBTF2. The first experiments revealed a major bottleneck in the parallel implementation of routine DGER from GotoBLAS when the matrix involved in the rank-1 update had a few columns (exactly the case that appears in (6)). In order

Table 2. Software employed in the evaluation

Platform	BLAS	Compiler	Optimization Flags	Operating System
XEON	GotoBLAS 1.00 MKL 8.0	`gcc 3.3.5`	`-O3`	Linux 2.4.27
ITANIUM	GotoBLAS 0.95mt MKL 8.0	`icc 9.0`	`-O3`	Linux 2.4.21

to overcome this problem, the code of these routine in the reference BLAS was manually inlined.

The results show that a notable reduction of the execution time was obtained by routine DGBTRF+M on ITANIUM for matrices of moderate bandwidth: approximately from $k_l = k_u$=50 to 150–200, depending on the BLAS implementation. The improvements of that routine on XEON were more modest and stable, around 5%. As expected, the efficacy of routine DGBTRF+C is slightly lower. The higher efficacy of the new codes for moderate bandwidth can be explained as follows: as the bandwidth increases, so does the optimal block size of the algorithm and thus, the updates of those blocks other than A_{22} can no longer be considered small operations (with respect to the number of threads used in the experiments) with no parallelism.

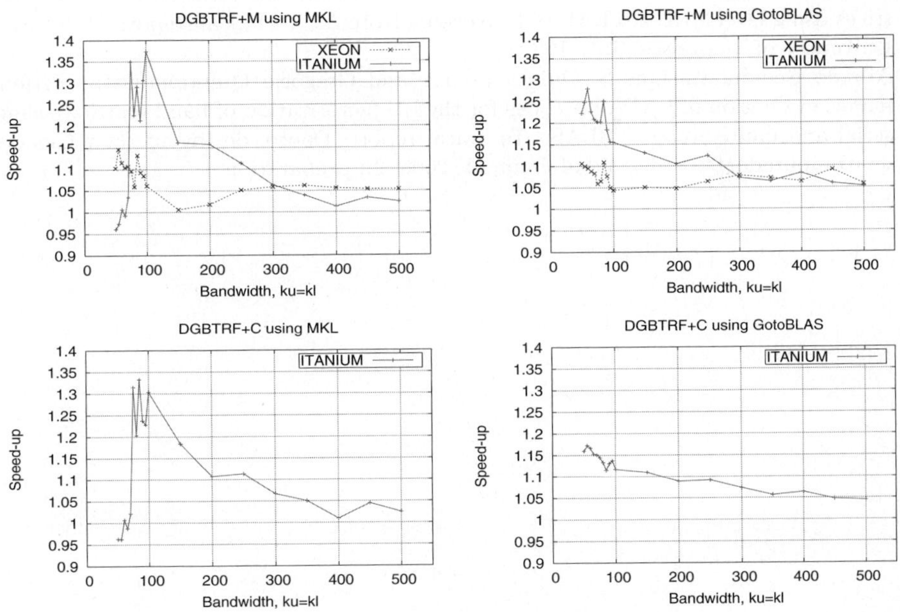

Fig. 2. Speed-up attained by routines DGBTRF+M (top) and DGBTRF+C (bottom) using multithreaded implementations of BLAS in MKL (left) and GotoBLAS (right)

5 Conclusions

We have presented two algorithms for computing the LU factorization of a band matrix that reduce the number of calls to BLAS per iteration so that coarser-grain parallelism is exposed. The routines are potentially better suited to explote the architecture of SMP systems. Although initial results seem to confirm our expectations, further experimentation, specially using architectures with more processors, is needed.

Several other (minor) conclusions have been extracted from our experience with band codes:

- BLAS does not support all the functionality that is needed for the factorization.
- The performance of BLAS-1 and BLAS-2 is much more important than in general for other dense routines.
- The optimal block size for the blocked routines needs careful tuning.

References

1. E. Anderson, Z. Bai, J. Demmel, J. E. Dongarra, J. DuCroz, A. Greenbaum, S. Hammarling, A. E. McKenney, S. Ostrouchov, and D. Sorensen. *LAPACK Users' Guide*. SIAM, Philadelphia, 1992.
2. Jeremy Du Croz, Peter Mayes, and Giuseppe Radicati. Factorization of band matrices using level 3 BLAS. LAPACK Working Note 21, Technical Report CS-90-109, University of Tennessee, July 1990.
3. Alfredo Remón, Enrique S. Quintana-Ortí, and Gregorio Quintana-Ortí. Performance evaluation of LAPACK codes for the LU factorization of band matrices using serial and multithreaded BLAS. Technical report, Depto. de Ingeniería y Ciencia de Computadores, Universidad Jaume I, 2006. In preparation.

A Tree-Based Distributed Model for BGP Route Processing*

Kun Wu, Jianping Wu, and Ke Xu

Department of Computer Science and Technology, Tsinghua University
Beijing 100084, P.R.China
{kunwu, xuke}@csnet1.cs.tsinghua.edu.cn, jianping@cernet.edu.cn

Abstract. The scalable architecture is one of the key issues of the next generation routers. The distributed routing protocol computing model is one of the most difficult challenges on the router control plane. This paper studies the route processing model of BGP, which has been deployed in backbone widely. A tree-based distributed computing model is discovered, based on the inherent parallel features of BGP. The model is described in two structures for two different system configurations. And the related algorithms are given. The performance promotion is analyzed theoretically. Finally, the promotion is proved by experiments.

1 Introduction

The router system can be divided into two functional planes inherently. They are the *data plane* and the *control plane*. The former is responsible for the route look-up and packet forwarding. The other covers the router control and management, including the routing information exchanges, route computing, the system maintenance, and so on. The scalability enhancements locate on both of the planes.

Plenty of researches[1][2] have been conducted on the scalable switching plane by far. Some commercial systems have been carried out under this concept [3][4][5]. Nevertheless, the centralized control is still the main stream structure in the control plane. The existing multi-control-board systems mainly focus on the redundant backup, other than the distributed control.

On the core routers in the backbone, the traditional centralized control model turns out to be a major potential bottleneck. The increasing route table requires more processing power and larger memory storage. Only the distributed control plane can match the performance requirements of the scalable router architecture. However, the inherent complexity of the control plane makes it very difficult to achieve an excellent decentralized structure. It is a major ongoing challenge to develop distributed routing protocol processing models, which are the most important components on the router.

* This work was supported by the National Key Fundamental Research Plan (973) of China (No. 2003CB314801) and the National Natural Science Foundation of China (No. 60473082).

M. Gerndt and D. Kranzlmüller (Eds.): HPCC 2006, LNCS 4208, pp. 119–128, 2006.

The main contribution of this work is a distributed BGP processing architecture. And the corresponding necessary algorithms are given to dispatch the sub-tasks efficiently under different system configurations. BGP is the default routing protocol in the Internet. The model in this paper can be well deployed on the core router in the scalable manner. The parallel computing and balanced storage can bring a promising performance improvement, which is verified by the experiments.

The rest of this paper is organized as follows. Section 2 gives an overview of the related systems and models. Section 3 analyzes the BGP route computing model, and proposes the tree-based distributed approach. Related algorithms are presented in this section. Section 4 evaluates the performance promotion theoretically, and conducts experiments to proof the improvements. Section 5 draws the conclusion and introduces the future work.

2 Related Work

The research on distributed control plane mainly focuses on modular software developing and the parallel routing protocol computing. Zebra is a typical modular routing software[6]. It implements the functions by a set of processes, including routing protocols and the route management. The the Vyatta[7], which comes from XORP, provides an open extensible router software reference model. [8] set up a separate server for the route computing. Furthermore, [9] moves the complexity to end systems, which require the routing information from route server. And, [10] is another proposal under this category. All of these researches enhance the router control plane structure in a distributed manner. However, the minimum granularity in these methods is one protocol, which is not fine enough to the scalable routers. [11]and[12] provide an abstract distributed model based on the routing behavior. This approach defines a set of database query language for the information exchange. It focuses on the overlay network much more than the single core router. [13][14][15] implement some flexible router systems. But, their main target is to develop an extensible router, which is very different from a scalable router.

The result in [16] can contribute to the distributed shortest path tree computing for OSPF. For BGP[17], which has been widely deployed in the backbone[18], the distributed route computing model is one of the key issues for the control plane in the scalable routers. [19] describes a distributed model for BGP. Agent technology is used in this model for the load balance. However, the load dispatching pattern is limited by the topology in this model.

3 Distributed BGP Route Computing

3.1 General Route Processing Model for BGP

BGP is illustrated as a path vector protocol[17]. The path vector is the basis for BGP route computing and route information exchanges. For the convenience of later discussion, some definitions are given. On a router running BGP, which is talking to n BGP neighbors, the path information to the destination r is denoted as $p_i(r)$. The path information about r collected from all of the n neighbors is written as $P(r) = \{p_i(r), i = 1, \ldots, n\}$.

The path compare is the fundamental operation during the shortest path competing. This process can be illustrated as a binary relation $\preceq$. While considering destination r, if the path from neighbor i is better than that from neighbor j, then $p_i(r) \preceq p_j(r)$. By the path selection rules defined in [17], $\preceq$ is *total order*, *linear order*, and *simple order* on the set $P(r)$. This means the following statements hold for $p_i(r)$, $p_j(r)$ and $p_l(r)$ in $P(r)$:

$$p_i(r) \preceq p_j(r) \quad \Rightarrow \quad i = j \tag{1}$$

$$p_i(r) \preceq p_l(r), p_l(r) \preceq p_j(r) \quad \Rightarrow \quad p_i(r) \preceq p_j(r) \tag{2}$$

$$p_i(r) \preceq p_j(r) \quad or \quad p_j(r) \preceq p_i(r) \tag{3}$$

Considering the equations above and the path selection rules in [17], the route computing for destination r is to find the optimal $\hat{p}(r)$ under the relation $\preceq$ at set $P(r)$. That is $\exists \hat{p}(r), \forall p_i(r) \in P(r), \hat{(p)}(r) \preceq p_i(r)$. In the runtime system, $P(r)$ reflect the network status at the current time. A set of all the possible $P(r)$ is denoted as $\mathbb{Q}(r)$, $\forall P(r) \in \mathbb{Q}(r)$. Hence, the BGP route computing result can be expressed as a function $\mathbb{R}$, which is defined on $\mathbb{Q}(r) \times P(r)$. That is $\mathbb{R} : \mathbb{Q}(r) \mapsto P(r)$, and $\mathbb{R}[P(r)] = \hat{p}(r)$.

3.2 The Parallel BGP Computing Framework

In this subsection, the conventional BGP computing model is to be improved in a distributed manner. For the purpose of discussion, assume $P(r) \neq \varnothing$. The infinite value is given as following.

Definition 1. *If the route updates received from neighbor i do not have information about route r, then the distance to r through i is assigned infinity.* $p_i(r) = \infty$.

For each neighbor i exporting the route to r, the path through i is *better* than the remainders.

$$\forall p(r) \in P(r), p(r) \neq \infty \quad \Rightarrow \quad p(r) \preceq \infty \tag{4}$$

If no route to r exists, $P(r) = \infty$, then no output should be given from $\mathbb{R}$. So, $\mathbb{R}$ can be extended more strictly to reflect the optimal path selection as $\mathscr{R}$, which is defined on $\mathbb{Q}(r) \times \{\{\hat{(p)}(r)\}, \varnothing\}$. $\mathscr{R} : \mathbb{Q}(r) \mapsto \{\{\hat{p}(r)\}, \varnothing\}$. $\mathscr{R}[P(r)] = \{p | p = \mathbb{R}[P(r)]\}$. That is

$$\mathscr{R}[P(r)] = \begin{cases} \{\hat{p}(r)\} & , P(r) \text{has member(s) better than} \infty \\ \varnothing & , P(r) \text{has} \infty \text{ only} \end{cases} \tag{5}$$

The computation structure of $\mathscr{R}[P(r)]$ can is dividable on $P(r)$ while taking the relation $\preceq$ into account.

Definition 2. *For $P(r)$'s subsets $P_1(r), P_2(r), \ldots, P_m(r)$, $m < \infty$, if $\forall i, j, i \neq j \Rightarrow P_i(r) \cap P_j(r) = \varnothing$, and $\bigcup_{i=1}^{m} P_i(r) = P(r)$, then the set $\{P_1(r), P_2(r), \ldots, P_m(r)\}$ is defined as a division of $P(r)$.*

Theorem 1. *For any division $\{P_1(r), P_2(r), \ldots, P_m(r)\}$ of $P(r)$, the following equation is hold.*

$$\mathscr{R}[P(r)] = \mathscr{R}[\bigcup_{i=1}^{m} \mathscr{R}[P_i(r)]] \tag{6}$$

Proof. If $\mathscr{R}[P(r)] = \varnothing$, from Definition 1 and Equation 4, $P(r) = \{p_i(r) | p_i(r) = \infty, \quad 1 \leq i \leq n\}$. Hence, any division of $P(r)$ is a set including ∞ only. $\mathscr{R}[P_i(r) = \varnothing, \forall i = 1, \ldots, m$. So, $\mathscr{R}[\bigcup_{i=1}^{m} \mathscr{R}[P_j(r)]] = \varnothing = \mathscr{R}[P(r)]$.
If $\mathscr{R}[P(r)] \neq \varnothing$, $\mathscr{R}[P_j(r)] = \hat{p}_j(r)$. Then, $\bigcup_{i=1}^{m} \mathscr{R}[P_j(r)] = \bigcup_{i=1}^{m} \{\hat{p}_j(r)\} = \{\hat{p}_1(r), \ldots, \hat{p}_m(r)\}$. For $\forall i = 1, \ldots, m$, we have $\hat{p}(r) \preceq \hat{p}_j(r)$. Otherwise, there must be at least a $\hat{p}_j(r)$ in $P(r)$, holding $\hat{p}_j(r) \preceq \hat{p}(r)$, and $\hat{p}_j(r) \neq \hat{p}(r)$. Assume $\hat{p}_J(r)$ is the *smallest* one. From the transitivity of $\preceq$ in Equation2, we have $\mathscr{R}[P(r)] = \{\hat{p}_J(r)\}$. However, $\hat{p}_J(r) \neq \hat{p}(r)$. This conflicts against the definition of $\mathscr{R}[P(r)]$. Hence the assumption is not valid. That is for $\forall i = 1, \ldots, m$, there is $\hat{p}(r) \preceq \hat{p}_j(r)$. Furthermore, We have $\mathscr{R}[\bigcup_{i=1}^{m} (R)[P_i(r)]] = \mathscr{R}[\{\hat{p}_1(r), \ldots, \hat{p}_m(r)\}] = \mathscr{R}[P(r)]$.

From Theorem 1, the optimal path computing based on relation $\preceq$ can be illustrated as a tree model. All the nodes are organized logically in the structure in Fig1.

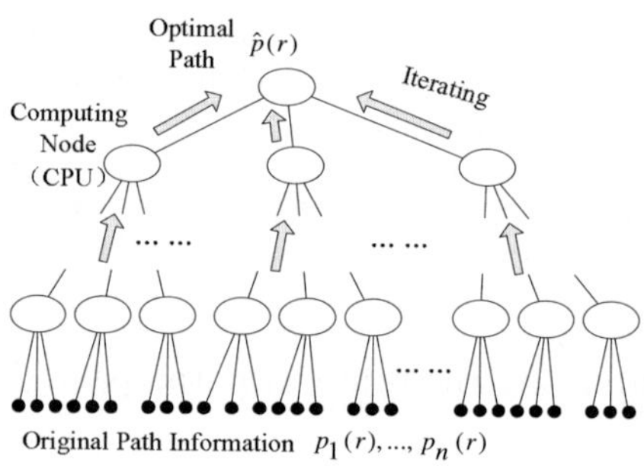

Fig. 1. The Tree-Like Computing Model

3.3 Models and Algorithms

Some definitions are given for the ease of discussion.

Definition 3. *The route information about r collected from some neighbors forms a set, which is defined as a* path group. *The best path dominated by relation $\preceq$ on a path group is called the* local optimal path. *The optimal one on all the paths is named as* global optimal path, *which is the $\hat{p}(r)$ above.*

Definition 4. *The bottom-up iterating model illustrated in Fig.1 is called a* iterating tree. *The leaf nodes are the original information from neighbors. The inner nodes are named* computing nodes.

There are two typical scalable router structure from the implementation view. They are named as the *Un-Bound Structure(UBS)*, and the *Load-Bound Structure(LBS)*. In UBS, All the computing nodes are equivalent, and the processing for the original information can be dispatched to every node freely. In LBS, the original information must be processed on the node, which the neighbors attached. In general UBS routers, every computing nodes have the same power. This is the k-ary iterating tree in this paper, and it has some useful features.

Definition 5. *If each inner node of the tree has k equivalent children nodes, the whole system is called a* k-ary iterating tree.

Theorem 2. *If the number of inner nodes on a k-ary iterating tree is u, and the number of leaf nodes is n, then $u \geq \frac{n-1}{k-1}$.*

Proof. Because of the definition of k-ary iterating tree, the children number of each inner node is no more than k. Every child node has an edge in the tree connected to it's father. So, the total number of the edges $v \leq uk$. These edges connect to all the nodes except the root node. That is $v = n + (u - 1)$, and $uk \geq v = n + (u - 1)$. In a real system, it is meaningful only if the inner node has more than one children. That is $k > 1$. So we have $u \geq \frac{n-1}{k-1}$.

A simple dispatching algorithm is given below.

Algorithm 1

1. Numbering u computing nodes randomly as $1, 2, \ldots, u$.
2. Numbering n neighbor randomly as $u + 1, u + 2, \ldots, u + n$.
3. Take node 1 as the root node.
4. Append the left $u + n - 1$ nodes to corresponding father orderly. If the child node number is x, then it's father's number is $\lfloor \frac{x+k-2}{k} \rfloor$

Algorithm 1 can be extended to support LBS. Before describing the solution, a few definitions and features are given.

Definition 6. *The father of a leaf node is defined as* last-to-leaf node, *which is denoted as ltl-node.*

Definition 7. *For an iterating tree, if the children number of all the inner nodes (ltl-nodes excluded) is k, and the children number of each ltl-nodes is no less than k. And, for the ltl-nodes whole children number is greater than k, the children are all neighbors, then this type of tree is named as* k-ary half-fixed iterating tree, *which is written as* k-ary hfi-tree.

If the number of neighbor attached is z. We define:

$$w = \begin{cases} k - z \,, k \geq z \\ 0 \quad\quad\;\, , k < z \end{cases} \tag{7}$$

Theorem 3. *For a k-ary hfi-tree,* $\sum_u w_i \geq u - 1$.

The proof is very similar with that of Theorem 2. It is not provided here for the limitation of the paper length. A width first dispatching algorithm extended from Algorithm 1 can be described as below.

Algorithm 2

1. Numbering all the node with $1, 2, \ldots, u$ in the decent resource order.
2. Set up an array $w[1 \ldots u]$ denoting the sequence in step 1. Extending $w[0] = 0$. Execute the following codes

```
for i = 1 to u
     w[i] := w[i] + w[i-1]
```

3. Take node 1 as the root node.
4. Appending reminder nodes to the correspoding father orderly. Assume the child node number is x, the father is y. Find an item from w, assuring $w[i-1] \leq x < w[i]$, then $y = i$.

4 Performance Evaluation

The UBS structure is more general model than LBS, and it is an ideal scalable framework for core router systems. For the limitation of the paper length, the performance evaluation is conducted mainly on the k-ary iterating tree, which is the solution for UBS. First of all, some often used symbols are given as following: n: the number of neighbors, u: the number of computing nodes, l: the depth of the inner nodes of the tree, k: the number of children for each non-leaf node, s: the BGP routing table size.

Considering the tree structure, we have the following statements.

$$u = \frac{n-1}{k-1} \tag{8}$$

$$n = k^l \tag{9}$$

Definition 8. *Under the concept described above, the ratio of* n *to* k *is called the key value of the system. It is written as below.*

$$E = \frac{n}{k} \tag{10}$$

4.1 Performance Analysis

Contribution to Computing. Assume the average time for one operation on two path under the relation $\preceq$ is $t_{\preceq}$. In the conventional models, the overall route table computing can be written as $t_0 = snt_{\preceq}$. In the k-ary iterating tree, the total transmission time can be reduced by pipeline the iterating cycles bottom up. Therefore, the overall route table buildup time $t_1 = (lkt_{\preceq} + (l-1)(t_t)) + skt_{\preceq}$, where the t_t is the transmission time for between two adjacent level. The speedup can be given as

$$S = \frac{snt_{\preceq}}{(lkt_{\preceq} + (l-1)t_t) + skt_{\preceq}}$$

In the core router, s is much greater than l and k, so we have $S \approx \frac{n}{k}$. The speedup of the k-ary iterating tree is the same as the key value E.

Contribution to Memory Storage. Assume the average memory cost for one path is α. In the traditional centralized models, the total memory cost on the node is $m_0 = \alpha n s_0$. However, the k-ary iterating tree reduce the cost on each node to $m_1 = \alpha k s_1$. Let $s_0 = s_1$, we have $R_m = \frac{m_0}{m_1} = \frac{n}{k}$, which is named as *memory reduction ratio.* On the other hand, let $m_0 = m_1$, then $R_s = \frac{s_1}{s_0} = \frac{n}{k}$, which is called *route storage ratio.*

Contribution to Availability. In the centralized model, the recovery time is the route rebuild time $t_0 = snt_{\preceq}$. In the k-ary iterating tree, assuming the crash probability of each node is the same, the rebuild time can be expressed as

$$t_1 = \sum_{i=1}^{l}((i-1)kt_{\preceq} + (i-2)t_t) + skt_{\preceq})\frac{k^{i-1}}{u}$$

The $(i-1)kt_{\preceq} + (i-2)t_t)$ can be ignored while s is big enough in the core router. So $t_1 = \frac{st_{\preceq}k(k^l-1)}{u(k-1)}$. From equations 8 and 9, we have the *recovery time ratio* $R_r = \frac{n}{k} = E$.

4.2 Experimental Verification

Some simulating experiments are conducted to verify the performance improvements of the k-ary iterating tree. The experiment environment is set up on a x86 hardware framework, carrying one PentiumIV 2.4GHz CPU, and 512M memory. The operating system is Linux Fedora 4 RC3, with compiler gcc-3.4.4. The core component of BGP protocol processing is implemented in a process. We statistic the time of the key processing of the route processing. The transmission time is parameterized as a relative value. The *Trans-Comp-Ratio (TCR)* is taken as the measurement of the ratio of the unit transmission time and the unit path comparing time. 27 neighbors are involved in the processing. k is set to 3. The key value $E = 9$.

Fig. 2 illustrates the speedup-to-route curves under different TCR. The speedup becomes stable to about the expected value 9 while route exceeds 40,000. The average speedup spread across $[10^4, 10^7)$ is listed in Tab.1. Under the condition having 10^5 routes, speedup is stable. For the system overhead introduced, the speedup observed is less than the expected value. However the overall error is not more than 10%.

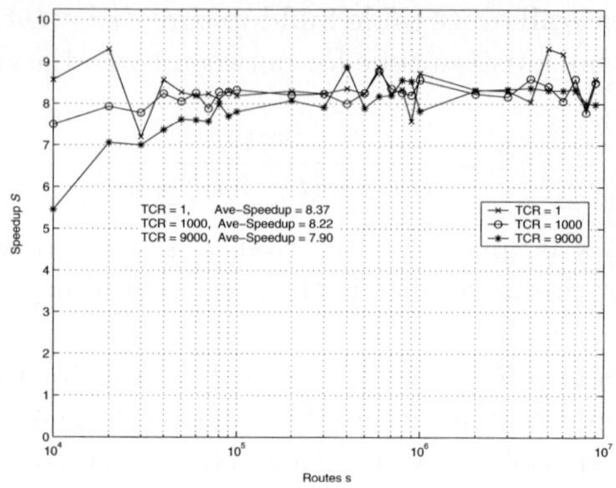

Fig. 2. Speedup-to-Route Curves

Table 1. Average Speedup Distribution 1

TCR	Ave-Speedup on route∈ $[10^4, 10^5)$	Ave-Speedup on route∈ $[10^5, 10^6)$	Ave-Speedup on route∈ $[10^6, 10^7)$
1	8.31	8.27	8.53
1,000	8.02	8.30	8.33
9,000	7.26	8.23	8.21

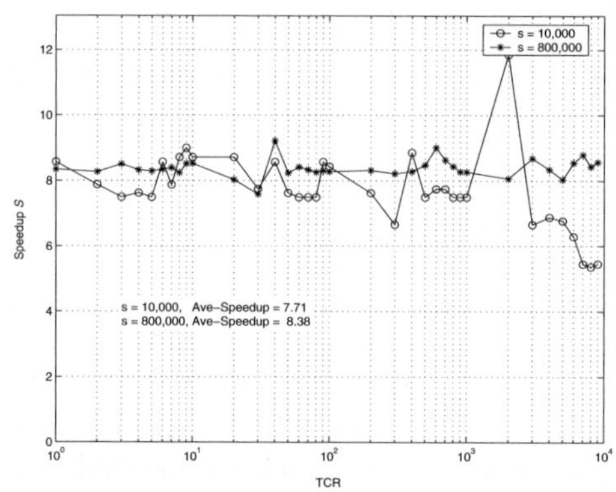

Fig. 3. Speedup-to-TCR Curves

Table 2. Average Speedup Distribution 2

routes	Ave-Speedup on TCR∈ $[10^0, 10^1)$	Ave-Speedup on TCR∈ $[10^1, 10^2)$	Ave-Speedup on TCR∈ $[10^2, 10^3)$	Ave-Speedup on TCR∈ $[10^3, 10^4)$
10,000	8.14	8.05	7.73	6.92
800,000	8.36	8.33	8.43	8.42

Fig. 3 depicts the speedup-to-TCR curves under different routes. The transmission forms the main overhead in the iterating algorithm. While the TCR is less than 10^3, speed up becomes stable. Tab.2 shows the average speedup trend against TCR. While number of route reaches 800,000, the error between observed value to the expected value (9) is no more than 8%.

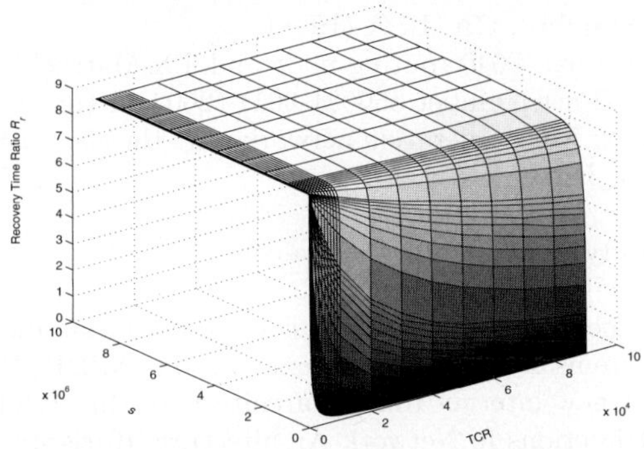

Fig. 4. Recovery Time Ratio

The failure probability of each node is assumed equal in the recovery verification experiments. The route table rebuilding time is taken as the recovery time. The routes varies from 10^4 to 10^7. The TCR varies from 10^0 to 10^4. The observed recovery time is illustrated in Fig. 4. The *recovery time ratio* increases rapidly while route number comes up. And it is very close to the expected value 9. The different configurations of TCR do not affect it very much. While routes exceeds 800,000, the error is less than 10% only if TCR¿1. This can be satisfied easily under a normal condition in the real circumstance.

5 Conclusion

Firstly, the BGP computing model is studied in this paper. The distributed architecture is given in a tree-based processing framework according to the parallel features. The k-ary iterating tree model and the k-ary hfi-tree model are given to deal with the UBS and LBS structures. And, the related load dispatching pattern algorithms are provided. The performance is improved by the tree-based model in there aspects, including the computing power, the memory, and the availability. The theoretical analysis shows that the enhancement is related to the ratio of neighbor number to k, which is defined as the *key value* of the model. Finally, some experiments are taken. The simulating results give a solid proof of performance promotion of the tree-based model proposed in this paper.

References

1. Iyer, S., McKeown, N.: Analysis of the parallel packet switch architecture. IEEE/ACM Transactions on Networking **11**(2) (2003.) 314–324
2. Chao, H.J., Deng, K., Jing, Z.: Petastar: A petabit photonic packet switch. IEEE Journal on Selected Areas in Communications **21**(7) (2003) 1096–1112
3. Dally, W.J., Carvey, P., Dennison, L.: The avici terabit switch/router. In: Hot Interconnects 6, Stanford, CA. USA (1998)
4. Juniper Networks, Inc.: T640 routing node and TX Matrix^{TM} platform: Architecture. white paper (Part Number 350031-001) (2004) http://www.juniper.net.
5. Cisco Systems, Inc.: Next generation networks and the cisco carrier routing system. white paper (2004) http://www.cisco.com.
6. (GNU Zebra) http://www.zebra.org.
7. (Vyatta Community) http://www.vyatta.com.
8. Caesar, M., Caldwell, D., Feamster, N., Rexford, J., Shaikh, A., van der Merwe, K.: Design and implementation of a routing control platform. In: Second Symposium on Networked Systems Design and Implementation (NSDI), Boston, USA (2005)
9. Yang, X.: Nira: A new internet routing architecture. In: ACM SIGCOMM Workshop on Future Directions in Network Architecture, Karlsruhe, Germany (2003)
10. Govindan, R., Alaettinoğlu, C., Varadhan, K., Estrin, D.: Route servers for inter-domain routing. Computer Networks and ISDN Systems **30**(12) (1998) 1157–1174
11. Loo, B.T., Hellerstein, J.M., Stoica, I., Ramakrishnan, R.: Declarative routing: Extensible routing with declarative queries. In: ACM Special Interest Group on Data Communication (SIGCOMM), Philadelphia, USA (2005)
12. Loo, B.T., Condie, T., Hellerstein, J.M., Maniatis, P., Roscoe, T., Stoica, I.: Implementing declarative overlays. In: 20th ACM Symposium on Operating Systems Principles (SOSP), Brighton, UK (2005)
13. Mosberger, D., Peterson, L.L.: Making paths explicit in the scout operating system. In: Second USENIX Symposium on Operating System Design and Implementation (OSDI), Seattle, USA (1996)
14. Kohler, E., Morris, R., Chen, B., Jannotti, J., Kaashoek, M.F.: The click modular router. ACM Transaction on Computer Systems **18**(3) (August 2000) 263–297
15. Decasper, D., Dittia, Z., Parulkar, G., Plattner, B.: Router plugins: A software architecture for next generation routers. IEEE/ACM Transactions on Networking **1**(2) (February 2000) 8
16. Xiao, X., Ni, L.M.: Parallel routing table computation for scalable ip routers. In: Second International Workshop on Network-Based Parallel Computing: Communication, Architecture, and Applications, Las Vegas, USA (1998)
17. Rekhter, Y., Li, T., Hares, S.: A border gateway protocol 4 (BGP-4). IETF:Requst for Comments 4271 (2006)
18. Huston, G.: Analyzing the internet's bgp routing table. Cisco Internet Protocol Journal **4**(1) (2001)
19. Zhang, X., Zhu, P., Lu, X.: Fully-distributed and highly-parallelized implementation model of bgp4 based on clustered routers. In: Networking - ICN 2005: 4th International Conference on Networking, Reunion Island, France (2005)

A Novel Scheme for the Parallel Computation of SVDs

Sanguthevar Rajasekaran and Mingjun Song

Computer Science and Engineering
University of Connecticut
Storrs CT 06269, USA
{rajasek, mjsong}@engr.uconn.edu

Abstract. The Singular Value Decomposition (SVD) is a vital problem that finds a place in numerous application domains in science and engineering. As an example, SVDs are used in processing voluminous data sets. Many sequential and parallel algorithms have been proposed to compute SVDs. The best known sequential algorithms take cubic time. This amount of time may not be acceptable especially when the data size is large. Thus parallel algorithms are desirable. In this paper, we present a novel technique for the parallel computation of SVDs. This technique yields impressive speedups.

We discuss implementation of our technique on parallel models of computing such as the mesh and the PRAM. We also present an experimental evaluation of our technique.

Keywords: Singular Value Decomposition, Two-sided Jacobi, One-sided Jacobi.

1 Introduction

The Singular Value Decomposition (SVD) is a vital problem with applications in many a domain. An important application of SVD is to reduce dimensionality in data mining and information retrieval fields. The well-known sequential bidiagonalization-based Golub-Kahan-Reinsch SVD algorithm [6] takes $O(mn^2)$ time (on an $m \times n$ matrix). For large values of m and n, this time could be prohibitive. With the advent of the internet and the subsequent data explosion, parallel techniques for computing SVDs have become increasingly more important.

The bidiagonalization-based SVD algorithm has been found to be difficult to parallelize and hence works on parallel SVD focus on Jacobi-based techniques. Both two-sided Jacobi and one-sided Jacobi techniques have been studied in this context. Brent and Luk [5] presented a parallel one-sided SVD algorithm using a linear array of $O(n)$ processors, with a run time of $O(mnS)$, where S is the number of sweeps. They also presented an $O(nS)$ time algorithm to compute the singular values of a symmetric matrix using an array of n^2 processors. Zhou and

M. Gerndt and D. Kranzlmüller (Eds.): HPCC 2006, LNCS 4208, pp. 129–137, 2006.

Brent [12] described an efficient parallel ring ordering algorithm for one-sided Jacobi.

Bečka and Vajteršic [2] presented a parallel block two-sided Jacobi algorithm on hypercubes and rings with a run time of $O(n^2S)$. They also gave an $O(nS)$ time algorithm on meshes [3]. Bečka *et al.* [1] proposed a dynamic ordering algorithm for a parallel two-sided block-Jacobi SVD with a run time of $O(nS)$. Okša and Vajteršic [10] designed a systolic two-sided block-Jacobi algorithm with a run time of $O(nS)$. Strumpen *et al.* [11] presented a stream algorithm for one-sided Jacobi that has a run time of $O(\frac{n^3}{p^2}S)$, where p is the number of processors (p being $O(\sqrt{n})$). They created parallelism by computing multiple Jacobi rotations independently and applying all the transformations thereafter. They show that each sweep of the Jacobi iteration algorithm can be parallelized on an $n \times n$ mesh in $O(nS)$ time. Clearly this algorithm is asymptotically optimal. Unfortunately their experimental results show that the value of S is much larger than what the sequential algorithm takes. In this paper, we employ their idea of separating rotation computations and transformations. We propose a novel algorithm for computing SVDs. This algorithm is fundamentally different from all the algorithms that have been proposed for SVD. It employs a specific "relaxation" of the Jacobi iteration. We call this *JRS iteration*. This algorithm is nicely parallelizable. For example, it enables the computation of all the rotations of a sweep in parallel such that the number of sweeps is reasonable.

We discuss the implementation of JRS iteration on various models of computing such as the mesh, the hypercube, and the PRAM. For example, on the CREW PRAM our algorithm has a run time of $O(S \log^2 n)$ (for a symmetrix $n \times n$ matrix).

The remainder of the paper is organized as follows. In Section 2, we introduce the sequential Jacobi-SVD algorithm. Section 3 describes our new JRS iteration algorithm. In Section 4, we show experimental results. Section 5 discusses parallel implementations of our new algorithms. Finally, we provide some concluding remarks in Section 6.

2 A Survey of the Basic Ideas

The SVD problem takes as input any $m \times n$ matrix A ($m \geq n$) and computes three matrices U, Σ, and V such that:

$$A = U\Sigma V^T,$$

where U is a $m \times n$ orthogonal matrix (i.e., $U^TU = I$), V is an $n \times n$ orthogonal matrix ($V^TV = I$), and Σ is an $n \times n$ diagonal matrix. If $\Sigma = diag(\sigma_1, \sigma_2, \ldots, \sigma_n)$, then these diagonal elements are the singular values of A. The column vectors of U are the left singular vectors of A, and the column vectors of V are the right singular vectors of A.

All the existing parallel algorithms use the Jacobi iteration as the basis. Jacobi iteration algorithm attempts to diagnolize the input matrix A by a series of *Jacobi rotations* where each rotation tries to zero-out an off-diagonal element. In particular, each Jacobi rotation involves premultiplying A by an orthogonal matrix and postmultiplying by another orthogonal matrix. We perform $(n^2 - n)/2$ rotations (in the case of a symmetric matrix) attempting to zero-out all the off-diagonal elements. These $(n^2 - n)/2$ transformations constitute a *sweep*. It can be shown that after each sweep the norm of the off-diagonal elements decreases and hence the algorithm converges. It is believed that the number S of sweeps needed for convergence of the sequential Jacobi iteration algorithm is $O(\log n)$ [6].

There are two varaints of the Jacobi iteration algorithm, namely, one-sided and two sided. Accordingly, there are two versions of our JRS iteration algorithm as well. Both the versions of JRS perform well in parallel.

2.1 Two-sided Jacobi SVD

The two-sided Jacobi iteration algorithm [9] transforms a symmetric matrix A into a diagonal matrix Σ by a sequence of Jacobi rotations (J):

$$\Sigma = \cdots J_3^T(J_2^T(J_1^T A J_1)J_2)J_3 \cdots = (J_1 J_2 J_3 \cdots)^T A (J_1 J_2 J_3 \cdots)$$

Each transform attempts to zero-out a given off-diagonal element of A. The Jacobi rotation, also called the Givens rotation[6], is defined as follows:

$$J(i,j,\theta) = \begin{pmatrix} 1 & \cdots & 0 & \cdots & 0 & \cdots & 0 \\ \vdots & \ddots & \vdots & & \vdots & & \vdots \\ 0 & \cdots & c & \cdots & s & \cdots & 0 \\ \vdots & & \vdots & \ddots & \vdots & & \vdots \\ 0 & \cdots & -s & \cdots & c & \cdots & 0 \\ \vdots & & \vdots & & \vdots & \ddots & \vdots \\ 0 & \cdots & 0 & \cdots & 0 & \cdots & 1 \end{pmatrix} \begin{matrix} \\ \\ i \\ \\ j \\ \\ \\ \end{matrix},$$

where (i,j) is an index pair of A to be zeroed, $c = \cos\theta$, $s = \sin\theta$, θ being called the rotation angle. It can be easily verified that $J^T J = I$, so the Jacobi rotation is an orthogonal transformation. The values of c and s are computed as follows. Consider one of the transformations: $B = J^T A J$. We choose c and s such that

$$\begin{pmatrix} b_{ii} & b_{ij} \\ b_{ji} & b_{jj} \end{pmatrix} = \begin{pmatrix} c & s \\ -s & c \end{pmatrix}^T \begin{pmatrix} a_{ii} & a_{ij} \\ a_{ji} & a_{jj} \end{pmatrix} \begin{pmatrix} c & s \\ -s & c \end{pmatrix}$$

is diagonal, *i.e.*, $b_{ij} = b_{ji} = 0$. By solving this equation and taking the smaller root [6], c and s are obtained by:

$$c = \frac{1}{\sqrt{1+t^2}}, \quad s = tc,$$

where

$$t = \frac{sign(\tau)}{|\tau| + \sqrt{\tau^2 + 1}},$$

and

$$\tau = \frac{a_{ii} - a_{jj}}{2a_{ij}}.$$

Depending on the order of choosing the element to be zeroed, there are classic Jacobi and cyclic Jacobi algorithms. In the classic Jacobi iteration algorithm, each transformation chooses the off-diagonal element of the largest absolute value. However, searching for this element requires expensive computations. Cyclic Jacobi algorithm sacrifices the convergence behavior and steps through all the off-diagonal elements in a row-by-row fashion. For example, if $n = 3$, the sequence of elements is $(1,2), (1,3), (2,3), (1,2), \ldots$. The computation is organized in sweeps such that in each sweep every off-diagonal element is zeroed once. Note that when an off-diagonal element is zeroed it may not continue to be zero when another off-diagonal element is zeroed. After each sweep, it can be shown that, the norm of the off-diagoal elements decreases monotonically. Thus the Jacobi algorithms converge.

2.2 One-sided Jacobi SVD

One-sided Jacobi algorithm, also called Hestenes-Jacobi algorithm [7][11], first produces a matrix B whose rows are orthogonal by premultiplying A with an orthogonal matrix U:

$$UA = B,$$

where rows of B satisfy:

$$b_i^T b_j = 0 \text{ for } i \neq j.$$

Followed by this B is normalized by:

$$V = S^{-1}B,$$

where $S = diag(s_1, s_2, \ldots, s_m)$, and $s_i = b_i^T b_i$. It can be easily shown that $A = U^T SV$, which is equivalent to the definition of SVD.

One-sided Jacobi is also realized by a series of Jacobi rotations, but on one side. For a given i and j, rows i and j are orthogonalized by $B = J^T A$ where $J = J(i, j, \theta)$ is the same matrix as in the two-sided Jacobi and:

$$\begin{pmatrix} b_i^T \\ b_j^T \end{pmatrix} = \begin{pmatrix} c & s \\ -s & c \end{pmatrix}^T \begin{pmatrix} a_i^T \\ a_j^T \end{pmatrix}.$$

Here c and s of J are chosen such that $b_i^T b_j = 0$. The solution of them is:

$$c = \frac{1}{\sqrt{1+t^2}}, \quad s = tc.$$

where

$$t = \frac{sign(\tau)}{|\tau| + \sqrt{\tau^2 + 1}},$$

and

$$\gamma = \frac{a_j^T a_j - a_i^T a_i}{2a_i^T a_j}.$$

As we could see, there is a close similarity between the one-sided and two-sided versions of the Jacobi algorithm.

3 JRS Iteration Algorithm

Since any rotation in the two-sided Jacobi algorithm changes only the corresponding (two) rows and (two) columns, and one-sided Jacobi algorithm changes only the corresponding (two) rows, there exists inherent parallelism in the Jacobi iteration algorithms. For example, the $n(n-1)/2$ rotations in any sweep can be grouped into $n-1$ non-conflicting rotation sets each of which containing $n/2$ rotations. For instance, if $n = 4$, there are 3 rotation sets: $\{(1,2),(3,4)\}$, $\{(1,3),(2,4)\}$, $\{(1,4),(2,3)\}$. Since each rotation can be performed in $O(n)$ time on a single machine, we can perform all the rotations in $O(n^2S)$ time on a ring of n processors [6]. The idea here is to perform each set of rotations in parallel.

We can think of the Jacobi algorithm as consisting of two phases. In the first phase we compute all the rotation matrices (there are $O(n^2)$ of them). In the second phase we multiply them out to get U and V. Consider any rotation operation. The values of s and c can be computed in $O(1)$ time sequentially. The algorithm of Strumpen *et al.* [11] performs all the $n(n-1)/2$ rotations of a sweep in parallel even though not all of these rotations are independent. Thus in their algorithm, all the rotation matrices can be constructed in $O(1)$ time using n^2 CREW PRAM processors. This will complete the first phase of the Jacobi algorithm. Followed by this the second phase has to be completed. This involves the multiplication of $O(n^2)$ rotation matrices. Since two $n \times n$ matrices can be multiplied in $O(\log n)$ time using n^3 CREW PRAM processors (see e.g., [4], [8]), a straight forward implementation of [11]'s algorithm runs in time $O(S \log^2 n)$ using n^5 CREW PRAM processors. In [11] an implementation on an $n \times n$ mesh has been given that runs in $O(nS)$ time. However, as has been pointed out before, the value of S is much larger than the corresponding value for the sequential Jacobi iteration algorithm.

Any parallel algorithm for SVD partitions the $n(n-1)/2$ rotations of a sweep into *rotation sets* where each rotation set consists of some number of rotations. All the rotations of a rotation set are performed in parallel. Most of the parallel SVD algorithms in the literature employ $(n-1)$ rotation sets each rotation set consisting of $n/2$ independent rotations. The algorithm of Strumpen et al. is an exception. We let multiple processors compute the rotation matrices of a rotation set (one matrix per processor), all the processors employing the **same** original matrix. In the sequential case, if A is the input matrix, computations will proceed as follows. $B_1 = J_1^T A J_1$; $B_2 = J_2^T B_1 J_2$; $B_3 = J_3^T B_2 J_3$; and so on. On the other hand, in parallel, computations will proceed as follows. $B_1 = J_1^T A J_1$; $B_2 = J_2^T A J_2$; $B_3 = J_3^T A J_3$; etc. The number of B_is computed in parallel will be decided by the number of available processors. Once this parallel computation is complete, all of the computed transformations will be applied to A to obtain a new matrix B. After this, again a parallel computation of rotation matrices will be done all with respect to B; B will be updated with the computed transformations; and so on.

In this paper we propose a fundamentally different algorithm for SVD. It is a specific "relaxation" of the Jacobi iteration algorithm that wel call the *JRS iteration algorithm.* Just like the Jacobi algorithm, there are two variants of the JRS iteration algorithm as well, namely, one-sided and two-sided. We provide details on these two variants in the next subsections.

3.1 Two-sided JRS Iteration Algorithm

The main idea behind the original two-sided Jacobi SVD is to systematically reduce the norm of the off-diagonal elements of a symmetric matrix A:

$$off(A) = \sqrt{\sum_{i=1}^{n} \sum_{i=1, j \neq i}^{n} a_{ij}^2}.$$

The convergence of the Jacobi algorithm is ensured by the fact that after each rotation, the norm of the off-diagonal elements decreases by twice the square of the element zeroed out in this rotation [6].

The JRS iteration algorithm also has sweeps and in each sweep we perform rotations (one corresponding to each off-diagonal element). The only difference is that in a given rotation we do not zero-out the targeted off-diagonal element but rather we decrease the value of this element by a fraction.

Let the element targeted in a given rotation be the (i, j)th element. Then we let

$$b_{ij} = a_{ij}(c^2 - s^2) + (a_{ii} - a_{jj})cs = \lambda a_{ij},$$

where λ is in the interval $[0, 1)$. When $\lambda = 0$, we get the original Jacobi iteration algorithm. We can solve for s and c as follows: If $a_{ij} = 0$, then set $c = 1$ and $s = 0$; Otherwise

$$\frac{a_{ii} - a_{jj}}{2a_{ij}} = \frac{c}{2s} - \frac{s}{2c} - \frac{\lambda}{2cs}.$$

Let $\tau = \frac{a_{ii} - a_{jj}}{2a_{ij}}$, $t = \frac{s}{c}$, then

$$(1 + \lambda)t^2 + 2\tau t + \lambda - 1 = 0.$$

According to [6], the smaller root should be chosen, so

$$t = \frac{sign(\tau)(1 - \lambda)}{|\tau| + \sqrt{\tau^2 + (1 - \lambda^2)}}.$$

Like in the regular Jacobi rotation, c and s can be computed as:

$$c = \frac{1}{\sqrt{1 + t^2}}, \quad s = tc.$$

We call the above algorithm JRS iteration algorithm.

3.2 One-sided JRS Iteration Algorithm

One-sided JRS algorithm is similar to the one-sided Jacobi algorithm. In each rotation, we let the norm of the corresponding two rows be reduced to a fraction of it. That is,

$$v_i^T v_j = \lambda u_i^T u_j.$$

The solution is similar to the two-sided Jacobi:

$$c = \frac{1}{\sqrt{1+t^2}}, \quad s = tc.$$

where

$$t = \frac{sign(\tau)(1-\lambda)}{|\tau| + \sqrt{\tau^2 + (1-\lambda^2)}},$$

and

$$\tau = \frac{u_j^T u_j - u_i^T u_i}{2u_i^T u_j}.$$

4 Experimental Results

We have implemented our JRS algorithms and tested them for convergence as well as performance. They have been compared with the regular Jacobi algorithms as well as the algorithms of [11]. We provide the experimental results in this subsection.

In this experiment, we have compared the number of sweeps taken by the different Jacobi approaches. We generated randomly several matrices of different sizes, including $10 \times 10, 50 \times 50, 100 \times 100, 200 \times 200, 500 \times 500$, and 1000×1000. The elements of the matrices were generated randomly to have a value in the interval $[1, 10]$. For each matrix size, we generated 10 matrices and for each algorithm we took the average number of sweeps. The convergence condition employed was on the norm of the off-diagonal elements. We used a norm value of 10^{-15}.

The results are shown in tables 1 and 2 for two-sided Jacobi and one-sided Jacobi algorithms, respectively. For two-sided Jacobi, we used symmetric matrices; for one-sided Jacobi, we generated unsymmetric matrices. In these tables, Independent Jacobi refers to the Jacobi algorithm where all the rotations in a sweep are done independently and in parallel. This is one of the algorithms employed in [11]. The values of the parameter λ used in JRS algorithm for matrices of different sizes are chosen experimentally, which are: 0.25, 0.5, 0.5, 0.75, 0.8, 0.85 respectively.

From tables 1 and 2, we see that the number of sweeps taken by the JRS is significantly less than that of Independent Jacobi of [11]. Also the number of sweeps taken by the JRS based algorithm is within a reasonable multiple of that of the sequential cyclic Jacobi algorithm.

Table 1. Experimental results for two-sided SVD

Matrix size	Cyclic Jacobi	Independent Jacobi [11]	JRS
10×10	5	11	16
50×50	7	6692	31
100×100	7	>300000	47
200×200	8	>300000	66
500×500	9	>300000	90
1000×1000	9	>300000	125

Table 2. Experimental results for one-sided SVD

Matrix size	Cyclic Jacobi	Independent Jacobi [11]	JRS
10×10	7	14	28
50×50	10	133	52
100×100	11	1037	62
200×200	12	193107	114
500×500	14	>300000	145
1000×1000	15	>300000	197

5 Parallelism

As our experimental results show, even when all the rotations in a sweep are done in parallel, JRS based algorithms converge fast. In particular, the number of sweeps is no more than a reasonable multiple of the number of sweeps taken by the sequential Jacobi algorithm. As a consequence, JRS based algorithms offer maximum parallelism. In fact most of the parallel algorithms that have been derived thus far (that employ Jacobi iterations) for SVDs can be readily translated into JRS based algorithms. We just mention a few of them below.

Based on the algorithms of [11] we get:

Theorem 1. *JRS algorithms run in time $O(nS)$ on an $n \times n$ mesh.*

The algorithm of [5] yields the following:

Theorem 2. *One-sided JRS algorithm can be implemented on a linear array of $O(n)$ processors to have a run time of $O(mnS)$.*

From our discussion in Section 3, we infer the following:

Theorem 3. *JRS algorithms can be implemented on a CREW PRAM to have a run time of $O(S \log^2 n)$.*

6 Conclusions

In this paper, we have proposed a novel algorithm (called JRS Iteration Algorithm) for computing SVDs. This algorithm enables one to perform all the rotations in a sweep independently and in parallel without increasing the number of sweeps significantly. Thus this algorithm can be implemented on a variety of parallel models of computing to obtain optimal speedups when the processor bound is $O(n^2)$. This method significantly decreases the number of sweeps over independent Jacobi proposed in [11]. Therefore, our method can be used in their stream algorithm to achieve a run time of $O(nS)$. Our algorithm can also be implemented on a CREW PRAM to have a run time of $O(S \log^2 n)$.

In the full version of this paper we provide additional experimental data, a value for λ (as a function of n) that results in the minimum number of sweeps, a convergence proof for JRS, a variant of JRS called *Group JRS*, etc.

Acknowledgments. This research has been supported in part by the NSF Grant ITR-0326155.

References

1. Bečka, M., Okša, G., Vajteršic, M., Dynamic ordering for a parallel block-Jacobi SVD algorithm, *Parallel Comp.*, 28, 243-262, 2002.
2. Bečka, M. and Vajteršic, M., Block-Jacobi SVD algorithms for distributed memory systems I: Hypercubes and rings, *Parallel Algorithms Appl.*, 13, 265-287, 1999.
3. Bečka, M. and Vajteršic, M., Block-Jacobi SVD algorithms for distributed memory systems II: Meshes, *Parallel Algorithms Appl.*, 14, 37-56, 1999.
4. Bini, D., and Pan, V., Polynomial and Matrix Computations, Vol.1, Fundamental Algorithms, Birkhäuser, Boston, 1994.
5. Brent, R.P., and Luk, F.T., The Solution of Singular-Value and Symmetric Eigenvalue Problems on Multiprocessor Arrays. *SIAM Journal on Scientific and Statistical Computing*, 6(1):69-84, 1985.
6. Golub, G.H., and Van Loan, C.F., *Matrix Computations*. John Hopkins Univer sity Press, Baltimore and London, 2nd edition, 1993.
7. Hestenes, M.R., Inversion of Matrices by Biorthogonalization and Related Results. *Journal of the Society for Industrial and Applied Mathematics*, 6(1):51-90, 1958.
8. Horowitz, E., Sahni, S., and Rajasekaran, S., *Computer Algorithms*, W.H. Freeman Press, 1998.
9. Jacobi, C.G.J., Über eine neue Auflösungsart der bei der Methode der kleinsten Quadrate vorkommenden linearen Gleichungen. *Astronomische Nachrichten*, 22, 1845.
10. Okša, G., and Vajteršic, M., A Systolic Block-Jacobi SVD Solver for Processor Meshes, *Parallel Algorithms and Applications*, 18(1-2), 49-70, 2003.
11. Strumpen, V., Hoffmann, H., Agarwal, A., A Stream Algorithm for the SVD, *Technical Memo 641*, Computer Science and Artificial Intelligence Laboratory, Massachusetts Institute of Technology, October 2003.
12. Zhou, B.B., and Brent, R.P., A Parallel Ring Ordering Algorithm for Efficient One-sided SVD Computations, *Journal of Parallel and Distributed Computing*, 42, 1-10, 1997.

Cache-Optimal Data-Structures for Hierarchical Methods on Adaptively Refined Space-Partitioning Grids

Miriam Mehl

Institut für Informatik, TU München
Boltzmannstraße 3, D-85748 Garching, Germany
mehl@in.tum.de
http://www5.in.tum.de/persons/mehl/

Abstract. The most efficient numerical methods for the solution of partial differential equations, multigrid methods on adaptively refined grids, imply several drawbacks from the point of view of memory-efficiency on high-performance computer architectures: First, we loose the trivial structure expressed by the simple i, j-indexing of grid points or cells. This effect is even worsened by the usage of hierarchical data and – if implemented in a naive way – leads to both increased storage requirements (neighbourhoodrelations possibly modified difference stencils) and a less efficient data access (worse locality of data and additional data dependencies), in addition. Our approach to overcome this quandary between numerical and hardware-efficiency relies on structured but still highly flexible adaptive grids, the so-called space-partitioning grids, cell-oriented operator evaluations, and the construction of very efficient data structures based on the concept of space-filling curves. The focus of this paper is in particular on the technical and algorithmical details concerning the interplay between data structures, space-partitioning grids and space-filling curves.

1 Introduction

As soon as we want to simulate real world applications described by (systems of) partial differential equations, we are faced with two increasingly important requirements: First, there is a rising need for accuracy and, second, the computing time should stay within manageable borders. These requirements force us to reduce both the number of unknowns – typically by sophisticated grid adaptivity – and the complexity of our solver – typically with the help of multigrid methods. Unfortunately, both methods – grid adaptivity and multigrid – come along with several drawbacks. Adaptively refined grids in general imply a substantial overhead in terms of storage requirements due to the need to store neighbourhoud relations and/or specialised stencils for the particular local adaptivity pattern. But also with respect to the efficiency of data access, adaptively refined grids cause considerable deficits since in an irregular data structure it is much more difficult to bring together data dependencies with data locality in the physical memory space. Multigrid methods even worsen this effect as they induce additional data dependencies between data of different grid levels.

M. Gerndt and D. Kranzlmüller (Eds.): HPCC 2006, LNCS 4208, pp. 138–147, 2006.

In the following, we present our approach to overcome the quandary between numerical efficiency and efficient memory usage described above. Sect. 2 introduces the used grids. Sect. 3 describes the algorithm of our solver and Sect. 4 demonstrates the construction of suitable memory efficient data structures. Finally, we present results on the applicability and efficiency of our concept in Sect. 5.

2 Space-Partitioning Grids

As a first towards hardware efficiency, we use so-called space-partitioning grids which offer both a high flexibility in term of the adaptivity pattern and a strict structuredness which is already inherently storage saving as the storage of dependencies of grid elements becomes obsolete. Space-partitioning grids, mostly referred to in the form of octrees, are widely used as a tool to solve different subproblems in the simulation context, for example grid generation, geometry description, visualisation, load distribution, and as a computational grid. This shows the variety of functionalities and advantages of the underlying concept making space-partitioning grids a very attractive choice – also as an overall integrating concept in particular in larger context such as coupled simulations [4].

Fig. 1 shows examples for space-partitioning Cartesian grids. The construction principle is the recursive refinement of grid cells into a fixed number of congruent subcells. The number of subcells can be varied. In the case of the well-known quadtree (two-dimensional) or the octree (three-dimensional), a partitioning into two in each coordinate direction is performed. We instead perform a partitioning into three in each direction for reasons we will explain later on in the context of the construction of data structures (Sect. 4).

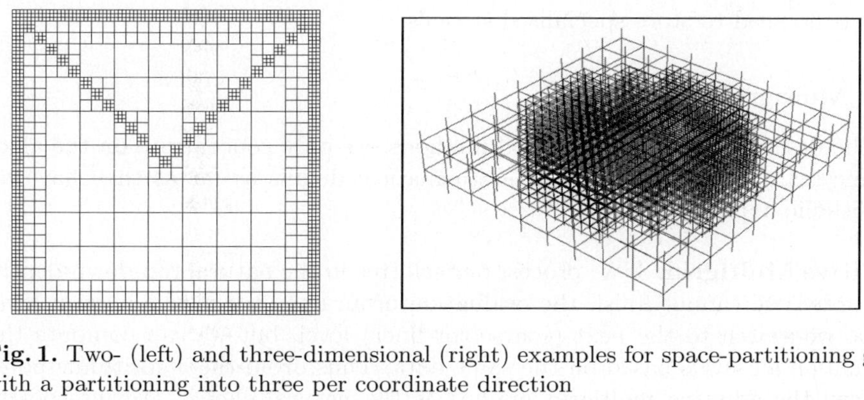

Fig. 1. Two- (left) and three-dimensional (right) examples for space-partitioning grids with a partitioning into three per coordinate direction

Arbitrarily local adaptive refinements can be mapped with space-partitioning grids[1]. The degrees of freedom are assigned to the vertices of the grid cells in our

[1] The only case we can not hadle without further efforts are unisotropic refinements. The generalisation of our concepts to unisotropic refinements is subject of our current work and in the implementation phase.

concept, but exceptions to this rule such as for the pressure in the Navier-Stokes equations [2,3] are possible. For reasons of clarity, we will restrict ourselves to vertex data in the following.

3 Algorithms

3.1 Operator Evaluation

Our main task – to achieve a high memory efficiency – requires the prevention of the storage of unnecessary informations and a high locality of data access (increasing the efficiency of the cache-usage). That is, we have to avoid the storage of neighbourhoodrelations and specialised difference stencils in dependence on the local adaptivity pattern, wich are, in our case, only caused by the vertex-oriented operator evaluation (processesing the grid vertex by vertex and using values at neighbouring vertices for operator evaluation) compute the value of the corresponding difference operator with the help of the values at neighbouring vertices. In addition, values at neighbouring verties might be arbitrarily far away from the current point in the physical memory space – and, therefore, are currently not in the cache with a high probability.

Thus, we switch to the so-called cell-oriented operator evaluation, a well-known concept in the finite element context [5]. It decomposes the operator into cell-parts which accumulate to the complete operator during a run over all cells contributing to the operator value. For the evaluation of the cell-part of an operator we allow only the usage of data owned to the cells, that is the degrees of freedom stored at the cells vertices. At boundaries between different refinements depths, the operator results from the accumulation of cell-parts of neighbouring cells appropriately combined with restriction and interpolation operators. Such, there is no need to store specialised stencils.

3.2 Multigrid

For the description of our multigrid methods, we only concentrate on the algorithmic realisation and leave out mathematical details as far as they have no algorithmic impacts.

Additive Multigrid. If we process our cell tree in the natural top-down-depth-first order, we cannot finish the evaluation of an operator at the current level before we switch to the next (coarser or finer) level, but we can compute the residual on all levels based on the same data. Thus, from the algorithmic point of view, the additive multigrid method is the natural choice: We interpolate all values to the finest level (dehierachisation) in the down-traversal (coarse to fine) of the cell tree, compute the cell-parts of the residual and smooth on the finest level, and restrict the fine grid residuals (those computed before fine-grid smoothing!) to the coarser grids during the up-traversal. Simultanously, we smooth on all grid levels as soon as the accumulation of residual parts (resulting from operator evaluation or restriction) is finished.

Multiplicative Multigrid. In contrast to additive multigrid methods, an iteration of a multiplicative multigrid method cannot be done within one top-down-depth-first sweep over the corresponding cell tree. We have to perform a run over the whole cell tree for each operation (smoothing, interpolation, restriction) but stop at the current grid level to prevent an overall cost of $O(N \log N)$ for a grid with N unknowns on the finest level. As a result, we have to swap out fine grid data to intermediate data structures whenever they are not needed.

The F-Cycle. If we now combine our multigrid methods with dynamical adaptivity, we end up with the so-called full-multigrid methods or F-cycles which start with an a priori defined preliminary and in general quite coarse grid, compute a first solution on this grid, apply some adaptivity criteria, refine or coarsen the grid according to these criteria, and, finally, compute the solution on the new grid. This results in an overall iterative process producing solutions on a sequence of incrementally enhanced grids.[2]

4 Data Structures

As mentioned above, our algorithm processes the grid in a top-down depth-first order and performs a cell-oriented evaluation of the discrete operators (difference stencils, restriction, interpolation). The cell-oriented view helps us to avoid the storage of unneccessary data (pointers to sons/fathers, specialised stencils at boundaries between different refinement depths) and, second, enhances the locality of data accesses if we additionally provide suitable data structures for vertex data[3].

4.1 Space-Filling Curves as an Ordering Mechanism

To achieve an optimal locality of data access and, thereby, a high cache-efficiency, we define a unique 'suitable' processing order of our grid cells and examine the resulting processing order of vertex data.

Space-filling curves [11] are used as an established tool for parallelisation and balanced load distribution [7,8,9,10] for PDE solvers on adaptive grids. The quasi-optimal locality of the class of self-similar recursively defined space-filling curves yields quasi-minimal communication costs [10]. Exactly this quasi-optimal locality is the property which is decisive for the optimisation of time locality of data usage in our algorithm, too. In general, self-similar recursively defined space-filling curves are given by a generating template connecting the *cells* of a first decomposition of the unit square (2D) or unit cube (3D) and a set of rules

[2] In the case of time-dependend problems, we have to adapt the grid to the temporal changes after each or after some time steps, in addition.

[3] Whenever we need cell-centered data such as the pressure in the Navier-Stokes equation, the allocation of the respective data structure is trivial: data are simply stored in a stream corresponding to the cell-stream resulting from our top-down-depth-first tree traversal.

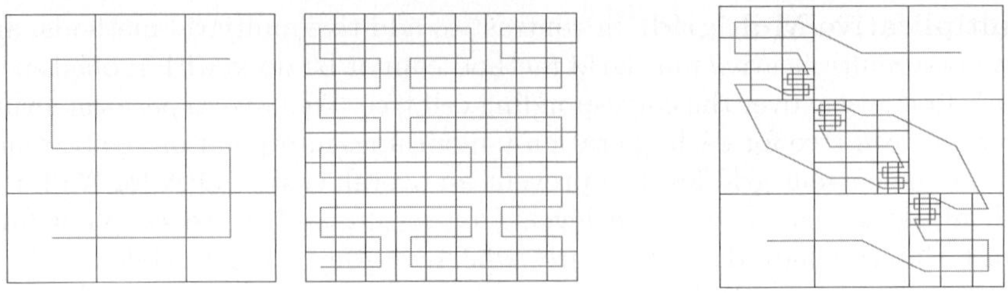

Fig. 2. Template, first two iterate and iterate for an adaptively refined space-partitioning grid of the the two-dimensional Peano curve

describing the recursive application of this template (possibly mirrored and/or rotated) on each of the subdomains in case of further refinement. Fig. 2 shows the generating template and the first iterate of the two-dimensional Peano curve as an example. The space-filling curves themselves are define as the limit of this recursive refinement process.

As we look for ordering mechanisms for the cells of a discrete grid, we are in fact not interested in the space-filling curves itself but in their iterates. Since the refinement rules of the curves are purely local, these iterates can be naturally generalised to adaptively refined space-partitioning grids. See Fig. 2 for an example. The high locality of the resulting oder of cells results from the fact that the curve visits all son (and grandson,...) cells of a father cell at once before they proceed to other cells not contained in the repective father cell. Such, all work to be done at one vertex of the grid (during the evaluation of operators) is finished within a short time period in the average case.

The resulting oder of grid cells can be easily generalised to an order of the cells of *all* refinement levels (needed hierarchical methods such as multigrid). Following the recursive definition of space-filling curves, we start with the coarsest cell (covering the whole computational domain), apply the curve's template in this cell, visit the son cells according to this template, their son cells (if existing) and so on. Thus, we end up with a particular – uniquely defined – top-down-depth-first odering as already presumed in Sect. 3.

4.2 Space-Filling Curves and Stacks

In the literature, we already find hints [12] on substantial improvements of the cache-preformance and, thus, the runtime of a PDE solver by the pure reordering of data according to their usage during the run along a space-filling curve. We go one step further and consequently use the properties of the Peano curve to construct data structures with optimal spatial locality perfectly supplementing the time locality described above. We will explain the underlying idea for two-dimensional regular grids with nodal data. Fig. 3 shows the Peano curve in two-dimensional regular grids. Moving along the curve, we see that the grid points at the left-hand side of the curves are processed in one direction during the first pass of the curve and exactly in the opposite direction during the second pass of the curve. The same holds for the points at the right-hand side of the curve.

This directly corresponds to a very simple type of data structures, the so-called stacks, which allow only two basic operations: put a datum on top of the stack and pop a datum from the top of the stack. In our examples, we can perform one iteration over all data with four data structures: one input stream containing all data in the order of their first usage, two stacks – corresponding to the left- and the right-hand side of the curve – on which we put data after the first pass of the curve and from which we pop them during the second pass, and one output stream collecting all vertex data after their last usage. The output stream can directly be used as an input stream for the next solver iteration if we process the grid along the same curve but in the opposite direction.

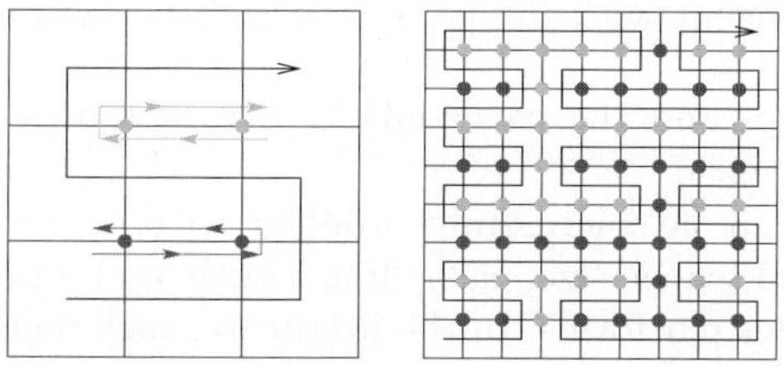

Fig. 3. Assignment of vertices and their data to two groups: left- and right-hand side of the Peano curve corresponding to two stacks used for the intermediate storage of data during solver iterations [4]

In the three-dimensional case, this concept works only if an analogue forward and backward processing of vertex data also holds on the two-dimensional faces of the cubic cells. We could not find any Hilbert curve fulfilling this requirement, whereas it is trivial to show for the Peano curve due to their dimension-recursive definition [13,14]. The generalisation to adaptively refined grids with hierarchical data is quite technical both in the two-dimensional and in the three-dimensional case. Therefore, we refer to other publications [15,13,14,16] for further details. We only state the results: we need eight intermediate stacks in the two-dimensional case and 27 stacks in the three-dimensional case, both independent of the current refinement depth of our grid.

5 Evaluation and Numerical Results

Before we present results on the hardware efficiency of our concept – measured by the cache-efficiency and the runtime – we would like to point out that our algorithms fulfill all numerical requirements, in particular offer a flexible and dynamical adaptivity and multigrid performance of the solvers. For detailed results see [15,13,14,6].

5.1 Cache-Efficiency and Storage Requirements

As described above, the storage of administrational data is almost obsolete due to the structueredness of the grid and the cell-oriented operator evaluation. Therefore, we end up with a very low storage requirement of only about five Byte per degree of freedom in the case of the three-dimensional Poisson equation (with unknowns stored as `float` variables), for example [13].

If we look at the cache-efficiency of our programs, we observe an extremely high hit-rate in the L2-cache of above 99.9% in two and in three dimensions [15,13]. Similar hit-rates could be achieved as well for all algorithmic extensions such as dynamical adaptivity [6], higher order discretisations [14], parallelization [17,18,19], and the Navier-Stokes solver [3,2]. In addition, we could show that the actual number of L2-cache-misses is only about 10% larger than the theoretical minimum given by the need to load data to the cache at least once per solver iteration [15,13][4]. Besides those absolute values, we would like to point out the robustness of our concept with the help of three observations:

- Both the storage- and the cache-efficieny do not depend of the degree of adaptivity of our grid.
- Our approach can be successfully applied without any knowledge of the cache parameters (cache size, cache line length, ...) and, thus, can be easily ported to different platforms. In the literature, such methods are denoted as cache-oblivious [22,23].
- The cache hit-rates scale perfectly with the number of degrees of freedom. That is, there are no size effects deteriorating the efficiency above a certain problem size.

5.2 Runtime and Parallel Efficiency

The runtime of a program is surely the last and most important criterion of efficiency. To give some indication of the potential of our program in comparison with other cache-optimized PDE solvers, we compare the runtimes with DiME [24], a cache- and runtime-optimised multigrid solver for partial differential equations on regular grids. Concretely, we compare one DiME-V-cycle with one pre- and one postsmoothing step to one iteration of our additive multigrid method for the solution of the two-dimensional Poisson equation on a regular grid. Table 1 shows the resulting runtimes. Our program is about five times slower than DiME, which is quite good if we take into account that we can handle fully adaptive grids, the runtime per degree of freedom is independent on the adaptivity pattern of the grid [15,13], the storage requirements of our program are very low (less than seven Bytes per degree of freedom even in the three-dimensional case [13] whereas DiME needs more than 27 Bytes per degree of freedom in the two-dimensional case), and there are still numerous further optimisation potentials for our program (see Sect. 6).

As mentioned above, space-filling curves are a well-known tool for balanced parallelisation of algorithms on adaptive grids. We stated in addition, that we can easily generalize our data structures to processes working on a part of the grid

[4] We measured and simulated the cache-efficieny with the help of the tools *cachegrind* [20] (simulation), *perfmon* [21] and *hpcmon* (hardware performance counter).

Table 1. Comparison of the runtimes per degree of freedom and multigrid iteration for our code (left hand side) and DiME [24] (right hand side) on an AMD Athlon XP 2400+ (1.9 GHz) processor with 256 KB cache and 1 GB RAM using the *gcc3.4* compiler with options `-O3 -Xw` (from [1])

grid res.	# deg. of freedom	runtime per it and dof
243	$5.95 \cdot 10^4$	$1.70 \cdot 10^{-6}$ sec
729	$5.33 \cdot 10^5$	$1.71 \cdot 10^{-6}$ sec
2187	$4.79 \cdot 10^6$	$1.72 \cdot 10^{-6}$ sec

grid res.	# deg. of freedom	runtime per it and dof
257	$6.60 \cdot 10^4$	$5.78 \cdot 10^{-7}$
513	$2.63 \cdot 10^5$	$4.23 \cdot 10^{-7}$
1025	$1.05 \cdot 10^6$	$3.85 \cdot 10^{-7}$
2049	$4.20 \cdot 10^6$	$3.74 \cdot 10^{-7}$

Table 2. Parallel speedup achieved for the solution of the three-dimensional Poisson equation on a spherical domain on an adaptive grid with 23, 118, 848 degrees of freedom. The computations were performed on a myrinet cluster consisting of eight dual Pentium III processors with 2 GByte RAM per node [17,25].

processes	1	2	4	8	16
speedup	1.00	1.95	3.73	6.85	12.93

only [17]. Table 2 shows the achieved parallel speedups for the three-dimensional Poisson equation on an adaptive grid on a spherical domain with 23, 118, 848 degrees of freedom [17].

6 Conclusions and Outlook

We presentet a new approach for the hardware- and, in particular, memory-efficient implementation of state-of-the-art numerical methods, that is dynamically adaptive multigrid solvers. Without loosing mathematical functionality and/or efficiency, we maintain an extremely high cache-efficiency and a very low storage requirement per degree of freedom. In terms of the runtime, the last and crucial criterion, we are about a factor of five to ten slower than highly optimised solvers working on regular grids. Although this is already a good result taking into account that we work on adaptive grids with arbitrarliy local refinements, this is a clear hint that there is still a potential for further improvements for example based on methods similar to those presented in [26,27] or with the help of streaming SIMD extensions (SSE). Furthermore, the simplification of stack definition and administration is a currently active – and promising – field of our research as it will reduce stalls of floating point operations due to administrational integer operations.

Besides the detailed examination of the applicability of these optimisation possibilities, an important focus of our future work is on the implementation of more complicated equations and systems of equations, in particular the three-dimensional Navier-Stokes equations. Hereby, we can already start from an existing implementation of the two-dimensional Navier-Stokes equations [2,3]. Such 'real' applications will also show the effective potential of our method in comparison to other (adaptive or regular) solvers.

References

1. M. Mehl, T. Weinzierl, Ch. Zenger. A cache-oblivious self-adaptive full multigrid method. To appear in: special issue Copper Mountain Conference on Multigrid Methods 2005, *Numerical Linear Algebra with Applications*, Wiley Interscience.
2. T. Neckel. *Einfache 2D-Fluid-Struktur-Wechselwirkungen mit einer cache-optimalen Finite-Element-Methode.* Diploma thesis, Institut für Informatik, Technische Universität München, (2005).
3. T. Weinzierl. *Eine cache-optimale Implementierung eines Navier-Stokes Lösers unter besonderer Berücksichtigung physikalischer Erhaltungssätze.* Diploma thesis, Institut für Informatik, Technische Universität München, (2005).
4. M. Brenk, H.-J. Bungartz, M. Mehl, and T. Neckel. Fluid-Structure Interaction on Cartesian Grids: Flow Simulation and Coupling Interface. In Bungartz and Schfer (eds.), *Fluid-Structure Interaction: Modelling, Simulation, Optimisation*, LNCSE series, Springer, to appear.
5. D. Braess. *Finite Elements. Theory, Fast Solvers and Applications in Solid Mechanics*, Cambridge University Press, (2001).
6. N. Dieminger. *Kriterien für die Selbstadaption cache-effizienter Mehrgitteralgorithmen.* Diplomarbeit, Institut für Informatik, Technische Universität München, (2005).
7. M. Griebel and G. Zumbusch. Hash based adaptive parallel multilevel methods with space-filling curves. In: Rollnik and Wolf (eds.), *NIC Series* **9**, (2002), 479-492.
8. A. K. Patra, J. Long, and A. Laszloff. Efficient Parallel Adaptive Finite Element Methods Using Self-Scheduling Data and Computations. In: Banerjee, Prasanna, and Sinha (eds.), *High Performance Computing – HiPC'99, 6th International Conference, Calcutta, India, December 17-20, 1999, Proceedings*, HiPC, Lecture Notes in Compter Science **1745**, (1999), 359-363.
9. S. Roberts, S. Klyanasundaram, M. Cardew-Hall, and W. Clarke. A key based parallel adaptive refinement technique for finite element methods. In: Noye, Teubner, and Gill (eds.), *Proceedings Computational Techniques and Applications: CTAC '97*, World Scientific, Singapore, (1998), 577-584.
10. G. W. Zumbusch. *Adaptive Parallel Multilevel Methods for Partial Differential Equations.* Habilitationsschrift, Universität Bonn, (2001).
11. H. Sagan. *Space-Filling Curves.* Springer, New York, (1994).
12. M. J. Aftosmis, M. J. Berger, and G. Adomavivius. *A Parallel Multilevel Method for Adaptively Refined Cartesian Grids with Embedded Boundaries*, American Institute of Aeronautics and Astronautics-2000-808, Aerospace Science Meeting and Exhibit, 38th, Reno, Nevada, Jan 10-13, (2000).
13. M. Pögl. *Entwicklung eines cache-optimalen 3D Finite-Element-Verfahrens für große Probleme.* Doctoral thesis, Institut für Informatik, TU München, (2004).
14. A. Krahnke. *Adaptive Verfahren höherer Ordnung auf cache-optimalen Datenstrukturen für dreidimensionale Probleme.* Doctoral thesis, Institut für Informatik, TU München, (2005).
15. F. Günther. *Eine cache-optimale Implementierung der Finite-Elemente-Methode.* Doctoral thesis, Institut für Informatik, TU München, (2004).
16. F. Günther, M. Mehl, M. Pögl, Ch. Zenger. A cache-aware algorithm for PDEs on hierarchical data structures based on space-filling curves. *SIAM Journal on Scientific Computing*, in review.

17. M. Langlotz. *Parallelisierung eines Cache-optimalen 3D Finite-Element-Verfahrens.* Diplomarbeit, Institut für Informatik, TU München, (2004).
18. W. Herder. *Lastverteilung und parallelisierte Erzeugung von Eingabedaten für ein paralleles Cache-optimales Finite-Element-Verfahren.* Diplomarbeit, Institut für Informatik, TU München, (2005).
19. F. Günther, A. Krahnke, M. Langlotz, M. Mehl, M. Pögl, and Ch. Zenger. On the Parallelization of a Cache-Optimal Iterative Solver for PDEs Based on Hierarchical Data Structures and Space-Filling Curves. In: *Recent Advances in Parallel Virtual Machine and Message Passing Interface: 11th European PVM/MPI Users Group Meeting Budapest, Hungary, September 19 - 22, 2004.* Proceedings, Lecture Notes in Computer Science, Vol. 3241/2004, Springer, Heidelberg, (2004).
20. J. Seward, N. Nethercote, and J. Fitzhardinge. *cachegrind: a cache-miss profiler*; `http://valgrind.kde.org/docs.html`
21. HP invent. *perfmon: create powerful performance analysis tools wich use the IA-54 Performance Monitoring Unit (PMU)*; `http://www.hpl.hp.com/research/linux/perfmon/index.php4`.
22. M. Frigo, C. E. Leierson, H. Prokop, and S. Ramchandran. Cache-oblivious algorithms. In: *Proceedings of the 40th Annual Sympoisium on Foundations of Computer Science*, New York, (1999), 285-297.
23. E. D. Demaine. Cache-Oblivious Algorithms and Data Structures. In: *Lecture Notes from the EEF Summer School on Massive Data Sets, University of Aarhus, Denmark, June 27-July 1*, Lecture Notes in Computer Science, Springer, (2002).
24. M. Kowarschik, C. Weiß. DiMEPACK – A Cache-Optimal Multigrid Library. In: Arabnia (ed.), *Proceedings of the International Conference on Parallel and Distributed Processing Techniques and Application (PDPTA 2001), Las Vegas, USA* **I**, (2001).
25. M. Langlotz, M. Mehl, T. Weinzierl, and C. Zenger. SkvG: Cache-Optimal Parallel Solution of PDEs on High Performance Computers Using Space-Trees and Space-Filling Curves. In: A. Bode und F. Durst (eds.), High Performance Computing in Science and Engineering, Garching 2004, Springer-Verlag, Berlin Heidelberg New York, (2005), 71-82.
26. C. C. Douglas, J. Hu, M. Kowarschik, U. Rüde, C. Weiß. Cache Optimization for Structured and Unstructured Grid Multigrid. *Electronic Transactions on Numerical Analysis* **10**, (2000), 21-40.
27. M. Kowarschik, C. Weiß. An Overview of Cache Optimization Techniques and Cache-Aware Numerical Algorithms. In: Meyer, Sanders, and Sibeyn (eds.), *Algorithms for Memory Hierarchies – Advanced Lectures*; Lecture Notes in Computer Science **2625**, Springer, (2003), 213-232.

CSTallocator: Call-Site Tracing Based Shared Memory Allocator for False Sharing Reduction in Page-Based DSM Systems

Jongwoo Lee[1], Sung-Dong Kim[2], Jae Won Lee[3], and Jangmin O[4]

[1] Dept. of Multimedia Science, Sookmyung Women's University, Seoul 140-742, Korea
bigrain@sookmyung.ac.kr
[2] Dept. of Computer Engineering, Hansung University, Seoul 136-792, Korea
sdkim@hansung.ac.kr
[3] School of Computer Science and Engineering, Sungshin Women's University, Seoul 136-742, Korea
jwlee@cs.sungshin.ac.kr
[4] NHN corp., 9th Fl. Venture Town Bldg. 25-1 Jungja-dong Bungdang-gu, Gyunggi-do, 463-844, Korea
jmoh@nhncorp.com

Abstract. False sharing is a result of co-location of unrelated data in the same unit of memory coherency, and is one source of unnecessary overhead being of no help to keep the memory coherency in multiprocessor systems. Moreover, the damage caused by false sharing becomes large in proportion to the granularity of memory coherency. To reduce false sharing in page-based DSM systems, it is necessary to allocate unrelated data objects that have different access patterns into the separate shared pages. In this paper we propose *call-site tracing-based shared memory allocator*, shortly *CSTallocator*. CSTallocator expects that the data objects requested from the different call-sites may have different access patterns in the future. So CSTallocator places each data object requested from the different call-sites into the separate shared pages, and consequently data objects that have the same call-site are likely to get together into the same shared pages. We use execution-driven simulation of real parallel applications to evaluate the effectiveness of our CSTallocator. Our observations show that our CSTallocator outperforms the existing dynamic shared memory allocator.

Keywords: False Sharing, Distributed Shared Memory, Dynamic Memory Allocation, Call Site Tracing.

1 Introduction

In distributed shared memory (DSM) systems, efficient data caching is critical to the entire system performance due to their non-uniform memory access time characteristics. Because the access to a remote memory is much slower than the access to a local memory, reducing the frequencies of the remote memory accesses with efficient caching can lead to decrease of the average cost of memory accesses, and subsequently improve the entire system performance [1]. A simple and widely

M. Gerndt and D. Kranzlmüller (Eds.): HPCC 2006, LNCS 4208, pp. 148 – 159, 2006.

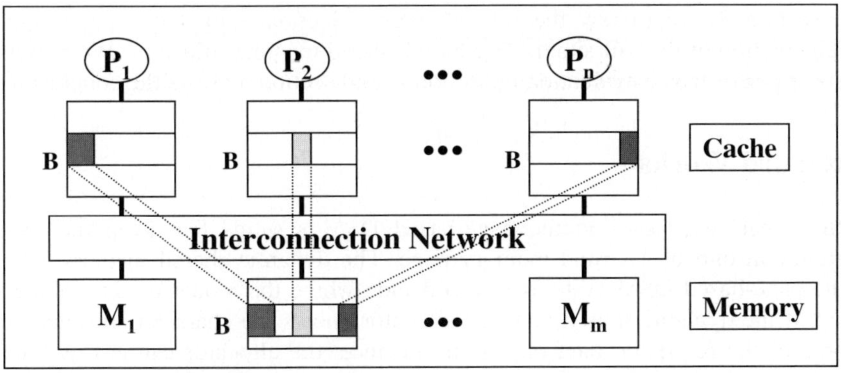

Fig. 1. Example of page replication in DSM systems

used mechanism for exploiting locality of reference is to replicate or migrate frequently used pages from remote to local memory [2]. But in case of page replication, the existence of multiple copies of the same page causes memory coherence problem (Fig. 1).

In DSM systems, ***false sharing*** happens when several independent objects, which may have different access patterns, share the memory coherency unit. Memory faults or misses caused by false sharing do not affect the correct executions of the parallel applications. As a result, we can say that false sharing is one major source of unnecessary overhead to keep the memory consistent [3, 4, 5, 6]. Especially, the problem becomes severe in PC-NOW DSM systems where the memory coherency unit is very large (generally, one virtual page). They say that the false sharing misses occupy 80% or so of the shared memory faults in page-based DSM systems [3, 4, 5, 6]. It means that the false sharing is the major obstacle for improving the memory performance in page-based DSM systems. In this paper, we present an efficient dynamic shared memory allocator for false sharing reduction in DSM system. The reasons why we chose to optimize dynamic shared memory allocator for reducing false sharing are that this approach is transparent to the application programmers, and almost all the false sharing misses happen in shared heap when multiple processes in a parallel application communicate with each other using shared memory allocated by dynamic shared memory allocator. The prediction of the future access patterns of each data object is necessary to reduce the false sharing misses caused by the data object. To accomplish this, we classify the data objects such that data objects requested at different locations in parallel program codes should not be allocated in the same shared page by tracing the call-site(object request location in parallel program codes). This is based on the idea that data objects requested at the different locations in program codes will show different access patterns in the future. Though the prediction technique of the access patterns we use is not always correct, we find out that our call-site tracing prediction technique could reduce the false sharing in comparison with other existing techniques**.** In order to measure the frequencies of page faults caused by false sharing(shortly false sharing misses), we use SPLASH and SPLASH II as a parallel application benchmark, and MINT as a multiprocessor architecture simulator.

In section 2, we review the related works. Section 3 explains the design and implementation of the call-site tracing-based shared memory allocator. We present the results of performance evaluation in section 4, and section 5 draws the conclusions.

2 Related Works

In this paper, we focus on the page-based DSM systems that keep the memory coherency in unit of a virtual memory page. The dynamic shared memory allocator for the page-based DSM systems has to decide where the requested data objects are placed. If the dynamic shared memory allocator knows the characteristics and access patterns of the requested data objects in advance, the allocator can easily place the data objects into the appropriate shared page with removing the causes of the false sharing. For example, the allocator can reduce the false sharing misses by placing the objects with much different access patterns to the different shared pages, or not placing non-related data objects into the same shared page. But, the dynamic shared memory allocator cannot know the characteristics and access patterns of the requested objects in advance. Therefore, the *typed allocation* is proposed in [7] where the clues provided by the programmers are used. In this typed allocation, the programmer must specify the memory access type through the allocation function arguments, such as Read-Only, Write-Mostly, and Lock types. Thus, the data objects with different types could be placed in the different shared pages. But, this scheme needs to additional overheads that user interfaces of the dynamic shared memory allocator have to be changed, and in turn the modification of the application source code is unavoidable. Moreover, it is not an easy job for the programmers to know in advance the access types of each shared data object. Our work assumes that there is no change in the API of the dynamic shared memory allocator.

Per-process allocation scheme assigns the different cache lines to the data objects requested by the different processes [3]. In this scheme, the data objects requested by the different processes are placed in the separate cache lines, so that it could reduce the possibility that data objects without relationships or with different access patterns are placed in the same cache line. This technique is effective where multiple processes request shared memory allocation evenly, but is likely to be ineffective where a dedicated process has the full responsibility of shared memory allocation [8]. In all the parallel applications used in our experiments, a dedicated process is also used for shared memory allocation, so it is inappropriate to compare this scheme with our approach.

Sized allocation scheme is proposed in [5, 6, 8], where the different-sized objects are prohibited from being placed in the same shared page. That is, by placing only the same-sized objects in the same shared page, this method tries to minimize the co-location of heterogeneous data objects. They say that, by using the object-size information for the prediction of the future access patterns, the transparency of the allocator API could be kept and the false sharing misses could be reduced simultaneously. Particularly according to [8], *allocation with separated tag* scheme and *minimizing the multi-page spanning* scheme could additionally reduce the false sharing misses. But this sized allocation is not enough to exactly predict the future access patterns of the shared data objects because the object size may not sufficiently represent the future access patterns. In our work, we will compare our call-site tracing technique with this sized allocation technique because these two methods have the similar assumptions that the interface of the dynamic shared memory allocator must not be modified and the shared memory allocation must be effective regardless of the existence of the allocation-dedicated process.

In [9], the hybrid allocation technique is proposed, which combines *per-processor allocation* and *minimizing the multi-page spanning* scheme. In this hybrid scheme, data objects requested by the different processors are placed in the different pages only when the object size is smaller than the page size. When the size of the data objects is bigger than the page size, on the other hand, they try to minimize the multi-page spanned allocations by prohibiting a shared object from being allocated in the page boundary. This technique could reduce the false sharing misses a little more by only combining the existing methods. But it is insufficient to accept this technique as a new prediction model of the future access patterns.

We find out by reviewing the previous works that the effective prediction of the future access patterns to be applied to the shared object allocations is an important factor to reduction of the false sharing misses. The shared objects which may have different access patterns must be placed in the different memory coherency unit. In this paper, we present ***call-site tracing based shared memory allocator***, shortly called ***CSTallocator***, where the future access patterns are predicted by the shared objects' request location(call-site) in the program codes. That is, the prediction is based on the instruction pointer from which the shared object allocation is requested. We hope that the objects with different call-sites may have the different access patterns in the future. By using the implicit information inherent in the program codes, our method not only keeps the API transparency, but also does not burden the programmers with the additional access type information. The call-site information of a shared object could be a useful clue for predicting the future access patterns because most parallel application programs call the allocation functions at different locations according to the object usage plans. Of course, the call-site tracing cost is more expensive than the cost of getting static information such as the allocation size passed via parameters or processor/process ID calling the function. Nevertheless, we can say that the call-site tracing overhead is not quite large because a call-site tracing procedure happens at a time only when the new call-site appears.

3 Design and Implementation of CSTallocator

With the information about objects' request locations in the program codes, we can infer the object's usage more accurately than with the object-size because multiple processes(or threads) cannot help to call the allocator at the different call-site according to the object's future usage. We expect that the future access patterns of the shared objects requested at different call-sites will be different even though the object sizes are the same. The only case that our expectation becomes wrong is when the usage of data objects requested at the same call-site changes abruptly and/or frequently. But it is difficult for the usage of the specific part of the program code to be dynamically changed, so we can use the object request call-sites as a clue for predicting the object's future access patterns.

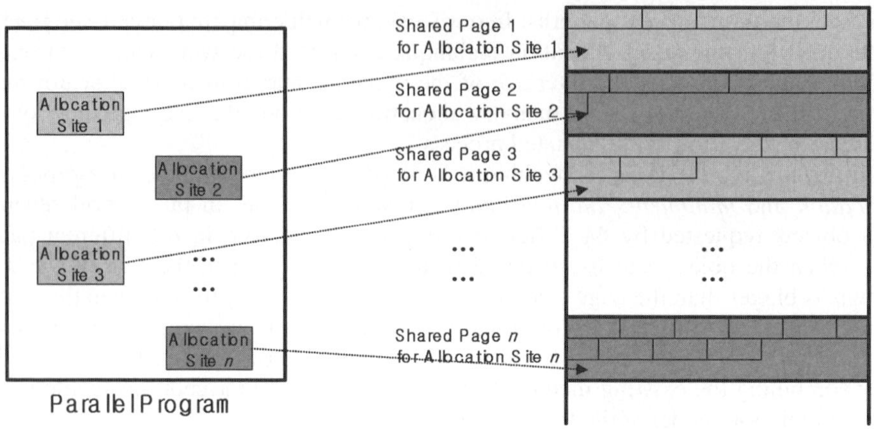

Fig. 2. Shared objects allocation example according to the call-sites in CSTallocator

Fig. 2 shows an example of the call-sites of each shared object in a parallel application program. In this figure, the shared objects are placed in the separate pages according to their allocator call-sites. The key idea is to prevent the shared objects requested at the different call-sites from being placed in the same shared page, while the different sized objects are allowed to be in the same page if the objects are requested at the same call-site. In our experiments, we intentionally allow this situation for the exact one-to-one comparison with the sized allocation scheme. In addition, we exclude the mixture scheme of call-site tracing and sized allocation for the accurate comparison of the two methods. Though the mixture technique considering both the call-site and object size is expected to show better performance, we do not discuss about the mixture technique here, and leave it as a future work.

3.1 Call-Site Tracing Technique

To accomplish the call-site tracing based allocation, firstly we have to identify the call-site where the shared memory allocation function is called in the program codes. The identification procedure must be done dynamically and transparently in the shared memory allocation function without additional formal parameters. For this purpose, we embed a module called *call path back tracker*, into the shared memory allocator. By back tracking the activation records accumulated in the process's (or thread's) stack, we could identify the call path from *main()*, the starting point of the program, to the current call-site. A return address has to be stored in the activation record for returning from the function call, and we could get this return address by identifying the size of local variables and the parameters used in the function. The stack back-tracking repeats till the *main()* function. For example, if we get "share_malloc() ← B() ← A() ← main()" from the stack back-tracking at a certain call-site, the ID of this call-site is represented as "A→B". The *share_malloc()* and *main()* functions are excluded in the call site ID representation because they always

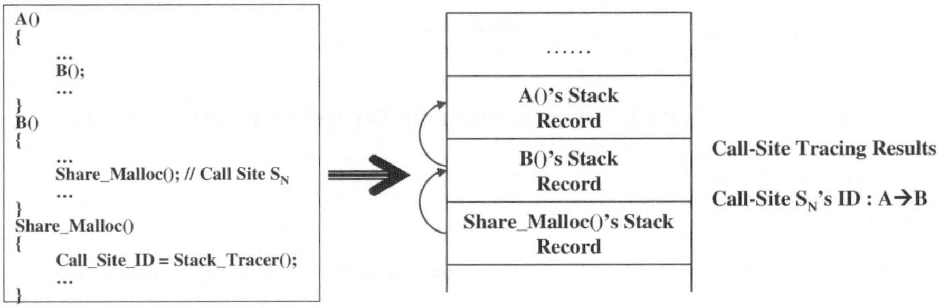

Fig. 3. An example of identifying call-site ID by call-site tracing

appear in every call-site ID. Fig. 3 shows an example of call-site tracing. In this figure, the call-site, S_N, is identified and registered with a call-site ID "A→B". And then, the shared objects with different call-site IDs are allocated in the different shared pages.

For the performance trade-offs of the stack back tracking, we have to consider the *back tracking depth of function call paths*. As a rule, a call-site ID can be defined after the back tracking to *main()* is completed. But in some parallel applications, we could identify all the call-sites without back tracking to *main()*. Therefore, we may decrease the overhead caused by the redundant stack back tracking if we could choose dynamically between the deep tracing and the shallow tracing. But the implementation of the dynamically depth-controlled back tracking is impossible because we cannot know the appropriate back tracking depth in advance to identify all the call-sites in a parallel application. So in our experiments, we will statically measure the effect of the back tracking depth adjustment on the performance. To do this, we define *length-N call chain*, which is the first *N* call paths from *share_malloc()* to *main()*. For example, length-1 call chain identifies only function *B()* which calls *share_malloc()*. In the same way, length-2 call chain includes function *B()*, which calls *share_malloc()*, and function *A()*, which calls *B()*, in the call-site ID. The longer the length of call chain, the deeper back tracking is to be done. In the prospect of the call chain length, we can expect that the possibility of false sharing would drop when using the longer length of call chain because it could identify the call-sites minutely.

3.2 Examples of Call-Site Tracing in Parallel Application Programs

Fig. 4 shows the call-site tracing results for the parallel applications used in our experiments. This figure summarizes all the paths from the *main()* to the call-sites identified during the applications' run-time. As we can see in this figure, cholesky(fig. 4(a)) and volend(fig. 4(d)) have a relatively large number of call-sites, on the other hand, mp3d(fig. 4(b)) and barnes(fig. 4(c)) have much smaller number of call-sites than the others. For convenience, we exclude the functions that have no call-site in this figure.

As we can see in this figure, most parallel application programs request shared memory at various locations in the program codes. In addition, we expect that the future access patterns to the shared objects allocated from the different call-sites are

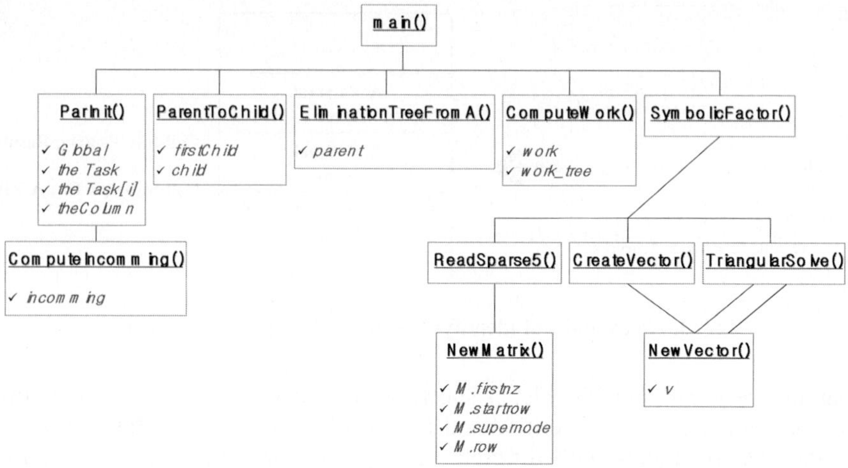

(a) Result of call-site tracing in Cholesky (maximum call chain length = 2).

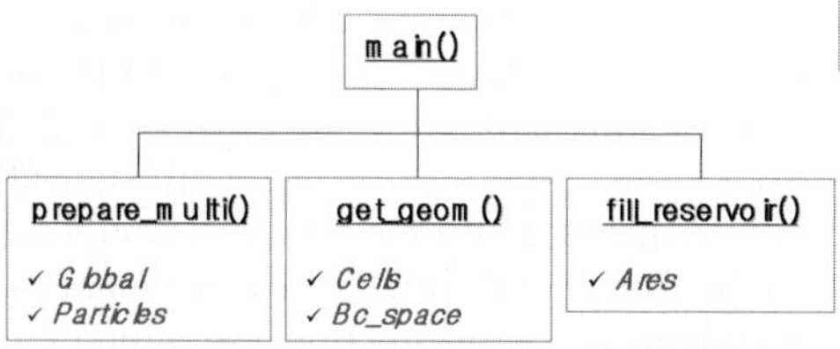

(b) Result of call-site tracing in Mp3d (maximum call chain length = 1)

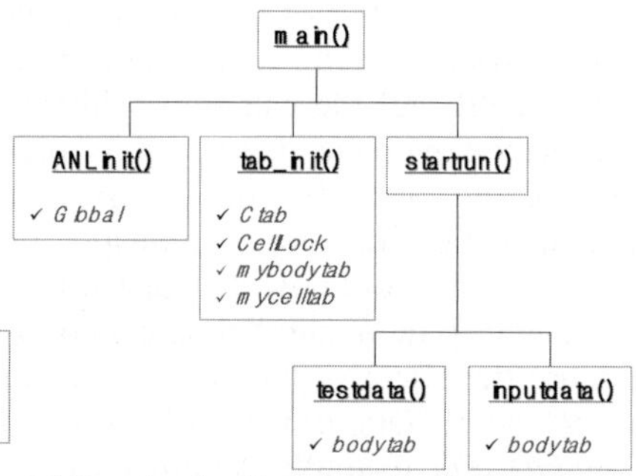

(c) Result of call-site tracing in Barnes (maximum call chain length = 1)

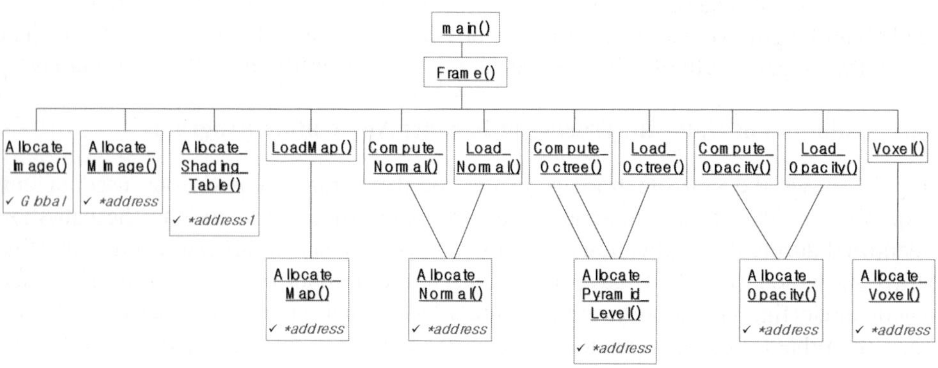

(d) Result of call-site tracing in Volrend (maximum call chain length = 2)

Fig. 4. Call-site tracing results in the four parallel application programs likely to differ from each other because the different call-sites mean the different programmer's intention. We are sure that our approach can reduce the false sharing misses because it prevents the shared objects with different access patterns from being allocated in the same shared page. Of course, the cost of stack back tracking increases in proportion to the length of call chain, but we can find out that the lengths of call chains are not altogether long in most applications as shown in fig. 4. The maximum length of call chain in the four parallel application programs used in our experiment is only 2, so we could identify all the call-sites by using shallow tracing only.

4 Performance Evaluation

This section explains the experimental environments and shows the results of the false sharing misses measurement, comparing with the performance of the two allocators, CSTallocator and the sized allocator.

4.1 Experimental Environments

We use the execution-driven technique to simulate a DSM system consisting of 16 nodes. The simulator consists of the front-end and the back-end simulators. The front-end simulator interprets the execution codes of the parallel application program binaries and simulates the executions of the processors. We use MINT(Mips INTerpreter) [10, 11] as a front-end simulator. The back-end simulator simulates the policies of the memory management system using MINT's outputs. MINT interprets the execution codes and calls functions provided by the back-end simulator in every memory reference. The back-end simulator implements the memory management policies and the memory coherence protocols to be simulated.

We use cholesky, mp3d, barnes, and volrend as parallel application program suites. These parallel applications are randomly selected from the Stanford's SPLASH [12] and SPLASHII [13]. We compare the number of false sharing misses when using the two allocation schemes, CSTallocator and sized allocation scheme. We also measure the effects of the length of call chain, *N*, on the number of false sharing misses when using CSTallocator.

4.2 Experimental Results

Table 1 shows how many false sharing misses are reduced in each parallel application when using our CSTallocator. The number of buckets in the second column is the number of the unique allocation slots found during the repeated shared memory allocation function calls. It represents the number of object sizes when using the sized allocation scheme, and the number of call-site IDs when using our CSTallocator respectively. Both the shared memory allocators manage the allocated objects as a linked list using the separate pointers for each bucket. The shared pages with the same bucket pointers are assigned to data objects with the same call-site ID or object size. Thus, the more buckets are found, the more sophisticated classification has been done. In general, the false sharing misses will decrease when the number of buckets increases.

Table 1. Results of performance comparison of CSTallocator and sized allocation (page size = 4KB, N = length of call chain)

(a) Cholesky

Allocator \ Measure	# of buckets	# of false sharing misses	Reduction rate(%)
Sized	10	44,717	
Call-Site-Tracing (N = 1)	15	40,921	8.5
Call-Site-Tracing (N = 2)	17	36,599	18.2

(b) Mp3d

Allocator \ Measure	# of buckets	# of false sharing misses	Reduction rate(%)
Sized	8	6,147,589	
Call-Site-Tracing (N = 1)	5	5,754,143	6.4

(c) Barnes

Allocator \ Measure	# of buckets	# of false sharing misses	Reduction rate(%)
Sized	27	5,805,705	
Call-Site-Tracing (N = 1)	7	5,104,413	12.1

(d) Volrend

Allocator \ Measure	# of buckets	# of false sharing misses	Reduction rate(%)
Sized	11	953	
Call-Site-Tracing (N = 1)	8	931	2.3
Call-Site-Tracing (N = 2)	12	883	7.3

From the result of table 1, we can see that our CSTallocator is much more effective for the false sharing reduction than the existing sized allocation scheme for all the parallel application programs used in our experiment. This observation indicates that the object request location in program codes, that is call-site, can be a better clue than the object size for predicting the objects' future access patterns. For example, we find out that the number of false sharing misses rather decreased for mp3d and barnes in which the sized allocation scheme uses more buckets. To our expectations, the false sharing misses reduction ratios of cholesky and volrend becomes larger in proportion to the length of call chain. This means that the future access patterns of the objects could be predicted more accurately with the fine-grained call-site identification. Moreover, the fact that the false sharing misses decrease even though the number of buckets decreases supports that our CSTallocator is also more effective in space efficiency than the sized allocation scheme.

4.3 Analysis of Space Efficiency

For the strict performance evaluation, we need to analyze space overheads caused by CSTallocator and the sized allocation scheme. The space overhead is the amount of memory used additionally by the proposed methods. For more accurate space efficiency analysis, we need to analyze the time efficiency in conjunction with space efficiency. But in our experiments, it is impossible to measure the actual execution time of the allocation functions because we use the simulation method instead of real executions. So we do not discuss about the time efficiency here, and leave it as a future work.

At first, we analyze the general shared memory allocator, which does not use the buckets such as object size or call-site ID. In the general shared memory allocator, the objects can be mixed up into the same shared page according to the sequence of requests. So in the general allocator, the allocation requests stream, S, is represented as:

$$S = \{s_1, s_2, \ldots, s_n\} \tag{1}$$

where s_i = requested size of i-th allocation ($1 \le i \le n$), n = total # of requests.

The number of pages needed to accept the above allocation requests stream is as follows:

$$\#\,\text{of pages required} = \left\lceil \frac{\sum_{i=1}^{n} s_i}{\text{page size}} \right\rceil \tag{2}$$

On the other hand, when using CSTallocator or sized allocation scheme, the allocation request stream can be represented as follows without considering the order of requests:

$$\begin{aligned} &S = \{S_{bucket_1}, S_{bucket_2}, \ldots, S_{bucket_k}\},\\ &S_{bucket_k} = \text{set of allocations with bucket ID } bucket_k,\\ &S_{bucket_1} \cap S_{bucket_2} \cap \ldots \cap S_{bucket_k} = \varnothing,\\ &BS = \{bucket_1, bucket_2, \ldots, bucket_k\}: \text{set of unique bucket IDs.} \end{aligned} \tag{3}$$

And the number of pages needed to accept the above stream is as follows:

$$\#\,\text{of pages required} = \sum_{bucket_k \in BS} \left\lceil \frac{|S_{bucket_k}| \times AvgSize_{bucket_k}}{\text{page size}} \right\rceil, \tag{4}$$

where $AvgSize_{bucket_k}$ is average size of each allocation request heading for $bucket_k$.

In comparison of the equation (2) with (4), the difference lies in the number of ceiling function. In equation (2), the ceiling function is applied at once, while it is applied as many as the size of the set BS (|BS|) in equation (4). This means that the maximum additional number of pages is limited to the number of the unique allocation sizes in sized allocation scheme and the number of call-site IDs in CSTallocator respectively. Thus, the following is valid:

$$\text{Space Overhead} = \left(\sum_{bucket_k \in BS} \left\lceil \frac{|S_{bucket_k}| \times AvgSize_{bucket_k}}{\text{page size}} \right\rceil \right) - \left\lceil \frac{\sum_{i=1}^{n} s_i}{\text{page size}} \right\rceil \leq |BS| \quad (5)$$

The obvious fact we can obtain from the above equations is that the shared page overhead is no more than the number of buckets regardless of which bucket classification methods are used. Table 2 shows the comparison results about the space efficiency measured by equation (5). As we can see in this table, the space efficiency of CSTallocator is better than that of the sized allocation scheme for mp3d and barnes, but is a little worse for cholesky and volrend. It means that the space efficiency gap between the two schemes is not quite large. Moreover, nowadays the space overhead such like the above can be surely tolerable if the memory specification of the current computer systems is taken into account.

Table 2. Space efficiencies of the two schemes (Page size = 4KB. In CSTallocator, maximum length of the call chain is used)

Parallel application programs (total # of pages needed in general allocation method)	# of additional pages (space overhead (%))	
	Sized allocation	CSTallocator
Cholesky (738)	10 (1.36)	17 (2.30)
Mp3d (553)	8 (1.45)	5 (0.90)
Barnes (308)	27(19.85)	7 (5.15)
Volrend (441)	11 (2.49)	12 (2.72)

5 Conclusions and Future Works

This paper presents an efficient shared memory allocation method for parallel applications which communicate via dynamically allocated shared memory in DSM systems. Without modifying the user interface of the shared memory allocator, the proposed call-site tracing-based allocator, called CSTallocator, can reduce the false sharing misses more effectively than the existing sized allocation scheme. Our CSTallocator prevents the shared objects with different call paths being allocated in the same shared page by tracing the object request location in the application program codes. We use the call-site as a clue for predicting the programmer's intention, and find out by simulation that the call-site help to predict the future access patterns of the shared objects more accurately than the existing sized allocation scheme. The CSTallocator additionally spends pages only as many as the number of unique call-sites in the applications. That is, our method could reduce more false sharing misses with a moderate space overhead than the sized allocation scheme. We are sure that our CSTallocator can contribute to both reduction of the false sharing misses and reduction of the cost on keeping the memory coherency in DSM systems.

In the future, we will verify how many false sharing misses can be reduced when using the mixture scheme of the sized allocator and our CSTallocator. And to measure the time efficiency as well as space efficiency, we will try to use the real DSM systems as a test bed instead of simulation environments.

References

[1] Andrew S. Tanenbaum. *Distributed Operating Systems*, chapter 6, pages 333-345. PRENTICE HALL, 1995.

[2] Jongwoo Lee, Yookun Cho. Page Replication Mechanism using Adjustable DELAY Counter in NUMA Multiprocessors. *Journal of the Korean Institute of Telematics and Electronics B*, 33B(6), pp.23-33, June 1996.

[3] Josep Torrellas, Monica S. Lam, and John L. Hennessy. Shared Data Placement Optimizations to Reduce Multiprocessor Cache Miss Rates. In *Proceedings of the 1990 International Conference on Parallel Processing*, volume II(Software), pages 266-270, August 1990.

[4] Susan J. Eggers and Tor E. Jeremiassen. Eliminating False Sharing. In *Proceedings of the 1991 International Conference on Parallel Processing*, volume I(Architecture), pages 377-381, August 1991.

[5] Jongwoo Lee, Yookun Cho. Shared Memory Allocation Mechanism for Reducing False Sharing in Non-Uniform Memory Access Multiprocessors. *Journal of Korean Information Science Society(A): Computer Systems and Theory*, 23(5), pp.487-497, May 1996.

[6] JongWoo Lee and Yookun Cho. An Effective Shared Memory Allocator for Reducing False Sharing in NUMA Multiprocessors. In *Proceedings of 1996 IEEE 2nd International Conference on Algorithms & Architectures for Parallel Processing(ICA^{3}PP '96)*, pages 373-382, June 1996.

[7] Roger L. Adema and Carla Schlatter Ellis. Memory Allocation Constructs to Complement NUMA Memory Management. In *Proceedings of the 3rd IEEE Symposium on Parallel and Distributed Processing*, December 1991.

[8] Jongwoo Lee, Moonhee Kim, Janghee Han, Daeku Ji, Jongwan Yoon, Jangseon Kim. Effects of Dynamic Shared Memory Allocation Techniques on False Sharing in DSM Systems. *Journal of Korean Information Science Society(A): Computer Systems and Theory*, 24(12), pp.1257-1269, December 1997.

[9] Boohyung Han, Seongje Cho, Yookun Cho. Techniques for Eliminating False Sharing and Reducing Communication Traffic in Distributed Shared Memory Systems. *Journal of Korean Information Science Society(A)*, 25(10), pp.1100-1108, October 1998.

[10] J. E. Veenstra. MINT Tutorial and User Manual. Technical Report TR452, Computer Science Department, University of Rochester, July 1993.

[11] J. E. Veenstra and R. J. Fowler. MINT: A Front End for Efficient Simulation of Shared-Memory Multiprocessors. In Proceedings of the Second International Workshop on Modeling, Analysis and Simulation of Computer and Telecommunication Systems(MASCOTS '94), pages 201-207, January-February 1994.

[12] J. P. Singh, W. Weber, and A. Gupta. SPLASH: Stanford Parallel Applications for Shared-Memory. *ACM SIGARCH Computer Architecture News*, 20(1):5-44, March 1992.

[13] Steven Cameron Woo, Moriyoshi Ohara, Evan Torrie, Jaswinder Pal Singh, and Anoop Gupta. The SPLASH2 Programs: Characterization and Methodological Considerations. In *Proceedings of the 22nd Annual International Symposium on Computer Architecture*, pages 24-36, June 1995.

Performance Evaluation of Storage Formats for Sparse Matrices in Fortran

Anila Usman[1], Mikel Luján[2], Len Freeman[2], and John R. Gurd[2]

[1] Department of Computer & Information Sciences,
PIEAS (Pakistan Institute of Engineering & Applied Science),
Islamabad, Pakistan
anila@pieas.edu.pk
[2] Centre for Novel Computing, The University of Manchester,
Oxford Road, Manchester M13 9PL, United Kingdom
{mlujan, len.freeman, john.r.gurd}@manchester.ac.uk

Abstract. Many storage formats have been proposed to represent sparse matrices. This paper extends to Fortran 95 the performance evaluation of sparse storage formats in Java presented at ICCS 2005, [7]. These experiments have the same set up (almost 200 sparse matrices and matrix-vector multiplication), but now consider the Fortran 95 Sparse BLAS reference implementation.

Keywords: Sparse matrix, storage format, Sparse BLAS, performance evaluation, JSA (Java Sparse Array).

1 Introduction

Sparse matrices (matrices with a substantial minority of nonzero elements, normally less than 10% nonzero elements) are pervasive in many mathematical and scientific applications. These matrices provide an opportunity to minimise storage and computational requirements by storing, and performing arithmetic with, only the nonzero elements. The many existing storage formats for sparse matrices are derived from different means of taking advantage of sparsity patterns in frequently occurring matrices.

The Basic Linear Algebra Subroutines (BLAS) standard includes for the first time a set of subroutines dedicated to operating with sparse matrices [3]. This part of the standard, hereafter referred to as the *Sparse BLAS*, primarily provides functionality for *iterative methods*. The Sparse BLAS does not state which storage formats must be supported, but ensures that the storage format used is completely transparent to the users. The Sparse BLAS leaves each specific hardware vendor the freedom to select the storage format (or formats) that performs best for its specific platforms.

In ICCS 2005 [7], the authors present a performance evaluation of different storage formats for sparse matrices in Java using the main kernel of iterative methods for linear systems: matrix-vector multiplication. The results show that a recently proposed storage format for Java, namely the Java Sparse Array (JSA)

M. Gerndt and D. Kranzlmüller (Eds.): HPCC 2006, LNCS 4208, pp. 160–169, 2006.

[6] performs similarly to, or better than, other long established storage formats. This paper concentrates on the performance of the same matrix operation and same storage formats with the same test matrices, but now their implementations are in Fortran 95 and belong to the reference implementation of the Sparse BLAS [4]. The JSA implementation has been translated to Fortran 95 and it is considered in the experiments.

The outline of the paper is as follows. Section 2 provides a brief description of some commonly used storage formats for sparse matrices and the JSA storage format. The Sparse BLAS reference implementations of the matrix-vector multiplication and other implementations used in the performance evaluation are described in Section 3. The performance evaluation (see Section 4) compares this operation using eight different storage formats. Around two hundred different symmetric/non-symmetric sparse matrices are considered in the performance study. Preliminary conclusions are given in Section 5.

2 Storage Formats for Sparse Matrices

There are many documented versions of different storage formats for sparse matrices. One of the most complete sources is the book by Duff *et al.* [2] (for a historical source see [8]). Some examples of these storage formats follow.

2.1 Compressed Sparse Row/Column Storage Formats (CSR/CSC)

CSR and CSC storage formats are not based on any particular matrix property and hence can be used to store any sparse matrix. In CSR, the nonzero values of every row in the matrix are stored together with their column number, consecutively in two parallel arrays, *Value* and *Col.* There is no particular order with respect to the column number, *Col.* The *Size* and *Pointer* for each row define the number of nonzero elements (nnze) in the row and point to the relative position of the first nonzero element of the row. Fig. 1 presents a sparse matrix stored in CSR.

The column based version, CSC, instead stores *Value* and *Row*, in two parallel arrays. *Size* and *Pointer* of each column allow each member of *Value* to be associated with a column as well as the row given in *Row*.

2.2 Block Entry Storage Formats

Block entry storage formats divide a matrix into blocks or submatrices (squares or rectangles) and define schemes to describe the memory position of a single block. If the block size remains fixed, for example, Block Sparse Row/Column (BSR/BSC) storage format can be obtained from CSR/CSC, respectively. Similarly, when the block size can vary the Variable Block Compressed Sparse Row/Column formats (VBR/VBC) are obtained.

$$A = \begin{pmatrix} & a_{12} & & & \\ & & a_{23} & & a_{25} \\ & & a_{33} & & a_{35} \\ a_{41} & a_{42} & & a_{44} & \\ & & & a_{54} & \end{pmatrix}$$

i	1	2	3	4	5
Size	1	2	2	3	1
Pointer	1	2	4	6	9

	1	2	3	4	5	6	7	8	9
Col	2	3	5	5	3	1	4	2	3
Value	a_{12}	a_{23}	a_{25}	a_{35}	a_{33}	a_{41}	a_{44}	a_{42}	a_{54}

Fig. 1. An example sparse matrix A stored using CSR

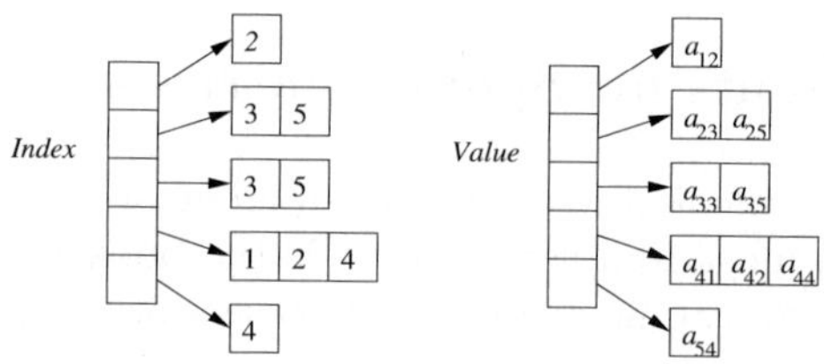

Fig. 2. The sparse matrix A stored using JSA

2.3 Java Sparse Array (JSA)

A recent storage format, designed particularly to suit Java, is JSA (Java Sparse Array), see Gundensen and Steihaug [6] for more details. JSA relies on being able to declare an array with individual elements being arrays – arrays of arrays. JSA is a row oriented storage format, similar to CSR. In Fortran 95, it is implemented as two arrays, each element of which is a pointer to an array. One of these arrays, *Value*, stores pointers to arrays which contain the matrix elements – each row in the matrix has its elements in a separate array. All the separate arrays can be reached through the pointers in the *Value* array; that is an array of pointers to arrays. The second array *Index* stores pointers to arrays which contain the column indices of the matrix, again one array per row. Fig. 2 shows the matrix A stored using JSA.

3 Fortran Implementations of Matrix-Vector Multiply

3.1 Reference Implementation of the Sparse BLAS

The Fortran 95 reference implementation of the Sparse BLAS was developed by CERFACS [4]. Although the standard does not indicate which storage formats must be supported, the Sparse BLAS reference implementation (SBF95) includes nine different storage formats (Coordinate or COO, CSR, CSC, diagonal or DIA, Block Coordinate or BCO, BSR, BSC, Block Diagonal or BDI and VBR[1] — see [2] for descriptions).

To use sparse matrices with the Sparse BLAS consist of three steps: (1) create a sparse matrix handle, (2) use this handle as a parameter in the Sparse BLAS routines, and (3) free any resources associated with the handle when it is no longer required.

For the first step and third step, the SBF95 uses a linked list to keep track of the different created/freed sparse matrices or handles. The handle is an integer used to access to the linked list which points to internal[2] data types implementing each of the mentioned storage formats. These hold specific information about the created matrix (such as the number of rows and columns, whether it is symmetric, etc.) and pointers to the arrays necessary for each of the different storage formats.

For the second step, the SBF95 transforms the handle into an internal pointer to the sparse matrix representation. Then several checks are performed for the correctness of the parameters (if these checks fail no part of the multiplication is performed and the execution stops immediately). For matrix-vector multiplication, 5 different checks are carried out, of which 4 involve subroutine calls to access data in the internal data types. After these checks, the implementations of matrix-vector multiplication present an if-then-else code structure separating special cases, where for example, the matrix is symmetric and only the elements in the upper triangular region are stored, from the general case. The actual implementation for each of the cases is part of the same subroutine; i.e. after the 5 checks no further subroutine call is made.

By default the SBF95 creates matrices in COO. Using internal utilities subroutines, users can transform from COO to the other storage formats.

3.2 Fortran Implementation of Java Sparse Array

The implementation of JSA is a separate and self-contained Fortran 95 module which defines the storage format as a data type. The data type holds the same information as the internal data types in the SBF95. The module has subroutines to allocate and assign the elements of a matrix, perform the matrix-vector multiplication and deallocate the memory used by a matrix.

Comparing the implementation of matrix-vector multiplication in JSA with those in the SBF95, the same checks and the same if-then-else structure are present. However the checks do not involve subroutine calls, and there is no handle to be translated into an internal representation.

4 Performance Evaluation

Matrix test data consists of 182 real, sparse symmetric and non-symmetric matrices available to download from the Matrix Market Collection [1] covering both

[1] The storage formats BDI and VBR are not used in the performance evaluation.

[2] Hereafter the word internal also means not part of the Sparse BLAS standard.

Table 1. Legend for the storage formats axis in Fig. 3

1	COO	7	BSC block size 2	13	BSC block size 8
2	CSR	8	BCO block size 4	14	BCO block size 16
3	CSC	9	BSR block size 4	15	BSR block size 16
4	DIA	10	BSC block size 4	16	BSC block size 16
5	BCO block size 2	11	BCO block size 8		
6	BSR block size 2	12	BSR block size 8		

systems of linear equations and eigenvalue problems, and representing many fields of Computational Science & Engineering.

The test program reads a matrix from file, multiplies a random vector by that matrix and records both the result vector and the time taken to calculate the result vector. The test program progresses in this way through the different implementations of the matrix operation. The test program checks the output results and we note that all the experiments produce the correct results.

The test machines are the same as those used in [7] to allow direct comparison. However, due to space limitations only results for one of the machines are reported; an Ultra Sparc 10 running Solaris and Sun's Fortran 95 compiler with optimisation flag `-fast`. In order to obtain meaningful times, the experiments report the execution times for each matrix-vector multiplication repeated 50 times. This not only gives timing results that are large enough compared with the accuracy of the timers (milliseconds), but is also a realistic simulation of the sequence of matrix-vector multiplications that dominates iterative methods in numerical linear algebra. The complete test program is run 10 times and the average of these times are reported.

4.1 SBF95 Performance Results

Fig. 3 gives the results of 8 different storage formats included in SBF95 for all matrices. Square block sizes between 2 and 16 are considered. No distinction is made between eigenvalue problems and systems of linear equations, nor between symmetric and non-symmetric matrices. The matrices in the Matrix Number axis are ordered by increasing nnze (the total number of non-zero elements).

Comparing these results with the results for the other machines not reported here, all the storage formats follow the same general pattern. For all of the storage formats there is a correlation between nnze and the observed execution times.

For the smaller problem sizes the block entry storage formats do not perform significantly differently for different block sizes. However, for the largest problems, the block entry storage formats, with block sizes of 4 and 8, perform better than the block formats with other block sizes. This suggests that there is a compromise between the gain resulting from more efficient use of the cache and the loss due to increases in the number of zero elements that are stored as the block sizes increase. For most of the test matrices, the point entry storage formats COO, CSR, CSC and DIA perform better than the block storage formats.

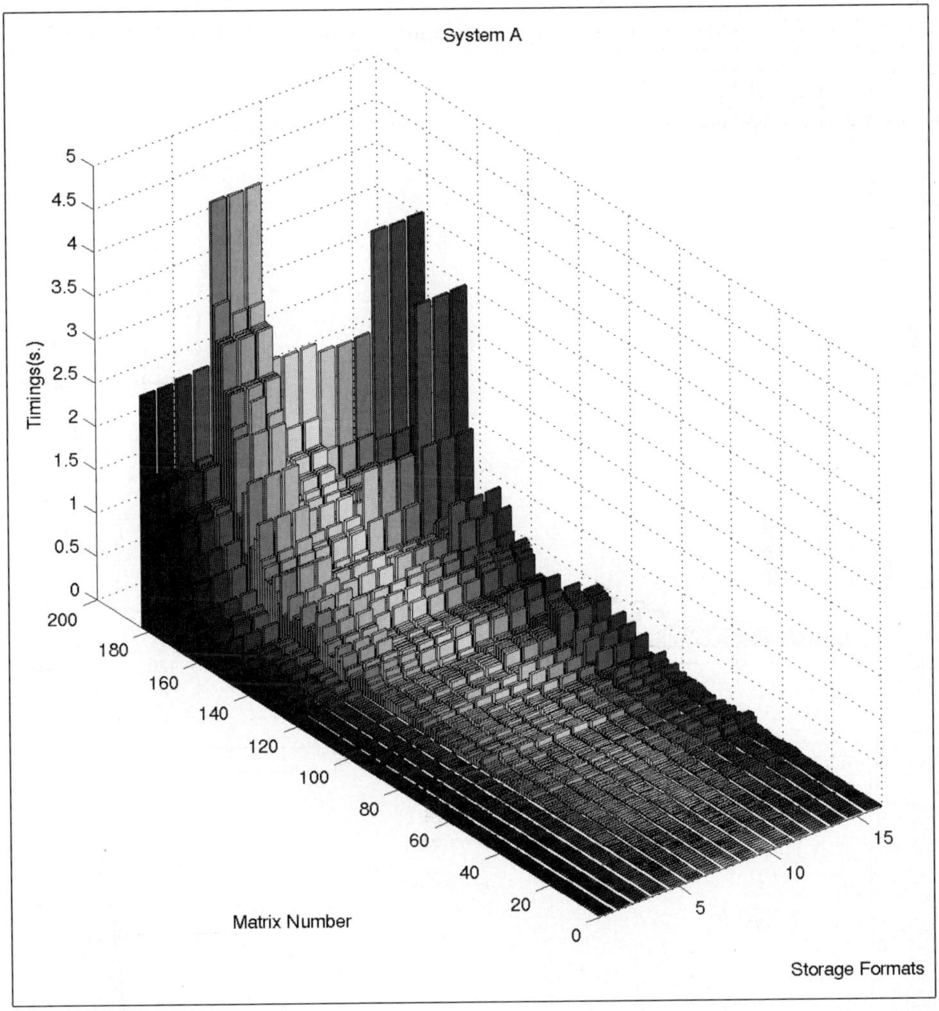

Fig. 3. Time results (seconds) for all matrices and storage formats. Table 1 presents the legend for the storage formats axis.

Fig. 4 takes a closer look at the results of CSR (as a typical representative of COO, CSC and DIA) and BSR with block sizes 4 and 8 (the fastest results among the block entry storage formats). The results are divided into two graphs (the upper graph covering matrices 1 to 140 and the lower graph covering matrices 141 to 182). For a few of the matrices with large numbers of nonzero elements (from matrix 173), the performance for BSR with block sizes of 4 and 8 is better than the performance of the point entry storage formats COO, CSR, CSC and DIA.

Performance Results for JSA

Fig. 5 presents the results of JSA and of the SBF95 implementation of CSR, referred to as CSR(SBF95) on the figure.

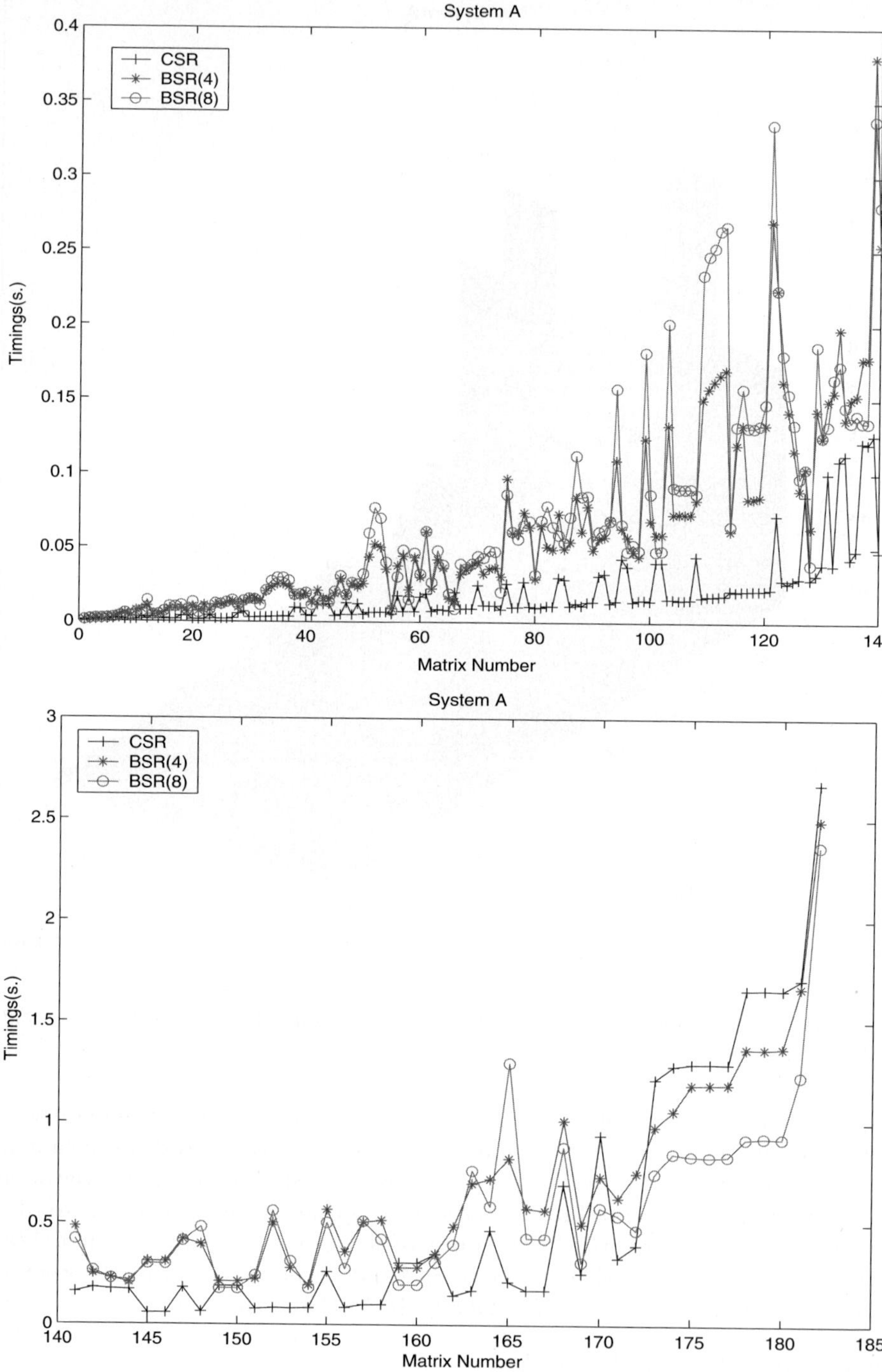

Fig. 4. Time results (seconds) for some selected storage formats

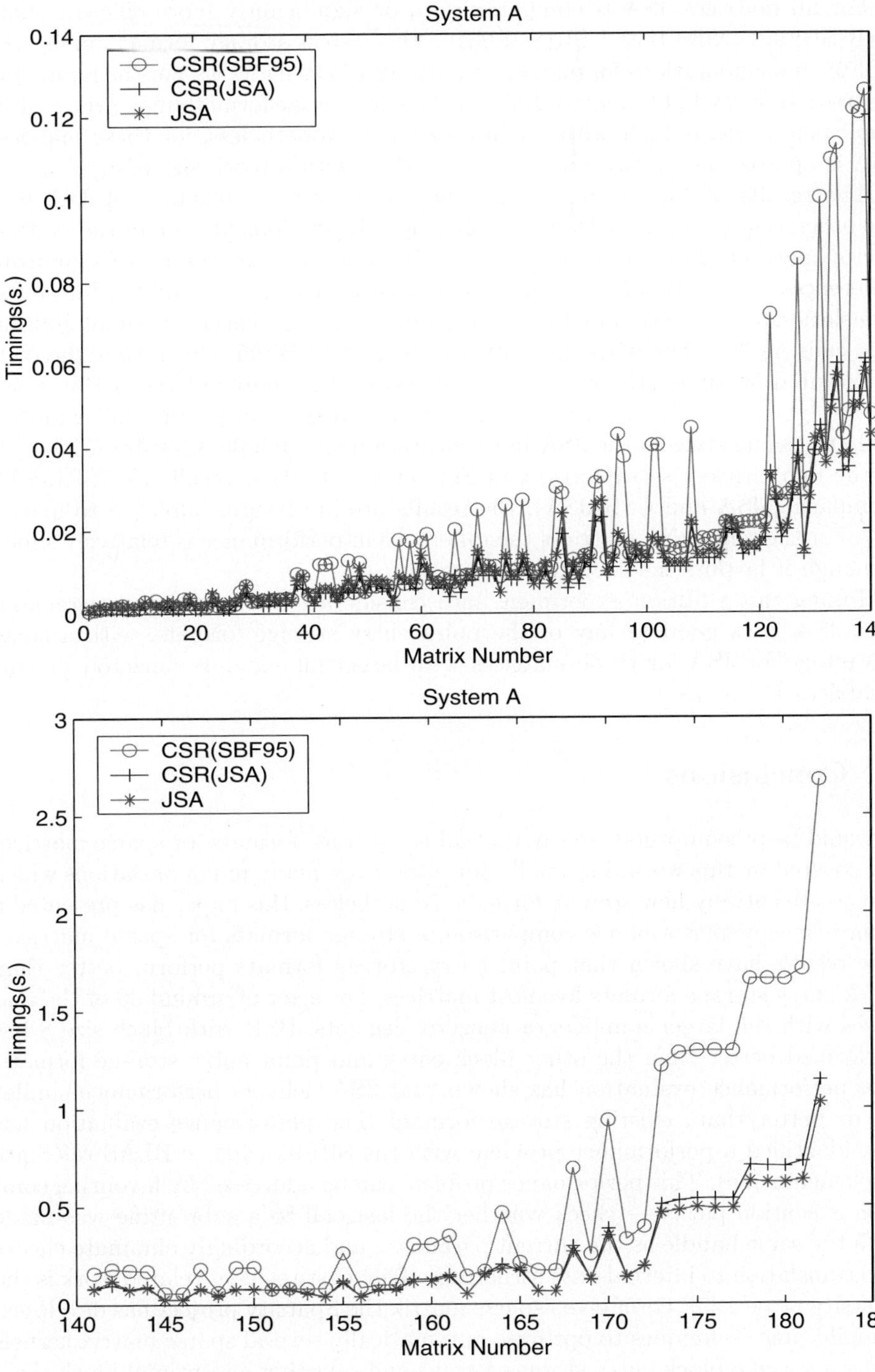

Fig. 5. Time results (seconds) for JSA

For all matrices JSA is comparable to, or significantly more efficient than, CSR(SBF95). Note that CSR(SBF95) is the fastest storage format among the SBF95 implementations for matrix 1 to matrix 173 (see Fig. 4). For the remaining matrices (matrix 174 to matrix 192), the fastest storage format among the SBF95 implementations is BSR with a block size of 8. Nonetheless, for these matrices JSA is approximately twice as efficient as BSR with a block size of 8.

The results of Fig. 5 give a clear indication of the advantage of JSA over the other storage formats. However, although the implementation of the matrix-vector multiplication for JSA and the SBF95 are similar, there are significant differences (see Section 3): (1) there is no need for the handle and the linked list translation or indirection in JSA; and (2) there are four checks, without subroutine calls in JSA, but with subroutine calls in the SBF95. Given that the code is executed 50 times, these differences could be the source of the performance advantage of JSA. To test this hypothesis, a CSR implementation following exactly the same style as the JSA implementation (no handle, no subroutine calls in the four checks) is developed and Fig. 5 presents these results as CSR(JSA). Comparing JSA and CSR(JSA), the results are indistinguishable up to matrix 177. For the remaining matrices the difference in performance is relatively small, although it favours the JSA implementation.

Taking this addition experiment into consideration, the conclusion remains that JSA is as good as any of the point entry storage formats, with a small advantage for JSA for those matrices with larger numbers of nonzero elements (matrices 177 to 182).

5 Conclusions

It would be presumptuous to say that all the storage formats for sparse matrices are covered by this work. Especially since there are many minor variations which can create entirely new storage formats. Nonetheless, this paper has presented a comprehensive performance comparison of storage formats for sparse matrices. The results have shown that point entry storage formats perform better than block entry storage formats for most matrices. For a set of around 30 of the matrices with the larger numbers of nonzero elements, BSR with block size 8 has performed better than the other block entry and point entry storage formats. The performance evaluation has shown that JSA delivers performance similar to, or better than, existing storage formats. The performance evaluation has also identified a performance problem with the SBF95 (Sparse BLAS reference implementation). This performance problem can be addressed by favouring common execution paths — check whether the last call to a subroutine was made with the same handle as the current parameter and accordingly eliminate checks and translation to internal data structures. The most relevant related work is the Sparsity project [5]. For a given sparse matrix the Sparsity project has developed compile time techniques to optimise automatically several sparse matrix kernels using a specific block entry storage format and selecting an optimal block size.

References

1. The matrix market. http://math.nist.gov/MatrixMarket/.
2. I. S. Duff, A. M. Erisman, and J. K. Reid. *Direct Methods for Sparse Matrices.* Oxford University Press, 1986.
3. I. S. Duff, M. A. Heroux, and R. Pozo. An overview of the sparse basic linear algebra subprograms: The new standard from the BLAS technical forum. *ACM Transactions on Mathematical Software*, 28(2):239–267, 2002.
4. I. S. Duff and C. Vömel. Algorithm 818: A reference model implementation of the Sparse BLAS in Fortran 95. *ACM Transactions on Mathematical Software*, 28(2):268–283, 2002.
5. Eun-Jin, K. A. Yelick, and R. Vuduc. SPARSITY: An optimization framework for sparse matrix kernels. *International Journal of High Performance Computing Applications*, 18(1):135–158, 2004.
6. G. Gundersen and T. Steihaug. Data structures in Java for matrix computations. *Concurrency and Computation: Practice and Experience*, 16(8):799–815, 2004.
7. M. Luján, A. Usman, P. Hardie, T. L. Freeman, and J. R. Gurd. Storage formats for sparse matrices in Java. In *Proceedings of the 5th International Conference on Computational Science – ICCS 2005, Part I*, volume 3514 of *Lecture Notes in Computer Science*, pages 364–371. Springer-Verlag, 2005.
8. U. W. Pooch and A. Nieder. A survey of indexing techniques for sparse matrices. *ACM Computing Surveys*, 5(2):109–133, 1973.

Performance Monitoring and Visualization of Grid Scientific Workflows in ASKALON*

Peter Brunner, Hong-Linh Truong**, and Thomas Fahringer

Institute of Computer Science, University of Innsbruck
{brunner, truong, tf}@dps.uibk.ac.at

Abstract. The execution of scientific workflows in Grids can imply complex interactions among various Grid applications and resources spanning multiple organizations. Failures and performance problems can easily occur during the execution. However, online monitoring and detecting failures and performance problems of scientific workflows in Grids is a nontrivial task. So far little effort has been spent on supporting performance monitoring and visualization of scientific workflows for the Grid. In this paper we present an online workflow performance monitoring and visualization tool for Grid scientific workflows that is capable to monitor the performance and to detect failures of Grid workflows. We also support sophisticated visualization of monitoring and performance result. Performance monitoring is conducted online and Grid infrastructure monitoring is integrated with workflow monitoring, thus fostering the chance to detect performance problems and being able to correlate performance metrics from different sources.

1 Introduction

Scientific workflows commonly compose several scientific tools and applications to perform complex experiments. The execution of scientific workflows in Grids frequently implies large amounts of complex interactions among various, diverse Grid applications and resources spanning multiple organizations. Due to the complexity and the diversity of both Grid workflows and resources, failures and performance problems can easily occur during runtime of Grid workflows. To monitor and detect failures and performance problems as early as possible at runtime is a key requirement from the scientific workflows community in the Grid.

However, until now performance monitoring and analysis tools for Grid scientific workflows are not well supported. Most research effort is dedicated to developing workflow languages and execution engines, as shown in [1]. Many existing Grid monitoring tools [2] do not support Grid scientific workflows. In this paper we present an online performance monitoring and visualization tool for Grid scientific workflows. The tool can monitor performance, detect failure of Grid scientific workflows, and present the monitoring and performance result through a sophisticated, but easy to use, visualization. Performance monitoring is conducted online, during the execution of the Grid

* The work described in this paper is supported by the European Union through the IST-2002-511385 project K-WfGrid.

** Corresponding author.

M. Gerndt and D. Kranzlmüller (Eds.): HPCC 2006, LNCS 4208, pp. 170–179, 2006.

workflows, and infrastructure monitoring is also integrated. The user can observe the performance of workflows as well as the Grid resources the workflow consumes by requesting and analyzing monitoring data of Grid infrastructure in parallel with application monitoring and analysis. As a result, the tool increases the probability to detect performance problems. Moreover it can correlate performance metrics from different sources. A prototype of this tool has been integrated into ASKALON environment [3] for developing scientific workflows in the Grid.

The rest of this paper is organized as follows: Section 2 outlines the ASKALON workflow system. In Section 3 we describe the architecture of our performance monitoring and visualization tool. Section 4 details techniques and features for performance monitoring, instrumentation and visualization of Grid scientific workflows. We illustrate experiments in Section 5. Related work is discussed in Section 6. Section 7 summarizes the paper and gives an outlook to the future work.

2 The ASKALON Environment

Our work on performance monitoring and visualization of scientific workflows in Grids is conducted in the framework of the ASKALON environment. ASKALON [3] is a Grid application development and computing environment which provides services for composing, scheduling and executing scientific workflows in the Grid. The main services in ASKALON are the *Resource Manager*, which is responsible for negotiation, reservation, allocation of resources and automatic deployment of services; the *Scheduler*, which determines effective mappings of workflows onto the Grid; and the distributed *Execution Engine* (EE), which is responsible for the reliable and fault tolerant execution of workflows. All ASKALON middleware services implement WSRF [4] by using Globus Toolkit 4.0 [5]. In ASKALON a user can compose Grid workflow applications using a UML-based workflow composition or can describe workflows using the XML-based Abstract Grid Workflow Language (AGWL). After the composition, the workflow is executed by the execution engine.

By using ASKALON a user designs a workflow, submits the workflow and observes the execution of the workflow on selected Grid sites. Performance bottlenecks and failures can occur at any time during the execution. A performance monitoring and visualization tool for scientific workflows enriches ASKALON by allowing users to monitor their running activities on selected Grid sites and to detect failure and abnormal behavior in Grid middleware.

3 Architecture of Workflow Performance Monitoring and Visualization Tool

Figure 1 depicts the architecture of the performance monitoring and visualization tool for Grid workflows within ASKALON. The performance monitoring and visualization tool relies on the SCALEA-G monitoring middleware [6] for collecting Grid infrastructure and application monitoring data. SCALEA-G services use peer-to-peer technology to communicate with each other and they can retrieve and store multiple types of monitoring data from diverse sensors.

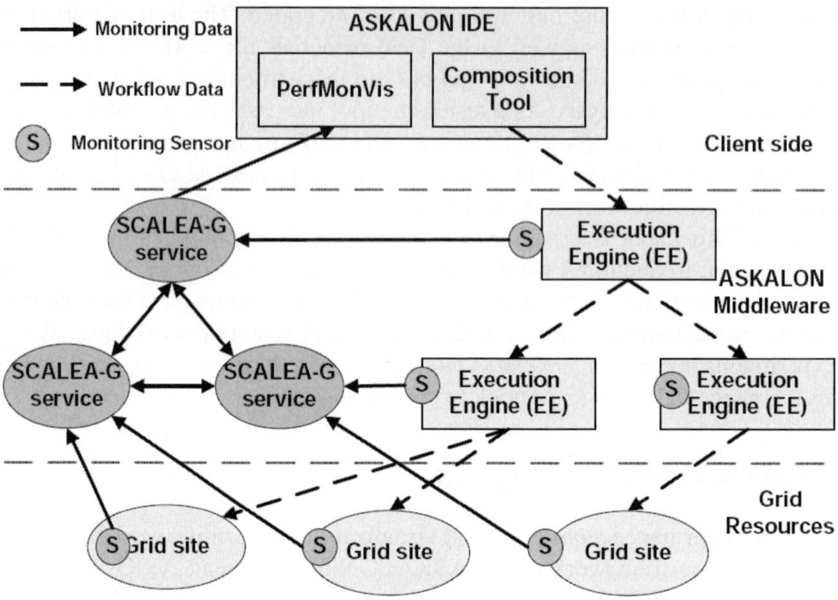

Fig. 1. Architecture of workflow performance monitoring and visualization tool

At the client side, the main GUI of the performance tool is integrated with the ASKALON IDE. From the IDE the user can compose workflows, submit them to the EE, and perform the performance monitoring and analysis of the workflows. Within distributed EEs, we have sensors capturing execution status of workflows. The sensors send monitoring data to SCALEA-G services which propagate the monitoring data to the main component (PerfMonVis) of the performance tool via subscription/query mechanism. On Grid sites where scheduled workflows are executed, sensors are also used to monitor Grid resources; these sensors provide the infrastructure information (e.g., machine and networks) of Grid sites. At the client side, monitoring data of Grid workflows and resources are received through data subscription and query. Monitoring data is analyzed and the performance results and failures are then visualized in the GUI.

4 Online Performance Monitoring and Visualization

4.1 Instrumentation and Monitoring Data

The execution of a Grid workflow is controlled by the EE under the guidance of the scheduler. However, invoked applications which perform the real work specified in workflow activities will be executed on distributed Grid sites by local Grid resource allocation and management (GRAM). To monitor the performance of Grid workflows, it is necessary to collect monitoring data not only within EEs but also at Grid sites.

To obtain workflow monitoring data we statically instrument EEs, which control job and data submissions to different Grid sites. Through the instrumentation, sensors are manually inserted into EEs to capture all the events associated with the workflow.

Table 1. Examples of workflow events

Event Name	Description
initialized	the activity has been initialized
queued	the activity gets in the EE queue
submitted	the activity has been submitted to the Grid site
active	the activity is active
suspended	the activity is suspended
completed	the activity is completed
failed	the activity is failed
canceled	the activity is canceled

Table 2. Examples of event attributes

Attribute Name	Description
ACTIVITY-NAME	original name of the activity
ACTIVITY-INSTANCE-ID	name of the activity instance
ACTIVITY-TYPE	type of activity
ACTIVITY-PARENT-NAME	parent name of the activity instance
COMPUTATIONALNODE	machine on which the activity is running
SOURCE-ID	source activity of a file transfer
DESTINATION-ID	destination activity of a file transfer

These events are then sent to the SCALEA-G middleware. Based on that they can be retrieved by the performance tool or any services or clients which are interested in obtaining workflow events. For instrumentation and monitoring at Grid sites, currently we have infrastructure sensors deployed in Grid sites. These sensors are used to monitor Grid resources and middleware services, e.g., capturing machine information, network bandwidth, and availability of GRAM and GridFTP. However, our previous work on instrumentation of Grid applications [7] has not been integrated into ASKALON.

We use a generic schema to describe various types of events. Basically, an event has an event name and an event time indicating the time at which the event occurs. Every event has event attributes that hold detailed monitoring data associated with the event. Event attribute can be used to store any interesting data, for example computational nodes on which an invoked application is executed, the source and the destination of a file transfer between activities, etc. Table 1 presents a few events captured in our system and Table 2 presents examples of event attributes.

Currently, all monitoring data collected is represented in XML form. Figure 2 presents examples of real events captured. In Figure 2(a) is an `initialized` event of a computational activity; the event consists of activity instance ID, the type of activity, the name of the parent instance. In Figure 2(b) is a `completed` event of a file transfer between two activity instances named `first-4` and `second-75`.

4.2 Online Monitoring and Visualization

Monitoring data collected from different sources, such as distributed EEs and Grid sites, is published into the SCALEA-G middleware. Each type of monitoring data is identified

```
<event>
  <eventname>activity_initialized
  </eventname>
  <eventtime>1142499781204</eventtime>
  <eventdata>
    <attrname>ACTIVITY_NAME</attrname>
    <attrvalue>first</attrvalue>
  </eventdata>
  <eventdata>
    <attrname>ACTIVITY_INSTANCE_ID
    </attrname>
    <attrvalue>first_4</attrvalue>
  </eventdata>
  <eventdata>
    <attrname>ACTIVITY_TYPE</attrname>
    <attrvalue>ControflowController
    </attrvalue>
  </eventdata>
  <eventdata>
    <attrname>ACTIVITY_PARENT_NAME
    </attrname>
    <attrvalue>whileBody_0_3</attrvalue>
  </eventdata>
</event>
```

(a)

```
<event>
  <eventname>activity_completed
  </eventname>
  <eventtime>1142499880862</eventtime>
  <eventdata>
    <attrname>ACTIVITY_INSTANCE_ID
    </attrname>
    <attrvalue>first/fVns1142499878780
    </attrvalue>
  </eventdata>
  <eventdata>
    <attrname>SOURCE_ID</attrname>
    <attrvalue>first_4</attrvalue>
  </eventdata>
  <eventdata>
    <attrname>DESTINATION_ID</attrname>
    <attrvalue>second_75</attrvalue>
  </eventdata>
  <eventdata>
    <attrname>ACTIVITY_TYPE</attrname>
    <attrvalue>FileTransfer</attrvalue>
  </eventdata>
</event>
```

(b)

Fig. 2. Examples of representation of events: (a) computational activity, (b) data transfer

by a tuple *(dataTypeID, resourceID)*, for example in case of workflow events the tuple `(wf.pma, 0c08d950-9979-11da-88a8-9ac1d10ab445)` indicates all events associated with the workflow whose UUID is `0c08d950-9979-11da-88a8-9ac1d10ab445`. The workflow performance monitoring and visualization tool can subscribe and/or query any monitoring data type of interest. In our tool, during the execution of a workflow, when the user starts to conduct the performance monitoring and analysis, the performance tool will subscribe all workflow events associated with the workflow. The tool will analyze events received and visualize performance results as well as failure detected in its GUI. Depending on the resulting analysis, the tool can subscribe/query other data types in order to find sources of problems. For example, if a workflow activity cannot be submitted to a Grid site, the user can query the monitoring data about the availability of local resource management on that Grid site.

The performance monitoring and visualization tool analyzes large amounts of diverse monitoring data and presents the result in an understandable way to the user. In ASKALON, our performance tool provides the following functionalities for online monitoring and visualization of Grid workflows:

- *Monitoring of execution phases of Grid workflows.* The user can monitor and analyze execution phases, e.g., queuing and processing, of Grid workflows during runtime. Detailed execution information of every single activities can be analyzed.
- *Monitoring of data transfer between Grid sites.* Monitoring of data transfer between Grid sites is an important feature. The user can observe the data transfer between activity instances through the visualization. Moreover, a specific data transfer can also be selected for analysis, for example to examine transfer time.
- *Interactive analysis of different activities.* The user can compare the execution of different running activities, for example examining different performance metrics

of the selected activities, load imbalances or overheads. This feature is particularly useful for analyzing parallel regions in scientific workflows.

- *Analysis and comparison of activity distribution and allocation for Grid sites.* How activities distributed to Grid sites as well as the utilization of Grid sites can be analyzed and compared. Allowing the user to select Grid sites that the user wants to analyze is a useful feature because if there are hundreds of Grid sites involved in execution of the workflow, the user can observe which sites are slower and find out why they are slow by using infrastructure monitoring.
- *Interactive querying of infrastructure monitoring during execution.* Infrastructure monitoring data, such as CPU load or network bandwidth between Grid sites can also be provided during the monitoring and analysis of Grid workflows. Grid workflows and resources consumed are analyzed in an integrated environment.

The monitoring and visualization tool can also help to develop the ASKALON middleware services. For example, performance results are used by the performance prediction service to provide estimated execution times of Grid workflows to support the scheduler. In ASKALON, the scheduling service can subscribe events indicating abnormal behaviors of Grid middleware and workflows so that it can reschedule the workflow.

Figure 3 shows the main GUI of the workflow performance tool. The top window (`Execution Trace`) visualizes the workflow representation together with detailed monitoring information. In the top-left pane is the static representation of the workflow, showing the hierarchical view of workflows. The hierarchical view explicitly defines, in detail, concepts and properties of workflows, including workflow, workflow regions and workflow activities, and their relationships based on the workflow performance ontology [8]. In the top-right pane is the execution trace of the workflow. The activities are sorted sequentially by the time they have been initialized. The example trace is for a workflow named `Wien2K`, which will be described in Section 5.1. We can monitor and visualize activities, workflow regions, data transfers, etc. By clicking an activity in the top-left pane, instances of that activity will be highlighted in the top-right pane. In the bottom-left pane is a tree representing performance information about the current selected activity, for example, the name of the activity, the parent activity of this activity, the machine where the activity is executed, queueing time, processing time, transfer time, source and destination of file transfers, etc. Detailed performance information such as average execution times per Grid sites and job distribution to Grid sites are also visualized.

5 Experiments

We have implemented a prototype of our workflow performance monitoring and visualization, and integrated it into ASKALON. Currently the prototype is based on Java 1.5, and JGraph [9] is employed for the visualization of the activities and of the workflow.

In this section, we present experiments illustrating the performance monitoring and visualization of real world Grid scientific workflows named `Wien2K` and `Invmod`. Our experiments are conducted within the Austrian Grid infrastructure [10], which connects several national Austrian Grid sites. For our experiments we selected two altix machines, including `altix1.uibk.ac.at` which is an altix 16 CPUs machine at the

University of Innsbruck and `altix1.jku.austriangrid.at` which is an altix 64 CPUs machine at the University of Linz. In all our experiments, we just specified the two Grid sites but the number of CPUs used for executing workflows is decided by the ASKALON scheduler.

5.1 Wien2K

`Wien2K` [11] is a program package for performing electronic structure calculations of solids using density functional theory, based on the full-potential (linearized) augmented planewave ((L)APW) and local orbital (lo) method. The problem size is 5.5 specifying the number of planewaves used, and the number of parallel k-points is 650. With this problem size, the two parallel sections - `pforLAPW1` and `pforLAPW2` - have 65 parallel iterations.

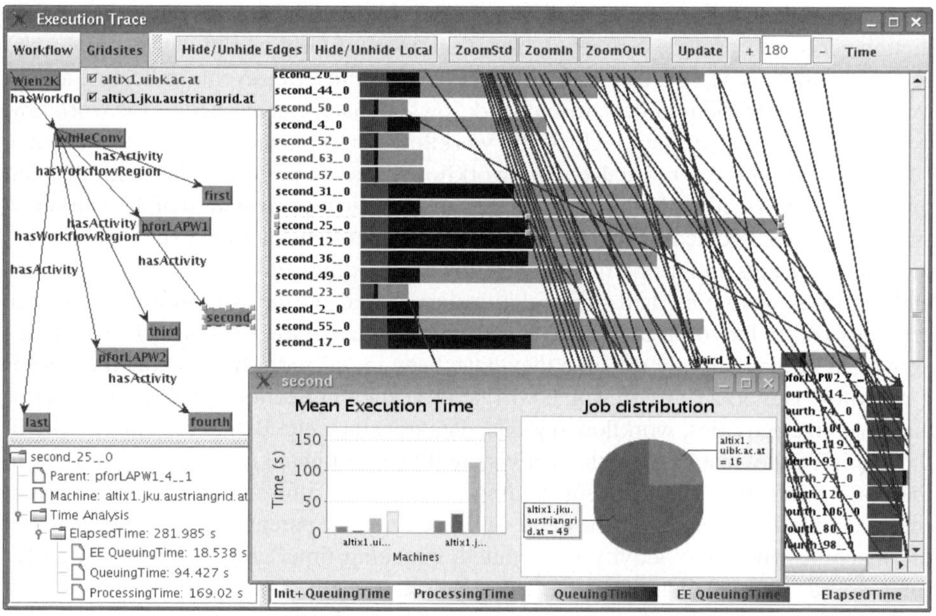

Fig. 3. Wien2K workflow experiment

This workflow has only one primary workflow region, named `pforLAPW1`, which takes a long time to finish. The rest of the workflow activities are not time-consuming. Through the trace execution in Figure 3, we observed that activities executed on `altix1.uibk` machine in Innsbruck are completed much faster than activities on `altix1.jku` machine in Linz. The window `second` shows performance information for activity `second`. The pane `Mean Execution Time` shows various average timing metrics (per number of instances) for activity `second` on different Grid sites whereas the pane `Job distribution` displays how instances of activity `second` are distributed on different Grid sites. Overall, timing metrics on `altix1.uibk` are better than that on `altix.jku`. Out of 65 parallel instances of activity `second`, the scheduler distributed 16 instances to `altix1.uibk`, which fully loads this 16-CPU machine. The remaining, 49 instances, are mapped to 64-CPU `altix1.jku` because the ASKALON performance prediction service indicates that an instance could be completed faster on `altix1.uibk` than on `altix1.jku`.

During our experimental work with `Wien2K`, we detected some errors made by the execution engine. For example, in one case, we did not get any `failed` event from the execution engine, but we observed in our visualization tool that one activity had been reinitialized. After analyzing the case, we found that the missing `failed` event was due to a bug in the execution engine.

5.2 Invmod

`Invmod` [12] is a hydrological application for river modelling which has been designed for inverse modelling calibration of the WaSiM-ETH program. `Invmod` has two levels of parallelism, one for the calibration of parameters that is calculated separately for each starting value using *parallel random runs*, the second for each optimization step represented by an inner loop iteration. In this experiment, the number of the parallel random runs is 7 and the parameter of the optimization step is 3.

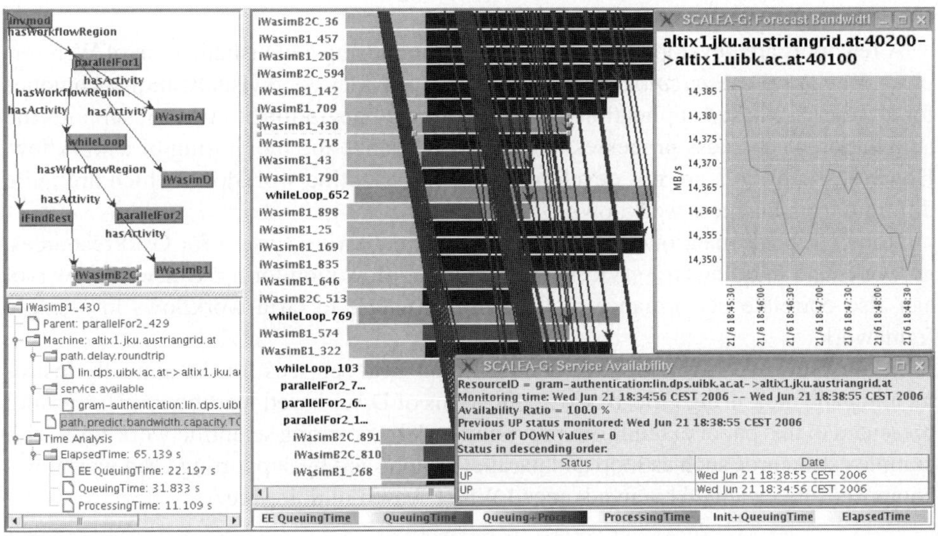

Fig. 4. Invmod workflow experiment

As shown in Figure 4, the `Invmod` workflow is more complex than `Wien2K` workflow. There are more parallel regions and data transfers, and a larger number of activities is executed in parallel. We observed the execution engine sends all the jobs, using GridFTP, from `altix1.jku` to itself, which is a performance overhead because this transfer is unnecessary as `altix1.jku` has a NFS (Network File System). In its current implementation, if a file must be sent to many activities which have different parameters, the execution engine just sends the file to every activity using GridFTP, without considering the underlying file system. By selecting data types in the tree in the bottom-left pane of Figure 4, infrastructure monitoring can also be invoked. For example, the window `Forecast Bandwidth` shows the network bandwidth between `altix1.jku` and `altix1.uibk` whereas the window `Service Availability` displays the availability of GRAM service on `altix1.jku`.

6 Related Work

While many tools support performance monitoring and visualization for scientific parallel applications, there is a lack of similar tools for Grid scientific workflows. P-GRADE [13] is a performance monitoring and visualization for Grid workflows. In contrast to our tool, P-GRADE does not support cyclic workflows. P-GRADE is based on Globus Toolkit 2 while our tool is based on advanced WSRF-based architecture using Globus Toolkit 4. Taverna Workbench [14] allows users to monitor the status of activities within Grid workflows, but visualizes information in a form of a simple table. Moreover, detailed execution phases are available only after the execution of the workflow. Our performance tool is more advanced because it can analyze monitoring data in detail at runtime, and dynamic calling relationship among activity instances can be examined online as well.

A few tools are developed for performance monitoring and visualization of Web services. *Web Service Navigator* [15], for example, provides good visualization techniques for Web service based applications. BEA WebLogic Integration Studio [16] supports the automation of business processes. It can also visualize the monitoring of a workflow. However, such tools are not developed for Grid scientific workflows which are quite different from business workflows.

None of above-mentioned tools integrates a monitoring system for Grid resources, networks and middleware services with Grid workflow monitoring. Our workflow tool can also correlate performance of Grid infrastructure to Grid workflows in a single framework.

In our previous work, we have demonstrated Grid services for dynamically instrumenting and measuring Grid-based applications of DAG-based workflows [7]. Our tool presented in this paper extends our previous work by covering scientific workflows with complex structures such as loop and parallel regions and by supporting Grid workflows composed and executed by an advanced WSRF-based middleware.

7 Conclusion and Future Work

Scientific workflows in Grids are complex and their execution commonly implies distributed, sophisticated interactions among various types of Grid applications and resources. We need performance monitoring and visualization tools that are capable to assist the user in monitoring these complex interactions and in detecting failures occurring during runtime. The contribution of this paper is a workflow performance monitoring and visualization tool that not only allows users to observe and analyze complex interactions during the execution of Grid workflows but also supports the correlation between the performance of Grid workflows and the underlying Grid infrastructure.

We are currently enhancing our tool with performance monitoring and visualization features that cover also invoked applications within workflow activities and code regions, according to our workflow performance ontology [8]. Moreover, we are integrating our tool into the K-WfGrid workflow system [17].

References

1. Yu, J., Buyya, R.: A taxonomy of scientific workflow systems for grid computing. SIGMOD Rec. **34** (2005) 44–49
2. Gerndt, M., Wismueller, R., Balaton, Z., Gombas, G., Kacsuk, P., Nemeth, Z., Podhorszki, N., Truong, H.L., Fahringer, T., Bubak, M., Laure, E., Margalef, T.: Performance Tools for the Grid: State of the Art and Future. Volume 30 of Research Report Series, Lehrstuhl fuer Rechnertechnik und Rechnerorganisation (LRR-TUM) Technische Universitaet Muenchen. Shaker Verlag (2004) ISBN 3-8322-2413-0.
3. Fahringer, T., Prodan, R., Duan, R., Nerieri, F., Podlipnig, S., Qin, J., Siddiqui, M., Truong, H.L., Villazon, A., Wieczorek, M.: ASKALON: A Grid Application Development and Computing Environment. In: 6th International Workshop on Grid Computing (Grid 2005), Seattle, USA, IEEE Computer Society Press (2005)
4. OASIS Web Services Resource Framework (WSRF) TC: http://www.oasis-open.org/committees/tc_home.php?wg_abbrev=wsrf (2006)
5. Globus Project: http://www.globus.org (2006)
6. Truong, H.L., Fahringer, T.: SCALEA-G: a Unified Monitoring and Performance Analysis System for the Grid. Scientific Programming **12** (2004) 225–237 IOS Press.
7. Truong, H.L., Fahringer, T., Dustdar, S.: Dynamic Instrumentation, Performance Monitoring and Analysis of Grid Scientific Workflows. Journal of Grid Computing **3** (2005) 1–18
8. Truong, H.L., Fahringer, T., Nerieri, F., Dustdar, S.: Performance Metrics and Ontology for Describing Performance Data of Grid Workflows. In: Proceedings of IEEE International Symposium on Cluster Computing and Grid 2005, 1st International Workshop on Grid Performability, Cardiff, UK, IEEE Computer Society Press (2005)
9. JGraph: http://www.jgraph.com/ (2006)
10. AustrianGrid: http://www.austriangrid.at/ (2006)
11. Blaha, P., Schwarz, K., Luitz, J.: WIEN97: A Full Potential Linearized Augmented Plane Wave Package for Calculating Crystal Properties. Institute of Physical and Theoretical Chemistry (2000)
12. Jasper, K.: Hydrological Modelling of Alpine River Catchments Using Output Variables from Atmospheric Models (2001)
13. Lovas, R., Kacsuk, P., Horvath, A., Horanyi, A.: Application of P-Grade Development Environment in Meteorology, http://www.lpds.sztaki.hu/pgrade (2003)
14. taverna: http://taverna.sourceforge.net/ (2006)
15. Pauw, W.D., Lei, M., Pring, E., Villard, L., Arnold, M., Morar, J.F.: Web services navigator: Visualizing the execution of web services, ibm systems journal, http://www.research.ibm.com/journal/sj/444/depauw.html (2005)
16. BEA Systems: Monitoring workflows, http://e-docs.bea.com/wli/docs70/studio/ch10.htm (2002)
17. K-WF Grid Project: http://www.kwfgrid.net (2006)

Exploring the Capacity of a Modern SMT Architecture to Deliver High Scientific Application Performance*

Evangelia Athanasaki, Nikos Anastopoulos, Kornilios Kourtis, and Nectarios Koziris

National Technical University of Athens
School of Electrical and Computer Engineering
{valia, anastop, kkourt, nkoziris}@cslab.ece.ntua.gr

Abstract. Simultaneous multithreading (SMT) has been proposed to improve system throughput by overlapping instructions from multiple threads on a single wide-issue processor. Recent studies have demonstrated that heterogeneity of simultaneously executed applications can bring up significant performance gains due to SMT. However, the speedup of a single application that is parallelized into multiple threads, is often sensitive to its inherent instruction level parallelism (ILP), as well as the efficiency of synchronization and communication mechanisms between its separate, but possibly dependent, threads. In this paper, we explore the performance limits by evaluating the tradeoffs between ILP and TLP for various kinds of instructions streams. We evaluate and contrast speculative precomputation (SPR) and thread-level parallelism (TLP) techniques for a series of scientific codes executed on an SMT processor. We also examine the effect of thread synchronization mechanisms on multithreaded parallel applications that are executed on a single SMT processor. In order to amplify this evaluation process, we also present results gathered from the performance monitoring hardware of the processor.

1 Introduction

Despite the efficiency of code optimization techniques and the continued advances in caches, memory latency still dominates the performance of many applications on modern processors. This CPU-memory gap seems difficult to be alleviated; on the one hand, CPU clock speeds continue to advance more rapidly than memory access times, on the other hand, the data working sets increase and complexity of conventional applications sets a limit on ILP.

One approach to maintain high throughput of processors despite the large relative memory latency has been Simultaneous Multithreading (SMT). SMT is a hardware technique that allows a processor to issue and execute instructions from multiple independent threads in the same cycle. The dynamic sharing of

* This research is supported by the Pythagoras II Project (EPEAEK II), co-founded by the European Social Fund (75%) and National Resources (25%).

M. Gerndt and D. Kranzlmüller (Eds.): HPCC 2006, LNCS 4208, pp. 180–189, 2006.

the functional units allows for the substantial increase of throughput, compensating for the two major impediments to processor utilization - long latencies and limited per-thread parallelism.

Thread-level parallelism (TLP) and speculative precomputation (SPR) have been proposed to utilize the multiple hardware contexts of the processors for improving performance of a single program. With TLP, sequential codes are parallelized so that the total amount of work is decomposed into independent parts which are assigned to a number of software threads for execution. In SPR, the execution of programs is facilitated with the introduction of additional threads, which speculatively prefetch data that is going to be used by the sibling computation threads in the near future, thus hiding memory latencies and reducing cache misses [13], [5], [10].

The benefit of multithreading on SMT architectures depends on the application and its level of tuning. In this paper we demonstrate that significant performance improvements are really difficult to achieve for optimized, fine-tuned parallel applications running on SMT processors. We tested two different configurations. Firstly, we balanced the computational workload of a parallel benchmark on two threads, statically partitioning the iteration space to minimize dynamic scheduling overhead. Secondly, we ran a main computation thread in parallel with a helper-prefetching thread. The latter was spawned to speculatively precompute L2 cache misses. Synchronization of the two threads is essential, in order to avoid the helper thread from running too far ahead, evicting useful data from the cache.

The rest of the paper is organized as follows. Section 2 describes related prior work. Section 3 deals with implementation aspects of software techniques to exploit hardware multithreading. Section 4 explores the performance limits and TLP-ILP tradeoffs, by considering a representative set of instruction streams. Section 5 describes the experimental framework, presents performance measurements obtained from each application, and discusses their evaluation. Finally, we conclude with section 6.

2 Related Work

SMT [12] is said to outperform previous execution models because it combines the multiple-instruction-issue features of modern superscalar architectures with the latency-hiding ability of multithreaded ones. However, the flexibility of SMT comes at a cost. When multiple threads are active, the static partitioning of resources (e.g., instruction queue, reorder buffer, store queue) affects codes with relative high instruction throughput. Static partitioning, in the case of identical thread-level instruction streams, limits performance, but mitigates significant slowdowns when non-similar streams of microinstructions are executed [11].

Cache prefetching is a technique that reduces the observed latency of memory accesses by bringing data into the cache before it is accessed by the CPU. Numerous thread-based prefetching schemes, either static or dynamic, have recently been proposed, including Collins et al., Speculative Precomputation [3], and Kim et al., Helper-Threads [5]. The key idea is to utilize otherwise idle hardware thread contexts to execute speculative threads on behalf of the main thread. These speculative threads attempt to trigger future cache-miss events far enough in advance of access by the non-speculative (main) thread, so that the memory miss latency can be masked. A common implementation pattern was used in these studies. A compiler identifies either statically or with the assistance of a profile the memory loads that are likely to cause cache misses with long latencies. Such load instructions, known as delinquent loads, may also be identified dynamically in hardware triggering speculative-helper threads [13]. SPR targets load instructions that exhibit unpredictable irregular, data-dependent or pointer chasing access patterns. Traditionally, these loads have been difficult to handle via either hardware or software prefetchers.

3 SPR and Synchronization Implementation Issues

There are two main issues that must be taken into account in order to effectively perform software prefetching using the multiple execution contexts of a hyper-threaded processor. First of all, the distance at which the precomputation thread runs ahead of the main computation thread, has to be sufficiently regulated. This requirement can be satisfied by imposing a specific upper bound on the amount of data to be prefetched. In our codes it ranges from $\frac{1}{A}$ ([10]) to $\frac{1}{2}$ of the L2 cache size, where A is the associativity of the cache (8 in our case). Whenever this upper bound is reached but the computation thread has not yet started using the prefetched data, the precomputation thread must stop its forward progress in order to prevent potential evictions of useful data from cache. It can only continue when it is signaled that the computation thread starts consuming the prefetched data. In our program codes, this scenario is implemented using synchronization barriers which enclose program regions (precomputation spans) whose memory footprint is equal to the upper bound we have imposed. In the general case, and considering their relatively lightweight workload, precomputation threads reach always first the barriers.

For codes whose access patterns were difficult to determine a-priori, we had to conduct memory profiling using the Valgrind simulator[8]. From the profiling results we were able to determine and isolate the instructions that caused the majority(92% to 96%) of L2 misses. In all cases, precomputation threads were constructed manually from the original code of the main computation threads, preserving only the memory loads that triggered the majority of L2 misses; all other instructions were eliminated.

Secondly, we must guarantee that the co-execution of the precomputation thread does not result in excessive consumption of shared resources that could be critical for the sibling computation thread. Despite the lightweight nature of the precomputation threads, significant processor resources can be consumed even when they are simply spinning on synchronization barriers.

The synchronization mechanisms have to be as lightweight as possible and for this purpose we have implemented lightweight spin-wait loops as the core of our synchronization primitives, embedding the `pause` instruction in the spin loop [4]. This instruction introduces a slight delay in the loop and de-pipelines its execution, preventing it from aggressively consuming valuable, dynamically shared, processor resources (e.g. execution units, branch predictors).

However, some other units (such as micro-ops queues, load/store queues and re-order buffers), are statically partitioned and are not released when a thread executes a `pause`. By using the privileged `halt` instruction, a logical processor can relinquish all of its statically partitioned resources, make them fully available to the other logical processor, and stop its execution going into a sleeping state. The `halt` instruction is primarily intended for use by the operating system scheduler. Multithreaded applications with threads intended to remain idle for a long period, could take advantage of this instruction to boost their execution. We implemented kernel extensions that allow from user space the execution of `halt` on a particular logical processor, and the wake-up of this processor by sending IPIs to it. By integrating these extensions in the spin-wait loops, we are able to construct long duration wait loops that do not consume significant processor resources. Excessive use of these primitives, however, in conjunction with the resultant multiple transitions into and out of the halt state of the processor, incur extra overhead in terms of processor cycles. This is a performance tradeoff that we took into consideration throughout our experiments.

4 Quantitative Analysis on the TLP and ILP Limits of the Processor

This section explores the ability and the limits of hyper-threading technology on interleaving and executing efficiently instructions from two independent threads. We constructed a series of homogeneous instruction streams, which include basic arithmetic operations (add,sub,mul,div), as well as memory operations (load, store), on integer and floating-point 32-bit scalars. For each of them, we tested different levels of instruction level parallelism.

In our experiments, we artificially increase(decrease) the ILP of the stream by keeping the source and target registers always disjoint, and at the same time expanding(shrinking) the target operands (T). We have considered three degrees of ILP for each instruction stream: minimum ($|T|$=1), medium ($|T|$=3), maximum ($|T|$=6).

4.1 Co-executing Streams of the Same Type

As a first step, we execute each instruction stream alone on a single logical processor, for all degrees of ILP (1thr columns of Table 1). In this way, all execution resources of the physical package are fully available to the thread executing that stream. As a second step, we co-execute within the same physical processor two independent instruction streams of the same ILP, each of which gets bound to a specific logical processor (2thr columns of Table 1). This gives us an indication on how various kinds of simultaneously executing streams of

Table 1. Average CPI for different TLP and ILP execution modes of some common instruction streams

	CPI					
	min ILP		*med ILP*		*max ILP*	
instr.	*1thr*	*2thr*	*1thr*	*2thr*	*1thr*	*2thr*
fadd	6.01	6.03	2.01	3.28	**1.00**	2.02
fmul	8.01	8.04	2.67	4.19	2.01	**3.99**
faddmul	7.01	7.03	2.34	3.83	1.15	**2.23**
fdiv	**45.06**	99.90	45.09	107.05	45.10	107.43
fload	1049.05	2012.62	1049.06	2012.43	1049.05	**2011.86**
fstore	1050.67	1982.99	1050.68	1983.07	1050.67	**1982.93**
iadd	1.01	**1.99**	1.01	2.02	1.00	2.02
imul	11.02	11.05	11.03	**11.05**	11.03	11.05
idiv	76.18	78.76	76.19	**78.71**	76.18	78.73
iload	2.46	4.00	2.46	3.99	2.46	**3.99**
istore	1.93	4.07	1.93	4.08	**1.93**	4.07

a specific ILP level, contend with each other for shared resources, and an estimation whether the transition from single-threaded mode of a specific ILP level to dual-threaded mode of a lower ILP level, can hinder or boost performance. For example, let's consider a scenario where, in single-threaded and maximum ILP mode, instruction A gives an average CPI of $C_{1thr-maxILP}$, while in dual-threaded and medium ILP mode the same instruction gives an average CPI of $C_{2thr-medILP} > 2 \times C_{1thr-maxILP}$. Because the second case involves half of the ILP of the first case, the above scenario prompts that we must probably not anticipate any speedup by parallelizing into multiple threads a program that uses extensively this instruction in the context of high ILP (e.g. unrolling). Bold elements of Table 1 indicate best case performance.

4.2 Co-executing Streams of Different Types

Table 2 presents the results from the co-execution of different pairs of streams (for the sake of completeness, results from the co-execution of a given stream with itself, are also presented). We examine pairs whose streams have the same ILP level. The slowdown factor represents the ratio of the CPI when two threads are running concurrently, to the CPI when the benchmark indicated in the fist column is being executed in single-threaded mode. Note that the throughput of integer streams is not affected by variations of ILP and for this reason we present only exact figures of medium ILP. Slowdown factors that vary less than 0.05 compared to the slowdown factor of the medium ILP case in a specific stream combination, are omitted. Bold elements indicate the most significant slowdown factors.

5 Experimental Framework and Results

We experimented on Intel Xeon processor enabled with HT technology, running at 2.8GHz. With the introduction of HT technology, the performance monitoring capabilities of the processor were extended, so that the performance counters

Table 2. Slowdown factors from the co-execution of various instruction streams

		Co-executed Instruction Streams				
	ILP	*fadd*	*fmul*	*fdiv*	*fload*	*fstore*
fadd	min:	1.004	1.004			
	med:	1.635	1.787	1.010	1.398	1.409
	max:	**2.016**	**2.801**	**2.023**	1.474	1.462
fmul	min:	1.002	1.004	1.006		
	med:	1.433	1.566	1.062	1.391	1.393
	max:	1.384	**1.988**			
fdiv	min:			**2.217**		
	med:	1.017	1.027	**2.374**	1.413	1.422
fload	min:	1.144	1.169			
	med:	1.286	1.255	1.153	**1.919**	**1.907**
	max:	1.684	1.358			
fstore	min:	1.134	1.133			
	med:	1.229	1.229	1.150	**1.897**	**1.887**
	max:	1.625	1.316			
	ILP	*iadd*	*imul*	*idiv*	*iload*	*istore*
iadd	med:	**2.014**	1.316	1.117	1.515	1.405
imul	med:	1.116	1.002	1.008	1.003	1.004
idiv	med:	1.042	1.019	1.033	1.003	1.003
iload	med:	**2.145**	0.941	0.934	1.621	1.331
istore	min:	4.072				
	med:	**4.299**	1.979	1.970	1.986	2.115
	max:	2.160	0.941	0.934	1.622	1.331

could be programmed to select events that are qualified by logical processor IDs, whenever that was possible. To use these performance monitoring capabilities, a simple custom library was developed. For each of the multithreaded execution modes presented in section 3 we present measurements taken for three events:

- **L2 Misses:** The number of 2nd level read misses as seen by the bus unit. For the TLP methods, including the prefetch hybrid method the L2 misses presented are the sum of the misses for both threads. For the pure software prefetch method, only the misses of the working thread are presented.
- **Resource stall cycles:** The number of clock cycles that a thread stalls in the processor allocator, waiting until store buffer entries are available. This performance metric is indicative of the contention that exists between hardware threads. For all cases, the results presented correspond to the sum of stall cycles on behalf of both logical processors.
- **μops retired:** The number of μops that were retired during the execution of the program. For all cases the μops number is the number of those retired for both threads.

We have used the NPTL library for the creation and manipulation of threads. Our operating system was Linux version 2.6.9. To force the threads to be scheduled on a particular logical processor within a physical package, we have used the `sched_setaffinity` system call. All user codes were compiled with gcc 3.3.5 compiler using the O2 optimization level, and linked against glibc 2.3.2.

We evaluated performance using two computational kernels, Matrix Multiplication and LU decomposition, and two NAS benchmarks, CG and BT. In MM and LU, we used 4096×4096 matrices, while in CG and BT we considered Class A problem sizes. In MM and LU, we applied tiling choosing tiles that completely fit in L1 cache, since this yielded the best performance. Furthermore, in MM we used blocked array layouts (non-linear layouts) with binary masks [2] and applied loop unrolling. The implementations of CG and BT were based on the OpenMP C versions of NPB suite version 2.3 provided by the Omni OpenMP Compiler Project [1]. We transformed these versions so that appropriate threading functions were used for work decomposition and synchronization, instead of OpenMP constructs. Both CG and BT are characterized by random memory access patterns, with the latter exhibiting somewhat better data locality.

The TLP versions of the codes are based on coarse-grained work partitioning schemes (**tlp-coarse**), where the total amount of work is statically balanced across the participant threads (e.g., different tiles assigned to different threads in MM and LU). The SPR versions use prefetching to tolerate cache misses, following the scheme we described in section 3. In the pure prefetching version (**spr**), the whole workload is executed by just one thread, while the second is just a helper thread that performs prefetching of the next data chunk in issue. In the hybrid prefetching version (**spr+work**), the workload is partitioned in a more fine-grained fashion with both threads performing computations on the same data chunk, while one of them takes on the prefetching of the next data chunk. This latter parallelization scheme was applicable only in MM and CG.

Figure 1 presents the experimental results for the aforementioned benchmarks. HT technology helped us to gain a speedup of $5\% - 6\%$ only in the case of NAS benchmarks, when applying the TLP scheme. In the SPR versions, although a significant reduction in L2 misses of the working thread was achieved in most cases, this was not followed by overall speedup. As Figure 1(d) depicts, in these cases, there was a noticable increase in the total number of μops, as well, due to the instructions required to implement prefetching. For LU and CG, specifically, the total μops were almost double than those of the *serial* case. Since these extra instructions could not be overlapped efficiently with those of the thread performing useful computations, as designated by the increased stall cycles in the *spr* case of all benchmarks compared to their *serial* versions, the reduction of L2 misses itself proved eventually not to be enough for performance improvement.

5.1 Further Analysis

Figure 2 presents the utilization of the busiest processor execution subunits, while running the reference applications. The first column (serial) contains results of the serial versions. The second column (tlp) presents the behavior of one of two threads for the TLP implementation (results for the other thread are identical). The third column (spr) presents statistics of the prefetching thread in the SPR versions. All percentages refer to the portion of the total instructions of each thread that used a specific subunit of the processor. The statistics

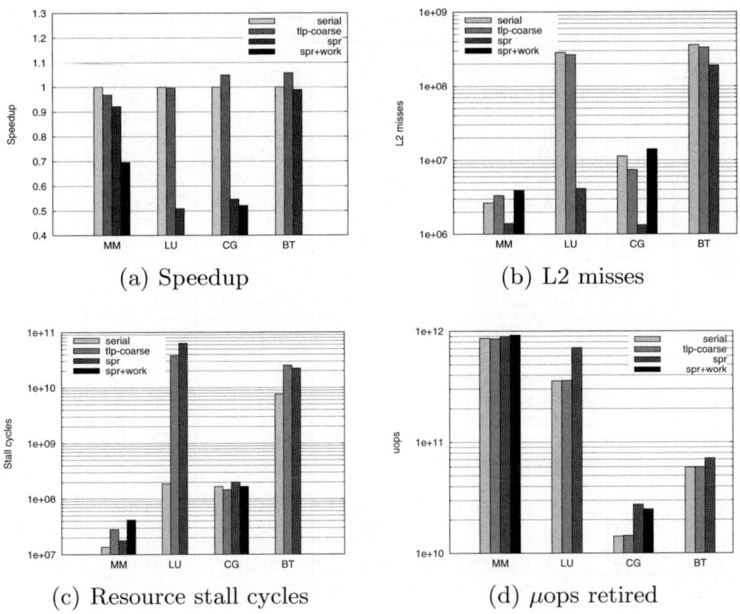

(a) Speedup (b) L2 misses

(c) Resource stall cycles (d) μops retired

Fig. 1. Experimental results

were generated by profiling the original application executables using the Pin binary instrumentation tool [6], and analyzing for each case the breakdown of the dynamic instruction mix, as recorded by the tool. Figure 3([4]) presents the main execution units of the processor, together with the issue ports that drive instructions into them. Our analysis examines the major bottlenecks that prevent multithreaded implementations from achieving some speedup.

Compared to the serial versions, TLP implementations do not generally change the mix for various instructions. Of course, this is not the case for SPR implementations. For the prefetcher thread, not only the dynamic mix, but also the total instruction count, differ from those of the worker thread. Additionally, different memory access patterns require incomparable effort for address calculations and data prefetching, and subsequently, different number of instructions.

In the MM benchmark the most specific characteristic is the large number of logical instructions used: at about 25% of total instructions in both serial and TLP versions. This is due to the implementation of blocked array layouts with binary masks that were employed for this benchmark. Although the out-of-order core of the Xeon processor possesses two ALU units (double speed), among them only ALU0 can handle logical operations. As a result, concurrent requests for this unit in the TLP case, will lead to serialization of corresponding instructions, without offering any speedup. In the SPR case of LU, the prefetcher executes at least the same number of instructions as the worker, and also puts the same pressure on ALUs. This is due to the non-optimal data locality, which leads prefetcher to execute a large number of instructions to compute the addresses

		Instrumented thread		
	EXECUTION UNIT	***serial***	***tlp***	***spr***
MM	ALU0+ALU1:	27.06%	26.26%	37.56%
	FP_ADD:	11.70%	11.82%	0.00%
	FP_MUL:	11.70%	11.82%	4.13%
	MEM_LOAD:	38.76%	27.00%	58.30%
	MEM_STORE:	12.07%	12.02%	20.75%
	Total instructions:	4590588278	2270133929	202876770
LU	ALU0+ALU1:	38.84%	38.84%	38.16%
	FP_ADD:	11.15%	11.15%	0.00%
	FP_MUL:	11.15%	11.15%	0.00%
	MEM_LOAD:	49.24%	49.24%	38.40%
	MEM_STORE:	11.24%	11.24%	22.78%
	Total instructions:	3205661399	1622610935	3264715031
CG	ALU0+ALU1:	28.04%	23.95%	49.93%
	FP_ADD:	8.83%	7.49%	0.00%
	FP_MUL:	8.86%	7.53%	0.00%
	FP_MOVE:	17.05%	14.05%	0.00%
	MEM_LOAD:	36.51%	45.71%	19.09%
	MEM_STORE:	9.50%	8.51%	9.54%
	Total instructions:	11934228188	7069734891	166842453
BT	ALU0+ALU1:	8.06%	8.06%	12.06%
	FP_ADD:	17.67%	17.67%	0.00%
	FP_MUL:	22.04%	22.04%	0.00%
	FP_MOVE:	10.51%	10.51%	0.00%
	MEM_LOAD:	42.70%	42.70%	44.70%
	MEM_STORE:	16.01%	16.01%	42.94%
	Total instructions:	44973276097	22486809710	8398026979

Fig. 2. Processor subunits utilization from the viewpoint of a specific thread

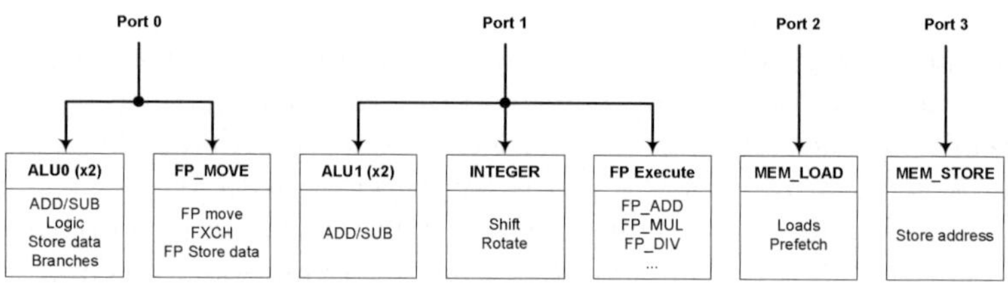

Fig. 3. Instruction issue ports and main execution units of the Xeon processor

of data to be brought in cache. These facts translate into major slowdowns for the SPR version of LU, despite any significant L2 misses reduction.

As can be seen in Figure 1, TLP mode of BT benchmark was one of few cases that gave us some speedup. The relatively low usage and thus contention on ALUs, in conjunction with non-harmful co-existence of *faddmul* streams (as Table 2 depicts) which dominate other instructions, and the perfect workload partitioning, are among the main reasons for this speedup.

6 Conclusions

This paper presents performance results for a SMT architecture, the Intel hyper-threaded microarchitecture. We examined scientific codes in which both TLP and SPR schemes were applied. Our evaluation was based on actual program execution, as well as simulation. The results gathered demonstrated the limits in achieving high performance for such applications.

SPR can achieve a fairly good reduction in L2 cache misses. However, in order to fine tune data prefetching, a considerable number of additional instructions have to be inserted into the pipeline. This increase in the number of μops, in combination with some kind of resource contention, harms performance in terms of execution time. Besides, optimized applications with a relatively high IPC (such as the tested microkernels), are really difficult to achieve even better performance without reducing the μops executed.

Coarse-grained work partitioning schemes do not have a significant impact on the number of μops executed (usually brings a slight increase). Total execution performance would be expected to be improved, especially in cases of L2 cache miss decrease. However, the two working threads, due to their symmetric profiles, compete for the same hardware resources. This contention constitutes in some cases a bottleneck to high performance.

References

1. Omni OpenMP Compiler Project. Released in the International Conference for High Performance Computing, Networking and Storage (SC'03), Nov 2003.
2. E. Athanasaki and N. Koziris. Fast Indexing for Blocked Array Layouts to Improve Multi-Level Cache Locality. In *Proc. of INTERACT'04*, Madrid, Spain.
3. J. Collins, H. Wang, D. Tullsen, C. Hughes, Y. Lee, D. Lavery, and J. Shen. Speculative Precomputation: Long-Range Prefetching of Delinquent Loads. In *Proc. of ISCA '01*, Göteborg, Sweden.
4. Intel Corporation. *IA-32 Intel Architecture Optimization.* Order Num: 248966-011.
5. D. Kim, S. Liao, P. Wang, J. Cuvillo, X. Tian, H. Wang, D. Yeung, M. Girkar, and J. Shen. Physical experimentation with prefetching helper threads on Intel's hyper-threaded processors. In *Proc. of IEEE/ACM CGO 2004*, San Jose, CA.
6. C. Luk, R. Cohn, R. Muth, H. Patil, A. Klauser, G. Lowney, S. Wallace, V. Reddi, and K. Hazelwood. Pin: Building Customized Program Analysis Tools with Dynamic Instrumentation. *SIGPLAN Not.*, 40(6):190–200, 2005.
7. D. Marr, F. Desktop, D. Hill, G. Hinton, D. Koufaty, J. Miller, and M. Upton. Hyper-Threading Technology Architecture and Microarchitecture. *ITJ*, Feb 2002.
8. N. Nethercote and J. Seward. Valgrind: A Program Supervision Framework. In *Proc. of RV'03*, Boulder, CO.
9. D. Patterson and J. Hennessy. *Computer Architecture. A Quantitative Approach*, pages 597–598. Morgan Kaufmann, 3rd edition, 2003.
10. F. Blagojevic T. Wang and D. Nikolopoulos. Runtime Support for Integrating Precomputation and Thread-Level Parallelism on Simultaneous Multithreaded Processors. In *Proc. of LCR'2004*, Houston, TX.
11. N. Tuck and D. Tullsen. Initial Observations of the Simultaneous Multithreading Pentium 4 Processor. In *Proc. of PACT '03*, New Orleans, LA.
12. D. Tullsen, S. Eggers, and H. Levy. Simultaneous Multithreading: Maximizing On-Chip Parallelism. In *Proc. of ISCA '95*, Santa Margherita Ligure, Italy.
13. H. Wang, P. Wang, R. Weldon, S. Ettinger, H. Saito, M. Girkar, S. Liao, and J. Shen. Speculative Precomputation: Exploring the Use of Multithreading for Latency. *ITJ*, Feb 2002.

A Statistical Approach to Traffic Management in Source Routed Loss-Less Networks*

Thomas Sødring[1], Raúl Martínez[2], and Geir Horn[1]

[1] Simula Research Laboratory
Martin Linges vei 17
1325 - Lysaker, Norway
{tsodring, geirho}@simula.no

[2] Departamento de Sistemas Informáticos
Universidad de Castilla-La Mancha
02071 - Albacete, Spain
raulmm@info-ab.uclm.es

Abstract. The evolution of high-performance networks has resulted in the development of new applications with Quality of Service (QoS) requirements. In this paper we review and evaluate the Simple Host (SH) traffic management mechanism that enables QoS provisioning within source-routed loss-less interconnect networks. SH provides statistical QoS guarantees for *brokered* multimedia traffic while supporting limited amounts of *unbrokered* background traffic. The results obtained from simulation show how to configure SH to provide statistical QoS based on jitter and latency requirements in an efficient manner.

Keywords: Networking protocol, Performance evaluation, Interconnection Networks, Traffic Management, Statistical Quality of Service, Advanced Switching.

1 Introduction

The evolution of high performance networks over recent years has been underpinned by the on-going development of interconnect networks such as InfiniBand [1], Advanced Switching (AS) [2] and Gigabit Ethernet [3]. The development of these interconnect platforms was necessitated by a need to support transactions of high volumes of data. This evolution has also seen the development of new applications providing multimedia entertainment to end-users that are served from high-performance networks that support large traffic volumes.

In the case of a multimedia content network serving a variety of video streams, there are strict quality of service (QoS) requirements that must be met. A video stream, for example, will have both latency and jitter requirements as a users' perception of the stream they are watching is bounded by its timely delivery. These requirements must

* This work was financed in part by the EU under the 6th framework program for IST as part of the SIVSS project. This work was partly supported by the Spanish CICYT under Grant TIC2003-08154-C06-02, by the Junta de Comunidades de Castilla-La Mancha under Grant PBC-05-005-1, and by the Spanish State Secretariat of Education and Universities under FPU grant.

M. Gerndt and D. Kranzlmüller (Eds.): HPCC 2006, LNCS 4208, pp. 190–199, 2006.

be met by the underlying network and necessitate the use of traffic management mechanisms to ensure that the traffic it is serving is not subjected to any unnecessary delays. To achieve this a network typically employs the use of a QoS provisioning mechanism that ensures the traffic receives certain levels of service.

In the Internet domain the IETF[1] have specified two mechanisms to deal with QoS: IntServ and DiffServ. The IntServ [4] protocol uses the Resource reSerVation Protocol (RSVP) [5] to reserve a set of resources along a path from source to destination and supports the use of per-flow admission control, classification and scheduling. IntServ has problems with scalability, which are related to the requirement of storing flow-state information in switches. Differentiated Services [6] (DiffServ) on the other hand moves the protocol complexity to the edge of the network and switches in the core merely implement a suite of buffering and scheduling techniques that are applied to a limited number of traffic classes. The switches at the edge of the network are responsible for classifying and policing the flows according to rules applicable to each traffic class.

The motivation behind this work is the development of a simple, efficient and effective traffic management scheme that provides statistical QoS guarantees in source-routed loss-less high performance networks. In particular, we focus on whether or not a network can meet the statistical delivery guarantees required by *brokered multimedia traffic* while supporting limited amounts of *un-brokered best-effort* traffic. In [7], we introduced a traffic management mechanism called Simple Host (SH) that works with source routed loss-less networks and detailed its applicability to Advanced Switching. In this article we refine and present results based on simulation relating to the idea presented in [7].

The remainder of this article is organised as follows. First we motivate the idea behind this work and review the SH protocol. Then the methodology we used to run a series of experiments is explained followed by an analysis of the results. We conclude with a discussion on the importance of our findings and a look at how this work could be taken further.

2 Simple Host

QoS is a topic that has been the focus of lot of attention and discussion over recent years. From a networking perspective the general QoS provisioning parameters are throughput, latency, jitter, and loss rate. Throughput is the effective number of data units transported per time unit, while latency (delay) is the time interval between the departure of a packet from the source to its arrival at the destination. Jitter represents the variance in latency and can be calculated as the difference between the latencies of consecutive packets belonging to a given flow. Figure 1 illustrates a typical probability density function for latency and shows that jitter is bound between the minimum and maximum latencies [8]. Finally, loss-rate is the percentage of packets that were not delivered to their destination and is usually represented as a“probability” of loss.

The QoS parameters an application may wish to specify can be expressed as the minimum throughput and maximum latency, jitter, and loss rate. For those applications with latency or jitter requirements, if a packet is unable to meet its deadline, its value

[1] The Internet Engineering Task Force http://www.ietf.org/

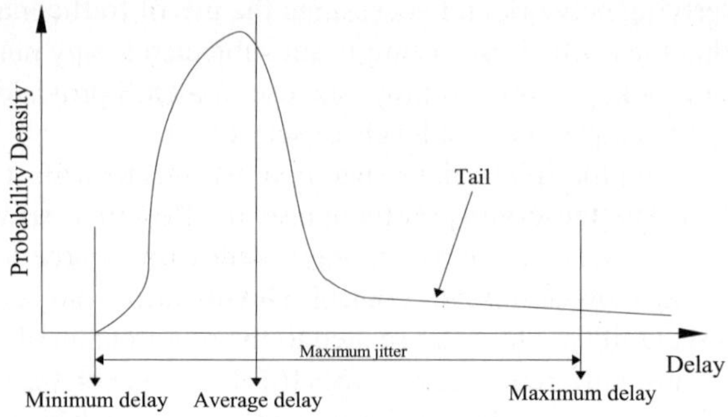

Fig. 1. Packet latency probability density

to the application may be greatly diminished or even worthless. In some cases, packets that miss their QoS deadlines can be considered lost.

The degree of tolerance or sensitivity to each of these parameters varies widely from one application to another. For example, multimedia applications are usually sensitive to latency and jitter, but many of them can tolerate packet losses to some extent. However, the severity of the effect of loss on video quality is also influenced by parameters such as the compression and encoding techniques used, loss pattern, transmission packet size, and the error recovery technique implemented [9]. For a further discussion about different applications and their requirements see [10].

In loss-less networks congested packets are not thrown away and as such the loss rate due to congestion is zero. As the packets are stored in network buffers the onset of congestion increases the latency experienced by packets traversing the point of congestion, which can have a knock-on effect on jitter as jitter is influenced by fluctuations in latency. In addition, this congestion can affect other flows that share common upstream links. However, if the congestion is not persistent, the congestion situation should dissipate after a short period of time and packets will reach their destinations. Depending on the latency introduced by the congestion a packet may or may not meet their QoS requirements. The goal of the SH protocol is to avoid persistent congestion situations and limit the transient congestion such that the brokered traffic meet their QoS requirements.

A common approach to achieve this is by using an admission control (AC) mechanism. The AC decides whether a new connection is accepted or rejected and ensures that the entry of additional traffic into a network does not disturb the QoS requirements of the existing traffic. Many AC schemes have been proposed. In [11] an implementation of both a probe and statistical approach to the problem is detailed. Probe-based algorithms are limited by a traffic awareness that is restricted to the traversal route while fluctuating traffic patterns, especially within a busy network, provide limited temporal information describing the network load. The collection and calculation of statistical data can be both costly to gather and process. Our work is related to [12] who proposed a framework that uses a distributed admission control scheme in order to provide QoS support within an IP network. Their protocol allows edge routers admitting traffic to provide QoS guarantees for the entire network with bandwidth-status updates circulating in a token. The main differences between our work and [12] is that their work is presented for the Internet domain and they report results where absolute guarantees on bandwidth requirements are provided. Moreover, no results on latency or jitter are presented.

Multimedia servers injecting traffic into a network will see that most of the bandwidth is consumed by packets belonging to persistent multimedia flows while a small amount of bandwidth is consumed by intermittent short-lived transmissions. Considering the short-lived transmissions as background traffic the network can save resources by only provisioning the multimedia traffic. The multimedia traffic has certain bandwidth, latency and jitter constraints that must be met so QoS is provisioned solely for this traffic.

SH is a traffic management mechanism that provides statistical QoS guarantees and is applicable when the traffic is clearly distinguishable as belonging to the category of either persistent multimedia flows or intermittent short-lived transmissions. It aims to minimize the complexity of providing QoS within a network by working at the edge of the network, allowing for a simple switch design within the network.

SH uses a bandwidth broker that has *a priori* knowledge of the bandwidth required by the flows. Central to the use of bandwidth broker is a network state graph that details the remaining bandwidth on each link in the network. When an endnode wishes to start injecting a flow, the routing algorithm is consulted to determine a set of routes that the flow can use while traversing the network. This information is sent to the traffic manager. Analysing each path to determine if there is available bandwidth, the traffic manager either grants or denies the endnode request to use one of the specified routes. If the bandwidth requirements are met the network state graph is updated consuming the bandwidth along the chosen path. Note that while there is no explicit requirement for multiple routes the use of multiple routes can lead to increased performance.

The traffic manager can be centralized in a given end node or distributed among the different end nodes. In [7] we proposed a distributed AC mechanism that could be applied to an Advanced Switching fabric. This work treated the entire network as a logical shared medium and exploited the use of multicast tables within Advanced Switching to distribute bandwidth updates.

As stated earlier, the SH framework supports both brokered and unbrokered traffic within a network at the same time. Using the network state graph, SH has a view of the bandwidth status of each link in the network. We define a parameter, β, that represents the proportion of bandwidth that each link reserves for multimedia traffic. The bandwidth broker makes sure that no link will be subjected to more than β multimedia traffic. However, the SH framework supports both brokered multimedia and unbrokered background traffic within a network at the same time. Therefore, we also consider a parameter, α, that represents the maximum amount of unbrokered traffic that is expected to be inserted into the network. Background traffic is injected on an intermittent basis and interleaved with the multimedia traffic, and as such interferes with the multimedia flows ability to meet its QoS requirements.

The focus of this work is to analyse the assignment of optimal values for the multimedia β parameter under certain loads of α background traffic. We achieve this observing how varying β for given loads of α effects the networks ability to meet the QoS requirements of the traffic it serves.

With this information, SH is able to provide statistical QoS guarantees limiting the fraction of traffic that exceeds specific end-to-end latency or jitter constraints below prescribed bounds.

3 Performance Evaluation

The experiments that are reported here are undertaken in order to determine what values of β (multimedia) traffic meet their QoS requirements under various loads of α (background traffic). We achieve this using three different values for α and varying β, and thus the amount of multimedia traffic. For each combination of α and β, we analyze the effect that the interaction of both types of traffic have on performance of the multimedia traffic, which is the traffic we provision QoS for. Specifically, we derive statistics regarding the probability that a packet belonging to a brokered flow will meet its latency and jitter requirements. These probabilities are dependent on the amount of α and β injected.

We developed a detailed simulator, capable of replicating the functionality of an AS fabric, however, the mechanism is applicable to any source-routed loss-less interconnection network technology. AS is a modern high performance networking architecture that evolved from the PCI Express protocol. It replaces the top layer of the PCI Express protocol stack in order to provide peer-to-peer communication through an AS fabric. PCI Express is itself an evolution of the PCI input/output bus of present day computers. Together, PCI Express and AS have the potential to become the next generation high performance interconnect [13].

3.1 Simulated Architecture

The topology used in all experiments is a fat-tree Bidirectional Multi-stage Interconnection Network (BMIN) with 16 end-points connected using 32 4-port switches (4 stages of 8 switches). AS supports the use of any topology, but we chose a BMIN as it is a commonly used interconnect topology in high-performance environments.

The switch model uses a combined input-output buffer architecture with a crossbar to connect the buffers. Virtual output queuing was implemented to minimize the head-of-line blocking problem at switch level. In our tests, the link bandwidth is 2.5 Gb/s but as links use the 8b/10b encoding scheme, the maximum effective bandwidth for data traffic is limited to 2 Gb/s. We assume an internal speed-up (x2) for the crossbar, as is the usual case for most commercial switches. As AS provides the freedom to use any crossbar scheduling algorithm, we chose to implement a first-come, first-serve (FCFS) scheduler.

A credit-based flow control protocol ensures that packets are only transmitted when there is enough buffer space at the other end to store them, making sure that no packets are dropped when congestion is present. The buffer capacity at the input and at the output ports of the switches is 16,384 bytes and in order to collect accurate latency statistics, we use infinite queues in the end nodes.

3.2 Traffic Model

As described earlier, α represents the maximum expected percentage of background load in the network. To simulate this each end node injects $2 \times \alpha/100$ Gb/s of traffic (this refers to the 2Gb/s link effective bandwidth) distributed uniformly among the other end nodes. The β parameter determines the maximum amount of multimedia traffic that the bandwidth broker allows overs a link. We use the β parameter as a basis to generate as many random flows as possible.

Both traffic types are simulated differently but have some common characteristics. Both are constant bit rate (CBR) and have a packet size equal to 2,048 bytes. The brokered traffic is comprised of 2 Mb/s point-to-point flows where all packets belonging to a particular brokered flow follow the same route through the network. The route of each background packet is chosen randomly among the possible routes available from the routing function. The injection time of the first packet belonging to a flow as well as when a node starts generating background traffic is calculated randomly.

3.3 Simulation Results

Simulations were run with the following values for α and β: $\alpha = 1, 7, 14$ and $\beta = 70, 75, 80, 85, 90, 95$. For each combination we ran 16 simulations. For each simulation we calculated and recorded the global average injection, the global average throughput, the average packet latency, and the average packet jitter. Figures 2 and 3 show the average values of these statistics for each of the 16 simulations. We also collected the latency and jitter of each packet belonging to a flow throughout the 16 simulations. Using this information we generated Figures 4, 5, 6, and 7. Note that the statistics are only shown for the multimedia traffic, which is the traffic with QoS requirements.

Figure 2 also shows the normalized aggregated average injection of the multimedia flows. This value is directly proportional to the β value as can be seen from the figure, but they are not the same. The reason for this is that the paths that endnodes use to communicate with each other are over shared links. In an ideal case, where links are not shared the theoretical maximum amount of injected multimedia traffic that the bandwidth broker can achieve is β. The figure shows this theoretical maximum and the actual amount of multimedia traffic that is able to be injected.

Figure 2 shows the normalized global throughput of the multimedia flows. The multimedia traffic manages to utilise all the bandwidth it injects except for the case where $\alpha = 14$ and $\beta = 90$ and 95.

Figure 3 shows the average values for both latency and jitter. We can see that both values grow gradually with the α and β parameters (the load of the network). This is the effect of the increasing congestion when the load increases. However, when the load is very high (the network is becoming saturated) the latency and the jitter grow exponentially. This happens when congestion is prevalent and congestion trees take form and start to grow, affecting not only the flows that traverse the point of congestion but also other upstream flows sharing common links. The effect of this congestion is evident from the latency and jitter values, but even more in the throughput results shown before, where some combinations of α and β were unable to utilise all the bandwidth that they inject.

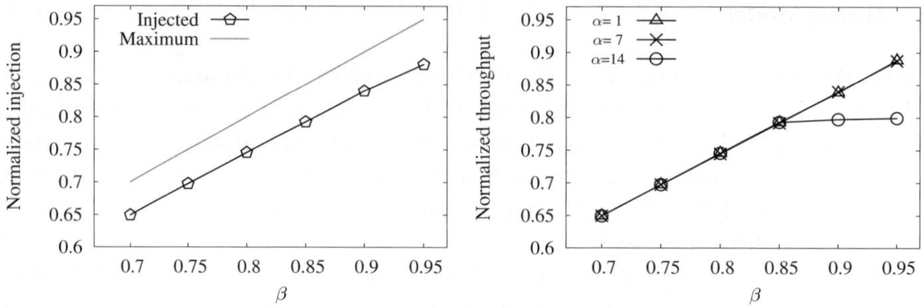

Fig. 2. Average injection and throughput

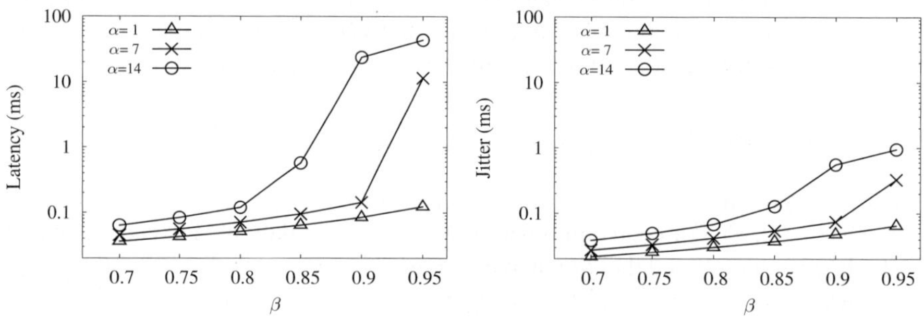

Fig. 3. Performance of average latency and jitter

We have shown the general performance of the network under different loads and presented results that show statistics relating to average values. However, there is little to gain from provisioning QoS based on average values. Our objective is to study the probability that the packets can meet certain latency and jitter deadlines under various combinatorial loads of multimedia and background traffic.

We generated a distribution histogram of the latency and jitter for each combination of α and β. To achieve this we calculated the latency and jitter values of each packet belonging to a multimedia flow across all simulations. We then divided the entire range of possible latencies into several intervals and calculated how many packets from the sequence fit within each interval. As a result we obtained distribution histograms similar to the ones shown in Figures 4(a) and 5(a) for $\alpha = 7$ and $\beta = 85$. The latency histogram provides a good representation of the networks performance. Using it, we can evaluate which latencies are probable and which are unlikely to be met. However, the information we wish to acquire is the fraction of traffic that exceeds a specific latency or jitter constraint, or looking at the same information from a different angle, what is the probability that a given packet will meet a certain latency or jitter requirement.

To compute these values, we acquired accumulated distribution histograms similar to those shown in Figures 4(b) and 5(b) from the distribution histograms. These figures show the proportion of packets that either meet or are below a certain latency (or jitter) constraint. From the accumulated histograms, we derived probability functions, again similar to the ones shown in Figure 4(c) and 5(c). Finally, from these functions we

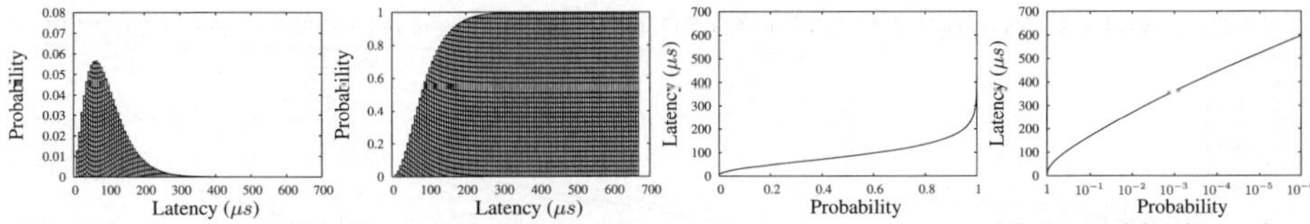

(a) Distribution histogram (b) Accumulated Distribution histogram (c) Probability (d) Logarithmic probability

Fig. 4. Latency performance for α=7 and β=85

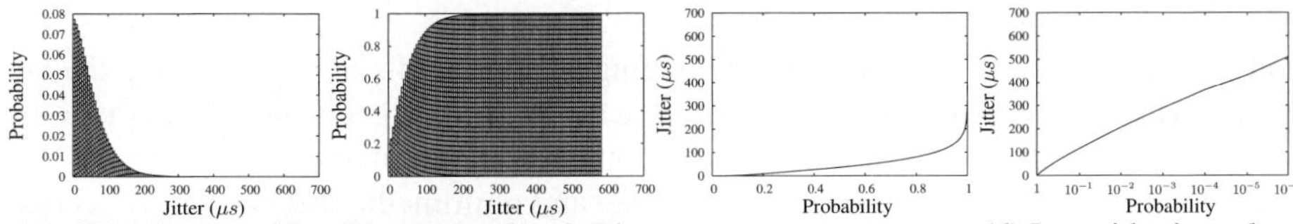

(a) Distribution histogram (b) Accumulated Distribution histogram (c) Probability (d) Logarithmic probability

Fig. 5. Jitter performance for α=7 and β=85

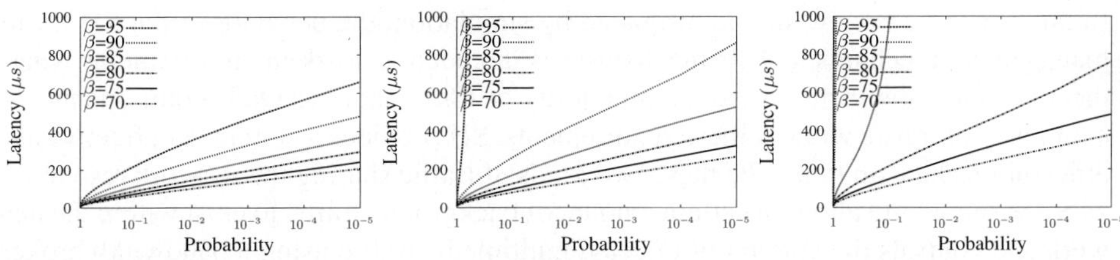

(a) α=1 (b) α=7 (c) α=14

Fig. 6. Probability that a packet experiences a certain (or higher) latency

calculated probability functions for meeting a given latency (or jitter). These are shown in Figures 4(d) and 5(d) with appropriate granularity.

Figures 6 and 7 show the probability that a packet will experience a certain latency or jitter while traversing the network. Given a precondition that a network is to have a background traffic load equal to or below a certain value the figures can be used to reason about an optimum β value that enables packets to meet certain QoS requirements.

As an example, consider the case where the QoS requirements of the multimedia traffic has both a latency and jitter deadline equal to 400 μs and the maximum fraction of traffic that is allowed to not meet the deadlines is 0.0001 (10^{-4}). In addition, the maximum expected background traffic is 7%. From Figure 7(b) we can see that we meet these jitter requirements with $\beta = 70, 75, 80$ and 85. Further we observe in Figure 6(b) that only $\beta = 70, 75$ and 80 meet the latency requirements. Therefore, to meet the above QoS requirements, the optimal β value is equal 80.

Summing up, our results show the probability of meeting the latency and jitter requirements under different traffic loads where the traffic load is characterized by the α and β parameters. We achieve this by generating histograms that detail the latency

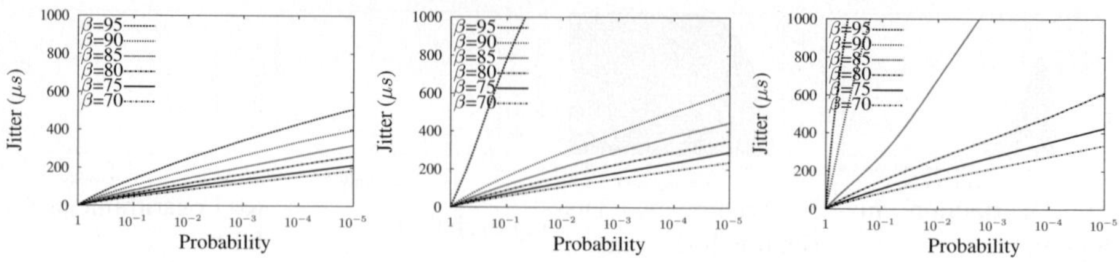

(a) α=1 (b) α=7 (c) α=14

Fig. 7. Probability that a packet experiences a certain (or higher) jitter

and jitter experienced by all packets belonging to multimedia flows traversing the network during simulation. The results provide a network manager, using SH as the traffic management mechanism, with the information necessary to make a decision regarding the configuration of the amount of background and multimedia that can safely co-exist within the network.

4 Conclusion

In this paper we reviewed, and evaluated by simulation, the behavior of the SH traffic management mechanism. This mechanism is intended to work in environments where there is a clear distinction between a majority of traffic that has QoS requirements and a minority of traffic without these requirements. SH provides a simple yet effective and efficient mechanism to handle these two kinds of traffic sharing the network resources.

SH assumes and allows a certain amount of background traffic to exist within the network and controls the amount of injected multimedia traffic using a bandwidth broker. This admission control mechanism limits the maximum amount of multimedia traffic that is allowed over each link in the network. We have denoted β as the configuration parameter that determines the amount of multimedia traffic.

The objective behind SH is to provide the multimedia traffic with statistical QoS guarantees. We considered these requirements as deadlines for the latency and jitter and the maximum fraction of traffic that is unable to meet them. If this scheme is to have the ability to allow unbrokered traffic into the network without disturbing the QoS requirements of the existing brokered multimedia flows, it must first make a decision regarding how much multimedia traffic to allow into the network.

We studied the performance of the network using three different loads for background traffic (α) and several possibilities for the multimedia load (β). The results obtained from simulations presented in this article show the effect that different loads of multimedia and background traffic have on latency and jitter and how this maps to the multimedia flows ability to meet its QoS requirements. Specifically, we derived statistics about the fraction of traffic that is unable to meet a certain latency or jitter constraint. With this information a network manager can choose an optimum value for the maximum multimedia traffic per link. This value should maximize the utilization of the network without breaking the specific QoS constraints of the multimedia traffic.

Future work includes undertaking simulations with variable bit rate traffic for multimedia flows and self similar traffic for background traffic which will create spatial and temporal hot spots throughout the network. Moreover, we also intend to study different possibilities for constraining the injection of background traffic without brokering it.

References

1. InfiniBand Trade Association: Infiniband architecture specification. 1.2 edn. (2004)
2. Advanced Switching Interconnect Special Interest Group: Advanced Switching Core Architecture Specification. revision 1.1 edn. (2004)
3. Sheifert, R.: Gigabit Ethernet. Addison-Wesley (1998)
4. Braden, R., Clark, D., Shenker, S.: RFC 1633 Integrated Services. IETF. (1994)
5. Braden, R., et al.: Resource ReSerVation Protocol (RSVP) – Version 1 Functional Specification. IETF. rfc2205 edn. (1997)
6. Blake, S., Black, D., Carlson, M., Davies, E., Wang, Z., Weiss, W.: RFC 2475 Differentiated Services. IETF. (1998)
7. Horn, G., Sødring, T.: SH: A simple distributed bandwidth broker for source-routed lossless networks. In Hamza, M., ed.: Proceedings of the IASTED International Conference on Computer, Networks and Information Security (CNIS), November 14-16, Phoenix, AZ, USA, ACTA Press, ISBN 0-88986-537-X (2005) 133–139
8. Wang, Z.: Internet QoS: Architecture and Mechanisms for Quality of Service. Morgan Kaufmann (2001)
9. Wang, Y., Zhu, Q.: Error control and concealment for video communication: A review. Proceedings of the IEEE **86**(5) (1998) 974–997
10. El-Gendy, M.A., Bose, A., Shin, K.G.: Evolution of the internet QoS and support for soft real-time applications. Proceedings of the IEEE **91**(7) (2003) 1086–1104
11. Reinemo, S.A., Sem-Jacobsen, F.O., Skeie, T., Lysne, O.: Admission control for DiffServ based Quality of Service in cut-through networks. In: Proceedings of the 10th International Conference on High Performance Computing (HiPC). (2003)
12. Bhatnagar, S., Vickers, B.: Providing quality of service guarantees using only edge routers. In: Proceedings of IEEE Globecom, IEEE (2001)
13. Mayhew, D., Krishnan, V.: PCI Express and Advanced Switching: Evolutionary path to building next generation interconnects. In: 11th Symposium on High Performance Interconnects, IEEE (2003)

Model-Based Relative Performance Diagnosis of Wavefront Parallel Computations

Li Li, Allen D. Malony, and Kevin Huck

Performance Research Laboratory
Department of Computer and Information Science
University of Oregon, Eugene, OR, USA
{lili, malony, khuck}@cs.uoregon.edu

Abstract. Parallel performance diagnosis can be improved with the use of performance knowledge about parallel computation models. The *Hercule* diagnosis system applies model-based methods to automate performance diagnosis processes and explain performance problems from high-level computation semantics. However, Hercule is limited by a single experiment view. Here we introduce the concept of *relative performance diagnosis* and show how it can be integrated in a model-based diagnosis framework. The paper demonstrates the effectiveness of Hercule's approach to relative diagnosis of the well-known Sweep3D application based on a Wavefront model. Relative diagnoses of Sweep3D performance anomalies in strong and weak scaling cases are given.

Keywords: Performance diagnosis, parallel models, wavefront, relative analysis.

1 Introduction

In recent years there has been growing interest in automating the process of parallel performance analysis, including the generation and running of experiments, the comparative analysis of performance results, the characterization of performance properties, and the diagnosis of performance problems. Performance diagnosis is a particularly challenging process to automate because it fundamentally is an intelligent system wherein we capture and apply *knowledge* about performance problems, how to detect them (i.e., their *symptoms*), and why they exist (i.e., their *causes*). In our work, we focus on performance knowledge engineering as the basis for building a framework to support automated performance diagnosis. The framework's function would be guided by expert strategies for problem discovery and for hypothesis testing, strategies that are captured and encoded in the knowledge base. We advocate looking to models of parallel computations as sources of performance knowledge, which present structural and communication patterns of a program. Models provide semantically rich descriptions that enable better interpretation and understanding of performance behavior. Here, performance knowledge can be engineered based on the model behavior descriptions so that bottom-up inference of performance

M. Gerndt and D. Kranzlmüller (Eds.): HPCC 2006, LNCS 4208, pp. 200–209, 2006.

causes is effectively supported. A diagnosis system then uses the knowledge for performance bug search and reasoning.

We developed the *Hercule* performance diagnosis system [6] to validate how performance knowledge derived from parallel models provides a sound basis for automating performance diagnosis processes and can explain performance loss from high-level computation semantics. This has been shown for several parallel models to date (e.g., master-worker, divide-and-conquer, and domain decomposition). However, we also realized that diagnosis of a single execution is incomplete as a comprehensive diagnosis process. Understanding of performance problems routinely involves *comparative* and *relative* interpretation. This paper reports our work to improve the Hercule methodology to support what we will term *relative performance diagnosis* (in the spirit of relative debugging [4]). The multi-experiment relative performance analysis is also addressed in [5] [7] etc.

In Section §2, we start with a description of the Hercule framework for diagnosis of the Wavefront model, and then discuss in Section §3 how Hercule is extended to allow relative execution diagnostic analysis. We show in Section §4 how Hercule explains certain performance anomalies in strong- and weak-scaling studies. Section §5 concludes with observations.

2 Hercule Automatic Performance Diagnosis Framework

Hercule is a prototype automatic performance diagnosis system that implements the model-based performance diagnosis approach. Hercule framework is displayed in figure 1. The Hercule system operates as an expert system within a parallel performance measurement and analysis toolkit, in this case, the TAU [3] performance system. Hercule includes a knowledge base that is composed of an abstract event library, performance metrics set, and performance factors for individual parallel models. The knowledge engineering approach that generates the knowledge base is discussed in [6]. Below, we use the Wavefront model as a driver example to describe Hercule's working mechanism.

Wavefront is a two-dimensional variant of a traditional pipeline pattern. Computation or data is partitioned and distributed on a two-dimensional process grid where every processor receives data from preceding processors and passes down data to successive processors in two orthogonal directions. Well-known pipeline performance problems include sensitivity to load imbalance, processor idleness when pipeline filling up and emptying, and so on. It is these types of problems that we want to find.

We use *abstract events* to describe behavioral and performance characteristics of the Wavefront model. The abstract event describing a Wavefront process node is shown in figure 2. An abstract event description includes constituent primitive events and their format and ordering, a related abstract event list, and associated performance attributes. Hercule implements the abstract event representation in a Java class library which provides a general programmatic means to capture model behaviors and allows for algorithm and implementation extension.

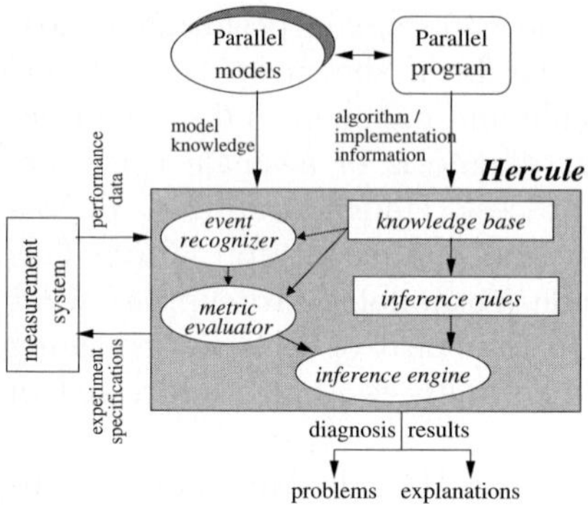

Fig. 1. Hercule diagnosis framework

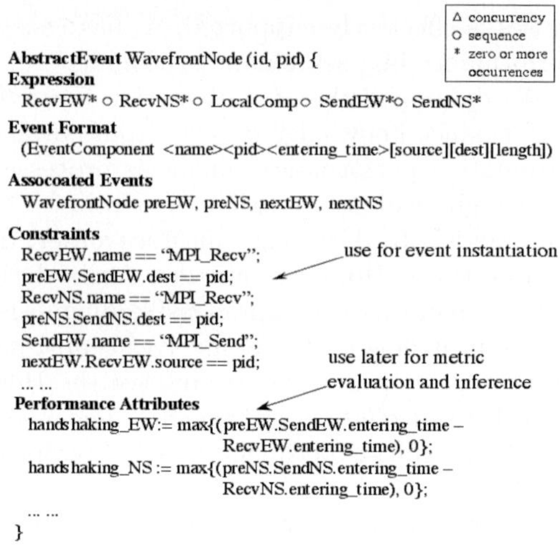

Fig. 2. Abstract event description of Wavefront

The *event recognizer* in Hercule fits event instances into abstract event descriptions as performance data stream flows through it. It then feeds the event instances into Hercule's performance *metric evaluator*, where performance attributes associated with the event instances are calculated. The metric evaluator also takes in model-specific metric evaluation rules from the knowledge database so it is able to produce metrics by synthesizing performance attributes of the related event instances. Thus, the performance metrics reflect model semantics. A metric named *pl_handshaking* in Wavefront model, for instance, refers to the performance penalty of waiting preceding processes in the Wavefront to pass down data. The model-specific metrics help in mapping low level performance information to a higher level of program abstraction.

Perhaps the most interesting part of Hercule is its cause inferencing system. The expert knowledge used to reason about performance symptoms can be structured as *inference trees* where the root is the symptom to be diagnosed, the branch nodes are intermediate observations obtained so far, and the leaf nodes are high-level performance factors that contribute to the root symptom. An inference tree for diagnosing symptom "low speedup" in Wavefront is presented in figure 3. An intermediate observation is obtained by evaluating a model-specific performance metric against the expected value (from performance modeling) or certain pre-set severity-tolerant threshold. In figure 3, for example, *pl_comm* means the communication cost associated with pipeline message passing. If it turns out to be significant comparing to the expected, the inference engine will continue to search for the node's child branches. The leaf nodes finally reached together compose an explanation of the root symptom.

We encode inference trees with production rules. Hercule makes use of syntax defined in the CLIPS [2] expert system building tool to describe production rules, and the CLIPS inference engine for operation. The inference engine pro-

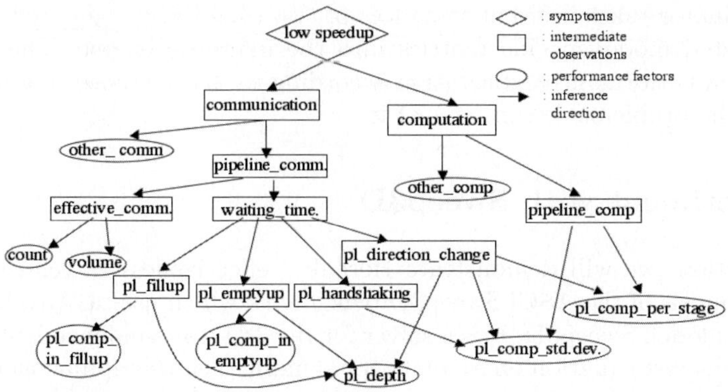

Fig. 3. An inference tree of Wavefront model that diagnoses low-speedup

vided in CLIPS is particularly helpful in performance diagnosis because it can repeatedly fire rules with original and derived performance information until no more new facts can be produced, thereby realizing automatic performance experiment generation and causal reasoning.

3 Hercule Extensions for Relative Performance Diagnosis

Understanding of performance problems routinely involves *comparative* and *relative* interpretation. Performance analysts often need to answer such questions in scalability analysis of a parallel application: what are most pronounced performance differences between two program executions with difference problem scales, which program design factors contribute to the differences, and what are magnitudes of their contributions?

Hercule's single execution diagnosis can be extended to support what we term *relative performance diagnosis* that is intended to answer the questions. To interpret what was happening at the performance anomalies with certain problem scale, we pick a performance reference run, in the family of scalability executions, which has comparatively normal performance and evaluate problematic runs against it. Relative performance diagnosis follows the same inference processes as presented in model-specific inference trees except for performance evaluation at branch nodes. Recall that cause inference in the inference trees is driven by performance evaluation, that is, to compare the model-specific metric with an expected value (from performance modeling) to decide on an intermediate observation. In relative performance diagnosis, we calculate the expected value based on model-specific metrics of the reference run to evaluate problem behaviors. Examples of relative diagnosis of anomalous Wavefront application executions will be presented in the next section.

Hercule extensions for supporting relative performance diagnosis manifest in the interfacing of the metric evaluator and the inference engine. To assert the performance observation associated with a branch node in the inference tree, the metric evaluator takes in event instances of two runs to be compared and feeds the calculated model-specific metrics into the inference engine. The inference engine sets a performance expectation according to the reference run metric and evaluates the problematic run against it.

4 Experiment with Sweep3D

In this section, we will demonstrate Hercule's effectiveness in relative performance diagnosis of the ASCI Sweep3D benchmark which uses a Wavefront computational model. Sweep3D [1] is a solver for the 3-D, time-independent, neutron particle transport equation on an orthogonal mesh. The three-dimensional space is partitioned on a two-dimensional processor grid, where each processor is assigned one columnar domain. Sweep3D exchanges messages between adjacent processors in the grid as wavefront propagates diagonally across the 3-D space in eight directions. Sweep3D is a well-researched parallel benchmark. Although parallelism overheads in Sweep3D have been minimized, for instance, by evenly distributing data across a process grid, leaving little room for performance tuning, Hercule can tell exactly how running time is spent in terms of model semantics, helping understand inherent performance losses of the model under an optimistic condition. Our performance study with Sweep3D focuses on overall scalability, looking at how well the application scales as the number of processors is increased (strong scaling) and as total problem size increases with the process count increase (weak scaling).

We ran tests on MCR, a linux cluster located at Lawrence Livermore national Laboratory. MCR has 1,152 nodes, each with two 2.4-GHz Pentium 4 Xeon processors and 4 GB of memory and has peak performance rating of 11.06 Tflop/s. The system interconnection is a customized 1024-port single rail QsNet network.

4.1 Case I: Diagnose Strong Scaling Performance Problems

Figure 4 shows the strong scaling behavior of Sweep3D with problem size 150^3, and angle blocking factor, *mmi*, equal to 3, k-blocking factor, *mk*, equal to 10. The application scales well in general, but at process count 32 the speedup drops and bounces up when process count increases to 36. We applied Hercule to contrast performance of run1 (with 32 processors) against run2 (with 36 processors) and diagnose performance anomaly cause. Hercule uses relative speedup (compared to two-processor run) to evaluate performance since there is no inter-processor communication in a sequential execution. The results that follow were generated in a completely automated manner.

Hercule first calculates speedup of run1 (with 32 processors), run2 (with 36 processors) relative to run3 (with 2 processors), and expected speedup of run1 based on run2 performance. It reaches a performance symptom of run1 that will be further explained.

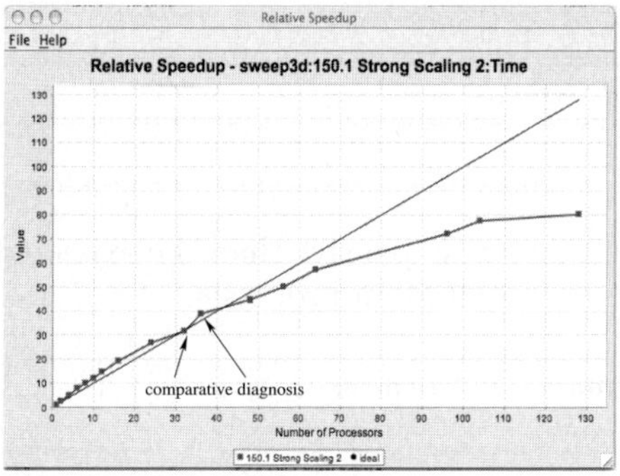

Fig. 4. Sweep3D strong scaling with problem size 150x150x150 (mmi=3, mk=10)

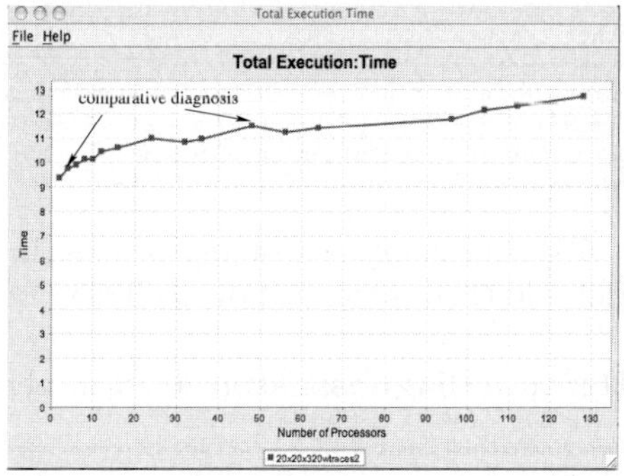

Fig. 5. Sweep3D weak scaling with problem size 20x20x320 (mmi=3, mk=10)

Hercule diagnosis step 1: find performance symptom

```
dyna6-166:~/PerfDiagnosis lili$ ./model_diag WF_speedup.clp 32pe.dup 36pe.dup 2pe.dup
Begin diagnosing ...
=========================================================================================
Speedup of run1 and run2 relative to run3
                run1           run2        expected run1
speedup        12.80          15.84        14.08
-----------------------------------------------------------------------------------------
run1 is slower than the expected value 14.08
-----------------------------------------------------------------------------------------
Next we look at the symptom low speedup.
=========================================================================================
```

Hercule then breaks runtime down into computation and communication, narrowing performance bug search.

Hercule diagnosis step 2: locate poorly performed functional groups

```
=========================================================================================
Level 1 experiment -- generate performance data with respect to comp. and comm..
-----------------------------------------------------------------------------------------
Relative speedup of functional groups in run1 and run2
                          run1           run2           expected run 1
computation:             16.035         19.906             17.694
communication:           1.115          1.172              1.042
-----------------------------------------------------------------------------------------
computation in run1 is longer than the expected.
-----------------------------------------------------------------------------------------
Next look at performance with respect to pipeline components.
=========================================================================================
```

As computation time per process stands out, Hercule further distinguishes pipeline-related computation and others.

Hercule diagnosis step 3: refine locating poorly performed functional groups

```
=========================================================================================
Level 2 experiment -- generate performance data with respect to pipeline components.
-----------------------------------------------------------------------------------------
Relative speedup of pipeline components in run1 and run2
                               run1         run2       expected run 1
computation in pipeline:     16.598       20.702       18.402
```

```
other computation:            10.452       12.405      11.03
-----------------------------------------------------------------------------------------------
computation in pipeline in run1 is slower than the expected most.
-----------------------------------------------------------------------------------------------
Next look at computation in pipeline.
===============================================================================================
```

Pipeline computation per process in run1 is more expensive than the expected. Hercule then looks at how well the load is distributed on processes.

Hercule diagnosis step 4: form performance hypothesis

```
===============================================================================================
                                        run1           run2           difference
computation in pipeline SDV (us):       236859         97548          139311
(w.r.t. processes)

-----------------------------------------------------------------------------------------------
Standard deviation of pipeline computation in run1 is significantly larger than run2, which
implies a load imbalance across processes.
-----------------------------------------------------------------------------------------------
Next testify the hypothesis load imbalance.
===============================================================================================
```

Hercule forms a load imbalance performance hypothesis based on the standard deviation of pipeline computations on all processes. It tests the hypothesis by looking at model-related overheads to which load imbalance possibly contributes most. It calculates and distinguishes performance impact of load imbalance on the overhead categories, and exemplifies occurrence of load imbalance with process behaviors in some specific computation step (iteration) and pipeline sweep. This way of explanation provides the users with both the nature of performance causes and evaluations of performance impact of the causes.

Hercule diagnosis step 5: test performance hypothesis

```
===============================================================================================
The impact of process load imbalance on performance manifests in pipeline-handshaking and
sweep-direction-change overhead.

Passing along data among successive pipeline stages (handshaking) takes 14.9% of pipeline
communication time. Pipeline handshake delay is unevenly distributed across processes.
std dev = 486463.75. process 31 involves the longest pipeline handshake cost.
-----------------------------------------------------------------------------------------------
Level 3 experiment for diagnosing handshaking related problems -- collect performance event
trace with respect to process 31
-----------------------------------------------------------------------------------------------
Pipeline HS delay is evenly distributed across iterations in the process 31. Next we look at
performance characteristics of iteration 3 which involves the longest pipeline HS.

Pipeline HS delay is evenly distributed across sweep in iteration 3 process 31, Next we look
at sweep 6 which involves the longest pipeline HS.

In iteration 3 sweep 6, computation are unevenly distributed across pipeline stages. For
example, in stage 4 process 4 spends 1964(us) doing computation, while in stage 10 process
31 spends 1590(us) computing.

In general, process 31 is assigned 23.6% less work load than process 4. Such discrepancy
causes process 31 idle for 29.5% of pipeline communication time..
-----------------------------------------------------------------------------------------------
When pipeline sweep direction change, processes may be idle waiting for successive pipeline
stages in previous sweep to finish, and for pipeline to fill up in a new sweep.  The sweep
direction changes comprise 34.6% of pipeline communication time. The delay is unevenly
```

```
distributed across processes. process 31 involves the longest pipeline direction change.

... ... (Due to the limit of space, we skip the interpretation of other overhead categories)
==========================================================================================
Diagnosing finished...
```

4.2 Case II: Diagnose Weak Scaling Performance Problems

The second experiment with Hercule demonstrates its capability of identifying and explaining parallelism overhead increases as both problem size and process count are increased in weak scaling study. Figure 5 shows the weak scaling behavior of Sweep3D with fixed problem size 20x20x320. We can see that runtime increases as more processors are used even though each process's computation load is kept the same. Hercule will compare 4-processor and 48-processor run and report and explain the performance difference. Again, the results that follow are generated in a completely automated fashion.

Hercule first calculates significance of performance difference and reaches a performance symptom, higher parallelism overhead.

Hercule diagnosis step 1: find performance symptom

```
dyna6-166:~/PerfDiagnosis lili$ ./model_diag WF_overhead.clp weak.48pe.dup weak.4pe.dup
Begin diagnosing ...
==========================================================================================
Runtime of run1 and run2 (in seconds)
                        run1          run2        difference%
runtime                11.489        9.815         17.055%
------------------------------------------------------------------------------------------
run1 is 17.055% slower than the run2.
------------------------------------------------------------------------------------------
Next we look at the symptom parallelism overhead.
==========================================================================================
```

Hercule then breaks runtime down into computation and communication, locating the functional group with most pronounced performance difference.

Hercule diagnosis step 2: locate poorly performed functional groups

```
==========================================================================================
Level 1 experiment -- generate performance data with respect to comp. and comm..
------------------------------------------------------------------------------------------
Runtime of functional groups in run1 and run2 (in seconds)
                            run1              run2              difference%
computation:                8.886             8.891             -5.624e-4
communication:              2.603             0.924             181.71%
------------------------------------------------------------------------------------------
communication cost in run1 is significantly higher than run2.
------------------------------------------------------------------------------------------
Next look at communication performance with respect to pipeline components.
==========================================================================================
```

Hercule further distinguishes pipeline-related communication and others.

Hercule diagnosis step 3: refine locating poorly performed functional groups

```
==========================================================================================
Level 2 experiment -- generate performance data with respect to pipeline components.
------------------------------------------------------------------------------------------
```

```
Runtime of pipeline in run1 and run2 (in seconds)
                                                    run1        run2        difference%
computation in pipeline:                            8.014       8.013       1.25e-4
other computation:                                   0.872       0.878        -6.83e-3

communication in pipeline:                          2.275       0.803       183.31%
         effective communication in pipeline:       0.943       0.571       65.15%
         waiting time in pipeline:                   1.332       0.231       476.62%
other communication:                                   0.328       0.121        171.07%

comm. count in pipeline (count/per process):        12288       12288          0
comm. volume in pipeline (byte/per process):        58982400    58982400       0
-----------------------------------------------------------------------------------------------
waiting time in pipeline in run1 is 476.62% higher than run2.
-----------------------------------------------------------------------------------------------
Next look at pipeline overheads.
===============================================================================================
```

Since waiting time in pipeline is significant, Hercule refines model-specific overhead categories and computes corresponding metrics.

Hercule diagnosis step 4: locate poorly performed pipeline components

```
===============================================================================================
Level 3 experiment -- generate performance data with respect to pipeline waiting time
-----------------------------------------------------------------------------------------------
Runtime of pipeline components in run1 and run2 (in seconds)
                                                run1              run2           difference%
 waiting time in pipeline                       1.332             0.231            476.62%
 pipeline fill-up:                              0.161             0.014            1050%
 pipeline empty-up:                             0.244             0.017            1335.29%
 pipeline handshaking:                          0.337             0.075            349.33%
 pipeline direction change:                     0.584             0.125            367.2%
===============================================================================================
```

There are increases in most overhead categories. We present below diagnosis results explaining two most pronounced categories, pipeline fill-up and empty-up.

Hercule diagnosis step 5: diagnose two most pronounced pipeline overheads

```
===============================================================================================
Diagnosing pipeline fill-up ... ...

In run1, pipeline fill-up delay is evenly distributed across iterations. We look at
performance characteristics of the iteration 0, which involves the longest pipeline fill-up.

In iteration 0, the depth of pipeline is 13. The pipeline tail, process 0 is being idle when
the pipeline is filling up by processes in preceding stages. The pipeline fill-up delay
comprises 335103us (20.8%) of process 0's total waiting time. The computations at preceding
pipeline stages together account for the long waiting time at the process. Reducing
computation load at preceding stages or pipeline depth will decrease filling up time.

In run2,  pipeline fill-up delay is evenly distributed across iterations. We look at
performance characteristics of the iteration 1, which involves the longest pipeline fill-up.

In iteration 1, the depth of pipeline is 3. The pipeline tail, process 0 is being idle while
the pipeline is filling up by processes in preceding stages. The pipeline fill-up delay
comprises 28707us (25.5%) of process 0's total waiting time.
-----------------------------------------------------------------------------------------------
Diagnosing pipeline empty-up ... ...

In run1, pipeline empty-up delay is evenly distributed across iterations.  Next we look at
performance characteristics of the iteration 4, which involves the longest pipeline empty-up.

In iteration 4, the depth of pipeline is 13. The pipeline head, process 0 is being idle when
the pipeline is emptying up by processes in successive stages. The pipeline empty-up
```

```
delay comprises 573,162us (35.5%) of process 0's total waiting time. The computations at
successive pipeline stages together account for the long waiting time at the process. Redu-
cing workload at successive pipeline stages or pipeline depth will decrease empty-up time.

In run2, pipeline empty-up delay is evenly distributed across iterations. We look at
performance characteristics of the iteration 1, which involves the longest pipeline empty-up.

In iteration 1, the depth of pipeline is 3. The pipeline head, process 0 is being idle when
the pipeline is emptying up by processes in successive stages. The pipeline empty-up
delay comprises 34858us (31.0%) of process 0's total waiting time.
==========================================================================================
```

As shown in the results, the increase of pipeline depth in run1 (48-processor run) is clearly the main cause of its overhead increase. Hercule illustrates and interprets performance impact of the pipeline depth with the behaviors of the process of the longest pipeline fill-up and empty-up. The pipeline depth also has a performance effect on sweep direction change. Due to limitation of space, we skip the interpretation of other overhead categories, event though Hercule is able to explain them equally well.

5 Conclusions

The use of relative performance diagnosis, where problems are discovered and explained in relation to other experiments, is important to support in diagnosis systems. For model-based diagnosis frameworks such as Hercule, we look to integrate relative analysis in the knowledge and inference engineering. In this paper, we report on how the Hercule framework was extended to enable relative diagnosis by adding an interface from the metric evaluator to the inference engine to analyze performance from different runs. In addition, decision rules were developed for problem identification and hypothesis testification. The paper demonstrates the effectiveness of the approach to relative diagnosis of the well-known Sweep3D application based on a wavefront model.

References

1. The ASCI sweep3d benchmark. `http://www.llnl.gov/asci_benchmarks/asci/limited/sweep3d/`.
2. CLIPS: A tool for building expert systems. `http://www.ghg.net/clips/CLIPS.html`.
3. TAU tuning and analysis utilities. `http://www.cs.uoregon.edu/research/paracomp/tau/tautools/`.
4. D. Abramson, I. Foster, J. Michalakes, and R. Sosic. Relative debugging: a new paradigm for debugging scientific applications. *The Communications of the Association for Computing Machinery*, 39(11):67–77, 1996.
5. K. Karavanic and B. Miller. A framework for multi-execution performance tuning. In *On-line Monitoring Systems and Computer Tool Interoperability*, pages 61–89. Nova Science Publishers, Inc., 2004.
6. L. Li and A. Malony. Knowledge engineering for automatic parallel performance diagnosis. Accepted by *Concurrency and Computation: Practice and Experience.*
7. F. Song, F. Wolf, N. Bhatia, J. Dongarra, and S. Moore. An algebra for cross-experiment performance analysis. In *ICPP 2004*, 2004.

Self-optimization of MPI Applications Within an Autonomic Framework

M. Iannotta[1], E. Mancini[1], M. Rak[2], and U. Villano[1]

[1] Università del Sannio, via Traiano, Benevento, Italy
massi.iannotta@gmail.com, {epmancini, villano}@unisannio.it
[2] Seconda Università di Napoli, Via Roma 29, 81031 Aversa (CE), Italy
massimiliano.rak@unina2.it

Abstract. An existing autonomic framework (MAWeS) can be used to provide run-time self-optimization for distributed applications. This paper introduces a new MAWeS Component that provides an interface for MPI applications. As case study, we will present the implementation of a dynamically-reconfigurable *n-body* solver, evaluating its obtained performance with and without the MAWeS framework under several different working load conditions.

1 Introduction

The high redundancy and logical complexity of present-day distributed computing systems have stimulated a great interest in autonomic computing systems [1,2], whose goal is to behave as the human autonomic system. Autonomic capabilities are usually classified in four different categories: *self-configuring*, in which the system can dynamically adapt to changing environments, *self-healing*, in which the system can discover, diagnose and react to disruptions, *self-optimizing*, in which the system can monitor and tune resources automatically, and *self-protecting*, in which the system can anticipate, detect, identify and protect itself against threats.

The first applications of autonomic systems [3,4,5,6] are mainly targeted to the configuration, management and optimization of large distributed systems running highly dynamic applications, such as web services. We have contributed to this research field with a prototype framework, MAWeS [7,8,9], which allows running applications to self-tune querying an optimization engine using a web services interface. Unlike most autonomic system existing today, MAWeS optimizations are not based on *reactive autonomicity*, i.e., on feedback control. MAWeS instead relies on *predictive autonomicity*, which uses feedforward control. The idea is to detect by means of external monitoring modules cyclic variations in parameters and their impact on performance, and to self-tune automatically the system, anticipating the need [10]. The MAWeS framework hinges on an existing description language and simulation environment (MetaPL and HeSSE, respectively [11,12,13,14]). The framework optimizations tools exploit the simulator to predict the system behavior in real, fictitious or future working conditions, anticipating the need for new configurations or tunings.

M. Gerndt and D. Kranzlmüller (Eds.): HPCC 2006, LNCS 4208, pp. 210–219, 2006.

In [15], we proposed the adoption of self-optimization techniques also in the context of MPI applications. However, the optimization engine presented in the mentioned paper was able to optimize applications only at start-up time, making it possible to adapt their computational and communications requirements to the load measured *at that time.* Successive load variations were systematically ignored. The objective in this paper is instead to execute MPI applications in the context of the MAWeS framework, thus exploiting its dynamical, predictive optimization capabilities. These are particularly suited to systems with a variable background CPU and network load, such as small-scale shared clusters. The objective of this experience is twofold: to evaluate the validity of the autonomic approach in the high-performance computing context, and to explore the feasibility of linking MPI code to a tool based on a service-oriented architecture through a web service interface.

In particular, we will present here a new MAWeS Component that acts as frontend for the framework. It provides a MAWeS interface for MPI applications. The component is made of a "standard" template that can be used to develop MPI applications interfaced to MAWeS, and of an interfacing daemon linked to a distributed load detection system. The latter is used to detect (or even to foresee) changed load conditions. The framework reacts distributing new configurations to the running application through the daemon interface.

The reminder of the paper is structured as follows. The next section sketches the structure of the MAWeS framework. Then, the design and the implementation of the new MAWeS frontend is presented. After that, the implementation of a reconfigurable MPI test application is described, presenting the performance results obtained. After an overview of related work on autonomic and self-optimization systems, the conclusions are drawn.

2 The MAWeS Framework and the MetaPL/HeSSE Methodology

The MAWeS Framework has been developed to support predictive autonomicity in web services based architectures. It is based on two existing technologies: the MetaPL language [12, 13] and the HeSSE simulation environment [11, 14]. The first is used to describe the software system and the interactions inside it; the second, to describe the system behavior and to predict its performance using simulation. The MAWeS framework uses MetaPL descriptions and HeSSE configuration files to run HeSSE simulations. Through the execution of multiple simulations, with different parameter values, it chooses the parameter set that optimizes the software execution.

The MAWeS framework partially hides the presence of a simulation environment exploited through a web service interface. It is structured in three layers (Figure 1), as follows:

Frontend made up of the software modules used by final users to access the MAWeS services. These modules aim to give high transparency to the tool, from the final user perspective;

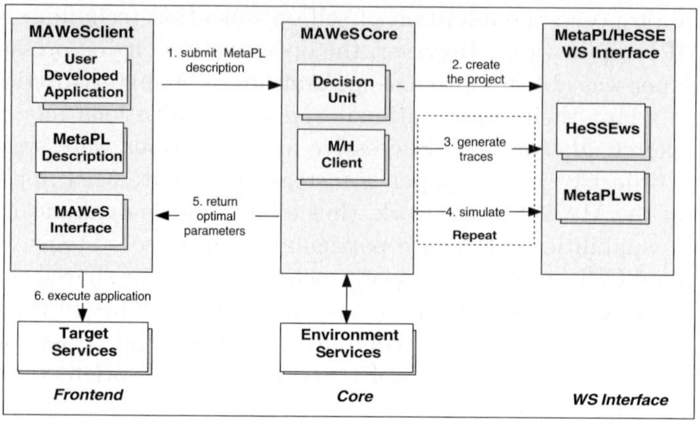

Fig. 1. MAWeS Framework logical schema

Core composed of the software and the services that manage MetaPL files and make optimization decisions;

WS Interface the set of Web Services used to obtain simulations and predictions through MetaPL and HeSSE.

The MAWeS Frontend provides a standard client application interface, `MAWeSclient`, which has to be extended by developers with their actual application code. The `MAWeSclient` client accepts as input a MetaPL file describing the application code. The MAWeS Core exploits environment services and the MetaPL/HeSSE Web Services interface using the application information contained in the MetaPL description, to find out optimal execution conditions. It is a software unit provided both as a web service and integrated into the `MAWeSclient`.

The sequence of events and calls that allow the execution of an application with optimal values of parameters is also shown in Figure 1. The MAWeS client submits the MetaPL application description to the MAWeS core (1). Then the services of the framework automatically find out the set of simulations needed, perform them (2,3,4), and return the set of optimal parameters for the target application (5). Finally, `MAWeSclient` starts the application code, passing to it the set of optimized parameters (6).

A critical issue is to point out the set of parameters that mostly affect performance, in order to find automatically the best application configuration. MetaPL descriptions consist essentially of code prototypes, enriched with task-to-processor mapping (`Mapping` tag). The Autonomic MetaPL extensions define additional language elements for this section. They introduce the `Autonomic` tag, included in `Mapping` element, which describes the target simulation configurations that can be used for application execution. Further details on MAWeS can be found in [7,8].

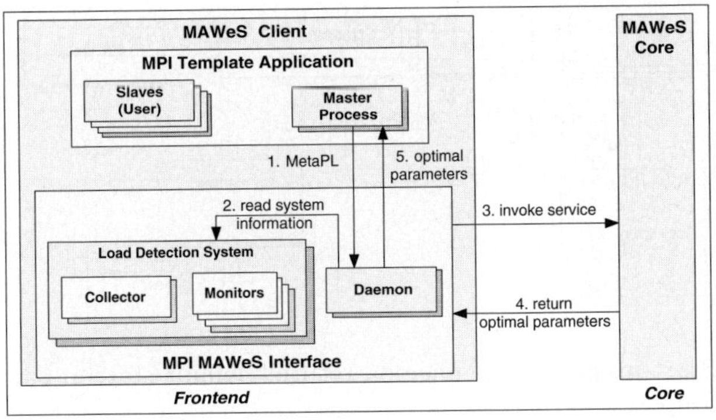

Fig. 2. Client-side of a Self-optimizing Application using MAWeS

3 Self-optimization of MPI Applications with MAWeS

In [15] we have shown the potential of the MetaPL/HeSSE methodology to self-tune scientific applications using predictive autonomicity. In the tests described in the mentioned paper, we optimized an MPI application performing a set of simulation *before* application start-up. In this paper, we aim to obtain an autonomic behavior performing the applications optimization while the application runs. It is important to point out that the MAWeS framework is based on a web services approach, whereas the target scientific application application is an MPI one. In order to interface the target application to the framework, a modification the original MAWeS Frontend is needed. As shown in Figure 2, the MPI-MAWeS frontend is made out of:

1. MPI Template Application (MTA)
2. MPI/MAWeS Interface (MMI)

The MTA provides a template that can be used to develop an MPI application interfaced to the MAWeS framework. In practice, the template allows the construction of master/worker codes. In addition to its "normal" coordination and (possibly) computational task, the master is the interface of the MPI code to the optimization framework. On the other hand, the workers contain no additional optimization logic and have only computational tasks. In our opinion, restricting our interest to master/worker codes is not great generality loss, as a "fake" master can in any case be introduced in different software design patterns. A second design choice of the template is to support only iterative codes, where each node performs its operations several times on a subset of the problem whole data set. The idea is to redistribute the data set whenever the optimization engine, due to the changed system load, points out a more profitable load sharing.

The MMI component provides a daemon, which is the direct interface with the MAWeS core, and a Load Detection System, useful to find changed system load conditions and to trigger the optimization engine in MAWeS. The MMI daemon module is essentially a gateway: it waits for UDP requests from the

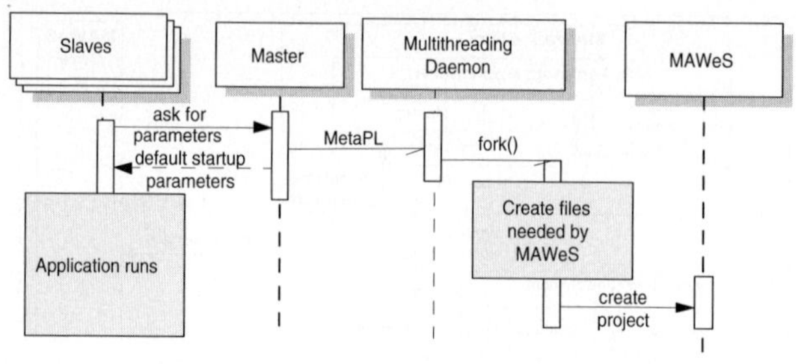

Fig. 3. Start-up phase: the daemon returns default startup parameters

MTA master, and then asks the MAWeS core for optimal parameters exploiting its web services interface.

The *Load Detection System* of MMI is designed with a master/worker approach. A *collector*, which usually resides on the same node as the MMI daemon and as the MTA master, queries a set of local monitors, which retrieve data about the state of each node. In the current implementation, the Load Detection System just takes into account CPU load.

3.1 The MTA-MMI Protocol

Figure 3 shows what happens when the application developed extending the MTA template is started. The application workers (which contain only user code) ask the master for the local data sets. This starts the MMI daemon and returns to the workers, in addition to actual data, a "default" set of parameters. In other words, the first iterations are not optimized. The MMI daemon generates the thread that will manage the optimizations. While the workers are working using the default configuration, MAWeS starts searching for the optimal configuration.

The workers compute, conclude their iteration, possibly return results or synchronize with the master, and, in the absence of any reconfiguration messages, start computing on the next iteration. As shown in Figure 4, the master asks the daemon to know if a better configuration is available. This happens with a frequency that depends on the application characteristics. In this case, a reconfiguration message is sent to the workers. The optimization logic is completely asynchronous with the user program. In the absence of reconfiguration messages, the MPI code runs at full speed without any introduced overhead. New configurations are broadcast to the workers that apply them as soon as possible (in practice, between one iteration and the next).

4 Case Study: N-Body

The objective of this section is to show how the framework can optimize the run-time behavior of a scientific MPI application. The example chosen is a code solving the n-body problem [16]. The universe of bodies is split in disjoint subsets

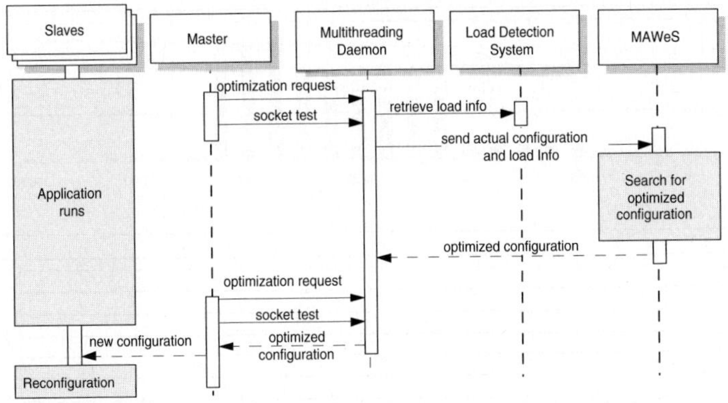

Fig. 4. Optimization phase: the daemon contacts the MAWeS framework to obtain an optimized configuration and returns it to the application

that are managed in a parallel way using a master-worker paradigm. The master initializes body positions and assigns distinct subsets of bodies to the workers. These, at every iteration, compute the gravitational forces between the bodies in their own subset against all bodies in the universe, and broadcast the updated bodies position and velocity, to be used in the next iteration.

As "global" iteration time is clearly the maximum of the iteration times of the different workers, the optimization strategy target is to level off any possible differences. This is fairly trivial in a homogeneous machine in the absence of external CPU load. In a system where the nodes have a variable fraction of CPU load due to other applications or services, achieving a good balance is instead a hard task.

As the single worker iteration times depend (besides on CPU load) on then number of bodies to be managed, MAWeS will control the latter parameter to obtain optimal execution conditions. In this context, "optimal" means balanced iteration times for the slaves and thus minimum overall execution time. To this end, MAWeS will occasionally ask for variations in the number of the bodies assigned to the slaves. As every slave knows the spatial coordinates of all bodies, not only of the locally-managed ones, reconfigurations can be easily obtained before starting a new iteration. It is clear that the algorithm proposed is highly non-optimized, due to the unnecessary data replications. However, our main goal here is to provide simply a proof-of-concept code. In an optimized code, where data are not replicated, bodies coordinates have to be transferred between workers when reconfigurations are performed. The effect of the overheads introduced can be easily taken into account by simulation in MAWeS.

4.1 Description of the Tests Performed

The environment used to test the application is the Fab4 Cluster, at the University of Sannio in Benevento, using MPICH version 1.2.7. During the tests, the MTA master, the MMI daemon and the Load Detection System Collector run on

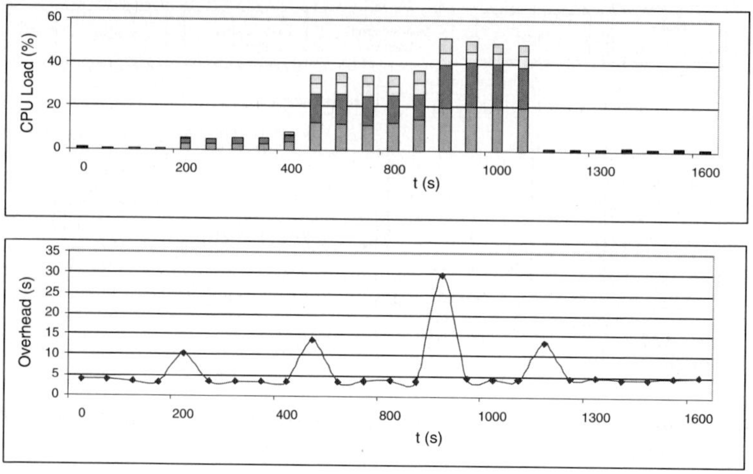

Fig. 5. Synthetic load time profile and corresponding MAWeS overhead

the cluster frontend, which hosts also the MAWeS Frontend Web Services and HeSSE. On each node (frontend and nodes) the Load Detection System is used to obtain periodically load information. In order to evaluate the tool behavior in a variety of conditions, we injected a synthetic CPU Load on some of the nodes during the target application execution. The MAWeS tool will adapt the application behavior to the system load, modifying the data sets of the workers.

Figure 5 shows in the upper diagram the synthetic additional CPU load injected in each node during the application run (the x axis reports the time from application start-up, the y axis the ratio of CPU time used by the synthetic load generator, in different colors for each node). The lower diagram shows the corresponding time for simulation, optimizations, ..., spent *in the MAWeS framework* (not in the application nodes, which simply receive pre-computed optimal configurations). The overhead is nearly negligible, except when, due to changes of system load, new configurations have to be compared, in order to find the optimal one for the future interval of time. It should be noted that this overhead has no direct effect on the application execution. The use of feedforward makes it possible to start the choice of a new configuration well before expected load changes. Should even MAWeS delay to communicate the new optimized configuration, the workers will continue to compute with the previous (likely, no longer optimal) one.

4.2 Experimental Results

As mentioned in the introduction, MAWeS is based on a feedforward approach. A predicted load profile is used, along with actual measurement data, to foresee changes of system load and to adapt proactively to the changes, anticipating the need for reconfigurations. However, the tool can be easily modified (this cannot be done on-the-fly in the current implementation) to adopt the more conservative

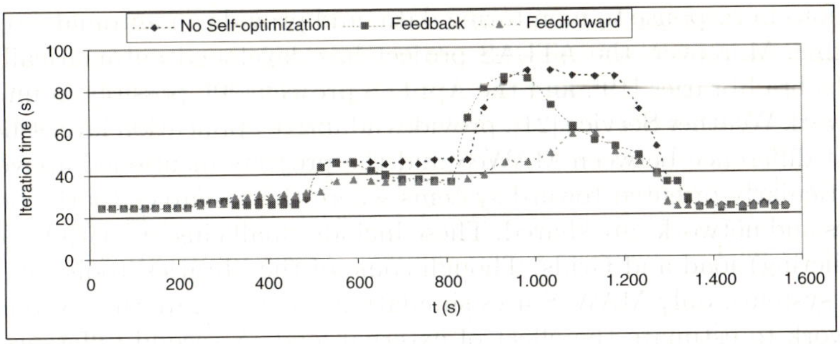

Fig. 6. System evolution without MAWeS, with feedback and with feedforward MAWeS

feedback approach. A comparison between the two strategies is indeed useful in the testing phase to understand the limits and the usefulness of the framework. It should be noted that, using the feedforward approach, the optimizations are carried out *before* the load is injected. With feedback, they are performed only *after* that a load variation is detected. As regards the previsions on system load, they can be obtained by historical data, and/or by making considerations on the type of host application load.

Figure 6 shows the evolution in time of the global iteration response times of the n-body application without any self-optimization (*No self-optimization*), and under MAWeS with feedback (*Feedback*) and feedforward strategy (*Feedforward*). In all the tests, the externally injected additional load is the one represented in the load profile in Figure 5. The optimization framework is supplied with perfectly accurate predictions of system load (in practice, with the actually-injected load profile). The comparison between the response times without and with MAWeS shows the validity of the self-optimizing technique implemented. Using MAWeS, it is clear that in the hypothesis of perfectly-accurate predictions the feedforward approach should lead to the best results. In fact, the feedback approach has a delay time linked to the optimization latency and overhead (see Figure 5). When the optimization is based on feedback, the system can optimize itself only after a short period of unoptimal performance, as it has to monitor the system to find unwanted situations. When this happens, the program can obtain the optimal parameters by MAWeS. These optimal parameters are obtained in advance if the feedforward approach is used.

5 Related Work

Even if the research efforts at the basis of MAWeS are placed in the context of autonomic computing, this paper is essentially concerned with parallel code self-optimization and tuning. This has been a very active research area in the last decade, and a wide body of literature has been produced on similar topics. In particular, the contributions more similar to our proposal are Active Harmony [17] and Autopilot [18]. Both of them allow the run-time configuration of

Discovery of Locality-Improving Refactorings by Reuse Path Analysis

Kristof Beyls and Erik H. D'Hollander

Department of Electronics and Information Systems (ELIS), Ghent University,
Sint-Pietersnieuwstraat 41, B-9000 Ghent, Belgium
{kristof.beyls, erik.dhollander}@elis.UGent.be

Abstract. Due to the huge speed gaps in the memory hierarchy of modern computer architectures, it is important that programs maintain a good data locality. Improving temporal locality implies reducing the distance of data reuses that are far apart. The best existing tools indicate locality bottlenecks by highlighting both the source locations generating the use and the subsequent cache-missing reuse. Even with this knowledge of the bottleneck locations in the source code, it often remains hard to find an effective code refactoring that improves temporal locality, due to the unclear interaction of function calls and loop iterations occurring between use and reuse.

The contributions in this paper are two-fold. First, the locality analysis is enhanced to not only pinpoint the cache bottlenecks, but to also suggest code refactorings that may resolve them. The refactorings are found by analyzing the dynamic hierarchy of function calls and loops on the code path between reuses, called *reuse paths.* Secondly, reservoir sampling of the reuse paths results in a significant reduction of the execution time and memory requirements during profiling, enabling the analysis of realistic programs.

An interactive GUI, called *SLO (Suggestions for Locality Optimizations)*, has been used to explore the most appropriate refactorings in a number of SPEC2000 programs. After refactoring, the execution time of the selected programs was halved, on the average.

1 Introduction

The memory access time is a growing bottleneck in high performance computer systems. A suitable memory hierarchy is one of the prominent answers, and its beneficial effect is proportional to a good data locality during program execution.

Poor temporal locality occurs when between consecutive reuses of the same data, a large amount of other data is accessed. Two consecutive reuses of the same data are called a *reuse pair.* In this paper, the distance of a reuse pair is defined as the number of accesses between use and reuse

The temporal locality of a reuse pair can only be increased by reordering memory accesses, so that its distance is reduced. The best existing analysis tools [1,2] highlight both the source locations generating the use and the reuse of long-distance reuse pairs. However, using this information, it often remains

M. Gerndt and D. Kranzlmüller (Eds.): HPCC 2006, LNCS 4208, pp. 220–229, 2006.

```
double inproduct(double*X,double*Y,int len) {
 int i; double result=0.0;
 for(i=0; i<len; i++)
  result += X[i]*Y[i]; //place of use
 return result; }
double sum(double *X, int len) {
 int i; double result=0.0;
 for(i=0; i<len; i++)
  result += X[i];        //place of reuse
 return result; }
```

(a) Reuse Pairs indicated by existing tools.

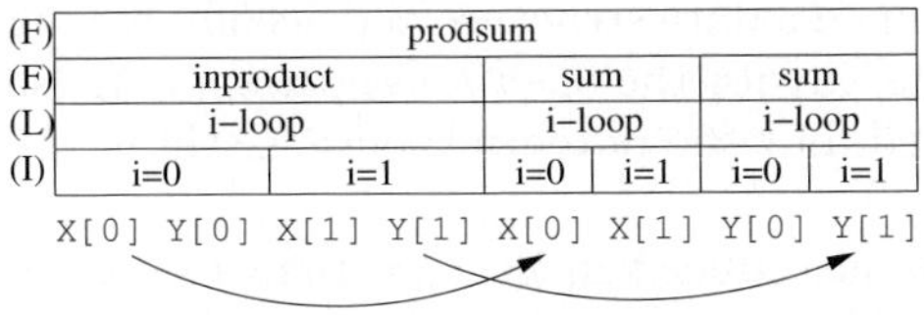

(b) Dynamic function-loop hierarchy, `len`=2.

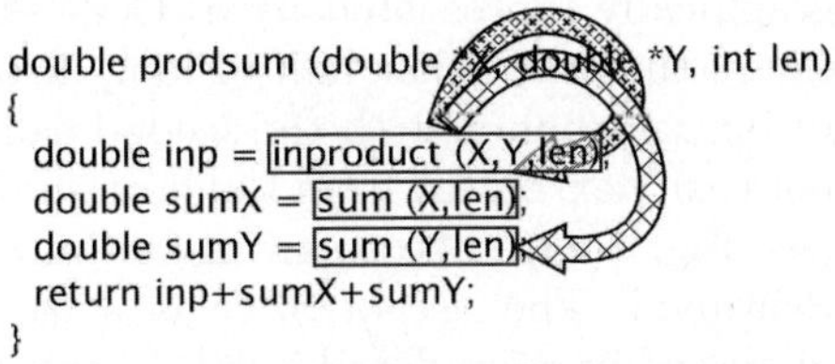

(c) Fusions indicated by SLO (converted from color to black-and-white by hand).

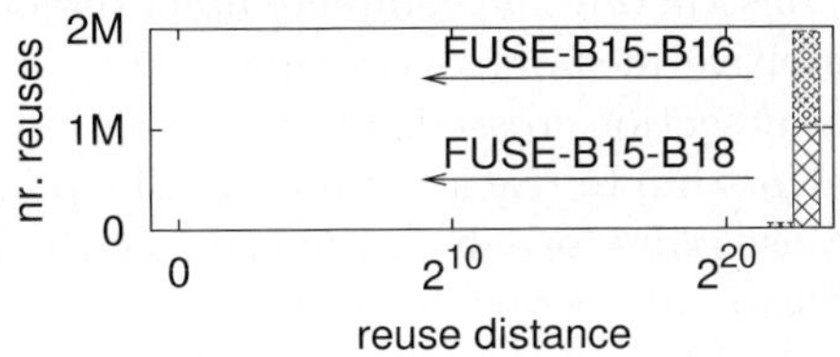

(d) The reuse distance histogram corresponding to (a), `len`=1000000.

Fig. 1. Example program containing long-distance reuses

hard to find a code refactoring that reduces the distance, due to the unclear interaction of function calls and loop iterations between the reuses.

Example 1. Consider the code in Fig. 1(a). The comments indicate the source location where the cache-missing reuse occurs and the location where the previous use of the data was. This view on the code does not reveal how to bring use and reuse closer together, since it is not clear from where the functions `inproduct` and `sum` are called.

In contrast, the presented method analyzes the *dynamic function-loop hierarchy*, which is the hierarchy of all function calls, loop executions and loop iterations at run-time. An example is given in Fig. 1(b), where (I) indicates a loop-iteration level, (L) a loop level and (F) a function-call level. Our method proceeds by analyzing the highest level in this hierarchy where "sections" are crossed between use and reuse. Significantly reducing the distance between use and reuse requires the merging of the corresponding sections. Merging these run-time sections by static source code transformations requires the following refactorings, depending on the type of sections that need to be merged:

(I) Iteration merging can be done by a *tiling-like* code transformation, so that less data is accessed in a given loop iteration.
(L) Loop merging can be done by fusing the corresponding loops.
(F) Function merging is done by fusing the corresponding functions.

Example 2. (continued) For the reuse of `X[0]` in Fig. 1(b), the highest-level sections that are crossed are function-call sections. Therefore, bringing use and reuse closer together requires fusing `inproduct` and `sum`. This refactoring is graphically represented by SLO as shown in Fig. 1(c). Furthermore, Fig. 1(d) shows that the other half of long-distance reuse pairs need to be optimized by fusing `inproduct` with the second call to `sum`. The need for these refactorings is not easily extracted from the bottleneck information shown by previous analyses (see Fig. 1(a)). The resulting optimized code runs 3 times faster on a 2.66Ghz Pentium4. The use of SLO as a tool to improve temporal locality has been discussed in more detail in [3]. In this paper, we present the underlying analysis.

Constructing the complete function-loop-hierarchy explicitly would result in prohibitive memory overheads, and prohibitive time overhead to find the highest-level section crossed. Therefore, we developed a data structure and accompanying algorithm to track the function-loop hierarchy for the *open reuse pairs*, i.e. those reuse pairs for which the use has occurred, but the potential reuse is still in the future, as presented in Sect. 2.

Furthermore, a sampling algorithm is introduced in Sect. 3, that speeds up the profiling of long-running programs. Previous sampling methods for cache measurement such as time and set sampling [1,4,5] only allow to control the sample rate, without guarantees about the resulting accuracy, since samples are taken systematically instead of randomly. In contrast, our algorithm takes samples randomly, which allows to derive a theoretically guaranteed accuracy.

The sampled profiling and analysis has been implemented in the GCC compiler. Using the data structures presented in Sect. 2, the analysis becomes doable for most SPEC2000 programs, albeit at a sometimes large memory overhead and a time overhead of a factor 1000 during profiling. As shown in Sect. 4, the sampling largely reduces memory overhead to an almost constant factor, and a time overhead of only a factor 5 for long-running programs compared to uninstrumented, fully-optimized execution. Using SLO, we were able to improve the locality of five already hand-optimized programs in SPEC2000, resulting in an average speedup of 2 on a number of different platforms. A comparison with related work is made in Sect. 5, indicating that the data structures and sampling techniques introduced in Sections 2 and 3 might also be profitable for a number of other program analyses. In Sect. 6, concluding remarks follow.

2 Compact Representation of the Function-Loop Hierarchy for Open Reuse Pairs

Pinpointing the refactoring for optimizing a given reuse pair is based on determining the highest level in the function-loop hierarchy where sections are crossed. Constructing the complete function-loop hierarchy would be prohibitively expensive. Therefore, a compact representation of the function-loop hierarchy for only the open reuse pairs has been developed. At each memory access, if the address has been accessed before, the highest-level sections crossed between the previous use and the current reuse is computed as follows.

First, the innermost function in the hierarchy that contains both use and reuse occur is determined. We call it the *Least Common Ancestor Function (LCAF)*

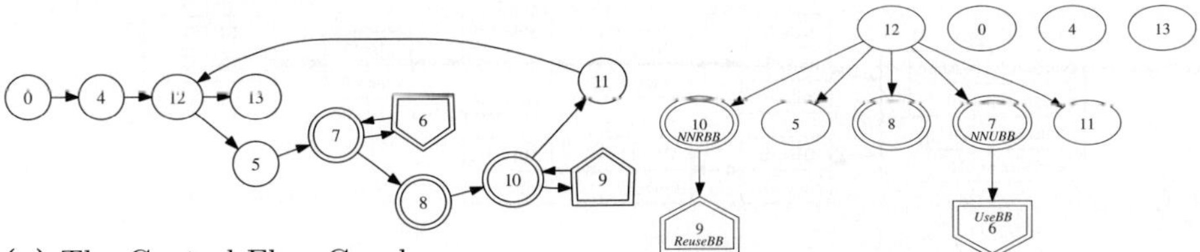

(a) The Control Flow Graph.

(b) The Nested Loop Forest.

Fig. 2. A Control Flow Graph and Nested Loop Forest. The basic blocks executed between use and reuse are indicated by double ellipses. For the considered reuse pair, the use occurs in basic block 6, while the reuse occurs in basic block 9.

since it is the least common ancestor of both use and reuse in the function-loop hierarchy. Then, the basic blocks executed between use and reuse are analyzed to find the outermost loops or loop iterations in the LCAF that are crossed between use and reuse, using the following definitions (see Fig. 2):

Definition 1. *The* ***Intermediately Executed Code (IEC)*** *of a given reuse pair is the set of basic blocks in the LCAF executed between use and reuse. The* ***UseBB*** *is the basic block in the LCAF containing the use; the* ***ReuseBB*** *contains the reuse.*

The ***Nested Loop Forest*** *of a function is a graph, with its basic blocks as nodes and edges going from loop headers to the basic blocks directly controlled by them. The* ***Outermost Executed Loop Header (OELH) of a basic block BB with respect to a given reuse pair*** *is the earliest ancestor of BB in the nested loop forest that has been executed between use and reuse.*

The ***Non-nested Use Basic Block (NNUBB)*** *is the OELH of UseBB. The* ***Non-nested Reuse Basic Block (NNRBB)*** *is the OELH of ReuseBB.*

When NNUBB=NNRBB, the highest level sections crossed are iterations of the loop for which the loop header is NNUBB. If NNUBB≠NNRBB, fusion of the loops or function calls associated with the NNUBB and NNRBB are required (Each function call is located in a separate basic block).

When a data reuse is detected, the algorithm in Fig. 4 is used to calculate the LCAF, NNUBB and NNRBB. The three steps in the algorithm are discussed below; the data structure for representing basic blocks executed between use and reuse is illustrated in Fig. 3:

1. On every memory access and basic block transition, the global time is increased. For each data address, the time of its last access is stored in a hash-table. At time of reuse, the time of use is retrieved from that hash-table.
2. At run-time, a stack is maintained that reflects the call stack of the program. The LCAF is simply the latest function on the call stack that was called before the time of use.
3. For each frame in the stack, the list of executed basic blocks is maintained sorted in Most Recently Used order, e.g. Fig. 3. At the time of reuse, the

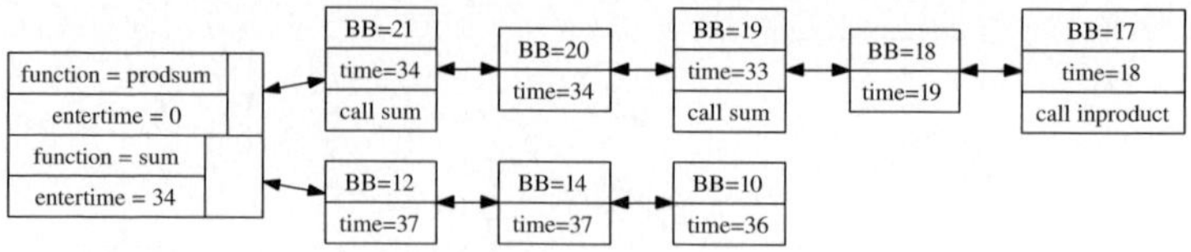

Fig. 3. Illustration of call stack, with an MRU list of basic blocks per call frame. The call stack for the code in Fig. 1 is presented. (current time=37, $t_u = 6, LCAF = prodsum, head(LCAF) = BB21, LAST(LCAF, t_u = 6) = BB17$).

```
t_u ← time of last use.
LCAF ← FirstFunctionBeforeTimeInStack(t_u)
head ← HEAD(LCAF); last ← LAST(LCAF, t_u); iec ← {head..last};
  NNRBB ← OELH(iec, head); NNUBB ← OELH(iec, last)
```

Fig. 4. Algorithm to find (NNUBB,NNRBB) for a reuse pair at time of reuse

ReuseBB is the head of the list of the LCAF, so NNRBB=OELH(head). The UseBB cannot be determined from this data structure. Also, it is not possible to record the UseBB at time of use, since the LCAF is unknown at that time. However, OELH(UseBB) is the same as the OELH of the last basic block in the list with an access time before the time of use, see lemma 1 below. Therefore, NNUBB=OELH(last).

Lemma 1. *The last basic block L in the MRU list with a time more recent than the time of use, has the same OELH as the UseBB.*

Proof. (a) If UseBB is executed once between use and reuse, then UseBB=L, so OELH(UseBB)=OELH(L).

(b) If UseBB is executed multiple times between use and reuse, then the trace of basic blocks between use and reuse has the following general form: ($UseBB_1 \cdots L \cdots UseBB_N \cdots ReuseBB$), where $UseBB_1$ is the dynamic instance of UseBB where the use occurs and $UseBB_N$ is the last instance of UseBB that is executed between use and reuse. L is executed between $UseBB_1$ and $UseBB_N$, so it must lay on some cyclic path from UseBB to itself. This cyclic path is part of a loop, and goes through the loop header which is part of the IEC. Since L and UseBB have a common loop header that is executed between use and reuse, they have a common loop header which is part of the IEC. As such, the OELH for both UseBB and L is the OELH of that common loop header. □

3 Overhead Reduction by Reservoir Sampling

While the data structures introduced in Sect. 2 enable to measure the data reuses and to pinpoint the required refactorings, it still has a large profiling overhead (factor 1000 on average) and a large memory overhead (over 2GB for some

SPEC2000 benchmarks), see Sect. 4. In this section, we introduce a sampling method to reduce both the time and memory overhead. The aim is to reduce the amount of reuse pairs for which the analysis in Sect. 2 is needed, while still guaranteeing the resulting error to be within a specified confidence interval.

3.1 Determining Number of Samples Needed for a Given Accuracy

Earlier sampling methods for memory traces, such as time sampling and set sampling [5] perform systematic sampling, i.e. the samples are not taken randomly. In contrast, we take uniform random samples of the reuse pairs in the program. This allows to perform the following analysis on the number of samples needed to obtain a given confidence interval.

The following assumptions are made: (1) A refactoring R is identified by the couple (NNUBB,NNRBB). (2) A refactoring R optimizes a given fraction R_f of all reuses for a given program run. (3) For a given reuse pair P, the probability that P is optimized by a refactoring R is R_P.

When the refactorings are determined for a sample of only n out of a total of N reuse pairs in a program run, the estimation for the fraction of reuse pairs that are optimized by refactoring R is $\hat{R}_P = c/n$, where c equals the number of reuse pairs in the sample that can be optimized by R. The corresponding large sample confidence interval [6] is, without assuming any distribution on R_P, $\hat{R}_P \pm z_{\alpha/2}\sqrt{\hat{R}_P(1-\hat{R}_P)/(n-1)}$. Therefore, the expected relative error within an α% confidence interval is $e = \left(z_{\alpha/2}\sqrt{\hat{R}_P(1-\hat{R}_P)/(n-1)}\right)/\hat{R}_P$ Rearranging this formula, the number of samples n that need to be taken to estimate R_P to within a certain error e for a given confidence interval α is: $n = \left(\left(z^2_{\alpha/2}(1-\hat{R}_P)\right)/e^2\hat{R}_P\right)+1$

In the experiments below, we wanted to find for each refactoring that optimizes at least 0.01% of all reuse pairs, the true fraction of reuses it optimizes with a maximum relative error of 10% ($e = 0.1$) with a 95% confidence interval ($z_{\alpha/2} = 1.96$). Substituting these values in the above formula, shows n must be at least 3.841.217. Therefore, a uniform random sample of at least 3841217 reuse pairs is needed to accurately pinpoint the refactorings.

3.2 Profiling Based on Reservoir Sampling

The goal of the sampling is to collect the (NNUBB, NNRBB) for n reuse pairs that have been uniform randomly selected. At the start of the program, it is unknown how many reuse pairs there will be in the program run, so it's impossible to select a fixed sampling rate (apart from the fact that a fixed sampling rate does not result in a completely random set of reuse pairs). Therefore, the implementation is based on reservoir sampling [7]. At the start of execution, each reuse pair detected is inserted in the sample, until there are n reuse pairs. From then on, the formula presented in [7, Alg. L] determines the number of reuse pairs to be skipped, before the next reuse pair is chosen to replace one of the pairs already present in the sample. As the program runs longer, the distance between reuse pairs that are sampled also grows larger. At the end of the execution, each reuse pair has an equal probability of being present in the sample [7].

In our implementation in the GCC compiler, every memory access, basic block and function call is instrumented. For each basic block and function call, a call to a run-time library is inserted that maintains the call stack data structure illustrated in Fig. 3. For each memory access, the following code is inserted:

```
++time;
if (HASH_ARRAY[addr & mask] != 0)  __full_check (addr, ref);
if (--accesses_to_next_traced == 0)  __track_address (addr, ref);
```

`time` is the global time. `HASH_ARRAY` is a large array that is indexed using the accessed address modulo the size of `HASH_ARRAY`. `HASH_ARRAY` contains non-zero entries for all addresses that are currently part of an open reuse pair that is sampled. `accesses_to_next_traced` is the number of accesses that need to be skipped until the next access that will be entered in the sample, according to [7, Alg. L]. When `HASH_ARRAY` is substantially larger than the number of samples n to be taken, and when the number of accesses already executed in the program run is substantially larger than n, both if tests are likely to be false. As a result, for most accesses in long running programs, the overhead of memory access instrumentation is just incrementing the `time` and `accesses_to_next_traced` variables and performing the two conditions in the if-test.

4 Experimental Results

We implemented the presented method in the GCC 4.1 compiler. The patch is available at `http://www.elis.ugent.be/~kbeyls/slo`. We used the SPEC2000 benchmarks, with all their available inputs, to evaluate the reduction in time and memory overhead due to the sampling. All profiling overhead tests were run on an AMD Opteron 1.6Ghz processor running Linux.

Fig. 5(a) shows the time overhead of profiling versus the original execution time of the non-profiled fully-optimized programs. It shows that profiling without sampling has a more or less constant overhead of about a factor 1000. Using sampling, the time overhead reduces to about 5 for long-running applications.

Fig. 5(b) shows the memory overhead. Using non-sampled profiling, a few runs lead to more than 2GB memory overhead, resulting in a program crash due to out of memory on the 32-bit platform. These results agree with those found by Fang et al. [8, Sect. 4.1] who instruments programs to measure the branch history between use and reuse: there are too many different paths executed between reuses to be able to store them in a reasonable amount of memory. Therefore, it seems that sampling techniques are required to practically measure the paths executed between reuses for arbitrary programs.

In contrast, the sampled profiling has an almost constant memory overhead of about 200MB, allowing to profile all SPEC2000 programs. Of those 200MB, about 150MB is consumed by the sample buffer containing 4 million reuse pairs, and the associated data. The slight fluctuation in memory overhead between

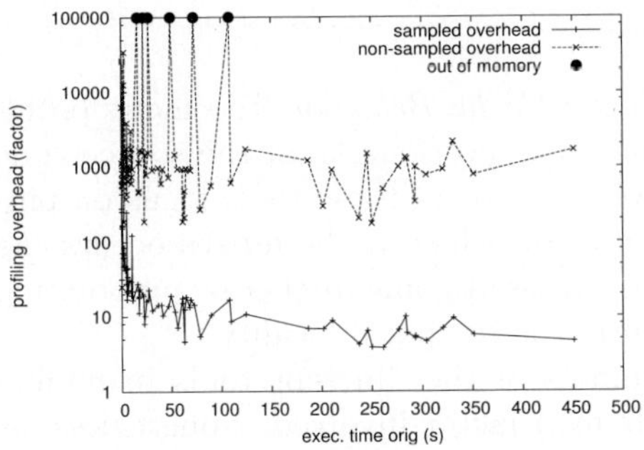

(a) Execution time overhead measured as a factor of the original execution time. (Vertical axis is logarithmic)

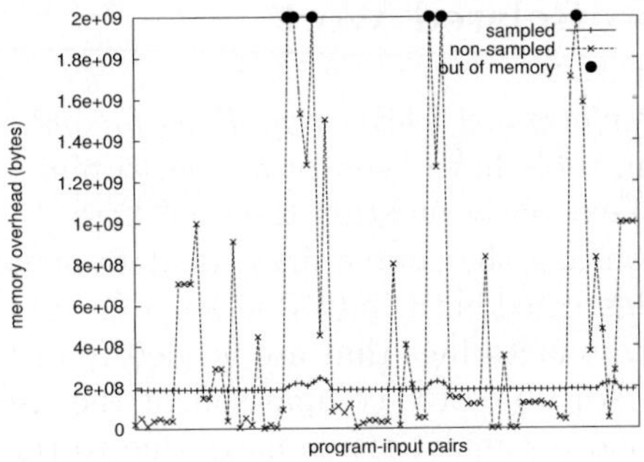

(b) Memory overhead.

Fig. 5. Time and memory overhead of sampled and non-sampled profiling for the SPEC2000 benchmarks

Table 1. Speedups on different platforms for five SPEC2000 programs after applying temporal locality optimizations based on suggestions made by SLO. The cache sizes of the largest cache level are indicated between parentheses for each platform.

	Speedup				**Analysis Time**	
	Pentium4 (512KB)	**Itanium (2MB)**	**Alpha (8MB)**	**Average**	**Non-Sampling**	**Sampling**
Art	4.11	1.54	1.16	2.39	12h13m	0h16m
Equake	1.10	2.93	3.09	2.30	59h33m	0h22m
VPR.route	1.51	1.40	1.41	1.44	18h32m	0h16m
Galgel	1.92	2.37	2.39	2.22	65h15m	0h24m
Applu	1.63	2.46	1.69	1.92	100h59m	0h28m

different program runs is due to the different number of basic block that are executed between use and reuse of those 4 million sampled reuse pairs.

We visually explored the analysis results using the SLO tool (available at `http://www.elis.ugent.be/~kbeyls/slo`) [3]. We optimized five SPEC2000 programs with a high cache miss rate by following the suggestions made by SLO. While some of the suggested refactorings could not be legally applied due to true data dependences, we still found a number of useful refactorings, resulting in an average speedup of 2 on a number of different platforms, see Tab. 1. The table also shows that thanks to sampling, the profiling time of these programs with reference input is reduced to less than 30 minutes for all programs.

Furthermore, SLO was also used to optimize the data locality of a video decoder that is being implemented in hardware, leading to a two-fold reduction of off-chip memory accesses. Since the video decoder is bandwidth-limited, this results in a doubled decoding frame rate [9].

5 Related Work

Performance Debugging Tools for optimizing Cache Behavior Too many profiling tools have been proposed to pinpoint cache bottlenecks in the source code to explicitly mention them all here. However, most of these tools focus on pinpointing the source lines, or data structures on which cache misses occurs. As illustrated in the introduction, our SLO tool aims at going further by pinpointing the refactorings that are needed to optimize the temporal locality.

An objective comparison of the usefulness of the different tools in finding good optimizations is hard, due to the human factor involved. Nonetheless, as far as we know, two other works have attempted to improve the data locality of Equake based on profiling results. In both [1] and [10], spatial locality is optimized by rearranging data layout, resulting in reported speedups of 1.4 and 1.24. In contrast, our method results in increased temporal locality, with an average speedup of 2.3. Both methods could be combined to further increase the overall locality and speedup. We are not aware of any earlier successful attempts of optimizing the temporal locality of the Art, VPR, Galgel and Applu programs.

Sampling Reuse Distance and Cache Simulation. The two most used methods to sample memory access traces for cache simulation are time sampling and set sampling [5]. In time sampling, a number of windows of consecutive memory accesses are traced, with large inter-window gaps of non-traced accesses. This method has its limitations for measuring long-distance reuses, since both the use and the reuse must be in the same window. Since there are billions of memory accesses between long-distance reuses in some applications, the windows should be billions of memory accesses large. In set sampling, only a subset of data addresses are sampled. For both time and set sampling, the accuracy cannot be determined theoretically, since the samples are not taken in a uniform random way. Typically, in order to reach less than 2% error, 10 to 20% of all accesses must be sampled [5]. While the calculation of error rates is not directly comparable, our sampling method allows to pinpoint all refactorings that optimize at least 0.01% of all reuses with at most 10% error by only sampling 0.09% of all accesses for the SPEC2000 programs.

Berg [1] uses the MMU in the processor to speed up the detection of reuses, resulting in a time overhead of only 40%. It is not clear how their method could be extended to also measure the code that is executed between reuses.

Other Program Optimization Strategies Based on Analysis of Data Reuses. In recent years, a number of techniques have been proposed that are based on profiling long-distance data reuse, e.g. inserting prefetch instructions selectively [8], inserting cache hints to improve replacement decisions [11], improving spatial locality [12], speculative memory disambiguation of memory instructions [13], optimizing the bandwidth usage in hardware implementations [9], predicting the execution time of programs [14], estimating energy consumption [15], detecting phases in program executions [16], etc. . The sampling proposed in Sect. 3 might be an interesting extension to these methods to reduce their profiling overhead.

6 Conclusion

Reuse path analysis is a new way to look at the locality behavior of a program. A method has been presented to suggest the required refactoring that improves temporal data locality for any given reuse pair. The algorithm and associated data structure make the implementation of the method practical for realistic applications. In addition, reservoir sampling of the reuse paths drastically reduces time and memory overhead. The techniques have been implemented in the GCC compiler and the visualizer SLO was developed to analyze the results. Using these tools, we optimized the temporal locality of five SPEC2000 programs, resulting in an average two-fold speedup on a number of different platforms.

References

1. Berg, E., Hagersten, E.: Fast data-locality profiling of native execution. In: SIGMETRICS. (2005) 169–180
2. Beyls, K., D'Hollander, E.H., Vandeputte, F.: RDVIS: A tool that visualizes the causes of low locality and hints program optimizations. In: ICCS. Volume 3515 of LNCS. (2005) 166–173
3. Beyls, K., D'Hollander, E.H.: Intermediately executed code is the key to find refactorings that improve temporal data locality. In: Computing Frontiers. (2006) 373–382
4. Martonosi, M., Gupta, A., Anderson, T.: Effectiveness of trace sampling for performance debugging tools. In: ACM SIGMETRICS. (1993)
5. Uhlig, R.A., Mudge, T.N.: Trace-driven memory simulation: a survey. ACM Comput. Surv. **29**(2) (1997) 128–170
6. Walpole, R., Myers, R.: Probability and Statistics for Engineers and Scientists. Prentice Hall (1993)
7. Li, K.H.: Reservoir-sampling algorithms of time complexity o(n(1 + log(n/n))). ACM Trans. Math. Softw. **20**(4) (1994) 481–493
8. Fang, C., Carr, S., Önder, S., Wang, Z.: Path-based reuse distance analysis. In: Compiler Construction. Volume 3923 of LNCS. (2006) 32–46
9. Devos, H., Beyls, K., Christiaens, M., Campenhout, J.V., D'Hollander, E.H., Stroobandt, D.: Finding and applying loop transformations for generating optimized FPGA implementations. (Transactions on HiPEAC) submitted.
10. Buck, B.R., Hollingsworth, J.K.: Data centric cache measurement on the intel itanium 2 processor. In: Proceedings of SuperComputing. (2004)
11. Beyls, K., D'Hollander, E.H.: Generating cache hints for improved program efficiency. J. of Systems Architecture **51**(4) (2005) 223–250
12. Zhang, C., Ding, C., Ogihara, M., Zhong, Y., Wu, Y.: A hierarchical model of data locality. In: POPL. (2006)
13. Fang, C., Carr, S., Onder, S., Wang, Z.: Instruction based memory distance analysis and its application to optimization. In: PACT. (2005)
14. Marin, G., Mellor-Crummey, J.: Cross-architecture performance predictions for scientific applications using parameterized models. In: SIGMETRICS. (2004)
15. VanderAa, T., Jayapala, M., Barat, F., Corporaal, H., Catthoor, F., Deconinck, G.: Instruction and data memory energy trade-off using a high-level model. In: ODES. (2004)
16. Shen, X., Zhong, Y., Ding, C.: Locality phase prediction. In: ASPLOS-XI. (2004) 165–176

Integrating TAU with Eclipse: A Performance Analysis System in an Integrated Development Environment

Wyatt Spear, Allen Malony, Alan Morris, and Sameer Shende

{wspear, malony, amorris, sameer}@cs.uoregon.edu

Abstract. The Eclipse platform offers Integrated Development Environment support for a diverse and growing array of programming applications and languages. There is an increasing call for programming tools to support various development tasks from within Eclipse. This includes tools for testing and analyzing program performance. We describe the high-level synthesis of the Eclipse platform with the TAU parallel performance analysis system. By leveraging Eclipse's modularity and extensibility with TAU's robust automated performance analysis mechanisms we produce an integrated, GUI controlled performance analysis system for Java, C/C++ and High Performance Computing development within Eclipse.

1 Introduction

IDEs (Integrated Development Environments) are increasing in popularity across many venues of software development. At the same time they are encompassing a larger set of roles within the software development process, such as advanced debugging, version control and software deployment. Many projects focus on developing new software tools to fill the growing requirements of IDE software systems. However, integration of preexisting command-line or otherwise standalone development tools into an IDE environment also offers several advantages. These include interoperability with preexisting data and tools that support the newly integrated components, user familiarity and, in many cases, the advantages derived from starting with a more mature, robust system.

The Eclipse platform supports a popular Java IDE with additional support available for for C and C++[3]. Support for Fortran and other languages is in development. High performance application development within Eclipse is also receiving increased attention[10]. However, performance analysis is an important component of the software development process that has received little integrated support in any of the Eclipse platform's IDEs, outside of the scope of Java development.

A number of specialized performance analysis systems exist. Although Eclipse projects may be able to make use of them, their interfaces are external to the platform. Most are command line based and not designed with IDE integration in mind. The complexity of a full, multi-language, performance analysis system

M. Gerndt and D. Kranzlmüller (Eds.): HPCC 2006, LNCS 4208, pp. 230–239, 2006.

makes implementing one specifically for a given IDE generally impractical. The difficulty is compounded further by issues of architectural and operating system support. Some performance analysis features, such as hardware event recording, may require platform specific configuration which would complicate deployment of any performance analysis solution internal to the IDE. A practical solution to these problems is to allow the IDE to wrap a consistent interface around an extant performance analysis system already configured and deployed on the desired platform.

TAU(Tuning and Analysis Utilities)[11] is a comprehensive performance analysis system which supports a wide array of architectures, operating systems and programming languages. Its features include automation of many performance analysis operations and utilities for converting its performance data output to a number of formats commonly used by other performance analysis tools. This makes it ideal for integration with the diversely deployed, extensible Eclipse platform.

The following two sections discuss Eclipse and TAU respectively, focusing on the features that facilitate and benefit from their integration. This is followed by a description of the Eclipse plugins that achieve the integration with TAU and an explanation of their implementation and future goals. Finally, the conclusion summarizes the success of the initial work.

2 Eclipse

Eclipse[4] is a popular software platform with support for customized IDE functionality. Its default set of plugins is designed for Java development but the Eclipse community is providing increasing support for other languages such as C/C++ and Fortran. Recently, support for high performance computing has also been provided via the Parallel Tools Platform(PTP)[10].

Two distinct advantages of the Eclipse platform are its portability and extensibility. The former is provided largely by Eclipse's Java-based implementation. Because it is implemented in Java it can be run consistently on Windows, Macintosh and many Unix based OSes. This can facilitate development of software for multiple architectures and helps provide a consistent interface if a user needs to migrate between different systems or operating systems during a project's development.

Because Eclipse is open source users are free to modify and extend its functionality as they see fit. This freedom is enhanced by Eclipse's fundamentally modular architecture. The result is a diverse array of enhancements and plugins made to increase Eclipse's functionality for software development, as well as other tasks.

2.1 The Eclipse Platform

Eclipse is essentially a platform for the support of plugins. The core of Eclipse provides the basic user interface and internal control mechanisms but virtually every useful activity that can be performed within Eclipse relies on a plugin.

Typically a single application, such as the Java IDE, is comprised of a set of plugins subdivided according to their individual functions. Eclipse's plugin API allows the inclusion of functionality from simple context menus to advanced program build managers.

Although Eclipse's default set of plugins are aimed at Java development they provide a significant level of general functionality and support for the expansion of its capabilities. The basic file controls, text editors and preferences widgets, for example, are plugin components that can be reused in many other applications.

Eclipse provides automated utilities for plugin installation and upgrading, but the manual procedure is as simple as moving the folder or jar file associated with a given plugin to the appropriate directory in the Eclipse installation. Additionally Eclipse allows control over which plugins are instantiated at runtime.

2.2 Eclipse Plugins

The Eclipse platform's facilities for extension and modification have resulted in a rich and growing supply of plugins for a wide variety of purposes. These serve both within Eclipse's original capacity as an IDE platform and in other more diverse applications. Performance analysis tools are by no means absent from this. The Test and Performance Tools Platform (TPTP)[14] is an example of a significant effort toward a fully integrated performance analysis system for Eclipse. Although the TPTP possesses a deliberately extensible API, presently it primarily supports analysis only within the JDT.

There are three Eclipse plugin collections, or projects, that lend themselves directly to the integration of the TAU performance analysis tools. The Java Development Tools (JDT), the C/C++ Development Tools (CDT) and the Parallel Tools Platform (PTP) all facilitate the development of programs and the use of programming paradigms that are supported by TAU and none include their own internal mechanisms for performance analysis.

JDT. The JDT[6] assists with Java development by providing a context sensitive source editor, project management and development control facilities, among other features. Most relevant to TAU integration is the JDT's run configuration management system. This grants control over the main method executed when a project is run from within Eclipse along with program arguments, environment variables and related options.

Because the JDT is a fundamental component of the default Eclipse package, it is the most mature of the projects to be integrated with TAU. Its API is fairly static and well documented, easing the development of plugins that add to its functionality.

CDT. Many of the CDT's features are comparable to those of the JDT. In particular the source editing features provide similar functionality such as syntax based colorization, search, replacement and refactoring tools and tree views of source structure. However, the build system of the CDT is naturally quite different. It supports both the use of external makefiles and an internally constructed “Managed” makefile system. In either case the compilation and linking of programs within the CDT is accomplished via user specified compilers and compiler options.

The run management systems of the JDT and CDT are similar in their user interfaces and internal APIs. In the case of a CDT, the run management system requires specification of a compiled binary file.

PTP. The PTP has made significant advances in support of parallel program development within Eclipse. While it is possible to program and compile parallel programs from within the CDT, it has no innate facilities to launch the resulting executables with a given parallel runtime. The PTP grants the ability to both launch and debug parallel programs from within Eclipse. Presently, the PTP only supports the OpenMPI[8] MPI implementation. However, parallel launching and debugging capabilities support for other MPI implementations and OpenMP are in development.

In many ways, the PTP acts as an extension of the CDT. This is especially evident in its compilation and execution system APIs. Thus, the creation of plugins that make use of these components is quite similar between the two projects.

3 TAU

TAU is a mature performance analysis system designed to operate on many different platforms, operating systems and programming languages. In addition to collecting a wide range of performance data it includes resources for performance data analysis and conversion to third party data formats.

Many of TAU's functions are closely bound to the underlying architecture of the system where the analysis takes place. Therefore, TAU is generally configured and compiled by the user to create custom libraries for use in performance analysis. In addition to generating system specific libraries, this configuration process allows specification of many performance analysis options allowing an extremely diverse range of performance experiments to be carried out with TAU. Each separate configuration operation produces a stub makefile and a library file that is used to compile an instrumented program for analysis.

3.1 Instrumentation

TAU's fundamental functionality is based on source code instrumentation. At the most basic level this consists of registering the entry and exit of methods within the program via calls to the performance analysis system. Performance analysis of a given program can be focused on a given set of functions or phases of the program's execution by adjusting which functions are instrumented. A common application of such selective instrumentation is to exclude small, frequently called routines to help reduce performance analysis overhead.

Manual instrumentation can be time consuming, especially for large programs. TAU includes utilities to perform automatic instrumentation of source code.

These utilities rely on the Program Database Toolkit, or PDT[7], to analyze the code so proper instrumentation points can be programmatically determined.

TAU also provides compiler scripts which act as wrappers of the compilers described at TAU's configuration. Use of these scripts in place of a conventional compiler results in fully instrumented binary files without modification to the original source.

3.2 Analysis

Depending on the configuration settings provided to TAU, it can generate a wide variety of performance data. As shown in fig 1, TAU includes utilities to convert both its profile and trace output to a diverse array of other performance data formats, allowing performance analysis and visualization in many third party performance analysis programs.

Additionally, TAU includes its own facilities for analysis of performance data. The ParaProf[2] profile analysis tool (seen in Figure 2), for example, provides a full set of graphical tools for evaluation of performance profile data.

4 Integration

Currently, three separate TAU plugins have been developed for Eclipse. Each allows performance analysis within the scope of a different Eclipse IDE implementation, one for the JDT, one for the CDT and one for the PTP. The TAU JDT plugin requires only the standard Eclipse SDK distribution and allows TAU analysis of Eclipse Java projects. The TAU CDT and TAU PTP plugins allow performance evaluation of C and C++ programs within the standard, sequential, C/C++ IDE implementation and the PTP's parallel implementation respectively. Both the TAU CDT and TAU PTP plugins support Fortran when the FDT plugin is installed.

All of the TAU Eclipse plugins share a similar user interface. As shown in Figure 1, a preferences screen allows the user to specify the location of TAU's installation. When this is done a TAU stub makefile can be selected. The makefile will determine which TAU libraries are included in the program's compilation. The preferences also include options for the output location of trace or profile data, the automatic deployment of the profile analysis tool ParaProf on profile output after program execution and other options relevant to the specific language or platform being analyzed.

TAU enabled compilation is invoked in the build system via a context menu in the project explorer, available for objects eligible for standard build operations. Similarly, executable objects are provided with a menu option that allows them to be run with TAU-relevant options automatically. Essentially, once the TAU plugins for the desired IDEs have been installed and configured, obtaining performance data from an Eclipse project is simplified to a sequence of mouse clicks.

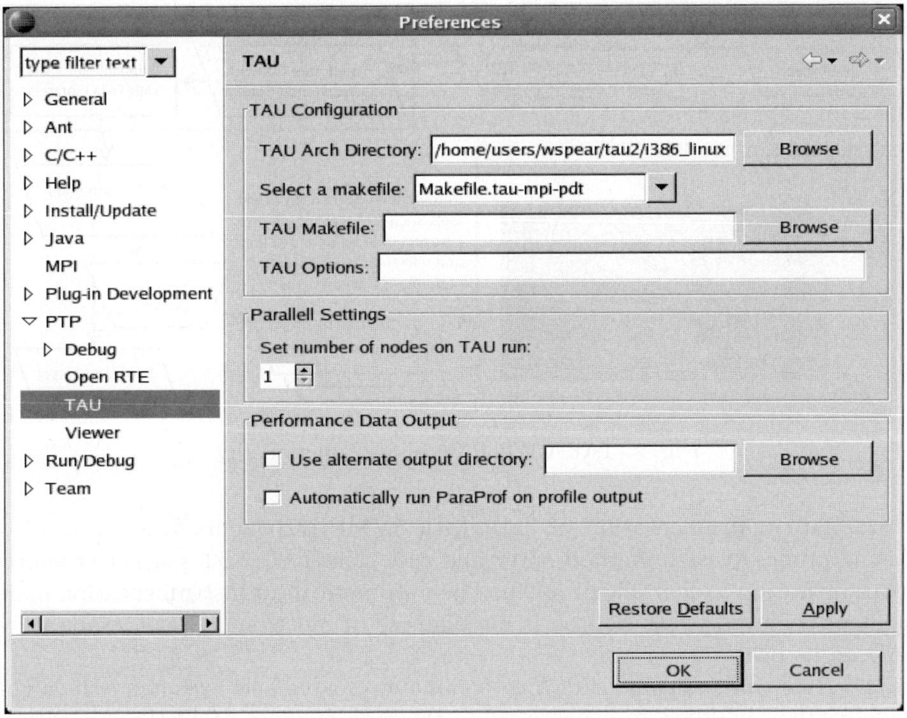

Fig. 1. The preferences screen for the TAU PTP plugin

4.1 The TAU JDT Plugin

The TAU plugin for the Java Development Tools (JDT) adds an *Instrument* command to the context menu of each Java project, package and source file. This command automatically instruments all of the Java source under the selected object. Thus, in addition to instrumenting an entire project, performance analysis can be done selectively at the source file or package level.

The plugin API of Eclipse's Java Development Tools (JDT) includes access to an Abstract Syntax Tree representation of Java source files. The TAU JDT plugin uses the AST to determine the proper instrumentation points within the source files.

The Java file, package and project context menus are also provided with an uninstrument Java command. This strips all automatically inserted TAU instrumentation from the selected Java file or files.

The TAU JDT plugin supports use of a selective instrumentation file, specified in the TAU JDT options panel. Files and methods can be specified for inclusion or exclusion from instrumentation. The selective instrumentation file supports wild card characters as well. The instrumentor simply accepts or rejects methods and files for instrumentation by referencing the names in the file against those returned by the AST.

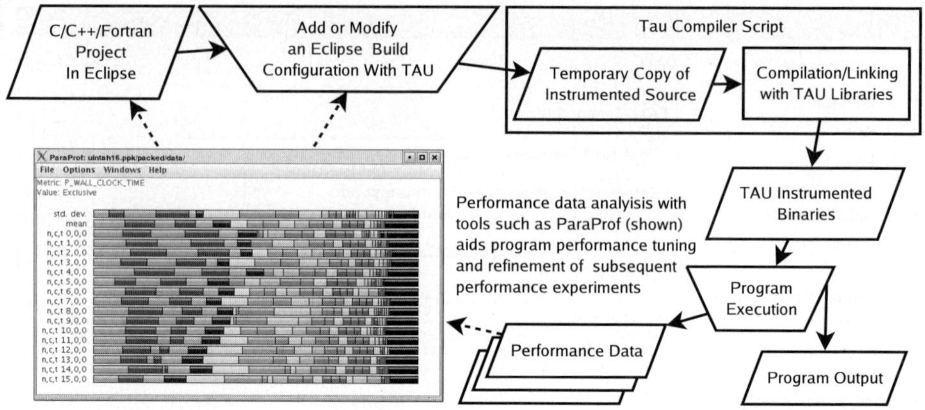

Fig. 2. TAU CDT/PTP plugin functionality

Standard instrumentation of Java with JVMPI[12] incurs the overhead of placing probes at each method entry and exit. The TAU JDT's selective source instrumentation avoids this problem. The automatic Java instrumentation management achieved using Eclipse is an example of functionality not available in TAU on its own.

The *Run Tau Instrumented Java* command creates and executes a Java run configuration modified to accommodate the requirements of TAU's invocation, specifically the inclusion of the TAU Java libraries. It builds a directory structure to contain the performance information generated by the instrumented program. It also generates and deploys the command line and environment arguments that specify the location of the required TAU libraries and the output location of the performance data.

4.2 The TAU CDT Plugin

The C/C++ Development Tools (CDT) project's interaction with TAU differs significantly from the JDT's. TAU's compiler script in conjunction with the TAU stub makefile of an appropriate configuration will automatically perform all of the necessary instrumentation, compilation and linking operations without affecting the source code of the project being analyzed.

The TAU CDT plugin adds a context menu option to Managed Make CDT projects, *Add TAU Configuration*, to create a new build configuration. When this is selected the plugin provides a list of existing build configurations to use as a template. The selected build configuration will be duplicated but with a TAU compiler script specified as the compiler and linker. The CDT plugin interacts with the Fortran Development Tools (FDT) identically to the CDT except that the TAU compiler script replacing the default compiler is Fortran specific. The compiler script is passed the stub makefile chosen in the TAU preferences screen. Because the compiler script recognizes compilation and linking commands as if it were a conventional compiler any preexisting customizations to the selected build configuration will be properly included in the new TAU build configuration. The compiler script's options, which may be set at the plugin preferences screen, include selective instrumentation similar to that described for the JDT.

The *Run Tau* command added to the binary context menu operates analogously to that provided by the JDT plugin. However, because the TAU libraries are included at compilation, the standard run system can be used without providing any TAU specific options. Figure 2 outlines the steps in obtaining performance data from an eclipse project via the plugin.

4.3 The TAU PTP Plugin

For the most part, the Parallel Tools Project (PTP) build system is equivalent to the CDT. Internally, the interfaces with the PTP build and run systems are similar to those of the CDT. The primary distinction is the necessity of setting a default number of processes with which to run the parallel code, as specified in the plugin's options.

As the PTP project develops, further enhancements of the TAU PTP plugin may be required. For example when the PTP implements remote execution of parallel computing jobs the TAU plugin will require a mechanism of obtaining the resulting performance data from the remote system.

Because program performance is an essential goal of high performance computing the TAU PTP plugin is likely to be the main focus of future development. Fortunately, its architectural similarity to the TAU CDT plugin may eventually allow the two to be combined, simplifying deployment and maintenance in the future.

4.4 Future Work

The present implementation of the TAU plugins allows access to most of TAU's capabilities from within Eclipse. However there are several avenues of integration and interface development that remain to be explored. Additionally there may be useful functions that can be developed within the Eclipse framework that have no parallel within the standalone TAU system.

One of the most immediate goals of future TAU plugin development is addition of persistent source instrumentation for projects within the CDT, FDT or PTP, similar to that already implemented in the JDT plugin. TAU's compiler script does offer some control over project instrumentation but allowing persistent instrumentation in the source code would greatly facilitate manual adjustment of that instrumentation. Manual instrumentation allows a finer level of user control which may be desirable in some performance analysis scenarios. Persistent source instrumentation would also necessitate some functionality to manage and allow switching between instrumented and uninstrumented code.

Another goal is to move the integration of TAU and Eclipse to a lower, more internalized level. Ideally, instead of calling on TAU's compiler scripts, the CDT build system will derive that functionality from the TAU plugin itself. Only the TAU makefiles and libraries would be required to build and run a TAU instrumented project once this is accomplished.

An important feature yet to be implemented is advanced management of performance data output. Ultimately the directory based solution will be replaced by a performance database solution within Eclipse analogous to the PerfDMF[5] system already included with the TAU.

In general, closer unification of TAU's peripheral utilities with Eclipse is a priority. For example allowing ParaProf to communicate with Eclipse could permit performance hot-spots depicted in ParaProf to programmatically link back to their source calls in Eclipse.

5 Conclusion

Growing interest in the use of IDEs for parallel computing raises questions regarding how parallel tools will be integrated. The nature of these tools, the variations specified in their software development and platform implementation, and the complexity of their function suggests that re-engineering of the tools to make them more compatible with the IDE framework would be a major undertaking. Instead, an approach that looks at the problem from a point of view of tool interoperability is more productive. We have described in this paper our approach for the integration of TAU with Eclipse following this strategy.

By providing a clean, simple interface between TAU and the Eclipse JDT, CDT and PTP projects we have created a relatively straightforward, unified mechanism of undertaking performance analysis within these respective Eclipse IDEs. High performance computing applications in particular require performance analysis and tuning during their development life-cycle to to achieve their implicit goal of efficient operation. Of course, conventional, sequential programs in more common venues also stand to benefit from IDE-aided performance analysis and optimization. In either case a consistent and robust mechanism for performance analysis is advantageous during the software development process.

Because TAU and Eclipse are only loosely coupled, significant changes in the functionality or interface of either of these actively developing programs is unlikely to disrupt of their integration. Maintenance and continued enhancement of the TAU plugins as described above will continue. TAU Eclipse integration will continue supporting IDE assisted performance analysis within Eclipse as the Eclipse platform and its IDE components continue to mature and as TAU's capabilities extend to new parallel analysis paradigms.

References

1. Amsden, J., "Levels Of Integration: Five ways you can integrate with the Eclipse Platform", http://www.eclipse.org/articles/Article-Levels-Of-Integration/levels-of-integration.html, March 2001
2. R. Bell, A. D. Malony, and S. Shende, "A Portable, Extensible, and Scalable Tool for Parallel Performance Profile Analysis", Proc. EUROPAR 2003 conference, LNCS 2790, Springer, Berlin, pp. 17-26, 2003
3. CDT - C/C++ Development Tools, http://www.eclipse.org/cdt

4. Eclipse, http://www.eclipse.org
5. K. Huck, A. Malony, R. Bell, L. Li, and A. Morris. PerfDMF: Design and implementation of a parallel performance data management framework. In Proc. International Conference on Parallel Processing (ICPP 2005). IEEE Computer Society, 2005
6. JDT - Java Development Tools, http://www.eclipse.org/jdt
7. K. A. Lindlan, J. Cuny, A. D. Malony, S. Shende, B. Mohr, R. Rivenburgh, C. Rasmussen. "A Tool Framework for Static and Dynamic Analysis of Object-Oriented Software with Templates." Proceedings of SC2000: High Performance Networking and Computing Conference, Dallas, November 2000.
8. Open MPI, http://www.open-mpi.org
9. Popescu, V. "Java Application Profiling using TPTP.", Eclipse Corner Article, http://www.eclipse.org/articles/Article-TPTP-Profiling-Tool/tptpProfilingArticle.html, Febuary 2006
10. PTP - Parallel Tools Platform, http://www.eclipse.org/ptp
11. S. Shende and A. D. Malony, "The TAU Parallel Performance System," International Journal of High Performance Computing Applications, ACTS Collection Special Issue, 2005
12. SUN Microsystems Inc., Java Virtual Machine Profiler Interface (JVMPI), http://java.sun.com/j2se/1.3/docs/guide/jvmpi/jvmpi.html
13. TAU - Tuning and Analysis Utilities, http://www.cs.uoregon.edu/research/tau/home.php
14. TPTP - Test and Performance Tools Platform, http://www.eclipse.org/tptp

Scalable Architecture for Allocation of Idle CPUs in a P2P Network

Javier Celaya and Unai Arronategui

University of Zaragoza,
Department of Computer Science and Systems Engineering
C/ María de Luna 1, Ed. Ada Byron, 50018 Zaragoza, Spain
{jcelaya, unai}@unizar.es

Abstract. In this paper we present a scalable, distributed architecture that allocates idle CPUs for task execution, where any node may request the execution of a group of tasks by other ones. A fast, scalable discovery protocol is an essential component. Also, up to date information about free nodes is efficiently managed in each node by an availability protocol. Both protocols exploit a tree-based peer-to-peer network that adds fault-tolerant capabilities. Results from experiments and simulation tests, using a simple allocation method, show discovery and allocation costs scaling logarithmically with the number of nodes, even with low communication overhead and little, bounded state in each node.

Keywords: Parallel and Distributed Architectures, Networking Protocols and Routing and Algorithms, Reliability and Fault-tolerance, Grid Computing, Peer-to-Peer Computing, Parallel and Distributed Algorithms.

1 Introduction

One of the most recent solutions in large scale computing, primarily oriented to embarrasingly parallel and computing-intensive problems, is the use of thousands or even millions of unreliable personal computers connected to the Internet. Projects like SETI@Home [1] and distributed.net [2] established a milestone in the field of Distributed Computing by harnessing the idle cycles of personal computers donated by volunteers to solve problems that could be split into independent parts. Each computer executes the processes associated to one or more parts, receiving the input data and returning the output results. Much work around this idea has been developed covering efficiency, throughput, fairness and security aspects, but the general structure still maintains some intrinsic disadvantages: only one entity in the network generates the workload, and the rest consume it, leading to centralized management and scheduling that negatively affect the scalability and fault-tolerance of the system.

To solve this problem, we propose an architecture where any participant of the network may need idle cycles to complete its tasks. When a node does not have enough computing power to finish its work in a certain amount of time, it may divide it into n independent tasks and query the network for n more idle

M. Gerndt and D. Kranzlmüller (Eds.): HPCC 2006, LNCS 4208, pp. 240–249, 2006.

machines that can each execute one of them. This is not a new idea, but there is little work covering this approximation. Projects with best results have been those who use peer-to-peer (P2P) networks and distributed protocols; with them, problems concerning scalability and fault-tolerance are drastically reduced. The use of an unstructured P2P network is simple and is based on already working ideas like Gnutella or Freenet, but does not allow the application of constraints to the idle CPU search.

For this reason we present a peer-to-peer network based on a balanced tree structure that finds the nearest free CPUs to the one that is demanding the execution of a number of tasks. At any time, any node of the network may request the execution of n tasks; this request is routed by neighbour nodes to those available ones that are closer to the originating client with a fast discovery protocol. The information about existing free nodes is dynamically managed by an availability protocol. These functionalities are obtained with little state in nodes, and low communication and CPU overhead. A simple allocation policy has been designed and implemented to evaluate the architecture behavior.

This paper takes some steps into a complete distributed computing solution, thus we will impose some restrictions to the environment: We assume that nodes execute batch tasks that do not communicate between them, so we won't be addressing the issues that arise from having dependencies. Also, we will suppose that there is low churn, that is, joins and leavings are not frequent. And finally, we will only consider a weak concept of fairness in the allocation of free nodes.

The rest of the article will be structured as follows: In Sect. 2 we will expose an overview of the system architecture and its behavior, and in Sects. 3 and 4 we will detail the protocols that allow the fast and scalable discovery of free nodes. Following, we will briefly present the hierarchical overlay topology and its management in Sect. 5. Finally, in Sect. 6 we will show the experimental results and in Sects. 7 and 8 we will explain what other work has been presented concerning distributed computing in peer-to-peer networks and the conclusions of this investigation.

2 System Architecture

The system has a three layer architecture:

- The first one defines the *connectivity protocol* that maintains the overlay links in the network. It conforms a tree-based network overlay, derived from the B-Tree [3], thus it is a balanced tree where each node can have between m and $2m$ children and the height is always a logarithm of the number of nodes N. The protocol states how nodes join and leave the network, how the tree is kept balanced, and how node failures are dealt with to rebuild the structure.
- The second layer is described by the *availability protocol*, that distributes information relative to the number of free nodes and computing power each time it changes. Every node of the network stores the global state of the branch that hangs below it, and communicates updates to its parent so it

can recompute the state of its own branch. This protocol uses a number of techniques that prevent the upper levels of the tree from being flooded with update notifications, while maintaining the information accurate enough to maximize the network use. Also, the conservative approach of notification updates yields to a more stable behavior of this protocol.

– Finally, the *discovery protocol* uses the information stored by each branch to route free nodes requests up and down the tree. It tries to find those free nodes that meet a trade-off between proximity to the client and computing power by distributing the requests among the appropriate branches at each level. Therefore the search is performed in a number of network hops that depends only on the height of the tree and, consequently, on the logarithm of the number of nodes of the network.

3 Discovery of Free CPUs

As it has been said, when a node has a number of tasks to be done, it requests the network to find that number of idle machines. This service is accomplished with the discovery protocol, that works as follows. By applying heuristic rules, it will try to allocate the fastest and nearest free nodes, so that tasks execution is efficiently done. To find them, the tree structure is exploited. Each inner node stores information about its descendants; not exhaustive information, but more general information about the branch as a whole. To be concrete, it knows the number of free CPUs of the branch, the maximum computing power and the minimum number of hops to a free node. This way, the management of this information becomes scalable as it does not depend directly on the number of nodes. How it is exactly managed will be explained in Sect. 4.

A node that receives a message with n pending tasks will first check if itself is ready to execute a new one. If so, it takes one of the tasks from the message. Then it distributes the remaining tasks between its child nodes according to the number of free nodes each branch has, calculated with the availability protocol, giving priority to the branches having more computing power or less hops to a free node. If it is not enough with the children branches to cope with all the tasks, then a new message with the last tasks will be sent to the father so that it can reach more distant branches. When the message arrives at the root of the tree and it cannot be sent to another branch, it is returned to the originating node meaning that there are no free nodes left in the network.

The worst case would be that of a leaf node that needs to allocate every node of the net. The request would have to go up to the root and then down to the rest of the tree; that is the longest path a request would traverse. As discovery of idle nodes is done concurrently in every branch, that would be the same as reaching one idle leaf node in the opposite side of the tree. This is done in $O(\log_m N)$ hops, being m the minimum number of children per node in the balanced tree and N the number of nodes in the network, thus making the discovery protocol highly scalable.

This is a best effort network. That is, the intermediate nodes make its best to route the message to the most suitable free nodes, but the reception is not guaranteed. In fact, the failure of nodes is frequent in a peer-to-peer network. For this reason, both the originating node and the allocated ones must avoid problems in the discovery phase and when sending the actual work to execute using timeout mechanisms, acknowledge messages and retransmissions.

4 CPU Availability Management

The information each node stores about its branch must be communicated to its parent so that it can efficiently route requests to the idle nodes it has under itself. Therefore, each node not only has information about its branch, but also about each of its sub-branches. The way this information propagates is critical, because it must be kept up to date without flooding the network with notification messages, specially near the root where there are less nodes per level. This propagation is performed by the availability protocol. Basically, when a node receives a notification of change from one of its child nodes, it must decide if it has to inform its parent, too.

With the maximum computing power and the minimum number of hops to a free node, the process is simple. The inner node has to calculate the new maximum and minimum values, respectively, between its child nodes and itself, and if it changes, route the new information to its parent immediately. When a notification goes up one level, the parent will compare these values with that of the other branches, and if it is not the new maximum or minimum then the notification will not propagate. So, it is less probable to jump up one more level as the number of nodes compared is greater, making the traffic self-limited and unlikely to reach the top levels.

The problems arise with the number of free nodes, because when a node gets ready (busy), the number of free nodes of each of its ancestors increases (decreases) by one. If the notification were sent with every change, the root would get informed of all of them, what leads to an unacceptably high traffic in the top levels. For this reason, we have designed a technique that delays the notification of the number of free nodes at each level of the tree, reducing the traffic routed up to the root. The basis of this method lies on sending a notification when the change is "important" enough. Actually, this means that the most significant bit set to one changes; that is, the number of free nodes crosses a boundary of power of two. For example, a notification would be sent if this value changes from 7 to 8 (111b and 1000b in binary) or from 32 to 31 (100000b and 11111b), but not when it changes from 23 to 24 (10111b and 11000b). Although this yields to a precision lack, there are three main reasons for using this technique:

1. Trying to provide optimality based on exact information is senseless when we are dealing with millions of nodes that are continually and concurrently changing state.
2. The traffic of notifications in the top levels is reduced because as a notification goes up the tree it is less probable of being routed to the next level. This depends also on the number of free nodes, as a high number has also less probabilities of being routed.

3. When the number of free nodes is low, the precision of this value is better. This is most relevant as the nodes of the network get busy, because they are correctly well-spent when there last only a few free nodes.

There are two policies deciding what availability value to take as reference for a branch when a child node notifies a change to its father: optimistic and conservative. An optimistic policy would use the same value sent by the child. On one hand, it has the advantage of having better precision in the information each node stores about its branch, but on the other hand the real number of free nodes of a branch could decrease and be less than the number its father is using, making top level nodes route requests to branches that cannot cope with them. A conservative policy would store a lower value, for example the higher power of two less than or equal to the notified value. With such a policy, the system has a better behavior against situations when there last very few free nodes, as it delays too big requests, although it does not make the most of the network.

We have decided to adopt a conservative approach. It forces a stabilization mechanism in the value of free CPUs each node contains, providing a gradual convergence in the occupation of the network.

5 B-Tree Based Topology

The overlay network topology is a hierarchy where every node of the network is mapped to a node of the tree. In our approximation we use a B-Tree [3] variant; it maintains the balance in every join and departure and allows more than two child nodes, thus reducing the tree height. But the main objective of the tree is grouping nearby nodes in the same branch. However, the concept of locality usually depends on many variables, so it is actually an approximation. We have decided to use the simple yet effective way of organizing the nodes in the tree by their physical address, their IP address actually. Based on the subnet partitioning of the IP address space and the studies on geographic locality of IP addresses [4], this method allows a fast and easy decision of where to insert a node in the tree when it joins the network, while maintaining good metrics between nodes of the same branch, specially near the leaves.

Our tree has some differences with the original B-Tree model. Each node holds only one value (its IP address) and forms part of a group of siblings; therefore, every node of the network participates in the management of the tree. There exist a constant m so that every node not being the root of the hierarchy always has between m and $2m$ siblings. If these limits are exceeded, the tree must be rebalanced by splitting or joining groups. Also, every node that is not a leaf has a pair of values that represent the interval of addresses of its descendants, including itself. These intervals are used to route messages along the tree, mainly in the operations of insertion and deletion of nodes.

Concerning *fault-tolerance*, every node knows the address of the k predecessors and k successors at the same level (they can be "brothers" or "cousins"). When a node fails, the tree structure can be repaired by its neighbours using these references, because they allow the communication between a node and the brother of its dead father. The value of k is an trade-off between fault-tolerance and an overload in the management of the tree.

5.1 Joining and Leaving the Network

The connectivity protocol consists in two operations that affect the structure of the network: joining and leaving. Joining is usually easier: when a node requests an insertion, the request message is routed up the hierarchy looking for a node whose interval contains the address of the new node, and then it goes down until it reaches the node with the nearest address to the new node's address. Finally they become brothers and the new node updates its references to its neighbours. Then the father node is notified, and it may request a group split to re-balance the tree if the number of its child nodes is greater than $2m$. When a node is added to the group of the root and it already has $2k$ nodes, a new root node will be created.

By leaving the network we assume, usually, a voluntary action, so the leaving node will supply its neighbours with the necessary information to maintain the network connectivity. First of all, a leaving node must check if it has any child. If so, it looks for a leaf node that becomes the new father of all of them, similarly to the creation of a new root node. Once done, or if it had no child nodes, it notifies its siblings and its father that it is going to leave and then they update their reference lists. Similarly to the joining, when the father node is notified of the node leaving, it must check if the number of child nodes is less than m. In that case, it will ask its predecessor or successor to send it child nodes, or to join into only one branch. One special case is when the father is the only one node in the root group. Then it will check if it has less than $2k$ child nodes, and if it has so, it will insert itself at the leafs, leaving its children as the new root group. In some cases nodes can fail and leave the network without notification. In this case, the fault-tolerance strategy presented above is applied.

6 Experimental Results

This architecture has been implemented as a simulated system with the OMNeT++ simulation framework. The allocation policy used in the tests has been a simple one, where as soon as nodes are discovered they are allocated.

Tests have been done aimed to measure free nodes discovery time, control messages traffic and CPU load. Every test has been issued with variations in the number of nodes, N, and the B-Tree parameter, m, to study the impact of the size and structure of the network in the performance of the protocols. The simulations have been performed with up to 50000 nodes and values of m from 6 to 10. Variations on the duration of the tasks and the size of the data have also been applied to recreate more realistic situations. There are three constants, though: the latency of the network connections has been established to 200 ms, the mean continental value for Internet, to simulate a very wide area network; 1

Mbps has been taken for the bandwidth, a conservative value for a home Internet connection; and the mean computing power of the nodes has been set to 2000 MIPS.

Time tests show that both the number of nodes and the number of child nodes per parent affect the discovery of free nodes. Just as expected, the last free node of the n requested is reached in $O(\log_m n)$ hops. For this reason, a network with a higher value of m performs better, while an increasing value of n is hardly appreciated. The results of the free nodes discovery time tests can be seen in Fig. 1 as a logarithmic growth. We can extrapolate the results to higher values of n. For example, we calculate that, for the test network, requesting the execution of 100,000 tasks would discover 100,000 free nodes in 2 seconds, 1,000,000 in 2.4 seconds, 10,000,000 in 2.8 seconds, and so on.

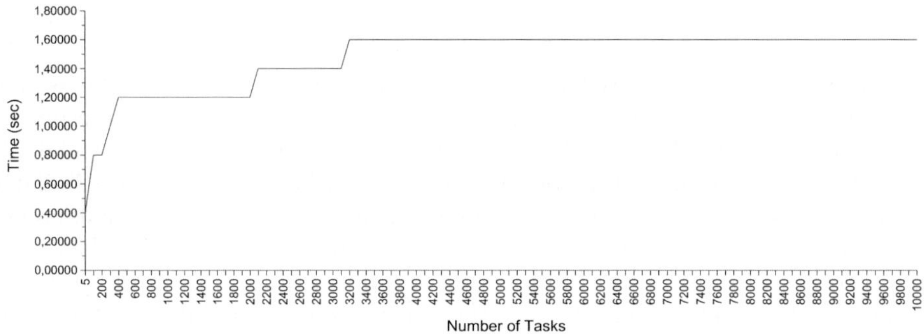

Fig. 1. Discovery time for as many free nodes as requested tasks. The test network has 50000 nodes and $m = 10$. One network hop is 200 ms.

Control traffic (traffic of non-data messages) and CPU load tests have been done under two situations: participants have a normal and high activity. Normal activity means that there are frequent requests from randomly chosen nodes, but the network does not get completely busy. On the other hand, under high activity, every node is busy and continuously receiving new requests, so we expect this to be the worst case. Traffic has been measured in bytes per second. CPU load is more difficult to measure in a simulation, but as every message is managed in constant time we have decided to express it in terms of messages per second. Tables 1 and 2 show the results of the normal and high-activity behavior. They present the value of m, the tree height and the mean and maximum values of CPU load and control traffic for the root and leaves of a network of 50000 nodes.

While the discovery protocol was positively affected by the value of m, the overall system load suffers when the tree is lower because each inner node has to deal with a greater number of child nodes, thus a trade-off is needed between them. Looking at each network variant it can be seen that, by using the availability protocol, under normal behavior both control traffic and CPU load is heavier at the leaves than at the root nodes. Also, control traffic hardly reaches

Table 1. CPU load and control traffic under normal activity. The net has 50000 nodes, with $m = 10$ and a bandwidth of 1Mbps.

		Root				Leaves			
	Tree	Load (msg/s)		Traffic (Bps)		Load (msg/s)		Traffic (Bps)	
m	height	mean	max	mean	max	mean	max	mean	max
4	7	0.08	2.87	24.10	519.98	3.56	4.21	1235.89	1343.51
6	6	0.09	5.64	24.98	1005.69	3.71	4.33	1293.18	1405.28
8	5	0.09	5.64	25.28	1005.35	3.85	4.56	1330.41	1436.15
10	5	0.09	5.62	25.43	999.48	4.12	5.77	1435.08	1485.60

Table 2. CPU load and control traffic under high activity. The net has also 50000 nodes, with $m = 10$ and a bandwidth of 1Mbps.

		Root				Leaves			
	Tree	Load (msg/s)		Traffic (Bps)		Load (msg/s)		Traffic (Bps)	
m	height	mean	max	mean	max	mean	max	mean	max
4	7	0.13	28.20	38.19	4980.56	39.68	41.62	14359.74	16031.69
6	6	0.13	28.46	39.53	5021.05	44.20	50.55	15198.24	16205.83
8	5	0.14	27.65	39.75	4526.82	54.15	59.85	17417.23	19256.67
10	5	0.14	28.44	40.05	5018.49	63.72	65.29	19566.90	21947.28

1KBps, what represents less than 1% of the total bandwidth. However, under high activity rate the root suffers waves of very high CPU load and control traffic.

Results are promising. As we can see in Tables 1 and 2, control overhead is very low. Under normal activity, the control traffic is only 1485 Bps and the CPU load only reaches 5.77 messages per second, in the worst case. And under heavy activity, the control traffic is 21947 Bps and the CPU load is 65.29 messages per second.

7 Related Work

As it has been pointed out in the introduction, the main approximation until now to a highly scalable distributed computing environment has been the harnessing of idle cycles donated by volunteers, as in SETI@Home project, the BOINC generic framework [5] and distributed.net. Those projects use the traditional client/server paradigm to schedule tasks and return results, what soon leads to scalability problems. For that reason, more elaborated network structures and distributed algorithms have been adopted. One example is Javelin++ [6], which extends the concepts of Javelin [7] replacing the broker that scheduled the tasks with a network of brokers. Recently, more strict peer-to-peer networks have been used to select the nodes which would execute the tasks. BOINC and similar projects adopt an application-driven perspective, in which the existence of an element that is generating all the workload determines the structure of the network and the management algorithms. Following a more general view, another family of projects, in which this paper is included, have proposed an architecture where every participant can generate the workload, which is better suited for this peer-to-peer philosophy, as every node is equal to the other ones.

CompuP2P [8] is one of the first works to use a decentralized peer-to-peer network to manage processor cycles as a shared resource. It arranges all the nodes in a Chord [9] ring and organizes them into 'compute markets', where idle cycles are traded with. However, it presents a scalability problem because it has no mechanism to limit the number of nodes in a market or to balance load between markets. G2-P2P [10] uses an object-oriented approach. It uses Pastry [11] to create a Distributed Hash Table (DHT) where computation objects are stored. Each object is assigned a random ID and stored in the Pastry node closer to that number. Using an uniform hashing function they claim to achieve a good load-balancing property, but there is no other criterion to select the most appropriate free node. In [12] the Pastry DHT is also used, but exploiting its locality awareness to discover near idle nodes. It then announces availability with controlled message floodings, what leads to inefficiency as a node surrounded by busy neighbours won't find a free node which is more distant than the maximum number of hops that a request is allowed to make. On the other hand, in [13] it is proposed the use of an unstructured overlay network, as it is easier to manage, and traverse it with random walks. Even though, that is also inefficient because there is no way of knowing if the next node of the walk is free or not.

For the overlay topology, other authors have proposed the use of a virtual tree on top of a DHT, where each node store only part of a tree index. Examples of this are P-Tree [14], P-Grid [15] and VBI-Tree [16]. However, they rely on a uniform distribution of the shared resource; for example, using a uniform hashing function for the DHT. For that reason, BATON [17] uses a balanced binary tree where each node of the network maintains one node of the tree. This type of organization is better suited for a non-uniform resource distribution because the tree gets balanced automatically when the insertions or deletions occur within the same zone. We adopt these ideas but with more than two children per node.

8 Conclusions and Future Work

In this paper, we have presented a network architecture that discovers the presence of idle machines with a scalable ($O(log_m N)$) and fast (1.8 seconds for 10000 requested tasks) method. It organizes the nodes in a balanced tree structure to efficiently distribute information about free nodes in a per-branch basis, that is eventually used to route the request from a client to the appropriate idle CPUs. The connectivity protocol, discovery protocol and availability protocol are all three designed in a totally distributed way, that provide high scalability and fault-tolerance. Moreover, the experimental simulation results show low overhead in the control traffic and CPU load.

We envision to validate these results with a real prototype to be implemented over the PlanetLab testbed. We believe this can be a valuable step to develop system support for high performance computing applications.

References

1. Anderson, D.P., Cobb, J., Korpela, E., Lebofsky, M., Werthimer, D.: Seti@home: an experiment in public-resource computing. Commun. ACM **45**(11) (2002) 56–61
2. Distributed.net: http://www.distributed.net (2000)
3. Bayer, R., McCreight, E.M.: Organization and maintenance of large ordered indexes. In: Record of the 1970 ACM SIGFIDET Workshop on Data Description and Access, November 15-16, 1970, Rice University, Houston, Texas, USA (Second Edition with an Appendix), ACM (1970) 107–141
4. Freedman, M., Vutukuru, M., Feamster, N., Balakrishnan, H.: Geographic Locality of IP Prefixes. In: Internet Measurement Conference (IMC) 2005, Berkeley, CA (2005)
5. Anderson, D.P.: Boinc: A system for public-resource computing and storage. In: GRID. (2004) 4–10
6. Neary, M.O., Brydon, S.P., Kmiec, P., Rollins, S., Cappello, P.: Javelin++: scalability issues in global computing. Concurrency: Practice and Experience **12**(8) (2000) 727–753
7. Christiansen, B.O., Cappello, P.R., Ionescu, M.F., Neary, M.O., Schauser, K.E., Wu, D.: Javelin: Internet-based parallel computing using java. Concurrency - Practice and Experience **9**(11) (1997) 1139–1160
8. Gupta, R., Somani, A.K.: Compup2p: An architecture for sharing of computing resources in peer-to-peer networks with selfish nodes. In: Online Proceedings of Second Workshop on the Economics of Peer-to-Peer Systems, Harvard University (2004)
9. Stoica, I., Morris, R., Liben-Nowell, D., Karger, D.R., Kaashoek, M.F., Dabek, F., Balakrishnan, H.: Chord: a scalable peer-to-peer lookup protocol for internet applications. IEEE/ACM Trans. Netw. **11**(1) (2003) 17–32
10. Mason, R., Kelly, W.: G2-p2p: A fully decentralised fault-tolerant cycle-stealing framework. In: ACSW Frontiers. (2005) 33–39
11. Rowstron, A.I.T., Druschel, P.: Pastry: Scalable, decentralized object location, and routing for large-scale peer-to-peer systems. In: Middleware. (2001) 329–350
12. Butt, A.R., Fang, X., Hu, Y.C., Midkiff, S.P.: Java, peer-to-peer, and accountability: Building blocks for distributed cycle sharing. In: Virtual Machine Research and Technology Symposium. (2004) 163–176
13. Awan, A., Ferreira, R.A., Jagannathan, S., Grama, A.: Unstructured peer-to-peer networks for sharing processor cycles. Journal Parallel Computing (PARCO) **32**(2) (2006) 115–135
14. Crainiceanu, A., Linga, P., Gehrke, J., Shanmugasundaram, J.: Querying peer-to-peer networks using p-trees. In: WebDB '04: Proceedings of the 7th International Workshop on the Web and Databases, New York, NY, USA, ACM Press (2004) 25–30
15. Aberer, K.: P-grid: A self-organizing access structure for p2p information systems. In: CooplS '01: Proceedings of the 9th International Conference on Cooperative Information Systems, London, UK, Springer-Verlag (2001) 179–194
16. Jagadish, H.V., Ooi, B., Vu, Q., Zhang, R., Zhou, A.: Vbi-tree: A peer-to-peer framework for supporting multi-dimensional indexing schemes. In: 22nd IEEE International Conference on Data Engineering (ICDE), 2006 (to appear). (2006)
17. Jagadish, H.V., Ooi, B.C., Vu, Q.H.: Baton: A balanced tree structure for peer-to-peer networks. In: VLDB. (2005) 661–672

A Proactive Secret Sharing for Server Assisted Threshold Signatures*

Jong-Phil Yang[1,**], Kyung Hyune Rhee[2], and Kouichi Sakurai[3]

[1] The Korean Intellectual Property Office, Government Complex Daejeon, Dunsan-Dong, Daejeon 302-701, Republic of Korea
jpyang@lisia21.net

[2] Division of Electronic, Computer and Telecommunication Engineering, Pukyong National University, 599-1, Daeyeon3-Dong, Nam-Gu, Pusan 608-737, Republic of Korea
khrhee@pknu.ac.kr

[3] Graduate School of Information Science and Electrical Engineering, Kyushu University, 6-10-1 Hakozaki, Higashi-ku, Fukuoka 812-0053, Japan
sakurai@itslab.csce.kyushu-u.ac.jp

Abstract. Threshold signature schemes distribute secret information to several servers and make the whole system that maintains the secret information fault-tolerant. Since threshold signature schemes typically assume that the shared signing function can only be activated by a quorum number of servers. If anyone has a power to activate the signing function of servers, he can easily compute valid signatures for a specific organization without knowing the private key. S. Xu et al. proposed a general construction to build threshold signature schemes (called as server assisted threshold signatures) which provide an organization (e.g., a user) with controllability for activating his private signing function in a certain enhanced way. In this paper, we newly propose proactive secret sharing schemes which are suitable for server-assisted threshold signatures.

Keywords: Secret sharing, Fault tolerant and Distributed Computing.

1 Introduction

A lot of organizations use their secret private keys to sign contracts or important electronic documents for secure transaction. Since these secret private keys for signing become main targets of malicious adversaries, the secrecy of private keys is one of the most import issues in secure electronic commerce. A promise solution to this problem is to distribute the signing function among multiple parties. Threshold signature schemes distribute secret information to

* This work was partially supported by Grant No. R01-2006-000-10260-0 from the Basic Research Program of KOSEF, and Strategic International Cooperative Program, Japan Science and Technology Agency (JST).

** This work has been done in part while at Kyushu University.

M. Gerndt and D. Kranzlmüller (Eds.): HPCC 2006, LNCS 4208, pp. 250–259, 2006.

several servers, and make the whole system that maintains the secret information fault-tolerant. However, to maintain a private key securely, it is too costly that each organization constructs a distributed system for securely managing its private key. Therefore, it is necessary to build a security system which stores securely secret private keys instead of organizations and performs securely their cryptographic functions. When a private key is securely shared among several servers, a threshold signature scheme typically assumes that the shared signing function can be only activated by a quorum number of servers. Therefore, it is inappropriate for settings where an organization employs some servers for a threshold protection of its private signing function. That is, if anyone has a power to activate the signing function of servers, he can easily compute valid signatures for a specific organization without knowing the private key. S. Xu et al. provided a general construction to build threshold signature schemes (called as server assisted threshold signatures) which provide an organization (e.g., a user) with *controllability* for activating his private signing function in a certain enhanced way [11]. In this paper, we design two types of proactive secret sharing schemes which are suitable for server-assisted threshold signatures.

2 Preliminaries

2.1 Model and Assumptions

In this paper, we use the following notations for description.

- Servers: There are n servers, $P_1, \ldots, P_n$, in the system. Every server keeps securely shares of each user's private key, and performs a signing function instead of each user.
- User: His secret private key is securely shared among n servers. At least t servers out of n servers perform signing function for the user to generate a valid signature, where $n \geq 2t - 1$. An organization, an individual or a certificate authority can act as a user.
- Communication: Servers have access to a dedicated broadcast channel; by dedicated we mean that if server P_i broadcasts a message; it is received by every other server and recognized that it came from P_i.
- Time: Time is divided into time periods which are determined by the common global clock. Each time period consists of a short refresh period, during which the players engage in an interactive refreshing protocol. The proactive secret sharing presented in this paper assumes a synchronous network.
- Adversary: We assume *short-term constrained adversary* introduced in [3] to characterize adversary; the adversary cannot break t or more servers during any time period. The adversary is also computationally bounded, implying that he cannot solve factoring and discrete logarithm problem.

Our proactive secret sharing schemes can be implemented via either threshold Schnorr in [9] or threshold RSA in [12]. Since the whole exponential operations in [12] are performed over integer, and the modules value in [9] are public because of obvious reasons, we use *abstract modulus* for the rest of this paper for generality.

2.2 Server Assisted Threshold Signatures

S. Xu et al. proposed a server assisted threshold signature scheme which the signing function is activated by a user [11]; we say that the server assisted threshold signature scheme provides *user controllability* for activating the user's signing function. It is based on hybrid threshold signature scheme, which is the combination of two-party signature schemes such as [7][8] and threshold signature schemes.

`Initialization:` A user(U) performs a 2-out-of-2 secret sharing to share his private key x;

$$x \overset{(2,2)}{\overbrace{\longrightarrow}} (x_U, x_P),$$

where x_U is U's share. U shares x_P among n servers by using an appropriate secret sharing scheme, namely

$$x_P \overset{(n,t)}{\overbrace{\longrightarrow}} (x_P^{(1)}, \cdots, x_P^{(n)}),$$

where each $x_P^{(i)}$ $(1 \leq i \leq n)$ is P_i's share.

`Signing:` Let g_1, g_1', g_2 and g_2' be appropriate algorithms which depend on the underlying ordinary signature scheme. To sign a message M, U generates a partial signature $\sigma_U = g_1'(M, x_U)$. Then, P_{i_j} contributes its partial signature $\sigma_P^{(i_j)} = g_1(M, x_P^{(i_j)})$, where $1 \leq i_j \leq n$ for $1 \leq j \leq t$. Given t valid partial signatures, anyone can compute $\sigma_P = g_2(\sigma_P^{(i_1)}, \cdots, \sigma_P^{(i_t)})$, where $1 \leq i_j \leq n$ for $1 \leq j \leq t$. Given σ_U and σ_P, anyone can compute a signature $\sigma = g_2'(\sigma_U, \sigma_P)$.

`Verification:` The verification algorithm is the same as that of the underlying signature scheme.

2.3 Combinatorial Secret Sharing

L. Zhou proposed combinatorial secret sharing which is based on additive secret sharing [5]. To avoid confusion, he used *share sets* to denote shares of a secret x by using a combinatorial secret sharing and used *shares* of x only for the values comprising a standard secret sharing. We use a function $CSS(n, t, x)$ for describing (n, t)-combinatorial secret sharing, where x is a secret.

1. Create $l = \binom{n}{t-1}$ different sets $T_1, \ldots, T_l$ of servers. These sets of servers represent the worst-case failure scenarios : sets of servers that could all fail under the assumption that at most $t-1$ servers are compromised.
2. Create l shares $\{x_1, \ldots, x_l\}$ using (l, l)-additive secret sharing. Associate share x_i with failure scenario T_i.
3. Include secret share x_i in X_P, the share set of a server P, if only if P is not in corresponding failure scenario. That is, for any server P, share set X_P equals $\{x_i \mid 1 \leq i \leq l \wedge P \notin T_i\}$.

Note that, without assigning x_i to any server in a failure scenario T_i, they ensure that servers in T_i do not have all together l shares to reconstruct the secret x. For any set Ω of servers, the constructed share sets satisfy the following conditions:

- Condition 1 : $\bigcup_{P \in \Omega} X_P = \{x_1, x_2, \dots, x_l\}$, where $|\Omega| \geq t$.
- Condition 2 : $\bigcup_{P \in \Omega} X_P \subset \{x_1, x_2, \dots, x_l\}$, where $|\Omega| \leq t - 1$.

2.4 Verifiable Secret Sharing

Verifiable secret sharing provides means for servers to check whether a set of shares constitute a sharing of a secret, so that erroneous shares from compromised servers can be detected and discarded. In this paper, we use generic functions to perform verifiable secret sharing for (l, l)-additive secret sharing in [5].

- $oneWay(x) = y : y$ is the validity check for secret x.
- $oneWay(x_i) = y_i$ for all $1 \leq i \leq l$: guarantees that $\{y_1, \dots, y_l\}$ are the validity checks for all shares.
- $vcConstr(y_1, \dots, y_l) = y$: ensures that y can be constructed from $\{y_1, \dots, y_l\}$ using $vcConstr$.

Each function mentioned above can be implemented by slight variation of Feldman's style of verifiable secret sharing [6]. Owing to self-explanatory for constructing additive versions, the concrete implementations are not introduced in this paper.

3 Conceptional Approach

In contrast to the ordinary PSS in [2][10], our PSS is a user-intervened method because of the nature of the server assisted threshold signature. Let x_U be the part held by the user, x_P be the part distributed among n servers, where $1 \leq i \leq n$. To periodically renew shares for x_P, we describe the concept of our PSS from (n, t)-polynomial secret sharing of view. When x_P is distributively stored as a value of $f(0) = x_P$ of a $t - 1$ degree polynomial $f(\cdot)$, we can update this polynomial by adding it to a $t - 1$ degree polynomial $g(\cdot)$, where $g(0) = \theta$, and θ is a *renewal value*, so that $f^{new}(0) = f(0) + g(0) = x_P + \theta$. We can renew shares for x_P thanks to the linearity of the polynomial evaluation operation:

$$f^{new}(\cdot) \leftarrow f(\cdot) + g(\cdot) \Longleftrightarrow \forall i \; f^{new}(i) = f(i) + g(i)$$

In above description, the value x_P shared among servers is changed. However, if the user updates his part as $x_U \circ \theta$, the user can successfully derive a real signature from σ_P which is generated by the newly changed and shared x_P[1]. In

[1] According to the implementation of g_2' in the section 2.2, the operator $\circ$ should be decided. Let us consider an example based on RSA without considering modulus operator. A user's private exponent (x) can be divided into two halves; $x = x_U - x_P$. Let m be a message to sign. In case of $g_2'(m^{x_U}, m^{x_P}) = m^{x_U}/m^{x_P}$, $\circ$ operator must be $+$.

this paper, we present two types of proactive secret sharing schemes, user original proactive secret sharing (shortly called as user origin PSS) and server original proactive secret shring (called as server origin PSS), according to someone who initiates the activity of proactive secret sharing: in user original PSS, the user initiates PSS, and the servers initiate PSS in the server original PSS.

4 User Origin PSS

In this section, we introduce user original PSS which consists of two protocols; *key renewal* and *key recovery*. To initialize the system, the user and n servers perform the following procedures. From now, we use x instead of x_P for easy description.

`(Step 0)`. A user performs initially $CSS(n,t,x)$. After that, the user sends securely X_{P_i} to P_i, and makes $oneWay(x) = y$ and $oneWay(x_i) = y_i$ public, where $1 \leq i \leq l$. A set of at least t servers of n can construct x. In this paper, we assume that this setup is securely performed.

4.1 Key Renewal

`(Step 1)`. The user generates randomly a *renewal secret* θ. The user performs $CSS(n,t,\theta)$ and generates additive shares $\{\theta_1, \theta_2, \dots, \theta_l\}$. Then, the user generates n *renewal sets*;

$$R_{P_i} = \{\theta_k \mid 1 \leq k \leq l \wedge P_i \notin T_k\}$$

After that, the user sends securely R_{P_i} to P_i, for each i. Then, he makes $oneWay(\theta)$ and $oneWay(\theta_k)$ public, where $1 \leq k \leq l$.

`(Step 2)`. Each P_i computes the validity checks for θ_i as many times as the number of receiving θ_i by itself, and compares them with the corresponding public values, $oneWay(\theta_i)$s, respectively. If the result is true, it checks

$$vcConstr(oneWay(\theta_1), \dots, oneWay(\theta_l))? = oneWay(\theta).$$

If the result is true, it performs the next steps. Otherwise, the procedure of key renewal is restarted `Step 1`. Each P_i computes a *new share set* $X_{P_i}^{new}$ by using X_{P_i} and R_{P_i}:

$$X_{P_i}^{new} = \{x_k + \theta_k \mid 1 \leq k \leq l \wedge P_i \notin T_k\}.$$

Finally, the new secret part shared by n servers, x^{new}, must be $x + \theta$. Then, each P_i deletes X_{P_i} and θ_k from memory. After that, Every P_i sends *Update success message* to the user. After receiving them, the user updates its own part as $x_U^{new} = x_U \circ \theta$, and deletes all secret values for key renewal from memory.

`(Step 3)`. All validity check values for x^{new} are publicly computed;

$$vcConstr(y_i, oneWay(\theta_i)) = y_i^{new}$$
$$vcConstr(y, oneWay(\theta)) = y^{new},$$

where $1 \leq i \leq l$.

4.2 Key Recovery

If a server P_m is corrupted, it is rebooted and initialized. The system initialization means that P_m is recovered from some Trojans or trapdoors.

(Step 1). After that, each $P_{i \neq m}$ sends securely necessary shares, $X_{P_i}^{new} \cap X_{P_m}^{new}$, to P_m.

(Step 2). After collecting necessary shares for recovery, P_m can verify the correctness of the received shares by using new validity check values at Step 3 in *Key Renewal.* Then, it can configure his share set at the same way as Step 2 in *Key Renewal.*

5 Server Origin PSS

Since server origin PSS performs both recovery and renewal at once, we call the procedure as *key update.*

(Step 0). The setup is the same as that of user origin PSS.

5.1 Configuration of Update Group

To configure a update group I of correct servers, a simple challenge, which can be originated by one of n servers, is broadcasted. Each server computes partial signatures as many times as the number of shares in share set and broadcasts them. Each server generates a list of reputations for $n-1$ servers except itself; If a server does not send partial signatures or sends faulty partial signatures, the reputation for the server is set to "Bad". Otherwise, the reputation is set to "Good". Each server broadcasts it. As a result, each server obtains a list of reputations which is generated by itself and $n-1$ received lists of reputations. Each server configures a candidate update group as follows:

- A server which has more than t "Bad" reputations is accused.
- A candidate update group consists of servers which have less than $t-1$ "Bad" reputations.

Note that n candidate update groups which are generated in different servers should be identical. A update group I consists of a chosen coalition of servers in the candidate update group. Since at most $t-1$ servers can be compromised in a time period, we can successfully configure a update group I.

5.2 Distribution of Key Update

Without loss of generality, we assume that the update group I consists of $\{P_1, P_2, \cdots, P_t\}$.

(Step 1). Every server P_i in I generates a random number θ_i, which is called as *update share* of P_i, where $1 \leq i \leq t$. A *update secret* θ will be equal to $\sum_{i \in I} \theta_i$, and no server know the value of θ. Each $P_i \in I$ performs $CSS(n, t, \theta_i)$, respectively.

(i) Let $\{\theta_i^1, \ldots, \theta_i^l\}$ be the result of (l,l)-additive secret sharing for θ_i. Then, we call an additive share of an update share (e.g., θ_i^j, where $1 \leq j \leq l$) as *share of update share*. Each P_i generates n *update share sets* for θ_i; $U_{P_i}^j = \{\theta_i^k \mid 1 \leq k \leq l \wedge P_i \notin T_k\}$, where $1 \leq i \leq t$ and $1 \leq j \leq n$.

(ii) Each P_i sends securely $U_{P_i}^j$ to $P_{j \neq i} \in I$, respectively. Then, each P_i makes $oneWay(\theta_i)$ and $oneWay(\theta_i^j)$ public, where $1 \leq j \leq l$. As a result, each P_i holds update share sets received from different servers.

`(Step 2).` For $P_i \in I$ to compute a *new share set* $X_{P_i}^{new}$, each P_i computes the validity checks for *share of update shares*, and compares the computed validity checks with the corresponding public values, respectively. If the result is true, it checks the follows;

$$vcConstr(oneWay(\theta_h^1), \ldots, oneWay(\theta_h^l)) \; ?= \; oneWay(\theta_h),$$

where $1 \leq h \leq t$. If the result is true, it performs the next procedures. Otherwise, it broadcasts an *accusation message* for any servers, which lead failures, to n servers, and the procedure for key update is restarted by choosing another update group from the candidate update group in *Configuration of update group*. Each $P_i \in I$ configures an *update set* Δ_{P_i} for update secret $\theta = \sum_{i \in I} \theta_i$ by using t update share sets received from t different servers:

$$\Delta_{P_i} = \{\alpha_k = \theta_1^k + \cdots + \theta_t^k \mid 1 \leq k \leq l \wedge P_i \notin T_k\}.$$

As a result, $\theta = \theta_1 + \cdots + \theta_t = \alpha_1 + \cdots + \alpha_l$. Then, each $P_i \in I$ computes a *new share set* $X_{P_i}^{new}$ as follows:

$$X_{P_i}^{new} = \{x_k + \alpha_k \mid 1 \leq k \leq l \wedge P_i \notin T_k\}.$$

Finally, x^{new} must be $x + \theta$. Then, each $P_i \in I$ deletes θ_i and t update share sets, which were used to configure Δ_{P_i}, from memory, respectively.

`(Step 3).` All validity check values for x^{new} are publicly computed as follows, where $1 \leq i \leq l$.

$$\begin{aligned}
oneWay(\alpha_i) &= vcConstr(oneWay(\theta_1^i), \ldots, oneWay(\theta_t^i)) \\
oneWay(\theta) &= vcConstr(oneWay(\theta_1), \ldots, oneWay(\theta_t)) \\
&= vcConstr(oneWay(\alpha_1), \ldots, oneWay(\alpha_l)) \\
vcConstr(y_i, oneWay(\alpha_i)) &= y_i^{new} \\
vcConstr(y, oneWay(\theta)) &= y^{new}
\end{aligned}$$

5.3 Recovery and Propagation

`(Step 1).` If a server $P_m \notin I$ is accused, P_m is rebooted and initialized. Otherwise, the system performs from `Step 3` to `Step 4`.

`(Step 2).` Each $P_i \in I$ sends necessary shares $X_{P_i} \cap X_{P_m}$ to P_m so that P_m recovers X_{P_m}. After collecting necessary shares for recovery, P_m can verify the correctness of the received shares by using previous validity check values in `Step 0`. Then, it can configure its share set.

(Step 3). Each $P_i \in I$ sends securely $U^j_{P_i}$ generated at Step 1 in *Distribution of key update* to $P_j \notin I$, respectively. Each $P_j \notin I$ holds therefore t update share sets for θ.

(Step 4). Each $P_j \notin I$ performs Step 2 in *Distribution of key update*, and also holds $X^{new}_{P_j}$ and Δ_{P_j}, respectively. Each $P_j \notin I$ deletes X_{P_j} and t update share sets from memory. Then, each $P_i \in I$ deletes X_{P_i} and n update share sets at Step 1 in *Distribution of key update*, but keeps update set Δ_{P_i} at Step 2 in *Distribution of key update*.

5.4 Delivery to User

(Step 1). As soon as the user connects to the system, each P_i sends securely Δ_{P_i} to the user, where $1 \leq i \leq n$. The user receives obviously necessary update sets for recovering θ from at least t servers. The user can verifiably recover θ by using the validity checking values at Step 3 in *Distribution of key update*. If the validity checking is successful, the user sends *update success message* to servers, and updates its own part as $x^{new}_U = x_U \circ \theta$. Then, the user deletes the received update sets from memory.

(Step 2). After receiving *update success message*, each server deletes its update sets from memory.

6 Evaluation

For the security of cryptographic schemes used to design our proactive schemes, refer to [5][9][11][12]. For our proactive secret sharing schemes to be fault-tolerant even in the presence of up to $t-1$ corrupted servers in every time period, they satisfy the following properties:

- *Independency:* New shares for the secret cannot be combined with old shares to reconstruct the secret.
- *Secrecy:* The secret remains unknown to adversaries.
- *Availability:* Correct servers have sufficient shares of the secret to reconstruct it.

Due to the lack of space, we omit the proof that our schemes guarantee the mentioned properties.

We implement our proactive secret sharing schemes by using JDK 1.4 and Bouncy castle package [1][4]. We simulate them under Pentium IV processors. The following notations are used to describe Fig. 1.

- UO-Renewal: The overall cost required to finish key renewal in user origin PSS.
- UO-User: The user's cost required to finish key renewal in user origin PSS.
- SO-Renewal: The overall cost required to finish key update without any recovery in server origin PSS.
- SO-User: The user's cost required to finish key update without any recovery in server origin PSS.

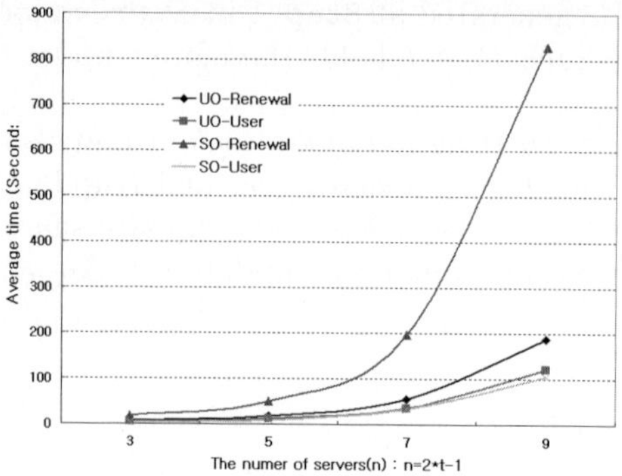

(a) Response times

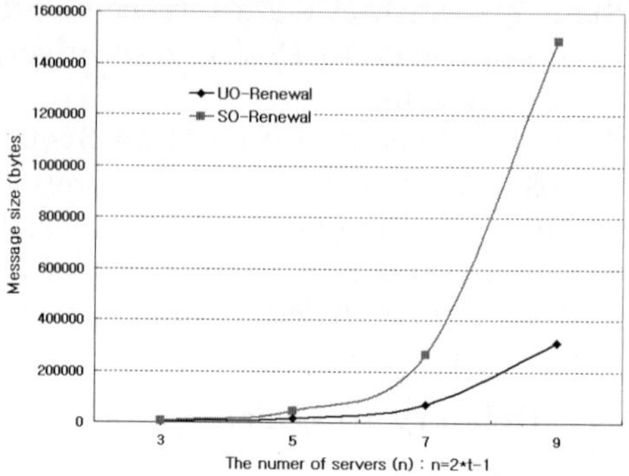

(b) Message traffics

Fig. 1. The costs for key renewal under 1024-bits keysize

Fig.1.(a) shows the measured response times under 1024-bits keysize according to the number of servers. As expected, user origin PSS is more efficient than server origin PSS about the response time for key renewal, and that is reasonable. From the view point of user's computation cost, the user can obtain computation benefit from the usage of server original PSS, because the user does not need to generate a secret value and shares it to servers in server origin PSS. We can see from Fig. 1.(a) that the user's burden of computation is about 70% of the overall cost in user origin PSS. In spite that the response time of user origin PSS is gradually increased according to the growth of the number of servers, the response time of server origin PSS is rapidly increased. Fig. 1.(b) shows the measured message traffics under 1024-bits keysize according to the number of servers. The message traffic of user origin PSS for key renewal is always less than that of server origin PSS. In particular, the message traffic of server origin PSS is very sensitive to the increment of the number of servers.

To sum up the result of simulations, server origin PSS requires higher computation and communication costs than user origin PSS. However, the user consumes less computation cost and hires conveniently the distributed system for threshold protection without taking care of renewal and recovery in server origin PSS. When a distributed system consisting of a large number of servers spreads on the low bandwidth network and the servers have low computing power, user origin PSS is a suitable strategy for deploying the distributed system. Otherwise, server origin PSS is suitable when a distributed system consisting of a small number of servers spreads on the high bandwidth network (e.g., LAN) and the servers have high computing power.

7 Conclusion

To the best of our knowledge, we newly provided the proactive secret sharing schemes for server assisted threshold signatures. According to the initiator, we designed two types of methods: user origin proactive secret sharing and server origin proactive secret sharing. Through performance simulations, we evaluated their computation and communication costs, respectively.

References

1. Bouncy Castle 1.24, https://www.bouncycastle.org.
2. C. Cachin, K. Kursawe, A. Lysyanskaya and A. R. Strobl, "Asynchronous verifiable secret sharing and proactive cryptosystems,", *In Proc. of 9th ACM CCS* (2002).
3. Haiyun Luo, Songwu Lu, "Ubiquitous and Robust Authentication Services for Ad Hoc Wireless Networks," UCLA Computer Science Technical Report 200030, Oct. (2000).
4. J. Garms and D. Somerfield , *Professional Java Security*, Wrox Press Ltd (2001).
5. Lidong Zhou, *Towards Fault-Tolerant and Secure On-line Services*, PhD Dissertation, Department of Computer Science, Cornell University, Ithaca, NY USA. April (2001).
6. P. Feldman, "A Pracitcal Scheme for Non-Interactive Verifiable Secret Sharing," *In Proc. of 28th FOCS* (1987).
7. P. MacKenzie, M. Reiter, "Networked Cryptographic Devices Resilient to Capture," *IEEE Security and Privacy'01*, May 14 - 16, (2001).
8. P. MacKenzie and M. Reiter. "Two-Party Generation of DSA Signatures," *Crypto'01*, LNCS 2139, pp.137-154, (2001).
9. R. Gennaro, S. Jarecki and H. Krawczyk, "Revisiting the Distributed Key Generation for Discrete-Log Based Cryptosystems," *RSA Security' 03*, (2003).
10. S. Jarecki, *Proactive Secret Sharing and Public Key Cryptosystems*, MIT Master of Engineering Thesis (1995).
11. S. Xu and R. Sandhu, "Two Efficient and Provably Secure Schemes for Server-Assisted Threshold Signatures," *CT-RSA*, (2003).
12. Tal Rabin, "A Simplified Approach to Threshold and Proactive RSA," *Advanced in Cryptology-CRYPTO 98*, LNCS 1462, pp.89-104, (1998).

An Efficient ID-Based Bilinear Key Predistribution Scheme for Distributed Sensor Networks*

Tran Thanh Dai, Cao Trong Hieu, and Choong Seon Hong**

Networking Lab, Department of Computer Engineering, Kyung Hee University
Giheung, Yongin, Gyeonggi, 449-701 Korea
daitt@networking.khu.ac.kr, hieuct@networking.khu.ac.kr,
cshong@khu.ac.k

Abstract. Security requirements are very pressing in distributed sensor networks due to exploitation purposes of these networks in human life, especially in military tasks. To obtain security in these sorts of networks, it is crucial to enable message encryption and authentication features among sensor nodes. This thing could be performed using keys agreed upon by communicating nodes. Nonetheless, acquiring such key agreement in distributed sensor networks becomes extremely intricate due to resource constraints. Up to now, there are many key agreement schemes proposed wired and wireless networks of which key predistribution schemes are considered to be the fittest solutions. Based on this observation, in this paper, we propose a key predistribution scheme relying on sensor nodes' unique identifiers. Our scheme exhibits several noteworthy properties: direct pairwise key establishment permission with explicit key authentication, high resiliency against information-theoretic security attack (node capture attack). We also present a detailed security and performance analysis of our scheme in terms of node capture attack, memory usage, communication overhead, and computational overhead.

1 Introduction

Advances in wireless communications and electronics over the last few years have sped up the development of networks of low-cost and multifunctional sensors. These sensors are tiny in size and able to sense, process data, and communicate with each other, typically over a radio frequency channel. They are usually deployed in a immense number and in the form of distributed networks to detect events or phenomena, collect and process data, and transmit sensed and processed information to interested users. Those distributed sensor networks are anticipated to be widely applied to many fields of human life ranging from civil applications to military applications.

In most of the applications, we truly need security measures to protect each sensor node in particular and the entire distributed sensor networks in general from malicious adversaries. According to typical approaches, security measures could be fulfilled

* This work was supported by MIC and ITRC projects.
** Corresponding author.

M. Gerndt and D. Kranzlmüller (Eds.): HPCC 2006, LNCS 4208, pp. 260 – 269, 2006.

based on efficient key agreement schemes. Nonetheless, sensor nodes typically operate in unattended conditions; have limited computational capabilities and memory, and battery-power capacity. Due to these resource limitations, the materialization of the efficient key agreement schemes in distributed sensor networks becomes a deeply intricate task. In fact, there are many key agreement schemes proposed for wired and wireless network environments which have been proved to be efficient and secure like trusted server schemes, public key based schemes, and key predistribution schemes. Nevertheless, constrained computation and energy resources of sensor nodes often make the first two schemes infeasible or too expensive for distributed sensor networks [8], [9], [10]. Recently, there are some attempts to solve the key agreement problem for sensor networks using elliptic curve cryptography (ECC) [11], [12]. However, the energy consumption of ECC is still expensive, especially compared to symmetric key based algorithms. Based on these analyses, it is straightforward to realize that key predistribution schemes seem to be the most feasible solution for the key agreement problem in distributed sensor networks.

A key predistribution scheme is a method to distribute off-line initial private pieces of information (keying materials) among a set of users, such that each group of a given size (in our scheme it is equals to two for the pairwise key generation purpose) can compute a common key for secure communication [13]. One branch of the key predistribution schemes is the ID-based key predistribution scheme. In that scheme, no previous communication is required and its key predistribution procedure consists of simple computations. Furthermore, in order to establish the key, each party should only input its partner's identifier to its secret key sharing function [14].

Due to those sorts of noteworthy properties, in this paper, we propose a highly resilient, resource-efficient and ID-based key predistribution scheme. Main contributions of our scheme are as follows:

1. Direct pairwise key establishment permission with explicit key authentication.
2. Substantially improved network resiliency against information-theoretic security attack (node capture attack).
3. Detailed theoretical analysis of security, memory usage, and communication and computation overhead.

The rest of the paper is organized as follows: section 2 mentions the related work; section 3 gives an overview of our building block; section 4 presents our proposed scheme; section 5 deals with the detailed security analysis; section 6 discusses performance analysis; section 7 concludes the paper.

2 Related Work

Recently, symmetric key cryptography has been received extensive studies to obtain various aspects of security in sensor networks. Perrig et al. [15] developed a security architecture for sensor networks which is comprised of two link layer protocols: SNEP and μTELSA. SNEP (Secure Network Encryption Protocol) provides data confidentiality, two-party authentication, and data freshness. μTELSA, the second part of SPINS, provides authenticated broadcast for sensor networks. Liu and Ning

[16] proposed a multi-level key chain method for the initial commitment distribution in μTESLA. Karlof, Sastry and Wagner [17] developed TinySec, the first fully implemented link layer security architecture for sensor networks. Eschenauer and Gligor [18] proposed a probabilistic key predistribution scheme recently for pairwise key establishment. The main idea is to let each sensor node randomly pick a set of keys from a key pool before deployment so any two sensor nodes have a certain probability of sharing at least one common key. Chan et al. [8] further extended this idea and developed three mechanisms for key establishment using the framework of predistributing a random set of keys to each node. The first one is q-composite keys scheme. This scheme is mainly based on [18]. The difference between this scheme and [18] is that q common keys, instead of just a single one, are needed to establish secure communication between a pair of nodes. By increasing the amount of key overlap required for key setup, the resiliency of the network is increased against node capture. The second one is multipath key reinforcement scheme applied in conjunction with [18] to yield greatly improved resilience against node capture attacks by trading off some network communication overhead. The main attractive feature of this scheme is that it can strengthen the security of an established link key by establishing the link key through multiple paths. The third one is random pairwise keys scheme. The purpose of this scheme is to allow node-to-node authentication between communicating nodes. Du et al. [19] proposed a method to improve [18] by exploiting a priori deployment knowledge. Specifically, by using node deployment knowledge and a wise key ring setup, the sensor networks get much higher probability of establishing a secure link between any pairwise of nodes. Zhu et al. [22] proposed a protocol suite named LEAP to help establish individual keys between sensors and a base station, pairwise keys between sensors, cluster keys within a local area, and a group key shared by all nodes.

3 Overview of Matsumoto-Imai's Key Predistribution Scheme

Matsumoto-Imai (MI) proposed a linear key predistribution scheme in [1] that allows distributing a common key to an arbitrary group of entities in a network without previous communications among the group nor accesses to any public key directory or whatsoever. In this section, we briefly describe how Matsumoto-Imai's key predistribution scheme works (MI scheme for short).

Let q be a prime power and m, l be positive integers. Let $\Psi = GF(q)$ and $\Psi^m = \{x \mid x = [x_1\ x_2\ \ldots\ x_m],\ x_i \in \Psi, i = \overline{1,m}\}$.

Suppose that each entity's (say, *entity i's*) identity y_i is a member of a set Υ and that $y_i \neq y_j, \forall i \neq j$.

And let Γ denote an one-way algorithm implementing an injection from Υ to Ψ^m.

The key setup server selects l (m, m) symmetric matrices $M^\tau s$ $(\tau = \overline{1,l})$ over Ψ randomly and independently from other entities.

The key setup server generates the *secret key sharing functions* Φ_i s:

$$\Phi_i(\omega) = \phi_i \Gamma(\omega)^T, \omega \in \Upsilon$$

for each $y_i \in \Upsilon$. Here, $\Gamma(\omega)^T$ is the transpose of $\Gamma(\omega)$ and ϕ_i is an (*l*, *m*) matrix defined by

$$\phi_i^T = \left[M_1 \Gamma(y_i)^T, ..., M_l \Gamma(y_i)^T \right]$$

Each entity *I* receives its own Φ_i from the center.

If entity A and entity B want to establish a pairwise cryptographic key, entity A computes $\Phi_A(y_B)$ and entity B computes $\Phi_B(y_A)$ independently. They are *l*-vectors over Ψ. It is easy to realize that both vectors are the same. This scheme could be used for key sharing among *n* entities by using symmetric *n*-linear mappings instead of the aforementioned symmetric bilinear mappings.

4 ID-Based Bilinear Key Predistribution Scheme

Since the purpose of MI scheme is to apply to the smart-card-based systems, not for distributed sensor networks, so we propose an ID-based bilinear key predistribution scheme inspired by MI scheme. We will later show that our scheme exhibit the fascinating properties satisfying security requirements due to specific characteristics of distributed sensor networks which have not been mentioned in MI scheme. Accordingly, our scheme consists of three phases, namely keying material predistribution, pairwise key establishment, and pairwise key reinforcement. The following are detailed description of these phases.

Keying material predistribution. Assume that each sensor node has a unique identification whose range is from 1 to *N* where *N* is the maximum number of sensor nodes that could be deployed during the entire lifespan of the sensor network. Each of the unique identifications is represented by $m = \log_2(N)$ bit effective ID in sensor nodes' memory. The keying material predistribution phase is to predistribute secret key sharing functions to each sensor nodes before deployment such that after deployment, neighboring sensor nodes can find a secret common key between them using these functions. It consists of the following steps:

1. Key setup server generates *l* $(m \times m)$ symmetric matrices $M^\tau s$ $(\tau = \overline{1,l})$ over finite field *GF(2)*. The M^τ s are private information and kept secret from both sensor nodes and adversaries. M^τ is used to generate the τth bit of a pairwise key between two neighboring sensor nodes, so *l* is the length of this key.

2. Key setup server computes secret key sharing function Φ_i for each sensor node S_i by first computing $\Phi_i^\tau = y_i M^\tau$ $(\tau = \overline{1,l})$ (1) and then generating Φ_i as

Φ_i as $\Phi_i = \begin{bmatrix} \Phi_i^1 \\ \Phi_i^2 \\ \cdots \\ \Phi_i^l \end{bmatrix}$ where $y_i(i = \overline{1, N})$ is the m-dimensional vector, the effective ID of sensor node S_i. This function is then distributed to each sensor node before node deployment.

Pairwise key establishment. After completing the keying material predistribution phase, each sensor node possesses a secret key sharing function. The object of this phase is to establish pairwise keys among neighboring sensor nodes using those functions. The procedure for establishing two neighboring sensor nodes S_i and S_j is described as follows with and added step to allow explicit key authentication.

1. After being deployed, S_i and S_j instantly broadcast their effective IDs y_i and y_j to their neighboring nodes. Since S_i and S_j are neighbors, S_i will get S_j's effective ID y_j and vice versa.
2. S_i computes the possible pairwise key $K_{ij} : K_{ij}^{\tau} = \Phi_i^{\tau} y_j^T (\tau = \overline{1, l})$ (2), where K_{ij}^{τ} indicates the τ th bit of the possible pairwise key K_{ij} between S_i and S_j. S_j carries out in the same way to get the possible pairwise key K_{ji}.
3. Up to this step, S_i / S_j needs to certify that the other has the same key as the one it computed. To do this, S_i / S_j has to show the other that it has the other's computed key by revealing secret information without revealing the computed key. As in [2], S_i / S_j generates a message M_i/M_j containing y_j / y_i, calculates the *message authentication code* (MAC) of M_i/M_j as a function of M_i/M_j and its computed key: $MAC = C_{K_{ij}}(M_i) / \ MAC = C_{K_{ji}}(M_j)$ and then send M_i/M_j plus MAC to the other (MAC can be calculate using a key-dependent one-way hash function such as HMAC [3]).

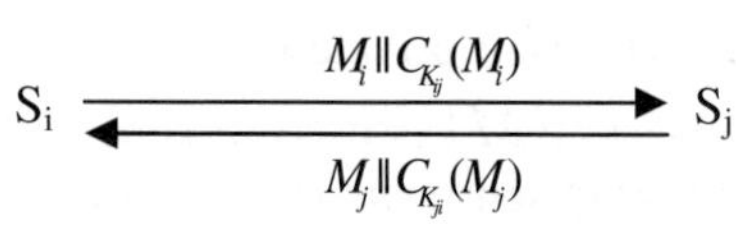

4. The recipient performs the same calculation on the received message, using its computed key, to generate a new *MAC*. The received *MAC* is compared to the calculated *MAC*. If the received *MAC* matches the calculated *MAC* then the receiver is assured that the message is from the alleged sender and its computed

key is exactly the same as that of the alleged sender. Since no one else knows the secret key, no one else could prepare a message with a proper *MAC*.

Up to this point, any two neighboring sensor nodes can establish a pairwise key to secure their communication link. However, as shown in [1], our proposed scheme is vulnerable to the *information-theoretic security attack* discussed later against the network resiliency. To prevent this sort of attack, there are two approaches. The first one is to allow two neighboring sensor nodes to take part in the *pairwise key reinforcement* phase. The second one will be discussed later on in security analysis section.

Pairwise key reinforcement (optional). This phase is aimed to reinforce a pairwise key between two neighboring sensor nodes S_i and S_j. It happens as follows: S_i and S_j randomly generate k_i and k_j respectively such that their lengths are equal to K_{ij} / K_{ji}. These keys are encrypted by K_{ij}, K_{ji} and transmitted to each end. Then, S_i / S_j computes a new pairwise key with S_j / S_i using the formula: $K = K_{ij} \oplus k_i \oplus k_j$ ($K = K_{ji} \oplus k_j \oplus k_i$) (3). In addition to the avoidance of information-theoretic attack, these formulas show that each node has the equal right to decide the value of the potential key *K*. It ensures that no node can get an advantage over the other from *K* selection.

This scheme substantially improves the security, resiliency and enable node to node authentication in the network. These features as well as other parameters will be thoroughly analyzed in the following sections of this paper.

5 Security Analysis

As already mentioned above, our scheme is vulnerable to information-theoretic security attack against network resiliency. Indeed, our scheme has a certain collusion threshold. As mentioned, $M^{\tau} (\tau = \overline{1,l})$ is a $(m \times m)$ matrix. By using *m* linearly independent secret Φ_i^{τ} s, M^{τ} can be easily revealed. Therefore, *m* is the value of the collusion threshold. In other words, an adversary only needs to compromise *m* sensor nodes to be able to compute any pairwise key of any two uncompromised neighboring sensor nodes using their effective IDs. It implies that with only *m* compromised sensor nodes, the adversary can compromise the entire network.

A straightforward solution to the attack is to increase the value of *m*. However, the increase in the value of *m* leads to the increase of memory size of sensor nodes needed to store Φ_i. The figure 1 show the relationship between *m* (number of compromised nodes), pairwise key length *l* and memory usage.

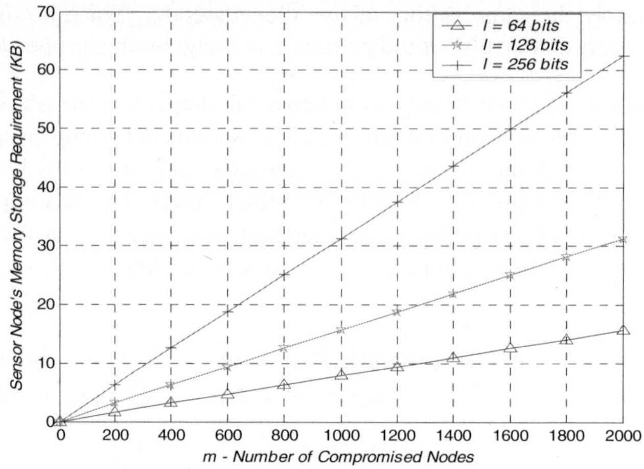

Fig. 1. Memory storage requirement against information-theoretic security attack

Assume that the key length of the pairwire key l = 128 bits, from the fig. 1, it is easy to realize that the solution can offer resistance to a collusion attack of up to 2000 compromised sensor nodes while using only about 32 KB of each sensor node's memory storage. This number of memory storage consumption is considered to be suitable for most sensor hardware platforms such as Berkeley Mica Motes with 128KB program memory [8]. Therefore, by increasing little amount of memory storage, the resiliency is significantly improved against *information-theoretic security attack*. This solution is considered to be acceptable in the sense that to successfully compromise the network, the adversary has to perform large scale attacks which are very expensive and more easily detectable.

The other solution has been briefly mentioned in section 3. This solution is partly inspired by an assumption in [4]. Accordingly, in this solution we assume that there exists a lower bound on the time interval T_{min} that is necessary for an adversary to compromise enough m sensor nodes, and that the time T_{est} for newly deployed sensor node to discover its immediate neighbors and establish initial pairwise keys with them is smaller than T_{min}. Taking advantage of the time interval T_{min}, two neighboring sensor nodes need to quickly exchange k_i and k_j to each other and then use (3) to change their initial pairwise key K_{ij} (K_{ji}) to the permanent pairwise key K. By doing in this way, we can eliminate the *information-theoretic security attack* from the entire network since the adversary could not compute the pairwise key K using (2).

In addition to information-theoretic security attack (node capture attack), our scheme also enable node to node authentication feature as already discussed. This feature, together with encryption techniques, is considered as a powerful tool to prevent some specific attacks carried out only in sensor networks such as sybil attack, sinkhole attack, hello flood attack, acknowledgement spoofing attack, etc [5], [6].

6 Performance Analysis

In this section, we analyze the performance of our scheme in term of memory usage, communication overhead and computational overhead.

As already analyzed in the aforementioned section, in our scheme, memory usage in each sensor node is in proportion to m given pairwise key length l and l given m. Increasing l, m or both result in the increase in security level (collusion threshold) but it implies more memory consumption to store that keying material in sensor nodes. The other approach to obtain higher security level (by eliminating collusion threshold attack) while the consumption of sensor nodes' memory storage could be significantly reduced is to include pairwise key reinforcement phase in the scheme. However, in this case, communication and computational overhead will be slightly increased. Thus, there must be trade-offs among security achievement, memory usage, communication overhead and computational overhead.

In our scheme, to establish the pairwise key, S_i and S_j need only to transmit three packets in case the pairwise key reinforcement phase is included. One packet is transmitted in the broadcast form. The other two packets are transmitted in the unicast form. These packets essentially contain the effective IDs of two nodes. Thus, the size of these packets is rather small. Therefore communication overhead of our scheme is rather low and can be acceptable in the distributed sensor network environment.

Considering computational overhead, it is easy to realize that our scheme is mainly based on multiplications of matrices $M^{\tau}(\tau = \overline{1,l})$ and sensor nodes' effective IDs over *GF(2)*. These multiplications essentially are exclusive-OR and AND bit operations. These multiplications consume much less computational time and require much less energy as well. The remaining computation constituting the overall computational overhead is *MAC* generation operations. These operations are considered as the least complex of the cryptographic algorithms and should intuitively incur the least energy cost [8]. For these reasons, the overall computational overhead of our scheme is not worth considering.

7 Conclusion

In this paper, we proposed a key predistribution scheme for distributed sensor networks inspired by the ID-based key predistribution scheme. Consequently, our scheme obviously inherits the noteworthy properties from that sort of scheme. First, the number of packets exchanged to establish a pairwise key between two sensor nodes which want to establish a secure communication channel is substantially minimized. Second, the key distribution procedure is composed of simple calculations so that computational costs are quite small and suitable for such computation limited devices as sensor nodes. Lastly, each sensor node has only to input its partner's identifier to its secret key sharing function to generate the desired key. Moreover, our schemes present two approaches which have been analyzed to cost sensor nodes much less their resource to tackle *information-theoretic security attack* inherited from ID-based key predistribution schemes. Our schemes also expose a technique that enable explicit key authentication which is expected to be the most effective solution to some sorts of attacks in distributed sensor networks. For all those reasons, there is no doubt that our scheme is an appropriate solution to the key agreement problem in distributed sensor networks.

References

1. Matsumoto, T., and Imai, H., "On the KEY PREDISTRIBUTION SYSTEM: A Practical Solution to the Key Distribution Problem", Advances in Cryptology - Crypto'87, Lecture Note in Computer Science, Vol. 293, 1988, pp. 185-193.
2. Stallings, W., "Cryptography and Network Security: Principles and Practice", Prentice Hall, 1998.
3. Rhee, M. Y., "Internet Security: Cryptographic Principles, Algorithms, and Protocols", Wiley, 2003.
4. Zhu, S., Setia, S., and Jajodia, S., "LEAP: Efficient Security Mechanisms for Large-Scale Distributed Sensor Networks", CCS'03, Washington, DC, USA, October 2003.
5. Wood, A. D., and Stankovic, J. A., "Denial of Service in Sensor Networks", IEEE Computer, Vol. 35, No. 10, October 2002, pp. 54-62.
6. Karlof, C., and Wagner, D., "Secure Routing in Wireless Sensor Networks: Attacks and Countermeasures", Proceedings of the First IEEE International Workshop on Sensor Network Protocols and Applications, May 2003, pp. 113 - 127.
7. Potlapally, N. R., Ravi, S., Raghunathan, A., and Jha, N. K., "A Study of the Energy Consumption Characteristics of Cryptographic Algorithms and Security Protocols", IEEE Transactions on Mobile Computing, Vol. 5, No. 2, February 2006.
8. Chan, H., Perrig, A., and Song, D., "Random key predistribution schemes for sensor networks", Proceedings 2003 IEEE Symposium on Security and Privacy, May 2003, pp. 197-213.
9. Du, W., Deng, J., Han, Y. S., Varshney, P. K., Katz, J., and Khalili, A., "A Pairwise Key Predistribution Scheme for Wireless Sensor Networks", ACM Transactions on Information and System Security, Vol. 8, No. 2, May 2005, pp. 228-258.
10. Liu, D., Ning, P., and Li, R., "Establishing Pairwise Keys in Distributed Sensor Networks", ACM Transactions on Information and System Security, Vol. 8. No. 1, February 2005, pp. 41-77.
11. Blaß, E.-O., and Zitterbart, M., "Towards Acceptable Public-Key Encryption in Sensor Networks", Proceedings of the 2nd International Workshop on Ubiquitous Computing, ACM SIGMIS, May 2005.
12. Wander, A. S., Gura, N., Eberle, H., Gupta, V., and Shantz, S. C., "Energy analysis of public-key cryptography for wireless sensor networks", Proceedings of the 3rd IEEE International Conference on Pervasive Computing and Communications, March 2005, pp. 324 - 328.
13. Blundo, C., Santis, A. D., Herzberg, A., Kutten, S., Vaccaro, U., and Yung, M. "Perfectly-secure key distribution for dynamic conferences", Advances in Cryptology - CRYPTO '92, Lecture Notes in Computer Science, Vol. 740, 1993, pp. 471-486.
14. Hanaoka, G., Nishioka, T., Zheng, Y., and Imai, H., "A Hierarchical Non-interactive Key-Sharing Scheme with Low Memory Size and High Resistance against Collusion Attacks", The Computer Journal, Vol. 45, No. 3, 2002.

4 Our Approach

We apply two stage operations for secure formation of clusters in the network.

Assumption 1. Once the sensors are deployed they remain relatively static in their respective positions.

Assumption 2. In a unit disk or transmission range of a sensor, all the neighboring sensors do not necessarily have a direct communication link among themselves. If two nodes i and j have a direct communication link, it is bidirectional; $\forall_{i,j}, (i, j) \in E \Rightarrow (j,i) \in E$ and it exists if and only if i and j have common keys.

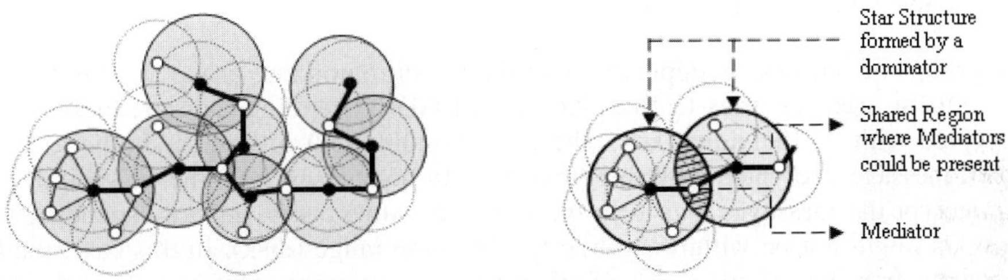

Fig. 4. (left) Dominators' coverage areas in WCDS (right) Mediator between two groups/stars

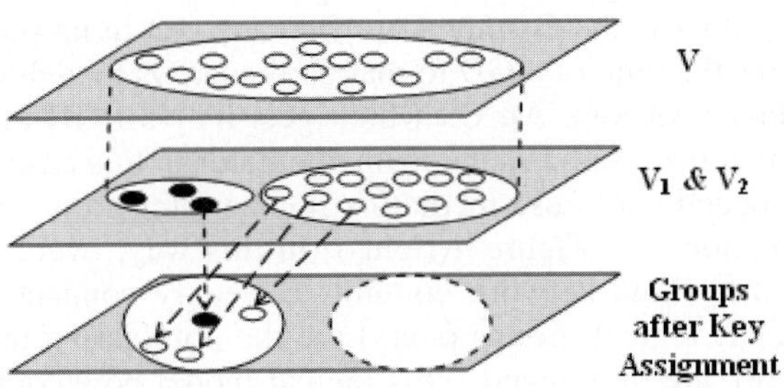

Fig. 5. Ranking of the sensors based on the key pre-distribution

4.1 Offline Rank Assignment

The sensors in the network are assigned their ranks based on the offline key-distribution. We divide the whole set of sensors V into two subsets, V_1 and V_2, where V_1 contains the probable group dominators (GD or cluster heads) and V_2 consists of ordinary sensors (Os). The set V_2 is further divided into several subsets $w_i \subset V_2$, i=1, 2, 3,….,N and N is the maximum number of possible proper subsets of V_2. Each w_i is assigned to one element in the set V_1. The sensors in the subset w_i (Os_1, Os_2,….Os_η) and corresponding one sensor from V_1 (let, $GD_{i,}$ i=1) are taken for group-wise key-predistribution (Figure 5). All the sensors in the set w_i are assigned two keys one of which is the group key shared by only the particular sensor and the GD_i. The GD_i contains all the individual keys of the sensors in its w_i and its own group key.

Assumption 3. All the sensors have same transmission range. Each node transmits within the transmission range isotropically (in all directions) so that each message sent is a local broadcast.

Assumption 4. The Base Station (BS) contains all the individual keys and group keys of the network.

Assumption 5. The number of Oss (value of η) in each group is decided on demand. It could be group specific or set to a common value for all the groups. η is actually the maximum degree (Δ (GD_i)) of a GD in a group.

4.2 Secure Cluster Formation

The groups of sensors are deployed over the target region one group at a time. After deployment, each Os tries to find out its own GD by sending a join request packet encrypted with its individual key. The corresponding GD in turn sends the join approval message encrypted with the group key. In both cases, both the GD and the Os can decrypt the messages and form the group. In some cases, the corresponding GD of an Os might not be within one-hop transmission range (disk). In this case, the Os detects the presence of other GDs of other groups in its surroundings, collects their ids and sends an error message to the base station (BS) with this information. The GDs within its one-hop transmission range also detects such erroneous Os and reports to the BS. The BS in turn assigns one of the neighboring GD as the adopter of the orphan Os. In the worst case, the Os might not find any GD in its surroundings. In this case, The BS assigns the rank of a GD to that particular Os though it does not contain any other sub-ordinate sensors. An Os which gets its own GD and another GD of another group in its transmission range is the mediator in this case. As stated earlier, all the stars thus shaped could use mediators for the inter-group (inter-star or inter-cluster) communication (see Figure 4(right)). In this way, eventually the resultant logical model of the whole network contains a weakly connected dominating set where the GDs of the logical groups (stars) are the dominating nodes and all other nodes in the network are dominated. This logical model now could be used for secure message delivery within the network (using the secret keys).The pseudo code for secure cluster (group) formation algorithm is presented in Figure 6.

All the groups of sensors could be deployed at a time or more groups could be deployed later based on demand. If it is needed, some sensors in a group could be deployed later. During the offline key pre-distribution, all the nodes are assigned the keys but all the nodes might not be deployed. When any of those remaining nodes is newly deployed, it follows the procedure of joining a group. If authorized by the access list of GD, it joins the group. Otherwise, GD forwards the id of this sensor to BS. BS informs GD about the individual key of that Os if it is a legitimate node. If authenticated by BS, GD generates a new group key and encrypts the new group key with the newly added node's individual key and sends it to that particular Os. All other nodes in the group know about the change of group key by a local broadcast by the

```
Let,
enc_i(.) - message encrypted by individual key of i
enc_iNOT(.) - message encrypted by an unknown individual key
enc_G(.) - message encrypted by the group key
enc_GNOT(.) - message encrypted by an unknown group key
Os_g - the set of Oss allowed under a group dominator g
locbr(.) - local broadcast within one hop transmission
range

for each s∈ V_Os
      locbr(enc_i(JOIN_REQ))
      if enc_G(JOIN_APRV) from any g∈ V_GD and hop(s,g)=1
            edge(s,g)
            dominator(s)←g
      else
            flood(enc_i(GD_ERR)) destined to BS
      end if
      if enc_GNOT(JOIN_APRV) from any g∈ V_GD and hop(s,g)=1
            neighbor_dominator(s)←g
      end if

for each g∈ V_GD
      if enc_i(JOIN_REQ) from any s∈ V_Os and s∈ Os_g
            send enc_G(JOIN_APRV)
            edge(s,g)
            sub-ordinate(g)←s
      end if
      if enc_iNOT(JOIN_REQ) from any s∈ V_Os
            mediator(g)←s
      end if
      if enc_i(GD_ERR) from any s∈ V_Os and hop(s,g)=1
            report enc_G(ORP_ERR) to BS
      end if
#In case of the BS:
if enc_i(GD_ERR) from any s∈ V_Os and enc_G(ORP_ERR) from any
g∈ V_GD
      if same id of s, issue command: Adopter_GD(s)←g
```

Fig. 6. Pseudo Code for Clustering Algorithm

GD of that group. In this case, the previous group key is used for encrypting the new group key. For leaving a group or cluster, the node simply leaves a message to inform the GD which in turn generates a new group key and multicasts it within the group members.

5 Performance Analysis and Comparison

We form a WCDS to cover almost all of the nodes in the network with minimum effort. The offline rank assignment reduces the burden of executing resource-hungry operations to form clusters like other clustering mechanisms. As shown in equation (1), WCDS requires less number (or equal to) of dominating nodes to cover the whole network than that of a CDS requires. Depending on the requirements we can increase or decrease the value of η (the expected degree of a GD in a group). In ideal case, the size of the dominating set created in our approach could be obtained by,

$$\begin{aligned} \textit{Size of Dominating Set} &= \frac{\textit{Number of vertices in the Graph}}{\eta+1} \\ &= \frac{\textit{Number of vertices in the Graph}}{\Delta(GD)+1} \end{aligned} \tag{2}$$

In our experiment, we generate random graphs of 20-200 and 40-200 nodes with expected average degree 6 and 12 respectively. To simulate the structure of the sensor network, we place the vertices randomly over a 2-D rectangular plane. The network size and density is set by changing the number of vertices and transmission ranges of the nodes. Applying our approach and two algorithms (I and II) of [13] we find that our approach generates much smaller number of group dominators or cluster heads. For a large number of sensors it works effectively. Figure 7 shows the size of dominating sets in comparison with that of Algorithm I and Algorithm II of [13]. The major advantage of our approach is the flexibility to set the value of η (expected maximum degree of a GD) according to the requirements.

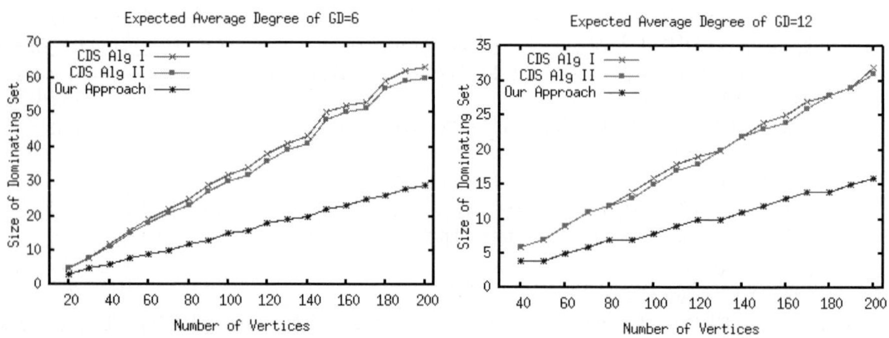

Fig. 7. (left) Size of the dominating set when expected average degree 6 (right) when expected average degree 12

We use the distinct group keys for each of the GDs and distinct individual keys for each Os. So, in general case, the number of distinct keys required for our network is equal to the number of sensors in the whole network.

Each group dominator (GD) in the network has to remember one group key and all the individual keys of the Oss of that particular group. So, the storage requirement for each GD in number of bits is,

$$\gamma_{GD} = (\eta + 1) \times k \tag{3}$$

and for each Os,

$$\gamma_{Os} = 2 \times k \tag{4}$$

where, η is the number of Oss in that particular group and k is the number of bits required for representing the key. As the value of η increases, the storage load for a GD increases. Hence, the value of η is set according to the requirements or a particular situation at hand. So, if initially we have α number of GDs and β number of Oss, the network wide storage usage for storing the keys is,

$$\begin{aligned}\Gamma_{network-wide} &= \alpha \times ((\eta+1)\times k) + \beta \times (2 \times k) \\ &= k \times (\alpha \times (\eta+1) + 2 \times \beta)\end{aligned} \tag{5}$$

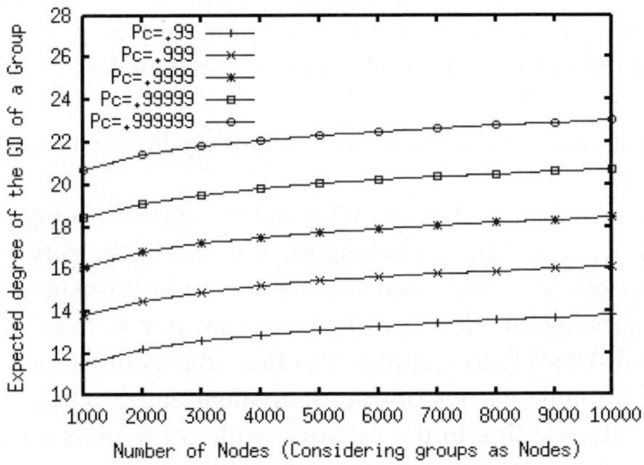

Fig. 8. Given a connectivity probability, expected degree of a GD from the high level view

After formation of clusters within the network, the mediators are used for communication among clusters. From the higher level view, we could consider the clusters (or groups) as nodes in a random graph G=(n, p), where n is the number of nodes (i.e. clusters in our case) for which the probability that an edge (i. e. communication link via mediator) exists between two nodes is p. p=0 when there is no edge and p=1 when the graph is fully connected. According to Erdös and Rényi [14], for monotone properties, there exists a value of p such that the property moves from "nonexistent" to "certainly true" in a very large random graph. The function defining p is called the threshold function of a property. Given a desired probability P_c for graph connectivity, the threshold function p is defined by,

$$P_c = \lim_{n\to\infty} P_r[G(n,p) \text{ is connected}] = e^{e^{-c}}, \text{ Where, } p = \frac{\ln(n) - \ln(-\ln(P_c))}{n}$$

Let, p be the probability that an edge (communication link via mediator) exists between two GDs of two clusters, n be the number of nodes (i.e. clusters/groups in the entire network in this case), and d be the expected degree of each GD, then,

$$d = p \times (n-1) = \frac{(n-1)(\ln(n) - \ln(-\ln(P_c)))}{n} \tag{6}$$

In our approach, the sensors could be added later on rather deploying all of them at a time. Sometimes the entire terrain info and deployment diagram could be available (consider a battlefield scenario where the sensors are deployed prior to the enemy forces' invasion). In this case, the extra sensors could be deployed within the range of its appropriate group or cluster. If the sensors are deployed randomly, in the worst case, all the extra or newly added sensors will not be within the range of their intended group dominator and even no other GD could be available in their surroundings. Hence, in the worst case, all the newly added sensors would be included in the dominating set which would increase the size of the dominating set. Still it could be less than the number of dominators needed in case of a connected dominating set (CDS) when the network size is very large. Keeping the size of the dominating set to a minimum is helpful as less number of dominators means less number of entry paths for the false information injected in the network and also it is cost effective if the expected dominators are considered to have relatively higher resources than those of the ordinary sensors.

Our scheme ensures that, each of the GDs and the corresponding Oss could directly form the groups (i. e. clusters) maintaining the security of the network from the bootstrapping state. As encryption is used for message-transmission within the network from the very beginning of the network formation, our scheme could successfully defend Hello Flood Attack [15] and most of other attacks in wireless sensor networks [16]. Again, as each node carries distinct individual and group keys, compromising one node affects only one link in the network while other links remain safe from the attacks by the adversaries. If the group key of a particular group is compromised, still the adversary needs valid individual keys of the Oss for decrypting the information sent from an Os. In case of the compromise of a GD, the base station gets involved for revoking the keys and even in this case, only one group is affected while others could still operate. The re-keying feature ensures robust security as with each addition of a new sensor, the group key is renewed. If considered as resource-exhaustive, the key renewal mechanism could be omitted. However, for military networks as security is the major issue, we could consider a slight increase of the usage of the resources in the sensors.

6 Conclusions and Future Works

This paper presents an efficient approach for secure clustering in distributed sensor networks based on key-predistribution and prior rank assignments. As the group dominators rule over all other sensors in the group for data transmission, the dominators could require more energy, processing and storage power. For this, a set of sensors with greater resources could be considered as dominators. As future works, we would like to deal with secure routing and an efficient method to prevent Denial-of-Service (DoS) and Distributed DoS (DDoS) attacks in distributed sensor networks using our approach. Due to the page limitations, we have shortened the details of some of the parts in this paper.

References

1. Carman, D. W., Kruss, P. S., and Matt, B. J., “Constraints and Approaches for Distributed Sensor Network Security”, NAI Labs Technical Report # 00-010, dated 1 September, 2000.
2. Nowak, R. D., "Distributed EM ALgorithms for Density Estimation and Clustering in Sensor Networks", IEEE Transactions on Signal Processing, Vol. 51, No. 8, August 2003, pp. 2245-2253.
3. Halgamuge, M.N., Guru, S.M., and Jennings, A., "Energy efficient cluster formation in wireless sensor networks", 10th International Conference on Telecommunications, 2003 (ICT 2003), Volume 2, 23 February-1 March 2003, pp. 1571-1576.
4. Lee, S., Yoo, J., and Chung, T., "Distance-based energy efficient clustering for wireless sensor networks", 29th Annual IEEE International Conference on Local Computer Networks, 2004, 16-18 Nov 2004, pp. 567-568.
5. Younis, O. and Fahmy, S., "Distributed Clustering in Ad-hoc Sensor Networks: A Hybrid, Energy-Efficient Approach", IEEE Transactions on Mobile Computing, 3(4), Oct-Dec 2004, pp. 366-379.
6. Ye, M., Li, C., Chen, G., and Wu, J., "EECS: an energy efficient clustering scheme in wireless sensor networks", 24th IEEE International Performance, Computing, and Communications Conference, (PCCC 2005), 7-9 April 2005, pp. 535-540.
7. Liu, J.-S. and Lin, C.-H.R., "Power-efficiency clustering method with power-limit constraint for sensor networks", Proceedings of the 2003 IEEE International Performance, Computing, and Communications Conference, 2003, 9-11 April 2003, pp. 129-136.
8. Tzevelekas, L., Ziviani, A., Amorim, M. D. D., Todorova, P., and Stavrakakis, I., "Towards potential-based clustering for wireless sensor networks", Proc. of the 2005 ACM conference on Emerging network experiment and technology, Toulouse, France, 2005, pp. 292-293.
9. I. Wokoma, L. S cks and I. Marshall, “Clustering n Sensor Networks using Quorum Sensing,” in the London Communications Symposium, University College London, 8th-9th September, 2003.
10. Banerjee, S. and Khuller, S., "A clustering scheme for hierarchical control in multi-hop wireless networks", Proc. of the IEEE INFOCOM 2001, Volume 2, 22-26 April 2001, pp. 1028 - 1037.
11. Clark, B. N., Colbourn, C. J., and Johnson, D. S., “Unit Disk Graphs”, Discrete Mathematics, 86: 165-177, 1990.
12. Garey , M. L. and Johnson, D. S., “Computers and Intractability: A Guide to the Theory of NP-Completeness”, W. H. Freeman, San Francisco, 1979.
13. Das, B. and Bharghavan, V., "Routing in ad-hoc networks using minumum connected dominating sets", Proc. IEEE International Conference on Communications (ICC'97), June 1997, pp. 376-380.
14. Erdos and Renyi, “On Random Graphs”, Publ. Math. Debrecen, Volume 6 (1959), pp. 290-297.
15. Karlof, C. and Wagner, D., “Secure routing in wireless sensor networks: Attacks and countermeasures”, Elsevier's Ad Hoc Network Journal, Special Issue on Sensor Network Applications and Protocols, September 2003, pp. 293-315.
16. Pathan, A-S. K., Lee, H-W., and Hong, C. S., "Security in Wireless Sensor Networks: Issues and Challenges", Proc. of the 8th IEEE ICACT 2006, Volume II, 20-22 February, Phoenix Park, Korea, 2006, pp. 1043-1048.
17. Gupta, G. and Younis, M., "Load-balanced clustering of wireless sensor networks", IEEE International Conference on Communications (ICC'03), Volume 3, 11-15 May 2003, pp.1848-1852.

Pairwise Key Setup and Authentication Utilizing Deployment Information for Secure Sensor Network*

Inshil Doh, Jung-Min Park, and Kijoon Chae

Dept. of Computer Science and Engineering, Ewha Womans University, Korea
isdoh@ewhain.net, pjm@kist.re.kr, kjchae@ewha.ac.kr

Abstract. For secure sensor network communications, pairwise key establishment between sensor nodes and data authentication are essential. In this paper, we propose cluster based key establishment mechanism in which clusterheads distribute key materials at the requests of their own members and an authentication mechanism using the pairwise keys established. Our proposal increases the degree of network connectivity and resilience against node capture by utilizing clusterheads which carry more key-related information. It also increases the authenticity of sensed data and further decreases unnecessary traffic by filtering false data.

1 Introduction

Sensor networks can be applied in various application area and draw a lot of attention these days. For reliable and efficient communication, security is a critical issue and key management and data authentication are the core of the security issues. However, because of basic constraints of sensor nodes, such as memory, computation, price, and energy, traditional approaches cannot be applied, and new security mechanisms for sensor network are needed[1]. Several key management mechanisms for sensor networks have been proposed, however, they still have various drawbacks. When sensor network field is clustered, most nodes need to establish pairwise keys with neighbor nodes located in the same clusters because when sensor nodes are dropped from a helicopter hanging above a deployment point, more nodes fall into the areas closer to the deployment point and it is seldom for a node to reside far away this point. For this reason, these nodes don't have to carry key materials for the pairwise keys with neighbor nodes in different clusters. Previous mechanisms did not consider this point and wasted storage. In this paper, we propose cluster-based key establishment mechanism in which clusterheads located near the deployment point in each cluster carry the key materials and distribute them at their own member nodes' requests. We also propose an authentication mechanism in which the clusterheads deployed on the routes verify and filter modified or forged data efficiently.

* This research was supported by the MIC(Ministry of Information and Communication), Korea, under the ITRC(Information Technology Research Center) support program supervised by the IITA(Institute of Information Technology Assessment).

M. Gerndt and D. Kranzlmüller (Eds.): HPCC 2006, LNCS 4208, pp. 280–289, 2006.

2 Related Work

Recently, many key establishment mechanisms have been proposed. Eschenauer and Gligor[3] proposed basic scheme in which each node randomly picks some secret keys from a large key pool. Chan et al.[4] extended the basic scheme to q-composite scheme requiring any two nodes share at least q (q > 1) common keys to establish a secure connection. Du et al.[5] presented a pairwise key pre-distribution scheme combining the basic scheme and Blom's key management scheme[2] together. Liu and Ning[6] proposed a similar pairwise key scheme based on Blundo[7]'s polynomial-based key distribution protocol. Du et al.[8] first employed sensor deployment knowledge in the key pre-distribution scheme. With the help of deployment knowledge, this scheme achieved a higher degree of connectivity and a stronger resilience against node capture attacks. However, it cannot keep the same performance when adversaries compromise nodes within a small local area. Yu and Guan[9] proposed group-based pairwise key establishment mechanism also using sensor deployment knowledge. This scheme achieves a much higher degree of connectivity with stronger resilience against node capture. However, this scheme still has drawbacks of security weakness and low rate of direct key establishment problems.

For authentication, Zhu et al.[10] proposed a mechanism which enables the base station to verify the authenticity of event reports when compromised nodes exist and also filters the false data packets injected into the network before they reach the base station. This mechanism has the overhead of forming associations between nodes and all the nodes should keep pairwise keys with their associated nodes which are *t*-hops away.

3 Network Model

3.1 Assumptions

We assume that there are no adversaries before sensor nodes are deployed. Clusterhead(CH)s are located in their own cluster areas and are more powerful in computation and transmission ability and have larger storage. We also assume that every node is static and knows the next node on the route to its own CH and the next node to the BS(Base station). All the routing paths from a CH to BS need to go through the CH when they need to cross a cluster.

3.2 Network Architecture

Sensor field is clustered in hexagonal shape and in every cluster a CH is located near the deployment point. We model this using a two-dimensional normal distribution(pdf), and assume that the deployment distribution for a node in a cluster follows a two-dimensional Gaussian distribution centered at a deployment point (x_i, y_i),

$$f_i(x,y) = \frac{1}{2\pi\sigma^2} e^{-[(x-x_i)^2+(y-y_i)^2]/2\sigma^2}$$

where σ is the standard deviation. The location of CH is assumed to be near the deployment point(x_i, y_i).

4 Key Establishment Mechanism

4.1 Blom's Symmetric Key Establishment Mechanism

Our proposal is based on Blom's symmetric key establishment mechanism. In Blom's scheme, a symmetric matrix $Kn \times n$ stores all pairwise keys of a group of n nodes, where each element k_{ij} is the key of node i used to secure its connection with node j. A = $(\mathrm{DG})^T\mathrm{G}$, where D = $(\lambda+1)\times(\lambda+1)$ is symmetric and G=$(\lambda+1)\times n$ is called public matrix. $(\mathrm{DG})^T$ is also called secret matrix. Each node i carries the $i'th$ row of secret matrix and the $i'th$ column of public matrix G. After deployment, each pair of nodes i and j can individually compute a pairwise key $k_{ij} = k_{ji}$ by only exchanging their columns in plain text because the key is the dot product of their own row and the column of the other's. Their rows are always kept secret. Blom's scheme has the property of λ-security, i.e. when more than λ rows are compromised, the secret matrix is derived and all keys can be computed by adversaries. So, λ is very critical for the security and should be properly determined.

4.2 Pairwise Key Establishment Mechanism

Based on Blom's scheme, our proposal establishes pairwise keys between every pair of neighboring nodes irrespective of cluster area.

Key material predistribution. Yu's mechanism[9] increased the security level by partitioning the network filed into hexagonal grids, and assigning two kinds of secret matrices under certain rules. However, it still has several shortcomings such as network connectivity and the possibility of disclosure of shared matrices.

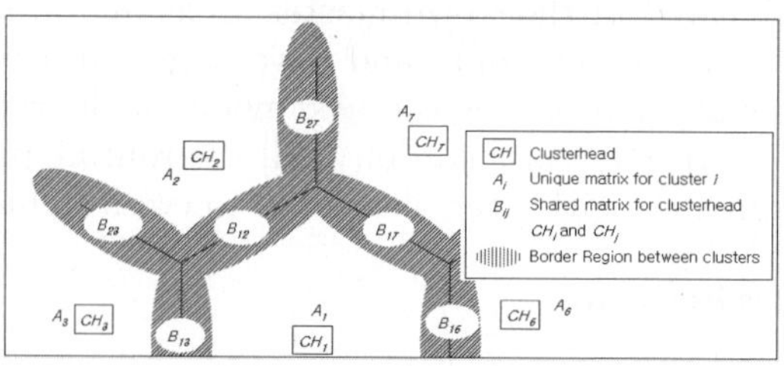

Fig. 1. Network clustering and Assignment of matrices

To overcome the shortcomings of previous schemes including Yu's mechanism, we locate CHs in every clusters and let the CHs distribute the key materials at the member nodes' requests after all the nodes being deployed in hexagonal clusters. Not all the nodes but only the ones which detect neighbor sensor nodes from

Table 1. Notations for our Proposal

Notation	Description
SN_{Xi}	Sensor node i deployed in cluster X
CH_Y	Clusterhead of cluster Y
BS	Base Station
A_X	Unique matrix A assigned to cluster X
B_{XY}	Shared matrix for neighboring sensor nodes deployed in different clusters X and Y
K_{ij}	pairwise key between nodes i and j
n	Number of clusters
m	Average number of sensor nodes in a cluster requiring inter-cluster communications

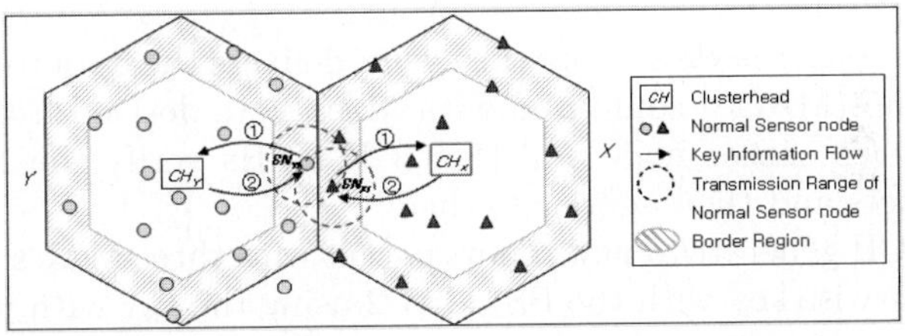

Fig. 2. Key information delivery after nodes deployment

different clusters get the shared key information. As in Fig. 1, Sensor network field is clustered as hexagonal shape, and for every sensor node in a cluster, a row from a unique matrix A assigned only for that cluster is predistributed. For inter-cluster communication, a shared matrix B is assigned for a pair of clusters. However, the information is not predistributed to sensor nodes, but their CHs carry the information from maximum six shared matrices B and distribute the information when their member nodes request the information. Every neighboring pair of CHs is predistributed their own pairwise keys before deployment, which means every individual CHs are predistributed at most six inter-clusterheads pairwise keys and one key with the BS depending on its location. Notations for our mechanism are shown in Table 1.

Pairwise key establishment. Pairwise keys are established through four steps.

Step 1. Sensor nodes deployment and Neighbor node detection : After all the nodes are deployed with key materials predistributed, all sensor nodes detect their own neighbor nodes through HELLO messages.
Step 2. Pairwise key establishment : Every pair of nodes in each cluster computes their own pairwise key using predistributed information.
Step 3. Request for key material to CHs : A sensor node detecting a neighbor node from different clusters requests corresponding information to its own

CH. As in Fig. 2, node SN_{Xi} and SN_{Yi} request the data to CH_X and CH_Y, respectively. The CHs deliver the encrypted information.

$$SN_{Xi} \rightarrow CH_X : Request(ClusterY, SN_{Xi})||MAC(Request(ClusterY, SN_{Xi}))$$
$$CH_X \rightarrow SN_{Xi} : Enc_{K_{CH_X - SN_{X_i}}}(\text{one row from } B_{XY}, \text{ one column from } G)$$

Step 4. Pairwise key computation: Sensor nodes compute pairwise keys with nodes belonging to different clusters using additionally distributed key materials from their own CHs.

5 Authentication Mechanism

When sensing an event, sensor nodes transmit the event report to BS in two phases.

Phase 1. Each sensor node sensing the event, delivers the data to their own CHs with MAC values(MAC0) computed with MAC keys derived from the pairwise keys between each sensor node and their CHs. CHs verify the MAC values of each data packets and then aggregate them.
Phase 2. The CH generates a new report packet with three MACs. MAC1 is computed using pairwise key with the BS, MAC2 using the key with next CH on the routing path, and MAC3 for the next normal node on the route to the BS.

Fig. 3 shows the two phase data delivery. In this way, not only outsider attacks, but also the insider attacks by compromised nodes can be detected. The outsider attacks are impossible because the adversaries don't know the pairwise keys and cannot generate legitimate MACs. Insider attackers, i.e. compromised nodes, can modify the data and generate fake MAC value, but this will be detected by the next CH, and even when the next CH cannot filter the false data, it can be finally detected by the BS. Security and overhead is further analyzed in section 6.

Table 2. MAC values

MACs	Used Keys	Description
MAC0	K_{CH-SN}	Verified by the CH
MAC1	K_{CH-BS}	Verified by the BS
MAC2	K_{CH-CH}	Verified by the next CH on the route
MAC3	$K_{CH-SN}, K_{SN-SN}, K_{SN-BS}$	Verified by the neighbor node

6 Performance Analyses

6.1 Security Analysis

Key establishment Security. Cluster size affects the degree of connectivity and the security level. When cluster size is big, relative number of nodes which need inter-cluster communication decreases. This lowers the number of rows from shared matrices and hence decreases the possibility of the disclosure

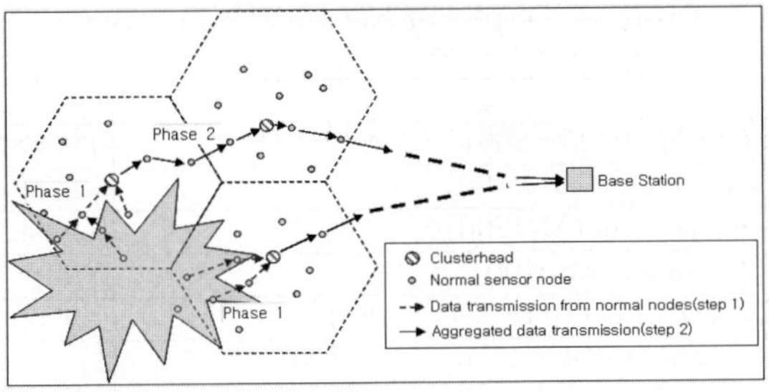

Fig. 3. Two phase message authentication

of shared matrices. However, with too big clusters, the degree of connectivity can be lowered and when the number of nodes sharing the dedicated matrix is more than λ, the matrix is reconstructed and the keys are calculated by the attackers. When cluster size is small, the number of clusters increases and the number of sensor nodes sharing dedicated matrices decreases. In this case, more nodes are related to inter-cluster communication and shared matrix can be in danger of disclosure. In addition, some nodes can be placed at different clusters from originally assigned one. When cluster size is too small, transmission range of a sensor node reaches non-neighboring clusters and causes that one-hop neighboring nodes cannot establish pairwise keys. According to Gaussian distribution and circumstances of Table 3, 99.87% of sensor nodes in a cluster are placed in a circle of radius 6σ, and if the distance of two nearest non-neighboring deployment points is greater than 6σ, we can say that most neighbors of each node are from its own cluster and the neighboring clusters[9].

Table 3. Parameters for performance

Notation	Value	Description
N	10^4	Number of sensor nodes
r	40 m	Transmission range of a normal sensor node
M	100 or 200	Memory size
σ	50 m	Standard deviation
S	$10^3 \times 10^3 m^2$	Sensor network field
Cn	200	Number of compromised nodes
l	$a \times \sigma$	Distance between the center of neighboring clusters

Table 4 lists the degree of connectivity of various mechanisms. Compared to other schemes given the memory requirement M = 100, our proposal achieves perfect degree of connectivity. Yu's scheme approaches perfect connectivity when w=4, which means that sensor nodes in one hexagon shares matrix B with neighbor sensor nodes from four neighbor hexagons. With w=4, Yu's mechanism achieves higher performance than the basic scheme or Du's deployment knowledge scheme. Our proposal can achieve perfect connectivity with the condition that all the matrix B is shared by only two hexagonal clusters, i.e. w=2.

Table 4. Degree of Network connectivity

Mechanisms	Pc
q-composite scheme(q=2, M=200)	0.9358
Eschenauer's basic scheme(M=100)	0.9867
Eschenauer's basic scheme(M=200)	0.9999
Du's deployment scheme(M=100)	0.9988
Yu's group based scheme(w=1∼3, M=100)	0.9969∼0.9999
Yu's group based scheme (w=4,5,6,7, M=100)	1
Our proposal(w=2, M $<$ 100)	1

1. Security for dedicated matrices

When cluster size becomes larger, the number of sensor nodes and the average number of compromised nodes in one cluster increase. It means that as the cluster size grows, the possibility that the dedicated matrix is disclosed also grows when there are compromised nodes in the network. With the number of compromised nodes 200 and M = 100, λ is 99 because M = (λ+1) for normal sensor nodes requiring only intra-cluster communications. This means no more than 99 compromised nodes in one cluster cannot reconstruct the dedicated matrix. This provides perfect resilience against node capture under the condition, M = 100, and it further lowers M of the normal sensor nodes.

2. Security for shared matrices

In our mechanism, required memory for pairwise key between sensor nodes from different clusters is M = (λ+1) $\times i$, (i =1,2). However, maximum number of clusters sharing a matrix is two, which means λ=49 when M = 100. In the worst case, the distance is 6σ, and more nodes share the same matrix. In this case, the number of clusters is fourteen, and with the number of compromised nodes 200, the number of compromised nodes sharing a matrix B is less than five even if all the compromised nodes are located in the border region. This is far from 49, and the matrix is perfectly safe from node capture attacks. Fig.4 shows the number of required matrices and the number of compromised nodes in one cluster as the distance between the deployment points of neighboring clusters increases. In (b), when total number of compromised nodes exceeds 400, the number of disclosed rows in matrix B can be more than λ, however, security for matrix A still shows perfect resistance because λ=99. As in Fig.5, our mechanism has much stronger resistance against node capture than the other mechanisms. It can be affected when the number of compromised nodes exceeds 400 and all the compromised nodes are in border region.

Authentication Security. An unauthorized node cannot inject false data or modify the data without being detected because of the hop-by-hop authentication. When normal nodes are compromised, they can modify the data, but it can be also detected by the next CH on the route because the CH can verify the MAC from the previous CH. Even if several compromised nodes collude, they can be detected by uncompromised CH or by the BS at the least.

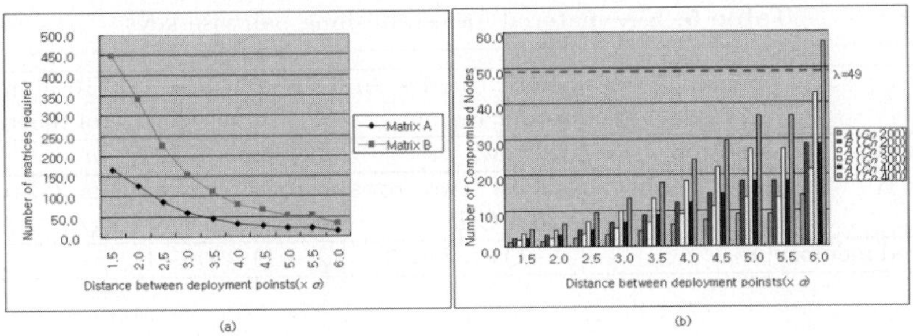

Fig. 4. (a) Number of matrices required (b) Number of compromised nodes (Cn=200,300,400)

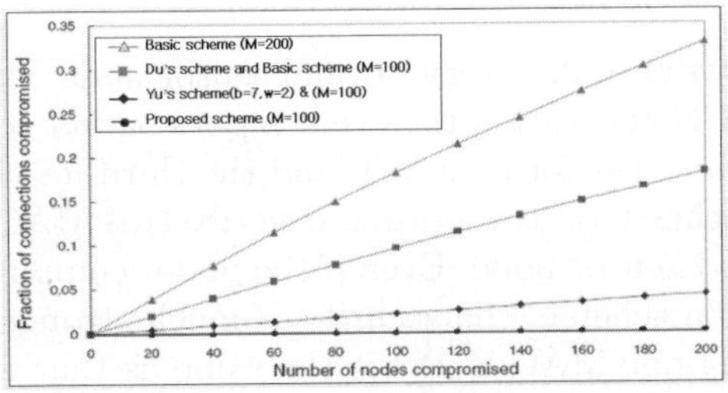

Fig. 5. Fraction of compromised connections

6.2 Overhead Analysis

Key establishment Overhead. Key related information used by the nodes are as in Table 5. By predistributing more information to CHs, normal sensor nodes have much lower memory requirements. Every CH requires $(\lambda+1)(m+1)$. Normal sensor nodes only require one row from the dedicated matrix, i.e, $\lambda+1$, and the nodes which need inter-cluster communications require $(\lambda+1)\times i$, where i=2,3, one row from matrix A, and one or two(maximum number of neighbor clusters) rows from matrix B. For public matrix G, only one seed number for one column is needed[5]. This can be neglected. For computation, all nodes including CHs require as many dot operations as the number of their respective neighbor nodes. The denser the network is, the more operations are required.

For sensor network, energy consumption is critical factor affecting on the lifetime of the network, and for communications, more energy is consumed than other aspects. In our proposal, for pairwise key establishment, normal sensor nodes don't require additional communication except the HELLO messages exchanges. Only the nodes deployed near the border region exchange with their CHs for key materials. Sensor nodes deployed in different cluster from their predefined locations need extra communications, however, it is quire rare. CHs need more communications than normal sensor nodes because they should distribute all the key materials at the requests from the member nodes. This can be tolerable because CHs have larger memory storage and are stronger in energy and

Table 5. Key material for establishing pairwise keys

	Sensor nodes in central area	Sensor nodes in border region	Clusterheads	Total number of matrices
Matrix A	one row	one row	one row	n
Matrix B	–	one or two rows (from CH)	m rows from six B matrices	$\simeq 3 \times n$ - $4 \times \sqrt{n} + 1$
Required memory	λ+1	$(\lambda+1)\times i$, i=2,3	$(\lambda + 1) \times (m+1)$	–

computation ability. Yu's mechanism require more communications for pairwise key setup because it needs path key establishment for neighbor nodes which cannot establish direct pairwise keys. Our mechanism not only lowers the communication overhead but also makes all the direct key establishment possible.

Authentication Overhead. Every normal sensor node needs to compute and verity one MAC. A CH generating the event report packet computes three packets, one for the BS, another for next CH, and the third for the next sensor node on the route. The other CHs compute and verify two MACs, one for next CH and the other for next sensor node. Even if the nodes compute more MACs than normal authentication scheme, the security is much stronger. And because the energy for computing one MAC is about the same as that for transmitting one byte, detecting insider and outsider attacks can eventually save more energy by eliminating unnecessary traffic.

6.3 Pairwise Key Establishment Probability with Neighbor Nodes

Yu's scheme offers a stronger resilience against node capture attacks with a higher degree of connectivity of the sensor network and a lower memory requirement compared to existing schemes. However, to prevent the shared matrices from being disclosed, the possibility of direct key establishment becomes low. To obtain the same degree of security, Yu's scheme needs to set the parameter b and w as 2 respectively, which means possibility of direct key establishment for inter-cluster communication is confined to one-third of our proposal. Among pairs of neighbor sensor nodes which require inter-cluster communications, more than 60% of the pairs should establish pairwise keys through path key establishment. To eliminate this overhead, it also impairs the efficiency of communications by setting up detours when routing. Our mechanism makes all pair of neighboring nodes establish their own pairwise keys with low memory requirement of normal sensor nodes and strong resilience against node capture.

7 Conclusion and Future Work

In this work, we proposed a pairwise key establishment mechanism which increases the security level and the possibility of direct key establishment. We also proposed an authentication mechanism using the pairwise keys to prevent the insider attacks. By clustering the network field into hexagonal shapes and deploy the clusterheads at the center of each cluster, we can make efficient use of clusterheads such as distributing key materials and filtering false data generated by compromised nodes on the routes. Experimental results show that our proposal requires lower memory requirement, increases the resilience against node capture, and guarantee the direct key establishment for every pair of neighboring nodes by making clusterheads carry more predistributed data than normal sensor nodes. It further decreases the amount of traffic generated by outsider and insider attackers by verifying the MAC values. We assume that clusterheads have larger memory storage and stronger in security. This assumption is quite reasonable in respect that in-network processing is necessary for sensor network communication. This architecture is more persuasive considering the efficient and secure data processing. For our future research, we will research on the secure routing based on the propsed architecture and possible attacks.

References

1. D. W. Carman, P. S. Kruus, and B. J. Matt, Constraints and approaches for distributed sensor network security, Technical report, NAI Labs, 2000.
2. R. Blom, An optimal class of symmetric key generation systems. Advances in Cryptology, Proc. of EUROCRYPT 84, LNCS 209, pp.335–338, 1985.
3. L. Eschenauer and V.D. Gligor, A key management scheme for distributed sensor networks, Proc. of the 9th ACM CCS'02, pp.41–47, 2002.
4. H. Chan, A. Perrig, and D. Song, Random key predistribution schemes for sensor networks, IEEE Symposium on Research in Security and Privacy, pp.197–213, 2003.
5. W. Du, J. Deng, Y. S. Han, and P. Varshney, A pairwise key predistribution scheme for wireless sensor networks, Proc. of 10th ACM CCS'03, 2003.
6. D. Liu and P. Ning, Establishing pairwise keys in distributed sensor networks, Proc. of 10th ACM CCS'03, pp.52–61, 2003.
7. C. Blundo, A. De Santis, Amir Herzberg, S. Kutten, U. Vaccaro, and M. Yung, Perfectly-secure key distribution for dynamic conferences, In Advances in Cryptology, CRYPTO'92, LNCS 740, pp.471–486, 1993.
8. W. Du, J. Deng, Y. S. Han, S. Chen, and P. Varshney, A key management scheme for wireless sensor networks using deployment knowledge, Proc. of IEEE INFOCOM, 2004.
9. Z. Yu and Y. Guan, A Robust Group-based Key Management Scheme for Wireless Sensor Networks, IEEE Communications Society, WCNC 2005.
10. S. Zhu, S. Setia, S. Jajodia, and P. Ning, An Interleaved Hop-by-Hop Authentication Scheme for Filtering of Injected False Data in Sensor Networks, Proc. of IEEE Symposium on Security and Privacy, 2004.

HAND: An Overlay Optimization Algorithm in Peer-to-Peer Systems*

Xiaoming Chen[1], Zhoujun Li [2], Yongzhen Zhuang[3], Jinsong Han[3], and Lei Chen[3]

[1] National Laboratory of Parallel and Distributed Processing, Changsha 410073
[2] School of Computer Science & Engineering, Beihang University, Beijing 10083
[3] Dept of Computer Science, Hong Kong University of Science & Technology, Hong Kong
chxmwp@163.com, lizj@buaa.edu.cn,
{cszyz, jasonhan, leichen}@cs.ust.hk

Abstract. Recently, peer-to-peer (P2P) model becomes popular due to its outstanding resource sharing ability in large-scale distributed systems. However, the most popular P2P system – unstructured P2P system, suffers greatly from the mismatching between the logical overlay and physical topology. This mismatching problem causes a large volume of redundant traffic, which makes peer-to-peer system far from scalable and degrades their search efficiency dramatically. In this paper we address the mismatching problem in an unstructured P2P architecture by developing a distributed overlay optimizing algorithm – HAND (Hops Adaptive Neighbor Discovery). Through comprehensive simulations, we show that HAND significantly outperforms previous approaches, especially in large-scale P2P systems.

Keywords: peer-to-peer, search efficiency, topology mismatching problem.

1 Introduction

Peer-to-peer computing, as a promising model to share and manage huge volumes of information in large-scale distributed systems, has gained more and more attentions in recent years [1-5]. In P2P systems, participating peers form a logical overlay on the top of the physical internet topology, so that they can provide computing and sharing services efficiently. In most of the overlays, neighbors of a peer are always chosen randomly. However this random neighbor choosing may cause severe *topology mismatching* problem, *i.e.* the logical topology does not match the underlying physical topology. In a mismatched overlay, the overlay paths are inefficient in the physical layer, leading to heavy traffic overheads in the search [6, 7]. Figure 1 is an example of topology mismatching. Figure 1(b) is the logical topology and Figure 1(a) is the underlying physical topology. A query message is delivered along the overlay path $A \rightarrow B \rightarrow C \rightarrow D$ in 1(b). However, in 1(a) we can observe that the query traverses twice on the several physical links. For example, the query is sent from A to B, and route through the same path back to A. Duplicated query messages may reach a certain peer

* This work is supported in part by the National Natural Science Foundation of China under Grants No. 60473057, 60573057 and 90604007.

M. Gerndt and D. Kranzlmüller (Eds.): HPCC 2006, LNCS 4208, pp. 290 – 299, 2006.

several times. In this example, node A, who initiates the query, afterwards receives its own query twice. This duplication is because that the overlay path is unaware of the underlying physical topology. The situation of mismatching in a real P2P system is even worse. According to Liu *et al*, topology mismatch problem incurs a large volume of unnecessary traffic, and more than 70% of overlay links are suffered [8]. Many previous studies have tried to address this problem by getting rid of the redundant traffic [1, 9-11]. However, the existing approaches have various drawbacks. Some of them, such as LTM [11], rely on clock synchronization, which is impractical and may lead to tremendous synchronization overheads and delays. Other approaches are based on distance probing. However these approaches bring additional traffic overheads, and work well only in static P2P systems.

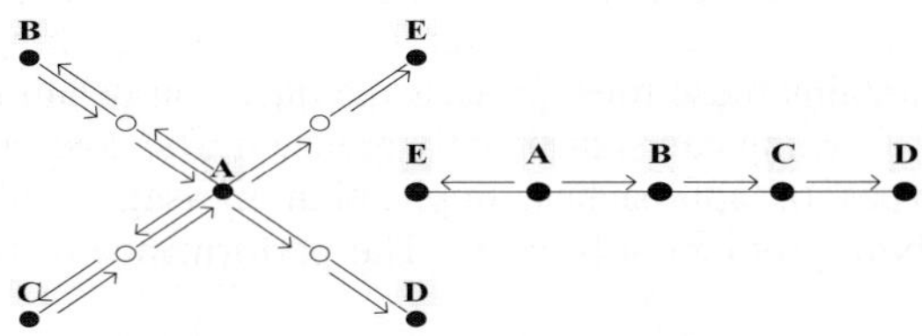

Fig. 1. An example of topology mismatch

We propose a novel algorithm, HAND (Hops Adaptive Neighbor Discovery), to address the topology mismatch problem. We first state that an optimal overlay is a logical overlay that matches the underlying physical topology and name it LCN – logical communication network. Here LCN is the target optimal overlay we want to achieve. LCN can effectively minimize the redundant traffic in the P2P searching. In HAND algorithm, we use a fully distributed triple hop adjustment strategy to adjust the mismatched overlay to LCN.

Our algorithm has several advantages comparing with previous ones. (a) HAND does not need any time synchronization. (b) The traffic overhead involving in the triple hop adjustment is very low. (c) It is applicable to the dynamic P2P environment. (d) HAND can maintain lower query response time.

The remainder of this paper is organized as follows. Section 2 introduces the related work. In Section 3, we present the HAND algorithm. In Section 4, we evaluation the performance of HAND and compare it with previous approaches. Finally, we conclude our work in Section 5.

2 Related Work

Many approaches have been proposed to reduce the large amount of unnecessary traffic caused by topology mismatching in decentralized unstructured P2P systems. They can be categorized into three types: forwarding-based, cache-based and overlay optimization approaches [11].

In forwarding-based approaches, a peer only selects a subset of its neighbors to re-broadcast query messages. When peers select neighbors, they use some metrics to determine what kind of neighbors should be selected. These metrics are some statistic information, including the number of query responses received from neighbors, the latency to connections to neighbors, and so on. Some algorithms use multiple metrics to select neighbors [10]. These forwarding-based approaches can improve the search efficiency. However, they shrink the search scope.

Cache-based approaches can be further divided into two sub-types: data index caching and content caching. Usually, researchers use two cache policies: the centralized caching policy and the local caching policy. Centralized P2P systems provide centralized index servers to keep indices of shared files of all peers. In the local caching policy, each peer maintains some information about other peers. These caching information can be the index of files, query strings and results, replicate data (file contents or query responses), and so on. For example, caching the index of files lets a peer, who maintains these files, process the query on behalf of all nodes within the given hops [12]. Caching can significantly reduce traffic costs and response time.

In the above types of approaches, duplication messages still incurred by the topology mismatching problem still exist. The performance of these approaches is limited.

A lot of overlay optimization algorithms have been proposed [8, 13-18]. HAND also belongs to this type. Here, we can divide all these overlay optimization algorithms into three sub-types: spanning tree based [14], cluster based [13, 15-17], minimum latency first approaches [8, 11]. The spanning tree based approaches first construct a rich graph; and then build a minimum spanning tree from that rich graph [14]. This kind of approaches incurs large traffic overheads to the system. Cluster based approaches use different techniques to identify physically closer nodes and connect them [13, 15-17]. These approaches may shrink the search scope significantly. Minimum latency first approaches use latency as the main metric to measure the distance between peers [8, 11]. However, most of the previous approaches require global latency information to conduct the optimization. In this paper, we also use latency as an optimization metric, but our algorithm HAND only requires local knowledge to perform the overall optimization.

3 HAND Algorithm

To overcome the drawbacks of the previous algorithms, we designed a HAND algorithm that meets the following criteria.

(1) The algorithm performs optimization without any clock synchronization.

(2) The algorithm is fully distributed. It is robust and reliable in decentralized P2P systems.

(3) The traffic overhead incurred in the optimization is trivial.

(4) The algorithm is tolerable to the dynamic P2P environment.

3.1 LCN

The goal of HAND is to adjust a mismatched overlay to a matched one, called a *LCN (logic-communication network)* in the context of this paper. LCN is an undirected graph $G=(V, E)$, which includes a vertex set V represents peers and an edge set E represents links of all nature neighbors.

Definition 1. If a pair of peers (v_1, v_2) satisfy the following condition it is called nature neighbors:

(I) (v_1, v_2) keeps a direct link in the overlay

(II) The shortest physical path between them does not include any other peer nodes.

That is to say if a peer sends a message to its nature neighbor, this message would not reach any other peer nodes. We call the link between the nature neighbors a *matched link.* If a pair of peer nodes satisfy only the second condition, they are named as *matched neighbors.* Note that matched neighbors are not necessary to be nature neighbors.

Definition 2. Let $G=(V, E)$ represent the topology graph of an overlay network, and $V=\{v_1, v_2 \dots v_n\}$ be the set of vertices, we define a n x n square matrix $A(G)=(a_{ij})$ where

$$a_{ij} = \begin{cases} 1 & \text{if } (v_i, v_j) \text{ is a pair of neighbor peers of overlay} \\ 0 & \textit{otherwise} \end{cases} \tag{1}$$

In addition, we define $B(G)=(b_{ij})$as a *matched adjacency matrix*, where

$$b_{ij} = \begin{cases} 1 & \text{if } (v_i, v_j) \text{ is a pair of nature neighbors} \\ 0 & \textit{otherwise} \end{cases} \tag{2}$$

Based on the above definition, we derive lemma 1, which can be proved obviously.

Lemma 1. The overlay network matches LCN *if* $A(G)=B(G)$.

3.2 Peer Sequence Characteristics

In practice, it is impossible to build a LCN using the centralized method, which assumes to have the global knowledge. However, a LCN implies the sequence relationship of peers, which lead to our decentralized adjustment algorithm - HAND.

3.2.1 Peer Sequence Mismatching

Definition 3. Let G* be a LCN of the network, and G' be the current overlay. A peer sequence $(v_1, \dots, v_k)$ of G' is matched if this sequence exists in G* in the same order, otherwise it is mismatched.

In practice, we use a *triple sequences* (v_1, v_2, v_3) to characterize overlay mismatching. A peer sequence of three adjacent nodes in a path is called a triple sequence. Figure 2 and Figure 3 is an example of peer sequence mismatching. Figure 2 represents a LCN and Figure 3 represents a P2P overlay network. The peer sequence *F-A-E* in Figure 3 has a different order in its LCN (*A-E-F* or *F-E-A* in Figure 2), so this sequence is mismatching. Another mismatching example is sequence *F-A-C* in Figure 3, which is split into two other sequence *F-E-A* and *E-A-C* in Figure 2.

Lemma 2. Let G* be LCN of the network, and G' be an overlay. G* and G' is matched if all the triple sequences of the two graph are matched.

3.2.2 Mismatching Detection

Based on Lemma 2, we propose a matching detection algorithm. Suppose a pair of probing messages is sent from a peer v_1 to its neighbors v_2 and v_3 to test the sequence v_2-v_1-v_3. Figure 4, 5 and 6 are used to illustrate the process. We suppose the delays of path (v_1, v_2) and (v_1, v_3) are x' and z' , respectively. When the probing message arrives at v_2, it is forwarded to v_3 directly. We therefore obtain the delay of physical path (v_2, v_3) and denote it as y'. Similarly, when the message reaches v_3, it is forwarded to v_2 directly to obtain the delay of physical path (v_3, v_2), which is also y'.

Lemma 3. If $y'=z'-x'$, then the sequence v_2-v_1-v_3 is mismatched and should be adjusted to v_1-v_2,-v_3.

Lemma 4. If $y'=x'-z'$, then the sequence v_2-v_1-v_3 is mismatched and should be adjusted to v_1-v_3,-v_2.

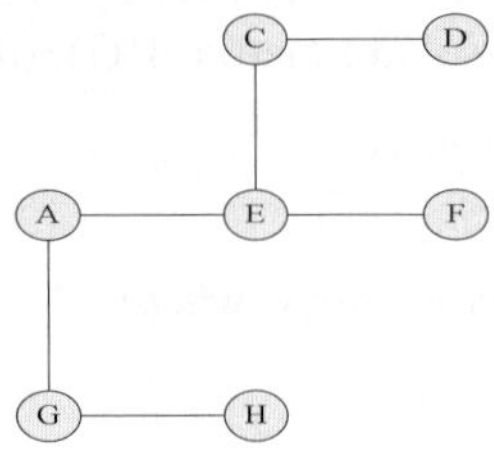

Fig. 2. LCN

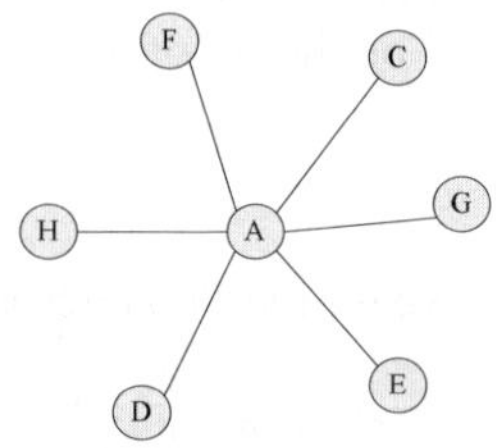

Fig. 3. An inefficient overlay

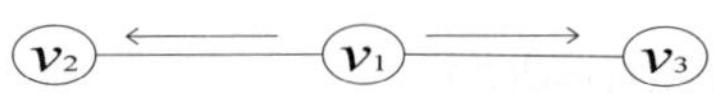

Fig. 4. The communication in overlay

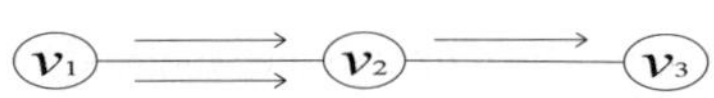

Fig. 5. The matching triple in Lemma 3

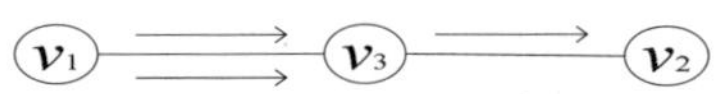

Fig. 6. The matching triple in Lemma 4

Table 1. Route table

Neighbor's IP	***Mark***	***Use-mark***
192.168.0.7	*A*	*B*
192.168.0.111	*B*	*A*

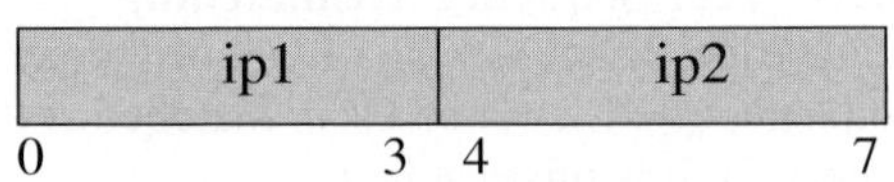

Fig. 7. Probing message body

If v_2-v_1,-v_3 is already matched, $y'=z'+x'$ is satisfied. However, if $y'=z'-x'$, it implies that the matched sequence should be v_1-v_2,-v_3 as in Figure 5, where the delay of path (v_1, v_2), (v_2, v_3) and (v_1, v_3) are x', y' and z' respectively. Similarly, if $y'=x'-z'$, it implies that the matched sequence should be v_1-v_3,-v_2 as in Figure 6. In practice, when the sequence is mismatched, the equation $y'=z'-x'$ and $y'=x'-z'$ can still be false due to error caused by forwarding delays and delay jitters. So we relieve the mismatching condition to be:

$$y' = z' - x' \pm \varepsilon,\ y' = x' - z' \pm \varepsilon \text{ and } y' = x' + z' \pm \varepsilon$$

where ε is a small positive real number.

3.3 Triple Hop Adjustment

In the previous section we describe the mismatching detection of a single sequence. In this section we will describe the mismatching detection of the whole P2P overlay. It includes designing the routing table, probing detect messages and adjusting mismatching sequences.

3.3.1 Routing Table and Probing Messages

Firstly, we describe our routing table and the probing message format. Table 1 shows the format of the routing table. Similar with Gnutella, it comprises a list of neighbor IPs. To guarantee that the pair of neighbors isn't probed repeatedly, we add two items in the routing table, *Mark* and *Use-mark*. *Mark* is a label given to each neighbor, the second node of a triple sequence. *Use-mark* records the candidates of the third node of the triple sequence. Therefore, *Mark* and *Use-mark* is a probing pair. When a peer adds a new neighbor, it must gives this neighbor a *Mark*; and then, this *Mark* should be added to the *Use-mark* of every other neighbor in the routing table; finally, the *Mark* of those old neighbors should be added to *Use-mark* of the new neighbor. When a neighbor leaves the network, we remove its record, and delete its appearance in the *Use-mark* of all other neighbors. Figure 7 is the format of probing message. The probing message has two segments, which are a pair of neighbor's IPs. For example, a probing message for sequence v_2-v_1-v_3 (Figure 4) consists of v_2's IP in the first segment and v_3 's IP in the second segment.

3.3.2 Probing Message Construction

Before we send detect messages, we must build the probing pair. For example in table 1, we find neighbor A, and select *B* from its *Use-mark*. Let A and B be a probing pair (A, B), and remove B from this *Use-mark* of A. We also delete A from the *Use-mark* of neighbor *B*. Finally, we encapsulate their IP in a probing message. In this example ip1 is 192.168.0.7, ip2 is 192.168.0.111. According to the above algorithm, each peer sends the probing message to a pair of neighboring peers. The peer receiving this message then re-sends it to the other peer. The probing continues until the *Use-mark* of the neighbors is empty.

3.3.3 Probing Algorithm

According to Lemma 3 and 4, a pair of probing messages can detect if the original sequence is matched or not and decide how to adjust the sequence to be a matched one. If the probing condition satisfies Lemma 3, edge (v_1, v_3) is deleted and a new edge (v_2, v_3) is added. If Lemma 4 is satisfied, edge (v_1, v_3) is deleted and a new edge (v_2, v_3) is added. In both situations, the routing tables of the corresponding peers are updated accordingly.

The probing message is issued by each node periodically. The HAND algorithm executed at each peer is as follows: for each triple sequence (*h-i-j*), we use *timestamp*(*i,h*) to denote the timestamp when the probing message arrives at *h* and *timestamp*(*i,j,h*) to denote when the probing message forwarded by *j* arrives at *h*.

4 Performance Evaluation

In this section, we present our simulation results. We first introduce the evaluation metrics and the simulation methodology, and then present the results comparing HAND with LTM.

4.1 Evaluation Metrics

We use two metrics to evaluate the performance of HAND - traffic cost and query response time. Traffic cost is the overhead in message transmission. In our simulation, we use comprehensive traffic cost, which is calculated from a function involving the consumed network bandwidth and the other related parameters. Response time of a query is the time period from when the query is issued until when the source peer receives a response result from the first responder.

In the comparison of HAND and LTM, we use three metrics: traffic cost reduction rate, response time reduction rate and traffic overhead, which are defined in [8]. The traffic cost reduction rate $R_c(*)$ and response time reduction rate $R_t(*)$ is calculated as follows:

$$R_c(*) = \frac{C_{(\text{Gnutella - like})} - C_{(\text{op - algorithm})}}{C_{(\text{Gnutella - like})}} \times 100\% \tag{3}$$

$$R_t(*) = \frac{T_{(\text{Gnutella - like})} - T_{(\text{op - algorithm})}}{T_{(\text{Gnutella - like})}} \times 100\% \tag{4}$$

where (*) represents a given mechanism that is used to search for all peers in a Gnutella-like overlay. Under this mechanism C(*Gnutella-like*) is the traffic cost incurred in a Gnutella-like overlay, C(*op-algorithm*) is the traffic cost incurred in an optimizing algorithm enabled Gnutella-like overlay. T(*Gnutella-like*) is the average response time of queries in a Gnutella-like overlay and T(*op-algorithm*) is that of an optimizing algorithm enabled Gnutella-like overlay. In this simulation we use blind flooding mechanism. Traffic overhead is the per minute traffic cost incurred by an optimizing algorithm on the P2P overlay.

4.2 Simulation Methodology

To accurately simulate the algorithms in P2P systems, we use trace driven simulation. First, we generate two topologies: one for physical application layer and the other for P2P overlay layer. We used a topology generator -BRITE[19] to generate a physical internet topology including 22,000 nodes. The logical topologies are obtained from real traces Clip2 [20]. The network size is ranging from 5,000 to 8,000 peers. The average number of neighbors in the P2P overlay ranges from 6 to 10. Both the physical topology and overlay topology follow the small world and power law properties [21].

We employed flooding search in our simulations. To accurately simulate flooding search, every node issues 0.3 queries per minute in our simulation, which is calculated from the observation in [22]. The file popularity follows a Zipf distribution [23].

In P2P systems, peers join or leave the overlay frequently. We simulate peer joining and leaving behaviors via tuning on/off logical peers. When a peer joins, a lifetime will be assigned to this peer. The lifetime is period the peer stay in the system, counted by seconds. The assignment of lifetime is defined according to the observation in. The lifetime will be decreased by each passing second. A peer will leave in next second when its lifetime reaches zero. During each second, there are a number of peers leaving the system. We then randomly turn on the same number of peers from the physical network to join the overlay.

4.3 Simulation Results

We first simulate our algorithm in the static environment where the peers do not join and leave frequently. In this simulation, we use the overlay topology with 8,000 peers on the top of the physic network with 22,000 nodes. Figures 8 and 9 plot the comparison of HAND and Gnutella. The results show that HAND can quickly converge to an optimal overlay. During this converging process, both the traffic cost and query response time are decreasing. Figure 8 and 9 show that our algorithm can effectively decrease the traffic cost by about 77% and shorten the query response time by about 49% in less than two minutes. In the simulation, each peer has 10 neighbors on average (*i.e.* the average connection of a peer is 10). The tradeoff between query traffic cost and response time is affected by the average connection. P2P systems with a higher average connection offer a faster search speed while increasing traffic. It has been proven that the more neighbor connections, the less query response time, but the more redundant traffic [17].

We then evaluate HAND in a dynamic environment with peer joining and leaving defined in Section 4.2. We evaluate HAND with different average neighbor connections and increasing network size in the dynamic environments.

First, we test two overlays of size 5,000 and 8,000. The average number of neighbors is always 10. Figure 10 shows that their traffic reduction is about 76% and 64% respectively. After converged, the average traffic costs of the two overlays are close. In addition, Figure 11 shows that HAND can effectively shorten the query response time by about 48%~54%. From the above analysis, we know the size of overlay network has a little impact on the effectiveness of our algorithm.

Second, we vary the average number of neighbors from 6 to 10 (6, 8, and 10). This means at the very beginning, the overlay network has 6, 8, and 10 average neighbor connections respectively. Figure 12 and Figure 13 plot the average traffic cost and average query response time running HAND algorithm on the three overlays. At the first few steps of the algorithm, the more neighbor connections, the higher the traffic cost and the shorter the response time. This is consistent with the relationship of neighbor connections, traffic cost and response time discussed in Section 4.3.1. After several steps, three topologies all converge to an optimal topology - LCN. So their average traffic costs and average query response time are almost the same. These figures also show that the reduction of traffic costs is about 42%~64% and query response time is about 40%~54%.

We then compare the performance of HAND with the performance of LTM. Figure 14 and Figure 15 plot traffic reduction rate and response time reduction rate respectively. From figure 14 we can see that HAND and LTM have almost the same traffic reduction rate. However, the response time reduction rate of HAND is higher than LTM in Figure 15.Their difference is about 4%. Figure 16 plots the traffic overhead of HAND and LTM. The traffic overhead of HAND is much less than that of LTM (including the traffic overhead of clock synchronization). On average, this traffic overhead reduction is about 55%.

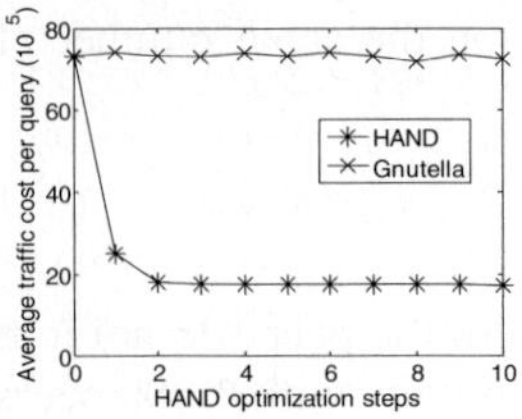

Fig. 8. Traffic reduction in static environments

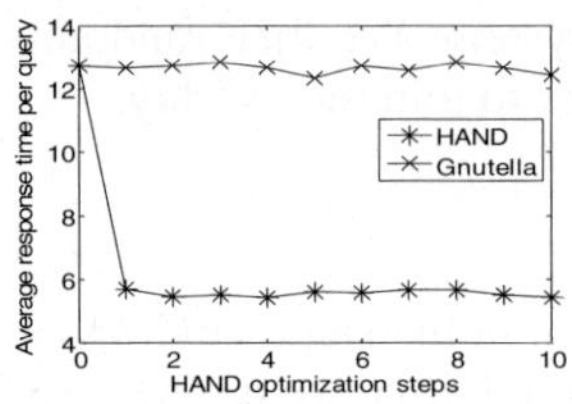

Fig. 9. Response time in static envrionments

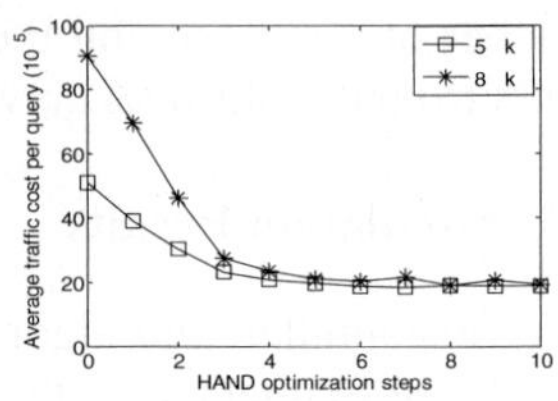

Fig. 10. Traffic cost of different topology size

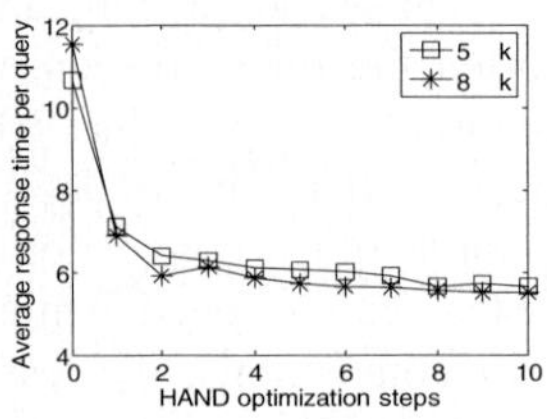

Fig. 11. Response time of different topology size

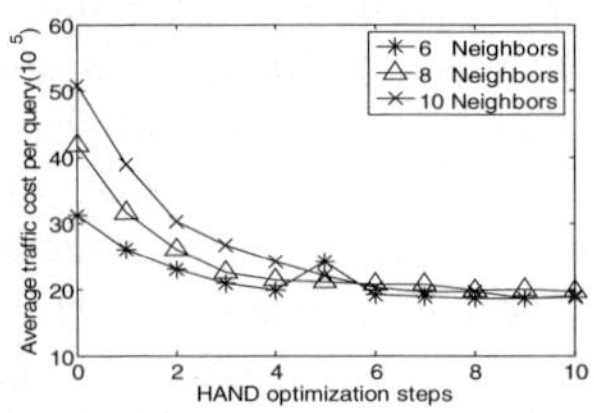

Fig. 12. Traffic cost of different average conn-ections

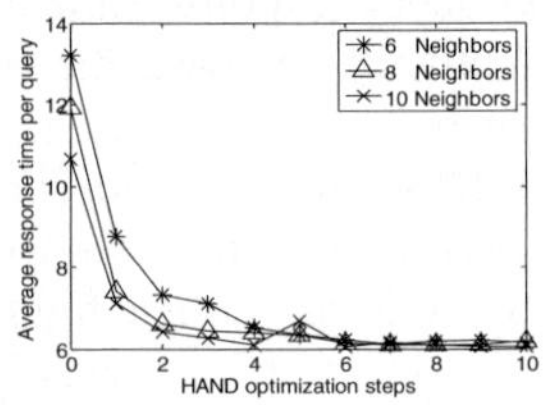

Fig. 13. Response time of different average connections

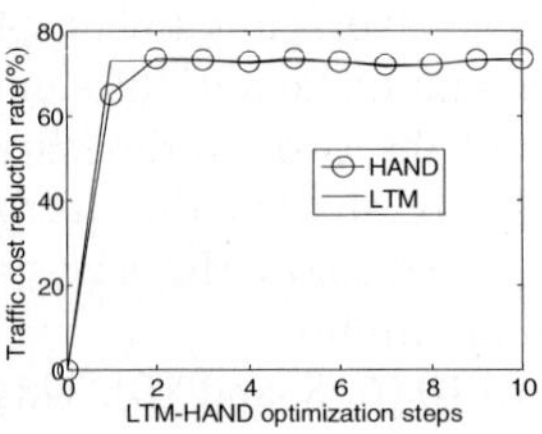

Fig. 14. Traffic reduction rate of HAND and LTM

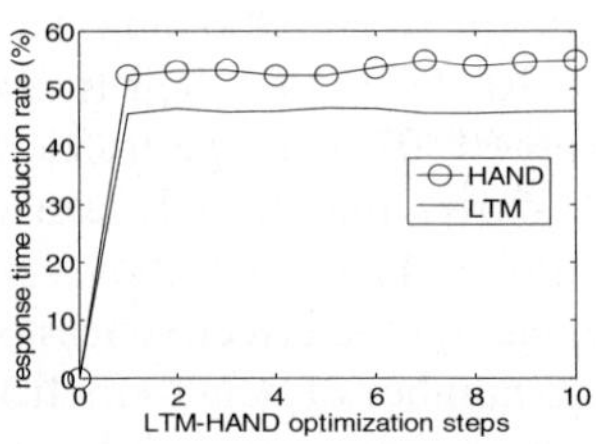

Fig. 15. Response time reduction rate of HAND and LTM

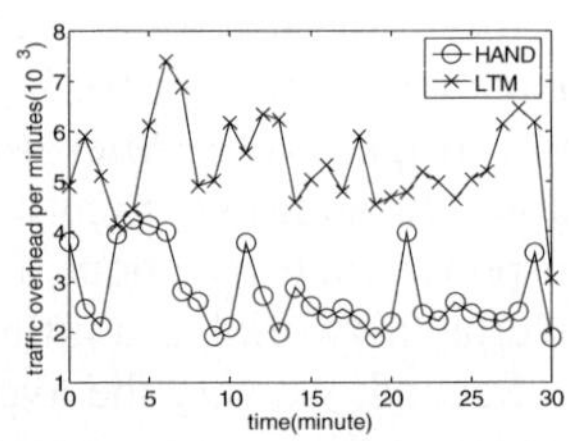

Fig. 16. Per minute traffic overhead of HAND and LTM

5 Conclusions and Future Work

In this paper, we propose a new solution for topology matching problem. The goal is to adjust the mismatching topology to an optimal LCN. We designed an algorithm HAND to optimize the overlay by considering the relation of LCN and the physical topology. This algorithm can be executed distributed by triple hops adjustment. In the future, we intend to deploy HAND into a real P2P streaming system.

References

[1] Y. Chawathe, S. Ratnasamy, L. Breslau, N. Lanham, and S. Shenker, "Making Gnutella-like P2P Systems Scalable," In Proceedings of ACM SIGCOMM, 2003.

[2] A. Nakao, L. Peterson, and A. Bavier, "A Routing Underlay for Overlay Networks," In Proceedings of ACM SIGCOMM, 2003.

[3] Y. Liu, Z. Zhuang, L. Xiao, and L. M. Ni, "A Distributed Approach to Solving Overlay Mismatching Problem," In Proceedings of IEEE ICDCS, 2004.

[4] X. Liao, H. Jin, Y. Liu, L. M. Ni, and D. Deng, "AnySee: Peer-to-Peer Live Streaming," In Proceedings of IEEE INFOCOM, 2006.

[5] L. Xiao, Z. Zhuang, and Y. Liu, "Dynamic Layer Management in Superpeer Architectures," In Proceedings of IEEE Transactions on Parallel and Distributed Systems (TPDS), 2005.

[6] I. Abraham, A. Badola, D. Bickson, D. Malkhi, S. Maloo, and S. Ron., "Practical locality-awareness for large scale information sharing," In Proceedings of IPTPS, 2005.

[7] Y. Liu, L. Xiao, and L. M. Ni, "Building a Scalable Bipartite P2P Overlay Network," In Proceedings of 18th International Parallel and Distributed Processing Symposium (IPDPS), 2004.

[8] Y. Liu, A.-H. Esfahanian, L. Xiao, and L. M. Ni, "Approaching Optimal Peer-to- Peer Overlays," In Proceedings of the 13th Annual Meeting of the IEEE International Symposium on Modeling, Analysis, and Simulation of Computer and Telecommunication Systems (MASCOTS), 2005.

[9] Z. Xu, C. Tang, and Z. Zhang, "Building Topology-aware Overlays Using Global Soft-state," In Proceedings of the 23rd International Conference on Distributed Computing Systems (ICDCS), 2003.

[10] Z. Zhuang, Y. Liu, L. Xiao, and L. M. NI, "Hybrid Periodical Flooding in Unstructured Peer-to-Peer Networks," In Proceedings of IEEE ICPP, 2003.

[11] Y. Liu, X. Liu, L. Xiao , L. M. Ni, and X. Zhang, "Location-Aware Topology Matching in P2P Systems", IEEE INFOCOM 2004, Hong Kong, China, March 2004.

[12] B. Yang and H. Garcia-Molina, "Efficient Search in Peer-to-Peer Networks," In Proceedings of the 22nd International Conference on Distributed Computing Systems (ICDCS), 2002.

[13] B. Krishnamurthy and J. Wang, "Topology Modeling via Cluster Graphs," In Proceedings of SIGCOMM Internet Measurement Workshop, 2001.

[14] Y. Chu, S. G. Rao, and H. Zhang, "A Case for End System Multicast," In Proceedings of ACM SIGMETRICS, 2000.

[15] V. N. Padmanabhan and L. Subramanian, "An Investigation of Geographic Mapping Techniques for Internet Hosts," In Proceedings of ACM SIGCOMM, 2001.

[16] S. Ratnasamy, M. Handley, R. Karp, and S. Shenker, "Topologically-Aware Overlay Construction and Server Selection," In Proceedings of IEEE INFOCOM, 2002.

[17] M. Ripeanu, A. Iamnitchi, and I. Foster, "Mapping the Gnutella Network," in *IEEE Internet Computing*, 2002.

[18] X. Liu, L. Xiao, A. Kreling, Y. Liu, "Optimizing Overlay Topology by Reducing Cut Vertices", In Proceedings of ACM NOSSDAV, 2006.

[19] BRITE, "http://www.cs.bu.edu/brite/."

[20] Clip2, " The Gnutella Protocol Specification v0.4, Document Revision 1.2."

[21] S. Saroiu, P. Gummadi, and S. Gribble, "A Measurement Study of Peer-to-Peer File Sharing Systems," In Proceedings of Multimedia Computing and Networking (MMCN), 2002.

[22] K. Sripanidkulchai, "The Popularity of Gnutella Queries and Its Implications on Scalability," 2005.

[23] L. Breslau, P. Cao, L. Fan, G.Phillips, and S. Shenker, "Web Caching and Zipf-Like Distributions: Evidence and Implications," In Proceedings of IEEE INFOCOM, 1999.

A High Performance Heterogeneous Architecture and Its Optimization Design

Jianjun Guo, Kui Dai, and Zhiying Wang

School of Computer, National University of Defense Technology,
410073 Changsha, Hunan, China
jjguo@tom.com

Abstract. The widely adoption of media processing applications provides great challenges to high performance embedded processor design. This paper studies a Data Parallel Coprocessor architecture based on SDTA and architecture decisions are made for the best performance/cost ratio. Experimental results on a prototype show that SDTA has high performance to run many embedded media processing applications. The simplicity and flexibility of SDTA encourages for further development for its reconfigurable functionality.

Keywords: Data Parallel, SDTA, ASIP.

1 Introduction

Nowadays, the developing media applications such as graphics, image and audio/video processing put forward great pressure on high performance embedded processor design. The increasing computational complexity and diversity of the applications make it hard for architecture designers to design an optimized architecture suitable anywhere. One typical embedded application field is from multimedia digital signal processing and security. It shows the same features, such as heavy computational workloads, large amounts of data, significant processing regularity, extensive data parallelism and real-time constraints [1]. Another typical application field is from scientific computing which shows the features of typically large number of arithmetic floating-point operations, typically large data and working sets, typically low temporal locality and high spatial locality depending on regularity of the applications [2]. All these application features make the traditional embedded microprocessors inadequate to meet the application requirements, thus high performance embedded processors are demanded.

As we all know, power wall, memory wall and frequency wall are great obstacles for high performance processor design. The energy consumption that accompanies the high performance is often beyond power budgets especially in embedded systems. Obviously, it has already become a bottleneck for the development of microprocessor today. With the announcement of several multi-core microprocessors from Intel, AMD, IBM and Sun Microsystems, the chip multiprocessors have recently expanded from an active area of research to a hot product area. In fact, multi-core technology is an efficient way to alleviate the pressure on power. For example, the fastest CELL processor adopts the heterogeneous multi-core architecture to attack the power wall [3][4].

M. Gerndt and D. Kranzlmüller (Eds.): HPCC 2006, LNCS 4208, pp. 300 – 309, 2006.

As for the memory wall, traditional cache-based design is still used but different variations have been proposed to overcome its limitations. Standard cache hierarchy is designed for general purpose applications which retain most of data in on-chip data cache/memory and stream instructions in from off-chip. That is to say, those applications with good temporal and spatial locality are ideally supported by cache-based memory hierarchies. However, for media and stream-based applications, they stream in much of their data in from off-chip and retain their instruction code in on-chip instruction cache/memory. They commonly spend 25-50% of execution time in memory stalls on standard cache-memory systems [5]. So poor temporal locality leads to compulsory misses in standard cache-memory hierarchies. Different variations of traditional cache system are proposed to solve this problem. Jouppi proposed the stream buffer [6], which stores the prefetched data in the stream buffer until needed. Palacharla and Kessler [7] enhanced the stream buffer by using a filter to limit unnecessary prefetches. Stride Prediction [8][9] overcomes the limitations of stream buffers in only prefetching successive cache lines and the prefetched data elements are placed directly into the cache. The performance of stride based prefetching is often significantly lower in smaller caches because the prefetched data pollutes the L1 cache. Stream Cache [10] separates from the L1 cache, which uses a Stride Prediction Table (SPT) to determine what data to prefetch, but stores the data in the stream cache instead of the L1 cache. These methods have comparatively good prefetching precision while the control mechanism is too complex and is not so suitable for high performance design.

All of the above are cache-based design, while the recently announced CELL processor adopts a three-level memory hierarchy to attack the memory wall [19]. That is Main Memory, Local Store and Registers. It uses local store and larger register files instead of the cache-based memory system. Its larger register file can also allow deeper pipelines to achieve higher frequency. The memory system of CELL introduces excellent performance while it needs fast DMA support to use local store efficiently and its SRAM needs full custom design which is beyond conventional design ability. It is also not transparent to programmers and offloads many works to the compiler or the programmer. So the memory system in our research is still cache-based but detailed research will be done to achieve acceptable performance.

High-end microprocessor designs have become increasingly more complex during the past decade with designers continuously pushing the limits of instruction level parallelism and speculative out-of-order execution. Although the complexity and diversity of the applications make it hard to design an optimized architecture suitable anywhere, it is possible to design high performance/cost ratio architecture especially for one or one kind of applications. Such an Application Specific Instruction Processor (ASIP) [11] is designed to perform certain specific tasks efficiently. Transport Triggered Architecture (TTA) can provide both flexibility and configurability during the design process [12]. In addition to the application specific instruction processor design, it also provides support for instruction level parallelism (ILP). Based on the flexibility and configurability of TTA, the high performance processor architecture can be specially designed according to the characteristics of specific applications.

This paper studies an embedded Data Parallel Coprocessor (DPC) based on Synchronous Data Triggered Architecture (SDTA). It can exploit ILP as much as possible with the support of custom function units for special operations and can match well with ASIP design process. The main work of this paper is to design a heterogeneous architecture composed of a LEON3 host processor and the DPC and also to make architecture decisions for performance concern. The rest of this paper is organized as follows. Section 2 briefly describes the transport triggered architecture. Section 3 describes the proposed architecture, while in section 4 architecture decisions are made and performance tests and results are given too. The last section concludes and discusses the future work.

2 Transport Triggered Architecture

As shown in Fig. 1, TTA proposed by Henk Corporaal et al [12] can be viewed as VLIW architecture. In TTA, there is only one data move instruction for programmers and the operation is triggered by a data transfer. This is different from traditional operation triggered architecture.

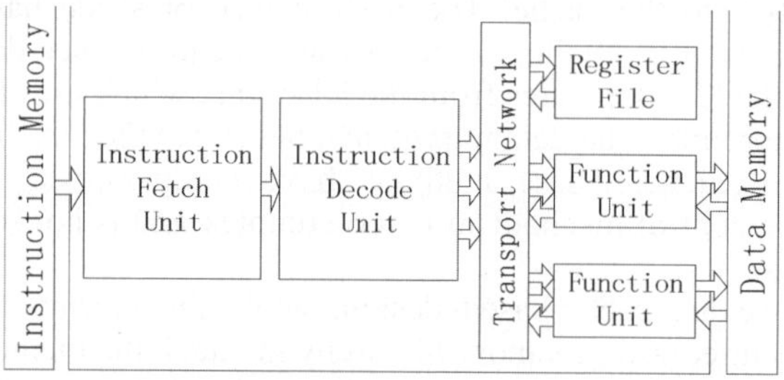

Fig. 1. Organization of Transport Triggered Architecture (TTA)

As shown in Fig. 2, the central part of TTA is very simple. Function units (FU) and register files (RF) are connected to buses by sockets. Every function unit has one or more operator registers (O), one trigger register (T) and result register (R). Data being transferred to the trigger register will trigger the function unit to do the operation using the data in the operator register and the trigger register as the source operands, and after certain cycles put the final result into the result register. One function unit can execute multiple operations (e.g. add, sub and mac) and the control code will determine what operation to execute through instruction decoding.

In TTA, all operations, including load/store, jump and branch, etc. are executed by a data move instruction. For example, a usual add operation can be represented by three data move operations:

$$add\ r_3, r_2, r_1 \quad \Rightarrow \quad r_1 \text{->} O_{add};\quad r_2 \text{->} T_{add};\quad R_{add} \text{->} r_3$$

First, the data in registers r_1 and r_2 are separately transferred to the operator register and trigger register of the adder unit, after some cycles (according to the latency of the adder unit) the result is sent from the result register to the register r_3.

In TTA, the number of the function units, register files and their ports, buses and bus connections are all flexible. This brings flexibility for the architecture design. In addition, one TTA instruction always contains several paralleled data move operations after code scheduling. The maximum number of parallel operations is the same as the number of the buses. Therefore, increasing bus number brings increase of performance. In addition, the architecture designer can customize special operations into the instruction set by designing special function units for some special operations [13]. Thus, some bottleneck operations in the applications can be implemented by special function units to meet performance requirement of the applications.

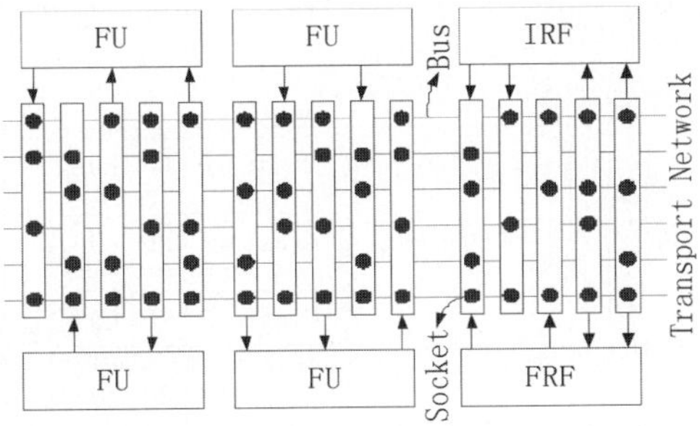

Fig. 2. General Structure of TTA. It is composed of function units and register files, all of which are connected by the transport network.

TTA with a synchronous control mechanism which is called Synchronous Data Triggered Architecture (SDTA) is used in our design. A single controller can be shared among multiple computational resources. It is an innovative architecture to support high performance computing efficiently. Due to the synchronous control, clock gating can be easily used to reduce power dissipation. The synchronous data triggered architecture programming paradigm (programming data transport) gives rise to undeveloped code scheduling degrees of freedom. The SIMD and MIMD programming paradigm, which combines large throughput with very low latency inter node communication, naturally fits into this synchronous data triggered architecture model. The major advantage of SDTA is that it can match well with the ASIP design process. When given a new application, we can first analyze the characteristics of the application, and then we can get the frequently used operations in the application. Special function units can be designed to accelerate the frequently used operations and these function units can be easily added to the SDTA. Therefore, the whole application can be accelerated. Synchronous data triggered architecture uses its registers and data transport capacity more efficiently for it reduces the coupling degree of the data transport and the operation execution. This becomes especially important when integrating intra-node parallelism onto a single chip.

3 Implementation

The whole chip is a heterogeneous multi-core architecture as shown in Fig. 3. It is composed of a LEON3 [14] host processor and the SDTA-based coprocessor. The coprocessor has eight clusters. The control tasks are accomplished by the LEON3 host processor while the computation intensive tasks are scheduled to the coprocessor by the host processor to satisfy the high performance requirement of the data intensive applications. Moreover, each cluster has eight parallel SIMD data paths and contains four integer units, two floating-point units, one integer divider, one floating-point divider, two integer compare units, one floating-point compare unit and one specific CORDIC [15] unit. The type and number of the function units derive from the analysis of applications to meet the performance requirements. In each cycle, all the function units in the SIMD data paths can transfer the specified data to trigger a new operation with the effective support of the 64-bits load/store unit. This suits for different embedded applications such as graphics and image processing, video and audio processing.

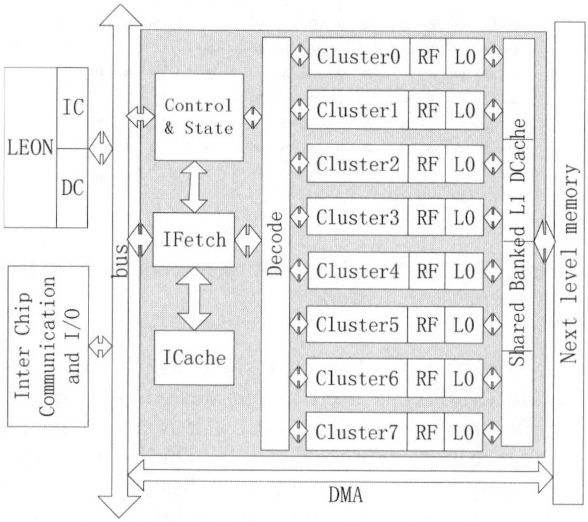

Fig. 3. Architecture Overview of the Whole Chip

3.1 Application Analysis and Compiler Work

When given a specific application, we first run it on the simulator to gather the necessary information such as the frequently used operations and the parallelism upper bound etc. by trace analysis. The type and number of the major operations in the application determine the type and number of the function units in the SDTA. According to the type of the operation, designer can quickly decide what function unit to implement; similarly, the number of the function unit is decided according to the proportion of the equivalent operations. The number of each function unit in our design is determined by the analysis of results from [1] to gain the maximum benefit. The number of buses determines how many operations can be executed at the same time, and so the parallelism upper bound means how many buses should be used. To make full use of the function units and also to save area, an 8-bus scheme is chosen. The number of the average active registers shows that how many registers should exist at the same time both to save the hardware cost and to keep the performance. The decision of register file structure will be described later.

The application program is written in C language. First the compiler translates it into serialized code in the form of data transport. Then the serialized code is scheduled and matched to different function units. The upper bound of simultaneous operations that can be scheduled is determined by the bus number in the cluster while the actual number of simultaneous operations that can be scheduled depends on the potential parallelism of the application and the registers available.

3.2 Function Units

The arithmetic clusters in the core adopt the SDTA architecture, where the operation is triggered by the data transfer. The computation is done by transferring values to the operator register and starting an operation implicitly via a move targeting the trigger register associated with the command. The operands may be the output of its own final stage or the output of their neighbor units.

The integer ALU performs the common operations including the arithmetical and logical ones. Many analyses such as [1] show that the data type used in many embedded applications are integer and data size less than 32bits is frequently used. To support all these features better, sub-word parallelism on byte or half-word is added. Bit operations like bit reverse are also added to support security applications. The multiplier supports 32-bit integer multiplication with 2 cycles' latency and it can also perform 2-way 16x16-bit or 4-way 8x8-bit multiplication.

The floating-point units perform operations on single and double precision floating-point operands. All operations are IEEE-754 compliant. The FPU is fully pipelined and a new operation can be started every clock cycle except that DIVF command which requires 20 cycles to complete the divide operation and is not pipelined.

The compare unit does the compare operation and returns a result, which can be used to predicate the conditional transfers by the condition codes. The predicates make the if-else conversion and conditional execution possible.

The CORDIC unit is based on the parallelization of the original CORDIC [15] algorithm and adopts a relatively simple prediction scheme through an efficient angle recoding. The proposed implementation, a pure combined logic without pipeline operation, has a delay of 33 clock cycles.

3.3 Memory Subsystem

The cache systems described in part one are mostly prefetch-based. Different prefetch methods are used to improve data fetch precision but complex prefetch mechanism makes the design complex too. Therefore, it is not so suitable for high frequency design. As illustrated in [16], when interconnect overheads are taken into account, sharing L2 cache among multiple cores is significantly less attractive than when the factors of delay and area are ignored. So only L1 cache and large register file system is chosen in our design. The L1 cache is multi-banked and high-bit address interleaved so that each bank has its own data space and each cluster can get its data from a bank to reduce collisions. The write-back policy and 256 bits long cache lines are selected to achieve higher performance. A compiler-managed L0 buffer is supported to store some uncacheable data. In addition, a DMA channel is used to allow the host processor and/or other peripherals such as I/O devices to access the processed data quickly. As for the register file, different structures are designed and tested. Finally, 8 banked and a total of 128 registers are used. Each bank has one read port and one write port. Experimental results in section 4.3 indicate this structure can achieve a best performance/cost ratio.

4 Experimental Results

A prototype chip with the host processor and one cluster is first implemented and tested to verify our design. Several typical kernel applications are selected to verify our design. By the analysis of different media processing applications, four kernel applications are finally chosen to be the representative set as shown in Table 1. These kernel applications are frequently used in many embedded applications. If they can be executed very fast, the execution time of the whole application will be much shorter.

Table 1. Typical digital signal processing application set

Name	Brief Description
fft	Fast Fourier Transform
fir	Finite Impulse Response filter
idct	Inverse Discrete Cosine Transform
matrix	Multiply of two matrix

The discrete fourier transform plays a very important role in representing the relationship between time field and frequency field of discrete signal. FFT is a fast implementation of discrete fourier transform, which is widely used in various DSP applications. The FFT application in this set is to transform an array containing 128 complex samples. Digital filter is the substitute of the traditional analog filter, which is used to change or modify the characteristics of the signal. FIR filter is one kind of the digital filter. In the FIR application of this set, the number of coefficients is 16, and the number of output samples is 40. IDCT is frequently used in video processing applications to decode the compressed image. The IDCT application in this set is to transform an 8x8 two dimensional image matrix. Then, MATRIX is an application which contains basic operations in different DSP applications, so accelerating these applications is a matter of great significance.

4.1 Silicon Implementation

Under the clock frequency of 230 MHz, the standard cell implementation of the prototype design utilizing a 0.18um CMOS technology resulted in an area of approximate 5.2mm^2 including the expected layout overhead with regard to the synthesis results with the Synopsys® Design Compiler®.

4.2 Cache Write Policy Selection

For L1 cache system, the performance of DPC under write-through and write-back policy is compared. Several 32bits applications are selected to do the experiment and the read and write miss ratio for three of them under write-back policy are shown in Table 2. Performance gains compared with write-through policy is also given in Table 2. From Table 2, it can be seen that write-back policy can achieve a better performance.

Table 2. The Read/Write miss ratio under write-back policy and its performance gains

Algorithm	Read miss ratio	Write miss ratio	Performance gains
32bits idct	7.50%	0.00%	2.23%
32bits fir	0.05%	2.09%	13.27%
32bits fft	4.76%	2.38%	4.35%

4.3 Optimal Register File Structure Selection

TTA structure depends on scheduling algorithm heavily, so the number of registers for compiler use is performance sensitive. Several typical applications in media signal processing are selected to determine the optimal register number and port number. The performance of these applications is compared between the LEON3 processor and the coprocessor with different register number and port number. The performance on LEON3 processor and DPC with 64 registers 2R/2W ports, 128 registers 1R/1W port and 256 registers 1R/1W port are compared. As shown in Fig. 4, the baseline performance is for LEON3 processor and 128 registers with 1R/1W port is the best performance/cost ratio point.

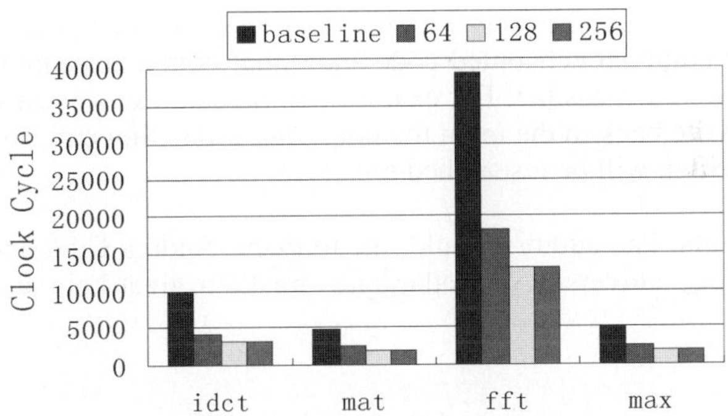

Fig. 4. Performance Comparison between LEON3 and DPC with different register file structure

4.4 Performance Comparison of the Whole Chip

The performance speedup of typical applications on DPC compared with LEON3 is shown in Table 3. The considerable speedup of FIR is due to that FIR uses sine operation frequently which can be accelerated by the CORDIC function unit in DPC. In addition, experiments show that the performance of DPC is better than Texas Instrument TMS320C64x series DSP processor [17] which is a fixed-point VLIW architecture comparable to DPC. For example, the speedup of 32bits IDCT and FIR is respectively 1.8 and 23.0. During the comparison, the applications on TMS320C64x are executed and evaluated in the CCS2.0 for C6000 [18].

Table 3. The typical algorithm performance comparison between LEON3 and DPC. The speedup is given here.

Algorithm	Speedup
8x8 32bits idct	2.9
128-dots 32bits fir	3795.1
radix-4 complex 128 16-bits fft	8.4
4x4 16-bits matrix	23.5

5 Conclusion

This paper studies a multi-core heterogeneous architecture that is very suitable for embedded applications for its flexibility and configurability. Through the application characteristics analysis and special function unit support, it is easy to modify the architecture to adapt to different applications. Traditional processor architecture is fixed, so the software must be adjusted to map to the hardware structure to gain best performance, whereas the DPC is configurable, so the hardware/software co-design method can be used to achieve a higher performance.

In the future work, the architecture will be further researched to use more efficient function units and simpler interconnection while gaining better performance and some low power mechanism will be implemented too. Software optimization such as compiler optimization will be researched further for the experiments show that handmade programs are 2-5 times faster than the compiler scheduled programs. Analyses of the compiler-scheduled code show that compiler cannot make fully use the registers as well as possible. For example, some data can stay in the registers and do not need to write back to the main memory after a function call, so a better register scheduling algorithm will be researched too.

Acknowledgments. The authors would like to thank Andrea Cilio and his colleagues in Delft technology university of Netherlands for their great help and support to our research work.

References

1. Jason E. Fritts, Frederick W. Steiling, Joseph A. Tucek. MediaBench II Video: Expediting the next generation of video systems research. Embedded Processors for Multimedia and Communications II. San Jose, California; March 8, 2005; ISBN / ISSN: 0-8194-5656-X; p. 79-93.
2. M. W. Berry. Scientific Workload Characterization By Loop-Based Analyses. SIGMETRICS Performance Evaluation Review, Volume 19, Issue 3, 1992, p.17-29.
3. J. A. Kahle, M. N. Day, H. P. Hofstee, C. R. Johns, T. R. Maeurer, D. Shippy. Introduction to the Cell multiprocessor. IBM Journal. Research & Development. Vol. 49 No. 4/5 July/September 2005.
4. Kevin Krewell. Cell moves into the limelight. Microprocessor Report. Feb. 14, 2005.
5. Jason Fritts. Multi-level Memory Prefetching for Media and Stream Processing. In Proc. of the IEEE International Conference on Multimedia and Expo(ICME2002) , August. 2002, p101–104.
6. N. P. Jouppi. Improving direct-mapped cache performance by the addition of a small fully-associative cache and prefetch buffers. In Proc. of the 17th Annual International Symposium on Computer Architecture, May 1990, p364–373.
7. S. Palacharla and R. Kessler. Evaluating stream buffers as a secondary cache replacement. In Proc. of the 21st Annual International Symposium on Computer Architecture, April 1994, p24–33.
8. J. W. C. Fu and J. H. Patel. Data prefetching in multi-processor vector cache memories. In Proc. of the 18th Annual International Symposium on Computer Architecture, May 1991, p54–63.

9. J. Fu, J. Patel, and B. Janssens. Stride directed prefetching in scalar processors. In Proc. of the 25th International Symposium on Microarchitecture, December 1992, p102–110.
10. Daniel Zucker, Michael Flynn and Ruby Lee. A Comparison of Hardware Prefetching Techniques For Multimedia Benchmarks. 3rd. IEEE International Conference on Multimedia Computing and Systems, Hiroshima, Japan, June, 1996, p236-244.
11. M.K. Jain, M. Balakrishnan, and Anshul Kumar. ASIP Design Methodologies : Survey and Issues. In Proc. of the 14th International Conference on VLSI Design(VLSID'01), Jan. 2001, p76-81.
12. Henk Corporaal and Hans Mulder. MOVE: A framework for high-performance processor design. In Supercomputing91, Albuquerque, November 1991, p692-701.
13. Jan Hoogerbrugge. Code generation for Transport Triggered Architectures. PhD thesis, Delft Univ.of Technology, February 1996. ISBN 90-9009002-9.
14. Leon3 Processor Introduction. http://www.gaisler.com/cms4_5_3/index.php?option=com_content&task=view&id=13&Itemid=53.
15. Jack E. Volder. The CORDIC trigonometric computing technique. IRE Transactions on Electronic Computers, vol. 8, 1959, pp. 330–334.
16. Terry Tao Ye. 0n-chip multiprocessor communication network design and analysis. PhD thesis, Stanford University, December 2003.
17. TMS320C64x CPU and Instruction Set Reference Guide. Texas Instruments, Inc, USA, 2000.
18. TMS320C64x DSP library programmer's reference. Texas Instruments, Inc, USA, 2003.
19. H. Peter Hofstee. Power Efficient Processor Architecture and The Cell Processor. In Proc. of the 11th International Symposium on High-Performance Computer Architecture (HPCA2005), San Francisco, CA, USA, February 2005, p258-262.

Development and Performance Study of a Zero-Copy File Transfer Mechanism for VIA-Based PC Cluster Systems*

Sejin Park[1], Sang-Hwa Chung[2], and In-Su Yoon[2]

[1] Telecommunication Network Business, Samsung Electronics Co.,LTD
Gumi, 730-350, Korea
sejini.park@samsung.com

[2] Department of Computer Engineering, Pusan National University
Pusan, 609-735, Korea
{shchung, isyoon}@pusan.ac.kr

Abstract. This paper presents the development and implementation of a zero-copy file transfer mechanism that improves the efficiency of file transfers for PC cluster systems using hardware-based VIA (Virtual Interface Architecture) network adapters. VIA is one of the representative user-level communication interfaces, but because there is no library for file transfer, one copy occurs between kernel buffer and user buffers. Our mechanism presents a file transfer primitive that does not require the file system to be modified and allows the NIC to transfer data from the kernel buffer to the remote node directly without copying. To do this, we have developed a hardware-based VIA network adapter, which supports the PCI 64bit/66MHz bus and Gigabit Ethernet, as a NIC, and implemented a zero-copy file transfer mechanism. The experimental results show that the overhead of data coy and context switching in the sender is greatly reduced and the CPU utilization of the sender is reduced to 30% ~ 40% of the VIA send/receive mechanism. We demonstrate the performance of the zero-copy file transfer mechanism experimentally, and compare the results with those from existing file transfer mechanisms.

Keywords: Zero-copy, File transfer, VIA, Gigabit Ethernet, Cluster system.

1 Introduction

As high-performance networks like Gigabit Ethernet have come into service, efforts have been made to improve the overall performance of cluster systems through reducing data copying between kernel and user buffers, decreasing context switching time, and minimizing the communication protocol overhead. User-level communication interfaces such as U-Net[4], VIA[5], M-VIA[6] and InfiniBand[7] have been developed to overcome these overheads for memory transactions. Another approach is a zero-copy file transfer mechanism such as the *sendfile* [1][2] system call based on TCP/IP. The *sendfile* system call is applied to web servers or cluster systems

* This work was supported by the Regional Research Centers Program(Research Center for Logistics Information Technology), granted by the Korean Ministry of Education & Human Resources Development.

M. Gerndt and D. Kranzlmüller (Eds.): HPCC 2006, LNCS 4208, pp. 310 – 319, 2006.

based on TCP/IP and is supported by various OSs such as Linux, Unix, and Solaris. However, there is no special mechanism or library for file transfers in the VIA-based cluster systems, so other methods are used to transfer files: the file can be read to a user buffer and sent using the VI, or transferred using extra sockets for file transfer.

Meanwhile, research into accessing files for NAS (Network Attached Storage) has progressed. NFS (Network File System) [10], based on TCP/IP, and DAFS (Direct Access File System) [8][9], based on VIA, are typical examples. They provide file access libraries to the client node, and the client node can access the file system using these libraries. In particular, DAFS offers better performance than NFS because DAFS uses the VIA protocol which is provided with user-level access, thereby reducing overheads such as data copy and context switching time [9]. However DAFS requires a dedicated file server and modification to the existing file system to implement zero-copy file transfer. In a cluster system, each node must act as client or server at the same time; in other words, any client node may be asked to transfer a local file to other nodes in the cluster system, just as for memory transactions. Therefore, if a general VIA-based cluster system can provide zero-copy file transfers between cluster nodes without modifying file system, it will be more convenient and efficient to construct cluster systems.

In this paper, we present a zero-copy file transfer mechanism based on VIA that has not been implemented previously. To do this, we developed HVIA-GE64, which is a second version of HVIA-GE (a Hardware implementation of VIA based on Gigabit Ethernet) [11], and used it as a NIC (Network Interface Card). We show the detailed implementation of the zero-copy file transfer mechanism and compare our experimental results with other mechanisms based on TCP/IP and VIA.

This paper is organized as follows. Section 2 briefly overviews existing file transfer mechanisms and related work, and Section 3 describes the implementation details of our zero-copy file transfer mechanism and HVIA-GE64. Section 4 discusses the experimental results of our mechanism, and finally Section 5 provides a brief conclusion.

2 Related Work

The most important factors in the file transfer performance are the number of data copies between user space and kernel space and the time consumed in context switching. Therefore, reducing these overheads will result in achieving better performance for file transfers. Figure 1 shows the three traditional methods for file transfer using TCP/IP on Linux cluster systems [1][3] and Figure 2 shows the number of data copies and context switches for those methods between user space and kernel space. As shown in Figure 1 (a), the *read* method reads the file data and copies it to a user buffer through a kernel buffer and then copies the data again to a socket buffer to transfer the data (Fig. 2 A→B). Of course, context switches between kernel space and user space also occur. In Figure 1 (b), the *mmap* method avoids one data copy from kernel buffer to user buffer that occurred in the *read* method of Figure 1(a) and copies data from kernel buffer to socket buffer directly (Fig. 2, C). However, because kernel buffer is shared with the user process without any copy, the number of context switches in executing the *mmap* method is the same as with the *read* method.

Figure 1 (c), the *sendfile* system call is a zero-copy file transfer mechanism based on TCP/IP. Because the *sendfile* system call in Linux kernel 2.1 does not use a user buffer, two context switches are avoided and only one data copy occurs (Fig. 2, C). Further, *sendfile* in kernel 2.4 does not copy file data from the kernel buffer to the socket buffer, but only appends the descriptor for the kernel buffer. NIC then reads the file data from the kernel buffer, referring to the descriptor in the socket buffer. (Fig. 2, D) Therefore, the *sendfile* system call implements a zero-copy file transfer.

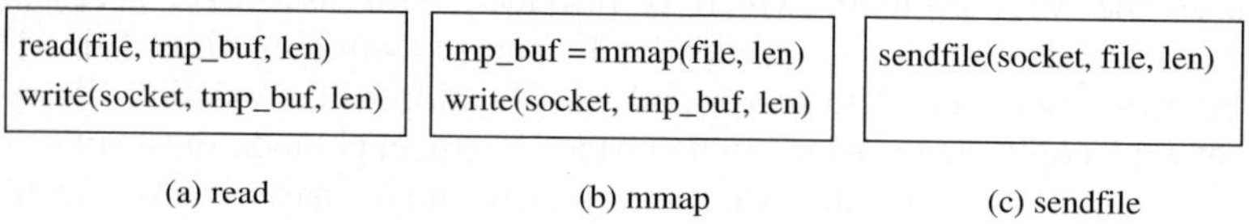

Fig. 1. File Transfer based on TCP/IP

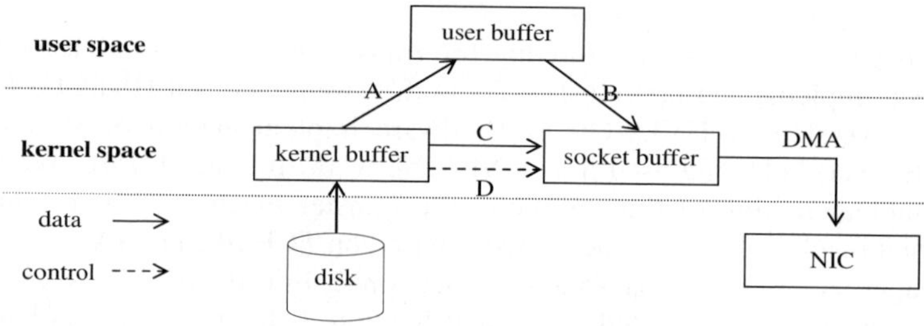

Fig. 2. Data copy based on TCP/IP

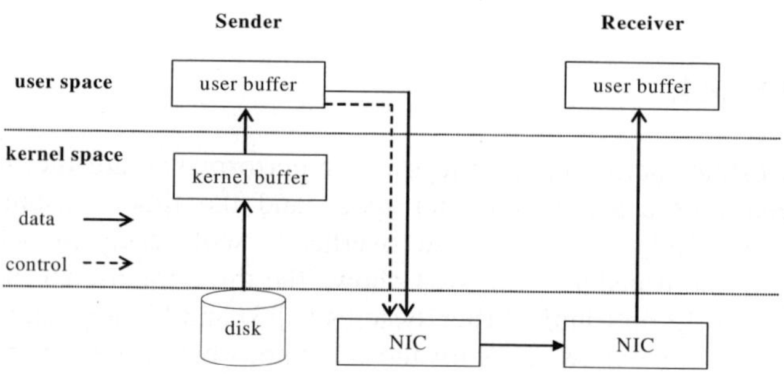

Fig. 3. File Transfer based on VIA

Figure 3 shows the general file transfer method in VIA cluster. As described in the introduction, VIA does not provide a file transfer library. Therefore, in a typical VIA cluster system, file data should be copied once from the kernel buffer to the user buffer, and then the data can be transferred using send/receive or RDMA.

3 Implementation of a Zero-Copy File Transfer Mechanism

3.1 A Hardware-VIA-Based Network Adapter

Figure 4 shows the block diagram of HVIA-GE64 based on the 64 bit/66 MHz PCI bus. HVIA-GE64 contains a Xilinx Virtex II Pro 30 FPGA for the VIA Protocol Engine and Intel's 82544EI Gigabit Ethernet chip for the physical network. This adapter can process address translation, doorbell and completion of RDMA write and send/receive in hardware. In particular, the address translation table (ATT) is stored on SDRAM in the adapter, allowing the VIA Protocol Engine to process address translation directly by accessing the table [11].

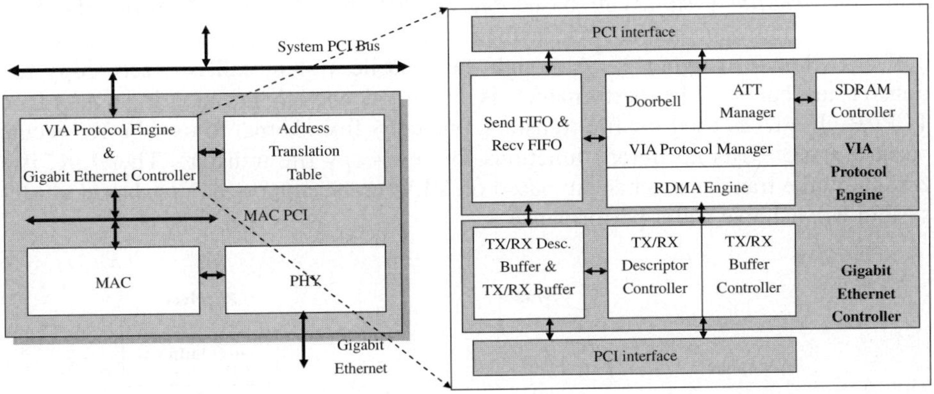

Fig. 4. HVIA-GE64

The right part of Figure 4 presents the VIA Protocol Engine and the Gigabit Ethernet Controller (GEC), which are the core modules of HVIA-GE64. The VIA Protocol Engine consists of Send/Receive FIFOs, ATT Manager, Protocol Manager, RDMA Engine, Doorbells, and local memory controller. It processes VIPL functions delivered to HVIA-GE64 through the PCI bus. In the case of *VipRegisterMem*, which is the VIPL function used to perform memory registration of a user buffer, the user buffer's virtual address, physical address, and size are sent to HVIA-GE64 as function parameters. The ATT manager receives information regarding the user buffer (i.e., virtual and physical addresses) and stores them on ATT.

When a send/receive request is posted to a send/receive queue, HVIA-GE64 is notified through the doorbell mechanism, and obtains the corresponding VI descriptor via DMA. Then, the VIA Protocol Manager reads the physical address of the user data through the ATT Manager. If the current transaction is a send, it initiates a DMA read operation for the user data in the host memory and transfers the data to the Tx buffer in the GEC via the Send FIFO. A send/receive transaction can also be implemented using RDMA, which enables a local CPU to read/write directly from/to the memory in a remote node without intervention of the remote CPU. An RDMA can be implemented as either RDMA read or RDMA write. If RDMA read is used, the local CPU must first send the request and then wait for the requested data to arrive from the remote node. Therefore, RDMA Engine in HVIA-GE64 is based on RDMA write, which is more advantageous in terms of latency.

Since HVIA-GE64 directly drives the Medium Access Control (MAC), GEC basically functions as a device driver for the MAC. GEC processes the initialization, transmit/receive, MAC management routines, and interfaces with the MAC using PCI. Although accessing the MAC directly complicates the design of the GEC and its internal buffers, the elimination of device driver access reduces the initialization overhead and the actual transmission time.

3.2 A Zero-Copy File Transfer Mechanism

In the VIA-based systems, there exist two mechanisms to transfer file data. One is using the *sendfile* system call based on TCP/IP, and the other is using the VIA send/receive mechanism after the file data is copied to the user buffer from the kernel buffer as shown in figure 3. Although the sendfile system call is a zero-copy file transfer mechanism, the performance is not good enough because it is based on TCP/IP. If VIA is used for file transfers, one copy from kernel to user buffer occurs because there exists no library functions for zero-copy file transfer. Therefore, if a zero-copy file transfer mechanism based on VIA can be supported, VIA-based cluster system will achieve better performance.

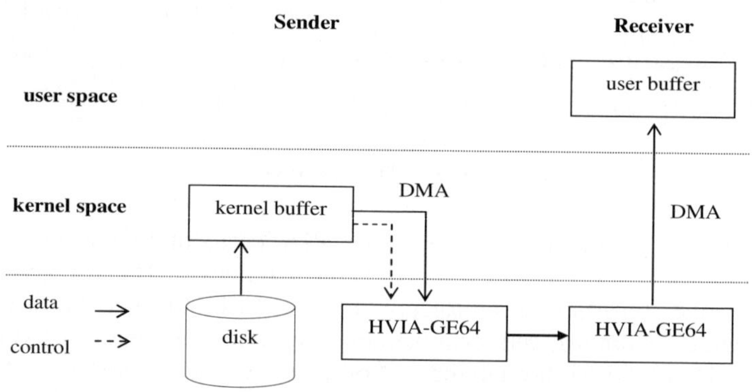

Fig. 5. A Zero-copy File Transfer Mechanism

Figure 5 shows the zero-copy file transfer mechanism for the VIA-based cluster system presented in this paper. The location of the file data that is read from HVIA-GE64 is the kernel buffer. The only action that occurs during a file transfer is that HVIA-GE64 reads data from kernel buffer. Therefore, when transmitting file data, the number of data copies between kernel space and user space is zero, and two context switches between kernel space and user space are eliminated. Two conditions must be satisfied to implement such zero-copy file transfer mechanism. First, the physical address of the kernel buffer must be supplied to HVIA-GE64 so it can read the kernel buffer directly. Second, HVIA-GE64 must be able to process a transport protocol to transmit data to remote nodes through the network. We implemented a zero-copy file transfer mechanism using HVIA-GE64 and the detailed implementation is explained below.

3.3 Detailed Implementation

Figure 6 shows our proposed zero-copy file transfer primitive. The *VipSendFile* primitive reads the file data from the disk to the kernel buffer, and delivers the physical address and length of the kernel buffer to HVIA-GE directly. The mechanism is similar to that for a *sendfile* system call, except that it uses a VI for transmission, not a socket. In the case of the *sendfile* system call, NIC must obtain the remote node address and the physical address and length of the kernel buffer from the corresponding socket. In our zero-copy file transfer mechanism, the sender can obtain the remote node address from *VIhandle* and the address information of the kernel buffer by executing the *VipSendFile* primitive. Then HVIA-GE64 transmits the kernel buffer's data using the address information by DMA.

```
VipSendFile( VIhandle     //VI handle
             fd,          //file descriptor
             offset,      //offset of the file
             len)         //length of the file
```

Fig. 6. VipSendFile: Our Zero-copy File Transfer Primitive

The scenario to transfer file data using VI in the sender is as follows. First, the VI is created and connection is established between the sender and receiver nodes. A user buffer must be registered in a typical send/receive, but the process of memory registration is omitted in the sender because the sender does not use a user buffer for file transfer. The user process then calls *VipSendFile* to transmit the file data.

At this time, the process between kernel space and HVIA-GE64 is as follows. When *VipSendFile* is called, the file data in the disk is read to the kernel buffer, the doorbell is rung in the HVIA-GE64 that notifies the file transfer, and the kernel buffer's information such as physical address and length is transmitted to HVIA-GE64. In a typical VIA-based system, the VI descriptor is posted to the Work Queue (WQ) to send/receive, but this process is omitted in this mechanism. Instead, HVIA-GE64 creates the virtual VI descriptor using *ViHandle* and the kernel buffer information, and then it reads the kernel buffer using DMA with a page unit. Once the file data has been read, it is transmitted to the remote node through the VI connection established previously. Therefore, this mechanism does not require the file system to be modified and allows the HVIA-GE64 to transfer data from the kernel buffer to the remote node directly without copying.

In a *sendfile* system call, there are no new functions in the receiver, because the work in the receiver is the same as for any typical receive. *VipSendFile* is the same as a *sendfile* system call. As shown in Figure 5, the receiver obtains the data using the *VipPostRecv* function, which is a typical receive function in VIA, and writes directly to the user buffer. Therefore, it is possible to use the receive function that VIA provides without modification.

4 Experimental Results

The performance of the HVIA-GE64 card was evaluated using two 2 GHz Opteron 246 PCs with 64-bit/66-MHz PCI buses. The PCs were running the Fedora core 1 for x86_64 with Linux kernel 2.4.22. In addition, for comparison purposes, the performances of TCP/IP, M-VIA, and *sendfile* system calls based on TCP/IP were measured using an Intel PRO/1000 MT Gigabit Ethernet card. This card includes Intel's Gigabit Ethernet MAC/PHY (82544EI), which is the same chip as the one in the HVIA-GE64. In experiments for file transfer, we omitted the performances of *read* and *mmap*, which are shown in Section 2, because the *sendfile* system call shows better performance. The performance of the M-VIA was measured using the vnettest program included with the M-VIA by the distributors. The performances of TCP/IP and *sendfile* system call were measured by modifying the vnettest program using the socket library.

4.1 Performance of a HVIA-GE64

Figures 7 and 8 show the latency and the bandwidth results of HVIA-GE64, M-VIA, and TCP/IP with an Ethernet MTU size of 1,514 bytes. The latency reported is one-half the round-trip time and the bandwidth is the total message size divided by the latency. The latency and bandwidth of the HVIA-GE64 were measured using send/receive on a single VI channel.

The minimum latency results of HVIA-GE64, M-VIA and TCP/IP are 8.2 μs 21.3 μs, and 27 μs, respectively. Therefore, the minimum latency of HVIA-GE64 is 2.8 and 3.3 times lower than M-VIA and TCP/IP, respectively. The maximum bandwidth results for 1 MB of user data are 112.1 MB/s, 90.5 MB/s, and 72.5 MB/s for HVIA-GE64, M-VIA, and TCP/IP, respectively. Therefore, HVIA-GE64 attains 24% and 55% higher bandwidth than M-VIA and TCP/IP, respectively.

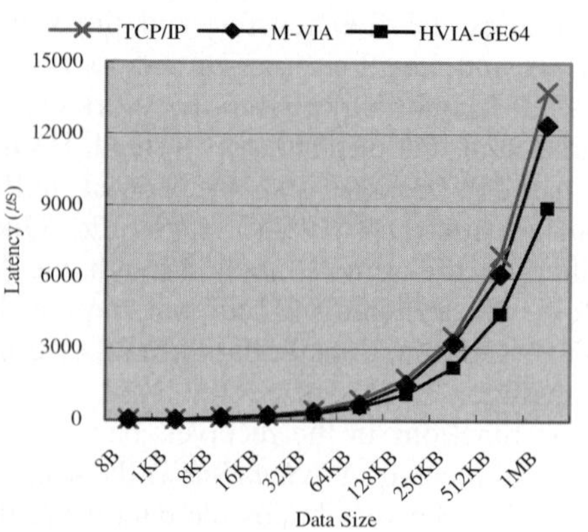

Fig. 7. Latency Comparison

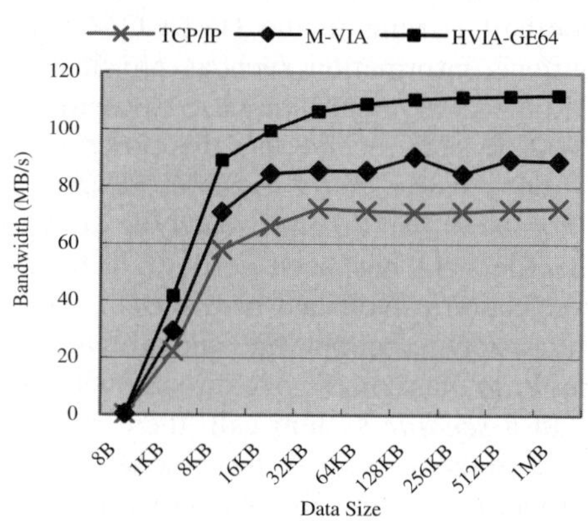

Fig. 8. Bandwidth Comparison

Figure 9 shows the CPU utilization results for HVIA-GE64, M-VIA and TCP/IP. The elements that influence CPU utilization are the CPU's jobs, such as data copy, context switching, system calls and other tasks for the CPU. CPU utilization for TCP/IP increases as data size grows, and becomes saturated at 55%-60%. In the case of M-VIA, CPU utilization is almost 100% when data size is below 4 KB, and saturated near 30%. When small size messages below 4KB is transmitted, the latency is the most important factor for the performance, therefore, M-VIA calls the polling function (*VipSendDone*) 50,000 times before suspending the send/receive process[14]. For HVIA-GE64, utilization decreases and saturates to below 1% when data size increases to about 10 KB. The only portions that require CPU cycles are posting send/receive on WQ and handling send/receive completion. Moreover, an RDMA write does not need an explicit *VipPostRecv* in the receiver, therefore an RDMA write is superior to a send/receive in terms of CPU utilization.

4.2 Performance of a Zero-Copy File Transfer

Figure 10 shows the total elapsed times of the three different mechanisms in the sender, and Figure 11 shows the CPU times of the three mechanisms in the sender. *TCP/IP-sendfile* represents a *sendfile* system call, *HVIA-send/receive* represents a typical send operation using VIA, and *HVIA-VipSendFile* is our zero-copy file transfer mechanism. The elapsed times are the times from the start of file read to the completion notified from the Gigabit Ethernet card or HVIA-GE64, and therefore include the total time for data copy, context switch and completion.

The *TCP/IP-sendfile* has more latency than the mechanisms based on HVIA-GE64 when the data size is larger. The *HVIA-VipSendFile* performs 47%–52% faster than the *TCP/IP-sendfile*. Although the *TCP/IP-sendfile* is a zero-copy protocol, the time to process the TCP/IP protocol is high. Between the two mechanisms based on HVIA-GE64, the *HVIA-VipSendFile* performs 3% faster than the *HVIA-send/receive* in the total elapsed time. The CPU time difference between the *HVIA-VipSendFile* and the *HVIA-send/receive* is 45 μs in sending 256 KB data as shown in Figure 11. Although the difference in the total elapsed time is not large, the difference in the CPU time is quite significant. The improvement is obtained by reducing one data copy and two context switches in executing the *HVIA-VipSendFile* compared with the *HVIA-send/receive*.

Figure 12 shows the CPU utilizations for the three mechanisms. The *TCP/IP-sendfile* causes 8% ~ 10% CPU utilization regardless of data size. In the case of the *HVIA-send/receive*, when the data size is below 6KB, CPU utilization is higher than that of the *TCP/IP-sendfile*. Although the total elapsed time of the *HVIA-send/receive* is lower than that of the *TCP/IP-sendfile*, the time to copy data between the kernel buffer and the user buffer is relatively long when the data size is below 6KB. However CPU utilizations of the *HVIA-send/receive* and the *HVIA-VipSendFile* decrease as data size increases, resembling the CPU utilization shown in Figure 9. Because HVIA-GE64 supports data transmission in hardware, the CPU utilizations are lower than that of the *TCP/IP-sendfile*. Moreover CPU utilization of the *HVIA-VipSendFile* is reduced to 30% ~ 40% of the *HVIA-send/receive*'s, because it removes two context switches and one data copy.

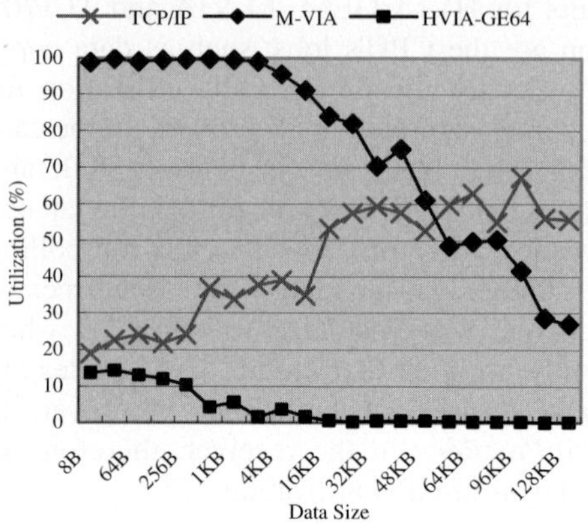

Fig. 9. CPU Utilization

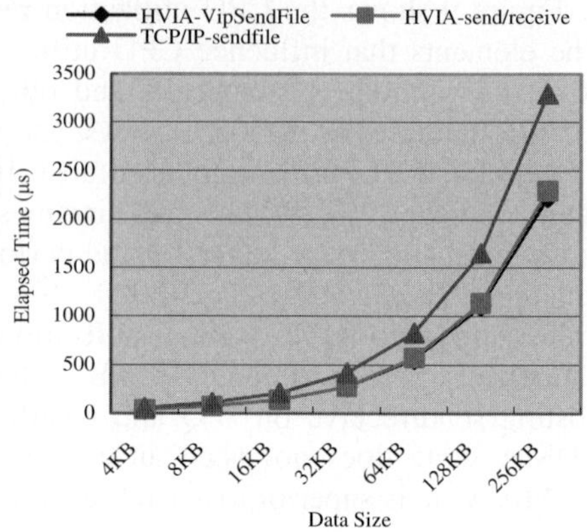

Fig. 10. Elapsed Time in the Sender

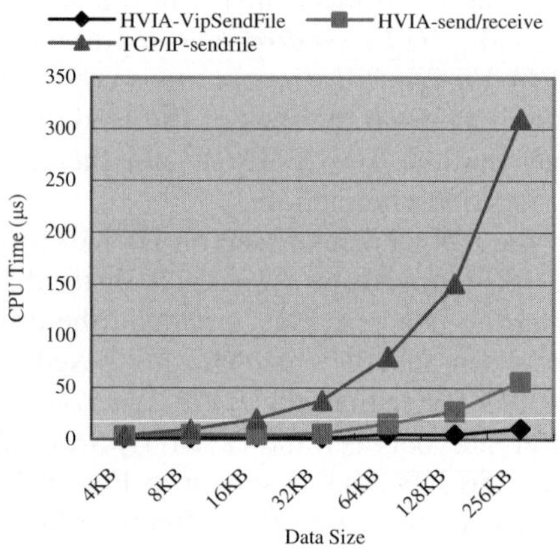

Fig. 11. CPU Time in the Sender

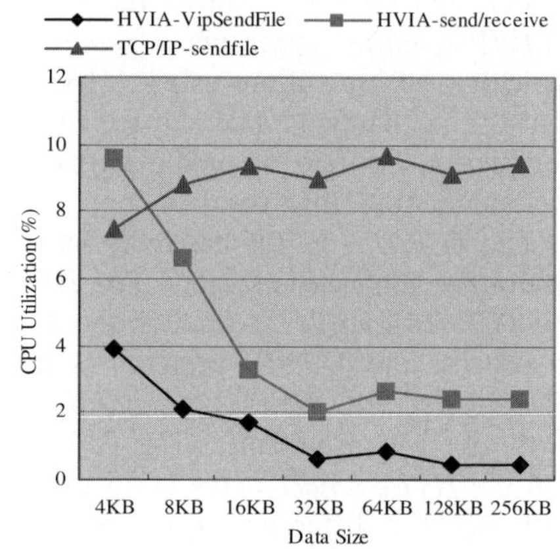

Fig. 12. CPU Utilization

5 Conclusions

This paper presents the development and implementation of a zero-copy file transfer mechanism that improves the efficiency of file transfers for VIA-based cluster systems based on HVIA-GE64. HVIA-GE64 supports the 64bit/66MHz PCI bus and Gigabit Ethernet and implements a hardware-based VIA protocol. This adapter can process virtual-to-physical address translation and doorbell in hardware. Our experiments with HVIA-GE64 showed a minimum latency of 8.2 μs and a maximum bandwidth 112.1 MB/s and much lower CPU utilization compared with TCP/IP and M-VIA. Our zero-copy file transfer mechanism provides a file transfer library that does not require file system modification and can transfer local files to remote nodes without data copying and eliminates two context switches between kernel and user space. As a result, our mechanism performs 47%–52% faster than the *sendfile* system call based on TCP/IP and the CPU utilization of the sender is reduced to 30% ~ 40% of the VIA send/receive mechanism.

References

[1] Dragan Stancevic, "Zero Copy I: User-Mode Perspective", Linux Journal, Volume 2003 Issue 105, January 2003
[2] Jeff Tranter, "Exploring The Sendfile System Call", Linux Gazette, Issue91, June 2003
[3] Oyetunde Fadele, "A Performance Comparison of read and mmap", http://developers.sun.com/solaris/articles/read_mmap.html
[4] T. von Eicken, A. Basu, V. Buch, and W. Vogels. "U-Net: A User-level Network Interface for Parallel and Distributed Computing", Proc, of the 15th ACM Symposium on Operating Systems Principles (SOSP), Colorado, December 3-6, 1995
[5] Virtual Interface Architecture Specification, http://www.viarch.org/
[6] P. Bozeman and B. Saphir, "A Modular High Performance implementation of the Virtual Interface Architecture", Proc. Of the 2nd Extreme Linux Workshop, June 1999
[7] InfiniBandTM Architecture, http://www.infinibandta.org/
[8] DAFS Collaborabive, Direct Access File System Protocol, Version 1.0, September 2001, http://www.dafscollaborative.org
[9] A. Feforova, M. Seltzer, K. Magoutis, S. Addetia, "Application performance on the Direct Access File System", ACM SIGSOFT Software Engineering Notes, Proc. of the 4th international workshop on Software and Performance, vol 29, January 2004
[10] Shepler, S., et. al. NFS version 4 Protocol, Internet Engineering Task Force RFC3010, December 2000
[11] Sejin Park, Sang-Hwa Chung, B. Lee, "Implementation and Performance Study of a Hardware-VIA-based Network Adapter on Gigabit Ethernet", Journal of Systems Architecture, Volume 51, October-November 2005
[12] M-VIA Core Release 1.2, http://old-www.nersc.gov/research/FTG/via/
[13] Hermann Hellwagner and Matthias Ohlenroth, “VI architecture communication features and performance on the Gigabit cluster LAN” Future Generation Computer Systems, Vol. 18, Issue 3, January 2002

DPCT: Distributed Parity Cache Table for Redundant Parallel File System

Sheng-Kai Hung and Yarsun Hsu

Department of Electrical Engineering,
National Tsing-Hua University, HsinChu 30055, Taiwan
{phinex, yshsu}@hpcc.ee.nthu.edu.tw

Abstract. Using parity information to protect data from loss in a parallel file system is a straightforward and cost-effective method. However, the "small-write" phenomenon can lead to poor write performance. This is still true in the distributed paradigm even when file system cache is used. The local file system knows nothing about a stripe and thus can not benefit from the related blocks of a stripe. We propose a distributed parity cache table (DPCT) which knows the related blocks of a stripe and can use them to improve the performance of parity calculation and parity updating. This high level cache can benefit from previous reads and can aggregate small writes to improve the overall performance. We implement this mechanism in our reliable parallel file system (RPFS). The experimental results show that both read and write performance can be improved with DPCT support. The improvement comes from the fact that we can reduce the number of disk accesses by DPCT. This matches our quantitative analysis which shows that the number of disk accesses can be reduced from N to $N(1-H)$, where N is the number of I/O nodes and H is the DPCT hit ratio.

1 Introduction

Commodity off the shelf computers have been used to form a Beowulf-like cluster[1] in recent years. Relying on clusters to solve computation-intensive problems (such as scientific applications) provides an economic solution. However, some data-intensive applications require a lot of data access from the I/O subsystem which has been the bottleneck[2] of a cluster system. A cluster system is an open architecture consisting of loosely coupled computers. Unlike commercial machines which have proprietary parallel file systems (such as PFS on the Intel Paragon[3], VESTA[4,5] on the IBM SP2 machine), commodity off the shelf clusters have no support for parallel file systems. Instead, they often use distributed fie system (such as NFS[6]) to provide an unified storage space. Distributed file systems usually provide an unified naming space but are less efficient for concurrent accesses. To address this issue, a number of parallel file systems[7] such as Parallel Virtual File System (PVFS) have been developed to improve the performance of concurrent accesses in the cluster environment. PVFS is a practical remedy for providing high performance I/O in the Linux cluster either through POSIX interfaces or MPI-IO[8]. However, it is usually used as a

M. Gerndt and D. Kranzlmüller (Eds.): HPCC 2006, LNCS 4208, pp. 320–329, 2006.

temporary storage for parallel computation. The reason is that PVFS does not store any redundant information in its striping structure. Using striping without keeping any redundancy information, the MTTF (Mean Time To Failure) of the system may be lower by $\frac{1}{N}$, where N is the number of I/O nodes in the cluster system. Things become even worse since not only disk failures must be considered but also other faults (power shortage, software crash, memory error...) may cause a node to fail. As a result, even if each node's disks use RAID (Redundant Array of Independent Disks)[9] to protect data from loss, it still doesn't help for the overall reliability.

We propose a reliable parallel file system (RPFS)[10] based on the original PVFS to protect data from loss. With a distributed parity cache table (DPCT), our RPFS can not only solve the small-write problem but also improve the performance of the original PVFS. The benefits come from the fact that we successfully reduce the number of disk accesses.

This paper is organized as follows. The related research topics are covered in section 2. Section 3 describes the system architecture of our RPFS along with DPCT. Finally, we will show the evaluation of our RPFS in section 4, and section 5 concludes our work.

2 Related Work

In this section, we present some file systems used in the cluster environment, either distributed or parallel ones. Sun Microsystems' NFS (Network File System) [6] is widely used in the traditional UNIX environment, but it lacks support for parallel semantics and the server node is always a single point of failure. It provides neither fault tolerance nor striping. A well known feature of NFS is its unified interface to user applications. However, its performance is notorious when serving many I/O clients. As we know, some of the clusters in the world still use NFS as their file system and depend on MPI-IO[8] for parallel accesses.

The concept of distributed RAID was first proposed by Stonebraker and Schloss[11] and then applied in Swift RAID distributed file system[12]. It supports the same redundancy schemes used in RAID level 0, 4, or 5. By the use of parity disks, an error can be recovered as long as it does not happen at the metadata server. Zebra[13] provides a similar technique as Swift but using logs for writing. xFS[14] decentralizes Zebra file system and makes the global file system cache available. GFS[15] and GPFS[16] connect disk storages and clients by interconnection switches. This makes the concept of servers disappeared and eliminates the failures caused by servers. However, these proprietary hardware costs more, and can not be applied in the commodity off the shelf clusters. OSM (Orthogonal Striping and Mirroring)[17] introduces a new concept, called RAID-x. It enhances write performance for parallel applications and improves scalability in the cluster environment. But it suffers from consuming network bandwidth when mirroring data.

CEFT-PVFS[18], like PIOUS[19] and Petal[20] provides mirroring to protect data from loss. CEFT-PVFS is based on PVFS and directly implement mirroring over PVFS. It can be regarded as a RAID-10 like parallel file system. However, it suffers from the data consistency problem between mirror nodes and working nodes. It also consumes too much network bandwidth when a lot of data needs to be written and mirrored. CSAR[21] uses the technique like HP AutoRAID[22], a hierarchical storage system supporting both RAID-1 and RAID-5. They both address the performance issue and devote their work to the small-write problem. CSAR is also based on PVFS and assumes that data must spread all over the I/O nodes. Its redundancy is built in the client library and cannot support POSIX interfaces.

Employing a cache to improve performance has been widely used either in hardware architecture or software infrastructure. Here, we propose a distributed parity cache table (DPCT) in our RPFS[10] to solve the small-write problem. Our DPCT is implemented in each of the I/O nodes and can also be regarded as a global server-side cache. This global server-side cache has the knowledge of a stripe, knows how to collect related blocks to form a stripe and is aware of its corresponding parity. With our novel cache replacement algorithm, we can greatly reduce the number of disk accesses and thus enhance the overall performance.

3 System Architecture

PVFS has three main components: mgr, iod and libpvfs. The mgr daemon should be run in the management node. It maintains the metadata of the system, including the mapping between a file name and its inode number, file striping size, I/O nodes used to construct a stripe, file permission sets and access rights...etc. There can be only one management node in the system. Iod runs on each of the I/O nodes, serving requests from clients and feeding them with the requested data. The more I/O nodes a system has, the more bandwidth it can provide. Libpvfs provides native calls of PVFS, allowing applications to gain the maximal I/O performance that could offer. Besides, a kernel module implements POSIX-complaint interfaces which allow traditional applications to run without any modification.

The system architecture of our reliable parallel file system (RPFS) is shown in Fig. 1. We add two functionalities which original PVFS does not provide. One is the redundacy mechanism, the other is the distributed parity cache table. These two functionalities are implemented without affecting the original striping structure of PVFS. We will describe both of them in more detail in the following subsections.

3.1 Distributed Parity Cache Table

As shown in Fig. 1, the distributed parity cache table (DPCT) is implemented in each of the I/O nodes. It lies between iod daemon and local file system and is used both as a read ahead and a write behind buffer. Each node has 4096 cache blocks with each block $(16K + 32)$ bytes in size. These blocks are divided into 1024 sets, with each set contains 4 blocks. These buffers are pre-allocated within

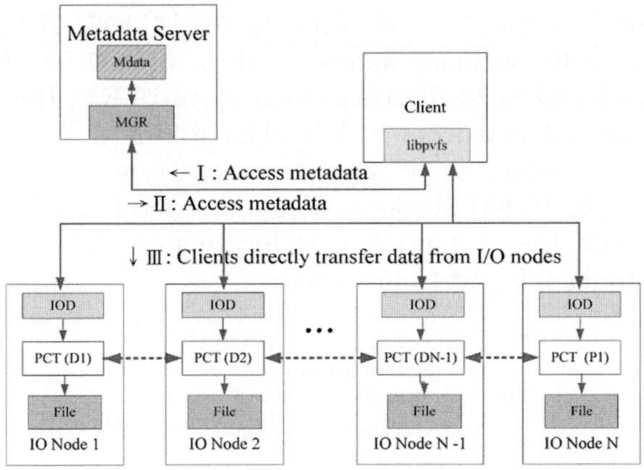

Fig. 1. System Architecture: A distributed parity cache table lies between the local file system and the iod daemon

each I/O node when the iod daemon is executed. Each cache block contains a 16-KB data region and a 32-bytes metadata. The 16-KB region is used to cache files, while the 32-bytes metadata contains many information used for cache replacement algorithm and parity updating. Fig. 2 shows detailed field names and their size of a cache block. DTag field describes the tag information which uniquely identifies the cache block itself. Different data segments which hash to the same set have different DTag values.

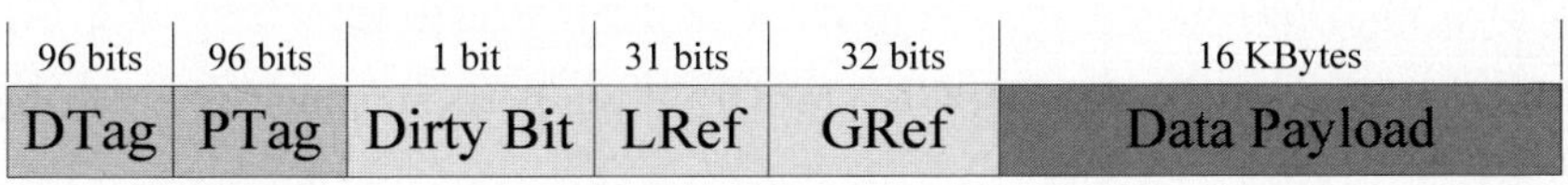

Fig. 2. Fields of a cache block: DTag is used to indicate whether this block contributes to a specific parity block described by PTag. If a write happens, the dirty bit would be set. LRef records the number of hits in this cache block. GRef indicates the number of related cache blocks being read or written within the same stripe.

If a write happens, the dirty bit of the cache block would be set. When the dirty block and its corresponding parity block have been written to disk, the corresponding dirty bit would be cleared. LRef field is used to record the number of hits in this cache block, which acts as the reference for cache replacement algorithm. GRef field monitors the related cache blocks (by the same DTag value) being read or written in DPCT. Cache blocks with the same DTag fields are used to construct a stripe and are XORed to form a parity block. Gref carries additional information when its value is bigger than one. The detail is to be explained in the next subsection. In our opinion, related blocks of a stripe are more important than unrelated blocks of the whole file system. A cache block may be reserved for a specific data segment when its PTag field is set. In this situation, if the identity of a data segment is the same with the PTag field of a cache block, the data segment always wins the cache block. Here, the identity of a data segment is its 64-bit inode number and its high 32-bit offset.

Each cache block can cache data segments up to 16 KB. However, users may specify striping size different from 16 KB when storing a file. To let our DPCT cache data segments regardless of their striping size, we need to split a bigger segment into multiple 16 KB blocks or combine several small segments to form a 16 KB cache block. Fig. 3 shows how to locate a cache block in DPCT when a file's striping size is different from 16 KB.

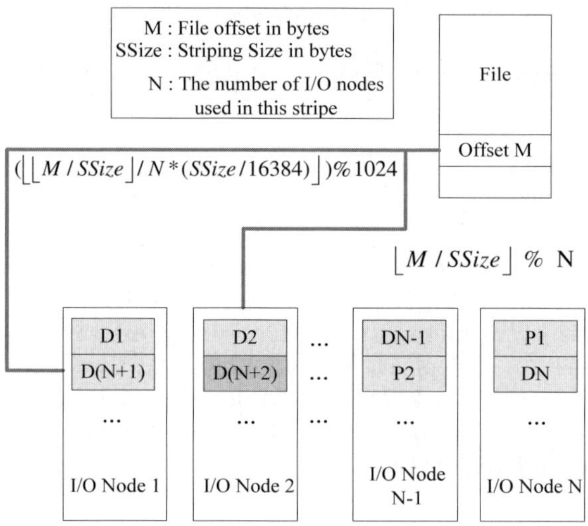

Fig. 3. Locate a cache block of DPCT: Our DPCT supports striping size either bigger or smaller than 16 KB

3.2 Cache Replacement Algorithm

With DPCT support, we need a novel cache replacement algorithm to solve the small-write problem and also improve write performance. When PTag of a cache block is null and its GRef field is less than two, LRef field is used solely to determine if this block should be replaced or not when there is a conflict. A conflict happens when two different data segments hash to the set. LRef is used to implement LRU (Least Recently Used) replacement algorithm.

Whenever a cache block is used for buffering write data, it will check its dirty bit first. If its dirty bit is set and there is a write conflict, the I/O node would write the dirty block to disk. At the same time, it would inform the I/O node holding the parity to write the corresponding parity block to disk. After these, the write request would be proceeded and cached. This can avoid data inconsistency. If the dirty bit is not set, the cache block is directly used to store the write data. After data has been written to cache, the PTag field of the cache block would be set to the same value as its DTag field. At the same time, a thread in the I/O node would communicate with other I/O nodes and update the PTag filed of the corresponding cache blocks belonging to the same stripe. This operation makes other I/O nodes to cache the required data segments, in order to efficiently calculate the parity. Here, the required data segments are those contributing to the same stripe and should be cached in I/O nodes. In this case, the required data segments have the highest priority. If other "not required" data segments hash to the same set and want to replace this block , the required data segment always wins.

There is a thread in each of the I/O nodes used to periodically monitor the status of a stripe. By checking the DTag fields of cache blocks, it knows how many data segments contributing to the same stripe are cached. The thread then updates the GRef fields of the corresponding cache blocks in DPCT. GRef field has higher priority than LRef in our cache replacement algorithm. This suggests that a stripe with more data segments cached should be kept in DPCT even if their LRef values are low.

The thread has another responsibility, that is to write the PTag fields of cache blocks. The PTag field of a cache block will be set under two situations. The first is the occurrence of writes, as described at the beginning of this subsection. The other is triggered by the condition that half of the data segments used to form a stripe are cached in DPCT.

If updating a parity block is needed, it will bring the required data segments either from cache blocks in DPCT or remote disks. It then uses these data blocks to calculate the new parity block to be written into disk.

3.3 Redundancy Mechanism

We store the parity blocks without disturbing the original striping structure of PVFS. We use RAID-5 to provide fault tolerance, the placement of parity blocks is shown in Fig. 4. Take $P1$ in Fig. 4 for example, if an application just writes block $D1$ or partial of $D1$, we need to fetch $D2$ and $D3$ blocks from disks to compute $P1$. This illustrates the problem of small writes. Whenever a small write happens, we require extra N disk accesses, where N is the number of I/O nodes. Using DPCT, we can quantify the number of required disk accesses in Eq. 1.

$$\sum_{x=1}^{N} x C_x^n (1-H)^x (H)^{N-x}, \text{ where } H \text{ is the hit ratio} \quad (1)$$

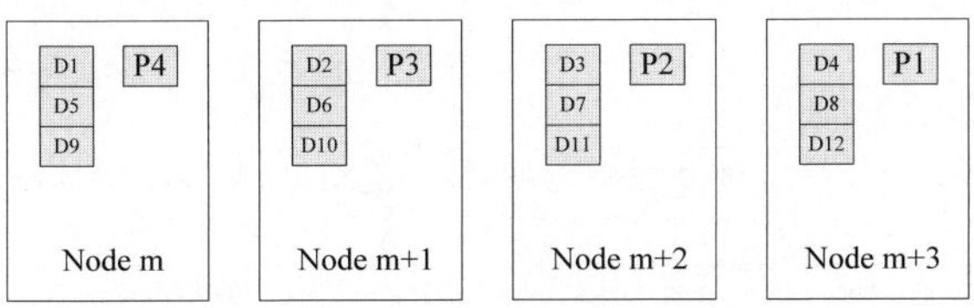

Fig. 4. Redundancy placement of RAID-5: $P1 = D1 \oplus D2 \oplus D3$, $P2 = D4 \oplus D5 \oplus D6$, $P3 = D7 \oplus D8 \oplus D9$, $P4 = D10 \oplus D11 \oplus D12$

Eq. 1 happens to be the mean value of binomial distribution. In other words, the number of disk accesses using DPCT is $N(1 - H)$. Where N is the number of I/O nodes in the system and H is the hit ratio of the DPCT. This means that we reduce the number of disk accesses from N to $N(1 - H)$ when "small-write" phenomenon happens. For example, assume that the number of I/O nodes is 16 and DPCT hit ratio is 0.7. We could reduce the number of disk accesses from 16 times to around 5 times.

4 Experiment Results

In this section, we will show the experimental results along with the evaluation environment used. Each node has a single AMD Athlon XP 2400+ CPU, except for the metadata server which has dual AMD Athlon XP 2400+ CPUs. Nine nodes are connected with a fast Ethernet switch to form a cluster, one metadata server and eight I/O servers. All nodes run Redhat Linux 7.3 with kernel version 2.4.20. We perform three kinds of tests and would describe them separately in the following subsections.

4.1 Throughput Test

To test the aggregate throughput, we use the pvfs_test.c utility accompanied with PVFS distribution. Pvfs_test.c is an MPI program which can be used to test the concurrent read and write of a single file or different files. In this test, each I/O node acts as a client too. Fig. 5 shows the results of read when eight clients are used. When reading, there is nothing to do with parity and thus incurring no overhead. This can be observed by the similar curves of PVFS and RPFS-NO-DPCT. The read performance of RPFS with DPCT is better than others because of the global server-side cache. Data segments are brought into cache blocks in the unit of 16 KB. This realizes the effect of prefetching.

Fig. 6 shows the results when using pvfs_test.c to perform write tests. Updating parity blocks has significant impact on RPFS because small write still needs

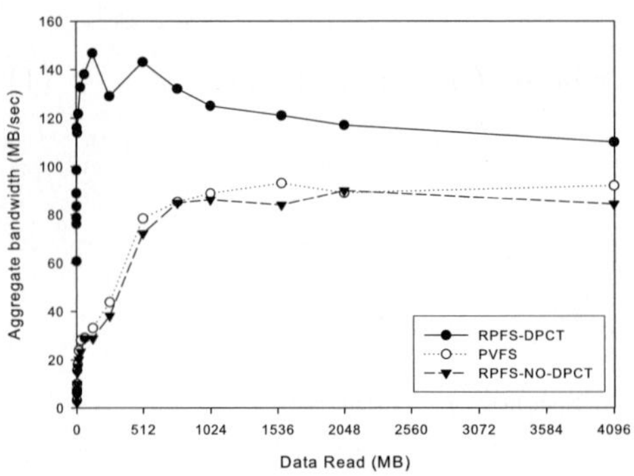

Fig. 5. Read Throughput

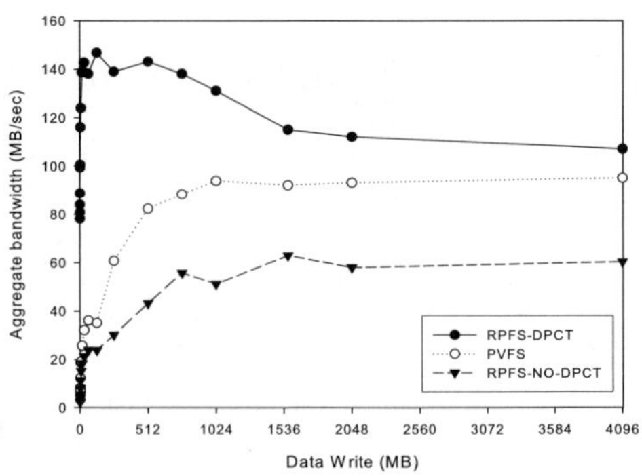

Fig. 6. Write Throughput

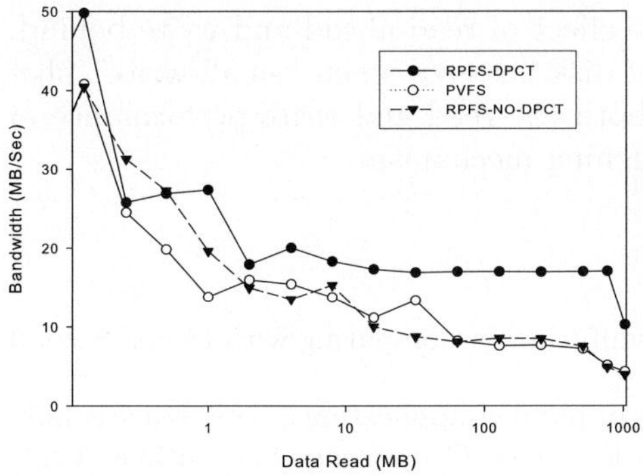

Fig. 7. Read Using Bonnie++

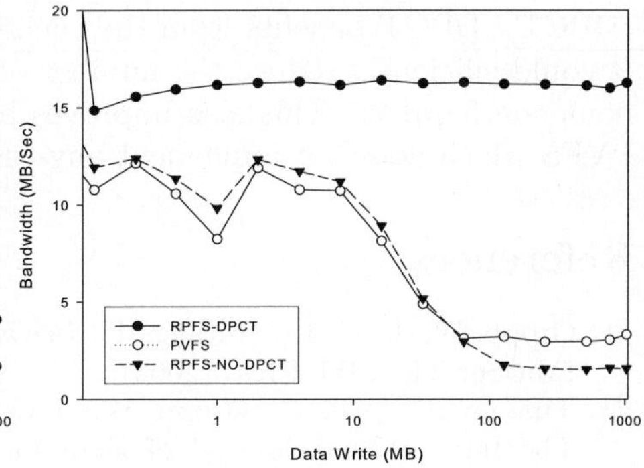

Fig. 8. Write Using Bonnie++

four operations : read the old striped blocks, read the old parity block, write the desired block and write the newly calculated parity block. The reason why every small write needs to update the corresponding parity block lies on the fact that PVFS does not implement cache mechanism. Without a global cache, we cannot aggregate small writes to form a big write. A big write not only can save the number of parity writing but also can reduce the needs to read old blocks again. Besides, the global cache can save request time because data may be accessed directly from DPCT instead of disk. Compared with Fig. 5, the write performance of PVFS is better than that of read. However, the write performance of RPFS-NO-DPCT is lowered by about 25%, due to parity writing. As for the RPFS-DPCT, parity writing has less impact due to the use of DPCT. Its write performance is only reduced by 3% when compared with its read performance.

4.2 POSIX-API Test

To help us understand the behavior of a traditional application which uses PVFS to access files, we use Bonnie++[23] for the test. In this test, a single client is used to run Bonnie++ program. Bonnie++ is a POSIX-compliant application and uses no native PVFS APIs to access files. We mount PVFS in the client's local directory so that it behaves like a local file system. All measurements are performed in the mounted directory. Fig. 7 and Fig. 8 show the results when using Bonnie++ to measure the bandwidth of read and write. Here, we use logarithmic scale in the X axis to clearly distinguish the performance of small data accesses. Again, we observe that RPFS-DPCT has performance benefits when compared with either PVFS or RPFS-NO-DPCT. In this test, the network bandwidth of fast Ethernet limits the throughput that a singe client could get from multiple I/O nodes. It is saturated at around 17 MB/sec.

5 Conclusion

Our reliable parallel file system (RPFS) extends PVFS with fault tolerance. The small-write problem can be alleviated by our distributed parity cache table

(DPCT). DPCT benefits from the cache effect of read ahead and write behind. It could efficiently reduce the number of disk accesses when "small-write" phenomenon happens. This also improves both the read and write performance of PVFS which does not implement any caching mechanism.

References

1. Gropp, W., Lusk, E., Sterling, T.: Beowulf Cluster Computing with Linux, Second Edition. The MIT Press (2003)
2. Thakur, R., Lusk, E., Gropp, W.: I/O in parallel applications: The weakest link. The International Journal of High Performance Computing Applications **12**(4) (Winter 1998) 389–395
3. Intel Supercomputer System Division: Paragon System User's Guide. (1995)
4. Corbett, P., Feitelson, D., Prost, J.P., almasi, G., Baylor, S., Bolmaricich, A., Hsu, Y., Satran, J., Sinr, M., Colao, R., B.Herr, Kavaky, J., Morgan, T., Zlotel, A.: Parallel file systems for IBM SP computers. IBM Systems Journal **34**(2) (1995) 222–248
5. Corbett, P.F., Feitelson, D.G.: The Vesta parallel file system. In: High Performance Mass Storage and Parallel I/O: Technologies and Applications. IEEE Computer Society Press and Wiley, New York, NY (2001) 285–308
6. Sandberg, R., Goldberg, D., Kleiman, S., Walsh, D., Lyon, B.: Design and implementation of the Sun Network File System. In: Proceedings Summer 1985 USENIX Conference. (1985) 119–130
7. Pâris, J.F.: A disk architecture for large clusters of workstations. In: Cluster Computing Conference, GA (1997) 317–327
8. Message Passing Interface Forum: MPI2: Extensions to the Message Passing Interface. (1997)
9. Patterson, D.A., Gibson, G.A., Katz, R.H.: A case for redundant arrays of inexpensive disks (RAID). In: Proceedings of the 1988 ACM SIGMOD International Conference on Management of Data, Chicago, Illinois. (1988) 109–116
10. Hung, S.K., Hsu, Y.: Modularized redundant parallel virtual file system. In: Asia-Pacific Computer Systems Architecture Conference 2005, Singapore (2005) 186–199
11. Stonebraker, M., Schloss, G.A.: Distributed RAID— A new multiple copy algorithm. In: Proceedings of 6th International Data Engineering Conference. (1990) 430–437
12. Long, D.D.E., Montague, B.R.: Swift/RAID: A distributed RAID system. Computing Systems **7**(3) (1994) 333–359
13. Hartman, J.H., Ousterhout, J.K.: The Zebra striped network file system. In: High Performance Mass Storage and Parallel I/O: Technologies and Applications. IEEE Computer Society Press and Wiley, New York, NY (2001) 309–329
14. Sweeney, A., Doucette, D., Hu, W., Anderson, C., Nishimoto, M., Peck, G.: Scalability in the xFS file system. In: Proceedings of the USENIX 1996 Technical Conference, San Diego, CA, USA (1996) 1–14
15. Soltis, S.R., Ruwart, T.M., O'Keefe, M.T.: The Global File System. In: Proceedings of the Fifth NASA Goddard Conference on Mass Storage Systems and Technologies, College Park, MD (1996) 319–342
16. Schmuck, F., Haskin, R.: GPFS: A shared-disk file system for large computing clusters. In: Proc. of the First Conference on File and Storage Technologies (FAST). (2002) 231–244

17. Hwang, K., Jin, H., Ho, R.S.: Orthogonal striping and mirroring in distributed RAID for I/O-centric cluster computing. IEEE Trans. Parallel Distrib. Syst. **13**(1) (2002) 26–44
18. Zhu, Y., Jiang, H., Qin, X., Feng, D., Swanson, D.R.: Design, implementation and performance evaluation of a Cost-Effective, Fault-Tolerance parallel virtual file system. In: International Workshop on Storage Network Architecture and Parallel I/Os, New Orleans, LA (2003)
19. Moyer, S.A., Sunderam, V.S.: PIOUS: a scalable parallel I/O system for distributed computing environments. In: Proceedings of the Scalable High-Performance Computing Conference. (1994) 71–78
20. K.Lee, E., Thekkath, C.A.: Petal: Distributed virtual disks. In: High Performance Mass Storage and Parallel I/O: Technologies and Applications. IEEE Computer Society Press and Wiley, New York, NY (2001) 420–430
21. Pillai, M., Lauria, M.: CSAR: Cluster storage with adaptive redundancy. In: Proceedings of the 2003 International Conference on Parallel Processing, Kaohsiung, Taiwan, ROC (2003) 223–230
22. Wilkes, J., Golding, R., Staelin, C., Sullivan, T.: The hp autoraid hierarchical storage system. ACM Transactions on Computer Systems (TOCS) **14**(3) (1996) 108–136
23. Coker, R.: Bonnie++ – file system benchmark. (http://www.coker.com.au/bonnie++/)

On High Performance Multicast Algorithms for Interconnection Networks

Ahmed Al-Dubai[1], Mohamed Ould-Khaoua[2], and Imed Romdhani[3]

[1,3] School of Computing, Napier University, 10 Colinton Road
Edinburgh, EH10 5DT, UK
{a.al-dubai, i.romdhani}@napier.ac.uk
[2] Department of Computing Science, University of Glasgow
Glasgow G12 8RZ, UK
mohamed@dcs.gla.ac.uk

Abstract. Although multicast communication, in interconnection networks has been a major avenue for a lot of research works found in literature, there are several key issues that should still be improved in order to meet the increasing demand for service quality in such systems. Apparently, most of the related works handle multicast communication within limited operating conditions such as low traffic load, specific network sizes and limited destination nodes. However, this paper investigates the multicast communication under different scenarios. It presents a comparison study of some well known multicast algorithms proposed for wormhole switched interconnection networks. Unlike the previous studies, this paper considers the multicast latency at both network and node levels. Performance evaluation results show that our proposed algorithm can greatly improve the performance of multicast operation.

Keywords: Wormhole Switched Networks, Path-based Multicast, Routing Algorithms.

1 Introduction

Multicast communication, in which a source node sends the same message to an arbitrary number of destination nodes in the network, is one of the most useful collective communication operations [1, 2, 3, 6]. Due to its extensive use, efficient multicast is critical to the overall performance of interconnection networks [1, 2, 3, 4, 14]. For instance, multicast is frequently used by many important applications such as parallel search and parallel graph algorithms [3, 14]. Furthermore, multicast is fundamental to the implementation of higher-level communication operations such as gossip, gather, and barrier synchronisation [1, 2, 4]. Ensuring a scalable implementation of a wide variety of parallel applications necessitates efficient implementation of multicast communication. In general, the literature outlines three main approaches to deal with the multicast problem: unicast-based [1, 3], tree-based [3, 13, 17] and path-based [2, 3, 8, 14, 15]. A number of studies have shown that *path-based* algorithms exhibit superior performance characteristics over their unicast-based and tree-based counterparts [2, 14, 15]. In path-based multicast, when the units (called flits in wormhole switched networks) of a message reach one of the

M. Gerndt and D. Kranzlmüller (Eds.): HPCC 2006, LNCS 4208, pp. 330 – 339, 2006.

destination nodes in the multicast group, they are copied to local memory while they continue to flow through the node to reach the other destinations [2, 3, 8]. The message is removed from the network when it reaches the last destination in the multicast group. Although many interconnection networks have been studied [3], and indeed deployed in practice, none has proved clearly superior in all roles, since the communication requirements of different applications vary widely. Nevertheless, n-dimensional *wormhole switched* meshes have undoubtedly been the most popular interconnection network used in practice [2, 3, 5, 6, 10, 11] due to their desirable topological properties including ease of implementation, modularity, low diameter, and ability to exploit locality exhibited by many parallel applications [3]. In wormhole switching, a message is divided into elementary units called flits, each of a few bytes for transmission and flow control. The *header* flit (containing routing information) governs the route and the remaining data flits follow it in a pipelined fashion. If a channel transmits the header of a message, it must transmit all the remaining flits of the same message before transmitting flits of another message. When the header is blocked the data flits are blocked in-situ. Meshes are suited to a variety of applications including matrix computation, image processing and problems whose task graphs can be embedded naturally into the topology [3, 6, 10]. Meshes have been used in a number of real parallel machines including the Intel Paragon, MIT J-machine, Cray T3D, T3E, Caltech Mosaic, Intel Touchstone Delta, Stanford DASH [3]. Recently, among commercial multicomputers and research prototypes, Alpha 21364's multiple processors network and IBM Blue Gene uses a 3D mesh. In addition, a mesh has been recently the topology of choice for many high-performance parallel systems and local area networks such as Myrinet-based LANs.

The rest of the paper is organised as follows. Section 2 outlines some related works; Section 3 accommodates the proposed multicast algorithm. Section 4 conducts extensive analysis and simulation experiments and Section 5 summarises this work.

2 Background and Motivation

In general, existing multicast communication algorithms rely on two main strategies. In view of the dominance of the start-up time in the overall multicast latency, algorithms in the first class try to reduce the number of start-ups required to perform multicast, but this has been shown to be inefficient under high traffic loads [8, 10, 14]. For instance, the Dual Path (DP) and Multi Path (MP) algorithms proposed in [10] use this strategy. Briefly, DP uses at most two copies of the multicast message to cover the destination nodes, which are grouped into two disjoint sub-groups. This may decrease the path length for some multicast messages. The MP algorithm attempts to reduce path lengths by using up to four copies (or $2n$ for the n-dimensional mesh) of the multicast message. As per the multi-path multicast algorithm, all the destinations of the multicast message are grouped into four disjoint subsets such that all the destinations in a subset are in one of the four quadrants when source is viewed as the origin. Copies of the message are routed using dual-path routing (see [10] for a complete description). Algorithms in the second class, on the other hand, tend to use shorter paths, but messages can then suffer from higher latencies due to the number of start-ups required [15]. Based on this strategy, for example, the Column Path (CP) algorithm presented in [15] partitions the set of destinations into at most $2k$ subsets (e.g. k is the number of columns in the mesh), such that there are at most two messages directed to each column.

Generally, most existing path-based algorithms incur high multicast latency. This is due to the use of long paths required to cover the groups serially like algorithms under the umbrella of the first multicast approach or those of the second category, in which an excessive number of start-ups is involved. In addition, a common problem associated with most existing multicast algorithms is that they can overload the selected multicast path and hence cause traffic congestion. This is mainly because most existing grouping schemes [8, 10, 15] do not consider the issue of load balancing during a multicast operation. More importantly, existing multicast algorithms have been designed with a consideration paid only to the multicast latency at the network level, resulting in an erratic variation of the message arrival times at the destination nodes. As a consequence, some parallel applications cannot be performed efficiently using these algorithms, especially those applications which are sensitive to variations in the message delivery times at the nodes involved in the multicast operation. Thus, our objective here is to propose a new multicast algorithm that can overcome the limitations of existing algorithms and thus leading to improve the performance of multicast communication in mesh networks. In a previous work [2], a new multicast scheme, the Qualified Group (QG) has been proposed for symmetric meshes. Such a scheme has been studied under restricted operating conditions, such as specific traffic load, fixed network sizes and a limited number of destination nodes [2]. In the context of the issues discussed above, this paper makes two major contributions. Firstly, the QG is generalised here with the aim of handling multicast communication in symmetric, asymmetric and different network sizes. Secondly, unlike many previous works, this study considers the issue of multicast latency at both the network and node levels across different traffic scenarios.

3 The Qualified Groups (QG) Algorithm

In an attempt to avoid the problems of existing multicast algorithms, this section presents the Qualified Group (QG) path-based multicast algorithm. The *QG* algorithm takes advantage of the partitionable structure of the mesh to divide the destination nodes into several groups of comparable sizes in order to balance the traffic load among these groups, which leads to avoid the congestion problem in the network. The groups, in turn, implement multicast independently in a parallel fashion, which results in reducing the overall communication latency. In general, the proposed algorithm is composed of four phases which are described below. For the sake of the present discussion and for illustration in the diagrams, we will assume that messages are routed inside the network according to dimension order routing [3, 10].

***Definition* 1.** *Consider a mesh (V, E), with node set V and edge set E, a multicast set is a couple $(p, Đ)$, where $p \in V$, $Đ = \{p_{1,} p_{2} ..., p_{k}\}$ and $p_i \in V, i = 1,...,k$. The node p is the source of the multicast message, and the k nodes in $Đ$ are the* *destinations. To perform a multicast operation, node* p *disseminates copies of the same message to all the destinations in* $Đ$.

We have adopted the dimension order routing due to the fact that this form of routing is simple and deadlock and livelock free, resulting in a faster and more compact router when the algorithm implemented in hardware, [3, 15]. However the *QG* algorithm can be used along any other underlying routing scheme, including the well-known Turn model and Duato's adaptive algorithms [3]. This is because the grouping scheme, as explained below, in *QG* can be implemented irrespective of the underlying routing scheme (in the algorithmic level), which is not the case in most existing multicast algorithms in which destination nodes are divided based on the underlying routing used (in the routing level) [8, 10, 15]. It is worth mentioning that such a research line will be investigated further in our future works.

***Phase* 1.** In this phase, a multicast area is defined as the smallest *n*-dimensional array that includes the source of the multicast message as well as the set of destinations. The purpose of defining this area is to confine a boundary of network resources that need to be employed during the multicast operation.

***Definition* 2.** *In the n-dimensional mesh with a multicast set* $(p, Đ)$, *a multicast area* G_{MA} *includes the source node* $p[d_1, d_2, \ldots d_n]$ *and destination nodes* $Đ\,[(d_1, d_2, \ldots d_n)]$ *such that* $\forall\, d_i \in \{d_1, d_2, \ldots, d_n\}$, *has two corners, upper corner* $u_{d_i} = \max(Đ[d_i], p[d_i])$ *and lower corner* $l_{d_i} = \min(Đ[d_i], p[d_i])$ *such that*

$$mid_{d_i} = \begin{cases} (l_{d_i} + u_{d_i})/2 \text{ if } (l_{d_i} + u_{d_i}) \text{ is even} \\ ((l_{d_i} + u_{d_i}) - 1)/2 \text{ if } (l_{d_i} + u_{d_i}) \text{ is odd} \end{cases}$$

***Phase* 2.** The multicast area G_{MA} is then divided into groups. The objective behind grouping the destination nodes is to distribute the traffic load over the multicast area in order to avoid traffic congestion, which contributes significantly to the blocking latency. Besides, grouping enables the destination nodes to receive the multicast message in comparable arrival times; i.e., this helps to keep the variance of the arrival times among the destination nodes to a minimum.

***Definition* 3.** *In an n-dimensional mesh with a multicast set* $(p, Đ)$, *a divisor dimension* Div_{d_i} *for* $Đ$ *satisfies the following condition*

$$Div_{d_i} = \min(N_{d_1}, N_{d_2}, \ldots, N_{d_n}),\; N_{d_i} = \left| Đ[d_i^{\Uparrow}] - Đ[d_i^{\Downarrow}] \right| :$$

$$Đ[d_i^{\Uparrow}] = \sum_{mid_{d_i}}^{u_{d_i}} Đ[d_i] \text{ and } Đ[d_i^{\Downarrow}] = \sum_{l_{d_i}}^{mid_{d_i}-1} Đ[d_i]$$

Notice that if $N_{d_1} = N_{d_2}$, d_1 is given a higher priority, i.e., a higher priority is given based on the ascending order of the dimensions. For instance, if $N_x = N_y = N_z$, X dimension will be considered as a divisor dimension. The divisor dimension is used as a major axis for the grouping scheme in this phase. The multicast area G_{MA} is then divided into a number of disjoint groups as formulated in the following definition.

***Definition* 4.** *Given an n-dimensional mesh with a multicast set* $(p, Đ)$ *and a multicast area* G_{MA}, $\forall G_i, G_j : G_i \subseteq G_{MA}$ *and* $G_j \subseteq G_{MA} \rightarrow G_i \cap G_j = \Phi$.

According to Definition 4, G_{MA} is divided into a number of primary groups as given in equation 1; where g_{pr} refers to the number of primary groups obtained after dividing the destination nodes over the division dimension, such that

$$g_{pr} = \begin{cases} p_t & \text{if } \exists G_i \subseteq G_{MA} : G_i = \Phi \\ 2^n & \text{otherwise} \end{cases} \tag{1}$$

where p_t is an integer, $1 \leq p_t < 2^n$

Phase 3. This phase is responsible for qualifying the groups already obtained in the preceding phase for a final grouping. Having obtained the primary groups, g_{pr}, we recursively find the multicast area for each group, $G_i \subseteq G_{MA}$, as defined in Definition 4, and determine the internal distance $Int(G_i)$ for each group G_i.

$$Int(G_i) = Dist(p_f(G_i), p_n(G_i)) + N_{G_i} \tag{2}$$

Where $Dist$ refers to the Manhattan distance in which the distance between tow nodes, for instance the distance between two nodes $(p1_x, p1_y)$ and $(p2_x, p2_y)$ is given by $Dist(p1, p2) = |(p1_x - p2_x) + (p1_y - p2_y)|$. While the first term, $Dist(p_f(G_i), p_n(G_i))$, in the above equation represents the distance between the farthest p_f and the nearest node p_n in a group G_i from/to the source node p, respectively, the second term, N_{G_i}, represents the number of destination nodes that belong to the relevant group $G_i \subseteq G_{MA}$. We then determine the external distance $Ext(G_i)$.

$$Ext(G_i) = Dist(p_n(G_i), p) \tag{3}$$

The minimum weight W_m for a group $G_i, 1 < i \leq g_{pr}$, where g_{pr} refers to the number of primary groups, is then calculated by

$$W_m(G_i) = Ext(G_i) + Int(G_i) \tag{4}$$

***Definition* 5.** *Given a multicast area* G_{MA} *and* $G_i \subseteq G_{MA}$, *where* $1 < i \leq g_{pr}$, *the average of the minimum weights* W_{av}, *for the multicast area* G_{MA}, *is given by*

$$W_{av} = \frac{\sum_{i=1}^{g_{pr}} W_m(G_i)}{g_{pr}} \tag{5}$$

***Definition* 6.** *Given a multicast area* G_{MA}, $G_i \subseteq G_{MA}$, *and* W_{av}, *the qualification point,* $QP(G_i)$, *for each group is calculated as follows*

$$QP(G_i) = \frac{(W_m(G_i) - W_{av})}{W_{av}} \tag{6}$$

The qualification point for each group is compared to an assumed threshold value TD, which is used to set a limit for the partitioning process.

***Definition* 7.** *Given a multicast area* G_{MA} *and* $G_i \subseteq G_{MA}$, *we say that* G_i *is a qualified group if and only if its minimum weight* $W_m(G_i) \leq W_{av}$ *or if its qualification point* $(QP(G_i)) \leq TD$.

For example, given that the threshold value is $TD = 0.5$, each qualified group must hold at least half of the total average weight W_{av} of the groups. Once a group $G_i \subseteq G_{MA}$ does not satisfy the condition formulated in Definition 7, it is treated as an unqualified group. In this case, this unqualified group is divided into two sub-groups based on its division dimension. If the new resulting groups are qualified the partitioning process is terminated. Otherwise, the unqualified group is divided into a number of sub-groups sb, where $2 \leq sb \leq 2^n$. For instance, for any unqualified group $G_i \subseteq G_{MA}$ in the 2D mesh, it can be divided into four groups at maximum, even if the new obtained groups are still larger than those which meet the qualification point. In fact, the partitioning process is terminated at this stage in order to reduce the number of comparisons during the qualifying phase. This helps to keep the algorithm simple and maintains a low preparation time.

***Phase* 4.** For each group resulting from Phase 3, the nodes which have the lowest communication cost, in terms of distance from the source node, are selected as the representative nodes of the qualified groups that can receive the multicast message from the source node. In other words, the nearest node for each qualified group is elected so that it could be sent the multicast message with a single start-up only. Concurrently, the representative nodes act as "source" nodes by delivering the message to the rest of the destination nodes in their own groups with one additional start-up time only. After qualifying all the groups, the source node sends the message to the representative nodes in the qualified groups. The source node performs this operation with a single start-up latency taking advantage of the multiple-port facility of the system by creating two disjoint paths in this step. Concurrently, every representative node in each group acts as a source node and, in turn, sends the message to the rest of the destinations in its own group.

4 Performance Evaluation

A number of simulation experiments have been conducted to analyse the performance of *QG* against *DP*, *MP* and *CP*. A simulation program has been developed to model the multicast operation in the mesh. The developed model has been added to a larger simulator called MultiSim [6], which has been designed to study the collective communication operations on multicomputers and has been widely used in the literature [2, 10, 12]. The simulation program was written in VC++ and built on top the event-driven CSIM-package [7]. We have used the 2D mesh with four injection channels and four ejection channels. Two unidirectional channels exist between each pair of neighbouring nodes. Each channel has a single queue of messages waiting for transmission. In our simulations, the start-up latency has been set at 33 cycles, the channel transmission time at 1 cycle and the threshold *TD* at 0.5. The network cycle time in the simulator is defined as the transmission time of a single flit across a channel The preparation time (which consists of dividing the destination nodes into appropriate subsets and creating multiple copies of the message as needed, depending on the underlying algorithm) of the DP, MP, CP and QG algorithms are set at 2, 2, 4 and 16 cycles, respectively. The preparation time was deliberately set higher in the QG algorithm to reflect the fact that our algorithm requires a longer time to divide the destinations into qualified groups. All simulations were executed using 95% confidence intervals (when confidence interval was smaller than 5% of the mean). The technique used to calculate confidence intervals is called batch means analysis. In batch means method, a long run is divided into a set of fixed size batches, computing a separate sample mean for each batch, and using these batches to compute the grand mean and the confidence interval. In our simulations, the grand means are obtained along with several values, including confidence interval and relative errors which are not shown in the figures. Like existing studies [1, 2, 3, 10, 15, 13], only the grand mean is shown in our figures.

4.1 Latency at the Node Level

This section presents the coefficient of variation of the multicast latency as a new performance metric in order to reflect the degree of parallelism achieved by the multicast algorithms. A set of simulation experiments have been conducted where the message inter-arrival times between two messages generated at a source node is set at 250 cycles. The message length is fixed at 64 flits and the number of destination nodes is varied from 20, 30, 40… to 60 nodes. The coefficient of variation (CV) is defined as SD/M_{nl}, where SD refers to the standard deviation of the multicast latency (which is also the message arrival times among the destination nodes) and M_{nl} is the mean multicast latency. The coefficient of variation of *QG* has been compared against that of *DP, MP* and *CP*. Table 1 contains performance results for the 16×16 mesh, which have been obtained by averaging values obtained from at least 40 experiments in each case. The $QG_{IMPR}\%$ in Table 1 refers to the percentage improvement obtained by *QG* over its *DP, MP* and *CP* competitors.

As shown in Table 1, *QG* achieves a significant improvement over *DP, MP* and *CP*. This is due firstly to the efficient grouping scheme adopted by *QG* which divides the destinations into groups of comparable sizes. Secondly, and more importantly, unlike in *DP, MP* and *CP*, the destination nodes for each qualified group in *QG* (except those selected in the first message-passing step) receive the multicast message in the second message-passing step, in parallel. This has the net effect of minimising the variance of the arrival times at the node level. In contrast, *DP, MP* and *CP* perform multicast with either longer paths as in *DP* and *MP* or in an excessive number of message-passing steps, as in *CP*.

Table 1. The coefficient of variation of the multicast latency in the DP, MP and CP algorithms with the improvement obtained by QG ($QG_{IMPR}\%$) in the 16×16 mesh

	#Destinations=20		# Destinations=40		# Destinations=60	
	CV	($QG_{IMPR}\%$)	CV	($QG_{IMPR}\%$)	CV	($QG_{IMPR}\%$)
DP	0.386	46.19	0.416	54.83	0.476	76.27
MP	0.326	23.48	0.365	35.69	0.420	55.56
CP	0.467	76.74	0.489	81.56	0.504	86.49
QG	CV= 0.2640		CV= 0.2695		CV= 0.27004	

4.2 Latency in the Presence of Multicast and Unicast Traffic

In some real parallel applications, a message may have to compete for network resources with other multicast messages or even with other unicast messages. To examine performance in such situation, results for the mean multicast latency have been gathered in the 10×10 mesh in the presence of both multicast (10%) and unicast (90%) traffic (similar studies are outlined in [8, 10, 15]). The message size is set at 64 flits and the number of destinations in a given multicast operation has been set to 10 and 20 nodes, respectively. The simulation results are provided in Figs. 1 and 2. Fig. 1 reports results for 10 destinations while Fig. 2 shows results for 20 destinations. Under light traffic, *QG, DP* and *MP* have comparable performance behaviour, with *MP* having a slightly lower latency. On the other hand, *CP* has a higher time.

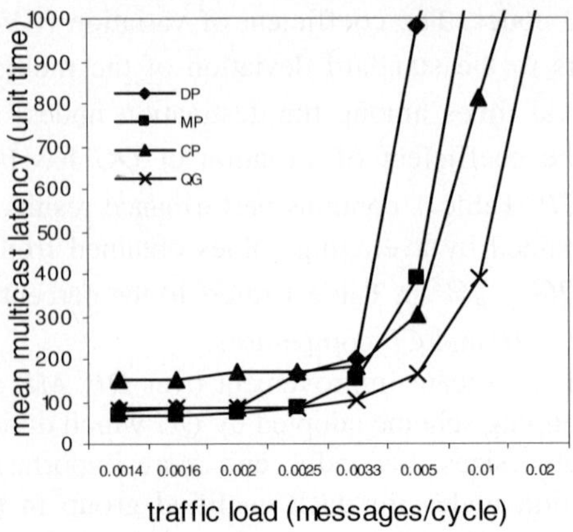

Fig. 1. Mean multicast latency in the 10×10 mesh. Message length is 64 flits, number of 10 destinations = 10 nodes, traffic consists of multicast (10%) and unicast (90%).

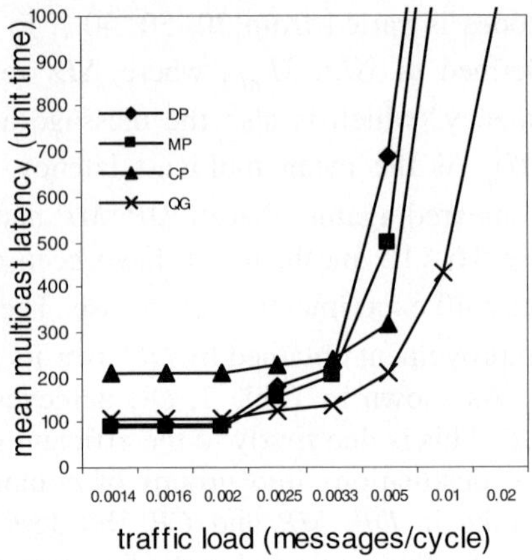

Fig. 2. Mean multicast latency in the 10×10 mesh. Message length is 64 flits, number of destination =20 nodes, traffic consists of multicast (10%) and unicast (90%).

This is mainly due to the dominating effect of the start-up latency in such a situation. However, under heavy traffic, an opposite behaviour is noticed in that *QG* performs the best in terms of both latency and throughput, followed by *CP*. More importantly, we can observe from Fig. 2 that as the number of destinations increases the performance advantage of *QG* becomes more noticeable over that of *CP*. This is mainly because *QG* alleviates significantly the congestion problem at the source node. In contrast, the source node in *CP* suffers from a higher load and as more destinations are involved in the multicast operation, the more severe this limitation becomes.

5 Conclusions and Future Directions

In this study, the QG multicast algorithm has been evaluated under different scenarios and conditions. Results from extensive simulations under different conditions have revealed that the *QG* algorithm exhibits superior performance over well-known algorithms, such as dual-path, multiple-path, and column-path algorithms. Unlike existing multicast algorithms, the *QG* algorithm can maintain a lower variance of message arrival times at the node level. Consequently, most of the destination nodes receive the multicast message in comparable arrival times. It would be interesting to further investigate the interaction between the important parameters that affect the performance of the *QG* algorithm, notably the grouping scheme, network size, threshold value, multicast group size, and traffic load, with the aim of proposing an analytical model that could predict, for example, the multicast latency given a particular grouping scheme, network size, multicast group size, and traffic load.

References

[1] Nen-Chung Wang, Cheng-Pang Yen and Chih-Ping Chu, Multicast communication in wormhole-routed symmetric networks with hamiltonian cycle model, Journal of Systems Architecture, vol.51, Issue 3 , March 2005, pp.165-183, 2005.

[2] A. Al-Dubai, M. Ould-Khaoua and L. Mackenzie, An efficient path-based multicast algorithm for mesh networks, *Proc. the 17th Int. Parallel and Distributed Processing Symposium (IEEE/ACM-IPDPS)*, Nice, France, 22 -26 April, pp. 283-290, 2003.

[3] J. Duato, C. Yalamanchili, L. Ni, Interconnection networks: an engineering approach, *Elsevier Science,*2003.

[4] Abderezak Touzene, Optimal all-ports collective communication algorithms for the *k*-ary *n*-cube interconnection networks , Journal of Systems Architecture, vol.50, Issue 4, pp. 169-236, 2004.

[5] Yuh-Shyan Chen, Chao-Yu Chiang and Che-Yi Chen, Multi-node broadcasting in all-ported 3-D wormhole-routed torus using an aggregation-then-distribution strategy, Journal of Systems Architecture, vol.50, Issue 9 , pp. *575-589*, 2004.

[6] P. K. McKinley, C. Trefftz, MultiSim: A simulation tool for the study of large-scale multiprocessors, *Proceedings of the Int. Symp. Modeling, Analysis and Simulation of Computer and Telecommunication Systems (MASCOTS' 1993)*, pp. 57-62. 1993.

[7] H. D. Schwetman, CSIM: A C-based, process-oriented simulation language, Tech. Rep. pp. 80-85, *Microelectronics and Computer Technology Corp*., 1985.

[8] E. Fleury, P. Fraigniaud, Strategies for path-based multicasting in wormhole-routed meshes, *J. Parallel & Distributed Computing*, vol. 60, pp. 26-62, 1998.

[9] Y.-C. Tseng, S.-Y. Wang, C.-W. Ho, Efficient broadcasting in wormhole-routed multicomputers: A network-partitioning approach, *IEEE Transactions on Parallel and Distributed Systems*, vol. 10, no. 1, pp. 44-61, 1999.

[10] X. Lin, P. McKinley, L.M. Ni, Deadlock-free multicast wormhole routing in 2D-mesh multicomputers, *IEEE Transactions on Parallel and Distributed Systems*, vol. 5, no. 8, pp. 793-804.1994.

[11] S. Cang, J. Wu, Time-step optimal broadcasting in 3-D meshes with minimal total communication distance, *J. Parallel & Distributed Computing*, vol. 60, pp. 966-997, 2000.

[12] D. F. Robinson, P. K. McKinley, C. Cheng, Path based multicast communication in wormhole routed unidirectional torus networks, *Journal of Parallel Distributed Computing*, vol. 45, 104 - 121, 1997.

[13] M. P. Malumbres, J. Duato, An efficient implementation of tree-based multicast routing for distributed shared-memory multiprocessors, *J. Systems Architecture*, vol. 46, 1019-1032, 2000.

[14] P. Mohapatra, V. Varavithya, A hardware multicast routing algorithm for two dimensional meshes, *the Eighth IEEE Symposium on Parallel and Distributed Processing*, pp. 198-205, News Orleans, October 1996.

[15] R. V. Boppana, S. Chalasani, C.S Raghavendra, Resource deadlock and performanceof wormhole multicast routing algorithms, *IEEE Transactions on Parallel and Distributed Systems*, vol. 9, no. 6, 535-549, 1998.

[16] S. Wang, Y. Tseng, C. Shiu , J, Sheu, Balancing traffic load for multi-node multicast in a wormhole 2D torus/mesh, the Computer Journal, vol. 44, no. 5, pp. 354-367, 2001.

[17] D. R. Kumar, W. A. Najjar, P. K. Srimani, *A new adaptive hadrdware tree-based multicast routing in k-ary n-cubes*, IEEE Computer, vol. 50, no. 7, pp. 647-659, 2001.

A Proactive Distributed QoS Control Framework for Cluster Web Site

Wang Xiaochuan and Jin Shiyao

School of Computer Science, National University of Defense Technology
410073 Changsha, China
{mrwangxc, syjin1937}@163.com

Abstract. In this paper, we present a distributed QoS control framework for cluster web applications. We describe the system structure and inner operations from the perspective of control theory. Controllers spread across all back-end servers, proactively controlling the actuators on the cluster entry. This can accommodate advanced but resource consuming control models and algorithms without central performance bottleneck problem in contrast to previous proposals. Our proposal also takes friendliness, flexibility and effectiveness of the control scheme into consideration for practical employment. We give an example implementation based on this framework. Experiments prove its feasibility and effectiveness.

1 Introduction

With the pervasive deployment of Internet, more organizations and enterprises extend their activity on WWW platform. Web applications face more challenges than ever before. The three main challenges are: highly diversified web application demands; highly variable user access patterns [1]; more demands on service availability, stability, cost and performance, referred as QoS (Quality of Service). Regarding the first two, QoS control may be the hardest one we have to conquer because it involves a highly dynamic process.

Based on review of state of art solutions and practical requirements, we claim that an advisable QoS control solution for cluster system, should satisfy three requirements:

1. Friendliness: It should work well with existing systems. Radical change should be avoided for cost and stability reasons. Further more, the deployment of such solution should not be complex or consuming too much.
2. Flexibility: It should work well with variation of the system under control with no or little manual operation or reconfiguration. Different QoS control types can be implemented expediently.
3. Effectiveness: Obviously, It's the basic requirement. It's essential for cluster under very high load. For example, it should be extensible.

In this paper, we mainly focus on basic architecture for adaptable and extensible QoS control of web cluster. In contrast to previous solutions, we adopt a distributed scheme to accommodate complex controllers designed using modern control theory

M. Gerndt and D. Kranzlmüller (Eds.): HPCC 2006, LNCS 4208, pp. 340 – 349, 2006.

without extensibility problem. In our proposal, QoS controllers reside on all servers working independently with each other. The architecture provides the foundation to meet aforementioned three criteria.

In section 2, we review state of the art QoS control mechanisms for web systems, especially web cluster. Section 3 presents the proactive distributed QoS control framework. Section 4 gives an example implementation based on the framework. Section 5 describes the experiments and results. In section 6, we summarize our conclusions and offer possible directions for future work.

2 Related Work

During the last decade, people have done a lot of work on web QoS control. They can be divided into two main categories: single node solution and cluster solution.

- *Single Node QoS Control:*

One classic approach to local QoS control is resource partition which imposes resource limits like maximum simultaneous threads on service entities to avoid over commitment and ensure QoS level [2,3]. The main problem for these methods is that it's difficult to determine the ideal resource limits under widely fluctuating loads. In addition, such limits don't relate to client side QoS metrics directly.

Another approach uses schedule strategy in favor of given request type such as shortest request, to achieve different QoS levels [4~6]. Although they have good performance for specific requests, they usually lack of quantitative metrics.

Admission control is another important method. By restricting the amount of requests entering the system, it can maintain the QoS level for some request types while giving up the others when the system is overloaded [7~9]. Like resource partition schemes, such proposals are usually based on inflexible fixed server side parameters like queue threshold, which ignore client perceived performance.

For some applications, service degradation is used when system is overloaded. By degrading image quality or page layout complexity, performance is maintained [10]. This method is very application specific.

In comparison with approaches based on intuition, analytical techniques are becoming popular in recent years. Mathematical tools and some modern theories like feedback control are used to design the QoS control subsystem quantitatively. [11~14] introduce different analytical models (queue theory, stochastic Petri net, productivity function, etc) to profile the web system and analyze its performance. With the success of feedback control theory in industrial fields, people introduce it to web QoS control [15~22]. The above approaches usually use a simplified and idealized system model. For example, analytical methods based on queue theory usually suppose a Poisson arrival. Feedback control often requires a linear model and uses traditional PI(D) controller. Unfortunately, Internet services tend to be highly nonlinear because of poorly understood traffic and internal resource demands. The mismatch between the model and actual system arouses doubt about the flexibility of the control mechanism in a changeful world. Moreover, a nontrivial fact is that such control system often needs a lot of preliminary work before deployment, such as offline system identification, control loop parameter configuration, etc. These hold back its wide use in practice. Although control theory based QoS control is a promising direction, there's still much to do.

Most of the above methods don't have direct relation with quantitative client perceived QoS metrics. Moreover, they often need mass modification of existing software (web server, OS, etc), even invent a new one from scratch. This could be especially expensive, ineffective and infeasible for practical application out of the lab. People usually want to protect their investment and operation experience. Smooth running without lots of disturbance is most desired. Further more, it makes system upgrade and transition very difficult. In a word, a black box control is preferable.

- *Cluster QoS Control:*

Similar with single node solutions, cluster QoS control also usually involves resource partition [23~26], preferential schedule [27, 30], admission control [28~30] , service degradation [30] and analytical control [30~32]. They usually have the same limitations like their single node counterparts.

It's notable that there're few solutions based on control theory for cluster. One important reason is extensibility. The control structure of a large cluster is rather complex and resource consuming. Floating computation is the daily grind. Models like matrix or artificial neural network may be involved. In a typical web cluster, QoS control is usually performed at the cluster entry called dispatcher. It takes charge of entire cluster system. Obviously, such control mechanism is a huge burden for dispatcher. Moreover, dispatcher codes usually work in OS kernel for performance reason. However many OS kernels don't allow floating operations at all, saying nothing of other useful tools. Development of control theory based schemes in dispatcher is rather expensive and complex. If a dedicated node is used for QoS control, the performance bottleneck and single point failure problem still exist.

3 Proactive Distributed QoS Control Framework

As Fig. 1 shows, a web cluster usually has an entry point called dispatcher between clients and server pool. In practice, the server pool may be divided into different functional layers. For simplicity, we just consider a single layer. Servers in this layer may provide different content. The dispatcher hides the inner structure of the cluster to clients. It's beneficial for application deployment, daily maintenance and system security. Our proposal also adopts such a structure. To eliminate single point failure, there may be a standby dispatcher, but this is not what this paper focuses on.

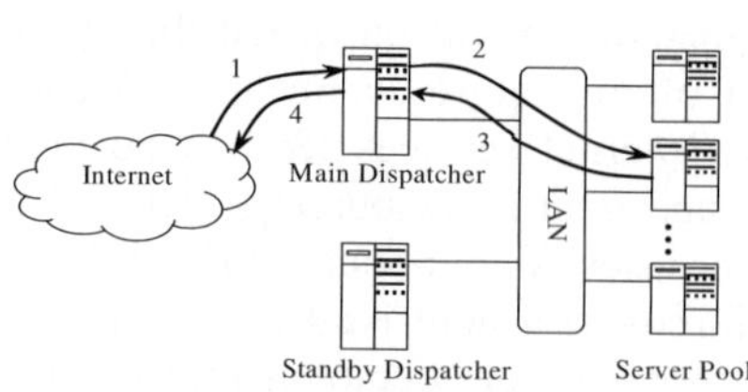

Fig. 1. Web cluster structure and processing diagram

For better QoS control, dispatcher acts as application gateway. On behalf of back-end web applications, it accepts client connections, reads in requests, and performs analysis and classification (arrow 1 in Fig. 1). This technique is usually called content-based schedule. It distinguishes between different request types or user types, preparing for further QoS control. After that, requests are distributed to back-end servers according to QoS control strategy (arrow 2 in Fig. 1). We'll describe this process later. Server responses are relayed back to clients by dispatcher (arrow 3, 4 in Fig. 1). Some techniques such as TCP handoff can send server response directly to clients, bypassing the dispatcher, but they usually need server OS modification. We still use dispatcher to relay server response for friendliness.

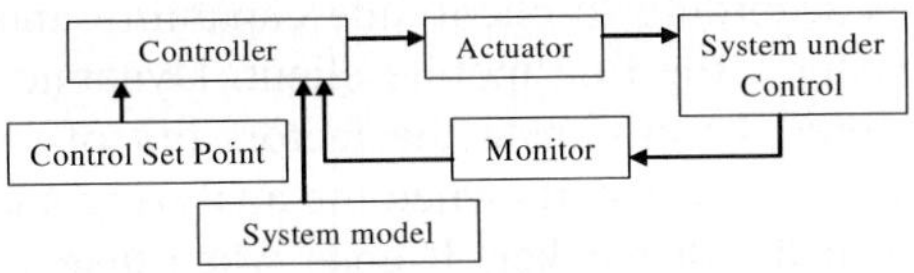

Fig. 2. A typical feedback control system

We'll describe QoS control from the perspective of control theory. A typical feedback control system is shown in Fig. 2. There are three types of basic element: monitor, controller and actuator. Monitor gathers run time information of the system under control. Controller deduces the control action according to the information collected by monitor, control objectives (set point) and specific control model and algorithm. Actuator executes the decisions of controller, influencing the object system directly. These three steps comprise the close control loop.

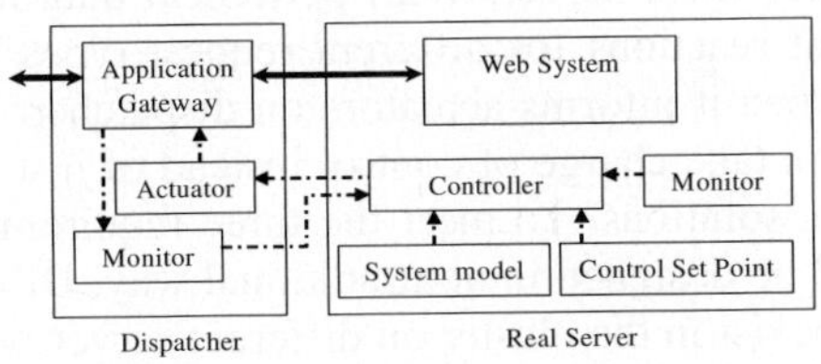

Fig. 3. Distributed QoS control framework

The distributed QoS control framework is shown in Fig. 3. For simplicity, we only give one server node. The dashed lines are control flows. The bold lines are application data flows. Monitors gather run time information and report to controllers periodically. Then controllers make decisions and inform actuator to execute them.

Adaptability and flexibility are guaranteed in this scheme. First, independent controllers are running on each server. Control expenses are apportioned among all nodes. Advanced but complex control methods can be used to achieve quantitative QoS control. Second, web application has extensive parallelism. Holistic cluster QoS objectives can be achieved by such parallel local solution. Third, it's convenient to customize control strategy, even use different control model for specific node or service. Last, all control elements are transparent to the web system. Server side software doesn't have to be modified or replaced at all.

Periodical control flow from server to dispatcher is natural heartbeat mechanism to detect failed server node, replacing additional server availability mechanism.

■ *Control Set Point*

Control set point is the object of the control action. For web QoS control, it's usually referred to as performance metrics. The most common are classified into two categories depending on their relationship to time: delay metrics and rate metrics. The former are directly proportional to time such as response times. The latter are inversely proportional to time such as data throughput. In this paper, we mainly consider two most common ones: response time and data throughput. The first is straightforward to users. The second is important for the service as a whole.

Response time varies according to client side capability and request type. It may take a long time to transfer a file to a mobile client. Dynamic content usually takes longer time than static one. To eliminate the factors out of server side control, we introduce Initial Response Time. It starts when the analyzing and classifying work are finished for the request at the dispatcher. It ends when dispatcher receives the first response fragment from the server. Thus, interactions with outside world are ignored while application characteristics are still taken into account.

■ *Monitor*

There're two types of monitor. The first type resides on real servers, collecting system resource usage information like CPU usage. The second type resides on dispatcher, collecting application performance data including Initial Response Time, data throughput, request arrival rate and leaving rate, etc. Performance data are collected for different {request type, server} combinations. Monitors work transparently to web applications, and send data to controllers on servers periodically.

■ *Controller*

Controller works only for local server. With periodical data fed by monitors, it deduces proper subsequent reactions for different request types based on QoS control model and algorithm. Then it informs actuators on dispatcher. We call this proactive approach because servers take charge of control instead of just accepting dispatcher's arrangement in previous solutions. To meet the three requirements remarked in Section 1, controller should be designed in nontraditional way. Different control methods may be designed and coexist in the cluster on different server without impact on other nodes. In Section 4, we'll present a controller based on fuzzy control theory.

■ *Actuator*

We choose Maximum Concurrent Request Number and Maximum Request Admitting Rate provided by controllers as the main control variables. They directly influence resource allocation on server. We also call them Service Capacity Index. The former is used to prevent resource over commitment; the latter is for QoS metrics control. Like monitors, they are also enforced by actuator on {request type, server} combinations. Thus, QoS control is achieved for different request type on different server.

4 Example Implementation

We implement a web cluster system (see Fig. 1) with QoS control mechanism based on the framework proposed in Section 3. The dispatcher is implemented as an HTTP gateway in Linux kernel 2.4. It supports HTTP 1.1 and classifies client requests based on the characteristic defined by the administrator.

Actuators and monitors are integrated in the HTTP gateway. Monitors collect performance statistics data and send them to controllers using UDP datagram every 2 seconds. The packet includes {concurrent request number, request admitting rate, average Initial Response Time / data throughput} triples for each request category managed by the controller. Actuators receive subsequent UDP messages from controllers in {Maximum Concurrent Request Number, Maximum Request Admitting Rate} format for each request category supported by the controller. The maximum communication load is (35 * request type number * server number) bytes every 2 seconds.

Actuator enforces controllers' decisions on the control variables. When thresholds of all candidate servers for a request category are exceeded, requests belonging to that category queue up at the gateway, until new resources are available or timeout. Controller message replaces traditional server health check mechanism. Three successive lose indicate the problem with the server or local network.

On real server, monitors collect system resource usage information periodically using system specific facilities. CPU usage monitor is implemented for the time being. Other resource monitors may be added in the future. Because CPU usage often fluctuates with peaks and valleys, we use a variation of DMC (Dynamic Matrix Control) algorithm to smooth it while keeping track of the main trend. The same work is done for concurrent request number and request admitting rate before they are fed to controller. Service health checking monitor is another important monitor type. It periodically examines service availability based on user customizable scripts for specific applications. When service dies, it sets Service Capacity Index to zero, thus dismiss the server from the cluster.

The controller has two jobs: quantitative QoS control for different request types (There're few such solutions before), preventing system resource over-commitment. In our implementation, we employs two independent fuzzy controllers for the above two purpose. The fuzzy controller performs a black box style control, satisfying friendliness flexibility and effectiveness requirements mentioned in Section 1. The fuzzy inference uses Single Fuzzifier, Product Inference Engine and Center Average Defuzzifier [33]. In fact, other control theories and techniques may be used in this framework. For the time being, all controllers and server side monitors are implemented for Windows platform and UNIX like OS such as Linux. Except the system resource usage monitor code, all other codes are platform independent.

- *Fuzzy Control Model for Quantitative Performance Control*

The model takes three fuzzy inputs:

Capacity Utilization (U): the utilization of capacity indicated by Maximum Request Admitting Rate. Let Cmax is Maximum Request Admitting Rate, C is request admitting rate at sampling time, then U = C / Cmax.

System Load (L): CPU usage percentage, preprocessed by DMC algorithm.

Relative Performance Error (E): error ratio to set point. E = (Preference – P) / Preference, where Preference is object performance metrics such as Initial Response Time, P is corresponding run time value preprocessed by DMC algorithm.

Output of the fuzzy inference is **Regulation Ratio (R)** of Maximum Request Admitting Rate. R = (Cnew – Cprevious) / Cprevious where Cprevious is previous value of Maximum Request Admitting Rate. We can get the new value Cnew from R and the above equation.

Based on the fuzzy sets defined for three inputs, we have 31 fuzzy inference rules.

- *Fuzzy Control Model for Preventing Resource Over-Commitment*

The model takes two fuzzy inputs:

Capacity Utilization (U): the utilization of capacity indicated by Maximum Concurrent Request Number. Let Cmax is Maximum Concurrent Request Number, C is concurrent request number at sampling time, then U = C / Cmax.

Relative Load Error (E): error ratio to set point. E = (Preference – P) / Preference, where Preference is object resource utilization (CPU usage), P is corresponding run time value preprocessed by DMC algorithm.

Output of the fuzzy inference is **Regulation Ratio (R)** of Maximum Concurrent Request Number. R = (Cnew – Cprevious) / Cprevious where Cprevious is previous value of Maximum Concurrent Request Number. We can get the new value Cnew from R and the above equation.

Based on the fuzzy sets defined for two inputs, we have 9 fuzzy inference rules.

5 Experiments and Results

The experiments are designed to validate the effectiveness of the QoS control framework, especially for quantitative QoS control. The test-bed consists of one dispatcher, two back-end servers, four client PCs. Each machine runs Linux kernel 2.4.8 with One 2.5 GHz Intel P4, 512 MB RAM and 1000 Mbps Ethernet network. Clients and servers belong to different subnets with dispatcher standing between them. Web servers are Apache 2.0.46. Maximum process number of Apache is 512; connection timeout value is 15 sec.

The first experiment is designed to validate the effectiveness of the QoS control framework under simple user load. We use sclient [34] to emulate 400 simultaneous users in 30 seconds, requesting same one image of 190K bytes. Object Initial Response Time is 100 ms. Initial Maximum Concurrent Request Number for all request types on each server is set to 50. Fig. 4 shows that the control process is smooth due to simple request pattern. The results converge to our objective.

The second experiment is to test the adaptability of the cooperative control framework to more realistic load. We use SURGE [35] to emulate 400 simultaneous users. Other settings are same with the first experiment. The results in Fig. 5 show larger fluctuations, but the control is still responsive with adequate precision.

To find the load produced by the controller, we ran the cluster without load and checked the load percentage with "top". That's around 2% every 2 seconds. The similar results are found on other platforms. Although it's negligible for server, dozens of such controller on single node will have serious impact, especially for a dispatcher implemented as application gateway. We configure five request types in our experiments. The total communication load is 350 bytes every 2 seconds.

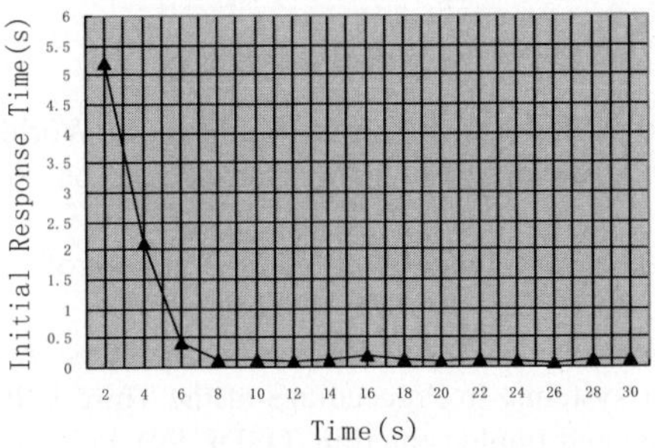

Fig. 4. Results under simple load pattern

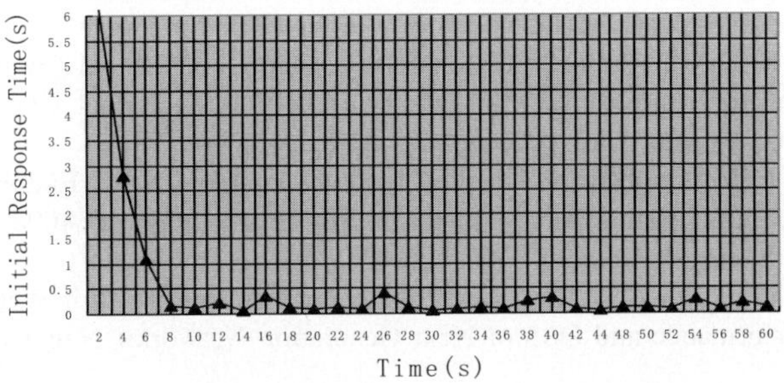

Fig. 5. Results under more realistic load pattern

From the above experiments results, we can see that our QoS control framework can perform effective quantitative QoS control with little burden on the whole system. There're few such results in previous research and implementation.

6 Conclusions and Future Work

Web cluster QoS control can improve client's experience and facilitate some valuable web site management modes. This paper has proposed a distributed QoS control framework based on back-end server's proactive control. It allows the use of more advanced but resource consuming control methods without single point performance problems in contrast to previous proposals. Thus quantitative QoS control can be achieved which is still a problem in previous solutions. Our proposal also takes friendliness, flexibility and effectiveness of the control scheme into account for practical employment. Under this framework, we developed a web cluster using fuzzy QoS controller which can perform quantitative QoS control. The experiments show its feasibility and effectiveness.

More controller types and other control objectives like QoS differentiation under this framework are worth further investigation. Distributed QoS control with multiple dispatchers is another interesting subject.

References

1. Mark E. Crovella and Azer Bestavros, Self-Similarity in World Wide Web Traffic: Evidence and Possible Causes in IEEE/ACM Transactions on Networking, 5(6):835--846, December 1997.
2. K. Li and S. Jamin. A measurement-based admission-controlled Web server. In Proceedings of IEEE Infocom 2000, Tel-Aviv, Israel, March 2000.
3. G. Banga, P. Druschel, and J. Mogul. Resource containers: A new facility for resource management in server systems. In Proceedings of the Third USENIX Symposium on Operating Systems Design and Implementation (OSDI '99), February 1999.
4. M. Harchol-Balter, M. Crovella, and S. Park. The case for SRPT scheduling in Web servers. Technical Report MIT-LCR-TR-767, MIT, October 1998.
5. B. Schroeder and M. Harchol-Balter. Web servers under overload: How scheduling can help. Technical Report CMU-CS-02-143, Carnegie-Mellon University, June 2002.
6. Jordi Guitart, David Carrera. Session-Based Adaptive Overload Control for Secure Dynamic Web Applications. In Proceedings of 34th International Conference on Parallel Processing (ICPP 2005), 14-17 June 2005, Oslo, Norway
7. L. Breslau, E.W. Knightly, S. Shenker, I. Stoica, and H. Zhang. Endpoint admission control: Architectural issues and performance. In Proceedings of ACM SIGCOMM 2000, Stockholm, Sweeden, October 2000.
8. R. Iyer, V. Tewari, and K. Kant. Overload control mechanisms for Web servers. In Workshop on Performance and QoS of Next Generation Networks, Nagoya, Japan, November 2000.
9. V. Kanodia and E. Knightly. Multi-class latency-bounded Web services. In Proceedings of IEEE/IFIP IWQoS 2000, Pittsburgh, PA, June 2000.
10. S. Chandra, C. S. Ellis, and A. Vahdat. Differentiated multimedia Web services using quality aware transcoding. In Proceedings of IEEE INFOCOM 2000, March 2000.
11. X. Chen, H. Chen, and P. Mohapatra. An admission control scheme for predictable server response time for Web accesses. In Proceedings of the 10th World Wide Web Conference, Hong Kong, May 2001.
12. Daniel A. Menasce, Daniel Barbara, Ronald Dodge. Preserving QoS of E-commerce Sites Through Self-Tuning: A Performance Model Approach. In the Proceedings of the 3rd ACM Conference on Electronic Commerce. Tampa, Florida, USA. Oct, 2001.
13. H. Chen and P. Mohapatra. Session-based overload control in QoS-aware Web servers. In Proceedings of IEEE INFOCOM 2002, New York, June 2002.
14. Zhangxi Tan, Chuang Lin, Hao Yin, Ye Hong. Approximate Performance Analysis of Web Services Flow Using Stochastic Petri Net. Accepted by GCC 2004, LNCS, October 2004, Wuhan China.
15. C. Lu, T. Abdelzaher, J. Stankovic, and S. Son. A feedback control approach for guaranteeing relative delays in Web servers. In IEEE Real-Time Technology and Applications Symposium, Taipei, Taiwan, June 2001.
16. Y. Diao, N. Gandhi, J. Hellerstein, S. Parekh, and D. Tilbury. Using MIMO feedback control to enforce policies for interrelated metrics with application to the Apache Web server. In Proceedings of the Network Operations and Management Symposium 2002, Florence, Italy, April 2002.
17. Matthew David Welsh. An Architecture for Highly Concurrent, Well-Conditioned Internet Services. Ph.D. thesis, U.C. Berkeley, September 2002.

18. Ronghua Zhang, Chenyang Lu, Tarek F. Abdelzaher, John A. Stankovic, ``ControlWare: A Middleware Architecture for Feedback Control of Software Performance," International Conference on Distributed Computing Systems, Vienna, Austria, July 2002.
19. Ronghua Zhang, Tarek F. Abdelzaher, and John A. Stankovic, ``Kernel Support for Open QoS-Aware Computing," Real-Time and Embedded Technology and Applications Symposium, Toronto, Canada, May 2003.
20. Ying Lu, Tarek F. Abdelzaher, Avneesh Saxena, ``Design, Implementation, and Evaluation of Differentiated Caching Services," IEEE Transactions on Parallel and Distributed Systems Vol. 15, No. 5, pp. 440-452, May 2004.
21. Chenyang Lu, Ying Lu, Tarek F. Abdelzaher, John A. Stankovic, Sang H. Son, ``Feedback Control Architecture and Design Methodology for Service Delay Guarantees in Web Servers," IEEE Transactions on Parallel and Distributed Systems, 2005.
22. Chengdu Huang and Tarek Abdelzaher, ``Bounded-Latency Content Distribution: Feasibility and Evaluation," *IEEE Transactions on Computers*, 2005.
23. 30 Vivek S,Mohit A. Locality-Aware Request Distribution in Cluster-based Network Servers. In:Proc. of ASPLOS-Ⅷ,ACM SIGPLAN,1998. 205~216
24. M. Aron, P. Druschel, and W. Zwaenepoel. Cluster reserves: A mechanism for resource management in cluster-based network servers. In Proceedings of the ACM SIGMETRICS Conference on Measurement and Modeling of Computer Systems, Santa Clara, CA, June 2000.
25. S. Ranjan, J. Rolia and E. Knightly, IWQoS 2002. QoS-Driven Server Migration for Internet Data Centers. In Proceedings of IEEE IWQoS 2002.
26. Bhuvan Urgaonkar, Prashant Shenoy. Dynamic Provisioning of Multi-tier Internet Applications Proceedings of the 2nd IEEE International Conference on Autonomic Computing (ICAC-05), Seattle, June 2005.
27. Xueyan Tang, Samuel T. Chanson, Huicheng Chi, and Chuang Lin. Session-Affinity Aware Request Allocation for Web Clusters. In Proceedings of the 24th International Conference on Distributed Computing Systems (ICDCS'04), IEEE Computer Society, 24-26 March 2004, Tokyo Japan, pp. 142-149.
28. A. Verma and S. Ghosal. On Admission Control for Profit Maximization of Networked Service Providers. In Proc. of 12th Int'l World Wide Web Conf. (WWW2003), Budapest, Hungary, May 2003.
29. Sameh Elnikety, Erich Nahum. A Method for Transparent Admission Control and Request Scheduling in E-Commerce Web Sites. In Proceedings of WWW2004, May 17–22, 2004, New York, New York, USA.
30. Bhuvan Urgaonkar, Prashant Shenoy. Cataclysm: Handling Extreme Overloads in Internet Applications. In Proceedings of the Fourteenth International World Wide Web Conference (WWW 2005), Chiba, Japan, May 2005.
31. Shan Z, Lin C, Marinescu D C, and Yang Y. QoS-aware load balancing in Web-server clusters: performance modeling and approximate analysis. Computer Networks Journal, September 2002, 40(2): 235-256.
32. Lin C and Marinescu D C. Stochastic high-level Petri nets and applications. IEEE Trans. on Computers, 1988, 37(7): 815-825.
33. Li-Xin Wang. A Course in Fuzzy Systems and Control. Prentice-Hall, Inc. 1997
34. Banga, G. and Druschel, P. Measuring the capacity of a web server. Usenix Symposium on Internet Technologies and Systems 1997.
35. P. Barford and M. E. Crovella, "Generating Representative Web Workloads for Network and Server Performance Evaluation," ACM SIGMETRICS '98, Madison WI, 1998.

Design and Implementation of Zero-Copy Data Path for Efficient File Transmission

Dong-Jae Kang[2], Young-Ho Kim[1], Gyu-Il Cha[1], Sung-In Jung[1],
Myung-Joon Kim[1], and Hae-Young Bae[2]

[1] Internet Server Group, Digital Home Division, ETRI,
161 Gajeong-dong, Yuseong-gu, Daejeon, Korea
{djkang, kyh05, gicha, sijung, joonkim}@etri.re.kr
[2] Dept. of Computer Science and Information Engineering, In-Ha University,
253 Yonghyeon-dong, Nam-gu, Incheon, Korea
hybae@inha.ac.kr

Abstract. Huge requests for file transmission by concurrent users cause excessive memory copy operations and system calls between kernel and user space, which are very expensive and require many CPU cycles for processing. And it has been a bottleneck that limits the number of serviceable requests and prevents CPU resources from being allocated to other processes. In this paper, we suppose the zero-copy data path for efficient file transmission to solve the upper described problems. To do that, we used existing system call interface, *sendfile*, for file transmission from one device to another and used the same buffer page to transmit a file. The supposed zero-copy data path reduces CPU cost per transaction by eliminating CPU copies between kernel and user space and by avoiding unnecessary system calls in user space. The improved CPU efficiency means that a larger number of requests can be serviced with the same CPU configuration.

1 Introduction

With the steep growth of multimedia and distributed computing technology, the size of requested data has been increased. And huge requests for file transmission by concurrent users cause excessive memory copy operations and system calls between kernel and user space which are very expensive operations in the remote storage server, shared backup server, ftp server and etc. Therefore, it has been a bottleneck that limits the number of serviceable requests and prevents CPU resources from being allocated to other processes in server. Generally, the operation which transfers file between network and disk needs execution of several system calls in user space and requires many CPU copies between kernel and user space. So, a lot of CPU cycles are consumed for processing them [1]. This means that the CPU resource may be unavailable during the huge copy operations [4] and context switching for each system call operation. But, in many cases, when the data don't need processing and handling in user space, so we don't have to pass the data through user area for transmitting it in server system, especially like as upper described server systems. It is possible to transfer data without copying to user area in many applications.

M. Gerndt and D. Kranzlmüller (Eds.): HPCC 2006, LNCS 4208, pp. 350–359, 2006.

In this paper, we design and implement the zero-copy data path for efficient file transmission to solve the upper described problems. Zero-copy means there is no CPU copy occurred between kernel and user space. Supposed zero-copy data path eliminates a lot of CPU copy operations and context switching between kernel and user space. To do that, we used existing interface, *sendfile*, for file transmission from one device to another. And we used same system buffer page allocated for getting data from source (disk or network) and putting it to destination (disk or network) device. Supposed zero-copy data path has several advantages. It reduces CPU cost per transaction by eliminating CPU copies between kernel and user space and by avoiding unnecessary system calls in user space. And it can improve the whole system performance by enhancement of CPU efficiency and it allows CPU resource to be allocated to another process and can service the larger number of requests with same CPU configuration.

2 Related Works

Recently, many researches have been performed to improve the system performance by eliminating CPU copy operations and system calls between kernel and user space. And the operations have been considered as bottlenecks in various server systems for a long time [1], [6], [10]. The zero-copy has been studied in two different ways, which use memory mapping technique, one is by using special hardware and the other by only software. The former is mainly related to data transmission using special network interface card (NIC), it can transmit data without CPU copy operations by mapping application buffer into memory on device [11], [14], [18]. The latter is related to memory mapping technique between user buffer and kernel memory for direct access to it without CPU copy operation [1], [2]. Several mechanisms supporting zero-copy have been supposed, including IOLite [9], and UVM virtual memory system [10] which uses some kind of page remapping between kernel and user space for data sharing. And it has been designed to create a fast in-kernel data path from one device to another, e.g., the data path between network and disk. These mechanisms do not pass the data between kernel and user space, but keep the data within the kernel. This means that applications do not manipulate data in any way, i.e., no data touching operations are performed by the application [1].

In this paper, we don't use memory mapping technique which needs memory initialization in startup time and memory management for allocation and de-allocation in operation time. So, we shared buffer page in kernel without using memory mapping between kernel and user space. Good example of such mechanism is the *sendfile* system call [2] in Linux operating system. Despite of the many supposed mechanisms, only a limited support for zero-copy, file transmission from disk to network using system buffer page, is provided in commodity operating systems like Linux. So, in this paper, we suppose the design and implementation of the in-kernel data path for file transmission supporting full functionalities for zero-copy and extended the function of *sendfile* system call interface in Linux.

3 Design and Implementation of Zero-Copy Data Path

In this section, we describe the details of design and implementation for zero-copy data path for file transmission.

3.1 Consideration for Design

We made several considerations for design of zero-copy data path, and it is reflected to our implementation. The considerations are as like bellow. First, zero-copy data path must be implemented in or upper VFS (Virtual File System) layer, because the implementation should be independent on specific sub systems. For example, it should be not dedicated function for specific file system. If it is implemented in VFS layer, it can be more general interface for all file system under VFS layer. And it will be able to operate correctly, regardless of any file systems. Second, its function is combination of several system call operations in kernel, especially, in VFS layer because the cost for system call should be minimized. The zero-copy data path should be supported by one interface and, regardless of data size, be called one time by application. System calls are very expensive operations because each system call requires two times of process context switching, and requires a lot of CPU copies between kernel and user space. Generally, like as *recv* and *write*, system calls are called repeatedly until requested all data are processed. It has been known as very time and resource consuming operation. If, regardless of data size, it is called one time by application, a number of data copy operations between kernel and user space will be eliminated. Additionally, process context switching for execution of system call will be reduced. Third, the implementation should have minimal dependence with other kernel subsystems, file subsystems, network subsystem and the other modules. This will make management and modification for it to be easy in future. Fourth, the CPU copy operations should be minimized. So, we use same system buffer page for data input and output. It eliminates the memory copy operations by CPU between kernel and user space. As a result, it will enhance system performance and CPU availability. The last, general common interfaces should be supported. By now, many researches support special interfaces to applications for usage of zero-copy. But, it may be obstacle against being used popularly. So, we used existing system call interface, *sendfile*, to adapt it.

3.2 Implementation of Zero-Copy Data Path

In this section, we describe the details of implementation for zero-copy data path for file transmission. First, we explain overview of it and next, explain the zero-copy operation and its interfaces in kernel.

3.2.1 Outline of Zero-Copy Data Path

Our zero-copy data path is the technique to eliminate the CPU copies between kernel and user space and to reduce the number of system call in user space by sharing the system buffer page. It is not implemented by memory remapping method between kernel and user space. So it doesn't have to create, manage and delete the pre-defined shared memory area. Application in user space only initiates the file transmission operation without manipulating the file and most of the operation is processed in kernel. The supposed zero-copy data path is consist of four kinds of modules as showed in Fig. 1 and disk and network adapter may be the same one or another. Disk read module and network receive module moves the data into the allocated buffer page from disk or network adapter, and they pass the pointer of the buffer page to other kernel module (disk write module or network send module) without passing the data into a user application.

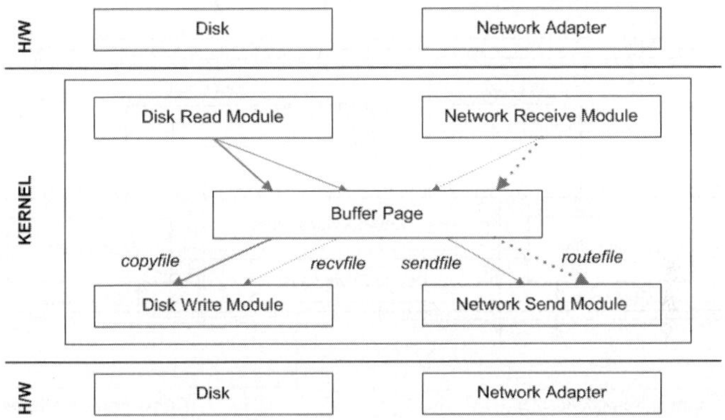

Fig. 1. Overview of zero-copy data path for file transmission. A solid line represents file transmission from disk, and a dotted line is from network. In the two cases, destination may be disk or network.

Zero-copy operations can be classified into *sendfile* (normal solid line), *recvfile* (normal dotted line), *copyfile* (bold solid line) and *routefile* (bold doted line) as like Fig.1. The *sendfile* is the operation from disk to network without CPU copy and context switching between kernel and user space. It can be used in media streaming, file uploading in ftp and so on. The *recvfile* is the reverse with sendfile, from network to disk. It can be used in file downloading, file saving in backup system and so on. The *copyfile* moves a file from disk to disk. Finally, the *routefile* is used for file transmission from network to network.

3.2.2 File Transmission Through Zero-Copy Data Path

Fig. 2 shows zero-copy data path (solid lines) and existing system call (dotted lines) operation for file transmission. In case of existing method, the file data has been copied at least four times per one execution, two DMA copies and two CPU memory copies as like Fig.2. And many user/kernel context switchings were required. The first copy is performed by the DMA engine, which copies the data from source device (disk or NIC) into buffer page in kernel address space. And then, the data is copied from the kernel buffer into the user space buffer and the *recv* or *read* system call returns. The return from this call caused a context switching from kernel back to user mode. Now the data is stored to a buffer in the user address space, and it can begin its way down again. The *write* or *send* system call causes a context switching from user to kernel mode. A third copy is performed to put the data into a buffer page in kernel address space again. In this time, the data is put into a different buffer page that is associated with a destination device. And the operation returns, creating our fourth context switch. Finally, the fourth copy happens as the DMA engine passes the data from the kernel buffer page to the destination device. As another method, *mmap* system call can be used. But, many system calls are still remained.

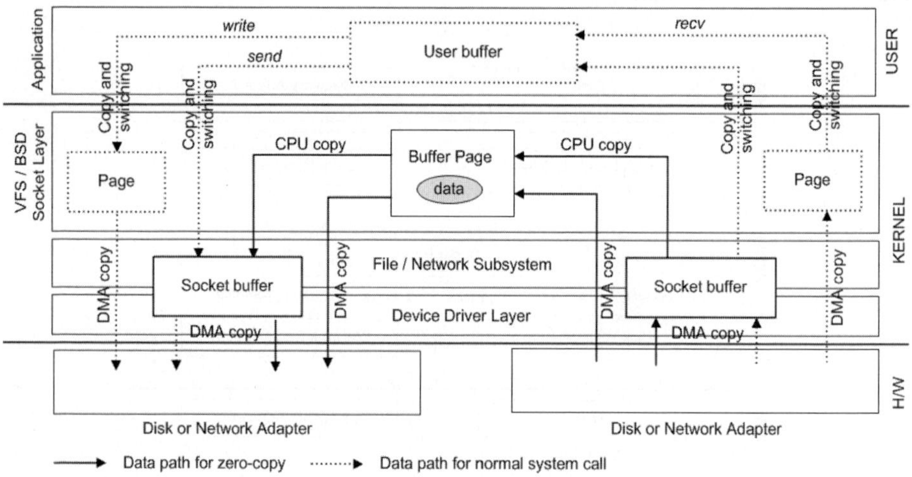

Fig. 2. Zero-copy data path for file transmission. A solid line represents file transmission flow for our implementation (zero-copy operations), and a dotted line is that of existing system call interface, *recv /send / read / write.*

But, zero-copy data path operations have no CPU memory copies between kernel and user space and user/kernel context switching anywhere as like Fig.2. We shares buffer page allocated for disk read module or network receive module with disk write module or network send module in Fig.1. Therefore, the zero-copy data path makes the data to be copied into a kernel buffer page from source device (NIC or disk) and then, it is passed from the same buffer page into a target device (NIC or disk).

As like Fig.2, when zero-copy operation is called, allocation of buffer page ***P*** for getting data from source device is performed in kernel. And allocated page is added to page cache. Then, the data from source device is taken into the buffer page ***P*** until it is full or all data are taken. In this time, if source device is network adapter, the data is taken by one CPU copy from socket buffer. If buffer page is full or all data are taken, the buffer page is put into the target device. If target device is disk, the buffer page is removed from page cache and it is modified with file information to be written. Then it is added to page cache again as page for output file. If the same buffer page is used for source and destination, it may break out serious problems in operating system. But, if destination device is network adapter, it is only removed from page cache, because data for network is not needed to be cached. Also, if destination device is network adapter, the data is move to target device through socket buffer with one CPU copy. Upper operations are repeated until all data are processed. Finally, it returns to user space. We remained one CPU copy in network subsystem, because zero-copy implementation should be independent on special subsystems as explained in section3.1.

3.2.3 Interface Generalization for Zero-Copy Data Path

We tried to adapt our zero-copy operations to existing *sendfile* system call interface in Linux 2.6. *sendfile* was a new feature in Linux 2.2. Originally, *sendfile* was intended to copy data between one file descriptor and another. Either or both of these file descriptor may refer to a disk or network. But, current *sendfile* system call interface only supports that input file descriptor should be a file descriptor opened for data reading from disk and output file descriptor should be a socket descriptor opened for data sending to network. So, we adapt our zero-copy operations to *sendfile* interface and support the full function of it.

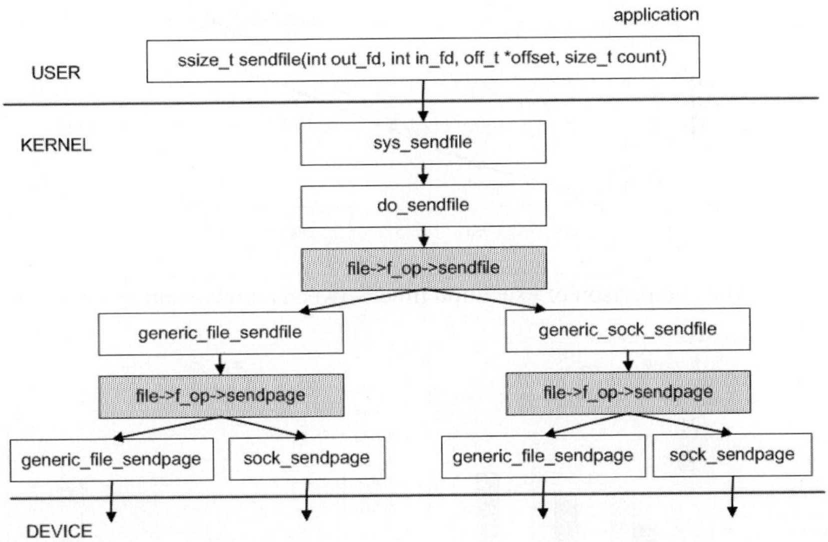

Fig. 3. Interfaces for zero-copy data path. Our zero-copy function is adapted to existing system call, *sendfile*, interface. And painted box supports generalized interface code for data input and output regardless of devices.

Fig.3 shows interface call sequence of zero-copy data path for file transmission using existing system call, *sendfile*. Described four zero-copy functions are identified by input arguments, *in_fd* and *out_fd*. In Fig.3, *generic_file_sendfile* performs disk read operaton, *generic_sock_sendfile* does network receive, *generic_file_sendpage* does disk write and *sock_sendpage* is network send module in Fig.1. *file->f_op->sendfile* gets the page-sized data into allocated buffer page from disk or network according to given *in_fd* argument. If input descriptor is file descriptor, *generic_file_sendfile* interface is called, if it is socket descriptor, *generic_sock_sendfile* is executed. And next, *file->f_op->sendfile* calls proper *file->f_op->sendpage* interface with given page pointer used for *file->f_op->sendfile*. That is, the buffer page for putting data to destination device (disk or NIC) is not allocated in *file->f_op->sendpage*. The buffer page is shared between *file->f_op->sendfile* and *file->f_op->sendpage* interfaces for data input and output. And then, *file->f_op->sendpage* calls the proper interface, *generic_file_sendpage* or *sock_sendpage*, for putting data into destination device (disk or NIC) according to given *out_fd* argument.

4 Experiments and Results

In this section, we tested and analyzed the performance of the supposed zero-copy data path. We leaved out the test for *sendfile* and *routefile* functions because their performance can be derived from that of recvfile and copyfile.

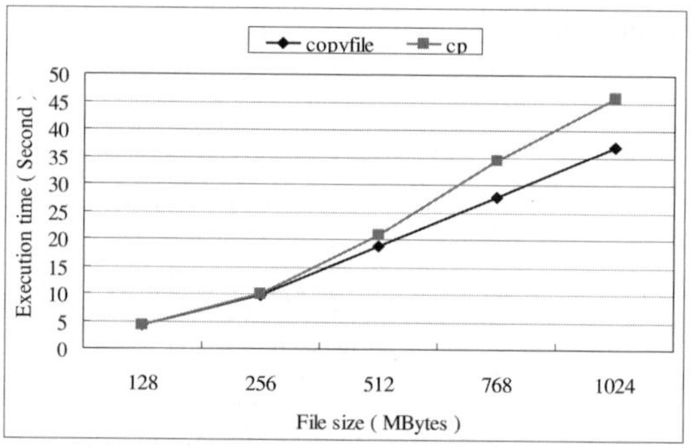

(a) The comparison of execution time between *copyfile* and *cp* operation

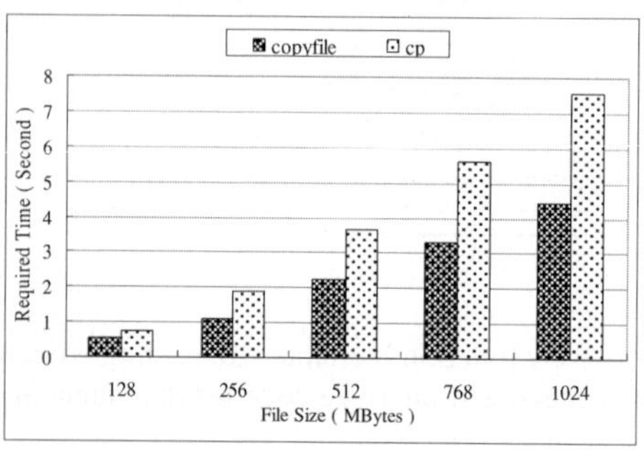

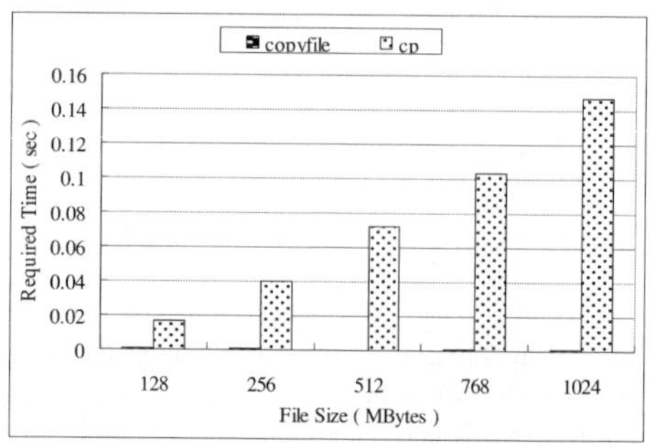

(b) CPU system time of *copyfile and cp* (c) CPU user time of *copyfile and cp*

Fig. 4. Comparison of *copyfile* and *cp* operation. (a) shows the execution time, (b) and (c) represents CPU system time and user time for each operation.

To test, we measured each overhead of the file transmission using zero-copy data path and existing system call under various file size and compared with each other. We used *Supermicro*^{TM} (Intel Xeon dual CPU, Hyper threading, 3.06GHz, 1 and 3G memory) system and PC (Intel Pentium 4 single CPU 2.60GHz, 512M memory) system. The systems were connected by private network to eliminate the unexpected effect by other computers in open network and TCP was used as network protocol. Required values were extracted by *getrusage*, *sar, time* and *vmstat* utilities under the various environment. *Booyo Linux* (linux-2.6.12.6), *Korea Linux Standard Spec,* developed by ETRI in Korea was used as the operating system for test machines.

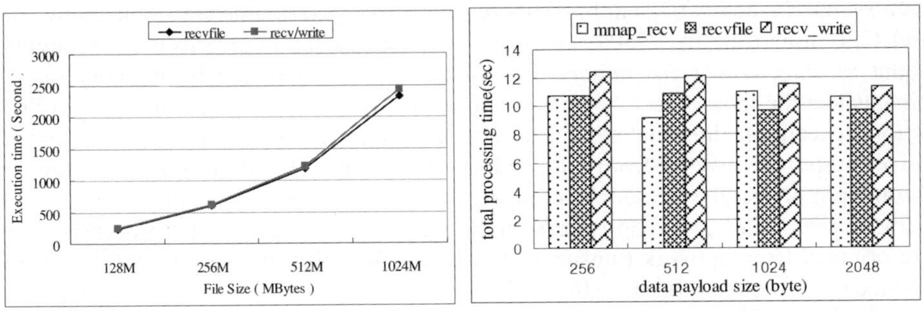

(a) execution time of *recvfile* and *recv/write* (b) execution time *recvfile, recv/write, mmap/recv*

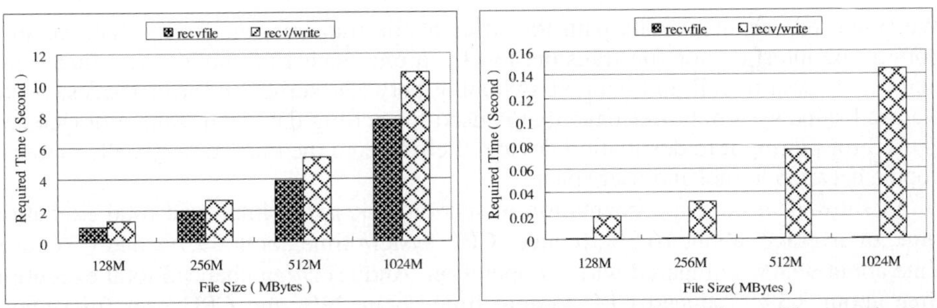

(c) CPU system time of *recvfile, recv/write* (d) CPU user time of *recvfile, recv/write*

Fig. 5. the comparison of *recvfile* and recv/write operation. (a) and (b) shows execution time, (c) and (d) represents the CPU system time and user time for each operation.

Fig.4 shows the comparison result of *copyfile* and *cp* operation and it was tested in single server system environment. In detail, (a) in Fig.4 represents total execution time according to various file size from 128M to 1G and it shows that file transmission by *copyfile* is faster than by *cp* operation, on average, about 16.5%. (b) and (c) in Fig.4 presents the CPU system time and user time of *copyfile* and *cp* operation. The gap of CPU usage rate between *copyfile* and *cp* operation is very big. In the case of CPU system time, *copyfile* reduced the cost by 39.1% compared with *cp* and in the case of CPU user time, it reduced the cost by 99.4%. In spite of huge reduction of CPU usage rate, the total execution time took a small improvement. The reason is that most total execution time, about 90%, is consumed at waiting for disk I/O and required memory. As a result, *copyfile* improved the execution time except waiting time, so improvement of total execution time is very smaller than that of CPU usage rate. But, it will enhance the CPU availability as much as reduction of CPU cost, and it will increase the number of serviceable requests. Fig.5 presents the comparison of *recvfile* and *recv/write* operation, especially, about execution time, CPU system time and CPU user time. (a) in Fig.5 is total execution time of *recvfile* and *recv/write* operation.

This was measured in concurrent user environment, 9 concurrent users from 3 systems transmitting the described size of data in figure, that is, the last case (1024M) means 9Gbyte was received. Total execution time was improved by 3.5% compared with *recv/write* and (c), (d) in Fig.5 shows that CPU system time was reduced by 28% and CPU user time by 99.4%. (b) in Fig.5 indicates the comparison of throughput about *recvfile*, *recv/write*, *mmap/recv*. mmap/recv is performed by memory remapping as like the most researches implementing existing zero-copy method. (b) in Fig.5 is total execution time that sender system transmits data to receiver system and receiver system receives it from NIC and saves to a disk. As showed in the figure, the throughput of *recvfile* is the best compared with other operations and has most stable performance. While, the gap is very slight, because the most of total processing time are oriented from network transmission and disk I/O. But, the system will be able to service more requests, because the CPU cost per operation is improved remarkably.

5 Discussion and Conclusion

We supposed zero-copy data path for efficient file transmission. To do that, we supported one interface for file transmission from one device to another. It reduced the number of system call and context switching between kernel and user spaces. Next, we used same system buffer page that is used for getting data from source device and is used for putting it to destination device. It eliminated the unnecessary CPU memory copies between kernel and user space.

As showed in the experimentation in section 4, *copyfile* improved total execution time, on average, about 16.5%, reduces CPU system time about 39.1% and CPU user time about 99.4% compared with *cp* operation. And *recvfile* enhanced total execution time about 3.5%, reduced CPU system time about 28% and CPU user time about 99.4% compared with *recv/write* operation. In case of CPU user time in *copyfile* and *recvfile*, the cost nearly came close to zero.

So, supposed zero-copy data path could enhance the CPU usage rate by removing CPU copies between kernel and user space and by avoiding unnecessary system calls. And it improved the whole system performance by bringing down the cost per operation. Additionally, it can allow CPU resource to be allocated to another process by improving CPU availability and can service the larger number of requests with same CPU cost.

References

1. Pal Halvorsen, Tom Anders Dalseng, Carsten Griwodz.: Assessment of Data Path Implementations for Download and Streaming. DMS'2005 (2005)
2. Dragan Stancevic.: Zero copy I: User-mode Perspective. Linux Journal, Issue 105 (2003)
3. Sotiropoulos, A., Tsoukalas, G., Koziris, N.: Efficient Utilization of Memory Mapped NICs onto Clusters using Pipelined Schedules. Cluster Computing and the Grid (2002) 238-238
4. Halvorsen, P., Jorde, E., Skevik, K.-A., Goebel, V., Plagemann, T.: Performance Tradeoffs for Static Allocation of Zero-copy Buffers. Euromicro Conference (2002) 138-143
5. Skevik, K.-A., Plagemann, T., Goebel, V., Halvorsen, P.: Evaluation of a Zero-copy Protocol Implementation. Euromicro Conference 2001, Proceedings (2001) 324-330

6. Pal Halvorsen, Thomas Plagemann, Vera Goebel.: Improving the I/O Performance of Intermediate Multimedia Storage Nodes. Multimedia Systems, Vol. 9, No 1 (2003) 56-67
7. Goldenberg, D., Kagan, M., Ravid, R., Tsirkin, M.S.: Zero-copy Sockets Direct Protocol over Infiniband-preliminary Implementation and Performance Analysis. High Performance Interconnects 2005, Proceedings (2005) 128-137
8. Amol Shukla, Lily Li, Anand Subramanian, Paul A. S. Ward, Tim Brecht.: Evaluating the Performance of User-space and Kernel-space Web Servers. Proceedings of the 2004 conference of the Centre for Advanced Studies on Collaborative research (2004)
9. Vivek, S., Pai, Peter Druschel., Willy, Zwaenepoel.: IO-Lite : A unified I/O Buffering and Caching System. ACM Transactions on Computer Systems, Vol.18, No.1 (2000) 37-66
10. 10 Charles, D., Cranor and Gurudatta M. Parulkar.: The UVM Virtual Memory System. USENIX Annual Technical Conference. Monterey CA USA (1999) 117-130
11. Piyush Shivam, Pete Wyckoff, Dhabaleswar Panda.: EMP: Zero-copy OS-bypass NIC-driven Gigabit Ethernet Message Passing. Proceedings of the 2001 ACM/IEEE conference on Supercomputing (2001)
12. Sejin Park, Sang-Hwa Chung, Bong-Sik Choi, Sang-Moon Kim.: Design and Implementation of an Improved Zero-Copy File Transfer Mechanism. PDCAT. Lecture Notes in Computer Science, Volume 3320 (2004) 446-450
13. Sotiropoulos, A., Tsoukalas, G., Koziris, N.: Efficient Utilization of Memory Mapped NICs onto Clusters using Pipelined Schedules. Cluster Computing and the Grid 2nd IEEE/ACM International Symposium (2002) 223-231
14. Yun-Chen Li, Mei-Ling Chiang.: LyraNET: A Zero-Copy TCP/IP Protocol Stack for Embedded Operating Systems. Embedded and Real-Time Computing Systems and Applications (2005) 123-128
15. Shinichi Yamagiwa, Keiichi Aoki, Koichi Wada.: Active zero-copy: A performance Study of Non-deterministic Messaging. Proceedings of the 5th International Symposium on Parallel and Distributed Computing (2005)
16. Xu Xiaofei, Ling Yi, Kang JiChang.: The Research on Zero-copy Receiving Method Based on Communication-page Pool. Parallel and Distributed Computing, Applications and Technologies (2003) 416 - 419
17. Tezuka, H., O'Carroll, F., Hori, A., Ishikawa, Y.: Pin-down Cache: A Virtual Memory Management Technique for Zero-copy Communication. IPPS/SPDP, Proceedings (1998) 308-314
18. Dong-Jae Kang, Kang-Ho Kim, Sung-In Jung, Hae-Young Bae.: TCP/IP Offload Engine Module Supporting Binary Compatibility for Standard Socket Interfaces. Grid and Cooperative Computing - GCC2005, LNCS, Volume 3795 (2005) 357-369
19. http://sourceforge.net/projects/zero-copy

Virtual Hierarchy Synthesis for Hybrid Mobile Ad Hoc Networks

Hyemee Park, Tae-Jin Lee, and Hyunseung Choo*

School of Information and Communication Engineering
Sungkyunkwan University 440-746, Suwon, Korea
Tel.: +82-31-290-7145
{hyemee, tjlee, choo}@ece.skku.ac.kr

Abstract. The interconnection of mobile ad hoc networks to fixed IP networks is one of the topics receiving more attention within the MANET working group of the IETF. In such integrated scenarios, commonly known as hybrid ad hoc networks, mobile nodes are witnessed as an easily deployable extension to the exiting infrastructure. Some ad hoc nodes act as *gateway* that can be used by other nodes to seamlessly communicate with hosts in the fixed network. Therefore, this research brings up several issues regarding Internet gateway discovery and address auto-configuration to be routable to the fixed Internet. In this paper, we focus on these elements to guarantee smooth interworking. The proposed Virtual Hierarchy Synthesis (VHS) scheme provides the efficient gateway discovery and optimal routing protocol are suitable in high mobility hybrid MANETs.

1 Introduction

The 1990s have seen a rapid growth of research interests in mobile ad hoc networking. The infrastructureless and the dynamic nature of these networks demands new set of networking strategies to be implemented in order to provide efficient end-to-end communication. This, along with the diverse application of these networks in many different scenarios such as battlefield and disaster recovery, have seen MANETs being researched by many different organizations and institutes. One interesting research area in MANET is routing. Routing in the MANETs is a challenging task and has received a tremendous amount of attention from researches. This has led to development of many different routing protocols for MANETs, however, most work has been concentrated on stand-alone ad hoc networks. Not much work has been done concerning the integration of ad hoc networks and the Internet. It is limited in supporting connectivity between communicating mobile nodes. Currently, users with portable devices in ad hoc networks want to access useful information on the Internet. In particular, the data collected from ad hoc devices is required to avail in central systems connected to the Internet for various purposes. Therefore, the interconnection of MANETs to fixed IP networks is increasingly important.

* Corresponding author.

M. Gerndt and D. Kranzlmüller (Eds.): HPCC 2006, LNCS 4208, pp. 360–369, 2006.

Mobility of user devices connecting to the Internet is of major interest in today's research in networking. The Mobile IPv6 [4] provides an important global mobility solution. Since nodes in ad hoc networks are inherently mobile, it seems inevitable that some of these nodes are likely to move between different ad hoc networks and to other parts of the Internet as well. Hence, we focus on connecting MANETs to global IPv6 networks, while supporting the mobility of ad hoc nodes by integrating Mobile IPv6 and MANET. In such integrated scenarios, commonly known as hybrid ad hoc networks, mobile nodes can easily be deployed and expanded through existing infrastructures. For seamless communications with Internet hosts, some ad hoc nodes act as gateways to be traversed by other nodes. The gateway discovery mechanism has an impact in terms of overall performance, and it is a key component in providing interoperability with fixed networks. Furthermore, connecting an ad hoc network to the Internet brings up several issues regarding routing and how to provide nodes in an ad hoc network with IP addresses that are routable to the fixed Internet. These are the most relevant elements to guarantee smooth interworking.

During the last year a number of schemes have been proposed to solve these challenges, however, they suffer from the limitation that a handoff overhead occurs from inefficient gateway discovery and addressing protocol in high mobility MANETs. In the proposed Virtual Hierarchy Synthesis (VHS) scheme, MANETs configure a virtual tree topology for reducing the overhead for the gateway discovery. To perform this procedure, the control message is propagated from the gateway to overall network and this message contains the network prefix information other than the gateway information. That is, ad hoc nodes auto-configure its IP address and discovers the gateway from this message. As the prefix delegation is applied, all nodes share a common global network prefix. It allows that routing overhead can be reduced due to avoid ingress filtering and unique address for each MN can be easily auto-configured. By integrating the addressing auto-configuration information into gateway discovery messages, the overall overhead is reduced. MANETs based on tree provides the optimal routing protocol for access to the Internet and reduces a handoff cost as well. By the comprehensive computer simulation, we compared the proposed scheme with existing schemes in terms of handoff delay and signaling overhead. Simulation results show that the newly proposed scheme has better performance than others.

The rest of this paper is organized as follows. In Section 2, related works are introduced with some discussions. Section 3 presents the proposed VHS scheme integrating Mobile IPv6 and MANET. The performance of the proposed scheme is evaluated in Section 4. Finally, we conclude in Section 5.

2 Related Works

There is a number of proposals in the literature to provide Internet connectivity for MANETs. One of the first proposals by Broch *et al.* [5] is based on integration of Mobile IP and MANETs employing a Dynamic Source Routing (DSR) [1]. Jonsson *et al.* [6] propose a method, called MIPMANET, to connect an ad hoc network to the Internet using Mobile IP with Foreign Agent (FA)'s

Care-of Address (CoA) and reverse tunneling. MIPMANET combines the use of Mobile IP protocol and Ad Hoc On-Demand Vector (AODV) [2]. Almmari *et al.* [7] analyzed the performance of mobile gateways in a MANET based on the Destination-Sequenced Distance Vector (DSDV) [3] routing protocol. Since these proposals are based on existing protocols, they are limited in supporting all routing protocols and do not provide scalability in heterogeneous network environments. The gateway discovery function of them is done proactively or reactively by using a routing protocol. It suffers from the flooding overhead for rediscovering a default route toward the gateway whenever handoff is performed. In addition, addressing protocol of these schemes cannot solve the ingress filtering problem in MANETs with multihop routing. It incurs the routing overhead by reverse tunneling, since the packet size is increased in each hop.

Hwang *et al.* [8] propose a self-organizing, self-addressing, and self-routing IPv6-enabled MANETs infrastructure. MANETs automatically organize nodes into tree architecture for self-organizing addressing protocol. When a new node joins the MANET tree, it receives a unique logical address, which represents its location of tree. The node uses the logical address as its 64 bit interface ID when configuring its global IP address. The proposed routing protocol efficiently reduces the flooding overhead by utilizing the default route between the parent and child in MANET tree topology. In order to make the routing efficient and to maintain the tree structure, each node regularly broadcasts control messages to its parent and children. This scheme allows mobile nodes to use the network prefix advertised by the gateway. However, it does not address how to propagate the network prefix of the gateway in MANETs. As mobile nodes configure its global IP using logical address of tree, the scheme does not provide an efficient mechanism to configure and manage a mobile node's address under high mobility. Although a node changes its point of attachment within the network, it need to reconfigure its address and perform Binding Update (BU) at its Home Agent (HA) and Corresponding Node (CN) as well. Furthermore, this scheme requires an additional function and overhead to organize and maintain the tree overlay.

3 Proposed Scheme

3.1 Virtual Hierarchy Synthesis (VHS) Scheme

In this paper, we consider the special property of the MANETs with the dynamic multihop topology and focus on the issue that the protocol overhead should be kept at minimum due to the scarcity of resources. To reduce the routing overhead by ingress filtering, the prefix delegation mechanism is proposed. This technique is also used to create virtual trees which are dynamically maintained and updated when unpredictable topology changes occur. In this section, we describe how to form and maintain a virtual tree with minimum overhead, and provide the optimal gateway discovery and efficient addressing mechanism based on the tree topology.

To propagate the prefix to overall networks, the gateway advertises the Router Advertisement (RA) message containing its global prefix. It is assumed that Access Router (AR) and MANET nodes operate under the IPv6 protocol, therefore all entities broadcast periodically the RA message to their neighbors. According to using the property, the solution uses the RA message, so that nodes can simply relay the prefix without a special control message or operation. Unfortunately, the multi-hop nature of an ad hoc network makes it impossible to use the advertisement message defined in IPv6. Therefore, the original RA message is extended to propagate over multiple hops through the ad hoc network and to share a common global network prefix using MANET nodes. Fig.1 presents the format of the modified prefix information option.

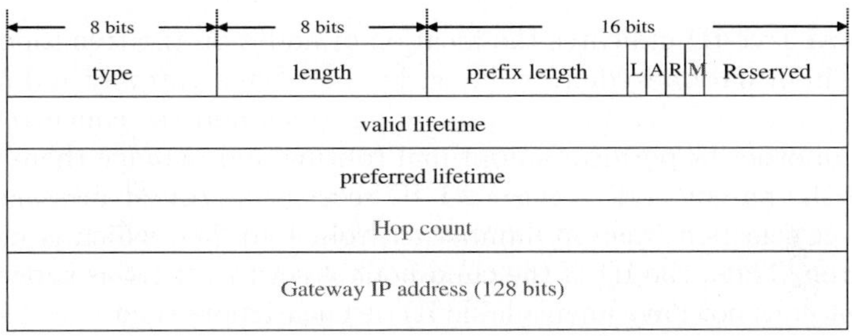

Fig. 1. Prefix information option

- The M flag means that this message is relayed over multi-hops.
- Hop count is the distance (in hops) from the gateway to the sender node.
- The IP address of the gateway is contained and the prefix length field represents the length (in bits) of the prefix part of this address.

And the initial distance transmitted by the gateway must be zero and set the M flag. When a node receives this message, it first generate a global address with the prefix advertised by a gateway and it's EUI-64 (Ethernet Unique Identifier) as the interface ID. Then, it increments the hop count one and forwards an updated version of the RA message to its neighbors. The gateway information contained in the RA message is therefore propagated in a hop-by-hop manner, until all nodes of the ad hoc network share the gateway IP and global network prefix. Prefix delegation is a method where all nodes of the MANET share the same prefix advertised from the AR, to reduce the tunneling overhead according to ingress filtering.

The proposed prefix propagation method leads to the creation of the virtual tree topology of the ad hoc network. Each virtual tree is rooted at a gateway, and it is formed by nodes using the global network prefix advertised by the gateway. In the proposed scheme, the AR acts as a gateway. We consider that a MANET connects to the Internet via an AR. The AR is a method of supporting Internet connectivity at the point of attachment between the Internet and MANET nodes. This overcomes the limitation that a MANET node with low power plays the role of a gateway.

In order to configure the virtual tree topology, each node tries to establish a parent-child association with its neighbor using 3-way handshake. Control packets of the procedure include the RA message disseminated for prefix delegation, PARENT REQUEST, and PARENT REQUEST ACK. When a node receives the RA message from a neighbor, it generates its global IP address with the prefix of the RA message. Then, it sends back a PARENT REQUEST to notify that it is selected as a parent node. When the parent receives the PARENT REQUEST from the child node, it selects the address which is not assigned yet in its child table and sends with the response message. As the parent node sends the ACK with the Tree ID as a logical address, parent-child association is established.

The logical Tree ID indicates the location of nodes on the tree topology. This is a routing information to deliver the packet between a gateway and nodes. The parent keeps the Tree ID and corresponding MN's address concurrently in its child table in order to perform an optimal routing and manage them easily. For example, if the parent node using 1.2.1 ID receives a request message from the child node, it selects a random number(x) from 1 to 255, which is not used by other children. Then, the ID of the child node is set to 1.2.1.x as shown in Fig.3. If the parent does not have an available ID, it must ignore the request so that the child chooses another parent. However, if the child moves away, it releases the resource and the Tree ID used by the child and deletes the entry of the child in its table. The reason why the Tree ID is not used as a global address of a node, is to consider the high mobility in MANETs. When an ad hoc node changes its point of attachment within the MANET tree, it must replace the global address based on Tree ID. It requires significant cost with regard to handoff. The Tree ID which is separated from the global IP addresses is proposed, in order to optimize routing in busy mobile environments.

A node may receive one or more RA messages from its neighbors. In this case, it selects the most appropriate node as a parent from one of the messages and sends a PARENT REQUEST back. In the proposed scheme, the desirable algorithm for selecting the parent node is that a node keeps its current prefix as long as it has neighbors with the same prefix, *i.e.*, until it cannot find a neighbor that uses the same network prefix. The main advantage of this algorithm is that they minimize the number of prefix changes. This greatly reduces the overhead induced by the sending of BU messages when a node changes its global address. And, a node chooses the parent node that advertises the shortest distance to a gateway, to maintain the shortest path. This ensures that one node does not concurrently generate multiple associations and avoids a tree loop. When a child node is established with a parent node, it ignores the RA message transmitted by other neighbors as long as it maintains the association.

In order to maintain the virtual tree topology, each node periodically checks its neighborhood. The nodes can detect the loss of its parent or child nodes using RA and Router Solicitation (RS) in the Neighbor Discovery Protocol (NDP) [9] of IPv6. If a child node does not receive the RA message from its parent within a predefined time, it assumes that its parent node has moved away. In this case, as the node transmits a RS message to its neighbors, it should restart a 3-way handshake procedure to select another parent. However, if a parent node does not receive the RA message from its child and the timer associated to it is expired, the parent node releases the resource and logical address assigned to the child. Therefore, in the proposed scheme, the RA message disseminated by a node allows its parent and children to simultaneously detect its existence.

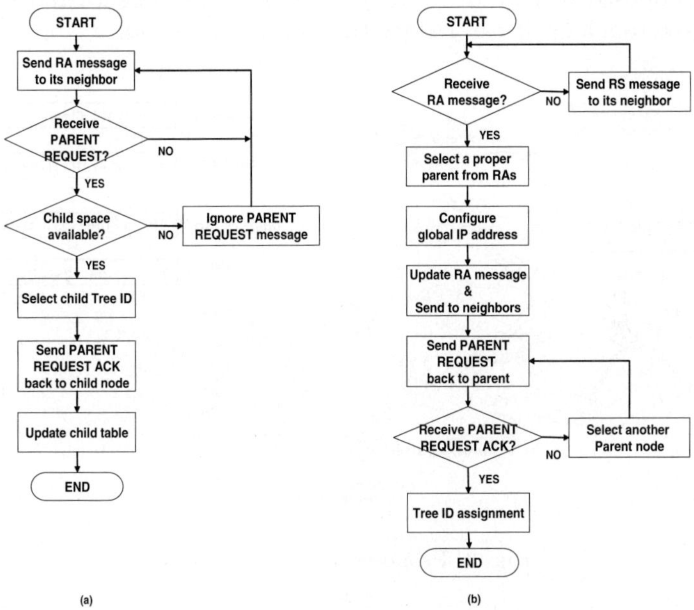

Fig. 2. Flow chart of VHS scheme(a) New node joins in network, and (b) Neighbors response to assign Tree ID

3.2 Integration with Internet Routing

We propose the routing protocol based on the virtual tree topology for communication with the Internet hosts. All traffic between ad hoc nodes and Internet hosts are forwarded along the path configured by parent-child association. When the MANET node wants to transmit the packet to the Internet host, it transmits the data packets to the gateway using the IPv6 routing header. The extended routing header contains the final destination address, *i.e.*, the address of the Internet host, and the destination field of the IPv6 header contains the gateway address. If intermediate nodes receive the packet with the destination field set to the gateway IP, they forward it to their parent node recursively. Only an ad hoc node with an IP address contained in the destination field of an IPv6 header can examine the routing header of this packet. Once the packet arrives at the gateway, the packet uses the address of the routing header as the final destination address. The modified packet is then forwarded to the Internet host.

In contrast, if the gateway receives the packet from the Internet host to the ad hoc MN, it adds the hop-by-hop option to the original IPv6 packet. It first searches its cache for the corresponding Tree ID of the destination address (MN's IP). Then, the Tree ID is inserted into the hop-by-hop option. As a result, intermediate nodes check this option for routing and deliver the packet to the child node selected by the longest prefix matching algorithm. This algorithm is used to determine how to forward a packet to its destination using the Tree ID. Since the proposed scheme uses the Tree ID and the parent-child relationship based on tree topology, it can reduce the routing overhead.

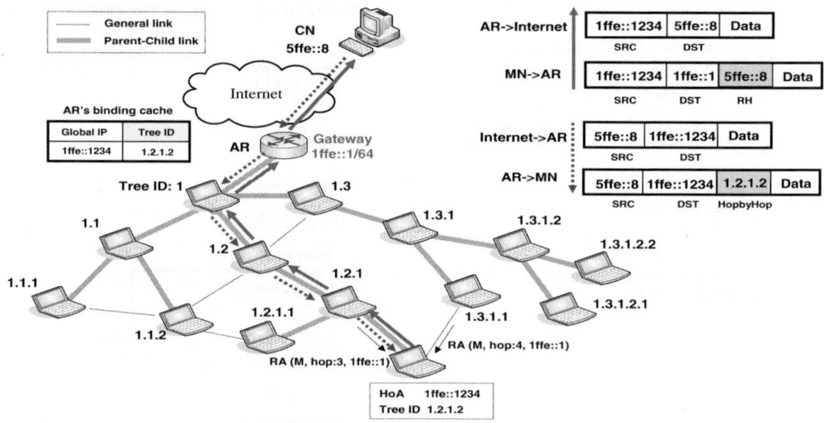

Fig. 3. Proposed routing protocol

3.3 Integration with Mobile IPv6

The proposed scheme supports seamless mobility of MANET nodes by integrating the Mobile IPv6 and ad hoc network. We consider that the global address acquired by an ad hoc node should be used as the Mobile IPv6 CoA of the node. Each change of global address in the ad hoc network will trigger the sending of at least one BU message. However, when the Tree ID changes, a new Local BU (LBU) between gateway and MN is used to immediately provide an optimal routing.

If the ad hoc MN moves from another MANET or changes its point of attachment within the network, it performs a 3-way handshake procedure to reconfigure the tree structure as it can no longer contact with its parent node any more. As a result, it has a new Tree ID. Then, it transmits the LBU message to the gateway in order to update the new ID. As the receiving gateway rewrites its cache entry, the gateway can insert the exact information in the hop-by-hop option when it is added to the packet transmitted by the CN. As it performs a light-weight BU procedure with the gateway, it does not require the route rediscovery mechanism of previous schemes between the gateway and the MN. Therefore it can reduce the flooding overhead of the proactive or reactive protocol.

4 Performance Evaluation

In this section, the performance of the proposed scheme is evaluated in comparison with integrating the Mobile IPv6 and existing routing protocol, AODV, DSR, DSDV, and MANETs based on tree topology. In order to evaluate these schemes, simulation implemented in C is conducted. Our simulation models a wireless network of 25 $\sim$ 100 mobile nodes placed randomly within a $150m \times 100m$ area. Radio propagation range for each node is $30m$. Using the simulation parameter in Table 1, we measure the total delay required when a node performs handoff and the overhead of the gateway discovery. Performance is evaluated under various mobility rates and numbers of nodes, to analyze the performance of mobility and node density effect.

Table 1. Simulation Parameters

Bit rates		**Processing time**	
Wire links	100 *Mbps*	MANET Nodes	0.01 *msec*
Wireless links	10 *Mbps*	Gateway, HA	0.01 *msec*
Propagation time		BU	0.01 *msec*
Wire links	500 μsec	NDP	0.01 *msec*
Wireless links	2 *msec*	**Average number of hops**	
Data size		Wire links	10
Message size	256 *bytes*	Wireless links	1

4.1 Impact of the MANET Node Mobility

The average total delay of the proposed and other schemes is first compared for various handoff rates. This simulation varies the handoff rates between 0.1 and 0.9. The network has 50 nodes. In addition, a node moves from another network or changes its point of attachment within the network, and it is placed randomly. The total delay is the summation of transmission, propagation and processing time in each hop between two end nodes. The metric calculates the delay of all procedures, such as NDP, BU, route discovery or routing table update, and tree reconfiguration *et al.*, occurring after handoff. When a node moves within a network, it first receives the RA message from the new neighbors. Therefore, it can detect its movement and perform the procedure to rediscover a default path toward the gateway. However, if a node moves from another network, the BU delay is added.

Fig. 4(a) presents the simulation results for the variation of handoff ratios. As presented in Fig. 4(a), our proposed scheme has better performance than other schemes. In the reactive protocol of AODV and DSR, the route discovery protocol is conducted other than NDP, BU procedures when a node performs handoff. Therefore, the RREQ message is forwarded by all nodes until the gateway is discovered and RREP is then delivered by the gateway. However, the proactive protocol of DSDV updates its routing table by adding a new node entry. Though the route discovery procedure is not required, it must transmit table information

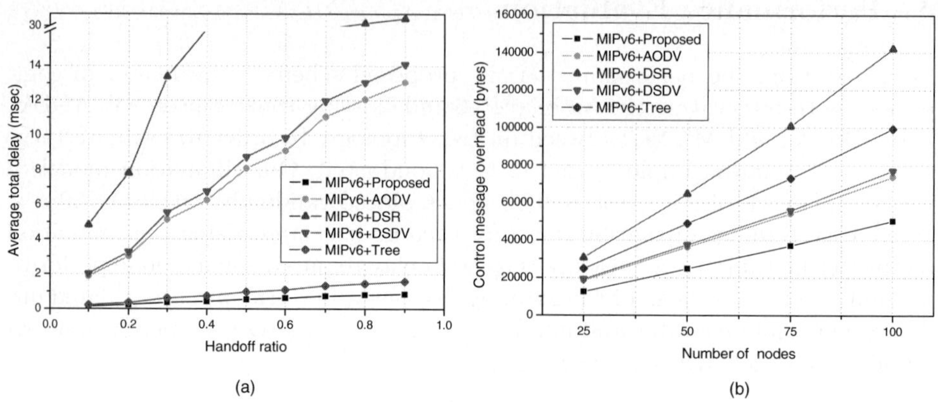

Fig. 4. Simulation results of four schemes (a)Handoff delay, and (b) Gateway discovery overhead

to all neighbors in order to notify a new node. The tree based schemes are not required to perform the route discovery or table update and only performs the procedure to relate with a new parent in tree topology. However, our scheme reduces the handoff delay than previous tree based scheme [8] as mobility rate is increasing, since it suffers from the BU storms by frequently changing of node's global address. Therefore, our proposed scheme is not significantly affected than other schemes, as the handoff is more frequent.

4.2 Impact of the Density of MANET

The overhead of the gateway discovery is evaluated for various network sizes in each scheme. In this simulation, the control message overhead is defined as the total bytes of the control message when a network is configured and initially attempts to connect with the Internet. That is, it is the sum of all control messages delivered until the gateway discovery is completed. We measure the overhead of 3-way handshake, route discovery and table update of these schemes as the network size is increasing. The simulation result is presented in Fig. 4(b). The result represents that the proposed scheme can reduce the control overhead in large scale networks. In particular, the previous work based on tree topology has higher control overhead than our scheme, since it requires an additional overhead to organize and maintain the tree overlay. Our proposal unicasts a PARENT REQUEST and ACK message of 3-way handshake by including the RA message, while previous scheme broadcasts most of the control messages. By integrating the addressing information into gateway discovery messages, the overall overhead of the proposed scheme is reduced.

5 Conclusion

In this paper, we present a protocol that builds a virtual trees, where each tree if formed by nodes that share a common global network prefix. By using the tree topology, an efficient gateway discovery and optimal routing is proposed in high mobility MANETs. In addition, this connectivity method is not dependent on a particular routing protocol. Therefore, it provides the optimized communication as well as scalability in heterogeneous network environments and backward compatibility with existing mechanisms.

Our simulation represents advantages of the proposed scheme compared to integrating MIPv6 and the existing routing protocols - AODV, DSR, DSDV, and tree based MANET. According to the simulation results, our newly proposed scheme shows up to about 50% of performance improvements in comparison with other schemes. Thus our solution provides Internet connectivity of ad hoc nodes with minimum routing overhead and delay in large-scale, high mobility hybrid MANETs.

Acknowledgment

This research was supported by Ministry of Information and Communication, Korea under ITRC IITA-2005-(C1090-0501-0019) and grant No. R01-2006-000-10402-0 from the Basic Research Program Korea Science and Engineering Foundation of Ministry of Science & Technology.

References

1. D. Johnson, D. Maltz, and J. Jetcheva, "The dynamic source routing protocol for mobile ad hoc networks," IETF, draft-ietf-manet-dsr-10.txt, July 2004.
2. C. E. Perkins, *et al.*, Ad hoc on-demand destance vector (AODV) routing," IETF, RFC 3561, July 2003.
3. C. E. Perkins, and T. J. Watson, "Highly dynamic destination-sequenced distance vector routing (DSDV) for mobile computers," ACM SIGCOMM'94, October 1994.
4. D. Johnson, C. E. Perkins, and J. Arkko, "Mobility Support in IPv6," IETF, RFC 3775, June 2004.
5. J. Broch, D. A. Maltz, and D. B. Johnson, "Supporting Hierarchy and Heterogenous Interfaces in Multi-Hop Wireless Ad Hoc Networks," IEEE International Symposium on Parallel Architectures, algorithms and Networks, June 1999.
6. U. Johnsson, *et al.*, "MIPMANET-Mobile IP for mobile ad hoc networks," IEEE Proc. of MobiHoc'00, August 2000.
7. H. Almmari and H. El-Rewini, "Performance Evaluation of Hybrid Environments with Mobile Gateways," 9th International Symposium on Computers and Communications, June 2004
8. R. H. Hwang, *et al.*, "Mobile IPv6-Based Ad Hoc Networks: Its Development and Application," IEEE JSAC, Vol. 23, No. 11, November 2005.
9. T. Narten, E. Nordmark, and W. Simpson "Neighbor Discovery for IP Version 6 (IPv6)," IETF, RFC 2461, December 1998.

Design and Analysis of High Performance TCP

TaeJoon Park[1], JaeYong Lee[2], and ByungChul Kim[2]

[1] Carrier Class Ethernet Team, ETRI,
161 Gajong-Dong, Yuseong-Gu, Daejeon, 305-350, Korea
tjpark@etri.re.kr

[2] Department of Infocom Engineering, Chungnam National University
220 Gung-Dong, Yuseong-Gu, Daejeon, 305-764, Korea
{jyl, byckim}@cnu.ac.kr

Abstract. Traditional TCP implementations have an under-utilization problem in high bandwidth delay product networks. This paper proposes a new congestion control mechanism, a high performance TCP (HP-TCP), to solve the under-utilization problem. The congestion avoidance period of the HP-TCP control is divided into linear and exponential growth phases, where the linear increase phase is similar to that of the legacy TCP; when there is no queueing delay in the linear increase phase, the congestion window grows exponentially to fill a large pipe quickly. The exponential increase phase can cause serious problems of overshooting the network capacity, which results in massive retransmissions and low bandwidth utilization. To solve this problem, the proposed algorithm uses the RTT status and the estimated bandwidth to prevent packet losses during the exponential growth phase. The simulation results show that the HP-TCP improves the convergence time and throughput performance of the TCP in high bandwidth delay product networks.

1 Introduction

In Grid environments, access to distributed data is very important, as well as access to distributed computational resources. Distributed scientific and engineering applications require the transfer of large amounts of data (terabytes or more) between storage systems in geographically different locations for analysis, visualization, and so on [1][2]. While the legacy TCP is the most commonly used and reliable transport protocol in the Internet, it is generally accepted that it is not suitable for massive data transfers such as Grid applications. Since the TCP congestion control algorithm is not dynamic enough for high bandwidth delay product (HBDP) networks, the packet drop rate required to fill a gigabit pipe using the current TCP protocol is beyond the limits of the currently achievable fiber optic error rates.

In this paper, we propose a modified TCP congestion control mechanism that can provide efficient data transfer in HBDP networks. The proposed congestion control mechanism has not only linear, but also exponential growth phases in the congestion window. When there is no queueing delay and it satisfies a certain condition, the congestion window grows exponentially during the congestion

M. Gerndt and D. Kranzlmüller (Eds.): HPCC 2006, LNCS 4208, pp. 370–379, 2006.

avoidance period. Otherwise, it maintains a linear growth in the congestion window similar to the legacy TCP congestion avoidance algorithm. Based on the relationship of the current and minimum round trip time (RTT), the proposed method selects between the linear and exponential growth phases of the congestion window update during the congestion avoidance period. To prevent packet loss during the exponential growth phase, our method uses not only the end-to-end delay information, but also the estimated bandwidth of the bottleneck node. In addition, the proposed method maintains a TCP-compatibility property in ordinary small BDP networks. The simulation results show that the proposed mechanism prevents overshooting during the exponential growth phase and improves the performance and fairness of the TCP in HBDP networks.

The remainder of the paper is organized as follows. Section 2 describes the related studies in this area; section 3 discusses the proposed mechanism. Section 4 discusses the analytic model for the HP-TCP. In Section 5, the simulation results are shown and discussed, and the paper is concluded in Section 6.

2 Related Work

The traditional TCP, i.e. TCP-reno, is compliant to the RFC 793 [3] and RFC 2581 [4] IETF standards. During the initial start up phase (slow start), the standard TCP exponentially increases the amount of transferred data until detecting either a packet loss by a timeout or triple duplicate ACKs. When a loss is detected, the TCP halves the congestion window (*cwnd*), which constrains the number of packets to be sent without acknowledgement, and moves into the congestion avoidance phase. During the congestion avoidance phase, the TCP increases the *cwnd* by one packet per window and halves the window for a packet drop. Thus, we call the TCP's congestion control algorithms Additive Increase Multiplicative Decrease (AIMD) algorithms. However, the AIMD control of the current TCP is not dynamic enough to fill a large pipe for large BDP networks. For example, for a standard TCP connection with 1500-byte packets and a 100 ms RTT, achieving a steady-state throughput of 10 Gbps would require an average congestion window of 83,333 segments and a maximum packet drop rate of one congestion event for every 5,000,000,000 packets [5]. The average bit error rate of (2×10^{-14}) is needed for full link utilization in this environment; however it is an unrealistic requirement for the current network technology.

Over the past few years, research has shown attempts to solve the under-utilization problem of the legacy TCP in HBDP networks [5][6] and can be classified into two main groups: loss-based and delay-based solutions. The loss-based solutions focus on a modification of the increase and decrease parameters of the TCP-AIMD algorithm, i.e. HS-TCP [5] . The loss-based solutions use packet loss as the only indication of congestion. However, these protocols have some drawbacks for deployment in terms of convergence times, compatibility, fairness, and so on [7][8]. The delay-based solutions propose the queueing delay as a congestion measure, i.e. FAST-TCP [6]. The delay-based solutions reduce or increase the transmission rate based only on RTT variations. However, when competing with other loss-based solutions, the delay-based solutions are penalized due to the aggressive nature of the loss-based solutions [9].

3 Congestion Control Mechanism of the High Performance TCP

The suggested TCP, HP-TCP, modifies the standard TCP's congestion control mechanism used in TCP connections with large congestion windows. The legacy TCP congestion control algorithm is composed of Slow Start (SS) and Congestion Avoidance (CA) phases. Since the linearly increasing TCP congestion avoidance algorithm is not dynamic enough, the packet drop rate required to fill a gigabit pipe using the current TCP protocol is beyond the limits of the currently achievable fiber optic error rates.

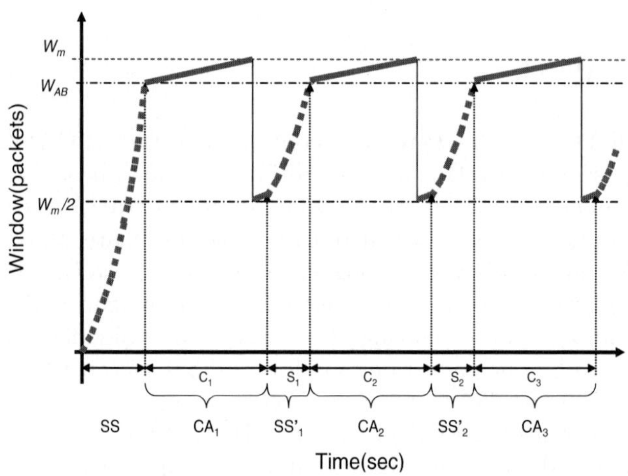

Fig. 1. Evolution example of window size including the startup phase in the proposed HP-TCP

The proposed congestion control mechanism can be divided into three phases: SS, CA, and SS'. The SS phase is the exponential growth phase; the CA phase is the additive increase phase; and the SS' phase is the exponential growth phase only activated in HBDP networks after the CA phase. Figure 1 shows an example evolution of the congestion window of the proposed HP-TCP.

The proposed congestion control mechanism has provided a solution to prevent overshooting in the SS phase and a fast bandwidth reserving solution in the CA phase to switch over to the SS' phase.

3.1 The SS Phase

The legacy TCP's SS phase is suitable for fast bandwidth reservations. To avoid a premature exit from the SS phase (Fig. 1) and to increase utilization, the initial *ssthresh* can be increased. However, a large initial *ssthresh* can cause slow start overshooting problems, multiple packet losses, and further reduce the utilization. The HP-TCP, like TCP-vegas, prevents overshooting during SS by only allowing exponential growth every other RTT. While TCP-vegas' congestion avoidance mechanism during the initial SS period is quite effective, it can still overshoot the available bandwidth, and it depends on sufficient buffering at the bottleneck router to prevent loss until realizing that it needs to slow down. To compensate for the weak points, we limited the maximum size of the *cwnd* in the SS phase by resetting *ssthresh* to the estimated bandwidth of the bottleneck node and adapting the pacing method to limit the burstness of sending packets. Limiting *ssthresh* to the available bandwidth can limit the fairness in TCPs that use losses for congestion verification. Therefore, in order to use a method that can estimate the bottleneck node's most available bandwidth, we use a bottleneck bandwidth estimation scheme to determine the *Pipesize*, which is defined as [10]:

$$Pipesize = \frac{delay \cdot bandwidth}{packet\_size} = \frac{delay}{packet\_spacing} \tag{1}$$

Usually, most problems in the delay-based congestion control methods, like TCP-vegas, are due to the reduction of *cwnd* depending on the delay. However, the proposed method only decides the *cwnd* switching point from exponential growth to additive increment and allows bandwidth competition using congestion control, as in the legacy TCP; thus, it minimizes problems in the legacy *cwnd* control that uses delay-based methods.

3.2 The CA Phase

In the TCP Reno, the congestion avoidance phase, CA (Fig. 1), starts when the congestion window exceeds the slow start threshold *ssthresh* or a packet loss is detected with triple duplicate ACKs. The start condition of the CA phase in the HP-TCP is the same as that of the TCP-reno. However, to manage the under-utilization problem during the CA phase in HBDP networks, we adopt an exponential growth of congestion window in the CA phase, called the SS' phase. In the HP-TCP, if the RTT does not increase for the predefined *waiting_time* and cwnd is smaller than the *Pipesize*, the CA phase switches to the SS' phase to drastically increase the congestion window and utilize the larger available bandwidth. The *waiting_time* is defined as:

$$waiting\_time = log_2(cwnd\_max) \times n \tag{2}$$

where *cwnd_max* is the maximum *cwnd* in the current CA phase.

As *waiting_time* is a function of *cwnd*, it can be used to control fairness; the larger the *cwnd*, the longer the *waiting_time*. Therefore, the connection of a larger *cwnd* delays the start of the SS' phase. Since the modified SS phase can quickly approach the achievable available bandwidth, the linear increasing rate of the *cwnd* does not need to be greater than that of the ordinary TCP. Hence, the increasing rate during the CA phase is the same as that of the legacy TCP.

The queueing duration, the CA period, is determined by the buffer capacity and the incoming/outgoing traffic rate. If the RTT exceeds the predefined threshold or the *cwnd* exceeds the estimated Pipesize during the SS' phase, the phase switches back to the CA phase again and the *cwnd* increases linearly until a packet loss is detected. In the event of a packet loss, the congestion widow size is halved for fairness; thus, the HP-TCP is compatible with the legacy TCP.

```
When an ACK for a new packet arrives;
If(cwnd > thresh_window){
   If (state == SS){
      // doubles its cwnd only every other RTT.
      If (RoundNumber == even)
         cwnd=cwnd +1;
      If ((RTT > BaseRTT) || (cwnd > Pipesize))
         state = CA;
   } else If (state == CA){
      cwnd = cwnd + 1/cwnd;
      If ((RTT ≤ BaseRTT) && (cwnd < Pipesize)) {
         waiting_time_counter ++;
         If(waiting_time== waiting_time_counter)
            state = SS';
      } else
         wait_time_counter = 0;
   } else If (state == SS'){
      If (ACK_count ≧ cwnd_base)
         cwnd=cwnd +1;
      If ((RTT > BaseRTT) || (cwnd > Pipesize)) {
         waiting_time = log2(cwnd_max)*n;
         state = CA;}
   }
}
When triple-duplicate ACKs arrive;
cwnd = cwnd / 2;
waiting_time = log2(cwnd_max)*n;
state = CA;
```

Fig. 2. Pseudo-code for the proposed congestion control

3.3 The SS' Phase

The SS' phase is similar to the initial slow start phase of the HP-TCP; however, the start time of the exponential growth is controlled by the combination of the *waiting_time*, RTT status, and estimated Pipesize of the connection. The start value of the exponential growth, *cwnd_base*, is the final *cwnd* in the CA phase and the increment, Incr, is given by (3).

$$Incr = \lfloor Max\{cwnd(i) - cwnd\_base, 0\}\rfloor / \lfloor cwnd(i) \rfloor \tag{3}$$

where i is the round number during the SS' phase. For example, when *cwnd_base*= K, the *cwnd* of the first round becomes $K+1$; it becomes $K+2$ in the second round, $K+4$ in the third, $K+8$ in the fourth, and so on.

The exit condition of the SS' phase in the HP-TCP is the same as that of the SS phase. Figure 2 shows the pseudo-code of the proposed mechanism. Only when the congestion window is larger than thresh_window, the proposed mechanism is used; otherwise, the normal TCP algorithm is used to maintain backward compatibility with the legacy TCP.

4 Analytic Model for the HP-TCP

In order to analyze the performance of the HP-TCP in a steady state, we used a stochastic model of TCP congestion control with parameters of loss rate (p) and average round trip time (RTT). We assumed that there was no ACK drop, and an unlimited amount of data to send in the sender; also, the congestion window size was not limited by the advertised receiver window size. The steady state TCP throughput $X(p)$ for the loss rate p is defined as the ratio of the total number of bits sent to the total transmission time. We define a "period" as the time between two adjacent packet drops. Y_i is defined as the number of packets sent in period $X_i(= C + S)$ for the ith period.

In the steady state, we can assume that the above evolution is a regenerative process, as shown in Fig. 1. The same assumption was used for modeling TCP throughput in [11]. Then, we can express the throughput as:

$$X(p) = \frac{E[Y]}{E[X]} \tag{4}$$

As shown in Fig. 1, the average value of X and Y (the subscript i is deleted hereafter) can be expressed as:

$$E[X] = RTT \cdot (E[C] + E[S]), \text{ and} \tag{5}$$

$$\begin{aligned} E[Y] &= E[Y_C] + E[Y_S]) \\ &= E(C)W_{AB} + \frac{W_m}{2}(k+1) + \sum_{i=0}^{k} 2^i \\ &= 1/p \end{aligned} \tag{6}$$

where Y_C and Y_S are defined as the number of packets sent in the CA phase and the SS' phase, respectively. $E[Y]$ is equal to $1/p$ because we assume the random packet drop probability p. W_{AB} is the maximum available bandwidth in units of packet; thus:

$$W_{AB} = \frac{AvailableBandwidth \cdot RTT}{MSS} \tag{7}$$

k is an integer satisfying (8) and $E[S]$ becomes $(k + 1)$ round.

$$\frac{W_m}{2} + 2^{k-1} < W_{AB} \leq \frac{W_m}{2} + 2^k \leq W_m \tag{8}$$

Hence, the average number of packets sent in each regenerative period, $E(Y)$, is given by:

$$\begin{aligned} E[Y] &\approx E[C]W_{AB} + \frac{W_m}{2}(k+1) + (2^{k+1}-1) \\ &\approx E[C]W_{AB} + (W_{AB} - 2^k)(k+1) + (2^{k+1}-1) \\ &\approx E[C]W_{AB} + (k+1)W_{AB} - (k-1)2^k - 1 \end{aligned} \tag{9}$$

From (9), it follows that

$$\begin{aligned} E[C] &\approx \frac{1/p - (k+1)W_{AB} - (k-1)2^k + 1}{W_{AB}} \\ &= \frac{1}{pW_{AB}} - (k+1) + \frac{(k-1)2^k + 1}{pW_{AB}} \end{aligned} \tag{10}$$

Therefore, from (5), (6), and (10), $X(p)$ is calculated as follows:

$$\begin{aligned} X[p] &\approx \frac{1/p}{RTT(E[C] + E[S])} \\ &= \frac{W_{AB}}{RTT\{1 + p((k-1)2^k + 1)\}} \end{aligned} \tag{11}$$

From (7), $2^{k+1} \leq W_m$ and $k = \lceil log_2(W_m) \rceil$. Thus, we can rewrite (11) as (12). For a small value of p, (12) can be approximated by (13):

$$X[p] \approx \frac{W_{AB}}{RTT\{1 + p((k-1)W_m/2 + 1)\}}\ , \text{ and} \tag{12}$$

$$\approx \frac{W_{AB}}{RTT} \qquad \text{if } pW_{AB} \ll 1\ . \tag{13}$$

5 Simulation Study

We performed simulations using the network simulator ns-2 [12] for a network with a bottleneck bandwidth of 800 Mbps and RTT values of 200 ms. The queue size at each bottleneck node is 50% of the bandwidth delay product of the bottleneck link. For convenience, the window size is measured in number of packets with a packet size of 1000 bytes. Drop tail queues are used in the routers.

Figure 3 compares the analytical results of equation (12) of section 5 and the simulation results. The environmental values used are as follows: a bottleneck bandwidth of 800 Mbps, an RTT value of 200 ms, a bottleneck buffer size of 10,000 packets, and a packet size of 1000 bytes. Drop tail queues were used in the routers. In the simulation, performance was measured by the packets delivered 100 seconds after the initial start for 1000 seconds. In comparison, the analytical results of equations (12) and (13) have been proved to be valid.

Figure 4 shows the congestion window variations of the three TCP variants with a loss rate of 10^{-6}. In spite of severe packet loss events, the congestion window of the HP-TCP quickly recovers the link capacity and the performance

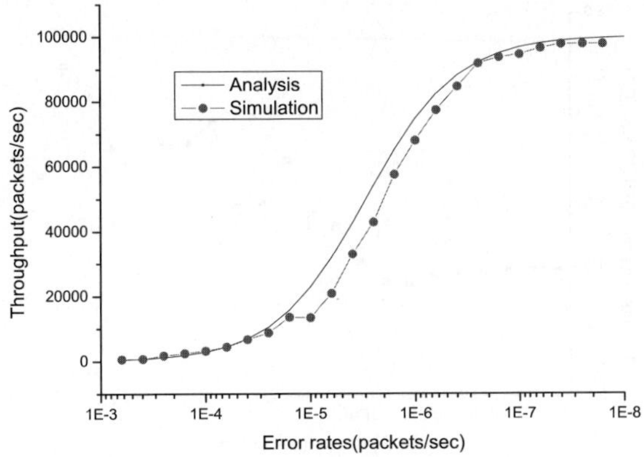

Fig. 3. Throughput comparison between analytical and simulation results

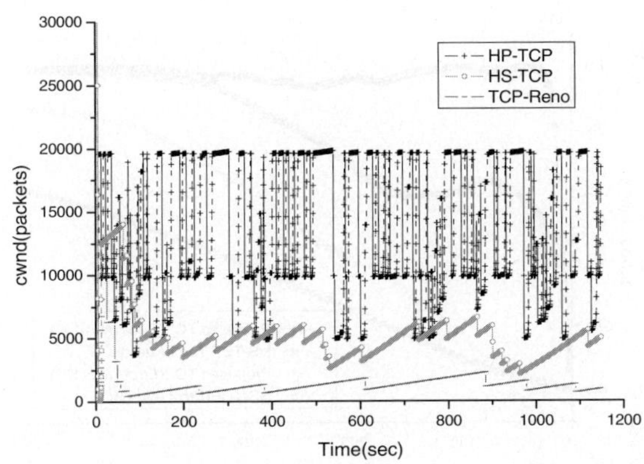

Fig. 4. Congestion window behavior of the HP-TCP, HS-TCP, and TCP Reno mechanisms with a loss rate of 10^{-6}

is almost maintained at the available bandwidth. However, the performances of the HS-TCP and TCP Reno are limited by the slow linear increase of the congestion window and the large loss probability.

The performance of the proposed HP-TCP and fairness has been compared with HS-TCP, which is known to be the representative TCP for HBDP networks. The parameters for the HS-TCP are set at 31 for low windows, 83,000 for high windows, and 10^{-7} for high p.

In Fig. 5, the performance of the HS-TCP is compared with that of the HP-TCP in terms of packet loss rate. In both cases, the difference is minimal with an almost zero drop rate (less than 10^{-7}), since both use 100% of the available bandwidth. However, with a reasonably high loss rate, greater than 3×10^{-6}, the throughput of the HP-TCP is 1.5 to 2.5 times better than that of the HS-TCP.

Figure 6 shows the fairness comparison between the HP-TCP and HS-TCP when the packet loss rate is 10^{-6} and 10^{-7} with two flows. To show the fair

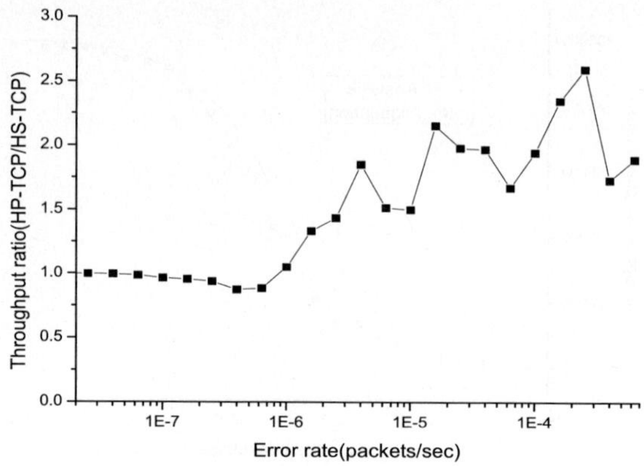

Fig. 5. Throughput ratio of the HP-TCP and HS-TCP mechanisms

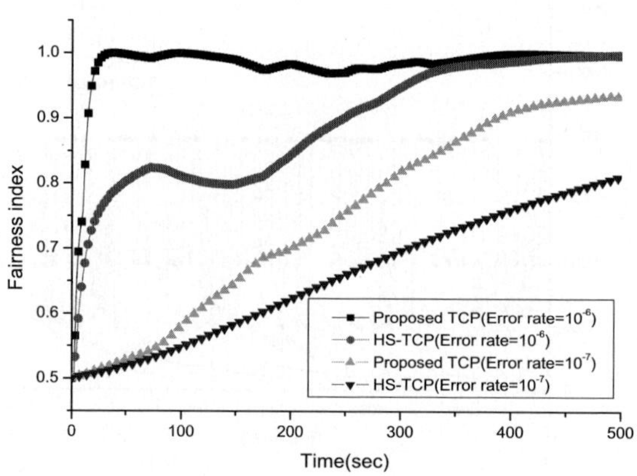

Fig. 6. Fairness comparison of the HP-TCP and HS-TCP mechanisms

share distribution across the connections, we use Jain's Fairness Index as defined in [13]:

$$FairnessIndex = \frac{\left(\sum_{i=1}^{n} b_i\right)^2}{n \sum_{i=1}^{n} b_i^2} \tag{14}$$

where b_i is the throughput of the ith flow and n is the total number of flows. When the throughputs of all flows are equal, the fairness index becomes 1. When the packet loss rate is 10^{-7}, the fairness index of HP-TCP increases faster than that of HS-TCP. When the packet loss rate increases to 10^{-6}, it takes 30 seconds to get the fair share for the HP-TCP, while it takes more than 300 seconds for the HS-TCP. In both cases, HP-TCP's fairness index converges to 1 faster than the HS-TCP. Hence, the HP-TCP shows improved fairness even for cases where the performance difference is minimal due to a low packet loss rate.

6 Conclusion

We proposed a modified TCP congestion control mechanism, the HP-TCP, with an exponential growth phase during the congestion avoidance state and a loss avoidance mechanism during the slow start state in HBDP networks. The RTT status and estimated bottleneck bandwidth are used to prevent overshooting during the exponential growth phase. In addition, we mathematically analyzed the performance of the proposed HP-TCP.

To obtain an analytical evaluation of the proposed HP-TCP, it was compared with the HS-TCP, which is the representative TCP for HBDP networks. We used our simulation results to verify the correctness of our mathematical analysis. In addition, the simulation results showed that the proposed mechanism improves fairness even when the performance difference is minimal due to a low loss rate. When the loss rate increases, the proposed method was proven to outperform other methods as well. The proposed HP-TCP can solve the TCP under-utilization problem in networks with a large available bandwidth. The proposed algorithm can be easily implemented in sender TCPs. Therefore, the proposed HP-TCP is a promising transport protocol for large data transfer applications in HBDP networks.

References

1. Allcock, W., et al: GridFTP: Protocol extensions to FTP for the Grid. GFD-R.020, (2004)
2. Kim, S., Park, S., Moon, J., Lee, H.: A low-crosstalk design of 1.25 Gbps optical triplexer module for FTTH systems. ETRI Journal, Vol. 28, no. 1. (2006) 9-16
3. IETF RFC 793: Transmission Control Protocol (1981)
4. IETF RFC 2581: TCP Congestion Control (1999)
5. IETF RFC 3649: HighSpeed TCP for large congestion windows (2003)
6. Jin, C., Wei, D.X., Low, S.H.: FAST TCP: Motivation, architecture, algorithms, performance. Proceeding of IEEE INFOCOM'04, Vol. 4. (2004) 2490-2501
7. Wang, R., Pau, G., Yamada, K., Sanadidi, M.Y., Gerla, M.: TCP startup performance in large bandwidth delay networks. Proceeding of IEEE INFOCOM 2004, Vol. 2. (2004) 796-805
8. Mascolo, S., Casetti, C., Gerla, M., Sanadidi, M.Y., Wang, R.: TCP Westwood: Bandwidth estimation for enhanced transport over wireless links. Proceeding of ACM/IEEE Mobi-Com 2001 (2001)
9. Mo, J., La, R.J., Anantharam, V., Walrand, J.: Analysis and comparison of TCP Reno and Vegas. Proceedings of IEEE INFOCOM (1999)
10. Giordano, S., Procissi, G., Russo, F., Secchi, R.: On the use of pipesize estimators to improve TCP transient behavior. Proceedings of ICC (2005)
11. Padhye, J., Firoiu, V., Towsley, d., Kurose, J.: Modeling tcp reno performance throughput: A simple model and its empirical validation. IEEE/ACM Transactions on Networking, Vol. 8. (2000)
12. The network simulator ns-2. Available: http://www.isi.edu/nsnam/ns/
13. Jain, R.: The art of computer systems performance analysis: techniques for experimental design, measurement, simulation and modeling. New York, John Wiley & Sons (1991)

On a NIC's Operating System, Schedulers and High-Performance Networking Applications

Yaron Weinsberg[1], Tal Anker[2], Danny Dolev[1], and Scott Kirkpatrick[1]

[1] The Hebrew University Of Jerusalem
{wyaron, dolev, kirk}@cs.huji.ac.il
[2] RADLAN - A Marvell Company, Israel
tala@marvell.com

Abstract. Two critical issues impact the overall performance of Linux clusters based on Intel servers: inter-process communication latency and data throughput. Underlying both of these performance challenges is the inefficient use of computational power and server CPU cycles to process the network protocols. Today's modern high-end Network Interface Cards (NICs) are equipped with an onboard CPU. In most cases, these CPU's are only used by the vendor and are operated by a proprietary OS, which makes them inaccessible to the HPC application developer. In this paper we present a design and implementation of a framework for building high-performance networking applications. The framework consists of an embedded NIC Operating System with a specialized scheduler. The main challenge in developing such a scheduler is the lack of a preemption mechanism in most high-end NICs. Our scheduler provides finer-grained schedules than the alternatives. We have implemented several network applications, and were able to increase their throughput while decreasing the host's CPU utilization.

1 Introduction

High Performance Computing (HPC) and supercomputers were synonymous for many years. But today, HPC clusters built from standards-based, common, off-the-shelf components, such as standard Intel based servers and TCP/Ethernet networking, are changing the economics, opportunities, and abilities of high-performance computing platforms. In an HPC application, latency is extremely important for inter-process communication among compute nodes. Additionally, computation tasks running on the compute nodes need all the CPU cycles that they can get.

Today's modern high-end NICs are equipped with a CPU onboard and line speeds reaching up to 10Gbps. Such high speeds pose a great challenge for today's CPUs and bus architectures. As a result, in order to better utilize the new high speed network infrastructure, the industry is moving towards an approach that offloads part of the hosts' protocol processing to the NIC (e.g. TCP Offload Engines (TOEs) [1]). In most cases, the NICs' CPUs are not fully utilized by the vendor and are usually running a proprietary OS that limit the applications ability to take advantage of them.

In this paper we propose a design and implementation framework that enables a developer to build a high-performance networking application. The developer can design tasks that will be executed at the NIC. The framework enables a developer to easily

M. Gerndt and D. Kranzlmüller (Eds.): HPCC 2006, LNCS 4208, pp. 380–389, 2006.

divide the application logic between the host and the NIC. The framework inherently facilitates the communication between the host and the NIC portions of the application. Another important contribution of this work is a modified NIC level scheduler. Most high-end NICs do not support preemption, thus when trying to schedule user tasks on the NIC, using a common real time scheduling algorithm, we found that there was a great inefficiency in the resulting schedule and the cluster throughput. Our new scheduling scheme (henceforth called: <Sched>++) is capable of extending any given non-preemptive scheduling algorithm with the ability to create finer-grained schedules.

2 Related Work

Today, HPC clusters can increase their achieved throughput by TOEs. However, to the best of our knowledge, there is no NIC oriented operating system that enables developers to design *applications* that can utilize the NIC's functionality. Although some vendors have developed a proprietary OS for their platforms, such an OS does not allow the application developer to integrate parts of the application code into the NIC. The ability to immigrate user tasks to the NIC can further increase the performance of HPC applications.

Spine – is a safe execution environment [2] that is appropriate for programmable NICs. Spine enables the installation of user handlers, written in Modula-3, at the NIC. Although Spine enables the extension of host applications to use NIC resources it has two major limitations. First, since all extensions are executed as a result of an event, building stand-alone applications at the NIC is difficult. Even for event-driven applications, the developer is forced to dissect the application logic to a set of handlers, thus confounding an object-oriented application design. Second, the Spine runtime is based on a virtual machine that offers many advantages such as safety and portability, but at a rather high price. The performance severely degrades and it is very unlikely that NIC manufacturers will agree on a single virtual machine as their device OS.

Arsenic – is a Gigabit Ethernet NIC that exports an extended interface to the host operating system [3]. Unlike conventional adaptors, it implements some of the protection and multiplexing functions traditionally performed by the operating system. This enables applications to directly access the NIC, thus bypassing the OS. The Ethernet Message Passing (EMP) [4] system, of the Ohio Supercomputer Center (OSC), is a zero-copy and OS-bypass messaging layer for Gigabit Ethernet. The EMP protocol processing is done at the NIC and a host application (usually MPI applications) can directly manipulate the NIC. Arsenic and EMP provide very low message latency and high throughput but lack the support for offloading.

TOE – is a technique used to move some of the TCP/IP network stack processing out of the main host and into a network card [1]. While TOE technology has been available for years and continues to gain popularity, it has been less than successful from a deployment standpoint. TOE only targets the TCP protocol, thus, user extensions are out of its scope.

3 Environment

Our programmable interface card is based on the Tigon chipset. The Tigon programmable Ethernet controller is used in a family of 3Com's Gigabit NICs. The Tigon controller supports a PCI host interface and a full-duplex Gigabit Ethernet interface. The Tigon has two 88 MHz MIPS R4000-based processors which share access to external SRAM. Each processor has a one-line (64-byte) instruction cache to capture spatial locality for instructions from the SRAM. Hardware DMA and MAC controllers enable the firmware to transfer data to and from the system's main memory and the network, respectively. The Tigon architecture doesn't contain an interrupt controller. The motivation is to increase the NIC's runtime performance by reducing the overhead imposed by interrupting the host's CPU each time a packet arrives or a DMA request is ready. Furthermore, on a single processor the need for synchronization and its associated overhead is eliminated.

4 NIC Operating System (NICOS)

This section presents the NICOS services. We start by describing the memory management service of NICOS, the NICOS task related APIs, the NICOS networking and the NICOS filtering APIs. Following that we provide a detailed description of the NICOS scheduling framework. We then conclude with several sample applications that use NICOS and we show the significant gain in the applications performance.

4.1 Memory Management

NICOS has to allocate memory each time a task, a queue or a packet is created. NICOS default memory allocation algorithm is based on the "boundary tag method" described in [5], which is suitable for most applications. Implementing a "generic" memory allocation mechanism is problematic: It takes up valuable code space, it is not thread safe and it is not deterministic. Since different realtime systems may have very different memory management requirements, a single memory allocation algorithm probably will not be appropriate. To get around this problem the memory allocation APIs provided in NICOS can be easily replaced by using the filtering APIs (see Section 4.4). A user's task can easily replace the default methods by installing a special kind of a filter. The registered method (i.e., the "filter" action) will be called instead of the default allocation routine. NICOS memory allocation APIs can also enable a developer to choose the *target* of the allocated memory. Memory consuming applications can allocate memory at the host. The memory is transparently accessed using DMA. This scheme is also suitable for developing OS bypass protocols, which removes the kernel from the critical path and hence reduces the end-to-end latency.

4.2 Task Management

NICOS provides several task management APIs that enable a developer to create/destroy tasks and to control their lifecycle state. The API enables a developer to create a periodic or non-periodic task, to yield, sleep, suspend, resume and kill a task. Although periodic tasks can be implemented by a developer on top of a sleep API, we added an explicit facility for periodic tasks so the OS is aware of them. Such a design allows the OS to minimize the ready-to-running[1] latency. Providing the timeliness guarantees required by NICOS has been a major challenge due to the non-preemptive architecture of these NICs.

4.3 Networking

The current networking API is very simple. NICOS provides only a single method that sends raw data.[2] The data is provided by the developer and includes all of the necessary protocol headers. NICOS supports synchronous and asynchronous send calls. The asynchronous ones are non-blocking. When using the synchronous mode, the execution is blocked until frame transmission is completed. Upon completion, the provided callback is called. Receiving a packet is currently done only via filter registration.

4.4 Filtering

When deciding which functionality is needed to be offloaded to the NIC, we looked for common building blocks in today's networking applications. We have found that the ability to inspect packets and to classify them according to specific header fields is such a building block. For instance, the classification capability is useful for firewall applications, applying QoS for certain traffic classes, statistics gathering, etc. Therefore we enhanced the NICOS services with a packet filtering (classification) capability, and the optional invocation of a user installed callback per packet match. In NICOS, a filter is a first class object - it can be introspected, modified and created at runtime.

The "Ping Drop" task (Program 1), which drops all ICMP packets, demonstrates the ease of use of the NICOS filtering API.

4.5 Scheduling

Schedulers for non-preemptive environments usually use an event-driven model. For example, the programmable NIC we are using for evaluating NICOS, provides a special hardware register whose bits indicate specific events. This event register is polled by a dispatcher loop that invokes the appropriate handler. Once the event handler runs to completion, the dispatcher loop resumes.

Providing timeliness guarantees for NIC based tasks can be beneficial for real-time and HPC applications. NICOS enables a developer to easily install a custom scheduler, implementing whatever scheduling policy is needed. NICOS provides several non-preemptive schedulers and an innovative scheme that can further improve their schedule. We have used this scheme to implement an enhanced version of the *Earliest Deadline First (EDF)* scheduler, which is described in details in the next section.

[1] The time from the moment a task becomes ready-to-run until it starts execution.

[2] In the future we plan to write a minimal networking stack for the NIC.

Program 1. Installing "Ping Drop" Filters

```
void registerPingDropFilters(void) {
  /* we would like to match ICMP packets */
  valueMask[0] = ICMP_PROTOCOL;
  bitMask[0] = 0x1; // match 1 byte
  /* start matching at ICMP_PROTOCOL_BYTE */
  pattern_filter.startIndex =ICMP_PROTOCOL_BYTE;
  pattern_filter.length = 1;
  pattern_filter.bitMask = bitMask;
  pattern_filter.numValues = 1;
  pattern_filter.valueMask = &valueMask;
  /* create the filter, add to Rx/Tx flows */
  pingDropFilter.filter_type = STATIC_PATTERN_FILTER;
  pingDropFilter.pattern_filter = &pattern_filter;
  nicosFilter_Add(&nicosTxFilters,&pingDropFilter,DROP,NULL,
                  GENERAL_PURPOSE_FILTERS_GROUP,&pingFilterTxId);
  nicosFilter_Add(&nicosRxFilters,&pingDropFilter,DROP,NULL,
                  GENERAL_PURPOSE_FILTERS_GROUP,&pingFilterRxId);
}
```

4.6 <Sched>++ Algorithm

Common Schedulers. The first scheduling algorithm we have implemented is the simple *Cyclic-Executive* scheduler [6]. The primary advantages of the Cyclic-Executive approach are: being simple to understand, easy to implement, efficient and predictable. Unfortunately, the deterministic nature of a Cyclic-Executive requires a lot of ad-hoc tweaking to produce deterministic timelines, which then must be tested thoroughly. The second scheduling algorithm we have implemented is the non-preemptive version of the EDF algorithm [7]. In EDF, the task with the earliest deadline is chosen for execution. In the non-preemptive version of EDF, the task runs to completion.

Both EDF and Cyclic-Executive are not optimal for a non-preemptive environment such as in our NIC architecture. For a set of scheduable tasks, the resulting task schedule meets the tasks' realtime requirements, however with a rather low CPU utilization. Therefore, we have devised a new task scheduling scheme (denoted as <Sched>++) which can be used to enhance any given non-preemptive scheduler. This scheme utilizes the **compiler** capabilities in order to create an optimized tasks schedule.

Related Definitions. A task is a sequence of operations to be scheduled by a scheduler. A task system $T = \{T_1, \cdots, T_n\}$, where each task T_i is released periodically, is called a *periodic task system.* Each task T_i is defined by a tuple (e_i, d_i, p_i, s_i), where e_i is the task's Worst Case Execution Time (WCET), s_i is the first time at which the task is ready to run (also known as the start time), d_i is the deadline to complete the tasks once it is ready to run, and p_i is the interval between two successive releases of the task. Thus, a task T_i is first released at s_i and periodically it is released every p_i. After each periodic release, at some time t, the task should be allocated e_i time units before deadline $t + d_i$. A *non-periodic* task is a task that is released occasionally, and at each invocation that task may require a different execution time. A *hybrid task system* is a system that contains both periodic and non-periodic tasks. To differentiate between the periodic and non-periodic tasks, a periodic task will be denoted as $\tilde{T}$.

<Sched>++ assumes a *hybrid task system*, where for each periodic task, $d_i = p_i$. To represent the runtime instance of a task, the notion of a *ticket* of a task is introduced. A ticket of a periodic task, $\tilde{T}_i$ is defined as the tuple (e_i, p_i, Pr_i), where e_i and p_i are the execution and the period of the task, and Pr_i is the task's priority. The ticket of a non-periodic task, T_j, is (e_j, Pr_j). This assumes that any type of task scheduler used by the OS can be extended using this ticket.

Algorithm Overview. <Sched>++ uses several compile-time techniques, which provide valuable information that can be used at runtime. The developer uses <Sched>++ specific compiler directives in order to define the system's tasks and tickets. The compiler uses these tickets as simple data structures in which it can store the calculated WCETs.

The compiler uses the generated control flow graph in order to calculate the WCET of the periodic and non-periodic tasks. Typical periodic tasks are comprised of a single calculated WCET, while non-periodic tasks may be comprised of a set of WCETs. In our context, a WCET is defined as the worst case execution time between two successive yields.

The ability of a compiler to modify the developer's code, at predefined places, is also utilized. By modifying the code, the ticket primitive is maintained automatically. The enhanced compiler updates the ticket with the task's next WCET prior to each *yield* invocation. This technique also eliminates the need to introduce a complicated runtime structure that contains all the WCETs of a given non-periodic task. A *single* ticket is recycled to represent the next task segment WCET at runtime.

<EDF>++ Algorithm. In order to implement the enhanced version of the EDF algorithm, the ticket of a periodic task $\tilde{T}$ is extended to be (e, p, Nr, Nd, Pr), where the additional fields Nr and Nd are the next release time and deadline of the task, respectively. Figure 1 presents the main logic behind the <EDF>++ algorithm, which is invoked by the *Yield()* function call. Part I of the algorithm starts with the classical EDF algorithm. The algorithm selects the next periodic task T_{next} that has the earliest deadline among all periodic tasks that are ready to run.

Part II of the algorithm is invoked when no periodic task is ready to run. The algorithm uses the tickets of the non-periodic tasks in order to select the next task to run. The chosen task should be able to run without jeopardizing the deadline of the next (earliest) periodic task. The scheduler considers the subset of non-periodical tasks that are ready to run, such that their next execution time is smaller than the slack time (the time until the next periodic task is ready). Among such tasks, the algorithm can use various criteria to pick the next task to be scheduled. For instance, one can use the algorithm in [8], which chooses a set of tasks that minimizes the remaining slack time. Any such algorithm would use the next execution time (WCET) of the tasks listed in their tickets. When there is no suitable task for execution, the IDLE task is invoked until the next periodic task is ready to run (part III).

Notice that the scheduling algorithm strives to schedule non-periodic tasks whenever there is an available time slot in the schedule. Available time slots may exist between

Yield() called from task T_k:

I: $T_{next} = \{\tilde{T}_i | \tilde{T}_i.Nd = \min(\tilde{T}_j.Nd | \tilde{T}_j.Nr \geq t)\}$;

`/* If no periodic task is ready, then choose from the non-periodic tasks */`

II: `if (`$T_{next} = NULL$`)`

  $SlackTime =$ `duration until next periodic task is ready;`

  `/* Pick the next non-periodic task that will run at most 'SlackTime' time units */`

  $T_{next} = PickNonPeriodicTask(SlackTime)$;

`/* if no task is ready, the Idle task will run for the time duration until the next periodic task is ready */`

III: `if (`$T_{next} = NULL$`)`

  $T_{next} = Idle\_Task(Timeout)$

`SwitchTo(`T_{next}`);`

Fig. 1. <EDF>++ Scheduler

periodic slots or whenever a task completes its execution ahead of time, which can only be determined at runtime.

<EDF>++ Evaluation. We have implemented an experimental system with both EDF and <EDF>++. Our task set includes twenty tasks where about half of them are periodic. We have executed the system with various periods and constraints. On average, for plain EDF, the IDLE task has been executed 28.6% of the time, yielding a CPU utilization of 71.4%. For the <EDF>++ algorithm, the IDLE task ran 14.7% of the time corresponding to 85.2% CPU utilization, an increase of 20% in the system's throughput.

We have also compared the response times of the tasks. Figure 2 shows a sequence of invocation times for a sample task measured from the system's start time. The x-axis shows the number of invocations, where the y-axis presents the time when the specific invocation occurred. The response times, in-between invocations, for the non-periodic

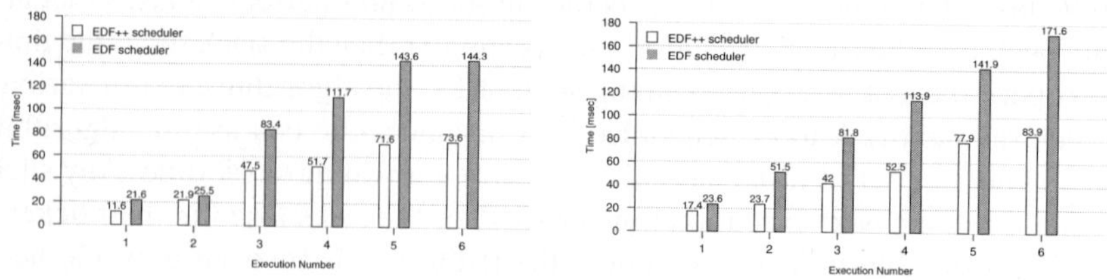

(a) Task A (b) Task B

Fig. 2. Invocation Times

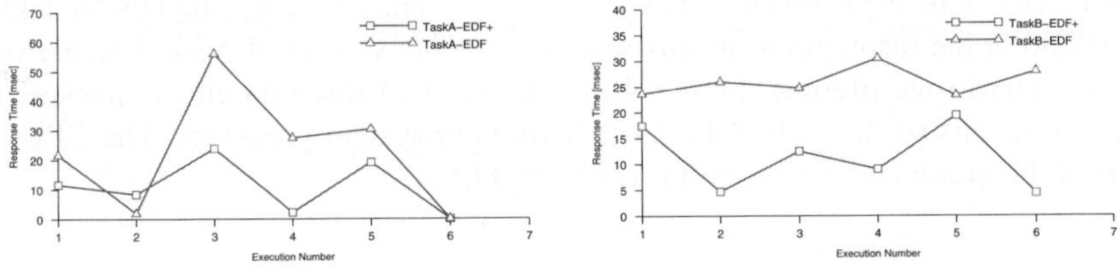

(a) Task A (b) Task B

Fig. 3. Response Times

tasks are presented in Figure 3. For example, the average response time for task A, using <EDF>++, is $10.83ms$ with standard deviation of $8.51ms$ versus $22.86ms$ and $18.87ms$ using EDF (a 53% decrease in the average waiting time). For task B the values are: $11.23ms$ and $5.78ms$ against $26.03ms$ and $2.54ms$ (57% decrease in the average waiting time). The graphs clearly show that the response times for the non-periodic tasks using the <EDF>++ scheduler are improved.

Regarding the response times for periodic tasks, the average response time is approximately the same ($1.37ms$ vs. $1.33ms$ with standard deviation of $2.49ms$ vs $2.39ms$). Thus, the improved response for the non-periodic tasks didn't affect the response time for the periodic tasks.

4.7 Sample Applications Using NICOS

A Firewall Application. An application of particular promise for offloading is a firewall application. Since a firewall is an application that filters packets according to a user defined security policy, earlier filtering (especially discarding packets) has a potential for significant improvements in performance. A firewall application on a NIC also has the additional advantage that it is harder for an adversary to modify than a software application running at the host.

We have designed and implemented a firewall application, called SCIRON [9], for a NIC. As presented in Section 4.4, NICOS provides a framework that enables a developer to install filters. Filters can be installed both at the firmware level and/or at the kernel level. SCIRON's firewall is implemented as a set of such firmware filters.

In order to simulate common kernel-based firewalls for performance evaluation, we have also installed filters at the driver layer. All comparisons shown below, compare the same firewall code (with the same filtering policy) between the driver based firewall and the NIC based firewall.[3]

Firewall Evaluation. This section compares the performance between host based and NIC based firewalls. Many parameters have an impact on the firewall's performance. For example, the number of rules, current CPU utilization, packet size, ratio of incoming to outgoing packets, total number of packets, number of packets accepted vs number rejected, can all potentially influence the performance.

[3] Currently, the firewall code is fully stateless, thus the state is not saved between successive filter action invocations.

Performance can be measured using two parameters. The first is the load on the CPU and the second is the throughput. In this section we discuss several typical scenarios. In the first scenario we present (Figure 4(a)), the firewall discards all the packets it receives. During this scenario the CPU is only running system processes. The CPU is on the left of the graph and throughput is on the right.[4]

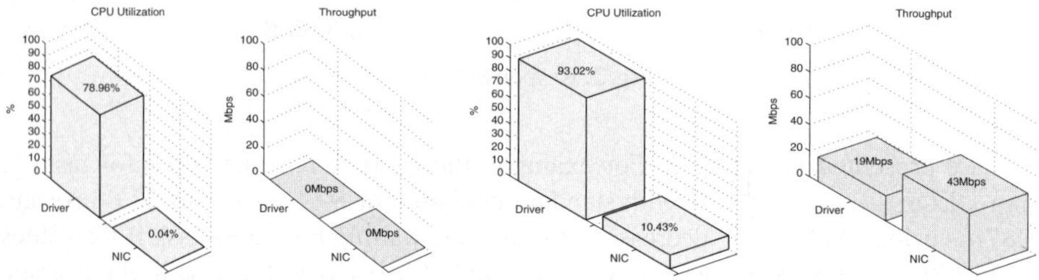

(a) Receiving - Discarding all packets (b) Receiving - 50% acceptance rate

Fig. 4. Firewall Performance

As expected, in this scenario the CPU utilization when using the firewall implemented on the NIC is approximately zero, whilst for the same firewall on the host it is quite high. The second scenario presented is given in Figure 4(b). This scenario is probably a more realistic behavior for a typical host machine. It is evident that the NIC based firewall has better performance both in CPU utilization and throughput.

STORM. Occasionally, clustered HPC applications need to synchronize the cluster's activities or to perform cluster-wide operations. Therefore, the cluster software usually needs to implement a basic locking and/or consensus algorithms that consumes a lot of bandwidth and degrades the cluster's performance.

A STORM cluster [10] consists of five nodes running Linux where each node is hosting a programmable NIC. As a proof of concept application we have implemented Lamport's Timestamp ordering algorithm [11] providing an agreed order on transmitted messages. This messaging system's performance is much better (latency of 84 us, and throughput of 768 Mb/s) than the host level implementation (latency of 200 us, and throughput of 490 Mb/s).

5 Conclusions

This paper presented a novel framework for building high-performance applications that can benefit from offload capabilities. The framework is comprised of a NIC Operating System and an innovative scheduler. We have implemented several HPC applications using this framework and have demonstrated increased application throughput. According to the International Technology Roadmap for Semiconductors (ITRS), by

[4] The benchmark ran on two hosts (Intel Pentium 4 CPU 2.4Ghz, 512MB) connected via 100Mbps ethernet.

2007, one million transistors will cost less than 20 cents. This current trend motivates hardware and embedded system designers to use programmable solutions in their products. We believe that programmable NICs will soon become widespread. The need for such a framework is apparent.

Acknowledgments. We would like to acknowledge Udi Weinsberg for his helpful suggestions regarding the scheduling algorithm. We would also like to acknowledge all of the NICOS project members: Maxim Grabarnik, Adamovsky Olga and Nir Sonnenschein.

References

1. Currid, A.: TCP offload to the rescue. Queue **2**(3) (2004) 58–65
2. Fiuczynski, M.E., Martin, R.P., Owa, T., Bershad, B.N.: Spine: a safe programmable and integrated network environment. In: EW 8: Proceedings of the 8th ACM SIGOPS European workshop on Support for composing distributed applications, New York, NY, USA, ACM Press (1998) 7–12
3. Pratt, I., Fraser, K.: Arsenic: A user-accessible gigabit ethernet interface. In: INFOCOM. (2001) 67–76
4. Shivam, P., Wyckoff, P., Panda, D.: Emp: zero-copy os-bypass nic-driven gigabit ethernet message passing. In: Supercomputing '01: Proceedings of the 2001 ACM/IEEE conference on Supercomputing (CDROM), New York, NY, USA, ACM Press (2001) 57–57
5. Wilson, P.R., Johnstone, M.S., Neely, M., Boles, D.: Dynamic storage allocation: A survey and critical review. In: Proc. Int. Workshop on Memory Management, Kinross Scotland (UK) (1995)
6. Cheng, S.C., Stankovic, J.A., Ramamritham, K.: Scheduling algorithms for hard real-time systems: a brief survey. (1989) 150–173
7. Liu, C.L., Layland, J.W.: Scheduling algorithms for multiprogramming in a hard-real-time environment. Journal of the ACM **20**(1) (1973) 46–61
8. Shmueli, E., Feitelson, D.G.: Backfilling with lookahead to optimize the performance of parallel job scheduling. In: Job Scheduling Strategies for Parallel Processing. (2003)
9. Weinsberg, Y., Pavlov, E., Amir, Y., Gat, G., Wulff, S.: Putting it on the NIC: A case study on application offloading to a Network Interface Card (NIC). In: IEEE CCNC. (2006)
10. (Super-fast Transport Over Replicated Machines (STORM)) Available at site: `http://www.cs.huji.ac.il/~wyaron`.
11. Lamport, L.: Time, clocks, and the ordering of events in a distributed system. Commun. ACM **21**(7) (1978) 558–565

A Microeconomics-Based Fuzzy QoS Unicast Routing Scheme in NGI*

Xingwei Wang, Qi Wang, Min Huang, and Nan Gao

College of Information Science and Engineering, Northeastern University, Shenyang, 110004, P.R. China
wangxw@mail.neu.edu.cn

Abstract. In this paper, a microeconomics-based fuzzy QoS unicast routing scheme is proposed and has been implemented by simulation. It attempts to make not only the user QoS requirements satisfied but also Pareto-optimum under Nash equilibrium on the network provider end-to-end utility and the user end-to-end utility achieved or approached along the found route by the proposed heuristic route selection algorithm based on the bi-directional Dijkstra algorithm and the intermediate list acceleration method. Simulation results have shown that the proposed scheme has better performance.

1 Introductions

NGI (Next-Generation Internet) is now becoming an integrated network, including terrestrial-based, space-based, sky-based, fixed and mobile sub-networks, supporting anywhere, anytime with any kind of information to communicate with anyone or even any object in fixed or mobile way [1]. End-to-end QoS (Quality of Service) should be supported in NGI. However, it is hard to describe the network status exactly and completely. With gradual commercialization of network operation, QoS pricing and accounting should be provided [2]. There exist conflicts on profits between the network providers and their users, and win-win should be attained. Support from QoS routing should be provided to help solve the above problems. In this paper, a microeconomics-based fuzzy QoS unicast routing scheme is proposed, trying to make Pareto-optimum under Nash equilibrium [3] on both the network provider and the user utilities achieved or approached along the found QoS route.

2 Problem Formulations

A network can be modeled as a graph $G(V,E)$, where V is node set and E is edge set. Just for algorithm description simplicity, the node parameters are merged into edge. Thus, the edge parameters become as follows: available bandwidth bw_{ij}, delay de_{ij}, delay jitter jt_{ij}, and error rate ls_{ij}. Suppose that the source node is $v_s \in V$ and the

* This work is supported by Program for New Century Excellent Talents in University; the National Natural Science Foundation of China under Grant No. 60473089; the National Development and Reform Commission under CNGI-04-13-2T, CNGI-04-6-2T and CNGI-04-15-7A.

M. Gerndt and D. Kranzlmüller (Eds.): HPCC 2006, LNCS 4208, pp. 390 – 399, 2006.

destination node is $v_d \in V$, look for the specific route l_{sd} between v_s and v_d, trying to make the network provider end-to-end utility TW and the user end-to-end utility TU achieve or approach Pareto-optimum under Nash equilibrium as much as possible with the following constraints satisfied: the available bottleneck bandwidth along l_{sd} is not smaller than the user bandwidth requirement bw_req; the delay along l_{sd} is not bigger than the user delay requirement de_req; the delay jitter along l_{sd} is not bigger than the user delay jitter requirement jt_req; the error rate along l_{sd} is not bigger than the user error rate requirement ls_req. The corresponding mathematical model is described as follows:

$$TW + TU \rightarrow \max\{TW + TU\} \tag{1}$$

$$TW \rightarrow \max\{TW\} \tag{2}$$

$$TU \rightarrow \max\{TU\} \tag{3}$$

$$s.t. \quad \min_{e_{ij} \in l_{sd}}\{bw_{ij}\} \geq bw\_req \tag{4}$$

$$\sum_{e_{ij} \in l_{sd}} de_{ij} \leq de\_req \tag{5}$$

$$\sum_{e_{ij} \in l_{sd}} jt_{ij} \leq jt\_req \tag{6}$$

$$1 - \prod_{e_{ij} \in l_{sd}} (1 - ls_{ij}) \leq ls\_req \tag{7}$$

TW and TU are calculated as follows:

$$TW = \sum_{e_{ij} \in l_{sd}} ws_{ij} \tag{8}$$

$$TU = \sum_{e_{ij} \in l_{sd}} us_{ij} \tag{9}$$

Among them, ws_{ij} and us_{ij} represent the network provider utility and the user utility on the edge e_{ij} respectively.

The above problem is NP-complete [4], and is resolved by the proposed scheme.

3 QoS Routing Scheme Descriptions

The proposed QoS routing scheme in this paper consists of three parts: edge evaluation, game analysis and route selection.

3.1 Edge Evaluation

Due to difficulty on exact expression of network status, the adaptability membership degree function is used to describe the adaptability of the candidate edge conditions to the user QoS requirements. The edge bandwidth adaptability membership degree function is defined as follows:

$$g_1(bw_{ij},bw\_req)=\begin{cases}0 & bw_{ij}<bw\_req\\ \left(\dfrac{bw_{ij}-bw\_req}{b-bw\_req}\right)^k+f_1(bw_{ij},bw\_req) & bw\_req\le bw_{ij}<b\\ 1 & bw_{ij}\ge b\end{cases} \tag{10}$$

$$f_1(bw_{ij},bw\_req)=\begin{cases}\varepsilon & bw_{ij}=bw\_req\\ 0 & otherwise\end{cases} \tag{11}$$

The edge delay adaptability membership degree function is defined as follows:

$$g_2(Jp,de_{ij},de\_req)=\begin{cases}0 & de_{ij}>de\_req\\ 1-e^{-\left(\frac{de\_req-de_{ij}}{\sigma_1}\right)^2}+f_2(Jp,de_{ij},de\_req) & de_{ij}\le de\_req\end{cases} \tag{12}$$

$$f_2(Jp,de_{ij},de\_req)=\begin{cases}\varepsilon & Jp=1\wedge de_{ij}=de\_req\\ 0 & otherwise\end{cases} \tag{13}$$

The edge delay jitter adaptability membership degree function is defined as follows:

$$g_3(Jp,jt_{ij},jt\_req)=\begin{cases}0 & jt_{ij}>jt\_req\\ 1-e^{-\left(\frac{jt\_req-jt_{ij}}{\sigma_2}\right)^2}+f_3(Jp,jt_{ij},jt\_req) & jt_{ij}\le jt\_req\end{cases} \tag{14}$$

$$f_3(Jp,jt_{ij},jt\_req)=\begin{cases}\varepsilon & Jp=1\wedge jt_{ij}=jt\_req\\ 0 & otherwise\end{cases} \tag{15}$$

The edge error rate adaptability membership degree function is defined as follows:

$$g_4(Jp,ls_{ij},ls\_req)=\begin{cases}0 & ls_{ij}>ls\_req\\ 1-e^{-\left(\frac{ls\_req-ls_{ij}}{\sigma_3}\right)^2}+f_4(Jp,ls_{ij},ls\_req) & ls_{ij}\le ls\_req\end{cases} \tag{16}$$

$$f_4(Jp,ls_{ij},ls\_req)=\begin{cases}\varepsilon & Jp=1\wedge ls_{ij}=ls\_req\\ 0 & otherwise\end{cases} \tag{17}$$

Formula (10), (12), (14) and (16) are Gaussian alike with smooth transition feature. $f_i(i=1,2,3,4)$ is used to deal with one hop route. Jp is a positive integer, representing the hop count of end-to-end route. ε is a positive pure decimal fraction and much smaller than 1. k, b, σ_1, σ_2 and σ_3 all are positive, $k>1$. An evaluation matrix $R=[g_1,g_2,g_3,g_4]^T$ of the candidate edge can be gotten from (10), (12), (14) and (16). According to the application nature, a weight matrix $B=[b_1,b_2,b_3,b_4]$ is given, b_1, b_2, b_3 and b_4 are the significance weights of bandwidth, delay, delay jitter and error rate on the application QoS respectively, $0<b_1,b_2,b_3,b_4<1$, $b_1+b_2+b_3+b_4=1$. The comprehensive evaluation value ω on the candidate edge conditions with regard to the user QoS requirements is computed as follows:

$$\omega = B \times R \tag{18}$$

Obviously, the bigger the value of ω is, the higher the adaptability of the candidate edge conditions to the user QoS requirements is.

3.2 Game Analysis

As shown in Fig.1, suppose there are n_u users in the network and are divided to two classes: high-end and low-end. There are s_u high-end users who are willing to pay more money for higher QoS, and the other $n_u - s_u$ users are low-end ones who can accept lower QoS with less cost. k_u is the number of users (including both high-end and low-end ones) who can be admitted by the network with the currently available resources. $n_u \geq k_u \geq s_u$. In Fig.1, Y-axis represents the price of the network for data transmission; X-axis represents the size of the data flow, and c represents the network cost. The producer surplus P_S and the customer surplus C_S are as follows [5]:

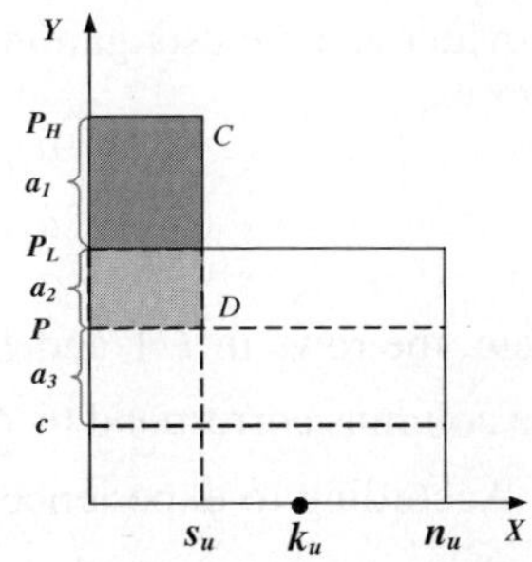

Fig. 1. Demand curve

$$C_S = a_1 \cdot s_u + a_2 \cdot (k_u - s_u) \tag{19}$$

$$P_S = (a_1 + a_2) \cdot s_u + a_3 \cdot (k_u - s_u) \tag{20}$$

Whether the available bandwidth of the candidate edge is abundant can be derived from the value of g_1. If $g_1 < h_1$ (h_1 is a constant and $0 < h_1 < 1$), the available bandwidth of the candidate edge is considered to be scarce, and the network provider can properly increase the price to suppress the user demands; if $h_1 \leq g_1 < h_2$ (h_2 is a constant and $0 < h_1 < h_2 < 1$), the available bandwidth of the candidate edge is considered to be moderate; if $g_1 \geq h_2$, the available bandwidth of the candidate edge is considered to be abundant, and the network provider can lower the price to attract the users. Thus, a tuning coefficient for the amount of bandwidth to be actually allocated to the user is introduced, and its definition is as follows [5]:

$$\rho = \begin{cases} \rho_1 & g_1 < h_1 \\ 1 & h_1 \le g_1 < h_2 \\ \rho_2 & g_1 \ge h_2 \end{cases} \tag{21}$$

$\rho_1 > 1$, $0 < \rho_2 < 1$, ρ_1 and ρ_2 are preset by experience. The actually allocated amount of bandwidth f_u to the user on the candidate edge is as follows:

$$f_u = \frac{u_c}{\rho P} \tag{22}$$

u_c denotes the cost paid by the user and P denotes the baseline price of the resource.

The network provider has two gaming strategies: s_1 and s_2, denoting whether it is willing to provide the bandwidth of the candidate edge to the user or not respectively; the user also has two gaming strategies: t_1 and t_2, denoting whether he is willing to accept the provided bandwidth of the candidate edge or not respectively. The network provider and the user gaming matrixes, $U1$ and $U2$, are defined as follows:

$$U1 = \begin{bmatrix} a_{11} & a_{12} \\ a_{21} & a_{22} \end{bmatrix} \qquad U2 = \begin{bmatrix} b_{11} & b_{12} \\ b_{21} & b_{22} \end{bmatrix}$$

Here, the rows in $U1$ and $U2$ correspond to s_1 and s_2 of the network provider, and the columns correspond to t_1 and t_2 of the user.

According to experience, a threshold value ω_0 is set. If $\omega > \omega_0$, the actual status of the candidate edge is considered better than that the user expected; if $\omega = \omega_0$, the actual status of the candidate edge is considered to be just what the user expected; if $\omega < \omega_0$, the actual status of the candidate edge is considered worse than that the user expected. Therefore, the element values of $U1$ and $U2$ are given as follows:

$$U1 = \begin{bmatrix} \dfrac{\left(P_S \dfrac{\omega}{\omega_0} - P_S\right)}{f_u} & \dfrac{\left(P_S \dfrac{\omega}{\omega_0} - P_S\right)}{f_u} \\ -\mu \dfrac{\left(P_S \dfrac{\omega}{\omega_0} - P_S\right)}{f_u} & -\dfrac{\left(P_S \dfrac{\omega}{\omega_0} - P_S\right)}{f_u} \end{bmatrix} \quad U2 = \begin{bmatrix} \dfrac{\left(C_S \dfrac{\omega}{\omega_0} - C_S\right)}{u_c} & -\mu \dfrac{\left(C_S \dfrac{\omega}{\omega_0} - C_S\right)}{u_c} \\ \dfrac{\left(C_S \dfrac{\omega}{\omega_0} - C_S\right)}{u_c} & -\dfrac{\left(C_S \dfrac{\omega}{\omega_0} - C_S\right)}{u_c} \end{bmatrix}$$

If the following inequations [6] are satisfied:

$$\begin{cases} a_{i^* j^*} \ge a_{ij^*} \\ b_{i^* j^*} \ge b_{i^* j} \end{cases} \qquad i, j = 1, 2 \ , \tag{23}$$

then $\{s_{i*}, t_{j*}\}$ represents the specific solution under Nash equilibrium. Here, $i*$ and $j*$ stand for some i and j, s_{i*} stands for s_1 or s_2, and t_{j*} stands for t_1 or t_2.

3.3 Route Selection

The basic idea is as follows: calculate a heuristic cost for each edge and use it as weight, and run the bi-directional Dijkstra algorithm and the intermediate list acceleration method [7] to select the route.

Heuristic cost. After the gaming result of the candidate edge e_{ij} is gotten, it is transformed into one kind of weight, denoted by Ω_{ij}, which is defined as follows:

$$\Omega_{ij} = \begin{cases} 1 & \textit{Nash Equilibrium} \\ >1 & \textit{non-Nash Equilibrium} \end{cases} \tag{24}$$

The heuristic cost $Tf_{ij}(\Omega_{ij}, ws_{ij}, us_{ij})$ of e_{ij} is defined as follows:

$$Tf_{ij}(\Omega_{ij}, ws_{ij}, us_{ij}) = \Omega_{ij}\left(q_1 \frac{1}{ws_{ij}} + q_2 \frac{1}{us_{ij}}\right) \tag{25}$$

In formula (25), Ω_{ij} represents the influence of Nash equilibrium on the route selection. q_1 and q_2 are the preference weights, representing whether and how much the network provider/user utility should be considered with priority when routing if needed. ws_{ij} and us_{ij} are the utilities which the network and the user actually get on the edge e_{ij} respectively. Obviously, the smaller the heuristic cost sum along the route, the nearer the network provider and the user utilities is to Pareto-optimum under Nash equilibrium. Thus, when routing, in the premise of meeting constraints (4)-(7), the objectives (1)-(3) are achieved or approached by means of minimizing the heuristic cost sum along the selected route, that is,

$$\text{minimize}\left\{\sum_{e_{ij}\in l_{sd}} Tf_{ij}(\Omega_{ij}, ws_{ij}, us_{ij})\right\} \tag{26}$$

Algorithm description. v_s and v_d are source and destination node respectively. Let pc_1 and Tc_1 denote pc label and Tc label of node v when searching direction is from v_s to v_d, and let pc_2 and Tc_2 denote pc label and Tc label of node v when searching direction is from v_d to v_s. $pc_1(v)$ is the minimum heuristic cost from v_s to v with the specific constraints satisfied, and $pc_2(v)$ is the minimum heuristic cost from v_d to v with the specific constraints satisfied. $Tc_1(v)$ is the upper bound of $pc_1(v)$, and $Tc_2(v)$ is the upper bound of $pc_2(v)$. S_i is the set of nodes with pc_1 label at Step i, and $S_i^{'}$ is the set of nodes with pc_2 label at Step i. To get the least-cost path from

v_s (or v_d) to current node at the same time as getting the least cost from v_s (or v_d) to current node, each node is given a λ_1 and a λ_2. $\lambda_1(v)$ represents the precedent node along the shortest path from v_s to v, and $\lambda_2(v)$ represents the precedent node along the shortest path from v_d to v .When the proposed algorithm ended, if $\lambda_1(v)=m$ (or $\lambda_2(v)=m$), the precedent node of v along the path from v_s (or v_d) with the minimum heuristic cost is v_m; if $\lambda_1(v)=m'$ (or $\lambda_2(v)=m'$), there does not exist the satisfied path from v_s (or v_d) to v; if $\lambda_1(v)=0$, $v=v_s$; if $\lambda_2(v)=0$, $v=v_d$. In addition, $\min bw_1(v_j)$ is the available bottleneck bandwidth from v_j to v_s, and $\min bw_2(v_j)$ is the available bottleneck bandwidth from v_j to v_d; $de_1(v_j)$ is the delay from v_j to v_s, and $de_2(v_j)$ is the delay from v_j to v_d; $jt_1(v_j)$ is the delay jitter from v_j to v_s, and $jt_2(v_j)$ is the delay jitter from v_j to v_d; $ls_1(v_j)$ is the error rate along the path from v_j to v_s, and $ls_2(v_j)$ is the error rate along the path from v_j to v_d. Tf_{ij}, bw_{ij}, de_{ij}, jt_{ij}, and ls_{ij} are the heuristic cost, the available bandwidth, the delay, the delay jitter, and the error rate of the edge e_{ij} respectively. MLS_1 is the middle list used to hold the sequence numbers of those temporary nodes when searching direction is from v_s to v_d, and MLS_2 is the middle list used to hold the sequence numbers of those temporary nodes when searching direction is from v_d to v_s.

Based on the algorithm described in [7], the following routing algorithm is proposed:

Step0. Initialization: $i=0$, $S_0=\{v_s\}$, $\lambda_1(v_s)=0$; $S_0'=\{v_d\}$, $\lambda_2(v_d)=0$. $\forall v\neq v_s$, $Tc_1(v)=+\infty$, $\lambda_1(v)=m'$, $k_1=s$; $\forall v\neq v_d$, $Tc_2(v)=+\infty$, $\lambda_2(v)=m'$, $k_2=d$. Do:

(1) $pc_1(v_{k_1})=0$, $pc_2(v_{k_2})=0$; $\min bw_1(v_{k_1})=+\infty$, $\min bw_2(v_{k_2})=+\infty$;

(2) $de_1(v_{k_1})=0$, $jt_1(v_{k_1})=0$, $ls_1(v_{k_1})=0$; $de_2(v_{k_2})=0$, $jt_2(v_{k_2})=0$, $ls_2(v_{k_2})=0$;

For each node v_j with $e_{k_1 j}\in E$ and $v_j\notin S_i$, add v_j to MLS_1 and compute $Tf_{k_1 j}$ according to (10)-(25); for each node $v_{j'}$ with $e_{k_2 j'}\in E$ and $v_{j'}\notin S_i'$, add $v_{j'}$ to MLS_2 and compute $Tf_{k_2 j'}$ according to (10)-(25).

Step1. Labeling procedure

1*st* labeling condition:

For each node $v_j\in MLS_1$, in order to achieve the objective of (26), if $Tc_1(v_j)>pc_1(v_{k_1})+Tf_{k_1 j}$, do:

(1) $pc_1'(v_j)=pc_1(v_j)+Tf_{k_1 j}$;

(2) $\min bw_1'(v_j)=\min\{\min bw_1(v_{k_1}), bw_{k_1 j}\}$;

(3) $de_1'(v_j)=de_1(v_{k_1})+de_{k_1 j}$, $jt_1'(v_j)=jt_1(v_{k_1})+jt_{k_1 j}$, $ls_1'(v_j)=1-(1-ls_1(v_{k_1}))(1-ls_{k_1 j})$;

For each node $v_{j'} \in MLS_2$, in order to achieve the objective of (26), if $Tc_2(v_{j'}) > pc_2(v_{k_2}) + Tf_{k_2 j'}$, do:

(1) $pc_2'(v_{j'}) = pc_2(v_{j'}) + Tf_{k_2 j'}$;

(2) $\min bw_2'(v_{j'}) = \min\{\min bw_2(v_{k_2}), bw_{k_2 j'}\}$;

(3) $de_2'(v_{j'}) = de_2(v_{k_2}) + de_{k_2 j'}$, $jt_2'(v_{j'}) = jt_2(v_{k_2}) + jt_{k_2 j'}$, $ls_2'(v_{j'}) = 1 - (1 - ls_2(v_{k_2}))(1 - ls_{k_2 j'})$;

2nd labeling condition:

In order to meet constraints (4)-(7), if

(1) $\min\{\min bw_1'(v_j), \min bw_2'(v_{j'})\} \geq bw\_req$;

(2) $de_1'(v_j) + de_2'(v_{j'}) \leq de\_req$, $jt_1'(v_j) + jt_2'(v_{j'}) \leq jt\_req$, $1 - (1 - ls_1'(v_j))(1 - ls_2'(v_{j'})) \leq ls\_req$;

then

(1) $Tc_1(v_j) = pc_1'(v_j)$, $Tc_2(v_{j'}) = pc_2'(v_{j'})$;

(2) $\min bw_1(v_j) = \min bw_1'(v_j)$, $\min bw_2(v_{j'}) = \min bw_2'(v_{j'})$;

(3) $de_1(v_j) = de_1'(v_j)$, $jt_1(v_j) = jt_1'(v_j)$, $ls_1(v_j) = ls_1'(v_j)$; $de_2(v_{j'}) = de_2'(v_{j'})$, $jt_2(v_{j'}) = jt_2'(v_{j'})$, $ls_2(v_{j'}) = ls_2'(v_{j'})$; $\lambda_1(v_j) = k_1$, $\lambda_2(v_{j'}) = k_2$;

go to Step2; otherwise, negotiate with user: if succeeded, go to Step2, otherwise the end.

Step2. Modification procedure

Step2.1. Find the currently working nodes.

Find the node v_{p_1} with the smallest $Tc_1(v)$ value (that is $\min_{v_j \in MLS_1} \{Tc_1(v_j)\}$) in MLS_1, let $pc_1(v_{p_1}) = Tc_1(v_{p_1})$, $S_{i+1} = S_i \cup \{v_{p_1}\}$, and $k_1 = p_1$; find the node v_{p_2} with the smallest $Tc_2(v)$ value (that is $\min_{v_{j'} \in MLS_2} \{Tc_2(v_{j'})\}$) in MLS_2, let $pc_2(v_{p_2}) = Tc_2(v_{p_2})$, $S_{i+1}' = S_i' \cup \{v_{p_2}\}$, and $k_2 = p_2$; $i = i + 1$;

Step2.2. Compute heuristic cost.

For each node v_j with $e_{p_1 j} \in E$ and $v_j \notin S_i$, compute Tf_{jp_1} according to (10)-(25); for each node $v_{j'}$ with $e_{p_2 j'} \in E$ and $v_{j'} \notin S_i'$, compute $Tf_{j' p_2}$ according to (10)-(25).

Step2.3. Modify the middlelists.

Remove v_{p_1} from MLS_1, and for each node v_j with $e_{p_1 j} \in E$ and $v_j \notin MLS_1$, add v_j to MLS_1; remove v_{p_2} from MLS_2, and for each node $v_{j'}$ with $e_{p_2 j'} \in E$ and $v_{j'} \notin MLS_2$, add $v_{j'}$ to MLS_2;

Step2.4. Find out whether there exists a converging node.

If $S_i \cup S_i^{'} = \phi$, go to Step1, otherwise let v_b represent the first node which is in both S_1 and S_2, and then go to Step 2.5.

Step2.5. Find out the final route.

If $MLS_1 \cap S_i^{'} = \phi$, concatenate the path from v_s to v_b and the path from v_b to v_d to form the final route l_{sd} from v_s to v_d. If l_{sd} meets constraints (4)-(7), output l_{sd}, the end; otherwise, negotiate with user: if succeeded, output l_{sd}, the end; otherwise, there is no feasible solution, the end.

If $MLS_1 \cap S_i^{'} \neq \phi$, let $H = MLS_1 \cap S_i^{'}$ and $|H| = q$. For each node $v_{j_i} (i = 1,2,..., q)$ in H, a path $l_i (i = 1,2,...q)$ from v_s to v_d is formed by concatenating the path from v_s to v_{j_i} and the path from v_{j_i} to v_d, so there are $q+1$ paths in total, that is: $l_1, l_2, ..., l_q$ and l_{sd} (l_{sd} is the path formed by concatenating the path from v_s to v_b and the path from v_b to v_d); find out the path l_e with the smallest heuristic cost sum from the $q+1$ paths formed above; if l_e meets constraints (4)-(7), output l_e, the end, otherwise negotiate with user: if succeeded, output l_e, the end; otherwise, there is no feasible solution, the end.

4 Performance Evaluations

Simulations have been done on NS (Network Simulator) 2 platforms. The scheme proposed in this paper, the fuzzy-tower-based QoS unicast routing scheme [8] and the Dijkstra-based unicast routing scheme have been performed over some actual and virtual network topologies. For simplicity, the above three schemes are denoted by M, F and D respectively. Comparison results on QoS unicast routing request succeeded rate(RSR), user utility(UU), network provider utility(NU), comprehensive utility(CU=UU+NU) achieved by M, F and D over CERNET2 topology(T1), CERNET topology(T2) and one virtual topology(T3, generated by Waxman2 [9] with average node degree 3.5) are shown in Table 1. From Table 1, it can be concluded that the proposed scheme has better performance.

Table 1. Performance Comparison

Scheme / Metric	M:F:D(T1)	M:F:D(T2)	M:F:D(T3)
RSR	1.0656:1.0555:1.0000	1.0043:1.0088:1.0000	1.0544:1.0434:1.0000
UU	1.0250:1.0192:1.0000	1.0029:1.0072:1.0000	1.0282:1.0212:1.0000
NU	1.1514:1.1313:1.0000	1.0214:1.0321:1.0000	1.1172:1.0888:1.0000
CU	1.1471:1.1282:1.0000	1.0224:1.0322:1.0000	1.1409:1.1250:1.0000

5 Conclusions

In this paper, a microeconomics-based fuzzy QoS unicast routing scheme is proposed. It can not only deal with the fuzziness of NGI network status, but also make both the network provider and the user utilities achieve or approach Pareto-optimum under Nash equilibrium along the found route with the user QoS requirements satisfied, supporting win-win from the aspect of routing. Simulation results have shown that the proposed scheme is effective and efficient.

References

1. Keiji T.: A Perspective on the Evolution of Mobile Communications. IEEE Communications Magazine. Vol. 41. No. 10 (2003) 66-73
2. Bob B., Vasilios D., Oliver H. et al: A Market Managed Multi-service Internet. Computer Communications. Vol. 26. No. 4 (2003) 404-414
3. Shi X. Q.: Microeconomics. Shanghai: Shang Hai University of Finance & Economics Press. (2000)
4. Wang Z., Crowcroft J.: Quality of Service Routing for Supporting Multimedia Applications. IEEE Journal on Selected Areas in Communications. Vol. 14. No. 7 (1996) 1288-1294
5. Hu Y. M., Zhang J. S.: Research on the Difference Pricing Strategy of Internet Congestion. Quantitative & Technical Economics. Vol. 35. No. 7 (2004) 81-85
6. Lin H., Matnsk C., Sajal K. D., Kalyan B.: ARC: an Integrated Admission and Rate Control Framework for CDMA Data Networks Based on Non-cooperative Games. In-ternational Conference on Mobile Computing and Networking. (2003) 326-338
7. Jin X. Q.: Bi-directional Dijkstra Algorithm and the Acceleration of Intermediate List. Computer Simulation. Vol. 21. No. 9 (2004) 78-81
8. Wang X. W., Yuan C. Q., Huang M.: A Fuzzy-tower-based QoS Unicast Routing Algorithm. Proc. of Embedded and Ubiquitous Computing (EUC'04), Aizu: Springer LNCS 3207. (2004) 923-930
9. Waxman B. M.: Routing of Multipoint Connections. IEEE Journal on Selected Areas in Communications. Vol. 6. No. 11(1988) 478-489

Adaptive Online Management for Congestion Control in QoS Sensitive Multimedia Services

Sungwook Kim[1] and Sungchun Kim[2]

[1] Department of Computer Science, Sogang University,
Shinsu-dong 1, Mapo-ku, Seoul, 121-742, South Korea
swkim01@sogang.ac.kr

[2] Department of Computer Science, Sogang University,
Shinsu-dong 1, Mapo-ku, Seoul, 121-742, South Korea
ksc@mail.sogang.ac.kr

Abstract. Over the years, the widespread proliferation of multimedia services has necessitated the development for an efficient network management. However, during network operations, traffic contention for limited network bandwidth may occur. Congestion is one of the most intense problems for more bandwidth consuming multimedia service. In this paper, we investigate the role of adaptive online pricing mechanism in order to manage effectively the network congestion problem. Our on-line approach is dynamic and flexible that responds to current network conditions. With a simulation study, we demonstrate that our proposed scheme enhances network performance and system efficiency simultaneously under widely diverse traffic load intensities.

1 Introduction

With the emerging more bandwidth consuming multimedia services, traffic congestion is one of the most intense problems in multimedia network management. In order to alleviate traffic overload condition, bandwidth migration and adaptation techniques [1]-[4] have been proposed. However, a major drawback of these approaches is the lack of ability to avoid congestion because they do not provide negative or positive incentives for users to control the unexpected growth of traffic.

Usually, users act independently and selfishly without considering the existence of other users. However, they are price sensitive. Therefore, the pricing strategy can have strong influence on the behavior of the users [5]. In this paper, we propose a new online price charging mechanism based on real time measurements. In order to control the network congestion for QoS sensitive multimedia service, adaptive price should reshape traffic congestion by providing negative incentives to users. Our approach combines adaptive bandwidth control concept with dynamic pricing concepts that can be achieved taking into account the economic considerations. Therefore, not only do we improve network performance, but we also achieve system efficiency. For the efficient bandwidth control, a well-balanced compromise between network performance and system efficiency is necessary.

For adaptive price management, we mainly concentrate on extending our previous bandwidth management research [1]-[4]. Based on our already proposed algorithms,

M. Gerndt and D. Kranzlmüller (Eds.): HPCC 2006, LNCS 4208, pp. 400 – 405, 2006.

our online price adjustment mechanism can exploit the adaptive nature of users including the interaction between user behavior and the network efficiency. Therefore, the service price in our scheme can vary dynamically based on the current traffic situations. This paper is organized as follows. Section II describes the proposed scheme in detail. In Section III, performance evaluation results are presented along with comparisons with schemes proposed in [1],[3]-[4]. Finally, concluding remarks are given in Section IV.

2 Dynamic Pricing Management

In this section, we describe the proposed adaptive online scheme in detail. In the first step, we propose a bandwidth demand estimation algorithm. In the second step, we develop an adaptive price control mechanism based on demand estimation and user behavior. According to this information, we finally derive each service price for different multimedia services.

Since exact service information in the future is not known, we are not able to exactly expect the bandwidth demand. Therefore, we estimate the expected bandwidth demand (B_e) by adaptive online manner. Our estimation algorithm can dynamically adjust the amount of B_e based on the current network conditions. For adaptive online adjustment, we provide a coordination paradigm for two different - uniformly distributed and non-uniformly distributed - traffic situations. When the traffic is uniformly distributed over time, the amount of traffics is close to stable. Therefore, traffic history is important information to estimate B_e. However, if traffic distributions are non-uniform, the traffic history cannot catch up with the temporal and spatial traffic fluctuations. At the time of temporary traffic fluctuations, B_e should be heavily influenced by current traffic changes. The main idea is to adjust B_e based on both the traffic history and recent traffic changes.

We define traffic window (W_{set_I}) to keep traffic history. By using traffic window, we can learn the pattern of coming service requests. In order to implement the traffic window management, we partition the time-axis into equal intervals of length *unit_time*. The value of D_n, which is defined as the average amount of requested bandwidths during traffic window, can be estimated as

$$D_n = \sum_{j \in W_{set\_I}} (B_j \times N_j) / T_u \qquad (1)$$

where N_i and B_i are the number of call requests during traffic window and the corresponding bandwidth of service type *i*, respectively. T_u is the length of the traffic window in terms of *unit_time*. However, the value of D_n is too slow to abrupt traffic changes. Therefore, in order to be more responsive to current traffic fluctuations, we adjust D_n every *unit_time* as a weighted average of recent request changes and traffic history as follows.

$$D_{n+1} = \alpha D_n + (1 - \alpha) C_b \qquad (2)$$

where C_b is the amount of requested bandwidth during the interval [t_c - *unit_time*, t_c], D_n is the expected reservation amount at the current time and D_{n+1} is the amount of our optimized bandwidth demand for the interval $[t_c, t_c + unit\_time]$. The parameter α controls the relative amount given to recent and past traffic request histories in our decision. Under diverse system environments, a fixed α value cannot effectively adapt to the changing request conditions. Our online algorithm also adaptively modifies α value each *unit_time* period. If current call blocking probability (CBP) is higher (lower) than predefined target probability (P_{target}), it means that the current requested service is larger (smaller) than the available bandwidth capacity, so the value of α should be decreased (increased). Based on this online approach, we can adaptively estimate the expected bandwidth demand (B_e) value according to the current network conditions. Every *unit_time*, the value of B_e is computed periodically as follows.

$$B_e = \begin{cases} \max\{D_n, D_{n+1}\} & \text{if } CBP > P_{target} \\ \min\{D_n, D_{n+1}\} & \text{if } CBP \le P_{target} \end{cases} \tag{3}$$

Our simple network economic model consists of two types of agents: users (consumers) and provider. Users require bandwidth to satisfy their QoS. Provider seeks to maximize network profit by selling bandwidth. In order to response varied multimedia traffic demands, the amount of bandwidth allocation is a discrete quantity in terms of the number of basic bandwidth units (BUs), where one BU is the basic amount of bandwidth allocation. Required bandwidth for different multimedia service is different number of BUs.

If the network is lightly loaded and all users are getting acceptable QoS, each BU is provided with fixed price (B_m). When a congestion occurs, that is, the demand D (requested bandwidth) exceeds the supply S (available bandwidth), our congestion-dependent price mechanism begins to impose the extra price charge, which forces the users to reduce their demand. Since users are very price sensitive, the increased price gives the users an incentive to back-off their requests. It means that the bandwidth demand and supply can be balanced dynamically. After the congestion situation is removed, extra charge falls to zero. Therefore, extra charge is iteratively adjusted up or down until the equilibrium of demand and supply.

Under our proposed model, the price adjustment process is performed at discrete intervals of time. The price adjustment interval is defined *adjustment_period* (A_P). Corresponding to the current network situations, we specify when this price adjustment procedure starts. To decide the starting time point (t_{s_p}), the conditions are proposed. At current time (t_c), if D is larger than S or current bandwidth utilization is lower than predefined target utilization (U_{target}), A_P is started ($t_c = t_{s\_p}$). During A_P, if estimated B_e is larger than S (or average bandwidth utilization is lower than U_{target}), our price adjust procedure starts. During k^{th} adjustment interval (A_P_k), the actual price per each BU ($P_{k+1}(BU)$) is obtained as

$$P_{k+1}(BU) = \max\{B_m, [\, P_k(BU) + F_p(B_{e\_k}, S_k)]\,\} \tag{4}$$

where B_m is minimum charge for one BU, B_{e_k} represent the expected bandwidth demand estimation during A_P_k, S_k is available bandwidth for new service and F_p is a function to calculate price adjustment. *we* assume that the price will not fall below a certain nonnegative minimum price (B_m). Function F_p, which estimates the price change in proportion to the difference between demand and supply, is defined as

$$F_p(B_{e\_k}, S_k) = (\delta_k \times \lceil (B_{e\_k} - S_k) / S_k \rceil) \quad (5)$$

where δ_k is a control factor used to adjust the convergence rate. If rapidly and continually bandwidth demand (or supply) increases, we can facilitate to bring the network to an equilibrium status by controlling the value of δ_k.

3 Simulation Experiments

In this paper, user reaction behavior (elasticity of dynamic price) is modeled based on different exponential demand functions. In these functions, the demand rate is exponential inverse proportional to the price increase [6]. We consider three forms for the demand function F in the example that follows. They are

$$F_1(x) = e^{-x^2}, \quad F_2(x) = \frac{e^{-x}}{1+x}, \quad F_3(x) = \frac{1}{1+x^4} \quad \text{for } x \geq 0. \quad (6)$$

In Fig.1 - Fig.2, we compare the network performance of our online price management scheme with existing schemes [1]-[4], which are designed for QoS sensitive multimedia service. Through performance evaluation of the communication/cellular network management, we can confirm that our online price strategy provides the ability to take care of the network congestion and encourages more efficient use of available bandwidth.

Fig.1 shows the performance comparison for multimedia cellular network. The curves indicate that the scheme based on adaptive price policy improves the performance of CBP and CDP significantly than a fixed price scheme. Fig.2 shows the performance comparison for wire-lined multimedia communication network. From the curves in Fig.2 we obtained, it can be seen that the proposed scheme achieves a better balance of economic and technical efficiencies and attains excellent performance from low to heavy traffic load distributions.

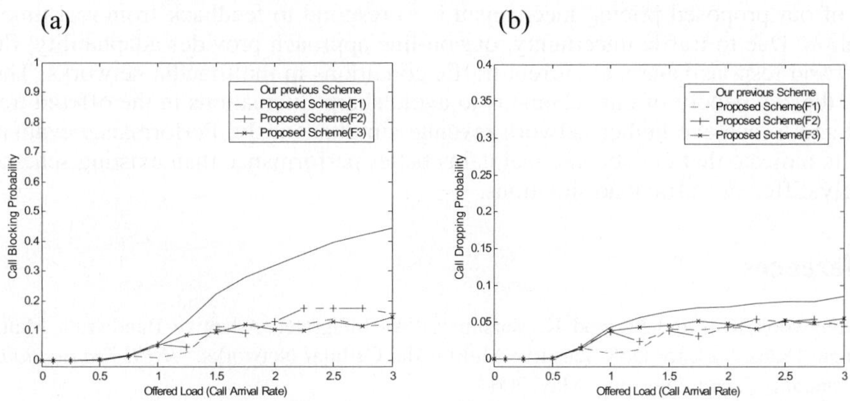

Fig. 1. Performance evaluation for multimedia cellular network (a) call blocking probabilities (b) call dropping probabilities

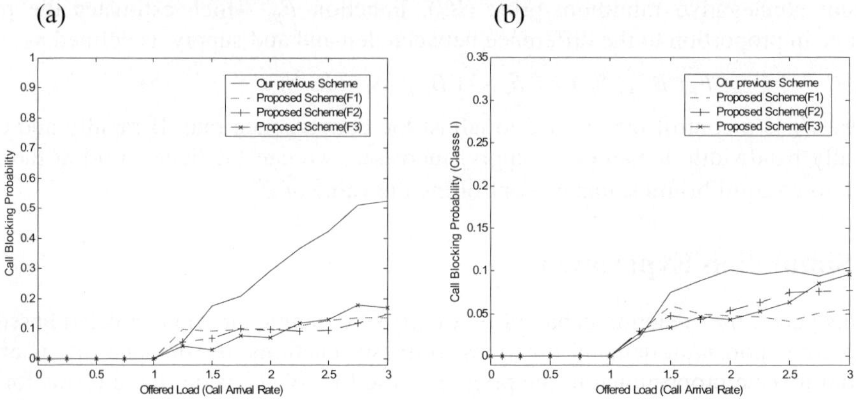

Fig. 2. Performance evaluation for multimedia communication network (a) call blocking probabilities (b) call blocking probabilities for *class I* data service

Simulation results have been concluded that dynamic and time-dependent pricing policy leads to a significant performance improvement and better network revenue. From a congestion management perspective, our adaptive price scheme achieves a quicker convergence to stable state. From an economic perspective, our scheme provides a higher system efficiency. Therefore, the scheme based on adaptive pricing policy can maintain well-balanced network performance between conflicting evaluation criteria in widely different traffic load situations while existing scheme [1]-[4] can not offer such an attractive trade off.

4 Summary and Conclusions

In this paper, online adaptive price management algorithm for multimedia networks is proposed. Our research efforts have focused on how to alleviate the congestion occurrences while maintaining a network system in the most efficient state. The main novelty of our proposed pricing mechanism is to respond to feedback from real time estimation. Due to traffic uncertainty, our on-line approach provides adaptability, flexibility and responsiveness to current traffic conditions in multimedia networks. Therefore, the key benefit of our scheme is to avoid sharp fluctuations in the offered traffic load and to achieve higher network revenue simultaneously. Performance evaluation results indicate that our scheme maintains better performance than existing scheme in widely different traffic load situations.

References

1. Sungwook Kim and Pramod K. Varshney, "An Integrated Adaptive Bandwidth Management Framework for QoS Sensitive Multimedia Cellular Networks," *IEEE Transactions on Vehicular Technology,* pp. , May, 2004.
2. Sungwook Kim and Pramod K. Varshney, "An Adaptive Bandwidth Reservation Algorithm for QoS Sensitive Multimedia Cellular Network", *IEEE Vehicular Technology Conference,* pp. 1475-1479, September, 2002.

3. Sungwook Kim and Pramod K. Varshney, “Adaptive Load Balancing with Preemption for Multimedia Cellular Networks”, *IEEE Wireless Communications and Networking Conference (WCNC),* March, 2003.
4. Sungwook Kim, “Adaptive On-line Bandwidth Management for QoS sensitive Multimedia Networks,” Ph.D dissertation, Syracuse University, Syracuse, NY, U.S, 2003.
5. L. Badia, M. Lindstrom, J. Zander, M. Zorzi, “Demand and Pricing Effects on the Radio Resource Allocation of Multimedia Communication System,” *Proc. IEEE GLOBECOM’03*, San Francisco, Dec. 2003.
6. P.C. Fishburn and A.M. Odlyzko, “Dynamic behavior of differential pricing and quality of service options for the Internet,” *Proceedings of the ACM First International Conference on Information and Computation Economics*, New York, pp.128–139, 1998.

BGPSep_D: An Improved Algorithm for Constructing Correct and Scalable IBGP Configurations Based on Vertexes Degree*

Feng Zhao, Xicheng Lu, Peidong Zhu, and Jinjing Zhao

School of Computer, National University of Defense Technology,
Changsha 410073, Hunan, China
fengzhao1980@tom.com

Abstract. IBGP configurations based on the route reflection may lead to forwarding loops and sub-optimal paths. Although the IBGP configuration generated by BGPSep guarantees three correctness properties of complete visibility, loop-free forwarding, robustness to IGP failures, and the number of IBGP sessions is smaller than in a full-mesh configuration, BGPSep does not reduce the number of IBGP sessions of its top level route reflectors. We improve BGPSep by removing some vertexes, whose degrees satisfy some conditions, from the IGP graph gradually. The improved algorithm is called BGPSep_D. We prove that BGPSep_D satisfies the three correctness properties. The performance of BGPSep_D is evaluated on several real-world backbone topologies. Experimental results indicate that BGPSep_D can generate an IBGP topology with much smaller maximum degree and a much smaller number of IBGP sessions than that produced by BGPSep.

1 Introduction

Border Gateway Protocol (BGP) [1] [2] is the widely used interdomain routing protocol. BGP can be divided into two parts: External BGP (EBGP) and Internal BGP (IBGP). As an alternative to full mesh IBGP, BGP route reflection [3] [4] is often used in the IBGP topology design.

However, IBGP configurations based on route reflection may lead to route oscillations, forwarding loops and sub-optimal paths [8][9][10]. These problems are hard to diagnose and debug, and networks with these problems are hard to manage.

Basu et al. [5] study the problem of route oscillations in IBGP with route reflection and prove deciding whether an IBGP configuration with route reflection can converge is NP-complete. They propose a modification to IBGP that guarantees convergence. Griffin and Wilfong study the conditions under which the BGP configuration converges to a stable path assignment [7], and also examine MED-induced oscillations [6]. Also they study the property of loop-free forwarding and prove that it is NP-hard to verify whether an arbitrary IBGP configuration is forwarding correct. They also

* This research was supported by the National Grand Fundamental Research 973 Program of China under Grant No. 2003CB314802 and the National High-Tech Research and Development Plan of China under Grant No. 2005AA121570.

M. Gerndt and D. Kranzlmüller (Eds.): HPCC 2006, LNCS 4208, pp. 406 – 415, 2006.

describe a set of sufficient conditions that an IBGP configuration is free of deflections and forwarding loops.

R. Musunuri et al. [11] present a selective path dissemination (SPD) protocol, which avoids IBGP anomalies specific to clustering. J.A. Cobb et al. [12] propose a complete solution that solves both clustering and MED induced anomalies in IBGP. However, these solutions require multiple path disseminations between route reflectors, so they may not be scalable. Also because they require changes to the IBGP messages, it is not so easy to deploy them.

M.Vutukuru et al. [13] present and analyze an algorithm, BGPSep, to construct an IBGP session configuration that is both correct and more scalable than a full mesh. They claim that to their knowledge, BGPSep is the first constructive algorithm to generate IBGP configurations with useful correctness guarantees, while scaling better than a full mesh. However, although the number of IBGP sessions is smaller than in a full-mesh configuration, BGPSep does not reduce the number of IBGP sessions of its top level route reflectors. That is, the maximum node degree of the IBGP topology remains the same as in a full-mesh configuration. A router should not maintain too much BGP sessions, because concurrent frequent updates on multiple peers may cause problems. For example, 53 peers generating 150 updates per second can cause a powerful router such as Cisco 12008 to miss sending a KEEPALIVE in time, thus resulting in a session reset [14]. So it is very meaningful to reduce the number of IBGP sessions a router needs to maintain.

We find that in the IGP topologies of some large ASes, there exist a lot of pendant vertexes, i.e. whose degree is one. Base on this property, we modify the BGPSep algorithm by removing some vertexes from the IGP graph gradually. We called the modified algorithm BGPSep_D. We evaluate the performance of BGPSep on several real-world IGP topologies. Experimental results show that the maximum degrees of the IBGP topologies generated by BGPSep_D for these IGP topologies can be reduced by about 9%-50% and the IBGP sessions can be reduced by about 10%-50%.

2 BGPSep

To facilitate the discussion of the BGPSep_D algorithm, we first describe BGPSep briefly.

According to [13], a full-mesh IBGP configuration satisfies the following desirable correctness properties:

P1 Complete visibility: The dissemination of routing information amongst the routers should be complete, in the sense that, for every external destination, each router picks the same route that it would have picked had it seen the best routes from every other EBGP router in the AS.

P2 Loop-free forwarding: After the dissemination of EBGP learned routes, the resulting routes picked by all routers and the subsequent forwarding paths of packets sent along those routes should be free of deflections and forwarding loops.

P3 Robustness to IGP failures: The route dissemination mechanism should be robust to node or link failures and IGP path cost changes-such changes should not result in a violation of the correctness property P2.

```
Algorithm 1. BGPSep
Input: IGP Graph G, set V of BGP routers
Output: Set I of IBGP sessions
if |V| = 1 then
    I = ∅;
else if |V| = 2 then
    {u,v} ← V ;
    I = {(u, v, peer)};
else
    /* Step 1: Choose a graph separator S ⊆ V . Routers
    in S are the route reflectors. */
    S =Graph-Separator(G) ;
    G_1,..., G_m ← components of V − S ;
    /* Step 2: Fully mesh the set of route reflectors
    */
    foreach u, v ∈ S , u ≠ v do
        I = I ∪ {(u, v, peer)};
    end
    foreach G_i do
        /* Step 3: Make every router in each component G_i a
        route reflector client of every route reflector
        */
        foreach u ∈ G_i, v ∈ S do
            I = I ∪ {(u, v, client)};
        end
        /* Step 4: Recursively apply BGPSep over each
           component */
        I_i = BGPSep(G_i);
        I = I ∪ I_i;
    end
end
return I;
```

Fig. 1. The BGPSep algorithm

An IBGP configuration generated by BGPSep guarantees properties P1, P2 and P3. As shown in Algorithm 1, BGPSep is based on the notion of a graph separator, a set of vertexes whose removal partitions a graph into roughly equal-sized connected components. BGPSep takes the graph $G = (V, E)$ formed by the BGP routers and produces the set I of IBGP sessions that must be established between the routers. Every element in I denotes an IBGP session and is of the form (u, v, t), where u and v are the routers between which the IBGP session is established and t is the type of the IBGP session. If t = "client", then the IBGP session between u and v is a client-route reflector session (with u being the client of route reflector v). If t = "peer", then the IBGP session between u and v is a normal non-client IBGP session. The recursion stops when the component has one or two routers. The algorithm uses a procedure Graph-Separator, which is a graph partitioning algorithm that takes a graph G and returns a graph separator S.

From the *Step 2* and *Step 3* of Algorithm 1, we can see that a top level route reflector has to establish IBGP sessions with all other BGP routers. For large ASes, the IBGP sessions number of a top level route reflector will be too many to influence the scalability.

```
Algorithm 2. BGPSep_D
Input: IGP Graph G, set V of BGP routers
Output: Set I of IBGP sessions
/* Step 1: removing the pendant vertexes gradually*/
I = ∅;
pending = true;
G' = G;
while pending == true do
   G = G';
   pending = false;
   foreach u ∈ G.V do
      if d_G(u) == 1 then /* d_G(u) is the degree of u in G */
         v = adj_G(u); /* v is the adjacent node of u */
         I = I ∪ {(u, v, client)}; /* let u be the client of v */
         G' = G' − {u};
         pending = true;
      end
   end
end
/* Step 2: Apply BGPSep over the remaining subgraph G' */
I_s = BGPSep(G');
I = I ∪ I_s;
return I;
```

Fig. 2. The BGPSep_D algorithm

3 The BGPSep_D Algorithm

We find that there exist a lot of pendant vertexes in the IGP topologies of some large ASes. So we modify the BGPSep algorithm by removing the pendant vertexes from the IGP graph gradually. If there is no more pendant vertex in the produced subgraph, then BGPSep is applied. We called the modified algorithm BGPSep_D, as shown in Algorithm 2.

The idea of BGPSep_D is simple. If the degree of node u is one in the IGP graph, then we let u be the client of its adjacent node v. If node v has complete visibility, then following the route reflection rules and the BGP route selection rules, node u will have complete visibility (this will be proved in the next section).

4 Proof of Correctness

In this subsection, we will prove that the IBGP configuration produced by BGPSep_D satisfies the properties P1, P2 and P3 described in the previous Section. To facilitate the discussion, we will borrow some related notations, definitions and lemmas used in [13].

Consider the IGP subgraph G induced by the BGP routers of a network in an AS. Let V denote the set of BGP routers. Let d denote any destination. For every router A ,let $Egress_d(A)$ denote the best egress router that A would have picked had it seen the best routes from every EBGP router in the AS.

Also we denote G_{sub} the subgraph G' that is generated after the *Step 1* of the BGPSep_D algorithm is finished.

Definition 1. A *signaling chain* between two routers A and B is defined as a set of routers $A(=R_0), R_1, R_2, ..., R_r, B=(R_{r+1})$ such that, for $i=1...r$, (i) R_i is a route reflector and (ii) at least one of R_{i+1} or R_{i-1} is a route reflector client of R_i.

Definition 2. A *signaling chain of monotone increase* between two routers A and B is a signaling chain $S: A(=R_0), R_1, R_2, ..., R_r, B(=R_{r+1})$ such that, for $i=1...r+1$, R_{i-1} is a route reflector client of R_i.

Definition 3. A *signaling chain of monotone decrease* between two routers A and B is a signaling chain $S: A(=R_0), R_1, R_2, ..., R_r, B(=R_{r+1})$ such that, for $i=1...r+1$, R_i is a route reflector client of R_{i-1}.

Definition 4. Given two signaling chains $S_1 = R_0, R_1, ..., R_k$ and $S_2 = R'_0, R'_1, ..., R'_l$, if $R_k = R'_0$ and $R_0, ..., R_k(=R'_0), R'_1, ..., R'_l$ is a signaling chain, then we called S_1 can *concatenate* with S_2. We denote the *concatenation* of S_1 and S_2 by $S_1 \| S_2$.

Definition 5. For a signaling chain $S: A(=R_0), R_1, R_2, ..., R_r, B(=R_{r+1})$, if there exists a shortest IGP path P from A to B such that for $i=0...r+1$, $R_i \in P$, then we say S is a *shortest signaling chain* between A and B.

Following the BGP route selection rule and the tie breaking rule, we can get the following claim:

Claim 1. For any destination d, if there exists a shortest signaling chain between router A and $Egress_d(A)$, then A learns of the best route via $Egress_d(A)$ for destination d.

Proof. The proof is in Appendix I.

4.1 Complete Visibility (P1)

For complete visibility to hold, we require that every router A chooses the route via egress $Egress_d(A)$ as its best route for destination d.

Lemma 1. In the IBGP configuration produced by BGPSep_D, for any destination d, there exists a shortest signaling chain between every router $A \in V$ and the egress router $Egress_d(A)$.

Proof. The proof is in Appendix II.

The following theorem follows from **Claim 1** and **Lemma 1**.

Theorem 1. The IBGP configuration output by BGPSep_D satisfies the property of complete visibility.

4.2 Loop-Free Forwarding (P2)

Theorem 2. The IBGP configuration output by BGPSep_D satisfies the property of loop-free forwarding.

The proof is omitted (it is similar to *Theorem 6* of [13]).

4.3 Robustness to IGP Changes (P3)

Because Algorithm 2 does not use IGP link costs in computing the IBGP configuration, we get the following Theorem.

Theorem 3. The IBGP configuration produced by BGPSep_D is not affected by changes in IGP link costs.

Theorem 4. The IBGP configuration produced by BGPSep_D satisfies the properties of loop-free forwarding and complete visibility in the face of IGP router and link failures.

The proof is omitted (it is similar to *Lemma 8* of [13]).

5 Implementation and Evaluation

We implemented the BGPSep_D algorithm in Matlab by extending BGPSep. The program reads the IGP graph from a file and writes the IBGP sessions to a file. We evaluate the performance of BGPSep_D on the backbone topologies of 5 ISPs annotated with inferred link costs from the Rocketfuel project [15]. (There is a very small connected component separated with the other part in the inferred topology of AS1221. And an IGP topology of an AS should be connected. So we modified the inferred topology of AS1221 slightly by removing the small connected component.) The used topologies are summarized in Table 1.

Table 1. ISP Topologies Used

AS	Name	Number of routers	Number of links	comments
1221	Telstra	104	302	Modified slightly
1239	Sprint	315	1944	
1755	Ebone	87	322	
3257	Tiscali	161	656	
3967	Exodus	79	294	

We compare the number of IBGP sessions produced by the BGPSep_D algorithm with those produced by BGPSep for these topologies. We assume conservatively that all the vertexes in the topology are external BGP routers, like [13].

The results are shown in Fig.3. We observe that the IBGP configuration produced by BGPSep_D results in a 10%-50% reduction in the number of IBGP sessions produced by BGPSep on these topologies.

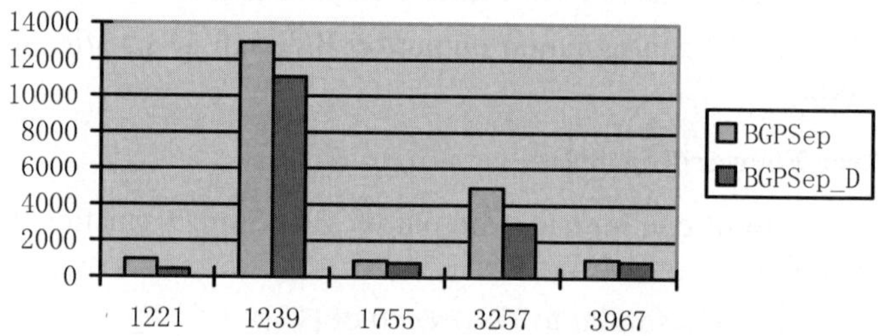

Fig. 3. Number of IBGP sessions: Rocketfuel ISP topologies

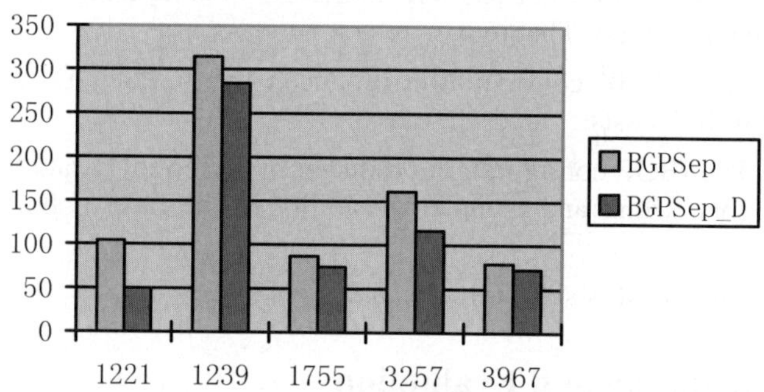

Fig. 4. The maximum degrees of the generated IBGP topologies: Rocketfuel ISP topologies

Compared with BGPSep, another key aspect of BGPSep_D scalability is the maximum degree of the generated IBGP topology. The results are shown in Fig.4. Also we observe that the IBGP configuration produced by BGPSep_D results in a 9%-50% reduction in the maximum degree of the generated IBGP topologies produced by BGPSep on these topologies.

6 Conclusion and Future Work

BGPSep is the first constructive algorithm to generate IBGP configurations with useful correctness guarantees, while scaling better than a full mesh. In this paper, we make a modification to the BGPSep algorithm. And we prove that the modified algorithm satisfies the correctness properties of complete visibility, loop-free forwarding and robustness to node and link failures. Although the modification is simple, an evaluation of the modified algorithm on real-world ISP topologies shows that it can achieve much better results than BGPSep.

References

1. Y. Rekhter and T. Li, Border Gateway Protocol 4, RFC 1771, SRI Network Information Center, July 1995.
2. Y. Rekhter, T. Li, and S. Hares, A Border Gateway Protocol 4 (BGP- 4). Internet Draft draft-ietf-idr-bgp4-26.txt, October 2004.
3. T. Bates, R. Chandra, and E. Chen, BGP Route Reflection – An Alternative to Full Mesh IBGP, RFC 2796,Network Working Group, April 2000.
4. T. Bates, R. Chandra, and E. Chen, BGP Route Reflection – An Alternative to Full Mesh IBGP, draft-ietf-idr-rfc2796bis-01.txt, Network Working Group, November 2004.
5. A. Basu, CH. Luke Ong, A. Rasala, F. Bruce Shepherd, and Gordon Wilfong, Route Oscillations in IBGP with Route Reflection. In Proc. ACM SIGCOMM, pages 235.247, Pittsburgh, PA, August 2002.
6. Timothy G. Griffin and Gordon Wilfong, Analysis of the MED Oscillation Problem in BGP. In Proc. 10th IEEE International Conference on Network Protocols, pages 90. 99, Paris, France, November 2002.
7. Timothy G. Griffin and Gordon Wilfong, On the correctness of IBGP configuration. In Proc. ACM SIGCOMM, pages 17.29, Pittsburgh, PA, August 2002.
8. D. McPherson, V. Gill, D. Walton, A. Retana. Border Gateway Protocol Persistent Route Oscillation Condition, RFC 3345, August 2002.
9. Nick Feamster, Proactive Techniques for Correct and Predictable Internet Routing. PhD thesis, Massachusetts Institute of Technology, September 2005.
10. Nick Feamster and Hari Balakrishnan, Detecting BGP Configuration Faults with Static Analysis. In Proc. 2nd Symp. On Networked Systems Design and Implementation (NSDI), Boston, MA, May 2005.
11. Ravi Musunuri, Jorge A. Cobb, Stable IBGP through Selective Path Dissemination, IASTED Parallel and Distributed Computing and Systems Conference (PDCS), Marina Del Ray, CA, Nov 3-5, 2003.
12. Ravi Musunuri, Jorge A. Cobb, A complete solution for IBGP stability, ICC 2004.
13. M. Vutukuru, P. Valiant, S. Kopparty, and H. Balakrishnan, How to Construct a Correct and Scalable IBGP Configuration, in Proceedings of IEEE INFOCOM, 2006.
14. A.Feldman et al, Measuring BGP Pass-Through times, in Proceedings of the PAM 2004 workshops.
15. Ratul Mahajan, Neil Spring, David Wetherall, and Tom Anderson. Inferring Link Weights Esing End-to-end Measurements. In Proc. 2nd ACM SIGCOMM Internet Measurement Workshop, pages 231.236, Marseille, France, 2002.

Appendix I: Proof of Claim 1

Let $S: A(=R_0), R_1, R_2, ..., R_k, B(= Egress_d(A))$ denotes the shortest signaling chain between router A and $Egress_d(A)$, and the best route selected by B for destination d is denoted by $r_d(B)$. Because R_k, B is a signaling chain, R_k will learn the route $r_d(B)$. If there is not another egress router for destination d, then R_k will choose $r_d(B)$ as its best route. Else let B' denote any other egress router for destination d.

Because B is the best egress router of A, for the egress router B' for destination d, the relation of $r_d(B')$ and $r_d(B)$ can be classified into two cases: (i) $r_d(B')$ has lower value of **local_pref,** longer **as_path** or a higher **med** value than $r_d(B)$ (if both has the same neighboring AS);(ii) the **local_pref** values, **as_path lengths** and **med** values of both routes are same.

In the case (i), R_k will choose $r_d(B)$ as its best route whether it has leant $r_d(B')$ or not.

In the case (ii), R_k is not an egress router for destination d, otherwise the best egress router of A will not be B because the IGP distance from A to R_k is smaller than that from A to B. The IGP distance from A to B' is not smaller than that from A to B, otherwise the best egress router of A will not be B. If the IGP distance from A to B' is greater than that from A to B, then the IGP distance from R_k to B' is greater than that from R_k to B and R_k will choose $r_d(B)$ as its best route. Else if the IGP distance from A to B' is the same as that from A to B, then the IGP distance from R_k to B' is the same as that from R_k to B, and R_k will choose $r_d(B)$ as its best route because the tie breaking rule is deterministic.

From the above discussion we know that R_k will choose $r_d(B)$ as its best route. Likewise, R_{k-1} will learn $r_d(B)$ from R_k further and select $r_d(B)$. On the analogy of this, at last A will learn $r_d(B)$ from R_1. ∎

Appendix II: Proof of Lemma 1

Let $B = Egress_d(A)$. We will analyze different cases as follows.

Case (1): $A \in G_{sub}$ and $B \in G_{sub}$. If A and B have an IBGP session with each other, then A and B belong to any shortest IGP path between A and B. Else consider a shortest path between A and B. It follows from the construction in BGPSep that this shortest path should pass through a set of recursively produced graph separators. Because the graph separators are configured as route reflectors and the routers in the components (separated by the separator) are all clients of these route reflectors, it follows that there exist router reflectors $R_1, ..., R_r$ on the shortest path such that $S: A, R_1, ..., R_r, B$ is a signaling chain. (Note that $R_1, ..., R_r$ need not be adjacent to each other on the shortest path). So S is a shortest signaling chain.

Case (2): $A \in G - G_{sub}$ and $B \in G_{sub}$. It follows from *step 1* in BGPSep_D that there exists a vertex $C \in G_{sub}$ such that there is single IGP path $P: A, R_1, ..., R_k, C$, $R_k \notin G_{sub}$. Also $S1: A, R_1, ..., R_k, C$ is a shortest signaling chain between A and C. From the discussion of Case (1), we know that there exists a shortest signaling chain $S2$ between router C and B. Because $S1$ is a signaling chain of monotone increase, $S1$ and $S2$ can be concatenate together to form a signaling chain $S = S_1 \| S_2$. Also because there is only an IGP path between A and C, S is a shortest signaling chain.

Case (3): $A \in G_{sub}$ and $B \in G - G_{sub}$. It follows from *step 1* in BGPSep_D that there exists a vertex $C \in G_{sub}$ such that there is single IGP path $P: B, R_1, ..., R_k, C$, $R_k \notin G_{sub}$. Also $S1: B, R_1, ..., R_k, C$ is a shortest signaling chain between B and C. From the discussion of Case (1), there exists a shortest signaling chain $S2$ between router C and A. Because $S1$ is a signaling chain of monotone increase, $S1$ and $S2$ can be concatenate together to form a signaling chain $S = S_1 \| S_2$. Also because there is only an IGP path between B and C, S is a shortest signaling chain.

Case (4): $A \in G - G_{sub}$ and $B \in G - G_{sub}$. There exists a vertex $C \in G_{sub}$ such that there is single IGP path $P: A, R_1, ..., R_k, C$, $R_k \notin G_{sub}$. Also $S1: A, R_1, ..., R_k, C$ is a shortest signaling chain between A and C. And there exists a vertex $D \in G_{sub}$ such that there is single IGP path $P': D, R_1', ..., R_l', B$, $R_1' \notin G_{sub}$. Also $S3: D, R_1', ..., R_l', B$ is a shortest signaling chain between D and B. From the discussion of Case (1), there exists a shortest signaling chain $S2$ between router C and D. Because $S1$ is a signaling chain of monotone increase, $S1$ and $S2$ can be concatenate together to form a signaling chain $S' = S_1 \| S_2$. Also $S3$ is a signaling chain of monotone decrease, S' and $S3$ can be concatenate together to form a signaling chain $S = S' \| S3$. S is a shortest signaling chain because there is only an IGP path from A to C and only an IGP path from D and B.

So in all cases, for any destination d, there exists a shortest signaling chain between every router $A \in V$ and the egress router $Egress_d(A)$. ■

DiffServ–Aware MPLS Scheme to Support Policy–Based End–to–End QoS Provision in Beyond 3G Networks

Kyungkoo Jun, Seokhoon Kang, and Byungjo Choi

Department of Multimedia Systems Engineering
University of Incheon, Korea
{kjun, hana, bjc97r}@incheon.ac.kr

Abstract. 3GPP proposes the policy-based network management to provide end–to–end QoS to services, and employs IntServ and DiffServ as policy enforcement means. Even if DiffServ is preferable to IntServ in scalability and configurability, it has the limitation of not being able to leverage multiple paths, hence undesirable in the network utilization aspect. In this paper, we propose DiffServ-aware Multiple Protocol Label Switching (MPLS) as a policy enforcement means. It enables to provide differentiated levels of QoS as well as to improve network utilization by splitting packets over multiple paths. It is also able to avoid the per–flow packet ordering problem occurring when splitting a flow into multiple paths by using a hashing–based scheme. We verify the effectiveness of our proposed scheme by the NS–2 based simulation.

1 Introduction

End–to–end QoS provision in packet–based cellular services is gaining more importance as there emerge more cases that UMTS networks interwork with external IP networks. Particularly, the end–to–end QoS provision is a critical issue when it concerns delay–sensitive realtime traffic of which volume will increase significantly when the UMTS networks start in near future the full–scale adoption of IP Multimedia Subsystem (IMS)[1] to allow cellular users to access external IP network based services.

3GPP[2] suggests adopting policy–based network management architecture in order to tackle the end–to–end QoS provision issues[3]. In the current proposed policy–based architecture, IntServ[4] and DiffServ[5] are employed as policy enforcement means and Common Open Policy Service (COPS)[6] protocol is used to carry policy–related control messages among network entities. Both IntServ and DiffServ, however, have limitations. IntServ has a drawback in scalability because it tries to reserve network resources per traffic flow, while DiffServ complementing IntServ by providing class–based differentiated QoS is unable to distribute packets over multiple routing paths existing between source and destination, i.e. IP packets flow to destination along only one path determined by shortest–path–only routing rule, discarding other paths having more hop counts.

M. Gerndt and D. Kranzlmüller (Eds.): HPCC 2006, LNCS 4208, pp. 416–425, 2006.

The lack of the multiple path support makes DiffServ improper in the network resource utilization aspect to be used in the environment in which networks connect together through multiple paths. Such inadequacy of DiffServ aggravates as more multiple paths are being established as a consequence of the evolution of cellular networks into All IP–based architecture which allows more external IP networks to connect to cellular networks with easy.

In this paper, we propose to use DiffServ–aware Multi Protocol Label Switching (MPLS)[7] as a policy enforcement means in the policy–based network management for Beyond 3G networks. DiffServ–aware MPLS complements DiffServ by enabling it to route packets with same destination over multiple paths. This paper presents the following contributions.

- We propose DiffServ–aware MPLS as a policy enforcement means in the policy–based network management in Beyond 3G networks in Section 3. In a case study in Section 4, we apply DiffServ–aware MPLS based policy to provide differentiated levels of QoS to traffic flows depending on their belonging 3GPP–traffic classes, which are defined in [3] as conversational, streaming, interactive, and background.
- We present a scheme to assign different bindings of DiffServ Code Point (DSCP) and Explicit Route Label Switched Path (ER–LSP) to traffic flows depending on the traffic characteristics in order to route them through separate paths in Section 4.
- We suggest a router architecture supporting the proposed DiffServ–aware MPLS in Section 3. One of the features is that it is able to avoid the per–flow packet ordering problem occurring when splitting a traffic flow into multiple paths. It is also able to limit the maximum allowed bandwidth per traffic class by using token buckets.

2 Policy–Based End–to–End QoS Provision

The policy–based network management architecture is proposed for the end–to–end QoS provision in [3]. Policy Decision Function (PDF) and IP Bearer Service (BS) Manager are two key entities to meet the end–to–end QoS requirements of application traffic. They are in the relationship of master and slave; PDF as a master determines which policy should be applied to satisfy QoS requirements stated in Service Level Agreement (SLA), while the IP BS Manager as a slave, acting as Policy Enforcement Point (PEP), carries the decided policy into action by tweaking network resource allocation of the UMTS domain as well as the external IP domain. PDF and the IP BS manager use Common Open Policy Service (COPS) [6] protocol to exchange policy enforcement related signaling messages: a set of query and corresponding response messages.

The IP BS manager provides the end-to-end QoS of services by managing both external IP network resources and the UMTS resources. To manage the external network resources, the IP BS Manager communicates to extern IP network entities by using a set of IETF standard IP mechanisms such as DiffServ and

IntServ. The management of the UMTS domain resources is, however, carried indirectly through the UMTS BS Manager. Due to the differences between the techniques used in the IP domain and the UMTS domain, the IP BS Manager communicates to the UMTS BS manager through the Translation function.

DiffServ, while providing a scalable policy enforcement means by differentiated packet scheduling according to DiffServ Code Point (DSCP) encoded in the packets, lacks the support of multiple paths when routing packets, as mentioned in Section 1.In the following section, we suggest using DiffServ–aware MPLS as a policy enforcement means in order to complement such shortcoming of DiffServ. Moreover, our scheme is able to avoid the packet ordering problem occurring when using multiple paths.

3 Policy Enforcement Using DiffServ–Aware MPLS

MPLS[8] is an IETF standard enabling high speed packet switching by circumventing CPU-intensive table lookup in the forwarding routing table necessary to determine the next hop router of an IP packet. Instead, it uses packet switching based on labels, which are assigned to each packet by evaluating the pair of source and destination IP addresses of packets. The path that a packet passes through by switching according to its label to reach its destination, is called Label Switched Path (LSP).

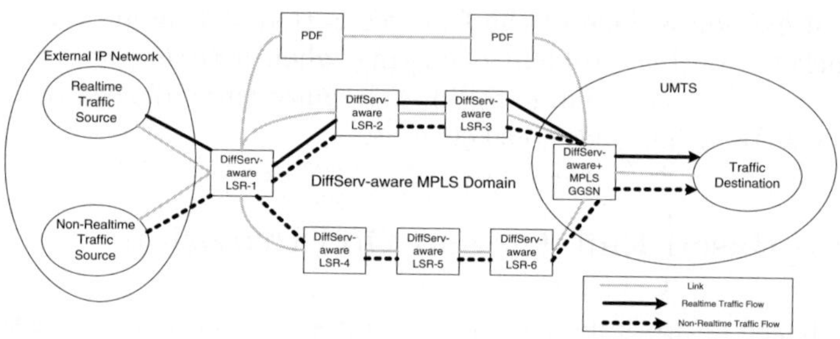

Fig. 1. The proposed DiffServ–aware MPLS based policy enforcement is capable of routing realtime and non–realtime traffic through multiple paths

DiffServ–aware MPLS[7] determines the label of a packet by considering not only the source and destination IP address pair but also the QoS class to which the packet belongs, which is encoded in DSCP. Therefore, it is feasible to route packets with the same source–destination IP address pair through different LSPs depending on their QoS requirements.

Figure 1 illustrates how the proposed DiffServ–aware MPLS based policy enforcement leverages multiple paths by routing the realtime and the non–realtime traffic through separate paths to destinations, as a consequence, being able to provide them each with differentiated QoS levels. The applied policy is to detour part of non–real traffic along a non–optimal path in order to protect the bandwidth on a shortest path for realtime traffic. As a result, realtime traffic and part of non–realtime traffic move along a shortest path consisting of Label Switch Router (LSR)–1, LSR–2, LSR–3, and GGSN, while the rest of non–realtime traffic along a one–hop longer path, LSR–1/LSR–4/LSR–5/LSR–6/GGSN.

The proposed traffic distribution of non–realtime traffic over multiple paths must be done carefully so that the packets from the same flow are not sent over different routes. Otherwise different delay of each path may cause disordered arrival of packets at a destination. It severely harms the TCP performance in particular; the disorder within a TCP flow can produce a false congestion signal and cause unnecessary throughput degradation [19]. A quick solution to preserve the per–flow packet ordering is to use either time stamp or sequence number to reorder the packets at intermediate routers or final destinations. It, however, is neither desirable nor practical when considering the complexity and the buffering necessary to store the following packets until the preceding packets arrive.

We suggest splitting traffic among multiple paths in a way that all the packets of a same flow are forwarded along a same path, preserving the packet ordering and at the same time leveraging the increased bandwidth because of the multiple paths. To differentiate the packets efficiently according to the flows, we propose a hashing scheme of 3–tuple or 5–tuple of the packet headers. The 3–tuple means the source and the destination IP address and the protocol, while the 5–tuple means the 3–tuple plus the source and the destination ports. In the simulation of Section 4, we use the 5–tuple hashing to provide more detailed level of flow classification. The 3–tuple hashing is more scalable though. We discuss the details of the suggested hashing scheme shortly.

Figure 2 shows the architecture of the LSRs to support the proposed DiffServ–aware MPLS based policy. We design the architecture to support the requirement that we split only the non–realtime traffic into multiple paths. It is because the time–sensitive characteristic makes the real–time traffic inappropriate to be distributed over multiple paths each of which has different delay, sometimes, exceeding the allowed maximum limit.

The router architecture consists of four parts from top to bottom: packet classifying queues, packet scheduler, labeler, and switching component. Firstly, incoming packets are classified according to their characteristics and, then stored in the corresponding packet classifying queues, which will be described in detail shortly. Secondly, the packet scheduler selects packets for transmission, and then the labeler marks the packets with labels for switching unless labeled yet. Finally, the switching component transmits the packets through one of outgoing network interfaces, which is determined by the label of the packets.

There are three types of the packet classifying queues, which are for the real-time traffic, for non–realtime traffic, and for DSCP–marked traffic respectively. The realtime traffic queues use a token bucket to limit the maximum allowed bandwidth for the realtime traffic. The realtime packets of which rate is below the maximum allowed bandwidth go into *in–profile* queue and are assigned EF, and otherwise into *out–profile* queue and AF. The non–realtime traffic queues are explained shortly. The DSCP–marked traffic queues stores the packets which are

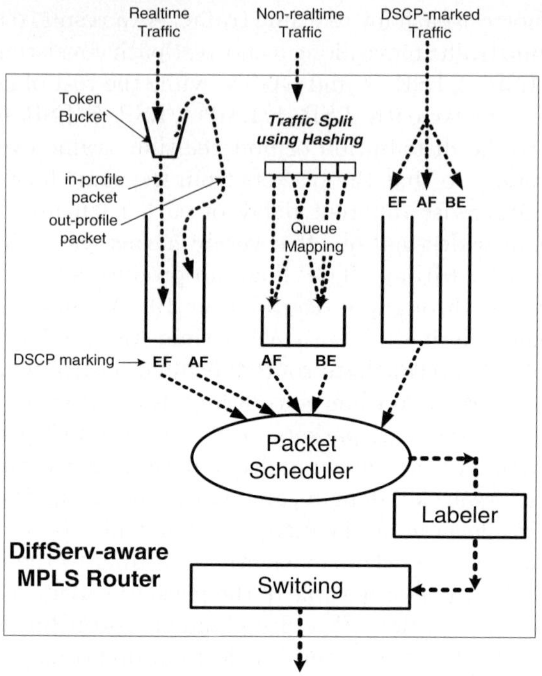

Fig. 2. The Label Switch Router structure supporting DiffServ–aware MPLS

already classified and marked accordingly by other edge DiffServ routers, thus three queues are named after the DSCP of the packets which they store: EF, AF, and BE.

The queuing process is as follows. Firstly, a hashing scheme splits non–realtime traffic stream into bins. The bins are then mapped to two queues based on a queue mapping strategy; a first fit strategy is used such that bins are mapped one by one to the first queue until the maximum allowed bandwidth for non–realtime traffic is reached. The rest of the bins are mapped to the second queue. Then, AF is assigned to the packets of the first queue, and BE to those of the second queue. As a consequence, the DiffServ marking is done in a per–flow based way; all the packets of a same flow have a same DSCP.

The packet scheduler selects the packets to transmit and the labeler determines the labels of the packets by considering the source/destination IP address pair and the allocated DSCP of the packets. Such label binding information is stored in Label Forwarding Information Base (LFIB), which is in turn referred when the switching component determines transmitting interface of outgoing packets according to their labels. The switching component operates in the same way as MPLS routers works.

We establish multiple Explicit Router(ER)–LSPs for which we are able to explicitly specify participating LSRs in order to route packets along different LSPs according to their QoS requirements. To set up such ER–LSPs, we use

Constraint–based Route Label Distribution Protocol (CR–LDP) [18], an extension of LDP which is a MPLS signaling protocol for LSRs to exchange with each other the binding information between labels and corresponding LSP. Those LDP message communicating LSRs are called *peer LSRs*; LSR–1 to 5 and GGSN are in the relationship of the peer LSRs.

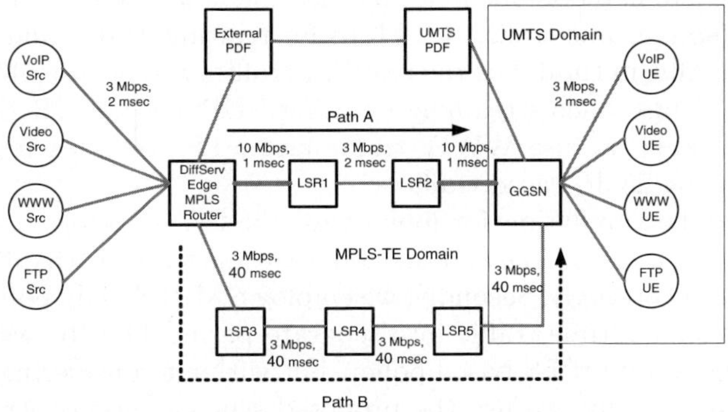

Fig. 3. The simulation setup consisting of an external IP network, a UMTS domain, and a set of connecting LSRs

Our proposed scheme adopts the E–LSP to combine the DiffServ and the MPLS information together. Since there are only four 3GPP–defined classes, 3–bit EXP field, which is able to represent eight levels of QoS, is sufficient to differentiate the 3G traffics properly, thus our proposed LSR structure is designed to treat the E–LSP accordingly. In our future research, we will investigate the necessary modification of the proposed LSR structure to accommodate the L–LSP scheme, which is expected to support more levels of QoS differentiation.

4 Simulation and Its Results

We perform a simulation to verify the performance and the effectiveness of our proposed DiffServ–aware MPLS based policy enforcement. We use NS–2 version 2.29 [9] to run the simulation. To enable routers to support the DiffServ–aware MPLS, we make necessary modification to MPLS–related modules of NS–2.

Figure 3 shows the simulation setup in which 3GPP–defined four types of traffic, i.e. conversational, streaming, interactive, and background, flow from an external IP network to a UMTS domain through a set of DiffServ–aware MPLS based LSRs. The simulation setup consists of three network parts: an external IP network hosting four traffic sources corresponding to 3GPP–defined service classes, a UMTS domain having traffic sinks, and a set of intermediate LSRs providing the DiffServ–aware MPLS capability. The LSRs connecting the external IP network with the UMTS domain provide two different paths. First, the path of Edge/LSR–1/LSR–2/GGSN, notated as *Path A* in the figure, is the shortest path taken as the default routing path. Second, the path of Edge/LSR–3/LSR–4/LSR–5/GGSN, *Path B*, is a non–optimal path with one additional hop. The maximum bandwidths of two paths are the same as 3 Mbps. It should be, however, be noted that path A and B have significantly different delays as 4 msec and 160 msec respectively, the delay of path B is 40 times bigger than that of path A.

We generate simulated traffic in a way that the non–real time traffic has more volume than the realtime traffic: 1.2 Mbps for realtime traffic and 4.8 Mbps for non–realtime traffic in total. For the realtime traffic, we generate 400 Kbps VoIP traffic and 800 Kbps video streaming by using UDP–based CBR traffic. For the non–realtime traffic, we use WWW traffic and FTP traffic each generating at 2.4 Mbps by using TCP–based traffic.

We perform the simulation for four cases. Firstly, we apply a DiffServ–only policy by marking the realtime and the non–realtime traffic with different DSCPs and scheduling accordingly. Secondly, we employ a MPLS–only policy by routing realtime and non–realtime traffic into separate paths. Thirdly, we use the proposed DiffServ–aware MPLS based policy, but without considering the per–flow packet ordering. Finally, we use the proposed scheme supporting the per–flow packet ordering. As the QoS metric to be compared among the cases, we measure the goodput per traffic classes.

Figure 4(a) shows the result of the DiffServ–only policy application case in which the realtime traffic bandwidth is ensured at the sacrifice of the non–realtime traffic. We observe that the goodputs of the VoIP and the streaming traffic are guaranteed to support the source transmission rate of 400 Kbps and 800 Kbps respectively but the goodputs of the WWW and the FTP traffic, which are around 900 Kbps, are largely dropped compared with the source transmission rates of 2.4 Mbps.

In the DiffServ–only case, the DiffServ Edge MPLS router of Figure 3 differentiate the realtime traffic from the non–realtime traffic by assigning the following DSCPs to incoming packets: EF to in–profile realtime traffic packets and AF to out–profile realtime traffic, while AF and BE for in–profile and out–profile non–realtime traffic respectively. To set up the maximum allowed bandwidth per traffic, we allocate a token rate of 1 Mbps each for VoIP and video streaming traffic, and 500 Kbps each for WWW and FTP traffic. As a consequence, most of the WWW and the FTP packets which are given the BE code are dropped , on the contrary all of the VoIP and the streaming packets are coded as EF, reaching the destination successfully.

Figure 4(b) shows the result of the MPLS–only policy application case in which the realtime traffic bandwidth is guaranteed by preventing the non–realtime traffic from sharing the same path, path A of Figure 3, instead the non–realtime traffic is allowed to use the path B exclusively. However, it has a drawback of low network utilization. We observe that the goodputs of the VoIP and the streaming traffic are ensured to support the source transmission rate of 400 Kbps and 800 Kbps each, consuming 40% of the maximum bandwidth of

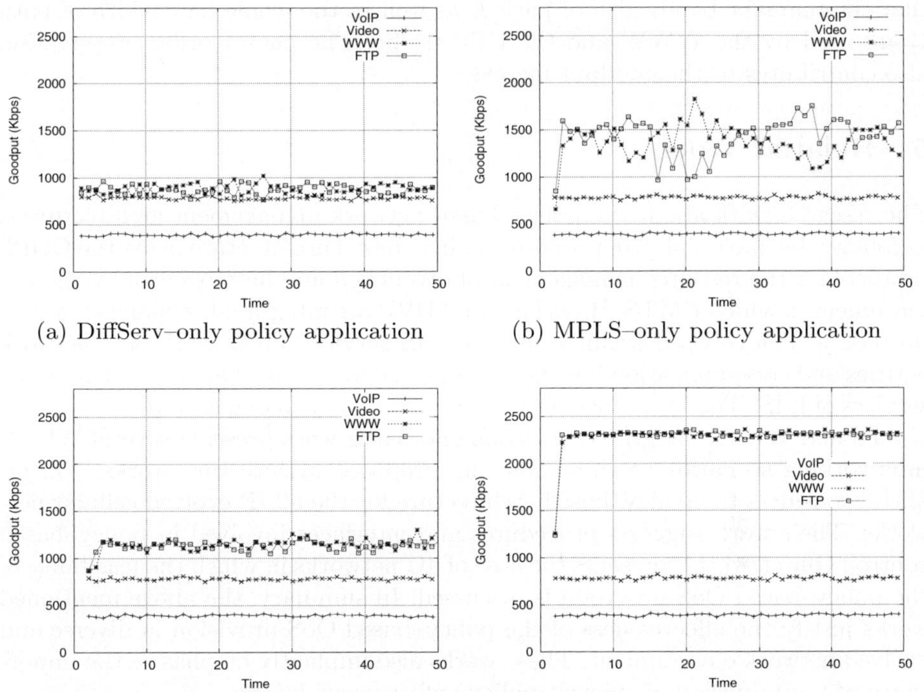

(a) DiffServ–only policy application

(b) MPLS–only policy application

(c) DiffServ–aware MPLS application without the packet ordering support

(d) DiffServ–aware MPLS application with the packet ordering support

Fig. 4. Goodput of traffic classes measured for four cases

the path A, and the goodputs of the WWW and the FTP traffic are average 1.4 Mbps each, which are little less than the maximum bandwidth of path B

Figure 4(c) shows the result of the DiffServ–aware MPLS based policy without the packet ordering support. The non–realtime packets are given AF and BE only depending on the packet rate, regardless of their belonging flow. Then, the AF packets are routed along path A, while the BE packets along path B, resulting in a case that the packets of a same flow are split into path A and B.

As a consequence, the goodputs of the WWW and the FTP traffic are even smaller than those of the MPLS–only case shown in Figure 4(b). The reason is that the disordering of TCP flow packets which are delivered through two delay–different paths, causes the false congestion signal, thus preventing the senders from raising the transmission rate. Nevertheless, the goodput of the realtime traffic is guaranteed.

Figure 4(d) shows the result of the proposed DiffServ–aware MPLS based policy application which supports the packet ordering. It splits the non–realtime traffic into path A and path B based on the flows. It not only guarantees the realtime traffic bandwidth but also increases the goodput of the non–realtime traffic as high as 2.3 Mbps each for the WWW and the FTP, which is much higher than 1.2 Mbps of the case without the packet ordering support. The reason is that the remnant bandwidth of path A as well as the whole bandwidth of path B are used by the WWW and the FTP traffic. The packet order preservation also contributes to the goodput increase.

5 Related Work

The research work about the policy–based network management architecture is as follows. Iacono et. al. [10] propose a policy–based architecture using the COPS protocol for the resource management of diversified and heterogeneous radio environment in which UMTS, HiperLan, and DVB are integrated. Zhuang et. al.[11] propose a policy–based architecture for IMS service. Their work lists network entities and communication interfaces necessary to support QoS of session–based services of IMS. The same authors[12] suggest a policy–based architecture for the UMTS and WLAN integrated environment. Their work presents several WLAN interworking scenarios to show how the proposed architecture works. Xin et. al.[13] considers the policy–based architecture for the all IP evolved cellular networks. Their work suggests procedures and interfaces involved in policy–based control. Gur et. al.[14] presents the case of 4G networks in which the feasibility of the policy–based QoS provision is discussed. In summary, the above mentioned works justify the effectiveness of the policy–based QoS provision in diverse and evolved network environment. These works also implicitly emphasize the importance of the provision of efficient policy enforcement means.

The research work about the policy enforcement schemes is as follows. Kim et. al. [15] presents the case study of using DiffServ as the policy enforcement means in 3G network connected to IP network through DiffServ–enabled routers. By allocating differentiated DSCP depending on traffic characteristics, it is able to ensure the QoS of realtime traffic. This scheme, however, is not able to utilize multiple paths as explained in Section 3. RFC 3564 [16] lists the scenario–based requirements that should be provided by the DiffServ–aware MPLS when it is used for traffic engineering purpose. We refer to its work to contemplate our simulation scenarios. One of the 3GPP technical specifications [17] suggests extending MPLS to be adopted as a policy enforcement means. Our work is founded on its work and develops the idea further.

6 Conclusions

In this paper, we propose to use the DiffServ–aware MPLS as a policy enforcement scheme to provide the required QoS to packet–based services in Beyond 3G networks interworking with external IP networks. The DiffServ–aware MPLS as the policy enforcement scheme is able to not only satisfy specified QoS requirements but also improve the network utilization by complementing the drawback of DiffServ, enabling to use multiple paths. Also, we design a router architecture to use a hashing scheme to split traffic stream based on flow to be able avoid the packet ordering problem occurring when using multiple paths. We verify the performance and the effectiveness of our proposed scheme by the NS–2 based simulation.

References

1. 3GPP Technical Specification 23.228 v.7.2.0 "IP Multimedia Subsystem (IMS) (R7)," Dec. 2005.
2. http://www.3gpp.org
3. 3GPP Technical Specification 23.207 v.6.6.0 "End–to–End QoS Concept and Architecture (R6)," Oct. 2005.
4. R. Braden et al., "Resource ReSerVation Protocol (RSVP) – Version 1 Functional Specification," RFC 2205, Sept. 1997.
5. S. Blake et al., "An Architecture for Differentiated Services," RFC 2475, Dec. 1998.
6. D. Durham et al., "The COPS (Common Open Policy Service) Protocol," RFC 2748, Jan. 2000.
7. L. Wu et al., "Multi–Protocol Label Switching (MPLS) Support of Differentiated Services," RFC 3270, May 2002.
8. E. Rosen et al., "Multi–Protocol Label Switching Architecture," RFC 3270, Jan. 2001.
9. http://www.isi.edu/nsnam/ns/
10. Iacono, S., Arneodo, F., Cardoso, K., Genet, M.G. and Zeghlache, D., "Policy based management for next generation mobile networks," in Proceedings of Wireless Communications and Networking 2003, Mar. 2003.
11. Zhuang, W., Gan, Y., Loh, K. and Chua, K., "Policy-based QoS Architecture in the IP Multimedia Subsystem of UMTS," IEEE Network, May/June 2003.
12. Zhuang, W., Gan, Y., Loh, K. and Chua, K., "Policy-based QoS Management Architecture in an Integrated UMTS and WLAN Environment," IEEE Communications Magazine, Nov. 2003.
13. Xin, Z., Xu, W., Fang, S., Yang, J., and Ping, Z., "Policy based end-to-end service control framework beyond 3G mobile network," in Proceedings of Vehicular Technology Conference 2004, Fall, 2004.
14. Gur, G., Alagoz, F., Tugcu, T., and AbdelHafez, M., "Exploring the Issues in Policy-Based Approaches for QoS Support in 3G+ Mobile Networks," In Proceedings of 2nd IFIP/IEEE International Conference on Wireless and Optical Communications Networks 2005, March 2005.
15. Kim, M., Nam, K., Lee, J., Hwang-Soo Lee, "A Case Study of Policy–based QoS Management in 3G Networks," in Proceedings of Vehicular Technology Conference, 2003, Spring, 2003.
16. Faucheur, F. et. al., "Requirements for Support of Differentiated Services–aware MPLS Traffic Engineering," RFC 3564, Jul. 2003.
17. 3GPP Technical Specification 23.802 v.6.6.0 "Architectural Enhancements for end–to–end Quality of Service (QoS) (R7)," Sep. 2005.
18. Jamoussi, B., "Constraint–Based LSP Setup using LDP," RFC 3212, Jan. 2002.
19. Wang, Z., "Internet QoS: Architectures and Mechanisms for Quality of Services," Morgan Kaufmann, 2001.

Effect of Flow Aggregation on the Maximum End-to-End Delay

Jinoo Joung[1], Byeong-Seog Choe[2], Hongkyu Jeong[3], and Hyunsurk Ryu[3]

[1] Sangmyung University, Seoul, Korea
jjoung@smu.ac.kr
[2] Dongkuk University, Seoul, Korea
[3] Samsung Advanced Institute of Technology, Kiheung, Korea

Abstract. We investigate the effect of flow aggregation on the end-to-end delay in large scale networks. We show that networks with Differentiated services (DiffServ) architectures, where packets are treated according to the class they belong, can guarantee the end-to-end delay for packets of the highest priority class, which are queued and scheduled with a strict priority, but without preemption. We then analyze the network with arbitrary flow aggregation and deaggregation, and again derive an upper bound on the end-to-end delay. Throughout the paper we use Latency-Rate ($\mathcal{LR}$) server model, and prove that FIFO, Strict Priority, and other rate-guaranteeing servers with aggregated flows are all $\mathcal{LR}$ servers to *individual flows* in certain conditions. We show that the delay bound of a flow that experiences aggregation and deaggregation, including the flows in DiffServ, depends on, among others, the burst sizes of the other flows within the aggregated flow and the number of the aggregations and the deaggregations.

1 Introduction

The problem of guaranteeing QoS within a packet switching network has been extensively studied and several solutions to this problem have been suggested. IETF has defined two services on IP networks which are collectively called Integrated services: the Controlled Load (CL) service and the Guaranteed Rate (GR) service [1,2]. The CL service defines a service that approximates the behavior of best-effort service under lightly utilized networks. The GR service, which we will refer as IntServ in this paper, guarantees the end-to-end QoS by means of reserving, allocating and providing an amount of predefined resource to each data traffic unit, which often is called a flow or a session, in every server. Let alone the resource reservation, managing flows in a network node means a lot of works. This complexity inhibits IntServ-type QoS architectures from being adopted in real networks. DiffServ [3] is another approach that has been proposed to solve the scalability problem of IntServ. It classifies packets, or the flows they belong, into a number of traffic classes. The packets are marked accordingly at the edge of a network. Therefore the hard works are necessary only at the edge nodes. Classes may be assigned with strict priorities, or a certain amount of bandwidth

M. Gerndt and D. Kranzlmüller (Eds.): HPCC 2006, LNCS 4208, pp. 426–435, 2006.

is provisioned for each class, as was the case with a flow in IntServ. With the support from a proper signaling scheme, DiffServ is an overly simplified version of IntServ, where many flows are aggregated into a single *class*, which is another name for an aggregated flow, and treated as a whole. We consider three networks with different QoS architectures. The first one is with IntServ-type architecture, where many flows are handled independently. The second network is with DiffServ-type architecture, where we consider the QoS characteristics of the highest priority class traffic with strict priority. The third network is with a hybrid architecture of aforementioned networks, where many flows may be aggregated and deaggregated arbitrarily. We focus on the queueing and scheduling behaviors of the flows or aggregated flows, and investigate the delay characteristics of them, in each networks.

QoS characteristics of the network with IntServ architecture have been well studied and understood by numerous researches in the past decade. Providing the allocated bandwidths, or service rates, or simply rates of an output link to multiple sharing flows plays a key role in this approach. Among a myriad of scheduling algorithms, we focus on the deficit round robin (DRR) [4], because the sorted priority scheduling algorithms, including Packetized Generalized Processor Sharing (PGPS), suffer from the complexity, which is $O(\log N)$ at best while N is the number of active flows in a scheduler [5]. The DRR, with many other rate-providing servers, is proved to be a Latency-Rate server [6], or simply $\mathcal{LR}$ server. All the work-conserving servers that guarantee rates exhibit this property and can therefore be modeled as $\mathcal{LR}$ servers. The behavior of an $\mathcal{LR}$ server is determined by two parameters, the latency and the allocated rate. The latency of an $\mathcal{LR}$ server may be considered as the worst-case delay seen by the first packet of the busy period of a flow. It was shown that the maximum end-to-end delay experienced by a packet in a network of $\mathcal{LR}$ servers can be calculated from only the latencies of the individual servers on the path of the flow, and the traffic parameters of the flow that generated the packet. More specifically for a leaky-bucket constrained flow,

$$D_i \le \frac{\sigma_i}{\rho_i} + \sum_{j=1}^{k} \Theta_i^{S_j}, \tag{1}$$

where D_i is the delay of flow i within a network, σ_i and ρ_i are the well known leaky bucket parameters, the maximum burst size and the average rate, respectively, and $\Theta_i^{S_j}$ is the latency of flow i at the server S_j. The second network with DiffServ architecture is then considered. In this network packets in a same class are enqueued to a single queue and scheduled in a FIFO manner within the class. The higher priority class may be served with a strict priority over the lower classes, or a certain amount of bandwidth may be assigned to each class. In the third network, flows may be aggregated with an $\mathcal{LR}$ server at any node in the network and deaggregated at some other node's input port. $\mathcal{LR}$ servers again schedule the aggregated flows and used throughout the network. There is a significant volume of researches for such networks with flow aggregation. End-to-end delay bounds with using *fair aggregator* was investigated [8]. It was shown that under condition that the scheduler is *fair*, the maximum delay of an aggregated flow is bounded. A fair scheduler, however, should be able to refrain itself from transmitting packets at full link capacity whenever one or more flows are not active, i.e. do not have packets to transmit at the moment. This mandates a non-work conserving type of scheduler behavior to bound the delay. Using Guaranteed Rate ($\mathcal{GR}$) servers [9] as *fair aggregator* was also investigated [10], and the maximum end-to-end delay was obtained. It was concluded that the aggregated scheduling provides even better delay performance than per-flow scheduling. Contrary to the work with $\mathcal{GR}$ servers [10], we still find the traditional per-flow scheduling performs better in general cases. This is because that the aggregated scheduling does not protect the flow under observation from other flows within the aggregate. If we have sufficiently large amount of burst from other flows through aggregation and deaggregation, then the aggregated scheduling performs quite poorly. This is obvious when we consider DiffServ as an extreme of the hybrid architecture where aggregation and deaggregation occur in *every* node. Finally we compare the end-to-end delays in each networks we analyzed.

2 Previous Works on $\mathcal{LR}$ Servers

We describe $\mathcal{LR}$ servers and its properties. The concept and the primary methodology for the analysis of $\mathcal{LR}$ servers are suggested by Stiliadis [7]. A *server* is a commonly used terminology which in convention means the combination of a scheduler and a transmitter that reside in a output port controller of a switch or a router. We assume a packet switch (router) where a set of flows share a common output link. We assume that the switches (routers) are store-and-forward devices. Let $A_i(\tau, t)$ denote the arrivals from flow i during the interval $(\tau, t]$ and $W_i(\tau, t)$ the amount of service received by flow i during the same interval. In the packet-by-packet model we assume that $A_i(\tau, t)$ increases only when the last bit of a packet is received by the server; likewise $W_i(\tau, t)$ is increased only when the last bit of the packet in service leaves the server. We further denote that a flow i is *backlogged* when one or more packets of i are waiting for service. In other words, if $A_i(0, t) - W_i(0, t)$ is larger than zero then the flow i is backlogged at t. A *server busy period* is a maximal interval of time during which the server is never idle. During a server busy period the server is always transmitting packets. A flow i *busy period* is a maximal interval of time $(\tau_1, \tau_2]$ such that for any time $t \in (\tau_1, \tau_2]$, packets of flow i arrive with rate greater than or equal to ρ_i or, $A_i(\tau_1, t) \geq \rho_i(t - \tau_1)$. Now we are ready for the definition and the primary characteristics of $\mathcal{LR}$ servers.

Definition 1. *A server $\mathcal{S}$ belongs in the class $\mathcal{LR}$ if and only if for all times t after time τ that the jth busy period started and until the packets that arrived during this period are serviced, $W_{i,j}^{S}(\tau, t) \geq \max\left(0, \rho_i(t - \tau - \Theta_i^S)\right)$. Θ_i^S is the minimum non-negative number that satisfies the above inequality.*

Lemma 1. *If $\mathcal{S}$ is an $\mathcal{LR}$ server, and flow i is leaky bucket constrained with parameters (σ_i, ρ_i), then the followings hold.*

1. *If $Q_i^S(t)$ is the backlog of flow i at time t, $Q_i^S(t) \leq \sigma_i + \rho_i \Theta_i^S$.*
2. *If D_i^S is the delay of any packet in flow i in server $\mathcal{S}$, $D_i^S \leq \sigma_i/\rho_i + \Theta_i^S$.*
3. *The output traffic of flow i from $\mathcal{S}$ conforms to the leaky bucket model with parameters $(\sigma_i + \rho_i \Theta_i^S, \rho_i)$.*

Proof. See the proof of theorem 3.1 of [7]. □

DRR is proved to be an $\mathcal{LR}$ server. For a detailed description on DRR, refer to the original paper [4]. The latency of the DRR server is given as $(3F - 2\phi_i)/r$, where ϕ_i is the quantum value (a relative weight given to a flow), F is the frame size, which is the sum of all ϕ_i over i, and r is the output link capacity.

3 Delay Bounds in DiffServ Architecture

Under a condition that there is no flow that violates the leaky bucket constraint specified for itself, with only a simple FIFO scheduler or with a strict priority scheduler, we will show that we can still guarantee the delay upper bound for each flows. In this environment, the sum of all flows that want to pass through a server is compared with the link capacity, and if it's less than the link capacity then delay upper bounds will be prescribed, as it can be calculated with the method explained in this section. We assume there is a maximum length for packets, and denote with L.

Lemma 2. *During a server busy period, the amount of service given by a FIFO server, $W^{\text{FIFO}}(\tau, t)$, is bounded by $W^{\text{FIFO}}(\tau, t) \geq \max\left(0, r(t - \tau - \frac{L}{r})\right)$, where τ and t are arbitrary instants within the server busy period, L is the maximum packet length, and r is the link capacity.*

Proof. Let τ be any instant in a server busy period. By definition during a server busy period the server is always serving packets at the rate r. Assume a packet was being served at τ. Assume the service for this packet is started at τ_0 and finished at τ_1. Let the length of this packet be L_1. Let us further denote by $\tau_2, \tau_3, \ldots$ the time instants the subsequent series of packets be served. Similarly we will denote by $L_2, L_3, \ldots$ the lengths of these packets. For the time instants at which the packet services are finished, $W(\tau, \tau_n) = \sum_{i=1}^{n} L_i = \sum_{i=1}^{n} r(\tau_i - \tau_{i-1}) \geq r(\tau_n - \tau) \geq \max\left(0, r(\tau_n - \tau - \frac{L}{r})\right)$. This inequality holds for any τ_n, including the service finish time of the last packet of the server busy period. For any time t in between the packet service instants, $\tau < t < \tau_n$, $W(\tau, t) = \sum_{i=1}^{n-1} L_i = \sum_{i=1}^{n-1} r(\tau_i - \tau_{i-1}) \geq r(\tau_{n-1} - \tau) \geq \max\left(0, r(t - \tau - \frac{L}{r})\right)$. The last inequality holds since $\tau_{n-1} > t - \frac{L_n}{r} \geq t - \frac{L}{r}$. From the above inequalities we conclude that $W(\tau, t) \geq \max\left(0, r(t - \tau - \frac{L}{r})\right)$, for any time between τ and the end of the server busy period. Moreover τ can be arbitrary, the lemma follows. □

If we consider a case where τ was chosen to be the starting time of the service of a maximum length packet, we can easily see that the upper bound given in lemma 2 is indeed tight. The identical logic derives the service lower bound for strict priority scheduler without preemption, as the following.

Lemma 3. *During a server busy period, the amount of service given by a strict priority server to the high priority traffic, $W^{SP}(\tau,t)$, is bounded by $W^{SP}(\tau,t) \geq \max\left(0, r\left(t-\tau-\frac{2L}{r}\right)\right)$, where τ and t are arbitrary instants within the server busy period.*

Proof. The proof of this lemma takes the same steps with the proof of lemma 2. The only difference is that with the strict priority scheduler even the highest priority queue has to wait for the completion of the service of the packet currently being served, since we assume a non-preemptive server. We omit the detail. □

Lemma 4. *Let K be the number of flows coming into server $\mathcal{S}$, a FIFO server or a strict priority server, and all the flows be leaky bucket constrained with parameters (σ_i, ρ_i), $1 \leq i \leq K$. The followings hold.*

1. *If $Q^S(t)$ is the backlog at time t of server $\mathcal{S}$, $Q^S(t) \leq \sigma + \rho\Theta^S$, where $\sigma = \sum_{i=1}^{K}\sigma_i$, $\rho = \sum_{i=1}^{K}\rho_i$, and $\Theta^S = L/r$ when $\mathcal{S}$ is FIFO, $2L/r$ when $\mathcal{S}$ is SP.*
2. *If D^S is the delay of any packet in server $\mathcal{S}$, $D^S \leq \sigma/r + \Theta^S$.*

Proof. The proof is similar to the proof of lemma 1. We omit the detail because of the space limitation. □

Lemma 5. *Under a condition that all the input flows are leaky-bucket constrained, during a flow i busy period a FIFO server or an SP server can provide service to flow i as the following: $W_i^S(T_0,t) \geq \max\left(0, \rho_i(t-T_0-(\sigma-\sigma_i)/r-\Theta^S)\right)$, where T_0 is the starting time of flow i busy period.*

Proof. By definition,

$$W_i^S(T_0,t) = Q_i^S(T_0) + A_i(T_0,t) - Q_i^S(t) \geq A_i(T_0,t) - Q_i^S(t). \tag{2}$$

Let us consider about the amount of backlogged packets of i at t, $Q_i^S(t)$. Again by definition the amount of the arrival during a flow i busy period is lower bounded as $A_i(T_0,t) \geq \rho_i(t-T_0)$. Let us first start with the assumption on arrival $A_i(T_0,t)$ as the following. $A_i(T_0,t) \leq \sigma_i^* + \rho_i(t-T_0)$, for some σ_i^*, $0 \leq \sigma_i^* \leq \sigma_i$. Then the maximum delay a packet from any flow will experience is bounded by lemma 4 with D^* as $D^* = \sigma^*/r + \Theta^S$, where $\sigma^* = \sum_{j=1}^{K}\sigma_j$, under condition that $\sigma_i = \sigma_i^*$. Since the maximum delay that a packet in any flow in a server is bounded, packets that arrived before $t - D^*$ must have been served at t. That is, for $t \geq T_0 + D^*$,

$$Q_i^S(t) \leq A_i(T_0,t) - A_i(T_0,t-D^*) \leq A_i(T_0,t) - \min_{\sigma_i^*} A_i(T_0,t-D^*). \tag{3}$$

Because $A_i(T_0,t-D^*) \leq \sigma_i^* + \rho_i(t-D^*-T_0) = \sigma_i^* + \rho_i(t-T_0-\sigma^*/r-\Theta^S) = \sigma_i^*(1-\rho_i/r) + \rho_i(t-T_0-(\sum_{j\neq i}\sigma_j)/r-\Theta^S)$, it is easy to see that $\arg\min_{\sigma_i^*} A_i(T_0,t-D^*) = 0$. Let us denote by D^0 the maximum delay with $\sigma_i^* = 0$, or equivalently $\min_{\sigma_i^*} A_i(T_0,t-D^*) = A_i(T_0,t-D^0) = \rho_i(t-T_0-D^0)$.

The last equality holds since $A_i(T_0, t-D^0) \geq \rho_i(t-T_0-D^0)$ by the definition of flow i busy period. Then from equation (2) and (3), for $t > T_0+D^0$, $W_i^S(T_0,t) \geq A_i(T_0,t-D^0) \geq \rho_i(t-T_0-D^0)$. For $t < T_0+D^0$, $W_i^S(T_0,t)$ can be zero, since the first packet from i busy period may experience the maximum delay. Therefore $W_i^S(T_0,t) \geq \max\left(0, \rho_i(t-T_0-(\sigma-\sigma_i)/r-\Theta^S)\right)$. □

The following theorem is a direct consequence from the definition of $\mathcal{LR}$ servers and lemma 5.

Theorem 1. *A FIFO server or an SP server, under conditions that all the input flows are leaky bucket constrained and the sum of average rates is less than the link capacity, is an $\mathcal{LR}$ server for individual flows with latency given as the following: $\Theta_i^S = (\sigma^S - \sigma_i^S)/r^S + \Theta^S$, where σ^S is the sum of all the σ_i^S within the server $\mathcal{S}$, r^S is the link capacity of $\mathcal{S}$, and $\Theta^S = L/r^S$ when $\mathcal{S}$ is FIFO, $2L/r^S$ when $\mathcal{S}$ is SP.*

From lemma 1 and theorem 1, the following corollary can be claimed, so that the maximum burst size at each server through the network can be calculated recursively.

Corollary 1. *The output traffic of flow i from a FIFO server or an SP server $\mathcal{S}$ conforms to the leaky bucket model with parameters $(\sigma_i^S + \rho_i\Theta_i^S, \rho_i)$, where σ_i^S is the maximum burst size of flow i into the server $\mathcal{S}$.*

The end-to-end delay of a network with DiffServ architecture with FIFO servers and/or Strict Priority servers therefore can be obtained by the following sets of equations: $D_i \leq \sigma_i/\rho_i + \sum_{n=1}^{N} \Theta_i^{Sn}$; $\Theta_i^{Sn} = (\sigma^{Sn} - \sigma_i^{Sn})/r^{Sn} + \Theta^{Sn}$; $\sigma_i^{Sn} = \sigma_i^{Sn-1} + \rho_i\Theta_i^{Sn-1}$, where $\mathcal{S}n$ is the nth server from the entrance of a network.

4 Delay Bounds in Networks with Arbitrary Flow Aggregations

A rationale for providing an amount of reserved service rate to an individual flow in IntServ architecture is to protect the flow from other data traffic from unpredictable sources that request best-effort service to the network, or malicious users that purposefully violate the input constraints. All the $\mathcal{LR}$ servers successfully achieve this mission, at the cost of the complexity of per flow scheduling and queueing. If we have a confidence in some of flows, however, that they never violate the promised leaky bucket parameters, or the network itself can shape the incoming traffic to conform to these parameters, then those trusted flows can be aggregated into a fatter flow while still be guaranteed for QoS, therefore we can greatly reduce the scheduling and queueing complexity in a server.

We consider a series of switches each with $\mathcal{LR}$ servers. The $(n-1)$th server from the network entrance, $\mathcal{S}(n-1)$, generates output traffic $I_{out}^{S(n-1)}$, or equivalently an aggregated flow of several elemental flows including i, which is the flow under observation. The next switch also has many output ports therefore many servers, including $\mathcal{S}n$ to which i is destined. Among the flows within $I_{out}^{S(n-1)}$, some of flows are switched to this server. Let us denote by I_{in}^{Sn} such a set of flows. Note that $I_{in}^{Sn} \subset I_{out}^{S(n-1)}$. In $\mathcal{S}n$, I_{in}^{Sn} is considered as a single flow, and queued into a single queue and served accordingly. There are other elemental flows or aggregated flows from the same or other input ports, which share the server $\mathcal{S}n$ with I_{in}^{Sn} and then aggregated with it, thus are consisted in I_{out}^{Sn}. Note that for the aggregation $\mathcal{S}n$ does nothing more than a normal $\mathcal{LR}$ server does. Also note that there may be other background flows that share the servers but are not aggregated into flow I_{out}^{Sn}. The amount of service given to the aggregated flow I_{in}^{Sn} is bounded as follows, because it is served by an $\mathcal{LR}$ server: $W_{I,in}^{Sn} \geq \max\left(0, \rho_{I,in}^{Sn}(t - T_0 - \Theta_{I,in}^{Sn})\right)$, where T_0 is the starting time of a flow I_{in}^{Sn} busy period, $\Theta_{I,in}^{Sn}$ is the latency of the aggregated flow I_{in}^{Sn} at $\mathcal{S}n$, and $\sigma_{I,in}^{Sn} = \sum_{k \in I_{in}^{Sn}} \sigma_k^{Sn}, \rho_{I,in}^{Sn} = \sum_{k \in I_{in}^{Sn}} \rho_k$, where σ_k^{Sn} is the maximum burst size of a flow k within I_{in}^{Sn}. The delay of a packet within flow I_{in}^{Sn} is also bounded by the lemma 1 as the following: $D_{I,in}^{Sn} \leq \sigma_{I,in}^{Sn}/\rho_{I,in}^{Sn} + \Theta_{I,in}^{Sn}$. We are interested in the output traffic characteristics of the flow i within I_{out}^{Sn}. Let $W_i^{Sn}(\tau, t)$ be the service given to the packets that belong to flow i, at the server $\mathcal{S}n$ during a time interval $[\tau, t)$. We argue the following.

Lemma 6. *Under a condition that all the input flows within I_{in}^{Sn} are leaky-bucket constrained, during a flow i busy period an $\mathcal{LR}$ server $\mathcal{S}n$ can provide service to flow i, within aggregated flow I_{in}^{Sn}, as the following: $W_i^{Sn}(T_0, t) \geq \max\left(0, \rho_i(t - T_0 - (\sigma_{I,in}^{Sn} - \sigma_i^{Sn})/\rho_{I,in}^{Sn} - \Theta_{I,in}^{Sn})\right)$, where T_0 is the starting time of flow i busy period.*

Proof. The proof takes the same steps with the proof of lemma 5. Since an aggregated flow is queued and scheduled as if they were in a FIFO server with some latency, lemma 5 can be considered as a special case of this lemma with the latency L/r and the allocated rate r. We omit the detail. □

The following theorem is a direct consequence from the definition of $\mathcal{LR}$ servers and lemma 6.

Theorem 2. *An $\mathcal{LR}$ server with an aggregated flow, under condition that all the input flows within the aggregated flow are leaky bucket constrained, is an $\mathcal{LR}$ server for individual flows with latency given as the following: $\Theta_i^S = (\sigma_{I,in}^S - \sigma_i^S)/\rho_{I,in}^S + \Theta_{I,in}^S$, where I_{in}^S is the aggregated flow.*

From lemma 1 and theorem 2, the following corollary can be claimed, so that the maximum burst size at each server through the network can be calculated recursively.

Corollary 2. *The output traffic of flow i within an aggregated flow I_{out}^S from an $\mathcal{LR}$ server $\mathcal{S}$ conforms to leaky bucket model with parameters $(\sigma_i^S + \rho_i\Theta_i^S, \rho_i)$, where σ_i^S is the maximum burst size of flow i into the server $\mathcal{S}$.*

The maximum end-to-end delay of a network of $\mathcal{LR}$ servers with aggregated flows can be obtained by the following sets of equations: $D_i \leq \sigma_i/\rho_i + \sum_{n=1}^{N} \Theta_i^{Sn}$; $\Theta_i^{Sn} = (\sigma_{I,in}^{Sn} - \sigma_i^{Sn})/\rho_{I,in}^{Sn} + \Theta_{I,in}^{Sn}$; $\sigma_{I,in}^{Sn} = \sum_{k \in I_{in}^{Sn}} \sigma_k^{Sn}$; $\sigma_i^{Sn} = \sigma_i^{Sn-1} + \rho_i\Theta_i^{Sn-1}$.

5 Numerical Results

We focus on a residential network environment, where the maximum number of hops and the number of flows are confined and predictable. Moreover in such networks the demand for real-time service is strong, especially for video and high quality audio applications. IEEE 802.3 Residential Ethernet Study Group [11] defines a bound for the end-to-end delay to be 2ms in a network of 7 hops for stringent audio and video applications [12]. We assume the 100Mbps Fast Ethernet links are used across the network.

Consider a network of arbitrary topology whose maximum radius is seven hops. We refer a hop by a switch, therefore a server. Consider the longest path where seven DRR servers are in series. In this longest path, at each server there are eight flows with the average rate of 10Mbps, including the flow under observation, i. The maximum burst size of i at the entrance of the network is 266 bytes. This maximum burst size, which is the same with the maximum packet size, is calculated by assuming MPEG-2 Transport Streams (TS) over IP over Ethernet. Note that the other flows do not have any burst size constraints. The flows other than i at different servers may or may not be the same ones. In this scenario the latencies at all the servers are identical and is 0.468ms, as given in equation (2). The end-to-end delay with seven servers is again obtained from equation (1) and is 3.490ms.

Next, consider a network of the same topology with the previous scenario, now with FIFO servers. Again the flow under observation, i, in the longest path of seven hops conforms to leaky bucket parameters (σ_i, ρ_i) at the entrance of the network, where σ_i is 266 bytes and ρ_i is 10Mbps. The maximum burst size at the nth server of i is denoted by $\sigma_i^{\mathcal{S}n}$. While i traverses through seven servers, it confronts with seven other flows, each with 10Mbps average rate, at each server. The other flows at nth server $\mathcal{S}n$ are assumed to have the maximum burst size that is same as the maximum burst size of the flow i at that server, $\sigma_i^{\mathcal{S}n}$. This assumption on the maximum burst size of other flows makes this scenario the worst case, because in a spanning tree network if i traverses the maximum length path then the flows confronted by i at any node cannot have traveled more hops than i. The maximum delay in this case for the flow under observation is 4.085ms. If SP servers instead of FIFO servers are used, the maximum network delay increases to 11.166ms.

Now consider a hybrid network with an aggregated flow which comprises flow 1 and 2. This aggregated flow traverses a network of LR servers in series. In every server there are six others flows, demanding 10Mbps per each flow. Again the flows 1 and 2 are constrained with leaky buckets at the entrance of the network with the parameters (266 bytes, 10Mbps). We define a *subnetwork* to be the collection of servers in series that the flow under observation belongs in an aggregated flow unaltered, therefore the subnetwork can be replaced with an equivalent virtual $\mathcal{LR}$ server. The maximum delay can be calculated from the maximum delay in the subnetwork 1 and the subnetwork 2. In this case the subnetwork 1 is the switch 1, and the subnetwork 2 is the collection of switches 2 through 7. The delay in the subnetwork 1 is 0.681ms, which is from lemma 1.

The maximum burst size of the aggregated flow into the subnetwork 2 is 10963 bits, and the maximum delay in the subnetwork 2 is 3.102ms. We obtain the maximum delay of the whole network by summing these two delays, which is 3.783ms. Finally we consider the case where the aggregated flow in the previous scenario is deaggregated into flows 1 and 2 at the input port of switch 7 to different output ports thus different servers, and each is confronted with other seven flows there. The total maximum network delay in this case is 6.501ms.

5.1 Summary

We compare a number of scenarios with various schemes of flow aggregations and examine their maximum delay performances. Table 1 summarizes the results. The performance of DiffServ depends heavily on the burst sizes of other

Table 1. Summary of performance comparison

Scenario	IntServ with DRR	DiffServ with FIFO	DiffServ with SP	Hybrid arch. with DRR with Flow Aggregation	Hybrid arch. with DRR with Flow Agg. and Deaggregation
Delay bound	3.490ms	4.085ms	11.166ms	3.783ms	6.501ms
Applicable to	Any case	Where leaky bucket conformance of all the flows are certain.	←	Where leaky bucket conformance of some flows are certain. Those trusted flows can be aggregated together.	←
Note	8 flows at each servers with leaky bucket parameters (266bytes, 10Mbps).	Other flows' bursts are set to be the same with flow *i*. EF class is the only traffic in the network.	EF class serviced with strict priority over other classes, but without preemption.	Flows 1&2 take the same path; Aggregation with DRR; Agg. only, at SW1.	Aggregation at SW1; Deaggregation at SW7.

flows that are aggregated with the flow under observation. IntServ, however, successfully protects the flows from each other's burst size variations, therefore is considered to have the predictable and robust performance. When there is only an aggregation at the entrance of the network, the hybrid architecture performs fine. When a deaggregation takes place in the middle of the path, however, the delay bound is very large due to the maximum burst size of the flow under observation that have increased to a significant level while traveling the path. The aggregation in the middle of a network will also lengthen the delay bound with the same reason. If we think of the DiffServ as an extreme of the hybrid architecture where flow aggregation and deaggregation occur in every node, then it is clear that the hybrid architecture cannot perform better than per-flow scheduling in general.

6 Conclusion

We have investigated how the maximum end-to-end delays can be guaranteed in networks where different QoS architectures are applied. Based on the analysis with $\mathcal{LR}$ server model we have confirmed that even a simplest class-based Strict Priority servers can guarantee a maximum end-to-end delay that is unsatisfactory but not excessive, especially in networks with spanning tree topologies. The hybrid networks with flow aggregation show better performances than DiffServ network, even with simple DRR servers. The maximum delays in such networks, however, depend heavily on how far the aggregation or deaggregation take places from the entrance of the network, since the farther the aggregation point the larger the maximum burst size, which can be interpreted as the degree of uncertainty. We suggest therefore to adopt the per-flow based IntServ architecture in moderate-sized networks. This study, however, does not reveal the statistical characteristics of the delay, e.g. the probability of delay violation or the correlation between the violated packets, which profoundly impact the AV stream performance. We suggest that the results shown in table 1 to be used as a performance index among different architectures and server choices.

References

1. R. Braden, D. Clark and S. Shenker, Integrated Services in the Internet Architecture: an Overview, IETF Request for Comments, RFC-1633. 1994.
2. P. P. White, RSVP and Integrated services in the Internet: A tutorial, IEEE Commun. Mag., vol. 35, pp. 100–106, May 1997.
3. S. Blake, D. Black, M. Carlson, E. Davies, Z. Wang, and W. Weiss, An architecture for Differentiated Services. IETF RFC-2475, 1998.
4. M. Shreedhar and G. Varghese, Efficient fair queueing using deficit round-robin, IEEE/ACM Trans. Networking, vol. 4, no. 3, pp. 375–385, June 1996.
5. S. Golestani, A Self-clocked Fair Queueing Scheme for Broadband Applications, in IEEE INFOCOM94, 1994.
6. D. Stiliadis and A. Varma, Latency-Rate servers: A general model for analysis of traffic scheduling algorithms, IEEE/ACM Trans. Networking, vol. 6, no. 5, Oct. 1998.
7. D. Stiliadis, Traffic Scheduling in Packet-Switched Networks: Analysis, Design and Implementation, Ph.D. Dissertation, U.C. Santa Cruz, June 1996.
8. J. A. Cobb, Preserving quality of service guarantees in spite of flow aggregation, IEEE/ACM Transactions on Networking, vol. 10, no. 1, pp. 43–53, Feb. 2002.
9. P. Goyal, S. S. Lam, and H. M. Vin, Determining end-to-end delay bounds in heterogeneous networks, In Proc. Workshop on Network and Operating Systems upport for Digital Audio and Video (NOSSDAV95), pages 287.298, Apr. 1995.
10. W. Sun and K. G. Shin, End-to-End Delay Bounds for Traffic Aggregates Under Guaranteed-Rate Scheduling Algorithms, IEEE/ACM Transactions on Networking, Vol. 13, No. 5, Oct. 2005
11. Residential Ethernet Study Group website, http://www.ieee802.org/3/re_study/index.html
12. F. Feng and G. M. Garner, Meeting Residential Ethernet Requirements: A Simulation Study, IEEE 802.3 Residential Ethernet Study Group, Sep. 2005

Heterogeneous QoS Multicast and Its Improvement on Edge-Based Overlay Networks*

Suogang Li, Jianping Wu, Ke Xu, and Ying Liu

Department of Computer Science and Technology,
Tsinghua University, Beijing, 100084 P.R. China
{lsg, xuke}@csnet1.cs.tsingha.edu.cn,
{jianping, liuying}@cernet.edu.cn

Abstract. Implementing traditional multicast in a network domain including core and edge routers is likely to indicate a huge burden of storage and forwarding for core routers. Supporting QoS aggravates the scalability problem of multicast. The recent multicasting on application layers is more scalable as well as increasing traffic load in network layers and end-to-end delivery delay. In the paper, we deploy the multicast supporting heterogeneous QoS on the overlay only comprising edge routers, and core routers may be unaware of multicast. With router resource limitation and QoS class constraint, we design an algorithm to build minimum cost trees. Also, we present a method to improve the tree performance through introducing some eligible non-member edge routers into multicast session. The simulation results demonstrate that the proposed multicast scheme is effective to decrease the total tree cost and satisfy the considered restriction factors. The scheme can offer an option to realize heterogeneous QoS multicast in DiffServ domains or MPLS VPN networks.

1 Introduction

Multicast is an efficient method to support multipoint communications. The native network-layer multicast is optimal in delivering data to a group of members, since it uses a delivery tree to forward data packets only once over each link. However, the network-layer multicast suffers from several problems. The first is the difficulty of deployment since it requires all the routers in the networks to implement multicast functions. The second is difficulty of supporting quality of service (QoS), whereas many multicast multimedia applications need various class services. The third problem is state scalability problem because each group session generates one entry of multicast routing table in each router. The last problem will be aggravated once heterogeneous QoS (i.e., different QoS classes) are provided in the multicast.

Recently, much study is focus on application-layer multicast among end system [1-4] and some of them have actually been realized with peer-to-peer (P2P) technology. The applications of file transfer and video on demand can be achieved on application layer, but the real time applications such as video-conferencing and remote

* This work is supported by the National Natural Science Foundation of China under Grant No. 60303006 and No. 60473082 and the National Grand Fundamental Research 973 Program of China under Grant No. 2003CB314801.

M. Gerndt and D. Kranzlmüller (Eds.): HPCC 2006, LNCS 4208, pp. 436 – 446, 2006.

experiment steering. Although no need for router support, the application-layer multicast depends on unicasting one or more data copy on network links, therefore it increases traffic in network.

In this paper we research overlay multicast in the network domain. A domain can be a normal Autonomous System (AS). We construct the overlay mesh network on the edge routers (ERs) of the domain and then deploy multicast on the overlay network. The core routers (CRs) will be unaware of multicast therefore achieving good scalability. The overlay is easy to realize since the ERs know the topology information of the network layer. Our multicast scheme establishes delivery trees with low cost as possible and considers two significant limitations, relating supported QoS classes and the resource of ERs. Besides, the tree performance is attempted to improve passing through the eligible non-member nodes.

The rest of the paper is organized as follows. The next section reviews related work. Section 3 gives a design overview of the multicast scheme. Section 4 proposes algorithms to construct multicast delivery trees satisfying limitations. Section 5 discusses the tree improving method. Section 6 studies the performance of the proposal through simulations. The last section takes a conclusion and future work.

2 Related Work

For implementing QoS multicast and solving scalability problem, several types of multicast were proposed recently.

(*a*) Application layer multicast (ALM). In ALM, end hosts accomplish member management and multicast forwarding and no router is needed to support multicast [1-4], etc. TAG [3] allows the hosts to learn network topology to optimize application multicast trees. MSN [4] considers interface bandwidth limitation in tree building. However, there is no ALM scheme supporting heterogeneous QoS requirements, which is supported in our proposed multicast scheme. AnySee [2] proposed inter-overlay optimizing scheme, where the nodes have no network layer topology and shortest path trees (SPTs) are built. It is different from our tree-improving scheme, as discussed in Section 5.1.

(*b*) Multicast in DiffServ. It is a facility to support QoS requested by members when multicasting in DiffServ domain. EBM [5] and M-DS [6] limit the multicast branching nodes in the ingress edge routers of the domain to scalability problem. This may result in that ingress routers become very busy and interface bandwidth may be not enough. Our proposal considers the resource limitation of the branching nodes.

(*c*) Aggregating multi-group on a single tree. Aggregated multicast [8] allows several groups which have similar member nodes on the domain to share the same tree, so the state maintained on core routers is decreased greatly. QMCT [9] employ the notion and design scalable QoS multicast scheme. In the paper, we implement multicast on the network edge and no multicast state on core routers at all. It is important to point out that the existing schemes and our proposal make the tradeoffs between optimality and scalability, in essence.

3 Design Overview

3.1 Relative Definitions

The current Internet includes many network domains. We firstly give the definitions of the network domains.

Definition 1. **Network domain:** Generally, a network domain comprises the edge and the interior, including ERs and CRs, respectively.

For example, the DiffServ network domain and the provider network domain specified in BGP/MPLS VPN. The CRs are only in charge of forwarding packets as quickly as possible, with little state information on packet flows, whereas the ERs offer necessary functions for flows. The functions include resource reservation, flow control, packet marking, managing VPNs, tunnels, multicast, etc.

Definition 2. **Member and non-member edge routers:** The ERs of a network domain are multicast member ER, called mER(g), when a group session g transits it. Correspondingly, the other ERs which the group session does not transit are called non-member ERs, i.e., nmER(g). The mER is called source ER or ingress ER where the session enters in the domain.

The proposed QoS multicast mechanism pushes the multicast function out to the network edges. The all ERs in the domain organize an overlay meshed network, called Edge Overlay Network (EON). On the EON, a tree with the minimum cost is constructed to connecting all the mERs related with the multicast group. The mechanism is called Edge Overlay QoS Multicast (EOQM). It takes advantage of edge routers of the domain and combines the merits of application-layer multicast and native network-layer multicast. Therefore, it becomes convenient for Internet service providers (ISPs) to deploy and manage QoS multicast services and utilize network resource efficiently. The CRs are completely free from multicast state maintaining.

In the routing in the EOQM, we assume that the ingress ER (ingERs) of each multicast session is designated for the session to compute the tree. The ingERs have full knowledge of other ERs in the session, and compute the tree according to its current conditions of the available bandwidth. The function components of on the ingER are showed in Fig.1. In this figure, the ingERs utilize directly the network layer topology information to compute multicast routing trees. The components of tree building and improvement will be discussed in the next two sections, respectively.

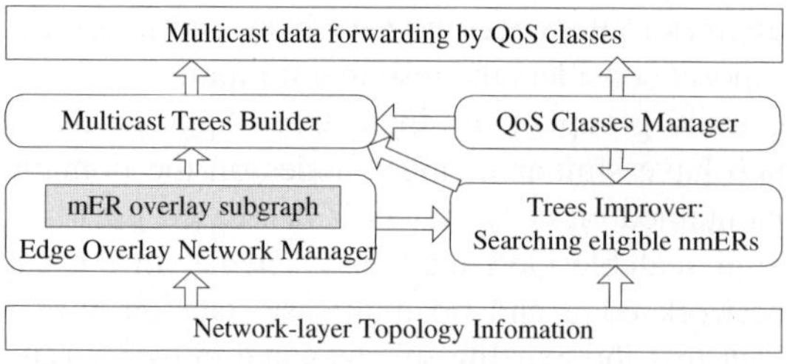

Fig. 1. EOQM function components on ingERs

As shown in Fig.1, one of advantages EOQM is no need for frequent state updates via broadcasting among all nodes; while the overlay multicast updates the state periodically since the participant nodes are end hosts. Therefore, routing performance of the overlay multicast may be deterred due to the imprecise routing information [4].

3.2 Design Targets

Four targets are achieved in EOQM which is listed as follows.

(*i*) The total cost of multicast tree spanning all mERs is as small as possible.
(*ii*) Heterogeneous QoS classes required by mERs are to be supported.
(*iii*) The mERs should be subjected to the constraint of resource limitation.
(*iv*) Eligible nmERs are introduced into the group to improve the tree performance.

Target (*iv*) will be discussed in Section 5. Target (*i*) is aimed to reduce the expenditure of multicast forwarding in the domain. The cost of the path between mERs is acquired by network layer information. For Target (*ii*), we assume all of the supported QoS classes are comparable [9], i.e., the resource satisfying the high class QoS must satisfy the low class QoS. This is means the node requiring high class QoS should not be placed downstream of the node requiring low class QoS. It is referred to as the *QoS Class Constraint* (*QCC*). The built tree in EOQM conforms to the QCC. The case of incomparable QoS classes will be studied in the future. Target (*iii*) indicates that another restriction is supposed to satisfy. That is named the *Member Resource Constraint* (*MRC*), which is related with the power of forwarding and the access capacity of interface in the mERs.

3.3 An Application Instance in BGP/MPLS VPN

The service of BGP/MPLS VPN is specified in [10]. ISP provides VPNs (Virtual Private Networks) to the customers using its network domain. The domain consists of P (Provider) routers and PE (Provider Edge) routers. The customers connect the PE through CE (Customer Edge) routers. The MPLS technology is used to implement VPNs in the P network domain. Herein, the PE router is an ER; the P router is a CR. Our proposal can be realized in the network model. The draft [11] specifies the multicast VPNs in the BGP/MPLS networks, which support several scalable multicast schemes. However, it cannot offer the heterogeneous QoS requirement. Again, the scheme tunneling the multicast packets from the ingress PE to all of egress PEs may make the ingress PE bear heavy load when many egress PEs are involved. These disadvantages can be compensated if applying the EOQM. It is practical and feasible.

4 Multicast Algorithm

In this section, we present the proposed solutions for obtaining good multicast trees. Above all, we should point out that the centralized route computation is used to increase routing efficiency and reduce message complexity. The centralized version does not create a single point of failure or even a performance bottleneck, as each session may select its ingress ER to perform the tree computation. This means that the overall computational load can be inherently distributed among different ERs for different sessions. We begin with the formulation of the edge overlay network model.

4.1 Problem Statements

An EON is a fully connected virtual network formed by edge routers which communicate with each other using links in the domain. The EON is modeled by a complete graph $G = (V, E)$, where V is a set of nodes representing ERs, and E is a set of edges. Each edge, being undirected, represents a unicast path between two ERs. Each edge $e(u, v) \in E$ is associated with a cost, $C(e)$, which represents the expenditure of passing the edge.

Multicast services are offered in the EON. All the mERs of a session g in a domain compose a node subset, $M(g) \subseteq V$. The ingress node is s. The multicast tree of g is a directed tree rooted at s, called $T(s \rightarrow M(g))$ or $T(s \rightarrow M)$, $T \subseteq G$. The total cost of T is $C(T) = \sum_e C(e)$. In the graph G, we called the overlay graph only formed by nodes in M a *pure* subgraph of g, denoted $G^P(g)$. Adding one or more non-member nodes into $G^P(g)$, the overlay graph is called a *hybrid* subgraph, called $G^H(g)$.

The multicast tree built is needed to satisfy the heterogeneous QoS requirement. The available QoS class set is Q, $|Q| = Q^N$. The multicast member $v \in M$ requests QoS class, $q(v) \in Q$. The constraint of QCC indicates the directed transfer from high class QoS node to low class QoS node. That is, the direction of edge e is $u \rightarrow v$ $(e{:}u \rightarrow v)$ if $q(u) > q(v)$. Obviously, the QoS class of source s is highest, i.e., $q(s) = 0$.

The limitation of MRC corresponds to the degree restriction of the nodes in the graph. Each node v in the graph is attached with a degree limitation $D^{Max}(v)$. The degree of node $v \in T$ is $d_T(v)$, $d_T(v) \leq D^{Max}(v)$.

The cost of the built multicast tree is expected to be as low as possible, as well as satisfying the constraints of QCC and MRC. We assume that the QoS requirement of each node can be met without considering resource reservation problem in our study. We refer to the problem as computing minimum cost trees with constraints of QoS class and node degree, QD-MCT for short. The definition is given as below.

Definition 3. **With constraints of QoS class and node degree, minimum cost directed tree problem (QD-MCT):** Given an undirected complete graph $G(V, E)$, a non-negative real cost $C(e)$ for each edge, a positive degree limitation $D^{Max}(v) \in Z^+$ and a QoS class $q(v) \in Q$ for each node v, where Q is the set of QoS classes supported, $Q = \{0, 1, \ldots\} \subset \mathrm{Z}$ with sort descending, $|Q| = Q^N$; find a spanning tree T of $M \subseteq V$ of minimum cost, subject to the constraints that the node u requiring high QoS class should not be placed downstream of the node v requiring low QoS class in T, for all u, $v \in M$ and that the node v degree in T is less than $D^{Max}(v)$ for all $v \in M$.

This problem can be formalized as

$$\begin{aligned} &\textit{Minimize} \quad C(T(s \rightarrow M)) = \textstyle\sum_{e \in T} C(e) \\ &\textit{Subject to} \quad (i)\ \forall e(u, v) \in E_T,\ e{:}u \rightarrow v, \text{ if } q(u) > q(v); \\ &\qquad\qquad\quad (ii)\ d_T(v) \leq D^{Max}(v). \end{aligned}$$

Solving the minimum cost tree of the node subset M in a graph is a well-known NP-complete (NPC) problem of Steiner Minimum Tree (SMT). The QD-MCT problem involves the constraints of QCC and MRC, which is corresponded to the problem computing directed SMT with degree limitations. Especially, the problem is reduced to *Traveling Salesman Problem* (*TSP*) when $D^{Max}(v) = 2$ for all $v \in M$. The QD-MCT problem is also NPC. The detailed process is omitted here for conciseness.

4.2 Heuristic Algorithm

We present the greedy heuristic algorithms to solve the QD-MCT problem for its complexity. Let T^{opt} denote the optimal QD-MCT tree and $C(T^{opt})$ denote the tree cost.

The heuristic algorithm tries to achieve the minimum cost spanning trees with the constraints of QCC and MRC in the overlay pure subgraph $G^P(g)$. The problem to compute Minimum Spanning Trees (MST) of a graph is polynomial solvable using the well-known Prim's algorithm. For the QD-MCT problem, the algorithm building a spanning tree of $G^P(g)$ incrementally, like Prim's algorithm.

In Prim's algorithm, a new edge with its connecting node off-tree is selected to join the current built tree if the edge cost is least in all candidate edge and results in no loop in the current tree. The step goes on until all nodes are contained. In our algorithm, we prefer to select the nodes with high class QoS rather than low class QoS nodes to join the current subtree due to the QCC. Therefore, the algorithm satisfies that the class QoS of every node on the current tree is not lower than that of any candidate node off the tree, and the direction of the newly selected edge is from the end on tree to the end off tree before joining. Finally, the gained tree is guaranteed to satisfy the QCC. In the initialization of the algorithm, the ingER must be the first to join the tree for it's default highest class QoS. For the MRC, assuming the degree limitation of each node, D^{Max} is identical, the algorithm should selects the new edge v to join the current tree which the degree $d_T(v) < D^{Max}(v)$. If not, another candidate edge with the second least cost is considered to join. See the algorithm as below.

Algorithm 1. **The heuristic for QD-MCT**

Input: The graph, $G^P(g) \subseteq G$; source node, s; node degree limitation, $D^{Max}(v)$ and required QoS class, $q(v)$, for $v \in V$, $q(s) = 0$. The QoS number, Q^N, the smaller q, the higher QoS class. Edge cost, $C(e(u, v))$ for $e(u, v) \in E$. *Output*: Tree T

(0) $T = \varnothing$ and push s into T.
(1) Classify the rest nodes into R_i, $i \in [0, Q^N\text{-}1]$, according to the QoS class of each node;
(2) for i from 0 to Q^N-1,
(3) while ($R_i \neq \varnothing$), do
(4) Select the node u in R_i and the edge $e(u, v)$ with v in T, with least $C(e)$ and $d_T(v) < D^{Max}(v)$;
(5) Join the node u and $e(u, v)$ in T and set $e{:}u \rightarrow v$
(6) Remove the u from R_i and update the degree d_T of u and v;

Node classifying in Step (2) of the algorithm needs to run in time $O(|V|)$. The later steps are like Prim's algorithm, being made to run in time $O(|E|\cdot\log|V|)$ using ordinary binary heaps. By using Fibonacci heaps [12], it can be sped up to run in time $O(|E|+|V|\cdot\log|V|)$, which is an improvement since $|V|$ is much smaller than $|E|=|V|\cdot(|V|-1)/2=O(|V|^2)$ in fully connected graph. So the total running time is polynomial.

5 Improving Multicast Through Non-member

We discuss how to improve the tree generated by Algorithm.1. The idea is to search some eligible nmER nodes, called *improving nodes*, to join into multicasting in order to reduce the total tree cost. We first illustrate the possibility of such nodes existing.

5.1 Possibility of Improving

A part of network topology is shown as Fig.2 (a). The overlay network shown in Fig.2 (b) comprises nodes o and $m_1 \sim m_n$. Node o connects $m_1 \sim m_n$ through x_u and then $y_1 \sim y_n$, respectively. The cost of path ox_u, x_uy_i and y_im_i is p_o, u_i and l_i, $1 \le i \le n$, respectively. The node r joins the overlay network as shown in Fig.2 (c), where r connects other nodes by x_v, the cost of path rx_v, x_ux_v and x_uy_i is p_r, p_x and v_i, respectively. We assume the cost of all paths is positive.

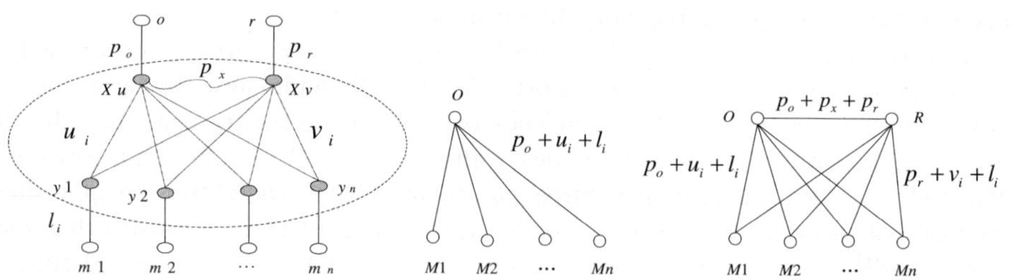

Fig. 2. An example for improving nodes

In the overlay network of Fig.2 (b), the cost of the multicast tree T_1 connecting o and $m_1 \sim m_n$ directly is

$$C(T_1) = \sum_{i=1}^{n} (p_o + u_i + l_i) = np_o + \sum_i u_i + \sum_i l_i = np_o + U + L \tag{1}$$

where $U = \sum_{i=1}^{n} u_i$ and $L = \sum_{i=1}^{n} l_i$.

In the overlay network of Fig.2 (c), the cost of the multicast tree T_2 connecting o and $m_1 \sim m_n$ passing r is

$$\begin{aligned} C(T_2) &= p_o + p_x + p_r + \sum_{i=1}^{n} (p_r + v_i + l_i) = p_o + p_x + (n+1)p_r + \sum_i v_i + \sum_i l_i \\ &= p_o + p_x + (n+1)p_r + V + L \end{aligned} \tag{2}$$

where $V = \sum_{i=1}^{n} v_i$.

Comparing Equ.(1) and (2), we see that $C(T_2) < C(T_1)$ as long as $(n+1)p_r + p_x + (V-U) < (n-1)p_o$. Let $\Delta = v\text{-}u$, then $(n+1)p_r < (n-1)p_o - p_x - \Delta$, i.e.,

$$p_r < \frac{(n-1)p_o - p_x - \Delta}{n+1} = \frac{n-1}{n+1} p_o - \frac{1}{n+1}(p_x + \Delta) \tag{3}$$

Considering the case where $p_x = 0$, i.e., $U = V$, the right part of Inequ.(3) must be positive if only $n \ge 2$. So we have

$$p_r < \frac{n-1}{n+1} p_o = (1 - \frac{2}{n+1}) p_o \tag{4}$$

where n denotes the children number of node o that is necessary to improvement occurring. When n is large, r is qualified to become the improving node as long as p_r is a little less than p_o.

Especially, when n=1, there is no improvement for the partial tree. Meanwhile, the partial tree is equivalent to unicast path between o and m_1. Therefore, it is impossible to improve the path $o \to m_1$ when the object is solving the shortest path in the case shown in Fig.2. This is just the case of AnySee [2], which differs from our scheme and also shows the different improving possibility. We point out that if routing metric in router x_u and x_v is identical to the cost of paths, then x_u will know path $x_u \to x_v \to y_1$ is more optimal than path $x_u \to y_1$ and no improvement exists. However, our scheme to optimize the total tree cost is always possible to find improvement.

5.2 Algorithm for Tree Improving

Now we give an improving algorithm how to search the improving node(s) among non-member routes for the group g, as shown in Algorithm.2

The algorithm searches improving node with smallest cost for the on-tree node having more than one child. In Step (6), we examine the degree limitation of current nmER for MRC constraint. The algorithm assumes the QoS class of improving node r is equal to the replaced node m$_i$. The exterior cycle of the algorithm runs no more than $|V_m|$ and the interior cycle runs no more than $|V_{nm}|$. Therefore, the algorithm is made to run in time $O(|V_m|\cdot|V_{nm}|) = O((|V_G|^2)/2) = O(|V_G|^2)$, for $|V_m|+|V_{nm}|=|V_G|$.

The tree cost after improved by Algorithm.2 is not more than that before improved definitely, since only the nmER(s) that is able to reduce the cost of the original tree is selected to join the tree. Assuming that the replaced node set is $\{m_i\} \subseteq M$ and the replacer of m_i is n_i, the reduced cost by Algorithm.2 is $\sum_i R^{min} = \sum_i [C(e(m_i \to n_i)) + \sum_k C(e(n_i \to h_k))]$, $h_k \in Ch(m_i)$.

Algorithm 2. **The improvement for QD-MCT**

Input: Group g, tree T, member node set $V_m=\{m_i\}$, non-member node set $V_{nm}=\{nm_j\}$, node degree limitation $D^{Max}(nm_j)$, $nm_j \in V_{nm}$. *Output*: improved tree T.

(0) Set variants: r, the current improving node, $R^{min} = \infty$, the current minimum cost related to r;
(1) for each m_i in T:
(2) Set $Ch_i = \{h_k\}$, the children set of m_i on T.
(3) If $|Ch_i| \leq 1$, then goto (1) to check next member node; otherwise, goto (4)
(4) Compute the cost sum P_i^m of the edges from m$_i$ to $h_k \in Ch_i$,; $P_i^m = \sum_k C(e(m_i \to h_k))$;
(5) for each nmER$_j$ in V_{nm}:
(6) If $D^{Max}(nm_j) < |Ch_i|$ then goto (5) to check next non-member;
(7) Compute the cost sum P_i^m from nm_j to $h_k \in Ch_i$,; $P_j^{nm} = C(e(m_i \to nm_j)) + \sum_k C(e(nm_j \to h_k))$;
(8) If $P_j^{nm} < P_i^m$ and $P_j^{nm} < R^{min}$, then $R^{min} = P_j^{nm}$, $r = nm_j$; otherwise goto(4) to check next node;
(9) If $R^{min} < \infty$, then add r and $e(m_i \to r)$ into T; add edges $e(r \to h_k)$ k into T and remove $e(m_i \to h_k)$;

6 Simulation Experiments

Firstly, we compute the minimum cost tree without any constraint, called T^M in the pure subgraph and the improved solution, T^{iM} in the hybrid subgraph, using the Prim's algorithm and the Algorithm.2, respectively. Next, we seek the minimum cost tree with constraints of QCC and MRC, called T^{CM} and T^{iCM} in the same mode.

(1) For tree performance of Algorithm.1

Computing the optimal solution, T^{opt} to *QC-MDT* is difficult since it is NP hard. We substitute T^M to T^{opt}. Obviously, $C(T^M) \leq C(T^{opt})$. Let the λ_1 and λ_2 denote the performance ratios before and after improved, respectively, as shown as follows.

$\lambda_1 = \frac{C(T^{CM}) - C(T^M)}{C(T^M)} \times 100\%$, $\lambda_2 = \frac{C(T^{iCM}) - C(T^{iM})}{C(T^{iM})} \times 100\%$. Smaller λ is meaning better performance.

(2) For improving effect of Algorithm.2

Let the μ_1 and μ_2 denote the improving effects for the trees with and without constraints, respectively, as shown as follows. Larger μ is meaning better improving.

$$\mu_1 = \frac{C(T^M) - C(T^{iM})}{C(T^M)} \times 100\% , \ \mu_2 = \frac{C(T^{CM}) - C(T^{iCM})}{C(T^{CM})} \times 100\% .$$

6.1 Setting of Experiments

We generate the experiment topologies using GT-ITM. Each node denotes a router in the router and total number of nodes are *N*. ER nodes are a half of the nodes. Random variant $\beta \in (0, 1)$ is the rate of the mER number to the number of all ERs, indicating the group size, $N_{mER}=\beta N/2$. The link cost between nodes is refered to as the corresponding edges delay, designated by the topology generator. The cost of a path connecting two ERs is the sum cost of all the links on the path. The degree limtation for each node is equal, $D^{Max}=10$ and QoS class number is $Q^N=3$ in simulations.

Two types of topology graph are generated: the sparse and the dense. The former holds more edges than the latter with the same number of nodes. It is implemented by setting the density parameter $p\notin(0, 1)$ that is the possibility that there exists an edge between any two nodes in the graph.

The simulation consists of four sets of experiments, setting $(N, p) = \{(1000, 0.8), (1000, 0.2), (500, 0.8), (500, 0.2)\}$, which denote the topology with many or few nodes and dense or sparse edges, respectively. In each group experiment, $\beta=\{0.05, 0.1, 0.2, \ldots, 0.9\}$, totaling ten. For the cases with or without QCC and MRC, the $(Q^N, D^{Max}) = \{(0, 0), (0, 10), (3, 0), (3, 10)\}$.

6.2 Simulation Results

The tree performance by Algorithm.1 is shown in Fig.3, where (a) for the unimproved trees and (b) for the improved trees. We can see that performance rates stand at 10% or so and not more than 20% when group size increasing in four networks topologies. It shows the good performance of the Algorithm.1.

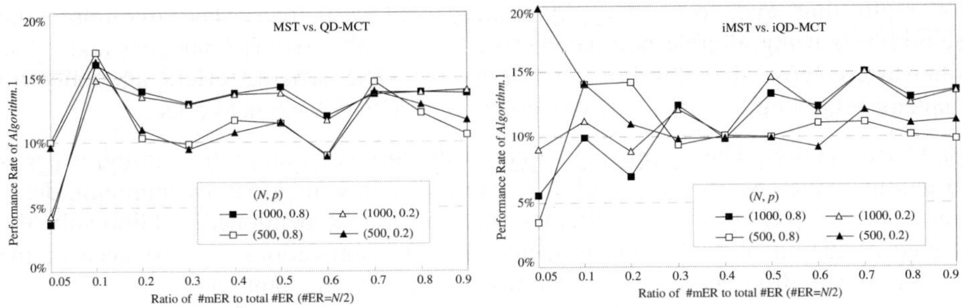

(a) (b)

Fig. 3. Performance of Algorithm 1

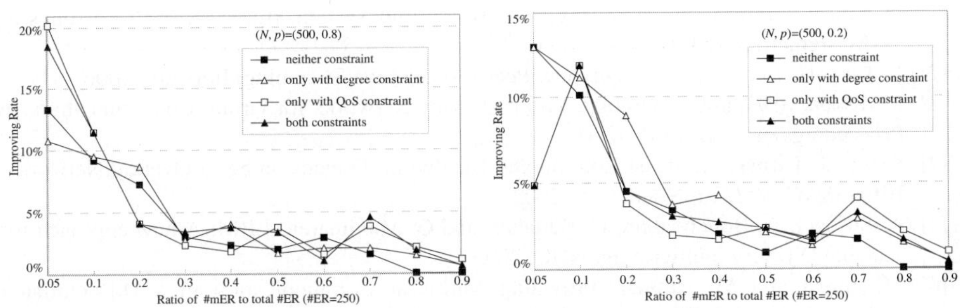

(a) (b)

Fig. 4. Improvement of Algorithm 2

The improving effects of Algorithm.2 are shown in Fig.4. The values of μ_1 and μ_2 are both about 5%. The summit of improving effect in QD-MCT is approach 18% when β=0.05 in Fig.4 (a).

We can obtain two facts from these curves. On one hand, the less β, the better the improvement is. The maximums of effects occur at β=0.1 in the four group curves, while there is nearly no improvement when β>0.8. The reason is that the little value of β means there more nmERs checked if they are eligible to improving the tree. On the other hand, improvement to the trees with constraints is larger than the one without any constraint. Under the constraint conditions, the performance of the trees by Algorithm.1 is worse than the trees without constraints. Therefore they have more opportunity to be improved.

7 Conclusions and Future Work

This paper presents an overlay multicast scheme supporting heterogeneous QoS in the general network domains. The scalability of the scheme is achieved since multicast is deployed on edge routers. The minimum cost trees are built to delivery traffic so as to increase total network performance. The tree-building algorithm supports different QoS requirements by members and considers the limited power of the member routers for replicating and forwarding in overlay multicast trees. The tree-improving algorithm is using eligible non-member nodes to replace some branching node. The simulations show the algorithms are effective. The scheme offers an option to implement QoS multicast in DiffServ domains or MPLS VPN networks.

FUTURE WORK. The optimizing object of the discussed algorithms in the paper is to minimize the cost of total trees. It is interesting how to solve the minimum delay tree or other types of the trees with these two constraints mentioned. Additionally, the degree limitation of each node is identical in the simulations. Performance of the algorithm with different degree limitations is studied in the future work.

References

[1] Y. Chu, S. G. Rao, and H. Zhang, A Case for End System Multicast, in Proceedings of ACM Sigmetrics, 2000.

[2] X. Liao, H. Jin, Y.Liu, etc., AnySee: Peer-to-Peer Live Streaming, Infocom, 2006

[3] M. Kwon, S. Fahmy, Topology-Aware Overlay Networks for Group Communication, In Proceedings of NOSSDAV, 2002

[4] S. Shi, J. Turner, Multicast Routing and Bandwidth Dimensioning in Overlay Networks, IEEE JSAC, Vol. 20, No. 8, Oct., 2002

[5] A. Striegel, A. Bouabdallah, H. Bettahar, and G. Manimaran, EBM: A new approach for scalable DiffServ multicasting, NGC, 2003

[6] M. Gupta and M. Ammar, Providing Multicast Communication in a Differentiated Services Network Using Limited Branching Techniques, IC, 2002

[7] M. Carlson, W. Weiss, S. Blake et al. An Architecture for Differentiated Services. IETF Internet RFC 2475, December 1998.

[8] A. Fei, J.-H. Cui, M. Gerla, and M. Faloutsos. Aggregated Multicast: an approach to reduce multicast state. Proceedings of Sixth Global Internet Symposium (GI2001), Nov., 2001.

[9] S. Li, J. Wu, K. Xu and Y. Liu, A Modularized QoS Multicasting Approach on Common Homogeneous Trees for Heterogeneous Members in DiffServ, WMSN, 2006

[10] [10] E. Rosen, Y. Rekhter, BGP/MPLS IP Virtual Private Networks (VPNs), IETF RFC 4364, Feb. 2006

[11] E. Rosen, R. Aggarwal, Multicast in MPLS/BGP IP VPNs, draft-ietf-l3vpn-2547bis-mcast-01, Dec. 2005

[12] T. H. Cormen, C. E. Leiserson, and R. L. Rivest, Introduction to Algorithms, McGraw Hill, 2000

On Multicasting Steiner Trees for Delay and Delay Variation Constraints*

Moonseong Kim[1], Young-Cheol Bang[2], and Hyunseung Choo[1]

[1] School of Information and Communication Engineering
Sungkyunkwan University 440-746, Suwon, Korea
Tel.: +82-31-290-7145
{moonseong, choo}@ece.skku.ac.kr
[2] Department of Computer Engineering
Korea Polytechnic University 429-793, Gyeonggi-Do, Korea
Tel.: +82-31-496-8292
ybang@kpu.ac.kr

Abstract. The objective of multicasting is to find a tree that has a minimum total cost, which called the **Steiner tree**. Multicast routing algorithms should support the required QoS. There are two important Quality of Service (QoS) parameters that need to be guaranteed in order to support the real time and multimedia applications. Firstly, we consider the delay parameter where, the data sent from source need to reach destinations within a certain time limit (delay bound). Secondly, in addition to the delay constraint, we add the delay variation constraint. The delay variation constraint is a bound on the delay difference between any two destinations. Our research subject is Delay and delay Variation Bounded Steiner Tree (**DVBST**) problem. The problem has been proved to NP-complete. In this paper, we propose efficient algorithm for DVBST. Simulations demonstrate that our algorithm is better in terms of tree cost as compared to the existing algorithms.

1 Introduction

New communication services involving multicast communications and real time multimedia applications are becoming prevalent. The general problem of multicasting is well studied in the area of computer networks and algorithmic network theory. In multicast communications, messages are sent to multiple destinations that belong to the same multicast group. These group applications demand a certain amount of reserved resources to satisfy their Quality of Service (QoS) requirements such as end-to-end delay, delay variation, cost, loss, throughput, etc.

For multicast communications such as a teleconference, it is very critical that the current speaker must be heard by all participants simultaneously, but otherwise the communication may lose the feeling of an interactive face-to-face discussion. Another similar dispute can be easily found in on-line video games. These are all related to the multicast delay variation problem [12][17].

* Dr. Choo is the corresponding author and Dr. Bang is the co-corresponding author.

M. Gerndt and D. Kranzlmüller (Eds.): HPCC 2006, LNCS 4208, pp. 447–456, 2006.

Moreover, network cost of the multicast routing tree is another factor that we have to consider, which is important in managing network resources. Varying levels of complexity may be present in a multicast problem depending upon the cost and/or criterion [21]. The Steiner tree is very useful in representing solution to multicast routing problems. The Steiner tree, studied extensively in network theory, deals with minimizing the cost of a multicast routing tree [7][23]. The Steiner tree problem is known to be NP-complete [6][9] and there are vast literatures [2][8][14][19].

Two kinds of delay and delay variation constrained multicast routing problems have been developed. One is Delay and delay Variation Bounded Multicast Tree (DVBMT) problem [10][13][17][18], without considering cost optimization. The other is Delay and delay Variation Bounded Steiner Tree (DVBST) problem. The two kinds of problems have been proved to NP-complete [17]. In this paper, our research subject is to construct the Steiner tree with delay and delay variation constraints. Simulations demonstrate that the proposed algorithms are better in terms of tree delay variation and tree cost as compared to the existing algorithms. Also, the time complexity is comparable to one of previous works.

The rest of the paper is organized as follows. In Section 2, we state the network model for the multicast routing, the problem formulations, and the previous algorithms. Section 3 presents the details of the proposed algorithms. Then, we evaluate the proposed algorithm by the computer simulation, in Section 4. Finally, Section 5 concludes this paper.

2 Preliminaries

2.1 Network Model

We consider that a computer network is represented by a directed graph $G = (V, E)$ with n nodes and l links where, V is a set of nodes and E is a set of links, respectively. Each link $e = (i, j) \in E$ is associated with two parameters, namely link cost $c(e) \geq 0$ and link delay $d(e) \geq 0$. The delay of a link, $d(e)$, is the sum of the perceived queueing delay, transmission delay, and propagation delay. We define a path as a sequence of links such that (u, i), (i, j), …, (k, v), belongs to E. Let $P(u, v) = \{(u, i),\ (i, j),\ \ldots,\ (k, v)\}$ denote the path from node u to node v. For given a source node $s \in V$ and a destination node $d \in V$, $(2^{s \to d}, \infty)$ is the set of all possible paths from s to d. *i.e.*,

$$(2^{s \to d}, \infty) = \{\ P_k(s, d) \mid \text{all possible paths from } s \text{ to } d,\ {}^{\forall}s,\ d \in V,\ {}^{\forall}k \in \Lambda\ \} \quad (1)$$

where, Λ is an index set. The path cost of P is given by $\phi_C(P) = \sum_{e \in P} c(e)$ and the path delay of P is defined by $\phi_D(P) = \sum_{e \in P} d(e)$. $(2^{s \to d}, \Delta)$ is the set of paths from s to d for which the end-to-end delay is bounded by Δ. Therefore $(2^{s \to d}, \Delta) \subseteq (2^{s \to d}, \infty)$.

For the multicast communications, messages need to be delivered to all receivers in the set $M \subseteq V \setminus \{s\}$ which is called the multicast group, where $|M| = m$. The path traversed by messages from the source s to a multicast receiver, m_i, is given by $P(s, m_i)$. Thus, multicast routing tree can be defined as

$T(s, M)$, the Steiner tree of $\bigcup_{m_i \in M} P(s, m_i)$, and the messages are sent from s to M through $T(s, M)$. The tree cost of tree $T(s, M)$ is given by $\phi_C(T) = \sum_{e \in T} c(e)$ and the tree delay is $\phi_D(T) = max\{ \phi_D(P(s, m_i)) \mid {}^{\forall}P(s, m_i) \subseteq T, {}^{\forall}m_i \in M \}$.

The multicast delay variation, $\phi_\delta(T)$, is the maximum difference between the end-to-end delays along the paths from the source to any two destination nodes.

$$\phi_\delta(T) = max\{|\phi_D(P(s, m_i)) - \phi_D(P(s, m_j))|, {}^{\forall} P \subseteq T, {}^{\forall} m_i, m_j \in M, i \neq j\} \quad (2)$$

Given a positive delay bound Δ and a positive delay variation bound δ, the objective of DVBST is to construct a minimum cost multicast tree $T(s, M)$ which spans s and M such that the delay and delay variation constraints are satisfied, *i.e.*,

$$P(s, m_i) \in (2^{s \rightarrow m_i}, \Delta), \ {}^{\forall}P(s, m_i) \subseteq T(s, M) \quad (3)$$

$$\phi_\delta(T) \leq \delta \quad (4)$$

$$\phi_C(T) \ is \ minimized. \quad (5)$$

2.2 Previous Algorithms for DVBMT

The DVBMT problem had been firstly introduced in [17]. Rouskas and Baldine proposed Delay Variation Multicast Algorithm (DVMA), finds a Multicast Tree spanning the set of multicast nodes. DVMA works on the principal of finding the kth shortest paths to the concerned nodes. If these paths do not satisfy delay variation bound δ, more longer paths are found. The complexity of DVMA is $O(klmn^4)$ where, k and l are the largest value among the numbers of the all appropriate paths between any two nodes under delay bound Δ, $|M| = m$, and $|V| = n$. DVMA is a high time complexity does not fit in modern high speed computer network environment.

Sheu and Chen proposed Delay and Delay Variation Constraint Algorithm (DDVCA) [18] based on Core Based Trees (CBT) [1]. Since DDVCA is meant to search as much as possible for a multicast tree with a smaller multicast delay variation under Δ, DDVCA in picking a core node has to inspect whether the core node will violate the Δ. In spite of DVMA's smart performance in terms of the multicast delay variation, its time complexity is very high. However, DDVCA has a much lower time complexity $O(mn^2)$ and has a satisfactory performance. Kim, *et al.* recently proposed a heuristic algorithm based on CBT like DDVCA, their algorithm hereafter referred to as KBC [10]. DDVCA overlooked a portion of the core selection. The selection of a core node over several candidates (possible core nodes) is randomly selected among candidate nodes. Meanwhile, KBC selects the good core node among candidate nodes as the same time complexity of DDVCA. KBC obtains the better minimum multicast delay variation than DDVCA.

However, if the above algorithms are used in DVBST then the minimum tree costs may not be guaranteed. For our goal DVBST, specific paths which are satisfied Δ have to be required. The paths should be obtained by Weighted Factor Algorithm (WFA) [11]. The WFA is introduced in next subsection. In this paper, the proposed algorithm constructs multicast tree using the paths because a multicast tree is union of paths for every multicast member. Furthermore, Kim, *et al.* recently proposed a heuristic algorithm based on WFA, their algorithm hereafter referred to as KBYC [13]. KBYC also uses expected multiple paths by WFA.

2.3 Weighted Factor Algorithm (WFA)

Kim, *et al.* recently proposed a unicast routing algorithm, their algorithm hereafter referred to as WFA [11], which is probabilistic combination of the cost and delay and its time complexity is $O(Wl + Wn\log n)$ where, $\{\omega_\alpha\}$ is set of weights and $|\{\omega_\alpha\}| = W$. Authors investigated the efficiency routing problem in point-to-point connection-oriented networks with QoS. They formulated the new weight parameter that simultaneously took into account both the cost and delay. In generally, the Least Delay path (P_{LD}) cost is relatively more expensive than the Least Cost path (P_{LC}) cost, and moreover, $\phi_D(P_{LC})$ is relatively higher than $\phi_D(P_{LD})$. The unicast routing algorithm, WFA, reinforces the properties with the weight ω, and then it is quite similar to a performance of kth shortest path algorithm. The weight ω plays on important role in combining the two the independent measures. If the ω is nearly to 0, then the path delay is low as in Fig. 1. Otherwise the path cost is low. Thus, the efficient routing path can be determined once ω is selected. Our proposed algorithm uses WFA for construction of multicast tree. Since a multicast tree is union of paths for every multicast member, the proposed algorithm has to select suitable weight for each member.

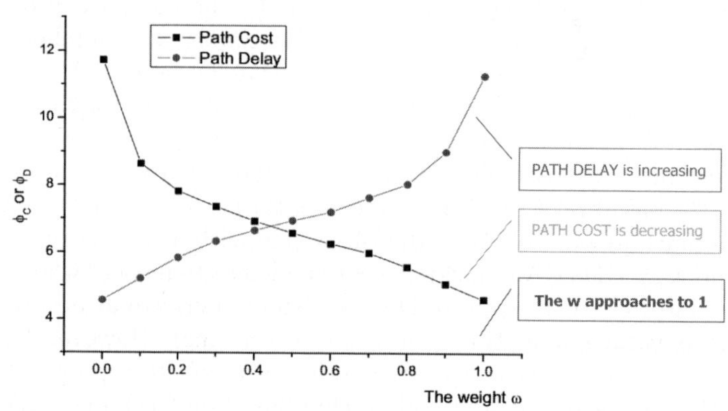

Fig. 1. Relationship between path delay and path cost from [11] (P_e: 0.3, $|V|$: 100)

3 The Proposed Algorithm

We now present algorithm to construct a multicast tree satisfying constraints (3) and (4) for the given values of the path delay Δ and the interdestination delay variation tolerance δ. Furthermore, the tree has the condition (5). We assume that complete information regarding the network topology is stored locally at source node s, making it possible to determine the multicast tree at the source node itself. This information may be collected and updated using an existing topology broadcast algorithm.

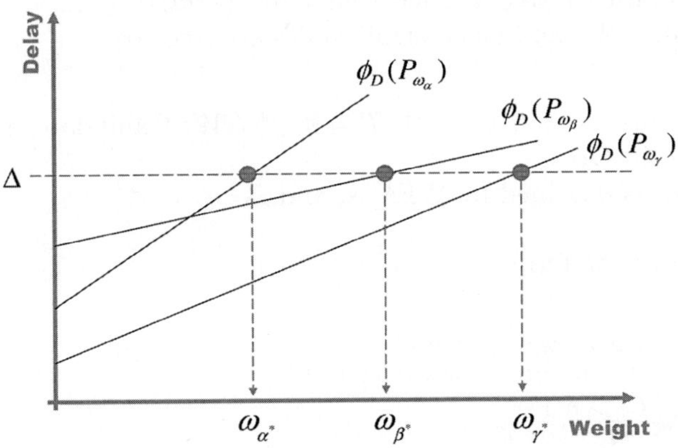

Fig. 2. Weight selection in proposed algorithm

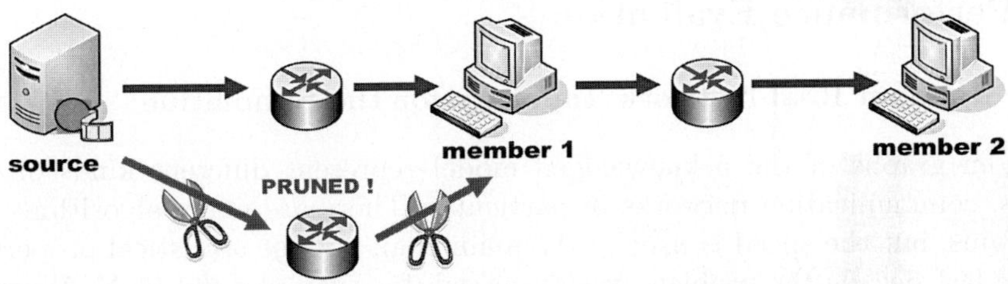

Fig. 3. Scenario illustrating loop creation in $G^* = P(s, m_1) \cup P(s, m_2)$

Our objective is to obtain the feasible tree of minimum cost for the given values of Δ and δ. Now, we introduce a heuristic algorithm for DVBST. WFA can check various paths with path delay or path cost for each ω as Fig. 1. For each multicast member, the proposed algorithm uses WFA and finds a pertinent weight considering with Δ and δ. Let $G = (V, E)$ be a given network and $M = \{ m_\alpha, m_\beta, m_\gamma \} \subset V$ be a multicast group. As indicated in Fig. 2, we select ω_{i^*} such that $min\{ \phi_C(P_{\omega_i}) \mid {}^\forall P_{\omega_i} \in (2^{s \to m_i}, \Delta) \ {}^\forall \omega_i \in W \}$. Since $\phi_C(P_{\omega_i})$ is decreasing as $\phi_D(P_{\omega_i})$ is increasing (see Fig. 1), we take ω_{i^*} such that $max\{ \omega_i \mid {}^\forall P_{\omega_i} \in (2^{s \to m_i}, \Delta) \}$. Let $G^* = \bigcup_{m_i \in M} P_{\omega_i^*}(s, m_i)$ be a connected subgraph of G. As shown in Fig. 3, G^* must be not tree. The proposed algorithm has to find the minimal spanning tree T from G^*. If there are several minimal spanning trees, pick an arbitrary one. And then, it constructs a multicast tree, $T(s, M)$, from G^* by deleting links in G^*, if necessary, so that all the leaves in $T(s, M)$ are multicast members. The proposed algorithm is described in detail as follows.

Proposed Algorithm $(G(V,E),\ M,\ s,\ \Delta,\ \delta)$

Input: A directed graph $G(V,E)$, M is the multicast group with $m = |M|$, a source node s, an end-to-end delay bound Δ, a delay variation bound δ.

Output: The multicast tree T such that $\phi_D(P(s,m_i)) \leq \Delta,\ \ ^\forall P(s,m_i) \subseteq T$, $\phi_\delta(T) \leq \delta,\ \ ^\forall m_i \in M$, and has a small multicast tree cost.

Begin
$CW[m_i] = \emptyset;\ \omega_{i^*} = \emptyset;\ G^* = \emptyset;\ T = \emptyset;$ /* CW: Candidates for Weight */
For $^\forall m_i \in M$ **Do**
    $CW[m_i]$ is obtained by WFA, s, and Δ
Do
    For $^\forall m_i \in M$ **Do**
        ω_{i^*} = maximum of $CW[m_i] \setminus \omega_{i^*}$
        $G^* = \bigcup_{m_i \in M} P_{\omega_i^*}(s, m_i)$
        T = Prune links for cycles in G^*
While($\phi_\delta(T) > \delta$)
Return T
End Algorithm.

4 Performance Evaluation

4.1 Random Real Network Topology for the Simulation

Random graphs of the acknowledged model represent different kinds of networks, communication networks in particular. There are many algorithms and programs, but the speed is usually the main goal, not the statistical properties. In the last decade the problem was discussed, for examples, by B. M. Waxman (1993) [22], M. Doar (1993, 1996) [4][5], C.-K. Toh (1993) [20], E. W. Zegura, K. L. Calvert, and S. Bhattacharjee (1996) [24], K. L. Calvert, M. Doar, and M. Doar (1997) [3], R. Kumar, P. Raghavan, S. Rajagopalan, D. Sivakumar, A. Tomkins, and E. Upfal (2000) [15]. They have presented fast algorithms that allow the generation of random graphs with different properties, in particular, these are similar to real communication networks. However, none of them have discussed the stochastic properties of generated random graphs. A. S. Rodionov and H. Choo [16] have formulated two major demands for the generators of random graph: attainability of all graphs with required properties and uniformity of distribution. If the second demand is sometimes difficult to prove theoretically, it is possible to check the distribution statistically. The method uses parameter P_e, the probability of link existence between any node pair. We use the method by Rodionov and Choo.

4.2 Simulation Results

We now describe some numerical results with which we compare the performance of the proposed scheme. We generate 100 different random real networks for

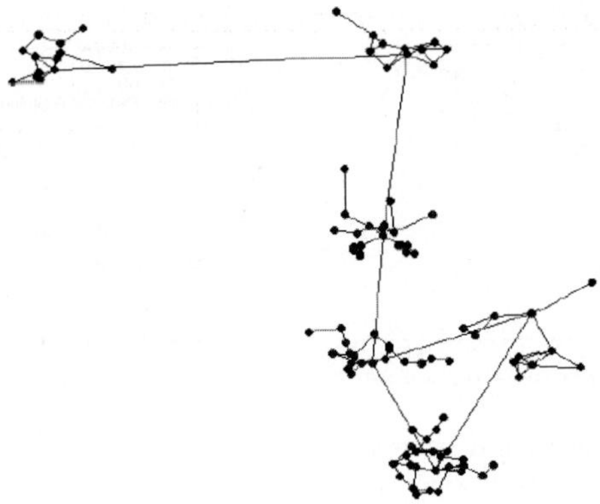

Fig. 4. Random graph generation by Rodionov and Choo [16]

each size of 50, 100, and 200. Each node in network has the average degree 4 or $P_e = 0.3$. The proposed algorithms are implemented in C. We randomly select a source node. The destination nodes are picked uniformly from the set of nodes in the network topology (excluding the nodes already selected for the destination). Moreover, the destination nodes in the multicast group, M, are occupied 10% $\sim$ 80% of the overall nodes on the network. The delay bound Δ value in our computer experiment is set to be 1.5 times the minimum delay between the source node and the farthest destination node [18]. We simulate 1000 times ($10 \times 100 = 1000$) for each $|V|$. For the performance comparison, we implement DDVCA, KBC, and KBYC in the same simulation environments. Since DDVCA outperforms DVMA [18] in terms of delay variation, we do not implement DVMA.

For $|V|$: 50, 100, 200 with P_e: 0.3 or average degree 4
  Do Generate random graph 100 times
    For $|D|$: 10%, 20%, 30%, ..., 80% of $|V|$
      Do Generate a source node and destination
        nodes 10 times, randomly uniform.
        - Run DDVCA and get the multicast tree cost.
        - Run KBC and get the multicast tree cost.
        - Run KBYC and get the multicast tree cost.
        - Run the proposed algorithm and get the multicast tree cost.

As indicated in Fig. 5, it is easily noticed that the proposed algorithm is always better than others. Since KBC and DDVCA use analogous core selection, their tree costs are similar. Because the proposed algorithm takes minimal cost paths which are bounded Δ, its tree cost cannot help but be good. The proposed algorithm enhancement is up to about 30.5% $\sim$ 38.7% ($|V| = 200$) in terms of tree cost for KBC. We also show that enhancement of KBYC is up to about 9.2% $\sim$ 19.7% ($|V| = 200$) in terms of tree cost for KBC.

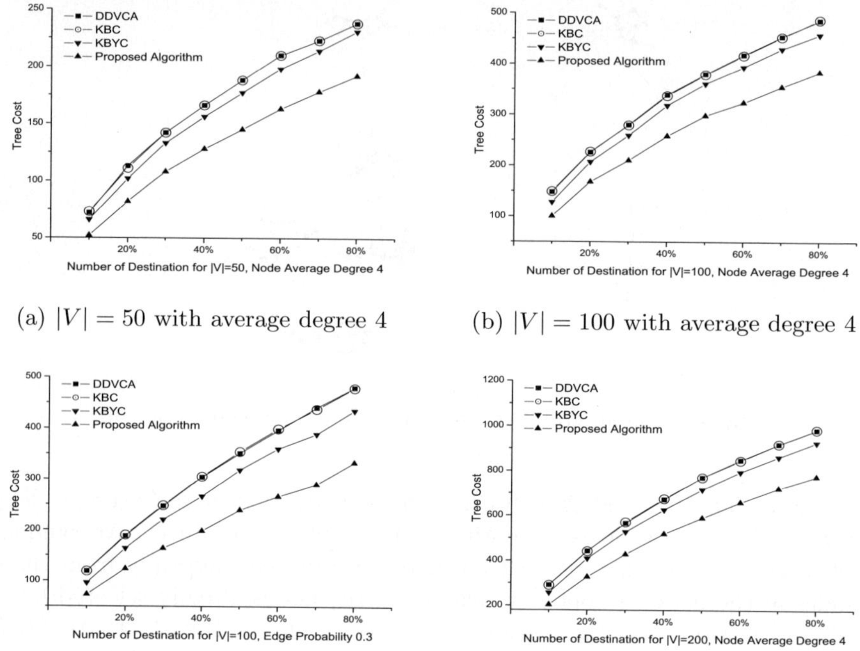

(a) $|V| = 50$ with average degree 4

(b) $|V| = 100$ with average degree 4

(c) $|V| = 100$ with $P_e = 0.3$

(d) $|V| = 200$ with average degree 4

Fig. 5. Tree costs

5 Conclusion

We have studied the problem of constructing minimum cost multicast trees that satisfy the end-to-end delay bound and delay variation bound, which is called as DVBST, and has been proved to be NP-complete [17]. We proposed algorithms using expected multiple paths. The expected multiple paths are obtained by WFA which is introduced in [11]. WFA is efficiently combining two independent measures, the cost and delay. The weight ω plays on important role in combining the two measures in WFA. Our proposed algorithm find suitable weight ω for each destination member, and it constructs the multicast tree for DVBST. The efficiency of our algorithms is verified through the performance evaluations and the enhancement is 30.5% $\sim$ 38.7% in terms of the multicast tree cost. Also, the time complexity is $O(Wml + Wmn\log n)$ which is comparable to one of previous works.

Acknowledgment

This research was supported by Ministry of Information and Communication, Korea under ITRC IITA-2005-(C1090-0501-0019).

References

1. A. Ballardie, B. Cain, and Z. Zhang, "Core Based Trees (CBT version 3) Multicast Routing protocol specification," Internet Draft, IETF, August 1998.
2. Y.-C. Bang, S.-T. Chung, M. Kim, and S.-S. Joo, "On Multicast Communications with Minimum Resources," Springer-Verlag Lecture Notes in Computer Science, vol. 3726, pp. 4-13, September 2005.
3. K. L. Calvert, M. Doar, and M. Doar, "Modelling Internet Topology," IEEE Communications Magazine, pp. 160-163, June 1997.
4. M. Doar, "Multicast in the ATM environment," Ph.D dissertation, Cambridge University, Computer Lab., September 1993.
5. M. Doar, "A Better Mode for Generating Test Networks," IEEE Proc. GLOBECOM'96, pp. 86-93, 1996.
6. M. R. Garey and D. S. Johnson, "Computers and Intractability: A Guide to the Theory of NP-Completeness," W. H. Freeman and Co., San Francisco, 1979.
7. F. K. Hwang and D. Richards, "Steiner Tree Problems," Networks, vol. 22, pp. 55-89, 1992.
8. G. Jho, M. Kim, and H. Choo, "Source-Based Minimum Cost Multicasting: Intermediate-Node Selection with Potentially Low Cost," Springer-Verlag Lecture Notes in Computer Science, vol. 3746, pp. 808-819, November 2005.
9. R. M. Karp, "Reducibility among combinatorial problems," Complexity of computer computations (R. E. Miller, J. W. Thather eds.), pp. 85-104, Newyork Plenum Press, 1972.
10. M. Kim, Y.-C. Bang, and H. Choo, "Efficient Algorithm for Reducing Delay Variation on Bounded Multicast Trees," Springer-Verlag Lecture Notes in Computer Science, vol. 3090, pp. 440-450, August 2004.
11. M. Kim, Y.-C. Bang, and H. Choo, "On Algorithm for Efficiently Combining Two Independent Measures in Routing Paths," Springer-Verlag Lecture Notes in Computer Science, vol. 3483, pp. 989-998, May 2005.
12. M. Kim, Y.-C. Bang, and H. Choo, "On Estimation for Reducing Multicast Delay Variation," Springer-Verlag Lecture Notes in Computer Science, vol. 3726, pp. 117-122, September 2005.
13. M. Kim, Y.-C. Bang, J. S. Yang, and H. Choo, "An Efficient Multicast Tree with Delay and Delay Variation Constraints," Springer-Verlag Lecture Notes in Computer Science, vol. 3982, pp. 1129-1136, May 2006.
14. L. Kou, G. Markowsky, and L. Berman, "A fast algorithm for Steiner trees," Acta Informatica, vol. 15, pp. 141-145, 1981.
15. R. Kumar, P. Raghavan, S. Rajagopalan, D Sivakumar, A. Tomkins, and E. Upfal, "Stochastic models for the Web graph," Proc. 41st Annual Symposium on Foundations of Computer Science, pp. 57-65, 2000.
16. A. S. Rodionov and H. Choo, "On Generating Random Network Structures: Connected Graphs," Springer-Verlag Lecture Notes in Computer Science, vol. 3090, pp. 483-491, August 2004.
17. G. N. Rouskas and I. Baldine, "Multicast routing with end-to-end delay and delay variation constraints," IEEE J-SAC, vol. 15, no. 3, pp. 346-356, April 1997.
18. P.-R. Sheu and S.-T. Chen, "A Fast and Efficient Heuristic Algorithm for the Delay- and Delay Variation-Bounded Multicast Tree Problem," Computer Communications, vol. 25, no. 8, pp. 825-833, 2002.
19. H. Takahashi and A. Matsuyama, "An Approximate Solution for the Steiner Problem in Graphs," Mathematica Japonica, vol. 24, no. 6, pp. 573-577, 1980.

20. C.-K. Toh, "Performance Evaluation of Crossover Switch Discovery Algorithms for Wireless ATM LANs," IEEE Proc. INFOCOM'96, pp. 1380-1387, 1996.
21. B. Wang and J. C. Hou, "Multicast Routing and Its QoS Extension: Problems, Algorithms, and Protocols," IEEE Network, vol. 14, no. 1, pp. 22-36, 2000.
22. B. W. Waxman, "Routing of multipoint connections," IEEE JSAC, vol. 6, no. 9, pp. 1617-1622, December 1988.
23. P. Winter, "Steiner Problem in Networks: A Survey," Networks, vol. 17, pp. 129-167, 1987.
24. E. W. Zegura, K. L. Calvert, and S. Bhattacharjee, "How to model an Internetwork," IEEE Proc. INFOCOM'96, pp. 594-602, 1996.

Periodic Message Scheduling on a Switched Ethernet for Hard Rcal-Time Communication

Myung Kyun Kim and Hee Chan Lee

University of Ulsan, Ulsan 680749, Korea
mkkim@ulsan.ac.kr

Abstract. This paper proposes a message transmission model for hard real-time communications of periodic messages on a switched Ethernet and also proposes an algorithm to schedule the messages to be transmitted within their deadlines. The proposed scheduling algorithm is a distributed one and is performed by the source and the destination nodes without the modification of the operational features of the standard Ethernet switch. When a new periodic message needs to be transmitted, it is first checked whether it can be scheduled on both the transmission and the reception links without affecting the already-schedlued messages, and a feasible schedule is made for the new message if it is schedulable. The proposed scheduling algorithm guarantees the transmission of periodic messages within their deadline and allows flexible message transmission on a hard real-time switched Ethernet.

1 Introduction

Real-time distributed control systems are becoming more widely used in industrial environment [1]. They are used to support a broad range of applications such as process control, factory automation, automotive, robotics, and so on. Switched Ethernet that are most widely used nowadays has many features for real-time communications such as providing a large amount of bandwidth, microsegmentation of network traffic, cut-through switching, and full-duplex links [2]. However, to provide hard real-time communications on the switched Ethernet, some real-time features must be added to both the end nodes and the switches to regulate the amount of traffic on the network in order not to overrun the output queue of the switch [3,4,5,6]. EtheReal switch [3] was built for real-time communication over a switched Ethernet, but it has no explicit support for real-time periodic traffic. Traffic shaping techniques have been proposed in [5,6] to provide a real-time communication by limiting the amount of traffic on the network, but their methods only show that the maximum delay on the network is bounded without considering the explicit deadlines of messages. Hoang et al. [4] have proposed hard real-time communication methods based on EDF (Earliest Deadline First) scheduling over the switched Ethernet. Their approach, however, assumes that both end nodes and the switch can schedule messages according to the EDF policy, which requires the addition of RT(Real-Time) layer to support the EDF scheduling above the MAC layer both on the nodes and on the switch.

M. Gerndt and D. Kranzlmüller (Eds.): HPCC 2006, LNCS 4208, pp. 457–466, 2006.

Pedreiras et al. [7] have proposed an elastic message transmission model called FTT-Ethernet to support dynamic real-time message requirements on Ethernet. Their method uses a synchronized message transmission based on a master-slave model. Their centralized scheme lowers the advantage of the distributed message transmission model of switched Ethernet (and also Ethernet) and the master becomes a single point of failure of the network.

This paper proposes a message scheduling algorithm of periodic messages for hard real-time communications over the switched Ethernet. The proposed scheduling algorithm is a distributed one and operates between source and destination nodes without requiring the modification of the operation of the standard switch. The switched Ethernet in this paper uses a synchronized message transmission model that is similar to the one proposed by Pedreiras et al. [7], but our model uses a distributed scheme to check the feasibility condition of a new message and to make a transmission schedule for the message. When a new periodic message needs to be transmitted, it is first checked whether it is feasible both on the transmission link and on the reception link without affecting the periodic messages that have already been scheduled, and a transmission schedule is made for the message if it is schedulable. For the admission control of new messages, two control messages, *Periodic_msg_req* and *Periodic_msg_rep* messages are exchanged between source and destination nodes. The proposed scheduling algorithm guarantees transmission of periodic messages within their deadline and allows flexible message transmission on a hard real-time switched Ethernet.

The rest of the paper is organized as follows. In section 2, the message transmission model on the switched Ethernet is discussed. In section 3, feasibility check and a message scheduling algorithm for a new periodic message on the switched Ethernet are described. An experiment of the proposed scheduling algorithm is described in section 4, and finally, conclusions and future works are discussed in section 5.

2 Message Transmission Model on the Switched Ethernet

Each node on a switched Ethernet is connected to a switch by a full-duplex link which consists of a transmission link (TL) and a reception link (RL) as shown in Fig. 1.

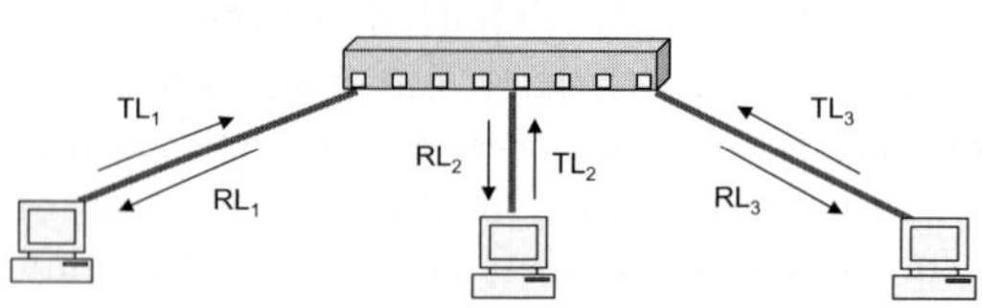

Fig. 1. A switched Ethernet example

Both transmission and reception links operate independently and the switch uses cut-through switching for fast forwarding of frames from the input ports to the output ports. In the cut-through switching, the switch determines the output port immediately after receiving the header of a frame and forwards the frame without waiting for the reception of the entire frame. When a switch operates in the cut-through switching mode, if a message SM_{ij} from node N_i to node N_j is transmitted at time t_0 from N_i, it arrives at N_j after $T_L = 2 * t_p + t_s$ amount of delay from t_0 if there is no collision at the output port of the switch to N_j and the output queue at the output port is empty. This is shown in Fig. 2-(a), where t_p is a propagation delay on a link and t_s is a switching delay in the switch. In Fig. 2-(a), RL_s and TL_s are the reception and transmission links of the switch, respectively. We assume that all of the nodes on a switched Ethernet are synchronized and both the transmission and the reception links are composed of a sequence of macro cycles (MCs), each of which is further divided into smaller basic message transmission units called elementary cycles (ECs) as shown in Fig. 2-(b). It is assumed that the switch uses cut-through switching, so the MC on the reception links starts after T_L amount of delay from the beginning of the MC on the transmission links. Each EC is further divided into two time periods: periodic cycle(PC) and aperiodic cycle(AC). PC is used for transmitting periodic messages and AC is for transmitting aperiodic messages. PL (Length of PC) and AL (Length of AC) are determined according to the ratio of the amount of message traffic of periodic messages and that of aperiodic messages.

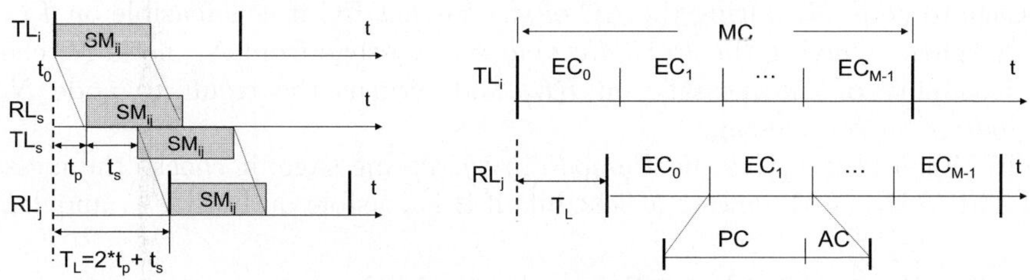

(a) Message transmission in cut-through switching (b) Message transmission model on a switch

Fig. 2. Message transmission model

All of the periodic messages have their strict message transmission deadlines, so in order for the messages to be delivered within their deadlines, the required network bandwidth must be allocated before transmitting the messages both on the transmission link and on the reception link. To reserve the required network bandwidth for periodic messages, an admission control process is performed between sender and destination nodes. During the admission control process, *Periodic_message_req* and *Periodic_message_rep* control messages are exchanged in the AC of the current EC between the sender and destination nodes. The control messages are given higher priorities than other normal aperiodic messages. In this paper, we only consider the scheduling of periodic messages. A periodic message SM_{ij} from node N_i to N_j has a real-time transmission requirement $\{P_{ij}, D_{ij}, C_{ij}\}$, where P_{ij}, D_{ij}, and C_{ij} denote the period, the deadline, and the length of the message, respectively. We assume that D_{ij} $= P_{ij} = k$ * EL (Length of EC) for some integer k. It is also assumed that the real-time transmission requirements of all of the periodic messages are known a priori, and the number of ECs in a MC is $M = \text{LCM}\{P_{ij}/EL\}$ for all i and j (LCM means least common multiple). A set of ECs from the $(k * p)$-th EC to the $((k+1)*p-1)$-th EC where $p = P_{ij}/EL$ and $0 \leq k \leq M/p-1$ is called the k-th *period boundary* of a periodic message with period P_{ij}.

3 Real-Time Scheduling of Periodic Messages on the Switched Ethernet

Each node on the switched Ethernet maintains a message schedule which contains an order of messages to be transmitted on each EC of a MC. When a source node wants to send a new periodic message during message transmission, an admission control process is carried out during the next AC between the source and destination nodes of the message. If the new message becomes feasible, the message is added to the message schedule and begins to be transmitted from the next period boundary of the message. The admission control process of a new periodic message SM_{ij} from node N_i to node N_j which has a real-time transmission requirement $\{P_{ij}, D_{ij}, C_{ij}\}$ is as follows.

(i) Node N_i checks the feasibility of SM_{ij} on TL_i and sends *Periodic_msg_req* message to node N_j during the AC of the current EC if it is feasible on TL_i.

(ii) When receiving the *Periodic_msg_req* message from N_i, node N_j checks the feasibility of the message on RL_j and returns the result to node N_i in *Periodic_msg_rep* message.

(iii) When N_i receives the *Periodic_msg_rep* message, it checks the message and adds SM_{ij} to the message schedule if it is feasible on both TL_i and RL_j.

3.1 Feasibility Check on Transmission Links

Scheduling of a message SM_{ij} with a real-time transmission requirement $\{P_{ij}, D_{ij}, C_{ij}\}$ is to find an ordered set of M/p ECs, $p = P_{ij}/EL$, in a MC, where the k-th EC, $0 \leq k \leq M/p-1$, belongs to the k-th period boundary of SM_{ij} and has available bandwidth greater than or equal to C_{ij} on both TL_i and RL_j. When a new message becomes feasible in an EC, the message is added to the last of the message schedule of the EC. For the feasibility check on the transmission links, each node keeps the current amount of traffic in each EC of its transmission link.

Fig. 3 shows an example of feasibility check on transmission link TL_3. In this example, a MC consists of 6 ECs and $EL = 1.2$ and $PL = 1.0$ $(AL = 0.2)$. Fig. 3-(a) shows the current message schedule of TL_3, where three messages SM_{32}, SM_{31}, and SM_{35} are being transmitted. In the figure, $T_3 = \{T_{3,i},\ 0 \leq i \leq 5\}$, where $T_{3,i}$ denotes the amount of current traffic in EC_i of TL_3. When node N_3 wants to send a new message SM_{34} that has a real-time requirement $\{2, 2, 0.4\}$ to node N_4, our algorithm first finds a set of 3 (M/P_{34}) ECs, where the k-th EC is the one that has a minimum current amount of traffic among the ECs in the k-th period boundary. In the case of the example of Fig. 3-(a), this corresponds

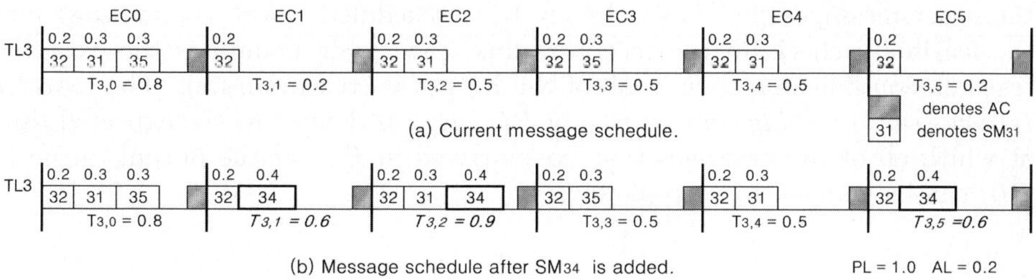

Fig. 3. An example of feasibility check of message SM_{34} on TL_3: real-time requirement of SM_{34} is {2, 2, 0.4}

to $\{EC_1, EC_2, EC_5\}$ (or $\{EC_1, EC_3, EC_5\}$). All of the ECs in the set have available bandwidths greater than 0.4 (C_{34}). So, SM_{34} is feasible on TL_3 and can be scheduled on the ECs in $\{EC_1, EC_2, EC_5\}$. Fig. 3-(b) shows the message schedule of TL_3 after the message SM_{34} is added.

After checking the feasibility on its transmission link, the source node transmits *Periodic_msg_req* message to the destination node. For the feasibility check on the reception link, the source node sends in the *Periodic_msg_req* message the real-time requirement for the message to be added and the current amount of traffic of each EC of the transmission link. The following is an algorithm to check the feasibility of a message SM_{ij} on TL_i.

// **Algorithm 1.** Feasibility check of SM_{ij} on TL_i
// Real-time requirement of SM_{ij} : $\{P_{ij}, D_{ij}, C_{ij}\}$
$p = P_{ij}/EL$;
for $(k = 0;\ k \le (M/p - 1);\ k++)$ {
  $T_{min} = \mathbf{min}_{0 \le m \le p-1}\ \{T_{i,k*p+m}\}$;
  if $(T_{min} + C_{ij} > PL)$ { // Not feasible
    Reject message transmission request of SM_{ij};
    return;
  }
}
Send *Periodic_msg_req*$[SM_{ij}, \{P_{ij}, D_{ij}, C_{ij}\}, T_i]$ to N_j;
return;

3.2 Feasibility Check on Reception Links

The messages that were transmitted in an EC by the source nodes arrive at each output queue of the switch to the destination nodes after $(t_p + t_s)$ amount of time delay, where t_p is a propagation delay and t_s is a switching delay in the switch. The messages that arrive at the same output queue are stored in the queue and transmitted one by one from the queue in FIFO(First-In, First-Out) order through the reception link of the destination node. In order that a new message should be feasible in an EC of the reception link, the new message must be able to be transmitted completely within that EC without affecting the other messages that have already been scheduled before the message. For the feasibility check on the reception link, each node maintains the expected transmission finishing time in all of the ECs of its reception link. The *expected transmission finishing time* in EC_i of RL_j, $R_{j,i}$, is defined as the expected time at which all of the messages that have arrived in EC_i at the output queue to RL_j can be transmitted completely.

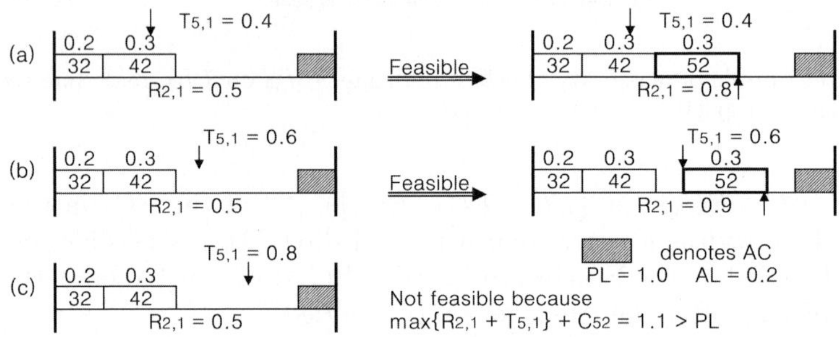

Fig. 4. Examples of feasibility check of SM_{52} in EC_1 of RL_2 where $C_{52} = 0.3$

When node N_j receives $Periodic\_msg\_req[SM_{ij}, \{P_{ij}, D_{ij}, C_{ij}\}, T_i]$ message from node N_i, it checks the feasibility of SM_{ij} in the ECs of RL_j using $R_j = \{R_{j,i}, 0 \leq i \leq M-1\}$, T_i, and C_{ij}. In order for SM_{ij} to be feasible in EC_k of RL_j, it must be satisfied that

$$\mathbf{max}\{R_{j,k}, T_{i,k}\} + C_{ij} \leq PL. \tag{1}$$

If the message SM_{ij} is scheduled to be transmitted in EC_k, the expected transmission finishing time $R_{j,k}$ is updated as follows:

$$R_{j,k} = \mathbf{max}\{R_{j,k}, T_{i,k}\} + C_{ij}. \tag{2}$$

Fig. 4 shows some examples of feasibility check of message SM_{52} and update of the expected transmission finishing time on RL_2. The cases (a) and (b) of the figure show the examples of feasible cases, but in case (c), SM_{52} is not feasible in EC_1 because $\mathbf{max}\{R_{2,1}, T_{5,1}\} + C_{52} = 0.8 + 0.3 = 1.1 > PL$.

As shown in Fig. 4-(b), there may be empty time periods on the reception link between adjacent massages when the arrival time of the new message at the reception link is greater than the current expected transmission finishing time of the reception link. When a source node has multiple messages to send in an EC, the waste of network bandwidth due to the empty time periods on the reception link can be reduced if each node sends the messages in the increasing order of message lengths. Thus, in our scheduling algorithm, each source node performs the admission control of messages in the increasing order of message lengths when it has multiple messages to add. The following is an algorithm to check feasibility on the reception link when a node receives *Periodic_msg_req* message from a source node.

```
// Algorithm 2. Feasibility check of SM_ij on RL_j
// N_j receives Periodic_msg_req[SM_ij, {P_ij, D_ij, C_ij}, T_i] from N_i
// next(S): function to return the next element in a set S and delete it
p = P_ij/EL;    S = ∅;
for (k = 0; k ≤ (M/p − 1); k++) {
    R_min = max{R_j,k*p, T_i,k*p};
    imin = k * p;    m = 1;
    while (m ≤ p − 1) {
        if (R_min > max{R_j,k*p+m, T_i,k*p+m}) {
            R_min = max{R_j,k*p+m, T_i,k*p+m};
            imin = k * p + m;
        }
        m = m + 1;
    }
    if (R_min + C_ij > PL) {              // Not feasible
        Send Periodic_msg_rep[SM_ij, NULL] to N_i;
        return;
    } else
        S = S ∪ {imin};
}
Send Periodic_msg_rep[SM_ij, S] to N_i;       // Feasible
while ((k = next(S)) ≠ NULL)          // update expected transmission
    R_j,k = max{R_j,k, T_i,k} + C_ij;     // finishing time R_j
return;
```

3.3 Update the Message Schedule

When node N_i receives *Periodic_msg_rep*$[SM_{ij}, S]$ message from node N_j, it examines the message. If S is NULL, SM_{ij} is not feasible on RL_j, so it rejects the transmission request of SM_{ij}. Otherwise, SM_{ij} is feasible on the ECs in S, thus, N_i updates the current amount of traffic of ECs in S and adds SM_{ij} to the message schedule. After that, the message SM_{ij} can be transmitted from the next period boundary of SM_{ij}. The following is an algorithm to update the message schedule when node N_i receives *Periodic_msg_rep* from node N_j.

```
// Algorithm 3. Updating the message schedule of TL_i
// N_i receives Periodic_msg_rep[SM_ij, S] from N_j
// next(S): return the next element in S and delete it
// Update_msg_schedule(SM_ij, k): add SM_ij to the message schedule of EC_k
if (S == NULL) {
    Reject the transmission request of message SM_ij;
    return;
}
p = P_ij/EL;
while ((k = next(S)) ≠ NULL) {
    T_i,k = T_i,k + C_ij;          // update T_i
    Update_msg_schedule(SM_ij, k);
}
return;
```

4 Experiment of the Proposed Scheduling Algorithm

We have implemented the proposed scheduling algorithm on systems using Linux 2.6.10 with high resolution POSIX timers [8] and experimented on a switched Ethernet which consists of a 10 Mbps switch (Cisco Catalyst WS-C1912C-EN switch) and 5 nodes. The switch operates in cut-through switching mode and the switching delay in the cut-through mode is 31 μs to 10 Mbps ports.

Table 1. The message set used in the experiment

Source	Generated messages
N_1	(N_4,{1,0.1}), (N_3,{3,0.1}), (N_5,{3,0.2}), (N_2,{2,0.3}), N_3,{2,0.3})
	(N_2,{2,0.4}), $(N_2, \{1, 0.4\})^*$, (N_5,{1,0.4}), $(N_3, \{3, 0.4\})^*$, (N_5,{3,0.4})
N_2	(N_5,{2,0.1}), (N_1,{2,0.2}), (N_3,{2,0.2}), (N_5,{1,0.2}), (N_3,{1,0.3})
	(N_3,{2,0.3}), (N_4,{2,0.3}), $(N_4, \{1, 0.4\})^*$, (N_5,{3,0.4}), $(N_4, \{3, 0.4\})^*$
N_3	(N_5,{3,0.1}), (N_1,{3,0.2}), (N_4,{1,0.2}), (N_2,{1,0.2}), (N_1,{3,0.2})
	(N_5,{3,0.2}), (N_2,{1,0.2}), (N_4,{3,0.3}), (N_1,{1,0.3}), $(N_4, \{2, 0.4\})^*$
N_4	(N_2,{3,0.1}), (N_2,{1,0.1}), (N_2,{2,0.1}), (N_3,{1,0.2}), (N_5,{2,0.2})
	(N_1,{3,0.2}), (N_3,{3,0.3}), (N_1,{2,0.3}), $(N_5, \{1, 0.3\})^*$, $(N_1, \{2, 0.4\})^*$
N_5	(N_3,{2,0.1}), (N_1,{3,0.1}), (N_3,{3,0.2}), (N_1,{2,0.3}), (N_4,{2,0.3})
	(N_1,{3,0.4}), (N_4,{1,0.4}), (N_3,{3,0.4}), $(N_4, \{2, 0.4\})^*$, $(N_2, \{3, 0.4\})^*$

cf. $(N_j, \{P_{ij}, C_{ij}\})^*$: N_j specifies a destination node, P_{ij} and C_{ij} specifies the period (deadline) and the message length of SM_{ij}, and * denotes the message that is not feasible

As discussed in section 2, a MC on the reception link begins after T_L amount of time from the beginning of the MC on the transmission link. We have used libnet library [9] to write messages to the network and libpcap packet capture library [10] to read messages from the network. There is a little more than 12 μs delay when copying a frame from the application to the frame buffer in the NIC and vice versa. The propagation delay on a link t_p is almost negligible, so $T_L = 2 * t_p + t_s = 31$ μs, but considering the delay due to copying between the application and the frame buffer in the NIC, we assumed that $T_L = 60$ μs, which is a little more than 55 μs (31 + 2*12). In this experiment, $EL = 2.0$ ms and $PL = 1.6$ ms.

For the experiment, we generated 50 messages randomly such that their periods are 1, 2, or 3 (M = 6) and their message lengths are within [0.1, 0.4] ms. Table 1 shows a set messages generated for the experiment. Each node sorts the messages according to their length and performs the admission control process in the increasing order of the message length. The experiment result shows that among the messages in the table, 41 messages are feasible according to the proposed scheduling algorithm. Fig. 5-(a) shows the response times of the messages whose periods are 1. As shown in the figure, all of the messages have been delivered within their deadline which is 1*EL (2 ms). Fig. 5-(b) and (c) show the response times of the messages with period 2 and 3, respectively. As shown in the figure, all of the messages have been delivered completely within their deadlines which are 2*EL (4 ms) and 3*EL (6 ms), respectively.

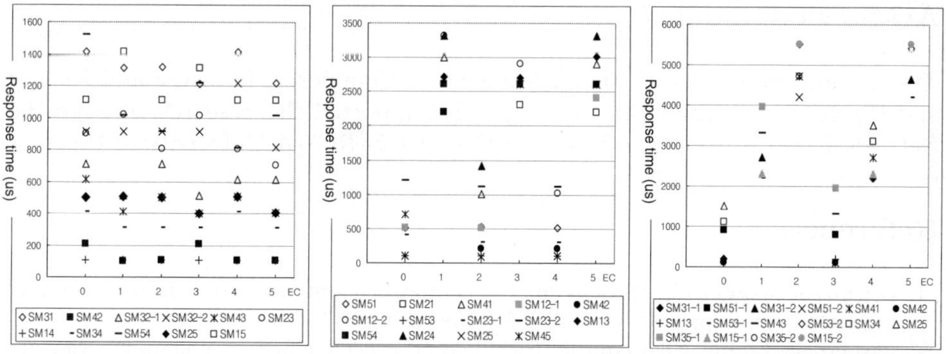

(a) Messages with period = 1 (b) Messages with period = 2 (c) Messages with period = 3

Fig. 5. Experiment results for messages in Table 1: response times of messages ($EL = 2.0$ms and $PL = 1.6$ms)

The proposed message scheduling algorithm guarantees transmission within their deadlines to the messages that have been added to the message schedule by the admission control process. The admission control process of our scheduling algorithm is a distributed one while FTT-Ethernet [7] uses a centralized admission control algorithm. Our scheduling algorithm operates between source and destination nodes without requiring modification to the standard Ethernet switch while the approach of Hoang et. al. [4] requires the modification of the switch to support EDF algorithm.

5 Conclusions and Future Works

This paper proposed a synchronized transmission model for hard real-time communications of periodic messages over the switched Ethernet and also proposed a dynamic message scheduling algorithm for the periodic messages. The proposed scheduling algorithm is a distributed one and is performed between the source and destination nodes without requiring the modification of the operational features of the standard Ethernet switch. Our message scheduling algorithm guarantees the periodic messages to be transmitted within their deadlines and also provides a flexible message transmission scheme for hard real-time communication among industrial applications.

The proposed message scheduling algorithm uses the exchange of control messages (*Periodic_msg_req* and *Periodic_msg_rep* messages) in an AC for the admission control of new periodic messages. We have assumed in this paper that the exchange of the control messages is performed within an AC, but some traffic restriction on asynchronous messages should be made to guarantee the transmission of the control messages within an AC. We continue to study on the traffic restrictions on asynchronous messages.

Acknowledgment

The authors would like to thank Ministry of Commerce, Industry and Energy and Ulsan Metropolitan City which partly supported this research through the Network-based Automation Research Center (NARC) at University of Ulsan.

References

1. Buttazzo Georgio C.: Hard real-time computing systems: Predictable cheduling algorithms and applications(2nd Ed.). Springer (2005)
2. Jasperneit J., and Neumann P.: Switched ethernet for factory communication. In Proc. of ETFA, Antibes, France (2001) 205–212
3. Varadarajan S. and Chiueh T.: EtheReal: A host-transparent real-time fast Ethernet switch. In Proc. of ICNP (1998) 12–21
4. Hoang H., Jonsson M., Hagstrom U., and Kallerdahl A.: Switched real-time ethernet with earliest deadline first scheduling- protocols and traffic handling. In Proc. of IPDPS, Fort Lauderdale, FL, USA (2002) 94–99
5. Loeser J. and Haertig H.: Low-latency hard real-time communication over switched Ethernet. In Proc. ECRTS, Catania, Italy (2004)
6. Kweon S. K., Shin K. G., and Workman G.: Achieving real-time communication over Ethernet with adaptive traffic shaping. In Proc. RTAS (2000) 90–100
7. Pedreiras P., Almeida L., and Gai P.: The FTT-Ethernet protocol: Merging flexibility, timeliness and efficiency. In Proc. ECRTS (2002) 134–142
8. Linux with high resolution POSIX timers: http://sourceforge.net/projects/high-res-timers
9. Libnet netwroking library: http://libnet.sourceforge.net
10. Libpacp packet capture library: http://ee.lbl.gov

Optical Traffic Grooming Based on Network Availability

Yeo-Ran Yoon, Tae-Jin Lee, Min Young Chung, and Hyunseung Choo*

Lambda Networking Center
School of Information and Communication Engineering
Sungkyunkwan University, Korea
{ookuma, tjlee, mychung, choo}@ece.skku.ac.kr

Abstract. As the bandwidth request for traffic streams grows excessively, traffic grooming on wavelength-division multiplexing (WDM) mesh networks becomes more important. In general it efficiently grooms low-capacity requests onto high-capacity lightpath. Thus network throughput and cost is improved. Also it affects the recent development of WDM passive optical network(PON) and the future fiber to the home(FTTH). Our objective is to improve the network throughput while maintaining the similar level of network cost. In this paper, we propose Dynamic Ordering on Lightpath Assignment (DOLA) algorithm, in which an order of lightpath assignment is determined according to the network availability. The comprehensive computer simulation results show that our proposed algorithm is up to about 14% superior to the well-known previous work [11].

1 Introduction

Recent progress in fiber optics and wavelength-division multiplexing (WDM) technology has lead the support for significant increase of traffic requests in the Internet. The WDM technology enables a considerable portion of the available fiber bandwidth to be used by allowing several signals with different wavelengths to be transmitted simultaneously through one fiber. In WDM networks, a lightpath provides a basic communication mechanism between two nodes and it passes through many fiber links and optical switches [3,4]

To satisfy the lightpath continuity constraint, the routing and assigning wavelength should be performed appropriately for each service request. This is well known as a routing and wavelength assignment (RWA) problem [4,5] or a lightpath-provisioning problem [1]. RWA is very important to efficiently utilize resources on optical networks consisting of WDM systems and optical switches, because it has significant effects on cost and performance of optical networks. Many solutions for the RWA discussed in the literature have assumed that the required bandwidth for connection request is the same as the WDM channel capacity [1,3,4,5,9]. However, the required bandwidth (OC-1, OC-3, or OC-12) is much smaller than the WDM channel capacity (OC-48, OC-192, or OC-768) in real environments.

If the entire bandwidth of a wavelength channel is allocated to a low-capacity connection, a large portion of the transmission capacity could be wasted. Thus,

* Corresponding author.

M. Gerndt and D. Kranzlmüller (Eds.): HPCC 2006, LNCS 4208, pp. 467–476, 2006.

traffic grooming schemes which are procedures for efficient multiplexing (demultiplexing) and switching low-capacity traffic streams onto (from) high-capacity lightpath are required to improve bandwidth utilization, to optimize network throughout, and to minimize network cost. Traffic grooming problems in static and dynamic environments are discussed in [11,12,13] and in [6,8,14], respectively. Especially in [11], Zhu and Mukherjee formulates a grooming problem as the Integer Linear Programing (ILP) to increase the network throughout. And they propose Maximizing Single-hop traffic (MST) and Maximizing Resource Utilization (MRU) algorithms based on the ILP solution.

This paper investigates traffic grooming problem in WDM mesh networks with the capability of single- and multi-hop grooming to support static traffic demands. Our objectives are to maximize the network throughput and to minimize the blocking probability. As we all know, maximizing of throughput guarantees the minimization on the blocking probability. For these, we propose a Dynamic Ordering on Lightpath Assignment (DOLA) algorithm. Based on the DOLA, the ordering on lightpath assignment is changed dynamically by using the current network resource utilization. Compared with the MRU algorithm, the DOLA achieves 14% improved performance in terms of the network throughput. The discussion with the MST along our own proposition is provided in another paper underway due to the space limitation.

The rest of the paper is organized as follows. Section 2 explains previous works on WDM SONET ring and mesh networks, single- and multi-hop grooming, and the general problem statement of the traffic grooming. We propose the DOLA algorithm for solving of the traffic grooming problem in mesh networks and describe it more in detail in Section 3. Section 4 evaluates the performance of the proposed one in terms of the network throughput and the blocking probability. Finally, we conclude this paper in Section 5.

2 Preliminaries

2.1 Related Works

Previous works in various environments: Most of the past traffic grooming research mainly focused on network design and optimization for SONET/WDM ring networks [2,7,10]. By employing wavelength add-drop multiplexers (W-ADMs), efficient wavelength assignment and time-slot assignment algorithms have been investigated to design a SONET/WDM ring network. The objective function in these studies is to minimize the total network cost in terms of the number of SONET add-drop multiplexers.

Optical transport networks keep evolving from interconnected SONET/WDM rings to optical WDM mesh networks in recent years. And the traffic grooming problem in optical WDM mesh networks becomes an important research area [6,8,11,12,13,14]. A generic graph model is proposed for provisioning multigranularity connections in multiwavelength optical WDM networks [12,13]. The authors present that different network optimization objectives and corresponding route selection schemes can be easily applied to this graph model. Based on the

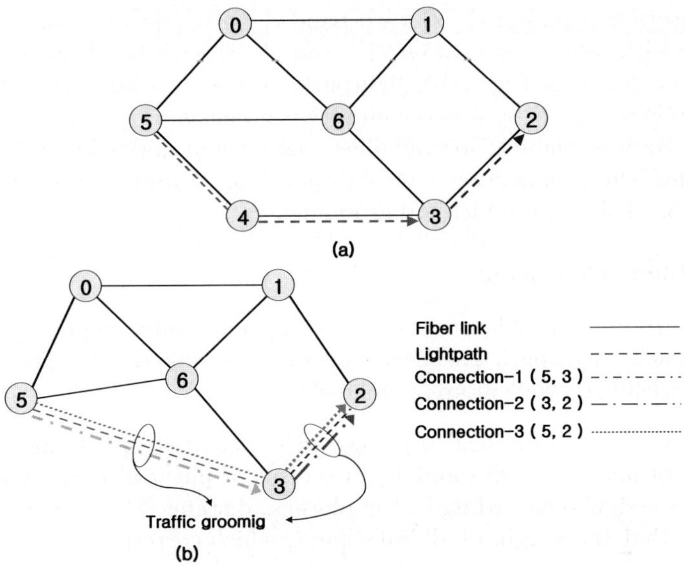

Fig. 1. Example of traffic grooming

model, they propose several grooming algorithms for static traffic demands and show that the performance of their algorithms is near optimal.

The authors in [6,8,14] discuss traffic grooming issues in dynamic traffic environment. They observe that, in a multigranularity WDM network, it accidentally blocks connections with high-capacity requirement more than those with low-capacity one [6]. This results in unfairness between connections on different capacities. Therefore, they propose a Call Admission Control (CAC) algorithm to implement the fairness. The authors in [8] and [14] study on-line provisioning mechanisms for connections with different capacities in a traffic groomable WDM network. Several on-line algorithms are proposed to optimize the network performance.

Single-hop and Multi-hop grooming: In general, traffic requests are carried through single-hop grooming and multi-hop grooming. Single-hop grooming is a method that allows a connection to transmit a single lightpath. This means that only end-to-end traffic grooming is allowed. On the other hand, multi-hop grooming is a method that allows a connection to traverse multiple lightpaths. In multi-hop grooming, a connection can be dropped at intermediate nodes and groomed with other low-capacity connections on different lightpaths before it reaches destination node. As expected, the multihop grooming leads to higher throughput than the single-hop one.

Fig. 1 shows an illustrative example of the traffic grooming in a WDM mesh network. Fig. 1(a) shows a physical topology with seven nodes and Fig. 1(b) shows a virtual topology constructed for given connection requests. Each fiber has two wavelength channels. The capacity given for each wavelength channel is OC-48 (2.5 Gb/s). Each node is equipped with a tunable transceiver. There are three connection requests: (5, 3) with bandwidth requirement of OC-12; (3, 2) with bandwidth requirement of OC-12; and (5, 2) with bandwidth requirement of OC-3. As shown in Fig. 1(b), lightpaths are set up for connection-1 (5, 3) and connection-2 (3, 2), and two connection requests are carried through these lightpaths. Because there is no available transceiver at nodes 5 and 2, connection-3 is groomed with connection-1 on lightpath (5, 3) and with connection-2 on lightpath (3, 2) through multi-hop grooming.

2.2 Problem Definition

The traffic grooming problem in which low-capacity traffic requests are groomed to high-capacity lightpath has been generally presented [11]. The parameters associated with the problem are as following.

- $G_p = (V, E_p)$: A physical topology with weighted directional graph. V is the set of network nodes and E_p is the set of physical links. The weight of a link is assigned according to the physical distance between two nodes. We assume that the weight of all links has 1, which corresponds to the fiber hop distance.
- C: Capacity of a wavelength
- W: Number of wavelengths in a fiber link
- D: A demand set. Each traffic request of the demand set has a capacity lower than C.
- Tr: Number of transceivers at each node.

We should determine the following two factors: 1) $G_v = (V, E_v)$: A virtual topology. V is corresponding to the set of nodes as in the physical topology. E_v is for the set of directional lightpaths assigned between nodes in a node pair; 2) Routing on traffic requests for the virtual topology. We focus on maximizing the network throughput. As discussed earlier, the blocking probability is calculated accordingly.

Traffic grooming usually can be divided into four subproblems [11]. These problems are not necessarily independent, but we deal with the four separately: 1) constructing the virtual topology that consists of lightpaths; 2) routing the lightpaths on the physical topology; 3) assigning wavelength to the lightpaths; 4) routing the traffic requests on the virtual topology. The virtual topology design and RWA problem is NP-complete, and traffic grooming in a mesh network is also an NP-complete problem.

There are lots of studies on RWA based on traffic grooming in the WDM networking field. The well-known routing approaches are fixed routing, fixed-alternate routing, and adaptive routing [9]. Fixed routing always routes the traffic through a predefined fixed route for a given source-destination pair. Fixed-alternate routing considers multiple fixed routes in the routing table, when a traffic request comes. And adaptive routing chooses the route between a source and a destination dynamically, depending on the current network state. This approach requires more computation and response time, but it is more flexible. The First-Fit (FF) method is the best known wavelength assignment approach.

In FF, all wavelengths are numbered and a lower numbered wavelength is considered when searching for an available wavelength. In our proposed heuristic algorithm, we use adaptive routing and FF wavelength assignment.

3 Dynamic Ordering on Lightpath Assignment (DOLA)

3.1 Motivation and MRU Algorithm

Our objective is to improve the network throughput by employing the network resource utilization as a decision criterion and by changing the order of lightpath assignments in the MRU algorithm [11]. Network resource utilization represents average traffic per wavelength link and it is an effective factor for traffic grooming. However, network resource utilization in the MRU algorithm does not reflect the current network status. Our proposed DOLA algorithm dynamically decides lightpath assignment order based on the practical resource utilization reflecting the current network status.

The MRU algorithm assigns a lightpath for a source-destination pair with the maximum network resource utilization value, subject to the constraints on the number of transceivers at two end nodes and the availability of wavelengths in the path connecting the two. Network resource utilization ($R(s,\ d)$) is defined as $T(s,\ d)/h(s,\ d)$, where $T(s,\ d)$ is the aggregate traffic capacity between s and d, which has not been successfully carried yet and $h(s,\ d)$ is the hop distance on the physical topology between s and d. This quantity shows how efficiently the network resource is utilized when carrying the traffic requests.

The connection requests between s and d will be carried on the new established single-hop lightpath as much as possible. If there is enough capacity in the network, every traffic of connection requests will traverse a single-hop lightpath (single-hop grooming). If no lightpath can be set up any more, the remaining traffic requests will be routed on the multi-hop lightpaths using currently available spare capacity at the virtual topology (multi-hop grooming). The details for single- and multi-hop grooming are explained in the subsection 2.1. This multihop grooming is performed based on the resource utilization $t(s,\ d)/h_v(s,\ d)$ for a traffic request $(s,\ d)$, where $t(s,\ d)$ denotes a capacity for each request, and $h_v(s,\ d)$ denotes the hop distance between s and d on the virtual topology constructed for existing lightpaths.

3.2 Detailed Description on DOLA Algorithm

The MRU algorithm sorts all of the node pairs s and d according to network resource utilization $R(s,\ d)$ between s and d, and puts them into a list L in a descending order, then attempts to assign lightpaths sequentially. But after a lightpath for one node pair is assigned, the network resource utilization for the other pairs is changed since the network status on the physical topology changes. The MRU algorithm never consider the change of the network resource, hence this does not reflect continuously the varying network resource utilization.

Accordingly, our proposed DOLA algorithm reflects varying network resource utilization by instant updates for other node pairs whenever a lightpath for one node pair is assigned. Then it finds the pair with the highest $R(s,\ d)$ and attempts to assign a lightpath for this node pair s and d. Because the DOLA algorithm assigns lightpaths dynamically based on the current network status, it uses the wavelengths and resources efficiently. Moreover, since the DOLA algorithm generates more lightpaths with the smaller number of hops than the MRU algorithm, connection requests which are not carried through single-hop grooming have more chance to be carried through multi-hop grooming.

$$T(s_i,\ d_i) = \sum_{\substack{j=1 \\ s_j=s_i \\ d_j=d_i}}^{|D|} c_j \tag{1}$$

$$R(s_i,\ d_i) = T(s_i,\ d_i)/h(s_i,\ d_i) \tag{2}$$

The proposed algorithm has two phases : 1) construction a virtual topology and 2) routing low-speed connections on the constructed virtual topology.

Step 1: Construct a virtual topology

- **1.1:** Sort all the node pairs $(s_i,\ d_i)$ according to network resource utilization $R(s_i,\ d_i)$ and put them into a list L in a descending order.
- **1.2:** Setup a lightpath for the node pair $(s_i,\ d_i)$ with the highest $R(s_i,\ d_i)$ in L using Dijkstra's shortest-path routing algorithm subject to the transceiver constraints. If it fails, delete $(s_i,\ d_i)$ from L; otherwise, let $T(s_i,\ d_i) = Max[T(s_i,\ d_i) - C,\ 0]$ and update $R(s_i,\ d_i)$. And the links of an assigned lightpath are deleted from the physical topology.
- **1.3:** Update $R(s_j,\ d_j)$ appropriately for each node pair $(s_j,\ d_j)$ whose shortest path contains the deleted links in **Step 1.2**.
- **1.4:** Go to **Step 1.2** until list L becomes empty.

Step 2: Route low-speed connections on the virtual topology in **Step 1**.

- **2.1:** Route all connection requests which can be carried through a single hop lightpath, and update the network status for the virtual topology.
- **2.2:** Route the remaining connection requests on the multi-hop lightpaths using currently available spare capacity of the virtual topology based on their connection resource utilization value $t(s_i,\ d_i)/h_v(s_i,\ d_i)$.

Step 3: Repeat **Steps 1** and **2** until entire wavelengths are fully exhausted.

As you see in the pseudocode for the DOLA algorithm(Fig. 3), the physical topology G_p, the number of wavelengths W, the capacity of each wavelength C, and the number of transceivers Tr are given. First of all, for each $(s_i,\ d_i)$ the algorithm finds the shortest path and its number of hops $h(s_i,\ d_i)$ (line 3). Then it aggregates capacities of the uncarried traffic requests for the same s_i and d_i (i.e. $T(s_i,\ d_i)$) in line 4. In line 5 we calculate the network resource utilization $R(s_i,\ d_i)$ and put them into a list L in a descending order (line 7). Even if the L is a list, we sometimes use the set notation for simplicity on L. In order to

$G_p(V, E_p)$	Physical Topology
$G_v(V, E_v)$	Virtual Topology constructed for existing lightpaths
W	The number of available wavelengths
C	Capacity per each wavelength
Tr	The number of Transceivers per each node
D	Demand set $D=\{de_1, de_2, \cdots, de_{\|D\|}\}$, where $de_i=(s_i, d_i, c_i)$
List L	Sorted demand set
$h(s_i, d_i)$	Number of hops in the shortest path for (s_i, d_i) in G_p
$T(s_i, d_i)$	Aggregate capacity for (s_i, d_i)
$R(s_i, d_i)$	Resource Utilization of (s_i, d_i) in G_p (= $T(s_i, d_i) / h(s_i, d_i)$)

Fig. 2. Notations used in the DOLA algorithm

```
1 Algorithm DOLA (Gp, W, C, Tr)
2 Begin
3   Find the shortest path for each (s_i, d_i) and its number of hops h(s_i, d_i)
4   Calculate T(s_i, d_i) = Σ_{j=1, s_j=s_i, d_j=d_i}^{|D|} c_j , i = 1, ··· ,|D|
5   Calculate R(s_i, d_i) = T(s_i, d_i) / h(s_i, d_i)
6   L = Ø
7   Sort (s_i, d_i)s in a descending order in terms of R(s_i, d_i) and L = L ∪ (s_i, d_i)
8   For w=1 to W
9   // Generate Gv (V, Ev)
10      While ( L ≠ Ø )
11        Select (s_i, d_i) with the maximum value of R(s_i, d_i) from L
12        If there exist available lightpaths and transceivers
13            Assign a lightpath to (s_i, d_i)
14            T(s_i, d_i) = Max[T(s_i, d_i)-C, 0]
15            If T(s_i, d_i) = 0
                  delete (s_i, d_i) from L
16            Delete the edges of the shortest lightpath associated with (s_i, d_i) from Gp
17            Find the new shortest path and the number of hops for each (s_i, d_i) whose
              shortest path overlapped with edges deleted
18            Recalculate R(s_i, d_i) = T(s_i, d_i) / h(s_i, d_i) for each (s_i, d_i)
19        Else
              Delete (s_i, d_i) from L and store it to a temporary list L'
20      EndWhile
21  // Routing demands in D on Gv
22      Single-hop grooming for the demands with small traffic requests
23      Multi-hop grooming for the remaining demands
24      Back to the original network status and restore list L = L ∪ L'
25  Endfor
26 End
```

Fig. 3. Pseudocode of the DOLA Algorithm

construct a virtual topology, it tries to assign a lightpath between a node pair (s_i, d_i) with the maximum $R(s_i, d_i)$ in L based on Dijkstra's shortest-path routing algorithm. If a lightpath is not assigned, (s_i, d_i) is deleted from L and store it to a temporary list L' (line 19). If there are available paths and transceivers, a lightpath between s_i and d_i is assigned (line 13) and $T(s_i, d_i)$ is updated to $Max[T(s_i, d_i) - C, 0]$ (line 14). At this moment, if $T(s_i, d_i)$ becomes zero, $(s_i,$

d_i) is deleted from L (line 15). The edges in the assigned lightpath are deleted from the physical topology (line 16).

For each node pair (s_i, d_i) whose shortest path traverses the deleted edges, we update $R(s_i, d_i)$ (lines 17 and 18). And the proposed altogithm selects a node pair (s_i, d_i) with the highest $R(s_i, d_i)$ in L and tries to setup a lightpath for the selected pair. Lines 10 $\sim$ 20 are repeated until the list L becomes empty, and the virtual topology is constructed. Next it carries all of the small traffic connection requests which can be carried through single-hop lightpaths, and updates the virtual topology status (line 22). Finally, if there are remaining connection requests, these are still carried by multi-hop lightpaths using currently available spare capacity on the virtual topology (line 23). Now the physical topology maintains the original network status again and the list L is restored by $L \cup L'$. Lines 8 $\sim$ 25 are repeated until entire wavelengths are exhausted.

4 Numerical Results

In this section, we compare our proposed DOLA algorithm with well-known MRU for the performance evaluate in terms of network throughput, which is successfully carried traffic capacity among traffic requests. We use the network topology of NSFNET as shown in Fig. 4(a) and random networks [?] for proper tests in various network topologies. We assume that the capacity of each wavelength is OC-48 and traffic demands are among OC-1, OC-3, and OC-12 as done in the previous work [11]. The traffic requests are randomly generated as follows: 1) the number of OC-1 connection requests for each node pair is uniformly distributed between 0 and 16; 2) the number of OC-3 connection requests is with uniform distribution over 0 through 8; 3) the number of OC-12 connection requests is selected uniformly for 0, 1, and 2. We simulate 1,000 times for each network topology. We observe the similar results for the network with bigger capacities.

We show the results of the MRU and DOLA algorithms on NSFNET in Fig. 4(b). We compare the network throughput as the number of wavelengths on each fiber link increases and each node is equipped with fifteen tunable transceivers. We observe that the DOLA algorithm demonstrates 6% $\sim$ 14% higher throughput than the MRU algorithm. Since the number of tunable transceivers at each node is limited to 15, increasing the number of wavelengths does not help to increase the network throughput when the number of wavelengths on each fiber link reaches 9.

Figures 4(c) and (d) show the results of the MRU and DOLA algorithm in random networks. To generate a random topology, we first generate a random graph with the number of nodes (N) and the edge probability for each node pair (P_e). Since a random graph must be connected, we generate a spanning tree with N nodes. As we know, $N-1$ tree edges create the spanning tree and guarantee a connected graph. Therefore, we calculate P_e again and determine whether there are other edges or not.

In Fig. 4(c), network throughput versus edge probability between node pairs (P_e) in random networks with fifteen nodes is shown. Each node is equipped with 15 tunable transceivers and carries up to 10 wavelengths. The results show

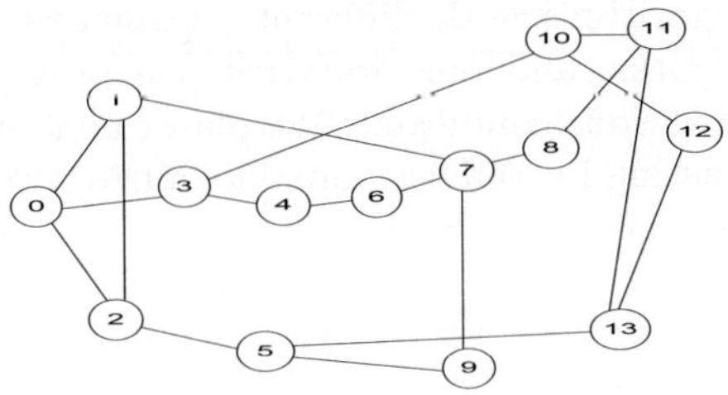

(a) NSFNET topology.

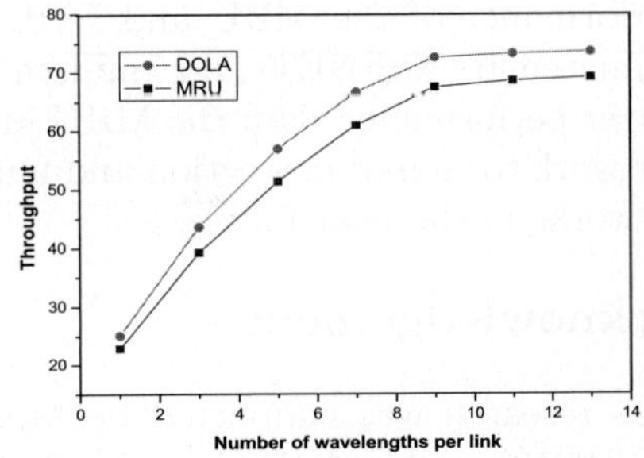

(b) Throughput comparison in NSFNET (Tr = 15).

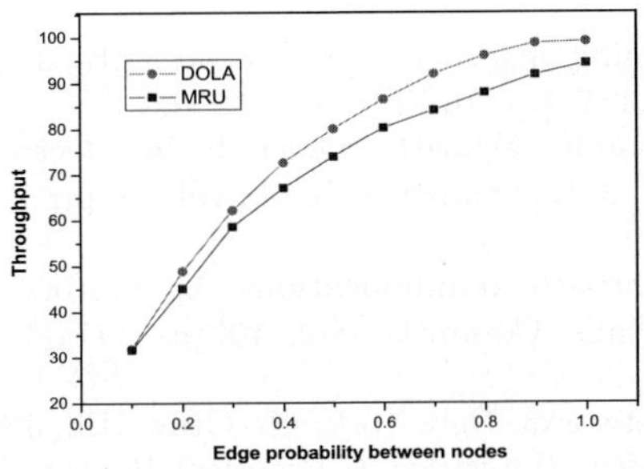

(c) Throughput in random networks (N = 15, W = 10, Tr = 15).

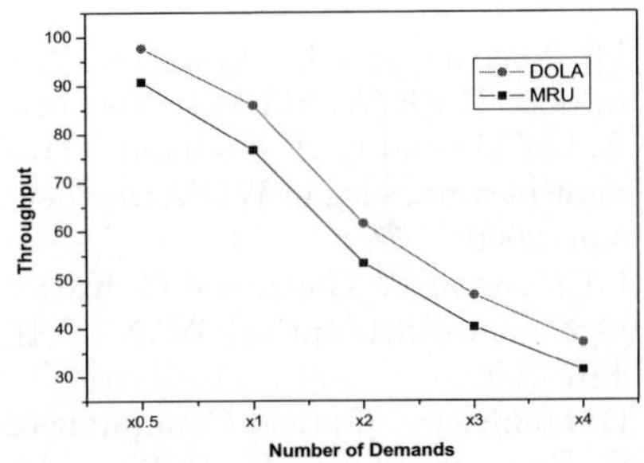

(d) Throughput in random networks (N = 15, P_e = 0.6, W = 10, Tr = 15).

Fig. 4. Throughput comparison

that the DOLA algorithm outperforms 6% $\sim$ 12% than the MRU in terms of network throughput. When the edge probabilities between node pairs (P_e) are 0.7 $\sim$ 0.9, it exhibits the highest throughput.

In Fig. 4(d), we compare the network throughput as the number of demands varies in random networks with fifteen nodes and P_e = 0.6. The condition on each node is same as in Fig. 6. We have adjusted the traffic by selecting the number range when deciding the number of demands at random. The DOLA algorithm accommodates 7% $\sim$ 18% more traffic demands than the MRU. And as the number of demands increases, the performance of the DOLA algorithm becomes better than that of MRU.

5 Conclusion

We have studied a traffic grooming problem in optical WDM mesh networks and proposed an improved DOLA algorithm with respect to network throughput based on the MRU. Our proposed algorithm updates network resource utilization dynamically based on the current network status and decides the process order on lightpath-assignment accordingly. It suits for changed topology frequency.

Performance of the MRU and DOLA algorithms with different parameters is evaluated in NSFNET and random networks, and our proposed one provide better performance than the MRU under various conditions. We plan to extend our work to shared protection and multicasting in traffic groomable WDM mesh networks in the near future.

Acknowledgement

This research was supported by MIC, Korea under ITRC IITA-2005-(C1090-0501-0019), and by KRFG funded by the Korean Government(MOEHRD) KRF-2005-042-D00248.

References

1. M. Alanyali and E. Ayanoglu, "Provisioning algorithms for WDM optical networks," IEEE/ACMTrans. Networking, vol. 7, pp. 767-778, Oct.1999.
2. A. L. Chiu and E. H. Modiano, "Traffic grooming algorithms for reducing electronic multiplexing costs in WDM ring networks," J. Lightwave Technol., vol. 18, pp. 2-12, Jan. 2000.
3. I. Chlamtac, A. Ganz, and G. Karmi, "Lightpath communications: An approach to high bandwidth optical WAN's," IEEE Trans. Commun., vol. 40, pp. 1171-1182, July 1992.
4. B. Mukherjee, Optical Communication Networks. New York: Mc-Graw-Hill, 1997.
5. R. Ramaswami and K. N. Sivarajan, Optical Networks: A Practical Perspective. San Francisco, CA: Morgan Kaufmann, 1998.
6. S. Thiagarajan and A. K. Somani, "Capacity fairness of WDM networks with grooming capability," Opt. Networks Mag., vol. 2, no. 3, pp. 24-31, May/June 2001.
7. J. Wang, W. Cho, V. R. Vemuri, and B. Mukherjee, "Improved approaches for cost-effective traffic grooming in WDM ring networks: ILP formulations and single-hop and multihop connections," J. Lightwave Technol., vol. 19, pp. 1645-53, Nov. 2001
8. C. Xin, Y. Ye, S. Dixit, and C. Qiao, "An integrated lightpath provisioning approach in mesh optical networks," in Proc. IEEE/OSA OFC '02, Mar. 2002, pp. 547-549.
9. H. Zang, J. P. Jue, and B. Mukherjee, "A review of routing and wavelength assignment approaches for wavelength-routed optical WDM networks,"Optical Network Mag. vol. 1, no. 1, pp. 47-60, Jan. 2000.
10. X. Zhang and C. Qiao, "On scheduling all-to-all personalized connections and cost-effective designs in WDM rings," IEEE/ACM Trans. Networking, vol. 7, pp. 435-443, June 1999.
11. K. Zhu and B. Mukherjee, "Traffic grooming in an optical WDM mesh network," IEEE J. Select. Areas Commun., vol. 20, pp. 122-133, Jan. 2002.
12. H. Zhu, H. Zang, K. Zhu, and B. Mukherjee, "A novel, generic graph model for traffic grooming in heterogeneous WDM mesh networks," IEEE/ACM Trans. Networking, vol. 11, pp. 285-299, Apr. 2003.
13. K. Zhu, H. Zang, and B. Mukherjee, "A Comprehensive Study on Next-Generation Optical Grooming Switches" IEEE J. Select. Areas Commun., vol. 21, No.7, Sep. 2003
14. K. Zhu and B. Mukherjee, "On-line provisioning connections of different bandwidth granularities in WDM mesh networks," in Proc. IEEE/OSA OFC '02, Mar. 2002, pp. 549-551.

Do We Really Need Dynamic Wavelength-Routed Optical Networks?

A. Zapata[1,2] and P. Bayvel[2]

[1] Telematics Group, Electronic Engineering Department, Universidad Técnica Federico Santa María, Valparaíso, Chile
[2] Optical Networks Group, Electronic & Electrical Department, University College London, UK

Abstract. It is widely believed that dynamic operation in wavelength-routed optical networks could overcome the inefficiencies of static allocation in wavelength usage. In this paper this hypothesis is reviewed by quantifying the wavelength requirements in dynamic wavelength-routed optical networks and the comparison of these to those of the static approach. To do so, new analytical and heuristic lower bounds for the wavelength requirements in dynamic networks are proposed. They are used to evaluate the optimality of existing algorithms whose wavelength requirements are evaluated by means of simulation. Results show that dynamic wavelength-routed optical networks can save wavelengths only at low traffic loads (< 0.4) and that the highest savings are achieved in sparsely physically connected networks.

1 Introduction

The ever increasing amount of data traffic, reportedly growing at a rate of about 60-100 % per year [1], and the emergence of networked multimedia applications impose high bandwidth demands (in the order of Tb/s) on transport networks. To satisfy the growing traffic demands, transmission systems and networks have migrated to an almost ubiquitous WDM (Wavelength Division Multiplexing) operation [2] in which data is transmitted using multiple carrier wavelengths (channels) over a fibre. Current WDM networks allow bandwidths in excess of 10 Tbp/s per fibre [3].

Currently WDM networks are operated statically. That is, lightpaths (where the lightpath is defined by a physical route and a unique wavelength assigned to that route) between network node pairs are allocated to accommodate the maximum expected traffic demand, known a priori. Before the network operation starts, lightpath allocation is performed and routers/switches are configured accordingly. Lightpath assignment is performed to avoid wavelength collisions in the same fibre and that the minimum number of wavelengths is used. Once network operation starts, data arriving at the electrical interface of an edge node is grouped by destination, converted into the optical domain and transmitted on the corresponding lightpath. These networks are now widely known as static WRONs (wavelength-routed optical networks) [4].

M. Gerndt and D. Kranzlmüller (Eds.): HPCC 2006, LNCS 4208, pp. 477–486, 2006.

Static WRONs are attractive because they are simple to operate and manage, it is possible to find the optimal (or near optimal) solution because there is no restriction on the processing time (due to the off-line nature of the lightpath allocation) and it has been found that the number of wavelengths required by the optimal allocation cannot be significantly further reduced using wavelength converters [4,5].

However, because the bandwidth allocation is carried out at wavelength granularity, allocated bandwidth (wavelengths) might not be used if the source remains idle for long periods of time. Given that that most of current networks operate at no more than 30% of their maximum capacity [6], under static operation a significant number of wavelengths channels will be inefficiently used. With the number of wavelengths mainly determining the cost of switches and physical impairment-compensating equipment, the network cost would thus be unnecessarily increased in static networks.

Although there are no conclusive research results to date, it is widely believed that the dynamic allocation of resources in optical networks would overcome the inefficiencies of the static allocation in wavelength usage, see for example [7,8,9,10]. The allocation of resources only when and where required with minimum delay would ensure that dynamic networks would provide the same service that static networks but at decreased cost, making them very attractive to network operators.

In this paper the potential wavelength savings of dynamic wavelength-routed optical networks with respect to the static approach are quantified by means of analysis and simulation. To do so, a new analytical lower bound for the wavelength requirements of dynamic networks is derived in Section 2. The analytical expression allows the study of the effect of the physical topology and the traffic load on wavelength requirements. An algorithmic (simulation-based) tighter lower bound is then proposed in Section 3. This algorithmic bound corresponds to the application of a near-optimal heuristic for the static case every time a new lightpath request arrives. Both bounds, analytical and algorithmic, allow the evaluation of the optimality of current proposals for dynamic lightpath allocation. This is carried out in Section 4 by quantifying the wavelength requirements of 3 different lightpath allocation algorithms by means of simulation for 7 real physical network topologies and comparing them to the wavelength requirements of the proposed bounds and static networks. A summary is given in Section 5.

2 Analytical Lower Bound for the Wavelength Requirements of Dynamic WRONs

(i) Network and Traffic Model

The network consists of N nodes arbitrarily connected by L unidirectional links (adjacent nodes are assumed connected by two fibres, one per direction). The physical connectivity of the network [4], α, is given by $L/N(N-1)$(there is a difference in a factor of 2 with respect to [4] because this paper considers unidirectional links whilst [4] considers bi-directional links). Traffic demand between each node pair is assumed governed by an ON-OFF process, independent for all node pairs. Traffic load ρ is given by $\mu_{ON}(\mu_{ON} + \mu_{OFF})$, where $\mu_{ON}(\mu_{OFF})$ is the mean duration of the ON (OFF) period.

(ii) Analytical Lower Bound
The lower bound for the wavelength requirements can be obtained by assuming that the set A of active connections (connections in the ON state; $0 \leq A \leq N(N-1)$) is routed using shortest paths (in number of hops) and fully re-utilising the wavelength space. This leads to the following mean number of wavelengths per link:

$$W_A = \left[|A| \cdot H_A/L\right] \tag{1}$$

where $|A|$ represents the cardinality of the set A and H_A the average path length (in number of hops) of the connections in the set A.

Different sets of active connections with cardinality $|A|$ have different values for H_A, which results in different values of W_A. The set A with the highest value for H_A determines the lower bound W_{LB} for the total wavelength requirement:

$$W_{LB} = \max_{\forall A} W_A = ||\hat{A}| \cdot H_{\hat{A}}/L| \tag{2}$$

where $\hat{A}$ corresponds to the set A of active connections with the longest routes and $H_{\hat{A}}$ corresponds to the average path length of the connections in the set $\hat{A}$.

By sorting all the possible $N(N-1)$ connections in decreasing path length (the path length of a connection corresponds to the number of hops of its shortest path) and letting h_i be the length of the i-th longest connection (thus, h_1 and $h_{N(N-1)}$ are the number of hops of the connections with the longest and the shortest paths, respectively), Eq.(2) can be re-written as follows:

$$W_{LB} = \left[\sum_{i=1}^{|\hat{A}|} h_i/L\right] \tag{3}$$

Although Eq.(3) represents a simple closed analytical expression for the lower bound on the wavelength requirement, it is difficult to evaluate because $|\hat{A}|$ depends in a non-trivial manner on the acceptable level of blocking B and the traffic load ρ. In the following an analytical approximation to evaluate $|\hat{A}|$ is given.

1. If the network is dimensioned to accommodate a maximum of $|\hat{A}|$ connections, B is equal to the probability of having more than $|\hat{A}|$ connections in ON state simultaneously, that is:

$$B = Pr\left\{n > |\hat{A}|\right\} \tag{4}$$

2. A connection is in the ON state with probability ρ (a Bernoulli random variable). Thus, the number of connections in ON state is a Binomial random variable:

$$Pr\{n = a\} = Bi(N(N-1), \rho) = \binom{N(N-1)}{a} \rho^a (1-\rho)^{N(N-1)-a} \tag{5}$$

3. Combining equations (4) and (5), the following expression for B is obtained:

$$B = \sum_{n=|\hat{A}|+1}^{N(N-1)} Bi(N(N-1), \rho) \tag{6}$$

Given the target acceptable blocking B, the traffic load ρ and the number of nodes N, the maximum number of active connections $|\hat{A}|$ can be numerically obtained from Eq.(6). However, the normal approximation of the binomial probability distribution [11] provides a simpler closed analytical expression for $|\hat{A}|$, as follows:

$$|\hat{A}| \approx \min\left\{N(N-1)\cdot\rho, N(N-1)\cdot\rho + \beta\sqrt{N(N-1)\cdot\rho\cdot(1-\rho)}\right\} \tag{7}$$

where β is such that the area under the normal distribution curve between $(-\infty, \beta)$ is equal to $(1-B)$ is equal to $(1-B)$ [12].

Eq.(7) is known to be accurate for $N(N-1)\rho(1-\rho) \geq 10$ [11], which means that N must be greater or equal to 11 for $\rho \in [0.1, 0.9]$, a typical condition in real networks.

To investigate the potential wavelength savings achieved by dynamically operating the network, the ratio R_W, defined as the ratio between the lower bounds for the wavelength requirements in the dynamic case and static cases, is given as:

$$R_W = \frac{\left\lceil \frac{|\hat{A}|H_{\hat{A}}}{L} \right\rceil}{\frac{N(N-1)H}{L}} = \frac{\left\lceil \frac{RH_{\hat{A}}}{\alpha} \right\rceil}{\frac{H}{\alpha}} \tag{8}$$

Where R, equal to $|\hat{A}|/(N(N-1))$, represents the fraction of connections which need to be accommodated in the network in the dynamic case compared to the static case. In the extreme case of $\alpha = 1$ (i.e., fully connected topology), $H = H_{\hat{A}} = 1$. Thus, $R_W = \lceil R \rceil = 1$ which means that fully connected networks do not benefit from dynamic operation in terms of wavelength savings, irrespective of the value of N, B and ρ.

For other values of α the analytical evaluation of Eq.(8) is problematic because of the lack of an expression for $H_{\hat{A}}$. However, the following equation - based on the assumption that the path length of the shortest paths is a Gaussian random variable with mean equal to H (as shown, for example, in [13]), is used to numerically estimate $H_{\hat{A}}$:

$$\phi\left(\frac{H_{\hat{A}} - H}{\sigma_h}\right) \approx 1 - \frac{|\hat{A}|}{N(N-1)} \tag{9}$$

where $\phi_{(Z)}$ corresponds to the value of the cumulative distribution function of the standard normal curve evaluated in z, H is the mean number of hops of the shortest paths (estimated from $\sqrt{(N-2)/\delta - 1}$ [14]) and σ_h is the standard deviation of the number of hops of the shortest paths (estimated from $\sqrt{\ln N}$ [13]).

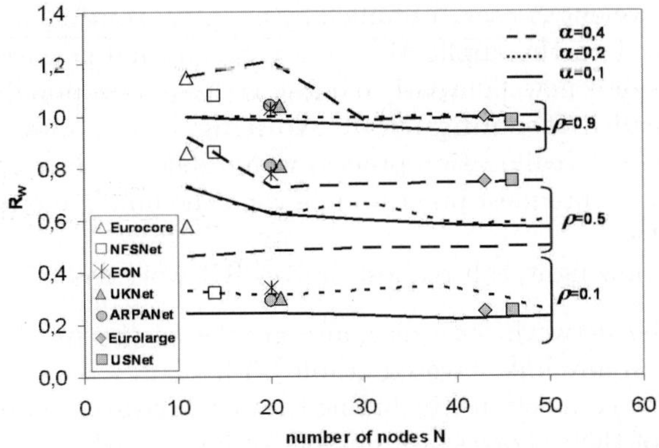

Fig. 1. Ratio R_W as a function of the number of nodes N

To study the effect that the number of nodes N, the physical connectivity α and the traffic load ρ have on the potential wavelength savings achieved in dynamic operation, R_W was calculated for a target blocking of $B = 10^{-6}$ and different values of N, α and ρ using the approximation provided by Eq.(9) for $H_{\hat{A}}$. The results are shown in Figure 1. R_W was also calculated for the following 7 real-world mesh topologies (for which the exact values of $H_{\hat{A}}$ can be calculated): **Eurocore** ($N = 11$, $L = 50$, $\alpha = 0.45$), **NSFNet** ($N = 14$, $L = 42$, $\alpha = 0.23$), **EON** ($N = 20$, $L = 78$, $\alpha = 0.2$), **UKNet** ($N = 21$, $L = 78$, $\alpha = 0.19$), **ARPANet** ($N = 20$, $L = 62$, $\alpha = 0.16$), **Eurolarge** ($N = 43$, $L = 180$, $\alpha = 0.1$) and **USNet** ($N = 46$, $L = 152$, $\alpha = 0.07$).

It can be seen that the number of nodes does not significantly affects the potential wavelength savings and that the ratio R_W decreases with ρ and α. That is, the highest benefits of dynamic operation are expected for low loads ($\rho \leq 0.5$) and sparsely connected networks ($\alpha \leq 0.2$). This is reasonable, as at low loads static networks are inefficiently used and highly-connected networks already require very low number of wavelengths per link in the static case (for example, 4 in Eurocore) making hard for dynamic operation to further decrease this requirement.

These results indicate that, for most values of the traffic load, dynamic operation has the potential of achieving significant savings in networks with a low physical connectivity ($\alpha \leq 0.2$). Increasing the levels of acceptable blocking slightly increases the percentage of savings as well, but the acceptable blocking level is usually determined by the applications rather than the network design process.

3 Algorithmic Lower Bound the Wavelength Requirements of Dynamic WRONs

The analytical lower bound derived in the previous section may be unachievable in practice, because adaptive routing does not necessarily use the shortest paths nor fully utilizes wavelength space. Due to the difficulty of modeling the length of paths and the wavelength usage obtained by adaptive routing, a tighter heuristic lower bound based on the application of a near-optimal heuristic for the static case [4] every time a new lightpath request arrives is proposed in this section. This heuristic (called **Reconfigurable Routing**, RR) however, would be impractical because the reallocation process would disrupt active connections and increase the lightpath request processing beyond the limits allowed by scalability considerations [15].

Every time a new lightpath request arrives RR works as follows:

1. Represent the network with as many graphs as the maximum number of wavelengths in any link (layered graph [15]).
2. Sort the active connections (including the new arrival) according to the number of hops of their shortest (in number of hops) paths (longest first)
3. Allocate lightpaths one by one, choosing connections according to the order established in step 1, as follows:
   i. Execute Dijkstra to find the shortest available path in every graph
   ii. Allocate the first path found which is at most e hops longer than the shortest path. If not such path is found on any of the graphs, block the request.

The parameter e in step 3.ii was varied between 0 and 3 depending on the traffic load, as higher values did not reduce the wavelength requirements in the studied networks.

4 Simulation Results

The following extreme (in terms of speed and blocking performance) dynamic lightpath allocation algorithms have been chosen to study their wavelength requirements in this paper:

- **Adaptive Unconstrained Routing - Exhaustive** (AUR-E). Shown to yield the lowest blocking to date, due to the online execution of the Dijkstra algorithm per request [16]. However, this way of operating makes AUR-E computationally intensive and thus, slow [15].
- **Shortest Path - First Fit** (SP-FF) [17] Shown to be the fastest algorithm available to date due to the use of pre-computed routes (only one per node pair) and the simplicity of the wavelength allocation algorithm. However, this algorithm exhibits higher blocking values than AUR-E.
- **k Alternate Paths using Shortest Path First Fit** (k-SP-FF). This algorithm attempts to achieve a good compromise between computational complexity and performance by applying alternate routing. Thus, the performance of fixed routing is improved without incurring in the high computational cost of Dijkstra-based AUR-E.

The wavelength requirements resulting from the application of the algorithms presented above was evaluated by means of simulation. Simulation details as follows.

The target blocking was set to a maximum value of 10^{-3} per node pair (thus, the network-wide blocking, denoted by B in previous sections, is also 10^{-3}).

ON and OFF periods were assumed identically and exponentially distributed for all node pairs. Lightpath requests were generated at the start of each ON period. To comply with the efficiency criteria [18] (that is, the transmission time of a burst should be at least as long as the overhead time) the mean ON period (μ_{ON}) was set to 5, 10 and 25 ms for the UK, European and US networks, respectively.

After eliminating transient simulation behaviour (first 10^3 lightpath requests per node pair), 10^4 lightpath requests per node pair were generated. To quantify the wavelength requirements, the number of wavelengths in each link was increased until no more than 10 requests generated per node pair was rejected (in this way, a maximum blocking of 10^{-3} is ensured).

The described simulation experiment was executed several times to obtain a confidence interval of 95% for the wavelength requirements of each link. For the SP-FF and k-SP-FF algorithms, 100 simulations were executed (for each network, for a specific value for the traffic load) and the confidence interval was in average 0.3% and 3.1% of the mean value for SP-FF and 3-SP-FF, respectively. For AUR-E and RR instead, only 15 simulations were executed, due to their high simulation time (as a way of illustration, the evaluation of the wavelength requirements of RR for the USNet topology for a unique value of the traffic load took more than 1 week in a Pentium 4 of 2.5 GHz and 256 MB RAM). The confidence interval was on average 3.5% and 3.4% of the mean value for AUR-E and RR, respectively.

The ratio R_W is plotted as a function of the traffic load in Figure 2 a)-g) for the SP-FF, 3-SP-FF (that is, up to 3 disjoint routes per node pair were used as higher values of k did not achieve better results in terms of wavelength requirements) and AUR-E algorithms. The ratio R_W obtained for the analytical and the heuristic lower bounds is also included for comparison.

From Figure 2a-g four main conclusions can be drawn:

- The heuristic lower bound is still significantly higher than the analytically obtained value: 46 % higher on average. This difference is due to the assumption of full wavelength reutilization and the use of shortest paths to obtain the analytical lower bound, which might not be fulfilled in the case of the heuristic lower bound. The heuristic lower bound predicts that potential wavelength savings can be achieved only at low/moderated traffic loads (0.3-0.5) and that sparsely connected networks experience the highest wavelength savings.
- SP-FF and 3-SP-FF require a much higher number of wavelengths than AUR-E to achieve the same blocking: 24% and 18% higher wavelength requirements in average, respectively. In networks of large size, as USNet for example, this difference might result in the installation of a few thousand extra wavelengths to offer the same service. Thus, in terms of resource utilization efficiency, SP-FF and k-SP-FF should not be considered for implementation in wavelength-routed networks.

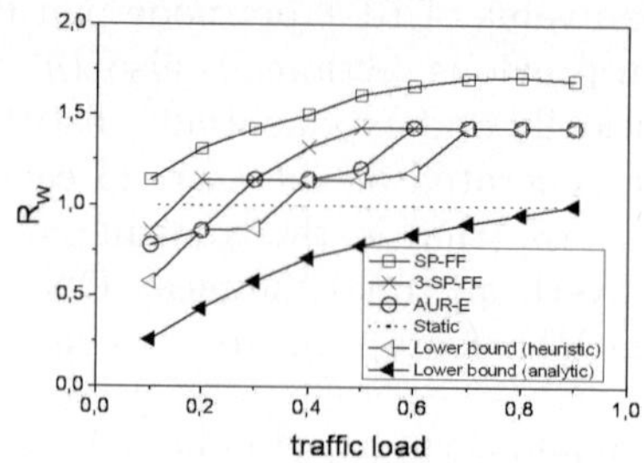

Figure 2.a R_w for Eurocore topology

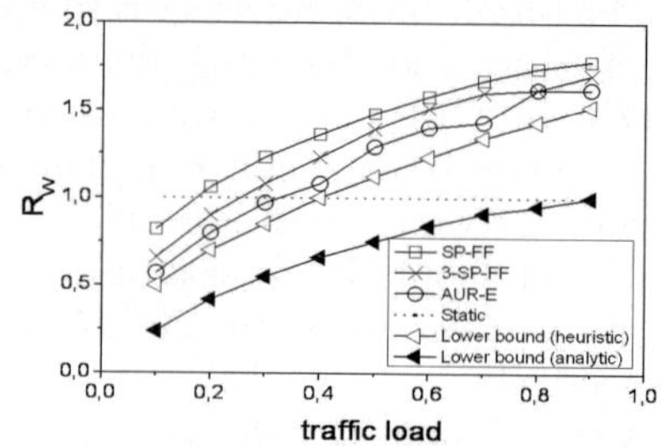

Figure 2.b R_w for NSFNet topology

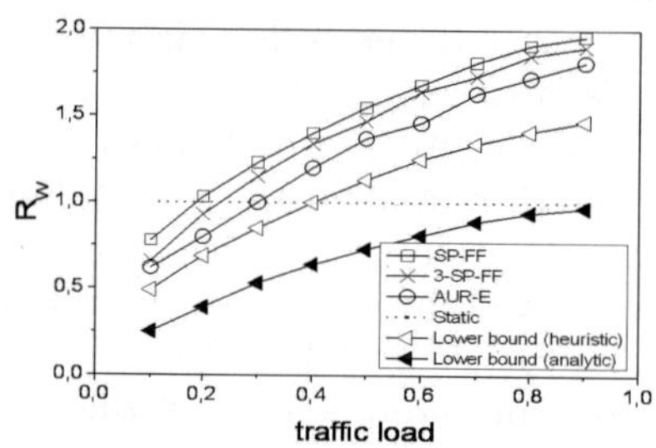

Figure 2.c R_w for EON topology

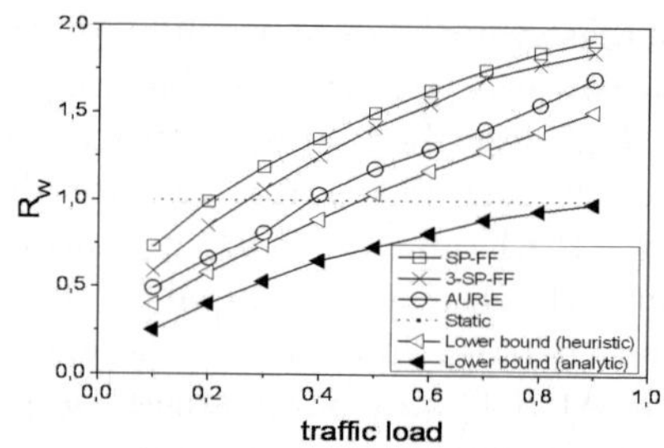

Figure 2.d R_w for UKNet topology

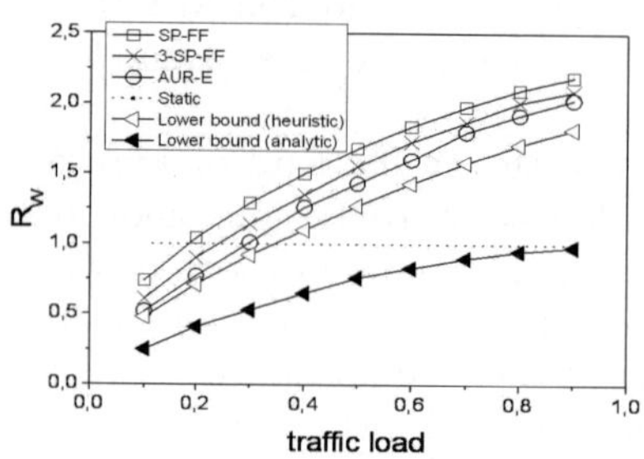

Figure 2.e R_w for ARPANet topology

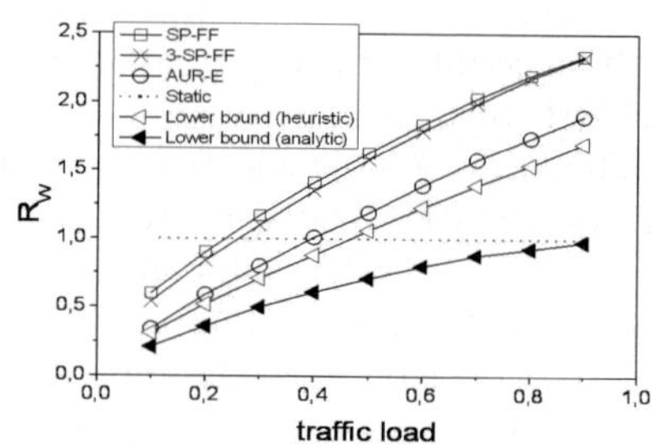

Figure 2.f R_w for Eurolarge topology

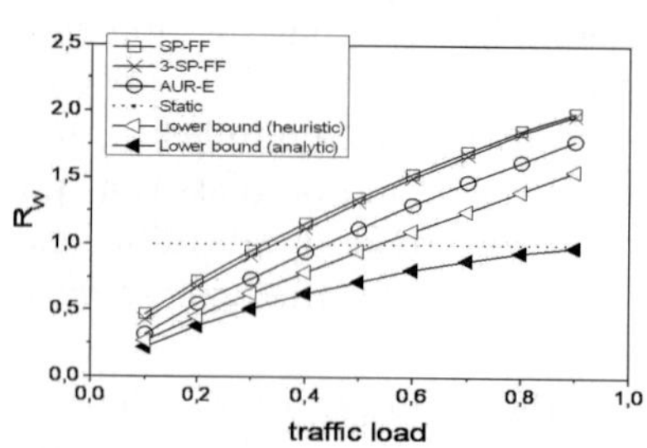

Figure 2.g R_w for USNet topology

- Although AUR-E corresponds to the algorithm requiring the lowest number of wavelengths, it requires on average, 14% higher number of wavelengths to achieve the same blocking performance than the heuristic lower bound. This percentage still represents a high number of additional wavelengths in networks of large size (about 2000 extra wavelengths in total for USNet). Therefore, the design of a lightpath allocation algorithm improving on the performance of AUR-E is still desirable. However, this is expected to be a very difficult task (since the proposal of AUR-E in 1996 there have been no better algorithms, despite significant activity in this research field).

- Considering the performance of AUR-E, the advantages of dynamic operation with respect to the static approach are observed only at low/moderate loads (< 0.4) and especially for sparsely connected networks: networks with physical connectivity, α, lower than 0.2 exhibit wavelength savings for traffic loads up to 0.4 whilst more connected networks (for example, Eurocore or NSFNet) exhibit savings only for loads up to 0.3. For loads in excess of 0.4 all the studied dynamic networks require more wavelengths than the corresponding static networks and thus, in that range of operation dynamic networks are not attractive in terms of wavelength requirements.

5 Summary

In this paper the question of whether dynamic operation in wavelength-routed optical networks brings benefits in terms of wavelength requirements compared to the static operation was addressed.

Through the derivation of an analytical and a heuristic lower bound for the wavelength requirements it was found that the traffic load and the physical connectivity were key factors affecting wavelength savings in dynamic networks and that dynamic operation could achieve wavelength savings only at low/moderated traffic loads (0.3-0.5, with the highest savings being exhibited by sparsely connected networks ($\alpha < 0.2$).

Both lower bounds were used as a benchmark for the wavelength requirements of dynamic lightpath allocation algorithms and it was found that the best performing algorithm (AUR-E) required 14% higher number of wavelengths than the heuristic lower bound. Considering such algorithm, dynamic wavelength-routed optical networks exhibited wavelength savings with respect to the static approach only at low/moderate traffic loads (0.3-0.4), with the highest savings achieved in sparsely connected networks ($\alpha < 0.2$).Wavelength-routed optical networks operating at higher traffic loads do not benefit from dynamic operation in terms of wavelength requirements. Such lower efficiency in wavelength usage of dynamic operation comes from the fact that lightpath allocation must be carried out in an on-line manner and thus, it cannot be optimized as in the static case. In addition, highly connected networks already require a low number of wavelengths in the static case, making difficult for dynamic operation to decrease further the wavelength requirement. These results are contradictory to the widely held view which says that savings will always occur under dynamic operation, and should encourage the research community to review the idea that dynamic operation of wavelength-routed optical networks is always desirable in terms of wavelength requirements.

Acknowledgements. Financial support from Fondecyt Project 1050361, USM Project 23.05.81, Fundación Andes and EU FP6 Integrated Project NOBEL/NOBELII is gratefully acknowledged.

References

1. K.G. Coffman and A.M. Odlyzko, "Growth of the Internet", in Optical Fiber Communication - vol. IV-B: Systems and Impairments, I.P. Kaminow and T. Li Eds., Academic Press, San Diego, 17-56, 2002
2. G. Agrawal, Fiber-Optic Communication Systems, 3rd Edition, Wiley-Interscience, Wiley & Sons, Chapter 8, 2002
3. Post-deadline papers in Proceedings of 31st European Conference on Optical Communications, ECOC 2005, Glasgow, Scotland, September 2005
4. S. Baroni and P. Bayvel, "Wavelength requirements in arbitrarily connected wavelength-routed optical networks", IEEE J. of Lightwave Technol., 15(2), 242-251, 1997
5. C. Assi et al., "Impact of wavelength converters on the performance of optical networks", Optical Networks Magazine, 3 (2), 22-30, 2002
6. A. Odlyzko, "Data networks are lightly utilized, and will stay that way" Rev. Network Economics 2, 210-237, 2003
7. O. Gerstel and H. Raza, "On the synergy between electrical and optical switching", IEEE Communications Magazine, 41 (4), 98-104, 2003
8. S. Sengupta, V. Kumar, D. Saha, "Switched optical backbone for cost-effective scalable core IP neworks", IEEE Communications Magazine 41 (6), 60-70, 2003
9. C. Assi, A. Shami, M. Ali, "Optical networking and real-time provisioning: an integrated vision for the next - generation Internet", IEEE Network, 15 (4), 36-45, 2001
10. M.J. O'Mahony, D. Simeonidou, D.K. Hunter, A. Tzanakaki, "The application of optical packet switching in future communications networks", IEEE Communications Magazine, 39 (3), 128-135, 2001
11. S. Ross, "A first course in Probability", 6th Edition, Prentice Hall, 2002, pp. 206
12. R. Yates and D. Goodman, "Probability and stochastic processes. A friendly introduction for Electrical and Computer Engineers", Wiley & Sons, 2nd Ed., 1999
13. S.N. Dorogovtsev, A.V. Goltsev, J.F. Mendes,"Pseudofractal sacle-free web", Physical Review E 65, 066122-1-066122-4, 2002
14. S. K. Korotky, "Network global expectation model: a statistical formalism for quickly quantifying network needs and costs", IEEE J. of Lightwave Technol. 22 (3), 703-722, 2004
15. M. Düser, A. Zapata, P. Bayvel, "Investigation of the Scalability of Dynamic Wavelength-Routed Optical Networks" J. of Optical Networking 3, 667-686, 2004
16. A. Mokhtar and M. Azizoglu, "Adaptive wavelength routing in all-optical networks", IEEE/ACM Transactions on Networking, 6(2), 197-206, 1998
17. I. Chlamtac, A. Ganz, G. Karmi, "Purely optical networks for terabit communication", in Proceedings of INFOCOM'89, v.3, 887-896, Ottawa, Canada, April 1989
18. M. Düser, P. Bayvel, "Analysis of a dynamically wavelength-routed optical burst switched network architecture", IEEE J. of Lightwave Technology 20(4), 574-585, 2002

Design and Implementation of Middleware and Context Server for Context-Awareness*

Jae-Woo Chang and Yong-Ki Kim

Research Center of Industrial Technology
Dept. of Computer Engineering, Chonbuk National University,
Chonju, Chonbuk 561-756, South Korea
jwchang@chonbuk.ac.kr, ykkim@dblab.chonbuk.ac.kr

Abstract. Context-awareness is a technology to facilitate information acquisition and execution by supporting interoperability between users and devices based on users' context. In this paper, we design and implement a middleware and a context server for dealing with context-aware applications in pervasive computing. The middleware plays an important role in recognizing a moving node with mobility by using a Bluetooth wireless communication technology as well as in executing an appropriate execution module according to the context acquired from a context server. In addition, the context server functions as a manager that efficiently stores into the database server context information, such as user's current status, physical environment, and resources of a computing system. To verify the usefulness of the middleware and the context server, we finally develop our context-aware application system which provides users with a music playing service in pervasive computing environment.

1 Introduction

In traditional computing environments, users actively choose to interact with computers. On the contrary, pervasive computing applications are embedded in the users' physical environments and integrate seamlessly with their everyday tasks [1]. Mark Wieser at Xerox Palo Alto Research Center identified the goal of future computing to be invisible computing [2]. An effective software infrastructure for running pervasive computing applications must be capable of finding, adapting, and delivering the appropriate applications to the user's computing environment based on the user's context. Thus, context-aware application systems determine which user tasks are most relevant to a user in a particular context. They may be determined based on history, preferences, or other knowledge of the user's behavior, as well as the environmental conditions. Once the user has selected a task from the list of relevant tasks, an application may have to move seamlessly from one device to another and from one environment to another based on the user's activity. The context-awareness is one of the most important technologies in pervasive computing, which facilitate information acquisition and execution by supporting interoperability between users and devices based on users' context.

* This work is financially supported by the Ministry of Education and Human Resources Development(MOE), the Ministry of Commerce, Industry and Energy(MOCIE) and the Ministry of Labor(MOLAB) though the fostering project of the Lab of Excellency.

M. Gerndt and D. Kranzlmüller (Eds.): HPCC 2006, LNCS 4208, pp. 487 – 494, 2006.

In this paper, we design and implement middleware and context server components for dealing with context-aware applications in pervasive computing. The middleware plays an important role in recognizing a moving node with mobility by using a Bluetooth wireless communication technology as well as in executing an appropriate execution module according to the context acquired from a context server. In addition, the context server functions as a manager that efficiently stores into database server con-text information, such as user's current status, physical environment, and resources of a computing system. In order to verify the usefulness of the middleware and the con-text server, we develop our context-aware application system which provides a music playing service in pervasive computing environment. The remainder of this paper is organized as follows. The next section discusses related work. In section 3, we de-scribe the overall architecture for context-aware application services. In section 4 and 5, we describe the design of our middleware and our context server for context-awareness. In section 6, we present the development of our context-aware application system using them. Finally, we draw our conclusions in section 7.

2 Related Work

In this section, we discuss the typical context-aware application systems. First, INRIA in France [3] proposed a general infrastructure based on contextual objects to design adaptive distributed information systems in order to keep the level of the delivered service despite environmental variations. The contextual objects (COs) were mainly motivated by the inadequacy of current paradigms for context-aware systems. The use of COs does not complicate a lot of development of an application, which may be developed as a collection of COs. They also presented a general framework for context-aware systems, which provides application developers with an architecture to design and implement adaptive systems and supports a wide variety of adaptations. Secondly, AT&T Laboratories Cambridge in U.K [4] presented a platform for context-aware computing which enables applications to follow mobile users as they move around a building. The platform is particularly suitable for richly equipped, networked environments. Users are required to carry a small sensor tag, which identifies them to the system and locates them accurately in three dimensions. Finally, Arizona State Univ. [5] presented Reconfigurable Context-Sensitive Middleware (RCSM), which made use of the contextual data of a device and its surrounding environment to initiate and manage ad hoc communication with other devices. The RCSM provided core middleware services by using dedicated reconfigurable Field Programmable Gate Arrays (FPGA), a context-based reflection and adaptation triggering mechanism, and an object request broker that are context-sensitive and invokes remote objects based on contextual and environmental factors, thereby facilitating autonomous exchange of information.

3 Overall Architecture for Context-Aware Application Services

Context is any information that can be used to characterize the situation of any entity [6]. An entity is a person, place, or object that is considered relevant to the interaction between a user and an application, including the user and applications themselves. In this section, we propose an overall architecture of context-adaptive system for supporting various context-aware application services, which is divided into three components, context server, middleware (fixed node), and moving node (client). First, the context server serves to insert remote objects into an object database and context information into a context database, as well as to retrieve them from the both databases. Secondly, a fixed node functions as a middleware to find, insert, and execute a remote object for context awareness. Finally, a moving object serves to execute a predefined built-in program according to the context information acquired from the middleware. Figure 1 shows the overall architecture for supporting various context-aware application services.

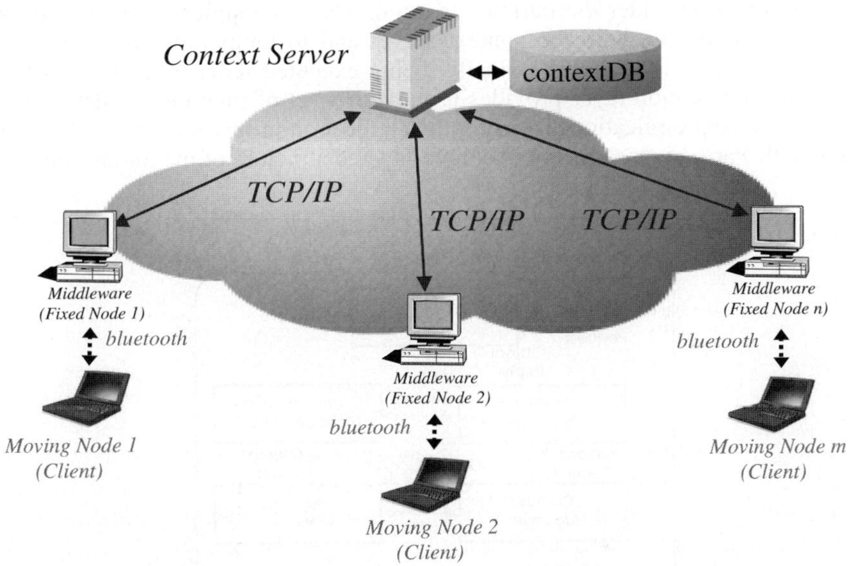

Fig. 1. Overall architecture for supporting context-aware application services

Because our architecture combines the advantage of the INRIA work with that of the AT&T work, it has a couple of powerful features. First, our middleware can define context objects describing context information as well as can keep track of a user's current location. Secondly, our context server can store context objects and their values depending on a variety of contexts as well as can mange users' current locations being acquired from a set of fixed nodes by using spatial indexing. Finally, our client can provide users with adaptive application services based on the context objects. Meanwhile, the context server communicates with a middleware by using a network based on TCP/IP, while a moving object communicates with a middleware using Bluetooth wireless communication [7].

4 Middleware for Context-Awareness

Our middleware for context-aware application services consists of three layers, such as detection/monitoring layer, context-awareness layer, and application layer. First, the detection/monitoring layer serves to monitor the locations of remote objects, network status, and computing resources, i.e., CPU usage, memory usage, bandwidth, and event information related with devices including Bluetooth. Secondly, the context-awareness layer functions as a middleware which is an essential part for handling context-aware application services. It can be divided into five managers, such as script processor, remote object manager, context manager, context selection manager, communication proxy manager. The script processor analyzes the content of context-aware definition script and executes its specified actions. The remote object manager man-ages a data structure for all the context objects used in application programs. The context manager manages context and environmental information including user preference and user location. The context selection manager chooses the most appropriate context information under the current situation. The communication proxy manager serves to communicate with the context server and to temporarily reserve data for retransmission in case of failure. Finally, being executed independently of the middleware, the application layer provides users with a set of functions to develop various context-aware applications by using the application programming interface (API) of the middleware. Figure 2 shows the three-layered structure of the middleware.

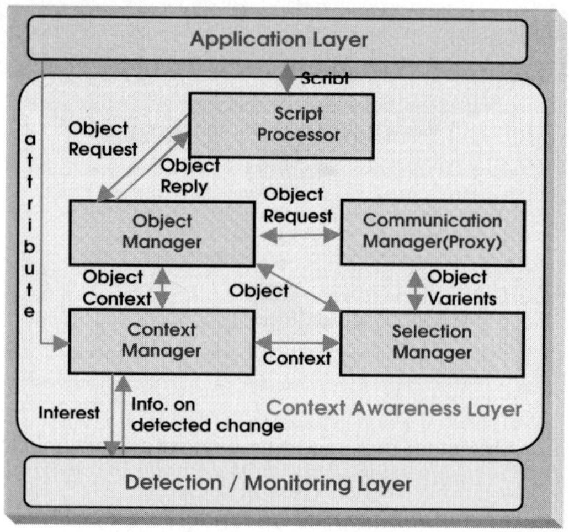

Fig. 2. Three-layered structure of the middleware

5 Context Server for Context-Awareness

For context-awareness, context server is required to store and retrieve remote objects and context information extracted from them. We design a context server which can efficiently manage both the remote object and the context information using a commercial DBMS called MySQL. This is because we can reduce the developing time, compared with using a storage system, and we can increase the reliability of the developed system. The designed context server analyzes and executes the content of packets delivered from the middleware. That is, the server determines whether the packet's content is contexts or context objects and stores them into the corresponding database. Figure 3 shows the structure of context server which is divided into four managers, such as communication manager (CM), packet analysis manager (PAM), context/object manager (COM), and mySQL query manager (SQM).

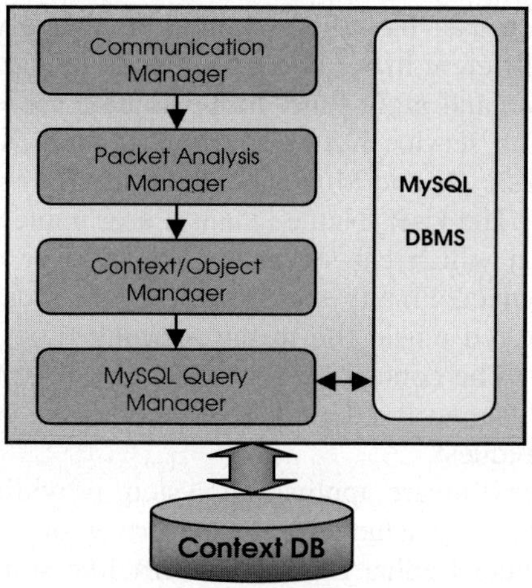

Fig. 3. Structure of context server

The CM serves to communicate between the context server and a middleware. The CM delivers to PAM the packets transferred from the middleware as well as to the middleware the result packets made from the server. The CM includes both file receiving module and TCP/IP socket module. The file receiving module is dependant on the TCP/IP socket module because it communicates with the middleware using TCP/IP socket. When the server is ready to communicate, it receives packets from the middleware. The PAM parses the packets from the CM and determines what action wants to be done currently. Based on the parsing, the PAM calls the most proper functions, i.e., context APIs, in the COM. The COM translates into SQL statements the content delivered from the PAM and delivers the SQL statements to SQM to execute them. The context APIs (application programming interfaces) for the COM is ContextDefine, ContextDestroy, ContextInsertTuple, ContextDeleteTuple, ContextSearch, ContextSearchTuples, and ContextCustom. The SQM executes the SQL statements from the COM using the MySQL DBMS and delivers the result to the middleware via the CM. The SQL includes the mySQL API module being implemented by using mysql C libraries.

6 Development of Context-Aware Application System

In this section, we first develop both our middleware and our context server which are designed for context awareness in the previous sections. For this, we implement them using GCC compiler 2.95.4 under Redhat Linux 7.3 (kernel version 2.3.20) with 1.7 GHz Pentium-IV CPU and 512 MB main memory. In order to show the efficiency of both our middleware and our context server implemented, we also develop a con-text-aware application system using them. The context-aware application system servers to provide users with a music playing service in pervasive computing environment. In the system, when a user belonging to a moving node approaches to a fixed node, the fixed node starts playing the user's music with his (her) preference according to his location. In general, each user has a list of his (her) music with his preference and even a user can have a different list of his (her) popular music depending on time, i.e., morning time, noon time, and night time. In the context server, a user record for the music playing application service has six attributes, such as User_ID, User_Name, Location, Music_M, Music_A, and Music_E. The User_ID serves as a primary key to identify a user uniquely. The User_Name means a user name and the Location means a user's current location which can be changed whenever a middleware finds the location of a moving object. Finally the Music_M, the Music_A, and the Music_E represent his (her) preferred music file in the morning time, the noon time, and the night time, respectively. The context server manages a list of music files for a user, processes queries given from a fixed node, and delivers the corresponding music file to the fixed node by its request.

We develop our context-aware application system providing a music playing service by using affix 2.0.2 as a Bluetooth device driver protocol and by using GCC 2.95.4 an a compiler, under Redhat Linux 7.3 (kernel version 2.4.20) with 866 MHz Pentium-III CPU and 64 MB main memory. In addition, the Bluetooth device follows the specification of Version1.1/Class1 and makes a connection to PCs using USB interfaces [8]. To determine whether or not our context-aware application system implemented works well, we test it by adopting a scenario used in Cricket [9], one of the MIT Oxygen project. For this, we test the execution of our context-aware application system in the following three cases; the first case when a user covered by a moving node approaches to a fixed node or move apart from it, the second case when two different users approaches to a fixed node, and the final case when a user approaches to a fixed node at different times. Among them, because the first case is the most general one, we will explain it in more detail. For our testing environment, we locate two fixed nodes in the database laboratory (DB Lab) and the media communication laboratory (Media Lab) of Chonbuk National University, respectively, where their middleware can detect a moving node by using Bluetooth wireless communication. There is a corridor between DB Lab and Media Lab and its distance is about 60 meter. We test the execution of our context-aware application system in a case when a user having a moving node moves from DB Lab to Media Lab or in a reverse direction. Figure 4 shows a testing case when a user having a moving node approaches to a fixed node. First, the fixed node receives a user name from the moving node as the moving node is approaching to it (①). Secondly, the fixed node determines whether or not the information of the user has already been stored into a server. If it does, the context server searches the music file belonging to the user in a current time and downloads the music file from the database (②). In case when the fixed node detects that a user is too far to communicate with it, the fixed node stops the process to play music and it re-moves the music playing process.

To analyze the performance of our context-aware application system, we measure an average time by adopting a boundary detection of beacons used in Cricket [9]. First, as a moving node is approaching to a fixed node, it takes 1.34 second for the fixed node to make a connection with the moving node. It means the time for the fixed node to detect the presence of a moving node. Secondly, it takes 0.5 second for the fixed node to start music playing service after making the connection between them. Finally, as a moving node is moving apart from a fixed node, it takes 1.45

Fig. 4. Testing case when a user is approaches to a fixed node

second for the fixed node to make a disconnection to the moving node. It means the time for the fixed node to detect the absence of the moving node. The time is relatively long because the kernel tries to communicate with the moving node even though the moving node is beyond the communication boundary of the fixed node. Therefore, it is very reasonable for the fixed node to set the time limit to two seconds. If it takes long time for a fixed node to establish a connection to a moving node and to detect a context from it, a user may consider the situation as a fault. Because the detection time for a context is less than two seconds, the context-aware application program is reasonable for the music playing application service.

7 Conclusions and Future Work

In this paper, we designed and implemented both our middleware and our context server for supporting a context-aware application system. The middleware played an important role in recognizing a moving node with mobility by using Bluetooth wireless communication as well as in executing an appropriate execution module according to the context acquired from a context server. In addition, the context server functions as a manager that efficiently stores into database server context information, such as user's current status, physical environment. To verify the usefulness of both the middleware and context server implemented, we developed our context-aware application system which provided users with a music playing service in pervasive computing environment. We tested it by adopting a scenario used in Cricket. It was shown that it took about 1.5 seconds for our context-aware application system to make a connection (or disconnection) between a fixed node and a moving node, thus being considered reasonable for our music playing application service. As future work, it is required to study on an inference engine to acquire new context information from the existing one.

References

1. K. Raatikainen, H. B. Christensen, and T. Nakajima, "Application Requirements for Middleware for Mobile and Pervasive systems", Mobile Computing and Communications Review, pp. 16-24, Vol. 6, No. 4.
2. M. Weiser, "The Computer for the Twenty-First Century", Scientific American, pp. 94-104, Sept. 1991.
3. P. Couderc, A. M. Kermarrec, "Improving Level of Service for Mobile Users Using Context-Awareness", Proc. of 18th IEEE Symposium on Reliable Distributed Systems, pp. 24-33, 1999.
4. A. Harter, A. Hopper, P. Steggles, A. Ward, P. Webster, "The anatomy of a Context-aware application", Wireless Networks Vol. 8, Issue 2/3, pp. 187-197, 2002.
5. S. S. Yau and F. Karim, "Context-sensitive Middleware for Real-time Software in Ubiquitous Computing Environments", Proc. of 4th IEEE Symposium on Object-oriented Real-time Distributed Computing, pp.163-170, 2001.
6. K. Cheverst, N. Davies, K. Mitchell, and A. Feiday, "Experiences of Developing and Deploying a Context-Aware Tourist Guide: The GUIDE Project", Proc. of 6th Int'l Conference on Mobile Computing and Networking, 2001.
7. Bluetooth Version 1.1 Profile, http://www.bluetooth.com.
8. Affix: Bluetooth Protocol Stack for Linux, http://affix.sourceforge.net.
9. N. B. Priyantha, A. Chakraborty, and H. Balakrishnan, "The Cricket Location Support System", 6th ACM/IEEE Int'l Conf. on Mobile Computing and Networking(MOBICOM), pp. 32-43, 2000.

Security and Privacy Analysis of RFID Systems Using Model Checking*

Hyun-Seok Kim[1], Il-Gon Kim[1], Keun-Hee Han[2], and Jin-Young Choi[1]

[1] Dept. of Computer Science and Engineering, Korea University, Seoul, Korea
{hskim, igkim, choi}@formal.korea.ac.kr

[2] e-Government Security Team, Ministry of Goverment Administration and Home Affairs, Seoul, Korea
keunhee@mogaha.go.kr

Abstract. Radio frequency identification (RFID) is expected to become an important and ubiquitous infrastructure technology. As RFID tags are affixed to all items, they may be used to support various useful services. However, this pervasive use of RFID tags opens up the possibility for various attacks violating user privacy and authentication among communication participants. Security mechanisms for RFID systems will be therefore of utmost important. In this paper, we describe problems of previous works on RFID security protocol and specify several known attacks with Casper, CSP and then verify their security properties such as secrecy and authentication using FDR model checking tool. Finally, we propose an RFID security protocol based on strong authenticaion that guarantees data privacy and authentication between a tag and a reader.
Keywords : RFID Security, Model Checking, Casper, CSP, FDR.

1 Introduction

Regarding RFID security, few issues are related to the data protection of the tags, message interception over the air channel, the eavesdropping within the interrogation zone of the RFID reader[1]. Among these, we will discuss two aspects on the risks posed to the passive party by RFID that has so far been dominated by the topics of data protection associated with data privacy and authentication between tag and reader. Firstly, the data privacy problem is that storing person-specific data in an RFID system can threaten the privacy of the passive party. This party might be, for example, a customer or an employee of the operator. The passive party uses tags or items that have been identified by tags, but the party has no control over the data which have been stored on the tags.

Secondly, the authentication will be carried out when the identity of a person or a program is checked. Then, on that basis, authorization takes place, i.e.

* This work was supported by Ministry of Commerce, Industry and Energy of Korea and the RFID Research Center of Korea University.

M. Gerndt and D. Kranzlmüller (Eds.): HPCC 2006, LNCS 4208, pp. 495–504, 2006.

rights, such as the right of access to data are granted. In the case of RFID systems, it is particularly important for tags to be authenticated by the reader and vice-versa. In addition, readers must also authenticate themselves to the backend, but in this case there are no RFID-specific security problems.

To satisfy above requirements, the most effective protective measure against an attack involving eavesdropping at the air interface is not to store any contents on the tag itself and instead to read only the ID of the tag. This measure, which is most often recommended in the technical literature and which is assumed by EPCglobal[2], offers the additional advantages that less expensive tags can be used, the memory for the associated data in the database is practically unlimited. For applications where relevant contents have to be stored on the tags themselves, only strong encryption procedures can provide reliable protection against eavesdropping. In this paper, we specify the hash[3] based RFID authentication protocols as the previous works which employs hash functions to secure the RFID communication using Casper[5], CSP[4]. Then we verify whether or not it satisfies security properties such as secrecy and authentication using FDR model checking tool[6]. After running FDR tool, we reconfirm the existence of known security flaws in this protocol and describe the problems of hash based technique. This paper is organized as follows. In brief, the related works on RFID security and authentication schemes associated with hash function will be described in Section 2. In Section 3, we outline the use of Casper, CSP, and FDR tool for analysing security protocols automatically. Our analyzed result of the protocol will be described in Section 4. The proposed security scheme associated with encryption are presented in Section 5. Finally, the conclusion and our future work are addressed in the last section.

2 Related Work

There have been many papers in the literature that attempt to address the security concerns raised by the use of RFID tags.

2.1 The Hash Lock Scheme

A reader defines a "Lock" value by computing lock = hash(key)[3] where the key is a random value. This lock value is sent to a tag and the tag will store this value into its reserved memory location (i.e. a metaID value), and automatically the tag enters into the locked state. To unlock the tag, the reader needs to send the original key value to the tag, and the tag will perform a hash function on that key to obtain the metaID value. The tag then has to compare the metaID with its current metaID value. If both of them are matched, the tag unlocks itself. Once the tag is in unlocked state, it can respond its identification number such as the Electronic Product Code (EPC)[2] to readers' queries in the forthcoming cycles. This approach is simple and straight forward to achieve data protection, i.e. EPC code stored in the tag is being protected. Only an authorized reader is able to unlock the tag and read it, then lock the tag again after reading the code. This scheme will be analyzed in this paper in detail.

2.2 The Randomized Hash Lock Scheme

This is an extension of hash lock[3] based on pseudorandom functions (PRFs). An additional pseudo-random number generator is required to embed into tags for this approach. Presently, tags respond to reader queries by a pair of values (r, hash(IDk || r)) where r is the random number generated by a tag, IDk is the ID of the k-th tag among a number of tags in ID1, ID2, . . ., IDk, . . ., IDn. For reader queries, the tag returns two values. One is the random number. The other is a computed hash value based on the concatenation(||) on its own IDk and r. Once the reader gets two values, it retrieves the current N number of ID (i.e. ID1, ID2, . . ., IDn) from the backend database. The reader will perform the above hash function on each ID from 1 to n with r until it finds a match. When the reader finds a match, the reader is able to identify that tag k is on its tag's ID list (i.e. tag authentication). The reader will then send the IDk value to the tag for unlocking it. Once the tag is in an unlocked state, the reader can get its EPC code in the next reading cycle.

3 Casper, FDR, and CSP

3.1 Casper, FDR

Over the last few years, a method for analyzing security protocol that first models communication security protocol using CSP[4], then verifies its secrecy, authentication and other properties using FDR(Failure-Divergence Refinement)[6]. In this method, the main difficulty is specifying the security protocol's behavior using CSP. Creating the description of the security model with CSP is a very error-prone and difficult task. To simplify the expression of the security protocol, and render this process more error free, Casper(A Compiler of Security Protocol Analyzer)[5] was developed by Gavin Lowe[7]. This tool enables a non-expert who is unfamiliar with CSP to express the security protocol's behavior more easily, without being familiar with the notation used by CSP notation, using various key types, messages, security properties and intruder knowledge descriptions contained in Casper. In brief, Casper is a compiler that translates a more simple and concise description of a security communication model into CSP code.

The Casper, FDR approach was chosen for the analysis of the security protocol because Casper provides not only simple notation but also formal semantics, which maps to the CSP language. In addition, FDR model checking is very good at verifying concurrency models such as communication protocols.

3.2 CSP

CSP(Communicating Sequential Processes)[4] is a language for process specification specially designed to describe communication processes, and it can describe both a pure parallelism and interleaving semantics. In CSP, the former(a pure parallelism) is expressed as " || "and the latter(interleaving semantics) as " |||".

The combination of a client, server and intruder are regarded as a process. The use of two different concurrency concepts is well suited to the description and analysis of network protocols. For example, security communication systems operated in distributed networks can be modeled briefly as follows.

```
SYSTEM = (CLIENT1 ||| CLIENT2 ||| SERVER)
|| INTRUDER
```

4 The Modelling and Analysis of the RFID Authentication Protocol Using Casper and FDR Tool

4.1 The Specification of Hash Unlocking Protocol

Firstly, we model the behavior of hash unlocking protocol at the hash lock scheme and attacker in Casper script.

Table 1. The Hash Lock Scheme Notation

T	RF tag's identity
R	RF reader's identity
DB	Back-end server's identity that has a database
Xkey	Session Key generated randomly from X
metaID	Key generated from reader using hash functioon
ID	Information value of tag
Xn	A random nonce generated by X
H	Hash function

Message 1.	R	− > T	: Query
Message 2.	T	− > R	: metaID
Message 3.	R	− > DB	: metaID
Message 4.	DB	− > R	: RKey, ID
Message 5.	R	− > T	: RKey
Message 6.	T	− > R	: ID

Fig. 1. The hash unclocking protocol

The general overview of above protocol(Fig.1) was already described in 2.1 section.

```
#Protocol description
0.        -> T  : R
1.  T   -> R  : (H(Rkey)) % metaID
2.  R   -> DB : metaID % (H(Rkey))
3.  DB -> R  : Rkey, Id
4.  R   -> T  : Rkey
5.  T   -> R  : Id
```

Before explaination of *# Protocol description*, we will describe % notation to show specific notation. The % notation is used so that the metaID can be forwarded to other participants. This is why a reader can not construct the metaID, since the other reader does not know the value of hash function where m is a message and v is a variable, denoting that the recipient of the message should not attempt to decrypt the message m, but should instead store it in the variable v. Similarly, $v \% m$ is written to indicate that the sender should send the message stored in the variable v, and the recipient should expect a message of the form given by m. Therefore, *metaID* is the certain not knowing result value of hash function for T. In *# Protocol description* header, to unlock the tag, at the first line, Message 0 means that T(Tag) must communicate with R(Reader). The reader needs to send query to the tag and the tag sends the metaID to authenticate with reader.(Message 1). The reader forwards this metaID to DataBase to be ensured his identity.(Message 2). The DataBase has to compare the metaID with its current metaID value and ,if both of them are matched, lets the reader know the key and Id of tag.(Message 3). The reader authenticates his identity with the tag sending key received by database. (Message 4). As a result, if both of them are matched, the tag unlocks itself. Once the tag is in unlocked state, it can respond its identification number(*Id*) to queries of readers in the forthcoming cycles.(Message 5). This approach is simple and straight forward to achieve data protection, i.e. EPC(Electronic Product Code)[2] stored in the tag is being protected. Only an authorized reader is able to unlock the tag and read it, then lock the tag again after reading the code.

```
#Specification
Secret(R, Rkey, [T])
Secret(R, Id, [T])
Agreement(T, R, [Id, Rkey])
```

In hash unlocking protocol Casper script, *#Specification* description represents secrecy and authentication properties. The line starting with *Secret* expresses *secrecy property* associated with data privacy in RFID system. For example, the first statement is interpreted as " R believes that Rkey is a secret which should be known only to R and T" and the second statement is " R believes that Id is a secret which should be known only to R and T". If R, T or DB is an intruder in this protocol, secret information will be leaked to him, in which case a man-in-the-middle attack is considered to have occurred. The line starting with *Agreement* define that *authentication property* associated with authentication between a tag and a reader. For example, the third line means that " T is authenticated to R with Id, Rkey."

4.2 Protocol Goals

Using CSP[4], we describe the properties, i.e. secrecy property associated with data privacy, authentication property associated with authentication between a tag and a reader.

Secrecy - also called concealment or confidentiality - has a number of different possible meanings; the designer of an application must decide which one is appropriate. The strongest interpretation would be: an intruder is not able to learn anything regarding communications between two participants of a system by observing or even tampering the communication lines. That is, the contents of message, sender and receiver, the message length, the time they were sent, and even the fact that a message was transmitted in the first place cannot be deduced.

The following predicate is implemented in CSP language.

```
SECRET_SPEC_0(s_) =
 signal.Claim_Secret?T_!s_?Rs_ ->
 (if member(Mallory,Rs_) then
 SECRET_SPEC_0(s_)
 else SECRET_SPEC_1(s_)) []leak.s_->
 SECRET_SPEC_0(s_)
```

The *SECRET _ SPECT _ 0* and *SECRET _ SPECT _ 1* represent secret property of above *#Specification* section meet in the system. Formally speaking, if T has completed a protocol run apparently with R(*signal.Claim _ Secret?T _ !s _ ?Rs _*), and R is honest and uncompromised, then the key accepted during that run by T is not known to anyone other than R(*SECRET _ SPECT _ 1*), otherwise the key is known by someone in the system(*leak.s _*). Similarly, if R has completed a run with the honest and uncompromised T, then the key accepted by R is not known to anyone other than T. That is, the end of a successful protocol run each principal(T, R) should ensure whether it or not same with matched key of the other principal. In addition, the participants of the protocol should be satisfied that the session key is a suitable key for communication. Each principal should also be able to confirm that the other protocol participant possesses the Session Key(RKey).

Authentication: A system provides strong authentication if the following property is satisfied: if a recipient R receives a message claiming to be from a specific sender S then S has sent exactly this message to R. For most applications this formulation must be weakened, since in most cases, communication channels are subject to both technical errors and tampering by attackers. A system provides weak authentication if the following property is satisfied: if a recipient R receives a message claiming to be from a specific sender S then either S has sent exactly this message to R or R unconditionally notices this is not the case.

```
AuthenticateINITIATORToRESPONDER
 Agreement_0(T) =
signal.Running1.INITIATOR_T.R ->
signal.Commit1.RESPONDER_R.T -> STOP
```

Formally speaking, the events of the form *Running1.INITIATOR _ T.R* in T's run of the protocol are introduced to mark the point that should have been reached by the time that R performs the *Commit1.RESPONDER _ C.M event*. Occurrence of *Running1.INITIATOR _ T.R* run means simply that Agent T is following a protocol run apparently with R. If a *Running1.INITIATOR _ T.R* event must always have occurred by the time that the *Commit1.RESPONDER _ R.T event* is performed, then authentication is achieved. That is, the protocol is checked against these specifications in order to determine if the properties of the protocol hold during subsequent runs of the protocol.

4.3 The Result of Verification

In this paper, we show verification results of the safety specification in hash unlocking scheme, we use traces refinement provided in FDR tool. If the trace events set of implementation model Q is a subset of the trace events set of specification model P, we can say that Q is a safe implementation. After running the FDR model checking tool, this protocol is not found to satisfy the Secret and Agreement requirements in Casper script. Through debugging the counter-example trace events, we reconfirm that hash unlocking protocol may be susceptible to a sniff and spoof attack by an intruder due to unsecured communication channel between reader and tag. A general attack scenario, which could be found in this protocol is described as below; *I _ Agent* means an intruder who can sniff messages and spoof his identity.

```
1.     Tag      ->  I_Reader   : H(RKey)
2. I_Mallory  ->  DataBase   : H(RKey)
3.  DataBase  -> I_Mallory   : RKey, Id
```

The notation *I _ x* represents the intruder I imitating some participant to fake or intercept a message. Through the man-in-the-middle attack of the hash unlocking protocol, an intruder masquerading as Reader in Message 1, 2 could forward the message *H(Rkey)*and in Message3, an intruder masquerading as Reader could intercept the *RKey, ID*.

That is, a hacker may obtain the current metaID value(H(RKey)) by querying a tag. The hacker replays the obtained metaID value and broadcasts it to any readers nearby, to get the specific random key for this MetaID value if any reader responds to his replay. Therefore, the hacker may have a chance to get the key to unlock the tag and obtain its data.

4.4 Analysis of the Hash-Based Protocols Using FDR

We can summarize verification results about above protocols using model checking.

Especially, the previous protocols are vulnerable to the replay attack, spoofing attack and can be tracked by an attacker.

The attacker performs the following attack.

1. Security against the spoofing attack : The attacker disguises as a right reader, then sends Query and reader to the tag. The attacker gets tag's response value due to not ensuring the response value of hash function from this attack.

2. Security against the replay attack : After the reader transmits Query to the tag, the attacker eavesdrops response value from tag.
3. Security against the traffic analysis and tracking: To receive responses, the attacker disguises the reader then transmits fixed Query and reader to the tag or overhears the information between the reader and the tag. Therefore, the attacker can analyzes the response of the tag.

5 The Proposed of the Strong Authentication Protocol for RFID Systems

5.1 The Modelling of the Strong Authentication Protocol Using Casper

In the previous schemes [3], they assumed that R is a TTP(Trusted Third Party) and the communication channel between R and B is secure. However, we assume that R is not a TTP and the communication channel is insecure like the current wireless network. We also assume that k is the secret session key shared between R and B, and R and B has enough capability to manage the symmetric-key crypto-system and sufficient computational power for encryption and decryption. The main idea of this framework is based on the security algorithm that employed in Yahalom protocol[8].

The proposed protocol must guaranty the secrecy of session key: in message 4, 5, the value of session key must be known only by the participants playing the roles of T, R. R, T must be also properly authentified to DB.

Message 1.	R	−> T	:	Query
Message 2.	T	−> R	:	Tn
Message 3.	R	−> DB	:	{ T, Tn, Rn } ServerKey(R)
Message 4.	DB	−> T	:	{ R, DBkey, Tn, Rn, Id} ServerKey(T)
Message 5.	DB	−> R	:	{ T, DBkey} ServerKey(R)
Message 6.	T	−> R	:	{ Id }DBkey

Fig. 2. The proposed strong authentication protocol

We describe this protocol by the next protocol description as follows:

```
#Protocol description
0.      -> T : R
1.   T -> R : Tn
2.   R -> DB: {T, Tn, Rn}{SKey(R)}
3a. DB-> T : {R, DBkey, Tn, Rn, Id}{SKey(T)}
3b. DB-> R : {T, DBkey}{SKey(R)}
4.   T -> R : {Id}{DBkey}
```

In this protocol description, we use the *Server Key* and *Tag's Nonce(Tn)* to minimize for burden of Tag and to ensure the authentication between Tag and

Reader. Functions could be defined that can take in an input parameter and return an output. It resembles a functional programming language in this aspect. The definition of a function called *SKey* that takes in the name of an Server and returns a *ServerKey* could be given as shared : *Agent −> ServerKey*. At the Message 1, we design that T makes random nonce Tn and sends R. Since it makes a simple challenge-response easily. Therefore, at the Message 2, through *T, Tn, Reader's Nonce(Rn), and Server Key*, R can be ensured the authentication from database. At the Message 3a, DB encrypts all of the *T, DBkey, Tn, Rn, and Id* received from R and sends these things to T to let R authenticate securely using server key. At the Message 3b, DB also sends *T, DBkey* to R to decrypt Tag's Id. At the last Message 4, T can send a his *Id* securely using DBkey received at the Message 3a.

In addition, we verify the proposed authentication protocol based encryption that establishes a secure channel between T and R, and confirm that the authentication protocol satisfies the secrecy and authentication property.

5.2 The Result of Verification

After running FDR tool, we confirm that our proposed protocol solves the security weakness in hash-based protocols.

– *Secrecy:* Spoofing, Replay Attack, Tracking, Eavesdropping on communication between tag and reader are attacks that threaten all participants. To protect from these attack, the countermeasures are therefore essentially identical in this protocol as follows.
  Firstly, shift all data except the ID to the backend. This is also to be recommended for reasons of data management. (i.e. the Id for the tag existing at the backend database will be shifted to protect spoofing, eavesdropping attacks securely to the tag through the database when the reader request.)
  Secondly, encode data transfer: we support encryption of the data transmission to ensure authorized access to the data that concern it and to protect replay attack and tracking.
– *Authentication:* When a tag receives a "get challenge(query)" command from a reader, it generates a random number Tn and sends it to the reader. The reader in turn generates a random number Rn and with it and the random number Tn generates an encrypted data block on the basis of an encryption algorithm and a server key(R). The data block is then returned to the database to authenticate the reader. Both reader and tag use the same encryption algorithm and since the server key is stored on the tag, the tag is capable of decrypting the server key(T). If the original random number Tn and the random number Tn', which has now been decrypted, are identical, then the authenticity of the tag vis-a-vis the reader has also been proved.

6 Discussion and Conclusions

Although hash-based scheme offers good reliability at low cost, an adversary can easily track the tag via its metaID or hash values. Furthermore, since the key K is sent in the clear, an adversary upon capturing the key can later spoof the tag. In this paper, we focus on the verification of the hash-unlocking protocol which is widely researched in RFID system and analysis of the vulnerabilities of the protocol using Casper ,CSP, and FDR. In verifying this protocol with FDR tool, we were able to reconfirm some of the known security vulnerabilities which are likely to occur in RFID system. Finally, we propose a strong authentication protocol based encryption algorithm against the man-in-the-middle, replay attacks and also verify its safety using FDR tool.

References

1. S. Sarma, S. Weis, D. Engels,: RFID systems and security and privacy implications. Workshop on Cryptographic Hardware and Embedded Systems (CHES) 2002, LNCS No(2523):(2003)454-469
2. EPCGLOBAL INC.: http://www.epcglobalinc.org.
3. S. Weis, S. Sarma, R. Rivest and D. Engels,: Security and Privacy Aspects of Low-Cost Radio Frequency Identification Systems. 1st Intern. Conference on Security in Pervasive Computing(SPC)(2003)
4. C.A.R. Hoare, Communicating Sequential Processes, Prentice-Hall, Englewood Cliffs, NJ(1985)
5. G. Lowe,: Casper: A compiler for the analysis of security protocols. Proceeding of the 1997 IEEE Computer Security Foundations Workshop X, IEEE Computer Society, Silver Spring, MD(1997)18-30
6. Formal Systems Ltd. FDR2 User Manual, Aug(1999)
7. P. Y. A. Ryan and S. A. Schneider. Modelling and Analysis of Security Protocols: the CSP Approach. Addison-Wesley(2001)
8. Lawrence C. Paulson.: Relations between secrets: Two formal analyses of the yahalom protocol, Journal of Computer Security(2001)

ITB: Intrusion-Tolerant Broadcast Protocol in Wireless Sensor Networks

Jin Wook Lee[1] and Yann-Hang Lee[2]

[1] Networking Technology Lab.,
Samsung Advanced Institute of Technology, P.O. 111, Suwon, Korea 440-600
[2] Real-Time Embedded Systems Lab.
Department of Computer Science and Engineering,
Arizona State University, Tempe, AZ 85287-8809, USA

Abstract. A large-scale wireless sensor network relies on broadcast protocol for data communication. This paper presents an idea for a base station to reliably and securely broadcast announcing messages to the network in an intrusion-tolerant way. Our Intrusion-Tolerant Broadcast Protocol based on Mutual Verification scheme guarantees that no forged information can be propagated to the network if an intruder is assumed to not compromise the base station. We show that our proposed protocol successfully runs and also demonstrate the effectiveness of the protocol by introducing a theorem and the proof of the theorem.

1 Introduction

A large-scale wireless sensor network may consist of thousands of nodes randomly deployed in an open environment. A typical communication in such a network would be a base station delivering commands to every sensor node in the network and then the sensor nodes sending sensed data from their close proximities to the base station for further analysis. For such a communication pattern, a broadcast protocol based on flooding would be an efficient solution. But there is a crucial issue concerning flooding-based broadcast protocols in wireless sensor networks - broadcast authentication.

It becomes critical that, once a message is sent out from the base station, there is a way for the node to guarantee that the message arrives at the recipient with its original content. A solution to this broadcast authentication is the private/public -key-based digital signature. TESLA-based protocols, proposed by Perrig et al. and Liu et al. as a broadcast authentication protocol, require a digital signature to distribute the initial parameters and all nodes to perform pre-setup procedures before undertaking a real task such as assigning the last key of a hash chain and synchronizing a global time clock [2] [3] [5]. Such pre-setup procedures create considerable overheads. We aim at an efficient and secure broadcast protocol for sensor networks. A novel broadcast protocol is proposed based on the notions of **intrusion-tolerance and mutual verification**.

2 Intrusion-Tolerant Broadcast Protocol

Intrusion-Tolerant Broadcast(ITB) Protocol we propose here introduces a mechanism to verify the behavior of neighbor nodes, detect and respond against threats during the

M. Gerndt and D. Kranzlmüller (Eds.): HPCC 2006, LNCS 4208, pp. 505–514, 2006.

broadcast. Our goal is to flood the broadcast messages to the network without allowing faked messages' propagation. We target a large-scale wireless sensor network consisting of a base station and a large number of sensor nodes. We use several terms to describe the protocol as defined in Table 1.

Table 1. Glossary

Glossary	*Description*
Level(l)	physical hop distance from the base station
Level Key(LK)	a value generated by one-way hash function and distinct according to level(l)
Shared Key(SK)	a symmetric key that is shared by all nodes. Pre-installed key.
Private Key($PrvK$)	a private key of the base station. It is not faked or generated by a node.
Level Synchronization	the protocol to assign the time duration for sending or receiving a message
Time interval	time duration for sending or receiving specific protocol messages
BRTS	Broadcasting Request To Send
BCTS	Broadcasting Clear To Send
BM	Broadcasting Message

2.1 Control Packet

The key idea of ITB protocol is that letting two-hop away nodes exchange the verification information of broadcast messages and allowing the nodes to verify each other before taking further actions. In order to realize the idea, we apply the concept of the typical RTS/CTS handshaking of 802.11 medium access control protocol [4], which is used to avoid collisions in wireless multi-hop networks. During RTS/CTS handshaking, a node is affected by two-hop away nodes' communication by receiving a RTS or a CTS control packet. We take advantage of this concept to deliver the verification information to two-hop away nodes in the routing path before delivering a data message. The protocol proposes three packets, BRTS, BCTS, and BM, in its broadcast handshaking mechanism. Consider a sensor node i which is l hops away from the base station and is in the radio range of a upstream node k of hop distance $l-1$. When a broadcast message is propagated from node k to node i, the packets that node i will send out in turn are:

$$BCTS_i^l : E_{SK}\{MAC(BM_i^l), LK^{l+1}\}$$

$$BRTS_i^l : E_{SK}\{MAC(BM_i^{l+1}), LK^{l+2}\}$$

$$BM_i^l : E_{SK}\{LK^l\} \mid E_{PrvK}\{Message\}$$

In the packet formats, $E_x\{M\}$ represents that message M is encrypted with a key x. Note that MAC of BRTS and BCTS is Message Authentication Code of BM rather than BRTS or BCTS themselves. Specially, the MAC in a BCTS that a node receives

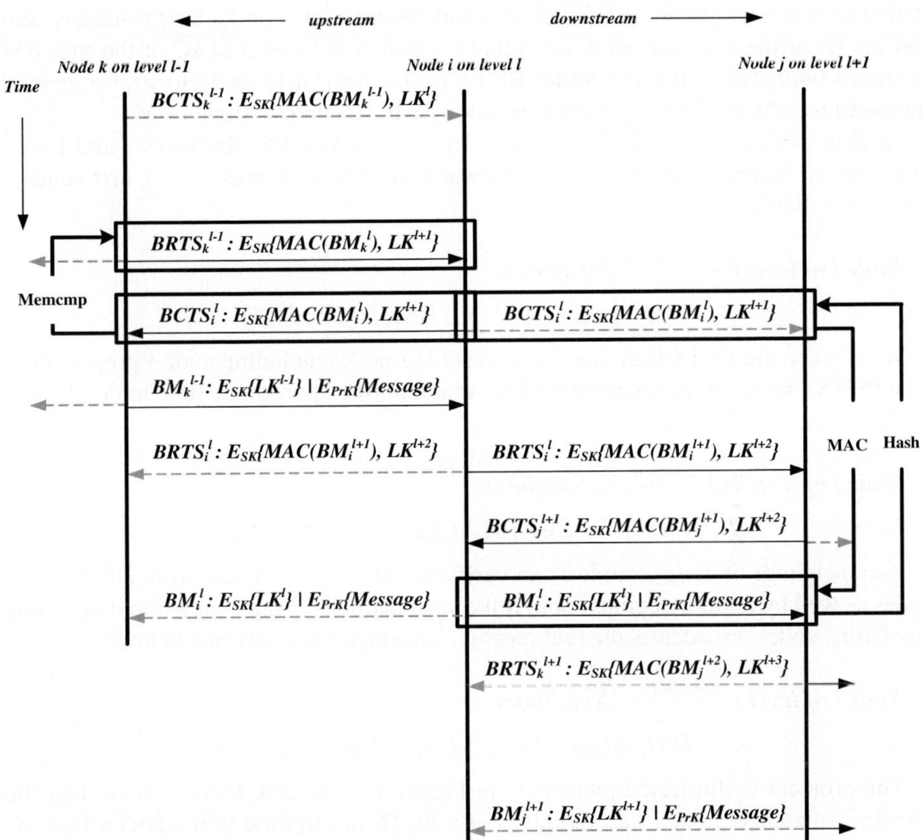

Fig. 1. Intrusion-Tolerant Broadcast Protocol in a single routing path

from upstream nodes will be used to verify a BM which the node will receive later. This allows two-hop away nodes to make an agreement with each other. For MAC, the use of collision-proof checksums is recommended for environment where a probabilistic chosen-plaintext attack represents a significant threat [7]. We apply MD5 as a MAC generation algorithm. Level Key(LK) is a verification key as an output of one-way hash function. LK of upstream nodes is not easily guessed due to its one-wayness.

2.2 Broadcast Step

The base station starts broadcasting a message by firstly sending a BCTS to its adjacent nodes. The reason that the base station firstly sends a BCTS is that each node starts its protocol operation on receiving a BCTS from one of parent-level nodes as shown in Figure 1. Originally CTS is a response of RTS. However, it is not possible for the neighbor nodes of the base station to receive a BCTS from the base station because the base station does not respond to any BRTS. Therefore, the protocol forces the base station to generates an initial $BCTS_0^0$ (Assume that the level of the base station is zero and the ID of the base station is zero) that contains LK^1 and a MAC of the first BM to start a broadcast. After the initial BCTS packet sent, it is ready to start a general forwarding cycle of sending a BRTS, receiving a BCTS and sending a BM.

Assume nodes i and j are two neighboring nodes with hop distances l and $l+1$, respectively. When node i is ready to forward the broadcast message, it first sends a BRTS as follows:

$$\textit{\textbf{Node i of level l}} \xrightarrow{broadcast} \textit{\textbf{Neighbors}} :$$

$$BRTS_i^l : E_{SK}\{MAC(BM_i^{l+1}), LK^{l+2}\}$$

After receiving BRTS from node i, all *level l+1* nodes including node j prepare their own BCTSs based on the received BRTS. Again, during appropriate time interval, node j transmit its BCTS as below:

$$\textit{\textbf{Node j of level l+1}} \xrightarrow{broadcast} \textit{\textbf{Neighbors}} :$$

$$BCTS_j^{l+1} : E_{SK}\{MAC(BM_i^{l+1}), LK^{l+2}\}$$

As a neighbor of node j, node i can verify the BCTS packet sent from all *level l+1* nodes as well by memory comparison. If the verification is completed without detecting any faults, node i broadcasts the real message enveloped in a BM like below:

$$\textit{\textbf{Node i of level l}} \xrightarrow{broadcast} \textit{\textbf{Neighbors}} :$$

$$BM_i^l : E_{SK}\{LK^l\} \mid E_{PrvK}\{Message\}$$

The protocol is illustrated pictorially in Figure 1. Note that, before forwarding the broadcast message, node i has already sent a BCTS in response to the BRTS from its parent node. Eventually node j hears this BCTS. Hence, node j of *level l+1* receives three packets from node i in the sequence of BCTS/BRTS/BM. Each packet embraces LK^{l+1}, LK^{l+2}, and LK^l respectively. From one-wayness of Level Key, LK^l is used to verify LK^{l+1} and LK^{l+2}. If the verification is completed without detecting any faults, the node officially accepts the broadcast message. Otherwise, the message is discarded and no forwarding operation is taken.

2.3 Two-Hop Away Communication

The protocol of each node is started up on receiving a BCTS sent by a parent-level node. However, a node does not have a way to verify this BCTS itself since the BCTS is originally destined to grandparent-level nodes (two-hop away nodes which are nearer to the base station, for instance, node j for node k in Figure 1) and the node does not have any information of this BCTS. Hence, the node stores this BCTS naively and keeps it unverified. The verification for a BCTS can be done by grandparent-level nodes. If this BCTS is not valid, grandparent-level nodes stop to broadcast the main message packet, BM and eventually a node does not receive the BM associated with the previously received BCTS. A BM has also a clue to verify the previously received BCTS. The LK of a BM is a key of verification for the received BCTS. Parent-level nodes cannot fake a BM and a BCTS without grandparent-level nodes' or child-leve nodes' detection.

3 Mutual Verification

In the protocol, each node receives three protocol packets from each of its upstream nodes and broadcasts out three protocol packets to its downstream nodes. The protocol provides a way to mutually verify each protocol packets. When the verification fails, a sender restrains itself from sending out the broadcast message or a receiver discards the received message. As a consequence, the broadcast operation stops along the broadcasting path that fails in the verification. The verification operation in the protocol relies on three computation actions, memory comparison, one-way hash function computation, MAC computations. Memory comparison is a lightweight action as compared to one-way hash function and MAC computation. Computational overhead of MAC computation is much more than that of one-way hash function. Therefore, a verification should be done in a way that the results of MAC and one-way hash function computation are stored to reduce the frequency of MAC and one-way hash function computation.

3.1 BCTS Verification

A BCTS packet is sent out on a recipient of a BRTS packet. There are three types of BCTS's that a node may receive. The first one is in an initial state that a node receives the BCTS's from its parent-level nodes. On receiving of the BCTS in the initial state, a node starts the protocol by storing the information of the BCTS coming from parent-level nodes. At this time the node dose not know whether the received information is legitimate or not. Only the source node of the BRTS can verify the BCTS since the payload of the BCTS should be identical to the payload of the corresponding BRTS. If the verification succeeds, the source takes the next action of sending a BM. Once a node receives the first BCTS, it can do the comparison to check out any BCTS received successively in the same broadcast window if there are multiple neighbor nodes. Subsequently, the node can receive a BRTS from a parent node in the upstream and enter a broadcast window of sending out a responding BCTS. There might be the second type of the BCTS that are originated from its sibling-level nodes. The verification for it does not require one-way hash function computation. A node compares its own BCTS with the ones from its sibling nodes. Lastly, there are a BCTS coming from child-level nodes as a response of its BRTS. It should have the same content as the BRTS. Memory comparison is good enough to verify them.

3.2 BRTS Verification

In the same way of the BCTS case, there are three types of BRTS packets that a node receives from its neighboring nodes. On receiving the first BRTS from one of its parent-level nodes, a node can verify the BRTS by comparing the LK field of the stored BCTS. The LK in the BRTS should be the output of one application of one-way hash function on the LK in the stored BCTS. If any verification is not succeeded, the node discards the stored BCTS and the received BRTS from the source node and registers the source node as a malicious node. Once a node succeeds in verification, the subsequent verifications are done by just comparison to reduce computational overhead. The other two types of BRTS packets are from the sibling nodes and the child nodes. The BRTS

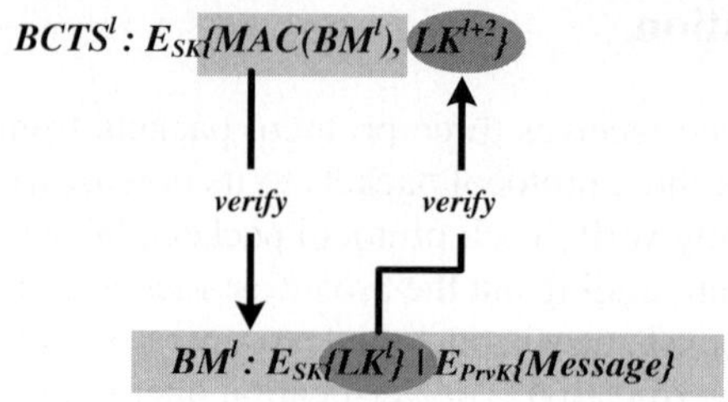

Fig. 2. Mutual Verification

from the sibling nodes should be identical to the one the node is sending out and the ones from the child nodes should have the correct LK (after one application of the hash function). It is possible to find any discrepancy in the comparison of these BRTS with the node's own BRTS. However, besides noting the possible malicious behavior in the neighboring nodes, the node with the verified BRTS can proceed to send out the broadcast message.

3.3 BM Verification

The verification of a received BM is done with the stored MAC and LK. When a node receives a BM from one of parent-level nodes for the first time, it computes the MAC for the whole message and decrypts the BM with the shared key, SK to extract LK. The computed MAC should be the same as the stored MAC which is included in the previously received BCTS. In addition, the node runs one-way hash function **twice** with the extracted LK and make a comparison with the stored LK of the received BCTS/BRTS. If this verification fails, a node discard all protocol packets coming from the same node. The notion of mutual verification is shown pictorially in Figure 2.

3.4 Effectiveness of the Protocol

We prove the effectiveness of the protocol through introducing a theorem based on three propositions.

Proposition 1 (Concurrency). *The neighbor nodes of node N receive the same message at the same time when node N transmits the message.*

Proposition 2 (Mutual Exclusion of Transmission). *The neighbor nodes of node N are not able to transmit any messages successfully at the same time when node N transmits a message.*

Proposition 3 (Watchdog). *The neighbor nodes of node N are able to monitor node N's transmission under any collision avoidance protocols.*

Theorem 1 (Mutual Verification). *A node cannot modify and forward the relaying messages successfully without being detected by one-hop away nodes if one-way function and MAC is secure.*

Proof. Assume there are three nodes in a routing path, which are node A, node B and node C. Let say that node A is a source node, node B is a relay node and node C is a destination node.

Suppose for contradiction that node B, an adversary that can modify a relaying message without being detected by node A and successfully forward the modified message to node C. As a result, node C receives a different message from the message that node A sent to node B. A proof of effectiveness of our protocol is divided by two parts according to the messages.

The behavior of node B on receiving a BCTS and a BM can be categorized as below;

Case I: On receiving a BCTS from node A,

- Scenario 1) Node B relays the received BCTS to node C
- Scenario 2) Node B modified the received BCTS and forwards it to node C

Case II: On receiving a BM from node A,

- Scenario 3) Node B relays the received BM to node C
- Scenario 4) Node B modified the received BM and forwards it to node C

Proof of Case I)

There are two scenarios for node B's action to avoid node A's detection in case that node B receives a BCTS from node A. One is that node B naively relays the received BCTS to node C. The other is that node B relays the received BCTS to node A and forwards a modified BCTS to node C at the same time. Scenario 1 is intuitively correct as the behavior of node B is correct. Let us consider Scenario 2. In order for node B to send a modified BCTS successfully to node C without node A's detection, it should be able to broadcast two different messages simultaneously, the original BCTS to node A and a modified BCTS to node C. However, this is contradiction of Proposition 1 and 2. Node A and node C exist within the radio range of node B, so node A receives the message from node B if node C receives it and vice versa. Therefore, node A is able to detect node B's misbehavior when node B modifies and forwards a BCTS. As a result of node A's detection, it does not forward a BM, which means node B should be able to create a BM that matches a modified BCTS in order to let node C trust node B.

Proof of Case II)

In proof of part I, we prove node B cannot forward a modified message without node A's detection. When node B relays a correct BCTS and receives a correct BM from node A, node B can choose the action like Scenario 3 and 4.

Scenario 3 is legitimate while Scenario 4 is needed to be proven. If node B receives a BM from node A, the transmission role of node A is finished at this moment. Hence, we can say node C has already received the correct BCTS from node B. If node B modifies the BM and forwards it, node C can detect the modification by using the previously received BCTS. Therefore, node B cannot disguise node A and node C at the same time (Proposition 1 and 2), which is contradiction.

4 Evaluation

We now are interested in evaluating the cost of the ITB protocol and demonstrating its applicability to the resource-constraint sensor devices. We quantify the computational overhead a node pays to securely broadcast a message and simulate energy consumption of the protocol. Communication and computational overheads will be considered and we justify our claim with a simulation result of energy consumption. Let n_a denote the average number of neighbor nodes for a node.

Communication Overhead. We define *Communication Overhead* to be the average amount of data a node is required to transmit and receive for a broadcast. Firstly, we discuss the number of communications a node is involved. In ITB protocol, the number of outgoing communications per each node is constant while the number of incoming communications is heavily dependent upon the number of neighbor nodes. Note that three outgoing transmissions(including BCTS, BRTS, and BM) are just necessary for a node to verify protocol packets and relay the broadcast message in ITB protocol. These transmissions are all broadcasts so retransmission is not applied. Hence, message complexity of outgoing communication is $O(1)$. The number of incoming communications is associated with the network density. A node receives one BCTS, one BRTS, and one BM packet from a single parent-level node. A node also receives the same sort of packets from its sibling-level nodes and child-level nodes. Therefore, the average number of incoming communications per node is calculated as $3 \cdot n_a$. Hence, message complexity of incoming communication is $O(n)$ where n is the average number of neighbor nodes.

From the standpoint of the number of communications, ITB protocol requires three more times number of outgoing and incoming communications than a simple flooding broadcast protocol does. However additional amount of data transfer for the protocol is only small bits excluding broadcast message body (Note that we assume the size of LK and MAC reasonably is 64 and 128 bits, respectively). In conclusion, ITB protocol is efficient in terms of communication overhead as the size of broadcast message increases.

Computational Overhead. *Computational Overhead* in ITB protocol is defined as the number of cpu cycle counts running security algorithms such as memory comparison, one-way hash function, and MAC. We analyze the expected number of such security algorithms runs for broadcasting a single 6-byte message. Thanks to the deterministic number of communications for a node, the computational overhead can be calculated through analyzing the each algorithm.

In order to verify incoming messages' validation a node makes a memory comparison between stored information and newly receiving information. Clearly the number of comparison is dependant upon the number of incoming messages since the comparison is used for verification. When a node is awaken by receiving a BCTS from a parent-level node, it just stores the content of the BCTS without any comparison. As a result of analysis of our algorithm, we obtain the total number of comparisons is less than $12 \cdot n_a - 5$, where n_a is the average number of neighbor nodes. The numbers of one-way hash function runs and MAC computations are obvious. They are counted four times and two times runs for a single broadcasting message respectively. Let C_c, C_h, and C_m denote the cost of computing a single comparison, the cost of computing a single hash function, and the cost of computing a single MAC computation, respectively. The total cost of computational overhead is derived as below:

$$(12 \cdot n_a - 5) \cdot C_c + 4 \cdot C_h + 2 \cdot C_m \quad (1)$$

The total computational cost is measured by the cpu cycle count of ITB algorithm. From PowerTOSSIM cpu cycle profiling [8], we get cpu cycle counts (34 cycles for memory comparison, 20228 cycles for one-way hash function, and 36437 cycles for MAC computation). As we see in Figure 3(a), the number of CPU cycles slightly increase as the number of neighbors increases. We claim that the computational cost of ITB protocol does not heavily rely on the number of neighbors nodes since the number of neighbor nodes affects only the number of memory comparisons. In conclusion, our ITB protocol is designed in consideration of scalability and efficiency.

To justify our claims of computational overhead, we implement the protocol and run it with MICA2 energy model (Refer to Table 2) to collect energy consumption data. We perform the PowerTOSSIM to simulate our protocol with real implementation code for MICA2 [8]. As can be seen in Figure 3(b), energy consumption caused by ITB protocol does not proportionally increase as the network density increases, which prove that our claims is reasonable.

Table 2. Energy Model for MICA2

model	energy(mA)
CPU active	8.93
CPU idle	3.2
CPU init	3.2
Radio RX	7.03
Radio TX	21.48
LED	2.2

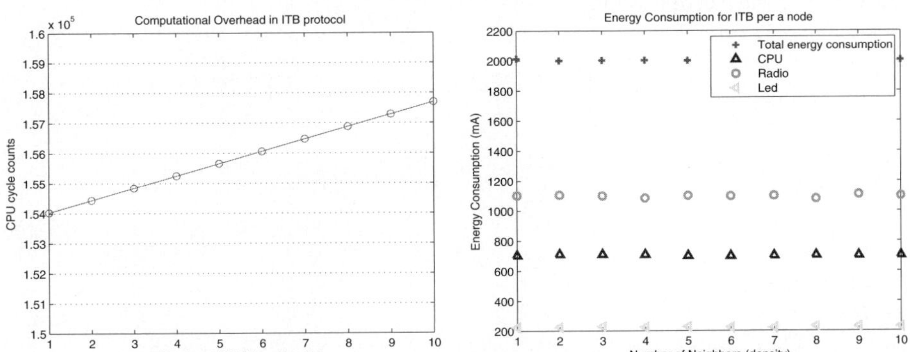

(a) Computational Overhead (b) Energy Consumption

Fig. 3. Average amount of power consumption for ITB protocol per node

5 Conclusion

In this paper we have proposed Intrusion-Tolerant Broadcast Protocol to realize secure broadcast via Mutual Verification mechanism. By using the proposed protocol, the base station is able to disseminate important messages to the network without propagating modified or garbage messages. Our protocol is secure in terms that any bad broadcasting messages modified by a compromised node are not propagated more than two-hop away from the compromised node. Our protocol makes any attacks of compromised node detectable, verifiable and responsible. Our analysis showed that ITB protocol is quite lightweight in a sense that communication and computational overhead are $O(n)$, where n is the average number of neighbor nodes(i.e., network density) respectively. Through discussing a theorem, we demonstrated the effectiveness of ITB protocol.

References

1. Y. H. Lee, A. Deshmukh, V. Phadke and J. W. Lee, *Key Management in Wireless Sensor Networks*, In Proc. of the First European Workshop (ESAS 2004), pages 190–204, 2004.
2. A. Perrig, R. Canetti, J. D. Tygar, D. Song, *Efficient and secure source authentication for multicast*, in Network and Distributed System Security Symposium (NDSS'01), 2001.
3. A. Perrig, R. Szewczyk, J. D. Tygar, V. Wen, D. Culler, *SPINS: Security Protocols for Sensor Networks*, Wireless Networks Journal (WINET), 8(5), pages 521–534, 2002.
4. A. E. F. Clementi, A. Monti and R. Silvestri, *802.11 Specification*, in ACM-SIAM SODA'01, 2001.
5. D. Liu and P. Ning, *Efficient Distribution of Key Chain Commitments for Broadcast Authentication in Distributed Sensor Networks*, In Proc. of the 10th Annual Network and Distributed System Security Symposium, pages 263–276, 2003.
6. A. Chakrabarti, A. Sabharwal and B. Aazhang, *Multi-Hop Communication is Order-Optimal for Homogeneous Sensor Networks*, in Proceedings of the third international symposium on Information processing in Sensor Networks, pages 178–185, 2004.
7. J. Kohl and C. Neuman, *The Kerberos Network Authentication Service (V5)*, in RFC 1510, 1993.
8. V. Shnayder, M. Hempstead, B.-R. Chen and M. Welsh, *PowerTOSSIM*, http://www.eecs.harvard.edu/ shnayder/ptossim/.

Authentication for Ubiquitous Multi Domain in Pervasive Computing Using PMI*

Deok Gyu Lee[1], Jong Sik Moon[1], Jong Hyuk Park[2], and Im Yeong Lee[1]

[1] Division of Information Technology Engineering, Soonchunhyang University, #646, Eupnae-ri, Shinchang-myun, Asan-si, Choongchungnam-do, Korea
{hbrhcdbr,comnik528,imylee}@sch.ac.kr
http://sec-cse.sch.ac.kr

[2] Center for Information Security Technologies, Korea University, 5-Ka, Anam-Dong, Sungbuk-Gu, Seoul, Korea
hyuks00@korea.ac.kr

Abstract. The Ubiquitous computer environment is thing which invisible computer that is not shown linked mutually through network so that user may use computer always is been pervasive. Intend computing environment that can use easily as user wants and it is the smart environment that user provides context awareness that is wanting computing environment. This Ubiquitous computing contains much especially weak side in security. Masquerade attack of that crawl that is quoted to user or server among device that is around user by that discrete various computing devices exist everywhere among them become possible. Hereupon, in this paper, proposed method that has following characteristic. Present authentication model through transfer or device. Suggest two method that realize authentication through device in case of moved to method (MD: Multi Domain) that realize authentication through device in case of moved user's direct path who device differs.

Keywords: Pervasive Computing, Multi Domain, Authentication, PMI.

1 Introduction

Ubiquitous computing aims at an environment in which invisible computers interconnected via the network exist. In this way, computers are smart enough to provide a user with context awareness, thus allowing the user to use the computers in the desired way. Ubiquitous computing has the following features: Firstly, a variety of distributed computing devices exist for specific users. Secondly, computing devices that are uninterruptedly connected via the network exist. Thirdly, a user sees only the personalized interface because the environment is invisible to him. Lastly, the environment exists in a real world space and not in a virtual one. However, the ubiquitous environment is weak in security. Since distributed

* This research was supported by the MIC (Ministry of Information and Communication), Korea, under the ITRC(Information Technology Research Center) support program supervised by the IITA (Institute of Information Technology Assessment).

M. Gerndt and D. Kranzlmüller (Eds.): HPCC 2006, LNCS 4208, pp. 515–524, 2006.

computing devices are spread out in the environment, it is possible to launch disguised attacks against the environment from a device authenticated by a user or a server. Also, although a user approves of the installation of only authenticated applications into the devices, there is a chance that a malicious code will be transmitted to surrounding devices that do not have computing capability. Since many ubiquitous computing devices do not provide efficient memory protection, the memory where user information (authentication information) is stored can easily be attacked. These problems can be resolved by using encryption or electronic signature for computing devices. But non-computing devices cannot be protected using encryption codes or electronic signatures, which bring up potential security issues. Also, when a device is moved out of the domain into a new user space, user authentication must be performed smoothly in the new space. This is so because a different device in the new user's space can be authenticated with the user authentication information in the new user space, and not with the previous user authentication information[3][4][7][9]. It is this paper's purpose to propose an authentication model through the movement of a smart device. The conceptualized model proposes two methods. One method is to implement the authentication through a device when an individual small device is moved into the multi domain. The other method is to implement the authentication through a device when another device is moved into the single domain. In this scheme, each device stores a different set of user authentication information. Minimum authentication information is stored in the smart device, and most authentication information is stored in the computing device. Since a smart device stores minimum authentication information, it transmits and receives authentication information from the Hub which first provides it with the authentication information. In this scheme, the Hub can be a user or a server[8]. This paper thus consists of five chapters. Chapter 1 covers this introduction. Chapter 2 covers requirements for authentication in ubiquitous computing. Chapter 3 covers ubiquitous computing research trends. Chapter 4 covers authentication methods using a smart device in the multi domain and the single domain. Chapter 5 compares existing ubiquitous researches and the proposed methods to verify the efficiency of the proposed methods. Finally, Chapter 6 concludes the research.

2 Related Works

In 1993, Mark Weiser of PARC (Palo Alto Research Center), a prominent figure in ubiquitous computing, took a look at the computer evolution in his paper and saw computer technology development from the point of view of changes in human relationships. The first wave was defined as mainframes, each shared by a lot of people. The second wave was defined as personal computers, each shared by one person. In his paper, the ubiquitous society was introduced. In society, a wide range of people use computers without being aware of various internal computers. This computer technology was defined as the third wave. His paper mentions a calm technology that permeates our daily life through invisible interfaces, which will be our major computer interface technology. He predicted that this technological innovation would bring ubiquitous computing to life[4][5]. Ubiquitous computing emphasizes the following researches. Since the ubiquitous computing network connects PCs, AV equipment on a server-oriented network, information electronic appliances, cellular phones, game consoles, control devices, and various other devices, core technologies such as miniaturization, cellular phone technology, technology for information electronic appliances, electronic control technology, and networking control technology have risen to the front. Among these, individual authentication technology and security technology have been named as the technologies that will allow users to utilize computers in a secure way. The research on authentication has been conducted as a national project in many countries. However, no research has been done regarding the provision of authentication in the multi-domain. The next part will cover general trends in the ubiquitous computing environment, and will describe existing methods and projects.

2.1 JARM Scheme

In 2002, Jalal's proposed a method that supports the user authentication level concept[3]. Different levels of user authentication information can be stored in different devices, which mean that minimum user information can even be stored in watches and smart rings. Medium-level user information can also be stored in a smart device like a PDA. With this method, if a device is moved from one user domain to another, the device can use the new user information in the new domain. However, the device cannot use the authentication information of the new domain, which restricts users who move from one domain to another from using the device. Therefore, with this method, all devices in one domain have authentication information, and a user can be authenticated through a device and can be authenticated against all devices using the level authentication information. This method cites multiple steps when it comes to the authentication through trust values for level authentication information. A device obtains a trust value by using the authentication protocol suitable for each device. The method that authenticates devices through trust values provides efficient authentication to a smart device, but the method often requires a high-level device to confirm the entire authentication or the smart authentication. If a middle-level device or a high-level device above the smart device is lost or located elsewhere, the entire authentication becomes impossible, thus requiring the redistribution of trust values to devices below that which was lost. The system is discussed in detail below.

1. The entire authentication information corresponds to the sum of trust values from device 0 to device N.
2. If a device is moved or lost, the entire authentication against devices below the lost device becomes impossible.

The JARM method can be described in detail as follows: In the ubiquitous computing environment, a user can be authenticated through various devices. A

user can be authenticated through one device, and little devices can be authenticated during multiple steps. During the multiple-step authentication process, authentication information is transmitted from higher-level devices to lower-level devices. The biggest concern in this process is how to trust devices. For instance, when a given password is used by a device, it is the choice of the device whether to trust the given password to authenticate trusted entities or not. Trust values can be transmitted to a device through its proper protocols. When a user wants to use one particular authentication method, trust values can be widely used. Examples for trust values in this method are shown below.
$C_{net} = 1 - (1 - C_1)(1 - C_2) \cdots (1 - C_n)$
C_{net} becomes here the trust value of the user. And $C_1 C_2 \cdots C_n$ becomes also a new appointment price of an each device. This method uses Kerberos, which was used as the authentication method for existing distributed systems. However, Kerberos has been adapted to suit the ubiquitous environment. Here, AD (Active Domain) means a domain for authentication, and is configured as Kerberos. This AD consists of three authentication components. The first component is AS (Authentication Server), which supports SSO within the active domain. The second component is TGS (Ticket-granting Server), which grants tickets that allow a user access to the active domain. The third component is the database, which stores all the information required for user authentication within the active domain.

3 Requirements for Ubiquitous Computing

With the advent of human-oriented ubiquitous computing, which is described as pervasive, or invisible computing, a user can concentrate on tasks without being aware that he is using computers. Despite the many benefits of the digital technology that ubiquitous computing utilizes, ubiquitous computing has unseen problems. Without addressing these problems, ubiquitous computing cannot be applied. Since a user uses many devices, user information can be copied in large volume and can be transmitted to unauthorized devices. This illegitimately collected user information can be used maliciously after changes on the network. These features and the environment of ubiquitous computing have allowed for a wide range of malicious attacks and uses, which are likely to become huge obstacles to the development of ubiquitous computing. Thus, to overcome these problems, the following requirements must be met when designing the ubiquitous computing system.

Mobility: A user's smart device that contains the authentication information must be mobile and be used for all services.

Entity Authentication: Even when a user with SM_A moves away from $Domain_A$, the user must be authenticated using the information of SM_A in $Domain_B$.

Corresponding Entity Authentication: When $Device_B$ is located in $Domain_A$, the corresponding entity authentication verifies that $Device_B$ and B are identical entities. This method implements the authentication for devices through the previous user's entity when several devices are connected to one domain. This method can provide a wide range of protection functions.

Data Outgoing Authentication: When the outgoing data authentication is provided by $Domain_A$, $Domain_A$ can confirm that $Device_A$ is the actual device in $Domain_B$ that requests the outgoing data authentication. This authentication method provides proof for the authentication data origin. However, this method does not provide protection for data duplication or alteration.

Connection/Non-connection Confidentiality: $Device_B$ in $Domain_A$ must provide connection confidentiality for the user data. $Domain_A$ receives B's information to obtain the final authentication from the higher-level device. Non-connection confidentiality means that $device_B$ must provide confidentiality for the user data prior to the connection to a specific domain.

4 Proposed Scheme

In the previous chapters, we have gone over the existing ubiquitous environment, JARM method, and PMI. Although many researches have been done regarding ubiquitous computing, the most active area of research is on communication rather than on security. Security is often researched only as part of the project, not as the main research topic. This paper has selected the JARM method as its research topic since the JARM method exclusively studies authentication in ubiquitous computing. In reviewing current existing researches, the researcher believes that several researches regarding security have been accomplished and published. At the time of research, the researcher discovered that ubiquitous computing must have mobility, entity authentication, corresponding entity authentication, outgoing data authentication, and connection/non-connection confidentiality as basic requirements. Thus, this paper proposes the adoption of PMI to meet the requirements listed above and to implement the authentication for the smart device. The ubiquitous computing devices lacked computing, storing, and other capabilities. But since a device must meet the requirements discussed earlier, applying PMI on top of the currently used encryption system will satisfy the device capabilities and requirements. Since all devices can carry out the authentication and access control with the PMI certificate, only activities authorized to the devices will be allowed. I will also propose a method that uses a PMI certificate for a device. The general system flow will be discussed after the consideration for the proposed method is reviewed.

4.1 Consideration for Proposed Scheme

The goal of the proposed method is to provide a device retaining the user information of the previous domain even when the device is moved into the multi domain. Thus, the following must be considered for the proposed method.

- A user device alone can be moved and this user device can be linked to other devices: This means that when a user's smart device is moved into the

multi domain, the smart device can be linked to devices in the multi domain to receive services. User information must therefore be extracted from the smart device, as other devices exist for services only.
- A user device is authenticated through the Hub where all device information is stored in the same space. This device is authenticated through the MDC (Multi Domain Center) when moved to a different space: When a user device is located in the user domain, the user device can be authenticated through the Hub, which connects all user spaces. When the user device moves away from the user space, the smart device can also be authenticated through the user's Hub. But when a smart device is moved, an authentication method other than the Hub must be used. That method is to use the MDC, which authenticates the smart device in the multi domain.
- Initial authentication information is granted to a smart device through a user's hub from MDC: During this process, an authorized user registers devices in the user's hub. If a user creates authentication information from the MDC for the first time, the authentication information is stored in the hub and to the smart device. At this point, higher-level MDC authenticates the smart device using the created authentication information given to the user.
- At this step, the privacy of the user location is not considered.

User registration and device registration must be done in advance in the single domain and the multi domain. These operations must be done in order for the initial MDC to grant a PMI certificate to a user. The next operation is to authenticate in the single and the multi domain. The authentication in the single domain is shown below. Let us assume that Bob has spaces where he can move around. Let us also assume that there are two active spaces for Bob: Bob1 and Bob2, and these two active spaces are located in $Domain_A$. When Bob's $Device_B$ moves away from Active Space 2 to Active Space 3, the single domain authentication occurs. Bob's SM_B in Active Space 2 notifies his Hub of movement, transmits the authentication information to $Device_B$ and the Hub, and requests the authentication from Active Space 3. At this point, Active Space 2 sends the SM_B authentication information to Active Space 3 for efficient authentication. Active Space 3 receives the request and compares the authentication information from Active Space 2 and the one from the Hub, and then authenticates SM_B in Active Space 3. All these steps complete the single domain authentication. In the multi-domain authentication, when user Bob moves to Alice's $Domain_A$, we can use Alice's $Device_A$ after getting SM_B authenticated. In this case, Bob's SM_B sends the movement signal to his Hub and requests the authentication after the move to Alice's Active Space. At this point, the authentication information is requested through Alice's Hub. Bob requests the authentication from MDC through Alice's Hub. If Bob's information does not exist in MDC, the authentication request is made against MDCM, which is a higher entity. In this way, although an entity is not systematically contained in or connected to the higher entity, the authentication using devices in other spaces can be completed through the Internet.

4.2 System Parameters

Next, system parameters used in this method are explained. Each parameter is distinguished according to its components. The components create and transmit the parameters.

- $*$: (SM: SMart device, D: Device, SD: Single Domain, MDC: Multi Domain Center, A: Alice, B: Bob, MDCM: MDC Manager, ASC: Active Space Center)
- $Cerr*$: public key of $*$ including Certification
- $PCert*$: public key of $*$ including PMI Certification
- n : PMI certification maximum issues number
- AP : Available Period
- i : user issued device
- pw : password
- $ID*$: $*$ of Identity
- $Hub*$: $*$ of Hub

r : user Hub generated random number
$E_*()$: $*$ key with Encryption
$R*$: $*$ of authority
$H()$: Secure Hash Function

4.3 Proposed Scheme

The detailed flow of these proposed methods is described below. In the first method, when a user moves to his domain with his smart device and attempts to use devices in the new domain, the user is authenticated using the smart device in which the user authentication information is stored. In the second method, when a user moves to the multi domain and attempts to use devices there, the user is authenticated using the smart device in which his user authentication information is stored.

User Registration and Device Registration. A user must have authentication information for all devices in the initial stage to use the devices in the single domain. A user receives a certificate from the MDC(Multi Domain Center), and his devices are granted a PMI(Privilege Management Infrastructure) certificate through a Hub. PMI certificates are granted according to the mutually agreed methods with the MDC. Granted PMI certificates are stored in a smart device.

Step 1. The following processes are required to create $Device_A$ authentication for User A. The MDC grants User A a certificate which allows User A to create n number of PMI certificates. A PMI certificate consists of the User A ID, privilege, and effective period of the certificate.
$MDC \rightarrow Hub_A : Cert_A[ID_A, R_A, n, AP]$

Step 2. User A grants PMI certificates to $Device_A$ and SM_A using the granted certificate. The certificate contains the path to a higher-level certificate.
$Hub_A \rightarrow SD_A\,(or Device) : PC_A = PCert_A[ID_A, H\,(Cert_A \| r)\,, i] \| AP$
$SD_A : E_{PK_D DC}[PC_A]$
$SD_A install : E_{pw(or PIN)}[E_{PK_D DC}[PC_A]]$

Step 3. User A informs the MDC of the certificate granted to his $Device_A$. Afterwards, User A's PMI certificate is used, and User A is authenticated using the PMI certificate path within SM_A.
$Hub_A \rightarrow MDC : E_{PK_D DC}[H\,(Cert_A \| r)\,, r, i]$

Authentication in the Multi Domain. When SM_A in $Domain_A$ moves to $Domain_B$ and uses User A's information to use $Domain_B$ and $Device_B$, SM_A uses User A's information as is.

Step 1. A movement signal is sent using $Device_A$ in $Domain_A$. If Hub_A receives the movement signal from SM_A, it removes itself from the space list.
$SD_A \rightarrow Hub_A : Signall\,(outgoing)$
$Hub_A : SD_DeviceList \rightarrow Delete[SD_A]$

Step 2. Hub_A notifies the MDC that it is moving out of $Domain_A$. If it moves to a different MDC, it notifies MDCM.
$Hub_A \rightarrow MDC : (ID_A, i)$
$MDC \rightarrow MDCM : (ID_A, i)$

Step 3. After notification that SM_A is finally located in $Domain_B$, it requests authentication from $Device_B$ in $Domain_B$.
$SD_A \rightarrow Hub_B : Signall\,(ongoing)$
$SD_A : E_pw[E_{PK_DDC}[PC_A]] = E_{PK_DDC}[PC_A]$
$SD_A \rightarrow Device_B : E_{PK_DDC}[PC_A]$
$Device_B \rightarrow Hub_B : E_{PK_{Hub_B}}[PC_B, E_{PK_DDC}[PC_A]]$

Step 4. Hub_B in $Domain_B$ verifies the authentication information from $Device_B$.
$Hub_B : E_{SK_{Hub_B}}[E_{PK_{Hub_B}}[PC_B, E_{PK_DDC}[PC_A]]] = PC_B, E_{PK_DDC}[PC_A]$
$Hub_B : PC'_B \doteq PC_B$

Step 5. If the $Device_B$ authentication information is passed, Hub_B transmits the authentication information to the MDC.
$Hub_B \rightarrow MDC : (ID_B, E_{PK_MDC}[E_{PK_DDC}[PC_A] \| ID_B])$

Step 6. The MDC verifies that the authentication information is generated from $Domain_B$ User. If confirmed, the MDC approves the authentication for SM_A.
$MDC : E_{PK_DDC}[PC_A] \| ID_B$
$MDC : PC'_A = PCert_A[ID_A, H\,(Cert_A \| r)\,, i]$

Step 7. In $Domain_B$, Hub_B accepts the received authentication for SM_A, and allows for the use of $Device_B$ in $Domain_B$.

4.4 Comparison with Proposed Scheme and JARM Scheme

This chapter will attempt to analyze the proposed protocol by classifying the user and device registration and the authentication in the multi domain, and compare the protocol to the existing method. In the existing method, SM_A is not authenticated by moving to AS1. Therefore, the research in this paper seeks to put emphasis on how to authenticate a user who wishes to use $Device_B$ by using user information A when SM_A is moved to $Domain_B$. The existing methods attempt to solve the problem by assigning different authentication information to the devices. But the weak point of this approach is to require all devices to be available when the entire authentication information is obtained. This method raises a problem in that authentication information cannot be obtained if a device is lost.

The proposed method and the existing method are compared below. The details of the proposed method will also be discussed below.

Mobility : A device containing authentication information that a user owns can use all services. In the proposed method, a PMI certificate is included in the device. Thus, the device can be moved away from the other devices, and can also be linked to a different device.

Entity Authentication : Although a smart device is moved away from its domain, the device in the multi domain can be authenticated using the previous user's information. Mobility is guaranteed because a user device uses its PMI certificate. However, if a certificate is not protected, it can be used by malicious users who can also be authenticated. To resolve this issue, the proposed method protects the PMI certificate by using device access and certificate access protection mechanisms such as the password and PIN.

Corresponding Entity Authentication : When a device is located in $Domain_A$, corresponding entity authentication is provided to verify that $Device_B$ and B are identical entities. This authentication method implements device authentication through the entity of the previous user when multiple devices are connected to one domain. This authentication can provide different levels of protection. Even when a smart device is moved, the authentication can be done using what is stored in the smart device. Also, a PMI certificate, which is an internal certificate and identical to the certificate from User A, can be used when performing the corresponding entity authentication.

Connection/Non-connection Confidentiality : $Device_B$ in $Domain_A$ must provide confidentiality for user data in both $Domain_A$ and $Domain_B$. $Domain_A$ receives information from B and receives the final authentication from the higher level. Non-connection confidentiality must provide data confidentiality before $Device_B$ connects to a specific domain.

Table 1. Comparison with Proposed Scheme and JARM Scheme (○ : offer, △ : part offer, × : non-offer)

item	Mobility	Entity Authentication	Corresponding Entity Authentication	Data Outgoing Authentication	Connection/Non-Connection Confidentiality
JARM Scheme	○	×	×	×	○
Proposed Scheme	○	○	○	△	○

4.5 Conclusion

Rapid expansion of the Internet has required a ubiquitous computing environment that can be accessed anytime anywhere. In this ubiquitous environment, a user ought to be given the same service regardless of connection type even though the user may not specify what he needs. Authenticated devices that connect user devices must be used regardless of location. If a device is moved to another user space from a previous user space, the authentication must be performed well in the transferred space. This is so because a device is not restricted to the previous authentication information, but can use new authentication information in the new space. This paper attempts to solve the problems discussed earlier by utilizing such entities as the Hub, ASC, and MDC in order to issue PMI certificates to devices that do not have computing capability. This provides higher-level devices the authentication information of the smart devices in order to authenticate the movement of these smart devices. With this proposed method, if a smart device requests the authentication after moving to the multi domain, the authentication is performed against the devices in the domain where the smart device belongs, and the smart device requests the authentication from the MDC. In the user domain, the authentication is performed through the Hub. However, the authentication is performed through MDC when a device is moved to the multi domain environment. This proposed method, therefore, attempts to solve the existing authentication problem. With regard to the topics of privacy protection, which is revealed due user movement key simplification (i.e., research on a key that can be used for a wide range of services), and the provision of smooth service for data requiring higher bandwidth, the researcher has reserved them for future researches.

References

1. A. Aresenault, S. Tuner, Internet X.509 Public Key Infrastructure, Internet Draft, 2000. 11
2. ITU-T, Draft ITU-T RECOMMANDATION X.509 version4, ITU-T Publications, 2001. 5.
3. Jalal Al-Muhtadi, Anand Ranganathan, Roy Campbell, and M. Dennis Mickunas,A Flexible, Privacy-Preserving Authentication Framework for Ubiquitous Computing Environments, ICDCSW '02, pp.771-776, 2002
4. Mark Weiser,"Hot Topics: Ubiquitous Computing," IEEE Computer, October 1993
5. M. Roman, and R. Campbell, GAIA: Enabling Active Spaces, 9th ACM SIGOPS European Workshop, September 17th-20th, 2000, Kolding, Denmark
6. S. Farrell, R. Housley, An Internet Attribute Certificate Profile for Authorization, Internet Draft, 2001.
7. anjay E. Sarma, Stephen A. Weis and Saniel W. Daniel, White Paper:RFID Systems, Security and Privacy Implications, AUTO-ID Center, MIT, Nov, 2002
8. Gen-Ho, Lee, Information Security for Ubiquitous Computing Environment, Symposium on Information Security 2003, KOREA, pp 629-651, 2003
9. Sung-Yong Lee and Hyun-Su Jung, Ubiquitous Research Trend and Future Works, Wolrdwide IT Vol. 3, No. 7, pp 1-12, 2002
10. Yun-Chol Lee, "Home Networks Technology and Market Trend", ITFIND Weeks Technology Trend(TIS-03-20) No. 1098, pp22-33, 2003
11. Aura Project home page. http://www-2.cs.cmu.edu/ aura/
12. CoolTown home page. http://www.cooltown.hp.com
13. Portolano home page. http://portolano.cs.washington.edu/
14. TRON Project home page. http://www.tron.org/index-e.html

Proxy-Based Service Discovery and Network Selection in 6LoWPAN

Shafique Ahmad Chaudhry, Won Do Jung,
Ali Hammad Akbar, and Ki-Hyung Kim*

Graduate School of Information and Communication
Ajou University, Suwon, Korea
{shafique, yarang, hammad, kkim86}@ajou.ac.kr

Abstract. Low Power Wireless Personal Area Networks (LoWPANs) have emerged as a catalyst technology for the realization of envisioned ubiquitous paragon. Considerable efforts are being carried on to integrate these LoWPANs with IP-based networks in order to make use of pervasive nature and existing infrastructure associated with IP-based technology. Provisioning of service discovery and network selection in such environments puts heavy communication and processing overhead. The access to closest services localizes the communication and increases the total network capacity. We introduce directory proxy agents to be deployed within LoWPANs in order to localize the service discovery communication. We also propose algorithms to make sure that service users are always connected to the closest proxy agent. The results show that our algorithms help finding the closest services and reduce the traffic overhead for service discovery considerably in LoWPANs.

1 Introduction

Low Power Wireless Personal Area Networks (LoWPANs) conform to the IEEE 802.15.4-2003 standard [1], which defines transmission and reception on the physical radio channel (PHY), the channel access, PAN maintenance, and reliable data transport (MAC). The IEEE 802.15.4 devices are characterized by low power, low bandwidth, short range, and low cost. While IEEE 802.15.4 provides specifications for physical and link layers, other alliances like ZigBee [2] are striving for defining the upper layers over IEEE 802.15.4 especially for sensor networks. The 6LoWPAN [3], a working group of the IETF [4], standardizes the use of IPv6 over IEEE 802.15.4. The motivation for IP connectivity, in fact, is manifold: a) The pervasive nature of IP networks allows use of existing infrastructure, b) IP based technologies, along with their diagnostics, management and commissioning tools, already exist, and are proven to be working, and c) IP based devices can more easily be connected to other IP networks, without the need for translation gateways etc.

Interworking of 6LoWPANs with IP networks brings in many challenges for service discovery and network selection. The direct deployment of IP-based service

* Corresponding author.

M. Gerndt and D. Kranzlmüller (Eds.): HPCC 2006, LNCS 4208, pp. 525 – 534, 2006.

discovery mechanisms on 6LoWPANs is not practical because of the drastic technological differences between both the technologies. The difference between packet sizes of 6LoWPAN and IPv6 is one of them; given that the maximum transmission unit for IPv6 is at least 1280 octets and it, therefore, cannot be mapped onto IEEE 802.15.4 frame which has 127 octets at physical layer. An IP-based service discovery mechanism, like Service Location Protocol (SLP) [5], message could easily be greater than available octets for application layer in IEEE 802.15.4. It means that a single message will be transmitted as multiple packets; therefore, causing more traffic load for bandwidth constrained 6LoWPANs.

The inherently broadcast based distributed service discovery mechanisms could overload the 6LoWPANs with service discovery overhead, especially in a network with a large number of nodes. The centralized service discovery mechanisms generate less service discovery traffic but need dedicated service coordinators, which are seldom available in the 6LoWPANs. This situation demands for an intuitive solution which can localize the service discovery communication without the use of dedicated service coordinators. This localization can be done by providing access to services which are closest to the user. This proximity-based services access localizes the communication, reduces the service discovery overhead and consequently increases the total network capacity.

Simple Service Location Protocol (SSLP) [6] has been proposed to provide service discovery mechanism in 6LoWPAN. It supports both the distributed as centralized service discovery models but does not address the issues related to the proximity of a service.

We propose a service discovery architecture that uses proxy agents for the directory agent in SSLP to localize the service discovery communication and at the same time help users to access the closer services. The introduction of proxy-agents softens the requirement of having a dedicated directory agent and at the same time localizes the service discovery communication. Directory Proxy Agents (DPAs) are inexpensive LoWPAN devices which act as proxy to the DAs in SSLP. These DPAs maintain and provide service information with the 6LoWPAN and IP-based networks. The connectivity between 6LoWPAN and IPv6-based networks enables the users to find and use local 6LoWPAN services as well as the services available in external IPv6-based networks. Our simulation results show that our architecture not only helps finding and using the closest services in inter-network environment but also considerably reduces the traffic overhead, as compared to other protocols, for service discovery.

The rest of the paper is as follows. Section 2 describes the service discovery architecture models and SSLP. In section 3, we review existing service discovery protocols and related work with an emphasis on 6LoWPANs. Interoperability of 6LoWPAN with IPv6 is thoroughly discussed in section 4. We describe our proposed service discovery architecture in section 5. In section 6 we present performance and evaluation of our scheme and section 7 concludes the paper.

2 Service Discovery Architectures

Service discovery architectures can be categorized in two broad classes i.e., centralized or directory-based model and distributed or non-directory-based model.

In the directory-based model a dedicated component, known as service-coordinator or Dirctory Agent (DA) maintains service information and process queries and announcements. The nodes, which offer a service, register their services with the DA by sending a unicast Service Registration (*SREG*) message to the DA. The clients, who need a service, search the service with the service coordinator by sending a unicast service request (*SREQ*) message. In case DA has a matching service registered with itself, it replies with a unicast service reply message, which also contains the address of service provider node. DA periodically broadcasts an advertisement message to notify other nodes about its existence. The directory-based architectures generate less service discovery overhead and are suitable for the environments with large number of services and network nodes.

The service discovery overhead has two major sources: a) Number of hops from the source node to the destination node, b) Number of neighbors of any node. Both these factors show more significance, when broadcast is used within the service discovery process. In a directory-based architecture most of the communication, except DA's advertisement broadcast, is unicast. The service coordinator's broadcast overhead can also be reduced by connecting all the nodes to their closest DAs.

The non-directory based architecture does not have any DA. The client can find a service using active or passive discovery mechanism. In the active discovery process, the clients broadcast or multicast the *SREQ*. Each active neighbor of the client receives the *SREQ*, sends a unicast service reply *SREP* if it offers the required service, and rebroadcasts the *SREQ*. Higher the number of neighbors, higher is the degree of broadcast. The number of *SREQ* broadcasts also depends on the scope of service discovery. In passive discovery mechanism, the servers broadcast their service advertisements periodically and clients compare these advertisements with their service needs. The non-directory based models show better performance in simple environments with few services.

3 Related Work

Service discovery is an actively researched area and has been studied largely in the context of wireless and mobile ad-hoc network systems. Universal Plug and Play (UPnP) [7], Jini [8], and Bluetooth [9] are some examples of existing industry standards. Each one of these architectures addresses different mixture of attributes, but most are mainly designed for traditional IP-based home or enterprise networks [10]. The resource and communication environment in LoWPANs is very different than traditional IP network making these architectures non-applicable to them.

The service discovery problem in sensor networks is different from traditional IP network in various respects. For sensor networks, their data centric embedded nature, relatively poor resources, emphasis on energy efficiency, and the lack of existing standards combine to render traditional resource discovery approaches ineffective. Dynamic Resource Discovery for Wireless Sensor Networks [11] defines the resource discovery problem in sensor networks and outlines the challenges involved in it. A peer to peer (P2P) service discovery architecture is proposed in [12]. Authors have proposed a distributed architecture which relies on dynamically identified location servers which provide a mapping between services and their locations. This work creates an overlay over the sensor networks, the implementation of which over 6LoWPAN is impractical due to the resource limitations.

After describing all these works, we state that currently there is no considerable service discovery architecture for 6LoWPANs except SSLP. SSLP supports both directory-based and non-directory-based discovery models. In SSLP, the client applications are called User Agents (UA), the nodes which advertise services are called Service Agent (SA) and service coordinator is called Directory Agent (DA). SSLP also offers interoperability with IP networks under SLP by introducing a translation agent (TA) which provides the translation of messages from SSLP to SLP and vice versa. Fig.1. shows this interoperability between SSLP and SLP. SSLP does not provide any mechanism to localize the service discovery communication by connecting to the nearest DA or by accessing the closest services.

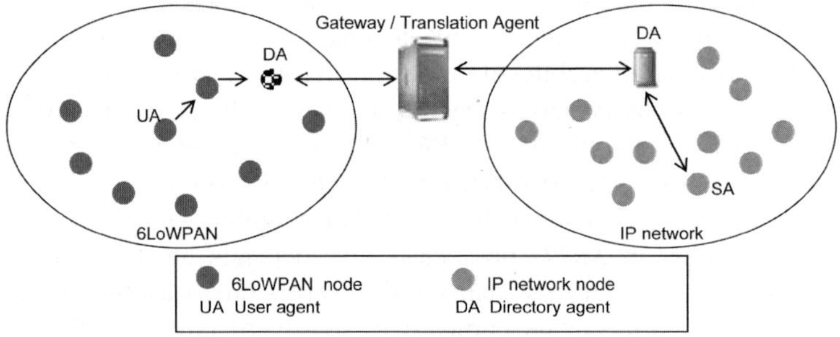

Fig. 1. Interworking of SSLP with SLP

4 Interoperability Between 6LoWPAN and IPv6

In this section we shall describe the possible scenarios for service discovery and network selection for a UA working in a 6LoWPAN, using interworked LoWPANs and IPv6 networks. We assume 6LoWPAN supports SSLP for service discovery whereas IP network supports SLPv2. To integrate a 6LoWPAN with IPv6 network a gateway is needed which connects both these networks. When a UA sends a *SREQ* for a service, if the service could be found in local network the SREP is sent locally, otherwise the SREQ is forwarded to the gateway in order to find a match from external networks. There are three possible scenarios to process a *SREQ*, generated by a UA in 6LoWPAN:

- **Both the networks support non-directory based architecture:** As the whole discovery mechanism is based on broadcast, huge amount of traffic is generated that puts heavy overhead on the network. This overhead is highly intolerable in 6LoWPANs, which already have low data rates.

- **Only one network has directory-based architecture:** This situation is burdensome as there is no direct route setup between the UA and SA which could exist in different networks. Moreover, an overhead of going through gateway is always involved.
- **Both the architectures support directory-based architecture:** The whole communication between two nodes in different networks can be done through gateway. This approach needs a dedicated node with sufficient resources to act as a DA for the 6LoWPAN. Unfortunately, 6LoWPAN nodes are characterized with limited resources, thus, lack the capability to work as a dedicated DA.

After discussing all the apparent scenarios with their respective advantages and disadvantages, we insist there is a need of a better architecture for service discovery and selection in 6LoWPANs. The resource limited nature of 6LoWPANs demands a service discovery architecture which can offer a trade-off between pure broadcast and putting dedicated DAs within a 6LoWPAN.

5 Proxy-Based Service Discovery

Considering the resource limitations with 6LoWPAN, we introduce Directory Proxy Agent (DPA) to be deployed in 6LoWPAN rather than placing a dedicated DA for each 6LoWPAN. The DPA acts as a proxy for the DA, localizes the service discovery communication, and resultantly, reduces the service discovery overhead.

We make use of the fact that 6LoWPAN nodes are inexpensive by deploying multiple DPAs, each responsible for a certain area, within a 6LoWPAN. All the DPAs maintain Local Cache (LC), for services in its local proximity, as well as External Network Cache (ENC) which represents the services with the neighbor IEEE 802.15.4 and IPv6 networks.

As shown in figure 2, three 6LoWPANs are connected to the external IPv6 network through a gateway. The scope of each 6LoWPAN is limited to a certain proximity and service directory services are offered by the DPA in that specified region. These DPAs could be arranged in a hierarchical way i.e. they can communicate with their peer DPAs as well as with the central DAs which might be the part of external IPv6-based network. All the SAs within the 6LoWPAN register themselves with the DPA. The DPA periodically broadcasts an advertisement message to notify its existence. As the DPA are deployed at specific locations, the service information maintained by them is also proximity-based.

Our architecture is independent of the underlying routing algorithms and can be implemented on any routing algorithm with minimal changes. We have evaluated its performance with AODV, however, we believe that availability of a hierarchical routing mechanism e.g. HiLoW [13] as underlying routing algorithm, will further improve its performance. The major strength of using HiLoW is that if 16-bit address of the service or destination node is known, it can be reached without using a routing table. Once a UA knows the 16-bit address of the DPA or SA, it can start communicating with that, without finding a route to the destination node.

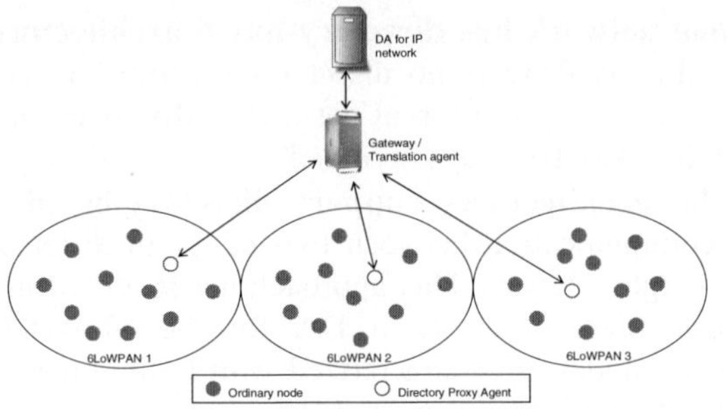

Fig. 2. DPA based service discovery architecture

Zigbee also supports hierarchical routing, especially in tree networks, routers move data and control messages through the network using a hierarchical routing strategy. In the same way a node without routing capacity can send the messages along the tree using hierarchical routing.

DPAs share their caching information with each other periodically. This sharing of information allows knowing about the services registered with the neighbor DPAs. This periodic sharing of information reduces flooding which, otherwise, will be required to find services from the neighboring networks. The connectivity of a DPA with the DA of IP network through the gateway facilitates to find services from IP networks. DPAs may exchange the services information with central DA as well, in order to maintain the information consistency.

5.1 Proximity-Based Service Access

Whenever a node wants to join a 6LoWPAN, it first tries to discover an existing 6LoWPAN. IEEE 802.15.4 specifies active and passive scanning procedures for this discovery operation. By following either one of the scanning procedures, the new device determines whether there is a 6LoWPAN in its personal operating space. Once a PAN is found, next step is to connect with the DPA. After getting the address of the DPA, the UA must find a route to DPA if an on-demand routing algorithm like AODV is being used. In case a hierarchical routing algorithm being used, knowing the address of DPA makes this UA capable of communicating with DPA. As the hierarchical routing is available, there is no need to explicitly find and maintain routes between the communicating nodes. If 16-bit short address of a node within 6LoWPAN is known, the path can be traversed by underlying routing mechanism. Though hierarchical routing algorithms eliminate the route finding process, they do not provide optimal routing path.

To access the closest services we make it essential that a UA is connected to the nearest DPA. This condition is required to ensure that the service request is sent to the closest DPA, which then replies with the nearest service's information. To realize this condition we propose neighbor assisted DPA discovery protocol. It uses DPA Discovery Request (*DDREQ*) and DPA Discovery Reply (*DDREP*) messages. The messages are used as follows:

- *DDREQ*: The request message is used to ask the neighbors about their respective DPAs. Initiated as a one hop-broadcast by the UA that needs to find the closest DPA.
- *DDREP*: This is the reply message in response to a *DDREQ***.** It contains the address of the DPA as well as distance to the DPA in terms of hop count.

The protocol works as follows. Whenever a UA needs to send a request to DPA, it checks with its single hop neighbors, by broadcasting *DDREQ* in one hop, the closest DPA in terms of hop count. The neighbors reply with *DDREP* that contains the address of closest DPA and distance to it in hop count. The nearest DPA, in terms of hop count, is considered as the closest DPA. Once the address of DPA is known, hierarchical routing makes it possible to send unicast messages between the UA and the DPA. Whenever a UA needs a service, it sends a unicast *SREQ* to the DPA. If the required service is registered with DPA, DPA responds with an *SREP*. UA then starts communicating with SA to use the service. Neighbor assisted DPA discovery algorithm helps in handling mobility of the UA as well. This protocol makes sure that the UA stays connected with the existing DPA as long as it is the nearest one, even when UA is moving. Fig. 3 illustrates the protocol.

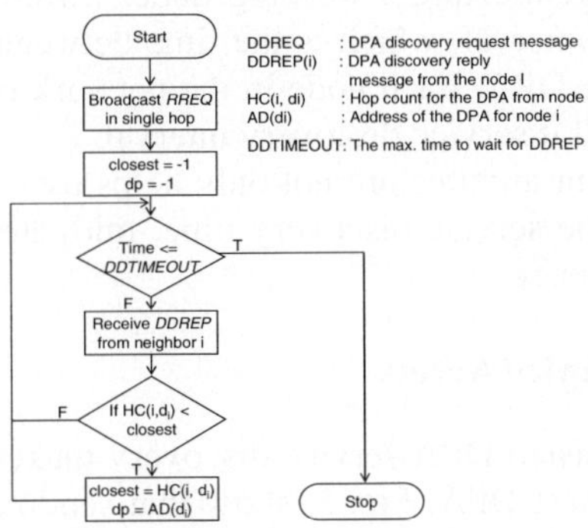

Fig. 3. Neighbor assisted discovery protocol

Legend: Findmatch(SREQ$_i$,LC) : The function which finds and returns a list of SAs which meet the requirements against SREQ i
list : The list of SAs providing the required service
HC : Hop count

```
begin procedure
    list ← Findmatch (SREQi,LC)
    best ← Findbestmatch(list)
    send a SREP with best as SA to UA which initiated SREQi
end procedure

begin procedure Findbestmatch (list)
    best ← first entry in the list
    for all other entries in list
        if (HC for current entry < HC for best)
            best ← current entry
        end if
    next
end procedure
```

Fig. 4. Algorithm to find the closest service

When a DPA receives a SREQ, it finds a match against user requirements, from LC. In case there are multiple matches the one with the minimum hop count is selected. However, in the situation when two services are at equal distance, the service with better performance is chosen. If the LC hit fails, the service is searched from the ENC and the SREP, for the closest SA, is sent to the UA. The algorithm is shown in fig.4.

6 Performance Evaluation

We have implemented our protocol in network simulator-2 (NS-2) by modifying the AODV implementation. We have modified AODV to evaluate our propagation schemes using ns-2. The simulation setup consists of 165 nodes, with a transmission range of 15 meters, spread over an area of 380m × 60m. Every simulation run is for 100 seconds, with Constant Bit Rate (CBR) being the traffic type. Inter-packet transmission delay varies between 0.05 and 0.5 seconds.

We examined various metrics including percentage of closest services found, cache effect, number of control packets generated, remaining energy and service discovery time. We evaluated our architecture's working under different scenarios by varying the DPA's advertisement interval, which is the time between two consecutive advertisement broadcasts by the DPA. Each node in the network tries to discover a service after a certain time; we call it service discovery interval.

The results show that our architecture not only helps user to access the closest services but also improves the service discovery time, mitigates the broadcasting overhead, saving the nodes' energy.

6.1 Proximity-Based Service Access

The usage of neighbor assisted DPA service discovery makes sure that UA is always making request to the closest DPA. Fig. 5. show that sending a SREQ to the nearest DPA considerably improves the access to the services which are in closest proximity. In the absence of proximity information, as in SSLP, the DA returns the IP address of a service which may not be the closest to the user. If the service is not registered with the DPA, the caching information will be used and UA is replied with the address of the service which is registered with the closest neighboring DPA.

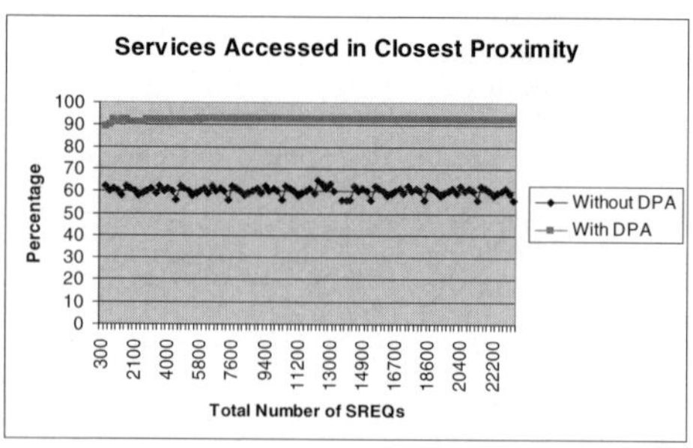

Fig. 5. Services accessed in closest proximity

6.2 DPA's Cache Effect

The role of cache is very essential and critical in the whole scenario. It is very important to find the optimal size of cache as well as the refresh rate. We tried different sizes and refresh rates. Fig. 6. shows that using the cache not only the local PAN services are accessible through DPA but also great fraction of services available in neighbors 6LoWPANs as well as with IP networks.

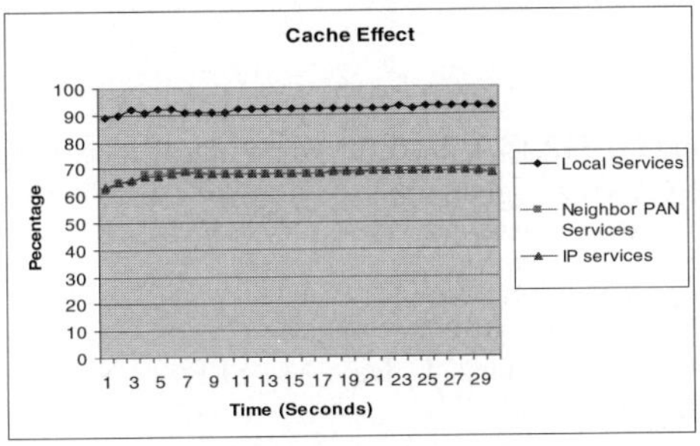

Fig. 6. DPA's cache effect

6.3 Number of Generated Control Packets

We define control packets as sum of total number of *SREQ*, *SREP* and DPA advertisement (DADV) messages. We varied the service discovery interval to examine its effect on control traffic. The results show that AODV generated more control traffic as compared to the situations where a DPA is available. This is mainly because of the fact that when service discovery interval is increased, the existing entries for the SAs are no more valid and services are searched again. Fig. 7 shows that the number of control packet generated, when DPA advertisement was sent every 5 seconds, is less than that of original AODV. The main reason is that when DPA advertises itself, the path to DPA is maintained and no additional message passing is needed.

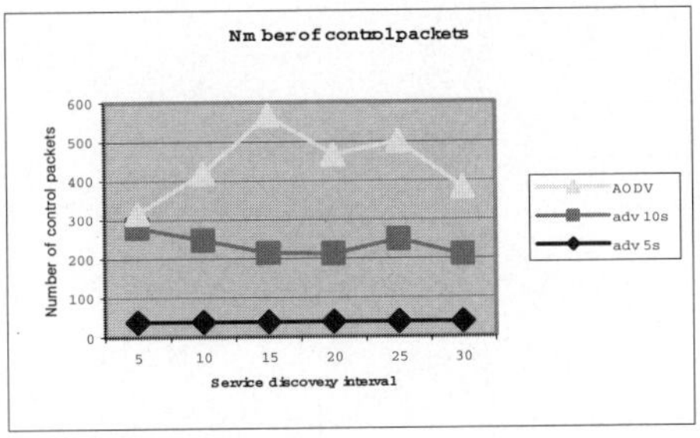

Fig. 7. Number of control packets

7 Conclusion

We introduce Directory Proxy Agents to be used in SSLP in order to relax the need of a dedicated DA. We propose algorithms to find the services which are closest to the user. The access to closest DPA and closest services minimize the service discovery overhead, and improves network capacity. The integration of 6LoWPAN with IPv6 means the services are made available from anywhere, yet closest to the user.

References

1. IEEE LoWPAN Standard 802.15.4-2003 http://standards.ieee.org/getieee802/-802.15.html
2. ZigBee Alliance http://www.zigbee.org
3. IPv6 over Low Power WPAN Working Group http://www.ietf.org/html-.charters/6lowpan-charter.html
4. Internet Engineering Task Force http://www.ietf.org/
5. Guttman, E., Perkins, C., Veizades, J., Day, M.: SLPv2 : Service Location Protocol, Version 2. RFC 2608, Jun. 1999
6. Kim, K., Yoo, S., Kim, H., Park, S.D., Lee, J.: Draft-daniel-6lowpan "Simple Service location Protocol (SSLP) for LoWPAN. draft-daniel-6lowpan-sslp-00 (work in progress) Jul. 9, 2005
7. Universal Plug and Play www.upnp.org/
8. www.jini.org/
9. http://www.bluetooth.com
10. Zhu, F., Mutka, W.M., Ni, L.M.: Service Discovery in Pervasive Computing Environments. IEEE Pervasive computing, (2005) page(s) 81-90
11. Tilak, S., Chiu, K., Abu-Ghazaleh, N.B., Fountain, T.: "Dynamic Resource Discovery for Wireless Sensor Networks" IFIP-NCUS (2005)
12. Sethom, K., Afifi, H.: A New Service Discovery Architecture for Sensor Networks. Wireless Telecommunications Symposium, WTS, California, (2005)
13. Kim K., Yoo, S., Park, J., Park, S.D., Lee, J.: Hierarchical Routing over 6LoWPAN (HiLow). draft-daniel-6lowpan-hilow-hierarchical-routing-00.txt (Work In Progress), (2005)

A Low-Power Hybrid ARQ Scheme for the RFID System*

Inwhee Joe

College of Information and Communications
Hanyang University
Seoul, Korea
iwjoe@hanyang.ac.kr

Abstract. The RFID (Radio-Frequency Identification) system consists of RFID tags and readers for contactless identification. Since RFID tags rely on limited energy, low-power design is very important for the RFID system. Also, the error rates tend to be high, because the link between the reader and the RF tag is wireless and RFID tags and readers could have mobility. To solve these two problems at the same time, we propose a hybrid ARQ (Automatic Repeat Request) scheme with energy efficiency for the RFID system. In our proposed scheme, the BCH channel coding is used together with the optimal packet length in order to minimize the energy consumption. From the results of performance evaluation, the proposed hybrid ARQ scheme can maximize energy efficiency compared to the ARQ scheme or channel coding alone, while satisfying the link reliability.

1 Introduction

The RFID system is a part of AIDC (Automatic Identification and Data Capture) technology, which enables to read data by RF communication without any contacts stored in tags, labels and cards with micro-chips built in them. The RFID technology is expected to replace the current bar-code system. Now, it is being discussed in the ISO (International Standard Organization) standard specifications [4], where the physical and logical requirements are defined for a passive RFID system operating in the 860 MHz - 960 MHz frequency range. The RFID system comprises RFID tags and readers. A reader transmits information to an RFID tag by modulating an RF signal. The tag receives both information and operating energy from this RF signal.

A low-power design is required for the RFID system because RFID tags rely on limited energy sources. Also, the error rates tend to be high, because the link between the reader and the RF tag is wireless and RFID tags or readers could

* This work was supported in part by grant No. IITA-2005-C1090-0501-0022 from the ITRC Support Program of the Ministry of Information and Communication, and in part by grant No. KRF-2005-003-D00185 from the Basic Research Promotion Fund of Korea Research Foundation.

M. Gerndt and D. Kranzlmüller (Eds.): HPCC 2006, LNCS 4208, pp. 535–541, 2006.

have mobility. According to the RFID standard, the CRC (Cyclic Redundancy Check) method is used for error detection and a data frame is retransmitted at the MAC (Medium Access Control) layer using the ARQ scheme, if there is any transmission error between the reader and the RFID tag. Since this ARQ scheme does not consider energy efficiency, it is not satisfactory for the RFID system.

In this paper, we propose a novel low-power hybrid ARQ scheme to improve the energy efficiency and link reliability at the same time for the RFID system operating in the 900 MHz range, by combining the energy-efficient BCH code and the ARQ scheme. In Section 2, we describe the proposed low-power hybrid ARQ scheme. In Section 3, we evaluate the performance of our scheme by mathematical analysis and computer simulation. Finally, we conclude the paper.

2 Low-Power Hybrid ARQ Scheme

Even if the ARQ scheme provides high reliability in the wired channel, the throughput drops rapidly in the wireless environment due to the increased frequency of retransmission. To counter this effect, a hybrid ARQ scheme is introduced by combining FEC (Forward Error Control) with ARQ schemes. However, when the hybrid ARQ scheme is applied to the RFID system, energy efficiency should be considered at the same time [2,3]. Here, We propose a low-power hybrid ARQ scheme using the BCH code and the optimal packet length. The BCH code is used because it consumes less decoding energy compared to other FEC schemes [1]. Also, the optimal packet length is used in terms of maximizing the energy efficiency [1]. Fig.1 shows the main process in our low-power hybrid ARQ scheme.

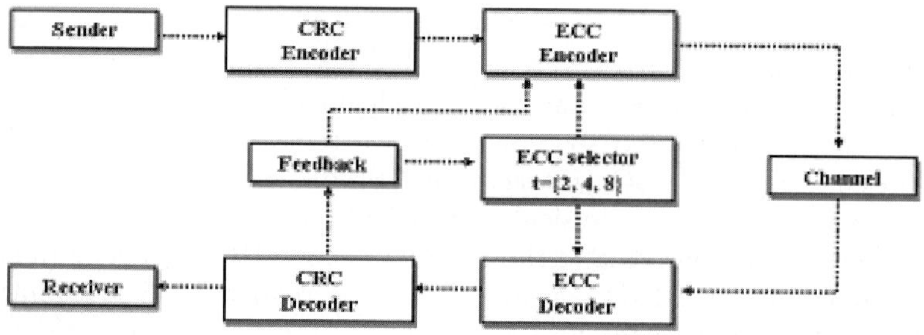

Fig. 1. Low-Power Hybrid ARQ Scheme

When data is transmitted first, it is encoded using the BCH code with the initial error correcting capability t of 2. At the receiver, the FEC portion in the proposed hybrid ARQ scheme first attempts to correct the transmission errors. If an uncorrectable error pattern is detected, then the receiver asks for a retransmission using ARQ. In this case, the data is encoded using the BCH code with the increased value of $t = 4$. After that, if there is any transmission error again, the error correcting capability t is increased to 8 for the last retransmission. The total transmission number is limited to 3, because the energy throughput drops rapidly according to the transmission number, which will be explained below.

If the error correcting capability t of the BCH code is 2, the packet length with the highest energy efficiency is 700bits, which is the optimal packet length in the sense of maximizing the energy efficiency. However, the closest value that can be provided by the BCH generator is 1023 bits, not the exact value of 700bits. When the BCH code is used with the error correcting capability of t, the packet error rate (PER) is given by

$$P_B = \sum_{i=t+1}^{n} \binom{n}{i} p_c^i (1 - p_c)^{n-i}, \tag{1}$$

where p_c is the channel BER.

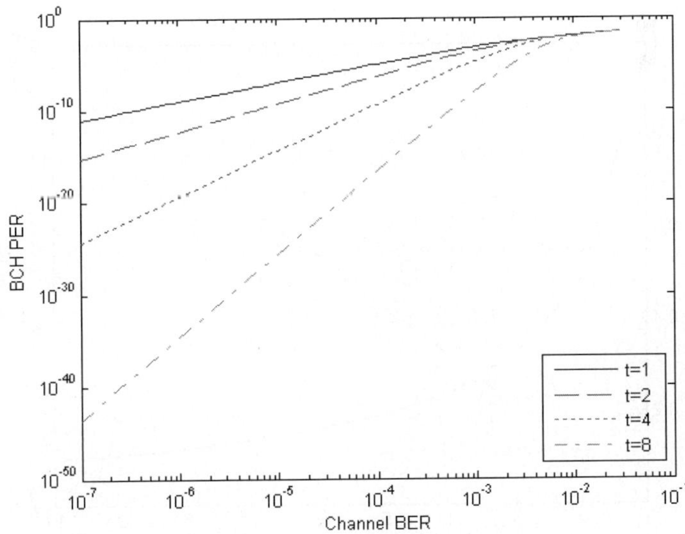

Fig. 2. Packet Error Rate vs. Channel BER as a function of BCH Code (t)

In the RFID system operating in the 900MHz, the radio range of the reader is up to 10m $\sim$ 20m, and the corresponding channel bit error rate (BER) is about $10^{-4} \sim 10^{-3}$ in this case. Evaluated from Eq. (1), the PER with the BCH code is shown in Fig. 2, if the optimal packet length is applied. In particular, when the channel BER is about $10^{-4} \sim 10^{-3}$ and $t = 2$, the PER with the BCH code is nearly 10^{-9}, which is low enough to meet the link reliability as an initial transmission condition. Therefore, the initial value of the BCH error correcting capability t is set to 2 from the analytical results above.

$$ETh_m = \frac{k_1(n - a - t)}{m(k_1 n + E_{dec} + k_2)} \tag{2}$$

$$Rel_m = (1 - P_B)P_B^{m-1} \tag{3}$$

k_1:Communication energy consumption ($k_1 = 1.8x10^{-6}$)
k_2: Start-up energy consumption ($k_2 = 24.86x10^{-6}$)
E_{dec}: Decoding energy (E_{dec}=0.18μm)
n: Packet length
a: Header length (a= 16bits)

Eqs. (2) & (3) present the energy throughput and reliability as a function of the transmission number m. For example, if the transmission number is m, it means that there are $m-1$ retransmissions plus one successful transmission. The evaluated results from Eqs. (2) & (3) are shown in Fig. 3 for the BCH code with t=2 and the channel BER 10^{-4}. The results present that the more transmission number, the more reliability is increased gradually, while the energy throughput is decreased dramatically.

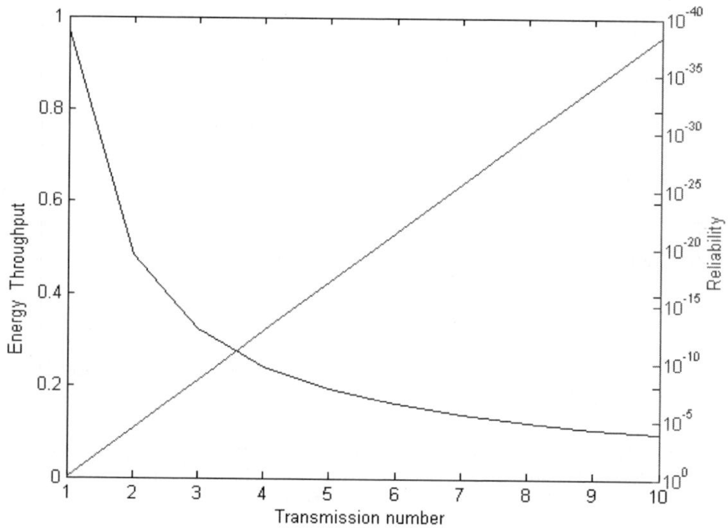

Fig. 3. Energy Throughput and Reliability as a function of Transmission Number

Based on the results, the retransmission should be minimized, because the energy throughput drops rapidly as the transmission number increases. Therefore, the total transmission number is limited to 3, as long as the link reliability is also concerned at the same time. The initial value of the BCH error correcting capability t is set to 2. If there is a transmission error, a data frame will be retransmitted with the increased value of t = 4 to improve the link reliability compared to the first transmission. Further, if there is an error again, the error correcting capability t of the BCH code is increased to 8, and the last data transmission will be performed.

3 Performance Analysis and Simulation

In this section, we compare the performance of the proposed hybrid ARQ scheme with the ARQ scheme or BCH coding alone. First, the energy efficiency is defined as:

$$\eta = EnergyThoughput \cdot Reliability \tag{4}$$

With this definition, the energy efficiency of the proposed hybrid ARQ scheme can be derived as a function of the channel BER to compare with that of the ARQ scheme or the BCH coding alone as follows:

$$\eta_{Hybrid} = \frac{k_1(n-a-t)}{m(k_1 n + E_{dec} + k_2)} \cdot (1 - p_{t2} \cdot p_{t4} \cdot p_{t8}) \tag{5}$$

$$\eta_{FEC} = \frac{k_1(n-a-t)}{m(k_1 n + E_{dec} + k_2)} \cdot \sum_{i=0}^{t} \binom{n}{i} p_c^i (1-p_c)^{n-i} \tag{6}$$

$$\eta_{ARQ} = \frac{k_1(n-a-t)}{m(k_1 n + E_{dec} + k_2)} \cdot (1 - p^3), \tag{7}$$

where $p = 1 - (1 - p_c)^n$.

Using Eqs. (5,6) and (7), the performance results can be plotted as shown in Fig. 4. Whereas the energy efficiency of the ARQ scheme alone is substantially lower than other two schemes, the energy efficiency of the proposed hybrid ARQ scheme is almost the same as that of the BCH coding alone until the channel BER is around 10^{-2}. However, when the channel BER becomes worse than that,

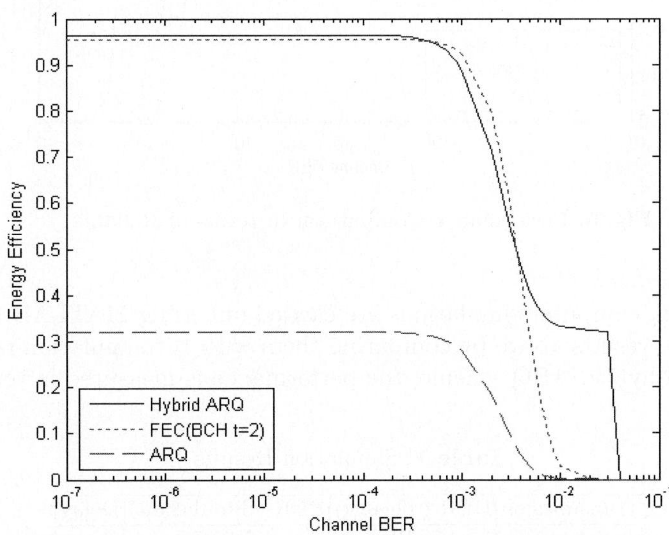

Fig. 4. Performance Comparison in terms of Energy Efficiency

the proposed hybrid ARQ scheme can improve the energy efficiency significantly compared to the BCH coding alone.

On the other hand, Fig. 5 compares the link reliability between the proposed hybrid ARQ scheme and other two schemes. Our hybrid ARQ scheme provides the most reliability all the time over the entire range of the channel BER. In particular, when the channel BER is $10^{-4} \sim 10^{-3}$ for the RFID system operating in the 900 MHz, the reliability of the proposed hybrid ARQ scheme is higher than any other schemes, even if the energy efficiency between the proposed hybrid ARQ scheme and the BCH coding alone is very similar. However, the proposed scheme is the only approach that can provide a reasonable energy efficiency even for the worst channel conditions. In summary, we can conclude that the proposed hybrid ARQ scheme is the best solution for maximizing the energy efficiency and the link reliability at the same time.

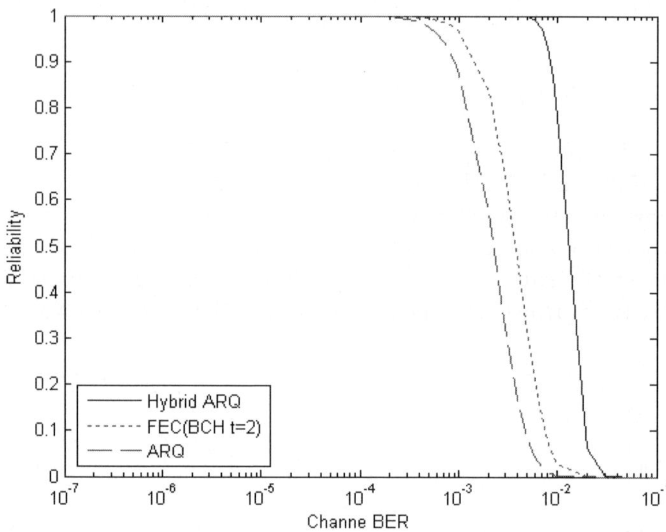

Fig. 5. Performance Comparison in terms of Reliability

In addition, computer simulations are carried out using MATLAB to confirm the analytical results above by comparing them with the simulation results. For the proposed hybrid ARQ scheme, the performance is measured in terms of the

Table 1. Simulation Results

Transmission	BER (Theory)	BER (Simulation)	Delay
1(t=2)	4.5x10^{-5}	4.7x10^{-5}	26ms
2(t=4)	8.7x10^{-7}	9.1x10^{-7}	51ms
3(t=8)	3.0x10^{-11}	0	77ms

BER error performance and the delay as a function of the transmission number for the packet length of 1023bits and the channel BER 10^{-4}. As shown in Table 1, we can find out that the simulation results agree with the analytical results.

4 Conclusions

For the RFID system, energy efficiency and link reliability are main design factors that should be considered. According to the RFID standard, the CRC method is used for error detection and a data frame is just retransmitted at the MAC layer regardless of energy efficiency, if there is any transmission error between the reader and the RF tag. In this paper, we have proposed a low-power hybrid ARQ scheme to deal with this issue. To improve the energy efficiency and reliability at the same time, the energy-efficient BCH code is combined with the ARQ scheme, leading to a novel hybrid ARQ scheme for the RFID system. Moreover, the optimal packet length is used for the energy efficiency. From the results of performance evaluation by analysis and simulation, we have shown that the proposed hybrid ARQ scheme can maximize energy efficiency compared to the ARQ scheme or channel coding alone, while satisfying the link reliability.

References

1. Inwhee Joe, "Optimal Packet Length with Energy Efficiency for Wireless Sensor Networks," Proceedings of IEEE International Symposium on Circuits and Systems (ISCAS), pp. 2955-2957, May 2005.
2. Z. SUN and X. JIA, "Energy Efficient Hybrid ARQ Scheme under Error Constraints," Wireless Personal Communications, Vol. 25, pp. 307-320, July 2003.
3. K. Sohrabi, J. Gao, V. Ailawadhi, and G.J. Pottie, "Protocols for Self-Organization of a Wireless Sensor Network," IEEE Personal Communications Magazine, Vol. 7, No. 5, pp. 16-27, October 2000.
4. ISO/IEC FDIS 18000-6, Radio-Frequency Identification for Item Management: Parameters for Air Interface Communications at 860 MHz to 960 MHz, July 2004.
5. Inwhee Joe, "A Novel Adaptive Hybrid ARQ Scheme for Wireless ATM Networks," ACM-Baltzer Wireless Networks Journal (WINET), Vol. 6, No. 3, pp. 211-219, June 2000.

Multi-Granularities Counting Bloom Filter

Zhou Mingzhong[1,2], Gong Jian[1,2], Ding Wei[1,2], and Cheng Guang[1,2]

[1] Dept. of Computer Science and Technology, Southeast Univ., Jiangsu, Nanjing 210096 China
[2] Jiangsu Province Key Laboratory of Computer Networking Technology
{mzzhou, jgong, wding, gcheng}@njnet.edu.cn

Abstract. Counting Bloom Filter is an efficient multi-hash algorithm based on Bloom Filter. It uses a space-efficient randomized data structure to represent a set with certain allowable errors, and allows membership and multiplicity queries over the set. Aiming at the set whose items frequencies following heavy-tailed distribution, this paper presents a novel algorithm called Multi-Granularities Counting Bloom Filter (MGCBF) based on Counting Bloom Filter. This algorithm applies hierarchical data structures through several counting bloom filters to store the items frequencies information in the set. The time and space complexities analysis of this algorithm illustrates that it can reduce the space needed dramatically with the cost of little additional compute-time. And the following experiments indicate this algorithm is more efficient than other algorithms with same errors probability when the items frequencies of the target set follow heavy-tailed distribution.

Keywords: heavy-tailed distribution; Bloom Filter; Counting Bloom Filter, Multi-Granularities Bloom Filter(MGCBF).

1 Introduction

With the pervasion of the computer and also the network, it becomes more and more serious to deal with the expanding data. Extracting useful information from very huge dataset efficiently is one of most important research directions. Bloom Filter is a space-efficient multi-hash algorithm created by B. Bloom in 1970's[1], and it is widely used in database and network applications [2]. But using original Bloom Filter to satisfy the needs of all kinds applications is becoming impossible because of the protean usages and quarries of data. Several algorithms based on original Bloom Filter come forth to fit the special needs of different applications [3][4][5].

In a great few of applications, some information of dataset need measuring is provided in advance (i.e. the size and the distribution). It can improve the performance of data disposal dramatically by taking advantage of those transcendental information. Heavy-tailed distributions (also known as power-law distributions) have been observed in many natural phenomena including both physical and sociological phenomena, for example, the size distribution of files transferring in network [6], the length distribution of flows intercepted in some router for a period [7], et. al. Heavy-tailed distribution means the most items frequencies are very small, while a very few items

M. Gerndt and D. Kranzlmüller (Eds.): HPCC 2006, LNCS 4208, pp. 542 – 551, 2006.

frequencies is so large that the total number of their tuples takes a large proportion of all tuples in the set. If the asymptotic shape of the distribution is hyperbolic, it is heavy-tailed regardless of the distribution for small values of the random variable.

For the dataset whose items frequencies following heavy-tailed distribution, this paper represents a novel algorithm based on Bloom Filter, called Multi-Granularities Counting Bloom Filter (MGCBF), allowing membership and multiplicity queries for individual items with quite few errors. Because the MGCBF takes advantage of the characteristic of heavy-tailed distribution given by the dataset, it can reduce the storage space (vs. CBF [3]) and calculating time (vs. SBF[4]), and improve the statistical precision (vs. SCBF [5]). This algorithm can be used in all kinds of dataset whose individual items frequencies following heavy-tailed distribution.

2 Previous Works

Bloom Filter are space efficient data structures whose size is m, allowing for membership queries over a given set $S=\{s_1,s_2,\ldots,s_n\}$. The Bloom Filter uses k hash functions, $h_1, h_2,\ldots, h_k$ to hash elements into an array V of size m. Initially, all positions of the array V are set to 0. For each element s, the bits at positions $h_1(s), h_2(s), \ldots, h_k(s)$ in the array are set to 1. Given an item q, its membership in the dataset can be checked by querying the bits at positions $h_1(q), h_2(q), \ldots, h_k(q)$. If and only if all those bits are set to 1, it is reported that $q\in S$. This algorithm can cause false positive error for it may be possible that not all bits at the positions $h_1(q), h_2(q), \ldots, h_k(q)$ are set by the item q. But no false negative error can be produced.

Counting Bloom Filter (CBF) is one of Bloom Filter's extensions. And it is introduced by L.Fan, P. Cao et. al. in [3]. This algorithm changes the bits in the array to counters, allowing estimates of the multiplicities of individual keys with a small error probability. When an element s want to insert, the counters at positions $h_1(s), h_2(s), \ldots, h_k(s)$ in the array add 1, And it means that CBF not only supports items insertions and queries like original Bloom Filter, but also can be used for items deletions. When an element sj want to be eliminated from the set, the counters at corresponding positions in the array subtract 1 if all these counters' values are bigger than or equal 1. Otherwise it is reported that $s_j\notin S$. The CBF inherits false positive errors from Bloom Filter, and also false negative errors are introduced if the counters space is not large enough. The method of counters space estimation is illustrated in [3], and the errors probability is also given.

S.Cohen and Y.Matias indicate Spectral Bloom Filter (SBF) in [4], which is another extension of original Bloom Filter. This algorithm is based on CBF, which uses a compact data structure to represent the counters array and a optimal method called recurring minimum to reduce the errors probability. But the maintenance of this compact data structure needs additional calculating time. And so the items frequencies distribution of the dataset can influence the time of data structure maintenance. To the dataset with same tuples number, the asymmetrical distribution of items frequencies need more compute time than symmetrical ones. When the frequencies follow heavy-tailed distribution, this situation is more deteriorated. The SBF does not give any optimizations for this situation for it does not fit the needs of those type datasets.

Space Code Bloom Filter (SCBF) is an algorithm for network flow length statistics proposed by S. Cohen and Y Matias in [5]. This algorithm can be used in other applications for the same requirements. The SCBF employs several Bloom Filters to estimate the packets number of individual flows. To make the data structure more efficiency, this algorithm is extended to a new algorithm called Multi-Resolution SCBF, applying sampling and multiplies resolution layers methods to express the flows whose packets number exceed some thresholds. The maximum likelihood estimation (MLE) and mean value estimation (MVE) methods are adopted to gauge all flows length information. Because of the real time requirement of network flow disposal, this algorithm uses the method of sampling. And this method cannot measure the items frequencies and their distribution accurately. It cannot satisfy the precise measurement needs of some special applications.

3 The Multi-Granularities Counting Bloom Filter

The MGCBF employs a serial of CBFs (MGCBF={cbf_0,cbf_1, …, cbf_{h-1}}), which use a set of different counter granularities (C={1,c_1,c_2, …, c_{h-1}}) to count the frequency of individual items in the dataset. The time and space complexity are main problems wanted to solve because of the long length of the sequence [2][3][4][5]. This algorithm supports related items insertions, queries and deletions, for frequencies statistical dataset whose frequencies following heavy-tailed distribution.

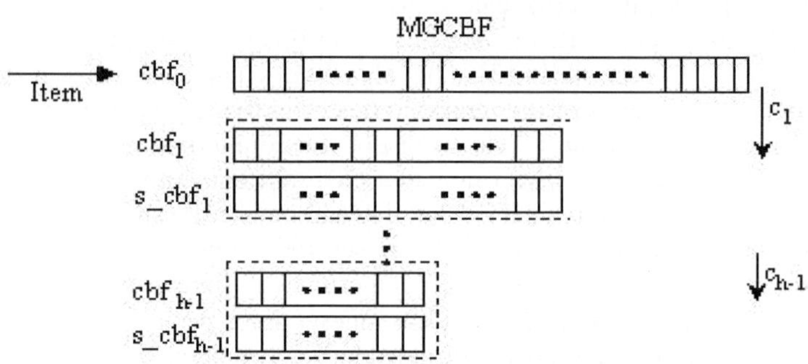

Fig. 1. The MGCBF data structure

The prototype of this algorithm is introduced as following:

(1) When an item x wanted to add into MGCBF, the counters at positions $h_1^0(x)$, $h_2^0(x)$, …, $h_{k0}^0(x)$ in the array V_0 add 1. (V_0 is the array structure of MGCBF's first CBF, cbf_0. h_1^0, h_2^0, …, h_{k0}^0 are the hash functions in cbf_0. Without the loss of generality, we suppose $h_1^0(x) \le h_2^0(x) \le \ldots \le h_{k0}^0(x)$);

(2) The second step is checking the value $h_1^0(x)$. If $h_1^0(x)=c_1$, the counters at positions $h_1^0(x)$, $h_2^0(x)$, …, $h_{k0}^0(x)$ decrease c_1, then the values in those counters are changed to 0, $h_2^0(x)-c_1$, …, $h_{k0}^0(x)-c_1$; At the same time, the counters add 1 at positions $h_1^1(x)$,$h_2^1(x)$, …, $h_{k1}^1(x)$ in V_1 which is the array of cbf_1, that means $h_1^1(x)+1$, $h_2^1(x)+1$, …, $h_{k1}^1(x)+1$;

(3) And then checking the value $h_1^1(x)$. If $h_1^1(x)=c_2$, we operate the same action as 2) in cbf_1 and cbf_2. This action keeps doing until cbf_{h-1} is checked. When an item *x* multiplicity query is need, we need get a minimum values in every layer CBF which make up of a set: $M(x)=\{min_0(x),min_1(x), \ldots, min_{h-1}(x)\}$, and then the frequency of *x* in S is reported as following:

$$\text{Counter}(x) = min_0(x)+min_1(x)*c_1+ \ldots + min_{h-1}(x)* \prod_{i=1}^{h-1} c_i \tag{1}$$

The MGCBF use different layers to analyze the items frequencies in the target dataset. The low frequently items only exist in low layers, while only very high frequent items can exhibit at every layer. When the items frequencies follow heavy-tailed distribution in the dataset, the arrays space reduces in power law as the layers increasing because only very small items are with very high frequencies. Because the counters in low layer CBFs' data structures should only maintain very small and controllable values, their space needed can reduce dramatically as the CBF. And multiply granularities makes counters values of every layer relative small, which can restrict the space needed in a small range. And so the storage resources needed in the MGCBF is relative smaller than those of the CBF, but this method will introduce additional time used for data structure maintenance and also the errors probability. Because the influences of errors caused by high-level CBFs are more important than those of low-level CBFs, the MGCBF applies an optimal method called recurring minimum introduced in [4] in high-level CBFs (cbf_1 and above) to reduce the errors probability dramatically by with the cost of a small storage resources and compute time. We will analyze the storage resource usages, calculating complexities and errors ratios in the following section.

```
insert (MGCBF,X)
  Initiate(MGCBF);   //set every counter of each granularity CBF to zero;
  for i:=1 to N, do
    cbf0.add(xi, 1);   //add the xi into the granularity 0 CBF
    for j:=1 to h, do
      if(minj-1(xi)= =cj)
        cbfj-1.remove(xi, cj);
        if cbfj.isminimum
          s_cbfj.add(xi, 1);
        else
          cbfj.add(xi, 1);
    return count(xi):= min0(xi)+min1(xi)*(c1 +)+ ... + minh-1(xi)* ∏_{k=1}^{k-1} ck;
  end_inserted
```

Fig. 2. Insertion algorithm in MGCBF

The pseudo-code of insertion items action in the MGCBF is illustrated in Fig.2. The parameters used in that can be described as following: $\mathbf{x_1,\ldots,x_N}$: the incoming packets sequence;$\mathbf{cbf_0, \ldots, cbf_{h-1}}$: the CBFs which construct the MGCBF, h is number of stages; $\mathbf{s\_cbf_1, \ldots, s\_cbf_{h-1}}$: the second CBFs used in high-level of the MGCBF; $\mathbf{C=\{c_0,c_1,c_2, \ldots, c_{h-1}\}}$: the space of counting unit in every level;

M(x$_i$)={min$_0$(x$_i$),min$_1$(x$_i$), …, min$_{h-1}$(x$_i$)}: the serial made up of minimum count of the items **x$_i$** in every stage; **Count(x$_i$):** the tuples number of **x$_i$** in the MGCBF.

4 Performance and Errors Analysis

The correct BibTeX entries for the Lecture Notes in Computer Science volumes can be found at the following Website shortly after the publication of the book: Because the Pareto distribution is one of widely used heavy-tailed distributions, this paper employs a set whose items frequencies following a Pareto distribution to analyze the performance and errors probability. And this paper also compares these characteristics with those of the other two algorithms (CBF and SBF). Firstly, we suppose the items frequencies of the target dataset ***S*** following a Pareto distribution whose cumulative distribution function is $F(x)=1-1/x^{\alpha}[\alpha>1]$, and the individual items number in ***S*** is set to *n*. Using the properties of the Pareto distribution [8], we can get the expectation $E(x)=\alpha/(\alpha-1)[\alpha>1]$. And so we can infer the tuples number in ***S*** is about $n\times E(x)=n\alpha/(\alpha-1)[\alpha>1]$.

4.1 Performance Analysis

In this section, we will analyze the performance in three different directions: the space complexity, the compute complexity and the error probability.

4.1.1 Space Complexity Analysis

When we suppose the most frequent item have P tuples in *S*, **Equation** (2) should be guaranteed if using the MGCBF:

$$P\le C_1+C_1\cdot C_2+\ldots+\prod_{i=1}^{h-1}C_i \tag{2}$$

And then the counters number n_i in the array of cbf_i can be estimated as the below equation in the MGCBF data structure according to properties of the Patero distribution:

$$n_i=F[x>(1\cdot C_1+C_1\cdot C_2+\ldots+\prod_{j=1}^{i-1}C_i)]\cdot n=(C_1+C_1\cdot C_2+\ldots+\prod_{j=1}^{i-1}C_i)^{-\alpha}\cdot n$$

Because we apply recurring minimum method to reduce the errors probabilities in high-level of the MGCBF, the space needed in level 2 and above is double of the original structure. With the supposition of *m*/*n*=*θ*, the total space needed in level i can be inferred by the above equation which we call m_i:

$$m_i\log_2(C_i)=\theta\cdot n\cdot\log_2(2C_i)\cdot(C_1+C_1\cdot C_2+\cdots+\prod_{j=1}^{i-1}C_j)^{-\alpha}$$

From this equation, we can calculate the all storage resource needed by the MGCBF to dispose the *S*:

$$M_{MGCBF}=\sum_{i=1}^{h} m_i \log_2(C_i)$$
$$=\theta\cdot n[\log_2 C_1 + C_1^{-\alpha}\cdot\log_2(2C_2)+\cdots+(C_1+C_1\cdot C_2+\cdots+\prod_{j=1}^{h-1}2C_j)^{-\alpha}\cdot\log_2(2C_h)] \quad (3)$$

4.1.2 Compute Complexity Analysis

Firstly, we on the assumption that the mean time of one tuple's insertion, query or deletion in individual layers of the MGCBF is following: $T_0,T_1,\ldots,T_{h-1}$. According to the cumulated distribution function (CDF) of the Pareto distribution, we can calculate the ratio of items exist in cbf_i :

$$F[x>(1\cdot C_1+C_1\cdot C_2+\ldots+\prod_{j=1}^{i-1}C_i)]=(C_1+C_1\cdot C_2+\ldots+\prod_{j=1}^{i-1}C_i)^{-\alpha}$$

For the item whose frequency is $f\in[C_1+C_1\cdot C_2+\ldots+\prod_{j=1}^{i-1}C_j, C_1+C_1\cdot C_2+\ldots+\prod_{j=1}^{i}C_j]$, the mean compute time of every item:

$$T(i)=T_1+\frac{T_2}{C_1}+\cdots+\frac{T_i}{\prod_{j=1}^{i-1}C_j}$$

And then we can calculate the mean compute time of one tuple at following:

$$T=(1-C_1^{-\alpha})\cdot T_1+[C_1^{-\alpha}-(C_1+C_1C_2)^{-\alpha}](T_1+\frac{T_2}{C_1})+\cdots+$$
$$[(C_1+C_1C_2+\cdots+\prod_{j=1}^{h-2}C_j)^{-\alpha}-(C_1+C_1C_2+\cdots+\prod_{j=1}^{h-1}C_j)^{-\alpha}](T_1+\frac{T_2}{C_1}+\cdots+\frac{T_i}{\prod_{j=1}^{i-1}C_j})$$
$$\Rightarrow T=T_1+C_1^{-\alpha-1}T_2+\cdots+(\prod_{j=1}^{h-1}C_j)^{-\alpha-1}T_h \quad (4)$$

4.1.3 Errors Probability Analysis

The errors caused by the MGCBF can be divided into two parts: (1) the inherent errors exist in the CBFs; (2) the errors caused by count numbers transferring in the layers.

About the errors of the Bloom Filter, B.Bloom analyzed it in detail when he introduced this algorithm [1]. When all n items of the target set S are inserted into the array V whose size is m, the error probability is $E=(1-e^{-kn/m})^k$. And then we can calculate the minimal value: k=ln2*m/n, that means $(1/2)^k=(0.6185)^{m/n}$. This error ratio can be controlled by modifying the values of k and n through estimating the value m :

θ=m/n=6 k=4 E=0.0561 θ=m/n=8 k=6 E=0.0215
θ=m/n=12 k=8 E=0.00314 θ=m/n=16 k=11 E=0.000458.

The detail analysis about CBF errors probability in [4][5] indicated that the inherent errors of the CBF are equivalent with the original Bloom Filter. The experiments of [4] illustrated that the error probability can be reduce to 1/18 if using the method of recurring minimum. And so the inherent errors probability can be controlled in a very small region in the high-level of the MGCBF.

The transferring of count numbers in levels of the MGCBF makes it important to reduce the error probability caused by this method. We can suppose the inherent errors in cbf_0, cbf_1, …, cbf_{h-1} are E_0, E_1, …, E_{h-1}. And the total errors in cbf_i can be regard as the iterative of its lower levels, it can be express as **Equation** (5) :

$$E_{i-1}' = 1-(1-E_0)(1-E_1)\cdots(1-E_{i-1}) \tag{5}$$

And we can get the actual errors of every CBF of the MGCBF:

$$(E_0', E_1', \ldots, E_{h-1}') = (E_0, 1(1-E_0)(1-E_1), \ldots, 1-(1-E_0)(1-E_1)\ldots(1-E_{h-1}))$$

From the equation above, we can conclude that the errors probabilities of items frequencies estimations increase as the CBF levels increasing in the MGCBF. But the errors probabilities of high-level are controlled in a very small region, and so the errors probabilities of high frequent items does not increase remarkably.

4.2 Comparison with Other Algorithms

The MGCBF is designed for the dataset whose items frequencies follow heavy-tailed distribution, and it can compactly use storage and compute resources. While the traditional algorithms do not use the known information, it can induce costs of statistical information maintenance. In this section, the MGCBF will be compared with the other widely used two algorithms: the CBF and the SBF in **performance** and errors probabilities. Because the SCBF uses sampling and MLE estimations, the application scope and statistical errors are absolutely different with the MGCBF, and so it is not considered when we operate the comparison.

Using the CBF, the space needed is about $m\log_2 P = \theta \cdot n \cdot \log_2 P$ for the correctness assurance [3][4]. And the SBF initializes the data structure with very little space, and adjusts the space of data structure to satisfy the needs in real time [4]. The SBF use the storage spaces more efficiently with the cost of compute time for maintaining the data structure.

We set the parameters of the MGCBF as following: $\theta=m/n=12$, $k=8$, $P=65536$ 、$C_1=C_2=\ldots=C_h=4$、$T_1=T_2=\ldots=T_h$ and$\alpha=3$. And we also set the mean compute time for reassigning storage resources to T_a in the SBF, and the ratio of the counters whose space should be assigned is β.

According to the **Equation** (1), we can calculate the minimal value of h is $h_{min}=8$. And so we can get the storage resources used by these three algorithms as **Equation** (6) while the space complexity equation of the SBF is according to [4]. Because the SBF uses table indexes to maintain the storage data structure, it is the most space efficient algorithm of these three ones.

$$\begin{aligned} CBF &: C = 12n * log_2(65536) = 12n \cdot 16 = 192n \\ SBF &: C = \frac{\alpha \cdot n}{\alpha - 1} + o(\frac{\alpha \cdot n}{\alpha - 1}) + O(n) = \frac{3n}{2} + o(\frac{3n}{2}) + O(n) < 20n \\ MGCBF &: C = \sum_{i=1}^{h} m_i log_2(2C_i) = 12n \cdot log_2 8 \cdot [1 + 4^{-\alpha} + \cdots + 21844^{-\alpha}] \approx 34.61n \end{aligned} \tag{6}$$

The MGCBF uses a hierarchical data structure to maintain the items information, which means it needs more compute time than the CBF, while the SBF needs more time to reassigned space for incoming items (that is to say $T_a >> T_1$). We compare the compute time of three algorithms in **Equation** (7). From these equations, we can indicate that compute time the MGCBF needs only a little more than that the CBF needed because of the heavy-tailed distribution of items frequencies in the target set. But compute time of the SBF is much more rely on the parameters T_a andβ. We can infer that the parameter $\beta \in [0.5,1)$, and so its compute time needed is far more than the others.

$$\begin{aligned} CBF &: T = T_1 \\ SBF &: T = T_1 + \beta \cdot T_a \\ MGCBF &: T = [1 + C_1^{-\alpha-1} + \cdots + (\prod_{j=1}^{h-1} C_j)^{-\alpha-1}]T_1 = 1.004T_1 \end{aligned} \tag{7}$$

We check the errors probability of the cbf_i in the MGCBF, and compare it with those of the other two algorithms in **Equation** (8). The errors probabilities of these three algorithms are very small because all of them use more space and hash functions to guarantee the precision. And it is verified the inference in § 4.1.3 that the errors probabilities may increase with the levels augment of the MGCBF but this increase is controlled in a small scope. If we want to make the algorithms more efficient by reducing the storage resources and hash functions, it will introduce more errors for all these algorithms. But the precision of the MGCBF will not reduce more rapid than the other twos.

$$\begin{aligned} CBF &: E = E_0 = 0.003 \\ SBF &: E \approx E_0/18 = 0.0002 \\ MGCBF &: E'_{i-1} = 1 - \prod_{j=0}^{i-1}(1 - E_j) = 1 - (1 - 4.6 \times 10^{-4})(1 - \frac{4.6 \times 10^{-4}}{18})^{i-1} \\ &= 1 - 0.997 \times 0.9998^{i-1} \end{aligned} \tag{8}$$

5 Experiments

This paper adopts five groups of datasets whose items frequencies following heavy-tailed distributions, and analyzes these datasets with three algorithms (the CBF, the SBF and the MGCBF) to compare and evaluate the performance and errors probabilities of these algorithms. The datasets are the packets coming from Abilene-III TRACE[9] and TRACE of the CERNET backbone. We apply the flow specification of 5-tuple to counting the flows by their lengths. **Fig** 3-a illustrates the distributions of five datasets: ***dif_num*** describes the individual items number in the datasets; ***mean*** indicates the mean value of individual items frequencies; ***P95*** depicts the frequency of item in 95 percentile; and ***P99*** depicts the frequency of item in 99 percentile; ***Max_val*** is used to characterize the tuples' number of the maximal frequent item in the dataset. And so from the **Fig**. 3-a , we can conclude that the individual items frequencies of all these five dataset follow heavy-tailed distribution though the items numbers of these datasets are far from the same ones.

The data of the other figures of **Fig.**3 are normalized for the convenience of comparison. **Fig.**3-b describes comparison of three algorithms' compute performance. The performance of the MGCBF is more efficient than the SBF, and near the CBF, The reason is that the SBF need more time for data structure maintenance, which is depicted in §4.1.3 in detail. The **Fig**.3-c indicates the MGCBF saves 60% storage resources comparing with the CBF. Though the SBF is more space efficient than the MGCBF, it needs more time for calculation. The error probability of the MGCBF is almost same with that of the CBF in the **Fig.** 3-d, which indicates that the recurring minimum method reduces the errors probabilities in the high-level CBFs dramatically. The errors probabilities of the SBF are smaller than those of the other ones, but it is not prefect as described in [4]. The reason may be that the experiments applied different datasets and different hash functions. In generally, the experiments results verify the correctness of conclusion inferred in §4.

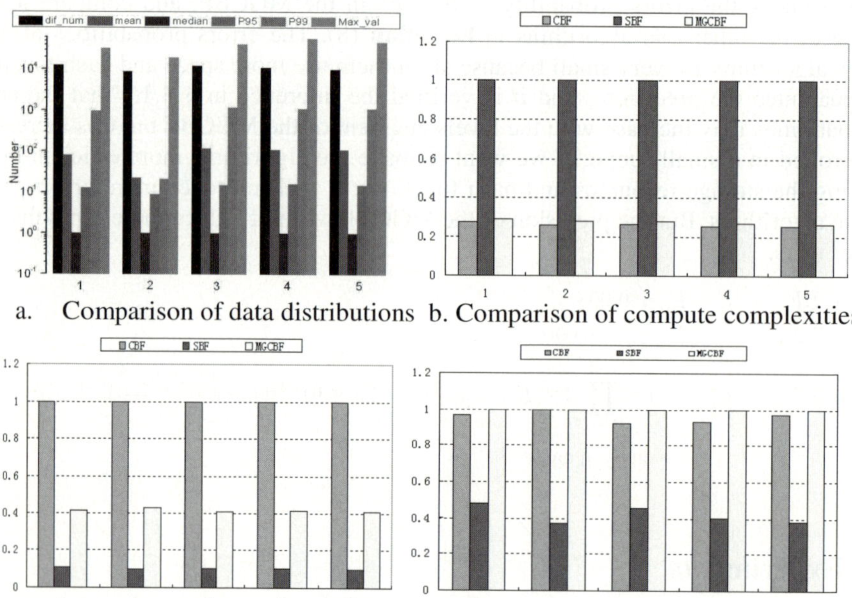

a. Comparison of data distributions b. Comparison of compute complexities

c. Comparison of storage resource d. Comparison of error probabilities

Fig. 3. Performance Comparison of the CBF, SBF, MGCBF using different datasets

6 Conclusion

This paper introduces a novel algorithm to count the number and multiplicities of individual items in a set, which is called Multi-Granularities Counting Bloom Filter (MGCBF). This algorithm is based on the Bloom Filter, and takes advantage of the fact that the items frequencies follow heavy-tailed distribution in this target set. This algorithm is more space efficient that traditional Counting Bloom Filter (CBF) with little compute time and few errors probabilities. And comparing to the Spectral Bloom Filter (SBF), this algorithm is more compute efficient with the cost of additional storage resources. The MGCBF not only support insertions and queries of items, but also support the items deletions, just like the SBF. And so the expansibility of the MGCBF is comparative fine. Further more, the more information about the dataset we can get, the subtler the parameters of the MGCBF we can set, and also the more precise results we can receive.

Acknowledgments. This research is partially support by the National Basic Research Program (called 973 Program), No. 2003CB314803; Jiangsu Province Key Laboratory of Network and Information Security BM2003201; the Key Project of Chinese Ministry of Education under Grant No.105084 and the National High-Tech Research and Development Plan of China No. 2005AA103011-1

References

1. B. Bloom, Space/Time trade-offs in hash coding with allowable errors. Commun.ACM. Vol. 13, no.7, pp. 422-426, July 1970.
2. A. Broder and M. Mitzenmacher. Network applications of bloom filters: A survey. In Proceedings of the 40th Annual Allerton Conference on Communication, Control, and Computing, 2002.
3. L. Fan, P. Cao, J. Almeida, and A. Z. Broder. Summary Cache: A Scalable Wide-Area Web Cache Sharing Protocol. IEEE/ACM Transactions on Networking, 8(3):281-293, 2000.
4. S. Cohen, Y. Matias. Spectral Bloom Filters. In Proceedings of the ACM SIGMOD 2003. San Diego, California, USA. June, 2003.
5. A. Kumar, J. Xu, et al. Space-Code Bloom Filter for Efficient Per-Flow Traffic Measurement. In IEEE Infocom 2004, Hongkong, China. March, 2004.
6. Zhang Y, Breslau L., V. Paxson and S. Shenker. On the Characteristics and Origins of Internet Flow Rates. In Proceedings of ACM SIGCOMM, Aug. 2002, Pittsburgh, PA.
7. A. Shaikh, J. Rexford and K. G. Shin. Load-Sensitive Routing of Long-Lived IP Flows. In Proceedings of the ACM SIGCOMM 1999.Cambridge, MA, USA. August , 1999.
8. The Pareto Distribution, http://www.ds.unifi.it/VL/VL_EN/special/special12.html, Dec, 2005.
9. The Abilene-III Trace Data – Illustrated, http://pma.nlanr.net/Special/ipls3.html, Dec, 2005.

Dynamic Execution Environments for Ubiquitous Computing Service

Seungkeun Lee

INRIA Rhône-Alpes
ZIRST 655 avenue de l'Europe, Montbonnot
38334 Saint Ismier cedex, France
sglee0901@gmail.com

Abstract. Ubiquitous service provides various functions to users by process of context acquisition and reasoning from ubiquitous environments. These services usually cooperate with other services for acquisition of context or providing service to users. In order to support the service execution, there are a lot of researches which are designed based on OSGi framework. OSGi framework supports the service execution environments and syntax based service matching. In previous studies, they have not been able to support the service mobility and interoperability effectively. This study proposes a dynamic execution environment for ubiquitous service which supports service mobility among service gateway and service matching based on semantic information. The proposed dynamic execution environment has advantages of interoperation and integrating the heterogeneous devices by applying these standard interface technologies and service discovery and dynamic service compositions using semantic information.

1 Introduction

Service Environments are dynamic configurations of services available in physical spaces for both collaborative and individual use. An important objective is to achieve service environments that are open, transparent, adaptive and degrade gracefully with respect to capabilities of the interactive devices available in the room. The term ubiquitous refers to the objective that device technology preferably should be designed in such a way that it melts into the periphery when not needed[1]. OSGi is an industry plan regarding the standards for allowing sensors, embedded computing devices and electronic appliances to access the Internet, and it provides a Java-based open standard programming interface for enabling communication and control between service providers and devices within home and small business networks[2,3]. Whereas previous technologies focused on interoperation among devices, OSGi places emphases on service delivery, allocation and management for the devices. Furthermore, linkage services concerning Jini or UPnP can be deployed or interacted based on development of OSGi-based applications. OSGi framework is already providing the service execution environment of ubiquitous services. An OSGi-based system has a structure for

M. Gerndt and D. Kranzlmüller (Eds.): HPCC 2006, LNCS 4208, pp. 552–561, 2006.

distributing new services, and the structure consists of only the elements on the local network, giving it relatively closed characteristics[3][4][5]. However, while service management and distribution can be dynamically executed within such single OSGi framework, there is insufficient support for applications with mobility among multiple frameworks. Therefore, there must be sufficient consideration for mobility of the users, devices and sensors among multiple OSGi frameworks in the expanded smart space, calling for research efforts in services supporting such mobility. An OSGi-based ubiquitous computing system has a structure for distributing new services, and the structure consists of only the elements on the local network, giving it relatively closed characteristics[6][7]. However, while service management and distribution can be dynamically executed within such single OSGi framework, there is insufficient support for applications with mobility among multiple frameworks[8]. And according to movements of services among multiple frameworks, more precise service matching mechanism is needed. This enables ubiquitous services to be accessed by semantic rather than by syntax. The semantic based service matching enables Services to be discovered, selected and composed automatically by other services. This study deals with designing an OSGi-based framework using a mobile agent technology that supports mobility and duplication with status information in the distribution environment. By supporting bundles in the form of a mobile agent, the designed framework also supports mobility of the bundles within multiple OSGi system environments. Therefore, it can support mobility of various elements such as services for specific components or users as well as device drivers. we also design a ontology for service description and the matching and composition methods. In implementation of this research, we use the OSGi framework's open source Knopflerfish 1.3.3[9][10].

2 OSGi (Open Services Gateway Initiative)

OSGi is a non-profit organization that defines standard specifications for delivering, allocating and managing services in the network environment. In its initial stage, OSGi was focused on the home service gateway, but it has recently expanded from a specific network environment to the ubiquitous environment. In turn, the objective of OSGi has become implementation of the service gateway for diverse embedded devices and their users[2][11].

The OSGi service framework is displayed in Fig. 1. The OSGi framework signifies the execution environment for such services and includes the minimum component model, management services for the components and the service registry. A service is an object registered in a framework used by other applications. For some services, functionality is defined by the interfaces they implement, allowing different applications to implement identical "service" types. The OSGi framework installs an OSGi component called a "bundle" and supports a programming model for service registration and execution. Furthermore, the framework itself is expressed with a bundle, which is referred to as a "system bundle". Bundles are service groups using services registered in the service registry as well as component units. Service implementation is delivered to and allocated in the framework through a bundle, which is a physical and logical unit. From the physical perspective, a bundle is distributed in a Java archive file format (JAR) that includes codes, resources and manifest files. The manifest file informs the framework of the bundle class execution path and declares Java packages that will be shared with other bundles.

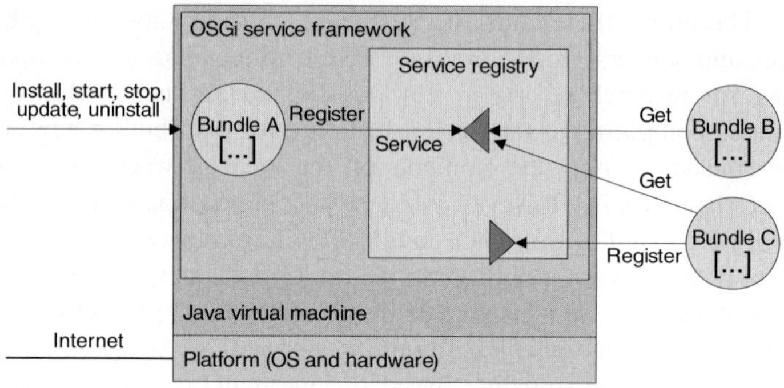

Fig. 1. OSGi Framework

The bundles can be registered in the service registry of OSGi framework. In this case, bundles can be found by syntax based query to service registry. If a bundle sent a query to service registry, service reference of required service is returned by syntax matching. Service registry is a set of key-values which are information composed of service interface name and service reference.

3 Semantic Based Service Description

Service Discovery can return a proper service reference in service registry based on a query of service requestor. Service is registered in service registry using semantic information. This semantic information is composed of service type, service input, service output and service context. Service matching is done by matching of these ontologies.

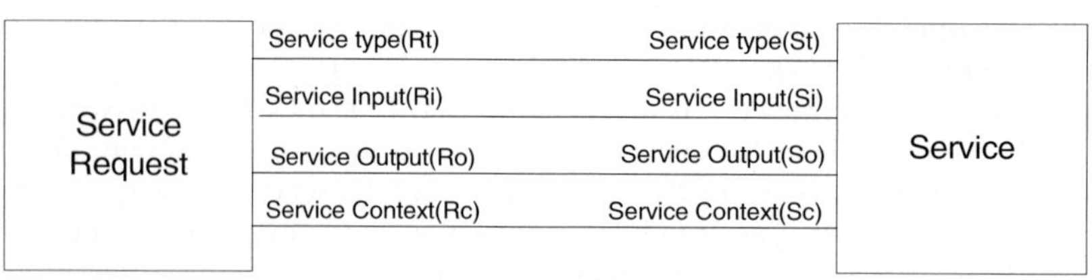

Fig. 2. Service Matching using Ontology

The service type ontology represents a type which service can be. This is different according to service domain. Service input/output is mapping into WSDL(Web Service Description Language) interface for calling service. This service can call the other service by sending of the parameters. These parameters include name, role, unit and data type. Service context is defined for searching service which is more usable to user. Service context include service location, approximation, openness. Figure 3 describes the service ontology for service description. For cxample, the homenetowk service wants to find a control service of a light in a home. If two services are matched using service type and grounding, the service discovery can provide a service which is located in more closely to user. Service QoS are used in the automatic service composition. This was introduced in another paper which was published in [12].

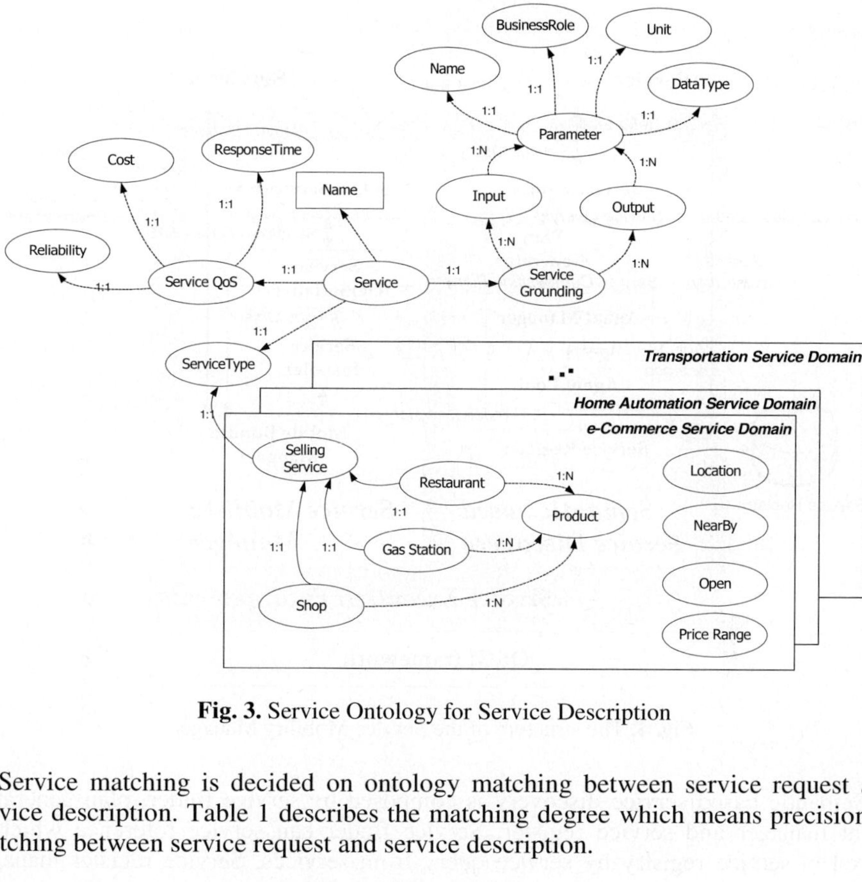

Fig. 3. Service Ontology for Service Description

Service matching is decided on ontology matching between service request and service description. Table 1 describes the matching degree which means precision of matching between service request and service description.

Table 1. The degree of ontology matching

Category	Case	Matching
"Exact"	Req = Srv	Ontology Matching
“PlugIn"	Req ⊂ Svr	
"Subsume"	Req ⊃ Svr	Approximate Matching
Intersection	Req ∩ Svr ≠ ∅	
Dispoint	Req ∩ Svr = ∅	Mismatch

4 System Design

The service execution environment consists of the service mobility manager and service discovery. Figure 4 describes the overall system. The service mobility manager provides the function of service mobility among execution environments. And, the service discovery provides the service matching and composition based on semantic based service description and ontology inference engine.

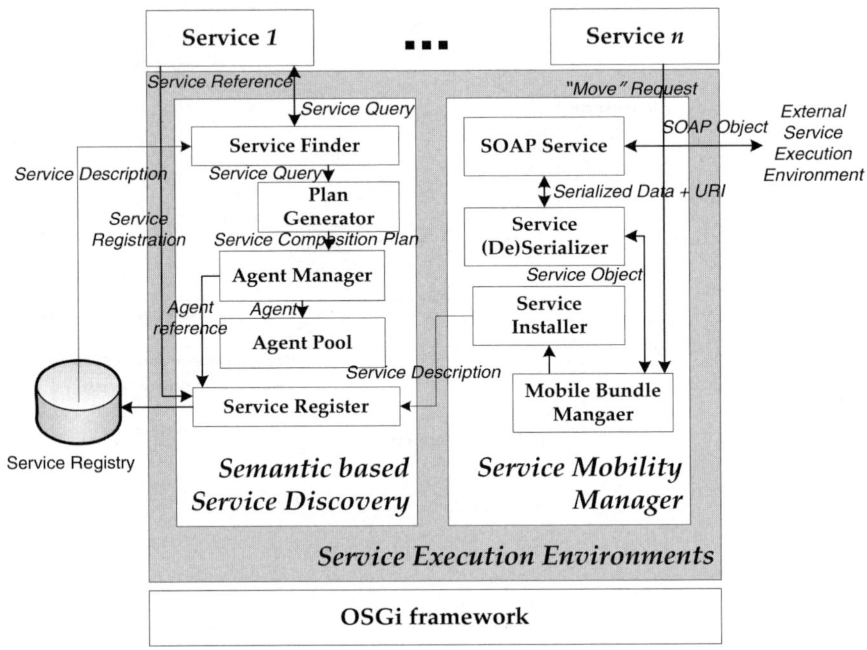

Fig. 4. The structure of the Service Mobility Manager

Semantic based service discovery is composed by service finder, plan generator, agent manager and service register. Service finder can service reference which is stored in service registry by service query from services. Service register manages service registry which stores service description. Plan generator is more complex component which can generate service composition plan. Agent manager make a agent which manage an execution of composition plan. Service wants other service with service request, service finder finds service reference with service description in service registry. If service is found, service reference is passed to service. If service is not found, this request is passed to plan generator. Plan generator makes a composition plan of service using service finder. A best composition plan is chosen among a few composition plans and makes an agent which manages service execution of service composition plan. This agent is registered in service register as a virtual service. Algorithm 1 describes the process of generation of composition plan.

Service Mobility Manager consists of Service Serializer/Deseializer, SOAP Manager. When the mobile bundle manager receives a mobility request from a bundle service, it manages the service bundle's lifecycle. The status information prior to mobility is marshalled into XML by the ServiceSerializer, and the SOAPClient delivers the SOAP message to the destination. The class file is installed through the destination BundleInstaller, and the ServiceDeserializer resumes the service by unmarshalling the SOAP message into an object.

Algorithm 1. Serviced Composition

```
public ServiceID compositeService(ServiceDescription
reqInput) {
    ....
ServicePlan[] servicePlan = new ServicePlan[100];
Input input = reqInput.getInput();
Output output = reqInput.getOutput();
//find service list which provides the requested result
ServiceID[] serviceID =
    serviceMatchaker.findService(output, reasoner);
for(int i = 0; i < serviceID.length; i ++ ) {
  //generation of service composition plan
  servicePlan[i] = serviceID[i];
  serviceChain[i] = makeServiceChain(servicePlan[i]);
}
//selection a best service plan
bestServicePlan =
   qosEvaluator.calculator(serviceChain);
//generation of agent which manage an execution service
composition paln
Agent agent =
   AgentFactory.createAgent(bestServicePlan);
//registration a virtual service as agent reference
agent.serviceID =
   viceRegister.registerService(reqInput);
....
}
```

We extend the status of bundle in OSGi which compose of 'Resolved', 'Starting', 'Active' and 'Stopping', adding 'DEAD', 'Movable', 'Moved'. 'Dead' status is different from 'Unistalled' which means that bundles are omitted automatically. 'DEAD' is used for checking information of service before it is received move request. The status of bundle is changed 'Movable' to 'Moved' for the service which is received move request. 'Move' status is used for the process which OSGi framework sends service to other OSGi framework. If OSGi framework would finish sending of service to other OSGi framework, service status is changed to 'Uninstalled', and this bundle is removed. Upon receipt of a mobility request, the service is switched to the mobility request state and execution suspension is requested. The service receiving the request returns after completing the action currently in process. If converted to the mobile state, the status information is serialized into the XML format using the Serializer, and service mobility is requested to the SOAPClient. The SOAPClient verifies the destination and generates an SOAP message to call SOAPService to the destination. If mobility is successful, the service currently being executed is deleted from the registry. At the destination, the SOAPService waiting for the SOAP message receives the URL information regarding the class location as well as the serialized data and delivers them to the MobileBundleManager. Prior to deserializing the received object, the MobileBundleManager installs the bundle from a remote location through the BundleInstaller. Upon successful installation, the object is deserialized and restored to the state prior to mobility. Finally, the service is converted to the RUNNABLE state and registered at the service registry.

5 Experiment

We propose a dynamic service execution environment for ubiquitous computing services. This environment is designed based on OSGi framework, and provide a semantic based service discovery and service mobility. In this chapter, we tested these two functions to prove the correctness of the environment. First, we tested semantic based service discovery with service registry which maintain a thousand of service descriptions. To view of results from execution environment, we implemented the diagnostic tool which can watch the status of the execution environments. Figure 5 shows the result which the execution environment can discover the correct services and the service composition plan.

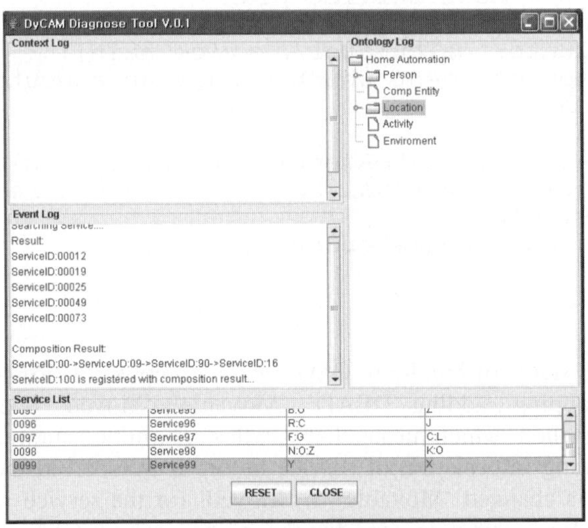

Fig. 5. The mobility of the MPlayer bundle

In order to test the service mobility proposed in this paper, an MPlayer bundle was developed for playing MP3 music files. JVM and OSGi mobile agent framework bundles were installed in a PDA and PC, respectively as an experiment environment as shown in Fig 6.

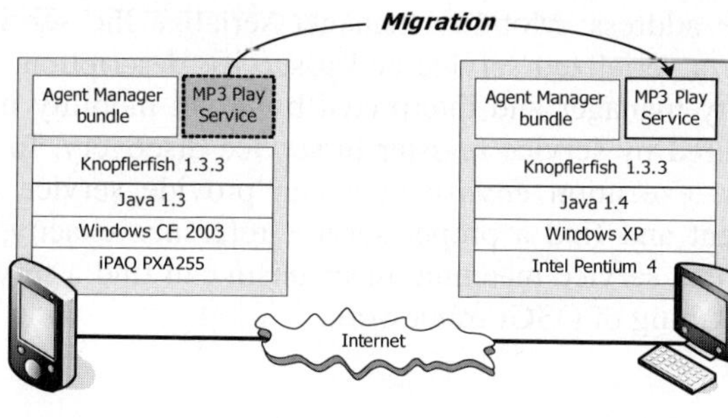

Fig. 6. The mobility of the MPlayer bundle

When a user listening to music from the MPlayer installed in the PDA moves into the space where the PC is located, the MPlayer service was moved to the PC and the file was resumed from the point where it had been playing from the PDA, providing a continuous MPlayer service to the user.

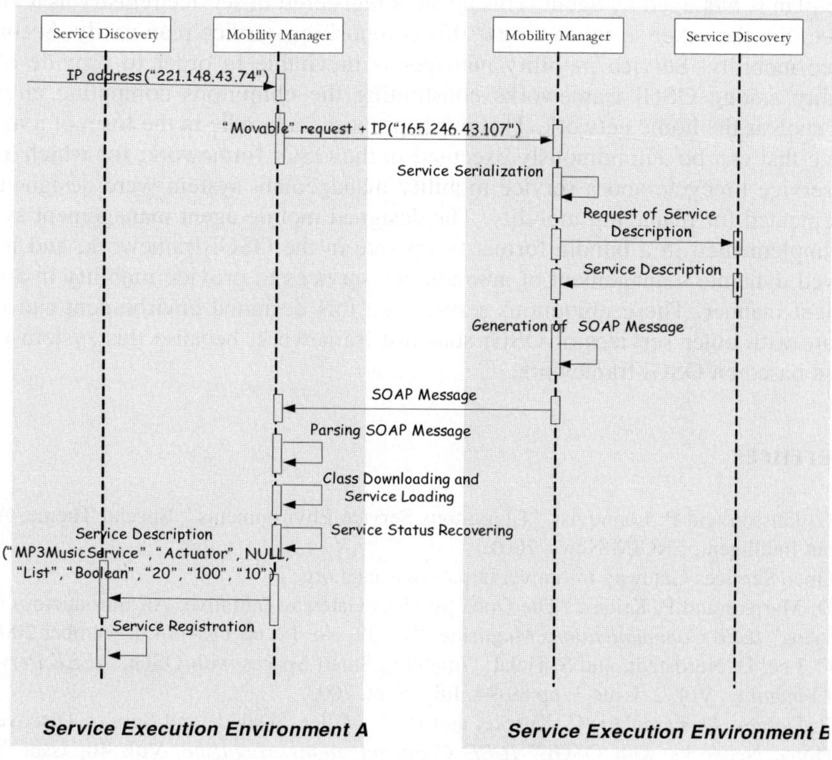

Fig. 7. Movement of the MPlayer service

Figure 7 describes the movement process of music service using the proposed service execution environment. If mobility manager received "move" request with IP address (221.148.43.74) of execution environment of service, this mobility manager send a request with own IP address. Mobility manager serialize the service and generate SOAP message with a serialized service and a service description. This message is sent to target mobility manager and interpreted by target mobility manager. Finally, this service is registered by service register in service discovery. In this experiment, the proposed service execution environment can provide service mobility among execution environment and find a proper service reference exactly. According to a few of experiments, this service matching mechanism can find a service exactly 76% more than syntax matching of OSGi framework.

6 Conclusion

We designed the dynamic service execution environment for dynamic interaction among services in ubiquitous environment. This system provides two important functions. First is a semantic based service discovery. Previous OSGi framework provides a syntax based service discovery. So, this environment provides a discovery of more correct service reference and service composition plan automatically. Service composition plan is managed by agent. This agent is registered in service registry as a virtual service. So, the other service can use this composition service plan easily. Second is service mobility. Service mobility manager is inevitable in order to provide object mobility among OSGi frameworks constituting the ubiquitous computing environment such as the home network. This paper proposed a bundle in the form of a mobile service that can be autonomously executed in the OSGi framework, for which a mobile service lifecycle and a service mobility management system were designed and implemented for managing mobility. The designed mobile agent management system was implemented in a bundle format to operate in the OSGi framework, and it also allowed dynamic management of autonomous services to provide mobility in a more efficient manner. These ubiquitous services on this designed environment can interoperate with other services of OSGi standard framework, because this system is designed based on OSGi framework.

References

1. G. Jansson and P. Lönnqvist, "Ubiquitous Service Environments", Special Theme: Ambient Intelligent, ERCIM News, 2001.
2. Open Services Gateway Initiative. http://www.osgi.org
3. D. Marples and P. Kriens, "The Open Services Gateway Initiative: An Introductory Overview," *IEEE Communications Magazine*, Vol. 39, No. 12, pp.110-114, December 2001
4. C. Lee, D. Nordstedt, and S. Helal, "Enabling Smart Spaces with OSGi," *IEEE Pervasive Computing*, Vol. 2, Issue 3, pp.89-94, July_Sept. 2003
5. P. Dobrev, D. Famolari, C. Kurzke, and B. A. Miller, "Device and Service Discovery in Home Networks with OSGi," *IEEE Communications Magazine*, Vol. 40, Issue 8, pp. 86-92, August 2002
6. K. Kang and J. Lee, "Implementation of Management Agents for an OSGi-based Residential Gateway," *The 6th International Conference on Advanced Communication Technology*, Vol. 2, pp.1103-1107, 2004

7. F. Yang, "Design and Implement of the Home Networking Service Agent Federation Using Open Service Gateway," *International Conference on Integration of Knowledge Intensive Multi-Agent Systems*, pp.628-633, Sept._Oct. 2003
8. R. S. Hall and H. Cervantes, "Challenges in Building Service-Oriented Applications for OSGi," *IEEE Communications Magazine*, Vol. 42, Issue 5, pp.144-149, May 2004.
9. K. Chen and L. Gong, Programming Open Service Gateways with Java Embedded Server^{TM} Technology, Addison Wesley, 2001
10. H. Zhang, F. Wang, and Y. Ai, "An OSGi and Agent Based Control System Architecture for Smart Home," *Proceedings of IEEE Networking, Sensing, and Control*, pp.13-18, March 2005
11. Knopflerfish. http://www.knopflerfish.org
12. Seungkeun Lee, Sehoon Lee, Kiwoon Rim, Jeonghyun Lee, "The Design on Webservices Framework Support Ontology Based Dynamic Service Composition, AIRS2005 (LNCS3689), 2005. 10.

A Dynamic Trust Model Based on Naive Bayes Classifier for Ubiquitous Environments

Weiwei Yuan, Donghai Guan, Sungyoung Lee*, and Youngkoo Lee

Department of Computer Engineering, Kyung Hee University, Korea
{weiwei, donghai, sylee}@oslab.khu.ac.kr, yklee@khu.ac.kr

Abstract. Computational models of trust have been proposed for use in ubiquitous computing environments to decide whether to provide services to requesters which are either unfamiliar with service providers or do not have enough access rights to certain services. Due to the highly dynamic and unpredictable characteristic of ubiquitous environments, the trust model should make trust decision dynamically. In this paper, we introduce a novel Naive Bayes classifier based trust model which can dynamically make trust decision in different situations. The trust evaluation is based on service provider's own prior knowledge in stead of assuming variable weights and pre-defined fixed thresholds. This model is also suitable to make decision when only limited information is available in ubiquitous environments. Finally we give the simulation results of our model and the comparison with the related works.

1 Introduction

Ubiquitous computing environment consists of a massively networked world supporting a population of diverse but cooperating mobile entities. The autonomous operation among the contributing units is necessary due to lack of central control [1]. Traditional authentication and access control are effective only in situations where the system knows in advance which users are going to access and what their access rights are. Later on, computational models of trust were proposed for ubiquitous computing environments which were capable of deciding on the runtime whether to provide services to service requesters which are either unfamiliar with service providers or do not have enough access rights to certain services. Access decision in ubiquitous computing environments has to rely on some kind of trust developed with past interactions.

Trust is the measure of willingness to believe in an entity based on its competence (e.g. goodness, strength, ability) and behavior within a specific context at a given time. Previous trust models used various time-consuming approaches to evaluate the trust value by considering different factors that may effect the trust decision. However, a common failing is that these models simply compared these painstaking gotten trust values with one or two fixed pre-defined thresholds to make the final trust decision, which is not suitable for the highly dynamic ubiquitous environments. For

* Corresponding author.

M. Gerndt and D. Kranzlmüller (Eds.): HPCC 2006, LNCS 4208, pp. 562 – 571, 2006.

example, in a ubiquitous supported smart office, the thresholds for different services providers to provide services may not be the same, e.g. the threshold for providing fax service may be higher than the threshold for enabling copy machine service. For the same service provider, its threshold to provide service may also change from time-to-time, e.g. the threshold for scanner may be raised since it has been frequently mis-operated by users recently. The change in threshold values is related to the changes in acceptance level of service providers to the whole ubiquitous environment. The raising of the scanner's threshold means that its acceptance level to the smart office has been decreased due to the previous unsuccessful interactions with the users. Hence we would dynamically make the decision due to the change in usage pattern.

The object of this paper is to propose a trust model in ubiquitous environments that can dynamically make trust decision based on different situations and different service providers. This paper sets the stage by introducing a novel Naive Bayes classifier based trust model, which makes decision based on each entity's own prior knowledge. The main advantage of our trust model is that it avoids using only one or two predefined fixed thresholds, and can dynamically update decisions according to each service provider's own judging standard. Moreover, our trust model can make use of limited information in decision making, which is usually the case in a real scenario.

The rest of the paper is organized as follows. We briefly introduce related work in Section 2. And we present the proposed trust model in detail in Section 3. Section 4 gives the simulation results. Finally, conclusions and future work are presented in Section 5.

2 Related Work

Since mid 90s the research on the key role of trust management models has been outlined in [2], [3], [4] to develop complex and dependable computer systems. In the field of ubiquitous computing, research has paid much more attention to build autonomous trust management as fundamental building block to design the future security framework, such as [6], [12], [13], [14], [15].

A general concept of dynamic trust model in ubiquitous computing environments had been given in [1]. In [5], the authors explained basic scenarios in ubiquitous computing and modeling requirements of trust. A solution to evaluate trust from the past experience was given in [7]. In [8], the authors proposed a role-based trust model in ubiquitous environment, where recommendations were used to make decision. Trust level, a measure of one's belief in the honesty, competence and dependability to a certain entity, was used to make decision in [9]. The trust was divided into 6 levels and operators such as time and distance were used to evaluate the trust level. In [10], the authors involved the concept of confidence, which reflects the communication frequency between two entities, in the trust evaluation. Trust value and confidence values were used to made the finally decision together. In [11], the authors proposed a novel Cloud-Based trust model to solve uncertain problem. These works involved great efforts to evaluate the trust values, however, when it comes to decision making based on these trust values, they just simply compare with one or two thresholds, which can not dynamically change due to the altering of the environments.

Our trust model provides improvement in earlier works by proposing a probabilistic model which involves precise computation to update the decisions dynamically. And the evaluation of the trust is also based on each entity's own situation which can better suit the ubiquitous environments.

3 Naive Bayes Classifier Based Trust Model

In our trust model, trust decision for unfamiliar service requesters is based on the recommendations from other entities in ubiquitous environments. One of the example scenarios is ubiquitous supported smart office as showed in Fig.1.

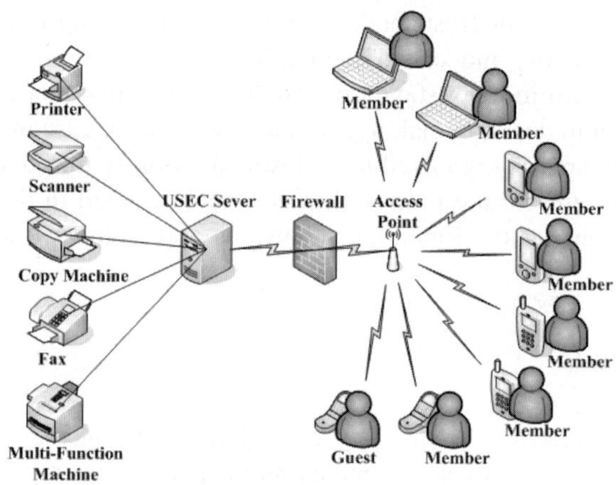

Fig. 1. Smart office supported by various ubiquitous units

The working procedure for the trust model to make decision is as follows: (1) Service requester (Sr_i) sends a request to USEC server to apply certain service. USEC server serves as service provider agent. (2) If Sr_i is not an acquaintance to service provider (Sp_j) or it does not have enough priority to access the service, USEC server will ask other entities who are now in a certain range of this smart office to give recommendations for Sr_i. (3) If entities who are requested to give recommendations have past interaction history with Sr_i, they will act as recommender (R_k) and give back recommendations to USEC server, (4) USEC server makes trust decision according to Sp_j's own judging standard based on the recommendations from the recommenders together with its own knowledge.

3.1 Factors Involved in Our Trust Model

There are totally five factors involved in our trust model.

Prior Probability. Prior probability reflects the acceptance level of certain service provider. It corresponds to the service provider's trusting beliefs for the whole ubiquitous environment. The lower the prior probability is, the more unbelieving the service provider is.

Definition 1. $P_{Sp_j}(y)$ and $P_{Sp_j}(n)$ are used to denote service provider Sp_j 's prior probability of acceptance and rejection respectively.

$$P_{Sp_j}(y) = \begin{cases} \frac{k}{m} & m \neq 0, \\ 0 & m = 0, \end{cases} \quad P_{Sp_j}(n) = 1 - P_{Sp_j}(y),$$

where $j, m, k \in N, k \leq m$ .Here m is the size of training sample; k is the size of acceptance sample. In case $P_{Sp_i}(y) \neq P_{Sp_j}(y)$, if $i = j$, it means we got one service provider in different situations. In this case, the same service provider has different acceptance levels for the environment due to the dynamic nature of service provider as well as the surroundings ubiquitous environment. Otherwise, if $i \neq j$, it means that they are different service providers. In case $P_{Sp_i}(y) > P_{Sp_j}(y)$ (i.e. Sp_i has a higher acceptance level when get same request), Sp_i is more likely to provide the service when requested. This situation is similar to our social society, Sp_i is easier to believe others comparing with Sp_j.

Trust Level. In our trust model, each entity is initially assigned a trust level according to its identity. If no information is available about the trustworthiness of an entity, it will be assigned as an unknown trust level. The trust level of an entity can be adjusted dynamically according to its behavior.

Definition 2. $Tl(S_k)$ is used to denote the trust level of entity S_k, where $k \in N$, $Tl(S_k) \in N$. Entity S_k may be a recommender or service requester.

If $Tl(S_k) > Tl(S_j)$ ($k, j \in N$), S_k is regarded as more reliable. However, in case $Tl(Sr_k)$ is unknown trust level, Sr_k may probably be provided services which are unavailable to the service requester who has a little bit higher trust level than him. This behavior also see parallels in our society, you may choose to trust an unfamiliar stranger that has never done harm to you instead of an acquaintance that had unpleasant interaction history with you.

Past Interaction History. Past interaction history is an entity's prior knowledge (this entity may be a recommender or service provider in our model) of acceptance to certain service requester.

Definition 3. $Pi(S_i, S_j)$ is used to denote the past interaction history between entities S_i and S_j. Entity S_i and S_j may be service requester, service provider or recommender.

$$Pi(S_i, S_j) = \begin{cases} \dfrac{n-(m-n)}{m} & m \neq 0, \\ 0 & m = 0, \end{cases} \quad \text{where } i, j, m, n \in N, i \neq j, n \leq m.$$

Here m and n denote the total communication times and successful communication times between S_i and S_j respectively. $Pi(S_i, S_j) \in [-1,1]$.

We suppose that past interaction history has Gaussian distribution. If S_i never communicate with S_j before, then $Pi(S_i, S_j) = 0$. If S_i and S_j had unpleasant interaction history, in previous work, $Pi(S_i, S_j)$ was set a positive small value. However, it means that the past interaction history for unknown entity is always worse even than the very malicious entity, which is obviously not correct. Hence our model set $Pi(S_i, S_j) \in [-1,0)$ for malicious entities, which is more convenient to differentiate unknown entities from malicious entities.

Time Based Evaluation. Intuitively, very old experiences of peers should have less effect in recommendation over new ones. Thus we take into account the time based evaluation.

Definition 4. $T(R_k, Sr_i)$ is used to denote the time based operator for recommender R_k to service requester Sr_i.

$$T(R_k, Sr_i) = \eta \frac{t_{R_k, Sr_i} - t_m}{\Delta\tau_0}, \text{ where } k, i \in N,$$

here t_{R_k, Sr_i} denotes the time when last communication between R_k and Sr_i happened. And η is time adapting operator. Suppose our measurement for time is based on a time window $[t_m, t_n]$, let $\Delta\tau_0 = t_m - t_n$.

Peer Recommendation. Peer recommendation is needed when service provider has no or not enough information to make decisions. Apparently if R_k had more interactions with Sr_i, the recommendation of R_k should be more importance for decision making, which introduces the notion of confidence.

Definition 5. $C(R_k, Sr_i)$ is used to denote the confidence for recommender R_k to service requester Sr_i.

$$C(R_k, Sr_i) = \frac{1}{std(M)\sqrt{2\pi}} \exp(-\frac{(m_k - mean(M))^2}{2std(M)^2}),\ i,k \in N,$$

where M is an array of communication times. $M[k] = m_k$, $k = 1, 2, ..., n$. Here m_k is the communication times between R_k and S_i. We suppose that M has Gaussian distribution.

We are now ready to use the above definitions to express the notion of peer recommendation.

Definition 6. $\Pr(R_k, Sr_i)$ is used to denote the peer recommendation from recommender R_k to service requester Sr_i.

$$\Pr(R_k, Sr_i) = \frac{Tl(R_k)}{Tl_N} * C(R_k, Sr_i) * \frac{n_k}{m_k} * \frac{T(R_k, Sr_i)}{\Delta\tau_0},\ \text{where } k, i \in N,$$

here m_k and n_k are the total communication times and successful communication times between R_k and Sr_i respectively. Tl_N is the total trust levels.

The final recommendation is the aggregate of all the peer recommendations.

Definition 7. $R(Sr_i)$ is used to denote the aggregate of recommendation for Sr_i from all the recommenders in the ubiquitous environment.

$$R(Sr_i) = \frac{\sum_{k=1}^{n} \Pr(R_k, Sr_i)}{n},$$

where $k, n, i \in N$, n is the number of the recommenders in the environment.

3.2 Trust Decision Making

Using the factors mentioned in section 3.1, our trust model uses Naive Bayes classifier twice to make the dynamic trust decision based on each service provider's acceptance level. Naive Bayes classifier is a technique for estimating probabilities of individual variable values, given a class, from training data and then to allow the use of these probabilities for classify new entities.

The decision is first made without recommendations, and it only depends on the service provider's own prior knowledge. Sometimes, the service provider may not be able to make the decision in the first decision, which means that the service requester is unfamiliar with the service provider or it does not have enough priority to access this service. Then recommendations given by other recommenders will be used to make the final decision together with service provider's own prior knowledge.

First Decision: When Sr_i gives a request to Sp_j, $h(Sr_i, Sp_j)$ is used to denote Sp_j's trust decision. Accept=1; Reject=0.

$$h(Sr_i, Sp_j) = \begin{cases} 1 & V_{NB|y} \geq V_{NB|n} \\ 0 & V_{NB|y} < V_{NB|n} \end{cases}, \tag{1}$$

where $V_{NB|y}$ and $V_{NB|n}$ are the acceptance and rejection value respectively.
Using Naive Bayes classier:

$$V_{NB} = \underset{v_m \in V}{\arg\max}\, P_q(v_m) \prod_n P(a_n|v_m) = \underset{v_m \in \{yes,no\}}{\arg\max}\, P_q(v_m) \prod_n P(a_n|v_m). \tag{2}$$

Definition 8. If attribute A has Gaussian distribution, we use $f_y(A)$ and $f_n(A)$ to denote the probability of A when given acceptance and rejection respectively.

$$f_y(A) = \frac{1}{std_y(A)\sqrt{2\pi}} \exp\left(-\frac{(A - mean_y(A))^2}{2std_y(A)^2}\right),$$

$$f_n(A) = \frac{1}{std_n(A)\sqrt{2\pi}} \exp\left(-\frac{(A - mean_n(A))^2}{2std_n(A)^2}\right),$$

where $mean_y(A)$ and $mean_n(A)$ denote the mean of A when given acceptance and rejection respectively. And $std_y(A)$ and $std_n(A)$ denote the standard deviation of A when given acceptance and rejection respectively.

$$V_{NB|y} = P_{Sp_j}(y) P(Tl(Sr_i)|y) f_y(Pi(Sr_i, Sp_j)), \tag{3}$$

$$V_{NB|n} = P_{Sp_j}(n) P(Tl(Sr_i)|n) f_n(Pi(Sr_i, Sp_j)), \tag{4}$$

where $P(Tl(Sr_i)|y)$ and $P(Tl(Sr_i)|n)$ are the probability of $Tl(Sr_i)$ when given acceptance and rejection respectively.

Final Decision: If $h(Sr_i, Sp_j)$ =0 in the first step of decision, Sp_j will use (1) to make trust decision again based on its own prior knowledge together with recommendations gotten from recommenders, that is, to add the factor of recommendation in(2).

$$V_{NB|y} = P_{Sp_j}(y) f_y(R_{Sr_i}) P(Tl(Sr_i)|y) f_y(Pi(Sr_i, Sp_j)), \tag{5}$$

$$V_{NB|n} = P_{Sp_j}(n) f_n(R_{Sr_i}) P(Tl(Sr_i)|n) f_n(Pi(Sr_i, Sp_j)). \tag{6}$$

As shown above, when making trust decision (both with and without recommendations), our trust model compares the value of $V_{NB|y}$ and $V_{NB|n}$. Since the calculation of $V_{NB|y}$ and $V_{NB|n}$ involves different factors as well as the prior probability, which reflects the current acceptance rate of Sp_j and varies from time-to-time, $V_{NB|y}$ and $V_{NB|n}$ will keep on changing according to different situations.

4 Simulation Result

Using the method mentioned in section 3, we got the simulation results as showed in Fig.2. For the brevity, we only give the simulation results of decision making for Sp_j with recommendations from recommenders. The result of decision making without recommendations is similar to this case.

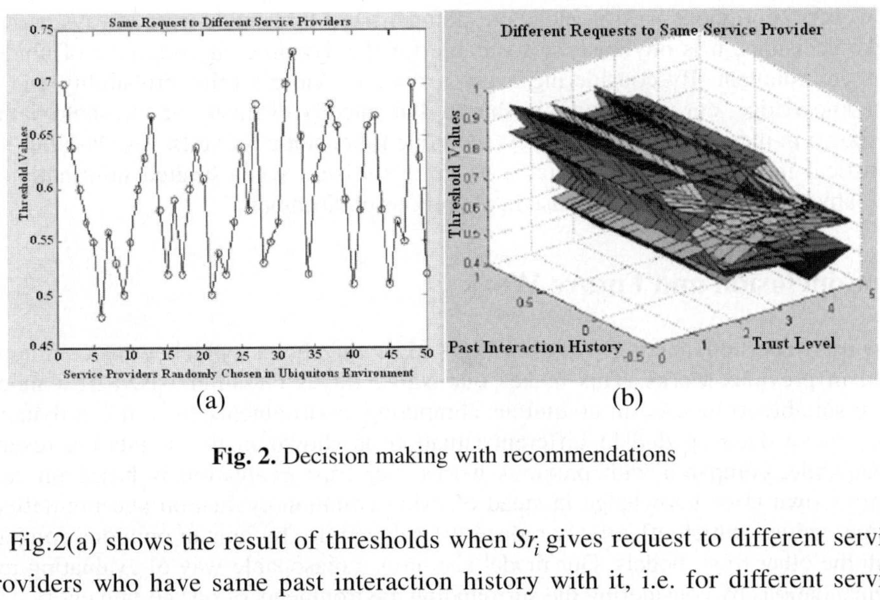

(a) (b)

Fig. 2. Decision making with recommendations

Fig.2(a) shows the result of thresholds when Sr_i gives request to different service providers who have same past interaction history with it, i.e. for different service provider Sp_m and Sp_n , $m, n \in N$, $Pi(Sr_i, Sp_m) = Pi(Sr_i, Sp_n)$. The thresholds here is the intersection of formulas (5) and (6), if the trust value is above threshold, our trust model will accept the request, otherwise, the request will be rejected. It is clear from the result of Fig.2(a) that when different service providers get the same request, even the past interaction histories between service provider and service requester are the same, the threshold is not a fixed value and it changes for different service providers. This is because the acceptance levels of different service providers are not the same.

Fig.2(b) gives the result of thresholds when different service requesters $Sr_1 \ldots Sr_k$, $k \in N$ give requests to same service provider Sp_j. Since all the requesters are given to the same service provider, Sp_j's acceptance level (i.e. prior probability) is same to all the service requesters. At the same time, our simulation set whole the recommendations given to different service requests to be the same, i.e., $R(Sp_m) = R(Sp_n)$ $1 \le m \ne n \le k$, $m, n \in N$ .However, Fig.2.(b) shows that the threshold keeps on changing. This is because of the variation in $Tl(Sr_m)$ and $Pi(Sr_m, Sp_j)$.

Our simulation results suggest that when requested by same service requester, different service providers or the same provider in different situations make different trust decisions. It is impossible to find a fixed threshold to make trust decision since the decision changed according to the entity's own prior knowledge. The results also give a look for the entity's dynamic trust decision with the variation of different factors. When one entity makes trust decision according to different service requesters, there is no fixed so-called threshold value for the service provider to make decision. However, in previous trust models, pre-defined fixed thresholds were always used to make decisions, it is obviously not suitable for the dynamic characteristic of ubiquitous environment. By considering every service provider's prior probability and its own knowledge, our trust model is able to dynamically evaluate the threshold values as shown in the simulation results. At the same time, since Naive Bayes classifier is a statistical method, it is also suitable to make decision when limited information is available, which is usually the case in ubiquitous environment.

5 Conclusion and Future Work

Our trust decision making avoids using simple thresholds, which were commonly used in previous works. This makes our Naive Bayes classifier based trust model more suitable to be used in ubiquitous computing environments since it can dynamically make decision due to different situation as shown in the simulation results. Meanwhile, compared with previous works, our trust evaluation is based on each entity's own prior knowledge in stead of using common evaluation and pre-defined weight values, which effectively reduce the subjectivity by human opinions compare with the other trust models. Our model also uses a reasonable way of evaluating recommendations by considering the surrounding environments of one certain entity.

We will add risk analysis in the coming work, since trust and risk always coupled tightly with each other. Other works like how to choose reliable recommenders to avoid unfair recommendations in ubiquitous trust model will also be involved in the coming work. We also propose to implement our trust model to be used in CAMUS, a middleware platform for a ubiquitous computing, in the future work.

Acknowledgment. This work is financially supported by the Ministry of Education and Human Resources Development (MOE), the Ministry of Commerce, Industry and Energy (MOCIE) and the Ministry of Labor (MOLAB) through the fostering project of the Lab of Excellency.

References

1. Marsh, S. P.: Formalising Trust as a Computational Concept. Ph.D. Thesis, University of Stirling (1994)
2. Blaze, M., Feigenbaum, J., Lacy, J.: Decentralized trust management. In Proc. of the 1996 IEEE Symposium on Security and Privacy, (1996) 164-173
3. Josang, A.: The right type of trust of distributed systems. In New security paradigms workshop, USA (1996) 119-131
4. English, C. and Nixon, P.: Dynamic Trust Models for Ubiquitous Computing Environments. In Proc. of the Fourth Annual Conference on Ubiquitous Computing, (2002)
5. Lamsal, P.: Requirements for modeling trust in ubiquitous computing and ad hoc networks. Ad Hoc Mobile Wireless Networks – Research Seminar on Telecommunications Software, (2002)
6. Ranganathan, K.: Trustworthy Pervasive Computing: The Hard Security Problems. In Proc. of IEEE Conference on Pervasive Computing and Communications, (2004)
7. Aime, M.D. and Lioy, A.: Incremental trust: building trust from past experience. In Proc. of IEEE International Symposium on a World of Wireless Mobile and Multimedia Networks, Italy (2005) 603-608
8. Shand, B., Dimmock, N., Bacon, J.: Trust for Ubiquitous, Transparent Collaboration. In Proc. of ACM: Special issue: Pervasive computing and communications, (2004) 711- 721
9. Liu, Z.Y., Jo, A.W., Thompson, R.A.:A Dynamic Trust Model for Mobile Ad Hoc NetWorks. In Proc. of 10th IEEE International Workshop on Future Trends of Distributed Computing Systems, (FTDCS'04) (2004) 80-85
10. Theodorakopoulos, G. and Baras, J.S.: Trust Evaluation in Ad-Hoc Networks. In Proc. of the 2004 ACM workshop on Wireless security, USA (2004) 1-10
11. He, R., Niu, J.W., Yuan, M.: A Novel Cloud-Based Trust Model for Pervasive Computing. In Proc. of the Fourth International Conference on Computer and Information Technology, (2004) 693-670
12. Guha, R., Kumar, R., Raghavan, P.: Propagation of trust and distrust. In Proc. of International Conference on World Wide Wed, USA (2004) 403-412
13. Michiardi, P. and Molva, R.: A collaborative reputation mechanism to enforce node cooperation in mobile ad-hoc networks. In Proc. of IFIP Communication and Multimedia Security Conference, Portoroz (2002) 107-121
14. Ganeriwal, S. and Srivastava, M.B.: Reputation-based framework for high integrity sensor networks. In Proc. of ACM Workshop on Security of ad-hoc and sensor networks, USA (2004) 66-77
15. Castelfranchi, C., Falcone, R., Pezzulo, G.: Trust in Information Sources as a source for Trust: A Fuzzy Approach. In Proc. of the second international joint conference on Autonomous agents and multiagent systems, (2003) 89-96

Context-Role Based Access Control for Context-Aware Application*

Seon-Ho Park[1], Young-Ju Han[1], and Tai-Myoung Chung[2]

[1] Internet Management Technology Laboratory,
Department of Computer Engineering,SungKyunKwan University,
440-746, Suwon, Korea +82-31-290-7222
{shpark, yjhan}@imtl.skku.ac.kr
[2] School of Information & Communication Engineering, Sungkyunkwan University,
440-746, Suwon, Korea +82-31-290-7222
tmchung@ece.skku.ac.kr

Abstract. The rapid growth of wireless network technology and the deployment of mobile computing devices have enabled the construction of pervasive computing environment. In pervasive computing environment, it is proliferated that many new applications that provide active and intelligent services by context information are collected by pervasive sensor devices. These new applications called context-aware applications must require new security mechanisms and policies different from typical ones. Specially, access control mechanism supports security policy that is based on context information, in order to provide automating context-aware services. So, this paper analyzes various access control mechanisms and proposes a context-role based access control mechanism for context-aware application.

1 Introduction

The development of computing and networking technology has changed existing computing environment into the ubiquitous computing environment. In existing computing environment, computing devices and applications are independent each other and are separated from human activities and environmental states. In ubiquitous computing environment, however, computing devices and applications can easily interoperate each other and interact with human activities by context-awareness. Context-awareness means that one is able to use context information. Context is any information that can be used to characterize the situation of an entity. An entity is a person, place, or object that is considered relevant to the interaction between a user and an application, including the user and application themselves.

From a security perspective, context-aware applications are a new challenge for exploring security policies that are intended to control access to sensitive

* This research was supported by the MIC(Ministry of Information and Communication), Korea, under the ITRC(Information Technology Research Center) support program supervised by the IITA(Institute of Information Technology Assessment).

M. Gerndt and D. Kranzlmüller (Eds.): HPCC 2006, LNCS 4208, pp. 572–580, 2006.

resources. Traditional security policy and access control mechanisms are based on only subject (users, processors) information. For example, in Access Control List (ACL) [1] that is a very commonly used access control mechanism, permission to access resources or services is controlled by checking for membership in the access control list associated with each object. However, this method is inadequate for context-aware applications of ubiquitous computing environment, because it does not consider context information. Granting a user to access without taking the user's current context into consideration can compromise security as the user's access privileges not only depend on the identifier of the user, but also on the user location and the users internal and environmental states. Thus, in the ubiquitous computing environment, to support automatically secure context-aware service and security policy based on varied context information, we must take context based access control into consideration. And also, the tool that manages efficiently complex security policy rule sets that contain various and large context information must be implemented [2,3].

In this paper, we present a new access control model that is suitable for context-aware applications in ubiquitous computing environment. The proposed model is Context-Role Based Access Control (CRBAC) that extends the role-based access control (RBAC) model. This model grants and applies permissions to users according to current context information. The context-role based access control model adds a notion of context-role to a traditional role based access control. The context-role represents environment state of the system by a mapping context-roles and context information. By using the uniform notion of a role to capture both user and context attributes, we will build an access control model that is easy to use and understand.

The reminder of this paper is organized as follows: Section 2 presents fundamental notion and model of traditional role-based access control and generalized role based access control mechanisms. Section 3 presents the proposed context-role based access control model illustrates formal definition and operation of context-role based access control. Section 4 presents a short evaluation and discussion about the model and its implementation. And section 6 concludes the paper.

2 Backgrounds and Related Works

2.1 Role Based Access Control

Role Based Access Control (RBAC) [4,5,6,7] is a form of non-discretionary access control that is designed for large, structured organizations. RBAC is an alternative to traditional discretionary access control (DAC) and mandatory access control (MAC). A major purpose of RBAC is to facilitate security administration and review. Many commercially successful access control systems for mainframes implement roles for security administration. A principle motivation behind RBAC is the ability to specify and enforce enterprise specific security policies in a way that maps naturally to an organization's structure. In RBAC, users are assigned to roles and roles are assigned to permissions. Because user-role associations change more frequently then role-permission associations, in most organizations, RBAC results in reduced administrative costs as compared to associating users directly with permissions.

To understand the various dimensions of RBAC, Sandhu et al [6,7] define a family of four conceptual models. First, $RBAC_0$ (the basic model) is indicating that it is the minimum requirement for any system that professes to support RBAC. $RBAC_1$ and $RBAC_2$ both include $RBAC_0$, but add independent features to it. $RBAC_1$ adds the concept of role hierarchies (situations where roles can inherit permissions from other roles). $RBAC_2$ adds constraints (which impose restrictions on acceptable configurations of the different components of RBAC). $RBAC_3$ includes $RBAC_1$ and $RBAC_2$ and, by transitivity, $RBAC_0$ [4,5,6,7].

RBAC is suited for large, structured organizations, for several reasons. First, roles and role hierarchies can reflect an organization's structure easily. So, RBAC encourages well-structured security policies that make sense in the context of the organization. Second, RBAC policies change very little over time, because permissions are assigned to roles, not users. Even though users frequently change jobs, the jobs them-selves seldom change [4,5,6,7]. But RBAC has a problem that its roles are inherently subject-centric. So, it cannot be used to capture security-relevant information from the environment which could have an impact on access decisions [8,9].

2.2 Generalized Role Based Access Control

Michael J. Convington et al have proposed the Generalized Role Based Access Control(GRBAC) model [8,9]. GRBAC is an extension of traditional Role-Based Access Control. It enhances traditional RBAC by incorporating the notion of object roles and environment roles, with the traditional notion of subject roles. An environment role can be based on any system state that the system can accurately collect. Object roles allow us to capture various commonalities among the objects in a system, and use these commonalities to classify the objects into roles.

But GRBAC has some problems. First, GRBAC is not suitable for large and complex organizations, because of defining too many roles in the system. Second, RBAC loses its advantage of data abstractions by object role. RBAC abstract user-level information from system level information by notion of permission is relationship between objects and operations. In GRBAC, because object role violates this relation-ship, the data abstraction could not be achieved. In addition, this problem violates user/role and role/permission associations. Finally, GRBAC has an unnecessary overlapping that environment roles and object roles, because certain physical environmental things can be also objects.

3 Context-Role Based Access Control Model

Traditional RBAC is very useful and offers an elegant solution to the problem of managing complex access control rule sets. But traditional RBAC has a problem that its roles are inherently subject-centric. So, it cannot be used to capture security-relevant information from the environment which could have an impact on access decisions. GRBAC can be used to capture security-relevant information from the environment. But, GRBAC has many problems as we mentioned in section 2.2.

Context-Role Based Access Control Model (CRBAC) supports the context based access control requirement of context-aware application in pervasive computing environment. CRBAC uses context-role in order to apply context information to access control decision. In this section, we first define Context-Role Based Access Control and then describe its operation.

3.1 CRBAC Definition

Based on the formalization of the traditional RBAC model, we present a formal definition of a CRBAC model. First of all, we describe following components of CRBAC:

- U(users): A user is an entity whose access is being controlled. U represents a set of user.
- C(context): C represents a set of context information in the system. C captures the context information that is used to define context role. Context information can be time, location, temperature, CPU usage, etc.
- R(roles): R represents a set of roles. A role has two roles that are user roles and context roles.
- UR(user roles): UR represents a set of user roles. It is equal to ROLE in traditional RBAC.
- CR(context roles): CR represents a set of context roles. The context role is used to capture security-relevant context information about the environment for use in CRBAC policies. The context role can contain time-related context role, location-related context role, etc.
- P(permissions): P represents a set of permissions. Permission is an approval to perform an operation on one or more CRBAC protected objects.
- S(sessions): S represents a set of sessions. A role is activated for user during each session. Activated role is a mapping between user roles and context roles.

CRBAC has three relations UA, PA, and CA that define the associations between user roles, user, permission assignments and context roles.

- UA: UA is the mapping that assigns a user role to a user.
- CA: CA is the mapping that assigns a context role to a context.
- PA: PA is the mapping that assigns permissions to a role.

The CRBAC model is illustrated in Figure 1. The CRBAC model shows relations and components of CRBAC model. Such relationships of components of CRBAC model is defined formally. Formal Definitions of CRBAC:

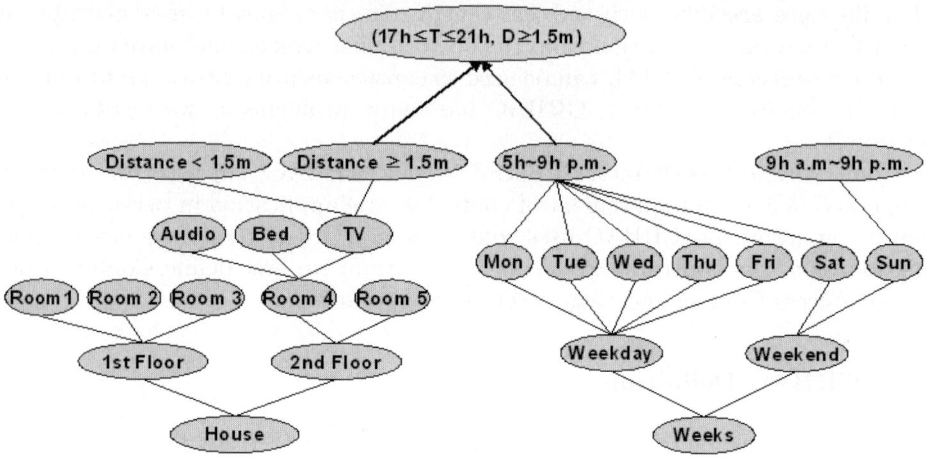

Fig. 1. Context-Role Based Access Control Model

- U, C, R, UR, CR, P, S(users, contexts, roles, user_roles, context_roles, permissions, sessions, respectively).
- $UA \subseteq U \times UR$, a many-to-many mapping user-to-user_role assignment relation.
- $assigned\_users(ur : UR) \rightarrow 2^U$, the mapping of user_role ur onto a set of users. Formally: $assigned\_users(ur) = u \in U|(u, ur) \in UA$
- $R \subseteq 2(UR \times CR)$, the set of roles.
- $PA \subseteq P \times R$, a many-to-many mapping permission-to-role assignment relation
- $assigned\_permissions(r : R) \rightarrow 2^P$, the mapping of role r onto a set of per-missions.
- $user\_sessions(u : U) \rightarrow 2^S$, the mapping of user u onto a set of sessions.
- $session\_roles(s : S) \rightarrow 2^R$, the mapping of session s onto a set of roles. Formally: $session\_roles(s_i) \subseteq r \in R|(session\_users(s_i), r) \in UA$

3.2 Context Roles

The context role is used to capture security-relevant context information in CRBAC policies. Context means situational information. Almost any information available at the time of an interaction can be seen as context information. Some examples [10] are:

- Identity
- Spatial information (e.g. location, orientation, speed, and acceleration)
- Temporal information (e.g. time of the day, date, and season of the year)
- Environmental information (e.g. temperature, air quality, and light or noise level)
- Social situation (e.g. who you are with, and people that are nearby)

- Resources that are nearby (e.g. accessible devices, and hosts)
- Availability of resources (e.g. battery, display, network, and bandwidth)
- Physiological measurements (e.g. blood pressure, heart rate, respiration rate, muscle activity, and tone of voice)
- Activity (e.g. talking, reading, walking, and running)
- Schedules and agendas

The context role shares many characteristics with user roles. So, context role has role activation, role revocation, and role hierarchies. Figure 2 show an example of role hierarchies of context role.

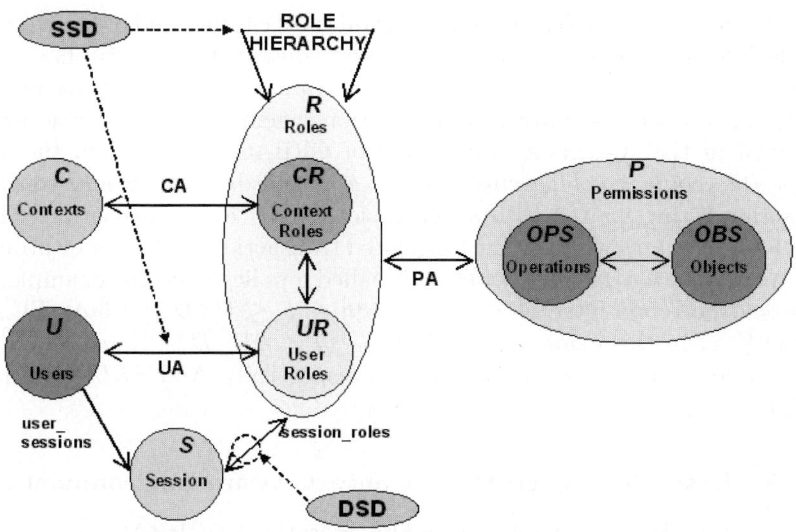

Fig. 2. Example of Context Role Hierarchies

In Figure 2,context role deal with two kind of context information that one is location-related context and the other is time-related context. Each context roles are hierarchically composed. And one context role may inherit from other kind of context roles. In Figure 2, context role $(18 \leq T \leq 21h, D \geq 1.5m)$ inherit from $(Distance \geq 1.5m)$ that is location-related context role and $(5h \sim 9h\ p.m.)$ that is time-related context role. Other context roles could be also hierarchically composed. For example, the temperature-related context role could have "body temperature" or "room temperature". Body temperature could also have "high temperature(over $37°C$)", "normal temperature(between $36°C$ and $37°C$), or low temperature(under $36°C$).

A request in traditional RBAC comes from a certain user who has a set of roles associated with her. This association is achieved via that define what roles U is allowed to take on based on the responsibilities and functions of U and that the set of roles are transferred to U and can subsequently be used during access requests. This is called role activation in RBAC. This role activation is also used for context roles. System administrator must define context roles and specify context variables and conditions that must maintain the values of the context variables for each context role. Then if an access request occurs, to mediate the access request, the system must gather context information and determine which of those context roles are active. Context role must require role revocation property. Activated context role may be invalid when related context state is changed. So policy rules for context role revocation should be carefully established.

3.3 Transactions of CRBAC

User roles and context roles provide powerful tools capture and organize security-relevant information about various users and context. But we still have not discussed how CRBAC is actually operated. In traditional RBAC, transactions are used to mediate access control. A transaction specifies a particular action to be performed in the system. A transaction of CRBAC is a tuple in the form of $< user\_role, context\_role, permission >$. A policy database would consist of a transaction listing, paired with a permission bit for each transaction. The permission bit indicates whether the associated transaction is allowed or prohibited. Each $< transaction, permissionbit >$ is called a policy rule. For example, policy rule would be represented as $\ll child, (18h \leq T \leq 1h, D \geq 1.5m), TV\_ON >$ $, ALLOW >$. In this example, $child, (18h \leq T \leq 21h, D \geq 1.5m)$ and TV_ON are user role, context role and permission, respectively. And $ALLOW$ is permission bit.

3.4 An Example of CRBAC in Context-Aware Environment

In this section, in order to illustrate the operation of CRBAC, an example of context-aware application in ubiquitous computing environment is illustrated. The environment of our example is an intelligent home that implements a home network and Context-Aware infrastructure. The intelligent home is equipped with various networked sensors/actuator devices such as cameras, microphones, RFID based location sensors, and so on. This intelligent home presents new and interesting security challenges. Given the sensitivity of context information that is generated and stored in the intelligent home, security policies can potentially be quite complex. A policy can restrict access to information or resources based on several factors, including user attributes, the specific resource or the context information. For example, users can be classified as "parent" or "child", "resident" or "guest", or even as "people" or "pet". Then, access rights can depend on the user's classification, as well as the associated identity. In addition, access control may be restricted based on the user's location, or based on environmental factors such as the time of day or temperature. For example, in the home, parents may restrict their children's access to the TV, allowing the watching of TV between 6:00 p.m. and 9:00 p.m. and only while they are 1.5 meters away from a TV. In this instance, the access control policy depends on context information composed of location and time information. So, certain CRBAC policy rule can be created for restriction of children's access to the TV. Security administrator must specify user role, context role and permission for composition of transaction. The transaction for this example is composed as $< child, (18h \leq T \leq 21h, D \geq 1.5m), TV\_ON >$. Because this transaction is a condition to allow, permission bit should be set to $ALLOW$. Consequently, policy rule is composed as $\ll child, (18h \leq T \leq 21h, D \geq 1.5m), TV\_ON >, ALLOW >$. This policy rule must be stored in security policy base and checked at each time the request of access control occurs.

4 Discussion and Evaluation

This paper introduced context roles and provided the context role based access control model. The major strength of CRBAC model is its ability to make access control decisions based on the context information. As we mentioned in section 2, traditional RBAC and Generalized RBAC does not appropriate for context-aware systems. Because traditional RBAC is inherently subject-centric, it cannot be used to capture security-relevant context from the environment which could have an impact on access decisions. And Generalized RBAC has three problems. The first problem is defining too many roles in the case of large system, the second problem is difficult to abstract data, and the third problem is unnecessary overlapping between object roles and environment roles. These problems make security policy is complicated and system overhead is increased, because too many security elements must be maintained. And because some object roles are also defined as certain environment roles, object role is unnecessary. Grouping objects may be more efficient than object roles. So Generalized RBAC does not also appropriate to context-aware applications. CRBAC meets requirements for the context-aware security service. Because role of CRBAC is composed subject roles and context roles, CRBAC can be used to capture security-relevant context from the environment which could have an impact on access decisions. And CRBAC maintains security principals that traditional RBAC originally possesses.

To successfully implement CRBAC in the context-aware environment, the several issues should be considered. First, the confidentiality of the context information must be guaranteed in access control process and transmission of context information. Compromised context information can cause a system make incorrect access control decisions and can comprise the security of the systems. Second, every time access control decision occurs, to check on activation of context role is too expensive. So the system should implement an efficient means of role entry testing for context roles.

5 Conclusion

In this paper, we introduced the Context-Role Based Access Control (CRBAC) model that provides context-based access control for context-aware applications in pervasive systems. And also, we explored traditional RBAC and Generalized RBAC and pointed out problems of these mechanisms. CRBAC has extended traditional role-based access control including the concept of context role. We are focused on solving the problem of secure and automatized service of context-aware applications in pervasive systems. Our work shows a formal definition of CRBAC, a notion and characters of context role, and a simple operation of CRBAC by using simple example of intelligent home.

References

1. Vinay Ahuja and Vijay Halaharvi : 'Network and Internet Security', IANE Publishing Company, 1999.
2. G. Zhang and M. Parashar: 'Context-aware dynamic access control for pervasive computing', In 2004 Communication Networks and Distributed Systems Modeling and Simulation Conference (CNDS'04), San Diego, California, USA, January 2004.
3. Michael J. Convington, Wende Long, et al.: 'Securing Context-Aware Applications Using Environment Roles', SACMAT'01, Chantilly, Virginia, USA, May 3-4, 2001.
4. David F. Ferraiolo, D. Richard Kuhn, Ramaswamy Chandramouli, ' Role-Based Access Control', Artech House, 2003.
5. Ravi S. Sandhu, Edward J. Coyne, et al.: 'Role-Based Access Control Models, IEEE Computer, Volume 29, Number 2, pages 38-47, Feb. 1996.
6. Ravi S. Sandhu : 'Role based access control' In Advances in Computers, volue 46. Aca-demic Press, 1998.
7. David F. Ferraiolo, Ravi Sandhu, et al. : 'Proposed NIST Standard for Role-Based Access Control', ACM Transac-tions on Information and System Security, Vol. 4, No. 3, Aug. 2001.
8. M. M. Moyer, M. J. Convington, et al. : 'Generalized role-based access control for securing future applications', In 23rd National Information Systems Security Conference. (NISSC 2000), Baltimore, Md, USA, Oct. 2000.
9. Matthew J. Moyer and Mustaque Ahamad : 'Generalized role based access control', Proceed-ings of the 2001 International Conference on Distributed Computing Systems (ICDCS), Mesa, AZ, Apr. 2001.
10. Mari Korkea-aho : 'Context-Aware Applications Survey', Internetworking Seminar, Apr. 2000.

Context Communication for Providing Context-Aware Application's Independency*

Kugsang Jeong, Deokjai Choi, and Gueesang Lee

Department of Computer Science, Chonnam National University,
300 Yongbong-dong, Buk-gu, Gwangju 500-757, Republic of Korea
handeum@iat.chonnam.ac.kr, {dchoi, gslee}@chonnam.ac.kr

Abstract. The behavior of context-aware applications depends on various aspects of the surrounding context. Most context-aware applications are dependent on middleware from the point of how to subscribe and query dynamically changing contexts. This dependency prevents application developers from easily developing context-aware applications because these applications either require knowledge of the internal architecture of the middleware or the help of an administrator. To relieve application dependency on middleware, we propose context communication between applications and middleware. An application communicates with middleware to register rules, which predefine interesting contexts, and to request context queries to acquire information. The context communication can allow application developers to spend time on developing application logic and less time on middleware architecture or its administration. We expect that our work will make it easier for developers to develop various context-aware applications.

1 Introduction

One of the key aspects of ubiquitous computing is context awareness. Ubiquitous computing in its early stage meant invisible interface computing with calm technology [1]. Ubiquitous computing has been evolved into more advanced computing giving rise to new concepts and various technologies. One such technology is context-aware computing. In context aware computing, applications are aware of contexts [2]. One example of context awareness is when an instructor enters a classroom to start a lecture, a context-aware application recognizes that he will start a lecture in the room and then projects a lecture file onto a screen. In the context-aware computing environment, users such as instructors can make use of services which automatically do tasks they would normally be required to do manually. Today context awareness is an important concept for ubiquitous computing and many context-aware applications are showing future possibilities for ubiquitous computing. With many more applications emerging, ubiquitous computing will soon become a common reality.

* This research was supported by the Program for the Training of Graduate Students in Regional Innovation which was conducted by the Ministry of Commerce Industry and of the Korean Government Regional Innovation which was conducted by the Ministry of Commerce Industry and Energy of the Korean Government.

M. Gerndt and D. Kranzlmüller (Eds.): HPCC 2006, LNCS 4208, pp. 581 – 590, 2006.

Context-aware applications adapt their behavior to dynamically changing contexts in environments. The middleware in ubiquitous computing infrastructures maintains context information which are sensed by sensors and predefined. Context-aware applications use context information to evaluate current contexts. There are two types for context-aware applications to adapt to current contexts: context trigger and context query. Context trigger applications have their own contexts and perform specific behaviors whenever a situation arrives at a predetermined context. Therefore context decision-making is important for triggering an application. The Labscape application of one.world infrastructure defines contexts as user's location changes in order to transfer experimental data to a computing device around a user whenever a user moves to a new location [3]. Context query applications acquire necessary context information from middleware during their runtime. The Context-Aware Mailing List application obtains a list of members who are in the building, and delivers messages only to them [2]. Context-aware computing architecture should support both context trigger and query function to allow applications to adapt their behavior to dynamically changing contexts.

In previous ubiquitous computing infrastructures supporting context-awareness context trigger and query functions are dependent on middleware's internal architecture or its administration. In earlier stage, applications received and requested context information from internal components that had context information which they are interested in [2]. Even if application developers use a discovery service, they needed to know the architecture of the middleware components. The next stage of infrastructures provided context information modeling which made it unnecessarily for application developers to know the middleware architecture [4]. To obtain context information, application developers should predefine rules based on context information models. The rule based context trigger and query separate application developers' concerns from the middleware architecture, but application developers still have dependencies on middleware because context information model is not a common model for sharing knowledge but a middleware-specific model and the registration of predefined rules can be performed by an administrator. Recent infrastructures support knowledge sharing context information model by using ontology [5, 6, 7, 8]. The rules for context trigger and query are based on ontology based context information model. They still need the help of administrators to register rules for context trigger into middleware. And they request middleware to query context information through remote invocation. The ways for context trigger and query described above may prevent context-aware applications from being developed easily, so the context-aware architecture should reduce the application's dependency.

To minimize the dependency of context-aware application on middleware and at the same time provide two types of context-aware applications, we propose context communication between context-aware applications and middleware to register context decision rules for context trigger and request context queries. Under our architecture, application developers compose rules for making a context decision and register them to the middleware through communication messages between applications and middleware. The middleware then generates a context object that evaluates the context decision rules and informs application of context information when the situation satisfies the registered rules. So when application developers develop applications, they do not have to modify middleware or know internal components at all. In case of context query, the application delivers a query statement to a relevant context object then the context object performs the query and transfers results to the application. All application developers need to do is to create context decision rules to register and to request a query to the context object and receive the results through communication messages. Therefore this work helps application developers create various context-aware applications easily, as needs arise to accelerate the evolution of ubiquitous computing environment.

Section 2 summarizes related works from the point of how dependent applications are on middleware. Then we describe our architecture and context communication in Section 3 and Section 4. We show application examples using our architecture in Section 5. Lastly Section 6 concludes our work.

2 Related Works

There are many infrastructures to support context-awareness for ubiquitous computing. In this section, we describe how they make a context decision for context trigger and query context from the point of applications' dependency on middleware.

Context Tool Kit provides context query and asynchronous events for context trigger [2]. After obtaining widget location through discovery service, applications subscribe to widget callback by using a specific method. Context decision is made by application's logic with all information issued from widgets. Applications can request context information by query method. So application developers are dependent on middleware's components because they should know in advance internal components like a location widget having context information which they are interested in. Also, they have to use middleware-specific methods for context query.

One.world provides context query and asynchronous events [3]. It maintains tuples based on data model to store and query information. Context query to tuples is performed by the structured IO operations and its own query language. Context decision is made by application's logic with all asynchronous events issued from internal components. Context trigger in one.world is also dependent on middleware's components. Context query uses internal components-independent query language based on tuples, though it uses middleware-dependent data model and query language.

GAIA provides the event manager to deliver contextual events to applications instead of explicit context decision function and the context service for queries with first order logic [4]. As applications should know event channels which provide information to make a context decision, and rules for query should be registered manually to the registry, applications are dependent on GAIA.

In SOCAM [7], application developers should create pre-defined rules in a file for context decision and pre-load them into context reasoning module in middleware. These steps can not be done without administrator's operation.

Some infrastructures can not provide both context trigger and context query while they provide one of the two functions independent on middleware. The Context Broker also supports context query function based on its own inference rules but does not support context decision function [5]. The Semantic Space provides context query engine based on RDF Data Query Language so that applications are independent of middleware. It does not, however, consider context decision function [6]. Context Management System has contributed to decoupling application composition from context acquisition [8]. To build a context-aware application in this architecture, an application developer composes an application by selecting building blocks and sets the context decision rules by using a tool. Context decision-making function is independent of middleware, but it does not provide context query function.

3 Context-Aware Computing Architecture

Our context-aware computing architecture, UTOPIA, is our initial work for realizing pervasive computing and is focusing on supporting not only application-defined context decision-making but context query [9]. The architecture of UTOPIA is shown in Figure 1. An application defines its application context as a context decision rule based on shared ontology and then registers the rule to the Context Register. The Context Register authenticates applications and accepts an application's context rules. The Context Object Generator generates an application-specific context object which evaluates the application's context rule. Whenever any situation in surrounding environment is changed, the application context object evaluates whether changed situation meets the rule. The Shared Ontology Base stores all context information in environments. The Event Provider signals context objects in order to re-evaluate the rule whenever context information, which context objects are interested in, changes. The Ontology Provider wraps sensed or predefined information into context information, and provides the Shared Ontology Base with the information.

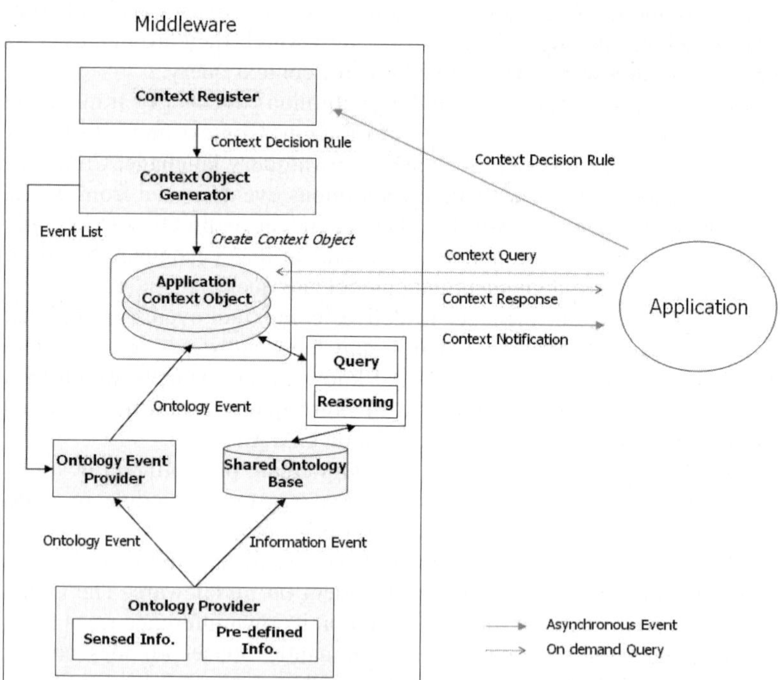

Fig. 1. UTOPIA Architecture

The context decision rule is a set of rules to describe the application context which a specific application is interested in. The context decision rule has the form of antecedent and consequent like *isIn(?x, ?y) ^ Instructor(?x) ^ LectureRoom(?y) => Lecture_start*. The context query is to acquire necessary context information from middleware. The statement for context query conforms to the syntax of Bossam which is one of an OWL query engine like *query q is isIn(?x, room_410) and Students(?x);* [10].

UTOPIA evaluates an application's context decision rules in order to determine whether the application is interested in the current situation and responds to context information query requested from the application. UTOPIA needs context communication between middleware and applications when applications register the rule into UTOPIA, request context query, and middleware notifies application context and responds to context query. The context communication described later in this paper is one of ongoing works to complete the implementation of UTOPIA.

4 Context Communication

4.1 CRDP (Context Register Discovery Protocol)

The context register waits for a request from an application and authenticates the application with an ID and password. The first step of an application is to contact the Context Register of infrastructure. To achieve this, we have implemented a simple Context Register Discovery Protocol.

The CRDP is a protocol that allows clients to dynamically obtain IP address and port number of context register. It operates similarly to Dynamic Host Configuration Protocol [11]. Application as a client broadcasts a discovery packet and context register as a CRDP server responds by sending an offer packet includes IP address and port number of Context Register.

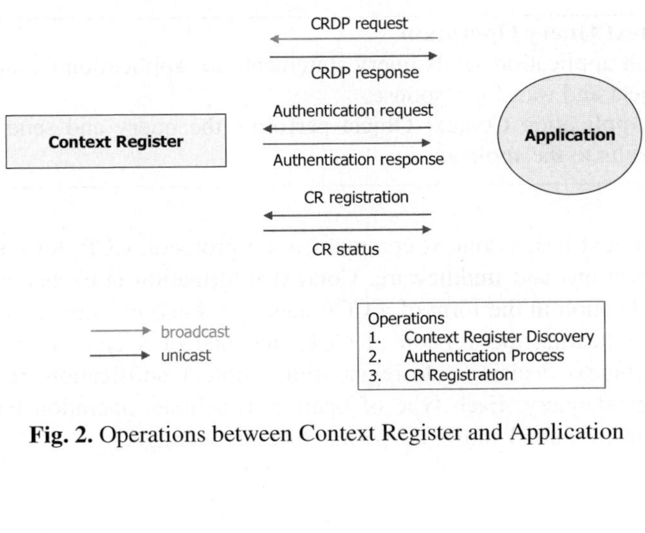

Fig. 2. Operations between Context Register and Application

After finding the location of Context Register, an application starts the authentication process. If the authentication is successful, an application can register its context rule to the Context Register. At authentication, the Context Register must figure out application's address and port number to use when the Application Context Object notifies context event to application. The Context Register then hands the context rule and application's location to the Context Object Generator. Figure 2 shows operations between the Context Register and a new application.

4.2 CCP (Context Communication Protocol)

The CCP supports both context trigger and context query. Asynchronous context notifications for context trigger are performed by subscribe/publish mechanism. A context-aware application registers context decision rule into context register for subscribing context. The application context object publishes contexts issued by making a context decision according to rules. The application subscribes contexts from the application context object. Context query is performed by request/respond mechanism. An application sends query statements to application context object to request a query. The application context object performs the query and sends the results to the application. Table 1 shows sequence of CCP operations

Table 1. Sequence of CCP operations

Context Trigger Operation
1. an application registers context decision rules to Context Register 2. CR verifies the syntax of rules and send acknowledgement to the application 3. application waits for contexts issued from Application Context Object 4. Application Context Object publishes the result of context decision-making
Context Query Operation
1. an application sends query statements to Application Context Object and wait for response 2. Application Context Object performs the query and send the results to the application

We propose text based context communication protocol, CCP, for communication between applications and middleware. Context information is exchanged between a co and an application in the form of a CCP message. Each message consists of a version number indicating the version of CCP, and one of 4 types of CCP operation which mean context decision rule registration, context notification, request and response to context query. Each type of operation includes operation type and type-specific contents. Figure 3 shows the format of each message.

version	operation type	operation message

a. CCP Message

register	register-id	decision rule 1	decision rule 2	...	decision rule *n*

b. Register Operation Message

notification	notification-id	application context	variable-value 1	variable-value 2	...	variable-value *n*

c. Notification Operation Message

request	request-id	query statement

d. Request Operation Message

response	request-id	variable-value 1	variable-value 2	...	variable-value *n*

e. Response Operation Message

Fig. 3. CCP Formats

We have implemented CCP by using socket and JMS (Java Message Service) [12]. To register context decision rule, we used socket based communication to register context decision rule and JMS to notify asynchronously contexts and request/respond to context query. Applications as socket clients request rule registration to Context Register which plays socket server's role. From the point of JMS, applications are subscribers for context which is published from an Application Context Object and requesters for context query.

The JMS provides subscribe/publish mechanism using topic connection factory in messaging system and request/response using queue connection factory. The Java Naming Directory Interface to find objects through names associated with objects can be used with the JMS. To communicate with middleware, applications need the name of JNDI (Java Naming and Directory Interface) factory [13], names of topic connection factory and topic for subscribe/publish service, and names of connection factory and queue for request/respond service. These information are predefined in middleware and the context register maintains them in a pool. They are dynamically assigned to applications and Application Context Objects by Context Register. When applications terminated, all information returned.

Context trigger is performed as followings. Applications obtain IP address and port number of Context Register through context register discovery protocol. The Context Register simply authenticates applications by id/password. After authentication, the application registers context decision rules to Context Register by *register* operation of CCP. The Context Register verifies the rules and sends applications acknowledgement piggybacking names of JNDI factory, topic connection factory and topic, and queue connection factory and queue for context query. Also the Context Register sends the rule with names of topic connection factory and topic to Context Object Generator. Then Context Object Generator generates an Application Context Object which makes a context decision according to rules and notifies asynchronously contexts through topic publisher of JMS. Applications install a topic subscriber and subscribe context notification from Application Context Object.

Context query is performed by *request* and *respond* operation after Application Context Object is generated. Applications send query statements to Application Context Object by *request* operation and wait for response by *respond* operation. If applications want to use only context query without context trigger, applications should register null rule so that it generates Application Context Object for performing context query.

5 Applications Using Context Communication

5.1 ULecture

ULecture is an application using both context trigger and context query communication. This application downloads a lecture material and projects it on the screen when a lecture starts. Application's context for ULecture is that a lecture starts at a lecture room. Context query is used to find the location of lecture material to download.

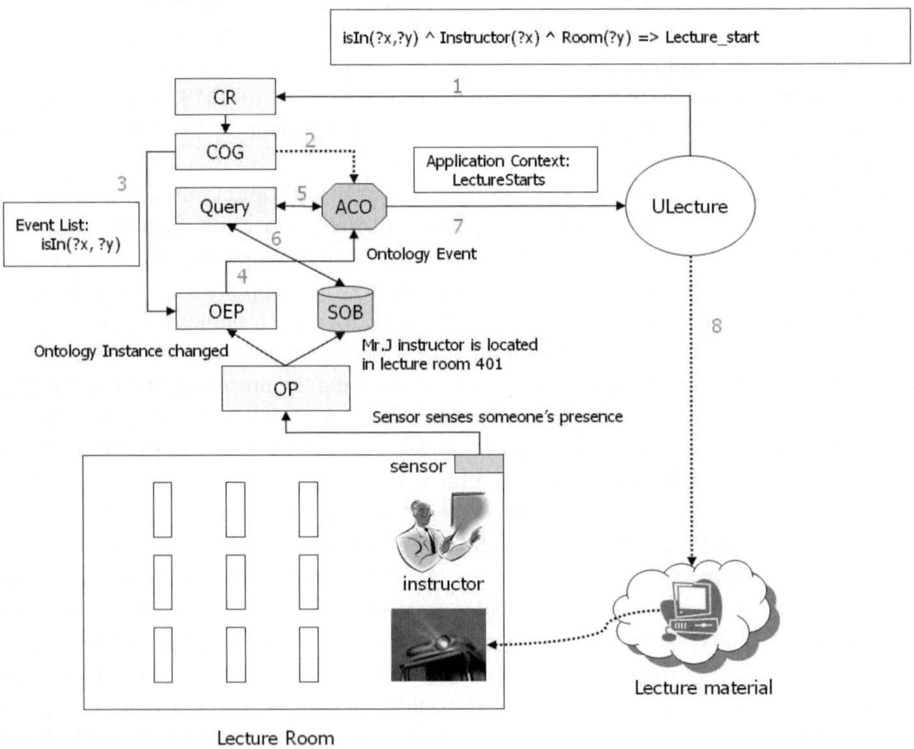

Fig. 4. ULecture Application

The context decision rule for ULecture is the same described in section 3. When being executed at first, ULecture registers the rule to Context Register with *register* operation of CCP. After registration, ULecture installs a topic subscriber and waits for context notification. In the middleware, an Application Context Object for ULecture is generated and an event list for Ulecture is registered into Ontology Event Provider. If a person enters a room, this context change is notified from sensor to Ontology Provider, ontology event provider and ULecture's Application Context Object in sequence. Then Application Context Object performs a context decision-making. If the person is an instructor and the room is a lecture room, context which the lecture is beginning is triggered. So Application Context Object publishes predefined context with instructor name and lecture room ID which are described as variables in the rules.

ULecture listens to context and request context query to find the location of lecture material with instructor's name. The query statement is *query q is teachLecture(K.Jeong, ?x) ^ hasLecture(Room_410, ?x) ^ hasLectureFileURL(?x, ?y);*. It means what the file location of lecture material which K.Jeong instructor teaches at room 410 is. The ULecture obtains the location of file from response of Application Context Object, and downloads it and turns on the projector in lecture room to project a file on the screen. We controlled a projector having infrared interface by using uATMEGA16 kit [14] which can record infrared control signals. Figure 4 shows the steps described above.

5.2 Context-Aware Mailing List

UAML is an application using context query, which is an upgraded version of one of the applications of Context Tool Kit. While original application delivers messages only to members of group who are currently in the building [2], our UAML can change dynamically recipients according to the result of context query. This application uses only context query, so it registers rules as a null. The Application Context Object for UAML is generated without context decision making for context trigger. CAML request context query according to user's operation. For example, when a user wants to deliver messages to members in room 410, the user just write query statement to get a list of members and their mail addresses: *query q isIn(?x, Room_410) ^ IAT(?x) ^ hasMailAddr(?x, ?y);*. Then UAML send this statement to CAML context object. Then CAML's Application Context Object performs and responds to context query. The UAML receives a list of the members and delivers messages to members in a list.

6 Conclusion

Context-aware applications in previous ubiquitous computing infrastructures are dependent on middleware. It means that applications should know middleware's internal components or need the help of an administrator to obtain context which they are interested in. Application dependency on middleware prevents application developers from easy development and deployment.

We proposed context communication supporting context trigger and context query to reduce application dependency. Context communication is performed by CRDP and CCP. The CRDP is to discover context register in middleware. The CCP is to deliver asynchronous context notification and request/respond to context query. The

CCP has 5 operations: register, notify, request and respond. Applications should find a Context Register by using CRDP to register context decision rules. After successful registration, applications wait for context notification from Application Context Object whenever context changes. Also applications can request context query to obtain necessary context information. We implemented ULecture and CAML with CRDP and CCP to show the effectiveness of context communication. Developers can develop easily applications which use context trigger and context query through context communication because they just predefine rules and query statements without regarding middleware's internal components and its operation.

References

1. Weiser, M.: The Computer for the 21st Century. Scientific American. 265, pp. 94-10.
2. Dey, A.K.: Providing Architectural Support for Building Context-Aware Applications. PhD Thesis, (2000)
3. Grimm, R.: System Support for Pervasive Applications. PhD Thesis, (2002)
4. Roman, M., Hess, C., Cerqueria R., Ranganathan, A., Campbell, R., Nahrsted, K.: Gaia: A Middleware Platform for Active Spaces. IEEE Pervasive Computing (2002) 74-83
5. Chen, H.: An Intelligent Broker Architecture for Pervasive Context-Aware Systems. PhD Thesis, (2004)
6. Wang, X., Jin, S.D., Chin, C.Y., Sanka, R.H., Zhang, D.: Semantic Space: An Infrastructure for Smart Spaces. PERVASIVE computing (2004)
7. Gu, T., Pung H.K., Zhang, D.: A Service-Oriented Middleware for Building Context-Aware Services. Journal of Network and Computer Applications (JNCA) Vol. 28. (2005) 1-18
8. Christonpoulou, E., Goumopoulos, C., Kameas, A.: An ontology-based context management and reasoning process for UbiComp applications. Joint conference on Smart objects and ambient intelligence: innovative context-aware services: usages and technologies. (2005)
9. Jeong, K., Choi, D., Lee, G., Kim, S.: A Middleware Architecture Determining Application Context using Shared Ontology. LNCS Vol. 3983 (2006)
10. Jang, M., Sohn, J.: Bossam: an extended rule engine for the web, Proceedings of RuleML, LNCS Vol. 3323 (2004)
11. Droms, R.: Dynamic Host Configuration Protocol. RFC 2131, IETF (1997)
12. Java Message Service, http://java.sun.com/
13. Java Naming and Directory Interface, http://java.sun.com/
14. uATMEGA16, http://www.digihobby.co.kr/

A Heterogeneous Embedded MPSoC for Multimedia Applications

Hong Yue, Zhiying Wang, and Kui Dai

College of Computer, National University of Defense Technology,
Changsha, Hunan, P.R.China 410073
yuehong_nudt@yahoo.com

Abstract. MPSoC are attractive candidate architectures for multimedia processing as multimedia schemes generally can be partitioned in control-oriented and data-dominated functions, which can all be processed in parallel on different cores. This paper presents a heterogeneous embedded MPSoC for a wide range of application fields with particularly high processing demands. It integrates three processor cores and various interfaces onto a single chip, all tied to a 32-bit AMBA AHB bus. The RISC core coordinates the system and performs some reactive tasks, and the cluster composed by two DSP cores perform transformational tasks with more deterministic and regular behaviors, such as the small and well-defined workloads in multimedia signal processing applications. The DSP cores are designed based on Transport Triggered Architecture (TTA) to reduce hardware complexity, get high flexibility and shorten market time. The processor is fabricated in 0.18um standard-cell technology, occupies about 21.4mm^2, and operates at 266MHz while consuming 870mW average power.

Keywords: Heterogeneous MPSoC, Transport Triggered Architecture, DSP, Embedded Processor.

1 Introduction

Embedded multimedia applications can be categorized into control-oriented and data-dominated, the computing platforms for multimedia processing always need to effectively handle both control-intensive and data-intensive tasks. Recent RISC architectures have been enhanced for data-intensive tasks by incorporating single-cycle multiply-accumulators, SIMD datapaths, or specific functional units [1] to have the ability to perform data-intensive tasks, but the performance is still far behind that of a DSP with similar computing resources [2]. This is because data-intensive tasks are very distinct from general-purpose computations. Similarly, DSP is enhanced to have some GPP features but it is not effective.

MPSoC are attractive candidate architectures for multimedia processing as multimedia schemes in general can be partitioned in control-oriented and data-dominated functions, which can all be processed in parallel on different cores. Each core can be adapted towards a specific class of algorithms, and individual tasks can be mapped efficiently to the most suitable core. The OMAP media processor of Texas Instruments company and the Tricore processor series of Infineon company are both this kind of processor [3][4].

M. Gerndt and D. Kranzlmüller (Eds.): HPCC 2006, LNCS 4208, pp. 591 – 600, 2006.

In general, different approaches are adapted to accelerate execution on embedded processors. In most cases, some kind of parallelization technique is employed on different levels [5]. Even combination of these techniques is used. Another very powerful means to accelerate multimedia processing is to adapt processors to specific algorithms by introducing specialized instructions for frequently occurring operations of higher complexity [6].

This paper presents a heterogeneous embedded MPSoC for both RISC and DSP tasks. It is an upgrade version of the processor illustrated in [17]. It integrates three processor cores and various interfaces onto a single chip, all tied to a 32-bit AMBA AHB bus. One core is a RISC core which coordinates the system and performs reactive tasks such as user interface; the other is a cluster composed by two novel VLIW DSP cores designed based on Transport Triggered Architecture (TTA) which performs data-intensive tasks with more deterministic and regular behaviors. Applications are mapped to the three cores according to the characteristics, small and well-defined DSP algorithms are executed on DSP cores, and the others are executed on RISC core.

This paper is organized as follows. In section 2, the proposed multi-core processor architecture is presented. The DSP cores designed based on TTA are described in section 3. In section 4, we will describe inter-processor cooperation in both the hardware and software way in detail. In section 5, the simulation and implementation results are reported. Section 6 concludes the paper.

2 SoC Architecture Overview

The SoC architecture, shown in Figure 1, comprises three cores that have been specifically optimized towards a particular class of tasks by employing different architectural strategies. The RISC processor is a 32bit processor compliant with the SPARC V8 architecture; it is optimized towards control-oriented tasks such as bitstream processing or global system control. The cluster composed by two TTA-DSP processor cores is particularly optimized towards high-throughput DSP-style processing, such as FFT, IDCT or filtering are both eight-issue VLIW architecture. The two cores are both designed based on TTA architecture; they offers high data level parallelism and high programming flexibility, more details are described in the next section.

A 32-bit AMBA AHB system bus [7] connects the three cores to off-chip SDRAM memory and flash memory via a 32-bit memory interface, to JTAG interface for debug, to SATA interface for access to the hard disk, and to the I/O bridge for access to different peripherals. The direct change of data between TTA-DSP Cache and external memory without placing a burden on the TTA-DSP cores is supported by the DMA tunnels. A 64kB Scratch-Pad memory exists to store some constant data that are used frequently in one procedure. If these data are always loaded from external memory, it is very time-consuming. Buffer between TTA-DSP DCache and system bus is to hide the memory access latency. The inter-processor communication is handled by IPCI (Inter-Processor Communication Interface), the detailed mechanism will be described in Section 4.

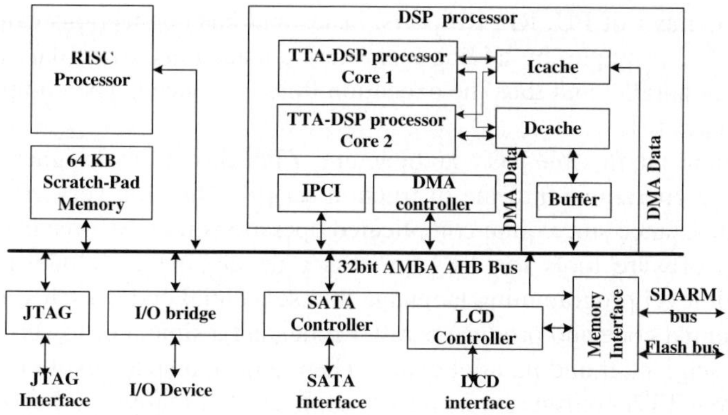

Fig. 1. Dual-core system-on-chip architecture

3 TTA-DSP Cluster

The two TTA-DSP cores are both highly parallel DSP cores with VLIW architecture. The processor core is designed based on TTA (Transport Triggered Architecture). Simple design flow, customized function units and flexible data level programmability make it very suitable for embedded DSP applications.

3.1 Transport Triggered Architecture

The main difference of TTA compared to traditional, operation triggered architecture is the way the operations are executed. Instead of triggering data transports, in TTA operations occur as a side effect of data transports, i.e., the execution begins when data is written to operand registers. This design implies that one data transport is needed to be explicitly appointed in the instruction. Figure 2 shows the basic TTA processor architecture [8].

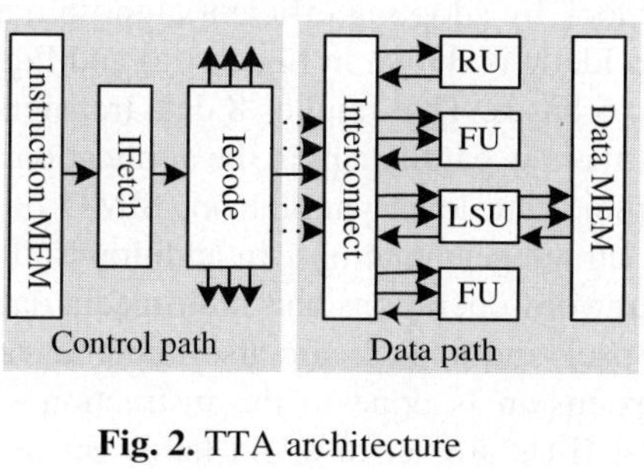

Fig. 2. TTA architecture

The data path of TTA is organized as a set of function units (FU) and register files (RF). The data transfers between these units are done through an interconnection network that consists of desired number of buses. The architecture is very flexible because the number of FU, RF, RF ports, buses and bus connections can be changed unlimitedly. For example, by adding buses and connections, more data transfers can be executed in parallel and thus the execution time is reduced. The complier supports all these changes [8].

In addition to the flexible basic architecture, TTA allows the designer to add application specific operations into the instruction set [9]. This is a very efficient way to improve performance since quite complicated operations may become trivial.

The user software tools for TTA processor development include a compiler to translate high-level programming language into sequential code, a scheduler to schedule the sequential code and produce parallel code, and a simulator to verify and evaluate both the sequential and parallel codes. Then, a design space explorer can be used to test different TTA configurations for the application. Finally, synthesizable Verilog code can be automatically generated for the chosen configuration using the Processor Generator.

3.2 TTA-DSP Cluster Architecture

The two TTA-DSP cores are both designed based on TTA according to the characteristics of multimedia signal processing applications. They compose a cluster to process data-oriented tasks. Using a multi-level multi-granularity instruction customization algorithm [19], the instruction sets of the two cores are decided. Therefore, different special function units are customized to execute some specific operations in two cores. That also means special instructions are added to the instruction set. Figure 3 shows the architecture of the two TTA-DSP cores. Both cores have 4 integer function units (IFU) and 2 floating-point function units (FFU), 2 load/store units (LSU), 2 integer register files (IRF) and 2 floating-point register files (FRF). Difference is that one core has a TriU unit and another has an IdctU. TriU is a function unit specific for trigonometric function calculation. It is designed based on CORDIC algorithm, so it can calculate many trigonometric functions in 32 cycles instead of other time-consuming or place-consuming ways [10]. And IdctU is an IDCT hardware engine to process the whole IDCT algorithm. Only less than 64 cycles are needed to finish the transform of an 8*8 block by adopting efficient algorithm and well-defined hardware architecture. TriU and IdctU is shown in Figure4(a) and Figure4(c). All these function units are connected to 8 buses. That implies 8 data transfers can be executed simultaneously provided that there is parallelism in the application.

To thoroughly exploit data level parallelism, SIMD data path is designed in the IFU. IFU integrates rich integer functions. In addition to the basic operations such as add, sub, multiply, subword operations and multimedia data manipulation operations such as mix, shuffle, pack and unpack, are also implemented in it. It means that multimedia instruction extension is done to the instruction set. Figure 4(b) shows the structure of an IFU. 4 IFUs are set to avoid hardware resources conflict. So at the same time, one IFU is executing multiply-accumulate, another IFU is executing add operation of 8 bytes (4 bytes packed in one 32bit register adds another 4 bytes packed in one 32bit register).

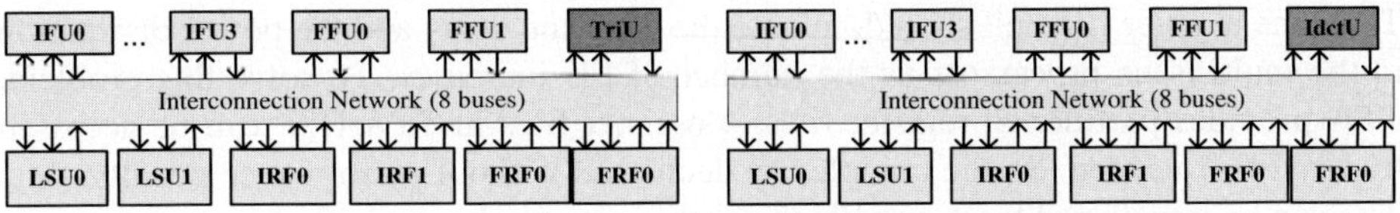

Fig. 3. TTA-DSP architecture

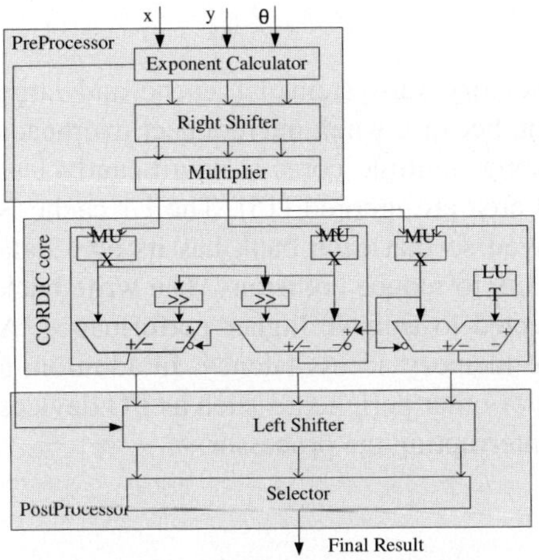

(a) TriU Structure

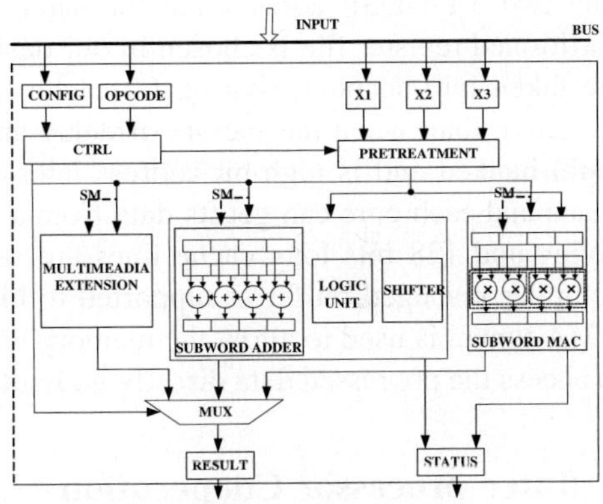

(b) IFU Structure

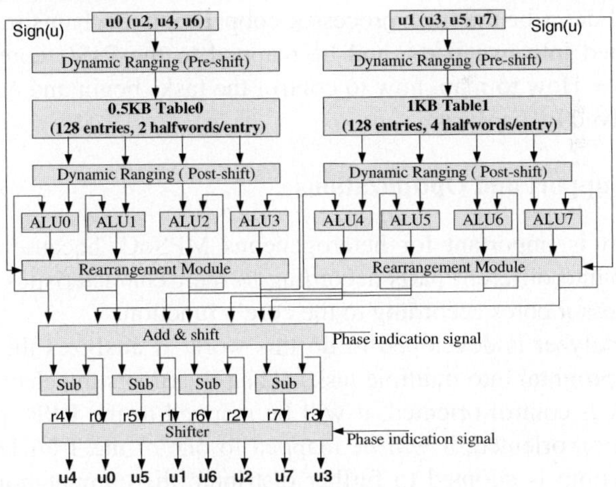

(c) IdctU structure

Fig. 4. Integer Function Unit Structure

In VLIW machines, the complexity of the register file grows rapidly as more and more FU are integrated on a chip. For N concurrent FU, the area of the centralized register files grows as N^3, the delay as $N^{3/2}$, and the power dissipation as N^3 [11].

Thus, the register file will soon dominate the area, the delay and the power dissipation in the multi-issue processors as the number of FU increases. To solve this problem, TTA provides partitioned register files. 4 register files in TTA-DSP can be accessed by any FU if needed. So the complexity decreased without performance penalty. And data can be transferred from one FU to another directly for each FU has its own registers, so lots of middle results saving are bypassed. Experimental results indicate that 4 banked and a total of 64 registers of each can achieve a best performance/cost ratio.

3.3 Memory Subsystem

The two TTA-DSP cores share the same memory subsystem. L1 cache and large partitioned register file is chosen in our design, because when interconnect overheads are taken into account, sharing L2 cache among multiple cores is significantly less attractive than when the factors of delay and area are ignored [15]. The L1 cache is multi-banked and is high-bit address interleaved so that each bank has its own data space and each core can get its data from a bank to reduce collisions. The write back policy and 128 bits long cache lines are selected to achieve higher performance. A compiler-managed buffer is supported to hide memory access latency. In addition, a DMA tunnel is used to allow the memory and/or other peripherals such as I/O devices to access the processed data directly without interrupting the processor.

4 Inter-processor Cooperation

This section will describe the inter-processor cooperation mechanism. The application will be partitioned into two parts, and be mapped to the RISC core and TTA-DSP cores by software. How to map, how to control the tasks begin and end, these control problems are solved by hardware.

4.1 Software Support and Optimizations

Software support is important for heterogeneous MPSoC, because one application must be divided into different parts according to their characteristics and be mapped to different processor cores according to the core's function.

A program analyzer is developed to do this work. It analyzes the input program, and divided the program into multiple tasks. Then it extracts the characteristics of the tasks. If the task is control-oriented, it will be mapped to the RISC processor; if the task is computation-oriented, it will be mapped to one of the TTA-DSP cores. MCP scheduling algorithm is adopted to further distribute the computation-oriented tasks on two DSP cores [19]. When the program is output, some assembly-written sentences have been added to distinguish these tasks. Hardware will identify the extra instructions and control the tasks mapping.

For TTA-DSP cores, optimizing complier is available that support its hardware features and perform code optimization, instruction scheduling, and VLIW parallelization. Based on the complier, C programs can be easily mapped to the TTA-DSP hardware architecture, whereas some computation-intensive core algorithms are typically written in assembly language and scheduled by hand for full exploitation of all data path features. The TTA-DSP compiler is being extended to support subword data types for SIMD.

During the architecture definition phase of TTA-DSP cores, cycle-accurate simulator have been developed for the optimization of architectural features and instruction sets. The simulator continues to be used for application development. But the integrated simulator for both RISC core and TTA-DSP is under way.

4.2 Hardware Inter-processor Communication Mechanism

A complete application begins to run on the RISC processor, for the control-oriented part is always first executed, and the DSP kernel algorithms will be called and run on the DSP cluster if needed. So the inter-processor communication is always happened during application running. The RISC processor and the DSP cluster communicate via an interrupt mechanism. This mechanism provides a flexible software protocol between the two parts.

There are 6 registers in IPCI used for the communication mechanism. Three are from RISC to DSPs, another three are from DSPs to RISC. The two sets almost have the same meaning. One register is command register, the interrupting processor can use this register to pass a command to the interrupted processor. The second register is data register, the interrupting processor can use this register to pass a data or an address or status information to the interrupted processor. The third register is a flag register. If the interrupt is responded, the flag will be cleared.

On writing to the command register, an interrupt is generated to the other processor and the flag register is set. The interrupted processor must acknowledge the interrupt by reading the data register (if necessary) and the command register for the associated interrupt. The interrupt is reset and the flag register is cleared when the command is read by the interrupted processor. If the interrupt is masked in the interrupt handler when the command register is written, no interrupt is generated and sent to the processor. However, the flag is set. If the interrupt is unmasked at a later time, an interrupt is generated to the processor.

5 Implementation Results

We have verified the proposed SoC from the micro-architecture design, to FPGA prototyping and the standard-cell based silicon implementation.

5.1 Performance Evaluation

To fully exploit the data-level parallelism of TTA-DSP cores, some DSP kernel algorithms in assembly code are hand written. The complete multimedia application simulation on both cores is really hard work, it is under way. So this section only presents the performance evaluation results of the TTA-DSP cores.

Figure 5 summarizes the performance comparisons between one TTA-DSP core, the state-of-the-art high-performance DSP processor TI TMS320C64xx and the general purpose embedded processor LEON3[14]. The results of TTA-DSP are get by our simulator, and the results of TMS320C64xx are get by TI development environment CCS2.0[12][13]. The results of LEON3 are get by LEON3 simulator[14]. All the 6 benchmarks are selected from TMS320C64x DSP Library [12]. The parameters of the benchmarks are listed in Table 1. They are measured in execution cycles.

The simulation results show that the performance of TTA-DSP core is better than that of the current popular DSP processors for various benchmarks, especially for the algorithms which contain trigonometric computations. The performance is also much better than that of the current popular general purpose embedded processors. The most special case is IDCT program. Because there is IdctU function unit in the TTA-DSP core, so the whole program is executed by one instruction. The performance is extremely high.

The performance of two cores cooperation is much better than that of one single core. The performance evaluation is in another paper [19].

Table 1. Benchmarks Description

No.	Name	Brief Description
1	FFT	16-bit 256 points radix 4 Fast Fourier Transform
2	FIR	16-bit samples for 16-sample 40-tap FIR filter
3	IDCT	Inverse Discrete Cosine Transform on 8*8 block of 16-bit pixel data
4	MAT_MUL	Multiply of two 4*4 matrix
5	IIR	IIR filter of 500 output samples
6	MAXIDX	Get the index of the maximum value of a vector

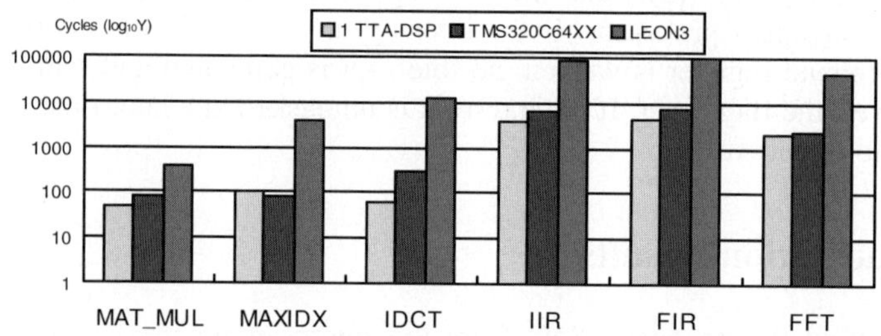

Fig. 5. Performance Comparison with TMS320C64xx

Table 2 gives the read and write miss ratio for three of them under write back policy. For L1 cache system, the performance of TTA-DSP under write through and write back policy are compared. Performance gains compared with write through policy is also given in Table 2. From Table 2, it can be seen that write back policy can achieve a better performance.

Table 2. The Read/Write miss ratio under write back policy and its performance gains

Benchmarks	Read miss ratio	Write miss ratio	Performance gains
MAT_MUL	5.42%	4.17%	3.25%
MAXIDX	3.33%	2.78%	4.36%
IDCT	7.50%	0.00%	2.23%
IIR	0.05%	2.09%	8.27%
FIR	0.05%	0.78%	13.27%
FFT	4.76%	2.38%	4.35%

5.2 Silicon Implementation

We have implemented the SoC in Verilog RTL code, which is verified with Modelsim. The design is synthesized using Design Compiler from Synopsys with the 0.18um standard-cell library . The net-lists are then placed and routed using SoC Encounter from Cadence for the 1P6M CMOS technology. Figure 6 gives the layout of the proposed SoC. The area is about 21.4mm^2 and it can operate at 266MHz and consume 870mW average power.

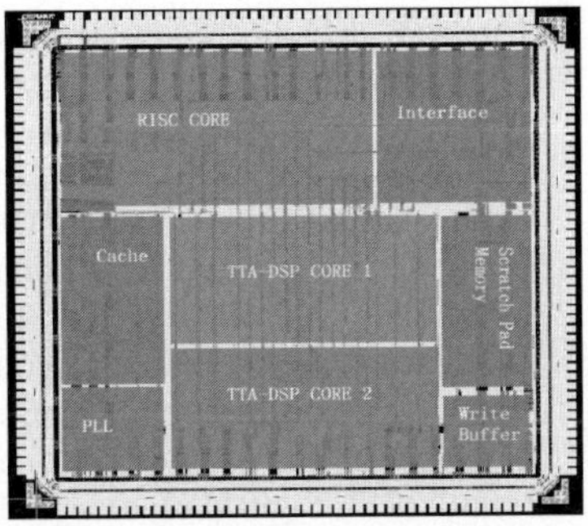

Fig. 6. Layout of the proposed SoC

6 Conclusions

This paper presents the design and the silicon implementation of the proposed heterogeneous embedded MPSoC for multimedia signal processing applications. It contains a RISC processor and two DSP processor cores. The RISC core coordinates the system and performs control-oriented tasks, and the DSP cores perform data-intensive tasks with more deterministic and regular behaviors, in the form of optimized assembly code. With effective mapping techniques, a wide range of multimedia signal processing applications can be fast executed on it. We are now studying the application partition techniques on the three processors and the code optimization techniques for TTA-DSP architecture. An integrated simulator and development environment for all the cores are developed.

Acknowledgements

This work was supported by Chinese NSF project No. 90407022 and some other funds. The authors would like to thank Andrea Cilio and his colleagues in Delft University of Technology in Netherlands for their great help and support to our research work.

References

1. M. Levy, "ARM picks up performance," Microprocessor Report, 4/7/03-01
2. R. A. Quinnell, "Logical combination? Convergence products need both RISC and DSP processors, but merging them may not be the answer," EDN, 1/23/2003

3. OMAP5910 Dual Core Processor – Technical Reference Manual, Texas Instruments, Jan,2003
4. Siemens, "TriCore Architecture". White Paper, 1998
5. Hans-Joachim Stolberg et al., "HiBRID-SoC: A Multi-Core System-on-Chip Architecture for Multimedia Signal Processing Applications," in Proc. DATE'03, pp.20008, March, 2003.
6. M. Berekovic, H.-J. Stolberg, M.B. Kulaczewski, P.Pirsch, et al., "Instruction set extensions for MPEG-4 video," J. VLSI Signal Processing Syst., vol. 23,pp. 27–50, October, 1999.
7. "ARM Ltd. AMBA Specification Rev. 2.0. [Online]," Available: www.arm.com
8. Corporaal H., "Microprocessor Architecture from VLIW to TTA," John Wiley & Sons Ltd., West Sussex, England, 1998
9. Hoogerbrugge J., "Code Generation for Transport Triggered Architecture," PhD Thesis, Delft University of Technology, Delft, The Netherlands, 1996
10. Hong Yue, Kui Dai, Zhiying Wang, "Key Technique of Float Point Function Unit Implementation based on CORDIC Algorithm", in Proc. Annual Conference of Engineer and Technology of China, pp.102-105, April 2005.
11. S. Rixner, W.J. Dally, B. Khailany, P. Mattson, et al., "Register organization for media processing," in Proc. HPCA-6, 2000, pp.375-386.
12. MS320C64x DSP Library Programmer's Reference, Texas Instruments Inc., Apr 2002
13. TMS320C55x DSP Programmer's Guide, Texas Instruments Inc., July 2000
14. T. Kumura, M. Ikekawa, M. Yoshida, and I. Kuroda, "VLIW DSP for mobile applications," IEEE Signal Processing Mag., pp.10-21, July 2002
15. Terry Tao Ye. 0n-chip multiprocessor communication network design and analysis. PhD thesis, Stanford University, December 2003.
16. TTay-Jyi Lin, Chie-Min Chao, "A Unified Processor Architecture for RISC&VLIW DSP", GLSVLSI'05, 2005.
17. Hong Yue, Kui Dai and Zhiying Wang, "A Dual-core Embedded System-on-Chip Architecture for Multimedia Signal Processing Applications", in Proc. ESA'06, June 2006.
18. M.-Y. Wu and D. D. Gajski, "Hypertool: A programming aid for message-passing systems," IEEE Trans. Parallel & Distrib. Systems, vol. 1, no. 3, pp. 330–343, July 1990.
19. Hong Yue, Kui Dai, Zhiying Wang and Xuemi Zhao, "An instruction customization algorithm for embedded application-specific heterogeneous multiprocessor", Journal of Computer Research and Development, being in reviewing

some elements. The structural logical blocks may contain either processes or other structural logical blocks (callcd *sub-blocks*). Structural logical blocks arc hicrarchically organized. At the top level of the hierarchy may exist only one structural logical block called the *system*. The processes represent the bottom level of the hierarchy. The *logical connection lines* are used to show the interconnection of logical blocks.

- *interfaces*. This part describes interfacing mechanisms of the system. Three different types of interfaces are introduced: *functional interfaces* (abstract sets of messages, which are accepted by the particular logical module), *data ports* (structural logical blocks, which are used to organize buffered data transfer between logical modules) and *trigger ports* (logical abstractness, which is used to define the execution order of two blocks).
- *requirements and restrictions*. The concept defines timing requirements, least memory requirements and scheduling restrictions.
- *test cases*. The concept also defines three types of test cases, which are used for automatic validation of the system's description and particular system implementations. The test cases are used to check the *system consistency (semantic and information consistency), description correctness* and *timing requirements*.

3.3 Platform Independent Program Representation

The PIPR application describes the aspects that the new concept defines. The PIPR application may be formally expressed by means of the mathematical language, but, in order to simplify the automatic analysis and code generation procedure a special form is used.

The PIPR application is represented as a set of different tuples. These tuples describe functions, processes, logical blocks, interfaces, timing requirements, test cases, etc. To simplify the data processing the tuples may be represented in a format, which is found to be the most suitable one, e.g. the XML format.

The most important part of the PIPR application is the definition of processes. The processes represent the bottom level of structural hierarchy and they actually reflect a lot of aspects (e.g. functionality, partitioning, interfaces, etc.) at the same time.

The formal definition of a process could be given as tuple <*ST, X, T, ...*>, where

ST - a finite set of the process's states
X - an input matrix of the process
T - a set of transition elements
... - other parameters, which are not important here

The set *T* stores tuples <*g, j, k*>. Every tuple defines a reference to a primitive (basic operation) *g*, a reference *j* to the next element in the set *T* and the alternative reference *k* to the next element in the same set. The matrix *X* defines the first tuple from the set *T*, which must be executed under the certain conditions.

So, such definition of a process is HW independent, because it is expressed in terms of primitives (basic operations), which, in their turn, are also HW independent.

As a result, the whole PIPR application is HW independent, since it is built-up of HW independent elements and it doesn't define any HW specific features.

3.4 HW Description File

The second input information of the introduced concept describes the HW platform, and it is used to adapt and optimize the generated implementation for this platform. It's necessary to notice, that this information is a description of HW from the programmer's point of view and not from the HW structural design point of view. HW structural descriptions provide a lot of detailed information, which is not required for the software generation process.

In contrast, the HW description file is a totally new description form, which explicitly provides information required for the software generation process. This information is given in a hierarchical form and it is split into two categories: the *HW architecture related information (HW model file)* and the *algorithm's implementation information (Implementation library file)*.

In the first category, the information about the number of processor cores, a vector processing unit type, a number of supported HW threads, etc. is stored (Fig. 3).

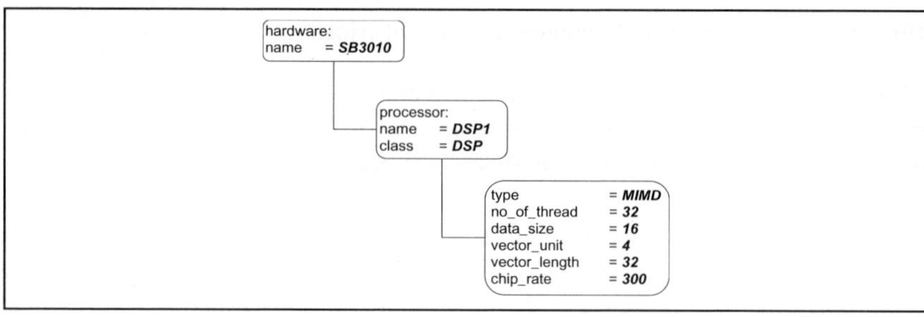

Fig. 3. A part of the HW model file

In the second category the information about execution times of every primitive, memory and thread requirements, etc. is given. It's also shown that some primitives have several alternative implementations and some non-primitive functions are already available as HW specific modules (Fig. 4).

Thus, the HW description file provides not a detailed description of a HW structure, but only that information, which is necessary for organizing of the efficient mapping of a PIPR application to a particular HW.

3.5 Starting Point

Before describing an application a decision about the programming (description) language was done.

The concept's description language was not developed that time and an idea of using common development tools was announced in the beginning.

There was a set of some well- and not-well known languages. All of these languages were intended to describe different things and had their own advantages and disadvantages, so, selection of a particular language was a scientific process, which required a definition of important criteria. The selected criteria were: language abstractness, standardization, wide spread, development tool availability.

And the following languages corresponded to these criteria: UML [4], SDL [5] and XML. The XML language was used to store the HW description file.

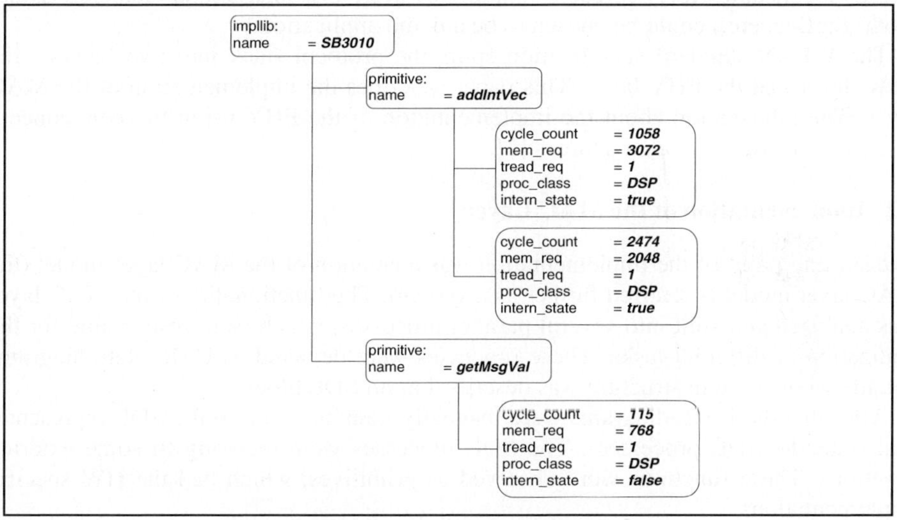

Fig. 4. A part of the implementation library file

This selection was only the starting point for the concept implementation and it was only one of many different implementation possibilities. The selected languages could not describe all the aspects of the concept (e.g. restrictions and requirements, test cases, etc.) and the additional information and additional processing were needed.

The additional processing was changing the descriptions represented in these languages. It was removing some irrelevant information and was using the additional information (e.g. restrictions and requirements, etc) to build the PIPR application.

3.6 Impacts

The new concept, one the one hand, simplifies and speed-ups the software development process, on another hand, it provides a long-lived and HW independent application specification, which can be reused for many hardware platforms and where porting to the hardware architectures can be captured in an automated process. The functionality of such an application can be easily upgraded or modified. Moreover, it reduces the development time and costs of a product.

4 Implementation of a Protocol Stack

4.1 Fundamentals

The usage of the new concept should be proved with the implementation of a protocol stack. Some aspects of the new development flow had to be determined during the implementation of a protocol stack's model.

The protocol stack 802.11a (WLAN) [6] was chosen as a demo application. This is already stable and widely spread standard, however, any other similar standard (Bluetooth, ZigBee, etc.) could be chosen to be a demo application.

The WLAN standard specification splits the protocol stack into two layers - the MAC layer and the PHY layer. This paper describes the implementation of the MAC layer. The information about the implementation of the PHY using the new concept layer may be found in other sources [3].

4.2 Implementation of the MAC Layer

The starting point of the implementation was a creation of the MAC layer model (the MAC layer model is referred further as a *system*). The functionality of the MAC layer was analyzed and split into several parallel processes, which were responsible for the realization of different tasks. These processes were depicted as UML state diagrams and the whole system structure was described as an SDL block.

After that the UML diagrams were manually transformed into the SDL representation (into the SDL processes). The SDL processes were referring to some external functions. These functions were observed as primitives, which had the HW specific implementations.

The definition of interfaces was the important step, because the MAC layer was cooperating with other elements (LLC, PHY and other layers) by means of the interfaces, which are, in general sense, connection points to the MAC layer's environment.

The MAC layer sent and received messages through these connection points. But the format of these messages inside the MAC layer and outside of it was different. The messages outside the MAC layer were HW specific, and, they were converted into a platform independent format while they were passing the connection points and vice versa. These message conversion functions had HW specific implementation and they were also observed as primitives.

The system was prepared for the first code generation procedure.

4.3 Code Generation

Originally one commercial tool was used for the code generation. The SDL description was passed to the input of the code generator (CG) and a C code was obtained at the output of it. This code was included into the simulation framework and different test cases were applied to it.

The generated code was able to provide a lot of nice options like graphical tracing within the tool's environment, automatic validation (with TTCN [7]), generation of message charts, etc.

However, the performance of the code generated by this CG was completely unsatisfying. It was totally impossible to fulfill time requirements defined in the standard, because the generated code was mainly intended for the simulation instead of a real-time execution of a system. It used a highly-dynamic message scheduler, a very flexible process scheduler, very complex implementations of simple data processing operations (e.g. assignment, addition of two numbers, etc.).

Even the use of the CG's plug-in for microcontrollers, which produced a highly-optimized C code, was not able to solve this problem. A creation of own CG had became a solution.

The main difference between the own developed CG and the CGs used in the beginning is that:

- the own developed CG observed the system not like a flexible SDL system, but as a system, which had a fixed structure. It allowed the CG to use optimized scheduler schemes, which could not be used with SDL systems;
- the own developed CG used a HW specific kernel;
- it avoided processing of different data types in the abstract way. Every data type had to be handled by a corresponding primitive.

The own CG receives conceptual information about the model, processes it and realizes a philosophy of the new software development concept.

5. Profiling Technique and Results

The generated code was approved on different HW platforms and in different simulation environments:

- PC (simulation in a framework, performance simulation, test cases)
- Infineon's MuSIC DSP environment (ARM part) [8]
- Sandbridge's SandBlaster 3010 DSP [9]

The real-time requirements were fulfilled during the performance simulation on the SandBlaster DSP.

A profiling was done over two different operations. The first operation (*OP1*) represents one of the use cases of the MAC layer when it sends an acknowledgement message in response to the incoming data frame. This operation must be accomplished within a certain time interval. The numerical values are defined in the communication standard specification [6].

The second operation (*OP2*) represents a simple message transfer from one parallel process to another. There are no time requirements for this procedure, but it's evident that these values must be much smaller than the processing time of the *OP1*.

The new concept assumes that it will be possible to perform the automatic profiling of some operation, without executing them. It requires the CG to know execution times of all primitives and kernel functions used by these operations. This information was not known that time, and it was not possible to perform automatic profiling. To profile these operations the time measurements were done during the code execution.

Two CGs provided an option to build-in a profiler into the generated code. But the profiling results couldn't be compared, since the CGs used different measuring methods. To make the profiling results comparable the measurement points were explicitly defined within the system description, i.e. they belonged to the design of a protocol stack.

The profiling results on the SandBlaster DSP are shown in the Table 1. Time information is given in processor's clock (kilo) cycles and in milliseconds.

Table 1. Profiling results

OPERATION	REQUIRED		conventional CG		own developed CG	
	k cycle	ms	k cycle	ms	k cycle	ms
OP1	~1,2	~0,016	~15 000	~200	~1,05	~0,014
OP2	-	-	~880	~12	~0,5	~0,006

It's possible to see a huge performance difference between two implementations, obtained with the help of the commercial tool and the own developed code generator. Long execution times of the code generated by the commercial tool are due to the high-flexibility and comprehensive facilities of the code generator.

6 Conclusions

The main result of the creation process is that the WLAN protocol stack described in a high-level language (PIPR) was converted to an executable application. Several HW specific implementations of the WLAN protocol stack were obtained and executed on different HW platforms and in different simulation environments.

References

1. W. Tuttlebee et al., "*Software Defined Radio: Enabling Technologies*", England, John Wiley, 2002
2. *Software Defined Radio Forum*, Version 2.1 of the Technical Report on "Architecture and elements of software defined radio systems", February 2000
3. R. Hossain, M. Wesseling, C. Leopold, "Virtual Radio Engine: a programming concept for separation of application specifications and hardware architectures", in *Proc. 14th IST Mobile and Wireless Communications Summit*, Dresden, June, 2005, in press.
4. ISO/IEC 19501, Unified Modeling Language (UML), ver. 1.4.2
5. Specification and Description Language (SDL), ITU Z.100 Standard
6. Wireless LAN, IEEE Std. 802.11a, 1999 Edition
7. Testing and test control notation version 3, TTCN-3, ITU Z.140 Standard
8. H.-M. Bluethgen et al., "Finding the optimum partitioning for multi-standard radio systems", *SDR Forum Technical Conference*, California, November 2005
9. SandBlaster DSP, http://www.sandbridgetech.com/sb_3000.htm

A New Address Mapping Scheme for High Parallelism MEMS-Based Storage Devices*

Soyoon Lee and Hyokyung Bahn

Dept. of Computer Engineering, Ewha University, Seoul 120-750, South Korea
sounie@ewhain.net, bahn@ewha.ac.kr

Abstract. MEMS-based storage is an emerging storage media that has several attractive features such as high-bandwidth, low-power consumption, and low cost. However, MEMS-based storage has vastly different physical characteristics compared to a traditional disk. First, MEMS-based storage has thousands of heads that can be activated simultaneously. Second, the media of MEMS-based storage is a square structure which is different from the rotation-based platter structure of disks. Third, the size of a sector in MEMS-based storage is smaller than 512 bytes of a conventional logical block. In this paper, we present a new address mapping scheme for MEMS storage that makes use of the aforementioned characteristics. This new scheme exploits the complete parallel feature of MEMS-based storage as well as the characteristics of the two dimensional square structure. Simulation studies show that the new scheme improves the performance of MEMS storage significantly by exploiting the high parallelism of MEMS storage.

1 Introduction

Storage capacity consumption is increasing at an enormous rate, and new technology and their by-products that try to satiate this storage appetite is being introduced everyday. At the server end of the storage consumer spectrum, disk capacity is increasing over 60% every year and cost of these devices seems to be coming down just as fast [11]. On the other end, as mobile consumer electronic products become an essential component in our everyday lives, the demand for increased storage capacity in these products is growing as well. Flash memory has been introduced into the market to serve such miniature consumer products. MEMS-based storage is a leading candidate to answer demands for a wide range of storage systems ranging from storage for small handheld devices to high capacity mass storage servers [4, 8, 12, 13, 17]. Since physical structure of MEMS-based storage is different from conventional disks, a new address mapping scheme appropriate for this media is needed. In this paper, we present a new address mapping scheme that exploits the high parallelism of a MEMS-based storage device.

Table 1 compares the characteristics of a conventional disk, Flash memory, and MEMS-based storage [11, 16]. A conventional disk represents storage for the server

* This work was supported by the Seoul R&BD program.

M. Gerndt and D. Kranzlmüller (Eds.): HPCC 2006, LNCS 4208, pp. 611 – 620, 2006.

Table 1. Characteristics comparison of a conventional disk, Flash memory, and MEMS-based storage. Though the characteristics of conventional disk and Flash memory are examined from the data sheet of current market products, the characteristics of MEMS storage are filled in from the expected model since there is no real product as yet. Hence, for some items, relative comparison may not be meaningful at this time.

	Conventional Disk	**NAND-type Flash Memory**	**MEMS-based Storage**
Size (mm)	101.6×147.0 × 26.1	12.0 × 17.0 × 1.0	10.0 × 10.0 × 2.0
Density (GB/cm^2)	0.14-0.24	0.49	1-10
Read access time (ms)	5-10	0.000025-0.025	0.56-0.80
Write access time (ms)	5-10	0.2-1.5	0.56-0.80
Shock resistance	Low	High	High
Cost ($/GB)	0.21	45	3-10
Bandwidth (MB/s)	17.3-25.2	100	75.9-320
Power consumption	High	Low	Low

end and Flash memory represents storage for the mobile end of the storage spectrum. Some of the advantages of MEMS-based storage can be summarized as follows.

- First, density of MEMS-based storage is very high. In 1 cm^2, shown in Figure 1, MEMS-based storage can hold more than 3GBs. Given the same dimension as conventional disks and/or Flash storage, the capacity of MEMS-based storage will be much higher. This implies MEMS-based storage may be used for a wide range of storage applications.
- Second, storage medium access time is stable and fast. Unlike Flash memory, read and writes to MEMS-based storage is stable at a few hundred microseconds. Access time for Flash memory, on the other hand, varies greatly depending on the operation. While reads are very fast, writes can be so slow as to be greater than a millisecond. Disks will continue to be limited to the ten's of milliseconds due to its mechanical limitation.
- Third, MEMS-based storage retains characteristics such as high bandwidth, shock resistance, low power consumption, and low cost that is superior to conventional disks and/or Flash memory making it a suitable storage medium for high capacity servers as well as for mobile consumer devices.

The advantages listed above make MEMS-based storage a leading candidate as tomorrow's storage medium. However, the mechanism by which MEMS-based storage operates has a couple of distinct characteristics [1, 2, 3, 4]. First, MEMS storage has thousands of heads that can be activated simultaneously instead of just a few as in disks. As shown in Figure 1, there is a single head associated with each region, and there are thousands of regions in each device. These heads may be active simultaneously. Second, the media of MEMS storage is a square structure, which is different from the rotation-based platter structure of disks. Again, this may be noted from Figure 1. The regions are where the data are stored, and note that these regions are square. Third, the size of a physical sector in MEMS storage is an order of magnitude different from that of a conventional disk. While the sector size of a disk is usually 512 bytes, the size of a physical sector in MEMS storage is 8 bytes.

We discuss the performance implications of these characteristics later in Section 3. Due to these differences in characteristics, new system software technologies appropriate for this media have been issues of recent research [2, 3, 15]. One of the most important management mechanisms for improving MEMS storage efficiency is the address mapping of logical blocks into physical sectors. In this paper, we present a new address mapping scheme that exploits the aforementioned characteristics of MEMS-based storage. We compare four address mapping schemes in terms of the average access time and the average request delay through trace driven simulation studies.

The remainder of this paper is organized as follows. Section 2 gives an overview of existing studies related to MEMS-based storage. We describe new address mapping schemes in Section 3, and show the experimental results in Section 4. Finally, Section 5 presents the conclusion of this paper.

2 Related Works

Several different MEMS storage projects are being conducted by major institutions such as Carnegie Mellon University [7], IBM Zurich Research Laboratory [8], and Hewlett Packard Laboratories [9]. Though substantial physical differences exist between MEMS storage models in these projects, they share the same basic design as shown in Figure 1. A MEMS-based storage device consists of the magnetic media (called *media sled*) that is divided into regions and groups of heads (called *probe tips*)

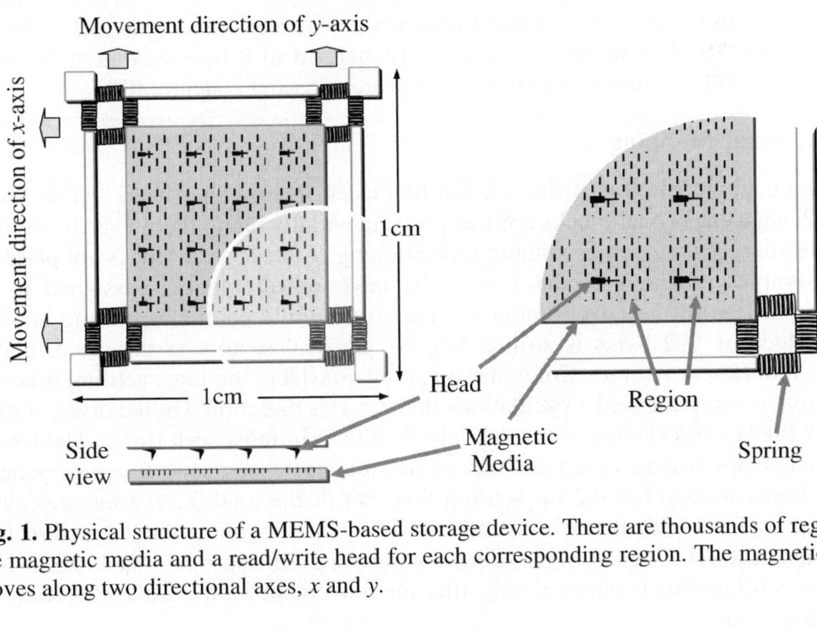

Fig. 1. Physical structure of a MEMS-based storage device. There are thousands of regions on the magnetic media and a read/write head for each corresponding region. The magnetic media moves along two directional axes, *x* and *y*.

used to access data on the corresponding region. To access data on a specific (x, y) location, MEMS-based storage suffers a substantial distance-dependent positioning time delay similar to disks. Unlike disks, however, the heads of MEMS-based storage are fixed and magnetic media itself moves to access data on a specific location. The movement of the media in the directions of x and y axes is independent and proceeds in parallel. Thus, the positioning time for a specific (x, y) location can be determined by the larger seek time of the x and y dimensions. In most current architectures, seek times on the x dimension $time_{seek\_x}$ is dominant over seek times on the y dimension $time_{seek\_y}$ because extra settling time must be included to $time_{seek\_x}$, but not to $time_{seek\_y}$. Settling time is the time needed for the oscillations of the magnetic media to damp out. This time is dependent on the construction of the magnetic media and the stiffness of the spring that sustains the magnetic media [4]. Since media access is performed in the direction of the y dimension after positioning, it requires constant media velocity in the y dimension and zero velocity in the x dimension. Hence, oscillation in the x dimension leads to off-track interference after seeking, while the same oscillation in the y dimension affects only the bit rate of the data transfer [6].

3 Address Mapping Schemes for MEMS-Based Storage

Today's storage interface abstracts a storage device as a linear array of fixed-size logical blocks [6]. Details of the mapping of the logical block address to the physical sector address are hidden. In the case of disks, the size of a logical block is 512 bytes while a physical sector is usually close to this value. However, in MEMS-based storage, the size of a physical sector is usually 8 bytes. Therefore, a new address mapping scheme is needed. To utilize the parallel access feature of MEMS devices, one logical block can be spread into multiple physical sectors across parallel-operating heads so as to be accessed concurrently rather than sequentially. In this paper, a typical logical block of 512 bytes is mapped to 8 byte sectors at the same relative position in 64 different regions, which are accessed concurrently.

3.1 Columnar Mapping

As a simple approach, logical blocks can be placed in columnar ways. As shown in Figure 2, adjacent logical blocks are mapped sequentially along the y-axis direction to allow for successive accesses without repositioning. After logical blocks are placed at the bottommost position of the y-axis, the next logical block is assigned to the topmost position of the next column. To exploit multiple parallel-operating heads, a logical block of 512 bytes is striped into 64 physical sectors at the same relative position in different regions. Columnar mapping considers the large settling time that is required to damp the sled's oscillations in the x-axis direction. Unlike disks, settling times of MEMS-based storage is relatively dominant in most seek times. Though the settling time for a disk is 0.5 ms out of the total 1-15 ms seek times in general, MEMS-based storage has 0.2 ms settling time out of the total 0.2-0.8 ms seek times [2]. Since adjacent logical blocks are allocated sequentially along the y-axis direction, accessing successive logical blocks does not need the settling time. (Refer to Section 2 that the settling time is required only after the movement for the x-axis direction and not for the y-axis.)

3.2 Fully-Parallel Columnar Mapping

Parallel access feature of MEMS-based storage should be maximally exploited to improve the system performance. Fully-parallel columnar mapping aims at maximizing the number of concurrent reads/writes of adjacent logical blocks. This can be done by mapping as many successive logical blocks as possible to the same relative (x, y) positions in different regions.

However, simultaneous activation of multiple heads in MEMS-based storage devices has some limitations. Due to power and heat considerations, it is unlikely that all the heads can be activated at the same time. For example, in the CMU MEMS storage model, only 1280 out of the total 6400 heads can be activated concurrently [2]. As a result, in order to fully utilize the parallel activity of 1280 heads, up to 20 successive logical blocks can be striped into the same relative position at different regions. This is calculated by 1280/64=20 because each logical block of 512 bytes should also be striped into 8 byte sectors of the same relative position at 64 different regions for parallel access. When this address mapping is used, the time of accessing adjacent 20 logical blocks can be reduced to a single block access time in the best case.

3.3 Snakelike Mapping

Like disks, the positioning time of heads in MEMS-based storage is a relatively large component of the total access times. When a request contains a large number of successive logical blocks, the head must read multiple columns, and this incurs additional positioning time delay. When the columnar mapping is used, to access the next block in the different column, the head should move from the bottommost position of the y-axis to the topmost position. To minimize this repositioning time delay, we use the snakelike address mapping scheme. In this scheme, odd columns and even columns place the logical blocks in the reverse order. As shown in Figure 2, the smallest logical block number of each odd column is placed at topmost and the largest number at bottommost. On the contrary, in the case of even columns, the smallest logical block number is placed at bottommost and the largest number at topmost. Through snakelike mapping, we can expect the small positioning time delay in case of large sequential accesses.

3.4 Fully-Parallel Snakelike Mapping

Fully-parallel snakelike mapping combines the ideas of fully-parallel mapping and snakelike mapping. First of all, a logical block of 512 bytes is striped across 64 different regions and 20 adjacent logical blocks are mapped to the same relative position at different regions. After this idea is adapted to the intra-position placement, the scheme uses the snakelike mapping in the inter-position placement. In odd columns, the smallest logical block number is placed at topmost and the largest number is placed at bottommost. Reverse order is employed for even columns. As shown in Figure 2, this scheme considers both minimizing the repositioning time delay and maximizing the parallel access features.

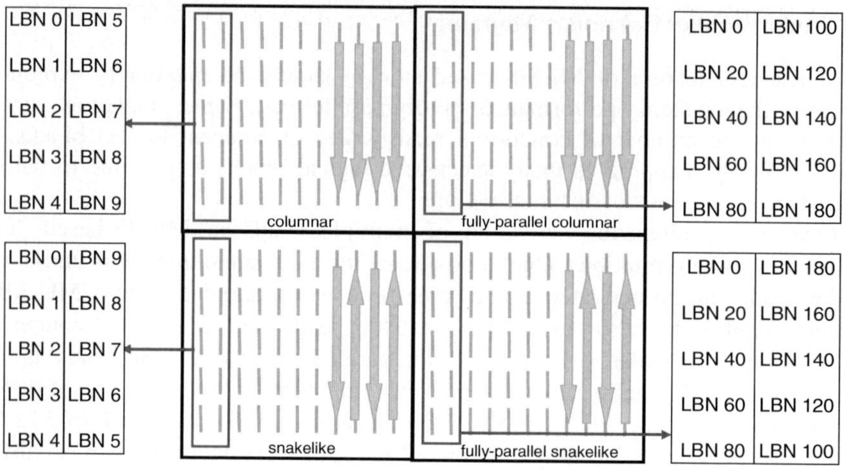

Fig. 2. Four different address mapping schemes. LBN denotes the Logical Block Number.

4 Experimental Results

In this section, we discuss the results from trace-driven simulations performed to assess the effectiveness of the four address mapping schemes. The parameters for the MEMS-based storage that we use for our experiments basically conform to the MEMS storage model presented by Griffin et al. [2, 4]. This MEMS-based storage device contains 6400 regions and each region has 2500 × 2440 bits in $x \times y$ dimensions. In this model, there are 2500 × 27 sectors in each region as the size of a sector is 80 bits (encoded data of 8 bytes) and *servo information* of 10 bits that identifies the sector information exists between adjacent sectors as well as at the top and bottom of the region as shown in Figure 3. Table 2 lists the detailed parameters of MEMS-based storage that was used in our experiments.

We use both a synthetically-generated trace and a real world disk trace. In the case of the synthetic trace, the inter-arrival times of requests conform to an exponential distribution for a range of mean arrival rates to simulate various workload conditions. The ratio of read and write operations is 67% and 33%, respectively, which are commonly used values [2]. The request size is also exponential with a mean of 4KB, and the logical block numbers of requests are uniformly distributed across the entire device.

For real world traces, because there are no traces that have been obtained directly from MEMS storage, we use the well-known Cello99 disk access trace. The Cello99 traces were collected from the disk activity of a time sharing server on the HP-UX operating system at Hewlett-Packard Laboratories. The Cello99 traces are available at [18]. To explore a range of workload intensities, we scale the traced inter-arrival times to produce a range of average inter-arrival times. For example, a scaling factor of two generates a workload that is two times more intense than the original trace.

Table 2. Experimental parameters of the MEMS-based storage device

Number of regions	6400
Number of heads	6400
Maximum concurrent heads	1280
Device capacity	3.2 GB
Physical sector size	8 bytes
Servo overhead	10 bits per sector
Bits per region	2500 × 2440
Settling time	0.22 ms
Average turnaround time	0.07 ms
Spring factor	75%
Media bit cell size	40 × 40 nm
Sled acceleration	803.6 m/s^2
Sled access speed	28mm/s
Per head data rate	0.7 Mbit/s
Command processing overhead	0.2 ms/request
On-board cache memory	0MB
Request scheduling algorithm	SPTF

When the total size of distinct blocks in the trace is larger than the 3.2 GB capacity of a single MEMS device, we used multiple media sleds. The sleds move simultaneously and their relative positions are unchanged. For request scheduling, we used the SPTF (Shortest Positioning Time First) algorithm because it is the most well-known scheduling algorithm for MEMS-based storage [2].

Figure 4 shows the average access times of the four address mapping schemes when the synthetic workload and the Cello99 trace are used. The fully-parallel snakelike mapping scheme outperforms the other three mapping schemes in terms of the average access time. For the Cello99 trace, the performance improvement of the fully-parallel snakelike mapping scheme against columnar mapping is as much as 80.5%. This is because concurrent read/write operations are maximized and at the

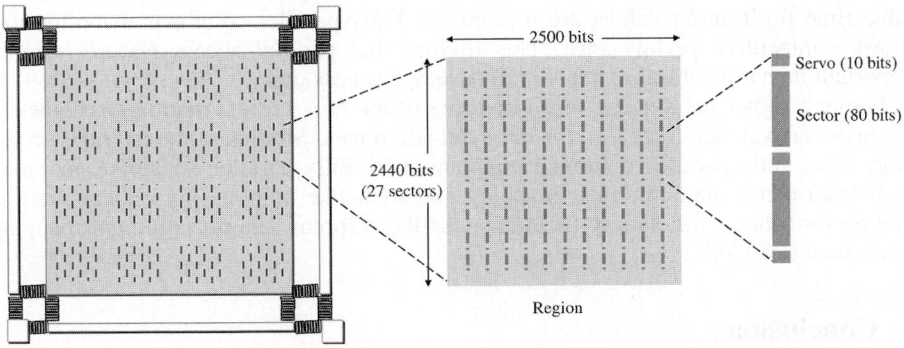

Fig. 3. Data organization of MEMS-based storage in our experiments. There are 6400 regions in a magnetic media sled and each region has 2500 × 2440 bits in $x \times y$ dimensions. The size of a sector is 80 bits (encoded data of 8 bytes) and "servo information" of 10 bits that identifies the sector information exists between adjacent sectors as well as at the top and bottom of the region.

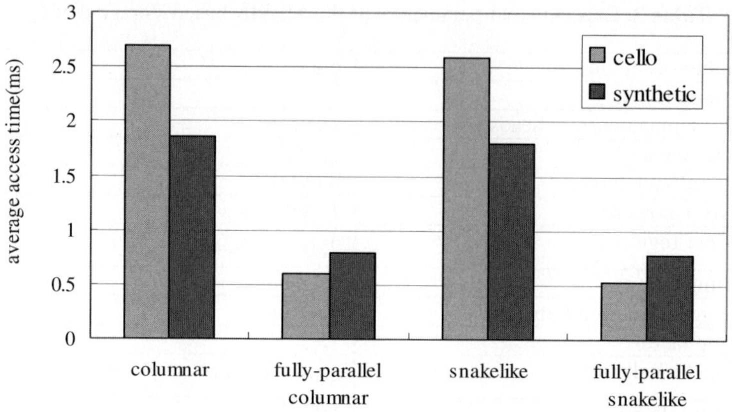

Fig. 4. Comparisons of average access time for the Cello99 and synthetic traces

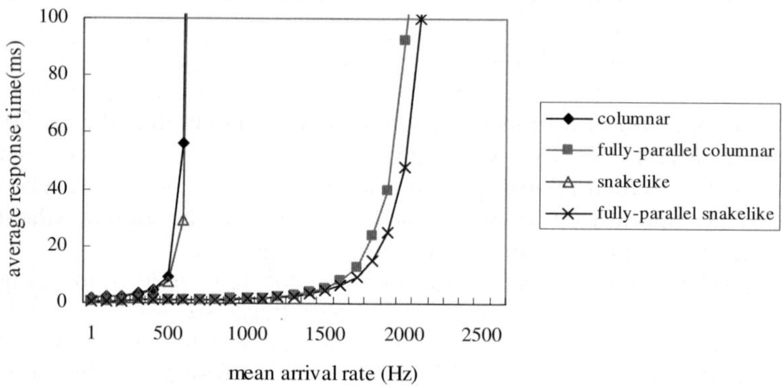

Fig. 5. Comparisons of average response time for the synthetic trace

same time positioning delays are minimized. Fully-parallel columnar mapping also shows competitive performance. This implies that parallel access feature is more important than repositioning delay in the average access time.

Figure 5 shows the average response times of the four address mapping schemes as the mean arrival rate increases for the synthetic traces. Similar to the average access time case, fully-parallel columnar mapping and fully-parallel snakelike mapping performed better than columnar mapping and snakelike mapping by a large margin. The performance gain of fully-parallel snakelike mapping against columnar mapping is as much as 93.7%.

5 Conclusion

MEMS-based storage is anticipated to be used for a wide range of applications from storage for small handheld devices to high capacity mass storage servers. In this paper, we presented a new address mapping scheme for MEMS-based storage that makes use of the physical characteristics of MEMS devices. The proposed scheme exploits the complete parallel feature of MEMS-based storage as well as the characteristics of the two dimensional square structure. Simulation studies have shown that the proposed scheme improves the performance of MEMS-based storage significantly in terms of the average access time and the average response time. Specifically, the performance improvement of the proposed scheme is 80.5% in terms of the average access time and 93.7% in terms of the average response time. We expect that the proposed address mapping scheme will be effective when employed to large server environments with heavy I/O requests such as multimedia and high performance scientific application servers.

References

1. B. Hong, S. A. Brandt, D.D.E. Long, E.L. Miller, K. A. Glocer, and Z.N.J. Peterson, "Zone-based Shortest Positioning Time First Scheduling for MEMS-based Storage Devices," *Proc. 11th IEEE/ACM International Symposium on Modeling, Analysis, and Simulation of Computer and Telecommunication Systems (MASCOTS'03)*, 2003.
2. J. L. Griffin, S. W. Schlosser, G. R. Ganger, and D. F. Nagle, "Operating system management of MEMS-based storage devices," *Proc. 4th Symposium on Operating Systems Design and Implementation (OSDI'00)*, pp. 227-242, 2000.
3. H. Yu, D. Agrawal, and A. E. Abbadi, "Towards optimal I/O scheduling for MEMS-based storage," *Proc. 20th IEEE/11th NASA Goddard Conference on Mass Storage Systems and Technologies (MSS'03)*, 2003.
4. J. L. Griffin, S. W. Schlosser, G. R. Ganger, and D. F. Nagle, "Modeling and performance of MEMS-based storage devices," *Proc. ACM SIGMETRICS Conference on Measurement and Modeling of Computer Systems*, pp. 56-65, 2000.
5. B. L. Worthington, G. R. Ganger, and Y. N. Patt, "Scheduling Algorithms for Modern Disk Drives," *Proc. ACM SIGMETRICS Conference on Measurement and Modeling of Computer Systems*, pp. 241-251, 1994.
6. S. W. Schlosser and G. R. Ganger, "MEMS-based storage devices and standard disk interfaces: A square peg in a round hole?" *Proc. 3rd USENIX Conference on File and Storage Technologies (FAST'04)*, 2004.
7. Center for Highly Integrated Information Processing and Storage Systems, Carnegie Mellon University, http://www.ece.cmu.edu/research/chips/
8. P. Vettiger, M. Despont, U. Drechsler, U. Dürig, W. Häberle, M. I. Lutwyche, H. E. Rothuizen, R. Stutz, R. Widmer, and G. K. Binnig, "The Millipede – More than one thousand tips for future AFM data storage," *IBM Journal of Research and Development*, Vol.44, No.3, pp.323-340, 2000.
9. Hewlett-Packard Laboratories Atomic Resolution Storage, http://www.hpl.hp.com/research/storage.html.
10. P. J. Denning, "Effects of scheduling on file memory operations," *Proc. AFIPS Spring Joint Computer Conference*, pp.9-21, 1967.
11. S. W. Schlosser, J. L. Griffin, D. F. Nagle, and G. R. Ganger, "Designing computer systems with MEMS-based storage," *Proc. 9th International Conference on Architectural Support for Programming Languages and Operating Systems*, Boston, Massachusetts, 2000.
12. B. Hong, "Exploring the usage of MEMS-based storage as metadata storage and disk cache in storage hierarchy," http://www.cse.ucsc.edu/~hongbo/publications/mems-metadata.pdf.

13. R. Rangaswami, Z. Dimitrijevic, E. Chang, and K. E. Schauser, "MEMS-based disk buffer for streaming media servers," *Proc. International Conference on Data Engineering (ICDE'03)*, Bangalore, India, 2003.
14. H. Yu, D. Agrawal, and A. E. Abbadi, "Tabular placement of relational data on MEMS-based storage devices," *Proc. International Conference on Very Large Databases (VLDB'03)*, pp. 680-693, 2003.
15. S. W. Schlosser, J. Schindler, A. Ailamaki, and G. R. Ganger, "Exposing and exploiting internal parallelism in MEMS-based storage," *Technical Report CMU-CS-03-125*, Carnegie Mellon University, Pittsburgh, PA, 2003.
16. Samsung Flash Memory, http://www.samsung.com/products/semiconductor/NANDFlash/SLC_LargeBlock/8Gbit/K9K8G08U1A/K9K8G08U1A.htm
17. L. R. Carley, J. A. Bain, G. K. Fedder, D. W. Greve, D. F. Guillou, M. S. C. Lu, T. Mukherjee, S. Santhanam, L. Abelmann, and S. Min, "Single-chip computers with microelectromechanical systems-based magnetic memory," *Journal of Applied Physics*, Vol.87, No.9, pp.6680-6685, 2000.
18. Public Software, Storage Systems Department at HP Labs, http://tesla.hpl.hp.com/public_software/
19. T. M. Madhyastha and K. P. Yang, "Physical modeling of probe-based storage," *Proc. IEEE Symposium on Mass Storage Systems*, 2001.
20. Dramaliev and T. M. Madhyastha, "Optimizing probe-based storage," *Proc. USENIX Conference on File and Storage Technologies (FAST'03)*, pp.103-114, 2003.

Practice and Experience of an Embedded Processor Core Modeling

Gi-Ho Park[1], Sung Woo Chung[2], Han-Jong Kim[1], Jung-Bin Im[1],
Jung-Wook Park[3], Shin-Dug Kim[3], and Sung-Bae Park[1]

[1] Processor Architecture Lab SOC R & D Center, System LSI Division, Semiconductor Business Yongin-City, Kyeonggi-Do, Korea
[2] Division of Computer and Communication Engineering, Korea University, Anam-Dong, Seongbuk-Gu, Seoul, Korea
[3] Supercomputing Lab. Yonsei University, 134, Shinchon-Dong, Seodaemun-Gu, Seoul, Korea

giho.park@samsung.com, swchung@korea.ac.kr,
{followjx, bin5000.im}@samsung.com, {pjppp, sdkim}@parallel.yonsei.ac.kr,
sung.park@samsung.com

Abstract. This paper presents our experience in developing an embedded processor core model for an SOC design. We developed an ARM1136 processor simulation environment based on the ARM's MaxCore tool and the SimpleScaclar simulator. A MaxCore ARM1136 instruction accurate (IA) model is developed to support application programmers for the writing application programs from the early design stage. The MaxCore ARM1136 processor model supports all ARMv4, ARMv5TE and ARM v6 instruction sets with 418 LISA instructions. This MaxCore IA Model can be integrated with the ARM's MaxSim system level design environment to develop application softwares and perform architecture explorations. A SimpleScalar ARM1136 cycle accurate (CA) model is also developed by enhancing the existing SimpleScalar-ARM version in the SimpleScalar 3.0. Most important micro-architectural features of ARM1136 processor are implemented in the enhanced SimpleScalar simulator. The accuracy of the developed SimpleScalar-ARM 1136 simulator is about 97% compared to ARM 1136 RTL simulation with the Dhrystone benchmark (100 iterations).

1 Introduction

As the advance of process technology, hundreds of millions of transistors can be integrated into a single chip. Architectural exploration with an accurate performance model is essential in designing these complex chips. Performance simulator enables the exploration of various design alternatives for the future complex SOC (System on a Chip) design. Especially, the processor core is the most important component in the SOC design.

We developed an ARM1136 processor simulation environment based on commercially available tools and simulators. Those are the ARM's MaxCore tool

M. Gerndt and D. Kranzlmüller (Eds.): HPCC 2006, LNCS 4208, pp. 621–630, 2006.

and the SimpleScalar simulator. The MaxCore tool is a processor core/IP modeling environment provided by ARM (formerly AXYS Design Automation Inc.) for the processor architecture exploration, software development and processor (re)configuration. The SimpleScalar simulator is one of most popular cycle-accurate simulator used in the computer architecture research community. ARM processor cores are the most popular processor cores in the SOC design these days.

MaxCore IA model has been implemented to support ARM v4, v5TE and v6 instruction sets based on the ARM's (formerly AXYS) ARM v4 instruction accurate model. The SimpleScalar simulator was enhanced to support ARM1136 processor architectural features. Those include the pipeline extension from 6 stages to 8 stages, the static branch prediction support, and the hit under miss feature. We checked the validity and accuracy of the simulator with ARMulator and ARM1136 RTL running test benchmarks. The Dhrystone benchmark [4] and the EEMBC benchmark suite [5] are used as the test programs. The accuracy of SimpleScalar based ARM1136 cycle accurate model is about 97% for the Dhrystone benchmark for the 100 iteration execution with respect to the actual ARM1136.

2 MaxCore IA Model Design

We use the ARM MaxCore processor modeling tool [1] for the ARM1136 IA model development. The MaxCore is based on the LISA (Language for Instruction Set Architecture) and its LISA+ variant. All ARMv6 instruction sets (236 instructions) were implemented with 418 LISA instruction sets. ARM32 v6 instructions and Thumb mode instruction was described as 346 and 72 LISA instructions respectively. The number of the LISA instructions is larger than actual number of ARM v6 instruction because different addressing modes for same instruction has been implemented as different instructions for the code reusability and the easy of debugging. Average simulation speed of the developed MaxCore IA model (Interactive model) is very fast, which is about 12 MIPS (Million Instructions per Second) on a SUN Blade 2000 workstation having an UltraSparc III 1.2GHz processor.

2.1 ARM11 MaxCore IA Modeling

ARM11 IA model was developed based on the ARM7 IA model (ARM v4 ISA support) provided by ARM (formerly AXYS). ARM10 instruction (ARM v5TE ISA) and ARM11 (ARMv6 ISA) are added to support complete ARMv6 ISA. We refer the ARM Architecture Reference Manual [2] and the ARM1136 Technical Reference Manual [3] to develop the IA Model.

The ARM11 IA simulator is implemented with the functional description of each instruction written in the LISA. The functional behavior of each instruction includes the behavioral model of the instruction, the instruction word decoding information, the assembler syntax and reference to other operations with LISA.

All 'operation's are related to each other in a hierarchical manner and the relation of the operations is described within the composition section of the each operation. Each instruction is described as a class of 'operation', so instruction set is described as hierarchically ordered tree structure.

2.2 MaxCore Instruction Accurate Model Validation Methods

When we add an instruction, a couple of hand-coded simple assembly test programs are executed to check the validity of the modeling for the added instruction. The full test has been done with the EEMBC benchmarks suite [5] after all instructions were modeled. The validation mechanism of MaxCore IA model is shown in Fig. 1.

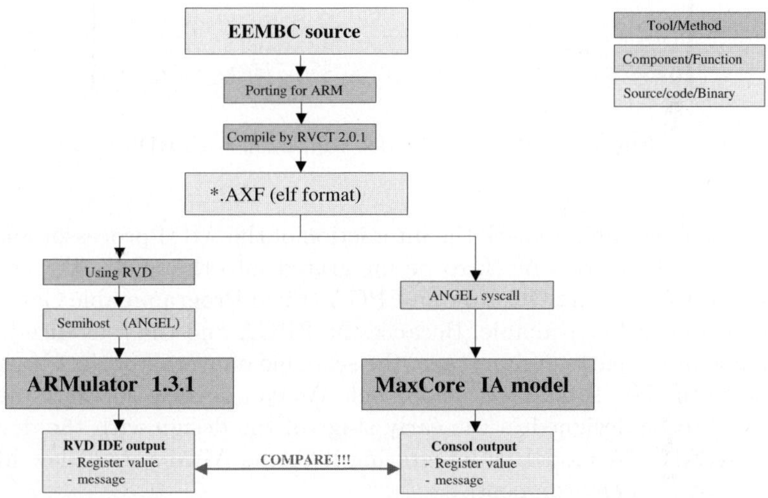

Fig. 1. MaxCore IA Test Flow

We use the ARM RealView Compilation Tool (RVCT) 2.0.1 to generate the ARM executable binary file (*.axf, elf format) for the execution in both the ARMulator and the MaxCore isim model, an interactive model of MaxCore ARM11 IA simulator. ARMulator is a functional emulator provided by ARM for the software development. The MaxCore ARM11 model was verified by comparing the value of the registers and the output messages generated by the test program as shown in Fig. 2. After completion of test program, we compared the last instruction executed, value of registers and output of test program in both the MaxCore ARM11 model (Upper part of Fig. 2) and the ARMulator ARM11 model (lower part of Fig. 2).

This MaxCore IA model can be integrated with the ARM's RealView MaxSim system level design environment to develop application software. Though ARMulator can be used to develop an ARM application programs for the ARM

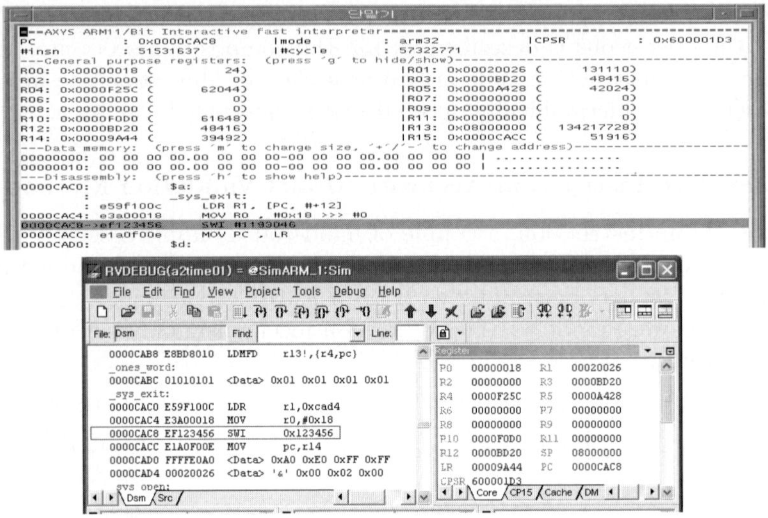

Fig. 2. MaxCore IA Model Validation with RDV

processor itself, we cannot check the interaction of the ARM processor and other IP (Intellectual Property) block to be integrated into the SOC. We should develop the software for an SOC after an FPGA (Field Programmable Gate Array) or a reference board is available. Because the FPGA and the reference board is not available in the early design stage, the schedule of overall project usually delayed due to the late software development. We can develop software programs for the SOC to be designed in the early stage of the design with the developed MaxCore ARM11 IA model by integrating it to the ARM's RealView MaxSim system level design environment.

3 SimpleScalar-CA Model Design

An ARM1136 cycle accurate (CA) model was developed based on the SimpleScalar 3.0 simulator. We enhanced the SimpleScalar-ARM version in the SimpleScalar 3.0, which is based on the ARM 7 processor. The SimpleScalar-ARM model is used as a baseline simulator to develop the ARM11 processor simulator.

However, we should modify the SimpleScalar-ARM model to reflect the differences of architectural features between ARM7 and ARM1136. Those features include the number of pipeline stages, the organization of execution units, write buffers.

The followings are important features modified from the original SimpleScalar-ARM to model the ARM1136 processor. We will call our developed ARM1136 simulator based on the SimpleScalar as SimpleScalar-ARM1136 in this paper. One of most important changes from the original SimpleScalar is related to the pipeline structure. ARM1136 has an 8-stage pipeline: fetch1, fetch2, decode, issue, execution1, execution2, execution3, and writeback as shown in Fig. 3. while the SimpleScalar simulator has only 6-stages, fetch, decode, issue, execution, writeback and commit.

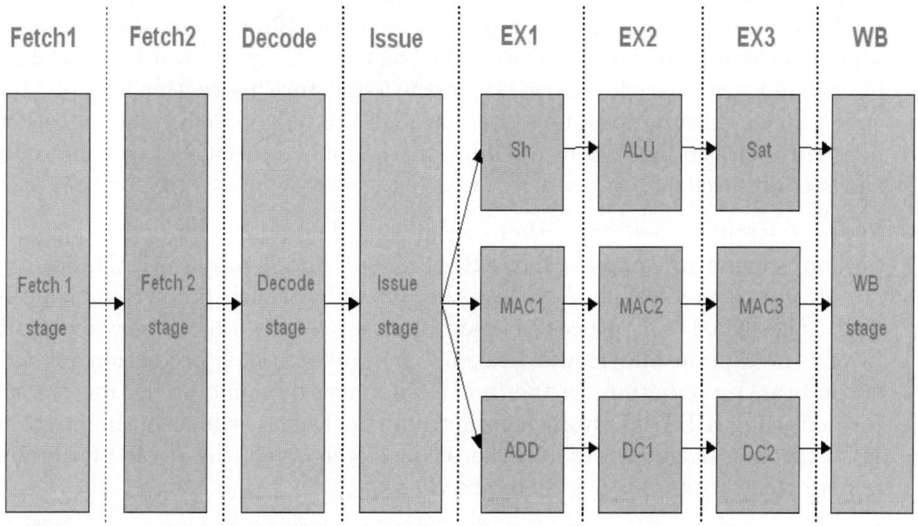

Fig. 3. Pipeline Stages of ARM1136 Processor

- Extend fetch stage from one two fetch stages : In SimpleScalar-ARM, there is only one fetch stage. The ARM1136 has two fetch stages, fetch1 and fetch2, to perform the instruction fetching and branch prediction. In SimpleScalar-ARM1136, fetch stage is extended to two stages based on the ARM1136 pipeline stage.
- Merging the commit stage into the writeback stage : In SimpleScalar-ARM 1136, commit stage is merged into writeback stage. We also adjust the branch misprediction resolution mechanism appropriately for the modified pipeline structure.
- Extend execution stage from one to three stages: In SimpleScalar-ARM, only one execution stage plays a role of multiple execution stages by setting execution latency and allowing multiple instructions in the execution unit. However this mechanism cannot support data forwarding in the execution stage. In SimpleScalar-ARM 1136, we explicitly divide one execution stage into three execution pipeline stages for three kinds of execution units (ALU pipeline, multiply pipeline and load/store pipeline). The ALU pipeline is composed of Shift, ALU and Saturation stage. If an instruction includes an operand to be shifted, it is performed in Shift stage. The ALU stage calculates the operands and the Saturation stage saturates the output from the ALU stage if necessary. The multiply pipeline and load/store pipeline

is implemented as MAC1, MAC2, and MAC3 pipeline stage and address calculation (ADD), DC1, and DC2 stage respectively. In the MAC1 stage and the MAC2 stage, two operands are multiplied. In the MAC3 stage, the output from the previous stage is accumulated. The Load/store pipeline has ADD, DC1, and DC2 stages. In the ADD stage, load/store address is calculated. If a shift operation is required to calculate the address in a load/store instruction, the shift operation in the Shift stage of the integer pipeline should precede load/store instruction. In other words, if a shift operation is necessary for a load/store, the load/store instruction is implicitly split into two instructions. One goes through the ALU pipeline and the other goes through the load/store pipeline. In the DC1 and DC2 stage, the data cache is accessed.

- Support static branch prediction mechanism: There is no static branch prediction scheme in SimpleScalar-ARM. In our SimpleScalar-ARM1136, the static branch prediction as well as the dynamic branch prediction is modeled. The ARM1136 processor uses bimodal prediction for the dynamic branch prediction and forward-taken/backward-untaken prediction for the static branch prediction. In the first fetch stage, dynamic prediction is performed using a BTAC (branch target address cache), which is similar to a BTB (branch target buffer). In the second fetch stage, the static prediction is performed if there is a miss in the BTAC.
- Interlock and data forwarding: In SimpleScalar-ARM(when in-order pipeline is set), an instruction is issued only when after all data dependencies of the previous instructions are resolved. In SimpleScalar-ARM1136, however, an instruction is issued regardless of data dependencies. Instead, required resolutions of data dependencies are checked in each execution stage. If the required data dependency has not been resolved, the instruction cannot proceed until the data dependency is resolved.
- Memory interface bandwidth: Mosts of modern processors have a write buffer between an L1 cache and memory to decouple stores from the pipeline. In SimpleScalar-ARM, the entry of the write buffer is not limited. We implemented an 8-entry write buffer, which may affect performance very much. The critical word first transfer feature is also modeled in SimpleScalar-ARM1136. We also implemented the 64-bit wide internal bus between the pipeline and the L1 cache, leading to two-word transfer at a time in case of LDM (LoaD Multiple) or STM (STore Multiple).
- HUM (Hit Under Miss) feature: When SimpleScalar-ARM is set to run in in-order, multiple data cache misses can go out to fetch the data from the lower level memory. When a cache miss occurs, subsequent data cache access can fetch data only if there is an empty entry in LSQ (load store queue). In our SimpleScalar-ARM1136, however, only one data cache miss can fetch data from the lower level memory to reflect the real ARM1136 processor architecture. If there are multiple cache misses, pipeline is stalled until only one cache miss remains.

- STORE operation in the execution stage : In SimpleScalar-ARM, a store operation is executed in writeback stage. In SimpleScalar-ARM 1136, a store operation is executed in the execution stage to reflect the pipeline stall in in-order pipeline with a write buffer.

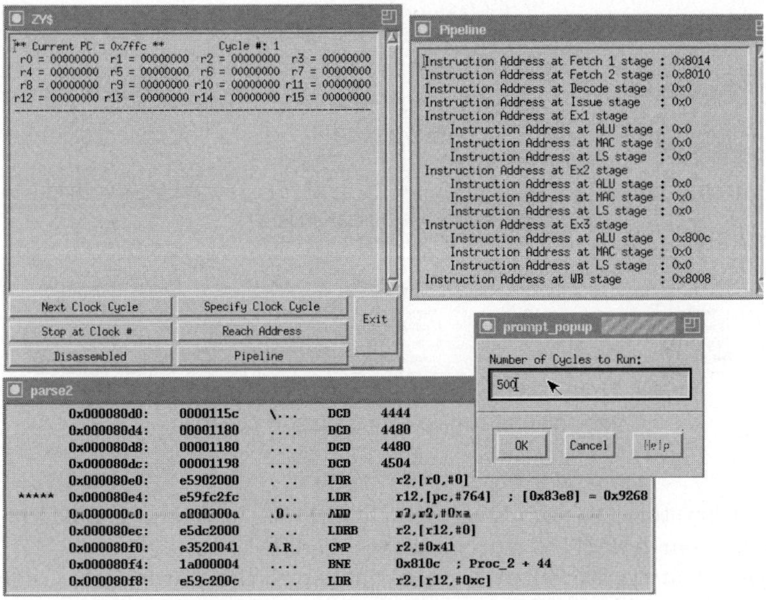

Fig. 4. Graphic User Interface of CA model Simulator

3.1 SimpleScalar-ARM1136 Cycle Accurate Model Validation Methods

We implemented a GUI (Graphic User Interface) for our SimpleScalar-ARM1136 simulator, as shown in Fig. 4 for efficient debugging and analysis of the SimpleScalar-ARM1136 simulator. The GUI shows register values, pipeline status and disassembled code to see the pipeline flows in the simulator.

To verify the accuracy, we compare the SimpleScalar-ARM 1136 with the ARM1136 RTL simulation. We use the ARM RealView Compilation Tool (RVCT) 2.0.1 to generate the ARM executable binary file for the execution. We execute the test program in ARM11 processor RTL simulation environment, SimpleScalar-ARM CA model and ARM RealView ARMulator 1.3.1 as shown in Fig. 5. We compare the output logs of these simulators to check the both functional correctness and cycle accuracy of SimpleScalar ARM1136 model.

Following steps are used to check the functional correctness and cycle accuracy of SimpleScalar-ARM1136 model.

1. Simple test programs with several instructions are written to verify the latencies in ARM1136 TRM [3]. These programs are run to verify the added features.

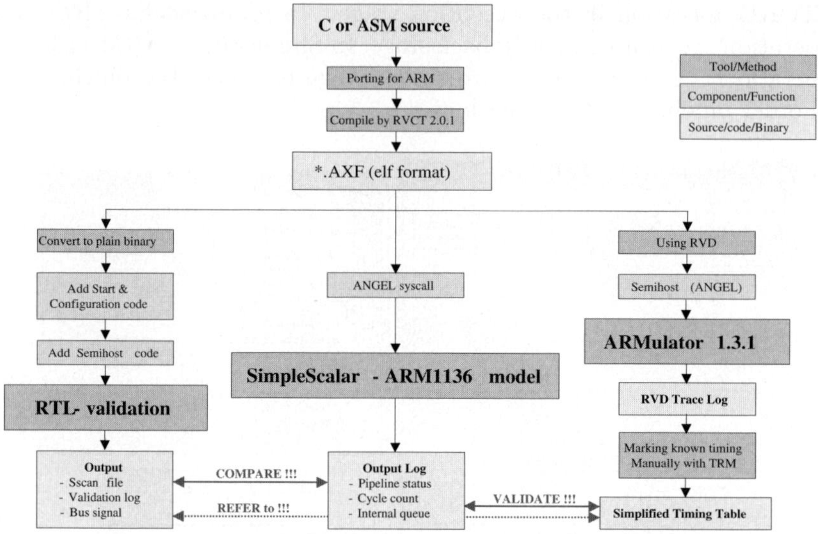

Fig. 5. SimpleScalar CA Model Test Flow

2. The Dhrystone (1-iteration) benchmark is run on ARM1136 RTL, SimpleScalar-ARM1136 and SimpleScalar-ARM.
3. When a feature is evaluated on a simulator, the feature is considered as stable if it provides similar improvement/degradation for all three simulation environments. In order to validation the stability of our simulators, we repeat procedure (2) by turning off the branch predictor.
4. Procedure (2) and (3) are repeatedly done on the Dhrystone benchmarks for 100-iteration execution.

Fig. 6 shows the cycle count comparison between our SimpleScalar- ARM1136 model and ARM1136 RTL when we run the Dhrystone benchmark with 100 iteration. As architectural features of ARM1136 explained in the previous section, are added to the original SimpleScalar-ARM simultor model, the accuracy is converging and stable to ARM1136 RTL simulation results even though there were some fluctuations. The left Y axis shows the number of cycles in our simulator to run the Dhrystone benchmark. The cycle count of the Dhrystone benchmark is 63,990 in ARM1136 RTL which has no error (0% error shown in the graph). The right Y axis shows the accuracy error which is calculated as Error = (# of execution cycles in simulator - 63,990) / 63,990. The accuracy of our final SimpleScalar-ARM1136 simulator is 96.9 % compared to ARM 1136 RTL with the Dhrystone benchmark (100 iterations).

3.2 Modeling Efforts and Accuracy

As we expected, the accuracy has been improved as we reflects the micro-architectural features to the simulator. It is much easier to develop the

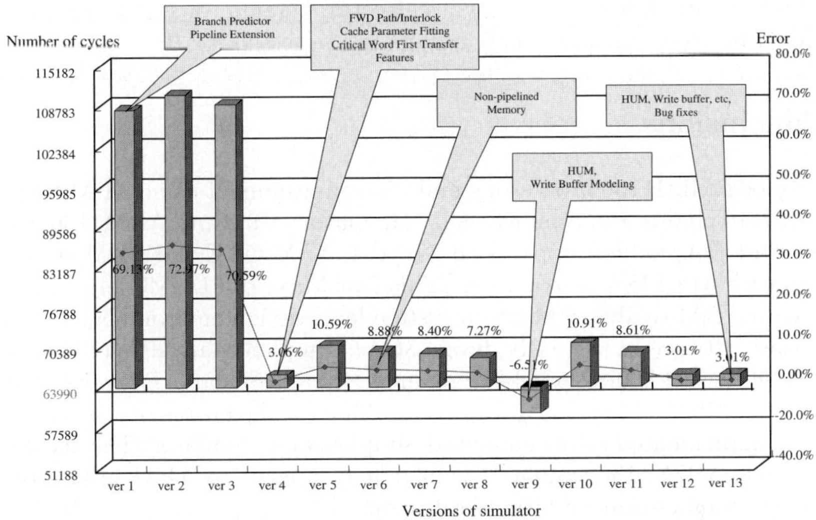

Fig. 6. Accuracy of SimpleScalar CA Model

instruction accurate (IA) model than the cycle accurate model. The MaxCore tool chain and the model provided from ARM (formerly AXYS) is very helpful for us to model it. The validation of the instruction accurate model can be done by simply checking the architectural status registers and output of benchmark programs. The golden reference, the ARM RealView ARMulator, is also very convenient to use for this purpose. We can find all information required for the modeling with the ARM architecture reference manual and ARM1136 technical reference manual.

Modeling for the cycle accurate (CA) model was much more difficult than that for the instruction accurate (IA) model. First of all, it is very difficult to have enough information for the ARM1136 micro-architectural features to achieve cycle accuracy of our model. The validation of the model is also very difficult for the CA model because the cycle count is not solely dependant on the processor core. We should have same system environment including memory system for the golden reference and our model. Adjusting our simulation environment including memory system architecture to existing RTL simulation environment requires much efforts for us. It is mainly because the RTL simulation environment is built for the verification purpose rather than any cycle accuracy checking purpose. If processor developing company, like ARM, provides a golden reference model designed for the cycle accuracy check to the architecture research society especially for the academia without any financial burden, it will be very helpful for the architecture researcher. It will be beneficial for themselves (the IP generating company) because they can get new and efficient micro-architecture ideas and simulation results based on their own processor/IP. Architect should consider the accuracy of the simulator especially for the features related to their target architecture before using any architectural simulator if it is not provided or certified by the company which designs the processor or IP.

4 Conclusions

We developed ARM1136 processor simulation environment using ARM RealView MaxCore tool chains and SimpleScaclar simulator. MaxCore ARM1136 instruction accurate (IA) model supports all ARMv4, ARMv5TE and ARM v6 instruction sets with 418 LISA instructions. This fast MaxCore IA Model can be integrated with ARM Realview MaxSim system level design environment to develop application software in the early design stage. SimpleScalar-ARM1136 cycle accurate (CA) model is developed by enhancing SimpleScalar-ARM version in the SimpleScalar 3.0. Most important micro-architectural features of ARM1136 processor are implemented in the enhanced SimpleScalar simulator. The accuracy of SimpleScalar-ARM 1136 simulator is 96.9 % compared to ARM 1136 processor with Dhrystone benchmark (100 iterations).

Based on our experience, it is very important to have a cycle accurate simulation model to perform the micro-architectural research. The simulation results can mislead researchers when the simulator is not very accurate. If it is difficult to secure the very accurate architectural simulator, researchers should consider very carefully for the accuracy of the simulator especially for the micro-architectural features related closely to their target architecture.

References

1. AXYS Design Automation Inc., MaxCore Tools Training Material, 2003. 10
2. ARM, ARM Architecture Reference Manual, ARM DDI 0100E, 2000. 6
3. ARM, ARM1136JF-S and ARM1136J-Stm Technical Reference Manual, r0p2, ARM DDI 0211D, 2003. 8
4. Reinhold P. Weicker, "Dhrystone Benchmark: Rationale for Version 2 and Measurement Rules," SIGPLAN Notices Vol. 23,No. 8, 1998. 8, pp. 49-62
5. "Embedded Microprocessor Benchmark Consortium" information available at www.eembc.org
6. D. Burger, T.M.Austin, "The SimpleScalar tool set, version 2.0," Technical Report TR-97-1342, University of Wisconsin Madison, 1997.

QoS Support for Video Transmission in High-Speed Interconnects

A. Martínez[1], G. Apostolopoulos[2], F. J. Alfaro[1], J. L. Sánchez[1], and J. Duato[3]

[1] Dept. de Sist. Inform., Univ. de Castilla-La Mancha, 02071 - Albacete, Spain
{alejandro, falfaro, jsanchez}@dsi.uclm.es
[2] CARV Lab., Inst. of Computer Science - FORTH, 71110 - Heraklion, Crete, Greece
georgeap@ics.forth.gr - Member of HiPEAC
[3] GAP - DISCA, Univ. Politécnica de Valencia, 46071 - Valencia, Spain
jduato@disca.upv.es - Member of HiPEAC

Abstract. Multimedia traffic presents some special requirements that are unattainable with a best-effort service. Current interconnect standards provide mechanisms to overcome the limitations of the best-effort model, but they do not suffice to satisfy the strict requirements of video transmissions. This problem has been extensively addressed at the general networking community. Several solutions have arisen, but they are too complex to be applied to high speed-interconnects. In this paper, we propose a network architecture that is at the same time compatible with the requirements of high-speed interconnects and provides video traffic with the QoS it demands.

Keywords: QoS, Switch Design, Scheduling, Virtual Channels, Clusters.

1 Introduction

The last decade has witnessed a vast increase in the amount of information and services available through the Internet. Clusters of PCs have emerged as a cost-effective platform to implement these services. They provide service to thousands or tens of thousands of concurrent users. These users usually demand specific quality of service (QoS) requirements [1].

In the next section, we will introduce the InfiniBand and PCI AS high-speed interconnect standards. These technologies provide mechanisms for QoS support that consist of the segregation of the traffic in traffic classes (TCs), virtual channels (VCs), and a mechanism to map the TCs to the VCs and then provide scheduling for the VCs. However, the scheduling algorithms proposed [2,3] are fairly simplistic and fail to provide certain kinds of traffic with the requirements they demand. For instance, video traffic is usually very concerned about jitter, and much less about latency [4].

There has been a very substantial body of work on mechanisms for providing QoS guarantees for packet switches[1]. Usually these works assume that the packet

[1] We will use the terms *packet switches* and *packet networks* to refer to general networking technologies.

M. Gerndt and D. Kranzlmüller (Eds.): HPCC 2006, LNCS 4208, pp. 631–641, 2006.

switch has significant amount of resources, in particular large random access buffers. Moreover, in packet networks packets may be dropped when buffering capacity is exceeded and latencies can be large. Thus, most of the scheduling policies focus on controlling packet losses and delays. On the other hand, in high-speed interconnects, switches are single-chip and, thus, have considerably more limited resources, latencies are very small due to the small geographical extent of the interconnect, and typically flow control is employed preventing packet drops. As a result, the interconnect environment requires special attention when developing QoS scheduling policies.

In this work, we introduce a QoS architecture that is tailored for the interconnect environment. By moving the complexity of the packet scheduling to the host network interfaces we are able to keep the packet processing in the switches very simple. We show how with a simple QoS architecture, we can support the QoS requirements of multiple different types of traffic: high-priority control traffic, medium priority video traffic, and low priority best-effort. All the types of traffic can coexist and receive the desired QoS without interfering with each other. At the same time, we are able to achieve high utilization of the interconnect and all these without requiring more than two queues per switch port. This allows us a significant reduction in the switch complexity, which is critical if we want to scale up the switch port densities.

The remainder of this paper is structured as follows. In the following section the related work is presented. In Section 3 we present our strategy to offer QoS support. Details on the experimental platform are in Section 4 and the performance evaluation is presented in Section 5. Finally, Section 6 summarizes the results of this study and identifies directions for future research.

2 Related Work

In this section, we will review the special characteristics of video traffic and its requirements. Next, we will analyze the two most recent technologies for high-speed interconnects (InfiniBand and PCI AS) and how they can provide QoS. Finally, we will review some algorithms for the provision of QoS to multimedia flows in general networking.

2.1 Video Traffic's Characteristics and Requirements

Video sequences are composed of a set of video frames that are generated at regular intervals. Compression algorithms produce frame patterns in which some frames are smaller than others. More specifically, there are intra-coded frames, which are basically normal pictures compressed with an algorithm like JPEG; besides, there are inter-coded frames, which only encode the differences with some neighbor frames. Therefore, frame size presents a lot of variability [5].

Ideally, the receiver should receive a complete frame exactly each inter-frame interval (usually 40 milliseconds). This is measured by *jitter*, the variation of the latency of two consecutive packets of the same flow [6]. This is important because if frames arrive too late they are obviously useless, but if they arrive too soon, they can overflow the reception buffer.

Furthermore, a latency of less than 100 milliseconds is desirable for interactive video [4]. This includes video-conference and video on demand, when the watcher has the ability to stop and peek through the sequence. Moreover, although there is some tolerance to packet loss, it should be very reduced.

2.2 QoS Support for Multimedia Traffic in Packet Networks

Over the last years there has been extensive work on how to schedule resources of a packet switch to provide guaranteed performance to traffic. The switch resources that need to be scheduled are buffer space (usually at the outgoing port) and link capacity. Both are managed through a *service discipline*. Typically, packet switch buffers are fairly large and support random access. When buffers become full, packets are dropped. Thus, general packet switches can introduce packet loss when their resources are oversubscribed. Performance guarantees usually include bounds on packet loss, delay, jitter, and transmission rate or throughput. A large number of service disciplines have been proposed (see [7] for an overview) each specifically targeted for providing certain types of guarantees.

The service disciplines operate at the flow level and consequently can provide different QoS guarantees to individual flows. An example of such flow oriented QoS architectures is the QoS architecture of ATM [8] and the Integrated-Services model [9] that was proposed for Internet QoS in mid-90s. Since such per-flow scheduling can prove a bottleneck as the number of flows grows, aggregate-QoS architectures have been proposed where QoS is provided collectively to all flows that belong to a certain class of service. There are only a few such classes of service, but flows now get only aggregate and not individual QoS. An example of such a QoS architecture is Differentiated Services [10], which is used to provide limited QoS in parts of the Internet today.

2.3 QoS Support in New High-Speed Interconnects

When compared with a generic packet switch, high-speed interconnect switches have some important differences mostly because of their much simpler and compact implementation. Firstly, flow control is commonly used to throttle the incoming traffic, and thus usually there are no packet drops due to running out of buffer space. Buffers themselves may be smaller than what one would expect from a generic packet switch. Furthermore, access to these buffers may be more restricted and random access may not be possible due to the strict time limitations. Similarly, the number of different queues may be limited.

InfiniBand was proposed in 1999 by the most important IT companies to provide present and future server systems with the required levels of reliability, availability, performance, scalability, and QoS [2]. Specifically, the InfiniBand Architecture (IBA) proposes three main mechanisms to provide the applications with QoS. These are traffic segregation with service levels, the use of VCs (IBA ports can have up to 16 VCs) and the arbitration at output ports according to an arbitration table. Although IBA does not specify how these mechanisms should be used, some proposals have been made to provide applications with QoS in InfiniBand networks [11].

On the other hand, PCI Express Advanced Switching (AS) architecture is the natural evolution of the traditional PCI bus [3]. It defines a switch fabric architecture that supports high availability, performance, reliability and QoS. AS ports incorporate up to 20 VCs (16 unicast and 4 multicast) that are scheduled according to some QoS criteria. Is is also possible to use a connection admission control implemented in the fabric management software.

These proposals, therefore, permit to use a significant number of VCs to provide QoS support. However, implementing a great number of VCs would require a significant fraction of silicon area and would make packet processing slower. Moreover, there is a trend of increasing the number of ports instead of increasing the number of VCs per port [12]. In general, the number of queues per port can have a significant effect on the overall complexity and cost of the interconnect switch. It is important to attempt to provide effective QoS with a number of queues as small as possible. Indeed, our proposal addresses this very effectively.

3 Architecture for QoS Support

Deadline[2]-based policies are among the most effective scheduling policies in packet networks. These policies operate as follows: each packet is labeled with a deadline and, thereafter, the switches solve all the scheduling and output conflicts choosing always the packet with the smallest deadline. This usually requires that all the packets in the buffers are taken into account for scheduling and, thus, random access buffers are needed. Another alternative is to set up heap buffers, that always keep at the top the packet with the lowest deadline [13,14]. However, these implementations are too expensive for high-speed interconnects. As far as we know, nobody has tried to adapt this kind of algorithms to this environment.

Note that in packet networks (like Internet) the deadline would be recomputed at each hop. This is not reasonable in this case. For high-speed networks, which span over a much shorter area, deadline would be computed once at the interfaces.

When traffic is regulated, the switches can avoid random access buffers and just take into account the first packet at each input buffer. The idea is that traffic coming from the interfaces has already been scheduled and is coming in descending order of deadlines. This being so, it is possible to just consider the first packet at each queue, being confident that packets coming afterward have higher deadlines.

[2] We will use the term *deadline* as a tag contained in the header of packets used for scheduling.

The behavior of the switch would be analogous to a sorting algorithm: if the switch has as input sorted chains of packets and has to produce at the output a sorted sequence, it only needs to look at the first packet of each input.

Let us have a look at the possible limitations of this algorithm:

- The traffic must be regulated. If a link is oversubscribed, we cannot guarantee that flows will get the demanded QoS. However, regulation on the traffic is always mandatory to provide strict guarantees on latency and throughput.
- The deadlines must not be recomputed. If we allowed the deadlines to change during the life of the packet, we would not be able to assure that the order established at the interfaces would be valid. However, in the high-speed interconnects environment latencies are expected to be very short and there is no need to recompute these deadlines.
- The packets are not coming always in order from the interfaces. The above scheduling policy assumes that packets arrive from interfaces ordered according to their deadlines. This may not be true all the time though. For example, immediately after a packet with large deadline has departed from the interface, a high priority small deadline packet arrives and is sent behind the large deadline packet. This will violate our assumptions and degrade the service offered to the high-priority packet. In order to limit the occurrences of these out-of-order packets, we use a non-work conserving deadline scheduling policy. For video traffic, an eligibility time (the minimum time when packets are allowed to leave) is also provided to reduce jitter. In this way, by bounding the cycle where packets are available for transmission, we can also bound the maximum distance between the deadlines of two out-of-order packets. In any case, as we will see in the evaluation section, the impact of out of order packets is rather limited mostly due to the low latency of the interconnect for regulated traffic.

In order to support control traffic, which is usually unregulated, we can safely assume that it will never congest any link by itself. This kind of traffic usually requires negligible bandwidth but demands low latencies. We can mix it with video traffic by providing small deadline tags.

Best-effort traffic must also be supported. In this case, high bandwidth is demanded and congestion may appear since this traffic is not regulated. Therefore, in order to not disturb regulated traffic, it requires a separate VC. Moreover, absolute priority should be given to regulated traffic over unregulated traffic.

Summing up, our proposal consists in a network architecture able to deal with three different classes of traffic: control traffic, which demands little bandwidth but low latency; video traffic, which demands guaranteed throughput, bounded latency and low jitter; and best-effort traffic, which demands as much bandwidth as possible. This is achieved with only two VCs and a feasible implementation on a single chip, as is usually the case in high-speed interconnects.

3.1 Generating Deadlines for Video Traffic

We propose a simple scheme to label video packets at the interfaces in order to provide these requirements. Each packet that belongs to a particular frame would receive a deadline covering the whole inter-frame period. This would smooth the burst along this period of time (see Figure 1 for an example). In other words, the first packet of the frame would receive a deadline near to the actual clock cycle, while the last one would receive a deadline equal to the clock plus the inter-frame time. The intermediate packets would be uniformly distributed between these.

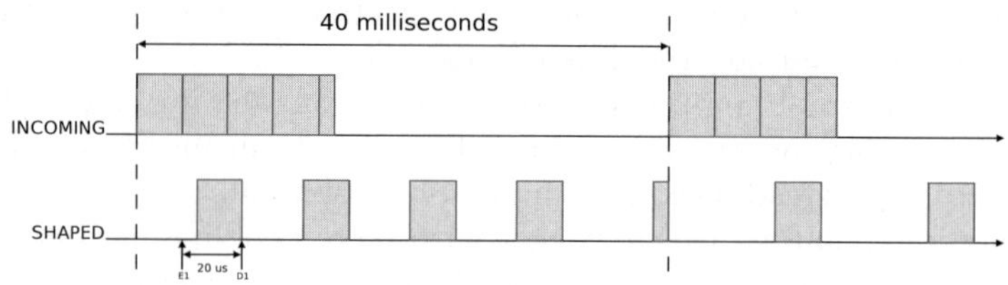

Fig. 1. Traffic shaping at network interfaces

In addition to deadlines, each packet would have an eligibility time. No packet would be allowed to leave the interface before this time has passed, in order to guarantee that the jitter would be as close to 0 as possible. We are computing the eligibility time of a packet as its deadline minus a constant value. We have found that 20 μs works well for this value.

With this strategy, buffers must have capacity for one whole frame per active video connection in the worst case. Note that this amount is also required in a work conserving alternative: the worst case is the same.

Therefore, the next packet to be chosen at a network interface would be the one with the lowest deadline from those which are eligible (the eligibility cycle has passed). This requires to keep two ordered queues at the interface, one with non-ready packets, in eligibility order, and another with eligible packets, in deadline order. This is affordable in these devices, since significant memory and processing capacity are available here.

4 Simulation Conditions

We have performed the tests considering three cases. First, we have tested the performance of our proposal, which uses 2 VCs at each switch port. It is referred to in the figures as *New 2 VCs*. We have also performed tests with switches using ideal, but impractically expensive random access buffers. In this case, it is referred to in the figures as *RAM buffers*. Note that we assume the same delays for the switch as in our proposal, which is not realistic, but serves us to examine which is the impact of order errors. Finally, we have also tested a traditional approach, based on the specifications of InfiniBand and PCI AS, with 4 VCs, noted in the figures as *Traditional 4 VCs*. In this case, there is a VC for each traffic class considered, both at the switches and at the network interfaces.

In the three cases, we have used 16 port switches, 8 Gbits/s links, and 8 Mbits of total buffering at each switch. To cope with the inefficiencies of the scheduler and packet segmentation overheads[3], the crossbar core operates twice as fast as the external lines (internal speed-up of 2.0).

The network used to test the proposals is a butterfly multi-stage interconnection network (MIN) with 64 end-points. The actual topology is a folded (bidirectional) perfect-shuffle. We have chosen a MIN because it is a usual topology for clusters. However, our proposal is valid for any network topology, including both direct networks and MINs. No packets are dropped at the switches because we use credit-based flow control at the VC level. However, if a video packet has to wait more than 100 milliseconds at the interface, it is completely useless and is, therefore, dropped.

In Table 1, the characteristics of the modeled traffic are included. Traffic consists in three categories: network control, video, and best-effort. The first category models short control messages that require short latency but demand negligible bandwidth. Video traffic is taken from actual MPEG-4 video sequences, which produce a video frame each 40 milliseconds, approximately. This is very bursty traffic and will heavily degrade the performance of the network. The required results for video traffic according to [4] are guaranteed bandwidth, latency below 100 milliseconds and jitter as short as possible. Finally, we have modeled two classes of best-effort traffic: *Best-effort* and *Background*. The former demands as much bandwidth as possible, while the latter would require the remaining bandwidth, if any.

Table 1. Traffic injected per host

TC	Name	% BW	Packet size	Notes
0	Network Control	1	[64,512] bytes	self-similar
1	Video	49	[64,2048] bytes	750 KByte/s MPEG-4 traces
2	Best-effort	25	[64,2048] bytes	self-similar, burst = 20
3	Background	25	[64,2048] bytes	self-similar, burst = 20

The self-similar traffic is composed of bursts of packets heading to the same destination. The packets' sizes are governed by a Pareto distribution, as recommended in [15]. In this way, many small size packets are generated, with an occasional large size packet. The periods between bursts are modeled with a Poisson distribution. If the burst size is long, it should show worst-case behavior because, at a given moment in time, many packets are grouped going to the same destination.

[3] Crossbars inherently operate on fixed size cells and thus external packets are traditionally converted to such internal cells.

5 Simulation Results

In this section, we show the performance of our proposal. We have considered three common QoS metrics for this performance evaluation: throughput, latency, and jitter. Note that packet loss is only possible at the interfaces for video packets, thereafter there is a credit-based flow control.

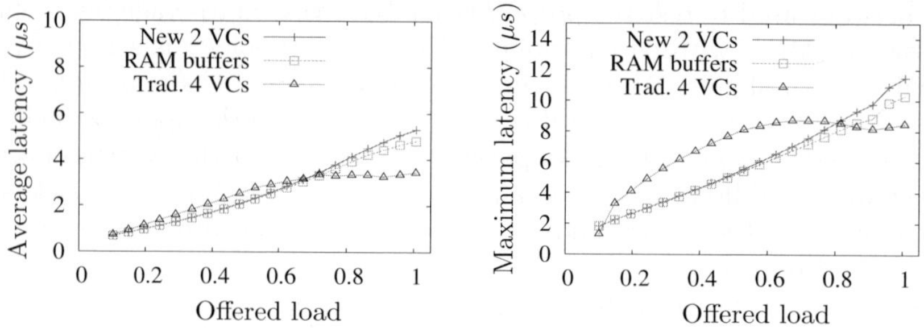

Fig. 2. Performance of *Network Control* traffic

The performance of *Network Control* traffic is shown in Figure 2. We can see that the three alternatives offer good latency results, both average and maximum. The differences between the *Traditional 4 VCs* case and the two based on deadlines come from the fact that in the latter ones only two VCs are used. Therefore, the *Network Control* traffic shares the VC with the *Video* traffic. We can conclude that this reduction of VCs is not causing problems, there is only a rather small performance loss at high load.

On the other hand, the small differences that can be observed between our proposal and the *RAM buffers* case are due to the order errors we discussed in Section 3. However, we see that the impact of this issue is very limited, even after mixing a few of high-priority packets with lots of low-priority video traffic.

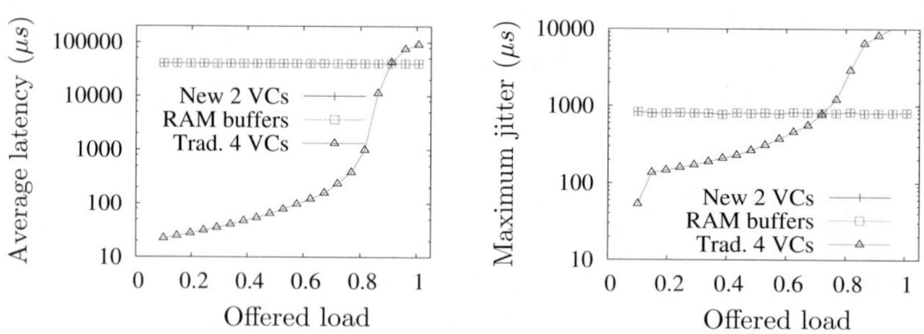

Fig. 3. Performance of *Video* traffic

Figure 3 shows the performance of video traffic in terms of average latency and maximum jitter. The deadline-based alternatives succeed in providing a

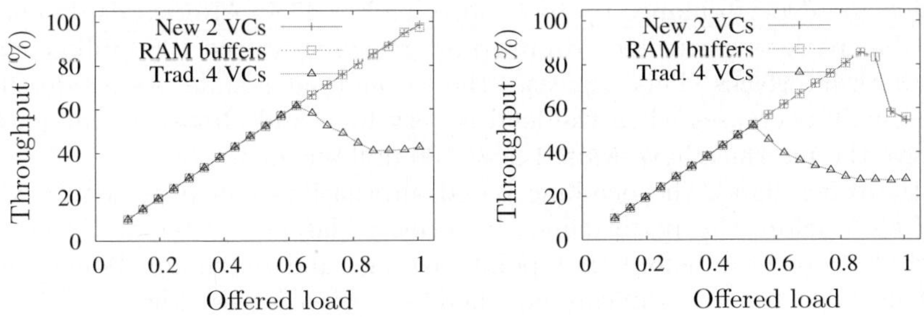

(a) Best-effort (b) Background

Fig. 4. Performance of best-effort traffic

constant latency of 40 milliseconds for all the video frames independently of the load. This is also reflected in jitter, which is low at all load levels. Note that this results are referred to the full video frames. Individual packets have much lower latency, of course. On the other hand, results for *Traditional 4 VCs* case are not so good, because latency varies with load and at high load the latency and jitter reach unacceptable values. Our two-VC scheme, with the traffic shaping based on deadlines, offers much better performance with less VCs.

To finish this part of the study, we will look at best-effort traffic results (Figure 4). We can see that the two deadline-based alternatives offer much better throughput to this kind of traffic than the *Traditional 4 VCs* architecture. Moreover, note that although we are using just one VC with our proposal for both best-effort TCs, there is QoS differentiation between the two classes of best-effort traffic: *Best-effort* traffic keeps good performance at high load, while *Background* decays.

We can conclude at this point that our proposal offers as good latency for *Control Traffic* as the other two options; offers as good jitter for video traffic as the unfeasible *RAM buffers* architecture due to the traffic shaping, while the *Traditional 4 VCs* case fails in this; and offers as good throughput as the other alternatives for bursty, unbalanced best-effort traffic.

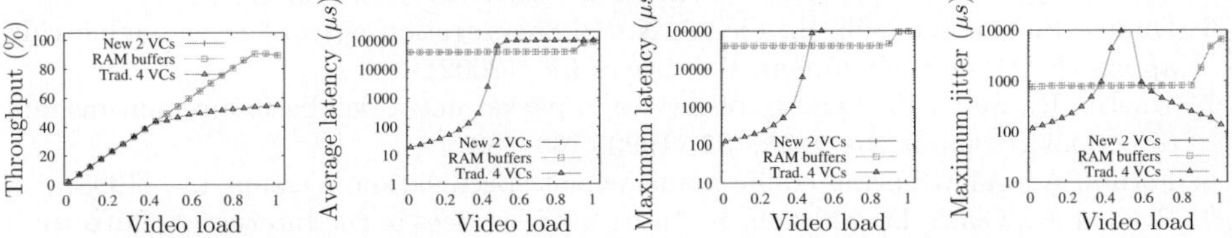

Fig. 5. Performance of full *Video* traffic injection

In the next experiment, we examine the capacity of the three alternatives to deal with video traffic. We vary the load from 0 to full input injection of video sequences, as can be seen in Figure 5. We observe that the *Traditional 4 VCs* case offers good performance up to a video load of 45%. Afterwards, latency is so bad that packets start being dropped after having waited 100 millisecond at the network interfaces. This decreases the throughput results. As a side effect, jitter actually decreases when the load is very high and almost all the packets that leave the interface have waited near 100 milliseconds.

On the other hand, the deadline based alternatives can cope with a video load of 85% before the performance in terms of latency, jitter or throughput is affected. Note that before that point, average and maximum latency is 40 milliseconds (for the full frames) and, therefore, jitter is very low.

6 Conclusions

In this paper, we propose a novel technique for supporting very efficiently deadline-based scheduling policies in a high-speed interconnect. Based on these policies, we are able to offer excellent performance compared with traditional solutions proposed in specifications like InfiniBand or PCI AS. Moreover, we show that the performance of our proposal is not far from what would be obtained using expensive random access buffers. We are able to use only 2 VCs per port, reducing considerably the cost and complexity of the interconnect switch. Even with only 2 VCs, we are able to provide QoS differentiation between multiple different classes of traffic and improve network utilization.

References

1. Miras, D.: A survey on network QoS needs of advanced internet applications. Technical report, Internet2 - QoS Working Group (2002)
2. InfiniBand Trade Association: InfiniBand architecture specification volume 1. Release 1.0. (2000)
3. Advanced switching core architecture specification. Technical report, (available online at `http://www.asi-sig.org/specifications` for ASI SIG members)
4. IEEE: 802.1D-2004: Standard for local and metropolitan area networks. `http://grouper.ieee.org/groups/802/1/` (2004)
5. Moving Picture Experts Group: Generic coding of moving pictures and associated audio. Rec. H.262. Draft Intl. Standard ISO/IEC 13818-2 (1994)
6. Duato, J., Yalamanchili, S., Lionel, N.: Interconnection networks. An engineering approach. Morgan Kaufmann Publishers Inc. (2002)
7. Guerin, R., Peris, V.: Quality-of-service in packet networks: basic mechanisms and directions. Comput. Networks **31** (1999) 169–189
8. Forum, A.: ATM Forum traffic management specification. Version 4.0. (1995)
9. Braden, R., Clark, D., Shenker, S.: Integrated Services in the Internet Architecture: an Overview. Internet Request for Comment RFC 1633, Internet Engineering Task Force (1994)
10. Blake, S., Back, D., Carlson, M., Davies, E., Wang, Z., Weiss, W.: An Architecture for Differentiated Services. Internet Request for Comment RFC 2475, Internet Engineering Task Force (1998)
11. Alfaro, F.J., Sánchez, J.L., Duato, J.: QoS in InfiniBand subnetworks. IEEE Transactions on Parallel Distributed Systems **15** (2004) 810–823

12. Minkenberg, C., Abel, F., Gusat, M., Luijten, R.P., Denzel, W.: Current issues in packet switch design. In: ACM SIGCOMM Computer Communication Review. (2003)
13. Ioannou, A., Katevenis, M.: Pipelined heap (priority queue) management for advanced scheduling in high speed networks. In: Proceedings of the IEEE International Conference on Communications (ICC'2001). (2001)
14. Yun, K.Y.: A terabit multiservice switch. IEEE Micro **21** (2001) 58–70
15. Jain, R.: The art of computer system performance analysis: techniques for experimental design, measurement, simulation and modeling. John Wiley and Sons, Inc. (1991)

Discrete Broadcasting Protocols for Video-on-Demand

Chao Peng[1,⋆], Hong Shen[1], Naixue Xiong[1], and Laurence T. Yang[2]

[1] Graduate School of Information Science
Japan Advanced Institute of Science and Technology
1-1 Tatsunokuchi, Ishikawa, 923-1292, Japan
p-chao@jaist.ac.jp
[2] Department of Computer Science,
St. Francis Xavier University, Antigonish, B2G 2W5, Canada

Abstract. The Video-on-demand (VOD) service allows users to view any video program from a server at the time of their choice. Broadcasting protocols can be used to improve the efficiency of a VOD system. The Harmonic Broadcasting Protocol has been proved to be bandwidth-optimal, but it is not efficient for the local storage. In this paper, we present the Discrete Broadcasting scheme, which can intelligently adjust its solution according to available bandwidth and local storage in order to achieve an ideal waiting time.

1 Introduction

In a Video-on-demand (VOD) system, a subscriber is expected to be able to watch his favorite video program in the server at the time of his choice. Usually such a system is implemented by a client-server architecture supported by certain transport networks such as CATV, telecom, or satellite networks. Clients use web browsers or set-top-boxes (STB) on their television sets. In a pure VOD system, each user is assigned a dedicated video channel so that they can watch the video they chose without delay and many VCR-like functions may be provided. But in this case the cost is too expensive, because it will quickly use up all of the available bandwidth on the VOD server when too many concurrent users are to be accommodated at the same time.

To reduce the tremendous bandwidth and I/O requirements, many alternatives have been proposed by sacrificing some VCR functions. Broadcasting is one of such techniques and is mostly appropriate for popular videos that are likely to be simultaneously watched by many viewers [9]. In this approach, the server uses multiple dedicated channels to execute frequent retransmissions of a "hot" video. Each client follows some reception rules to grab and store data from appropriate channels so as to play the whole video continuously. What distinguishes broadcasting from other VOD distribution methods is that the server's

⋆ Supported by "Fostering Talent in Emergent Research Fields" program in Special Coordination Funds for promoting Science and Technology by Ministry of Education, Culture, Sports, Science and Technology.

M. Gerndt and D. Kranzlmüller (Eds.): HPCC 2006, LNCS 4208, pp. 642–652, 2006.

broadcasting activity is independent of the number of viewers. The bandwidth savings can be considerable since a few (10 or 20) very popular videos are likely to account for nearly 80% of demands [2].

The simplest solution is to periodically broadcast the video on several channels, each differentiated by some time. By this method the server need at least K channels in order to keep the waiting time below L/K (here L is the length of the whole video). To enhance the efficiency in channel usage, many schemes [2,4,5,6,7,8,10] have been proposed by imposing a large enough client receiving bandwidth and an extra buffering space at the client side.

In this paper, we present the *Discrete Broadcasting Protocol*(DB), which can intelligently adjust its solution according to available resources such as available channels and local storage. It can reduce the average waiting time of a 120 minutes video to 3 minutes if we allocate a bandwidth of 4 times the consumption rate. It can also be modified for VOD service even when the local storage is very small. Some of our results have already been realized in industrial application and have got a good performance.

2 Model and Analysis of VOD Broadcasting Protocols

The broadcasting problem in the VoD service can be described as the following: In a VOD system, given a video of size S (Mb) and consumption rate c (Mb/s), if the available bandwidth on the VOD server is B, the endurable delay for the client is D and the available storage size of the client is M, we should find a broadcasting scheme which can satisfy these tree constraints.

Usually we will bound the storage m and the delay d but minimize the bandwidth b. We can also bound m and b but minimize d, but the method is the same. What is different is the storage issue, yet it is often assumed to be unlimited and been neglected. The following are the parameters we need to consider in a VOD system.

The parameters of a given video	
L	The length of a video program, in seconds.
S	The size of a video program, $S = L \cdot c$, in Mb.
c	The consumption rate, in Mb/s.
Performance parameters of the system	
d	The max delay for any client, in seconds.
b	The bandwidth needed for the server, in Mb/s.
m	The maximum storage used by any client, in Mb.
Constraint parameters of the system	
D	The endurable maximum delay , in seconds.
B	The available bandwidth of the server, in Mb/s.
M	The minimum local storage size, in Mb.

Figure 1 illustrates the basic ideas of VOD. Here we use a single tape with length L and width c to denote a whole video which is Constant Bit Rate (CBR)

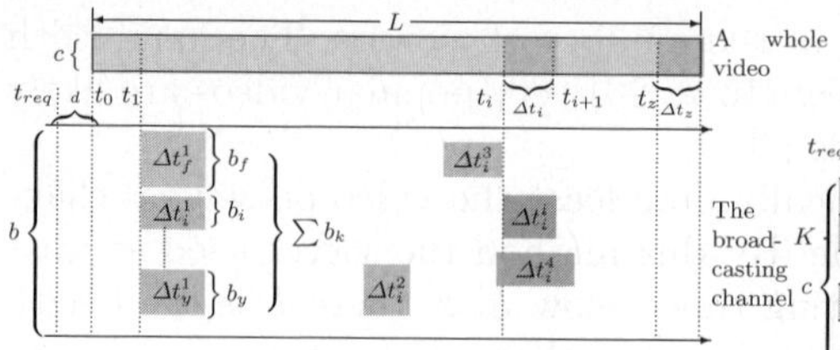

Fig. 1. A general framework for VOD

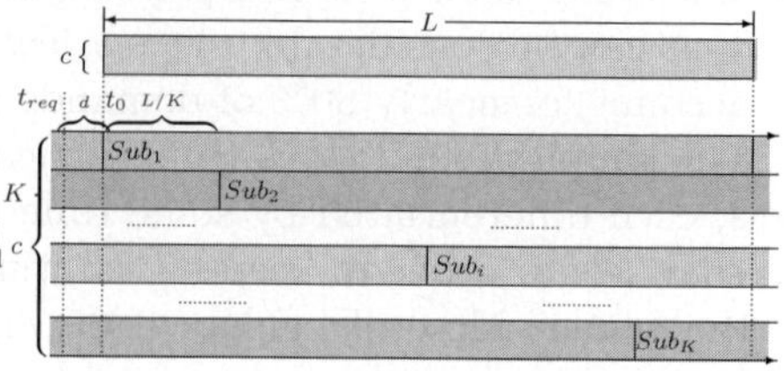

Fig. 2. The SBP Protocol

encoded, so its size is $S = L * c$. At the server side, we will use a channel of bandwidth b to broadcast this video. Suppose the client sends a request for this video at time t_{req}, and starts to consume this video at time t_0, then the period between this two points is the delay $t_0 - t_{req}$ of this user. The maximum delay experienced by any client of a video is the viewing delay d of this video.

To improve the efficiency, we can divide a whole video into small segments (the segment starts at time t_i is denoted as Δt_i) and arrange these segments into the broadcasting channel according to a certain schedule. To guarantee that a client can watch the video smoothly, the required segment Δt_i must be already in the storage of the client's STB at time $t_{req} + d + t_i$.

Thus we need to make sure that the segment Δt_i can be downloaded during the period from t_{req} to $t_{req} + d + t_i$. Sometimes there may be more than one such segments during this period, for example, there are three Δt_i during the period from t_{req} to $t_{req} + d + t_i$ in Figure 1. The client can choose to download the last appearance Δt_i^3 if he knows the schedule of all segments at the time he starts to download, for this may in some cases decrease the storage requirement. But in most cases he doesn't have such knowledge, then he has to download it at its first appearance Δt_i^1 and store it for future consumption. For a certain video, we can calculate the storage requirement. Refer to the shadow tape of the video in Figure 1, at time t_i, all segments at the left side have been consumed and can be cleared from the storage, but some segments at the right side may have been downloaded or partially downloaded and they will stay in the storage until they are consumed. The total volume of these segments will reach a maximum value at some time, so it will be the storage requirement for users enter at t_{req}. Then the largest value among all users enter at different time is the minimum storage requirement m for this video.

As an example, the *Staggered Broadcasting Protocol* (SBP) in [1] rebroadcasts the whole video on $b/c = K$ distinct channels (each with bandwidth c) at equal time intervals L/K, and thus the maximum viewing delay will be L/K (Figure 2). So for a $7200sec$ video, we need 12 such channels to guarantee a $600sec = 10min$ viewing delay, which is not so efficient for bandwidth.

3 The Discrete Broadcasting Protocol for VOD

In our *Discrete Broadcasting Protocol* model, we first assume that the bandwidth b allocated for the broadcasting channel will be a multiple of the consumption rate c. Our second assumption is that a CBR video of length L will be divided into n segments with same size L/n. Let's arrange them by time order and use $S_i(1 \leq i \leq n)$ to denote the ith segment.

In this simplified model, the maximum delay depends on the maximum distance between the beginning time of any two neighboring S_1 segments. Since all segments are equally sized, we can assume that this distance is $k * (L/n)$, $k \in N$. Thus the maximum delay is near $k*(L/n)$ in the case that the client just misses the first frame of the video when he starts to download. If the minimum distance of any two neighboring S_1 segments is also $k * (L/n)$, and the arriving times of the clients are uniformly distributed, then the average delay will be $\int_{0^+}^{\frac{kL}{n}} (\frac{kL}{n} - x)dx / \frac{kL}{n} = \frac{kL}{2n}$.

To satisfy the smooth watching requirement, we need to make sure that the client's STB can find an S_i during any period of length $L*i/n$ which starts from the begin of an S_1. Then we can make sure that after a client start consume S_1, he can download a S_i before $t_{req} + d + L*(i-1)/n$ (or find S_i at then) when he should start to consume S_i. If we fix the distance between any two neighboring S_1 segments to be (L/n), then we need to put an S_i in any period of length $L*i/n$ and we can use a single channel of $1*c$ to broadcast S_1.

Now let's analyze the bandwidth requirement. Since we need to put an S_i in any period of length $L*i/n$, then S_1 will occupy $1*c$ bandwidth, S_2 will occupy $c*1/2$ and S_i will occupy $c*1/i$, we have that $b \geq \frac{c}{1} + \frac{c}{2} + ...\frac{c}{i} + ... + \frac{c}{n} = \sum_{i=1}^{n} \frac{c}{i} = c * H_n$.

Based on this analysis, we designed Algorithm1. Figure3 is the first 52 columns of one result scheduling table of Algorithm1 when the available bandwidth is $4*c$. We can see in Figure3 that the whole video is divided into 26 equal-size segments. But according the analysis, the number of segments should be $max\{n|\lceil H_n \rceil = 4\} = 30$. The gap lies in that every segments are equally-sized and are broadcasted using the same bandwidth $1*c$, so the scheduling table is a discrete table. In such a discrete case, we cannot make sure that the distance between any two neighboring S_is is exactly i. The reason is because there will be a confliction when you try to put one segment at time t yet all 4 sub-channels at that time slot are already occupied.

Algorithm 1. The Discrete Broadcasting Algorithm
SERVER:

Divide the whole video into n equal-size segments;
Put all S_1 segments into the first sub-channel;
For $i = 2$ *to n* **do**
  $t_{cur} = t_{next} = 0$;
  While *the video is not finished* **do**
    Calculate the next time t_{next} to put S_i;
    If *find a vacancy in $(t_{cur}, t_{next}]$* **then** *put S_i;*
      Else *report error and exit;*
    $t_{cur} = t_{next}$;
  End while loop
End for loop

CONSUMER:

```
Start downloading all segments;
If find segment S_1 then start viewing S_1;
For i = 2 to n do
    If find segment S_i in the local storage then
        start viewing S_i;
      Else report error and exit;
    For all segments S_k in the broadcast channel;
      If k > i and S_k is not in the local storage then
        Download S_k into the local storage;
End for loop
```

For example, see S_{14} in Figure3, its first appearance is at t_{10}, so its next appearance should be at t_{24} for the most efficient case. But we find that all 4 sub-channels at t_{24} slot are already occupied, thus we can only put it at t_{23}. Such kind of conflictions will happen more often as n increases. So $L/max\{n|\lceil H_n \rceil = b/c\}$ is a theoretic optimal lower bound for maximum waiting time and cannot be achieved in most cases. Notice that there are some blank positions at the head, but we cannot use them to accommodate more segments. Because if we put an S_{27} there, we cannot find a vacancy for the next S_{27} since the columns from t_5 to t_{52} are already full.

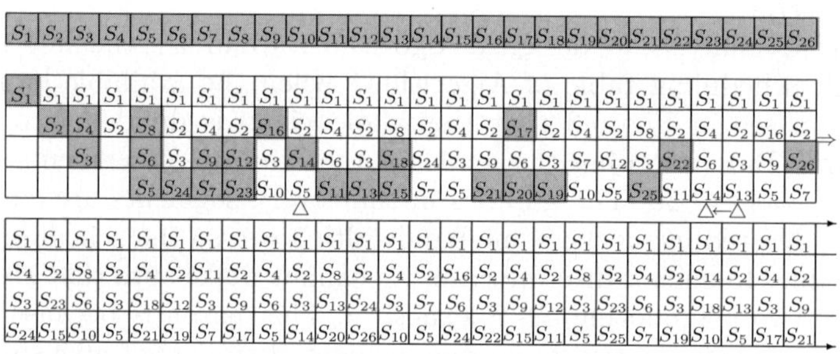

Fig. 3. The Full Discrete Broadcasting Protocol for VOD

Suppose the length of the video program to be broadcasted is $120min = 7200sec$, then the length of each segment is $277sec$ if we adopt the table in Figure3. Thus the maximum delay is $277sec = 4.6min$ and the average delay is $2.3min$. If the format is high quality MPEG-II-compressed NTSC video at about $10Mbps$, then we have $r = 4 * c = 40Mbps$ while $m = 7200 * 10 * 8 * 9/26 \approx 3115Mbytes \approx 3.1Gbytes$, the ration $9/26$ is calculated from the table which means that a user need to store at most 9 segments any time.

4 The Block Discrete Broadcasting Protocol

The scheduling table of DB will be very complex when n is large. So we present the *Block Discrete Broadcasting Protocol*(BDB), which arranges a short BLOCK table and then repeat broadcasting the segments according this BLOCK table. Notice that there are some blank positions at the head of the table in Figure3, we can utilize them by using the BLOCK table method. And we need not to change the algorithm at the client side.

Algorithm 2. The Discrete Block Broadcasting Algorithm
SERVER:

Divide the whole video into n *equal-size segments;*
Put all S_1 *into the first sub-channel of the BLOCK;*
For $i = 2$ *to* n **do**
  $t_{cur} = t_{next} = 0;$
  While $t_{cur} < l_{BLOCK}$ **do**
   Calculate the next time t_{next} *to put* S_i*;*
   If $t_{next} < l_{BLOCK}$ **then**
    If *find a vacancy in* $(t_{cur}, t_{next}]$ **then** *put* S_i*;*
     Else *report error and exit;*
   Else
    If *find a vacancy in* $(t_{cur}, l_{BLOCK}] \cup [0, t_{next}]$
     then *put a segment* S_i *there;*
     Else *report error and exit;*
  $t_{cur} = t_{next};$
  End while loop
End for loop
Repeat broadcasting BLOCK.

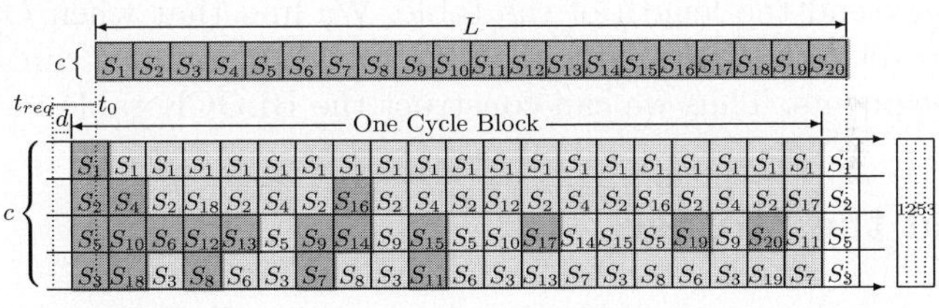

Fig. 4. The Scheduling Table of the Discrete Broadcasting Protocol

Suppose the length of a BLOCK table is l columns. To satisfy the smooth watching requirement, we need to make sure that the client's STB can find an S_i during any period of length $L * i/n$ which starts from an S_1. This means that we have to put an S_i in any i consecutive columns. Thus the space occupied by S_i will be no less than $\lceil \frac{l}{i} \rceil$ since this table will be repeatedly broadcasted. And the lower bound of the number of segments in a BLOCK table with l columns of Algorithm 2 will be: $b \geq l + \lceil \frac{l}{2} \rceil + \lceil \frac{l}{3} \rceil + ... \lceil \frac{l}{i} \rceil + ... + \lceil \frac{l}{n} \rceil = \sum_{i=1}^{n} \lceil \frac{l}{i} \rceil > l * H_n$.

Figure 4 is one output BLOCK table of Algorithm 2. We divide the whole video into $n = 20$ equal-size segments in this example and the length of the table is also $l = 20$ columns. The deep-grey shadowed segments in the BLOCK table show all those segments need to be downloaded for a client who sent a request at time t_{req}. Since $H_{20} \approx 3.60$ and $\sum_{i=1}^{20} \lceil \frac{20}{i} \rceil = 80$, the table has achieved the lower bound. For a $120min = 7200sec$ video with $c = 10Mbps$, the maximum delay is $7200/20sec = 6min$ and the average delay is $3min$. We can also calculate from the table that $r = 4 * c = 40Mbps$ while $m = 7200sec * 10Mbps * 9/(20 * 8) = 4050Mbytes = 4.05Gbytes$. Here we set $l = n = 20$, but this does not mean that we should always set $l = n$. In contrary, we can adjust the value of l to find the most efficient schema. Consider an example when $b = 3 * c$, since $max\{n|\lceil H_n \rceil = 3\} = 10$, we cannot divide the video into more than 10 segments and load them into a scheduling table.

We start from $n = 8$, let $l = 8$ and we will have $\sum_{i=1}^{8} \lceil \frac{8}{i} \rceil = 24$. Fortunately a table with 3 lines and 8 columns can accommodate exactly 24 segments. Using Algorithm2, we construct the table in Figure 5. Now let's try $n = 9$. If we

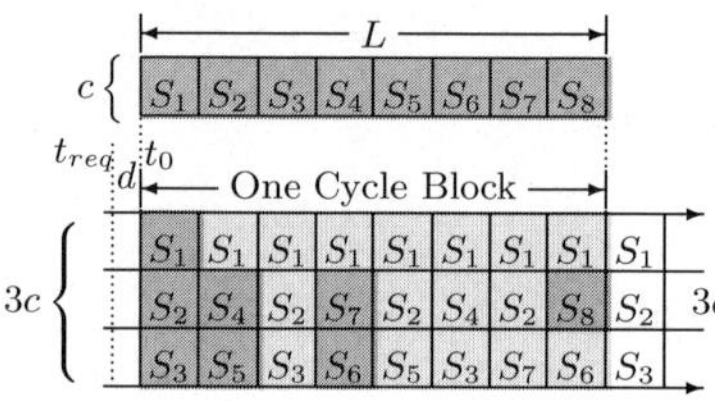

Fig. 5. A Block of BDB when n=8

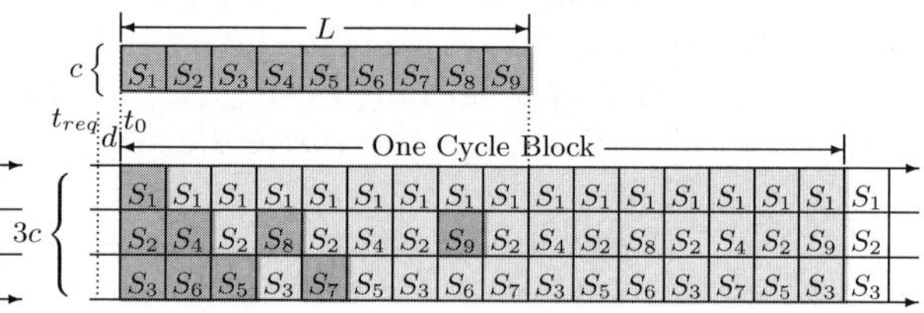

Fig. 6. A Block of BDB when n=9

let $l = 9$ then $\sum_{i=1}^{9} \lceil \frac{9}{i} \rceil = 29$. Yet a table with 3 lines and 9 columns can accommodate only 27 segments. So we cannot construct such a BLOCK table. Then let us extend the length of the table. We find that when $l = 16$ we have $\sum_{i=1}^{9} \lceil \frac{16}{i} \rceil = 48$. While a table with 3 lines and 16 columns can accommodate exactly 48 segments. Thus we can construct the BLOCK table in Figure 6.

5 Bound the Storage

We assume that there are enough local storage in all above schemes. And the true local storage requirements of them are all above 35% and around 40%. But sometimes in real applications we will encounter the problem of local storage shortage. Suppose there is a STB whose storage size is $2Gbytes$, which can satisfy the requirement for any video with $c \leq 5Mbps$ and $L \leq 7200sec$ since $m \leq 7200 * 5 * 40\%/8 = 1800Mbytes$. Now if there is a video with $c = 10Mbps$ and $L = 7200sec$, then $L * c = 9000Mbytes$ and $m = 3150Mbytes > 2Gbytes$.

If will still divide the video into n equal-size segments, then we can store $s = \lfloor n * 2/9 \rfloor$ segments by using $2Gbytes$. To guarantee the smooth watching of this video, we need to make sure that the client can find an S_i from the local storage at time t_i. We find that in most cases, the longer the distance between any two neighboring S_is, the longer the time S_i will be stored in the local storage. If too many segments need to be stored, then the limited local storage will be crammed sooner or later. The result is some segments must be discarded and the video cannot be smoothly consumed.

So one method to depress the local storage requirement is trying to reduce the distance between any two neighboring S_is when i is large. Let's reconsider about the BLOCK table, one characteristic of this table is that the maximum storage required will never exceed $S * l/n$. Since at a certain time when you are watching S_i, you need only to find the segments from S_{i+1} to S_{i+l-1}. You need not to consider about S_{i+l} because it can always be found at the period from when you start to watch S_{i+1} to when you finish S_{i+l}, which is a whole cycle BLOCK. If you find S_{i+l} at exactly the time when is should be consumed, you can watch it directly because its broadcasting rate is the same as its consumption rate. Else you just load it into your local storage. Thus at anytime you need only to store the current segment and the next $l-1$ segments at most.

Of course this property is nonsense when $l \geq n$, but it does make sense when $l \ll n$. The following example in Figure7 show the point. In this example, we divide the video into $n = 19$ equal-size segments and arrange them into a $6 * 4$ BLOCK table. Since the length is only $l = 4$ and $\sum_{i=1}^{19} \lceil \frac{4}{i} \rceil = 24$, we have to use $6 * c$ bandwidth so that the table can accommodate more segments. Now according to the above analysis, the local storage requirement will be $9000 * 4/19 \approx 1895Mbytes < 2000Mbytes$. Since $l \ll n$ in this BLOCK table and the

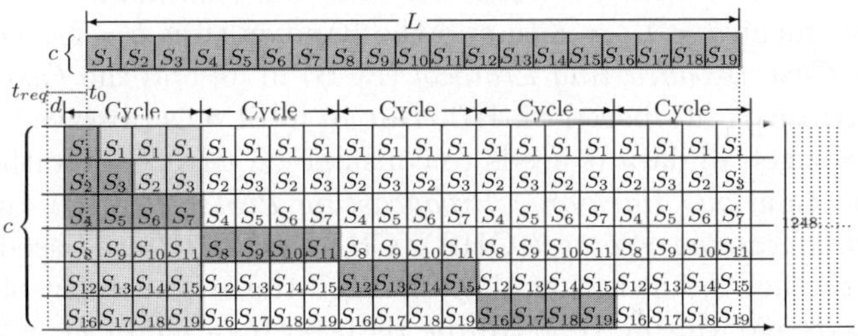

Fig. 7. The Storage Efficient Scheme for VOD

local storage is limited, we need only to slightly adapt the algorithms at the client side to satisfy the new requirement.

The deep-grey shadowed segments in the BLOCK table in Figure7 show all those segments need to be downloaded for a client who sent a request at time t_{req} , the maximum delay is $7200/19sec = 379sec = 6.3min$ and the average delay is $3.2min$.

6 Performance Analysis and Comparison

The *Pyramid Broadcasting Protocol*(PB)[10] can provide shorter waiting time than previous schemes with the same available bandwidth. This protocol partitions an L-length video into n sequential segments of geometrical series increasing sizes and multiplexes M different videos into each logical channel.

A user should download the first segment at the first occurrence and start playing, then he will download the subsequent segment at the earliest possible time. To ensure smooth watching, each channel need plenty of bandwidth and the I/O rate is very high, while local storage requirement can reach more than 70% of the whole video. To address these issues, the authors of [2] proposed the *Permutation-based Pyramid Broadcasting Protocol*(PPB). In PPB, each channel multiplexes its segments into p periodic bit streams and transmits them in $1/p$ times lower rate. But the local storage requirement of PPB is still high since the exponentially increasing speed may cause the last segment to be as large as 50% of the whole video, and the synchronization mechanism in it is very difficult to implement. So authors of [4] proposed the *Skyscraper Broadcasting Protocol* (SB). In SB, fixed bandwidth c is assigned to each logical channel. A video will be divided into n segments, the length of segments are [1, 2, 2, 5, 5, 12, 12, 25, 25, 52...]. Each of these segments will be repeatedly broadcasted on its dedicated channel at the consumption rate c. SB uses a value W as an upper bound to control the maximum size of each segment.

A significant progress was achieved by [5], in which the *Harmonic Broadcasting Protocol*(HB) was proposed. HB equally divides a video into n segments $[S = S_1 \uplus S_2 \uplus \cdots \uplus S_n]$, and each segment S_i will be divided into i equal-size subsegments $[S_i = S_i^1 \uplus S_i^2 \uplus \cdots \uplus S_i^i]$. Then HB allocates a single channel with bandwidth $C_i = c/i$ for each segment S_i. Thus the maximum delay is the length of the first segment $d = L_1 = L/n$ and the bandwidth is $b = H_n \cdot c$.

The *Stair Case Broadcasting Protocol* (SCB) in [6] and the *Fast Broadcasting Protocol* (FB) in [7] are based on HB. But in [8] it was observed that the user in HB may not get all data it needs to consume on time. The authors of [8] also proposed the *Cautious Harmonic Broadcasting Protocol*(CHB) and the *Quasi-Harmonic Broadcasting Protocol*(QHB). CHB and QHB are based on the same idea of HB, the new point is that they changed the arrangement of subsegments so that the average waiting time can be reduced by a factor near 2. But for the same Maximum Waiting Time, HB is still the best. Later they are proved to be Bandwidth-Optimal by Engebretsen and Sudan in [3].

The *Discrete Broadcasting Protocol* (DB) we propose in Section 3 is similar to HB. It can be deemed as the discrete version of HB. BDB is a simple version of DB, it's delay will be reduced when the BLOCK length is prolonged. But its performance is already quite good even when $l = n$. The Figure 8 shows the number of segments we can reach in DB, BDB and HB by using the same Bandwidth. In Figure 9 we compare the best maximum delay we can achieve in PB, SB, BDB, HB and the mean-delay of BDB by using the same Bandwidth. In both figures we set $l = n$ for BDB and we choose the best α value for PB. In Section 4 we have discussed the solution of DB when the local storage is very small compare with the video size. For common BDB which uses a BLOCK of $l = n$, the maximum local storage requirement is around 40% of the whole video. But if we shorten the length of the broadcasting BLOCK, we can decrease the local storage requirement by the cost of increasing the bandwidth. Figure 10 shows the local storage usage situation for two examples of the above two schemes. The data are calculated from the scheduling BLOCK Tables from Figure4 and Figure7 respectively. The following table compares the performance of these

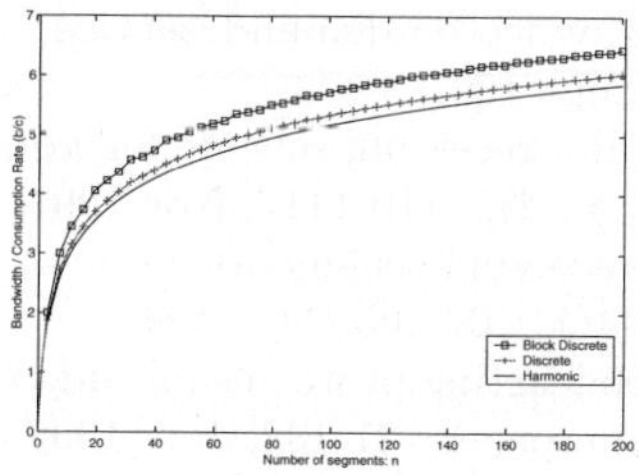

Fig. 8. Bandwidth Ratio v.s. Number of segments

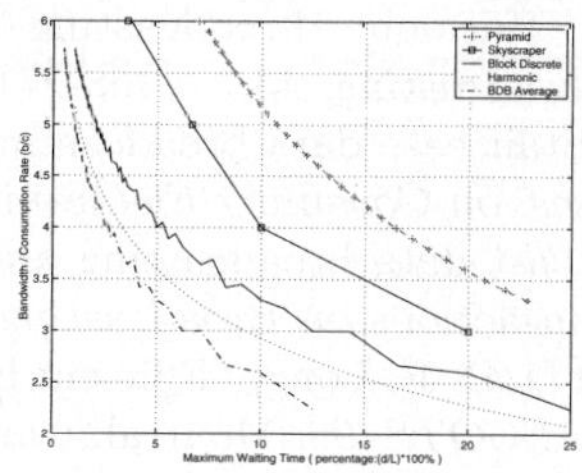

Fig. 9. The b/c Ratio versus the max-delay

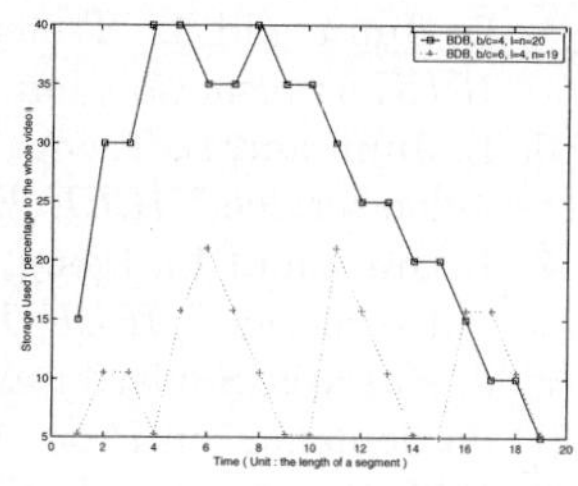

Fig. 10. Storage used in Figure4 and 7

Table 1. Performance and resources requirements comparison

	HB	SCB	FB	DB	BDB	PB	PPB
Maximum Delay	4min	8min	8min	4.6min	6min	20min	30min
Average Delay	4min	4min	4min	2.3min	3min	10min	15min
Local Storage	3.4GB	2.1GB	4.2GB	3.8GB	3.6GB	6GB	6.75GB
Disk I/O rate	30Mbps	20Mbps	40Mbps	40Mbps	40Mbps	50Mbps	50Mbps

protocols. The sample video is a $120min$-long MPEG-II-compressed NTSC video at a consumption rate of about $c = 10Mbps$. The bandwidth is $b = 40Mbps$.

7 Conclusion

In this paper we present the efficient *Discrete Broadcasting Protocol* for VOD service. We also work out the *Block Discrete Broadcasting Protocol* as an extension of DB. Both DB and BDB can achieve lower average delay than the *Harmonic Protocol* with the same bandwidth. Furthermore, HB cannot work when the local storage is less than 37% of the whole video size. BDB can achieve a 6.3-minute max delay when the local storage is around 20% of the whole video size by using a bandwidth of 6 times the consumption rate. The discrete characterisc also makes our protocols more flexible and easy to implement.

References

1. K. C. Almeroth and M. H. Ammar, "The use of multicast delivery to provide a scalable and interactive video-on-demand service," *IEEE Journal on Selected Areas in Communications*, 14(5):1110-1122, Aug 1996.
2. C. C. Aggarwal, J. L. Wolf, and P. S. Yu, "A permutation-based pyramid broadcasting scheme for video-on-demand systems," in *Proc. International Conference on Multimedia Computing and Systems*, pages 118-26, June 1996.
3. L. Engebretsen and M. Sudan, "Harmonic broadcasting is optimal," in *Proc. 13th annual ACM-SIAM SODA*, San Francisco, California, Pages: 431-432, Jan 2002.
4. K. A. Hua and S. Sheu, "Skyscraper broadcasting: a new broadcasting scheme for metropolitan video-on-demand systems," in *Proc. ACM SIGCOMM '97 Conference*, Cannes, France, pages 89-100, Sept 1997.

5. L. Juhn and L. Tseng, "Harmonic broadcasting for video-on-demand service," *IEEE Transactions on Broadcasting*, 43(3): 268-271, Sept 1997.
6. L. Juhn and L. Tseng, "Stair case data broadcasting and receiving scheme for hot video service," *IEEE Trans. on Consumer Electronics*, 43(4), 1110-1117, Nov 1997.
7. L. Juhn and L. Tseng, "Fast data broadcasting and receiving scheme for popular video service," *IEEE Transactions on Broadcasting*, 44(1):100-105, Mar 1998.
8. J.-F. Paris, S. Carter and D. D. E. Long, "Efficient broadcasting protocols for video on demand," in *Proc. MASCOTS '98*, Montral, Canada, pages 127-132, July 1998.
9. P.M. Smithson, J.T. Slader, D.F. Smith and M. Tomlinson, "The development of an operational satellite internet service provision," in *Proc. IEEE GlobalCom'97*, pp. 1147-1151, Nov 1997.
10. S. Viswanathan and T. Imielinski, "Metropolitan area video-on-demand service using pyramid broadcasting," *Multimedia Systems*, 4(4):197-208, 1996.

Multistage Authentication Scheme for Mobile Ad-Hoc Network Using Clustering Mechanism*

Hyewon K. Lee[1] and Jaeyoung Choi[2]

[1] Korea Information Security Agency, Seoul, Korea
hkl@kisa.or.kr
[2] School of Computing, Soongsil University, Seoul, Korea
choi@ssu.ac.kr

Abstract. While applicable fields of ad-hoc network become extensive, security gets more attention, especially authentication. Threshold cryptography [7] allows ad-hoc nodes to authenticate each other by cooperation of nodes. There are many authentication schemes based on threshold cryptography, but these assume pre-key distribution before network configuration. This paper proposes a reliable multistage authentication scheme for clustering based wireless ad-hoc networks. The secured key distribution mechanism is considered in contrast to current schemes. Every node in networks is required to authenticate another before data transmission. The effectiveness of the proposed scheme is verified by means of simulations

1 Introduction

As nodes in ad-hoc network join or leave network dynamically and unconsciously, all of them are exposed to critical security-related problems, especially authentication. For authentication in ad-hoc network, [7] has proposed threshold cryptography algorithm, which divides secret into n shares and distributes each share to individual node. The secret is restored only when k or more than k shares are known, and attackers will be able to threat network's security system only when more than k genuine shares are scraped. Several authentication schemes have been proposed [1-4] based on threshold cryptography, but these do not consider key (or share) distribution and assume that key is already learned before network configuration.

In this paper, we propose multi-level authentication strategy for cluster-based wireless ad-hoc network using threshold cryptography. In clustering ad-hoc network, some nodes are elected as cluster-heads, and each cluster-head controls its cluster from others independently. Our proposed scheme is composed of 4 authentication stages as follows: key manager selection and authentication between key manager and cluster-heads, authentication between cluster-heads, authentication between key cluster-heads and common nodes, and authentication between corresponding nodes. When two common nodes want to initiate data transmission, authentication between them should go first. Especially, ID of each node is used on communication, which allows powerful digital signature and blocks non-reputation.

* This work was supported by Korea Research Foundation Grant (KRF-2004-005-D00172).

M. Gerndt and D. Kranzlmüller (Eds.): HPCC 2006, LNCS 4208, pp. 653 – 661, 2006.

This paper is organized as follows: section 2, some authentication schemes are described and analyzed. Section 3 investigates a novel proposed scheme which exploits authentication scheme for wireless ad-hoc nodes and digital signature. Performance analysis and simulation results are presented in section 4. Section 5 draws the conclusions.

2 Related Works

Ad-hoc network is a multi-hop network without any prepared base station. It is capable of building a mobile network automatically without any help from DHCP servers or routers to forward or route messages. Applications of wireless ad-hoc network are extensively wide, such as battle field or any unwired place; however, these are exposed to critical problems related to network management, node's capability, and security because of absence of centralized controls, restricted usage on network resources, and vulnerability of nodes which results from the special wireless ad-hoc network's character, shared wireless media. These problems induce ad-hoc network to be weak from security attacks from eavesdropping to DoS. To guarantee secure authentication is the main part of security service in ad-hoc network because networks without secure authentication are exposed to exterior attacks.

2.1 Threshold Cryptography

In threshold cryptography algorithm, secret D is divided into n shares (D_1, D_2, ⋯, D_n) using (1) and (2) [7]. It has following characteristics:

- With k or more than k D_i shares, D can be computable.
- With k-1 or less than k-1 D_i shares, D cannot be computable.

This threshold cryptography algorithm is based on polynomial interpolation given k points in the 2-dimensional plane $(x_1, y_1), (x_2, y_2), \cdots,(x_n, y_n)$ with different x_is. Then, there will be only one polynomial $q(x)$ of degree k-1 which for all $q(x_i) = y_i$. For example, k-1 polynomial is generated as (1) and each D_i is evaluated as (2). The threshold cryptography provides strong key management to security system because the secret is still safe even if k-1 systems are attacked.

$$q(x)= a_0 + a_1x + a_2x^2 + \cdots + a_nx^n. \tag{1}$$

$$D_1 =q(1), \cdots, D_i =q(i), \cdots, D_n =q(n). \tag{2}$$

Several kinds of key distribution schemes for secured communication have been devised to provide authentication [1-2]. These schemes employ distributed authentication mechanism [7] via cooperation of nodes, while they miss confidence of the nodes which participate in authentication process. Also, the authors in [2] assume that public and service keys are pre-distributed into certificate authority (CA) nodes on initial network configuration. However, the pre-key-distribution scheme has problems in that network should interoperate with infrastructure based networks and network scalability is degraded.

2.2 CGSR (Clusterhead Gateway Switch Routing)

Many routing protocols for ad-hoc networks are based on flat-tiered networks with no central controls. However, flat-tiered configuration has bad efficiency when we think of code division, channel access, routing and bandwidth allocation. In CGSR (Clusterhead Gateway Switch Routing) [8] protocol, 1-hop distant nodes organize a cluster, and one cluster-head manages common nodes within the cluster. A common node residing on two different clusters works as a gateway. All packets are forwarded via cluster-head and gateway.

CGSR employs DSDV (Destination Sequence Distance Vector) [12] protocol and inherits DSDV's character for routing. Each node keeps cluster member table, which contains information about cluster-heads for destination node. Cluster member tables are periodically broadcasted to networks using DSDV. Once update message is received, ad-hoc nodes update their cluster member table, respectively. When a node receives any message from others, the node forwards it to a proper cluster-head using its cluster member table and routing table.

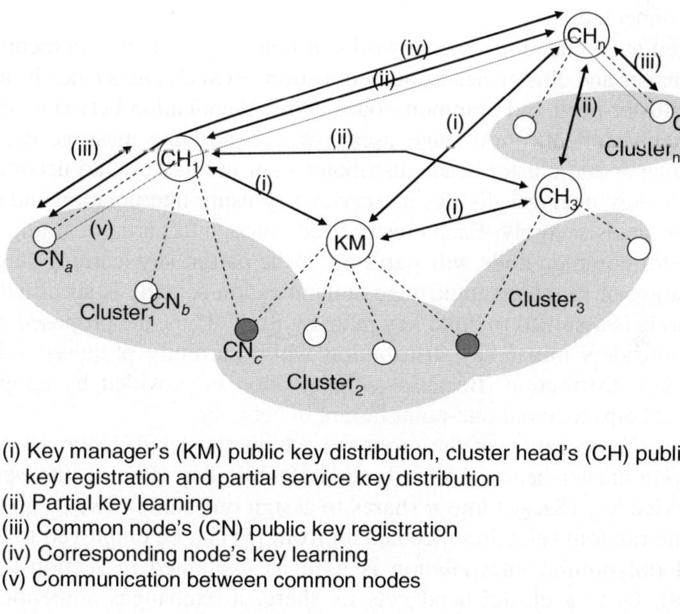

(i) Key manager's (KM) public key distribution, cluster head's (CH) public key registration and partial service key distribution
(ii) Partial key learning
(iii) Common node's (CN) public key registration
(iv) Corresponding node's key learning
(v) Communication between common nodes

Fig. 1. Configuration of ad-hoc network using clustering mechanism

3 Design of Securing Model for Mobile Ad-Hoc Network Using Clustering Mechanism

In order to provide authentication irrespective of infrastructure and in order to resolve pre-key-distribution problem, we consider multistage authentication scheme, which is independent of current public networks. The proposed scheme adopts public key infrastructure and threshold cryptography mechanism [7]. In public key system, a pair of public and private keys is generated by each node. Public keys are distributed to networks, while private keys are kept in each individual node. In threshold cryptography, a secret is divided into *n* nodes and at least *k* valid shares are required to build the genuine secret. Based on this simple idea, network service key is distributed to ad-hoc networks, and more than *k* nodes will participate in the authentication process.

Fig. 1 illustrates the hierarchical configuration of ad-hoc network for proposed authentication scheme. As explained above, a network is divided into clusters that each cluster is composed of a cluster-head and common nodes in cluster-based network. A cluster may be composed of only one node. For an instance, cluster_1 has one cluster-head (CH_1) and three common nodes (CN_a, CN_b and CN_c) in Fig. 1. A common node residing on two different clusters works as a gateway to connect two individual clusters. In Fig. 1, CN_c and CN_d operate as a gateway. A cluster-head helps packet routing, and packets will be routed via cluster-head and (or) gateway to reach proper destination. A cluster-head may allocate address for common nodes within its cluster and guarantee the uniqueness of address allocation. A common node may select one of the nearest cluster-heads and belong to the cluster. The arrow lines in Fig. 1 specify signal message exchange between ad-hoc nodes, and the dotted lines specify that nodes are logically connected.

The proposed authentication scheme works in four stages, such as authentication between key manager and cluster-heads, authentication between cluster-heads, authentication between cluster-head and common nodes and authentication between communication parties. Each authentication stage uses 3-way handshake message exchange. At first, a key manager and cluster heads distributes their public key into network, respectively. Then, the key manager divides its service key using interpolation and distributes share to cluster heads securely. Each cluster head enters into partial learning stage, and authentication for common node will start only if the partial key learning stage is done. Detailed operation of proposed multistage authentication scheme is specified in Fig. 2, where every node is assumed to build key pairs by itself. Thus, the proposed authentication scheme considers initial key distribution while currently proposed schemes are based on pre-key-distribution. Because authentication is provided by cooperation of several nodes, it helps to avoid one-point failure in network.

Fig. 2 shows the operation of the proposed scheme in detail. Upon a key manager is elected among cluster-heads, it distributes its public key (PK_{KM}) into network and divides its service key (SK_{KM}) into *n* shares to assign one share ($SK_{KM\_CH[i]}$) to a cluster-head[*i*]. The random selection mechanism from [5] can be employed to elect a key manager, and polynomial interpolation is used as explained in section 2.1 for key distribution (a). Once a cluster-head gets its share, it exchanges authentication mes sages with the other cluster-heads to verify each other. As more than k ($0 \le k < n$) authentication messages are got, the cluster-head will get proper signature (b). Now, the cluster-head broadcasts its public key into its cluster, which encourages a common node to start authentication phase with the cluster-head. A common node exchanges authentication messages with its cluster-head, and the cluster-head exchanges authentication messages with other cluster-heads to get proper signature for the common node (c). Only a common node verified by a cluster-head is able to initiate communication with others. To initiate communication with another party, authentication between end-nodes should go first, which gives destination's public key to a source via a cluster-head (or cluster-heads) (d).

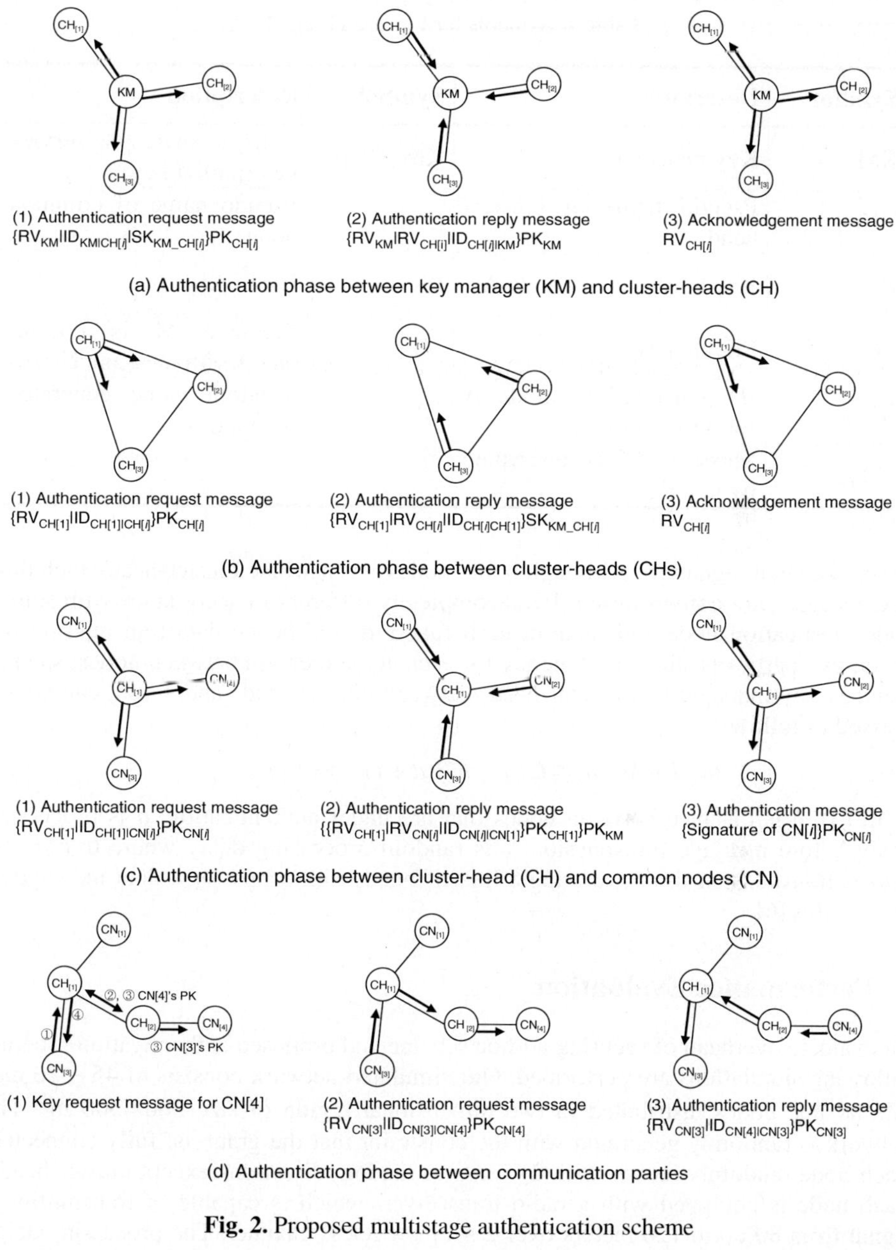

Fig. 2. Proposed multistage authentication scheme

Whenever an authenticated node sends data packet, it attaches ID signed by destination's public key, which allows the destination node to identify that the packet is from genuine source node. As shown in Fig. 2, the ID learned from multistage authentication, is generated by using destination node's IP address and random hash function, and so each ID is definitely different from node to node, which allows much

Table 1. Symbols for Fig. 1 and Fig. 2

Symbol	Description	Symbol	Description
KM	Key manager	KM_CH[*i*]	CH[*i*]'s share of service key (partial key)
CH[*i*]	Pseudo name of cluster-head	CN[*i*]	Pseudo name of common node
K_A	Key generated by node A	PK	Public key
SK	Private key	M|N	Message M is cocatenated with message N
$ID_{A\|B}$	ID generated by node A for destination B	RV_A	Random value generated by node A
$\{M\}_K$	Message M is encrypted by key K		

stronger digital signature. Although some nodes, even worse cluster-heads including key manager, are compromised, ID is completely different in accordance with source node, destination node and random hash function, and hence data transmission between two parties is still safe. Latency for each authentication between nodes, such as between key manager and cluster-head, between cluster-heads and so on, can be expressed as follows:

$$h(3d + 8s / b) \le C_{\text{auth}} \le h(3(d + r) + 8s / b). \tag{3}$$

In (3), *h* is distance between nodes that are under authentication, *d* is processing delay before message transmission, *r* is random processing delay where $0 \le r \le 1$ and $8s/b$ is transmission time over single link of bandwidth *b* bps for a data message of size *s* bytes [6].

4 Performance Evaluation

To evaluate overhead of securing ad-hoc scheme and proposed authentication scheme, following simulations are performed. Our simulated network consists of 45 ~ 70 mobile ad hoc nodes distributed in two dimensional region of size 500×500 m^2. The network is randomly generated with the constraint that the graph be fully connected. Each node randomly moves with 2~5 m/s to random directions except cluster-heads. Each node is equipped with a radio transceiver, which is capable of transmitting a signal from 80 up to 120 meters over 2 Mbps wireless channel. The processing delay for transmitting a message is randomly chosen between 5 ms and 10 ms. The propagation delay is 500 ms. The IEEE 802.11b MAC protocol is used for the data link layer, while IP runs for the network layer. For clustering mechanism, we employ CGSR protocol [8]. Each node is assumed to be supplied by a battery with enough power to at least make it able to carry out a complete operation. Table 2 specifies network characteristics. Total number of exchanged packet and total latency of authentication are measured in both schemes.

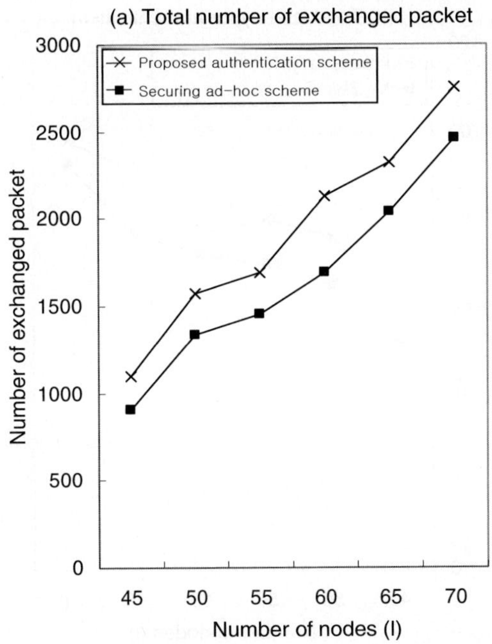

Fig. 3. Total number of exchanged packet for multi-stage authentication

Total number of exchanged packet and latency for multi-stage authentication are illustrated in Fig. 3 and Fig. 4. As shown in Fig. 3 and Fig. 4, number of exchanged packet and latency increases when the number of nodes in network increases, but the increasing rate is not so aggravated. The proposed authentication scheme requires slightly more messages, and it is somewhat slow in authentication. However, merely 10-second difference between securing ad-hoc scheme in [1] and proposed scheme is found from the convergence rate in Fig. 5. The result of simulations shows that proposed scheme provides more proper secure communication to mobile nodes than current scheme with negligible expense.

Table 2. Characteristics of mobile ad-hoc test network

Parameter	Value
Network size	500 * 500 m^2
Number of nodes (l)	45, 50, 55, 60, 65, 70
Number of cluster-heads (n)	2/3 * l
Node's speed	2-5 m/s
Node's transmission range	80~120 m
Packet propagation time	500 ms

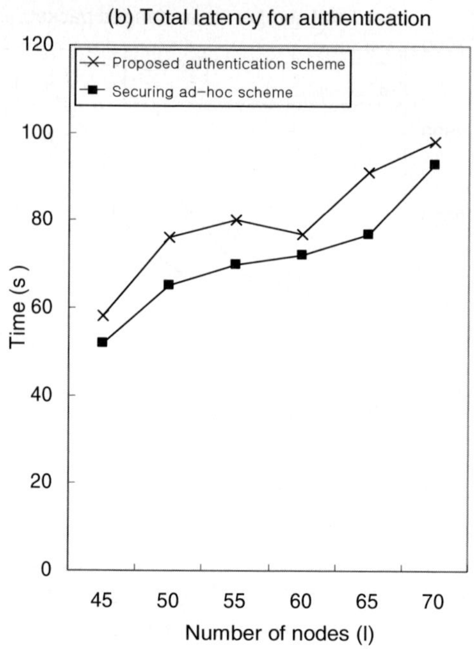

Fig. 4. Latency for multi-stage authentication

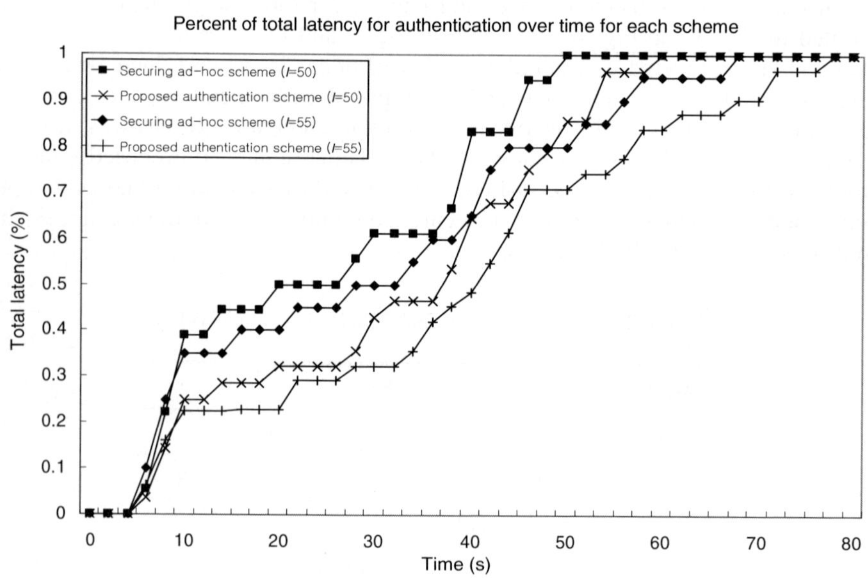

Fig. 5. Convergence rate of total latency for authentication over time for each scheme

5 Conclusions

This paper proposes an efficient multistage authentication scheme for clustering based ad-hoc network. Our proposed scheme is composed of 4 authentication stages as follows: key manager selection and authentication between key manager and cluster-heads, authentication between cluster-heads, authentication between key cluster-heads and common nodes, and authentication between corresponding nodes. When two common nodes want to initiate data transmission, authentication between them should go first. Especially, ID of each node is used on communication, which allows powerful digital signature and blocks non-reputation.

The result of simulations shows that the proposed scheme requires just a little more packets and delay before data transmission than securing ad hoc scheme [1]; however, these additional augmentations in delay become negligible when we think of advantages from digital signature and non-reputation.

References

1. Zhou, L. and Hass, Z. J. : 'Securing Ad-Hoc Networks'. IEEE Network Magazine, Vol. 13, No. 6, pp. 24–30 (1999)
2. Kong. J., Zerfos. P., Luo. H., Lu. S., and Zhang. L. : 'Providing Robust and Ubiquitous Security Support for Mobile Ad-Hoc Networks'. ICNP'01, pp. 251–260 (2001)
3. Frankel. Y., and Desmedt., Y. G. : 'Parallel Reliable Threshold Multisignature'. Tech. Report TR-92-04-02, Univ. of Wisconsin-Milwaukee (2002)
4. Desmedt. Y. and Frankel. Y. : 'Threshold Cryptosystems'. LNCS 435, pp. 307–315 (1990)
5. Heinzelman. W. R., Chandrakasan. A., and Balakrishnon. H. : 'Energy Efficient Communication in Protocol for Wireless Microsensor Networks'. IEEE Proc. Hawaii Int'l Conf. Sys. SCI., pp. 1–10 (2000)
6. Kulik. J., Heinzelman. W. R., and Balakrishnon. H. : 'Negotiation-Based Protocols for Disseminating Information Dissemination in Wireless Sensor Networks'. Selected Papers from Mobicom '99, pp. 169–185 (2002)
7. Shamir. A. : 'How to Share a Secret'. Massachusetts Institute of Technology Communication of the ACM, 22(11), pp612–613, (1979)
8. Chiang, C., Wu, H., Liu, W. and Gerla, M. : 'Routing in Clustered Multihop, Mobile wireless Networks with Fading Channel'. Proc. IEEE Singapore International Conference on Networks, Singapore (1997)
9. Jordan, R. and Abdallah, C.T. : 'Wireless Communications and Networking: an Overview'. Antennas and Propagation Magazine, IEEE (2002)
10. Yang, H., Luo, H., Ye, F., Lu, S. and Zhang, L. : 'Security in Mobile Ad Hoc Networks: Challenge and Solution'. IEEE Wireless Communications (2004)
11. Mishra, A., Nadkarni, K., Patcha, A. and Tech, V. : 'Intrusion Detection in Wireless Ad Hoc Networks'. IEEE Wireless Communications (2004)
12. Perkins, C. and Bhagwat, P. : 'Highly Dynamic Destination Sequence Distance Vector Routing (DSDV) for Mobile Computers'. Proceedings of ACM SIGCOMM'94, London, U.K., pp.234–244 (1994)

Fast and Memory-Efficient NN Search in Wireless Data Broadcast

Myong-Soo Lee and SangKeun Lee*

Korea University, Seoul, South Korea
{lms9711, yalphy}@korea.ac.kr

Abstract. It is observed, surprisingly, that existing nearest neighbor search methods may not work well on mobile clients with very limited memory space. To resolve this problem, a novel method for nearest neighbor search is introduced in the context of a representative of indexes in wireless data broadcast. In the proposed scheme, a mobile client performs the nearest neighbor search in memory-efficient manner by making a sequential access to index packets according to their broadcast order over a wireless channel. The performance evaluation demonstrates that our approach outperforms existing ones considerably in terms of access time without compromising tuning time.

1 Introduction

Nearest neighbor (NN) query is one of the most important queries in location-based services. The example of an NN query is "Find the nearest neighbor hotel." [13]. Much literature is available on methods of solving NN query in spatial databases. Recently, research has been presented on air indexes for NN query in wireless data broadcast environments (e.g., R-tree for wireless broadcast [18], D-tree [15], Hilbert curve index [17], and grid-partition index [18] etc.). In particular, the grid-partition index partitions the search space for NN queries into grid cells and indexes all the objects that are potential nearest neighbors of a query point in each grid cell. It is reported [18] that the grid-partition index outperforms the other indexes in terms of energy conservation.

NN search method in the grid-partition index is proposed at [18] with the implicit assumption that a mobile client has local storage space enough to hold all of the potential nearest neighbors' index packets to compute the distance between a query point and each object during NN detection phase. Mobile clients, however, may suffer from intrinsic limitation such as small user interface, limited storage space, and scarce battery power. In this paper, we observe that the original NN search methods with R-tree or the grid-partition index presented in [18] may lead to long access time for NN queries with very limited space of local storage, which is mainly caused by a lot of backtracking during NN detection.

* Corresponding author. This work was supported in part by MIC & IITA through IT Leading R&D Support Project.

M. Gerndt and D. Kranzlmüller (Eds.): HPCC 2006, LNCS 4208, pp. 662–671, 2006.

This backtracking incurs significant access latency because the client must wait for next broadcast cycles to obtain potential objects.

Motivated from this observation, a new NN search method in wireless data broadcast is proposed in the context of the grid-partition index [18]. The basic idea is for a client to retrieve index packets for NN search according to their broadcast order on the air. A similar philosophy and its impact on wireless transaction processing was reported at our early work [11]. In the current work, we examine the impact of sequential access on fast and memory-efficient NN search in wireless data broadcast. The proposed method shows the following nice evaluation results:

- Access efficiency is considerably improved by eliminating backtracking during NN detection, and
- Energy conservation is comparable (or marginally superior) to the existing one.

The subsequent sections of this paper are organized as follows. The proposed methodology is explained in Section 2. Section 3 presents the performance evaluation. Section 4 concludes this paper.

2 Proposed Approach

This section focuses on supporting the NN search in wireless broadcast environments, in which clients are responsible for retrieving data by listening to the wireless broadcast channel. To handle the issue of energy conservation of mobile devices, a well-known data and index organization for air indexing technique, called (1,m) indexing technique [8], is widely accepted. Throughout this paper, thus, we employ the (1,1) indexing scheme for simplicity.

2.1 Background and Motivation

Grid-partition index is built on the pre-computed solution space that can be represented by *Voronoi Diagrams* (VDs)[2]. Let $O = \{O_1, O_2, \cdots, O_n\}$ be a set of points, the *Voronoi Cell* (VC) for O_i is defined as the set of points q in the space such that $dist(q, O_i) < dist(q, O_j), \forall j \neq i$. The VD for the running example is depicted in Figure 1(a), where C_1, C_2, C_3, C_4, C_5 and C_6 denote the VCs for the six objects, O_1, O_2, O_3, O_4, O_5 and O_6, respectively.

In the fixed grid-partition index (FP) [18], the search space is divided into fixed-size grid cells. Figure 1(a) presents an example of the FP. The whole space is divided into four grid cells, G_1, G_2, G_3 and G_4. Parameters, g_x and g_y, are the fixed width and height of a grid cell, while S_x and S_y are the scales of the x- and y-dimensions in the original space. The original space is thus divided into $\frac{S_x}{g_x} \cdot \frac{S_y}{g_y}$ grid cells. Figure 1(b) presents the index structure for the FP (in this example, S_x is omitted because of S_x=g_x). The FP consists of two levels: the upper-level index, which is built upon the grid cells, and the lower-level index,

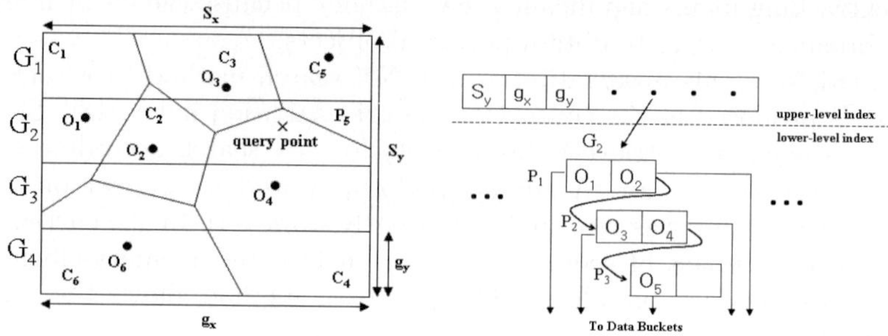

(a) A Running Example (b) Index Structure

Fig. 1. Example of the Fixed Grid-Partition Index

which is built upon the objects associated with each grid cell. The upper-level index maps a query point to the corresponding grid cell, while the lower-level index facilitates access to the objects associated with each grid cell. One packet of the lower-level index (i.e., P_1, P_2 and P_3) may have several pointers to data buckets and coordinates.

With the fixed grid-partition index, an NN query is answered by executing the following three steps: 1) *locating the grid cell*, 2) *detecting NN*, and 3) *retrieving data*. The first step locates the grid cell in which the query point lies. We access the upper-level index and get parameters, i.e., S_x, S_y, g_x, g_y. Then, given a query point (q_x, q_y), we use a mapping function, $adr(q_x, q_y) = \lfloor \frac{q_y}{g_y} \rfloor \cdot \lceil \frac{S_x}{g_x} \rceil + \lfloor \frac{q_x}{g_x} \rfloor$, to calculate the address of the grid cell. In the second step, all the objects associated with a grid cell are obtained, and the nearest neighbor is detected by comparing their distances to the query point. As objects in the grid cell and the query point are shown in Figure 1(a), the grid cell G_2 is associated with objects, O_1, O_2, O_3, O_4 and O_5.

In a grid cell, the objects are sorted according to one dimension. Here, the sorted sequence of the object is O_1, O_2, O_3, O_4 and O_5. For an NN query, the checking process continues in the increasing order of the distance between the current object and the query point in the sorting dimension (*dis_sd*). Initially, the current shortest distance between objects and the query point *min_dis* is set to infinite. At each checking step, *min_dis* is updated, if the distance between the object being checked and the query point, *cur_dis*, is shorter than *min_dis*. Then, the process continues until *dis_sd* is longer than *min_dis*. For this method, the NN query is searched without evaluating all associated objects. In a grid cell of grid-partition index, the objects are broken into two lists according to the query point in the sorting dimension. The first list consists of the objects with coordinates smaller than the query point, and the rest form the second list. The two lists are checked alternatively, for searching the NN in a grid cell. Table 1 presents the process for the running example to detect NN. Note that the NN search reads 7 (redundant) index packets: *root* (i.e., upper-level index), P_1, P_2, P_3, P_1, P_2 and lastly P_1.

Table 1. Processing NN search in the fixed grid-partition index

Step	Object	*dis_sd*	*cur_dis*	*min_dis*
1	O_4	0.2	1.5	1.5
2	O_5	1	1.8	1.5
3	O_3	1.2	1.7	1.5
4	O_2	2.5	Stop	.

Notice that, in the NN search method of the grid-partition index, mobile clients have to store two lists which contain the information (e.g., ID, coordinate and pointer to the data bucket) of all objects within a grid cell. Since, however, a mobile client may not have enough memory space, these two lists may not be stored at the client's side. In this case, NN search may not be performed effectively.

2.2 Our Approach

In order to alleviate the local memory constraint observed from [18], a novel method for searching the NN is proposed in the context of the fixed grid-partition index. The basic idea is to perform the NN search in order of the sorted sequence in the grid cell rather than in order of the closest distance in the sorted dimension. Since clients compute the distance between objects in packet and the query point just by reading incoming packets, the proposed method can be used even with very limited memory space of mobile clients.

Table 2. Processing NN search with our approach

Step	Object	*dis_sd*	*cur_dis*	*min_dis*
1	O_1	-3.5	3.7	3.7
2	O_2	-2.5	2.7	2.7
3	O_3	-1.2	1.7	1.7
4	O_4	-0.2	1.5	1.5
5	O_5	1	1.8	1.5

(* Notation (-) indicates that the object locates the left-side of the query point.)

Table 2 presents an example of the proposed method using Figure 1. As shown in the table, the objects are searched sequentially in order of the position in a grid cell. The search process continues until $dis\_sd > min\_dis$. In this example, the objects are searched for in the order of O_1, O_2, O_3, O_4 and O_5. Thus, an NN search can be complete within the current broadcast cycle. Although the proposed method reads most of index packets in a grid cell, as it will turn out, the energy consumption is found to be "comparable" since it removes the duplicate visits of packets during the search process (the performance evaluation in Section 3 confirms this argument). With the running example in Figure 1, the proposed method reads 4 index packets (the sequence is *root*, P_1, P_2, and P_3).

The detailed proposed algorithm is presented in Algorithm 1, which illustrates the second "detecting NN" step that obtains objects associated with a grid cell, and detects the NN by comparing their distances to the query point.

Algorithm 1. Nearest-Neighbor Search

Input: the query point $p(x, y)$
Output:the pointer to the data object
Procedure:

```
ptr := the pointer to the grid cell that contains the query point;
min_dis := ∞ ;
flag := false;
for each packet pkt in the current grid cell do
  Get the object point set set from pkt;
  for each object point o(x, y) of set do
    cur_dis := the distance between o(x, y) and p(x, y);
    dis_sd := o.x - p.x;
    if dis_sd > min_dis then
      flag := true;
      break;
    if min_dis > cur_dis then
      min_dis := cur_dis;
      ptr := the pointer to the data object of o(x, y);
  if flag == true then break;
return ptr;
```

3 Performance Evaluation

This section evaluates the performance of the proposed NN search method by comparing it with the limited memory version of existing search methods in R-tree and the fixed grid-partition (FP) [18]. This paper assumes mobile clients are equipped with very limited memory space, therefore the modified NN search methods in R-tree (BASE in R-tree) and the FP (BASE in FP) are presented here for the comparison purpose.

- **BASE in R-tree:** As in the work [18], R-tree is broadcasted in width-first order on air. For query processing, no matter where the query point is located, the minimum bounding rectangles are accessed sequentially, while impossible branches are pruned similarly in the original algorithm [13]. The width-first order, however, asks the client to maintain the distance information between all the nodes at the same level and the query point. In order to emulate NN search on R-tree with limited memory space, the base memory with the size of a single packet is assigned to mobile clients.
- **BASE in FP:** In order to emulate the existing search method in FP, the client only stores the order of object's ID to search an NN instead of maintaining the information (ID, coordinate and pointer) of all objects. Consequently, to evaluate the distance between an object and the query point, the

client must read the index packet with the object. In order to emulate NN search on in FP with limited memory space, the base memory with the size of a single packet is assigned to mobile clients.

In the data broadcast scenario, for simplicity, a *flat* broadcast scheduler is employed (i.e., each data object is broadcasted once per cycle in the ascending order of x-dimension values). To multiplex the index and data on the broadcast channel, we employ the (1,1) interleaving technique [8] where a complete index is broadcasted at the beginning of the broadcast cycle, the period of time when the complete set of data objects is broadcasted.

In the performance evaluation, two datasets, UNIFORM and SKEWED, are used. In the UNIFORM dataset, 5,929 points are uniformly generated in a square Euclidean space. The SKEWED dataset contains 5,922 cities and villages in Greece. These are extracted from the point dataset available from [1]. The system model in the simulation consists of a base station, a number of mobile clients, and a broadcast channel. The packet size is varied from 64 bytes to 512 bytes. In each packet, two bytes are allocated for the packet id. Two bytes are used for one pointer and four bytes are for one coordinate. The size of a data object is set at 1K bytes. These system parameters are very similar to those in the work [18] except the memory space of mobile clients.

Two performance metrics are typically used to measure access efficiency and energy conservation: access time and tuning time. They are defined as follows[3][8].

- **Access time:** the period of time elapsed from the moment a mobile client issues a query to the moment the requested data is received by the client.
- **Tuning time:** the period of time spent by a mobile client staying active in order to obtain the requested data.

3.1 Impact of Grid Size

This subsection evaluates the performance of the proposed method by varying the number of grids in order to determine the best grid partition. Figure 2 shows the performance for UNIFORM and SKEWED datasets while a packet has the size of 128 bytes.

Figure 2(a) and (b) demonstrate the tuning time of compared search methods. Although not obviously shown in the figure, in most cases, the larger the number of grids is, the better is the tuning time. As the number of grids becomes larger, the number of index packets related to each grid cell becomes smaller and the upper-level index becomes larger. In the FP, the number of packet accesses of upper-level index is one or two regardless of the size of upper-level index. Consequently, the decreased number of index packets that should be visited, results in better tuning time. The proposed method behavior is similar to the BASE in FP in most cases except for the skewed data of a small number of grids. In the BASE in FP, when the number of grids is smaller than 144, the number of index packets of a grid cell becomes larger. Therefore, the probability of searching the NN in the current broadcast cycle reduces. In this case, the tuning time increases

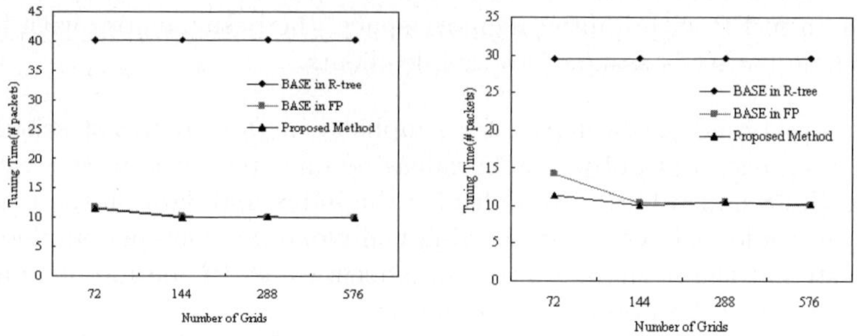

(a) UNIFORM (b) SKEWED

Fig. 2. Tuning Time vs. Number of Grids

because index packets should be repeatedly visited during each broadcast cycle. In these experiments, the best cell partition is found to be 144.

3.2 Impact of Packet Size

Figure 3 presents the access time for all methods. In this figure, the performance is evaluated with different size of packets varying from 64 bytes to 512 bytes. It can be seen that the BASE in R-tree suffers the worst performance, and in most cases, the proposed method has superior performance. Since very limited memory incurs many backtracking in R-tree, the BASE in R-tree has long access latency.

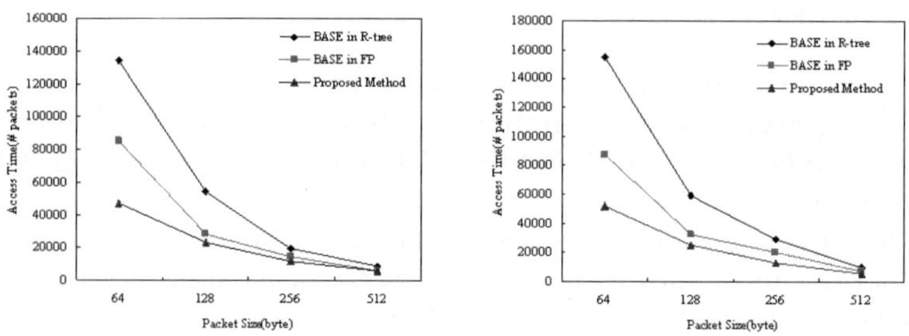

(a) UNIFORM (b) SKEWED

Fig. 3. Access Time vs. Size of Data Packets

The improvement of the proposed method over the BASE in FP is 47% for 64 bytes and 9% for 512 bytes on average. In the BASE in FP, to compute the distance between the object and the query point, the client must read the index packet with the object. Thus, while searching an NN, the object of previous packet may be searched after searching the object of later packet in broadcasting order. Sincc packets passed in the broadcast channel cannot be accessed, the next broadcast cycle must be waited for. However, the proposed method is highly likely to search the NN within the current broadcast cycle because clients evaluate the distance as soon as they read the packet from the broadcast channel. In particular, as the packet capacity becomes smaller, the BASE in FP requires the more number of backtracking due to the increased number of packets in the grid cell. Thus, the performance gap between the proposed method and the BASE in FP increases.

Figure 4 presents the tuning time for all methods. Unexpectedly, as presented in the figure, the proposed method and the BASE in FP have similar performance in both datasets. In the BASE in FP, objects should be repeatedly visited to compute the distance between objects and the query point in each broadcast cycle. Hence, the larger broadcast cycle, the worse the tuning time. Therefore, although the proposed method reads most of index packets in a grid cell, the performance is comparable.

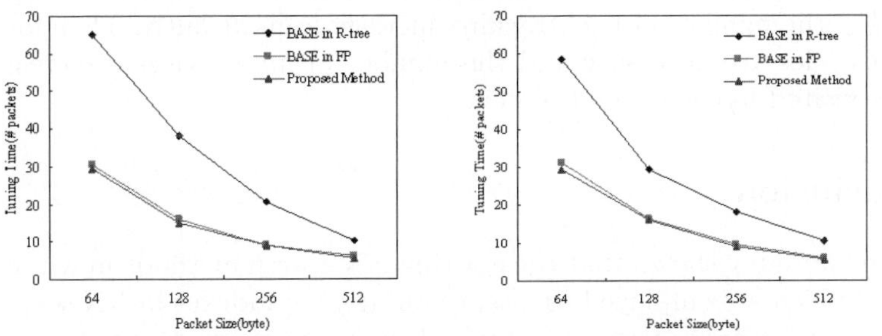

(a) UNIFORM (b) SKEWED

Fig. 4. Tuning Time vs. Size of Data Packets

3.3 Impact of Scalability

Figure 5 presents the performance scalability of all methods to the number of data objects. The access and tuning time of methods are measured by fixing the packet size to 128 bytes. The data are all uniformly distributed and the number of grids is set to 100.

Figure 5(a) presents the access time, where the proposed method has superior performance in most cases. The improvement of the proposed method over the BASE in FP is 3% for 1,000 objects and 48% for 50,000 objects on average. As the dataset becomes larger, the gap between the proposed method and the BASE in FP increases. This is mainly because the BASE in FP requires more backtracking for the large index packets related to each grid cell.

Figure 5(b) demonstrates the tuning time, where the proposed method has superior performance in most cases. As the number of objects is lower than 10,000,

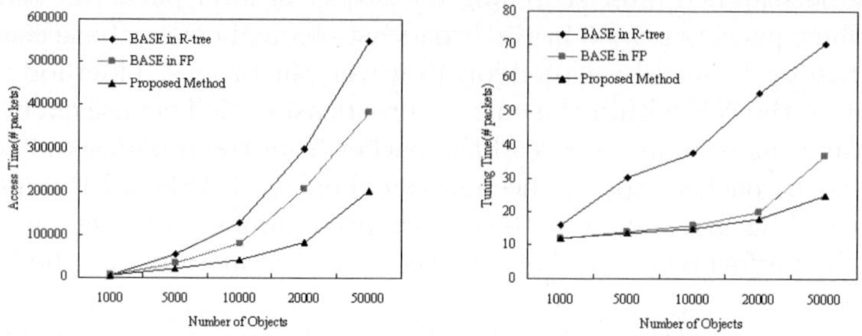

(a) Access Time (b) Tuning Time

Fig. 5. Performance vs. Size of Dataset

the proposed method have comparable performance to the BASE in FP. As the number of objects becomes larger than 10,000, however, the proposed method outperforms the BASE in FP. In the BASE in FP, as the number of objects become larger, the number of backtracking increases significantly. Therefore, more broadcast cycles are necessary and the number of index packets increases that should be visited in each broadcast cycle.

4 Conclusion

We have remarkably shown that the existing NN search methods in wireless data broadcast, which are employed as a representative of indexes in NN search problem, may be unsuitable with very limited memory space. Motivated from this observation, a novel NN search method has been presented in the context of grid-partition index. Our approach benefits from the sequential access to incoming index packets according to the order they appear over a wireless channel.

The conducted performance evaluation confirms the superiority of the sequential access manner. It is demonstrated that access time of the proposed method outperforms the existing methodologies significantly. Tuning time of the proposed method, which was expected to be bad, is found to be "comparable" to the emulated version of an existing method, which is rather counter-intuitive. We are currently performing further experiments with *depth-first order* broadcasting of R-tree and optimal m on $(1,m)$ indexing technique.

References

1. Spatial dataset. http://www.rtreeportal.org/datasets/spatial/greece/cities_loc.zip.
2. M. Berg, M. Kreveld, M. Overmars, and O. Schwarzkopf. *Computational Geometry: Algorithms and Applications*. Springer-Verlag, 1996.
3. A. Celik, P. Ding, , and J. Holliday. Data broadcasting with data item locality and client mobility. In *Proceedings of the IEEE International Conference on Mobile Data Management*, page 166, 2004.

4. M.-S. Chen, K.-L. Wu, and S. Yu. Optimizing index allocation for sequential data broadcasting in wireless mobile computing. *IEEE Transactions on Knowledge and Data Engineering*, 15(1):161–173, 2003.
5. B. Gedik, A. Singh, and L. Liu. Energy efficient exact kNN search in wireless broadcast environments. In *Proceedings of the Annual ACM International Workshop on Geographic Information Systems*, pages 137–146, 2004.
6. A. Guttman. R-trees: A dynamic index structure for spatial searching. In *Proceedings of the ACM SIGMOD International Conference on Management of Data*, pages 47–54, 1984.
7. H. Hu, J. Xu, and D. L. Lee. A generic framework for monitoring continuous spatial queries over moving objects. In *Proceedings of the ACM SIGMOD International Conference on Management of Data*, pages 479–490, 2005.
8. T. Imielinski, S. Viswanathan, and B. Badrinath. Data on air: Organization and access. *IEEE Transactions on Knowledge and Data Engineering*, 9(3):353–372, 1997.
9. R. Kravets and P. Krishnan. Power management techniques for mobile communication. In *Proceedings of ACM/IEEE International Conference on Mobile Computing and Networking*, pages 157–168, 1998.
10. D. L. Lee, W.-C. Lee, J. Xu, and B. Zheng. Data management in location dependent information services. *IEEE Pervasive Computing*, 1(3):65–72, 2002.
11. S. Lee, C.-S. Hwang, and M. Kitsuregawa. Using predeclaration for efficient read-only transaction processing in wireless data broadcast. *IEEE Transactions on Knowledge and Data Engineering*, 15(6):1579–1583, 2003.
12. W.-C. Lee and B. Zheng. DSI: A fully distributed spatial index for location-based wireless broadcast services. In *Proceedings of the IEEE International Conference on Distributed Computing Systems*, pages 349–358, 2005.
13. N. Roussopoulos, S. Kelley, and F. Vincent. Nearest neighbor queriess. In *Proceedings of the ACM SIGMOD International Conference on Management of Data*, pages 71–79, 1995.
14. J. Xu, B. Zheng, W.-C. Lee, and D. L. Lee. Energy efficient index for querying location-dependent data in mobile broadcast environments. In *Proceedings of the IEEE International Conference on Data Engineering*, pages 239–250, 2003.
15. J. Xu, B. Zheng, W.-C. Lee, and D. L. Lee. The D-tree: An index structure for planar point queries in location-based wireless services. *IEEE Transactions on Knowledge and Data Engineering*, 16(12):1526–1542, 2004.
16. B. Zheng, W.-C. Lee, and D. L. Lee. Spatial queries in wireless broadcast systems. *Wireless Networks*, 10(6):723–736, 2004.
17. B. Zheng, J. Xu, W.-C. Lee, and D. L. Lee. Spatial index on air. In *Proceedings of the IEEE International Conference on Pervasive Computing and Communications*, pages 297–304, 2003.
18. B. Zheng, J. Xu, W.-C. Lee, and D. L. Lee. Energy-conserving air indexes for nearest neighbor search. In *Proceedings of the International Conference on Extending Database Technology*, pages 48–66, 2004.

A New Proposal of TCP for IEEE 802.11 Wireless Networks

Fábio C. M. Araújo[1], Cláudia J. Barenco Abbas[2], and L. Javier García Villalba[3,*]

[1] Electrical Engineering Department, Faculty of Technology
University of Brasilia - UnB
70900 Brasília, Brazil
ufa@ieee.org

[2] Departamento de Computación y Tecnología de la Información
Universidad Simón Bolívar - USB
Oficina MYS 213-B, Apartado Postal 89.000
Caracas, 1080 Venezuela
barenco@ldc.usb.ve

[3] Grupo de Análisis, Seguridad y Sistemas (GASS)
Departamento de Sistemas Informáticos y Programación (DSIP)
Facultad de Informática, Despacho 431
Universidad Complutense de Madrid - UCM
C/ Profesor José García Santesmases s/n, Ciudad Universitaria
28040 Madrid, Spain
javiergv@sip.ucm.es

Abstract. The TCP was designed to work in an environment that is almost link error free. So it supposes that any error during receiving data is due to congestion. This assumption is not always true, especially in a wireless environment, which has a high BER when compared to other medium like fiber and twisted pair. This work describes a simple and backward compatible way to increase the performance of TCP in wireless links. This solution is evaluated in a simulated environment using the NS2 simulator and is based in informing the TCP whenever the link is under high BER, so it can trigger the proposed algorithm that will hold the transmission and save the congestion window status for late recovery from the disconnection.

Keywords: TCP, 802.11, link error, wireless networking.

1 Introduction

Wireless environments is becoming widely used, as they have more flexibility in use and can be more easily deployed in a already built place. With the increasing use of wireless networks, people expect that the performance on these kind os networks be the same as in conventional wired ones.

The main protocol used to control flow data and to give some reliability in transmission over the Internet is the TCP [12], whenever the network is wired or wireless.

* Supported by the Spanish Ministry of Education and Science under Project TSI2005-00986.

M. Gerndt and D. Kranzlmüller (Eds.): HPCC 2006, LNCS 4208, pp. 672 – 681, 2006.

The TCP was designed to work on an environment that is almost link-error free, so it supposes that any error during receiving data is due to congestion. This assumption is not always true, especially in a wireless environment, which has a high BER when compared to other medium like fiber and twisted pair.

With this assumption, TCP reacts wrongly in presence of a high error link, shrinking its congestion window. This procedure leads to a poor throughput performance, as the congestion window is decreased when it shouldn't, because there are still bandwidth left to transmit.

In other words, the TCP error control is mainly centered in congestions losses, and it does not take into account the transient random errors or temporary "black-outs", so typical in a wireless environment. Even though sometimes the congestion results in burst drops, the recovery in case of congestion (smoothed windows increases to avoid more congestion) is not an optimal solution in case of link errors.

This work describes a simple and backward compatible way to increase the performance of TCP in wireless links. This solution will be evaluated in a simulated environment using the NS-2 simulator [8] and is based in informing the TCP whenever the link is under high BER, so it can trigger the algorithm that will hold the transmission and save the congestion window status for late recovery from the disconnection.

The rest of this paper is organized as follows: In section 2, the TCP RENO [5] error control is described and its real effectiveness in wireless link. Section 3 explains how the new approach for TCP, called TCP_Ufa, works. Simulation results and comparisons with TCP RENO are shown in section 4. Conclusion and future work are described in section 5.

2 TCP RENO

The TCP RENO was chosen for this work because it is the most used TCP version nowadays is based on it [7] [11] [1]. It basically uses the following error control mechanisms:

- **Slow Start:** This algorithm controls the flow of the transmission using the congestion window. It operates by observing that the rate at which new packets should be injected into the network is the rate at which the acknowledgments are returned by the other end. Each time an ACK is received, the congestion window is increased by one segment. The sender can transmit up to the minimum of the congestion window and the receiver window. When that ACK is received, the congestion window is incremented from one to two, and two segments can be sent. When each of those two segments is acknowledged, the congestion window is increased to four. This provides an exponential growth (it is not exactly exponential because the receiver may delay its ACKs, typically sending one ACK for every two segments that it receives).
- **Congestion Avoidance:** Congestion avoidance is a way to deal with lost packets. This mechanism makes use of a threshold, usually called Slow Start Threshold, or sshthresh. When congestion occurs (timeout or the reception of duplicate ACKs), one-half of the current window size (the minimum between congestion window and the receiver's window, but at least two segments) is

saved in ssthresh. Additionally, if a timeout occurs, congestion window is configured to one segment (slow start). The transmission restarts in the exponential growth phase (cwnd = cwnd * 2) until the sshthesh is reached. Once this occurs, the congestion window size is increased linearly (cwnd = cwnd + 1). When segment acknowledgments are not received, the slow-start threshold is set to half of the current congestion window size, and the algorithm restarts.

- **Fast Retransmit:** In this mechanism, the duplicate acknowledgment is used to verify if the segment was lost or if it was delivered out of order. When an out-of-order segment is received, TCP generates a DUPACK. The purpose of this duplicate ACK is to let the other end know that a segment was received out of order, and to tell it what sequence number is expected. TCP does not know whether a duplicate ACK is caused by a lost segment or just a reordering of segments, so it waits for a small number of duplicate ACKs to be received. It is assumed that if there is just a reordering of the segments, there will be only one or two duplicate ACKs before the reordered segment is processed, which will then generate a new ACK. If three or more duplicate ACKs are received in a row, it is a strong indication that a segment has been lost. TCP then performs a retransmission of what appears to be the missing segment, without waiting for a retransmission timer to expire.
- **Fast Recovery:** When Fast Retransmit indicates that a segment is missing (three or more duplicate ACKs), the congestion avoidance is performed, instead of the slow start. In this situation, the ssthresh set to one-half the current congestion window. The segment is retransmitted, but the congestion window is configured to ssthresh plus 3 segments. This inflates the congestion window (cwnd) by the number of segments that have left the network and which the other end has cached. Each time a duplicate ACK is received, the cwnd is increment by the segment size. This inflates the congestion window for the additional segment that has left the network. When a ACK of new data is received (acknowledgment of the retransmission) the cwnd is set to ssthresh value (half of the value it was before the loss). If there are duplicate ACKs, the cwnd growth is linear.

These algorithms had shown themselves very useful in conventional networks, but in wireless environment, they had some issues. Fast Retransmit makes use of slow start, which in turn shrinks the window when a loss occur. If this loss was originated from a error link, probably there is no need to decrease the transmit rate, since most errors in wireless links are due to random transient or temporary "black-outs". The Fast Recovery algorithm can be efficient when just one segment gets lost in a transmission window, but it is not optimized when multiple losses happen in a single window. In respect to these problems, many proposals were made. Ramakrishnam and Floyd [9] proposed a Explicit Congestion Notification (ECN) at the IP layer in order to trigger the TCP congestion control. Hence, TCP performance can be enhanced by means of avoiding losses of data windows due to limited buffer space at the bottleneck router, and congestive collapse can be avoided also. However, by not receiving an explicit notification the TCP sender will not be able to safely assume that a detected drop was not caused due to congestion. Even more, this solution adds complexity, as the routers in the route and the TCP itself must be ECN-enabled.

There are other kind of proposals, in transport layer, as the Indirect TCP (I-TCP) [2]. In this solution, the TCP connection is split in two: one between the mobile station and the base station, and other between the base station and the fixed station. This way, the errors in wireless link are hidden from the source TCP, as the base station in on charge of the retransmission. Even more, the transmission rates are increased due the low Round Trip Time (the proximity of the base station and the mobile host), recovering faster from congestion. However, splitting the connection makes the end-to-end TCP semantics: the data sent by the fixed station must be acknwolegded before it is delivered. If these packets do not reach the destination (for instance, the base station's buffers overran), they will not be retransmitted. Moreover, the I-TCP also suffer from erroneous wireless links, as the TCP does.

Another approach for TCP in wireless environment is Freeze-TCP [6]. Unlike the I-TCP, Freeze TCP is a true end-to-end mechanism, which does not require any intermediaries, neither change in TCP code is required on the sender side. The modifications are restricted in mobile client TCP. It handles the task of identifying an imminent disconnection due to a potential hand off, fading signal strength, or any other problem arising due to wireless media and notifies the sender of any impending disconnection by advertising a zero window size (ZWA- zero window advertisement) and prevents the sender from entering into congestion avoidance phase. By getting the advertised window as zero, the sender enters the persist mode and locks (freezes) all the timers related to the session. And periodically sends the ZWP (Zero Window Probes) until the receiver's window opens up. Yet, Freeze-TCP is only useful, if a disconnection occurs while the data is being transferred. It is not effective, in case of a disconnection, when there is no transaction going on between sender and receiver.

WTCP [10] is another approach where the end-to-end semantics are preserved. In this approach, the base station have a double stack (TCP+WTCP), and monitors the connection. When the base station receives data segments from the transmitter, it maintains an array with the sequence number of the data segments and their arrival time. Each time that a segment is sent to the mobile station, the WTCP adds the time that the segment lost in its buffer in the timestamp of the segment. This way, the source will believe this segment has been sent after it really was. "Lying" to the source is justified by the effect it would differently have on the RTT (round-trip time) measurement and especially its variance: a packet that would be successfully transmitted only after several retransmissions attempts would give a huge RTT sample. As TCP uses a smoothed RTT estimation, it may take a long period to resettle down. The base station acknowledges a segment to the fixed host only after the mobile host actually receives and acknowledges this same segment, what tends to maintain the end-to-end semantics.

Considering duplicate ACKs and timeouts events, the base station manages local retransmission of lost segments. In case of timeout, the transmission (congestion) window is reduced to one because the lost packet can be the the first of a burst loss, typical of wireless links. On the other hand, for duplicate ACKs, the packet is simply retransmitted without further process assuming that the receipt of duplicate ACKs is a proof of the acceptable link quality. However, "Lying" maybe disadvantageous. As the Retransmission Time Out (RTO) value is close linked to RTT estimatives, packets could stay a long time in the buffers of the base station because of the momentary bad quality link. This scenario can lead the source to time out more often. The throughput would deteriorate due to this wrong RTT estimation. Generally, split solutions often suffer from high software overhead in case of duplication of protocol stack and often require a lot of buffering capacity.

There are many other TCPs approaches for wireless links. More information on the solutions mentioned here and new ones, refer to [13] and [3].

3 TCP_Ufa

The proposed solution is based on TCP RENO and operates in a simple way. Basically it verifies the quality of the link, and if it is not propicious for transmission, an algorithm is triggered to avoid unnecessary changes on the congestion window.

Initially, the protocol would have two algorithms: one for long disconnections and another for short disconnections, as the characteristics of different disconnection times may affect the performance of the protocol.

When the probable interference time is high, the algorithm forces TCP to enter Slow Start (cwnd = 1). The value of the present cwnd is stored. So, the TCP does not need to wait for the RTO to enter Slow Start. Additionally, it avoids data to be inserted in a medium with a high error probability. When is sensed that the disconnection was end, TCP gets back with normal operation. The Slow Start algorithm was chosen because it is a mandatory mechanism in modern TCPs [4].

In case the interference is short, the algorithm forces TCP to half the cwnd, as just a burst of data would be lost, and Fast Recovery makes the data flow return to the previous state relatively quickly.

If the medium is sensed to be relatively propitious to data traffic, TCP_Ufa will not interfere in transmission in any way. The flowchart of TCP_Ufa is showed in figure 3.

Before the simulations were executed, it was necessary to determine at which point the disconnection is considered short or long. So, a prior simulation was run to determine this point, comparing with TCP RENO. In the first run, TCP_Ufa was configured to operate with only the short disconnection algorithm, and the disconnection time increasing in every run. The results were depicted in figure 1. The same simulations were run with TCP_Ufa configured only with the long disconnection algorithm (figure 2).

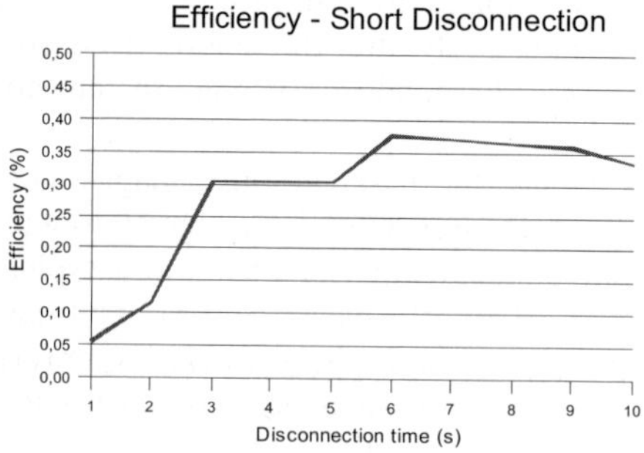

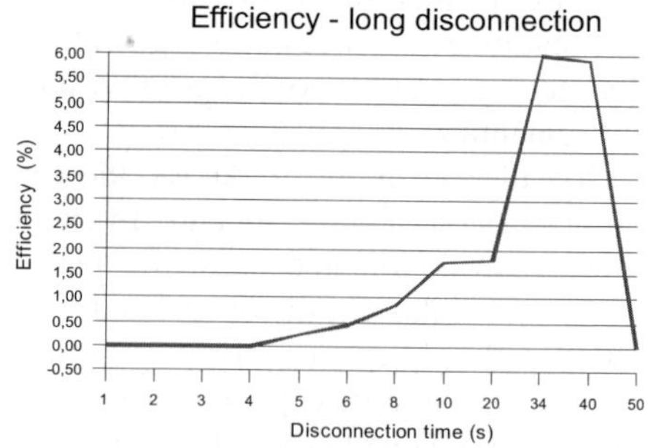

Fig. 1 and 2. Short and long algorithm performance when compared with TCP RENO

Up to 5 seconds of disconnection, the short disconnection algorithm shows a better efficiency when compared with the long disconnection algorithm. From 6 seconds and above, the results shows that long disconnection algorithm outperforms the short one. So, the optimal point where the disconnection is considered long is from 6 seconds and beyond.

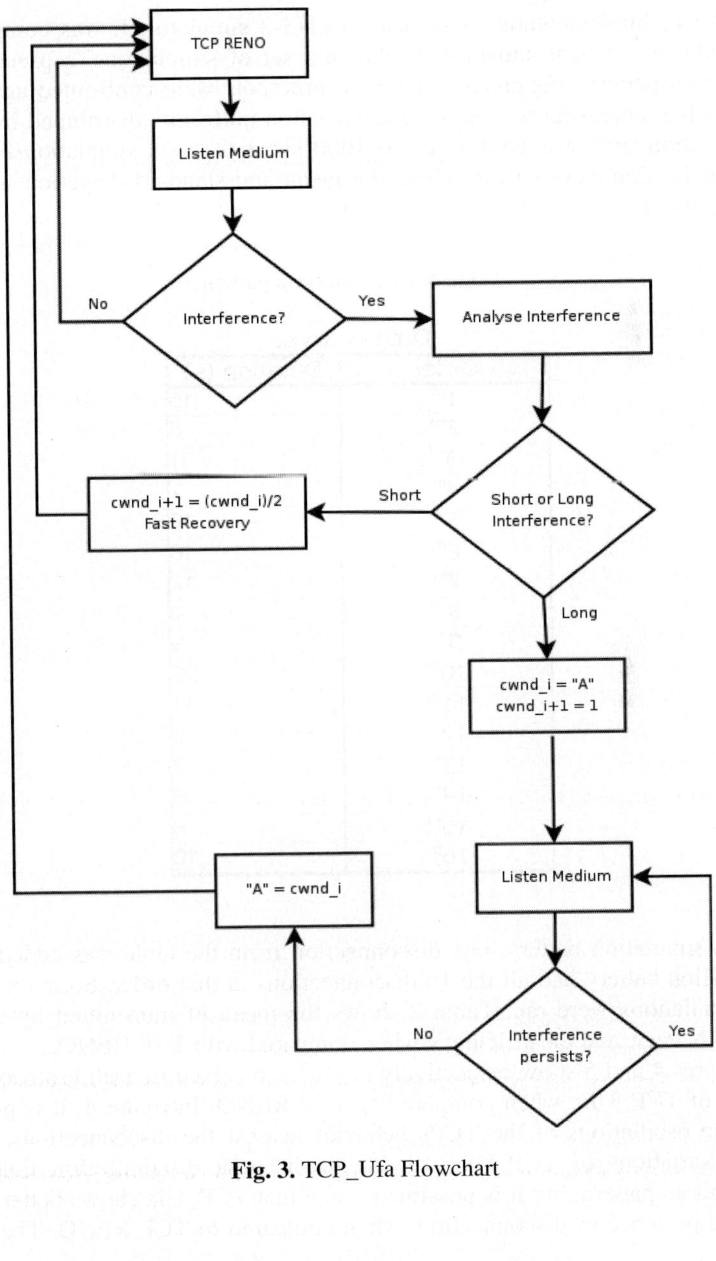

Fig. 3. TCP_Ufa Flowchart

The interference prediction can be archived with an Auto-Regressive Model (AR), with a model to predict the fading signal, using spectral estimation, linear prediction and interpolation, like the one described in [4].

4 Simulation Results and Analysis

The TCP_Ufa implementation was made in NS-2 simulator. It was compared with TCP RENO in identical situation. In the first set of simulations, a pseudo-random disconnection pattern was created. The two protocols were confronted against these pattern, with a increasing number of disconnection uniformly distributed in time. The total simulation time was configured to 1000 seconds. Each simulation battery was made with 10 simulations from where the mean and standard deviation were calculated. The pseudo-random disconnection pattern are in table 1:

Table 1. Disconnection pattern

Disconnections	
Order	Duration (s)
1st	15
2nd	8
3rd	31
4th	7
5th	19
6th	16
7th	12
8th	4
9th	9
10th	28
11th	11
12th	6
13th	7
14th	9
15th	8
16th	10

In each simulation battery, one disconnection from the table was added until the last simulation battery had all the 16 disconnections in that order. So, a total number of 160 simulations were ran. Table 2 shows the mean of transmitted bytes in each simulation battery, and the efficiency when compared with TCP RENO.

The figures 4 and 5 show respectively the bytes received in each protocol, and the efficiency of TCP_Ufa, when compared to TCP RENO. In figure 4, it is possible to observe the oscillations of the TCPs behavior against the disconnections. The performance variations of TCP_Ufa are due to different disconnection times in the pseudo-random pattern, but it is possible to note that TCP_Ufa shows better performance in the presence of disconnection, when compared to TCP RENO. These results also show in figure 5 that TCP_Ufa have peaks of efficiency at large disconnections (3rd, 5th and 10th).

Table 2. Simulation Results

Disconnection number in 1000 seconds	Accumulated Disconnection Time (s)	TCP RENO (μ)	Standard Deviation (σ)	TCP_Ufa (μ)	Standard Deviation (σ)	Efficiency (%)
1	15	4,624,879	1035	4,698,674	1456	1.595609
2	23	4,287,687	986	4,359,877	1235	1.683658
3	54	3,987,417	924	4,093,658	1756	2.664407
4	61	3,589,731	1054	3,674,598	1356	2.36416
5	80	3,498,745	1108	3,609,745	1875	3.172566
6	96	3,400,547	1047	3,494,898	1235	2.774583
7	108	3,356,981	983	3,435,687	1356	2.344547
8	112	3,331,568	854	3,411,751	1249	2.406765
9	121	3,286,546	905	3,366,999	1307	2.44795
10	149	3,146,874	1096	3,275,217	1568	4.078428
11	160	3,093,498	937	3,205,879	1147	3.632813
12	166	3,052,476	928	3,125,871	1409	2.404442
13	173	3,009,986	901	3,076,544	1286	2.21124
14	182	2,975,654	895	3,039,546	1253	2.147158
15	190	2,933,658	915	2,994,796	1352	2.084019
16	200	2,889,854	938	2,950,487	1292	2.098134

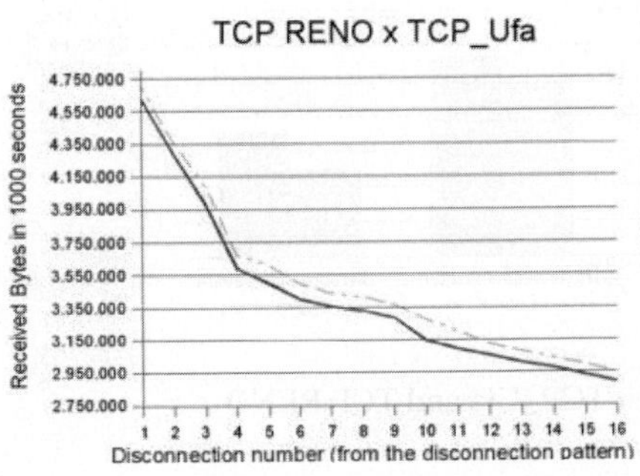

Fig. 4. Received bytes x number of disconnections from the pattern

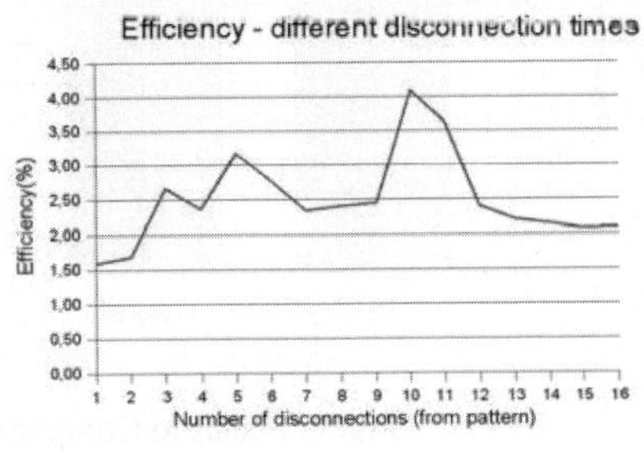

Fig. 5. TCP_Ufa efficiency x number of disconnections

In the second set of simulation, the protocols passed in the same disconnection sequence, but this time with a non-controlled traffic, which generated congestion, besides disconnections. These congestions were made raising the transmission rate above the channel capacity, twice in the 1000 seconds simulation. The first congestion was programed to happen between disconnections, and the second occurred just after a disconnection. Again, a simulation battery consisted in a 10 equal 1000 seconds simulations, and the mean and the standard deviation were calculated based on the results of the battery. Initially, an disconnection/congestion-free environment were simulated for comparison basis. So, a 25 seconds long congestion was added to the above scenario. Another identical congestion was added again, configuring two congestions during simulation, also with no disconnection. After that, a disconnection pattern was added to the one-congestion environment. This congestion occurred on the first half of the simulation, exactly between two disconnections (between 4th and 5th disconnection). Finally, another congestion was added, but this time, it would occur just before (5 seconds) in the 10th disconnection. The results obtained are shown in table 3. The figure 6 illustrates the efficiency of the TCP_Ufa for each scenario.

Table 3. Results for scenarios with congestion

Scenario	TCP RENO (□)	Standard Deviation (□)	TCP_Ufa (□)	Standard Deviation(□)	Efficiency (%)
Free	9.026.987	979	9.021.976	1152	-0,06
1 Congestion	8.842.684	1.068	8.840.595	1.201	-0,02
2 Congestions	8.598.479	1.073	8.599.653	1.456	0,01
Disconnection pattern + 1 Congestion	7.219.877	1.108	7.412.583	1.254	2,67
Disconnection pattern+ 2 Congestions	6.628.246	1.085	6.884.656	1.230	3,87

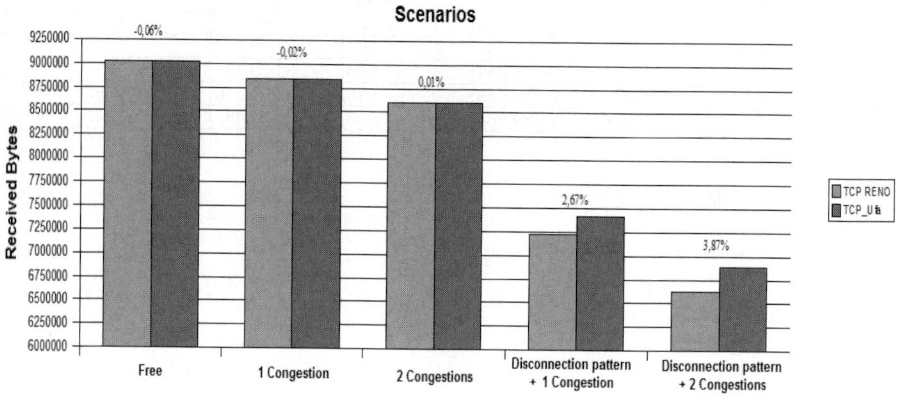

Fig. 6. Scenario Comparison between TCP_Ufa and TCP RENO

In the first 3 scenarios, both of the protocols behave in a very similar way. TCP_Ufa is set to not interfere with TCP RENO congestion control in absence of disconnections. With the introduction of the disconnection pattern, some efficiency differentiation starts to appear. In an one-congestion plus disconnections scenario, there is a better performance of TCP_Ufa when compared to TCP RENO.

Finally, in the last scenario, there is a congestion that starts just before a disconnection. In this moment, both of the protocols starts the congestion process (transmission restricted to congestion window and ssthresh halves). However, TCP_Ufa starts its disconnection algorithm, shutting the transmission and storing the cwnd value, while TCP RENO continue to shrink the window, thinking there is still a congestion in the network. TCP_Ufa resume the transmission as it was in the moment of disconnection, but TCP RENO begins to transmit with the ssthresh too low, delaying the achievement of a reasonable transmission rate. This explain the better performance of TCP_Ufa, when compared with TCP RENO in these situations.

5 Conclusion and Future Work

This paper shows a new proposal for TCP RENO for a better performance in wireless environments. This proposal is done in way of minimizing the modifications in TCP, keeping the changes on the sender side, and using a computing low cost algorithm. The presented protocol were designed on top of TCP RENO, in NS-2. One of the major advantages of this new implementation is the compatibility with modern and older TCP design. Just under severe interference, TCP_Ufa enters in action and even then, just in sender side. This permit full legacy compatibility, as the end-to-end semantics are preserved and there is no need of intermediate nodes. As a future work, TCP_Ufa have to be tested in other situations over wireless links, like multimedia scenarios (delay sensitive), bursty traffics, and multi-hop environments. Mobility should also be verified. Another suggestion is to test TCP_Ufa with new kinds of applications, as in this work only CBR was used. Finally, it is important to note that the TCP performance issue on wireless links is not yet solved, and there is no IETF defined standard to permanently deal with this situation. A considerably number of solutions have been proposed but none consolidated as a consense.

References

1. Bagal, P. : Comparative study of RED, ECN and TCP Rate Control. Tech. Report, 1999.
2. Bakre, A.; Badrinath, B.: Implementation and Performance Evaluation of Indirect TCP, IEEE Transactions on Computers, vol. 46, no. 3, pp. 260-278, 1997.
3. Balakrishnan, H. et al: A Comparison of Mechanisms for Improving TCP Performance over Wireless Links - ACM/IEEE Transactions on Networking, 1997.
4. Eyceoz, T. et al: Prediction of Fast Fading Parameters by Resolving the Interference Pattern- Asilomar Conf. on Signals, Systems and Computers, 167-171, California, 1997.
5. Floyd, S.; Henderson, T.: The NewReno Modification to TCP's Fast Recovery Algorithm - IETF RFC 2582, 1999.
6. Goff, T. et al: A True End-to-End Enhancement Mechanism for Mobile Environments, INFOCOM 2000.
7. MacDonald, D. et al: Microsoft Windows 2000 TCP/IP Implementation Details.
8. Network Simulator 2 (NS-2). http://www.isi.edu/nsnam/ns/.
9. Ramakrishnan, K. et al: A Proposal to Add Explicit Congestion Notification (ECN) to IP. RFC 2481, 1999.
10. Ratnam, K.; Matta, I.: WTCP: An Efficient Mechanism for Improving TCP Performance over Wireless Links, Proc. IEEE Symposium on Computers and Communications, 1998.
11. Salim, J.; Ahmed, U.: Performance Evaluation of Explicit Congestion Notification (ECN) in IP Networks – IETF RFC 2884, 2000.
12. Stevens, W.: TCP Slow Start, Congestion Avoidance, Fast Retransmission, and Fast Recovery Algorithms - RFC 2001, 1997.
13. Tsaoussidis, V. et al: Open issues on TCP for Mobile Computing, Journal of Wireless Communications and Mobile Computing, John Wiley & Sons, Issue 1, Vol. 2, 2002.

Gradient-Based Autoconfiguration for Hybrid Mobile Ad Hoc Networks

Longjiang Li[1], Xiaoming Xu[2,3], and Yunze Cai[2]

[1] Department of Computer Science and Engineering, Shanghai Jiaotong University, Shanghai 200030, P.R. of China
[2] Department of Automation, Shanghai Jiaotong University, Shanghai 200030, P.R. of China
[3] University of Shanghai for Science and Technology, Shanghai 200093, P.R. of China
{e_llj, xmxu, yzcai}@sjtu.edu.cn

Abstract. Autoconfiguration for mobile nodes is an important problem in hybrid mobile ad hoc networks (MANETs) where pure MANETs are interconnected to the external Internet. The Dynamic Configuration and Distribution Protocol (DCDP) is an efficient approach, which extends DHCP to a stateless autoconfiguration protocol for wired and wireless networks. However, the distribution randomness in the DCDP results in an unbalanced distribution of address resources, which wastes the limited network address space and, subsequently, incurs scalability problems. In this paper, we propose a new approach, which can implement a more balanced distribution of address resources than DCDP. As the address distribution process continues, almost every node holds address resources. Thus, a new node can obtain its address almost immediately, when it joins the network. Finally, the simulation shows that the proposed scheme outperforms the existing algorithms.

1 Introduction

In many practical scenarios [1] [2], mobile ad hoc networks (MANETs) are increasingly seen as "extensions" of infrastructure networks where pure MANETs need to be connected to external wired network via the gateways. When a gateway is available, global scope address is needed.

Many autoconfiguration protocols have been proposed in recent years, but most [3, 4, 5, 6, 7, 8, 9, 10, 11] of them focus on pure MANETs [12] [13]. Although the methods for pure MANETs can be applicable to hybrid MANETs with a little modification, the performance is usually not too good. In hybrid MANETs, a gateway can act as the starting point of address allocation. Multiple gateways also can improve the performance of networks through the coordination via wired network, which is not available in pure MANETs. The Dynamic Configuration and Distribution Protocol (DCDP) [14] is a effective autoconfiguration protocol, which extends Dynamic Host Configuration Protocol(DHCP) [15] to a stateless autoconfiguration protocol for wired and wireless networks. It does not depend on multi-hop broadcasting, but the distribution randomness in the DCDP results in an unbalance distribution of address resources, which wastes the limited network address space and, subsequently, incurs scalability problems.

M. Gerndt and D. Kranzlmüller (Eds.): HPCC 2006, LNCS 4208, pp. 682–691, 2006.

In order to overcome these drawbacks, we propose a new approach , which tries to make the address distribution more balanced to improve the performance. We also use the address pool as the basic unit of address resources just similar as in [16]. In addition, a gradient distribution function is introduced to control the distribution of address resources. A configured node can determine whether to distribute address resources to other nodes using the gradient distribution function without relying on global information. Since the address distribution process does not rely on multi-hop broadcast, much communication overhead can be saved. As the address distribution process continues, almost every node holds address resources. Therefore, a new node can obtain its address almost immediately, when it joins the network. The address resources also can be utilized more economically. When a node need leave the MANET, the node can transfer its address capsulated in an address message to any one of neighboring nodes for reuse.

This paper is structured as follows. Next section introduces the basic idea of the proposed approach, namely gradient-based address distribution protocol(GADP). Section 3 gives a succinct comparison between GADP and several known approaches, which depicts the superiority of GADP over others. Section 4 presents some simulation results, which is equal to our analysis. Section 5 concludes the paper.

2 Gradient-Based Address Distribution

2.1 Basic Idea

The basic idea behind the proposed scheme is to set up a gradient to distribute address resources over all nodes in a MANET. We also use the address pool as the basic unit of address resources just similar as [16]. If a node has at least an address pool, it can configure the first element in the address pool as its address. If a node has been configured address, we say that it is a configured node, otherwise an unconfigured node. An unconfigured node has no address resource.If the node has a neighboring node which has redundant address resources, it may send a request to the neighbor to obtain a free address as its address. If not, it can broadcast queries within its one-hop scope periodically until one configured neighbor replies or just wait for some other configured node emerging. For a configured node, if it distributes part of its address resources to other nodes, its address resources should decrease. In order to compensate the loss of its address resources, the node also can ask a neighboring node, which is assumed to have more amount of address resources, to distribute address resources to it. The amount of address resources that are distributed between two neighboring nodes may have the different values. In the binary splitting method, a configured node can distribute half of its address resources to the requester, which is simplest. Thus the binary splitting method is also adopted in this paper.

2.2 Binary Splitting Method

We use a 3-tuple: (*prefix*, *start*, *level*) to represent an address pool, where *prefix* is the prefix of the address pool, *start* is the first address in the address pool and *level* indicates how many times the address pool can be split any more. It is easy to see that the length of an address pool is equal to 2^{level}. The scheme can be compatible easily with IPv4 or IPv6 by customizing different prefix, which is constant during the address allocation. When we only consider the case where all nodes will be configured from the same node, because all nodes have the same *prefix*, for convenience, we may neglect *prefix* and only use a 2-tuble: (*start*, *level*) to represent an address pool.

A node can have several address pools, which are all recorded in a variable, denoted by *list*. A configured node uses the *start* value of the first element in the *list*, i.e. $list[1].start$, as its address.

Suppose the length of the whole available address space for the MANET is K and then the address pool of the first node should be (0, K). In order to be consistent, we use (-1,-1) to express that a node has not been configured yet. Without loss of generality, we use a simple rule as followed to give an instantiated version of the binary splitting idea.

Definition 1 (Binary Splitting Rule). $(start, level) \rightarrow (start, level - 1) \vee (start|2^{(K-level)}, level - 1)$.

That is to say, that an address pool,$(start, level)$, may be split into two address pools, $(start, level - 1)$ and $(start|2^{(K-level)}, level - 1)$.

Note that *start* is coded in binary code and "|" is a bitwise OR operation, e.g. for a network with $K = 3, (010, 1) \rightarrow (010, 0) \vee (110, 0)$. An address pool can be split, if only its level is greater than zero.

2.3 Gradient Distribution Function

DCDP leads to an unbalance distribution of address resources, because it adopts opportunistic distribution of the address resources. Our scheme tries to minimize this randomness.

Consider an ideal MANET. There is no loss of packet. It can be expressed as a connected undirected graph, where all links are bidirectional. Each bi-directional link is considered as the union of two simplex unidirectional links. We denote the MANET by $G(V, E)$ where V is the set of nodes and E is the set of edges. Each node can have several address pools. A variable *list* is used to record all its address pools. When a node receives an address request from anther node, it must decide whether to distribute address resources to the requester. Consider two neighboring nodes, B and A. Assume that A has sent a request to B for an address allocation, i.e., A acts as a requester and B a server. B should run a decision function $f(B, A)$. If $f(B, A)$ returns *true*, B will try to distribute address resources to A. Otherwise, B will not.

Definition 2 (Gradient Distribution Decision Function). *Node B runs the decision function, $f : V \times V \rightarrow [true, false]$, to decide whether to distribute address resources to node A. Here,*

$$f(B, A) = \begin{cases} true \ if \ F^{+}(B) > F^{-}(A), \\ false \ otherwise. \end{cases}$$

where $F^+ : V \mapsto \mathbf{R}$ and $F^- : V \mapsto \mathbf{R}$ is two metrics to evaluate the amount of address resources of a node. $F^+(B)$ is used when node B acts as a server, and $F^-(B)$ is used when node B acts as a requester.

The definition 2 means that one node distributes some of its free address resources to another only when its $F^+(.)$ is greater than $F^-(.)$ of the opponent. Different implementations of $F^+(.)$ and $F^-(.)$ can lead to different solutions, which may have different performances. Now we give two instances of them as followed.

B1) $F^+(B) = F^+(B) = \sum_{u \in B.list} u.level$;
B2) $F^+(B) = B.list(1).level$,
$F^-(B) = \sum_{u \in B.list} u.level$.

Note that *B.list* holds all address pools of B. In B1), *level* values of B' address pools is summed up to get the metric. Although other solutions may be possible, we do not use the total number of addresses that the node holds as the metric, because it seems to be too large to handle. Fig. 1 shows an example where both node A and B have two address pools and $B.list(2).level > A.list(2).level > B.list(1).level > A.list(1).level$. In Fig. 1a, B1) is applied and $f(B, A)$ return *true*, so the address distribution from B to A is allowed. In contrast, B2) is applied and $f(B, A)$ return *false* in Fig. 1b, so the address distribution is disallowed. Since B1) often leads to the "oscillating" phenomenon [18], we use the scheme derived from B2) as the standard version of GADP. Due to space limitation, the discussion of the "oscillating" phenomenon is not shown.

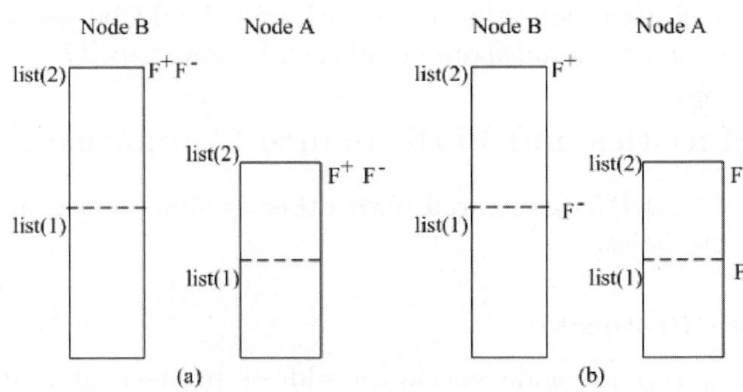

Fig. 1. Two instances of the decision function a) It derives from B1; b) It derives from B2.

2.4 Address Distribution Path

Given a distribution decision function, a node can decide whether to distribute address resources to its neighbors. Now we can set out to discuss the whole behavior of distribution process.

As stated before, if an unconfigured node,say A, knows that one of its neighboring node, say B, has been configured, the node can send requests to the neighbor for a free address pool. B can run the distribution decision function to distribute address resources to A and the B's address resources may probably thus decrease. If B has some neighbor, say C, which has more redundant address resources, B also can ask C to compensate its loss. In fact, we propose that the node chosen by B should have most address resources in B's neighborhood and at least more address resources than B. C can work in the same way.The process continues until some node can not find any neighbor that has more address resources than itself.

2.5 Mechanism for Network Partition and Merger

There are common two kinds of scenarios about network partition and merger.

The first scenario is that a MANET partitions and then the partitions merge again. Since all nodes get their addresses from the process starting with the same allocation initiator, all addresses are different from each other and there is no conflict if the partitions become merged later.

The second scenario is that two or more separately configured MANETs merge. The allocation may start with different allocation initiators. In hybrid mobile ad hoc network, pure MANETs need to be connected to external wired network via the gateways. If gateways are used to act as allocation initiator, multiple gateways can obtain the different prefixes through the coordination via wired network. If each node may attach the prefix as a part of its address, there will be no duplicate address in all MANETs. Therefore, GADP is not sensitive to network partition and merger.

In addition, our scheme also can be extended to the application in pure MANETS. In such situation, allocation initiators should be generated with a pseudo-random algorithm and more details can be found in [17].

3 Characteristics and Performance Comparison

In this section, GADP is compared with other several typical approaches to stress its characteristics.

3.1 Perkins's Protocol

In this protocol [7], each node selects an address by itself and employs DAD mechanism (query-based DAD) to determine whether the selected address is valid or not. No state is maintained. No address reclamation is needed. However, broadcast adopted in conflict detection leads to high communication overhead, high latency, and small scalability.

3.2 MANETconf

The protocol [4] maintains a global allocation state, so a new node can obtain a free address to join the subnet. However, the state management and synchronization to maintain global state incur high complexity, high communication overhead, high latency, and low scalability.

3.3 PACMAN Protocol

PACMAN [10] follows a hybrid approach and uses cross-layer information from ongoing routing protocol traffic to provide an address assignment and carry out the duplication address detection passively. A node running PACMAN assigns an address to itself using a probabilistic algorithm. The protocol can support frequent network partitioning and merging. The node configuration time is always fixed (in the order of milliseconds), but the address conflict can take place with a certain probability. Therefore, this protocol can be suitable for some applications where address conflicts can be tolerated to some extent.

3.4 Our Scheme

In contrast, GADP can generate global unique addresses, if only the allocation starts with the unique address prefix. Since the multi-hop broadcast is not needed, the communication overhead can be reduced very much. The allocation process tends to balance the distribution of address resource, so almost every node holds address resources at last. When a new node joins the MANET, it can be expected to obtain its address almost immediately. When a node need leave the MANET, the node can transfer its address capsulated in an address message to any one of neighboring nodes for reuse. Finally, GADP does not rely on routing protocols or certain topologies, so it can be extensible and suitable for hybrid MANETs with unpredictable mobility of mobile nodes.

4 Evaluation

We implement query-based DAD [7], DCDP [14] together with GADP to compare their performance. The simulation is based on Scalable Wireless Ad hoc Network Simulator (SWANS version 1.0.6) [19]. Since topologies in MANETs are dynamic and unpredicted, using an uniformly random topologies is more appropriate.

Due to the scarcity of resources in MANETs such as bandwidth and energy, the communication overhead should be kept at a minimum. Because every successfully received packet or sent packet, either unicast packet or broadcast packet, must have consumed bandwidth (and power as well), we use the average number of packets on each node, including what a single node receives or sends, as the evaluation metric for communication overhead.

A node must be configured an unique address before participating in the network function, so the configuration latency also should be kept at a minimum. In query-based DAD, nodes participating in the allocation try a maximum of three times for broadcast of duplicate address detection packet. The configuration latency is related with the network size. While in GADP, except domain initiators, every node tries infinitely to broadcast query packets until it receives a reply from one of its configured neighbors. Since address resources need spend time to reach each nodes, what we care most is how long a new node can be configured after the network has operated for some time. In order to be easy to compare, the retry intervals for both are set to be the same and we use the configuration latency of the last node, i.e.,the total configuration time minus the time when the last node joins, as the evaluation metric for configuration latency.

4.1 Simulation Parameters

The random waypoint mobility model [20] is adopted in the simulation. Node speeds are randomly distributed between zero and the maximum speed 10 m/s and the pause time is consistently 30 seconds. The retry period is randomly between 5 and 10 seconds. We select a node at random and configure its address (0, 0, 13). All other nodes join the MANET one by one. When all nodes are configured, the simulation program stops running and prints the result. Each simulation is repeated 10 times using different random-number generator initializations and the graphs show the average value.

4.2 Simulation Results

Different Node Numbers. We vary the number of nodes, from 200 to 300, within a fixed field (7000meter × 7000 meter). Fig. 2(a) shows the average number of packets, including what a single node receives or sends, with different node numbers. The number of packets generated in query-based DAD is about 5 times of that of GADP on average. And as the node density increases, the communication overhead of query-based DAD increases because the link number increases. However, the communication overhead of GADP and DCDP decreases because the neighboring nodes become more so that the number of retries to obtain addresses decreases. The communication overhead of GADP is less than that of DCDP, because the distribution of address resources in GADP is evener than that in DCDP.

Fig. 2(b) shows the relationship of the latency and the node number. As the node number increases, the node density increases but the configuration time in

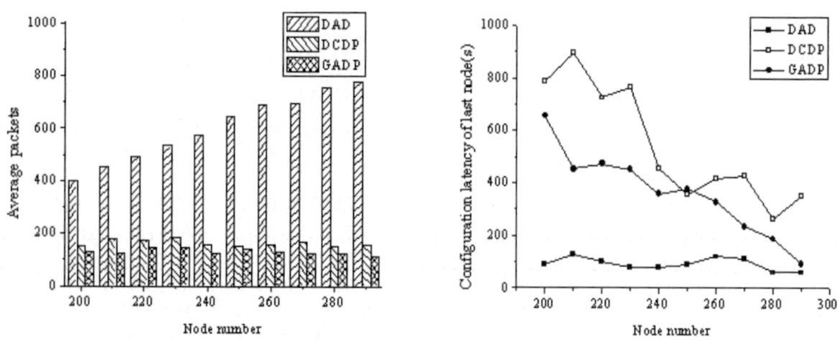

(a) Communication overhead (b) Configuration latency

Fig. 2. A simulation for different node numbers

both GADP and DCDP decreases. The configuration time in GADP is less than in DCDP and the curve of the configuration time in GADP is also smoother than that in DCDP.

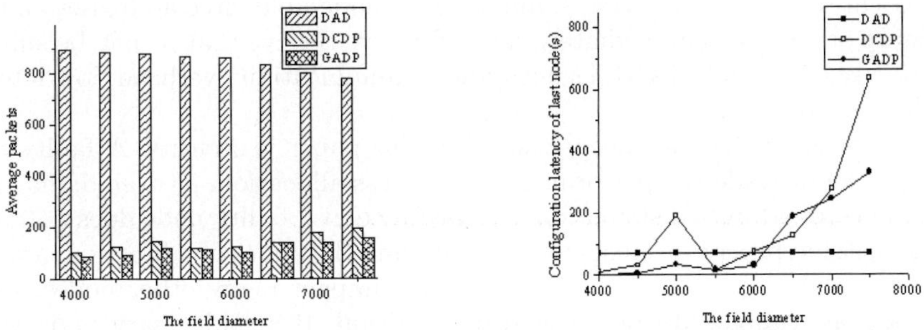

(a) Communication overhead (b) Configuration latency

Fig. 3. A simulation for different field sizes

Different Node Densities. In order to investigate the impact of different node densities, we vary the diameter of the field from 4000 meter to 7500 meter and keep the number of nodes to be 300. Fig. 3(a) shows the communication overhead within different sizes of fields. As the density increases, the communication overhead in both GADP and DCDP increases because the chance that a node has a configured neighboring node decreases.

Fig. 3(b) shows the configuration latency within different sizes of fields. As the density decreases, the latency in both GADP and DCDP also increases because the neighboring nodes may become fewer and most nodes need move farther to meet a configured node. Moreover, the latency in GADP increases slower than that in DCDP. That is to say, a node can be configured faster in GADP than in DCDP.

From the results in Fig. 2(b) and Fig. 3(b), we can arrive at a conclusion that the latency in GADP depends on the node density but it is not sensitive to the node number while the latency in query-based DAD increases for large scale MANETs. GADP also has a better performance than DCDP because the distribution of addresses in GADP is more balanced than in DCDP.

5 Conclusion and Future Work

We have presented an address autoconfiguration solution, abbreviated to GADP, for hybrid mobile ad hoc networks, where some nodes play the role of gateways and can be treated as the start point of the address allocation. GADP sets up a gradient to distribute address resources over all nodes in a MANET. GADP inherits the merit of DCDP, and increases the balance of address resources. Unlike query-based DAD algorithm, GADP does not rely on the multi-hop flooding, so it can reduce the overhead of address configuration significantly . Each node runs locally a decision process to dispense address resources to other nodes, which may lead to the balanced distribution of address resources over the whole network. GADP is not sensitive to the network partitioning and merging. As the allocation process progresses, almost each node holds free addresses and can configure other nodes immediately. Therefore, we believe that it may be suitable for large scale hybrid MANETs with low communication overhead, low latency, and high scalability.

A major issue that has been ignored in this paper is security. A faulty node perhaps can degrade the performance of address allocation, or even damage the consistency of address distribution. For instance, if a faulty node does not abide by the decision process or misbehaves in running the allocation function, a wrong address possibly generated by the node may impact many other nodes, which increases the probability of the address collision. If an adversary node which knows the state of the system misbehaves deliberately, the robustness of GADP may be suspicious in the worse case. Those problems illustrated above need more research.

Acknowledgments

This paper is supported by the National Key Fundamental Research Program (2002cb312200) and the National Natural Science Foundation of China (60575036).

References

1. I.K. Park, Y.H. Kim, and S.S. Lee, "IPv6 Address Allocation in Hybrid Mobile Ad-Hoc Networks," The 2nd IEEE Workshop on Software Technologies for Embedded and Ubiquitous Computing Systems, May 2004, pp:58-62.
2. R. Wakikawa, J. Malinen, C. Perkins, A. Nilsson, and A. Tuominen, "Global connectivity for IPv6 mobile ad hoc networks," IETF Internet Draft, draft-wakikawa-manet-globalv6-04.txt, Oct. 2005.
3. M. Gunes and J. Reibel, "An IP Address Configuration Algorithm for Zeroconf Mobile Multihop Ad Hoc Networks," Proc. Int'l. Wksp. Broadband Wireless Ad Hoc Networks and Services, Sophia Antipolis, France, Sep. 2002.
4. S. Nesargi and R. Prakash, "MANETconf: Configuration of Hosts in a Mobile Ad Hoc Network," Proc. IEEE INFOCOM 2002, New York, NY, Jun. 2002.
5. J.Boleng, "Efficient Network Layer Addressing for Mobile Ad Hoc Networks," Proc. Int'l. Conf. Wireless Networks, Las Vegas, NV, Jun. 2002, pp: 271-77.
6. H. Zhou, L. M. Ni, M. W. Mutka, "Prophet Address Allocation for Large Scale Manets," Proc. IEEE INFOCOM 2003, San Francisco, CA, Mar. 2003.
7. C. Perkins, J. T. Malinen, R. Wakikawa, E. M. Belding-Royer, and Y. Sun, "IP address autoconfiguration for ad hoc networks," IETF Draft, 2001.
8. N. H. Vaidya, "Weak Duplicate Address Detection in Mobile Ad Hoc Networks," Proc. ACM MobiHoc 2002, Lausanne, Switzerland, Jun. 2002, pp: 206-16.

9. K. Weniger, "Passive Duplicate Address Detection in Mobile Ad Hoc Networks," Proc. IEEE WCNC 2003, New Orleans, LA, Mar. 2003.
10. K. Weniger, "PACMAN: Passive Autoconfiguration for Mobile Ad Hoc Networks," IEEE JSAC, Special Issue on Wireless Ad Hoc Networks, Mar. 2005.
11. Y. Sun and E. M. Belding-Royer, "Dynamic Address Configuration in Mobile Ad Hoc Networks," UCSB tech. rep. 2003-11, Santa Barbara, CA, Jun. 2003.
12. K. Weniger and Z. Martina, "Mobile Ad Hoc Networks - Current Approaches and Future Directions", IEEE Network, Karlsruhe Univ., Germany, July 2004.
13. J. Jeong, J. Park, H. Kim, H. Jeong, D. Kim, "Ad Hoc IP Address Autoconfiguration," draft-jeong-adhoc-ip-ADDR-autoconf-04.txt, Jul. 2006.
14. A. Misra, S. Das, A. McAulley, and S. K. Das, "Autoconfiguration, Registration, and Mobility Management for Pervasive Computing, " IEEE Personal Communications, Volume 8, Issue 4, Aug. 2001, pp: 24-31.
15. R.Droms, J. Bound., B. Volz, T. Lemon, C. Perkins, M. Carney, "Dynamic Host Configuration Protocol for IPv6 (DHCPv6)," Network Working Group RFC 3315, Jul. 2003.
16. A. J. McAulley, and K. Manousakis, "Self-Configuring Networks," MILCOM 2000, 21st Century Military Communications Conference Proceedings, Volume 1, 2000, pp:315-319.
17. L. Li, X. Xu, "Optimistic Dynamic Address Allocation for Large Scale MANETs," Lecture Notes in Computer Science, Volume 3794, Dec. 2005, pp:794-803.
18. Z. Hu; B. Li, "ZAL: Zero-Maintenance Address Allocation in Mobile Wireless Ad Hoc Networks," in: Proceedings of 25th IEEE International Conference on Distributed Computing Systems, Jun. 2005, pp:103-112.
19. R. Barr, JiST-Java in Simulation Time: User Guide and Tutorial. Mar. 2004.
20. J. Broch,J. Broch, D. Maltz, D. Johnson, Y. Hu, J. Jetcheva, "A performance comparison of multi-hop wireless ad hoc routing protocols," in: Proceedings of the Fourth Annual ACM/IEEE Inter-national Conference on Mobile Computing and Networking, Oct. 1998, pp: 85-97.

Model-Aided Data Collecting for Wireless Sensor Networks

Chongqing Zhang, Minglu Li, and Min-You Wu

Department of Computer Science and Engineering,
Shanghai Jiaotong University, Shanghai, China
zhangchongqing@sjtu.edu.cn

Abstract. In this paper, we address the problem of collecting data from sensor nodes using a model-aided approach. In our approach, a model is maintained by a node and a replica of the model is maintained the base station. The base station uses the replica model to estimate the actual measurement data of the sensor node in usual time, and an actual measurement datum is sent to the base station only when the error of the model's corresponding estimation exceeds allowable error bound. In such a way, energy can be saved by reducing the transmission of actual measurement data. Experimental results show the effectiveness of our approach.

Keywords: Wireless Sensor Networks, Model-aided, Data Collecting, Data Fitting.

1 Introduction

Wireless sensor networks can offer us revolutionary new methods of accessing data from real environment [1]. However, because of the limited power of sensor nodes, collecting data is still a challenging work. For example, a Berkeley mote is only powered by two alkaline AA batteries [2]. Furthermore, it is infeasible to replenish the energy of sensor nodes by replacing the batteries in many applications [1]. Therefore, data collecting approaches of high energy-efficiency are strongly needed.

Motivated by the need of extending the network lifetime of energy-constrained wireless sensor networks, there has been considerable research in the area of energy-efficient data collecting in sensor networks and many techniques [3, 4, 5, 6, 7, 8, 9, 10, 14] have been proposed and developed. Among these techniques in-network aggregation and compression are two noticeable techniques. Although the measures they take are different, they are both trying to save energy by reducing the total amount of data transmitted.

Aggregation [3] is an in-network query processing technique for wireless sensor networks. By such a technique, for an aggregation query (e.g., the average rainfall of the monitored area), sensor readings are accumulated into partial results that are combined as messages propagate toward the base station. TinyDB [4] and Cougar [5] are two examples of utilizing aggregation to reduce energy consumption. On the other hand, compression attempts to take advantage of the correlation in the data and exploit coding techniques to reduce the size of data transmitted. For example, in [6],

M. Gerndt and D. Kranzlmüller (Eds.): HPCC 2006, LNCS 4208, pp. 692 – 699, 2006.

Ganesan et al. used wavelet based approach to compress the sensor data; while in [7] Chou et al. used distributed source coding to reduce the redundancy of the data to be transmitted to the sink.

All sensor nodes are still needed to transmit their data both in aggregation and compression. In [10], a model-aided approach was proposed to overcome this problem. However, the models adopted in this approach are fixed and not adaptive to the phenomenal changes. In this paper, adaptive models are adopted to improve the model-aided approach. In our approach, a predictive model M_i is induced by a sensor node N_i using data fitting [12] and an identical model M_i' is sent to the base station. The base station utilizes M_i' to estimate the actual measurement data of node N_i. At the same time, M_i is used by node N_i to judge how the estimations of model M_i agree with the actual measurement data. A measurement datum will be reported to the base station only when the error of corresponding estimative figure exceeds allowable error bound. In such a way, communication cost can be reduced and measurement error can be controlled in an allowable range.

The rest of this paper is organized as follows. In section II, We give the WSN model on which our research are based and present an overview of our approach. In section III, we discuss our approach in detail. In section IV, the implementation issues are discussed. Experimental results are presented in section V to show the effectiveness of our approach. We conclude in section VI.

2 Overview of Approach

We give an overview of our approach using an example of monitoring the blood pressure of patients in a hospital. Fig. 1 gives how the blood pressure of hypertension patients changes in 24 hours [11]. DBP and SBP denote diastolic blood pressure and systolic blood pressure, while SH and EH represent secondary hypertension and essential hypertension. As the figure shows, the blood pressure does not change desultorily. On the contrary, it fluctuates cyclically (with a period of 24 hours). The blood pressure changes continuously and it reaches its highest and lowest points at approximately 8 o'clock AM and 2 o'clock PM. Our approach uses these rules to achieve its energy-efficiency.

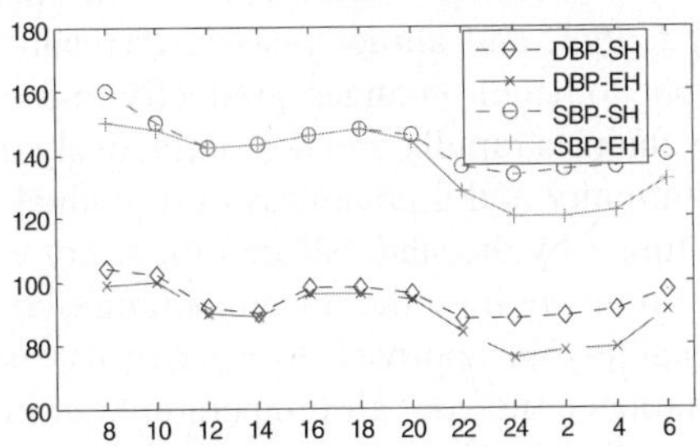

Fig. 1. Blood pressure curve in 24 hours [11]

Fig. 2 gives an overview of our approach. The base station uses commands to request the sensor nodes to sense the patients' blood pressure for a period of time *T* and at a certain frequency *f*, with an error bound ε is allowed. As the figure shows, a pair of models is maintained, with one model M_i distributed on node N_i and the other M_i' on *BS*. M_i and M_i' are always kept in synchronization. Model M_i is induced by a light-weight algorithm running on N_i from the measurement data set. Assume at a time instant *t*, a copy M_i' of M_i is sent to *BS*. Then at next time instant $t + 1$, *BS* can utilize model M_i' to estimate the actual measurement data of the sensor node N_i. At the same time, N_i still measures the blood pressure and compares the estimation E_i^{t+1} of M_i with the actual measurement data X_i^{t+1}. If $|E_i^{t+1} - X_i^{t+1}| \le \varepsilon$ (ε is the allowable error bound), the measurement data X_i^{t+1} is not reported to *BS*, otherwise X_i^{t+1} is reported to *BS*.

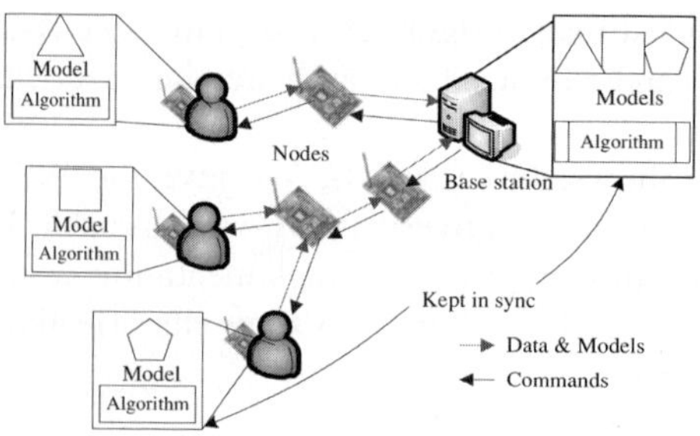

Fig. 2. Overview of Approach

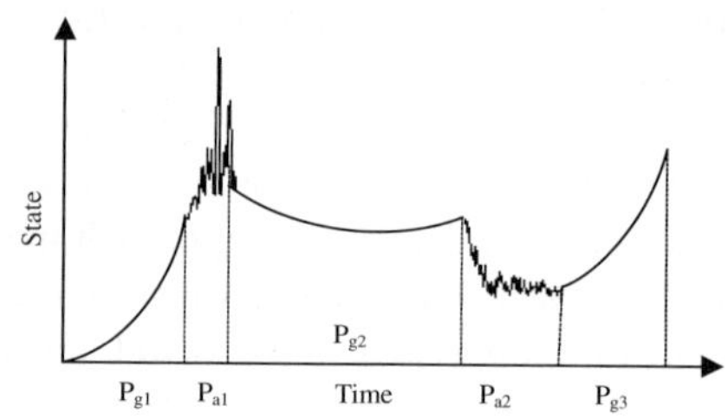

Fig. 3. Gradual changes versus abrupt changes

3 Principle of Approach

3.1 Predictability of Phenomena

The changes of natural phenomena follow some temporal and spatial rules. In this paper we focus on the temporal rules of the changing processes of physical phenomena. One temporal rule is that the changing process of a phenomenon may consist of gradual phases and abrupt phases. An example is the air temperature in a garden. The air temperature may change rapidly and violently in a short time, yet in most of the time it changes slowly and smoothly. Figure 3 shows a changing process that has gradual phases: Pg1, Pg2, Pg3, and abrupt phases: Pa1 and Pa2. During a gradual phase, the state of the phenomenon changes gradually and continuously; while the state of the phenomenon changes rapidly and discontinuously during an abrupt phase.

In many cases, the continuity and gradualness of a gradual phase make it possible to predict the state after time t by the state before time t. For example, we can predict the air temperature in an hour by how the air temperature changes before now. It is this predictability that enables our approach to achieve its energy-efficiency. Examples of such kind of phenomena include: air temperature, air humidity, earth temperature, soil fertility, soil humidity, body temperature, blood pressure, health of machines or buildings, pH value of lake water, concentration of pollutant, diffusion of contaminants, etc.

As for some phenomena, predicting the future state by previous state is very hard, sometimes even impossible. For example, the irregularity of the noise in a workshop makes it difficult to predict its intensity in the future. Our approach is not applicable for monitoring such kind of phenomena.

3.2 Models

Problem Definition. For a sensor node N_i, given the measurement data $\{X_i^0, X_i^1, \ldots, X_i^t\}$ before time instant t, an error bound ε, conclude a predictive model M that minimizes $Num(E_i^{t+a}: |E_i^{t+a} - X_i^{t+a}| \leq \varepsilon)$, where $a \geq 1$.

However, at time instant t, $\{X_i^{t+a}, X_i^{t+a+1} \ldots\}$ are unknown. As a result, these data cannot help us to figure out model M. What we can depend on is the data set $\{X_i^0, X_i^1, \ldots, X_i t\}$. So what we should do is to derive a proper model M from {Xi0, Xi1, …, Xit} and hope the prediction of M will agree with the actual future measurement data.

The continuity and gradualness of a gradual phase make it can be represented as a unitary function or several unitary functions with time as the independent variables. Based on this, unitary functions with time as the independent variables are adopted as models depicting how the monitored phenomenon changes in a gradual phase. Assume a unitary function f(x) for a phase Pg is derived at time instant t and sent to BS, then BS can use f(x) to estimate the actual state after time instant t.

From above analysis, it can be seen the key problem of our approach is to derive the function f(x) from limited measurement data. This problem can be viewed as a data fitting problem [12]. There are generally three problems to solve: 1) identifying a target function with unknown parameters, 2) identifying a proper data set and 3) determining the unknown parameters of the target function. Problem 2 and 3 are answered in following sections. Here we answer how to solve problem 1.

What target functions should be adopted is strongly application-dependent. As for different applications, the target functions that should be adopted may be quite different. If the change of the monitored phenomenon follows an obvious function type, then we have an obvious choice. Otherwise, if the function $f(x)$ is continuous and has $n + 1$continuous derivatives on an open interval (a, b), then according to Taylor Theorem [13], for $x \in (a, b)$, $f(x)$ can be represented as the summation of a polynomial of $(x - x_0)$ and a remainder:

$$f(x) = f(x_0) + f'(x_0)(x - x_0) + f''(x_0)\frac{(x - x_0)^2}{2!} + \cdots + f^n(x_0)\frac{(x - x_0)^n}{n!} + R_n(x) \tag{1}$$

where $x_0 \in (a, b)$. Based on this, polynomial functions can be adopted to fit the measurement data if there is not an obvious choice.

Note that time also affects the selection of target functions. For example, it can be seen from Fig.1 that the blood pressure curve takes on different shapes in different time phases of a day. As a consequence, using only one function to model the blood pressure curve of a whole day is not appropriate. The proper way is that different functions should be taken to model the blood pressure curve in different time phases. By storing multiple target functions for different time phases on nodes, proper functions can be selected according to the time when the model is building.

Although unitary functions are adopted in this paper to represent how the monitored phenomenon changes, this does not mean unitary functions are the only models that can be used. We think adopting what modeling tool is highly application-dependent.

4 Implementation of Approach

4.1 Building the Models

Identifying the data set concerns how to determine the number and location of the data. At a time instant *t*, what is known to node N_i is the measurement data $\{X_i^0, X_i^1, \ldots, X_i^t\}$ before *t*, and what we need to do is to derive a function that can predict the measurement data after *t*. Generally speaking, data that are adjacent in time are more correlated [15]. Based on this, we use the data measured in an interval before time t to fit the target function. As for the length of the interval, on the premise of that the target models can be constructed successfully; the length should be as small as possible. In the experiments presented in section V, an interval contains only 5 data, yet the result is very satisfactory.

Suppose the chosen data set is (x_i, y_i), $i = 1, 2, \ldots, n$, $x_i \neq x_j$ if $i \neq j$. Let *f(t)* be the target function and *f(t)* is expressed as:

$$f(t) = a_1 r_1(t) + a_2 r_2(t) + \ldots + a_m r_m(t) \tag{2}$$

where $\{r_1(t) \ldots r_m(t)\}$ is a group of predefined functions. $\{a_1 \ldots a_m\}$ is the coefficients that need to be determined. Least square fitting [12] is adopted by *MADG* to determine $\{a_1 \ldots a_m\}$. To do this, we need to make expression (2) get the least value.

$$J(a_1, \ldots, a_m) = \sum_{i=1}^{n} [f(x_i) - y_i]^2 \tag{3}$$

According to the necessary conditions for an extremum: $\partial J / \partial a_k = 0$, where $k = 1, 2, \ldots, m$, following group equation is derived:

$$\begin{cases} \sum_{i=1}^{n} r_1(x_i)[\sum_{k=1}^{m} a_k r_k(x_i) - y_i] = 0 \\ \vdots \qquad \vdots \qquad \vdots \\ \sum_{i=1}^{n} r_m(x_i)[\sum_{k=1}^{m} a_k r_k(x_i) - y_i] = 0 \end{cases} \tag{4}$$

Group equation can be expressed as:

$$R^T R A = R^T Y \tag{5}$$

where

$$R = \begin{bmatrix} r_1(x_1) & r_2(x_1)\cdots r_m(x_1) \\ r_1(x_2) & r_2(x_2)\cdots r_m(x_2) \\ \vdots & \vdots \quad \vdots \\ r_1(x_n) & r_2(x_n)\cdots r_m(x_n) \end{bmatrix}, \ A = (a_1, \ldots a_m)^T \text{ and } Y = (y_1, \ldots y_n). \tag{6}$$

By solving group equation (5), $\{a_1 \ldots a_m\}$ can be derived. If $r_1(t) \ldots r_m(t)$ are linearly independent, then $R^T R$ is reversible, and equation (5) has sole solution. In our approach, this is guaranteed by always using polynomial functions as the target functions.

4.2 Algorithm

In our approach, most of the work is done on sensor nodes. The base station simply uses the models to compute the expected value. A node N_i performs operations shown in Fig. 4 when a data is measured.

In the algorithm, data fitting is done only when there are enough data in the data set. Lines 13 to 18 are used to guarantee the effectiveness of the model. This can ensure the gain in performance if a correct model is used, and also the performance does not degrade if there is not an accurate model.

```
1: while measures a data X do {
2:    Add measurement X to DataSet;
3:    If X is the first data then {
4:       set y = X as the model  and send model to BS;
5:       continue; }
6:    E = estimate of model;
7:    If |E - X| < ε then continue;
8:    If Num(DataSet) < DataNum then {
9:       set y = X as the model and send model to BS;
10:      continue; }
11:   f(x) = data fitting result;
12:   bool = false;
13:   For each data d in DataSet do {
14:      E = estimate of f(x);
15:      If | E - d | > ε then {
16:         bool = true;
17:         set y = X as the model and send model to BS;
18:         break; } }
19:   If bool then continue;
20:   Set f(x) as model and send model to BS; }
```

Fig. 4. Algorithm running on a node

5 Experimental Results

The performance of our approach (denoted as Model-Aided) was tested against other two approaches. In one approach (denoted as Simple), all data are sent to the base station. In the other approach (denoted as Cache), the latest measurement data X_l of a node is cached by the node and the base station. A measurement data X_c is sent to the base station only when the absolute value of the difference between X_l and X_c is bigger than error bound ε.

MicaZ motes [16] are used to test the performances of all approaches. Four motes are deployed to monitor the air temperature in the garden outside of our laboratory. We monitored the temperature for 5 days. The monitoring results of four motes are quite similar. Fig. 5 shows the air temperature curve that is drawn from the data collected by a node using approach Simple in 40 hours.

The number of transmitted packets is adopted to evaluate the performances of three approaches. For sake of simplicity, a measurement datum or a model is both regarded as a data packet. A node measures the air temperature every 1 minute. The allowable error bound is 0.05 Celsius. Only linear functions are taken to fit the temperature data. A dataset includes 5 data, i.e. data measured in 5 minutes.

Fig. 6 presents the comparative results of three approaches in cost which is evaluated by the number of sent packets. From the figure, it can be seen that even the performance of Cache is much better than Simple. By adopting models, our approach achieves better performance further than Cache.

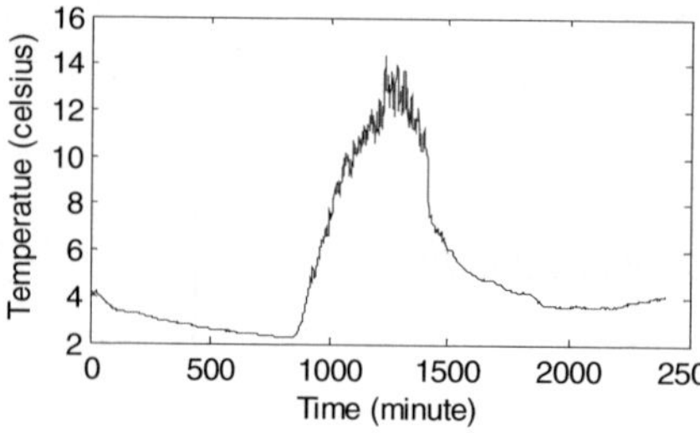

Fig. 5. Air Temperature Curve

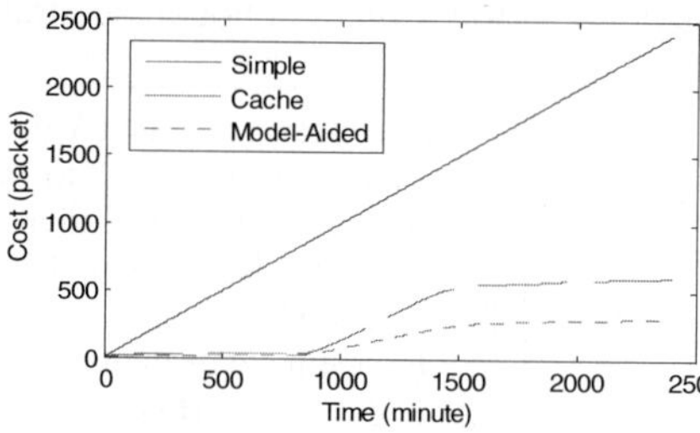

Fig. 6. Packets Sent of Three Approaches

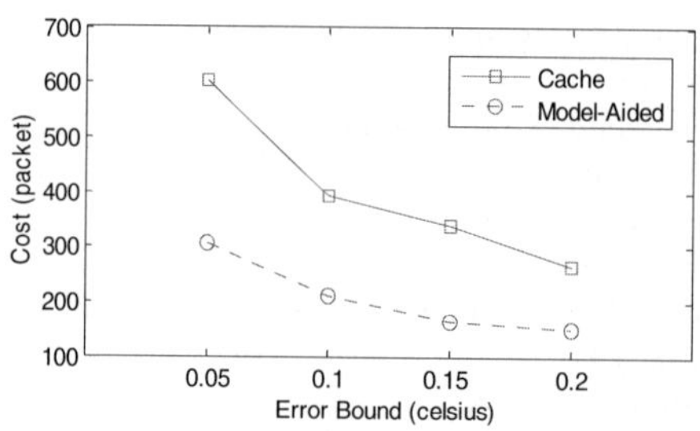

Fig. 7. Error Bounds versus Cost

Fig. 7 reveals the comparative results of two approaches, Cache and Model-Aided, against the error bound. Approach Simple is not compared because the costs of Simple are identical under different error bounds. In both Cache and Model-Aided, the number of sent packets drops as the error bound increases. It can also be observed that Model-Aided outscores Cache under all error bounds.

6 Conclusion

In this paper, temporal rules of the changing processes of natural phenomena are exploited to improve the energy-efficiency of wireless sensor networks. By maintaining replicated models on sensor nodes and the base station, energy can be saved by reducing the data transmitted to the base station. In the next step, we plan to integrate model-aided approach with spatial rules of natural phenomena and make full use of spatio-temporal rules to heighten the energy-efficiency of wireless sensor network.

References

[1] I. F. Akyildiz, W. Su, Y. Sankarasubramaniam, and E. Cayirci, "Wireless Sensor Networks: A Survey," Computer Networks, Vol. 38, No. 4, pp. 393-422, March 2002.

[2] M. Horton, D. Culler, K. PIster, J. Hill, R. Szewczyk, and A. Woo, "MICA, The Commercialization of Microsensor Motes," Sensors, Vol. 19, No. 4, pp 40-48, April 2002.

[3] B. Krishnamachari, D. Estrin and S. Wicker, "The Impact of Data Aggregation in Wireless Sensor Networks," DEBS, 2002.

[4] S. Madden, M. J. Franklin, J. M. Hellerstein, and W. Hong, "Tag: A tiny aggregation service for ad hoc sensor networks," OSDI, 2002.

[5] Y. Yao and J. Gehrke, "Query processing in sensor networks," CIDR, 2003.

[6] D. Ganesan, B. Greenstein, D. Perelyubskiy, D. Estrin, and J. Heidemann. An Evaluation of Multi-resolution Search and Storage in Resource-constrained Sensor Networks. ACM SenSys 2003.

[7] J. Chou, D. Petrovic, and K. Ramchandran, "A distributed and adaptive signal processing approach to reducing energy consumption in sensor networks," INFOCOM, 2003.

[8] B. Babcock and C. Olston, "Distributed Top-K Monitoring," ACM SIGMOD 2003.

[9] A. Deligiannakis, Y. Kotidis and N. Roussopoulos, "Hierarchical In-Network Data Aggregation with Quality Guarantees," EDBT 2004.

[10] D. Chu, A. Deshpande, J. M. Hellerstein and W. Hong, "Approximate Data Collection in Sensor Networks using Probabilistic Models," ICDE 2006.

[11] C. Jin, Z. Qian, L. Chen, X. Wang, "Ambulatory blood pressure monitoring in secondary hypertension," Chinese Journal of Cardiology, Vol 27, No 3, pp. 50-53, May 1999.

[12] R. L. Burden, J. D. Faires, "Numerical Analysis," Brooks Cole publishing company, December 2000.

[13] D. Hughes-Hallett, A. M. Gleason, P. F. Lock, D. E. Flath, et al, "Applied Calculus," Wiley, April 2002.

[14] C. Intanagonwiwat, R. Govindan and D. Estrin, "Directed diffusion: A scalable and robust communication paradigm for sensor networks," MOBICOM 2000.

[15] I. F. Akyildiz, M. C. Vuran, O. B. Akan, "On Exploiting Spatial and Temporal Correlation in Wireless Sensor Networks," WiOpt 2004.

[16] Crossbow, Inc. Wireless sensor networks. http://www. xbow.com.

Low Latency and Cost Effective Handoff Based on PBF Scheme in Hierarchical Mobile IPv6

Jongpil Jeong, Min Young Chung, and Hyunseung Choo*

Intelligent HCI Convergence Research Center
Sungkyunkwan University
440-746, Suwon, Korea
Tel.: +82-31-290-7145
{jpjeong, mychung, choo}@ece.skku.ac.kr

Abstract. In this paper, an efficient neighbor AR (Access Router) discovery scheme and handoff procedure using neighbor information, called the Pre-Binding Update and Forwarding (PBF) scheme, are proposed. It allows each AR and Mobility Anchor Point (MAP) to know neighboring ARs and MAPs, and therefore, a mobile node (MN) can perform the handoff process in advance, using the proposed handoff mechanism. It is important to note that the Inter-MAP domain handoff improvement of the proposed scheme is up to about 57% and 33% for handoff latency in comparison of the Hierarchical Mobile IPv6 (HMIPv6) and the Hierarchical Mobile IPv6 with Fast handover (F-HMIPv6), respectively. In HMIPv6, the total signaling cost rapidly increases in proportion to the number of Correspondent Nodes (CNs) communicating with the MN. Therefore this is combined with the Inter-MAP forwarding scheme, which operates differently to HMIPv6. It does not transmit the Binding Update (BU) message to the CNs and Home Agent (HA) when the MN moves among adjacent MAPs, after the Pre-BU process for Inter-MAP handoff. The proposed scheme demonstrates superior performance, which the total signaling cost is smaller than that of the HMIPv6 scheme, until 8 forwarding steps. Therefore, it is sufficiently satisfied with the requirements of real-time applications, and seamless communication is expected.

1 Introduction

Eventually, Mobile IPv6 (MIPv6) [1,2] will become an essential part of Mobile Internet. In addition, the Internet Engineering Task Force (IETF) considers Fast Mobile IPv6 (FMIPv6) [3] and HMIPv6 [4], which enhances handoff performance (latency and data loss), to provide localized mobility management for standard Mobile IPv6. Mobile IPv6 handoff incurs high handoff latency, data loss, and global signaling. Basically, FMIPv6 reduces handoff latency by link layer (L2) triggers and prevents data loss by creating a bi-directional tunnel between a mobile node's previous subnet's access router (oAR) and next subnet's access

* Corresponding author.

M. Gerndt and D. Kranzlmüller (Eds.): HPCC 2006, LNCS 4208, pp. 700–709, 2006.

router (nAR). HMIPv6 prevents global handoff signaling by appointing a MAP that acts like a local HA. In Mobile IPv6 and HMIPv6, no information is exchanged among ARs. Therefore, only after completing L2 handoff, an MN can receive information regarding the AR, to which the MN will handoff via an agent advertisement message. On-going communication sessions with other hosts are impossible before completion of this handoff process, which is the major portion of overall handoff latency [5,6].

In the proposed mechanism, an AR can learn information regarding its geographically adjacent ARs - typically, global address, L2 identifier, and the prefix information of ARs that are currently being advertised. The current AR that the MN is visiting would be able to inform the MN of the prefix information of ARs to which the MN would likely handoff. If this is possible, the MN can start the AA process for On-link Care of Address (LCoA) in the case of Intra-MAP handoff, and LCoA and Regional Care of Address (RCoA) in the case of Inter-MAP handoff. After completion of the Address Auto-configuration (AA) process, the MN transmits an incomplete binding update message to a MAP, and then the MAP performs a Duplicate Address Detection (DAD) process using this message. Through this signaling flow, there is a remarkable decrease in handoff latency.

This paper is organized as follows. In Section 2, the previous schemes are discussed for the analysis of handoff latency. The motivation of this work and the new scheme are presented in Section 3, based on HMIPv6. In Section 4, the performance of the proposed scheme is evaluated. Finally, this paper is concluded in Section 5, presenting future directions.

2 Related Works

This section provides a brief overview of the differences to be taken into account for the various approaches to reduce the handoff latency. The latency due to a handoff using *basic MIPv6* is directly proportional to the minimum round-trip time required for a BU to reach either the HA, the CN or the oAR in case forwarding from oAR is allowed. The interruption time starts the moment the MN stops listening to the oAR and finishes when the first packet arrives via the new route, from either the HA, CN or oAR [2].

Using the *anticipated FMIPv6*, handoff is prepared in advance. Assuming the fast binding acknowledgement (F-BAck) is received via the oAR, i.e, the overlapping area is based on the mobile speed to make it possible, and then handoff is performed, the latency is proportional to the difference between receiving the F-BAck and the reception of the first packet forwarded to the nAR [3]. Fig. 1 depicts the FMIPv6. The handoff latency in HMIPv6 is the same as in the case of MIPv6. However, instead of the proportional minimum round-trip time between the MN and the HA or the CN and the oAR, when forwarding from previous access router is enabled, it is proportional to the round-trip time between the MN and the MAP or the oAR [4].

The combination of both FMIPv6 and HMIPv6, named *Hierarchical Mobile IPv6 with Fast handover (F-HMIPv6)*, introduces the difference in sending the

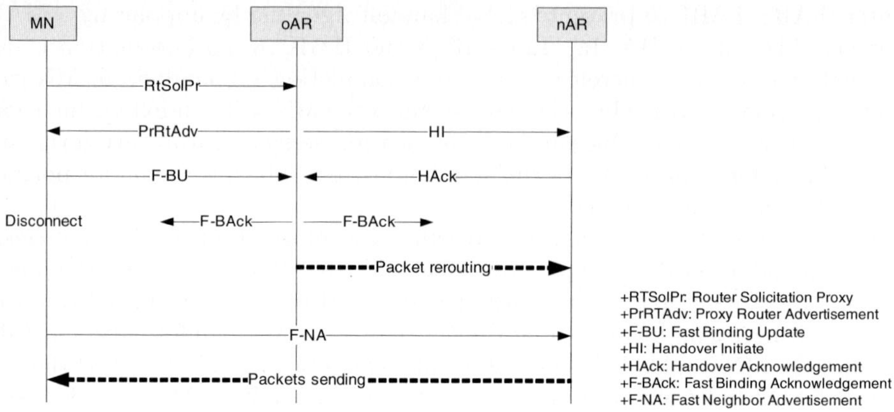

Fig. 1. Fast handover for Mobile IPv6 operation

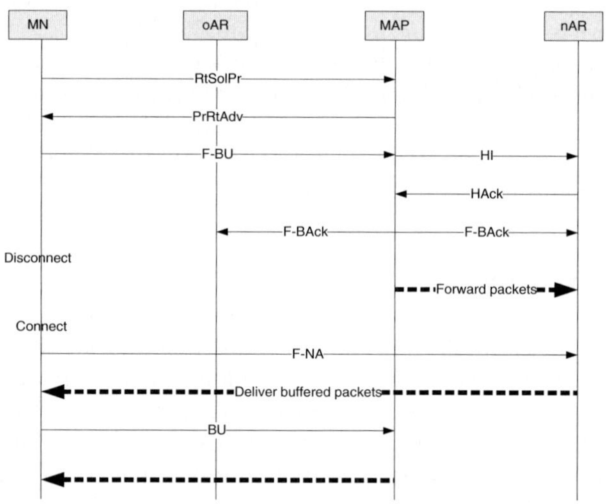

Fig. 2. Hierarchical Mobile IPv6 with Fast handover (F-HMIPv6) operation

Fast Binding Update (F-BU) to the MAP instead to the oAR. This means that the proxy router advertisement (PrRtAdv) and the Handover Initiate (HI) cannot be transmitted simultaneously. The introduction of the hierarchy results in the forwarding which is performed by the MAP; in the case of a symmetric topology, this will result in an advantage [3,4,12]. Fig. 2 depicts the F-HMIPv6.

Another protocol integrating FMIPv6 and HMIPv6 [13], namely *FMIPv6 for HMIPv6 (FF-HMIPv6)*, is introduced. When MN moves within a MAP domain, FF-HMIPv6 works almost identical to FMIPv6. The only difference is that the MN sends a binding update (BU) to MAP (instead of HA in FMIPv6) after arriving at nAR. When the MN moves beyond the MAP domain, it sends a fast BU

(F-BU) to its previous MAP while arriving at nAR. Then, a bi-directional tunnel between previous MAP and nAR is created and packets destined to the MN are forwarded via the tunnel until the MN completes the HMIPv6 handover at its new MAP domain. FF-HMIPv6 differs from F-HMIPv6 [12]. F-HMIPv6 only supports intra-MAP handovers. FF-HMIPv6 intra-MAP handover signaling is between oAR and nAR as inherited from FMIPv6 while F-HMIPv6 is between the oAR and MAP, which requires a new message. On the other hand, FF-HMIPv6 imposes both FMIPv6 and HMIPv6 tunneling overheads during a handover.

3 The Proposed Protocol

3.1 PBF (Pre-binding Update and Forwarding)

In Mobile IPv6, after an MN moves from one subnet to another and performs the AA process, the MN informs the current network of its global address. The AR receiving this message verifies whether the MN's address can be used by the DAD. The ARs and MAPs know their respective neighboring AR's address information using the neighbor discovery scheme. When the MN senses that it will perform handoff, it transmits a ***handoff solicitation message*** to the current AR, and the AR transmits an ***advertisement message*** with options. When the MN receives this message, it performs an AA process to the new CoA in advance before the actual handoff, and transmits a ***Pre-BU message*** to the AR, to which the MN will handoff. The ***Pre-BU message*** is the same as the ***typical BU message*** with an additional reserved field of the mobility header. The AR (AR and MAP in case of Inter-MAP domain handoff) performs the DAD and records the address as an incomplete state of the MN's address, meaning that it will not be used for routing of the MN until the MN transmits the ***real BU message*** after the actual handoff. After L2 handoff, the MN transmits the ***BU message*** to the new MAP via the new AR. The new AR and MAP finish the handoff process and directly progress routing by changing the MN's address in an incomplete state into a complete one, without the DAD process.

When the MN enters another regional network, it sends the BU to the first AR of the subnet. The AR relays it to the MAP2 (new MAP), and the MAP2 sends it back to the MAP1 (old MAP). When the MAP2 receives its message, it compares the one to the MAP list and finds the MN's field. And it updates the current MAP address of the MN. In other words, the MAP2 of the MN serves as the HA at the home network and handles movements of the MN directly through the registration to the HA and CNs until the maximum number of forwarding link allowed (q). The packet encapsulation is implemented for the communication steps as follows. First, the CN transmits packets to the registered MAP1 of the MN. The MAP1 receives them and retrieves its next MAP address of the MN. Second, MAP1 encapsulates packets and transmits them to the MAP2. The MAP2 receives them and resolves the address of the next MAP. The MAP2 transmits them to the resolved address of the next MAP. These steps are iteratively done in the last MAP (MAPq). Finally, the MAPq decapsulates packets and relays them to the registered AR of the MN. The AR receives those packets and transmits them directly to the MN. These steps have a slight burden in terms of the tunneling cost per data packet. These steps are considered as factors in performance analysis. In this work, the maximum number of forwarding links permitted between MAPs is not fixed but optimized for each MN, in order to minimize total signaling cost. The optimal number is obtained based on the operational difference between the existing and proposed scheme.

3.2 Signal Flows

Fig. 3 represent the signal flows of the proposed handoff process. Since the signal flow patterns in both Intra-MAP and Inter-MAP domain handoff cases are similar, only the Inter-MAP domain handoff is described in detail.

1. The MN receives a ***beacon message*** from the nAR.
2. The MN transmits a ***(proxy) agent solicitation message*** that includes the L2 identifier information of nAR to the current AR (oAR) to request the nAR's information (the address prefix and global address of the nAR).
3. In response to the ***agent solicitation message***, the oAR verifies the neighboring AR information using the L2 identifier in the message. If it retrieves the requested information, the oAR transmits the MN a ***(proxy) agent advertisement message*** that includes the requested information (address prefix and global addresses of nAR and MAP2).
4. The MN performs the AA process and acquires the RCoA and LCoA.
5. The MN transmits a ***Pre-BU message*** to the nAR.

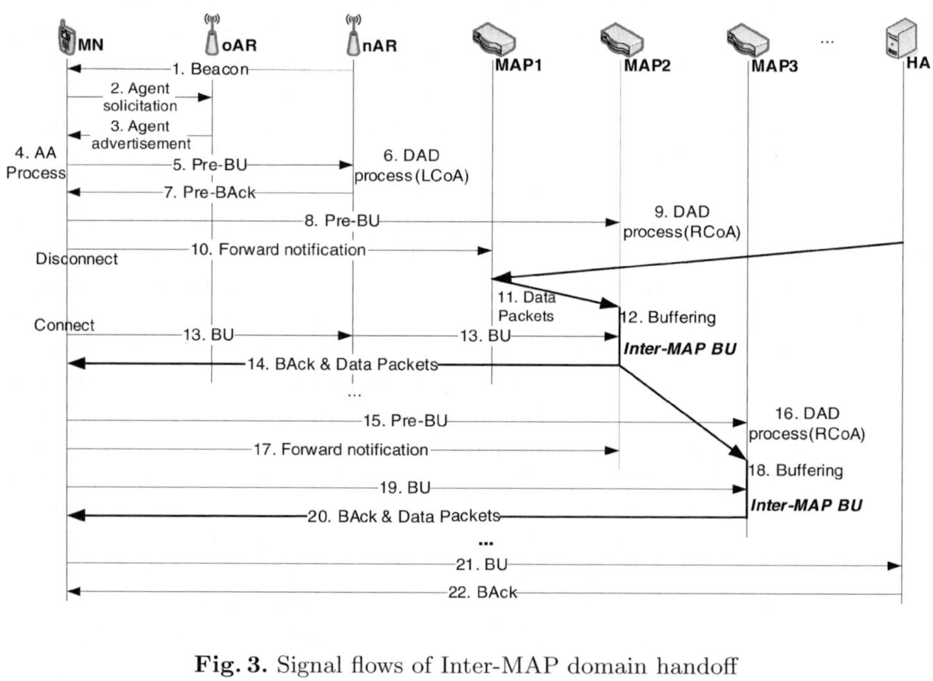

Fig. 3. Signal flows of Inter-MAP domain handoff

6. The nAR performs the DAD processing of the LCoA. The nAR receiving the ***Pre-BU message*** records the MN's address as incomplete.
7. The nAR transmits the ***Pre-BAck message*** to the MN.
8. The MN also transmits a ***Pre-BU message*** to the MAP2 for the DAD processing of RCoA.
9. Upon receiving it, the MAP2 performs the DAD process for RCoA.
10. The MN also transmits a ***forward notification message*** to the MAP1.
11. It requests the MAP1 to forward future packets arriving at the MAP1 to the MAP2. The MN performs the L2 handoff.
12. The MAP2 is ready to buffer the packet destined to the MN. During the AA and DAD processes, the MN still can receive packets from the oAR.
13. As soon as the completion of the L2 handoff, the MN transmits a BU to the MAP2 via nAR.
14. The nAR and MAP2 receive the message and changes the MN's address state to completed one. The MAP2 generates a MN list and transmits its RCoA to the MAP1 after using this ***BU message***. Using this message the MAP1 records MAP2's RCoA to MN's list. Therefore, when the HA transmits the message, it is delivered by this route. At this step, a MN's location information does not change with the HA and CNs. In addition, the MAP2 transmits the buffered packet destined to the MN with a ***BAck message***.
15. The MN transmits a ***Pre-BU message*** to the MAP3 for the DAD processing of RCoA.
16. Upon receiving it, the MAP3 performs the DAD process for RCoA.
17. And then, the MN transmits a ***forward notification message*** to the MAP2.
18. It requests the MAP2 to forward future packets arriving at the MAP2 to the MAP3. The MAP3 is ready to buffer the packet destined to the MN.
19. The MN transmits a BU to the MAP3 via nAR.
20. The nAR and MAP3 receive the message and changes the MN's address state to completed one. In addition, the MAP3 transmits the buffered packet destined to the MN with a ***BAck message***.
21. Until 8 forwarding steps ($q \leq 8$) among MAPs, the MN receives the data and transmits a ***BU message*** to the HA and CNs for normal routing.
22. The HA and CNs receive the message and transmits the ***BAck message*** to the MN.

4 Performance Evaluation

4.1 Handoff Latency

In this section five schemes are compared, MIPv6, FMIPv6, HMIPv6, HMIPv6 with fast handover (F-HMIPv6) and the proposed scheme. A simulation study is performed using ns-2 and its extensions [7,8], in order to evaluate the performance of the proposed mechanism. Fig. 4 illustrates a network topology of our simulation study [14,15]. The total simulation duration is 80 sec and a MN moves with 2 m/sec crossing contiguous cells. It is assumed that each cell is an 802.11 network. It is

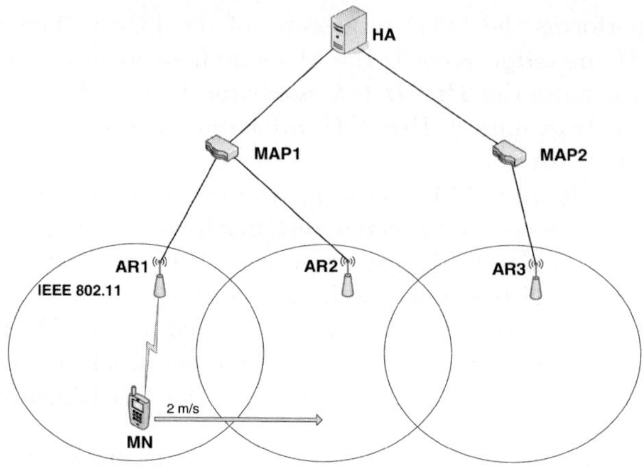

Fig. 4. Network topology for the simulation

also assumed that traffic is UDP packets, the packet interval is 10 msec, and the packet size is 256 bytes. The CN begins transmitting packets to the MN at 10 sec after the simulation start time and finishes transmitting packets when simulation is completed. For L2 trigger time, AA, and DAD, it is assumed that the L2 handoff latency and address resolution time are 200 msec and 300 msec, respectively, and the agent advertisement period is set to 1 sec. At $t = 25$ sec, an Intra-MAP domain handoff occurs from AR1 and AR2, and at $t = 60$ sec, Inter-MAP domain handoff occurs from AR2 within MAP1 to AR3 within MAP2.

First, the handoff latency is studied - the sum of the L2 and L3 handoff latency. The handoff latency of basic Mobile IP is defined as the period between the disconnection of the MN's wireless link and reception of AR's binding acknowledgement by MN. It is used for Intra-MAP domain handoff. In Inter-MAP domain handoff, handoff latency is the time from when MN triggers link-down in the current network to when the MN receives HA's first binding acknowledgement after handoff.

As presented in Fig. 5, the basic MIPv6 demonstrates the largest Intra-MAP domain handoff latency. The relationship between the FMIPv6 and the F-HMIPv6 is the same trend as it is between the MIPv6 and HMIPv6. The difference is a reduction of address resolution time in handoff latency. When using the fast handover scheme, a MN progresses the AA process in advance to reduce the latency related to address configuration. The proposed scheme presents the minimum latency. For Inter-MAP domain handoff, HMIPv6 presents the largest handoff latency. As in Intra-MAP domain handoff, the proposed scheme also presents the minimum latency. Note that the Inter-MAP domain handoff improvement of the proposed scheme is very large. This is due to the fact that the proposed scheme performs the AA and DAD processes in advance. MIPv6 presents many packet losses due to the large handoff latency, and HMIPv6 presents decreased packet losses than the base MIPv6 through the advantages of the MAP entity. The proposed scheme presents superior performance without

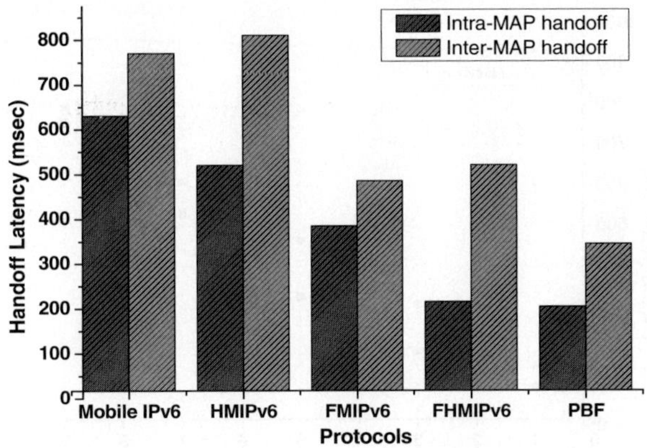

Fig. 5. The comparison of the handoff latency

packet loss. In the proposed scheme, the MAPs buffer and forward packets during the handoff period, this improves overall performance. The reason that the transmission time in the buffered packet decreases is that the transmission rate of a buffer is different from that of the source. This can be incurred through the use of an optimized buffering algorithm. Compared with MIPv6 and HMIPv6, the proposed scheme also demonstrates superior performance. It is sufficiently satisfied with the requirements of real-time application and seamless communication is expected when using the proposed scheme in L2 information exchange.

4.2 Modeling Total Signaling Cost

In this subsection, the performance of the proposed scheme based on HMIPv6 and HMIPv6 is compared and evaluated. There is a total signaling cost, which influences performance, consisting of a location update and packet delivery (Tunneling Cost) in the element. From analysis [9], the overall average signaling cost function in the proposed scheme is obtained in the HMIPv6. First of all, the HMIPv6's total signaling cost is represented as $C^{HMIPv6}(k, \lambda_a, T_f) = C^{HMIPv6}_{Location\_Update} + C^{HMIPv6}_{Packet\_Delivery}$. where, k is the number of ARs under the MAP, λ_a is the average packet arrival rate for each MN, T_f is the average time each MN stays in each subnet before making a movement. It is assumed that the MN moves from the MAP*0* to the MAPq, and therefore, the total signaling cost is calculated as $C^{HMIPv6}_{TOTAL}(k, \lambda_a, T_f, q) = q \cdot (C^{HMIPv6}_{Location\_Update} + C^{HMIPv6}_{Packet\_Delivery})$.

The total signaling cost in the proposed scheme is acquired as follows.

$$C^{PBF}(k, \lambda_a, T_f) = C^{PBF}_{Location\_Update} + C^{PBF}_{Packet\_Delivery}$$

Then, the total signaling cost is calculated as shown below.

$$C^{PBF}_{TOTAL}(k, \lambda_a, T_f, q) = C^{PBF}_{Location\_Update} + \sum_{i=1}^{q} C^{HMIPv6}_{Packet\_Delivery}$$

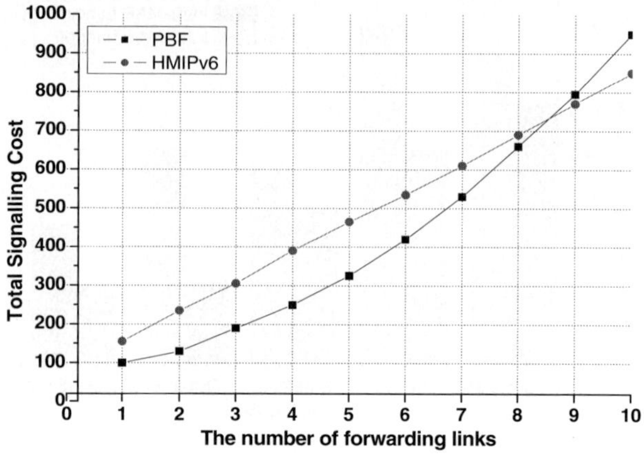

Fig. 6. The comparison of the total signaling costs

In Fig. 6, it is assumed that $T_f = 4$, $\lambda_a = 0.3$, and $\gamma = 5$, where, γ is the number of CNs. In the proposed scheme, the MN calculates the total signaling cost and compares it with the HMIPv6 total signaling cost. If the cost in the proposed scheme is smaller than that of the HMIPv6 scheme, the MN does not behave in home registration and CN registration but extends the forwarding links to other MAPs. These steps are iteratively done until the proposed scheme cost is greater than that of the HMIPv6 cost. The total signaling cost of the proposed scheme is smaller than that of the HMIPv6 scheme, until 8 forwarding steps ($q \leq 8$). However when $q \geq 9$, the proposed cost becomes greater than that of the HMIPv6 cost, the MN transmits the registration message to the HA and the CNs, and removes all the previous links among MAPs.

5 Conclusion

In this paper, an efficient neighbor AR discovery scheme and the handoff procedure using the neighbor information, called the PBF scheme, are proposed. The proposed neighbor ARs discovery scheme allows each AR and MAP to know its neighboring ARs and MAPs, and therefore the MN can perform the handoff process in advance. According to the simulation study, the proposed handoff mechanism demonstrates improved performance over existing mechanisms, due to the fact that the proposed scheme performs AA and DAD processes in advance.

In HMIPv6, the signaling cost rapidly increases in proportion to the number of CNs communicating with the MN when the MN moves among regional networks. Therefore the proposed scheme is combined with the Inter-MAP forwarding scheme, the proposed scheme does not transmit the BU to the CNs and the HA when the MN moves among adjacent MAPs. Instead, the current location of the MN is informed by transferring the modified BU to the previous MAP. According to the results of the performance analysis, the number of forwarding links is determined, allowing up to approximately 8 regional networks without BU information.

In a future study, detailed performance evaluation of the proposed approach is expected. Also, by comparing numerical results and optimizing handoff latency, we will study the method. And we will research the solution about the Layer 2 part and the heterogeneous network.

Acknowledgment

This research was supported by Ministry of Information and Communication, Korea under ITRC IITA-2005-(C1090-0501-0019) and grant No. R01-2006-000-10402-0 from the Basic Research Program Korea Science and Engineering Foundation of Ministry of Science & Technology.

References

1. T. Narten et al., "Neighbor Discovery for IPv6," Internet-Draft, October 2005.
2. D. Johnson, C. Perkins, and J. Arkko, "Mobility Support in IPv6," IETF RFC 3775, June 2004.
3. R. Koodli, "Fast Handovers for Mobile IPv6," RFC 4068, July 2005.
4. H. Soliman and K. E1-Malki, "Hierarchical Mobile IPv6 mobility management(HMIPv6)," RFC 4140, August 2005.
5. K. Omae, M. Inoue, I. Okajima and N. Umeda, "Performance Evaluation of Hierarchical Mobile IPv6 Using Buffering and Fast Handover," Technical Reports of IEICE, IN2002-152, December 2002.
6. K. Omae, et al., "Hierarchical Mobile IPv6 Extension for IP-based Mobile Communication System," Technical Reports of IEICE, IN2001-178, February 2002.
7. The Network Simulator - ns-2, http://www.isi.edu/nsnam/ns
8. Website, http://mobqos.ee.unsw.edu.au/ robert
9. D. Choi and H. Choo, "Cost Effective Location Management Scheme in Hierarchical Mobile IPvv6," Springer-Verlag Lecture Notes in Computer Science, vol. 2668. pp. 144-154, May 2003.
10. J. Jeong, et al., "Improved Location Management Scheme Based on Autoconfigured Logical Topology in HMIPv6," ICCSA 2005, vol. 3480, pp. 291-300, May 2005.
11. S. Pack et al., "An Adaptive Mobility Anchor Point Selection Scheme in Hierarchical Mobile IPv6 Networks," Technical Report, Seoul National University, 2004.
12. H. Jung et al., "A Scheme for Supporting Fast Handover in Hierarchical Mobile IPv6 Networks," ETRI Journal, vol. 27, Number 6, December 2005.
13. Y. Gwon et al., "Scalability and Robustness Analysis of Mobile IPv6, Fast Mobile IPv6, Hierarchical Mobile IPv6, and Hybrid Mobile IPv6 Mobility Protocols Using a Large-scale Simulation," 2004 IEEE International Conference on Communications, vol. 7, pp 4087-4091, June 2004.
14. R. Hsieh et al., "Performance analysis on Hierarchical Mobile IPv6 with Fast-handoff over End-to-End TCP," In Proceeding of GLOBECOM, November 2002.
15. R. Hsieh et al., "A Comparison of Mechanisms for Improving Mobile IP Handoff Latency for End-to-End TCP," Proceedings of the Ninth Annual International Conference on Mobile Computing and Networking, MOBICOM 2003, September 2003.

Distributed Classification of Textual Documents on the Grid

Ivan Janciak[1], Martin Sarnovsky[2], A Min Tjoa[3], and Peter Brezany[1]

[1] Institute of Scientific Computing, University of Vienna, Nordbergstrasse 15/C/3 A-1090 Vienna, Austria
janciak@par.univie.ac.at, brezany@par.univie.ac.at

[2] Department of Cybernetics and Artificial Intelligence, Technical University of Kosice, Letna 9, Kosice, Slovakia
martin.sarnovsky@tuke.sk

[3] Institute of Software Technology and Interactive Systems, Vienna University of Technology, Favoritenstrasse 9-11/E188, A-1040 Vienna, Austria
tjoa@ifs.tuwien.ac.at

Abstract. Efficient access to information and integration of information from various sources and leveraging this information to knowledge are currently major challenges in life science research. However, a large fraction of this information is only available from scientific articles that are stored in huge document databases in free text format or from the Web, where it is available in semi-structured format.

Text mining provides some methods (e.g., classification, clustering, etc.) able to automatically extract relevant knowledge patterns contained in the free text data. The inclusion of the Grid text-mining services into a Grid-based knowledge discovery system can significantly support problem solving processes based on such a system.

Motivation for the research effort presented in this paper is to use the Grid computational, storage, and data access capabilities for text mining tasks and text classification in particular. Text classification mining methods are time-consuming and utilizing the Grid infrastructure can bring significant benefits. Implementation of text mining techniques in distributed environment allows us to access different geographically distributed data collections and perform text mining tasks in parallel/distributed fashion.

Keywords: Text Mining, Multi Label Text Categorization, Distributed Text Mining, Grid Computing, JBOWL, GridMiner.

1 Introduction

The process of data mining is one of the most important topics in scientific and business problems. There is a huge amount of data that can help to solve many of these problems. However, data are often geographically distributed in various locations. While text is still premier source of information on the web, the role of text mining is increasing.

M. Gerndt and D. Kranzlmüller (Eds.): HPCC 2006, LNCS 4208, pp. 710–718, 2006.

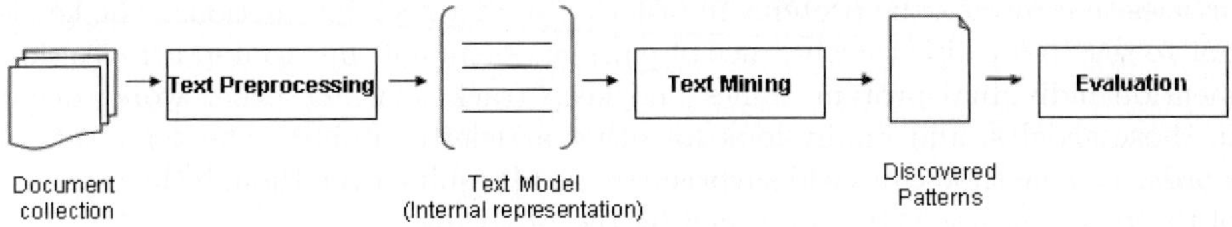

Fig. 1. Process of knowledge discovery in textual documents

Nowadays, the information overload means a big problem, so text mining algorithms working on very large document collections take very long times on conventional computers to get results. One approach to face this problem is distributed computing - distributed text mining algorithms can offer an effective way to mine extremely large document collections.

Motivation of this work is to use the Grid computational capabilities to solve text mining tasks. Some of the methods are time-consuming and use of the Grid infrastructure can bring significant benefits. Implementation of text mining techniques in distributed environment allows us to perform text mining tasks, such as text classification, in parallel/distributed fashion.

Knowledge discovery in texts is a variation on a field called knowledge discovery in databases, that tries to find interesting patterns in data. It is a process of semiautomatic non-trivial extraction of previously unknown, potentially useful and non-explicit information from large textual document collection, as depicted on Figure 1. A key element of text mining is to link extracted information together to form new facts or new hypotheses to be explored further by more conventional means of experimentation. While regular data mining extracts the patterns from structured databases of facts, text mining deals with problem of natural language processing. The biggest difference between data mining and text mining is in the preprocessing phase. Preprocessing of text documents is completely different, in general, it is necessary to find a suitable way to transform the text into an appropriate internal representation, which the mining algorithms can work on. One of the most common internal representations of document collection is the *Vector Space Model* [6]. Text mining phase is the core process of knowledge discovery in text documents. There are several types of text mining tasks as follows:

- Text categorization : assigning the documents into the pre-defined categories
- Text clustering : descriptive activity, which groups similar documents together
- Information retrieval : retrieving the documents relevant to the user's query
- Information extraction : question answering.

Nowadays, text mining plays important role in the area of processing biomedical databases that contain huge volume of textual documents. For example, one of the current questions in genomics is to inspect which proteins interact with others. There has been notable success in looking at which words co-occur in articles that discuss the proteins in order to predict such interactions. The key is not to search for direct occurrence of pairs in document, but to find articles that mention individual protein names and keep track of which other words occur in those articles, and finally look for other articles containing the same sets of words. This method can yield surprisingly good results, even though the meaning of the texts are not being discerned by the programs.

The structure of the rest of the paper is organized as follows. Section 2 discusses the classification of documents using multi-label classification. Section 3 describes the design of sequential and distributed versions of the text mining service implemented using the knowledge discovery framework - GridMiner. Experimental performance results are discussed in Section 4. Related work is presented in Section 5 and we briefly conclude in Section 6.

2 Text Classification Based on a Multi-label Algorithm

Text Classification is the problem of assigning a text document into one or more topic categories or classes based on document's content. Traditional approaches to classification usually consider only the unilabel classification problem. It means that each document in collection has associated one unique class label.

This approach is typical for data mining classification tasks, but in a number of real-world text mining applications, we face the problem of assigning the document into more than one single category. One sample can be labeled with a set of classes, so techniques for the multi-label problem have to be explored. Especially in text mining tasks, it is likely that data belongs to multiple classes, for example in context of medical diagnosis, a disease may belong to multiple categories, genes may have multiple functions, etc.

In general, there are many ways to solve this problem. One approach is to use a multinomial classifier such as the Naive Bayes probabilistic classifier [4], that is able to handle multi-class data. But most of common used classifiers (including decision trees) cannot handle multi-class data, so some modifications are needed. Most frequently used approach to deal with multi-label classification problem is to treat each category as a separate binary classification problem, which involves learning a number of different binary classifiers and use an output of these binary classifiers to determine the labels of a new example. In other words, each such problem answers the question, whether a sample should be assigned to a particular class or not.

In the work reported in this paper, we used the decision trees algorithm based on the Quinlan's C4.5 [7]. A decision tree classifier is a tree with internal nodes labeled by attributes (words), branches are labeled by weight of the attribute in a document, and leafs represent the categories [1]. Decision tree classifies a sample by recursively testing of the weights in the internal nodes until a leaf is reached.

While this algorithm isn't suitable to perform multi-label classification itself, we use the approach of constructing different binary trees for each category.

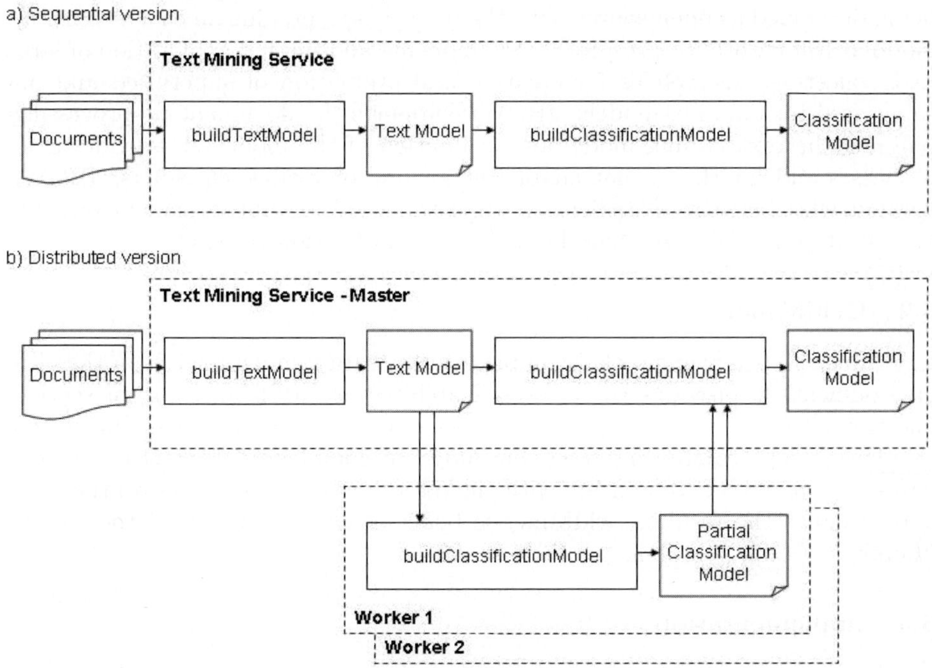

Fig. 2. Design of sequential and distributed text mining services

The process of building many binary trees can be very time consuming when running sequentially, especially on huge document collections. Due to the fact that these binary classifiers are independent on each other, it is natural to find a suitable way how to parallelize the whole process. Growing of these binary trees is ideal for parallel execution on a set of distributed computing devices. Such a distribution might be desirable for extremely large textual document collections or large number of categories, which e.g. can be associated with a large number of binary classifiers.

3 Architecture and Implementation of the Classification Service

Building distributed text mining services is an inherently difficult and complex task. To reduce the complexity, the first goal was to design a sequential version of Text Classification Service based on the multi-label algorithm implemented in the JBOWL library, which is discussed below.

3.1 JBOWL

JBOWL - (Java Bag-of-Words Library) [2] is an original software system developed in Java to support information retrieval and text mining. The system is being developed as open source with the intention to provide an easy extensible, modular framework for pre-processing, indexing and further exploration of large text collections, as well as for creation and evaluation of supervised and unsupervised text-mining models. JBOWL supports the document preprocessing, building the text mining model and evaluation of the model. It provides a set of classes and interfaces that enable integration of various classifiers. JBOWL distinguishes between classification algorithms (SVM, linear perceptron) and classification models (rule based classifiers, classification trees, etc.).

3.2 GridMiner

GridMiner [3] is a framework for implementing data mining services in the Grid environment. It provides three layered architecture utilizing a set of services and web applications to support all phases of data mining process. The system provides a graphical user interface that hides the complexity of the Grid, but still offers the possibility to interfere with the data mining process, control the tasks and visualize the results. GridMiner is being developed on top of the Globus Toolkit[1].

3.3 Implementation

The interface of the sequential and distributed versions of the service defines two main methods needed to build final model: *BuildTextModel* and *BuildClassificationModel*. While the first one is implemented as a pure sequential method, the second one can build the final model distributing the partial binary classifiers. This behavior of the service depends on its configuration. A simplified architecture of both versions is depicted in Figure 2. Moreover, other methods were implemented to provide term reduction and model evaluation, but these methods were not used during the performance evaluation experiments discussed in Section 4.

1. **BuildTextModel** - This method creates the *Text Model* from the documents in the collection. The model contains a document-term matrix created using TF-IDF weighting [8], which interprets local and global aspects of the terms in collection. The input of the method is a parameter specifying the text model properties and the location of the input collection.
2. **BuildClassificationModel** - The *Classification Model*, as the result of the decision tree classifier, is a set of decision trees or decision rules for each category. This service method creates such a model from the document-term matrix created in the previous method. The sequential version builds the model for all categories and stores it in one file. The process of building the model iterates over a list of categories and for each of them creates a binary decision tree. The distributed version performs the same, but it distributes the work of building individual trees onto other services, so called workers, where partial models containing only trees of dedicated categories

[1] http://www.globus.org

are created. These partial models are collected and merged into the final classification model by the master node and stored in the binary file, which can be passed to a visualization service.

4 Experiments

In this section, we present experiments performed on the local area network of the Institute of Scientific Computing in Vienna. As the experimental test bed, we used five workstations Sun Blade 1500, 1062MHz Sparc CPU, 1.5 GB RAM connected by a 100MBit network.

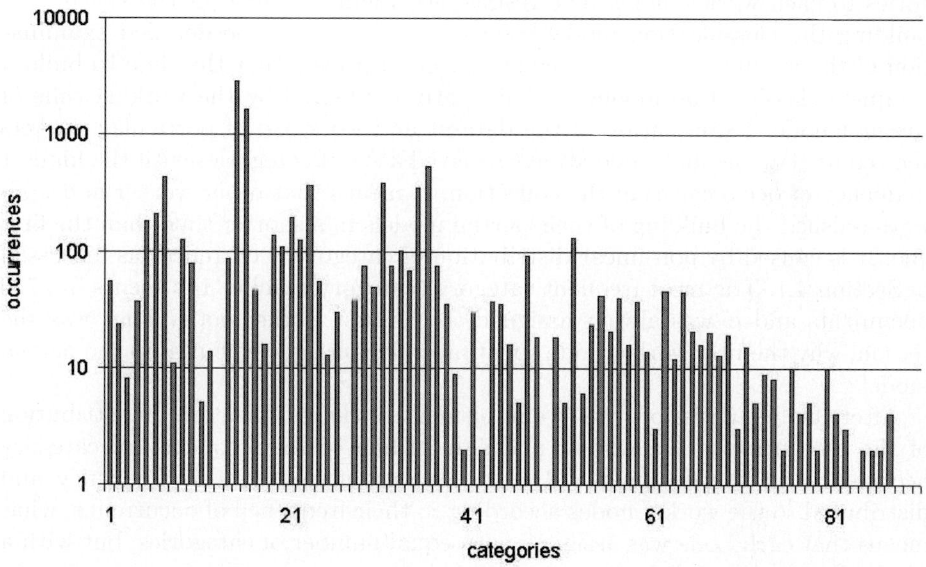

Fig. 3. Logarithmic distribution of categories frequency in the Reuters-21578 dataset

4.1 The Training Dataset

The Reuters-21578 [5] document collection was used as the training data in our experiments. It is de facto a standard dataset for text classification. Its modification, ModApte split [1], consisting out of 9603 training documents in 90 categories and formatted in XML, was used in all tests. Figure 3 depicts the logarithmic distribution of category frequencies in the Reuters-21578 collection. It shows that there are only two categories that occur more than 1000 times in the collection, while there is a lot of categories with frequency of occurrence less than 10. The time needed to build binary classifiers for categories with highest frequency of occurrence is significantly longer than the building time for the rest of the categories. This is a key factor for tasks distribution and optimization. In our case it is a decision how to assign partial categories and associated binary decision trees construction to the worker nodes.

4.2 Performance Results

The main goal of the experiments was to prove, that the distribution of processes mentioned above, can reduce the time needed to construct the classification model. We started the experiments using the sequential version of the service, in order to compare the sequential version with the distributed one. The time to build the final classification model on a single machine using the ModApte dataset was measured three times and its mean value was 32,5 minutes. Then we performed the first series of the distributed service tests without using any optimization of distribution of categories to the worker nodes. According to the number of worker nodes, the master node assigned the equal number of categories to each worker node. The results, see Figure 4, show us the speedup of building the classification model using multiple nodes. The detailed examination of the results and the document collection proved that the time to build a complete classification model is significantly influenced by the working time of the first node. Examination of the dataset and workload of particular workers showed us that the first node always received a set of categories with the highest frequency of occurrences in the collection. It means that other worker nodes always finished the building of their partial models in a shorter time than the first one. It is caused by non-linear distribution of category occurrences as discussed in Section 4.1. The most frequent category (category number 14) occurs in 2780 documents and it was always assigned to the first worker node. That was the reason, why the first worker node spent much longer time to build-up the partial model.

After the first series of tests, we implemented the optimization of distribution of the categories to the worker nodes according to the frequency of category occurrences in the documents. Categories were sorted by this frequency and distributed to the worker nodes according to their frequency of occurrence, what means that each node was assigned with equal number of categories, but with a similar frequency of their occurrences.

We run the same set of the experiments as in the first series and the results showed us more significant speedup using less worker nodes, see optimized bars in Figure 4. The best performance results were achieved using optimized distribution on 5 worker nodes (5.425 minutes), which was comparing to single machine computing time (32.5 minutes) almost 6 times faster. The minimal time to complete classification model is limited by the time of processing of the most frequent category - if this is assigned to a single worker node.

5 Related Work

In this section, we describe projects utilizing the Grid to perform advanced knowledge discovery in textual documents. DiscoveryNet[2] provides a service oriented computing model for knowledge discovery, allowing the user to connect to an use data analysis software as well as document collection that are made

[2] http://www.discovery-on-the.net

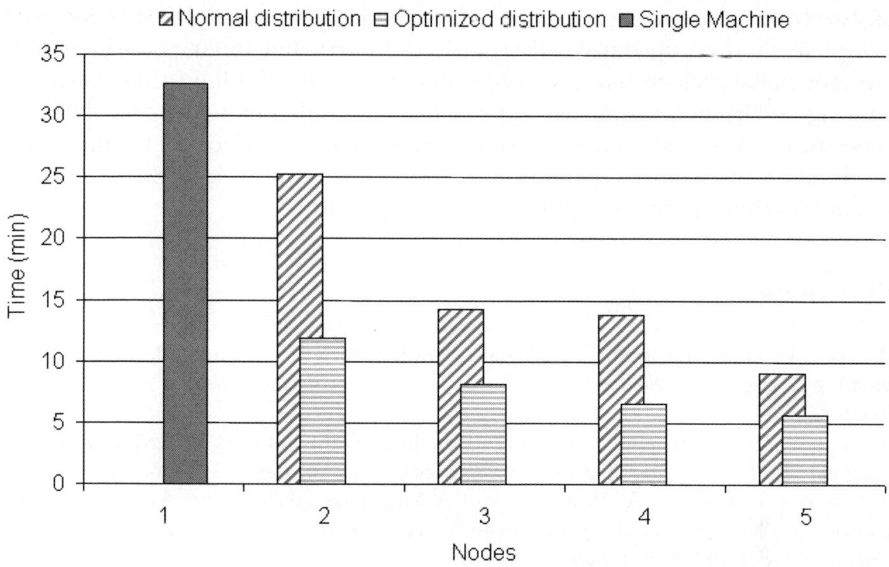

Fig. 4. Performance results of the normal and optimized distribution of nodes workloads

available online by third parties. The aim of this project is to develop a unified real-time e-Science text-mining infrastructure that leverages the technologies and methods developed by the DiscoveryNet and myGrid[3] projects. Both projects have already developed complimentary methods that enable the analysis and mining of information extracted from biomedical text data sources using Grid infrastructures, with myGrid developing methods based on linguistic analysis and DiscoveryNet developing methods based on data mining and statistical analysis. National Centre for Text Mining[4] is also involved in research activities covering the Grid based text mining. Primary goal of this project is also focused to develop an infrastructure for text mining, a framework comprised of high-performance database systems, text and data mining tools, and parallel computing.

6 Conclusions and Future Work

In this paper we presented a comparative study of sequential and distributed versions of classifiers based on decision trees. We proposed an idea how to distribute the process of building a multi-label classification model in the Grid environment by splitting the set of particular binary classifiers, needed to construct the final models into the workpackages, that are computed in distributed fashion. The results proved that the distributed version can bring significant benefits and

[3] http://www.mygrid.org.uk
[4] http://www.nactem.ac.uk

helps to reduce computing time needed to build the classification model. We also implemented an optimized distribution of particular binary classifiers onto the worker nodes, which had inconsiderable impact on the final time reduction comparing to the non-optimized approach. On a different real-world datasets, the speedup of the distributed version may be more significant. In our future research effort, we plan to explore and extend this approach of distribution of text classification services to other text mining tasks.

References

1. C. Apte, F. Damerau, and S. M. Weiss. Towards language independent automated learning of text categorisation models. In *Research and Development in Information Retrieval*, pages 23–30, 1994.
2. P. Bednar, P. Butka, and J. Paralic. Java library for support of text mining and retrieval. In *Proceedings of Znalosti 2005, Stara Lesna*, pages 162–169, 2005.
3. P. Brezany, I. Janciak, A. Woehrer, and A Min Tjoa. Gridminer: A framework for knowledge discovery on the grid - from a vision to design and implementation. In *Cracow Grid Workshop*, Cracow, December 2004.
4. Pedro Domingos and Michael J. Pazzani. On the optimality of the simple bayesian classifier under zero-one loss. *Machine Learning*, 29(2-3):103–130, 1997.
5. D. D. Lewis. Reuters-21578 text categorization test collection distribution 1.0. http://www.research.att.com/ lewis, 1999.
6. H. P. Luhn. A statistical approach to mechanized encoding and searching of literary information. *IBM Journal of Research and Developement*, 4:309–317, 1957.
7. J. R. Quinlan. Learning first-order definitions of functions. *Journal of Artificial Intelligence Research*, 5:139–161, 1996.
8. G. Salton and C. Buckley. Term-weighting approaches in automatic text retrieval. *Information Processing and Management*, 24:513–523, 1988.

Towards Job Accounting in Existing Resource Schedulers: Weaknesses and Improvements

Herbert Rosmanith, Peter Praxmarer, Dieter Kranzlmüller, and Jens Volkert

GUP, Joh. Kepler University Linz
Altenbergerstr. 69, A-4040 Linz, Austria/Europe
rosmanith@gup.uni-linz.ac.at

Abstract. The vision of having access to tremendous amounts of computation and storage resources on demand, together with access to special devices, similar to the availability of today's power grids has been formulated by Ian Foster and Carl Kesselman in [1] in 1997 and since then has been known by the term Grid computing.

As this vision slowly became reality and we're now at the verge to having Grids production ready not only for scientific communities but also for industrial partners security, accounting and billing are now major concerns that need to be reflected and further improved.

This paper analyzes two of the major local resource managers, Condor [2] and Torque[3], that are being used as local resource managers in the major grid middlewares Globus [4,5,6,7,8] as well as in the gLite and LCG [9,10] software stack with respect of being able to track malicious jobs and enforce a site policy.

As weaknesses have been found we also present an approach that is capable of truly tracking any kind of job.

1 Introduction

Local resource management systems (LRMS) such as Condor [2], Torque[3] (based on OpenPBS), LSF[11], SGE [12], and others are responsible for scheduling and co-allocating cluster resources to execute jobs in parallel on a number of nodes. Furthermore they gather accounting information and have to enforce a site policy. Site policies can be e.g. limiting a jobs runtime, its number of forked processes etc.

For this to work it is essential that the LRMS is capable of monitoring a job during all its lifetime and take appropriate actions if a job is violating the policy.

2 The Problem

We examined Torque [3] and Condor [2] with respect to monitoring a malicious process that detaches from the control of the LRMS and thus circumvents any accounting and billing mechanism applied by the LRMS. A scenario like this allows the malicious process to consume more of the computational resources than it was supposed to consume. In fact, this practically renders the accounting

M. Gerndt and D. Kranzlmüller (Eds.): HPCC 2006, LNCS 4208, pp. 719–726, 2006.

and billing system useless, since it is not capable of gathering appropriate data which is especially of interest when industry provides computational resources where using it has to be paid for.

2.1 Torque

We found that Torque[1] [3], and it's predecessor OpenPBS [13] track a job by means of its *session id* [14]. Thus, detaching a job from the control of the *pbs_mom* process can easily be done by changing the process's session id.

The following code illustrates the steps necessary to accomplish it:

```
int fire_n_forget() {
  switch(fork()) {
  case -1:
      exit(-1);
  case 0:
      close(0);
      close(1);
      close(2);
      setsid();
      while(1) {
          // consume cpu
      }
  default:
      break;
  }
}
```

The above presented code closes all file descriptors, and forks a new process. The resulting child process acquires a new *session id* by calling *setsid()* and continues with consuming the cpu-time. The parent process which has been actually tracked by the LRMS terminates and vanishes from accounting.

The result is that Torque considers the job being terminated although a child process is still running. It is worth noting that calling *fork()* and *setsid()* are unprivileged system calls and can be considered normal operations throughout program execution.

With respect to the so-called *jobmanagers* that are part of the Grid middleware and provide the interface between the *gatekeeper* and the local resource management system it can be said that they also fail in detecting what has been going on and thus report a successful job termination.

2.2 Condor

Condor is another important resource management system that has been developed at the University of Wisconsin [2,15]. Its features are extensive, allowing

[1] Version 1.2.0p6.

distributed submission, job priorities, user priorities, submission of workflows modeled as a directed acyclic graph of interdependent jobs. It supports PVM and MPI jobs, as well as job checkpointing and migration for certain types of jobs. Its functionality is provided through six so-called *universes*:

- Vanilla
- MPI
- PVM
- Globus
- Scheduler
- Standard

A detailed description of each universe can be found in [15]. In our further examination we focus on the *standard universe*, which extends serial jobs by the use of the following services [15]:

- Transparent process *checkpoint* and *restart*
- Transparent process migration
- Remote System Calls
- Configurable File-I/O buffering
- On-the-fly file compression/decompression

Within the *standard universe* a program is statically linked against the condor library which provides a customized implementation of the Unix C-library. It is customized in that way that most of the system calls are blocked or re-routed. In particular, a program linked to this library is unable to perform a *fork()*, or *setsid()* call. This makes Condor seemingly invulnerable to the attack performed against the Torque batch scheduler as it is described above.

Nevertheless it is possible to make a program running in the *standard universe* execute system calls. The idea is shown in the following listing:

```
int my_waitpid(int p1,void *p2,int p3) {
  register int rc;
  asm("\
      mov %2,%%ebx\n\
      mov %3,%%ecx\n\
      mov %4,%%edx\n\
      mov %1,%%eax\n\
      int $128"
      : "=r"(rc)
      : "i"(__NR_waitpid), "m"(p1), "m"(p2), "m"(p3));
  return rc_errno(rc);
}
```

The code presented above performs the *waitpid()* system call directly in assembler. Since the *my_waitpid()* function is not defined in the Condor C library it is not replaced by a Condor approved implementation and thus goes into the application code without modification.

Obviously, all other system calls can be implemented in a similar way, and thus the mechanism provided by Condor also fails in preventing system calls.

In order to prove that it is really possible to leave the Condor monitoring system we implemented an application that starts a child process on a Condor-controlled cluster node which provides an interactive session while the parent process has terminated. With the termination of the parent process the job has terminated for Condor too, while the child process still consumes the CPU-time unharmed.

3 Proposed Solution

As we have seen that the approaches undertaken by Torque as well as Condor fail in preventing a malicious user to circumvent their job monitoring systems a different approach is needed. While the mechanism provided by Torque can be circumvented with rather small effort, circumventing Condor's mechanism requires more effort, but is still possible.

We draw the conclusion that it is not enough to provide a *user space* based approach but instead need a facility running in *kernel space* that supports the monitoring facility. As proof of concept, we implemented a loadable kernel module for the Linux kernel, version 2.6.15, that redirects arbitrary interrupt vectors to its own routines. Of special interest is software interrupt number 128 which is called whenever a system call is performed. When the kernel module is loaded it registers itself and redirects interrupt number 128 to its own implementation which checks and records the requested system call and finally calls the original implementation.

3.1 Implementation Notes

The implementation of this kernel module has been done in assembler for the IA32 architecture. A port to the AMD64 architecture is in progress. With the *sidt (store interrupt descriptor table)* machine instruction a pointer to the *IDT (interrupt descriptor table)* is acquired. An *ISR (interrupt service routine)* can be redirected by overwriting the appropriate offset with the address of the new ISR (see [16] for a description of the Linux kernel architecture). In its current implementation the new ISR logs the requested system call and calls the original ISR.

It also would have been feasible to modify the *syscall* implementation within the Linux source directly. A disadvantage of this approach is that it requires patching and recompiling the Linux kernel. This would certainly reduce the acceptance and thus has been avoided. Instead our solution can be dynamically loaded as a kernel extension without modifying any of the existing kernel source code.

3.2 Monitoring System Calls

The above presented kernel module can be used to monitor system calls from user space. Its use is illustrated in Figure 1.

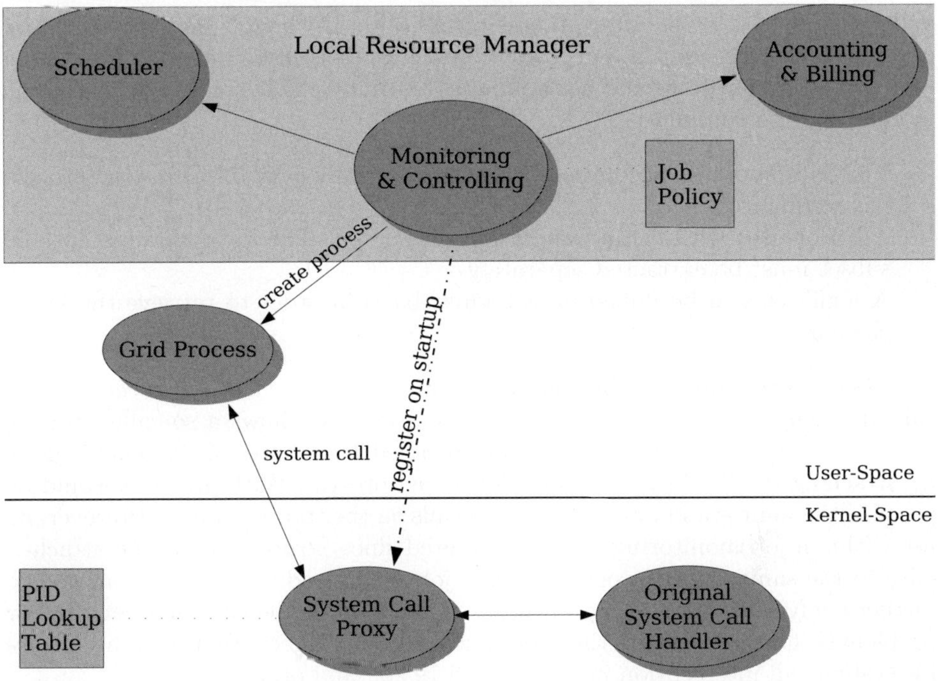

Fig. 1. Systemcall interaction

When the operating system boots, the kernel module is loaded and attaches itself as *system call proxy (syscpx)*. Upon start of a new job the jobmanager subscribes itself at the syscpx using a character device. After the subscription every process within a job created by the jobmanager is monitored while other system processes are ignored. Furthermore it is possible to instruct the syscpx to monitor only a defined subset of system calls (e.g. only *setsid()* and *fork()*). In order to prevent a malicious jobmanager from intervening with other processes the syscpx only allows a jobmanager to monitor and control processes which have been created and forked by itself.

The syscpx monitors the activities of the registered job by *notifying* the monitoring and controlling facility of the local resource manager about requested system calls. In addition, the syscpx can be instructed to send a *query* to the controlling facility to ask whether a process is allowed to perform the requested system call. In this way the monitoring and controlling facility can enforce a policy regarding the execution of system calls.

4 Related Work

With regard to accounting and monitoring most of the Unix systems, as well as Linux, provide a *process accounting* [14] system. This facility collects various data about the process id, memory use, execution time etc. of a running process and stores it to a logfile after process termination. Although this data could be quite useful for accounting purposes of grid jobs, it is more intended for system administration purposes and has a number of disadvantages regarding the online monitoring of a running job:

- The data about an accounted process is only fully available *after* the process has terminated.
- All processes within the system are accounted. The data about a specific subset must be extracted separately.
- A logfile has to be polled on a regular basis in order to retrieve the latest changes.

A different tool provided by Unix-based operating systems is the *ptrace* system call. It provides tracing and debugging facilities and allows a so-called *tracing process* to control another process, the so-called *traced process.* Tracing is done by attaching the tracing process to the traced process. With *ptrace* it would be possible to monitor and intercept system calls on the traced process. However, its use within a job monitoring system is limited since *ptrace* can only be attached once to the same traced process. Thus a job monitored in this way, can not be further analyzed for debugging purposes on demand by the user. A second major problem is that *ptrace* slows the process execution. Projects that facilitate *ptrace* for system call interception include [17,18,19,20], and others.

An interesting work is *systrace* [21], which similar to our work uses a module in kernel space to intercept system calls and allows or disallows those calls according to a policy. Furthermore it allows the creation of audit logs to support forensic purposes and intrusion detection. The process of policy creation is also facilitated by a graphical user interface.

While the approach is similar, systrace has been developed with enforcing system security in mind, our tool - in comparison - is intended to support accounting and billing systems of workload management systems.

As monitoring and enforcing policies within operating systems is an important topic for implementing trusted systems, the SELinux project [22], and the subsequent Linux Security Module [23] have been developed with this in mind. The SELinux project implements ideas which have been previously investigated within the academic Flask kernel [24], and allows the implementation of highly sophisticated system policies on a Linux system. Since Linux 2.6, auditing support is included in the kernel, too. It has to be investigated whether the auditing system can provide enough information in a mode suitable for LRMS requirements.

5 Conclusion

This paper describes how to simply circumvent the monitoring facilities of two major workload management systems. It was found that current workload management systems fail in tracking this type of job mainly because their monitoring mechanism solely operates in the user space.

The proposed solution thus extends the Linux kernel by a loadable kernel module that gathers reliable data in the kernel space which subsequently can be used by the workload management system for appropriate monitoring and accounting of running jobs. The presented approach is also able to not only monitor system calls, but also to intercept them, thus adding an additional layer of control.

Acknowledgments

We thank our colleagues, especially Paul Heinzlreiter and Martin Polak, for their support and fruitful discussions.

Availability

All source code is available on request from the authors.

References

1. Foster, I., Kesselman, C.: Globus: A metacomputing infrastructure toolkit. The International Journal of Supercomputer Applications and High Performance Computing **11**(2) (1997) 115–128
2. Litzkow, M., Livny, M., Mutka, M.: Condor - a hunter of idle workstations. In: Proceedings of the 8th International Conference of Distributed Computing Systems. (1988)
3. Torque resource manager 2.0. (http://www.clusterresources.com/pages/products/)
4. Foster, I., Kesselman, C., Tuecke, S.: The anatomy of the grid: Enabling scalable virtual organizations. International J. Supercomputer Applications **15**(3) (2001)
5. Foster, I., Kesselman, C.: Globus: A metacomputing infrastructure toolkit. The International Journal of Supercomputer Applications and High Performance Computing **11**(2) (1997) 115–128
6. Foster, I., Kesselman, C., eds.: The grid: blueprint for a new computing infrastructure. Morgan Kaufmann Publishers Inc., San Francisco, CA, USA (1999)
7. Foster, I., Kesselman, C.: Computational grids. In: VECPAR. (2000) 3–37
8. Foster, I., Kesselman, C., Nick, J., Tuecke, S.: The physiology of the grid: An open grid services architecture for distributed systems integration (2002)
9. gLite - Lightweight Middleware for Grid Computing. (http://www.glite.org/)
10. LCG - LHC Computing Grid Project. (http://lcg.web.cern.ch/LCG/)
11. LSF - Load Sharing Facility. (http://accl.grc.nasa.gov/lsf/)
12. Microsystems), W.G.S.: Sun grid engine: Towards creating a compute power grid. In: CCGRID '01: Proceedings of the 1st International Symposium on Cluster Computing and the Grid, Washington, DC, USA, IEEE Computer Society (2001) 35
13. OpenPBS - The Portable Batch System Software. (http://www.altair.com/software/)
14. Stevens, W.R.: Advanced programming in the UNIX environment. Addison Wesley Longman Publishing Co., Inc., Redwood City, CA, USA (1992)
15. Tannenbaum, T., Wright, D., Miller, K., Livny, M.: Condor – a distributed job scheduler. In Sterling, T., ed.: Beowulf Cluster Computing with Linux. MIT Press (2001)

16. Mauerer, W.: Linux Kernelarchitektur. Hanser Fachbuchverlag (2003)
17. Goldberg, I., Wagner, D., Thomas, R., Brewer, E.A.: A secure environment for untrusted helper applications. In: Proceedings of the 6th Usenix Security Symposium, San Jose, CA, USA (1996)
18. Wagner, D.A.: Janus: an approach for confinement of untrusted applications. Technical Report UCB/CSD-99-1056, EECS Department, University of California, Berkeley (1999)
19. Garfinkel, T.: Traps and pitfalls: Practical problems in system call interposition based security tools. In: Proc. Network and Distributed Systems Security Symposium. (2003)
20. Jain, K., Sekar, R.: User-level infrastructure for system call interposition: A platform for intrusion detection and confinement. In: NDSS, The Internet Society (2000)
21. Provos, N.: Improving host security with system call policies. Proceedings of the 12th USENIX Security Symposium (2003)
22. Loscocco, P., Smalley, S.: Integrating flexible support for security policies into the linux operating system. In: Proceedings of the FREENIX Track: 2001 USENIX Annual Technical Conference, Berkeley, CA, USA, USENIX Association (2001) 29–42
23. Smalley, S., Fraser, T., Vance, C.: Linux security modules: General security hooks for Linux (2003) `http://lsm.immunix.org/docs/overview/linuxsecuritymodule.html`.
24. Spencer, R., Smalley, S., Loscocco, P., Hibler, M., Andersen, D., Lepreau, J.: The Flask Security Architecture: System Support for Diverse Security Policies. In: Proc. 8th USENIX Security Symposium, Washington, DC (1999)

Mapping Heavy Communication Workflows onto Grid Resources Within an SLA Context

Dang Minh Quan

Paderborn Center of Parallel Computing, University of Paderborn, Germany

Abstract. Service Level Agreements (SLAs) are currently one of the major research topics in Grid Computing. Among many system components for supporting SLA-aware Grid jobs, the SLA mapping mechanism receives an important position. It is responsible for assigning sub-jobs of the workflow to Grid resources in a way that meets the user's deadline and as cheap as possible. With the distinguished workload and resource characteristics, mapping a heavy communication workflow within SLA context defines new problem and needs new method to be solved. This paper presents the mapping algorithm, which can cope with the problem. Performance measurements deliver evaluation results on the quality and efficiency of the method.

1 Introduction

Mapping and running jobs on suitable resources are the core tasks in Grid Computing. With the case of Grid-based workflows, where a single job is divided into several sub-jobs, the majority of efforts for this issue concentrate on finding a mapping solution in best effort manner[1,2,3]. In the SLA (Service Level Agreement) context, where resources are reserved to ensure the Quality of Service (QoS), mapping a workflow requires different mechanism. The literature recorded some proposed solutions for this problem in [4,5,6]. Most of the proposed mechanisms suppose a workflow including many sub-jobs, which are sequent programs, and a Grid service having ability to handle one sub-job at a time. This is not sufficient enough as sub-jobs in many existed workflows [7,8,9] are parallel programs, and many High Performance Computing Centers (HPCCs) provide computing service under single Grid service [10]. It is obvious that a HPCC can handle many sub-jobs, which can be either sequent programs or parallel programs, at a time. Moreover, all of them did not consider the case of having heavy communication among sub-jobs in the workflow. This paper, which is a continuous work in a series of efforts supporting SLA for the Grid-based workflow [12,13,14], will present a mechanism to handle all stated drawbacks.

1.1 Workflow Model

Like many popular systems handling Grid-based workflow [1,2,3], we also suppose Directed Acyclic Graph (DAG) form of the workflow. User describes the specification about the required resources to run sub-jobs, data transfer among

M. Gerndt and D. Kranzlmüller (Eds.): HPCC 2006, LNCS 4208, pp. 727–736, 2006.

sub-jobs, the estimated runtime of sub-jobs and impose the expected runtime of the whole workflow. User wants the system to finish running the whole workflow in time. In the scope of this paper, the time is computed in slot. Each slot equals with a specific period of real time. Figure 1 presents a concrete Grid workflow. Each sub-job of the workflow has different resource requirements as described in table 1.

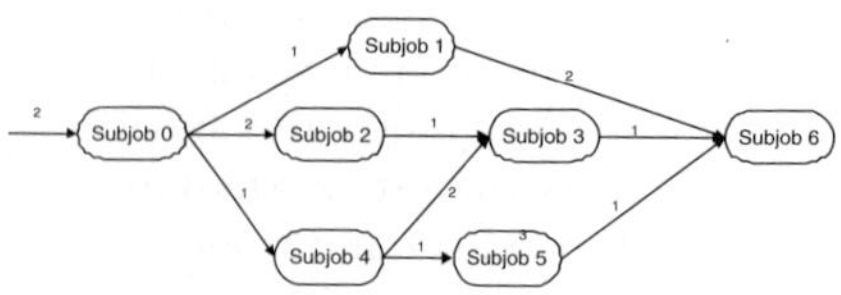

Fig. 1. A sample workflow

Table 1. Resource requirements for sub-jobs

Sj_ID	CPU	Storage	exp	runtime
0	18	59	1	6
1	16	130	3	8
2	20	142	4	5
3	22	113	4	8
4	18	174	2	9
5	20	97	3	14
6	16	118	1	4

It is noted that a sub-job of the workflow can be either a single program or a parallel program and the data to be transferred among sub-jobs is very large, usually in GB scale. The case of light communication among sub-jobs of the workflow was handled in [14]

1.2 Grid Service Model

The computational Grid includes many High Performance Computing Centers (HPCCs). We believe that only HPCCs have enough conditions to support SLA for sub-jobs of the workflow. The resources in each HPCC are managed by software called local Resource Management System (RMS). In this paper, the acronym RMS is used to represent the HPCC as well as the Grid service of that HPCC. Each RMS has its own resource configuration and this configuration is usually different from other RMSs. Those differences include number of CPU, number of memory, storage capacity, software, expert, service price, etc. To ensure that the sub-job can be executed within a dedicated time period, the RMS must support advance resource reservation, for example CCS [10]. Figure 2 depicts a sample CPU reservation profile in such RMS. Queuing-based RMSs are not suitable for our requirement, as no information about the starting time is provided. In our system, we reserve three main types of resource: CPUs, storages and experts. An extension to other devices is straightforward.

If two sequent sub-jobs are executed in the same RMS, it is not necessary to do data transfer task and the time used for this task equal to 0. Otherwise, the data transfer task must be performed. To make sure that a specific amount of data will be transferred within a specific period of time, the bandwidth must also be reserved. Unfortunately, up to now, there is no mechanism responsible for that task in the worldwide network. Here, to overcome that elimination, we use central broker mechanism. The link bandwidth between two local RMSs is determined

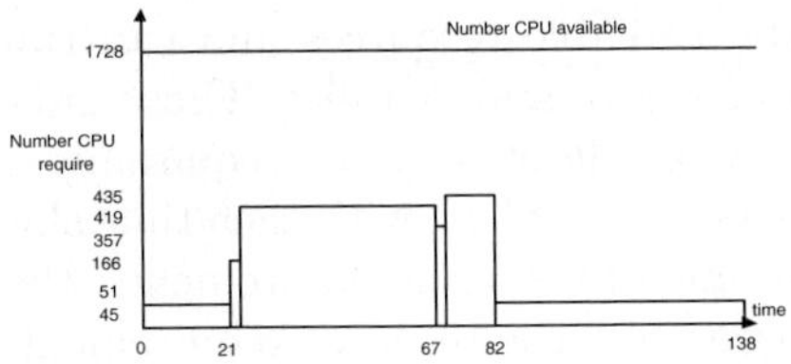

Fig. 2. A sample CPU reservation profile of a local RMS

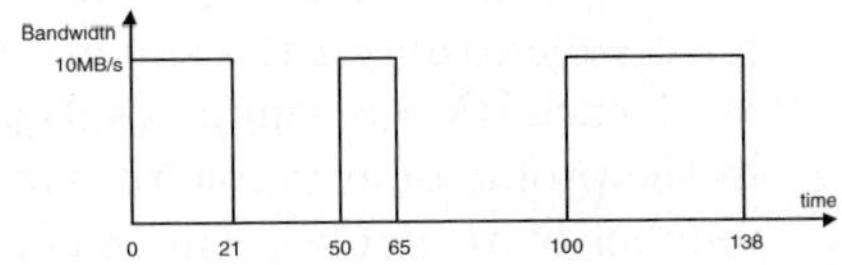

Fig. 3. A sample bandwidth reservation profile of a link between two local RMSs

as the average bandwidth between two sites in the network. Whenever having a data transfer task on a link, the SLA broker will determine which time slot is available for that task. During that specified period, the task can use the whole bandwidth and other tasks must wait. Using this principal, the bandwidth reservation profile of a link will look similar to the one as depicted in Figure 3. A more correctly model with bandwidth estimation [11] can be used to determine the bandwidth within a specific time period instead of the average value. In both cases, the main mechanism is unchanged.

1.3 Mapping Mechanism Requirement

The formal specification of the described problem includes following elements:

- Let R be the set of Grid RMSs. This set includes a finite number of RMSs, which provide static information about controlled resources and the current reservations/assignments.
- Let S be the set of sub-jobs in a given workflow including all sub-jobs with the current resource and deadline requirements.
- Let E be the set of data transfer in the workflow, which express the dependency between the sub-jobs and the necessity for data transfers between the sub-jobs.
- Let K_i be the set of resource candidates of sub-job s_i. This set includes all RMSs, which can run sub-job s_i, $K_i \subset R$.

Based on the given input, a feasible and possibly optimal solution is sought, which allows the most efficient mapping of the workflow in a Grid environment with respect to the given global deadline. The required solution is a set defined as

$$M = \{(s_i, r_j, start\_slot) | s_i \in S, r_j \in K_i\} \quad (1)$$

A feasible solution must satisfy following conditions:

- The total runtime period of the workflow must be within the expected period given by user.
- All $K_i \neq \emptyset$. There is at least one RMS in the candidate set of each sub-job.
- The dependencies of the sub-jobs are resolved and the execution order remains unchanged.

– Each RMS provides a profile of currently available resources and can run many sub-jobs of a single flow both sequentially and parallel. Those sub-jobs, which run on the same RMS, form a profile of resource requirement. With each RMS r_j running sub-jobs of the Grid workflow, with each time slot in the profile of available resources and profile of resource requirements, the number of available resources must be larger than the resource requirement.

In the next phase the feasible solution with the lowest cost is sought. The cost of a Grid workflow is defined as a sum of four factors: money for using CPU, money for using storage, cost of using experts knowledge and finally money for transferring data between the involved resources. If two sequent subjobs run on the same RMS, the cost of transferring data from the previous subjob to the later subjob is neglected. It can be shown easily that the optimal mapping of the workflow to Grid RMS with cost optimizing is a NP hard problem.

2 Related Work

In two separated works [5,6], Zeng et al and Iwona et al built systems to support QoS features for Grid-based workflow. In their work, a workflow includes many sub-jobs, which are sequent programs, and a Grid service has ability to handle one sub-job at a time. To map the workflow on to the Grid services, they used Integer Programming method. Applying Integer Programming to our problem faces many difficulties. The first is the flexibility in runtime of the data transfer task. The time to complete data transfer task depends on the bandwidth and the reservation profile of the link, which varies from link to link. The variety in completion time of data transfer task makes the constraints presentation very complicated. The second is that a RMS can handle many parallel programs at a time. Thus, presenting the constraints of profile resource requirement and profile of resource available in Integer Programming is very difficult to perform.

With the same resource reservation and workflow model, we proposed an algorithm which mapping a light communication workflow to Grid resources in [14]. The proposed algorithm uses Tabu search to find the best possible assignment of sub-jobs to resources. In order to shorten the computation time caused by the high number of resource profiles to be analyzed and by the flexibility while determining start and end times for the sub-jobs, several techniques for reducing the search space are introduced. However, these techniques cannot be applied to solve the problem in this paper because of different workload context.

Metaheuristics such as GA, Simulated Annealing [15], etc were proved to be very effective in mapping, scheduling problems. McGough et al also use them in their system [4]. However, in our problem, with the appearance of resource profiles, the evaluation at each step of the search is very hard. If the problem is big with highly flexible variable, the classical searching algorithms need very long time to find a good solution. In the scope of this paper, we apply several standard Metaheuristics to our problem as means of comparing.

3 Planning Algorithm for Heavy Communication Workflows

The input of the mapping procedure includes information about workflow and information about RMSs. Information about workflow is provided in a file describing sub-jobs and a file describing the dependence. Information about RMSs is stored in a relational database. They include the description of the resource configuration in each RMS, the resource reservation profile of each RMS and the bandwidth reservation profile of each link. The information is collected from RMSs by the monitoring module. Based on this information, the system will do mapping. The overall mapping mechanism, which is called H-Map, is presented in Figure 4.

1. Determine candidate RMSs for each sub-job.
2. Determine assigning sequence for all sub-jobs of the workflow
3. Generate reference solution set
4. With each solution in reference set
Use specific procedure to improve the solution as far as possible
5. Pick the solution with best result

Fig. 4. Mapping mechanism overview

3.1 Determining Candidate RMSs for Each Sub-job

Each sub-job has different resource requirement about type of RMS, type of CPU, etc. There are a lot of RMSs with different resource configuration. This phase finds among those heterogeneous RMSs the suitable RMSs, which can meet the requirement of each sub-job. Each resource parameter of an RMS is represented by number value and is stored in a separate column in the database table. For example, with the parameter Operating System, Linux, Sun, Window, Unix are represented by value number 1, 2, 3, 4 respectively. The co-relative resource requirement parameter of a sub-job is also represented by number value in the same manner. Thus, the matching between sub-job's resource requirement and RMS's resource configuration is done by several logic checking conditions in the WHERE clause of the SQL SELECT command.

3.2 Determining the Assigning Sequence of the Workflow

When the RMS to execute each sub-job, the bandwidth among sub-jobs was determined, the next task is determining time slot to run sub-job in the specified RMS. At this point, the assigning sequence of the workflow becomes important. The sequence of determining runtime for sub-jobs of the workflow in RMS can also affect the total runtime especially in the case of having many sub-jobs in the same RMS.

In general, to ensure the integrity of the workflow, sub-jobs in the workflow must be assigned basing on the sequence of the data processing. However, that principle does not cover the case of a set of sub-jobs, which have the same priority in data sequence and do not depend on each other. To examine the problem, we determine the earliest and the latest start time of each sub-jobs of the workflow in ideal condition. The time period to do data transfer among sub-jobs is computed by dividing the amount of data to a fix bandwidth. The earliest and latest start, stop time for each sub-job and data transfer depends only to the workflow topology and the runtime of sub-jobs but not the resources context. Those parameters can be determined by using conventional graph algorithms. A sample of those data for the workflow in Figure 1, in which the number above each link represents number of time slots to do data transfer, is presented in Table 2.

```
assign_number of each candidate RMS =0
While m_size < max_size {
  clear similar set
  foreach sub-job in the workflow {
    foreach RMS in the candidate list {
      foreach solution in similar set {
          if solution contains sub-job:RMS
            num_sim++
          store tuple (sub-job, RMS, num_sim) in
          a list }}
    sort the list
    pick the best result
    assign_number++
    If assign_number > 1
      find defined solution having the same
      sub-job:RMS and put to similar set
}}
```

Fig. 5. The algorithm generate reference set

Table 2. Valid start time for sub-jobs of workflow in Figure 1

Sub-job	Earliest start	Latest start
0	0	0
1	7	22
2	8	17
3	18	23
4	7	7
5	17	17
6	32	32

The ability of finding a suitable resource slot to run a sub-job depends on number of resource free during the valid running period. From the graph, we can see sub-job 1 and sub-job 4 having the same priority in data sequence. However, from the data in table 2, sub-job 1 can start at max time slot 22 while sub-job 4 can start at max time slot 7 without affecting the finished time of workflow. Suppose that two sub-jobs are mapped to run in the same RMS and the RMS can run one sub-job at a time. If sub-job 4 is assigned first at time slot 7, sub-job 1 will be run from time slot 16 thus the workflow will not be late. If sub-job 1 is assigned first, in the worse case at time slot 22, sub-job 4 can be run at time slot 30 and the workflow will late 23 time slots. Here we can see, the latest time factor is the main parameter to evaluate the full affection of the sequence assignment decision. It can be seen through the affection, mapping the sub-job having smaller latest start time first will make the latency smaller. Thus, the latest start time value determined as above can be used to determine the assigning sequence. The sub-job having smaller latest start time will be assigned earlier.

3.3 Generating Reference Solution Set

A solution is found by determining each sub-job of the workflow run by which RMS. We do not consider time factor in this phase so a reference solution is defined as a set of map sub-job:RMS with all sub-jobs in the workflow. Each solution in the reference solutions set can be thought as the starting point for local search so it should be spread as wide as possible in the searching space. To satisfy the space spreading requirement, number of the same map sub-job:RMS between two solutions must be as small as possible. The number of member in the reference set depends on the number of available RMSs and number of sub-jobs. During the process of generating reference solution set, each candidate RMS of a sub-job has a co-relative *assign_number* to count the times that RMS is assigned to the sub-job. During the process of building a reference solution, we use a similar set to store all defined solution having at least a map sub-job:RMS similar to one in the creating solution. The algorithm is defined in Figure 5.

While building a solution, with each sub-job in the workflow, we select the RMS in the set of candidate RMSs, which creates minimal number of similar sub-job:RMS with other solutions in the similar set. After that, we increase the *assign_number* of the selected RMS. If this value larger than 1, which means that the RMS were assigned to the sub-job more than one time, there must exist solutions that contains the same sub-job:RMS and thus satisfying the similar condition. We search those solutions in the reference set, which have not been in the similar set, and then add them to similar set. When finished, the solution is put to the reference set. After all reference solutions are defined, we use a specific procedure to refine each of the solution as far as possible.

3.4 Improving Solution Quality Algorithm

Before improving the quality of the solution, we have to determine specific runtime period for each sub-job and each data transfer task as well as the makespan of the present solution. The start time of a data transfer task depends on the finish time of the source sub-job and the state of the link's reservation profile. We use *min_st_tran* variable to present the dependence on the finish time of the source sub-job. The start time of a sub-job depends on the latest finish time of the related data transfer tasks and the state of the RMS's reservation profile. We use *min_sj_tran* variable to present the dependence on the latest finish time of the related data transfer tasks. The task to determine timetable for the workflow is done with the procedure in Figure 6.

For each sub-job of the workflow in the assigning sequence, firstly, we find all the runtime period of data transfer task from previous sub-jobs to current sub-job. This period must be later than the finish time of the source sub-job. Note that with each different link the transfer time is different because of different bandwidth. Then, we determine the runtime period of the sub-job itself. This period must be later than the latest finish time of previous related data transfer task. The whole procedure is not so complicate but time consuming. The time consuming steps are the searching reservation profiles, and they make the whole procedure long time consuming.

```
foreach sub-job k following the assign sequence {
  foreach link from determined sub-jobs to k{
    min_st_tran=end_time of source sub-job
    search reservation profile of link the
    start_tran > min_st_tran
    end_tran = start_tran+num_data/bandwidth
    store end_tran in a list
  }
  min_st_sj=max (end_tran)
  search in reservation profile of RMS running
  k the start_job > min_st_sj
  end_job= start_job + runtime
}
```

Fig. 6. Algorithm determine timetable for workflow

```
while (num_loop < max_loop) {
  foreach subjob in the workflow {
    foreach RMS in the candidate list {
      if cheaper then put (sjid, RMS id, improve_value)
      to a list }}
  sort the list according to improve_value
  from the begin of the list{
    Compute time table to get the finished time
    If finished time < limit
      break
  }
  Store the result
  num_loop ++;
}
```

Fig. 7. Procedure to improve the solution quality

The overall of solution quality improvement procedure is described in Figure 7. In one iteration, we can move only one sub-job to one RMS with the hope to decrease the cost. So we only consider the move, which can decrease the cost. With each solution we compute the time table, if it satisfies the deadline then update the result.

4 Performance Evaluation

Performance experiment is done with simulation to check for the quality of the mapping algorithms. The hardware and software used in the experiments is

Table 3. Experiment results of the H-Map algorithm

	H-Map		SA		ILS		GLS		GA		EDA	
Wf	Rt	Cost	Rt	Cost	Rt	Cost	Rt	Cost	Rt	Cost	Rt	Cost
1	1	632.58	105	632.58	21	632.58	14	632.58	25	632.58	62	632.58
2	0.5	756.96	84	756.90	37	756.96	19	756.90	22	756.96	88	756.90
3	1	780.42	114	780.42	54	780.42	29	780.34	27	780.34	126	780.34
4	1	836.18	78	836.18	81	836.18	46	836.18	28	836.18	178	836.18
5	1	892.02	105	892.02	114	892.02	59	892.02	29	892.21	241	892.02
6	2	948.10	90	948.10	147	948.10	80	948.10	36	1005.27	390	947.86
7	1	1003.7	78	1003.99	201	1003.7	98	1003.99	36	1075.19	462	1003.7
8	2	1059.89	121	1059.89	250	1059.89	127	1059.89	32	1059.89	558	1059.89
9	2	1184.21	130	1184.21	307	1184.21	167	1184.21	44	1248.86	659	1183.92
10	2	1308.53	146	1332.53	398	1308.53	187	1308.53	47	1383.53	680	1308.53
11	3	1364.14	124	1376.42	502	1364.14	222	1377.63	52	1440.81	956	1364.42
12	2	1488.12	184	1551.74	462	1521.74	303	1501.95	51	1569.39	854	1536.74
13	7	1512.26	174	1512.26	620	1512.26	354	1566.09	56	1620.17	1136	1512.26
14	3	1567.74	162	1631.15	815	1567.74	392	1568.15	56	1663.81	1255	1601.15
15	6	1591.12	161	1675.67	876	1591.12	524	1621.67	70	1764.18	1663	1621.67
16	5	1786.56	180	1871.81	1394	1840.31	763	1843.55	85	1914.91	2845	1830.87
17	7	1889.78	197	1960.87	1695	1892.27	1258	1936.83	93	2028.06	4170	1961.30
18	10	2217.34	272	2276.33	2046	2283.67	1623	2256.53	1953	2406.97	10976	2276.33

rather standard and simple (Pentium 4 2,8Ghz, 2GB RAM, Linux Redhat 9.0, MySQL). The whole simulation program is implemented in C/C++. The goal of the experiment is to measure the feasibility, the quality of the solution and the time needed for the computation. To do the experiment, 18 workflows with different topologies, number of sub-jobs, sub-job specifications, amount of data transferring were generated and mapped to 20 RMSs with different resource configuration and different resource reservation context by 6 algorithms H-Map, Simulated Annealing (SA), Guided Local Search (GLS), Iterated Local Search (ILS), Genetic Algorithm (GA), Estimation of Distribution Algorithm (EDA) [15]. The implementation of those algorithms is described in [16]. The final result of the experiment is presented in table 3 with column Wf (Workflow) presents the id of workflows, column Rt (Runtime) and Cost record the cost and runtime of solutions generated by each algorithm correlative with each workflow respectively.

The experiment results show that H-Map algorithm finds out higher quality solution with much shorter runtime than other algorithms in most cases. Some of the metaheuristics such as ILS, GLS, EDA find out equal results with small problems. But with big problem, they have exponent runtime and find out unsatisfied results.

5 Conclusion

This paper has presented a method, which performs an efficient and precise assignment of heavy communication workflow to Grid resources with respect to SLAs defined deadlines and cost optimization. In our work, the distinguished character is that a sub-job of the workflow can be a sequent or parallel program and a Grid service can handle many sub-jobs at a time. The performance evaluation showed that the proposed algorithm creates solution of equal or better quality than most standard metaheuristics and needs significantly shorter computation time. The latter is a decisive factor for the applicability of the method in real environments, because large-scale workflows can be planned and assigned efficiently.

References

1. E. Deelman, J. Blythe, Y. Gil, C. Kesselman, G. Mehta, S. Patil, M. Su, K. Vahi, and M. Livny, "Pegasus : Mapping Scientific Workflows onto the Grid", Proceedings of the 2nd European Across Grids Conference, Nicosia, Cyprus, January 28-30, 2004.
2. D. P. Spooner, S. A. Jarvis, J. Cao, S. Saini and G. R. Nudd, "Local Grid Scheduling Techniques Using Performance Prediction", IEEE Proceedings - Computers and Digital Techniques, 150(2), pp. 87–96, 2003.
3. R. Lovas, G. Dzsa, P. Kacsuk, N. Podhorszki, D. Drtos, "Workflow Support for Complex Grid Applications: Integrated and Portal Solutions", Proceedings of 2nd European Across Grids Conference, Nicosia, Cyprus, 2004.

4. S. McGough, A. Afzal, J. Darlington, N. Furmento, A. Mayer, and L. Young ,Making the Grid Predictable through Reservations and Performance Modelling, The Computer Journal, v.48 n.3, pp. 358–368, 2005
5. L. Zeng, B. Benatallah, A. Ngu, M. Dumas, J. Kalagnanam, H. Chang, QoS-Aware Middleware for Web Services Composition, IEEE Transactions on Software Engineering, v.30 n.5, pp. 311–327, may 2004
6. I. Brandic and S. Benkner and G. Engelbrecht and R. Schmidt, QoS Support for Time-Critical Grid Workflow Applications, Proceedings of e-Science 2005
7. S. Ludtke, P. Baldwin, and W. Chiu, "EMAN: Semiautomated Software for High-Resolution Single-Particle Reconstructio" , Journal of Structure Biology, v. 128, 1999.
8. G. B. Berriman, J. C. Good, A. C. Laity, "Montage: a Grid Enabled Image Mosaic Service for the National Virtual Observatory" , ADASS, v. 13, 2003.
9. L. Richter, "Workflow Support for Complex Grid Applications: Integrated and Portal Solutions", Proceedings of the 2nd European Across Grids Conference, 2004.
10. L. Burchard, M. Hovestadt, O. Kao, A. Keller, and B. Linnert, "The Virtual Resource Manager: An Architecture for SLA-aware Resource Management", Proceedings of the IEEE CCGrid 2004, IEEE Press, pp. 126–133, 2004.
11. R. Wolski, "Experiences with Predicting Resource Performance On-line in Computational Grid Settings", ACM SIGMETRICS Performance Evaluation Review, v. 30 n. 4, pp. 41–49, 2003.
12. D.M. Quan, O. Kao, "SLA negotiation protocol for Grid-based workflows", Proceedings of the International Conference on High Performance Computing and Communications (HPPC-05), LNCS 3726, pp. 505–510, 2005.
13. D.M. Quan, O. Kao, "On Architecture for an SLA-aware Job Flows in Grid Environments", Proceedings of the 19th IEEE International Conference on Advanced Information Networking and Applications (AINA 2005) , IEEE Press , pp. 287–292, 2005.
14. D.M. Quan, O. Kao, "Mapping Grid job flows to Grid resources within SLA context", Proceedings of the European Grid Conference,(EGC 2005), LNCS 3470, pp. 1107–1116, 2005.
15. C. Blum, A. Roli, "Metaheuristics in combinatorial optimization: Overview and conceptual comparison", ACM Computing Surveys, v. 35 n.3, pp. 268–308, 2003
16. D.M. Quan, "A Framework for SLA-aware execution of Grid-based workflows", PhD thesis, University of Paderborn - Germany, 2006.

The SLA-Compatible Fault Management Model for Differentiated Fault Recovery

Keping Long[1], Xiaolong Yang[1,2,*], Sheng Huang[2], Xin Yang[2], and Yujun Kuang[1]

[1] Research Centre for Optical Internet and Mobile Information Networks, University of Electronic Science and Technology of China, Chengdu 610054, China
[2] Chongqing University of Posts and Telecommunications, Chongqing 400065, China
{longkp, yangxl, huangs}@cqupt.edu.cn

Abstract. The paper proposes a SLA-compatible fault management model for differentiated fault recovery after introducing SLA, service-differentiated and existing fault recovery techniques, and gives the basic ideas, detailed processing schemes and consideration of this model.

Keywords: Service Level Agreement(SLA), Fault Recovery, Quality of Recovery.

1 Introduction

With the emergence of multifarious new Internet applications and the increasing of diversified requirements of various customers, the capability to offer differentiated services is considered as one of the key faculties of ISP networks. In addition, service differentiation is a valuable opportunity for operators to increase their income from their infrastructure, by selling high added-value services. So, how to provide and manage differentiated services in single network becomes more and more significant, the appropriate and effective model and schemes for service management are very interesting. At this point, service level agreement (SLA) based differentiated service management is a good candidate.

A service level agreement (SLA) [1, 2] is a formal contract between service provider and its subscriber that contains the rights, obligation and punishment of both sides during service provisioning, it has became to the most prevalent evaluation criterion for telecommunication services in recent years. In a SLA, performance specifications are the most important components, and these performance specifications are described by service level specifications (SLSs). An SLS is a set of specific performance parameters and their values that together define the service offered to a traffic stream in a network. These performance parameters refer to service availability, reliability, stability, fault recovery and so on. The performance commitments of fault recovery are indispensable in every SLA. So, differentiated fault recovery for different customers according to their demands (SLA) is very

[*] Contact Author: Xiaolong Yang (E-mail: yxl@uestc.edu.cn, Tel: +86-28-8320-7895, Fax: +86-28-8320-7885) is currently with University of Electronic Science and Technology of China, Chengdu 610054, China.

M. Gerndt and D. Kranzlmüller (Eds.): HPCC 2006, LNCS 4208, pp. 737 – 742, 2006.

necessary and significant. In this article, we propose a SLA-compatible fault management model for differentiated fault recovery, describe its detailed mechanisms and procedures, and give some operations and concepts for the management and maintenance of network resources and information databases.

2 Fault Recovery Schemes

In multi-layer networks, there have two types of existing fault recovery mechanisms: single-layer recovery and multi-layer recovery, according to the cooperative relationship among independent layers. Single-layer fault recovery schemes are shown in Fig.1, it has two main types - protection and restoration. According to back-up resources' sharing relationship and recovery granularity, protection schemes can be classified as dedicated protection, shared protection, link protection, path and sub-path protection. And link, sub-path and path restoration are general approaches of restoration schemes [3].

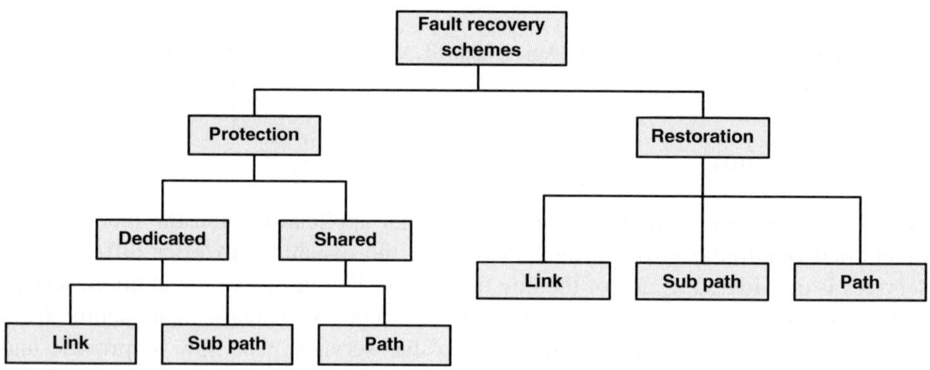

Fig. 1. Single-layer fault recovery schemes

In existing multi-layer recovery mechanisms, sequential recovery and integration recovery are the two primary recovery approaches. The sequential multi-layer recovery has three main schemes: bottom-up, top-down and diagnostic [4, 5]. Hold-off timer and recovery token are the two typical recovery schemes of bottom-up recovery mechanism.

In practical networks, each above fault recovery scheme has its own advantages and disadvantages in recovery time, recovery success rate or optimal utilization of resources.

3 The SLA-Compatible Differentiated Fault Management Model

Failure is unavoidable in any networks, so fault recovery is required. How to satisfy various recovery demands of customers while optimizing the use of network resources in current network infrastructure is the most concerned thing of service providers. In this paper, we will propose a SLA-compatible fault management model and mechanisms to support differentiated fault recovery. The main ideas of our fault management model are: 1) Differentiated fault recovery treatment, select appropriate recovery schemes for each interrupted service flow according to its SLA, these selected recovery schemes can be single-layer or multi-layer. 2) The recovery performance is measurable and controllable, and each recovery processing is monitored on-line. The matched recovery schemes of each service can be adjusted or re-matched if the actual measured performance can't satisfy the committed one. 3) Differentiated resources manage and assignment, differentiated resources assignment schemes (like preemption) will be used when there has a competition of recovery resources. 4) All SLA and recovery mechanism are administrable and operable, including add, delete, modify and mapping. Fig.2. shows the architecture and main components of our model. There are two planes: service plane and manage plane. Service plane is consists of ISP service networks and various clients and their equipments, this plane in charge of the path provisioning and transmission for customer's traffic. Manage plane has several important components including SLA and Fault management processor, Manage information databases (SLA and recovery schemes), performance measurer and monitor. These components work together to provide an intelligent platform for the management and control of SLA, fault recovery schemes and recovery processing.

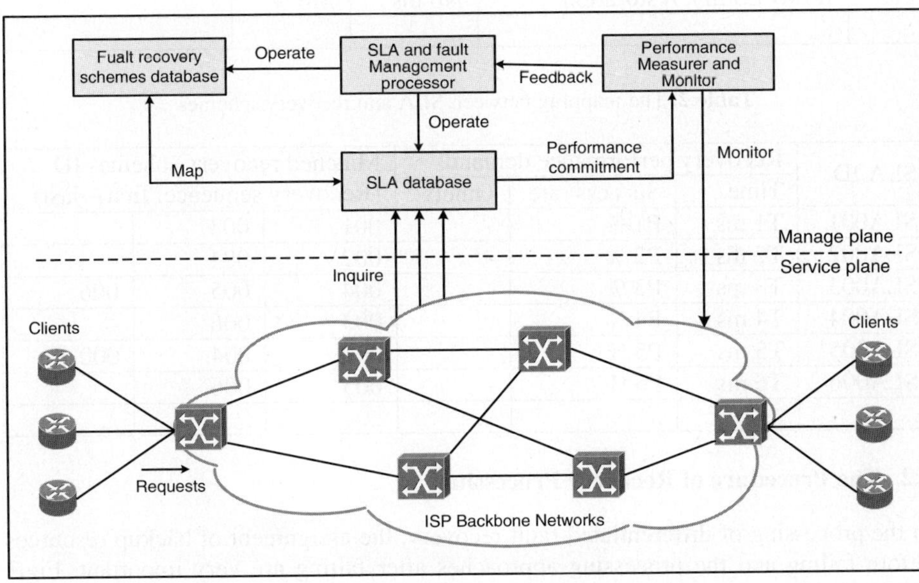

Fig. 2. Architecture of proposed fault management model

3.1 The Mapping of SLA and Fault Recovery Schemes

In order to provide differentiated fault recovery based on SLA demands, there must has mapping relationships between SLAs and fault recovery schemes. In our fault management model, statistic performances of recovery schemes and are maintained in fault manage database, like Table 1 shows (N, N1, … Nn > 0; 0 < M, M1, ... Mn <

100). According to the performance commitments of each SLA, one or more appropriate fault recovery schemes will be selected to match this SLA, and these mapping information are maintained in SLA database. When a failure occurs, the interrupted service will be recovered by using the corresponding matched recovery schemes. A typical mapping relationship between SLAs and fault recovery schemes are shown in Table 2, "Matched recovery schemes ID" matches the recovery schemes' number in table 1.

Table 1. Performances of various recovery schemes

No.	Name	Recovery time	Success rate	Backup resources
000	Best-effort restoration	N ms	M %	-
001	Optical 1:1 dedicated protection	N1 ms	M1 %	100%
002	Optical M:N shared protection	N2 ms	M2 %	(100N/M)%
003	IP/MPLS 1:1 path protection	N3 ms	M3 %	100%
004	IP/MPLS path restoration	N4 ms	M4 %	-
005	IP/MPLS sub-path restoration	N5 ms	M5 %	-
006	IP/MPLS link restoration	N6 ms	M6 %	-
...	...	...	...	...

Table 2. The mapping between SLA and recovery schemes

SLA ID	Recovery performance demands			Matched recovery schemes ID (Recovery sequence: first→last)		
	Time	Success rate	Others			
SLA001	T1 ms	P1 %	-	001	003	-
SLA002	T2 ms	P2 %	-	002	004	-
SLA003	T3 ms	P3 %	-	004	005	006
SLA004	T4 ms	P4 %	-	003	006	-
SLA005	T5 ms	P5 %	-	001	004.	000
SLA006	T6 ms	P6 %	-	005	006	-
...	...	...	...	...	...	...

3.2 The Procedure of Recovery Processing

In the processing of differentiated fault recovery, the assignment of backup resources before failing and the processing approaches after failing are very important. Fig.3 shows the flow chart of path provisioning before failing and Fig.4 shows the flow chart of fault recovery after failing.

3.3 Differentiated Assignment of Recovery Resources

Resources competition may occur during the processing of path provisioning and fault recovery while network's traffic load is serious. In this situation, our model provides a mechanism try to achieve differentiated use and assignment of recovery resources.

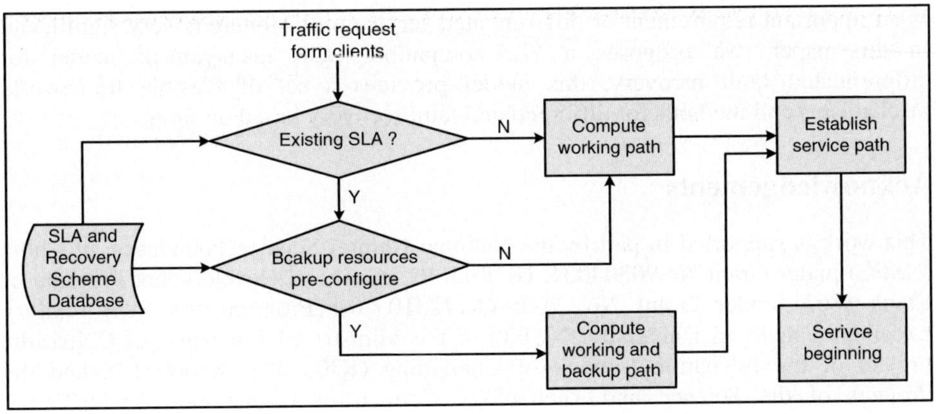

Fig. 3. The flow chart of path provisioning

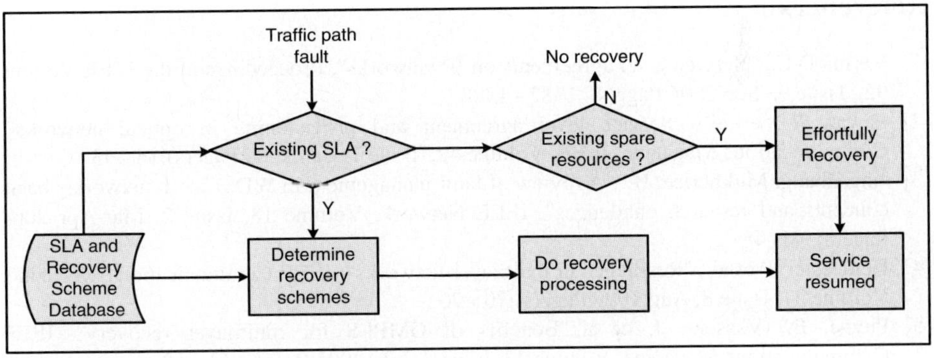

Fig. 4. The flow chart of fault recovery

The main ideas of this mechanism are: 1) The interrupted service whose SLA has more strict performance demands and higher price has the priority of using resources. 2) If recovery resources is deficient, the high-priority interrupted service can preempt the resources belongs to the lower-priority one.

3.4 The Management and Maintenance of Databases

There are two databases in our model: SLA database and recovery schemes database. SLA's attributes, performance specifications and mapping relationships with recovery schemes are kept in SLA database. And recovery performances of various fault recovery schemes are stored in recovery schemes database. These two databases are administrable and operable, manual or intelligent.

4 Conclusions

Nowadays, models and approaches for differentiated service provisioning are widely concerned by research institutes and service providers. Differentiated fault recovery as an important requirement of differentiated service provisioning is very significant. In this paper, we proposes a SLA-compatible fault management model for differentiated fault recovery, this model provides a set of feasible framework, mechanisms and methods for differentiated fault recovery based on SLA.

Acknowledgements

This work is supported in part by the National Natural Science Foundation of China (NSFC) under Grant No.90304004, Hi-Tech Research and Development Program of China (863) under Grant No. 2005AA122310, the Program for New Century Excellent Talents in University (NCET) of the Ministry of Education of China, the Project of the Education Council of Chongqing (KJ060508, KJ060513), and the Projects of the Science and Technology Council of Chongqing (2005BB2062, 2005AC2089).

References

[1] Verma D.C.; "Service level agreements on IP networks", Proceedings of the IEEE Volume 92, Issue 9, Sep 2004 Page(s): 1382 - 1388

[2] Fawaz, W. et al., "Service level agreement and provisioning in optical networks", Communications Magazine, IEEE, Volume 42, Issue 1, Jan 2004 Page(s):36 - 43

[3] Jing Zhang; Mukheriee, B.; "A review of fault management in WDM mesh networks: basic concepts and research challenges", IEEE Network, Volume 18, Issue 2, Mar-Apr 2004 Page(s): 41- 48

[4] Demeester, P et al., "Resilience in multilayer networks", IEEE Communications Magazine, Volume 37, Issue 8, Aug.1999 Page(s):70 - 76

[5] Puype, B., Vasseur, J. et al.,"Benefits of GMPLS for multilayer recovery", IEEE Communications Magazine, Volume 43, Issue 7, July 2005 Page(s):51 - 59

Towards SLA-Supported Resource Management

Peer Hasselmeyer[1], Bastian Koller[2], Lutz Schubert[2], and Philipp Wieder[3]

[1] C&C Research Laboratories, NEC Europe Ltd., 53757 Sankt Augustin, Germany
hasselmeyer@ccrl-nece.de
[2] Höchstleistungsrechenzentrum Stuttgart,
Allmandring 30, 70550 Stuttgart, Germany
{koller, schubert}@hlrs.de
[3] Research Centre Jülich, 52415 Jülich, Germany
ph.wieder@fz-juelich.de

Abstract. Achievements and experiences in projects with focus on resource management have shown that the goals and needs of High Performance Computing service providers have not or only inadequately been taken into account in Grid research and development. Mapping real-life business behaviour and workflows within the service provider domain to the electronic level implies focusing on the business rules of the provider as well as on the complexity of the jobs and the current state of the HPC system. This paper describes an architectural approach towards a business-oriented and Service Level Agreement-supported resource management, valuable for High Performance Computing providers to offer and sell their services. With the introduction of a Conversion Factory the authors present a component that is able to combine the Service Level Agreement, the system status, and all business objectives of the provider in order to address the business needs of service providers in the Grid.

1 Introduction

Current solutions for resource management of High Performance Computing (HPC) environments were mainly developed neglecting the business needs and goals of service providers. The introduction of Service Level Agreements (SLAs) [1,9] provided an instrument to express business-related terms, but it became clear that configuring the system only according to an SLA does not solve the problem of automatic resource configuration at all.

The main purpose of SLAs is to define certain Quality of Service (QoS) parameters in a way that appropriate service levels can be maintained during interaction with customers. SLAs are therefore an important tool for automatic business enactment and QoS management, in particular monitoring the performance of services and detecting violations. By using SLAs, the customer has a document which states certain quality properties, in most cases bound to penalties for the service provider failing to deliver this quality. But the creation of the SLA, as well as the configuration of the system, is bound to different aspects. The authors aim specifically at developing an architecture for SLA-supported

M. Gerndt and D. Kranzlmüller (Eds.): HPCC 2006, LNCS 4208, pp. 743–752, 2006.

resource management which takes three major aspects into account: (a) The complexity of the job which is bound to (b) the availability of resources on the service provider's system, and all based on (c) the respective *Business Level Objectives* (BLOs) [7,10] of the provider.

Nowadays, the usage of SLAs requires translation of SLA terms. Those terms are mostly stated as high level technical to low level technical terms and they can be used to derive the configuration of the system. SLA-specific configurations have to be based on the Business Level Objectives of the service provider while at the same time they have to respect the current state of the system resources (e.g. the load of the system). This state information is referred to as *infrastructure knowledge* throughout the paper.

A number of approaches towards this goal have already been proposed. Inter alia projects like TrustCoM [12] and NextGRID [11] use SLAs within the service provisioning process, the latter e.g. for automatic generation of SLAs with respect to Business Level Objectives which are translated into policies. Based on this idea, we describe how infrastructure configuration can be automated taking into consideration SLAs as well as the service provider's environment.

The paper starts by describing a business scenario and presents an overview how SLAs are currently handled at a service provider. Section 3 details the relationship between SLA contents and resource configuration. The architectural approach proposed in this paper is presented in Section 4, introducing the different components of this system as well as explaining a generic mapping process. Section 5 provides details on the usage of this approach for Service Level Agreement negotiation as well as for system configuration purposes. Finally, the concluding section discusses limitations and advantages of the proposed architecture, specifically related to potential upcoming implementations of the architecture introduced here. Additionally, references to ongoing research are provided for areas that are not in the focus of our work, but which are of essential importance once the proposed architecture is implemented.

2 Business Scenario

2.1 Description

The approach presented in this paper is based on actual business requirements which we shall exemplify in the context of a scenario taken from the TrustCoM project. In this scenario, an aircraft manufacturer intends to re-design the seats of an existing air plane. As this task involves complex calculations, a third-party HPC service is contracted to take over task-specific computations (i.e. to execute a job). In the given example, the HPC provider allows customers to run computational jobs at a specific Quality of Service. For this kind of providers, configuration and maintenance of the underlying HPC infrastructure is a particularly complex issue – specifically when performed autonomously and when the respective task is unknown to the service provider, i.e. no a priori configuration information exists that could be reused.

2.2 Limitations of Existing Solutions

Existing solutions commonly rely on the HPC provider to negotiate a contract with a consumer, in case of the example with the aircraft manufacturer. If the provider has no experience with that particular job, he has to perform a complexity analysis of the task manually. The authors of this paper are aware of the fact that such complexity analyses are hard to automate. We therefore only sketch an idea of "semi"-automation of such analyses by using a database that stores already calculated complexities (see Section 4). From the complexity information, an administrator has to derive the resource requirements for this task – taking into consideration not only complexity but also provider-specific knowledge about previous executions of a job (if available).

Having the complexity received from an *Analyst*, the provider's *Administrator* can usually (manually) map the requirements of the job to the provider's resources. As an extension to our example of re-designing the air plane's seats, let us assume that the job belongs to a (provider-defined) complexity class C. The provider would use his knowledge of jobs with this complexity and calculate the requirement of having, say, 64 nodes available for 24 hours to execute this kind of job.

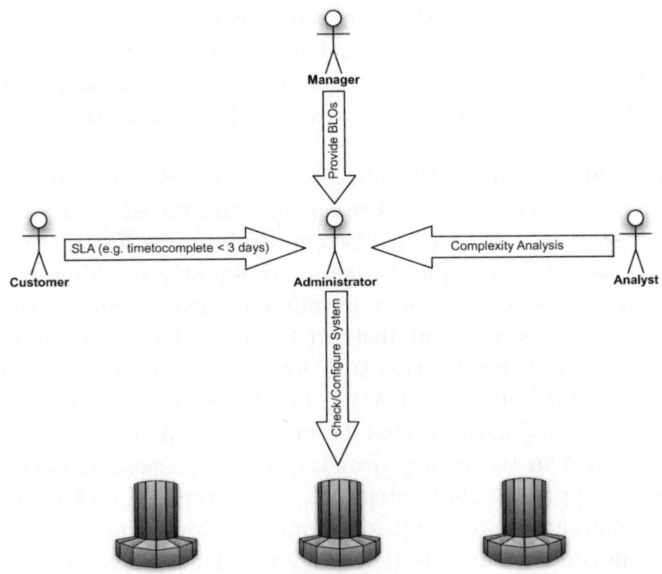

Fig. 1. A human-centric approach to resource configuration at service providers

It is important to mention that this kind of calculation reflects only the theoretical approach towards the manual mapping. In real business cases, such a calculation is also influenced by the HPC providers policies (mainly the Business Level Objectives). Business Level Objectives are abstract goal definitions of the respective business party, generally defined by a *Manager* before any business is conducted. Having an HPC provider, a BLO could for instance be "*maximise workload on this machine*". The BLOs influence translation from SLAs to configuration parameters, as in addition do information about the system and potential constrains from the customer, who may, for the sake of the example, request the job to be finished within three days. With this information the Administrator can determine whether the "*64 nodes*" requirement can really be fulfilled within the valid time period for this job or, depending on the provider's Business Level Objectives, whether he actually wants to fulfil it. Based on his calculations, the Administrator can then configure the resources.

Current solutions require involvement of different parties to collect the necessary information to execute the job according to the requirements of a customer. Fig. 1 shows the different parties and tasks they carry out as described in this section. It can be seen that this approach lacks automatic mechanisms to carry out the previously outlined tasks.

3 Dependency of Resource Configuration on SLAs

In order to configure his resources in a sensible and efficient way, a service provider has to take different facts into account. One set of aspects is defined by Service Level Agreements. But SLAs are only one part of the whole picture. The configuration of a system depends on other aspects as well: the Business Level Objectives of the service provider, the complexity of the job and the current system status. In the following we discuss how far SLAs influence the configuration of the system.

Terms within SLAs can usually not be used directly for configuring the affected resources. A Service Level Agreement can consist of a set of abstract terms which, unfortunately, mean different things to different providers. The term "performance", for example, is defined differently by different parties and it is therefore calculated and provided in different ways depending on the respective infrastructure. By going from abstract terms to the infrastructure layer, we need a mapping of high level terms to a low (technical) level (resulting in an *Operational Level Agreement* (OLA)). This mapping is inevitable as the service provider has to understand what he needs to provide in order to fulfil the conditions of an SLA in terms of processing power, processing time, etc.

We foresee an approach that integrates infrastructure and business-specific knowledge to allow automatic and autonomous conversion of SLA terms into configuration information. As discussed below, such a conversion process may also support the management of the infrastructure during operation.

SLAs have strong impact on the configuration system, in particular if service provision has to be dynamic and on-demand. TrustCoM and NextGRID currently examine how Service Level Agreements can be created and used in a way that is adequate for both customers and service providers. In these projects SLAs are not negotiable. This simplifies selection and creation of SLAs, by using e.g. a protocol and SLA representation like WS-Agreement [1], and enables the use of databases to store (static) configuration information. However, studies of real business use cases have shown that pre-defined configurations can only be used in a limited set of scenarios, as the configuration of the service provider's system does not only depend on the Service Level Agreements, but also on the job type(s) and the current workload.

Although, at the time of writing, Business Level Objectives like "*utilise my resources a hundred percent*" or "*get the maximum profit, while spending as little money as possible*", are not supported by available resource management systems, current research activities to solve this problem are on-going in NextGRID, TrustCoM, and the upcoming project BREIN [2]. For the purposes of this paper, we assume that BLOs are represented as documents in a database and that they can be retrieved easily. Additionally, a certain degree of knowledge of the service provider's infrastructure is required, e.g. whether and how the service provider can execute the requested jobs in time. In summary, a configuration system that builds on a "base" configuration represented by BLOs and the complexity analysis of a job seems like a promising approach towards SLA-supported resource management.

4 Mapping Service Level to Operational Level Agreements

The outlined scenario has high demands on the HPC provider's capability of configuring its resources on-the-fly. Related to the discussion in the previous section, this one presents our architectural approach to mapping Service Level Agreements to Operational Level Agreements.

4.1 An Architecture for SLA-Supported Resource Management

Current research in the Service Level Agreement area deals mainly with high-level application-specific terms as SLA content (e.g. "time-to-complete" as used in NextGRID). These terms are used since the customer should not know and in most cases does not want to know the details of the service provider's configuration. It is therefore mandatory to map those high level-terms to the low-level technical layer usable for calculating the requirements on the system.

Our architecture (see Fig. 2) is an approach to automate the process of configuring a service provider's system based on:

- the BLOs of the service provider,
- the *complexity* of the job,
- the *configuration knowledge* of former job runs (stating something like: "*for this job, I need the following system resources*"), and
- the Service Level Agreement itself.

The resulting set of OLAs will be the basis for the system configuration. The translation of OLAs to actual system configurations is performed by the *Execution Management System* (EMS). The EMS is not part of our proposed

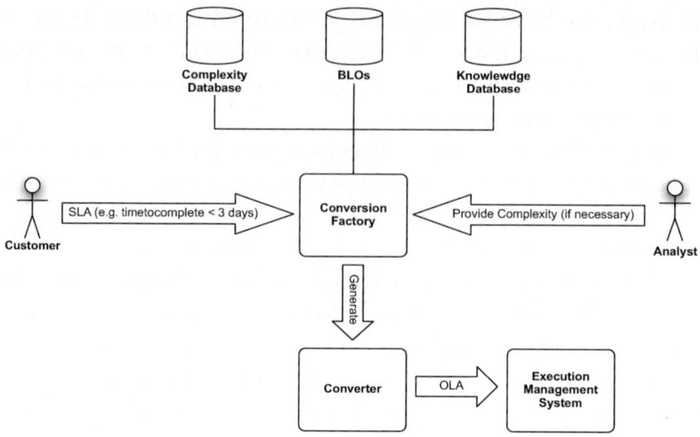

Fig. 2. An architecture for SLA-supported resource management

architecture but the entity responsible for consuming and interpreting the Operational Level Agreements.[1]

The proposed architecture consists of the *Conversion Factory*, the *BLO Database*, the *Knowledge Database*, the *Complexity Database*, and the *Converter.* The main component of our proposed architecture is the Converter. The Converter instance is created based on the Business Level Objectives of the service provider, the complexity of the job and the infrastructure knowledge. Business Level Objectives represent the goals of the service provider. In an ideal world, the service provider's main goal would be to satisfy the customer's needs. But in the business world, both customers as well as service providers have additional internal goals. All goals need to be observed when negotiating and executing a business transaction.

The function of the Converter is to provide the information to (a) check resource availability, (b) configure the system, and (c) monitor the status. Thus, depending on changes of the provider's infrastructure, the Converter needs to perform basically the same task multiple times in order to optimally support a business transaction. In particular monitoring is dependent on the business relation (as it is different for each individual SLA).

We therefore introduce a "Conversion Factory" that instantiates individual Converters. Each Converter is specific to the particular needs of a business relationship. For this purpose the Conversion Factory queries the different databases to retrieve information about potentially available, pre-calculated complexities or previous business transactions. In particular we see the need for a Knowledge Database, a Complexity Database, and a BLO Database. The Knowledge Database stores information about system requirements for different complexity classes, generated by running "similar" jobs in the past. In addition, the Knowl-

[1] The function of the EMS is analogous to that of the OGSA Execution Management Services which "are concerned with problems of instantiating and managing, to completion, units of work" [5].

edge Database in our approach represents an information service which also provides data about the current status of the system. Realisations of the architecture described may obviously choose to implement such a service separately or to exploit the respective service of the middleware in use. The Complexity Database stores previously calculated complexities of jobs; it could, as far as our scenario is concerned, contain the assignment of the complexity class C to the aircraft manufacturer's job. The BLO Database stores the Business Level Objectives of the service provider.

Obviously, the Complexity Database does not contain all possible job-complexity pairs. Complexity calculation of so far "unknown" jobs is therefore needed. Generally, it is impossible to predict the behaviour (and hence complexity) of an unknown application. Potential solutions include letting the job run with a reduced quantity of resources to estimate the time requirements (relative to the system's capabilities) – however, such an approach neglects for example the relationship between input data complexity and time requirements. Manual processing will therefore be needed in most cases. This does nevertheless not diminish the improvements brought about by our Converter.

While the Knowledge Database and the Complexity Database are exploited to reduce the mapping complexity and thus speed up this process, the purpose of the BLO Database is to provide information on the service provider's goals and their relative importance. As mentioned before, BLOs represent the policies to be taken into consideration when generating configuration information. Such priorities could directly influence configurations, e.g. with statements like "*jobs from companies X,Y,Z get higher/lower priority*".

Please note that, as opposed to the Knowledge and Complexity Databases which are ideally populated automatically, the service provider wants to retain control of the definition of its own BLOs. Automatic definition of BLOs does therefore not seem to be a sensible option.

As described, the Converter is created based on the SLA, the Complexity, and the infrastructure knowledge. It provides three interfaces:

1. **Availability.** This interface is used to check the availability of an offered service for negotiation purposes.
2. **Status.** This interface provides monitoring data from the EMS converted to SLA level terms to check the status of the service.
3. **Configuration.** Once an SLA is negotiated, this interface is used to convert the SLA-specific parameters into valid configuration information usable as input to the EMS based on the pre-configured conversion mechanisms.

Once instantiated, the Converter is available until either the negotiation process fails or the service provisioning period expired.

4.2 The Mapping Process

This section gives an overview of the processes which are executed by the Conversion Factory to create an appropriate Converter instance. The different processes executed during the mapping process are:

- **Complexity analysis of the job**. When an SLA (respectively an "offer") is sent to the Conversion Factory, the complexity of the job is retrieved. This can either be done by querying the Complexity Database or a (human) analyst.
- **Knowledge Database lookup**. The Knowledge Database contains information about the infrastructure and the available resources. In addition it provides information about previous configurations of the system with respect to different job complexities.
- **Application of BLOs to the SLA (template)**. As our approach concentrates on the service provider domain, the customer is only implicitly relevant (through the SLA – which to maintain is also in the service provider's interest). The process of creating the Converter instance is influenced by the Business Level Objectives of the service provider.

5 Usage

The proposed architecture can be used to support SLA negotiation as well as configuration of the service provider's system. This chapter will give a short overview of the capabilities of the Conversion Factory / Converter approach and its expected use within a business interaction.

1. **Negotiation.** During the negotiation of a Service Level Agreement, the service provider has to figure out whether it can fulfil a service request according to the SLA terms. Our architecture supports this by checking the availability of the requested resources according to SLA and infrastructure status. The mapping and Converter creation process takes place as described in Section 4.2. Using the Converter, the negotiation component of the service provider can check the status of the system and thus the availability of the resources with respect to the SLA, by using the provided interface (see Section 4.1).

   Coming back to our example scenario: The aircraft manufacturer asked for re-designing the air plane's seats within three days. We assume that the Complexity Database has an entry for the respective job, telling the Conversion Factory that the job's complexity class is C. A Knowledge Database query then translates C into a resource requirement of 64 nodes for 24 hours. At this stage, the BLOs and the system's status have to be taken into consideration. As "time-to-complete" is set to three days, and given that the respective HPC resources are available, the BLOs now determine the level of service provided and communicated to the customer as the job is accepted. In case the system's schedule (or any BLO) does not allow the execution of the job, the provider has to reject the request or create a counter-offer.
2. **Configuration**. Having made an agreement, the service provider has to configure its system according to the Service Level Agreement, the BLOs, and the job complexity. In case the Converter was already instantiated during negotiation, it can be used to map the SLA terms to the infrastructure-specific (technical) parameters, which will be sent to the Execution Management System in

order to configure the systems. This configuration is subject to the BLOs of the provider. Imagine the aircraft manufacturer being a "high priority" business partner and the objective of the service provider is to provide "*minimal response time for high priority business partners*". This BLO would cause the EMS to run the job as soon possible. In case of the customer being a "low priority" business partner (and assuming that the job is not a "rigid" [4] one), less than 64 nodes may be allocated for more than 24 hours to keep resources in reserve for higher priority customers.

3. **Execution (Monitoring).** With the introduction of the status interface, we enable the service provider to use the Converter for monitoring during the execution of the service. Getting the data from the EMS, the Converter delivers the data mapped to SLA terms to the service provider's management system.

   In case of high priority jobs, monitoring could be used for violation prevention. If some of the allocated nodes fail, the EMS could be notified of an impeding SLA violation. With that warning, the EMS can adjust the system accordingly, e.g. by migrating (parts of) the job, to ensure the fulfilment of the SLA.

6 Conclusions

In this paper we introduce an architecture for SLA-supported, half automatic resource management. The architecture reflects the business needs of the service provider and aims at a more automatic and autonomous resource configuration through the introduction of a Conversion Factory. To achieve this, Service Level Agreements are mapped to provider-specific Operational Level Agreements with the aid of formalised business goals (Business Level Objectives), complexity analyses, and knowledge of previous configurations.

We are aware that the approach presented in this paper implies the availability of sophisticated methods and algorithms to realise the logic of components like the Conversion Factory and the Converter. Also the formalisation of SLAs, OLAs, and BLOs is still subject of various research activities of varying maturity.

With respect to the classification of jobs the Γ model described in [6] and the related scheduling architecture [3] may serve as a starting point towards the realisation of the complexity analysis as well as the resource assignment and configuration knowledge provision capabilities needed. Furthermore, our approach calls for services to map SLAs to OLAs while observing the provider's Business Level Objectives. In this case the authors are confident that semantic techniques as e.g. presented in [8] will help to provide appropriate solutions.

Acknowledgements

This work has been supported by the NextGRID project and has been partly funded by the European Commission's IST activity of the 6th Framework Programme under contract number 511563. This paper expresses the opinions of the authors and not necessarily those of the European Commission. The European Commission is not liable for any use that may be made of the information contained in this paper.

The developments presented in this paper were also partly funded by the European Commission through the IST programme under Framework 6 grant 001945 to the TrustCoM Integrated Project.

References

1. A. Andrieux, K. Czajkowski, A. Dan, K. Keahey, H. Ludwig, T. Nakata, J. Pruyne, J. Rofrano, S. Tuecke, and M. Xu. Web Services Agreement Specification (WS-Agreement) Version 2005/09, September 2005.
2. The BREIN project. Website, `http://www.gridsforbusiness.eu`.
3. K. Cristiano, R. Gruber, V. Keller, P. Kuonen, S. Maffioletti, N. Nellari, M.-Ch. Sawley, M. Spada, T.-M. Tran, Ph. Wieder, and Wolfgang Ziegler. Integration of ISS into the VIOLA Meta-scheduling Environment. In S. Gorlach and M. Danelutto, editors, *Proc. of the Integrated Research in Grid Computing Workshop*, pages 357–366. Università di Pisa, November 28–30, 2005.
4. D. G. Feitelson and L. Rudolph. Toward convergence in job schedulers for parallel supercomputers. In D. G. Feitelson and L. Rudolph, editors, *Job Scheduling Strategies for Parallel Processing*, volume 1162 of *LNCS*, pages 1–26. Springer, 1996.
5. I. Foster, H. Kishimoto, A. Savva, et al. The Open Grid Services Architecture, Version 1.0, January 2005. `http://www.ggf.org/documents/GFD.30.pdf`.
6. R. Gruber, P. Volgers, A. De Vita, M. Stengel, and T.-M. Tran. Parameterisation to tailor commodity clusters to applications. *Future Generation Comp. Syst.*, 19(1):111–120, 2003.
7. J. O. Kephart and D. M. Chess. The Vision of autonomic computing. *IEEE Computer*, 36(1):41–50, 2003.
8. L. Li and I. Horrocks. A Software Framework for Matchmaking Based on Semantic Web Technology. In *Proc. of the Twelfth International World Wide Web Conference (WWW 2003)*, Budapest, Hungary, May 20–24, 2003. ACM.
9. H. Ludwig, A. Keller, A. Dan, R. P. King, and R. Franck. Web Service Level Agreement (WSLA) Language Specification, 2003. `http://www.research.ibm.com/wsla/WSLASpecV1-20030128.pdf`.
10. Ph. Masche, P. Mckee, and B. Mitchell. The Increasing Role of Service Level Agreements in B2B Systems. In *Proc. of WEBIST 2006 – 2nd International Conference on Web Information Systems and Technologies*, Setúbal, Portugal, April 11–13, 2006. To appear.
11. The NextGRID project. Website, 24 June 2006 `http://www.nextgrid.org/`.
12. The TrustCoM project. Website, 24 June 2006 `http://www.eu-trustcom.com/`.

Reliable Orchestration of Resources Using WS-Agreement

Heiko Ludwig[1], Toshiyuki Nakata[2], Oliver Wäldrich[3], Philipp Wieder[4], and Wolfgang Ziegler[3]

[1] IBM TJ Watson Research Center, Hawthorne, NY, USA
hludwig@us.ibm.com

[2] NEC, Central Research Laboratory, Tokyo, Japan
t-nakata@cw.jp.nec.com

[3] Fraunhofer Institute SCAI, Department of Bioinformatics, 53754 Sankt Augustin, Germany
{oliver.waeldrich, wolfgang.ziegler}@scai.fraunhofer.de

[4] Research Centre Jülich, 52425 Jülich, Germany
ph.wieder@fz-juelich.de

Abstract. Co-ordinated usage of resources in a Grid environment is a challenging task impeded by the nature of resource usage and provision: Resources reside in different geographic locations, are managed by different organisations, and the provision of reliable access to these resource usually has to be negotiated and agreed upon in advance. These prerequisites have to be taken into account providing solutions for the orchestration of Grid resources. In this document we describe the use of WS-Agreement for Service Level Agreements paving the way for using multiple distributed resources to satisfy a single service request. WS-Agreement is about to be released as a draft recommendation of the Global Grid Forum and has already been implemented in a number of projects, two of which we will presented in this paper.

1 Introduction

1.1 Motivation to Use WS-Agreement

Today, the access to and use of services offered by a provider, e.g. use of resources in a Grid environment, are usually governed by static agreements made beforehand between the parties. The parties thus already have to be in contact before the actual (business) interaction takes place. However, in a large Grid environment with a large number of providers of different services and an even larger number of service consumers this static approach is not feasible.

In contrast WS-Agreement [1] offers a reliable mechanism to create solid electronic agreements between different parties interested in setting up dynamic collaborations that include mutual obligations, e.g. between service provider and service consumer. WS-Agreement thus fills the existing gap when trying to establish a dependable relation between these parties (that must not have established a formal contractual relationship beforehand) enabling them e.g. to provide a

M. Gerndt and D. Kranzlmüller (Eds.): HPCC 2006, LNCS 4208, pp. 753–762, 2006.

service on one side and to consume this service at the other side while the formal framework of this exchange is governed by a dynamic contract. WS-Agreement empowers both parties to establish a dynamic electronic contract that governs their ad hoc business connection set up to meet an actual need of one party with a service provided by the other party. Once both sides have fulfilled their obligations specified in the agreement, the agreement is completed and both partners have no longer mutual obligations from this agreement.

WS-Agreement is now in the state of a proposed recommendation which is the last phase before becoming a recommendation of the Global Grid Forum (GGF [5]). Nevertheless, it is already used in a number of projects - two of them being presented in this paper - and has attracted attention from several large communities like the telecommunications and IT industries organised in the IPsphere Forum [7] or the partners in the European project EGEE [4].

1.2 WS-Agreement

The objective of the WS-Agreement draft specification, defined by the GRAAP Working Group [6] of the GGF, is to provide a domain-independent and standard way to establish and monitor Service Level Agreements. The specification comprises three major elements: (i) A description format for agreement templates and agreements; (ii) a basic protocol for establishing agreements, and (iii) an interface specification to monitor agreements at runtime.

Agreements according to the specification are bilateral and set up between two roles, the *Agreement Initiator* and the *Agreement Responder*. These roles are independent of the roles of service provider, the domain that performs jobs or other services, and service consumer. An agreement defines a dynamically-established and dynamically-managed relationship between these parties. The object of the relationship is the delivery of a service, e.g. the execution of a job or the availability of compute resources, by one of the parties within the context of the agreement. The management of this delivery is achieved by agreeing on the respective roles, rights and obligations of the parties. The agreement may specify not only functional properties for identification or creation of the service, but also non-functional properties of the service such as performance or availability.

Fig. 1 outlines the main concepts and interfaces of the WS-Agreement specification. In the chosen example, the Agreement Responder is a service provider, the Agreement Initiator the service consumer. An Agreement Responder exposes an interface of an *Agreement Factory*, which offers an operation to create an agreement and an operation to retrieve a set of agreement templates proposed by the agreement provider. Agreement templates are agreements with fields to be filled in. Templates help an Agreement Initiator to create agreements that the agreement provider can understand and accept. The `createAgreement` operation returns accept or reject, if a synchronous reply is expected. Otherwise, in case of a longer decision-making process, the service responder can convey the decision to an `AgreementResponse` interface that the Initiator exposes. If the `createAgreement` operation succeeds, an agreement instance is created. The agreement instance exposes the terms of the agreement as properties that can be

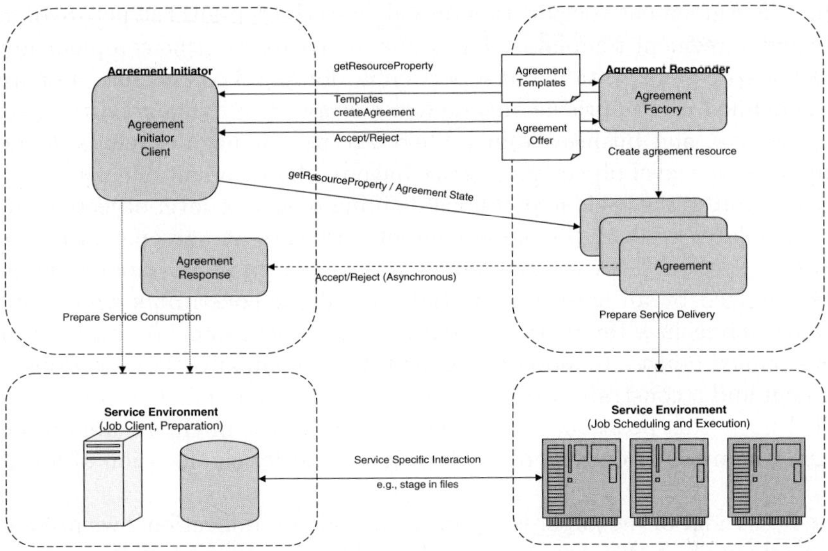

Fig. 1. Concepts and interfaces of WS-Agreement

queried. In addition, runtime states for the agreement as a whole and its individual terms can be inspected by the Initiator. All interfaces exposed by the parties, *Agreement Factory*, *Agreement* and *AgreementResponse* are resources according to the Web Services Resource Framework (WSRF) [13]. Upon acceptance of an agreement, both service provider and service consumer have to prepare for the service, which typically depends on the kind of service subject to the agreement. For example, a service provider schedules a job that is defined in the agreement. A service consumer will make the stage-in files available as defined in the agreement. Further service specific interaction may take place between the parties governed by the agreement.

The WS-Agreement draft specification defines the content model of both, agreements and agreement templates as an XML-based language. Structurally, an agreement consists of a name, a context section, and the agreement terms. The agreement context contains definitorial content such as the definition of the parties and their roles in the agreement. The agreement terms represent contractual obligations and include a description of the service as well as the specific guarantees given. A service description term (SDT) can be a reference to an existing service, a domain specific description of a service (e.g. a job using the Job Submission Description Language (JSDL [2], a data service using Data Access and Integration Services, etc.), or a set of observable properties of the service. Multiple SDTs can describe different aspects of the same service. A guarantee term on the other hand specifies non-functional characteristics in service level objectives as an expression over properties of the service, an optional qualifying condition under which objectives are to be met, and an associated business value specifying the importance of meeting these objectives.

The WS-Agreement specification only defines the top-level structure of agreements and agreement templates. This outer structure must be complemented by means of expressions suitable for a particular domain. For example, a guarantee term is defined as comprising the elements scope, qualifying condition, service level objective, and business value. There are no language elements defined to specify a service level objective. Parties have to choose a suitable condition language to express the logic expressions defining a service level objective. Agreement Templates contain the same content structure as agreements but add a constraints section. This section defines which content of a template agreement can be changed by an agreement initiation and the constraints which must be met when filling in a template to create an agreement offer. A constraint comprises a named pointer to an XML element in the context or term sections of the agreement and a constraint expression defining the set of eligible values that can be filled in at this position. For example, an Initiator might be able to choose between a number of options of a job or must specify the location of a stage in file.

The remainder of the paper is organised as follows. In Section 2 we present two implementations of WS-Agreement, followed by a summary of the experiences made (Section 3). Based on these experiences we present the discussion of requirements for negotiation of Service Level Agreements in Section 4. An overview of further developments in the GRAAP-WG related to this paper concludes the paper.

2 Implementations

2.1 Wide Area Business Grid

In this section, we describe a system using WS-Agreement in combination with JSDL to express SDTs for complex, wide-area jobs. This system was developed within the framework of the Japanese Business Grid Project [3] (2003-2005). The project's mission was to develop a business Grid middleware realising a next generation business application infrastructure.

Overview of the system. The main objective in designing and implementing the system has been to make it easy to deploy and run business applications like a 3-tier Web application in a data centre. The main characteristics of the system are:

- Job submission is realised by a job description using WS-Agreement and JSDL, and the Application Contents Service (ACS).
- Brokering is used to allocate the necessary resources.
- Automatic deployment and configuration of programmes and data (including the necessary preparation of hosting environments) is realised by a language similar to the Configuration Description Language (CDL [8]), and the ACS.
- Management of heterogeneous resources is realised through Grid middleware agents providing a common interface (resource virtualisation).

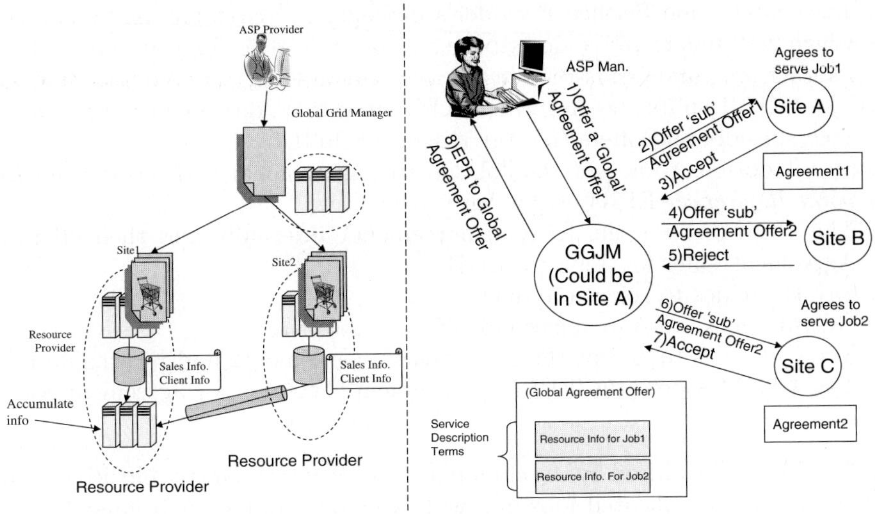

Fig. 2. Wide Area Business Grid

Use of WS-Agreement and JSDL. We chose JSDL with extensions as the domain-specific language used within the Agreement. The link between JSDL constructs and WS-Agreement is realised through a convention specifying that the content of the JSDL element `JobName` corresponds to WS-Agreement's `ServiceName`. We also extended the JSDL constructs in order to make the description of more complicated resource requirements easier. This resulted in two types of resource description information for each job: (i) "Global" resource information for describing the type of the job or providing general characteristics of the job (e.g. whether to allow automatic load control or not) and (ii) "local" resource information which describes the resource needed for each of the components that make up a pool to carry out the job. Several jobs which are related - such as a 3-tier Web application for a Web shop and a batch job which calculates the sales info every weekend - can be included in a single agreement.

Implementation of the Wide Area Business Grid. The agreement template as described above is utilized in order to realise a wide-area business Grid. The aim here is to make it possible to share IT resources based on the contract/agreement among (i) distributed centres in an enterprise and (ii) trusted partners' data centres (resource providers), thus making it possible for an Application Service Provider (ASP) to dispatch a complex job from a single portal, as depicted on the left side of Fig. 2. The right side of this figure pictures a scenario where an ASP wants to dispatch a complex job consisting of two 3-tier Web applications spanning over two sites (resource providers). The sequence of events which occur in this scenario is as follows:

1. The Application Service Provider's manager prepares an *Agreement Offer* which contains resource descriptions of both Job1 for one site and Job2 for another site, and sends the offer to the Global Grid Job Manager (GGJM).
2. The GGJM splits the Agreement Offer into two Agreement Offers, one for each sub-job, and offers the Agreement for Job1 to site A.
3. Site A decides to accept the Job, creates agreement1 and returns the *Endpoint Reference* (EPR) of the Agreement.
4. The GGJM stores the EPR of agreement1 internally and then offers the Agreement Offer for Job2 to site B.
5. Site B decides to reject the offer.
6. GGJM offers the Agreement Offer for Job2 to site C.
7. Site C decides to accept the offer, creates agreement2 and returns its EPR.
8. The GGJM stores the EPR of agreement2, creates an agreement for the complete job and returns it to the ASP manager.

The flexibility introduced by the combination of WS-Agreement and JSDL allows to handle very complicated jobs in a wide-area distributed environment.

2.2 VIOLA Meta-scheduling Environment

The German VIOLA project [11] develops among other components a meta-scheduling environment providing resource reservations based on WS-Agreement. The immediate objective of this development is to co-allocate computational and network resources in a UNICORE-based Grid, but we designed the environment to support arbitrary types of resources and to be integrated into different Grid middleware systems. The main integration effort to access other middleware, like e.g. Globus, is to implement the client-side interface of the Meta-Scheduling Service. Since it is beyond the scope of this paper to explain the system in detail we refer to [12] for a complete architectural description of it and to [10] for the definition of the negotiation protocol currently implemented.

Fig. 3 sketches the basic architecture of the meta-scheduling environment and its integration into an arbitrary Grid middleware. The VIOLA Meta-Scheduling Service communicates with a client application using WS-Agreement, it receives a list of resources (pre-selected by a resource selection service which is not pictured here), and it returns reservations for some or all of these resources. To interact with varying types of scheduling systems we use the adapter pattern approach. The role of an Adapter is to provide a single interface to the Meta-Scheduling Service by encapsulating the specific interfaces of the different local scheduling systems. Thus the Meta-Scheduling Service can negotiate resource usage by exploiting a single interface independent of the underlying resource type. To achieve this, the Meta-Scheduling Service first queries local scheduling systems for the availability of the requested resources and then negotiates the reservations across all local scheduling systems which, in order to participate in the negotiation process, have to be capable and willing to

- let the Meta-Scheduling Service reserve resources in advance by offering data about job execution start and stop times, and

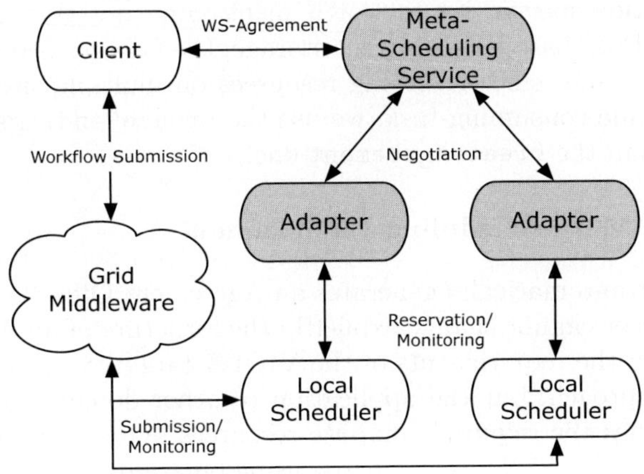

Fig. 3. High-level meta-scheduling architecture

- provide at least partial access to their local schedules, e.g. by publishing information about available free time slots.

A prototype of the architecture shown in Fig. 3 has been implemented using the UNICORE Grid middleware and this prototype is used within the VIOLA testbed to demonstrate the execution of MPI jobs supported by the co-allocation capabilities of the Meta-Scheduling Service. Furthermore a network resource management system supporting advance reservation is integrated to enable the reservation and scheduling of end-to-end network connections with a guaranteed Quality of Service level.

3 Experience with Implementing WS-Agreement

By implementing WS-Agreement-based resource orchestration frameworks in the two different projects we gained substantial experience which we continuously feed into the standardisation process. In this section we examine some issues that we think are of value for potential implementers of WS-Agreement.

3.1 Wide Area Business Grid

One of the objectives of Business Grid was to make the system dynamically adapt to workload fluctuation. Specifying a range of CPU resources together with SLA conditions using Guarantee Terms is one candidate to meet our goals, but we wanted to let the user be able to specify the controlling of the SLA in a flexible manner. To meet our goals, we use a combination of a statically allocated resource range specified by an Agreement together with policies specified by ASP managers. As a by-product, we currently do not use the Guarantee Terms. In addition, some of the domain-specific terms are specified outside the Agreement.

Implementation-specific issues worth mentioning are that the system uses J2EE RI with Pointbase for persistent storage, the Globus Toolkit 4 for WSRF-related services, and, since reserving resources on multiple sites can be a very complex and time consuming task, we use the `createPendingAgreement` operation rather than the `createAgreement` one.

3.2 VIOLA Meta-scheduling Environment

Since the client automatically generates an Agreement Offer based on the input of a user, the user cannot change explicitly the structure of an Agreement Offer. With respect to the requirements of the VIOLA target domain (MPI jobs) this is a feasible approach, but the application to other domains may change this. An Offer specifies the required compute resources by using Service Description Terms and the required Network QoS by using Guaranty Terms. On reception of an Agreement Offer the Meta-Scheduling Service tries to co-allocate all required resources and only in case of success an Agreement is created.

The negotiation process between the Meta-Scheduling Service and the Adapters is implemented based on a proprietary protocol. This is done because the WS-Agreement protocol does neither specify means to change an existing agreement nor to cancel an agreement once it was created. This point is covered again in Section 4. The creation of an agreement is done in a synchronous manner. In our case it is feasible because one important goal in the development of the Meta-Scheduling Service was to make the negotiation process between the subsystems as effective as possible and therefore minimise the overhead of the negotiation as much as possible.

4 Requirements for Negotiation of Service Level Agreements

In this subsection, we describe our wish-list with respect to the current draft. A smaller part is under consideration for the current WS-Agreement recommendation while the bigger part will be addressed in subsequent versions of WS-Agreement. A **cancellation mechanism** is required to meet a needs of multi-site allocation use case for Business Grid. This is being integrated into the current specification. With regards to **re-negotiation**, we feel that there is a need to distinguish between two types of re-negotiation: (i) Re-negotiation is done in the initial phase where the two parities negotiate on the Agreement terms and (ii) re-negotiation is needed after an Agreement has reached the *Observed* state.

Here we would like to discuss two issues for the latter re-negotiation type, with the assumption that the current Agreement is in the Observed state during the re-negotiation. There are two major features we think are necessary to make WS-Agreement an even more useful specification for SLAs: a modifiable expiration time and a modifiable list of resources. The first requirement is able to satisfy requests arising from dynamic business conditions, where one party may want to extend or shorten the time that an Agreement will be effective. The second requirement addresses situations - like the one described before in the Business Grid example - where resources will be allocated within a certain range specified in order to meet the SLAs. However, due to increasing resource demand there may be cases in which the Agreement Initiator would like to plan for enhancements to the current system. This could be done in the following two alternative ways: Re-negotiation of the original Agreement or creation of a new Agreement with references to the original Agreement.

The discussion of the items of the wish-list has recently been started in the GRAAP working group. As a result of this discussion several small subgroups ("design-teams") have already been installed or will be installed soon. These sub-groups focus on

- the further development of WS-Agreement to include e.g. cancellation and re-negotiation,
- a protocol for a multi-step negotiation of agreements as used for example in the VIOLA project,
- provision of other term languages, e.g. including network related terms, and
- terms related to reservation, e.g. start, duration, strategy.

5 Future Perspectives

Even before the WS-Agreement specification has become a proposed recommendation of the GGF early adopters have already started using WS-Agreement prototypes in several projects, e.g. Cremona [9]. It has also attracted attention from several large communities, as described above, and we expect a broader uptake once the specification is published as proposed recommendation.

The next major activity to be carried out is the demonstration of interoperability between different implementations of WS-Agreement. In parallel the GRAAP Working Group will start to work on a negotiation protocol based on WS-Agreement, where it is planned to include most of the features described in Section 4. We already started discussion and work on the protocol for the multi-step negotiation between resource provider and resource consumer. Finally, the third activity started recently is focusing on new term languages to support a broader range of usage scenarios for WS-Agreement, e.g. expressing domain-specific service level objectives for network or security.

Acknowledgements

Some of the work reported in this paper is funded by the German Federal Ministry of Education and Research through the VIOLA project under grant #01AK605L. This paper also includes work carried out jointly within the CoreGRID Network of Excellence funded by the European Commission's IST programme under grant #004265. In addition, part of this work had been funded by the Ministry of Economy, Trade, and Industry (METI), Japan with administrative support by Information-Technology Promotion Agency (IPA), Japan. The authors would like to thank Hiro Kishimoto, Andreas Savva, Nobutoshi Sagawa, Shinya Miyakawa for various discussions.

References

1. A. Andrieux, K. Czajkowski, A. Dan, K. Keahey, H. Ludwig, T. Nakata, J. Pruyne, J. Rofrano, S. Tuecke, and M. Xu. Web Services Agreement Specification (WS-Agreement), September 2005. 3 Apr 2006 <https://forge.gridforum.org/projects/graap-wg/document/WS-AgreementSpecificationDraft.doc/en/26>.
2. A. Anjomshoaa, F. Brisard, M. Drescher, D. Fellows, A. Ly, S. McGough, D. Pulsipher, and A. Savva. Job Submission Description Language (JSDL) Specification v1.0. Grid Forum Document GFD.56, Global Grid Forum, November 2005.
3. The Business Grid Project. 30 Mar 2006 <http://www.ipa.go.jp/english/softdev/sixth.html>.
4. EGEE – Enabling Grids for E-sciencE. 3 Apr 2006 <http://public.eu-egee.org/>.
5. GGF – The Global Grid Forum. 3 Apr 2006 <http://www.ggf.org>.
6. Grid Resource Allocation Agreement Protocol Working Group. 3 Apr 2006 <https://forge.gridforum.org/projects/graap-wg/>.
7. The IPsphere FORUM. 3 Apr 2006 <http://www.ipsphereforum.org/home>.
8. S. Loughran et al. Configuration Description, Deployment, and Lifecycle Management (CDDLM) Deployment API. Grid Forum Document GFD.69, Global Grid Forum, March 2006.
9. H. Ludwig, A. Dan, and B. Kearney. Cremona: An Architecture and Library for Creation and Monitoring of WS-Agreements. In M. Aiello, M. Aoyama, F. Curbera, and M. Papazoglou, editors, *Proc. of the 2nd International Conference on Service Oriented Computing (ICSOC 2004)*, pages 65–74. ACM Press, 2004.
10. A. Streit, O. Wäldrich, Ph. Wieder, and W. Ziegler. On Scheduling in UNICORE – Extending the Web Services Agreement based Resource Management Framework. In *Proc. of Parallel Computing 2005 (ParCo 2005)*, Malaga, Spain, September 13–16, 2005. To appear.
11. VIOLA – Vertically Integrated Optical Testbed for Large Application in DFN. 29 Mar 2006 <http://www.viola-testbed.de/>.
12. O. Wäldrich, Ph.Wieder, and W. Ziegler. A Meta-scheduling Service for Co-allocating Arbitrary Types of Resources. In R. Wyrzykowski, J. Dongarra, N. Meyer, and J. Wasniewski, editors, *Proc. of the 2nd Grid Resource Management Workshop (GRMWS'05) in conjunction with PPAM 2005*, volume 3911 of *LNCS*, pages 782–791. Springer, 2006.
13. OASIS Web Services Resource Framework (WSRF) TC. 3 Apr 2006 <http://www.oasis-open.org/committees/wsrf>.

Dynamically Scheduling Divisible Load for Grid Computing

Salah-Salim Boutammine, Daniel Millot, and Christian Parrot

GET / INT
Département Informatique
91011 Évry, France
{Salah-Salim.Boutammine, Daniel.Millot, Christian.Parrot}@int-evry.fr

Abstract. This paper presents an improved framework for an existing adaptive method for scheduling divisible load. The proposed framework delimits in a new way the application field of the method. This method can be used to schedule parallel applications whose total workload is unknown a priori, and can deal as well with the unpredictable execution conditions commonly encountered on grids. Some parameters which quantify the dynamic nature of these varying execution conditions are discussed in the paper.

Keywords: scheduling, divisible load, parallel application, grid, master-worker, on-line, multi-round, heterogeneity, dynamicity.

1 Introduction

In many application domains, such as data mining or customer records processing for instance, data from huge collections can be processed independently. Such computation load can be arbitrarily divided among multiple computers and is then called "divisible load". For the execution of the corresponding applications, the use of multiple computers is quite straightforward when these computers have the same CPU and communication link speeds. On the contrary, scheduling the computation is much more intricate when the execution platform is heterogeneous. The Divisible Load Theory (DLT [1, 2, 3, 4]) has proved to be able to provide useful schedules in such a case.

In this paper, we consider applications that process a finite –but a priori unknown– amount of data independently; the total workload is supposed arbitrarily divisible in chunks. Such applications are suitable for the master-worker programming model. The master collects the data in an input buffer and until it gets the last data item, it iteratively distributes chunks to the workers, then collects the corresponding results from them. Clearly, for such a programming model to be useful, the processing cost for a chunk by a worker must dominate the corresponding communication costs between master and worker.

It has to be noted that although we consider "divisible load", DLT cannot be straightforwardly applied, as we suppose that the total workload of the application is not known a priori. For this reason, only runtime algorithms can be used.

M. Gerndt and D. Kranzlmüller (Eds.): HPCC 2006, LNCS 4208, pp. 763–772, 2006.

We want such an application to terminate as soon as possible on a set of grid resources, but we do not consider the fundamental problem of choosing the resources to be used and the order in which they are served. We adopt a one-port communication model without contention [5], and consider both computation and communication latencies, due to the context of grid computing. Execution parameters on a grid, such as available computing power or network bandwidth, vary both in space (heterogeneity) and time (dynamicity). We assume that we know all past values of these parameters and are unaware of the future ones. As the makespan minimization problem is NP-complete when latencies are considered [6], it can only be heuristically dealt with.

In [7], we have presented the On-Line Multi-Round method (denoted "OLMR" thereafter), an adaptive method that can be used for scheduling parallel applications when the total workload of the application is not known a priori and in the dynamic context of grids. Assuming that a set of resources has been identified, the method aims at distributing the tasks of a parallel application on this set of resources optimally, in order to minimize the makespan of the application. It focuses on scheduling during the “steady-state” phase [8], which lasts from the time instant when each worker has received a first chunk up to the instant when the master gets the final data item to be processed. From this time instant, begins the “clean-up” phase. The problem of scheduling the remaining load is suitable for DLT, as now the total workload is known: namely the amount of data still present in the master input buffer. So, according to the optimality principle, we can try and minimize the makespan by synchronizing the termination of the computation of all the workers. This, being possible only if the master does not overload any worker too much during the “steady-state” phase, which would cause a discrepancy too large for late workers to catch up during the “clean-up” phase, thus preventing a synchronous termination of all workers. This is precisely the aim of the OLMR method.

This paper is organized as follows. Section 2 presents related work. Section 3 gives an overview of the OLMR method. Section 4 successively details the new framework for the scheduling algorithm and studies the conditions for the method to succeed. Section 5 concludes the paper and outlines future work.

2 Related Work

The divisible load model (DLT [1, 2, 3, 4]) has been largely studied; it is the first model that enables optimality proofs for scheduling methods in the context we have chosen [1, 3, 9, 10]. This model is well suited to the problem of scheduling the “clean-up” phase. On the contrary, it is not suited to scheduling the “steady-state” phase, as the total workload is not known during this phase.

Several multi-round methods have been proposed in the literature [11, 12, 13]. On the one hand the iterated distributions of multi-round methods have the advantage (when using a one-port model without contention) of making the nodes active earlier, and on the other hand they have the drawback of increasing the time wasted by latencies (due to affine cost). Several strategies for distributing the load to slaves have already been studied. First of all, some strategies fix the number of rounds arbitrarily (multi-installment methods [11]). They are well suited to a model with linear costs for homogeneous platforms. For heterogeneous platforms, other strategies have been proposed, which are able to take account of the affinity of costs when determining the load of the nodes for each round. For instance, the workload which is delivered at each round by the UMR method [12] follows a geometric progression whose common ratio is adjusted so that all the nodes work for the same duration in each round and so that computation overlaps communication exactly. It is proved [12] that this method minimizes the total execution time; provided that we know how to select the best set of nodes to be used. The PMR method [13] introduces periodicity for the rounds (without imposing any particular value for the period) and requires that all nodes work during the whole period and that computation overlaps communication exactly. It is proved [13] that this method maximizes the amount of load processed by time unit.

Unfortunately none of these methods can be used when the total workload is not known a priori or if the execution parameters vary in time. On the contrary, the On-Line method presented by Drozdowski in [14] can deal with this situation. It proceeds incrementally and adjusts the size of the chunk to be sent to a worker for each new round in order to try and maintain a constant duration for the different rounds; so doing, it avoids contention at the master and limits discrepancy between workers. But this method has the drawback that computation never overlaps communication in any worker node, as the emission of the chunk of the next round is at best triggered by the return of the result of the previous one, with no possible anticipation.

The OLMR method presented in [7] improves on Drozdowski's On-Line method it is based on by avoiding idle time of the computing resources. [7] compares both methods in a static context. Under appropriate hypotheses, which are met when execution parameters are stable (cf Lemma 6.1 in [14]), rounds are asymptotically periodic (as for the PMR method). Be additionnal hypotheses satisfied, the OLMR method definitely minimizes the makespan.

3 Overview of the OLMR Method

When the total load is important compared to the available bandwidth between master and workers, the workload should be delivered in multiple rounds [11, 12, 13]; and this is what OLMR does. It proceeds incrementally, computing the size $\alpha_{i,j}$ of the chunk to be sent to a worker N_i for each new round j, in order to try and maintain a constant duration τ for the different rounds.

OLMR determines $\alpha_{i,j}$ so as to make the distribution asymptotically periodic with period τ, an arbitrarily fixed value, for all the workers. For worker N_i, let $\sigma_{i,j-1}$ be the elapsed time between the begining of the emission of the chunk of its $(j-1)^{th}$ round and the end of the reception of the result corresponding to this chunk. OLMR determines the value of $\alpha_{i,j}$ as Drozdowski's method does, i.e.

$$\alpha_{i,j} = \alpha_{i,j-1} \cdot \frac{\tau}{\sigma_{i,j-1}}. \quad (1)$$

That is it allocates comparatively bigger (resp. smaller) chunks to workers with higher (resp. lower) performance. Hence, this method can take the heterogeneous nature of computing and communication resources into account, without explicit knowledge of execution parameters.

[14] shows that, in a static context, with affine cost models for communication, the way $\alpha_{i,j}$ is computed using equation (1) ensures the convergence of $\sigma_{i,j}$ to τ when j increases indefinitely. Being an estimation of the asymptotic period used for task distribution, τ is also an upper-bound on the discrepancy between workers. Being able to control this bound makes it possible to minimize the makespan during the "clean-up" phase.

The following notations are used throughout the rest of the paper:

- N number of workers,
- γ_i start-up time for a computation by worker N_i,
- $w_{i,j}$ computation cost for a chunk of size 1 of the jth round by worker N_i,
- β_i (resp. β_i') start-up time for a communication from the master to N_i (resp. from N_i to the master),
- $c_{i,j}$ (resp.$c_{i,j}'$) transfer cost for a data (resp. result) chunk of size 1 of the jth round from the master to worker N_i (resp. from N_i to the master).

As suggested in section 1, the values of the execution parameters of any worker N_i — here $w_{i,j}$, $c_{i,j}$ and $c_{i,j}'$ — depend on the round.

Chunks contain only data to be processed. For a chunk of size α, we denote $\alpha'(\alpha)$ the size of the corresponding result, where function $\alpha'()$ is supposed to be increasing.

We assume that costs are roundwise affine in the size of chunks. Hence, for a chunk of strictly positive size α (i.e. $\alpha \in \mathbb{R}^{+*}$) of the jth round, we define the cost of:

- sending the chunk to worker $\mathrm{N_i}$ $\quad \alpha \cdot c_{i,j} \quad + \beta_i,$
- processing the chunk on worker $\mathrm{N_i}$ $\quad \alpha \cdot w_{i,j} \quad + \gamma_i,$
- receiving the corresponding result from worker $\mathrm{N_i}$ $\quad \alpha'(\alpha) \cdot c_{i,j}' + \beta_i'.$

We indicated in section 1 that the processing cost for a chunk should dominate its communication costs. We choose to formulate this assumption as:

$\forall \alpha \in \mathbb{R}^{+*}$,

$$\alpha \cdot \min_{j \in \mathbb{N}^*} w_{i,j} + \gamma_i \geq \left(\alpha \cdot \max_{j \in \mathbb{N}^*} c_{i,j} + \beta_i\right) + \left(\alpha'(\alpha) \cdot \max_{j \in \mathbb{N}^*} c_{i,j}' + \beta_i'\right) \qquad (2)$$

$for\ i = 1, N.$

Equation (2) ensures that sending chunks of any size α to a worker N_i and receiving the corresponding results of size $\alpha'(\alpha)$ costs less than processing these chunks.

For each round j, OLMR divides the chunk sent to N_i into two subchunks "I" and "II" of respective sizes $\overline{\alpha}_{i,j}$ and $\alpha_{i,j} - \overline{\alpha}_{i,j}$. Dividing the chunks in two parts

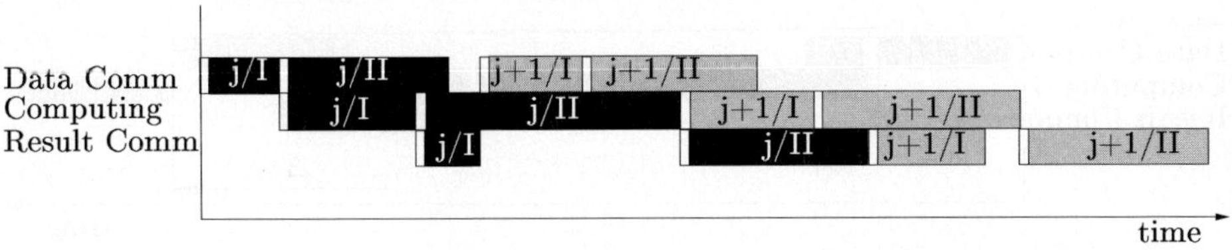

Fig. 1. Overlapping between communication and computation with OLMR

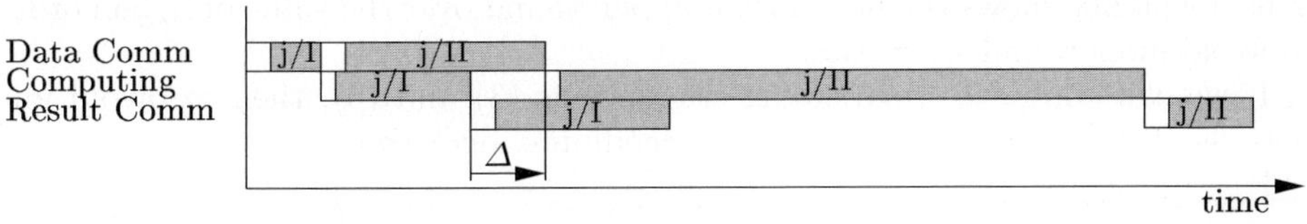

Fig. 2. Example of intra-round starvation with OLMR

is enough in order to avoid idle time of the computing resources; as can be seen in Fig.1, the division allows the computation to overlap the communications. In order to compute $\alpha_{i,j}$, we use a value of $\sigma_{i,j-1}$ derived from the measurement of the elapsed time (including both communications and computation) for subchunk I of the previous round: $\overline{\sigma}_{i,j-1}$. We will show that, thanks to this anticipation in the computation of $\alpha_{i,j}$, we can avoid the inter-round starvation.

Unfortunately, while attempting to deal with inter-round starvations , there is a risk of creating intra-round starvation between subchunks I and II (see the idle period Δ on Fig.2). We explain below how to prevent both risks.

As we assume that (2) holds, intra-round starvation can be avoided if $\overline{\alpha}_{i,j}$ is large enough for the processing of subchunk I to overlap the sending of subchunk II of size $\alpha_{i,j} - \overline{\alpha}_{i,j}$. Thus, there is no intra-round starvation if and only if

$$\overline{\alpha}_{i,j} \geq \frac{\alpha_{i,j} \cdot c_{i,j} + \beta_i - \gamma_i}{w_{i,j} + c_{i,j}}. \tag{3}$$

Inter-round starvation between the jth and $(j+1)^{th}$ rounds of N_i could occur if, for round j, subchunk I happens to be too large compared to subchunk II (see Fig.3).

Let $\nu_{i,j}$ be some real number dominating $\overline{\alpha}_{i,j}$ and $\overline{\alpha}_{i,j+1}$:

$$\nu_{i,j} \geq max\left(\overline{\alpha}_{i,j}, \overline{\alpha}_{i,j+1}\right). \tag{4}$$

There is no inter-round starvation between the jth and $(j+1)^{th}$ rounds of N_i if

$$\overline{\alpha}_{i,j} \cdot w_{i,j} + \alpha'\left(\overline{\alpha}_{i,j}\right) \cdot c'_{i,j} \leq \alpha_{i,j} \cdot w_{i,j} - \nu_{i,j} \cdot c_{i,j+1} + \gamma_i - \left(\beta'_i + \beta_i\right). \tag{5}$$

If $\left(w_{i,j} \cdot Id + c'_{i,j} \cdot \alpha'\right)$ has an inverse function, then (5) can be rewritten

$$\overline{\alpha}_{i,j} \leq \left(w_{i,j} \cdot Id + c'_{i,j} \cdot \alpha'\right)^{-1} \left(\alpha_{i,j} \cdot w_{i,j} - \nu_{i,j} \cdot c_{i,j+1} + \gamma_i - \left(\beta'_i + \beta_i\right)\right).$$

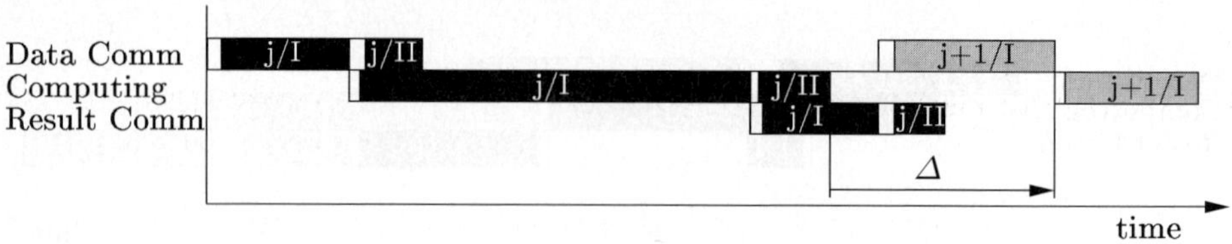

Fig. 3. Example of inter-round starvation with OLMR

This inequality shows the need for an upper bound over the value of $\overline{\alpha}_{i,j}$ in order to avoid inter-round starvation.

If we can choose $\overline{\alpha}_{i,j}$ satisfying inequations (3) and (5), then we avoid idle periods of N_i. Fig.4 gives the OLMR scheduling algorithm.

while (the last data item has not been acquired) **do**
  if (Reception from N_i of the result of subchunk I of its $(j-1)^{th}$ round) **then**
  - Get $\overline{\sigma}_{i,j-1}$, $\omega_{i,j-1}$, $c'_{i,j-1}$ (and γ_i for the first result from N_i)
  - Compute $\sigma_{i,j-1}$ .... (cf. (6))
  - Compute $\alpha_{i,j}$ .... (cf. (1))
  - Compute $\overline{\alpha}_{i,j}$ .... (cf. (8))
  - Send a subchunk of size $\overline{\alpha}_{i,j}$ to N_i as subchunk I of its jth round
  - Send a subchunk of size $(\alpha_{i,j} - \overline{\alpha}_{i,j})$ to N_i as subchunk II of its jth round

  end if
end while

Fig. 4. OLMR scheduler

Thanks to equation (1), the OLMR scheduler computes $\alpha_{i,j}$. Finally, nothing remains but to determine $\overline{\alpha}_{i,j}$ and $\sigma_{i,j-1}$.

4 A Closer View of the Method

In order to determine the size of the chunk to be sent for the next round without waiting for the result of the currently processed chunk, it suffices to replace the measured value $\sigma_{i,j-1}$ in expression (1) by some computed value derived from $\overline{\sigma}_{i,j-1}$. But we only know the values of the execution parameters for the data whose result have been received by the master. [7] shows that the value of $\sigma_{i,j-1}$ can be fixed as:

$$\sigma_{i,j-1} = \overline{\sigma}_{i,j-1} + (\alpha_{i,j-1} - \overline{\alpha}_{i,j-1})\, w_{i,j-1} + (\alpha_{i,j-1} - 2\cdot\overline{\alpha}_{i,j-1})\, c'_{i,j-1} + \gamma_i. \quad (6)$$

In this paper, we do not reconsider this result, but we improve the choice of the value of $\overline{\alpha}_{i,j}$.

The following theorem proposes a way to set the value of $\overline{\alpha}_{i,j}$ so that constraints (3) and (5) are both satisfied.

Theorem 1. *Given $\alpha_{i,j}$, if γ_i, $w_{i,j}$, $c_{i,j}$, β_i, $c'_{i,j}$ and β'_i satisfy (2) and*

$$(\alpha_{i,j} - \nu_{ij}) \cdot w_{i,j} + \gamma_i \geq \nu_{ij} \cdot c_{i,j} + \beta_i \tag{7}$$

for i=1,N.
Then, taking

$$\overline{\alpha}_{i,j} = \alpha_{i,j} - \nu_{i,j}, \tag{8}$$

where ν_{ij} satisfies (4), the workers will compute without any idle period during the steady-state phase.

Proof. Thanks to (7), we have

$$(\alpha_{i,j} - \nu_{i,j}) \cdot (w_{i,j} + c_{i,j}) \geq \alpha_{ij} \cdot c_{ij} + \beta_i - \gamma_i$$
$$\alpha_{i,j} - \nu_{i,j} \geq \frac{\alpha_{ij} \cdot c_{ij} + \beta_i - \gamma_i}{w_{i,j} + c_{i,j}}.$$

Then using definition (8) of $\overline{\alpha}_{i,j}$, we have

$$\overline{\alpha}_{i,j} \geq \frac{\alpha_{i,j} \cdot c_{i,j} + \beta_i - \gamma_i}{w_{i,j} + c_{i,j}}.$$

So constraint (3) is satisfied.
By definition (8), we have

$$\nu_{i,j} \cdot w_{i,j} = (\alpha_{i,j} - \overline{\alpha}_{i,j}) \cdot w_{i,j}.$$

By hypothesis (2), the last inequality can be rewritten as:

$$(\nu_{i,j} \cdot c_{i,j+1} + \beta_i) + \left(\alpha'\left(\nu_{i,j}\right) \cdot c'_{i,j} + \beta'_i\right) - \gamma_i \leq (\alpha_{i,j} - \overline{\alpha}_{i,j}) \cdot w_{i,j},$$
$$\overline{\alpha}_{i,j} \cdot w_{i,j} + \alpha'\left(\nu_{i,j}\right) \cdot c'_{i,j} \leq \alpha_{i,j} \cdot w_{i,j} - \nu_{i,j} \cdot c_{i,j+1} + \gamma_i - (\beta'_i + \beta_i).$$

As $\alpha'()$ is increasing and ν_{ij} satisfies (4), we have

$$\overline{\alpha}_{i,j} \cdot w_{i,j} + \alpha'\left(\overline{\alpha}_{i,j}\right) \cdot c'_{i,j} \leq \alpha_{i,j} \cdot w_{i,j} - \nu_{i,j} \cdot c_{i,j+1} + \gamma_i - (\beta'_i + \beta_i).$$

That is, inequality (5) is satisfied.

Although different, hypotheses (2) and (7) both make the assumption that processing should dominate communications. We can show that if

$$\alpha_{ij} \geq 2 \cdot \nu_{ij}, \tag{9}$$

then hypothesis (7) of Theorem 1 is satisfied; provided inequality (2) be satisfied.

In order to fix the value of $\overline{\alpha}_{i,j}$ according to constraint (5), we need a value for $\nu_{i,j}$. We can use statistical characteristics of the random variable $\overline{\alpha}_i$ which has the value $\overline{\alpha}_{ik}$ for the kth round of N_i. Let's limit ourselves to order 1 and 2 moments of the probability distribution of $\overline{\alpha}_i$, respectively its mean $\underset{k=j-j_0,j-1}{m}(\overline{\alpha}_{i,k})$ and standard deviation $\sigma_{k=j-j_0,j-1}(\overline{\alpha}_{i,k})$ over its values for rounds $j-j_0, j-j_0+1, j-j_0+2, \ldots, j-1$; for some j_0.

We can choose

$$\nu_{i,j} = m_{k=j-j_0,j-1}(\overline{\alpha}_{i,k}) + \lambda_i \cdot \sigma_{k=j-j_0,j-1}(\overline{\alpha}_{i,k}) \qquad \forall j \in \mathbb{N}^*. \tag{10}$$

As the amount of data processed during the steady-state phase is finite, there necessarily exists a real number λ_i such that inequation (4) is true (for each N_i).

Proposition 1. *Let us consider hypotheses of Theorem 1 are satisfied. If we choose $\overline{\alpha}_{i,j}$ and ν_{ij} according to (8) and (10) respectively, then the probability that the workers will compute without any idle period during the steady-state phase is greater than*

$$1 - \frac{1}{{\lambda_i}^2}.$$

The proof of this result is a direct consequence of the Bienaymé-Chebyshev's inequality.

When the values of the order 1 and 2 moments of the random variable $\overline{\alpha}_i$ are fixed, the sufficient condition (9) can be verified provided sufficiently small values of λ_i are used; precisely those that increase the probability to avoid inter-round starvation.

Proposition 1 requires the knowledge of $(\lambda_i)_{i=1,N}$. Even an heuristic estimate of a convenient value of λ_i allows the use of this theorem. Nevertheless, OLMR may still be used when these values (which characterize the dynamicity of execution parameters) are not known. Starting with arbitrary values (e.g. $\lambda_i = 0$ corresponding to a stability assumption), the scheduler could, if necessary, adjust λ_i values at any round according to information provided by the workers. Actually an inappropriate value of λ_i used for some round will lead to an intra- or inter-round starvation observable by the corresponding worker. The scheduler could then adjust this value for the next round, according to the type of starvation observed by the worker.

Parameters τ and λ_i are characteristic of the evolution of the execution parameters. On the one hand, τ characterizes their speed of evolution. Practically, it is the period that should be used for reconsidering their value. Parameter τ can be adjusted according to the finest time scale characterizing the evolution of the execution parameters. So doing, this evolution is taken account of (in average) over the duration of a round. On the other hand, λ_i measures the amplitude of their variations on such a period. The obvious dependence between τ and λ_i can take on the most varied forms. For instance, we can have rapid variations (small τ) with little consequence on the scheduling of the application (λ_i close to 0), or on the contrary slow variations (large τ) with important consequences on the scheduling (λ_i far from 0).

5 Conclusion

This paper reminds of the principle of the OLMR method presented in [7] and proposes an improved framework which delimits the application field of the method in a new way. OLMR method optimizes the workload distribution in order to minimize the makespan when executing parallel applications on shared resources such as those of a grid. It can be used when the information that scheduling algorithms traditionally need is lacking; so it can deal with the heterogeneity and dynamicity of the grid. Sufficient conditions have been stated for full usage of the computing resources by means of avoiding idle time. In order to design the OLMR method, we had to consider the characterization of the dynamicity of the execution conditions. This led us to define $N+1$ parameters: τ and $(\lambda_i)_{i=1,N}$.

As for PMR method [13], the use of the OLMR method must be coupled with some mechanism able to optimally select the resources to be used. Such resource selection can rely on heuristics, e.g. a greedy algorithm over the set of nodes ordered according to decreasing bandwidth (bandwidth-centric allocation [15]). It can be done at each round or according to the dynamicity of the execution parameters. OLMR method is susceptible to numerous developments, either tending to confirm the results of this paper or aiming at enlarging its potentialities. First of all, it is useful to check experimentally that, under the hypotheses of our model, the method gives the expected results. For that, we are currently developping simulation programs, using the SimGrid toolkit [16] in order to study OLMR behavior in various conditions and make comparisons with other methods. Furthermore, OLMR could be adapted in different ways: in this paper, τ and λ_i have implicitly been considered as constant throughout all the rounds, but this hypothesis restricts the degree of approximation (order one) of the dynamicity that the scheduler takes into account. From one round to the next, the value of τ could be adapted in order to take further account of the evolution of heterogeneity and dynamicity that would be noticed.

References

[1] V. Bharadwaj, D. Ghose, V. Mani, and T.G Robertazzi. Scheduling divisible loads in parallel and distributed systems. *IEEE Computing Society Press*, 1996.

[2] T.G. Robertazzi, J. Sohn, and S. Luryi. Load sharing controller for optimizing monetary cost, March 30 1999. US patent # 5,889,989.

[3] T.G. Robertazzi. Ten reasons to use divisible load theory. *IEEE Computer*, 36(5)(63-68), 2003.

[4] K. van der Raadt and H. Casanova Y. Yang. Practical divisible load scheduling on grid platforms with apst-dv. In *Proceeding of the 19th International Parallel and Distributed Processing Symposium (IPDPS'05)*, volume 1, page 29b, IEEE Computing Society Press, April 2005.

[5] O. Beaumont. Nouvelles méthodes pour l'ordonnancement sur plates-formes hétérogènes. Habilitation à diriger des recherches, Université de Bordeaux 1 (France), December 2004.

[6] A. Legrand, Y. Yang, and H. Casanova. Np-completeness of the divisible load scheduling problem on heterogeneous star platforms with affine costs. Technical Report CS2005-0818, UCSD/CSE, March 2005.

[7] S. Boutammine, D. Millot, and C. Parrot. An adaptative scheduling method for grid computing. In *Proceedings of the 11th international Euro-Par Conference (Euro-Par 2006)*, Springer-Verlag, 2006.

[8] L. Marchal, Y. Yang, H. Casanova, and Y. Robert. Steady-state scheduling of multiple divisible load applications on wide-area distributed computing platforms. *Int. Journal of High Performance Computing Applications*, 2006, to appear.
[9] J. Sohn, T.G. Robertazzi, and S. Luryi. Optimizing computing costs using divisible load analysis. *IEEE Transactions on Parallel and Distributed Systems*, 9(3), March 1998.
[10] T.G. Robertazzi. Divisible load scheduling. http://www.ece.sunysb.edu/˜tom/dlt.html.
[11] V. Bharadwaj, D. Ghose, and V. Mani. Multi-installment load distribution in tree networks with delays. *IEEE Transactions on Aerospace and Electronic Systems*, 31(2):555–567, 1995.
[12] Y. Yang and H. Casanova. *UMR: A Multi-Round Algorithm for Scheduling Divisible Workloads.* IEEE Computing Society Press, April 2003.
[13] O. Beaumont, A. Legrand, and Y. Robert. Optimal algorithms for scheduling divisible workloads on heterogeneous systems. Technical Report 4595, INRIA, Le Chesnay(France), October 2002.
[14] M. Drozdowski. *Selected problems of scheduling tasks in multiprocessor computing systems.* PhD thesis, Instytut Informatyki Politechnika Poznanska, Poznan, 1997.
[15] O. Beaumont, L. Carter, J. Ferrante, A. Legrand, and Y. Robert. Bandwidth-centric allocation of independent task on heterogeneous platform. Technical Report 4210, INRIA, Rhône-Alpes, Grenoble(France), June 2001.
[16] H. Casanova, A. Legrand, and L. Marchal. Scheduling distributed applications: the simgrid simulation framework. In *Proceedings of the 3th International Symposium on Cluster Computing and the Grid (CCGrid03)*, IEEE Computing Society Press, 2003.

Computational Efficiency and Practical Implications for a Client Grid

Nianjun Zhou and Richard Alimi

IBM
150 Kettletown Road, Southbury, Connecticut 06488–2685, USA
{jzhou, ralimi}@us.ibm.com

Abstract. Client grid computing models based on participation of non-dedicated clients have been popular for computationally intensive tasks. Two fundamental requirements of these models are efficiency and accuracy. Common implementations use 1) checkpointing mechanisms for higher efficiency and 2) redundancy to achieve accurate results. In this paper, we formulate client grid computation using stochastic models and analyze the effects of checkpointing and redundancy in relation to performance. We first quantify the computation times required for a task with and without checkpointing, then the relationship between result accuracy and redundancy. Finally, we give a sensitivity analysis for parameters relating to client availability, checkpointing, and redundancy to provide guidelines on design and implementation of client grid systems.

1 Introduction

Grid computing is a well-developed computational model used for resource-intensive or distributed computations. One such model, termed Public-Resource Computing [1], uses commonly available machines as part of a grid for research-oriented projects. People can "donate" the spare resources on their personal computers. Notable implementations include BOINC [2], which runs projects including SETI@home [3], and World Community Grid [4]. United Devices [5] markets Grid MP which provides a similar grid platform and is also employed by World Community Grid. We term these computational models *client grids* to differentiate them from models composed of dedicated machines. The client grid uses a single *management center*, and many clients which perform work on behalf of the grid, but are not dedicated to it. There are computational *jobs* that must be completed, each having its own priority. Each job is divided into a set of *tasks* which are executed by clients. Tasks running on clients typically run at low priority so they will not interfere with the client's normal work. Since clients are not dedicated and their reliability and characteristics are generally unknown few performance and accuracy guarantees can be made by the grid. The management center must also be cautious of the results collected from clients.

This computational model's performance has not, to the authors' knowledge, been explored analytically. Designers and administrators would benefit from

M. Gerndt and D. Kranzlmüller (Eds.): HPCC 2006, LNCS 4208, pp. 773–782, 2006.

modelling performance based on configuration parameters and client availability and reliability. Understanding these parameters and the factors that dictate performance are important to an efficient client grid solution.

Resource allocation and incentives in grid environments have been extensively studied. Many schemes and implementations for resource allocation [6, 7, 8] as well as studies concerning economics and incentives [9, 10, 11, 12] are available. Our model does not yet account for these issues since it is well known from existing client grid implementations [2, 4] that people do contribute spare resources. In the future, however, our model could be extended to include these considerations and analyze the degree to which people donate resources.

Checkpointing, an implementation of distributed snapshots [13], is important to many grid environments. It has been widely implemented and deployed [5, 4, 2, 3, 6]. Performance of individual checkpointing mechanisms has been evaluated [14], but their benefit in the context of general client grid computation has not been analyzed.

Result verification is one aspect that has been explored. Because clients cannot be trusted, the management center must verify results returned to it. Some projects deployed on BOINC use a technique called homogeneous redundancy [15]. This method distributes redundant copies of a task to "numerically equivalent" clients, thus giving a higher probability that results for divergent computations will match. Other methods [16] apply to *fault-tolerant* computations in which results for a set of tasks are combined to produce a single result within a certain accuracy. Our paper assumes the strict case which requires a certain number of results for a single task to match exactly.

This paper contributes the first attempt known by the authors to analyze client grid performance as a function of task characteristics and both availability and reliability of clients. Although the work itself is theoretical, it is grounded in the authors' experiences in client grid design, implementation, and knowledge of the World Community Grid project. The results can provide guidelines for application designers to effectively port applications to the grid.

We begin with an explanation of the problem and assumptions in Sects. 2 and 3. Important characteristics of the grid architecture are captured, with some simplifying assumptions. Our analysis is presented in Sects. 4 and 5. Sect. 4 is devoted to analyzing client task processing, while Sect. 5 considers task redundancy at the management center. Sect. 6 provides a practical analysis of adjustable or observable parameters and how they relate to performance and accuracy. Finally, Sect. 7 summarizes our contributions and gives directions for future analysis in client grid computing.

2 Problem Formulation

When a client is available to do work, it queries the management center. The management center assigns a task to the client which executes it. Upon completing the task, the client returns the result to the management center. There are factors which make implementation and analysis challenging. First, clients are not dependable since they can be powered off, busy with other work, or disconnected from the network. Tasks can save state periodically by a mechanism known as checkpointing. Second, results returned from clients may not be blindly trusted for reasons such as network transmission errors, faulty or overclocked hardware, or even manual tampering [15]. They are verified by redundant execution.

We begin by analyzing the time required by a client to complete a single task. We then determine the total computational time required for a task as it relates to task redundancy.

3 Stochastic Models

3.1 Client Availability

Each client has an *up* state in which it is available, and *down* state in which it is idle. We assume that upon entering the *up* state, the time before the client returns to the *down* state is distributed according to an exponential distribution,

$$p(t) = \frac{1}{T_{\text{up}}} e^{-\frac{t}{T_{\text{up}}}} \quad \text{for } t \in [0, \infty) \ .$$

We denote the average time the client spends in the *up* state as T_{up} and the average time spent in the *down* state as T_{down}. The exponential distribution is used here because it is both commonly used and simple. We assume i.i.d distributions for client availabilities, and that all clients have the same computational power.

Clients execute tasks in the *up* state. When the client enters the *down* state, it immediately stops executing its current task and does nothing until it enters the *up* state again. Each client can only store a single task at a time. That is, the client must request a task, execute it, then return the result before requesting another task. For this reason, we assume that a client is always connected to the network in the *up* state. Allowing a client to queue tasks then provides no advantage. Actual implementations will benefit by allowing the client to queue downloaded tasks, but analysis of this capability is left as future work.

A task requires a fixed amount of computation time. A *checkpointing* mechanism can also be implemented where the state of the computation of the task is saved at fixed intervals and the execution of the task can resume from the last saved state when the client re-enters the *up* state. Without a checkpointing mechanism, the task must be restarted from scratch.

4 Grid Client Analysis

4.1 Accountable Time for a Single Task

The term *accountable time* denotes the computational time that the client actually spends on this task while in the *up* state. Note that this is not equivalent to computational time required to complete the task. The client may have had to restart the task from scratch because it did not complete successfully the previous time. If the task supports checkpointing, then the task will be restarted from the last saved state. We begin by analyzing the case where the task does not implement checkpointing, then extend the results to handle the latter situation.

Without Checkpointing. Without a checkpointing mechanism, the client must be in the *up* state long enough to complete the entire task. The task takes t_{d} computational time to complete. We also define the ratio $\alpha = \frac{t_{\mathrm{d}}}{T_{\mathrm{up}}}$. The probability that a single execution of the task fails is then

$$p_{\mathrm{fail}} = \int_0^{t_{\mathrm{d}}} \frac{1}{T_{\mathrm{up}}} e^{-\frac{t}{T_{\mathrm{up}}}} \, dt = 1 - e^{-\alpha} \ . \tag{1}$$

We now determine the expected accountable time spent during a failed execution. The probability of entering the *down* state at time τ given that the execution of the task fails is a conditional probability distribution $p_{\mathrm{down}}(\tau) = \frac{e^{-\frac{\tau}{T_{\mathrm{up}}}}}{T_{\mathrm{up}}(1-e^{-\alpha})}$ for $\tau \in [0, t_{\mathrm{d}})$ (Considering that the probability of being in the *up* state for duration shorter than t_{d} is $1 - e^{-\alpha}$). The expected accountable time used for the failed execution is then

$$t_{\mathrm{fail}} = \int_0^{t_{\mathrm{d}}} \tau p_{\mathrm{down}}(\tau) \, d\tau = T_{\mathrm{up}} \left(\frac{1 - e^{-\alpha} - \alpha e^{-\alpha}}{1 - e^{-\alpha}} \right) \ . \tag{2}$$

Theorem 1. *The average accountable time required to complete a single task without checkpointing is*

$$t_{\mathrm{acct}} = t_{\mathrm{d}} \left(\frac{e^{\alpha} - 1}{\alpha} \right) \ . \tag{3}$$

Proof. Let i be the value of a random variable I denoting the first successful execution after $i - 1$ failures. I is then distributed geometrically with probabilities $p_{\mathrm{succ}}(i) = p_{\mathrm{fail}}^{i-1} (1 - p_{\mathrm{fail}})$ for $i \in [1, \infty)$. The accountable time is given by the function $t_{\mathrm{d}} + (i - 1)\, t_{\mathrm{fail}}$. We take the expected value, then substitute in (1) and (2). □

With Checkpointing. With a checkpointing mechanism, the task can be restarted from the last saved state rather than from scratch. The time interval between checkpoints is denoted as t_{c} and we require $\lceil \frac{t_{\mathrm{d}}}{t_{\mathrm{c}}} \rceil$ successful executions. We define γ to be the checkpoint frequency, $\gamma = \frac{t_{\mathrm{c}}}{t_{\mathrm{d}}}$.

Theorem 2. *The average accountable time required to complete a single task with checkpointing is*

$$t_{\mathrm{acct}} \approx t_{\mathrm{d}} \left(\frac{e^{\gamma\alpha} - 1}{\gamma\alpha} \right) \ . \tag{4}$$

Proof. Because the exponential distribution is memoryless, we can replace t_{d} in (3) with t_{c} to get the accountable time required for each chunk. Multiplying by $\lceil \frac{t_{\mathrm{d}}}{t_{\mathrm{c}}} \rceil$ gives $t_{\mathrm{c}} \left(\frac{e^{\gamma\alpha}-1}{\gamma\alpha} \right) \lceil \frac{t_{\mathrm{d}}}{t_{\mathrm{c}}} \rceil$ as the total accountable time. Approximating $\frac{t_{\mathrm{d}}}{t_{\mathrm{c}}} \approx \lceil \frac{t_{\mathrm{d}}}{t_{\mathrm{c}}} \rceil$ if $t_{\mathrm{d}} >> t_{\mathrm{c}}$, then produces the desired result. □

4.2 Average Client Time for a Single Task

We have calculated the time spent on the task while the client is in the *up* state, but ignored the *down* state. If we denote the percentage of time the client spends in the *down* state by s (thus, $s = \frac{T_{\text{down}}}{T_{\text{down}}+T_{\text{up}}}$) then the expected time required to complete a task after being assigned to a client becomes $t_{\text{client}} = \frac{t_{\text{acct}}}{1-s}$. Without checkpointing, this expression is

$$t_{\text{client}} = \frac{t_{\text{d}}}{1-s}\left(\frac{e^{\alpha}-1}{\alpha}\right)$$

and with checkpointing it becomes

$$t_{\text{client}} = \frac{t_{\text{d}}}{1-s}\left(\frac{e^{\gamma\alpha}-1}{\gamma\alpha}\right) .$$

5 Management Center Analysis

5.1 Task Redundancy and Computational Time

We begin by analyzing the redundancy required for an individual task. The target is a relationship between task result quality and completion time. Though there are many types of corruption that can occur, we only concentrate on two. The first type is corruption caused by hardware or network transmission errors, and the second is sabotage or some other correlated corruption. In either case, the management center will accept client results for a task until q of them match, at which point it will decide on a final result for the task. We seek a value for the task redundancy q such that the probability the task result is correct is sufficiently high. We assume a task is not assigned to the same client multiple times.

Random Corruption. The first type of corruption, hardware and network transmission errors, is simple to analyze. The number of possible errors is virtually infinite, and we assume the probability of each occurring is equally likely. The probability that two individual client results actually match is then essentially zero. Formally, there is a single correct result R and a set of corrupt results $\{R'_1, R'_2, \ldots, R'_N\}$ where N is virtually infinite. We assert that under the assumption that only random hardware and transmission errors occur, the required redundancy for a single task is $q = 2$ and the probability that the selected task result is corrupt is $p_{\text{corrupt}} = 0$. The expected number of redundant copies sent out is then $c = \frac{2}{1-\epsilon}$ where ϵ represents the proportion of corrupt results to correct results.

Correlated Corruption. We now consider the second type of corruption. We assume that only two distinct results R and R' are returned for a single task. Such instances have been attributed [16] to people attempting to improve their "score" by modifying their clients to process tasks faster. A certain percentage $\delta \in [0, 1)$ of the clients return the corrupt result R', while $(1-\delta)$ of the clients return the correct result R. At the point when it decides on a final task result, the management center will have received either q of result R or q of result R'. It will select that result and the task will be marked complete. We now relate the probability of choosing the corrupt result R' to parameters δ and q.

Theorem 3. *Given task redundancy q, the probability that the management center chooses a corrupt client result for a task is*

$$p_{\text{corrupt}}(q) = \delta^q \sum_{i=0}^{q-1} \binom{i+q-1}{i} (1-\delta)^i \ . \tag{5}$$

Proof. The probability of completing the task after receiving $(i+q)$ client results is

$$\binom{i+q-1}{i} \left\{ \delta^i (1-\delta)^q + \delta^q (1-\delta)^i \right\} \text{ for } i \in [0, q-1] \ . \tag{6}$$

We extract from this expression the probability that the result R' is selected after $(i+q)$ are returned. Then, $p_{\text{corrupt}}(q, i) = \delta^q \binom{i+q-1}{i} (1-\delta)^i$ for $i \in [0, q-1]$. Summing this expression over $i \in [0, q-1]$ gives the total probability the corrupt result R' is selected as the task result. □

Eq. (5) provides a way to choose the task redundancy q such that $p_{\text{corrupt}}(q)$ is sufficiently low. We can now calculate the expected number of copies of the task sent out using (6):

$$c(q) = q + \sum_{i=0}^{q-1} i \binom{i+q-1}{i} \left\{ \delta^i (1-\delta)^q + \delta^q (1-\delta)^i \right\} \ . \tag{7}$$

Computational Time. We can now quantify the total average client time required to complete a single task

$$t_{\text{task}} = c(q) * t_{\text{client}} \ . \tag{8}$$

Note that this is the amount of client computational time, not the wall clock time. This distinction is important since tasks are executed in parallel. From (7) and (5), we can first relate a single task completion time with the redundancy selected, and furthermore, relate to the accuracy required for the accepted result. More discussion can be found discussion of Fig. 2 in Sect. 6.

6 Practical Implications

We now look at the main results and what they mean to client grid implementations. We begin by looking at the effects that checkpointing and client availability

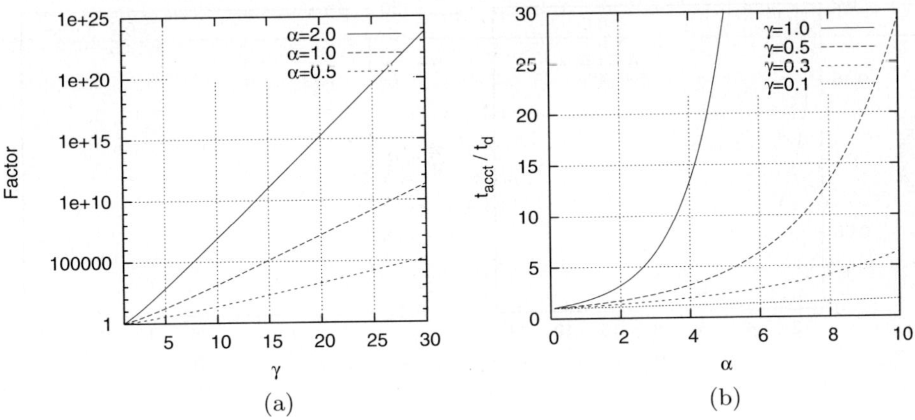

Fig. 1. (a) checkpoint improvement and (b) client availability

have on task completion time, then move to task redundancy. From our working knowledge, existing client grid implementations typically do not select values for checkpointing and redundancy parameters based on any formal arguments. The following analyses are intended to help designers and administrators make decisions in selecting parameters that give both performance and accuracy.

6.1 Checkpointing and Client Availability

Calculating the ratio of t_{acct} with checkpointing to t_{acct} without checkpointing gives us the improvement factor $\frac{e^{\gamma\alpha}-1}{\gamma(e^{\alpha}-1)}$ shown in Fig. 1 (a). As task size t_{d} grows larger in relation to the client's average *up*-time T_{up}, it is increasingly more beneficial to implement a checkpointing mechanism that checkpoints as often as possible. One must ensure, however, that the checkpointing mechanism does not require so much overhead that it becomes wasteful. Based on these results, one can balance the improvement factor with the checkpointing frequency γ.

It is also useful to look at accountable time as a function of the client's availability. From (4) we derive the ratio $\frac{t_{\text{acct}}}{t_{\text{d}}} = \frac{e^{\gamma\alpha}-1}{\gamma\alpha}$ shown in Fig. 1 (b). The effect on accountable time is drastic in the case of no checkpointing ($\gamma = 1$), but drops significantly as checkpointing increases. This ratio gives an idea for how frequently one might want to checkpoint a task based on typical client availability statistics.

6.2 Task Redundancy

Result integrity is typically of high importance, but it would be useful to obtain high integrity without unnecessarily impacting performance. We consider here the case of correlated corruption as discussed in Sect. 5.1.

In order to specify the task redundancy q, one must understand the probability of corruption given candidate values for q. Eq. (5) is shown in Fig. 2 (a). One extremely nice feature is that probability of corruption decreases extremely fast

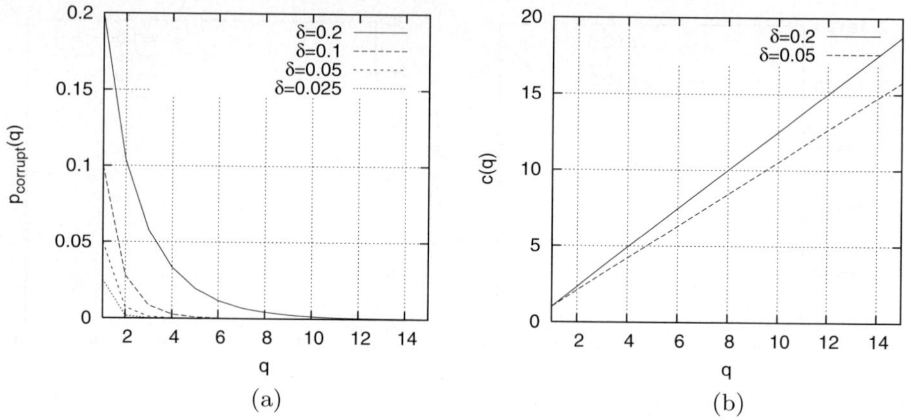

Fig. 2. (a) task redundancy and (b) expected copies sent out

as redundancy increases. The plot indicates that $q = 2$ or $q = 3$ are ususally sufficient as long as at least $\frac{9}{10}$ (thus, $\delta \leq \frac{1}{10}$) of the clients are trustworthy. Looking at (7) shown in Fig. 2 (b), we also see that the expected number of tasks actually sent out appears to scale linearly. Further, the number of extra tasks sent out is not too much more than q, even with relatively large (for a typical client grid environment) values of δ.

7 Summary and Future Work

We have established the basis for stochastically analyzing client grid efficiency and given practical results that can be applied to actual implementations. Our results quantify average computational time required by clients for individual tasks in relation to client availability. We have also quantified the reliability that can be expected from client grid computations under two types of corruption. As stated, our work establishes a basis for analysis of client grid systems. However, it can be extended to account for other common occurrences and trends.

An exponential distribution is used to model times between state changes in the client. This may not, however, be the actual behavior. A more accurate probability distribution might be obtainable only from an existing implementation, or an analysis could be done for arbitrary distributions.

Another extension is to allow heterogeneous clients each classified by power and availability. This might be easily analyzed as a set of independent client grids, each containing clients of similar classification.

If we allow for types of tasks other than computational tasks (for example, network-intensive tasks), then it would also be useful to look at other types of resources such as network bandwidth or storage.

Client task queuing features are becoming more popular in client grid implementations. This capability allows a client to download a set of tasks (up to some threshold) when possible so the client is not necessarily starved for work if disconnected from the network for a long period of time. The *up* state is split into two: *up-connected* and *up-disconnected*. An appropriate stochastic model needs to be developed for switching between these three states and determining times spent in each state.

Job completion times are extremely important for client grid environments. Multiple jobs compete for grid resources, and it is useful to quantify their individual completion times. We have preliminary results for job completion times as a function of job parameters and client behavior, but would like to extend them to determine feasibility and assignment of job parameters given completion deadlines. These results will appear in a future publication.

The management center is responsible for storing task input files and client result files until a final task result can be generated. With a large number of tasks, the storage requirement can be quite large. If the management center were to use a scheduling scheme that completed individual tasks sooner, task input files and extraneous client result files could be purged sooner as well. It would therefore be useful to quantify the disk usage for various scheduling schemes in relation to client resource usage.

The authors are involved in development of a client grid infrastructure within IBM, while other team members are actively involved in World Community Grid [4]. With these unique perspectives and the considerations listed in this section, we aim to improve our model and ensure the accuracy of our assumptions in practical implementations.

References

1. Anderson, D.P., Cobb, J., Korpela, E., Lebofsky, M., Werthimer, D.: Seti@home: An experiment in public-resource computing. Commun. ACM **45**(11) (2002) 56–61
2. Berkeley Open Infrastructure for Network Computing (BOINC): (`http://boinc.ssl.berkeley.edu/`)
3. SETI@home: (`http://setiathome.ssl.berkeley.edu/`)
4. World Community Grid: (`http://www.worldcommunitygrid.org/`)
5. United Devices: (`http://www.ud.com/`)
6. Litzkow, M., Livny, M., Mutka, M.: Condor - a hunter of idle workstations. In: Proceedings of the 8th International Conference of Distributed Computing Systems. (1988)
7. Foster, I., Kesselman, C.: The globus toolkit. (1999) 259–278
8. Foster, I., Kesselman, C., Lee, C., Lindell, R., Nahrstedt, K., Roy, A.: A distributed resource management architecture that supports advance reservations and co-allocation. In: Proceedings of the International Workshop on Quality of Service. (1999)
9. Buyya, R.: Economic-based Distributed Resource Management and Scheduling for Grid Computing. PhD thesis, Monash University, Melbourne, Australia (2002)
10. Clearwater, S.H., ed.: Market-based control: a paradigm for distributed resource allocation. World Scientific Publishing Co., Inc., River Edge, NJ, USA (1996)
11. Wolski, R., Plank, J.S., Brevik, J., Bryan, T.: Analyzing market-based resource allocation strategies for the computational grid. International Journal of High Performance Computing Applications **15**(3) (2001) 258–281

12. Wolski, R., Brevik, J., Plank, J.S., Bryan, T.: Grid resource allocation and control using computational economies. In Berman, F., Fox, G., Hey, A., eds.: Grid Computing: Making The Global Infrastructure a Reality. John Wiley & Sons (2003)
13. Chandy, K.M., Lamport, L.: Distributed snapshots: determining global states of distributed systems. ACM Trans. Comput. Syst. **3**(1) (1985) 63–75
14. Pruyne, J., Livny, M.: Managing checkpoints for parallel programs. In: Workshop on Job Scheduling Strategies for Parallel Processing (IPPS '96), Honolulu, HI (1996)
15. Taufer, M., Anderson, D., Cicotti, P., III, C.L.B.: Homogeneous redundancy: a technique to ensure integrity of molecular simulation results using public computing. In: IPDPS. (2005)
16. Germain-Renaud, C., Playez, N.: Result checking in global computing systems. In: ICS '03: Proceedings of the 17th Annual International Conference on Supercomputing, New York, NY, USA, ACM Press (2003) 226–233

Developing a Consistent Data Sharing Service over Grid Computing Environments*

Chang Won Park[1], Jaechun No[2], and Sung Soon Park[3]

[1] Intelligent IT System Research Center
Korea Electronics Technology Institute
Bundang-gu, Seongnam-si, Korea
[2] Dept. of Computer Software
College of Electronics and Information Engineering
Sejong University, Seoul, Korea
[3] Dept. of Computer Science & Engineering
College of Science and Engineering
Anyang University, Anyang, Korea

Abstract. Data replication is an important issue in computational grid environments where many data-intensive scientific applications require high-performance data accesses to remote storages. However, providing the consistent data replication service for computational grid environments is not an easy task because it requires a sophisticated technique to minimize I/O and communication latencies incurred during data copy steps to the remote clients. We have developed a replication architecture that is combined with data compression to reduce I/O and network overheads, with the help of data base. In this paper we describe the design and implementation of our replication architecture and present performance results on Linux clusters.

1 Introduction

The consistent data replication mechanism is a critical issue in computational grid environments where a large amount of data sets generated from large-scale, data-intensive scientific experiments and simulations are frequently shared among geographically distributed scientists [1,2,3]. In such an environment, the data replication technique significantly affects performance by minimizing the remote data access time. However, developing an efficient data replication technique is not an easy task because it may occur lots of I/O and network overheads during data copy steps to the remote clients. Furthermore, most of data replication techniques do not provide the consistent data replication service.

The usual way of managing consistent data replicas between distributed sites is to periodically update the remotely located data replicas, as implemented in Globus toolkit [4,5]. This method is implemented under the assumption that the data being replicated is read-only so that once it has been generated would not it be modified in any grid site. This assumption is no longer true in such

* This work was supported in part by a Seoul R&BD program.

M. Gerndt and D. Kranzlmüller (Eds.): HPCC 2006, LNCS 4208, pp. 783–792, 2006.

a case that a remote site may modify or update the data replicated to its local storage. If another remote site tries to use the same data replicated on its storage before the data is updated with the new one, then the data consistency between distributed sites would fail and thus the scientist would get wrong results.

We have developed a replication architecture that enables the geographically distributed scientists to safely share large-scale data sets. Using the metadata stored in database, our architecture provides a high-level, easy to use API (Application Programming Interface) for the data retrieval. Also, the metadata allows to combine a high-performance parallel I/O, called MPI-IO [6], to data compression to minimize I/O and network overheads, while hiding the detailed structure to users.

In order to support the consistent data replication service, our architecture supports two kinds of data replication methods, called owner-initiated data replication and client-initiated data replication. In the owner-initiated data replication, the data owner who owns the application data sets starts the data replication to share the data sets with remote clients. In the client-initiated data replication, the client who needs the data sets starts the data replication by connecting the data owner. Since, instead of being written to files, all the necessary data attributes and compression-related metadata are stored in database, our data replication architecture can easily be ported to any grid computing environments.

The rest of this paper is organized as follows. In Section 2, we discuss an overview of our software architecture. In Section 3, we present the design and implementation of our compression and replication methods. Performance results on the Linux cluster located at Sejong University are presented in Section 4. We conclude in Section 5.

2 Software Architecture for Data Replication

We describe the design and implementation of our replication architecture. We group the remote clients, according to the data sets to be shared, in order to minimize the effort in integrating the data replicas spread over the remote clients.

2.1 Client Grouping

Figure 1 shows an overview of our replication architecture. Our replication architecture supports the client grouping to eliminate unnecessary communication overheads. By making client groups, our architecture can easily detect the other clients who share the same data replicas and can let them take the new copy of the modified data, without affecting other clients.

The client grouping is performed by using the metadata stored in data base located at the data owner. The data base tables are organized to seven tables; *application_registry_table*, *run_registry_table*, *data_registry_table*, *file_registry_table*, *process_registry_table*, *client_registry_table*, and *replication_registry_table*.

When a client wants to receive the data replicas from the data owner at the first time, he should register to the metadata database table by clicking on the run_registry_table, data_registry_table, and file_registry_table to select the data

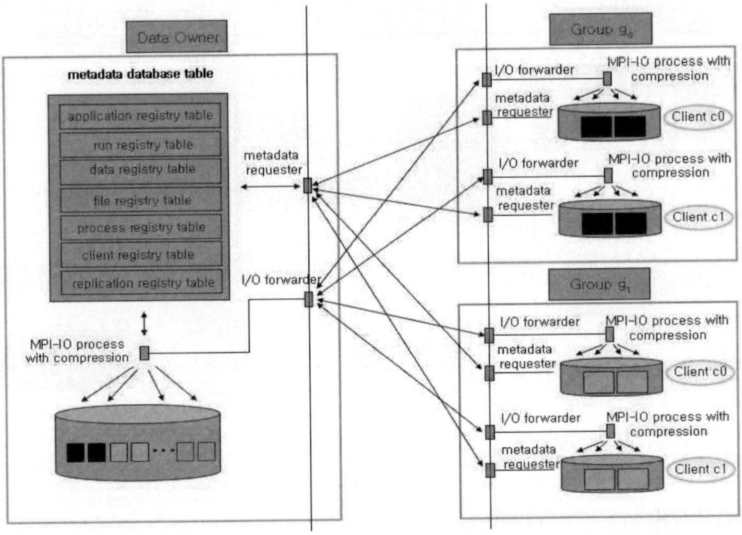

Fig. 1. Client grouping

sets of interest. If there is a group who has already registered to share the same data sets, the new client will then be a member of the existing group. Otherwise, a new group where the new client belongs to is created by taking the new group ID in the client_registry_table. Also, the replication_registry_table is appended to reflect the new client registration.

Figure 1 shows two remote client groups created based on the data replicas. Each client in a group is identified with groupID and clientID, such as (g_0, c_0) for the first client and (g_0, c_1) for the second client in Group g_0.

The communications for the metadata and real data are separately performed through the different communication ports. I/O forwarder is responsible for the real data communication and invokes the MPI-IO process to perform I/O. The metadata requester is responsible for transferring the metadata accessed from the metadata database table located at the data owner.

3 Data Compression and Replication

In order to achieve high-performance I/O and communication bandwidth, we use MPI-IO and data compression. The consistency between the distributed data replicas is achieved by applying either the client-initiated data replication method, or the owner-initiated data replication method.

3.1 Data Compression

With the aim of reducing I/O and communication overheads, our replication architecture combines MPI-IO with data compression using lzrw3 algorithm. After

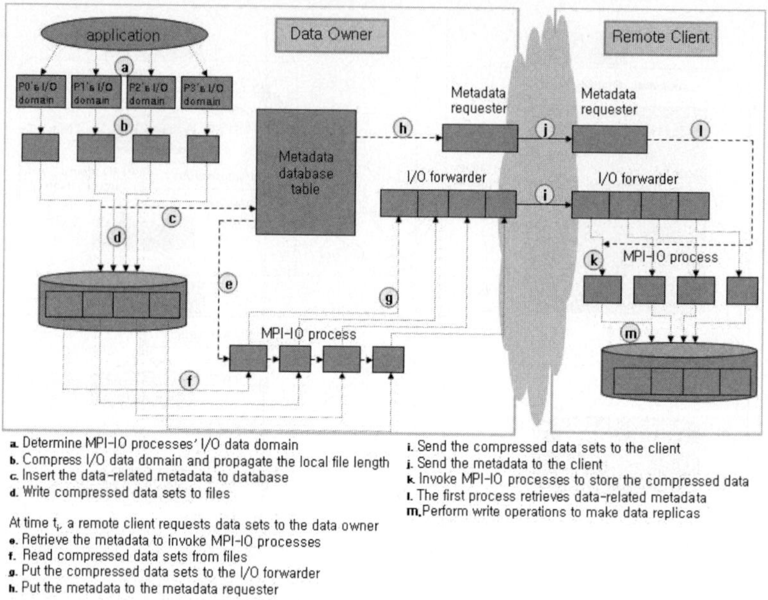

Fig. 2. Data replication combined with compression

merging the I/O data domain in MPI-IO, if the data compression is applied, the compression library is then linked to the replication architecture. Otherwise, the native file system I/O call is used. The compression-related metadata, such as the compressed global size, number of processes involved in the compression, local compressed file length, and starting file offset of each process, are stored in the data_registry_table, file_registry_table, and process_registry_table, respectively.

Figure 2 describes the data compression steps taken in replicating the data sets to the remote clients belonging to two groups, g_0 and g_1. After determining the I/O domain of each process at step a, the data sets are compressed at step b. Since the compressed file length of each process becomes unpredictable, it should be propagated among the processes to calculate the starting local file offset. The data attributes and compression-related metadata are stored into the metadata database table at step c and the compressed real data are stored in the file at step d.

Suppose that a remote client requests the data sets from the data owner at time t_i. If the data sets being requested have already been compressed, the data sets are then retrieved from the file using the compression-related metadata selected from the metadata data base table at steps e and f.

The compressed real data sets and the corresponding metadata are transferred to the requesting client, by I/O forwarder and metadata requester, at steps g through j. After receiving the compressed data sets, the I/O forwarder invokes MPI-IO processes at step k, using the metadata received by the metadata requester(step l), and the compressed data sets are stored in the local storage of the requesting client at step m.

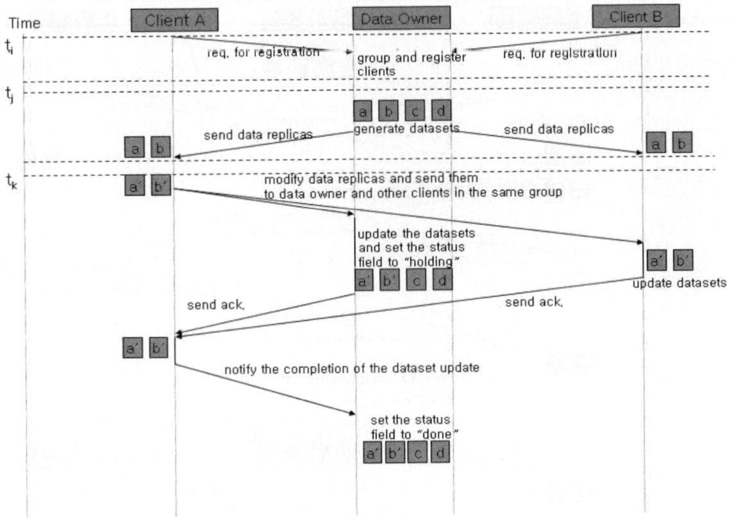

Fig. 3. Owner-initiated replication

3.2 Owner-Initiated Data Replication

In the owner-initiated data replication, when user generates data sets at the data owner, our architecture replicates them to the remote clients to share the data sets with the data owner. Also, when a remote client changes the data replicas stored in its local storage, it broadcasts the modifications to the members in the same group and to the data owner for replica consistency. Figure 3 shows the steps taken in the owner-initiated replication.

At time t_i, since the client A and client B want to receive the same data sets, a and b, from the data owner, they are grouped into the same client group. When an application generates the data sets at time t_j, a and b are replicated to the client A and client B.

Suppose that the client A modifies the data replicas at time t_k. The client A requests for the IP address of other members in the same group, sends the modified data sets to them, and waits for the acknowledgements.

When the data owner receives the modified data, it updates them to the local storage and sets the status field of the replication_registry_table to "holding" to prevent another client from accessing the data sets while being updated. When the data owner receives the notification signal from the client who initiated the data modification, it sets the status field to "done", allowing another client to use the data replica.

The owner-initiated data replication approach allows remote clients to share the data replicas safely, provided that they find out the corresponding status field is set to "done". Moreover, even though a remote client crashes, it doesn't affect the data consistency since as soon as the data replicas are modified the change is immediately reflected to the data owner and to the other clients in

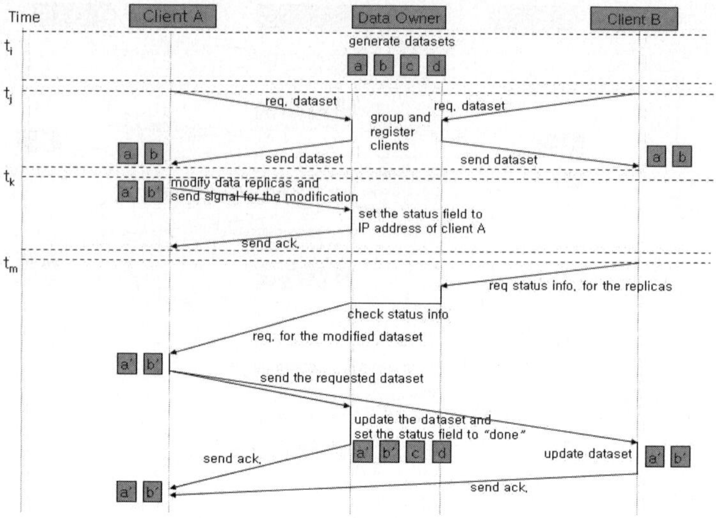

Fig. 4. Client-initiated replication

the same group. However, if the data modification to the data replicas happens frequently, the heavy communication bottleneck then incurs even if no one else would use the data sets modified.

3.3 Client-Initiated Data Replication

Figure 4 shows the client-initiated data replication where only when the modified data replicas are needed by users are those data replicas sent to the requesting client and to the data owner. Unlike in the owner-initiated data replication, there is no data communication when users on the data owner produce the application data sets. If a client needs to access the remote data sets stored in the data owner, he will then get the data replica while registering to our architecture.

In Figure 4, after the application generates the data sets at time t_i, the client A and client B request for the data sets, a and b, to the data owner at time t_j. Because they want to receive the same data sets, they become members of the same client group.

Suppose that client A modifies the data replica at time t_k. He just sends a signal to the data owner to update the corresponding status field of the data set with the IP address of client A.

At time t_m, suppose that the client B accesses the data replica stored in its local storage but not having been updated by client A. In order to check the replica consistency, client B first requests the status information of the data replica from the data owner. The data owner finds out that the data set has been modified by client A and requests the data from client A. Client A sends the modified data to the data owner and to the client B, and then waits for the acknowledgements from both. After the data owner updates the modified data set and sets the status field to "done", it sends back an acknowledgement to the client A.

In the client-initiated data replication approach, the data replicas are sent to the remote clients only when they actually need the data sets. Therefore, unlike the owner-initiated data replication approach, the client-initiated data replication does not cause unnecessary data communication. However, if a client who keeps the modification of the data replica crashes before the data modification is updated to the data owner and to the other members of the same group, a significant data loss will then happen.

4 Performance Evaluation

In order to measure the performance of replicating costs to remote clients, we used two Linux clusters, located at Sejong university. Each cluster consists of eight nodes having Pentium3 866MHz CPU, 256 MB of RAM, and 100Mbps of Fast Ethernet each. The operating system installed on those machines was RedHat 9.0 with Linux kernel 2.4.20-8.

The performance results were obtained using the template implemented based on the three-dimensional astrophysics application, developed at the University of Chicago. The total data size generated was about 520MB and among them, 400MB of data were generated for data analysis and for data restart, and then

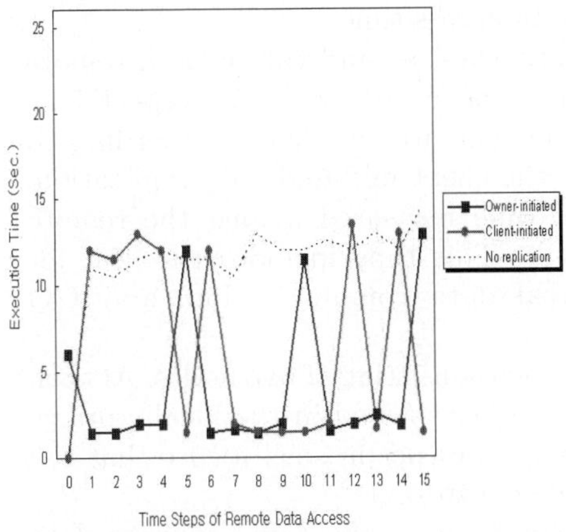

Fig. 5. Execution time for replicating visualization data on the remote clients as a function of time steps for accessing remote data sets. Two client groups were created, each consisting of two nodes.

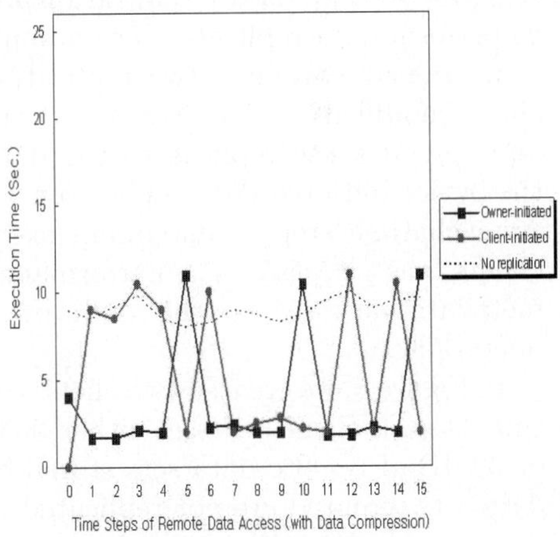

Fig. 6. Execution time for replicating compressed visualization data on the remote clients as a function of time steps for accessing remote data sets. Two client groups were created, each consisting of two nodes.

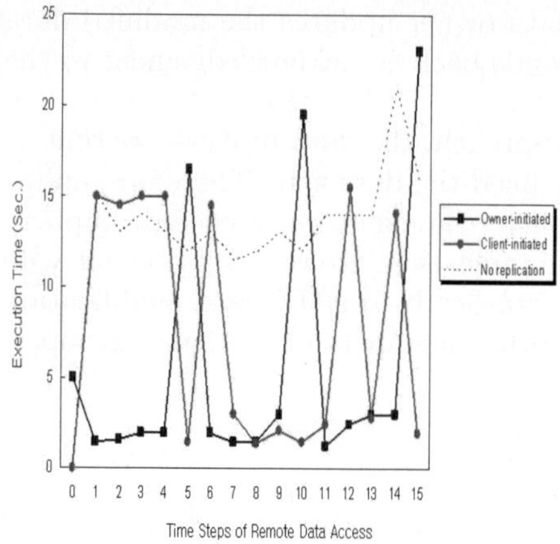

Fig. 7. Execution time for replicating visualization data on the remote clients as a function of time steps for accessing remote data sets. Two client groups were created, each consisting of eight nodes.

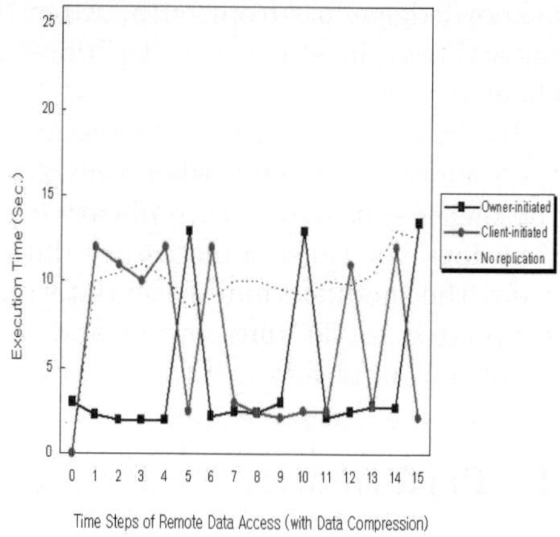

Fig. 8. Execution time for replicating compressed visualization data on the remote clients as a function of time steps for accessing remote data sets. Two client groups were created, each consisting of eight nodes.

the remaining 120MB of data were generated for data visualization. The data sets produced for data visualization are used by the remote clients, thus requiring to perform data replications to minimize data access time.

In order to evaluate two replication approaches, a randomly chosen remote client modified 30MB of replicas at time steps 5, 10, and 15, respectively, and spread those replicas to the data owner and to the clients, according to the owner-initiated data replication and to the client-initiated data replication. At each time step, a maximum execution time measured among the remote clients was selected as a performance result. This time includes the cost for metadata accesses to the data owner, real data communication, and I/O operations.

In Figure 5, we created two client groups, each consisting of two nodes. At each time step, a client accesses either 30MB of replicas stored in the local storage, or 30MB of remote data sets stored on the data owner in such a case that the data sets required are not replicated to the local storage.

In the owner-initiated replication, as soon as an application produces data sets at time step 0, all the remote clients receive the necessary visualization data sets to replicate them to the local storage. These replicas are used until the modification to the replicas happens at time steps 5, 10, and 15, respectively.

When the replicas stored in the local storage are used, the execution time for accessing visualization data sets drops to almost 3 seconds needed for communicating the corresponding data properties, such as file name, data type, and status information to check replica consistency, with the data owner.

If the modification to the replicas happens at time steps 5, 10, and 15, the modified replicas are then broadcast to the data owner and to the clients in the sample group, thereby increasing the execution time for accessing remote data sets.

In the client-initiated replication, since there is no replica stored in the client side until time step 4, each remote client should communicate with the data owner to receive the data sets needed. From time step 5, since each client can use the data replicas stored in its local storage, the execution time for accessing data sets dramatically drops to almost 3 seconds.

When the replicas are modified at time steps 5, 10, and 15, the client-initiated approach just sends to the data owner the IP address of the client modifying the replicas, and thus it takes no more than 3 seconds to access metadata. However, in Figure 5 we can see that, at time steps 6, 12, and 14, another client tries to access the modified replicas, thus incurring the data communication and I/O costs to update the replicas to the requesting client and to the data owner.

Without data replication, the data communication for accessing the remote data sets is consistently carried out at the clients, affecting the performance.

In Figure 6, we replicated the compressed data sets on the two client groups, each consisting of two nodes. The total data size being transferred was about 20MB, showing almost 30% of reduction in the data size. Due to the data compression, the I/O times for retrieving the compressed data sets from the data owner and for writing them to the remote clients are significantly reduced. We asynchronously propagated the compressed file length between I/O processes to calculate the starting file offset of the next process. We found out that it took about 100msec, thus not significantly affecting the I/O time.

The communication time for transferring the compressed data sets to the remote client is also decreased. However, because we use a single I/O forwarder to merge the compressed data sets from multiple MPI-IO processes, the reduction in the communication time is not as large as we expected. As a future work, we plan to develop a striped communication in which multiple I/O forwarders can connect the remote clients to transfer the evenly divided data sets.

In Figure 7, we increased the number of nodes in each group to eight. Using the owner-initiated data replication, we can see that as the number of nodes in each group increases the execution time for replicating remote data sets also goes up due to the increment in the communication overhead to broadcast the replicas to the data owner and to the other clients.

On the other hand, as can be shown in Figure 7, the client-initiated data replication shows not much difference in the execution time to receive the modified data sets because less number of nodes than in the owner-initiated data replication is involved in the communication. In Figure 8, we performed the same experiments as described in Figure 7, using the compressed data sets. As shown in Figure 8, both replication methods clearly show the reduced execution time with data compression.

5 Conclusion

We have developed a data replication architecture to build a consistent data sharing environment over the geographically distributed scientists. In order to reduce the I/O and communication costs in replicating the data sets on the remote clients, we used MPI-IO and data compression, and then they showed the clear benefits in the execution time. Also, in order to maintain the consistency in case of replica modifications or updates, we developed two kinds of data replication methods. In the owner-initiated data replication, the replication occurs when applications generate the data sets in the data owner location. In the client-initiated data replication, only when the data sets are needed by a remote client are the necessary data sets replicated to the requesting client. Due to the data broadcast, the owner-initiated data replication approach shows the increased communication overhead when the number of nodes per group becomes large. On the other hand, the client-initiated data replication shows the constant communication cost even with the increased number of nodes per client group.

References

1. B. Allcock, I. Foster, V. Nefedova, A. Chervenak, E. Deelman, C. Kesselman, J. Leigh, A. Sim, A. Shoshani, B. Drach, and D. Williams. High-Performance Remote Access to Climate Simulation Data: A Challenge Problem for Data Grid Technologies. SC2001, November 2001
2. R. Moore, A. Rajasekar. Data and Metadata Collections for Scientific Applications. High Performance Computing and Networking (HPCN 2001), Amsterdam, NL, June 2001
3. A. Chervenak, E. Deelman, C. Kesselman, L. Pearlman, and G. Singh. A Metadata Catalog Service for Data Intensive Applications. GriPhyN technical report, 2002
4. A. Chervenak, I. Foster, C. Kesselman, C. Salisbury, S. Tuecke. The Data Grid: Towards an Architecture for the Distributed Management and Analysis of Large Scientific Datasets. Journal of Network and Computer Applications, 23:187-200, 2001
5. A.Chervenak, R. Schuler, C. Kesselman, S. Koranda, B. Moe. Wide Area Data Replication for Scientific Collaborations. Proceedings of 6th IEEE/ACM International Workshop on Grid Computing (Grid2005), November 2005.
6. R. Thakur and W. Gropp. Improving the Performance of Collective Operations in MPICH. In Proceedings of the 10th European PVM/MPI Users' Group Conference (Euro PVM/MPI 2003), September 2003

Analysis of Interoperability Issues Between EGEE and VEGA Grid Infrastructures

Bartosz Kryza[1], Łukasz Skitał[1], Jacek Kitowski[1,2], Maozhen Li[3], and Takebumi Itagaki[3]

[1] Academic Computer Center CYFRONET-AGH, Nawojki 11, Cracow, Poland
{bkryza, lskital, kito}@agh.edu.pl

[2] Institute of Computer Science, AGH University of Science and Technology, Mickiewicza 30, Cracow, Poland

[3] School of Engineering and Design, Brunel University, Uxbridge, UB8 3PH, UK
{Maozhen.Li, T.Itagaki}@brunel.ac.uk

Abstract. Recent Grid initiatives brought a few large Grid infrastructures deployed over the world. However most of them are not easily interoperable and creation of Virtual Organizations that could include resources from a couple of different Grid systems is practically impossible.

In this paper, we present an analysis of issues that need to be resolved when providing interoperability between heterogenous Grid environments. The analysis is based on the example of EGEE and VEGA Grid systems middleware integration. A universal interoperability layer is proposed – called Grid Abstraction Layer – that will enable integration of these two Grid systems as well as other Grid environments and allow for creation of multi-Grid Virtual Organizations.

1 Introduction

Nowadays, Grid systems are being applied to a variety of domains of human life from scientific computations, biology and medicine to business. As usually at the beginning of new technology emergence the ongoing activities are diversified into different directions, making adoption of Grid systems in real life applications both fuzzy and difficult. The recent standardization efforts undertaken by Global Grid Forum [1] are expected to make it more convenient and less detail driven. Standards such as Open Grid Services Architecture (OGSA) and Web Services Resource Framework (WSRF) have been proposed to facilitate various Grids interoperable by employing service-oriented architectures.

One of the obstacles identified at present is the lack of unified way of referring to resources, both those from the Grid and from the application domains, addressing problems of semantic interoperability. This issue is included into the research area by several European projects related to Semantic Grid, like InteliGrid [2], OntoGrid [3] and K-Wf Grid [4], among others. The main idea of these attempts is to provide a Virtual Organization [5] with a formalism for describing the domain and its relation with the Grid resources as well as a knowledge base that would manage these descriptions. Often ontologies are proposed as a way of describing the content of Virtual Organizations and use of knowledge bases that manage these

M. Gerndt and D. Kranzlmüller (Eds.): HPCC 2006, LNCS 4208, pp. 793–802, 2006.

descriptions – thus reusing the results of Semantic Web [6] and tailoring them to the peculiarities of grid environments. One of the good examples of using intelligence techniques to capture and manage knowledge resources for next generation Web and decision making in a cooperative framework is Knowledge Grid ([7,8,9]), which incorporates grid computing ideas to offer intelligent services.

Another difficulty found by the users is variety of grid environments, that are being developed rather independently. They can be treated as interfaces for the users, like for example Globus [10,11] or Unicore [12,13], each of them offering a kind of system interoperability. Their dissimilar foundations result in different features and functional discrepancy, with important distinctions between subsequent versions. Nevertheless, the environments are starting to be widely used in scientific and business projects built upon a selected grid environment, like LCG [14], EGEE [15], TeraGrid [16], CoreGrid [17], BalticGrid [18], China National Grid (CNGrid) [19], and many others. Despite of the standardization efforts, a need for syntactic and structural interoperabilities between the existing grid environments have been strongly required in order to benefit seamlessly from distributed resources and services provided by the computing environments.

The purpose of the paper is to identify and discuss interoperability issues arising while dealing with different grid technologies with focus on gLite for EGEE and Vega Grid Operating System (GOS) for CNGrid. The organization of the paper is as follows: first a short overview of the related work is given, followed by gLite and CNGrid features description. Then a proposal of their interoperability is introduced with several possible realizations defined. Finally a summary and overview of the ongoing work is given.

2 Grid Interoperability Efforts

Interoperability between different Grid systems has been an issue for some time now, mostly due to several large Grid infrastructures which were managed by incompatible Grid middleware software. A notable attempt has been subject of an EU-IST GRIP project [20], concerned with integration of two major Grid middleware frameworks – Globus [11] and Unicore [13] . The proposed solution included translation of job requests from UNICORE to Globus and mapping of UNICORE user certificates to Globus proxy certificates [21]. The implemented translation layer did not require any changes to the middleware components themselves. The problem of assigning resources to particular requests (e.g. limited by a particular Virtual Organization) has been solved by wrapping a whole Globus Virtual Organization concept into a UNICORE Usite concept. In Snelling et al. [22] an interesting list of a minimal set of Grid services that is necessary for interoperability namely: authentication services, job brokering and resource discovery is presented. It omits from this set such issues as data management and performance monitoring. In Stockinger et al. [23] a discussion of integration of data management systems between CrossGrid and DataGrid projects is described. The work included integration of replica location services on both global Grid and local storage levels as well as integration of data access cost prediction at local storage systems. Malawski et al. [24] describes approaches to integrating CrossGrid middleware and applications based on Globus Toolkit 2.4 with the service oriented OGSA architecture. The similar approach, concerning LCG 2.6 and Globus Toolkit 4 interoperability, is reported in [25], for potential application for complex biomedical studies. Some existing approaches to grids interoperability are limited to user level [26] in order to make selection of the requested grid environment. The abstraction model for a grid execution framework, published recently [27] is useful for both making the grids interoperable at different level and practical interoperability implementations.

The ongoing EGEE project aims to provide the biggest Grid infrastructure in Europe with production quality, spanning 39 countries with over 200 sites including over 20,000 CPU and on average 10,000 concurrent jobs per day [28], to a bunch of various scientific applications. The gLite grid middleware is being developed by EGEE as a successor of the CERN's LCG grid middleware [14]. LCG has proved its utility value in several European projects, like CrossGrid, providing scientists with a production testbed [29] and European DataGrid [30]. Its follow up - EGEE2 - started on April 2006 will continue the effort on gLite development and infrastructure evolution.

On the other hand, the ongoing CNGrid project, supported by the National High-Tech R&D Program (the 863 program), aims to provide the biggest Grid infrastructure in China. Up to now, there are totally nine high performance computing centers having been built with a total computing capacity of 17 Tflops. These CNGrid nodes are interconnected by network backbones (CERNET, which is the China Education and Research Network, and CSTNet). Hosts in these centers include Dawning 2000/3000, Galaxy 3, Sunway, Tsinghua Tongfang PC-Cluster, HKU Gideon 300 cluster, etc. [31]

The issue on the interoperation of EGEE and CNGrid has received some attention from the Grid communities of the two sites. For example, the recently funded EUChinaGrid project [32,33] is carrying out some work in this area. However, the interoperability work to be fulfilled by EUChinaGrid will be limited due to the instrument nature of the project. Another, very recent, effort on Grid interoperation was the establishment of Grid Interoperability Now (GIN) including representatives from EGEE, PRAGMA and TeraGrid communities [34]. GIN concentrates on preparing a show-case of interoperation between these Grids by providing a testbed for evaluating the interoperability issues between the mentioned Grids. This includes VO authorization and authentication, job submission, data management and monitoring. In the context of Virtual Organizations the assumption for now is that all interoperating Grids must use Virtual Organization Management Service (VOMS) [35].

3 Overview of EGEE and VEGA Grid Middleware Platforms

The EGEE (Enabling Grid for E-Science in Europe) and VEGA (Versatile services, Enabling intelligence, Global uniformity and Autonomous control) are by far the biggest Grid infrastructures in Europe and China respectively.

The EGEE middleware, called gLite [36], is based on LCG-2 (Large Hadron Collider Grid) suite developed at CERN. gLite is essentialy based on the pre-WS Globus Toolkit 2.4. The architecture of gLite is divided into 6 classes of services and components:

- *Grid Access* - including Command Line and Application Programming interfaces,
- *Job Management Services* - which provide several services for job submission, and accounting,
- *Data Services* - provide the users and application with data discovery and transfer functionality,
- *Security Services* - which cover the functionality of authentication, authorization and auditing,
- *Information and Monitoring Services* - includes monitoring of network, hardware resources and job execution as well as service discovery,
- *Helper Services* - serves as a place for adding additional services like Network Bandwidth Allocation service or Agreement Service.

On the other hand, the VEGA middleware [37], especially its core component called Grid Operating System (GOS), has been implemented according to the OGSA specification and using Globus Toolkit 3 as the basis of the system. It emerged from the network of Dawning superservers deployed in the most important Chinese supercomputing centers [38]. In VEGA, the architecture reflects the OGSA three layers that is: Application Layer, Platform Layer and OGSI Layer. The lowest, OGSI Layer is based on GT3 and Web Services. The top OGSA layers contain such components as:

- *User interface and programming tools* - this includes GSML based portals and Abacus programming language,
- *Grid-level Task Management* - is based on the concept of *grip* (GRId Process), and provides framework for executing and scheduling of such jobs,
- *Global Data Management* - provides abstraction over GT3 data services GASS and GridFTP,
- *Address Space Management* - provides mapping between abstract concepts from VEGA GOS as grip or agora to physical grid resources,
- *Agora Management* - management of Virtual Organizations registered in the Grid.

4 Interoperability Between EGEE and VEGA

The main motivation for providing the interoperability between VEGA and EGEE is to enable creation of inter-Grid Virtual Organizations that would reuse the vast computational and storage resources of the EGEE and VEGA Grid infrastructures.

The major requirement here, is to enable creation and support for basic Grid use cases in a inter-Grid Virtual Organization. In gLite, the VOMS (Virtual

Table 1. The equivalence of key functional components of EGEE and VEGA

EGEE component	VEGA component	Functionality
Computing Element	GRAM	job submission
Workload Management System	Grip Management	resource brokering
R-GMA	MDS-2	information services
FiReMan and Data Movement	VegaFS	data management
VOMS	Agora	VO management
Cmd line interface	GSML Portal	user interface
gLite Security	GSI	security

Organization Manangement System) stores some additional information on each user or group such as roles and rights. This information could be extended to provide information on roles and rights to access other Grid resources - and still remain transparent to the end user. In VEGA the Agora management framework [39] is also very flexible, as it is based on a theoretical EVP model (Effective, Virtual, Physical) where every community (Virtual Organization) can be defined using 4-tuples of the form: (Subject, Object, Context, Policy).

Another important assumption is to allow users of the two Grid environments to use their native user interfaces in a way they are used to. However in order to provide this functionality, some intervention into the respective middleware layers is necessary. Basically, in the case of any two grid environments the set of components whose functionality must be mapped in between consists of: authentication, authorization and Virtual Organization mapping; job submission; information services; data discovery and transfer; and monitoring.

The core components of EGEE and VEGA that will need to be mapped in order to achieve the interoperability are depicted in Table 1.

The proposed idea for interoperability between the mentioned two Grid environments is in creating an intermediate abstraction layer, called Grid Abstraction Layer (GAL), that will act as an intermediary for all kinds of requests that can be made between the Grids. The interoperability layer will provide virtual equivalents of all basic Grid components that should allow reusing the shared Grid resources by users of the native Grids in the most invisible way. It is planned that GAL layer will reuse existing standards and recommendations such as JSDL [40] for job description or CIM [41] for resource description. GAL will also make integration and creation of multi-Grid Virtual Organizations easier for other Grid environments. The GAL layer will maintain mappings of all kinds of Grid resources that are reused between the Grid infrastructures. Most importantly, that includes mapping of jobs and their descriptions that are being executed in the other Grid environment, mapping of files stored in both Grid's storage space and user authentication and authorization mapping. It is also envisioned that ontologies will be used for description of hardware resource mappings and for matchmaking purposes. Of course, in order to provide the end users with full transparency of the two Grids, some extension or modification of their middleware is necessary. In the case of gLite this can be achieved to some extent

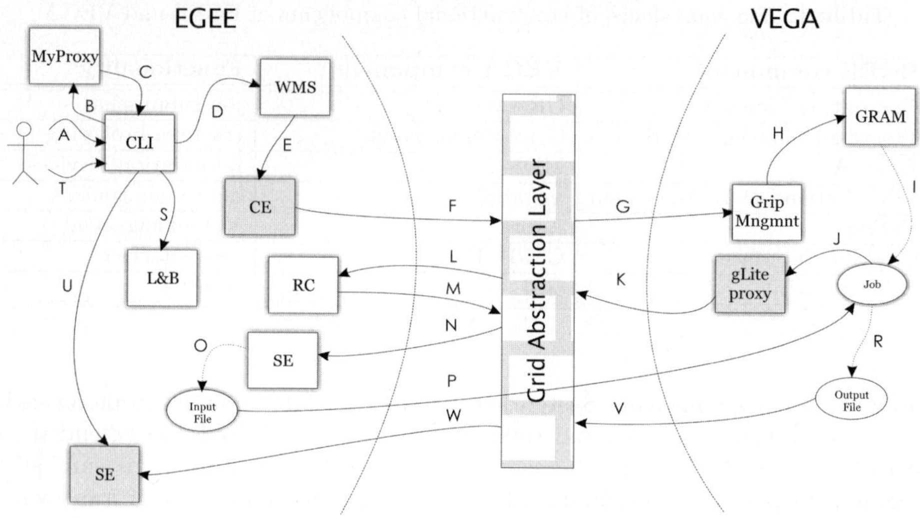

Fig. 1. A job submission scenario from EGEE to VEGA

by provision of fake Computing and Storage Elements that will appear to the gLite core components as usual services, although they will interoperate with the GAL layer in order to forward requests to the VEGA Grid system. Another option would be to extend existing core components of gLite such as Workload Management System (WMS) with the capability of submitting the jobs to GAL layer, however, since WMS is a central point of the gLite middleware this could become difficult due to administrative issues.

5 Example Scenario Analysis

Let's try to analyze a seemingly simple case of submitting a job by an EGEE user, that will be scheduled for execution on a VEGA Grid node (Figure 1).

5.1 Login and Proxy Generation

The first step the gLite user has to make in order to perform any action is to login to some Command Line Interface (CLI) node (A). Afterwards he has to create a temporary proxy certificate using voms-proxy-init (B,C), which will be used for requests performed on users behalf. The proxy will also be used to authorize users requests at several points of the job submission with Virtual Organization Management System (VOMS). In case of EGEE and VEGA the same certificates can be used, since both these environments depend on the Globus Security Infrastructure (GSI), provided that the sites trust each other.

5.2 Job Submission

The next step is submission of the job to the Workload Management System (WMS) (D), using command glite-job-submit. The WMS finds the best matching Computing Element for execution of the job (E). Let's now assume that in this case, WMS chooses a fake Computing Element (CE) which forwards all jobs to the GAL layer for execution in the VEGA Grid environment (F). At this point GAL will convert the job request from gLite's JDL to a GGF standard Job Submission Description Language (JSDL) [40]. This step is motivated by the idea that the GAL should also enable in the future integration of other Grid environments than gLite and VEGA. The GAL figures out, based on the parameters of the job such as users VO membership and required resources, that it will be submitted to the VEGA Grid. The job is then submitted to the VEGA resource broker or translated directly to Resource Specification Language (RSL) [42] and submitted to some particular VEGA Grid node - by means of the local Grid Resource Allocation Manager (GRAM) interface. Now, while the job is being executed (I), it requests the Input File. In gLite, a job can request input data in two basic ways. One is to use a script, that executed just before the actual job execution, transfers the file using gLite CLI command glite-get. The other way is to use the gLite I/O API directly in the job code. In either case, the VEGA Grid site, that belongs to the inter-Grid Virtual Organization - should provide wrappers of these command in order for jobs to be executed fully transparently. Such library, when called by the job, will contact the Grid Abstraction Layer, and ask it to provide the requested file - in this case by indicating the proper lfn file name. In this case, GAL will contact gLite's Replica Catalogue (L) to obtain the location of the best replica (M) and then contact the proper Storage Element (N) to transfer the file to the local storage where the job is being executed in the VEGA Grid (P). Finally the job is executed and output file is produced to some local storage of VEGA Grid (R). Of course, at any time during the job execution, the user should be able to monitor the status of the job using the command glite-job-status. In such case the CLI contacts the Logging&Bookkeeping (L&B) database (S), that in gLite stores all information about past and current jobs, which should be kept up to date with the information about the job being executed in the VEGA Grid by the fake Computing Element.

5.3 Obtaining Job Results

After the job execution is finished, the user requests the output produced by the job by issuing command glite-job-output (T) and the CLI connects to the L&B component in order to get the location of the data produced by the particular job. The L&B returns to the CLI the location of the output data generated by the job and CLI transfers the data to the users' local storage using GridFTP protocol. Afterwards, the fake CE element is called by the GAL layer and informs the WMS that the job has been completed and how to access the produced data. More complicated case is when the job executed in VEGA Grid would produce a file that would be registered somewhere in the VegaFS [43]. This file should then be made accessible to the gLite user through the fake Storage Element component (U). In order to provide this functionality such fake Storage Element would have to be informed about the creation and registration of files in the VEGA Grid file system through the GAL layer (V, W). Since the SE node provides a Local Replica Catalogue of files that are stored in it, it should be possible to implement it in such a way that it would provide a view on the files accessible in the VEGA Grid storage. Alternatively, in the case of job execution, the job could also use the gLite Proxy wrapper for the glite-put command and thus store the file directly in some gLite Storage Element through the GAL layer.

6 Conclusions and Future Work

In this paper we have presented an idea for interoperability of two Grid infrastructures, namely EGEE and VEGA, at the middleware level. The identification of components that need to be mapped between the two Grid environments was performed and specified. Discussion of different approaches to integration of the EGEE and VEGA was presented, including an outline of possible challenges and drawbacks. An example of a job submission scenario from EGEE to VEGA Grid was sketched, along with detailed identification of potential problems and challenges, such as authorization, authentication and data management . Additionally, a Grid Abstraction Layer component, acting as an intermediary between the Grid environments was introduced. The envisioned solution will be generic enough to enable other Grid environments to reuse the results of this work in future Grid interoperability efforts.

The future work will include careful design and implementation of the GAL layer along with any missing abstraction models of Grid concepts - such as for instance the concept of Virtual Organization or data registered in the Grid. Also, several wrapper libraries and services for both gLite and VEGA that will enable the actual interoperation of the resources in the EGEE and VEGA grid systems will be developed. In particular, implementation of Computing and Storage Elements for the purpose of forwarding the Grid requests from gLite to VEGA Grid through the Grid Abstraction Layer will be provided.

Acknowledgements

This research has been done in the framework of EU IST-034363 Gredia project. AGH University of Science and Technology grant is also acknowledged.

References

1. http://www.ggf.org/
2. Intelligrid project homepage, http://www.inteligrid.com/
3. Ontogrid project homepage, http://www.ontogrid.org.ontogrid/
4. K-Wf Grid project homepage, http://www.kwfgrid.net/

5. Foster, I., Kesselman, C., and Tuecke, S., The Anatomy of the Grid: Enabling Scalable Virtual Organizations, Int.J. Supercomputer Applications, 15(3)(2001).
6. http://www.w3.org/2001/sw/
7. Zhuge, H., The Knowledge Grid, World Scientific Publishing Co., Singapore, 2004.
8. Zhuge, H., The Future Interconnection Environment, IEEE Computer, 38 (4)(2005) 27-33.
9. Zhuge, H., China's E-Science Knowledge Grid Environment, IEEE Intelligent Systems, 19 (1) (2004) 13-17.
10. Foster, I., and Kesselman, C., The Globus project: a status report, Proc. IPPS/SPDP'98 Heterogeneous Computing Workshop, 4-18, 1998.
11. GLOBUS Project website, http://www.globus.org
12. Erwin, D., (ed.) Joint Project Report for the BMBF Project Unicore Plus, ISBN 3-00-011592-7, 2003.
13. UNICORE Project website, http://www.unicore.org
14. The LHC Computing Grid, http://lcg.web.cern.ch/
15. The Enabling Grids for E-sciencE (EGEE), http://public.eu-egee.org/
16. http://www.teragrid.org/
17. The European Research Network on Foundations, Software Infrastructures and Applications for large scale distributed, GRID and Peer-to-Peer Technologies (CoreGrid), http://www.coregrid.net/
18. The BalticGrid Project, http://www.balticgrid.org/
19. China National Grid (CNGrid), http://www.cngrid.org/en_index.htm
20. GRIP Project website, http://www.grid-interoperability.org
21. Rambadt, M., and Wieder, P., UNICORE - Globus: Interoperability of Grid Infrastructures, Proc. Cray User Group Summit 2002.
22. Snelling, D., van den Berghe, S., von Laszewski, G., Wieder, P., MacLaren, J., Brooke, J., Nicole, D., and Hoppe, H.-C., A UNICORE Globus Interoperability Layer, Online, http://www.grid-interoperability.org/D4.1b_draft.pdf
23. Stockinger, K., Stockinger, H., Dutka, L., Slota, R., Nikolow, D., Kitowski, J., Access Cost Estimation for Unified Grid Storage Systems, 4th International Workshop on Grid Computing (Grid2003), Phoenix, AZ, USA, November 17, 2003, IEEE Computer Society Press, 2003.
24. Malawski, M. Bubak, M., Zajac, K., Integration of the CrossGrid Services into the OGSA Model, in: Bubak, M., et al. (Eds.), Proc. Cracow'02 Grid Workshop, Dec.11-14, 2002, Cracow, Poland, ACC Cyfronet AGH, 2003, Cracow, pp. 140-147.
25. Tirado-Ramos, A., Groen, D., Sloot, P.M.A., Exploring OGSA Interoperability with LCG-based Production Grid for Biomedical Applications, in: Bubak, M., (eds.) Proc. Cracow'05 Grid Workshop, Nov.20-23, 2005, Cracow, Poland, ACC Cyfronet AGH, 2006, Cracow, in press.
26. P. Lindner, E. Gabriel and M. Resch, GCM: a grid configuration manager for heterogeneous grid environments, Int.J.Grid and Utility Computing, 1(1)(2005) 4-12.
27. Amin, K., von Laszewski, G., Hategan, M., Al-Ali R., Rana, O., and Walker, D., An Abstraction model for a Grid Execution Framework, Euromicro J. Systems Architecture, 2005, in press.
28. Laure E., Production Grids Infrastructure Enabling eScience, CESNET Conf., March 6-8, 2006, Prague, Czech Repulic
29. Development of a Grid Environment for Interactive Applications (CrossGrid), http://www.crossgrid.org/
30. The European DataGrid (DataGrid), http://www.eu-datagrid.org/

31. Guangwen Yang, Hai Jin, Minglu Li, Nong Xiao, Wei Li, Zhaohui Wu,Yongwei Wu, Feilong Tang, Grid Computing in China. Journal of Grid Computing 2(2): 193-206 (2004).
32. EUChinaGrid Project homepage, http://www.euchinagrid.org/.
33. EUChinaGrid Project Deliverable D5.1: Project Presentation, http://www.euchinagrid.org/docs/EUChinaGRID-Del5.1v1.3.pdf
34. Catlett C., Grid Interoperation Now DRAFT Charter, http://forge.ggf.org/sf/go/doc3216?nav=1
35. Alfieri R., Cecchini R., Ciaschini V., dell'Agnello L., Frohner A., Gianoli A., Lorentey K. and Spataro F., VOMS, an Authorization System for Virtual Organizations., In Proc. of European Across Grids Conference, 2003, Santiago de Compostela, Spain, pp. 33-40, LNCS 2970, Springer, 2004
36. EGEE Consortium, EGEE Middleware Architecture, EU Deliverable DJRA1.4, Available at https://edms.cern.ch/document/594689
37. Zhiwei Xu, Wei Li, Li Zha, Haiyan Yu and Donghua Liu, Vega: A Computer Systems Approach to Grid Computing, In Journal of Grid Computing: GCCO3: A Spotlight of Grid Computing, June 2004, pp. 109-120, Volume 2, Number 2, Springer-Verlag
38. Xu Z., Sun N., Meng D. and Li W., Cluster and Grid Superservers: The Dawning Experiences in China., In Proc. 2001 IEEE Intrnl. Conf. on Cluster Computing (CLUSTER 2001) 8-11 October 2001, Newport Beach, CA, USA,
39. Wang H, Xu Z., Gong Y. and Li W., Agora: Grid Community in Vega Grid, In Proc. of Second International Workshop on Grid and Cooperative Computing, GCC 2003, Shanghai, China, December 7-10, 2003, Part 1, pp. 685-691, 2003, Springer
40. Anjomshoaa et. al., Job submission Description Language, Version 1.0, GFD-R.056, Available at http://forge.gridforum.org/projects/jsdl-wg
41. Quirolgico S., Assis P., Westerinen A., Baskey M. and Ellen Stokes, Toward a Formal Common Information Model Ontology., WISE Workshops, pp. 11-21, 2004, Springer
42. The Globus Resource Specification Language RSL v1.0., Online, http://www-fp.globus.org/gram/rsl spec1.html
43. Wei Li, Jianmin Liang and Zhiwei Xu, VegaFS: A Prototype for File-Sharing Crossing Multiple Administrative Domains., In Proceedings of IEEE International Conference on Cluster Computing, 1-4 December 2003, Kowloon, Hong Kong, China, IEEE Computer Society, pp. 224-231, 2003

Temporal Storage Space for Grids*

Yonny Cardenas, Jean-Marc Pierson, and Lionel Brunie

LIRIS CNRS UMR 5205, INSA de Lyon
{yonny.cardenas, jean-marc.pierson, lionel.brunie}@liris.cnrs.fr

Abstract. A distributed system like grid can be viewed as hypothetical infinite storage system for storing large data entities. Frequently these large data entities can not be stored in one unique location. In this paper we propose a temporal storage space managed as a collaborative cache system where clients have a perception of this space as an unlimited temporal storage capacity. They use the temporal storage space to share and to reuse large data entities in a distributed way. Temporal storage space is composed of a set of autonomous caches that work cooperatively with individual caches are generic cache services deployed inside organizations and a specialized service to coordinate global operations that facilitate the cooperation between caches. Implementation issues of a prototype in Globus Toolkit 4 are discussed.

Keywords: grid caching, collaborative cache, grid data access, temporal storage.

1 Introduction

Distributed systems support different models of distributed computations. Frequently, a significant quantity of data must be reused and shared between different locations. An important proportion of data must thus be moved from different places but typically these data are used only for a limited period of time. This data movement, without global coordination, tends to use inefficiently an important quantity of computing and network resources. In this respect an advanced system is necessary to support temporal data dissemination in distributed systems.

Distributed system like grid [1] can be viewed as an hypothetical infinite storage system that allows to store large data entities (terabytes and more). Since these data are in constant growth, it is necessary to manage the data volume in a dynamic way. The distributed data management system proposed in peer-to-peer systems where the size of data objects is relatively small and data copy proliferation expected or tolerated is undesirable or expensive. Grid data management solutions such as Storage Resource Broker SRB [2] that provides uniform access to distributed data and Data Replication Service DRS [3] that integrates data services for replica operations do not support automatic management of distributed temporal data.

* This work is supported by the Région Rhône-Alpes project RAGTIME and the French Ministry for Research (ACI GGM).

M. Gerndt and D. Kranzlmüller (Eds.): HPCC 2006, LNCS 4208, pp. 803–812, 2006.

In this work, the focus is put on the automatic operation and organization of large volumes of temporal data for reuse and sharing. We propose a temporal data management system for distributed environments based on a new collaborative caching approach. Contrary to traditional collaborative caching [4] where the cooperation is limited to data resolution, the collaborative cache capacities are extended to manage distributed temporal data in our approach. Several caches cooperate to place and manage large data entities. In this work we assume that datasets updates and consistency mechanisms are not required since most scientific data entities are accessed in read-only manner [3].

The contributions of this paper include: 1) a generalization of the functionality required for reuse and share of distributed temporal data in the grid; 2) a description of an approach to grid caching based on an extension of collaborative cache capabilities; 3) a description of a Temporal Storage System (TSS) to operate and coordinate temporal data which is a practical implementation of our grid caching approach; and 4) a description of an implementation prototype of the TSS in the Globus Toolkit 4 [5] (GT4) environment.

2 Grid Caching Approach

Caching is recognized as one of the most effective techniques to manage temporary data. A cache is a place where data can be stored temporary so that they are easily and quickly available in the near future. Caching includes the replacement strategies for determining which data should be kept in the cache. In this respect cache is a system that controls automatically data objects contained for a limited period of time in a storage resource [6].

The main objective of cache is to reduce the need to get data from original sources. Similarly, cache permits to reuse and share data by different clients, thus data resources are used more efficiently. To scale cache capabilities, several collective cache schemes have been proposed for distributed systems [4]. These collaborative cache approaches have been developed for web traffic on Internet which has different characteristics in contrast to traffic in a grid. The main difference is related with the size of data objects which are relatively small in web, hence the data copy proliferation is expected or tolerated. Several peer-to-peer data management mechanisms are based on this characteristic [7]. In contrast to web environment, grid may manage large data entities, currently terabytes and soon petabytes: These data entities are expensive to store and transfer, data copy proliferation in grid is thus undesirable.

In traditional web collaborative caching, groups of caches work together for retrieving a particular data entity, the potential cooperative scalability is reduced to data resolution. Grid supports distributed models of computation that need more capabilities such as storing working data anywhere in the distributed system.

Dynamic interaction requirements in distributed systems also raise an important question about distributed caching; how to deal with caches which do not collaborate ? Organizations have diverse and sophisticated temporal storage mechanisms but they usually work isolated. Grid technology needs to integrate these specialized mechanisms to scale cache capabilities. A fundamental assumption of conventional collaborative cache approaches is that the deployment and dynamic interaction of caches across administrative domains is not constrained. Grid technology is the result of agreements to provide basic interoperability standards to help cooperative cache systems to be deployed in multi administrative distributed systems.

A new collaborative cache approach is necessary to adapt caching to requirements of temporary data management in grid. We propose a grid caching approach based on three principles: Autonomy, Collaboration, Coordination.

The **autonomy principle** aims at gathering different and specialized temporal storage mechanisms that can be managed with cache techniques which are not originally designed to work collaboratively. This principle establishes that each cache is autonomous and may apply particular technologies, control techniques, and policies for internal data resources management.

The autonomy principle has two dimensions: technical and administrative. Technical because collaborative cache approach aggregates high capacity of temporal data storage from heterogeneous and specialized mechanisms. And administrative, because these mechanisms are administrated by different corporations. Thus each corporation shares storage resources keeping the internal control and functionality.

Our approach aggregates independent entities or caches. Each entity has specialized capacities and dexterity that it shares with others caches. We build a distributed temporary data system as a federation of autonomous caches.

The **collaboration principle** aims at extending the cache capabilities in two dimensions: internal and external in reference to cache location in the distributed system. Internally, temporary data sharing level between corporative applications is extended introducing cache service concept. Cache service is an external mechanism that operates and manages temporal data that are used by wide variety of clients inside a corporation. Cache service provides standard operations to reuse and share temporal data between applications inside an organization. Externally, the collaborative cache approach does not restrict its interactions to data resolution. Since it interacts to execute other data management operations.

The cache service works as a temporary data gateway between corporative location and the external distributed system. Thus a cache service exposes standard cache operations outside the corporation, to other cache services. Therefore collaborative cache increases the cooperation capabilities expanding cache interactions to all cache operations. Cache service concept is the result of collaboration principle, a generic abstraction of the cache mechanism. We define a cache service as an independent and generic entity that offers temporal storage to internal and external applications and services [8].

The **coordination principle** operates and controls interactions between caches to manage the collaborative cache scheme [4]. Cache interaction capabilities are organized to implement a collaborative cache system. Similarly to the collaboration principle, the coordination principle is supported by exposed operations in each cache service. With this principle each cache must support coordination operations that permit to control and operate the inter-cache actions. In this way it is possible to build the collective mechanisms that federate distributed caches as a uniform cache system [9].

To support the collective coordination mechanisms each cache service implements special monitor and control operations. To monitor, each cache service provides information for other services to support inter-cache data sharing, e.g. description of cache data content. Additionally cache services provide information on their own configuration and behavior (which replacement policy is in use, how full is the cache, ...). To control, each cache service adjusts dynamically its configurable parameters for interaction with other caches. Similarly, control mechanism establishes collective configuration parameters such as cache information exchange, collective data resolution or data global placement.

3 Temporal Storage Space (TSS)

We propose the Temporal Storage Space (TSS) which is a data management system that dynamically operates and coordinates distributed temporal data. TSS is used for sharing and reusing data between members of the distributed system. Clients have the perception that TSS has an unlimited capacity for storing temporal data: These clients delegate temporal data distribution and administration to TSS. The Temporal Storage Space (TSS) is designed following the principles of autonomy, collaboration and coordination (described previously in section 2). Thus, TSS is a distributed virtual space that is built as a collaborative cache system where a group of caches work together to gather temporal storage resources dynamically.

TSS is formed from a group of Local Cache Services (LCS) (described in section 3.1) and one Collective Coordinator Service (CCS) (described in section 3.2) which are deployed at multiple organization domains. LCSs come together to build the temporal storage space as illustrated in figure 1.

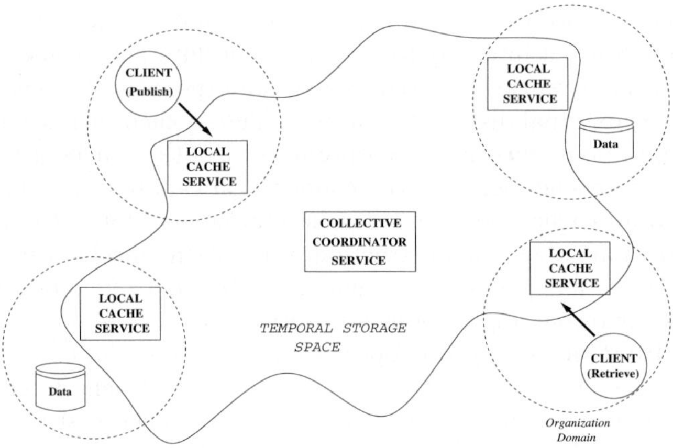

Fig. 1. Temporal Storage Space with LCS and CCS

3.1 Local Cache Service (LCS)

The Local Cache Service (LCS) is the component that implements the cache service locally inside each organization. LCS exposes local cache functionalities to external system and works as a gateway for data access between the local organization and the external system. It resolves data queries from applications using local internal catalog which is updated periodically with summaries from other LCSs.

LCS translates external data access operations to internal cache operations and vice versa. It includes basic metadata in its catalog using a database system, which is used to extend internal cache query resolution capacity. LCS implements conventional interfaces inside an organization and standard interfaces for external interactions. LCS offers functionalities inside organizations where it is deployed, e.g. a health center deploys a LCS that is used for the health center internal applications. On the other hand, the same LCS offers cache functionalities to external organizations in a grid virtual community e.g. the same LCS deployed in the health center will offer cache functionalities to LCSs of other organizations in the same healthgrid virtual community.

Working in a collaborative mode, the LCS requests remote LCSs to execute publish and retrieve operations(3.3). Regularly each LCS sends reports of its data content to all LCSs. LCS sends periodically cache information reports to CCS.

3.2 Collective Coordinator Service (CCS)

The Collective Coordinator Service (CCS) is a specialized service that coordinates the interactions between LCSs. In this way, it implements TSS specific coordination functionalities. CCS implements operations to indicate locations to place data between different sites of the distributed system. CCS uses metrics to evaluate which location will be selected to place the Data Entity. Sophisticated placement algorithms can base the placement selection on multiple metrics, combining them in a single or hybrid metric. Placement algorithms permit to affect the operation and objective of TSS, for example, efficient resources use, increasing access performance, load balancing, data copy proliferation, etc. A discussion about these sophisticated placement algorithms are beyond the scope of this paper. In this work we use a simple mechanism for placement determination that is based on cache storage capacity. Cache storage capacity refers to the level of availability of the individual LCS storage resource. It can be calculated in a variety of ways, including number and size of data entities and access per second. Storage capacity is calculated from cache report. CCS updates its collective internal catalog with cache information provided by LCS. The collective catalog consistency is supported with periodic reports from LCS.

The CCS resolves LCS requests for distributed placement of particular data entities. It checks the information in its collective internal catalog in order to extract information useful to make a decision about distributed data placement in TSS.

3.3 LCS and CCS Operations

This section describes how the LCS and CCS support three main operations defined for the temporal storage space: data publish, retrieve, and placement. The operations described in this section make references to grid implementation using GT4 middleware.

Data Publish Operation permits clients to put data into the TSS. Figure 2(a) illustrates the exchange protocol behind the publish operation. Before the client prepares the optional metadata to registry with data in LCS, it uses basic and extensible XML schema to register the metadata. The client sends a request to LCS for data placement into virtual storage space giving an explicit description of the data (metadata)(1). The LCS executes its placement policy to establish if it can store data locally or if it must use any external storage (2). If the data must be stored externally, the LCS sends a message to the CCS asking for global distributed placement (3). The CCS executes the global placement policy based on its collective internal catalog and it returns the selected LCS where to place the data(4). The LCS sends a request to remote LCS for external data placement (5). The LCS starts the data transfer process using the GridFTP service (6). The GridFTP servers perform the data transfer between sites(7). The remote LCS checks periodically the status of each data transfer request to ensure that each file was transferred successfully (8). LCS periodically builds a data content report, then it sends this report to CCS and all LCSs. Each LCS updates its catalog with this external cache information for future data resolutions (9).

Data Retrieve Operation permits clients to get data from TSS. Figure 2(b) illustrates the protocol of the retrieve operation. The client sends a data request to LCS (1). The LCS queries its local internal catalog to establish the data location. If the data are stored in remote LCS, LCS creates a data requests to remote LCS where data are localized (2). The remote LCS starts the transfer process using the GridFTP service (3). The GridFTP transfer service performs the data transfer between sites(4). In the local site, LCS checks periodically the status of each transfer request to ensure that each file is transferred successfully (5). Finally, the LCS delivers data to client (6).

Data Placement Operation is executed by CCS to decide the distributed placement of data. Each LCS builds periodically a data content report then the LCS sends this report to CCS. This way, CCS collects cache information of all LCSs(1). CCS uses this information to determine global data placement. To support this process CCS maintains the collective internal catalog, which contains cache information about all LCSs.

3.4 TSS Information

The TSS needs information for collaborative cache operation, describing the main elements of collaborative cache such as storage, data, cache and transfers. Cache information provides a description of each element itself and its behavior. We classify these information as element entity and element activity respectively. Element entity information includes static information about elements such as

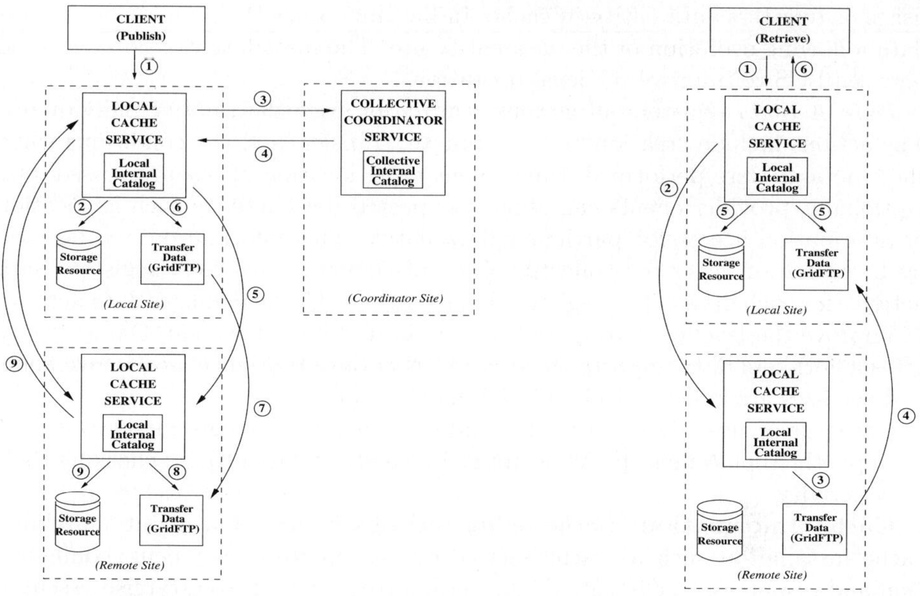

(a) Publish operation

(b) Retrieve Operation

Fig. 2. LCS and CCS operations

an identification and configurable attributes or parameters. Element activity information includes dynamic information such as behavior and actions.

TSS manages information about the following elements: storage, data, cache and transfers.

Storage Information: It is information about storage resource. *Storage entity* information includes characteristics such as type of device, storage capacity, rate transfer, delay, filesystem. *Storage activity* information includes information about use of storage resources: storage in use, available space.

Data Information: *Data entity* information describes individually each data object, this includes identification and description structure of data registered. Data entity specifies permitted actions on data content. An unique identifier is established for each registered data entity which can be based on Logical File Name (LFN) or Universal Resource Identifier (URI). Cache service clients use unique identifier as key to make operations related with data entities. *Data entity* information is used for managing individual data object: access type, share level, authorization, This information is established at data registry operation. Data entity information may include optional information such as metadata. Metadata represent semantic content of data entity, including references to external specific and detailed metadata or metadata services. Cache service registers essential metadata pieces for general semantic classification: target applications, data provenance, software version, algorithms, parameters and short annotations. This metadata can be used for data discovery mechanism to improve retrieval operation. Initial metadata is supplied by the provider client who publishes or registers data entity in cache. In life time other clients add more metadata following evolution of the data entity use. The metadata can be exchanged between LCS to improve retrieval operations.

Data Activity registers all actions realized on each data entity individually. The actions include creation, access, update, transfer, and the conditions when these actions were performed: time, client type, location of resources used, etc. Optionally, provider clients can register expected data activity such as location of information access for particular data entity. This information can be used by coordination layer for collective data placement. Each LCS registers data activity for each data entity registered in its service. Cache manages data activity to improve the level of sharing and reusing in distributed system. Data activity gives a notion of data importance because those data that have more movements and accesses are believed to be more important [9].

Cache layer uses data entity and data activity to manage data locally, for example the replacement policy is applied on a set of data objects checking their data activity.

Cache Information: *Cache entity* registers information about individual cache installation such as cache service reference, site deployment, configuration and resources. Cache entity information permits CCS to establish essential characteristics of group caches.

Cache Activity registers all actions realized by each LCS individually. This is information about dynamic cache service behavior. Cache activity shows individual and dynamic cache service status and performance. This information includes: storage capacity in use, cache hit and miss ratio, replacements ratio, number of intercache data request generated and served, etc. Cache activity information is extracted from monitoring data activity of individual LCS about its data entities. It is registered by each LCS in internal catalog and it is exchanged for collective coordination actions.

Transfer information is related mainly with time expected to deliver data objects between cache services. *Transfer entity* describes logical interconnection end-to-end between caches, this includes achievable transfer bandwidths, cost and availability. *Transfer activity* includes dynamic information such as performance and latency. Transfer activity is established from specialized services that provide information about resource performance in the distributed environment. Alternatively transfer activity can be obtained by the cache service by monitoring their own activity. Historical information concerning data transfer rates is used as a predictor of future transfer times.

4 Prototype

We have developed a prototype of the LCS and CCS that support the main functionalities described in section 3. Both are implemented as independent grid services. LCS is a generic and configurable cache that manages local storage resource at location where it is deployed, configurable local parameters are: storage capacity available, data entity time to life, replacement method, etc. LCS supports dynamic replacement of cache method (LRU, LFU and Size)[4].

LCS prototype manages cache information using a database management system. It stores minimal optional metadata about data entity supplied by data provider using a pre-established XML format. It coordinates the data transfers with GridFTP invocation and delegation service. As grid service, LCS exposes its operations to receive requests from other LCS to implement the extended cache collaboration. LCS prototype can be invoked directly by user applications, LCSs or other grid services to publish or retrieve data.

The first CCS prototype is a basic implementation of grid service described in section 3.2 It registers periodically cache information reports such as used cache capacity from LCS. Optionally CCS can monitor the capacity of remote LCS. CCS registers the cache information in internal catalog. It processes synchronously the requests of LCS for collective data placement. In this version CCS uses a simple algorithm to select LCS to place the specific data entity: It selects the LCS with the most available storage capacity. Algorithms and more complex metrics can be implemented that use CCS catalog cache information.

LCS prototype is deployed in three French laboratories of GGM Project [10]: LIRIS, IRIT, and LIFL located in Lyon, Toulouse and Lille respectively. The CCS prototype is deployed in Lyon.

The prototype is designed based on the Web Services Resources Framework (WSRF) [11] standards to support interoperability. It is implemented using middleware Globus Tool Kit version 4.0 (GT4) [5]. Currently we are testing the operational functionality of the prototype. We plan to make performance measurements for large files transfer and increased number of data file placement requests between LCS and CCS.

5 Related Work

Our goal is to develop a general-purpose system that provides a flexible cache service at different level. By contrast, several other data systems take a different approach and provide higher-level data management capabilities using highly integrated functionality. These systems include Storage Resource Broker (SRB) [2]. SRB provides distributed clients with uniform access to diverse storage resources. It has notion of uniformity but has not autonomous administration for replicas placement, copies proliferation control, and internal resource optimization.

Data Replication Service(DRS) [3] is a high level data management service. It is a service that coordinates others services such as GridFTP, RLS, RFT and Delegation. It is designed as a grid service and implemented with WSRF standards. It can be used with different metadata catalog services. It coordinates the operations with data replicas, but it does not perform global data management, does not optimize placement and data copies resolution, nor it guarantees data consistency. Transparency is not achieved since clients are aware of the replication, and must supply all details for replica operations.

Distributed Parallel Storage System (DPSS) [12] is a mechanism for network data cache. It is used to access massive amounts of data. It is a temporal storage system that works as a virtual disk for remote clients. DPSS reduces the impedance or contention between different data storage systems. Internally it works as a large buffer and supports concurrent data access. DPSS can work as federative system for different storage systems, but it does not manage data distribution between different sites. DPSS does not establish a global view of data in the distributed system and it does not have any notion of collaboration.

6 Conclusion

We have described a temporal storage space (TSS) used to share and to reuse large data entities in distributed system like grid. This virtual space gives the perception of unlimited storage capacity. Users in distributed system delegate temporal data administration to this storage service. We presented the main components of TSS proposal: a group of distributed caches, collective coordination mechanism, cache information interchange, and cache operations. We described how the distributed data placement management is supported by TSS using these components. We also presented the implementation of the system based on interaction of two specialized services: Local Cache Service (LCS) and Collective Cache Service (CCS). We made a detailed description of the main operations for access data (publish and retrieve), and coordination operation for distributed data placement.

References

[1] Foster, I., Kesselman, C., Nick, J., Tuecke, S.: The physiology of the grid: An open grid services architecture for distributed systems integration (2002)

[2] San Diego Supercomputer Center (SDSC): Storage resource broker (srb). http://www.sdsc.edu/srb (2003)

[3] A. Chervenak R. Schuler, C. Kesselman, S.K.B.M.: Wide area data replication for scientific collaboration. In: 6th IEEE/ACM International Workshop on Grid Computing - Grid 2005, Seattle, Washington, USA, IEEE ACM (2005)

[4] Wang, J.: A survey of web caching schemes for the internet. ACM Computer Communication Review **25** (1999) 36–46

[5] Globus Project: The globus project. http://www.globus.org (2002)

[6] Barish, G., Obraczka, K.: World wide web caching: Trends and techniques (2000)

[7] Androutsellis-Theotokis, S., Spinellis, D.: A survey of peer-to-peer content distribution technologies. ACM Comput. Surv. **36** (2004) 335–371

[8] Y. Cardenas, J. Pierson, L.B.: Service de cache pour les grilles de calcul. In: Extraction des connaissances: Etat et perspectives, Toulouse, France, Editions Cepadues (2005) 199–202

[9] Y. Cardenas, J. Pierson, L.B.: Uniform distributed cache service for grid computing. In: Sixteenth International Workshop on Database and Expert Systems Applications (DEXA 2005), Copenhagen, Denmark, IEEE Computer Society (2005) 351–355

[10] Grid for Geno-Medicine Project: Grid for geno-medicine. http://liris.cnrs.fr/projets/liris/projets/ggm (2004)

[11] Global Grid Forum (GGF): Web services resource framework (wsrf). http://www.globus.org/wsrf (2003)

[12] Tierney, B., Johnston, W., Lee, J.: A cache-based data intensive distributed computing architecture (2000)

e-AIRS: An e-Science Collaboration Portal for Aerospace Applications

Yoonhee Kim[1], Eun-kyung Kim[1], Jeu Young Kim[1], Jung-hyun Cho[1], Chongam Kim[2], and Kum Won Cho[3]

[1] Dept. of Computer Science, Sookmyung Women's University, Seoul, Korea
{yulan, kimek, wldud5, abaekho}@sookmyung.ac.kr

[2] School of Mechanical and Aerospace Engineering, Seoul National University, Seoul, Korea
chongam@snu.ac.kr

[3] Supercomputing Application Technology Dept. KISTI Supercomputing Center, Daejon, Korea
ckw@kisti.re.kr

Abstract. Computational simulation in fluid dynamics often requires high performance and massive data intensive process. In addition, advanced expertise to manage complicated simulation processes with various environmental conditions in order to obtain reasonable results is essential but not easy to achieve. This paper presents an e-Science Aerospace Integrated Research System (e-AIRS) that aims to use Grid technology to establish an integrated and collaborative environment to enable distributed computational simulation and remote experiments for remote scientists in aerospace area. e-AIRS provides easy-to-use portal services to manage workflow-based experiment process from application design, execution, monitoring, and visualization for Grid applications.

Keywords: aerospace, CFD, collaboration, portal.

1 Introduction

Most research activities related to e-Science[13] in many countries often include massive, computationally expensive and data intensive processes in astrophysics, particle physics, biology, chemistry, engineering application, environmental engineering and medical science. Space Technology (ST) field which is a vigorous area among engineering application parts in the inside and outside of the country, with the purpose of offering an environment that allows aerospace researchers reduces efforts and avoid overlapping investment by connecting numerical researchers and experimental data in the aerospace field as well as achieves remote visualization. Due to lack of appropriate and automatic collaborative methods in aerospace, it takes time on obtaining adequate tools and performing experiments using limited experimental equipments. Moreover, geometrical separation among research institutions and their unskilled co-works frequently result in duplicate investments. Therefore, it is urgent to arrange an infrastructure for aerospace researchers which permits not only work

M. Gerndt and D. Kranzlmüller (Eds.): HPCC 2006, LNCS 4208, pp. 813 – 822, 2006.

under the integrated environment and share results but also continue working and analyze results remotely. To offer an integrated environment that links nationwide experiment tools with controlling computers and makes it possible for ST researchers to cooperate with each other as well as carry out remote numerical experiments, this activity is going to sum up related state-of-the-art technologies and focuses on developing required core software.

What scientists in the area of engineering and scientific computing mainly need is a grid-enabled portal [18], by which they are able to deploy applications easily without concern with the complex details of grid technologies. Specifically, a portal can be viewed as a Grid-based problem solving environment that allows scientists to access distributed resources, and to monitor and execute distributed Grid application from a web browser. With graphical user interfaces, the portal is an integrated development environment based on grid services. Since our e-AIRS is based on portal system for enabling aerospace researchers to securely and remotely access and collaborate on the analysis of aerospace vehicle design data, primarily the results of wind-tunnel testing and numeric (e.g., computational fluid-dynamics: CFD) [5] model executions. This system provides an environment for Aerospace research scientists to make as easy as possible to use. They are actively involved from the beginning when we design overall architecture and design user interface of each step.

This paper describes the e-AIRS portal, which serve as a web interface to computational grids to access remote data files and remote services. We describe an e-Science Aerospace Integrated Research Systems (e-AIRS). It provides a development environment for geometrical mesh generation, computation, monitoring and visualization. The main feature of the e-AIRS is effective integration of several computational tools into graphical user interfaces implemented in the web portal.

2 Related Work

DAME [11][16] is an e-Science pilot project, demonstrating the use of the GRID [14] to implement a distributed decision support system for deployment in maintenance applications and environments. DAME will develop a generic test bed for Distributed diagnostics that will be built upon grid-enabled technologies and web services. The generic framework will be deployed in a proof of concept demonstrator in the context of maintenance applications for civil aerospace engines. The project will draw together a number of advanced core technologies, within an integrated web services system:

DAME will address a number of problems associated with the design and implementation of On-line decision support systems. The most significant of these are access to remote resources (experts, computing, knowledge bases etc), communications between key personnel and actors in the system, control of information flow and data quality, and the integration of data from diverse global sources within a strategic decision support system.

The new web services model for information brokerage on the Internet offers an inherently pragmatic framework within which to address these issues. DAME will exploit emerging open standard web service architectures over a GRID network to demonstrate how the data management aspects of maintenance support systems can be handled within a unified knowledge broker model.

DAME is an EPSRC[8] funded e-Science pilot project. It will demonstrate how Grid technology can facilitate the design and development of decision support systems for diagnosis and maintenance, in which geographically distributed resources, actors and data are combined within a virtual organization.

The DAME project exploits the emerging OGSI/OGSA[5][6][12] Grid service architecture to demonstrate how the data management aspects of maintenance support systems can be handled within a unified knowledge broker model. A proof of concept demonstrator is being built around the business scenario of a distributed aircraft engine maintenance environment, motivated by the needs of Rolls-Royce and its information system partner Data Systems and Solutions.

The GECEM[9] project aims to use and develop Grid technology to enable large-scale and globally-distributed scientific and engineering research. The focus of the project is collaborative numerical simulation and visualization between the UK and Singapore from which experience will be gained in the use of Grid technology to support research in the context of an 'extended enterprise'. In addition to these high-level objectives, the project also looks to develop Grid-enabled capability and products for use by the wider community. This paper reports on the current status of the GECEM project and discusses a prototype Grid integrating resources at the BAE Systems Advanced Technology Centre near Bristol, the Welsh e-Science Centre (WeSC)[15] in Cardiff, and the University of Wales, Swansea (UWS). This Grid is capable of taking a model geometry generated at BAE Systems, transferring it to UWS where it is meshed. The meshed geometry is then transferred to WeSC where it is used to solve a computational electromagnetic problem and visualized.

GEODISE[10] will provide Grid-based seamless access to an intelligent knowledge repository, a state of the art collection of optimization and search tools, industrial strength analysis codes, and distributed computing and data resources.

3 e-AIRS Portal

The e-AIRS portal, released for the first time in Jan 2006, builds upon the core features in the GridSphere [1] portal framework to provide developers with a framework for developing Grid-enabled portlets. The GridSphere portal framework provides an open-source portlet based Web portal. Indeed, the overall view of development within the GridSphere framework has been generally positive. It provides common portlets as standard, and these can be extended or added to in the form of new portlets. The portlets are implemented in Java and built upon reusable Java Server Pages (JSP) based user interface components, called action components.

Our main goal is to establish the powerful and user friendly collaboration environment to aerodynamic researchers. The e-AIRS portal provides the main interface through which services are accessed. User can access all e-AIRS software and hardware resources will be via a portal. This will hide the complexity of the system from the end-user. In addition, it also provides an interface to mesh generation, calculation and monitoring of such applications on remote resources, and collaborative visualization, exploration, and analysis of the application results.

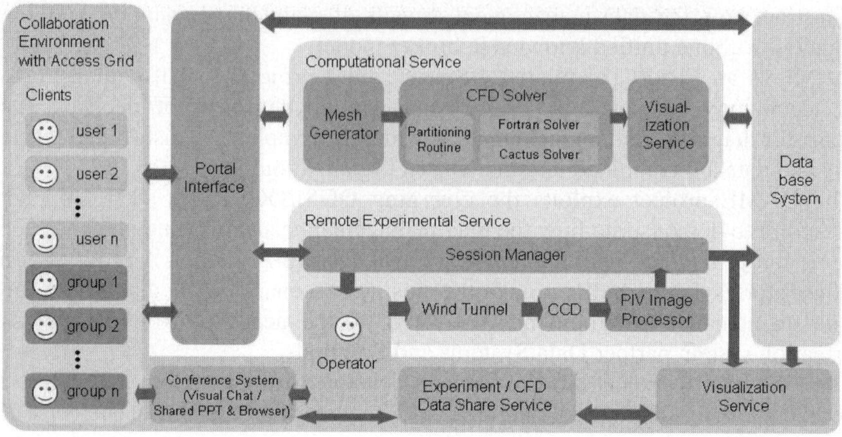

Fig. 1. An e-AIRS architecture

Our portal contains two major portlets. One is the "Computational Service" (CS Service) portlet. Like Fig. 1 shows, CS Service portlet has three main components: "Mesh Generation Service", "CFD solver service", and "Monitoring and Visualization Service". This portlet provides service to select data files and computation resources, submit jobs to remote machines and transfer files for mesh generation and CFD simulation. All geometries, meshes, solvers and post-processing instructions are prescribed either interactively using the graphical user interface. The CS service offers pre-processor (e-AIRSmesh) and post-processor (e-AIRSview) as necessary steps for numerical calculation, and enables numerical analysis with high performance numerical tools. In that e-AIRSmesh and e-AIRSview were developed in java applet forms, this service is feasible without installing any other programs only if internet is accessible. Moreover, user can check process of calculation and result with portal. The CS service supports a problem-solving environment the enables a user to perform unstructured mesh generation in platforms, and to perform the visualization of grid data and visual steering of the solution.

Fluid dynamics studies now affect engineering and design wherever the flow of gases or liquids is significant. For that reason, CFD has become one of the most dynamic and innovative sectors of technical computing market segment. And because it requires large compute resources, executing CFD applications in parallel has helped the growth of supercomputing resources over recent years.

The other major portlet is "Remote Wind Tunnel Service" portlet. This portlet is performing a wind tunnel experiment through portal interface without visiting an institute and keeping an experiment process. Users of this actual experimental service can use web portal as interface communicating with operator. Detailed information on each subject will be discussed in following sections.

3.1 Mesh Generation Service: e-AIRSmesh

The Mesh Generation Service is to do the pre-process which is the first phase of numerical analysis process for CFD. The generation of a grid or mesh about any geometry is often the most difficult phase of the CFD process and is anticipated to be the most time consuming part of the entire process for new users. Since our mesh generation tool, an e-AIRSmesh, is based on portal system as Java applet which provides user-friendly interface. It brings together most of the portal's preprocessing technologies in one environment. Its purpose is to make a geometry object, a mesh object, and to control boundary conditions of aerodynamic form. The approach adopted for parallel mesh generation is based upon a geometrical partitioning of the domain. The complete domain is divided into a set of smaller sub-domains and a mesh is generated independently in each sub-domain. The combination of the sub-domain meshes produces the mesh for the complete domain.

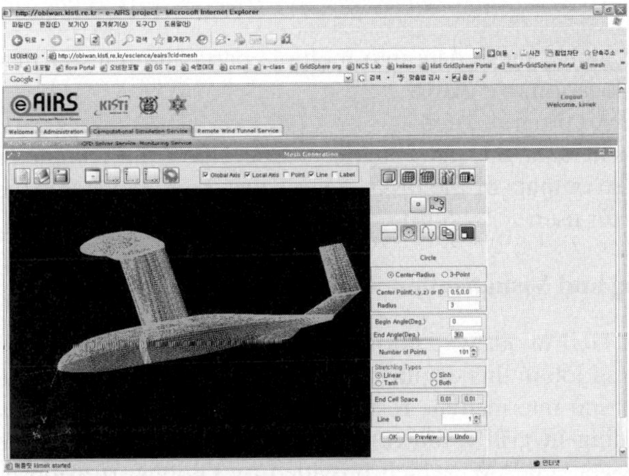

Fig. 2. e-AIRSmesh

The aerodynamic shape optimization problem solved in this work can be understood as a problem of obtaining an airfoil shape - represented by a set of parameters known as design variables - that will minimize/maximize a performance index subject to a set of specified constraints. An example of an objective might be the minimization of drag; an example of a constraint might be to satisfy a certain lift.

3.2 CFD Solver Service

CFD is one of the main research strategies for fluid dynamics. This experience includes the solution of unusual and difficult problems such as the prediction of the unsteady forces generated by passing vehicles, the generation of wind-noise, and the flow of blood in medical devices. Two CFD solvers were developed, Fortran-based and Cactus-based, while numerical calculation service uses the solver based on Fortran. The Cactus-based solver is prepared as a prerequisite work for the extension of this project to workflow environment. The solver module calculates the results in terms of displacements and stresses for each load case. To provide access to a Grid execution environment to which e-AIRS applications can be submitted. This execution environment is responsible for scheduling the application and returning the results to the portal. The simulations require the user of large meshes and the use of significant computational resources. We can provide a complete solution or help you staff obtain the tools and expertise to bring the solution process in-house.

The forms that are used to select data files and resources, and initiate execution are, in fact, a set of portlets. The portlet has three main components: file selection , resource discovery, and job submission. This portlet will commit the resources and the files selected to access the CFD solver service. The CFD simulation codes are submitted using the Grid Resource Allocation and Management (GRAM) by Globus Toolkit. The resource list allows a user to select a machine. It procedure as follows:

1. Divide the global mesh data into multiple parts and transfer them to distributed compute power location in e-AIRS;
2. Run the solver program; and
3. Collect result data set and combine them. Upon the completion of a job, the server collects the outputs of the tasks, aggregates them and stores the final results in the data storage for user.

3.3 Monitoring and Visualization Service

Fig. 3 shows the GUI in our web portal of job monitoring. A user can monitor a latest status of submitted job in this display. Users are easily able to check the current status from the system and intermediate result of currently running job. Basically, status of jobs that is running in grid resources will be monitored by using Globus Resource Allocation Manager (GRAM) which provided by Globus. If a job is submitted to a computing server it's in the 'PREPARE' state while sitting in the queue waiting to be executed. In case of normal completion the job status is 'DONE', otherwise the job is 'INRPT', 'ERR' or 'FAILED'.

The user can also view a convergence graph file of a job. These features allow greater control on the job executions. A user can monitor the progress of a job accurately; hence the user can make a decision whether to stop the current execution. Such interactive exchanges can close the loop on simulation experiments by providing visualization of intermediate results and then allowing scientists to respond by manipulating the simulation, while it is running.

One of the most important steps in simulation is how to display results to users. We have e-AIRSview (Fig. 4) system to display result of simulation. An integrated post-processor program, e-AIRSview, comes packaged with the e-AIRS portal. Fig. 4 is an example of the graphics produced by the e-AIRSview. It was developed as stand-alone application for aerospace simulation result viewing software. When the computational simulation finishes (on remote machines), all output data will be transferred to a storage server. In this stage, location of these files will again be recorded in DB. Thus for each calculation, input parameters, location of output data with some extra information, e.g., owner, date, are all recorded in DB. User can download the results of the finite element analysis graphically for each completed job.

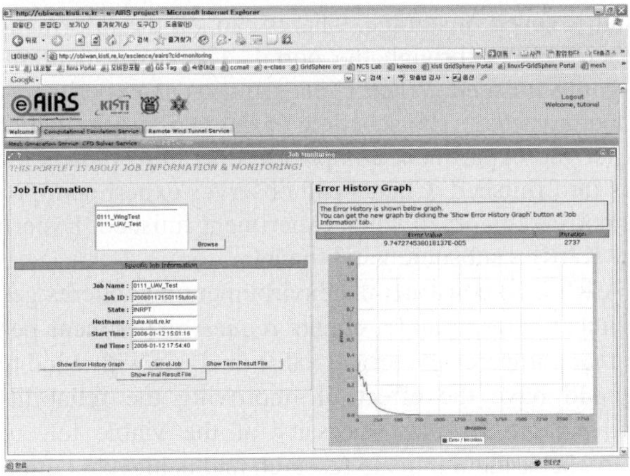

Fig. 3. Monitoring Service UI in e-AIRS

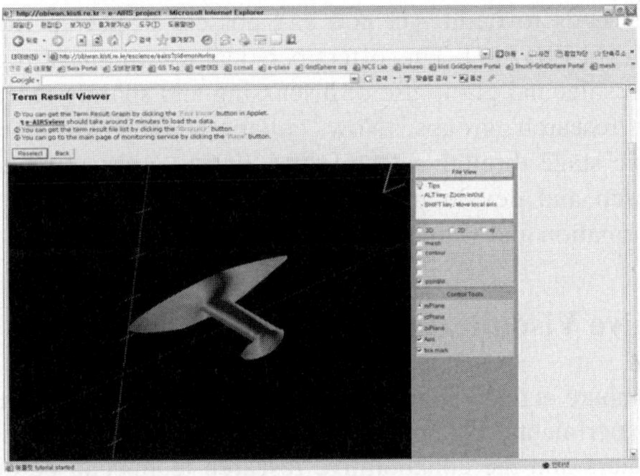

Fig. 4. e-AIRSview

Currently, e-AIRSview is downloading image from the storage (execution server) to portal (client) and loading all data in to the memory before it actually display image. To resolve this problem we are trying to developed new display system that is suitable in web.

3.4 Remote Experimental Service

Remote Wind Tunnel Service is performing a wind tunnel experiment through portal interface without visiting an institute and keeping an experiment process. The client of the wind tunnel service can use web portal as interface communicating with operator. The client enters conditions of an experiment on a portal interface. Operator then performs the experiment based on these. The particle images made by CCD camera are transported to the PIV server. The PIV server outcomes vector field of velocity. The result is saved in storage. Clients can get the information of performing experiments or their results through the portal. The experiment session have three statuses, new/on-going/finished. When client requests an experiment, the status of the session is new. Operator enters the schedule of the experiment and changes the status into on-going. After the experiment completed, operator uploads the result files and changes the status into finished. Clients can observes experiment process at real time through portal, is notified by email when experiment finished. Experiment is based on PIV, supported by KARI's subsonic wind tunnel.

The goal of e-AIRS is to construct the environment that clients get numerical value analysis, if experiments are possible, clients request experiment performing and are offered the result that compares numerical value analysis with wind tunnel experiment result. Clients would have the effect of improving the reliability by comparing through this environment. Besides necessity of the viable teleconference gathers strength in order to check the duplicate research and achieve a collaborative research. This means the system should offer the collaboration environment to research groups for separate portal use. This paper set sights on constructing system that is offered the collaboration environment based on access grid system. To construct access grid each research institute needs to build AG nod and possess it, set up from a conference room nod to PC level PIG [17]. This offers personal use of analyzing and comparing with experiment results and group use of discussing and sharing research results with other researchers/research groups. Now collaboration environment construction approaches a basic stage through a simple installation and test by using AG toolkit [3]. Future research and development will construct collaboration environment of sharing files, information and viable tools.

4 Collaborative Visualization Environment

Users want to compare actual experimental data with simulated output or to compare between actual experimental data. The simulations of aerospace science run on very large supercomputers and the collaborative research teams can be distributed across both intra and inter-continental networks. Support for multiple concurrent users is also an important aim. We show how distributed groups can view simultaneously a visualization of results of simulation and can steer the application into a single seamless collaboration environment. Fig. 5 shows an example of using AG. This utilizes the power of Access Grid (AG) in being able to coordinate multiple channels of communication within a virtual space. The AG infrastructure is one of the most widely used technologies for collaborative computing in the scientific community. It provides for a rich communication environment that goes beyond teleconferencing and standard desktop-to-desktop videoconferencing to enable group-to-group interaction and collaboration. The AG architecture is open, extensible, scalable, and leverages established standards. Each AG node can be equipped with multiple video cameras for video transmission and multiple microphones for audio transmission. AG nodes meet in 'AG venues' to simplify the complicated software interactions that bring the separate nodes together. In the venue, participants can share applications, files, desktops, browsers, and video for large teams to exchange data across sites [19].

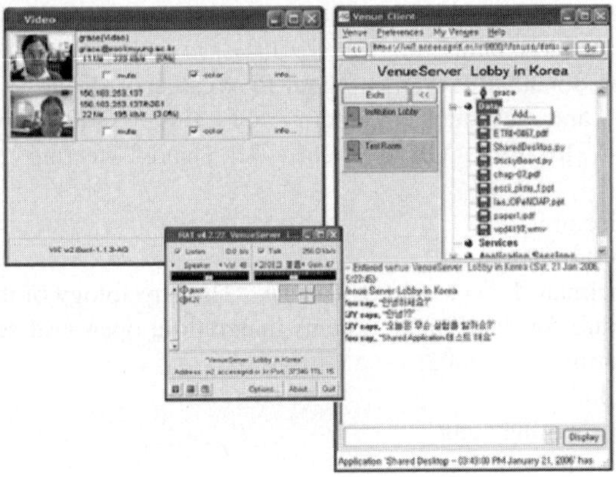

Fig. 5. Snapshots of Video & Audio Collaboration Service using AG venue client

5 Conclusion

The e-AIRS is still in the early stages of development. However, a simple prototype of the e-AIRS described in this paper has been implemented and demonstrated. We have presented the e-AIRS portal based on a combination of GridSphere and our own e-AIRS portlet applications. It serves a convenient research environment independent of time and space. In addition, our portal will make it easy to access research devices. This will make the non-experts produce their own research data more conveniently.

We have also presented main modules of our framework, which are designing, job steering, jobs and resource monitoring and visualization. Job steering submits an application to resource for its execution, that is, dispatches the jobs to suitable resources and aggregates the jobs outputs. Monitoring handles access control to Grid resources and monitors jobs over allocated resource. The e-AIRS eases a user from the need to understand the complexity of a grid system.

For future work of this research, we are going to develop parametric control engine service, which helps expedite establishment of various experimental environment using e-AIRS. We are also tracking the development of Web Services Resource Framework (WSRF) and related web service standards. Our approach appears less well suited to situations in which arbitrary web services need to be composed together to create applications on-the-fly.

References

1. http://www.gridsphere.org/gridsphere/gridsphere
2. http://www.kari.re.kr/
3. http://www.accessgrid.org/
4. http://www.cfd-online.com/
5. GGF http://www.ggf.org/documents/GFD.53.pdf
6. I. Foster, H. Kishimoto, A. Savva, D. Berry, A. Djaoui, A. Grimshaw, B. Horn, F. Maciel, F. Siebenlist, R. Subramaniam, J. Treadwell, J. Von Reich. The Open Grid Services Architecture, Version 1.0. Informational Document, Global Grid Forum (GGF), January 29, 2005.

7. http://www.dantecdynamics.com/piv/princip/index.html
8. http://www.epsrc.ac.uk/default.htm
9. David W. Walker, Jonathan P. Giddy, Nigel P. Weatherill, Jason W. Jones, Alan Gould, David Rowse, and Michael Turner. "GECEM: Grid-Enabled Computational Electromagnetics", In: Proc. UK e-Science All Hands Meeting 2003, page 436-443, September 2003.
10. http://www.geodise.org/
11. http://www.cs.york.ac.uk/dame/
12. I. Foster, C. Kesselman, J. Nick, and S. Tuecke. "The physiology of the grid: An open grid services architecture for distributed systems integration, open grid service infrastructure" wg, global grid forum, June 2002.
13. http://escience.or.kr/
14. http://www.gridcomputing.com/
15. http://www.wesc.ac.uk/
16. T. Jackson, J. Austin, M. Fletcher, M. Jessop. "Delivering a Grid enabled Distributed Aircraft Maintenance Environment (DAME)", In: Proceedings of the UK e-Science All Hands Meeting 2003.
17. http://agcentral.org/help/glossary/pig
18. http://developers.sun.com/prodtech/portalserver/reference/techart/jsr168/pb_whitepaper.pdf
19. Shared Applications Programmers Manual http://www-unix.mcs.anl.gov/ fl/research/accessgrid/documentation/SHARED_APPLICATIONS_MANUAL/ProgrammersManual_SharedApplicationsHTML.htm

A Parallel Plug-In Programming Paradigm*

Ronald Baumann[1,2], Christian Engelmann[1,2], and Al Geist[2]

[1] Department of Computer Science
The University of Reading, Reading, RG6 6AH, UK
[2] Computer Science and Mathematics Division,
Oak Ridge National Laboratory, Oak Ridge, TN 37831-6164, USA
{r.baumann, c.engelmann}@reading.ac.uk,
{baumannr, engelmannc, gst}@ornl.gov
http://www.csm.ornl.gov

Abstract. Software component architectures allow assembly of applications from individual software modules based on clearly defined programming interfaces, thus improving the reuse of existing solutions and simplifying application development. Furthermore, the plug-in programming paradigm additionally enables runtime reconfigurability, making it possible to adapt to changing application needs, such as different application phases, and system properties, like resource availability, by loading/unloading appropriate software modules. Similar to parallel programs, parallel plug-ins are an abstraction for a set of cooperating individual plug-ins within a parallel application utilizing a software component architecture. Parallel programming paradigms apply to parallel plug-ins in the same way they apply to parallel programs. The research presented in this paper targets the clear definition of parallel plug-ins and the development of a parallel plug-in programming paradigm.

1 Introduction

Today, parallel and distributed scientific computing is a tool that enables researchers world-wide to solve large-scale problems in many different research areas, such as climate, nanotechnology, quantum chemistry, nuclear fusion, and astrophysics. Scientific high-end computing (HEC) utilizing tens-to-hundreds of thousands of processors enables new scientific breakthroughs in these areas using computational simulations (simulated experiments) of real-world problems. HEC exploits multi-processor parallelism of scientific algorithms on a large scale using common parallel programming paradigms, such as single program multiple data (SPMD) and multiple program multiple data (MPMD). The scientific computation is performed by a set of cooperating individual processes or tasks communicating via message passing and/or remote method invocation as part of a parallel scientific application.

* This research is sponsored by the Mathematical, Information, and Computational Sciences Division; Office of Advanced Scientific Computing Research; U.S. Department of Energy. The work was performed at the Oak Ridge National Laboratory, which is managed by UT-Battelle, LLC under Contract No. De-AC05-00OR22725.

M. Gerndt and D. Kranzlmüller (Eds.): HPCC 2006, LNCS 4208, pp. 823–832, 2006.

Software component architectures, such as Harness [1,2] and the Common Component Architecture (CCA) [3,4], allow assembly of scientific applications from individual software modules based on clearly defined programming interfaces, thus improving the reuse of existing solutions and simplifying application development. Furthermore, the plug-in programming paradigm additionally enables runtime reconfigurability making it possible to adapt to changing application needs, such as different application phases, and system properties, like resource availability, by loading/unloading appropriate software modules.

Component architectures and plug-in programming are well understood technologies for non-parallel system architectures. HEC applications are inherently parallel requiring adaptation of these technologies to parallel and distributed system architectures. Past research focused on the following two approaches:

- In Harness, the component framework itself runs in a distributed virtual machine (DVM) fashion, where all processes cooperate in a virtual machine environment. This approach enables assembly of SPMD and MPMD applications from components, but requires the DVM to setup all components based on locality. It is not very scalable, but applications are easy to implement due to the support for global component setup. Fault tolerance is addressed through plug-in checkpoint/restart mechanisms using the DVM as a highly available backbone for management and storage.
- In CCA, the component framework itself runs in a parallel (SPMD) fashion, where each process individually is able to manage its own set of components. This approach also enables assembly of SPMD and MPMD applications from components, but requires each component framework instance to setup its components based on locality. It is very scalable, but applications are difficult to implement due to the required local component setup. Fault tolerance is addressed trough application checkpoint/restart mechanisms.

The notion of *parallel plug-ins* evolved from research within the Harness project. Similar to parallel programs, parallel plug-ins are an abstraction for a set of cooperating individual plug-ins communicating via message passing and/or remote method invocation as part of a parallel scientific application. Parallel programming paradigms (SPMD/MPMD) apply to plug-ins in the same way they apply to programs. Parallel plug-ins within Harness are effectively realized by utilizing the distributed virtual machine. Within CCA, parallel plug-ins are effectively realized by implementing parallel programs consisting of plug-ins. Up until now, both approaches avoided a clear definition of parallel plug-ins as it was not well understood.

The research presented in this paper targets the clear definition of parallel plug-ins and the development of a parallel plug-in programming paradigm that combines both approaches in order to further improve reuse of existing solutions and to simplify application development. Harness is being used as a proof-of-concept research vehicle to prototype parallel plug-in management and experimental parallel plug-ins.

In the following sections, we first briefly discuss past and ongoing related research and development efforts. Secondly, we present a clear definition of parallel plug-ins, their programming models (types) and their programming requirements (communication, coordination, and fault tolerance). We continue with a description of our prototype implementation for two distinct scientific application scenarios. This paper concludes with a short summary of the presented research and a discussion future work.

2 Related Work

The research in Harness [1,2] is a collaborative effort among Oak Ridge National Laboratory (ORNL), University of Tennessee, Knoxville, and Emory University focusing on the design and development of technologies for flexible, adaptable, reconfigurable, lightweight environments for heterogeneous parallel and distributed scientific metacomputing.

As part of the Harness project, a variety of experiments and system prototypes were developed to explore lightweight pluggable runtime environments, assembly of scientific applications from software modules, highly available DVMs, fault-tolerant message passing, fine-grain security mechanisms, and heterogeneous reconfigurable communication frameworks.

Currently, there are three different Harness system prototypes, each concentrating on different research issues. The teams at ORNL [5,6,7,8] and at the University of Tennessee [9,10,11,12] provide different C variants, while the team at Emory University [13,14,15,16] maintains a Java-based alternative.

Conceptually, the Harness software architecture consists of two major parts: a runtime environment (RTE) and a set of plug-in software modules. The multi-threaded RTE manages the set of dynamically loadable plug-ins. While the RTE provides only basic functions, plug-ins may provide a wide variety of services needed in fault-tolerant parallel and distributed scientific computing, such as messaging, scientific algorithms, and resource management. Multiple RTE instances can be aggregated into a DVM.

Our research in parallel plug-ins focuses on the C-based lightweight Harness RTE from ORNL [7] using its dynamic, heterogeneous, reconfigurable communication framework (RMIX) [6] plug-in for fault-tolerant message passing and remote method invocation.

RMIX allows software components to communicate using various remote method invocation (RMI) and remote procedure call (RPC) protocols, such as ONC RPC, by facilitating dynamically loadable provider plug-ins to supply different protocol stacks. While the RMIX base library contains functions that are common to all protocol stacks, like networking and thread management, RMIX provider plug-ins contain protocol stack specific functions for connection management, message formats, and data encoding. Since it is up to the provider plug-ins to reuse RMIX base library functions, implementations may range from lightweight to heavyweight. Moreover, client- and server-side object stubs are very lightweight and protocol independent as they only perform an adaptation to the RMIX system. In addition to standard synchronous RMI/RPC mechanisms, RMIX also offers advanced RMI/RPC invocation semantics, such as asynchronous and one-way. RMIX is not a high-performance message passing system. Its RMI/RPC mechanisms are designed for loosely-coupled systems.

The Harness-RMIX plug-in contains the RMIX base library as well as client- and server-side object stubs of the Harness RTE. Stubs for Harness plug-ins are implemented as separate plug-ins. Since the Harness RTE supports plug-in dependencies, a plug-in requiring RMIX automatically loads its stub plug-in(s), which subsequently loads the RMIX plug-in.

The already mentioned Common Component Architecture (CCA) [3,4] is a component-based approach targeted at the needs of large-scale, complex, high-end, scientific simulations. CCA relies on a standardized component framework model for scientific applications based on an interface description language for component interfaces, a port model for unified component interaction, core component framework services, a framework configuration API and a framework repository API. Several CCA framework implementations exist that serve different areas of interest for scientific applications. Furthermore, a set of CCA components exist as well as a number of CCA-based scientific applications.

It is our hope that the work presented in this paper will be eventually incorporated in some form into production-type component architectures for scientific HEC, such as the Common Component Architecture, to further improve reuse of existing solutions and to simplify application development.

Other related past and ongoing research and development efforts include lightweight plug-in design patterns [17], a various number of pluggable component frameworks (e.g. the Open CORBA Component Model Platform [18,19]), as well as recent accomplishments in RTEs for parallel and distributed system architectures, such as Open RTE [20].

3 Parallel Plug-Ins

The intent of our research presented in this paper is to merge plug-in programming technologies with pluggable component frameworks for parallel architectures using common parallel programming models and appropriate design patterns in order to provide better reuse of existing solutions as well as easier scientific application development.

A *parallel plug-in* can be defined as a set of individual plug-ins cooperating in a parallel architecture to perform a common task, such as solving a computation or providing a service. Participating individual plug-ins may be located on the same or on distributed computational resources, and communicate with each other for task control and data transfer purposes. Similar to a parallel program, a parallel plug-in is a parallel programming abstraction, where the same parallel programming models, such as SPMD and MPMD, apply.

While the execution environment of a parallel program is a parallel or distributed operating system, parallel plug-ins reside within a component framework for parallel architectures, *i.e.,* within a parallel program. They provide a componentized approach for building parallel applications by offering parallel building blocks with clearly defined interfaces.

In the past, parallel plug-ins have been effectively realized by by utilizing a pluggable DVM environment or by implementing parallel programs consisting of individual plug-ins. Both approaches dealt with the necessary coordination of individual cooperating plug-ins by using either distributed control [8] or localized control [3], while avoiding a clear definition of parallel plug-ins. In both cases, it is up to the plug-in component programmer to take care of task control and data transfer without access to appropriate design patterns.

3.1 Parallel Plug-In Types

A parallel plug-in consists of one or more individual plug-ins that add features and capabilities to the runtime environment of a component framework as part of a parallel application. The following parallel plug-in types (see Figure 1) can be defined based on number, location, and purpose of involved individual plug-ins:

- The singleton (or service) plug-in is a special case of parallel plug-in as it involves only one individual plug-in at one node within the context of a parallel application.
- The SPMD (or replicated) plug-in follows the known SPMD parallel programming model and involves more than one node within the context of a parallel application. The same plug-in code is replicated to different nodes and applied to different data.
- The MPMD (or distributed) plug-in follows the known MPMD parallel programming model and involves more than one node within the context of a parallel application. Different plug-in codes are distributed to different nodes and applied to different data.

3.2 Parallel Plug-In Communication

Communication is essential for parallel plug-ins. Similar to parallel programs, parallel plug-ins communicate using message passing and/or RMI/RPC in order to coordinate individual tasks and to transfer necessary data. In order to identify individual collaborating plug-ins within a parallel plug-in, naming and message/invocation routing is needed.

Message passing systems, such as PVM [21] and MPI [22], typically only provide naming and routing mechanisms for individual processes of a parallel application and not for individual plug-ins. Message tags may be used to identify individual plug-ins within an individual process. However, a separate plug-in naming service is needed in order to support dynamic adaptation, *i.e.*, dynamic loading and unloading of parallel plug-ins.

RMI/RPC systems typically provide a naming and routing mechanism for exported objects or program parts. Individual plug-ins that are part of a parallel plug-in may be exported as objects or program parts using a RMI/RPC system for naming and routing of invocations. Furthermore, local services of individual component framework processes may be exported as objects or program parts as well to enable remote access to basic component framework services, such as loading and unloading of plug-ins.

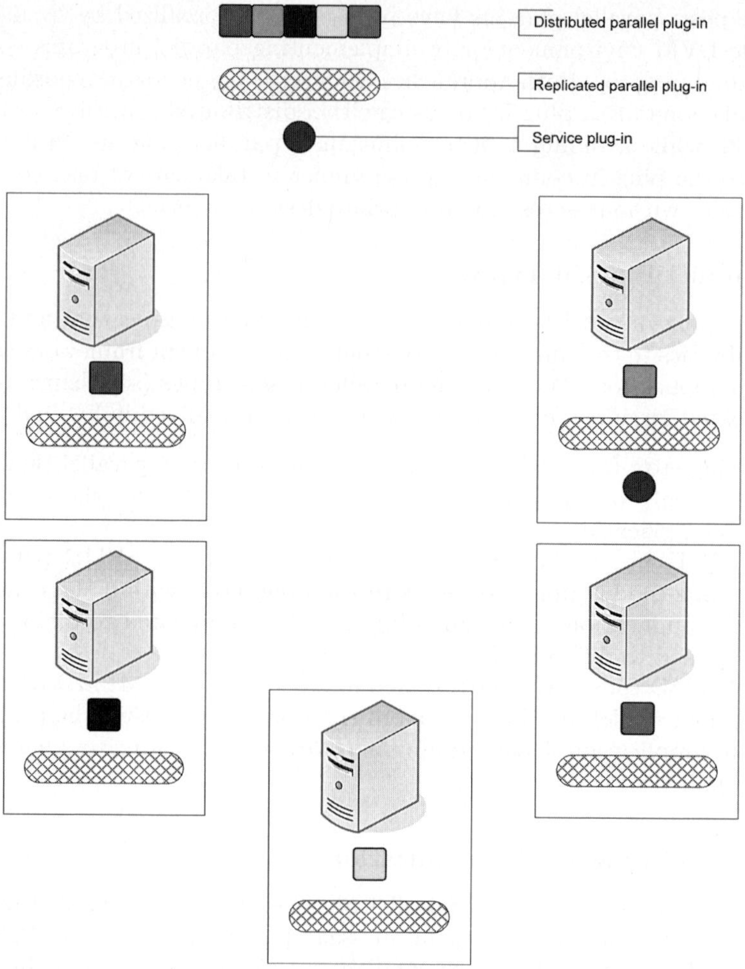

Fig. 1. Parallel Plug-in Types

3.3 Parallel Plug-In Coordination

Coordination of individual plug-ins that are part of a parallel plug-in is needed to perform loading, unloading, task and data distribution, and fault tolerance mechanisms. This coordination is accomplished via separate coordination (service) plug-ins and via the parallel plug-in itself using the communication subsystem with its plug-in naming scheme to address individual plug-ins.

In order to load a parallel plug-in, the used component framework itself or a separate service plug-in contacts individual component framework processes and loads the appropriate individual plug-in based on the parallel plug-in type. Unloading a parallel plug-in is implemented in the same way.

Task and data distribution can be performed by the parallel plug-in itself if it supports self configuration. However, in order to simplify parallel plug-in and application development, a separate service plug-in may be used to execute task and data distribution based on the parallel plug-ins programming model, *i.e.*, by automatically partitioning data and/or assigning tasks.

Similarly, fault tolerance mechanisms, such as checkpoint/restart, may be performed by the parallel plug-in itself if supported, but may also be coordinated by a separate service plug-in based on a design pattern that matches the fault behavior and fault tolerance requirements of the parallel plug-in.

3.4 Parallel Plug-In Fault Tolerance

Fault tolerance is typically realized in three steps: detection, notification, and reconfiguration. In parallel computing, detection and notification are typically performed by the communication system using timeouts to detect faults and the naming scheme to identify faulty communication endpoints.

A parallel plug-in may be reconfigured using the same techniques as for parallel programs, with the exception that reconfiguration takes place within the parallel plug-in programming scope. Parallel plug-in state may be regularly stored on stable storage in the same way a parallel program checkpoint is stored. Upon failure, the entire parallel plug-in or parts of it (individual plug-ins) may be restarted on different nodes. Loss of state due to failures may be ignored if the application is able to continue without extensive reconfiguration [23].

As already mentioned earlier, fault tolerance mechanisms for parallel plug-ins may be encapsulated into a separate service plug-in based on a design pattern that matches the fault behavior and fault tolerance requirements of the parallel plug-in. Furthermore, such a service plug-in may directly interface to existing fault tolerance technologies, such as checkpoint/restart layers.

4 Prototype Implementation

A proof-of-concept prototype has been implemented as part of a Master's thesis [24] using the Harness RTE as a research vehicle.

The Harness RTE offers a lightweight backbone to load and unload individual plug-ins into a multi-threaded process residing on a single node. Multiple Harness RTEs located on the same or different nodes are used as a lightweight backbone to load and unload parallel plug-ins utilizing a parallel parallel plug-in manager for coordination and the Harness-RMIX plug-in for communication.

In the following, we describe the developed parallel plug-in manager and our efforts in implementing a parallel plug-in design pattern for two distinct scientific application scenarios.

4.1 Parallel Plug-In Manager

The parallel plug-in manager (PPM) is itself a service plug-in and provides parallel plug-in management services for both parallel plug-in programming models,

SPMD and MPMD. These services include: loading, unloading, task and data distribution, and fault tolerance support.

The PPM loads a parallel plug-in by starting and contacting a set of Harness RTEs via remote method invocation to load a specific individual Harness plug-in. The set of available nodes is given by the user. The number of involved Harness RTEs must be equivalent or more than the number of individual plug-ins needed. A round robin schedule may be used in a more sophisticated solution to allow oversubscribtion of nodes.

Loading a parallel plug-in introduces the problem of partial success, *i.e.*, not all individual plug-ins were loaded due to unavailability of resources (plug-in, Harness RTE, or node). If there are more nodes available than needed, the PPM retries to load an individual plug-in at a different location.

The parallel plug-in loading fails if not all required individual plug-ins were loaded. This decision entirely depends on the parallel plug-in to load and is guided by the user by configuring the PPM appropriately.

Task and data distribution depending on the parallel plug-in programming model may be performed as part of the parallel plug-in loading procedure. Furthermore, fault tolerance mechanisms are supported via the PPM by offering restart of failed parallel plug-in parts.

4.2 Monte Carlo Integration

The first proof-of-concept parallel plug-in has been implemented using the SPMD programming model performing a Monte Carlo integration algorithm in a bag-of-tasks fashion. The developed parallel plug-in consists of a Monte Carlo integration algorithm, where each individual plug-in performs an equal share of the overall computation. The PPM loads the parallel plug-in on all available nodes, while accepting partial loading success.

Fault tolerance has been implemented using a separate service plug-in to reload failed plug-ins upon notification. The Monte Carlo integration share of a failed plug-in is repeated entirely. The degree of fault tolerance is $n-1$, *i.e.*, $n-1$ out of n Harness RTEs may fail.

4.3 Image Processing

The second proof-of-concept parallel plug-in has been implemented using the MPMD programming model performing a sequence of image processing algorithms in a pipeline fashion. The developed parallel plug-in consists of a set of individual plug-ins, each performing a different computation and forwarding its result to the next plug-in. The PPM loads the parallel plug-in on the necessary number of nodes. It does not accept partial loading success.

Fault tolerance has been implemented using a separate service plug-in to reload failed plug-ins and to reconfigure the pipeline upon notification. Each plug-in stores its results temporarily until completion has been acknowledged by the next plug-in in the pipeline. The image processing algorithm of a failed plug-in is repeated for all unacknowledged results. The degree of fault tolerance is 1, *i.e.*, 1 Harness RTE may fail. The degree may be increased significantly using stable storage for intermediate results.

5 Conclusion

With this paper, we presented results of our recent research in a parallel plug-in programming paradigm for software component architectures in parallel and distributed scientific high-end computing. We defined the parallel plug-in abstraction, associated programming models, and resulting programming requirements. We demonstrated similarities and differences between parallel plug-ins and programs with regards to their programming models and execution environments. We described a Harness-based proof-of-concept prototype of a parallel plug-in manager and of parallel plug-ins for two distinct scientific application scenarios. Further implementation details have been published in a Master's thesis [24].

Our research indicates that the parallel plug-in programming paradigm presented in this paper is an appropriate design template for software component architectures in parallel and distributed scientific high-end computing.

It is our hope that the work presented in this paper will be eventually incorporated in some form into production-type component architectures, such as the Common Component Architecture, to further improve reuse of existing solutions and to simplify application development.

References

1. Geist, G.A., Kohl, J.A., Scott, S.L., Papadopoulos, P.M.: HARNESS: Adaptable virtual machine environment for heterogeneous clusters. Parallel Processing Letters **9**(2) (1999) 253–273
2. Beck, M., Dongarra, J.J., Fagg, G.E., Geist, G.A., Gray, P., Kohl, J.A., Migliardi, M., Moore, K., Moore, T., Papadopoulous, P., Scott, S.L., Sunderam, V.: HARNESS: A next generation distributed virtual machine. Future Generation Computer Systems **15**(5–6) (1999) 571–582
3. Common Component Architecture Forum: Home Page. Available at http://www.cca-forum.org (2006)
4. SciDAC Center for Component Technology for Terascale Simulation Software (CCTTSS): High-Performance Scientific Component Research: Accomplishments and Future Directions. Available at http://www.cca-forum.org/db/news/documentation/whitepaper05.pdf (2005)
5. Oak Ridge National Laboratory, Oak Ridge, TN, USA: Harness project. Available at http://www.csm. ornl.gov/harness (2006)
6. Engelmann, C., Geist, G.A.: RMIX: A dynamic, heterogeneous, reconfigurable communication framework. In: Lecture Notes in Computer Science: Proceedings of International Conference on Computational Science (ICCS) 2006, Part II. Volume 3992., Reading, UK (May 28-31, 2006) 573–580
7. Engelmann, C., Geist, G.A.: A lightweight kernel for the Harness metacomputing framework. In: Proceedings of 14th Heterogeneous Computing Workshop (HCW) 2005, Denver, CO, USA (April 4, 2005)

8. Engelmann, C., Scott, S.L., Geist, G.A.: Distributed peer-to-peer control in Harness. In: Lecture Notes in Computer Science: Proceedings of International Conference on Computational Science (ICCS) 2002. Volume 2330., Amsterdam, The Netherlands (April 21-24, 2002) 720–727
9. University of Tennessee, Knoxville, TN, USA: Harness project. Available at http://icl.cs.utk.edu/harness (2006)
10. University of Tennessee, Knoxville, TN, USA: FT-MPI project. Available at http://icl.cs.utk.edu/ftmpi (2006)
11. Fagg, G.E., Bukovsky, A., Vadhiyar, S., Dongarra, J.J.: Fault-tolerant MPI for the Harness metacomputing system. In: Lecture Notes in Computer Science: Proceedings of International Conference on Computational Science (ICCS) 2001. Volume 2073. (2001) 355–366
12. Fagg, G.E., Bukovsky, A., Dongarra, J.J.: Harness and fault tolerant MPI. Parallel Computing **27**(11) (2001) 1479–1495
13. Emory University, Atlanta, GA, USA: Harness project. Available at http://www.mathcs.emory.edu/harness (2006)
14. Kurzyniec, D., Wrzosek, T., Sunderam, V., Slominski, A.: RMIX: A multiprotocol RMI framework for Java. In: Proceedings of IEEE International Parallel and Distributed Processing Symposium (IPDPS) 2003, Nice, France (April 22-26, 2003) 140–145
15. Kurzyniec, D., Sunderam, V.S., Migliardi, M.: PVM emulation in the Harness metacomputing framework - Design and performance evaluation. In: Proceedings of International Symposium on Cluster Computing and the Grid (CCGRID) 2002, Berlin, Germany (May 21-24, 2002) 282–283
16. Sunderam, V., Kurzyniec, D.: Lightweight self-organizing frameworks for metacomputing. In: Proceedings of IEEE International Symposium on High Performance Distributed Computing (HPDC) 2002, Edinburgh, Scotland (July 24-26, 2002) 113–124
17. Mayer, J., Melzer, I., Schweiggert, F.: Lightweight plug-in-based application development. In: Lecture Notes In Computer Science: Revised Papers from the International Conference NetObjectDays on Objects, Components, Architectures, Services, and Applications for a Networked World (NODe'02). Volume 2591., Erfurt, Germany (October 7-10, 2002) 87–102
18. Object Management Group, Inc: CORBA Component Model. Available at http://www.omg.org/technology/documents/formal/components.htm/ (2006)
19. ObjectWeb Consortium: OpenCCM - The Open CORBA Component Model Platform. Available at http://openccm.objectweb.org/ (2006)
20. OpenRTE Team: OpenRTE project. Available at http://www.open-rte.org (2006)
21. Geist, G.A., Beguelin, A., Dongarra, J.J., Jiang, W., Manchek, R., Sunderam, V.S.: PVM: Parallel Virtual Machine: A Users' Guide and Tutorial for Networked Parallel Computing. MIT Press, Cambridge, MA, USA (1994)
22. Snir, M., Otto, S., Huss-Lederman, S., Walker, D., Dongarra, J.: MPI: The Complete Reference. MIT Press, Cambridge, MA, USA (1996)
23. Engelmann, C., Geist, G.A.: Super-scalable algorithms for computing on 100,000 processors. In: Lecture Notes in Computer Science: Proceedings of International Conference on Computational Science (ICCS) 2005. Volume 3514., Atlanta, GA, USA (May 22-25, 2005) 313–320
24. Baumann, R.: Design and development of prototype components for the Harness high-performance computing workbench. Master's thesis, Department of Computer Science, University of Reading, UK (March 6, 2006)

Hybrid MPI-Thread Implementation on a Cluster of SMP Nodes of a Parallel Simulator for the Propagation of Powdery Mildew in a Vineyard

Gaël Tessier[1], Jean Roman[1], and Guillaume Latu[2]

[1] INRIA Futurs and LaBRI UMR 5800, ScAlApplix project
Université Bordeaux 1 and ENSEIRB, 33405 Talence Cedex - France
http://www.labri.fr/projet/scalapplix
[2] LSIIT UMR 7005, Université Strasbourg 1
67412 Illkirch Cedex - France

Abstract. This paper describes a new hybrid MPI-thread implementation of a parallel simulator for the propagation of a parasite in a vineyard. The model also considers the structure, the growth and the susceptibility of the plant. Two spatial scales are distinguished for the dispersal of the fungus. A realistic discrete model was used for the local and neighbouring dispersal, and a stochastic model integrates distribution laws for the long-range dispersal over the parcel. An algorithmic description of the new parallel simulator is given and real life numerical experiments on an IBM SP5 are provided, using up to 128 processors.

1 Introduction

This paper deals with the simulation of a biological host-parasite system: *powdery mildew* is a fungus parasite of grapevine. The propagation of the parasite involves a large number of multiscale mechanisms between the environment, the host and the pathogen. Many epidemiological studies have been performed on this topic, using dispersal models based on differential calculus such as in [7] and [9]. Yet, the dynamics of the spread of epidemics is still not well known.

The main purpose of this work is to integrate knowledge obtained during experiments and to reproduce the events that interact in this complex system, so as to understand better its dynamics and to have a more effective control of epidemics. Our original approach consists in modeling both the architecture and the growth of the vinestocks, and the dispersal of the parasite over the vineyard. By coupling these two models, we produce realistic simulations of the biological system. Initially, a sequential simulator considering only one grapevine has been developed. Since the simulation requires a large amount of calculations, we have designed a parallel version able to simulate the spread of epidemics over a parcel; its characteristics and performances were previously described in [2], [8].

Nowadays, massively parallel high performance computers are generally designed as networks of SMP nodes. Each SMP node consists of a set of processors that share the same physical memory, to fully exploit shared memory advantages, a relevant approach is to use an hybrid MPI-thread implementation. The

M. Gerndt and D. Kranzlmüller (Eds.): HPCC 2006, LNCS 4208, pp. 833–842, 2006.

rationale that motived this new hybrid implementation was that the communications within a SMP node can be advantageously substituted by direct accesses to shared memory between the processors in the SMP node using threads. As a consequence, the MPI communications are only used between processors that host threads from different MPI processes. We achieve to reduce the global amount of data exchanged over the network by using shared memory and by modifying the original algorithm.

The remainder of the paper is organized as follows: Sect. 2 describes a brief review of the biological system and of the biomathematical model, Sect. 3 describes our hybrid implementation and Sect. 4 provides a performance analysis of the simulation. Finally, Sect. 5 gives some biological results from a real life simulation and some conclusions.

This interdisciplinary work is a collaboration between the INRIA Futurs ScAlApplix project and the LSIIT UMR 7005 for the computer science field, the INRA UMR Santé Végétale in Villenave d'Ornon for the biological investigations and the MAB UMR 5466 for the mathematical models.

2 Description of the Biological Model

2.1 Biological Model

We consider the simulation of a single season, from January to the beginning of September with a time step of one day. Location and onset of primary infection are some input parameters of the simulation.

Vinestocks are hierarchical plants modeled by binary trees. Each node of the tree represents an element of the plant, and contains information on its spatial and biological state. Powdery mildew (see [1]) is a fungus that spreads thanks to microscopic airborne spores. Its biological cycle consists of: the infection of a leaf or of a cluster by spores; a latency period during which the rising colony is only growing; a sporulation phase during which spores are released by wind.

As for the dispersal of spores, which is a key step of computations, two scales have been distinguished:

neighbouring dispersal - in a local domain englobing the source stock and its direct neighbours, the spores are spread within *dispersal cones* thrown from each sporulating colony. If some spores reach one edge of the rectangular parallelepiped delimiting the volume of the grapevine, they either fall on the ground, or are transmitted to the contiguous grapevine, or are dispersed over the vineyard depending on the exit side;
long range dispersal - at long range, dispersal is computed stochastically. Random drawings following distribution laws yield displacements to spread spores over all the stocks of the parcel.

One iteration of the simulation on a single grapevine consists of several steps: fetching of the variables describing the environmental conditions, vineyard management practices, growth and apparition of organs, primary infection at the given date and dispersal of spores within the plant. The dispersal of spores turns out to be the most costly part of computations. From each sporulating colony, spores are spreaded within a *dispersal cone*. For efficiency, the volume delimiting the grapevine was cut out with a discrete mesh of parallelepipeds called *voxels* [6]. This avoids to consider each leaf of the binary tree to determine the set of leaves that intercept the cone as shown in Fig. 1.

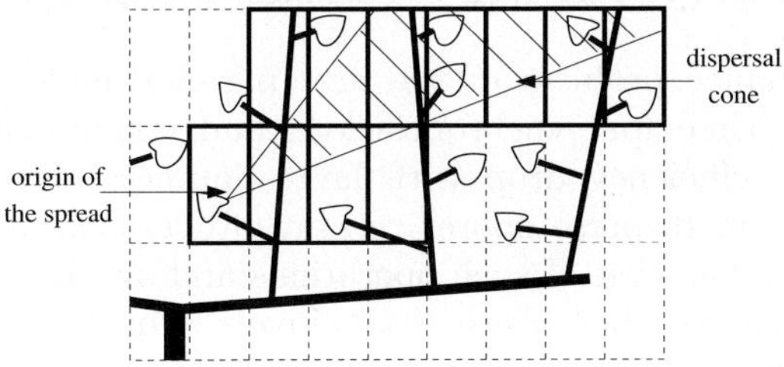

Fig. 1. Dispersal cone in a grapevine; only bold voxels intercept the cone

In comparison with the previous biological model used in [2], some modifications and improvements have been provided; an important one is the possibility for leaves to carry multiples lesions. This significantly increases the number of dispersal cones, hence the amount of computations and the volume of data exchanged during the neighbouring dispersal. The algorithmic complexity of the local dispersal within a vinestock grows as $O(\#leaves \cdot \log(\#leaves))$, where $\#leaves$ is the total number of leaves in the vinestock.

2.2 Previous Parallel Simulator

The parallelization was based on an SPMD [3] approach. The field was modelled as a 2D mesh of vinestocks and the set of stocks attributed to a processor, called *local_ stocks*, was determined according to a 2D-cyclic distribution [3]. Then, an iteration of the simulation was split into three important phases as in Fig. 2.

```
for vine in local_stocks do
  vine_computations
end for
neighbouring_dispersal
long_range_dispersal
```

Fig. 2. Algorithm of a parallel iteration after budding

`vine_computations` corresponds to the operations performed on a single grapevine, except that dispersal cones that exit the vine volume by lateral sides are added to the adequate cone lists for the neighbouring dispersal. Other exiting spores are accumulated in the *vine* data structure for later long-range dispersal.

The number of cones or spores cannot be predicted; algorithmic complexities of `neighbouring_dispersal` and `long_rang_dispersal` could only be evaluated dynamically. Also, since a process does not calculate the list of processes that have contiguous stocks with its own, `MPI_Alltoall` communications are performed during these both dispersals: each process exchanges with other ones the dispersal cones that are transmitted between contiguous stocks and the spores that are spreaded at long range over the whole field.

3 Using of an Hybrid MPI-Thread Programming

One main performance problem of the first parallel implementation was the increasing part of time for synchronizations and communications, as shown in [2], making the efficiency drop with large numbers of processors. Indeed, as any grapevine can disperse spores on any other one, communications are `MPI_Alltoall`-like. The cost of such operations and synchronizations increases highly along with the number of processors involved in the simulation.

To overcome this problem, an hybrid approach mixing processes and threads has been implemented. Some other works have integrated OpenMP threads in parallel MPI implementations ([5]) but far less used POSIX threads which are more flexible though.

The idea is to benefit from the high speed of shared memory accesses by replacing n monothreaded processes in the previous simulator by n/p processes, each one containing p simulation threads that compute the growth of stocks and colonies and the dispersal of spores, and one master thread responsible for inter-process MPI [4] communications. On the one hand, *the number of processes involved in global communications will be smaller*, on the other hand, threads in a same process can exchange data via the shared memory, *avoiding MPI communications.*

3.1 Design of the Parallel Algorithm

Both master and simulation thread codes are iterative over the time steps. During each iteration, synchronizations are necessary: in each process, the master thread has to wait for the end of some calculations in order to send local data to other processes. Similarly, simulation threads have to wait for the completeness of communications in order to perform calculations on received data. Listings 1.1 and 1.2 present the general structure of an iteration code for communication and simulation threads.

Before explaining in details the different steps of computations and communications, we describe briefly the variables and the data structures in each process. Each process knows the number of processes `n_procs` involved in the simulation, and each one runs one communication thread and `n_threads` simulation ones. `local_vines` is the set of stocks assigned to the process and each simulation thread is attributed a private subset of that one, called `thread_vines`.

`neighbour_cones` is an array of `n_procs` arrays of dispersal cones (source, direction, number of spores). Its kth entry refers to the list of cones from grapevines in `local_vines` that have to be spreaded over the stocks of the MPI process of rank k. Also `recv_cones` is an array of `n_procs` -1 arrays where cones from other processes are received.

`ldist_spores` is an array of `n_threads` arrays. Its kth entry is an array with the field dimensions: it has as many lines as rows in the parcel and as many columns as stocks in the rows. Each one of its elements is associated to the stock of same coordinates in the field. An element in this array is the number of spores received by the corresponding stock during the long range dispersal from all others in the set `thread_vines` of the simulation thread k in the current process.

Listing 1.1. Communication thread

```
  get_paramaters()
barrier_wait(GET_PARAM)
barrier_wait(CALC_LOC)
  comm_disp_neighbour()
barrier_wait(COMM_NEIGHBOUR)
barrier_wait(CALC_NEIGHBOUR)
  comm_disp_longdist()
barrier_wait(COMM_LONGDIST)
barrier_wait(CALC_LONGDIST)
  update_system_variables()
```

Listing 1.2. Simulation thread

```
barrier_wait(GET_PARAM)
  for vine in thread_vines do
    evolution(vine)
    local_dispersal(vine)
  end for
barrier_wait(CALC_LOC)
  calc_disp_neighbour_local()
barrier_wait(COMM_NEIGHBOUR)
  calc_disp_neighbour()

  spread_spores_longdist()
barrier_wait(CALC_NEIGHBOUR)

barrier_wait(COMM_LONGDIST)
  calc_disp_longdist()
barrier_wait(CALC_LONGDIST)
```

Hereafter, we will describe each function calls of Listings 1.1 and 1.2.

get_parameters consists in reading input files to fetch global data describing the environment, such as temperature, wind speed and direction...

evolution performs to the management practices of viticulture system (shoot topping, leave cutting for example) and the growth and the apparition of organs in grapevines.

local_dispersal computes the dispersal of spores from each sporulating colony within dispersal cones.

When a cone reaches a lateral side of the delimiting volume of the stock, it has to continue its way in the neighbour stock. The MPI process to which it will be sent, is determined. Then the characteristics of the cone are written in the corresponding array in `neighbour_cones` data structure.

Moreover, when the cone exits the volume of the stock by the above edge, its spores are accumulated in a counter of the stock to be spread through long range dispersal.

comm_disp_neighbour sends to and receives from each other process the cones in the arrays `neighbour_cones` and `recv_cones`.

calc_disp_neighbour_local is called during the communication of neighbour dispersal cones. Each simulation thread traverses the entry of `neighbour_cones` for its own process. This entry corresponds to the cones that were transmitted between neighbour stocks that are both allocated to the same process. In this way, it allows at least a partial overlap of the communications by computation.

Note that a dispersal cone can be transmitted from one stock to a contiguous one but not again to another neighbour. If the cone passes through two grapevines, its spores are finally accumulated for long range dispersal.

calc_disp_neighbour does the same as the previous calculation step, except that input cones are in the arrays `recv_cones` received from other processes.

spread_spores_longdist considers each stock in `thread_vines`. For each one, the corresponding simulation thread performs two Gaussian random drawings that yield a displacement in the referential of the field. It either identifies a target stock or the outside of the field. In the first case, the spores accumulated in the source stock are written in `ldist_spores`.

comm_disp_longdist merges the arrays in `ldist_spores` in the first entry, by adding the numbers of spores that correspond to the same target stock. The merge result is sent to all other processes and the arrays received from them are merged. At the end of the communication, an element in the first array of `ldist_spores` is the number of spores received by the corresponding stock from all others in the parcel.

calc_disp_longdist takes as input the merged array constructed during the communication. Each simulation thread computes the dispersal of spores on each stock in its `thread_vines` array.

update_variables eventually updates some global variables, increments the current day or indicates the end of the simulation to the simulation threads.

3.2 Data Distribution

As in the initial simulator, stocks are allocated to processes and we have kept a 2D block-cyclic distribution. A set of stocks almost uniformely distributed over the parcel is attributed to each process. As a region in the neighbourhood of a primary foci of infection generates more dispersal cones and then computations, this distribution tends to balance the computations between the processes if the blocks are small.

On the other hand, to benefit from the overlap of communications in the function `calc_disp_neighbour_local`, it is important to have as many contiguous stocks as possible allocated to a process, that means big blocks. The chosen size is square blocks of four stocks.

Inside a process, a simple static 1D block distribution has been implemented between threads for the moment.

The load-balancing was first completely static and rested upon the quality of the initial distribution. The need of a dynamic load-balancing between threads appeared during the performance evaluations. Its implementation will be discussed in the next section.

4 Performance Analysis

The hybrid parallel simulator is implemented with MPI [4] and POSIX threads.

A platform located at Université Bordeaux 1 (Talence, FRANCE) was used for the numerical experiments. This high performance parallel cluster consists of 13 nodes of 8 IBM P575 Power 5 DualCore processors connected by a Federation switch. Up to 128 processors were used for the simulations. Several aspects were evaluated during our benchmarks:

- the load-balancing revealing the need of a dynamic strategy;
- the influence of the ratio `n_threads/n_procs` confirming the efficiency of the hybrid programming;
- the scalability, which is a key point for the performance of a parallel program.

4.1 Load-Balancing

Like in the initial simulator, the evolution of stocks is very fast and costs about the same for every grapevine. So it is rather well-balanced between processes and threads, and counts for a little part of computation time.

On the contrary, `local_dispersal` cost highly depends on the severity of disease. Preleminary tests underlined the load-imbalance that was responsible for the disappointing initial performance. Indeed, due to synchronization barriers, the length of a computation step is the one of the simulation thread that carries out most of the calculations. To illustrate this point, a simulation was run with four processes, each one with four simulation threads. Elapsed times in the different computation steps were cumulated over the whole simulation in each thread. Table 1 shows the minimum and maximum cumulated times for these steps in one of the processes, the variation between theses times and in the last column, the importance of this variation relative to the total simulation time.

Bold times hilight the most costly computation steps: local and neighbouring dispersal. Their load-imbalance is very important, particularly relative to the total time. Because of that, this configuration with four processes of four threads was less efficient that the one with sixteen monothreaded processes.

So, a dynamic load-balancing mechanism was implemented for the local dispersal. Each simulation thread gets exclusively one vinestock in which it

Table 1. Minimum and maximum times for computation steps between threads of one process during the simulation (242 days) of a 32×32 parcel

Computation step	min time (s)	max time (s)	(max-min)/total
Evolution	13.3	21.9 (+65 %)	1.5 %
Local_dispersal	**127.4**	**264.5 (+108 %)**	**23 %**
Calc_disp_neighbour_local	**141.6**	**224.0 (+58 %)**	**14 %**
Calc_disp_neighbour	**49.4**	**108.2 (+119 %)**	**10 %**
Spread_spores_longdist	0.018	0.0193 (+7 %)	0 %
Calc_disp_longdist	1.68	2.36 (+40 %)	0.1 %
Total simulation	584		

computes the local dispersal, and so on until every stocks in `local_vines` has been considered. As a result, minimum and maximum times for the local dispersal in the same simulation became 212.5 s and 220.9 s (compared with 127.4 s and 264.5 s). Furthermore, time was also gained at the following barrier of synchronization, hence reducing times for the `calc_disp_neighbour_local` to between 49.8 s and 104.1 s. However, this step remains unbalanced. Its dynamic load-balancing will be studied in a future version, and it will require some important changes in data structures for the neighbouring dispersal. Finally, the total simulation time was reduced to 497.6 s instead of 584 s.

4.2 Influence of the Ratio n_threads/n_procs

A series of simulations were performed on a 32×32 parcel with different configurations from 1 to 16 threads per process and with 16 to 128 processors. Table 2 reports the total simulation times for these different configurations. We do consider here a number of processors equal to `n_procs` × `n_threads`.

Bold times are the best performances for different number of processors. These performances are better from 12 % to 23 % than the mono-threaded ones and from 20 % to 47 % than the 16-threaded ones. It appears that best results are obtained with "intermediate" number of threads: generally four threads per process which means four processes per nodes.

The examination of the refined profilings explain theses results. Indeed, the cumulated times over the simulation for the local dispersal in each simulation thread follow a similar evolution to the total simulation time, as shown in Tab 3.

It is about the same evolution for the cumulated times of the two steps `calc_disp_neighbour_local` and `calc_disp_neighbour`.

As presented in Sect. 4.1, an effective load-balancing mechanism has been added to the local dispersal. So, these increasing times for high numbers of threads per process are not only due to bad distribution or load-imbalance.

Table 2. Total simulation times for different configurations on a 32×32 parcel

Number of processors	`n_threads` per process				
	1	2	4	8	16
16	567.6	515.8	**497.6**	567	626.8
32	349.5	293.6	**280.5**	467.2	485.6
64	221.1	177.9	**154.9**	206.7	300.3
128	137.1	**105.4**	126.2	168.9	200.4

Table 3. Maximum cumulated times for the local dispersal for the simulation of a 32×32 parcel with 16 processors

`n_threads` per process				
1	2	4	8	16
263.6	231.6	220.9	256.7	296.7

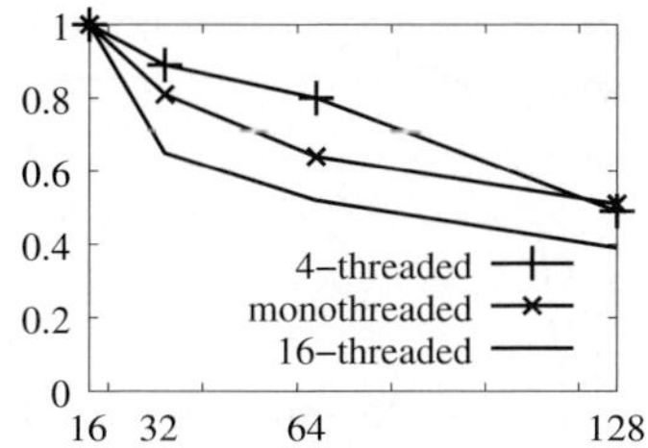

Fig. 3. Comparison of relative efficiency between the monothreaded, 4-threaded and 16-threaded parallel simulators

In fact, during this computational step, the `n_threads` simulation threads successively reserves stocks until all local stocks have been considered. While computing the dispersal, they write in the `n_procs` arrays of `neighbour_cones` the dispersal cones that are transmitted between neighbour stocks. To avoid write concurrency problems in the arrays, each thread reserves a line in the `n_procs` arrays. Both reservations of stocks and lines are implemented with mutual exclusion mechanisms. As the number of threads per process increases, the concurrency does so and the exclusion mechanisms are more often used.

Further development will include finding a solution to our concurrent accesses problem.

4.3 Scalability

Scalability was the main performance issue of the previous parallel simulator with only MPI-communications. It has been improved with the hybrid MPI-threads simulator. By comparing relative efficiency of the monothreaded and four-threaded parallel simulators, results turn out to be quite good. Figure 3 represents the relative efficiency of these two configurations of the simulator.

However, it should be noticed that, due to the modifications of the biomathematical model, in particular the ability of leaves to carry multiple lesions, computation time represents an higher part of total simulation time. Transitional phase during which the epidemics is developing can present important load-imbalance between processes, making now the computation times not inversely proportional to the number of processors. Finally, the attribution of primary infection stocks and the load-imbalance between threads during the neighbouring dispersal explain the drop of efficiency with some configurations (see Sect. 4.1).

5 Biological Results and Conclusion

In terms of performances, the results are quite good. The improvements to communication, load-balancing and scalability have been observed during the experiments thanks to the modifications brought to the simulator.

Nevertheless, they are not perfect. The main improvement that seems to be necessary at the moment, is a good load-balancing of computations. It includes two points: a full dynamic load-balancing between threads of a same process for

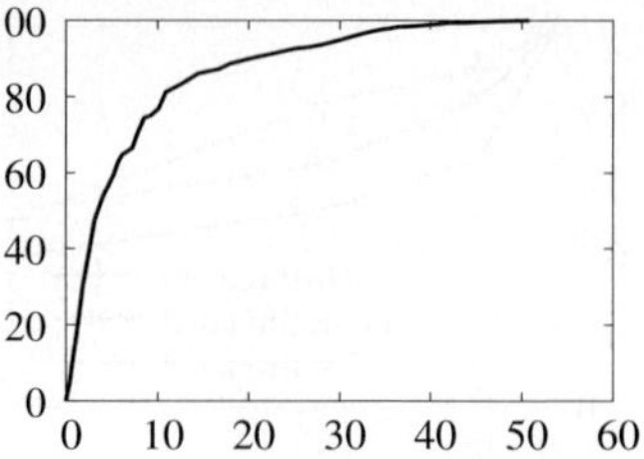

Fig. 4. Percentage of infected stocks in function of the percentage of infected leaves in the parcel during the simulation of a 32×32 parcel

all computation steps; and a better static distribution between processes, that takes into account the amount of calculations induced by a primari foci and its neighbouring region.

As for the biological results, calibration of the model has been carried on further. Figure 4 shows the evolution of the percentage of infected stocks in function of the percentage of infected leaves in the parcel during a real simulation.

Such results characterize the dispersal degree of the epidemics and correspond to field data: when 15 % of the leaves in the field are infected, in average 80 % of the stocks are infected, as we can observe on this simulation output.

References

1. Bulit (J.) et Lafon (R.). – Powdery mildew of the vine. *In: The powdery mildews*, éd. par Academic press, London. – DM Spencer, 1978.
2. Calonnec (A.), Latu (G.), Naulin (J.-M.), Roman (J.) et Tessier (G.). – Parallel Simulation of the Propagation of Powdery Mildew in a Vineyard. *In: 11th International Euro-Par Conference*. pp. 1254 – 1263. – Springer Verlag, sept 2005.
3. Grana (Ananth), Gupta (Anshul), Karypis (George) et Kumar (Vipin). – *Introduction to Parallel Computing*. – Addison Wesley, 2003, Second Edition édition. ISBN 2-201-64865-2.
4. MPI Forum. – *Message Passing Interface MPI Forum Home Page*. Available from http://www.mpi-forum.org/.
5. Nikolaos Drosinos and Nectarios Koziris. – Performance Comparison of Pure MPI vs Hybrid MPI-OpenMP Parallelization Models on SMP Clusters. *In: IPDPS*. – IEEE Computer Society, 2004. ISBN 0-7695-2132-0.
6. Samet (Hanan). – *The Design and Analysis of Spatial Data Structures*. – University of Maryland, Addison-Wesley Publishing Company, 1989. ISBN 0-201-50255-0.
7. Shigesada (Nanaka) et Kawasaki (Kohkichi). – Invasion and the range expansion of species: effects of long-distance dispersal. *In: Proceedings of BES Annual Symposium 2001 'Dispersal'*, chap. 17, pp. 350–373. – Blackwell Science (in press), 2002.
8. Tessier (G.). – Simulation parallèle de la propagation de l'oïdium dans une parcelle de vigne. *In: RenPar'16*. – avril 2005.
9. Zawolek (M. W.) et Zadocks (J. C.). – Studies in Focus Development: An Optimum for the Dual Dispersal of Plant Pathogens. *Phytopathology*, vol. 82, n 11, 1992, pp. 1288–1297.

Exploring Unexpected Behavior in MPI*

Martin Schulz[1], Dieter Kranzlmüller[2], and Bronis R. de Supinski[1]

[1] Center for Applied Scientific Computing
Lawrence Livermore National Laboratory
P.O. Box 808, L-560, Livermore, CA 94551, USA
schulzm@llnl.gov, bronis@llnl.gov
[2] Institute of Graphics and Parallel Processing (GUP)
Joh. Kepler University Linz
A-4040 Linz, Altenbergerstr. 69, Austria/Europe
dk@gup.jku.at

Abstract. MPI has become the dominant programming paradigm in high performance computing partly due to its portability: an MPI application can run on a wide range of architectures. Unfortunately, portability in MPI is only guaranteed for compiling codes; it does not necessarily mean that an MPI program will actually result in the same behavior on different platforms. The MPITEST suite provides a series of micro kernels to test MPI implementations across different systems. All codes of MPITEST conform to the MPI standard; however, their behavior is implementation dependent, potentially leading to unexpected results. In this paper we introduce MPITEST and present examples from the test suite along with their surprising results and consequences on a series of platforms. The goal of this work is to demonstrate this problem in general and to raise awareness in the MPI user community.

1 Motivation

The Message Passing Interface (MPI) [6] is the most widely used programming paradigm for parallel and high performance computing. It provides programmers with a comprehensive interface for exchanging messages between application tasks, yet can be used with only a few primitives. An important feature for the programmer is its portability, which promises the use of a single MPI application across a wide range of architectures.

Yet, in practice the portability of MPI is only guaranteed during compilation of the code, and not during its execution. Even though the MPI standard defines the runtime behavior of the individual primitives, it does not necessarily mean that an MPI program will actually result in the same behavior on different platforms. This can lead to unexpected results on new platforms or the manifestation of previously dormant bugs. In particular, less experienced users are vulnerable to these traps.

* Part of work was performed under the auspices of the U.S. Department of Energy by University of California Lawrence Livermore National Laboratory under contract No. W-7405-Eng-48. UCRL-PROC-222368.

M. Gerndt and D. Kranzlmüller (Eds.): HPCC 2006, LNCS 4208, pp. 843–852, 2006.

We present MPITEST (`http://www.gup.uni-linz.ac.at/mpitest/`), a benchmark suite that helps users to explore such potential problems and compare the impact of individual constructs across different machines. It contains a series of micro kernels that conform to the MPI standard, but may lead to unexpected behavior, which is not portable in the sense of execution behavior across different platforms. In this paper we present a number of surprising results and potential consequences of examples from the MPITEST suite on a series of platforms. Our primary goal is to raise awareness of this problem in the MPI user community.

We begin with a general overview of MPITEST and describe our experimental setup in Section 2. We then introduce three examples of MPITEST benchmarks in Sections 3, 4, and 5. We conclude with a discussion of the impact that these issues have for code portability.

2 The MPITEST Suite: Overview and Setup

Many benchmark suites exist to help users evaluate MPI performance across platforms. These are either sets of micro benchmarks, like SKAMPI [7] or Sphinx [3,2], or sets of application kernels, like the NAS parallel [1] or the ASCI Purple [5] benchmarks. However, aside from basic MPI functionality correctness microbenchmarks, no comprehensive benchmark suite exists that covers non-performance related characteristics of MPI implementations.

We close this gap with MPITEST, a test suite that targets corner cases in the MPI standard and portability issues in MPI implementations. It currently contains codes that investigate nondeterminism, receive ordering, and asynchronous communication. In the remainder of this paper we present one representative code from each of these classes. For maximum flexibility, each of the test codes can be customized using additional compile-time and runtime parameters.

To ease the development and to ensure consistent parameter handling, startup, and output formats, each benchmark is based on a common library developed as part of MPITEST. This library implements the actual main routine; reads and evaluates runtime parameters; is responsible for printing headers that include machine configuration, benchmark parameters, and overall timing; and provides common services like timing routines.

For an easier and more homogeneous execution of MPITEST benchmarks across a heterogeneous set of platforms, we also developed a test harness as part of MPITEST that detects the target platform and automatically creates the necessary execution environment. This harness is integrated with the MPITEST build setup and includes the necessary scripts for batch submission and interactive startup, handling of all command line parameters, and a uniform and archivable result file storage. Combined with the uniform output format in the MPITEST library we ensure a long-term storage of the all relevant information necessary for later result evaluation and reproduction.

MPITEST is designed to be very portable and we have implemented it on a large number of systems. For the experiments in the following sections we used

Table 1. Machine configurations

	Berg	*Thunder*	*BG/L*	*SGI*
CPU Type	Power4	Itanium-2	PowerPC 440	MIPS R12000
Clock freq.	1.3 GHz	1.4 GHz	700 MHz	400 MHz
CPUs/Node	8	4	1 or 2	128 (CC-NUMA)
# Nodes	4	1024	65536	1
Memory/Node	256 GB	8 GB	512 MB	64 GB
Interconnect	Federation Switch	Quadrics Elan–4	Custom Networks	NUMAflex

four machines at LLNL and GUP/JKU: *Berg*, an IBM Power–4 cluster; *Thunder*, an Itanium-2 cluster; *BG/L*, the Blue Gene/L system; and *SGI*, an SGI Origin 3800 installation. The exact system parameters are listed in Table 1.

3 Wildcard Receives

As a first example and typical case of unexpected behavior, we have chosen so-called wild card receives. This kind of receives, where the receiving task accepts a message from any sender, are a natural way to introduce nondeterminism into a program and quite common in today's MPI applications. A typical example of such a scenario is shown in Figure 1. Both tasks Q and R send a message to task P, while task P posts two wildcard receives accepting messages from any task. Depending on the timing of tasks Q and R, the latencies in the network, as well as the implementation of the MPI library, the first receive in P can match either the message sent by Q or by R. Both scenarios are legal executions.

However, the degree of nondeterminism and the number of actually observed permutations depends on the characteristics of the hardware and its usage. Thus, codes debugged on only one platform might always reveal only a certain subset of executions. Specific situations, e.g. those leading to incorrect behavior, may not occur during testing. However, such sporadic errors may suddenly occur, particularly when slightly changing the code (e.g., by disabling debugging statements) or if the code is ported to another machine.

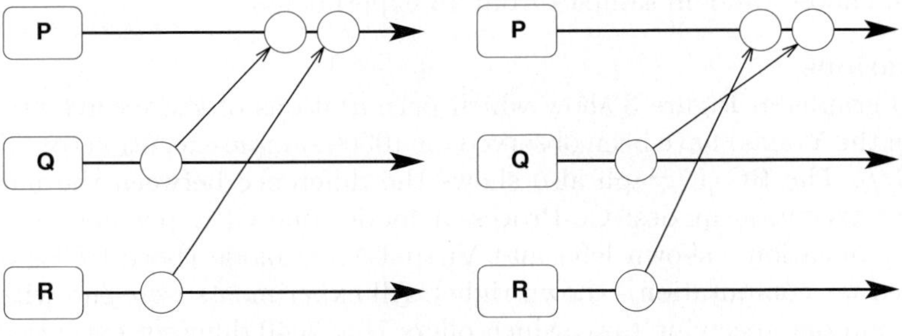

Fig. 1. Nondeterministic behavior at wildcard receives

```
Wait for constant time
Buffer[0]=rank
MPI_Send(Buffer,receive task)
```

N-1 Sender

```
for i = [ 0,N [
{
   MPI_Recv(MPI_ANY_SRC)
   Add sender ID to string
}
Print string with receive order
```

1 Receiver

Fig. 2. Pseudo code for nondeterministic receives

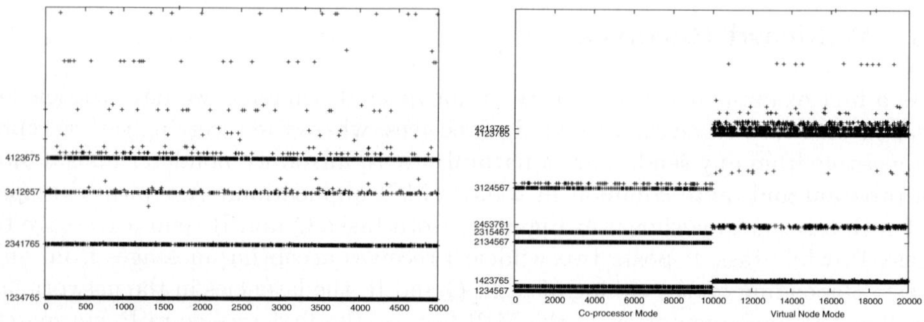

Fig. 3. Nondeterministic behavior of portable programs on different hardware architectures: *Thunder* (left), BG/L (right)

Implementation

Figure 2 illustrates the MPITEST program that focuses on nondeterministic receives [4]. All but one task send a message to a fixed receiver after waiting for a random time. That one task receives all incoming messages and records the order in which they appear. We repeat this process for a large number of times and then choose random samples from all experiments.

Observations

The two graphs in Figure 3 show which permutations of the receive order (encoded on the Y-axis) have been observed for 10000 sample executions on *Thunder* and *BG/L*. The BG/L graph also shows the difference between the machine's two main execution modes: Co-Processor mode (one CPU per node dedicated for communication - shown left) and Virtual-Node mode (both CPUs on each node used for computation - shown right). All experiments were run with seven sending and one receiving task, which offers $7! = 5040$ different executions.

These graphs show that the observed receive orders on these two platforms are substantially different. On *Thunder* four distinct receive orders dominate, while on *BG/L* a larger number of (different) orders can be observed (especially in Virtual-Node mode). Further, on *BG/L* a significant difference can also be seen even on the same machine when comparing the execution under the two different modes. Using Virtual-Node mode, the diversity is much larger. Considering that codes are often developed on clusters or on BG/L using the simpler (default) Co-Processor mode before performing scaling experiments in Virtual-Node mode, bugs due to race conditions may easily escape detection.

4 Message Ordering and Overtaking Messages

Another example of unexpected behavior occurs with message overtaking. In MPI, the message ordering is guaranteed based on the following statement in the standard's specification [6]:

Section 3.5, Semantics of point-to-point communication
"Messages are non-overtaking: If a sender sends two messages in succession to the same destination, and both match the same receive, then this operation cannot receive the second message if the first one is still pending. If a receiver posts two receives in succession, and both match the same message, then the second receive operation cannot be satisfied by this message, if the first one is still pending. This requirement facilitates matching of sends to receives. It guarantees that message-passing code is deterministic, if tasks are single-threaded and the wildcard *MPI_ANY_SOURCE* is not used in receives."[1]

It is clear that if this rule would not be enforced, programs would be nondeterministic even without wildcard receives (see Section 3). In this case, any program that sends two messages from one task to the other (with the same message tag) would lead to nondeterministic behavior, such that the execution delivers different results with the same input data. This would introduce all the problems of nondeterminism to any program that does not tag each message uniquely, making program development difficult and program execution unpredictable and unreliable.

For this reason, MPI introduced the non-overtaking rule, which needs to be guaranteed by the underlying MPI implementation. Unfortunately, keeping the ordering intact also introduces a certain overhead and may affect performance. Therefore, vendors of MPI implementations have to be careful when implementing this feature and it is not clear, if every implementation actually guarantees non-overtaking in all cases.

Implementation
The following test tries to verify the non-overtaking feature of a particular MPI implementation. The code is illustrated in Figure 4, and obviously at least two

[1] However, these semantics don't prevent indirect overtaking, e.g. when a task P sends two messages, one first to task Q and another one to task R, while task Q forwards the message upon reception to task R as well. In this case, it is possible that either the message from task P or the message from task Q arrives first at task R.

```
for i = [ 0, 5 [
  MPI_Isend(long_message,0)
for i = [ 0, 5 [
  MPI_Isend(short_message,0)
MPI_Waitsome(N-1, requests)
Verify send order
```

1 Sender

```
MPI_Recv(MPI_ANY_SRC)
Check arrival order
```

1 Receiver

Fig. 4. Pseudo code for message overtake experiments

tasks are needed. Task 1 implements a sender, which transfers messages of different length to task 0. The messages include a variable, which contains the ordering number of the particular message. In order to check, whether the order is disrupted, messages of different length (very long messages of 1 MByte and short messages of 1 Byte) are interleaved, such that a series of long messages may still be in progress when the short messages are issued.

Messages are issued using the non-blocking *MPI_Isend* operation since it allows messages to be transmitted in an order different from their issue order. Assuming the MPI implementation is optimized for latency hiding, it could transfer shorter messages, while longer messages are still in the queue. As a result, the ordering of messages at *MPI_Isend* may be disrupted. Consequently, for our test case, the receiver side would have to correct the ordering such that the general non-overtaking rule of MPI is not violated.

Observations

The observations for this example are divided into two cases. In case 1, we observed the order of the messages at the receiver in order to verify, if the MPI specification is violated. In all our experiments, this has never been the case, which indicates that ordering is implemented correctly on all tested systems.

In case 2, we verified the order of outgoing messages at the sending task. When using only 2 tasks, where each message has to use exactly the same route from sender to receiver, we did not observe any out of order messages at the sender. However, by scaling up the number of tasks, out-of-order message transfer took place. As an example on the *SGI*, we used 16 MPI tasks with a message length of 1 MByte, and we observed 29% of out-of-order send operations. This is a clear indication that SGI's MPI implementation applies some kind of optimization to the sending task, while still enforcing the correct order at the receiver.

5 Large Windows for Asynchronous Receives

The MPI standard guarantees in–order message delivery for any communication channel, i.e., messages sent between two tasks using the same communicator and

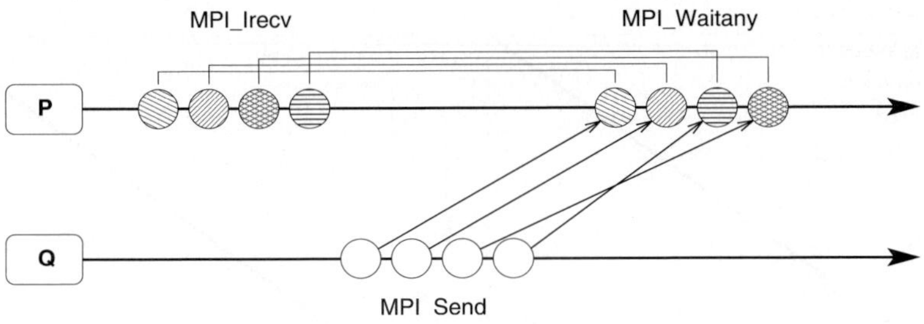

Fig. 5. Multiple outstanding receives

```
for i= [ 0, N [
{
   Buffer[0]=i
   MPI_Send(Buffer, recv. task)
}
```

1 Sender

```
for i = [ 0,M [
   req[i]=MPI_Irecv()
for i = [ 0,N [
{
   MPI_Waitany(req array)
   Print(Recveived value, i)
   req[recvd]=MPI_IRecv()
}
```

1 Receiver

Fig. 6. Pseudo code for varying receive windows

message tag. However, this guarantee only refers to the time when the receive is posted, not when messages are actually received by the application. While this is not relevant for blocking receives, it can cause problems for asynchronous non-blocking operations when multiple receives are concurrently active.

Figure 5 shows an example. Task P posts four asynchronous receives that are matched with four sends from task Q. According to the MPI standard, the order of the messages is preserved with respect to the order in which the receives are posted. The order in which the user receives the messages, i.e., the corresponding *MPI_Waitany* finishes, however, can vary.

Implementation

To stress the above scenario, MPITEST contains a micro kernel that can vary the number of concurrent receives. The pseudo code is shown in Figure 6. A sending task sends N messages to a receiver with a message ID in the body. The receiver firsts posts M asynchronous receives and then receives all N messages from the sender. For this, the receiver uses *MPI_Wait_any* to query the next

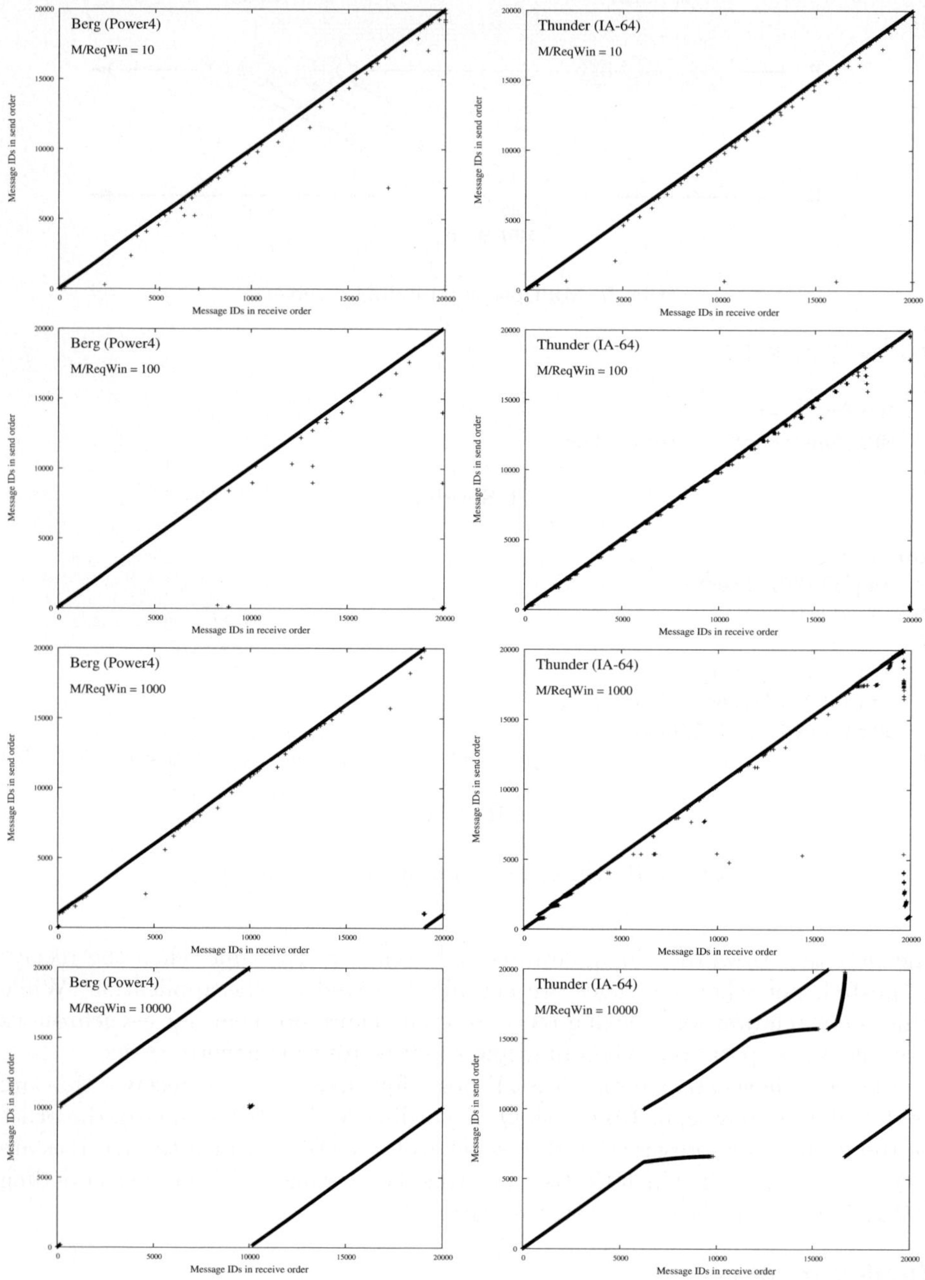

Fig. 7. Send vs. receive order on *Berg* (left) and *Thunder* (right) with varying receive window width: 10, 100, 1000, and 10000 (from top to bottom)

available message, records the message ID contained in the message body, and then reposts the corresponding receive operation.

Observations
For this micro kernel, we plot the message order in which the message was sent vs. the order in which it was received. In the ideal case, i.e., with a complete in-order delivery, the result should be a straight line. Figure 7 shows the results of the code on two representative platforms: *Berg* and *Thunder*.

For small receive windows, both platforms behave similarly and close to the expected in-order case. There are, however, several messages that get completed out-of-order. With an increasing message window, the behavior of the two platforms differs significantly. On *Berg*, the graph gets shifted upward by the size of the message window. This means that MPI always completes the last request in the window first, and since this request is restarted once completed, all other receives are stalled. Only once all messages have been transmitted, MPI receives the old messages in the remainder of the request array. On *Thunder*, on the other hand, larger request windows up to 1000 requests behave similar to the small window case. For very large windows, however, the behavior changes drastically into a seemingly chaotic, but reproducible state.

6 Conclusions

The utilization of MPI for many parallel applications has led to a wide acceptance of MPI in the community. However, the MPI standard leaves some freedom for implementations on particular platforms. While this may not cause any problems in the majority of cases, there are some pitfalls with the possibility for substantial effects on correctness and reliability.

The MPITEST suite is a first step towards methodically testing MPI implementations for these issues. With a combination of test cases and experimental setups, users can evaluate the behavior of MPI implementations. This test suite should increase the awareness of the possibility of such unexpected behavior and provide feedback to implementors about the compliance (or lack thereof) to the MPI standard and/or user expectations.

The issues explored by MPITEST also indicate that user application testing on a small set of platforms is unlikely to cover the set of possible message orderings, regardless of how many times the application tests are performed. Thus, subtle errors can persist in applications for extended periods. Additional work can complement MPITEST to provide users with greater assurance of the correctness of their applications by increasing and randomizing the message orders covered [8].

We are continuing to expand the test suite with new test cases being developed over time or provided by the community. The package itself will eventually be available for download from the project webpage at http://www.gup.uni-linz.ac.at/mpitest/, such that interested users can easily check their own MPI implementations or contribute to MPITEST. Feedback from the application users community is clearly needed and welcome at `mpitest@gup.jku.at`.

References

1. D. Bailey, T. Harris, W. Saphir, R. V. der Wijngaart, A. Woo, , and M. Yarrow. The NAS parallel benchmarks 2.0. Report NAS-95-020, NASA Ames Research Center, Moffett Field, CA, Dec. 1995.
2. B. de Supinski. The ASCI PSE Milepost: Run-Time Systems Performance Tests. In *Proceedings of the International Conference on Parallel and Distributed Processing Techniques and Applications (PDPTA)*, June 2001.
3. B. de Supinski and N. Karonis. Accurately Measuring Broadcasts in a Computational Grid. In *Proceedings of the 8th IEEE International Symposium on High-Performance Distributed Computing (HPDC)*, pages 29–37, 1999.
4. D. Kranzlmüller and M. Schulz. Notes on Nondeterminism in Message Passing Programs. In *Proceedings of the 9th European PVM/MPI Users' Group Meeting*, pages 357–367, Sept. 2002.
5. Lawrence Livermore National Laboratory. The ASCI purple benchmark codes. http://www.llnl.gov/asci/purple/benchmarks/limited/code_list.html, Oct. 2002.
6. Message Passing Interface Forum (MPIF). MPI: A Message-Passing Interface Standard. Technical Report, University of Tennessee, Knoxville, June 1995. http://www.mpi-forum.org/.
7. R. Reussner, P. Sanders, L. Prechelt, and M. Müller. SKaMPI: A Detailed, Accurate MPI Benchmark. In *Proceedings of the 5th European PVM/MPI Users' Group Meeting*, pages 52–59, Sept. 1998.
8. R. Vuduc, M. Schulz, D. Quinlan, B. de Supinski, and A. Sæbjørnsen. Improving Distributed Memory Applications Testing By Message Perturbation. In *Proceedings of Parallel and Distributed Systems: Testing and Debugging (PADTAD)*, July 2006.

Zone-Oriented Byzantine Agreement on Zone-Based Wireless Ad-Hoc Network

Chien-Fu Cheng[1], Shu-Ching Wang[2,*], and Tyne Liang[1]

[1] Department of Computer Science, National Chiao Tung University, 1001, Ta Hsueh Rd., Hsinchu, Taiwan 300, R.O.C.
{cfcheng, tliang}@cs.nctu.edu.tw

[2] Department of Information Management, Chaoyang University of Technology, 168, Jifong E. Rd., Wufeng, Taichung County, Taiwan 413, R.O.C.
scwang@cyut.edu.tw

Abstract. A wireless ad-hoc network system may suffer from various types of hardware failure. In order to enhance the fault-tolerance and reliability of the wireless ad-hoc networks, we revisit the Byzantine Agreement problem in the zone-based wireless ad-hoc network in this paper. The proposed protocol is called as the Zone-Oriented Agreement Protocol (ZOAP) which can make each fault-free mobile processor reach an agreement value to cope with the faulty component in the zone-based wireless ad-hoc network.

Keywords: Byzantine agreement, fault-tolerance, distributed system, zone-based wireless ad-hoc network, malicious fault.

1 Introduction

The wireless ad-hoc networks have attracted significant attentions recently due to its features of infrastructure less, quick deployment and automatic adaptation to changes in topology. Therefore, wireless ad-hoc network suits for military communication, emergency disaster rescue operation, and law enforcement [2].

We know that the reliability of the mobile processor is one of the most important aspects in wireless ad-hoc networks. In order to provide a reliable environment in a wireless ad-hoc network, we need a mechanism to allow a set of mobile processors to agree on an agreement value [7]. The Byzantine Agreement (BA) problem [4], [5], [6], [8], [9], [10] is one of the most fundamental problems to reach an agreement value in a distributed system.

The BA requirement can be satisfied when the following constraints are met:

(Agreement): All fault-free processors agree on an agreement value.
(Validity): If the source processor is fault-free, then all fault-free processors agree on the initial value sends by the source processor.

The traditional BA problem was focused on the fixed and well-defined network [4], [5], [6], [8], [9], [10]. However, the network structure of wireless ad-hoc network is not fixed, and it can change its topology at any time by the feature of mobility.

* Corresponding author.

M. Gerndt and D. Kranzlmüller (Eds.): HPCC 2006, LNCS 4208, pp. 853 – 862, 2006.

Hence, the tradition solutions for the BA problem were not suited for the wireless ad-hoc network.

In this study, we re-visit the BA problem in the zone-based wireless ad-hoc network. The detailed description of network structure will be presented in section 2.2. The proposed protocol is called as the Zone-Oriented Agreement Protocol (ZOAP). ZOAP can make each fault-free mobile processor in the zone-based wireless ad-hoc network reach an agreement value.

The rest of this paper is organized as follows. Section 2 will serve to introduce the definitions and conditions of the ZOAP in a zone-based wireless ad-hoc network. Then, in Section 3, we shall introduce the BA protocol ZOAP. The new BA protocol ZOAP will be brought up and illustrated in Section 4. Finally, in Section 5, we shall present the conclusion.

2 The Definitions and Conditions for BA Problem

The detailed definitions and conditions for BA problem are shown in this section. They are the failure type of a fallible mobile processor, network model, number of rounds required in the "message-exchange phase" and the constraints.

2.1 The Failure Type of a Fallible Mobile Processor

The symptoms of mobile processor failure can be classified into two categories: dormant fault and malicious fault (also called as Byzantine fault) [5]. The dormant faults of a fallible mobile processor are crashes and omission. A crash fault happens when a processor is broken. An omission fault takes place when a processor fails to transmit or receive a message on time or at all. However, the behavior of a mobile processor with malicious fault is unpredictable. If the BA problem can be solved with mobile processor with malicious fault, the BA problem can be also solved with mobile processor with dormant fault.

2.2 The Network Model

In this study, the BA problem is discussed in a zone-based wireless ad-hoc network, such as the network structure in Joa-Ng and Lu's [3]. An example of zone-based wireless ad-hoc network is shown in Fig. 1. Each mobile processor in the network can be identified as unique, and can get its zone id through the Global Positioning System (GPS). Let N be the set of all mobile processors in the network and |N|= n, where n is the number of mobile processor in the network. Let Z be the set of all zones in the network and |Z|= z, where z is the number of zones in the underlying network and $z \geq 4$. If there are at least $\lceil \mu_i /2 \rceil$ malicious faulty mobile processors in Z_i, then Z_i will be the malicious faulty zone. Here, Z_i is the i-th zone, and μ_i is the number of mobile processors in Z_i, $0 \leq i \leq z$. If the gate way processor of the zone Z_i is faulty, then Z_i will also be the malicious faulty zone. Let Z_m be the maximum number of malicious faulty zones allowed. If there is no mobile processor in the i-th zone, then Z_i is the away zone.Let Z_a be the maximum number of away zone.

2.3 The Number of Rounds of Message Exchange Required by ZOAP

The term "round" denotes the interval of message exchange between any pair of processors [1], [4], [5], [8], [9], [10]. Fischer and Lynch [1] pointed out that t +1 ($t=\lfloor(n-1)/3\rfloor$) rounds are the minimum number of rounds to get enough messages to achieve BA. The unit of Fischer and Lynch [1] is processor, but the unit of the zone-based wireless ad-hoc network is zone. So that, the number of required rounds of message exchange in the zone-based wireless ad-hoc network is Zm +1 ($Zm=\lfloor(z-1)/3)\rfloor$). The detailed description of the "message-exchange phase" will be presented in section 3.1.

2.4 The Constraints of ZOAP

The number of processors with malicious fault allowed in the network depends on the total number of processors. For example, in Lamport et al. [4], the assumption of failure type of the fallible processor is malicious and the unit of the network is processor. So, the constraint of Lamport et al. [11] is $n>3f_m$ (f_m is the maximum number of malicious faulty processors allowed). However, Siu et al. [8] find that the correct constraint on number of processors required should be $n>(n-1)/3+2f_m$. Due to the proposed protocol ZOAP is designed for the zone-based wireless ad-hoc network with fallible processors, and the unit of the network is zone. Hence, the constraints of the ZOAP is $z>\lfloor(z-1)/3\rfloor+2Zm+Z_a$.

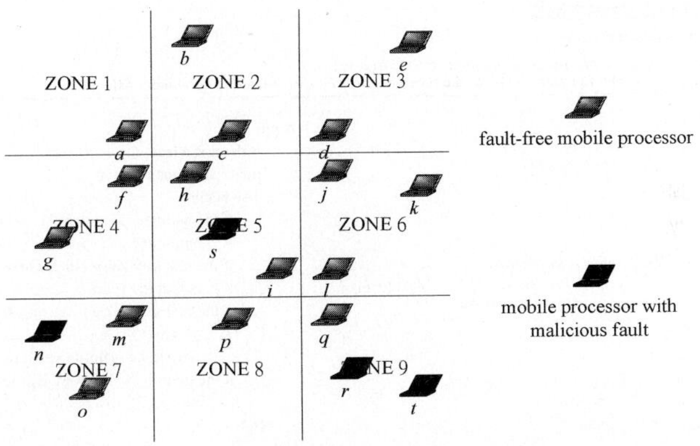

Fig. 1. An example of zone-based wireless ad-hoc network

3 Zone-Oriented Agreement Protocol (ZOAP)

In this section, the detailed description of our proposed protocol Zone-Oriented Agreement Protocol (ZOAP) is shown here.

There are three phases in our protocol ZOAP, and they are the "message-exchange phase", "decision-making phase", and the "extension-agreement phase". The number of rounds required for running ZOAP is Z_m +1 ($Z_m=\lfloor(z-1)/3)\rfloor$). And, ZOAP can tolerate Z_m malicious faulty zones, where $z>\lfloor(z-1)/3\rfloor+2Z_m+Z_a$. The protocol ZOAP is shown in Fig. 2.

Zone-Oriented Agreement Protocol (ZOAP)

Definition:

Each mobile processor can communicate to other mobile processor in the same zone directly.
Each mobile processor can communicate to other mobile processor in the different zone through the gate way processor.
Three mobile processor roles: sender processors, relay processor and receiver processor, depending on the message flow. A message sent from a sender processor to a receiver processor may be passed through some inter-mediate (relay processor, gate way processor).
A relay processor cannot falsify a message from a sender processor to a receiver processor. This is achieved using encryption technology (such like RSA).
The return processor, a mobile processor that moves away from the network in the "message-exchange phase", and returns to the network before the beginning of "extension-agreement phase".
Each mobile processor receives other mobile processors' messages at each round of "message-exchange phase". If no message is received, the value is replaced with δ^0.

Message-Exchange phase:

Compute the number of rounds required γ : $\gamma = Z_m+1$, where $Z_m=(z-1)/3$

If r=1, then:

The source processor sends its initial value v_s using encryption technology to all other mobile processors.
Each mobile processor stores the value from the source processor in the root s of its mg-tree.

For r > 1, do:

Without a source processor, each mobile processor transmits the values at level r-1 in its mg-tree using encryption technology to all other mobile processors. If the value at level r-1 is δ^i, the value δ^{i+1} is replaced as the transmitted value, where 1 i $(z-1)/3$ -1
Each receiver processor applies function RMAJ on it received messages and stores the function RMAJ values in the corresponding vertices at level r of its mg-tree.

Decision-Making phase:

Step 1: Delete vertices with repeated name of the mg-tree.
Step 2: Uses the function VOTE to process root s of each mobile processor's mg-tree and obtain the agreement value VOTE(s).
Step 3: Transmit the VOTE(s) to the return processors if there is any return processor in the network.

Extension-Agreement Phase:

For the return processor only

Step 1: Receive other mobile processors' VOTE(s) values.
Step 2: Uses the function VOTE to the received messages to obtain the agreement value.

RMAJ(V) =
1. The majority value in the vector $V_i = [v_1, \ldots, v_{i-1}, v_i]$, if it exists.
2. A default value is chosen, otherwise.

VOTE(a) =

```
begin
    if the a is a leaf
    then output the value a                                /* condition 1*/
    else begin
        if the number of value δ^0 is 3*(Z_m-?+1)+[(z-1) mod 3]
            output the value a                             /* condition 2*/
        if the majority value is δ^i, where 1≦ i ≦ (z-1)/3
            output the value δ^(i-1)                       /* condition 3*/
        if the majority value is the non-δ^j value,
            where 0≦ j ≦ (z-1)/3, m  {0,1}
            output the majority value m                    /* condition 4*/
        if the majority value does not exist
            output the default value                       /* condition 5*/
    end
end.
```

Fig. 2. The BA Protocol Zone-Oriented Agreement Protocol (ZOAP)

3.1 Message-Exchange Phase

At the first round of "message-exchange phase" (r=1), the source processor sends its initial value v_s using encryption technology to all other mobile processors. Hence, each mobile processor in the source mobile processor's zone can receive the initial value v_s from source mobile processor directly. For other mobile processors which are not in the source mobile processor's zone, the source processor can send its initial value by relaying through the gateway processor. Hence, the mobile processors in the different zone can also receive the messages from the source processor.

After the first round of message exchange (r>1), each mobile processor without source processor sends the values at level r-1 in its mg-tree (the detailed description of the mg-tree is presented in Appendix) to each zone's processors, if the value at level r-1 is δ^i , then replace the value δ^{i+1} as the transmitted value, where $1 \leqq i \leqq Z_m - 1$ (The value δ^i is used to report the absent value.). At the end of each round, the receiver processor uses function RMAJ (Function RMAJ is shown in Fig. 2) on it received values which from the same zone to get a single value. And each receiver processor stores the received messages and function RMAJ value in its mg-tree.

3.2 Decision-Making Phase

In the "decision-making phase", each processor without source processor removing the vertices with repeated names in order to avoid the influence from a faulty processor from repeating in an mg-tree. Then, using function VOTE on the root s of each processor's mg-tree. The agreement value VOTE(s) is obtained. The function VOTE only counts the non-value δ^0 (excluding the last level of the mg-tree) for all vertexes at the r-th level of an ic-tree, where $1 \leqq r \leqq Zm+1$. The condition 1, condition 4, and condition 5 in the function VOTE are similar to conventional majority vote [5]. The condition 2 is used to remove the influence by a malicious faulty processor. The condition 3 is used to describe the existence of an absentee. The detail descriptions of function VOTE is shown in Fig. 2.

3.3 Extension-Agreement Phase

The goal of the "extension-agreement phase" is to allow return processor to compute an agreement value which is the same as other fault-free mobile processors' agreement value. In the "extension-agreement phase", each return processor receives the agreement values from other mobile processors which join whole the "message-exchange phase". The VOTE function is used on the values received to obtain the agreement value. The return processor can then obtain the same agreement value as other fault-free mobile processors. The reason is that each fault-free mobile processor can reach an agreement value if $z > \lfloor (z-1)/3 \rfloor + 2Zm + Z_a$. Hence, there are at least $z - \lfloor (z-1-Z_a)/3 \rfloor - Z_a$ zones that are fault-free and have the same agreement value. That is, in the worst case, a return processor can receive $z - \lfloor (z-1-Z_a)/3 \rfloor - Z_a$ copies of the same value, which is large then $\lfloor (z-1-Z_a)/3 \rfloor$. Therefore, a return processor can then decide the agreement value bye using the function VOTE.

4 An Example of Executing ZOAP

In this section, an example of executing the proposed protocol ZOAP is shown here. A zone-based wireless ad-hoc network is shown in Fig. 1.

The worst case of the BA problem is that the source processor with malicious faults. For example, suppose P_s is the source processor, which means P_s may transmit different values to different zones. In order to reach an agreement value among fault-free mobile processors in our example, ZOAP needs 3 (⌊(9-1)/3)⌋+1) rounds of message exchange in the "message-exchange phase".

4.1 Message-Exchange Phase

At the first round of the "message-exchange phase", the source processor P_s sends its initial value to all mobile processors in the zone-based wireless ad-hoc network. The message stored by each zone's fault-free processors at the first round of "message-exchange phase" is shown in Fig. 3. At the *r*-th (*r*>1) round of message exchange, except for the source processor, each processor sends it's RMAJ values at the (*r*-1)th level in its mg-tree to all other mobile processors without source processor. Then, each receiver processor applies RMAJ to its received messages from the same zone and stores the received values and the RMAJ values at the corresponding vertices at level r of its mg-tree. However, some of processors may move away from the network in this phase. So, other mobile processors in the network could not receive the messages from the mobile processors which are not in the network. As shown in Fig. 4, any mobile processor in the network would not receive the message from the mobile processors P_f, P_g, and P_k. The mg-trees of fault-free mobile processor P_a at the second and third round in the "message-exchange phase" are shown in Fig. 5 and Fig. 6.

	Level 1 root *s*
Z_1's fault-free mobile processors	1
Z_2's fault-free mobile processors	0
Z_3's fault-free mobile processors	1
Z_4's fault-free mobile processors	0
Z_5's fault-free mobile processors	0
Z_6's fault-free mobile processors	0
Z_7's fault-free mobile processors	0
Z_8's fault-free mobile processors	0
Z_9's fault-free mobile processors	1

(mg-trees)

Fig. 3. The mg-tree of each processor at the first round

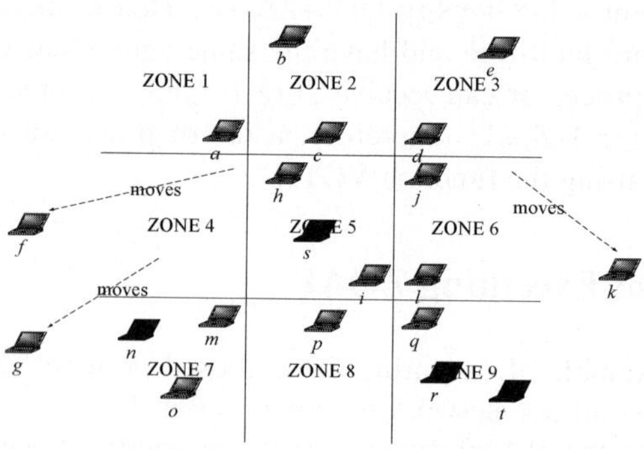

Fig. 4. P_f, P_g ,and P_k move away from the network in the "message-exchange phase"

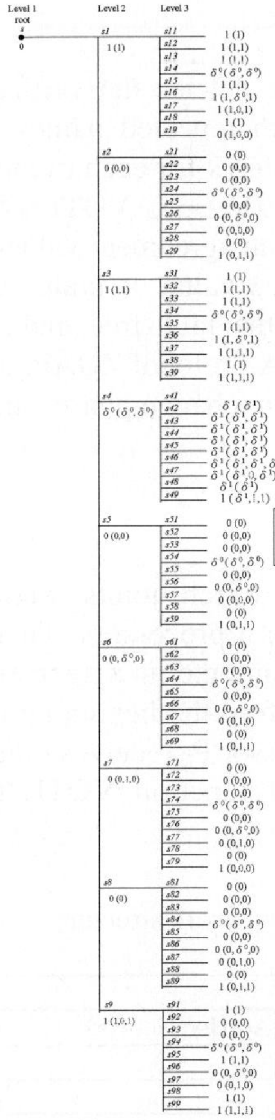

Fig. 6. The final mg-tree of mobile processor P_a

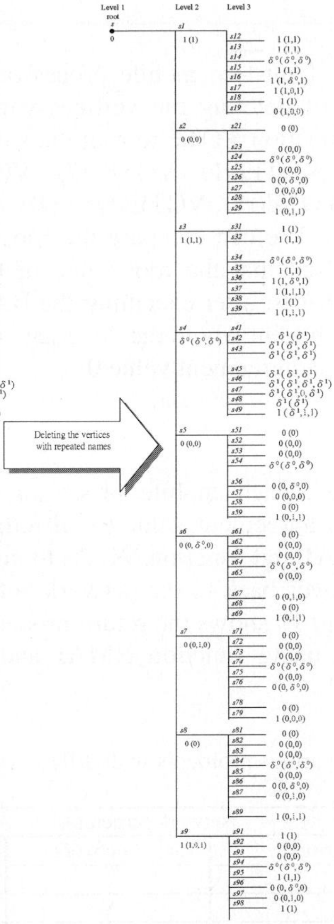

Fig. 7. Mg-tree without repeated vertices

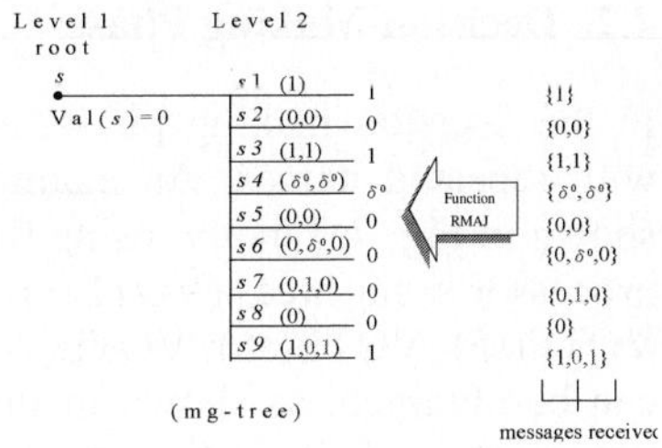

Fig. 5. The mg-tree of processor P_a at the second round of the "message-exchange phase"

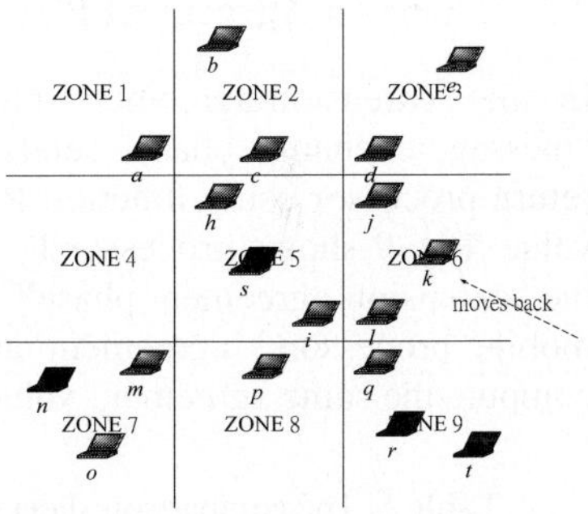

Fig. 9. The mobile processor P_k return to the network before the beginning of "extension-agreement phase"

◆ (VOTE($s2$) = 0,0, δ^0, 0, 0, 0, 0, 1) = 0 by the condition 3 in the function VC

◆ (VOTE($s3$) = 1,1, δ^0, 1, 1, 1, 1, 1) = 1 by the condition 3 in the function VC

◆ (VOTE($s4$) = δ^1, δ^1, δ^1, δ^1, δ^1, δ^1, δ^1, 1) = δ^0 by the condition 2 in the function VC

◆ (VOTE($s5$) = 0,0,0, δ^0, 0, 0, 0, 1) = 0 by the condition 3 in the function VC

◆ (VOTE($s6$) = 0,0,0, δ^0, 0, 0, 0, 1) =0 by the condition 3 in the function VC

◆ (VOTE($s7$) = 0,0,0, δ^0, 0, 0, 0, 0) = 0 by the condition 3 in the function VC

◆ (VOTE($s8$) = 0,0,0, δ^0,0,0,0,0) = 0 by the condition 3 in the function V

◆ (VOTE($s9$) = 1,0,0, δ^0,1,0,0,1) = 0 by the condition 3 in the function V

■ (VOTE(s) = 1,0,1, δ^0,0,0,0,0,0) = 0 by the condition 3 in the function V

Fig. 8. Mobile processor P_a uses function VOTE to compute its agreement value 0

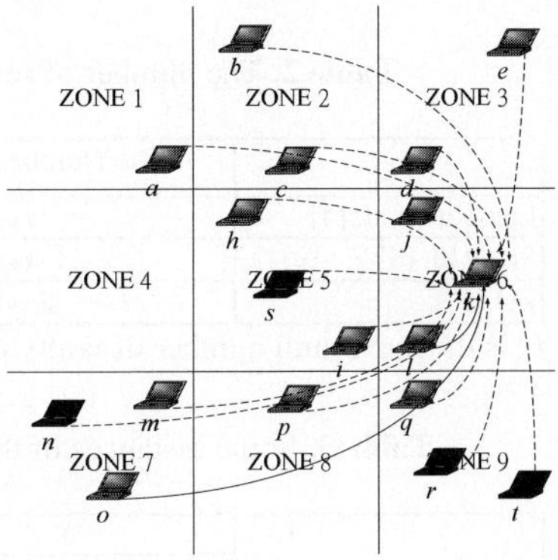

(VOTE(s) = 0,0,0,δ^0,0,0,0,0,1) = 0

Fig. 10. Return processor P_k uses Function RMAJ and Function VOTE to compute the agreement value

4.2 Decision-Making Phase

In the decision making phase, each fault-free mobile processor deletes the vertices with repeated names. An example of deleting the vertices with repeated names is shown in Fig. 7. Finally, using the function VOTE to root the value *s* for each mobile processor's mg-tree (VOTE(*s*)= (VOTE(*s*1), VOTE(*s*2), VOTE(*s*3), VOTE(*s*4), VOTE(*s*5), VOTE(*s*6), VOTE(*s*7), VOTE(*s*8), VOTE(*s*9)) =0), an agreement value 0 can be obtained, as shown in Fig. 8. We can compare the root *s* value of fault-free processor in Z_1 in Fig. 3, we can find that the root value of the fault-free mobile processors in Z_1, is replaced by 0. That is, after executing the BA protocol ZOAP, all the fault-free mobile processors which joint whole the "message-exchange phase" and decision making phase can agree on an agreement value 0.

4.3 Extension-Agreement Phase

In the "extension-agreement phase", each mobile processor which joints whole "message-exchange phase" sends its agreement value to all return processors. Then, return processor using function RMAJ and function VOTE to compute its agreement value. Fig. 9 shows processor P_k moves back to the network before the beginning of the "extension-agreement phase". Fig. 10 shows the return processor P_k receives other mobile processors' agreement and using function RMAJ and function VOTE to compute the same agreement value 0.

Table 1. The comparison sheet of various protocols with different network structure

	Network Structures		
	Fully connected network	General network	Ad-Hoc Network
Lamport et al. [4]	V		
Siu et al. [8]	V	V	
ZOAP	V	V	V

Table 2. The number of rounds required in the "message-exchange phase"

	The Number of Processor Required	Constraint
Lamport et al. [4]	t+1, t=$\lfloor(n\text{-}1)/3\rfloor$	$n > 3f_m$
Siu et al. [8]	t+1, t=$\lfloor(n\text{-}1)/3\rfloor$	$n >\lfloor(n\text{-}1)/3\rfloor+2f_m$
ZOAP	Z_m+1, Z_m=$\lfloor(z\text{-}1)/3\rfloor$	$z >\lfloor(z\text{-}1)/3\rfloor+2Z_m+Z_a$

f_m is the maximum number of faulty processors in the network.

Table 3. Some instances of the number of rounds required in the fixed network

	The number of rounds required		
	n=512, z=64	n=512, z=32	n=512, z=16
Lamport et al. [4]	171	171	171
Siu et al. [8]	171	171	171
ZOAP	22	11	6

Fixed network: the topology of the network is not changeable.

5 Conclusion

In this paper, we revisit the BA problem in a zone-based wireless ad-hoc network with fallible processor. Since the traditional fixed network such as fully connected network, broadcasting network, and general network are all special case of the zone-based network. If the protocol can solve the BA problem in a zone-based wireless ad-hoc network, the protocol also can solve the BA problem in the fully connected network, broadcasting network, and general network. Table 1 shows the comparison sheet of various protocols with different network structures.

For performance, due to the traditional BA protocol [4], [5], [6], [8], [9], [10] could not solve the BA problem in the zone-based wireless ad-hoc network. Therefore, we compare the performance of our protocol with pervious protocols in the fixed network. For example, if there 512 processors fall into 64 zones. The protocols proposed by Lamport et al. [4], and Siu et al. [8] need 171 ((n-1)/3 +1) rounds of message exchange to reach an agreement value in the network too. However, ZOAP only needs 22 ((z-1)/3 +1) rounds of message exchange to reach an agreement value. Therefore, the performance of ZOAP is more efficient than previous results [4], [5], [6], [8], [9], [10] when the network is divided into zones. Table 2 shows the number of rounds required of some previous protocols and Table 3 shows some instances of the number of rounds required for various protocols.

References

1. M. Fischer, and N. Lynch: A Lower Bound for the Assure Interactive Consistency. Information Processing Letters, Vol. 14. No. 4. (1982) 183-186.
2. X. Hong, K. Xu, and M. Gerla: Scalable routing Protocols for Mobile Ad Hoc Networks. IEEE Network, Vol. 16. No. 4. (2002) 11-21.
3. M. Joa-Ng and I.T. Lu: A Peer-to-Peer Zone-Based Two-Level Link State Routing for Mobile Ad Hoc Networks. IEEE Journal on Selected Areas in Communications, Vol. 17. No. 8. (1999) 1415-1425.
4. L. Lamport, R. Shostak, and M. Pease: The Byzantine Generals Problem. ACM Trans. Programming Language Systems, Vol. 4. No. 3. (1982) 382-401.
5. F.J. Meyer and D.K. Pradhan: Consensus with Dual Failure Modes. IEEE Trans. Parallel and Distributed Systems, Vol. 2. No. 2. (1991) 214-222.
6. G. Neiger: Distributed consensus revisited. Information Processing Letter, Vol. 49. (1994) 195-201.
7. Silberschatz, P.B. Galvin, G. Gagne: Operating System Concepts. 6th. Ed., John Wiley & Sons, Inc., (2002).
8. H.S. Siu, Y.H. Chin, and W.P. Yang: A Note on Consensus on Dual Failure Modes. IEEE Trans. Parallel and Distributed System, Vol. 7. No. 3. (1996) 225-230.
9. S.C. Wang, Y.H.Chin and K.Q. Yan: Byzantine Agreement in a Generalized Connected Network. IEEE Trans. Parallel and Distributed Systems, Vol. 6. No. 4. (1995) 420-427.
10. K.Q. Yan, Y.H. Chin and S.C. Wang: Optimal Agreement Protocol in Malicious Faulty Processors and Faulty Links. IEEE Trans. Knowledge and Data Engineering, Vol. 4. No. 3. (1992) 266-280.

Appendix

When a fault-free mobile processor receives the message sent from the source processor, it stores the received value, denoted as val(s), at the root of its mg-tree as shown in Fig. 3. In the second round, each processor transmits the root value of its mg-tree to itself and all the other zones. If mobile processors in Z_1 send message val(s) to Z_2, then mobile processors in Z_2 store the function RMAJ value of the received messages from zone Z_1, denoted as val($s1$), in vertex $s1$ of its mg-tree. Generally, message val($s1….z$), stored in the vertex s1…z of an mg-tree, implies that the message just received was sent through the source processor, the processors of zone Z_1,…, the processors of Z_z; and the processors in Z_z is the latest processors to pass the message. An example of mg-tree with two levels and three levels are shown in Fig. 5 and Fig. 6.

Priority-Enabled Optimization of Resource Utilization in Fault-Tolerant Optical Transport Networks*

Jacek Rak

Gdansk University of Technology, Narutowicza 11/12
80-952 Gdansk, Poland
jrak@pg.gda.pl

Abstract. In this paper, a novel *SCMTO* heuristic algorithm of establishing survivable connections in wide-area networks, that jointly optimizes the level of resource utilization as well as the time of connection restoration, is proposed. The algorithm is dedicated to multipriority traffic. Two service classes are assumed, referred to as low and high priority, respectively. Unlike typical optimization methods, *SCMTO* algorithm also guarantees fast service restoration by processing the high-priority connections with the author's *a posteriori* optimization, based on the Largest-First graph coloring heuristics.

The algorithm was evaluated for the US Long-Distance Network. The results show that with only little capacity utilization degradation, fast restoration can be achieved. Compared to the results of the typical *a priori* optimization, the obtained average values of connection restoration time were up to 20% shorter (and even up to 41% for the high-priority traffic only). They were achieved together with reduction in resource utilization ratio up to 48%.

Presented solutions are dedicated to WDM-based high-performance optical communications network architectures.

1 Introduction

Nowadays we observe an increasing dependency of society on large-scale complex networked systems. This amplifies the importance of assuring the network survivability, defined as *the capability of a system to fulfill its mission in a timely manner, in the presence of attacks, failures or accidents* [2, 8]. This paper investigates the survivability issues of end-to-end connections in optical wide-area networks, where, due to wavelength division multiplexing (WDM), each network link consists of a set of channels (wavelengths), each one capable of transmitting the data independently at peak electronic speed of a few Gbps. A failure of any network element may thus lead to large data and revenue losses [9]. Survivability is achieved by introducing redundancy. It means that for the main path of a connection, referred to as *active path*, there are additional (redundant) paths, called *backup paths* (or shortly *backups*), to protect the connection in case of a certain failure scenario [5, 6, 10, 13, 14]. The most common is

* This work is partially supported by the Ministry of Education and Science, Poland, under the grant 3 T11D 001 30.

M. Gerndt and D. Kranzlmüller (Eds.): HPCC 2006, LNCS 4208, pp. 863 – 873, 2006.

the protection against a single network element failure (i.e. a single link or a single node). Survivable networks are based either on dedicating backup resources in advance (referred to as *protection* or *preplanned* scheme) or on dynamic *restoration* [10, 13, 14]. Concerning the scope of protection/restoration, *path*, *region* or *link protection/restoration* is typically used [5, 6, 11, 13].

However, assuring survivability by establishing backup paths results in an excessive level of resource (link capacity) utilization. This in turn limits the number of connections that may be established. Typical technique of reducing the link capacity utilization ratio, here referred to as the *a priori* approach, is based on sharing the backup paths [3, 6, 13]. However, this capacity-effective optimization finds backups that are often not the shortest ones, since, under such sharing, the costs of links do not reflect well the real link lengths. Long backups make in turn the task of connection restoration time-consuming [14]. Nowadays, optimization that is capacity-effective, but causes the restoration process take much time, is often not acceptable. Network operators are interested in assuring fast connection restoration (especially with regard to important connections), even for the price of worse link capacity utilization ratio.

Although the issues of assuring fast restoration of broken connections as well as the optimization of link capacity utilization have been separately investigated by many research groups, they still have not been extensively studied jointly. In general, it is not possible to achieve minima for both the functions. Depending on the importance of the criteria, one always has to find a compromise.

The objective of the paper is to propose the *SCMTO* algorithm optimizing the network resource utilization which simultaneously provides fast restoration of important connections. The model is dedicated to multipriority traffic (two service classes are assumed). Based on the demanded guarantee on fast service restorability, connections are assigned low or high priorities, respectively. According to the proposed algorithm, utilization of backup path resources, allocated for the high-priority connections, is optimized using the author's *a posteriori* technique, that is based on the Largest-First graph coloring heuristics. This optimization, described in Section 2.2, provides fast connection restoration together with medium level of link capacity utilization. In order to better utilize the backup path link capacities, the proposed *SCMTO* algorithm uses the typical capacity-effective *a priori* optimization routine, described in Section 2.1, for all the low-priority connections.

To the best knowledge of the author, this is the first paper to propose the algorithm, dedicated to the multipriority traffic, that jointly optimizes the average restoration time values as well as the ratio of resource utilization. The results show that, when using the proposed *a posteriori* optimization of resource utilization, with only little capacity utilization degradation, fast restoration can be achieved and the resource utilization kept at the low level.

The rest of the paper is organized as follows. Section 2 shows the principles of optimizing the backup path resource utilization. The author's *a posteriori* technique is described in detail. The proposed *SCMTO* algorithm is given in Section 3. Section 4 presents the modeling assumptions. The results, obtained for the US Long-Distance Network, are described in Section 5 and include: the lengths of backup paths, the values of connection restoration time and the level of link capacity utilization.

2 Optimization of Link Capacity Utilization

Typical methods optimizing the ratio of link capacity utilization are generally based on the concept of sharing the link capacities that are reserved for backup paths. If, for each connection, provisioning 100% of the requested capacity after a failure is required, then *capacity of a link channel, reserved for backup purposes, may be shared among different backups, if the respective protected parts of active paths are mutually disjoint*[1] [4, 6]. Indeed, backup capacities may be shared, if they are to protect against different failure scenarios (which are not expected to occur simultaneously).

Considering the optimization strength, there are three main variants [6]. *Intra-demand sharing* implies sharing the link capacities of backup paths that protect disjoint parts of an active path of the same connection. *Inter-demand sharing* denotes sharing the backup paths that protect disjoint parts of active paths of different connections. *Parallel intra- and inter-demand sharing* is the most capacity efficient optimization, combining the features of both intra- and inter-demand sharing.

2.1 *A Priori* Optimization of Resource Utilization

A typical optimization technique, presented in [4, 6], here referred to as the *a priori* approach, is performed before finding a given backup path. It is applied when calculating the cost ξ_{ij} of each network link l_{ij}. If the backup path for the kth connection of capacity $r^{(k)}$ is to be found, then the cost ξ_{ij} of each link l_{ij} is determined as:

$$\xi_{ij} = \begin{cases} 0 & \text{if } r^{(k)} \le m_{ij}^{(k)} \\ (r^{(k)} - m_{ij}^{(k)}) \cdot s_{ij} & \text{if } r^{(k)} > m_{ij}^{(k)} \text{ and } f_{ij} \ge r^{(k)} - m_{ij}^{(k)} \\ \infty & \text{otherwise} \end{cases} \tag{1}$$

where: $r^{(k)}$ is the requested capacity; f_{ij} is the unused capacity of a link l_{ij};
s_{ij} is the length of a link l_{ij}; $m_{ij}^{(k)}$ is the capacity reserved so far at l_{ij} that may be shared.

If, for a link l_{ij}, the requested capacity $r^{(k)}$ is less or equal to the capacity already reserved for backups and may be shared, then the link cost ξ_{ij} is set to 0. However, if the demanded capacity $r^{(k)}$ is greater then the capacity $m_{ij}^{(k)}$ that may be shared, but there is enough unused capacity f_{ij}, then the link cost ξ_{ij} is determined by the extra capacity that is to be reserved at l_{ij} and also often reflects the length s_{ij} of this link. Otherwise it is set to infinity. After having calculated the costs of all the network links, the backup path is then found as the cheapest one (regarding the aggregated sum of costs of the used links), with help of any algorithm of path finding (e.g. Dijsktra's [1]).

Since, under the *a priori* optimization, the obtained costs ξ_{ij} of backup path links (Eq. 1) do not reflect well the real lengths of links, the established backups are often not the shortest ones. Summarizing, the *a priori* resource utilization optimization is capacity-efficient, but, due to increasing the length of backup paths, makes the process of connection restoration take much time. This in turn confines its applicability to connections of low priority (i.e. with low guarantee on fast service restorability).

---

[1] Depending on the kind of protection (either against a single link, or a single node failure) these parts of active paths must be mutually link- or node-disjoint, respectively.

2.2 The Proposed A *Posteriori* Optimization of Resource Utilization

The author's heuristic algorithm of establishing Survivable Connections with the *a POsteriori* optimization of link capacity utilization (*SCPO*), shown in Fig. 1, simultaneously provides fast connection restoration. Fast restoration of connections is achieved here due to not increasing the length of backup paths. After performing the optimization, backups remain unchanged due to parallel:

- performing the *a posteriori* optimization after the active and backup paths of all the connections in the network are tried to be found and installed,
- confining the scope of optimization to the single network link.

Input: A set K of demands to establish survivable connections, each of capacity $r^{(k)}$

1 Establish connections by sequentially processing the demands from K as follows:
For each demand k:
  1.1 Find* and install its active path
  1.2 Find* and install its backup path(-s)

2 After having processed all the demands from K, apply the *a posteriori* optimization of backup path link capacity utilization, as follows:
For each network link l_{ij}:
  2.1 Divide the set B_{ij} of backup paths, installed on channels of a link l_{ij}, into the subsets B_{ij}^s such that:
    – each subset contains backups that may share capacity one another**
    – the number of subsets B_{ij}^s for a link l_{ij} is minimized.
  2.2 For each subset B_{ij}^s:
    2.2.1 delete link channel allocations for the backups of B_{ij}^s, made in Step 1.2
    2.2.2 apply sharing by allocating one common channel for all the backups of a given B_{ij}^s (allow sharing with regard to all the backups of the subset B_{ij}^s).

* Perform the following:
a) calculate the matrix Ξ of link costs (i.e. for each link l_{ij} set its cost ξ_{ij} to the link length s_{ij}, if the amount of unused capacity f_{ij} is not less than $r^{(k)}$; otherwise set ξ_{ij} to infinity). If a backup path is to be found then, in order to assure that the backup path is link-disjoint with the respective active path, set the costs ξ_{ij} of links, used by the active path, to infinity
b) find the path, using any algorithm of path finding (e.g. Dijkstra's [1])
c) if finding the path is infeasible due to the lack of resources, then reject the demand and delete all the previously installed paths of the demand, else install the found path

** Any type of sharing is allowed. Inter- or intra-demand sharing is used most commonly.

Fig. 1. *SCPO* algorithm

The problem of optimally dividing the set B_{ij} of backup paths of each link l_{ij} into the subsets B_{ij}^s (Step 2.1 of the *SCPO* algorithm) is NP-hard [7], as it is equivalent to the vertex coloring problem of an induced graph of conflicts G_{ij}. In such a graph:

- the vertices denote backup paths that run through a link l_{ij},
- there is an edge between a given pair of vertices p and r of G_{ij}, if and only if the respective backup paths cannot share a common channel (i.e. when parts of the protected active paths are not disjoint each other) or when sharing a channel between such backups is not the aim of optimization (for instance an edge, denoting backups protecting the same (different) connection(s), must be added, if intra-(inter-)demand sharing is not needed, respectively).

Due to the computational complexity reasons, during experiments the *Largest-First* (*LF*) heuristics [7], having thc polynomial computational complexity, was used to color the vertices of each induced graph G_{ij}. Using the vertex coloring algorithm, any two neighboring vertices in G_{ij} must be assigned different colors. After applying the coloring routine to G_{ij}, all the vertices that obtained the same color, represent backup paths that belong to one particular subset B_{ij}^{s}. They protect mutually disjoint parts of active paths and thus may share a common channel at l_{ij}. The total number of colors used for G_{ij} determines the number of channels of a link l_{ij} that will become allocated for backup paths after applying the Step 2.2 of the *SCPO* algorithm.

Summarizing, the proposed *a posteriori* optimization provides fast connection restoration, but, due to confining the optimization scope to the single link, is less capacity-effective, compared to the *a priori* technique. It is thus the most suitable for optimizing the backup resources, utilized for connections requiring fast restoration.

3 *SCMTO* Algorithm of Establishing Survivable Connections

The proposed algorithm of establishing Survivable Connections for Multipriority Traffic with Optimization (*SCMTO*), shown in Fig. 2, uses both the described optimization approaches. Based on the demanded guarantee on fast service restorability, connections are assigned low or high priorities, respectively. The main objective of the algorithm is to optimize the network resource utilization. However, for high-priority connections it also provides fast connection restoration. That is why the *SCMTO* algorithm performs the *a priori* optimization for low-priority connections and the *a posteriori* optimization for connections of high priority.

Input:	A set D of demands to establish survivable connections, each of capacity $r^{(d)}$
1	Divide the set D of demands into two subsets: K and L such that subset K contains the high-priority demands and subset L contains the low-priority demands
2	Try to establish connections for all the demands from K by applying the *SCPO* algorithm
3	For each demand from L try to establish a connection with simultaneously applying the *a priori* optimization (for backup path finding only)

Fig. 2. *SCMTO* algorithm

It is worth mentioning that the *SCMTO* algorithm applies the *a posteriori* optimization to all the established backups of high-priority connections at once at each network link, but processes the low-priority connections with the *a priori* optimization sequentially (one after another). The proposed algorithm is dedicated to the preplanned survivability scheme and may be used for all cases of protection scope as well as for all types of failures. However, here only protection against a single link failure is considered. In order to simplify the comparisons, the standard metrics of distance as well as the common Dijkstra's shortest path algorithm [1] are used in the paper for all cases of path finding. However, the proposed method may be easily applied for other metrics and algorithms of path finding. The traffic is assumed to be *static* here, meaning that the set of connection requests is given in advance, but this technique may be also applied for the case of *dynamic traffic* as well.

4 Modeling Assumptions

The experiment was performed for the U.S. Long-Distance Network, presented in Fig. 3, consisting of 28 nodes and 45 bidirectional links. Due to the size of the network, only computations with the use of *SCMTO* heuristic algorithm were feasible[2]. Simulations, performed using the author's network simulator, dedicated to optical networks and implemented in *C++*, were to measure the link capacity utilization ratio and the values of connection restoration time. Connection restoration time was measured according to [10, 14, 15] and comprised: time to detect a failure, link propagation delay, time to configure backup path transit nodes and time of message processing at network nodes (including queuing delay). All the links were assumed to have 32 channels. Channel capacity unit was considered to be equal for all the network links. Network nodes were assumed to have a full channel (wavelength) conversion capability.

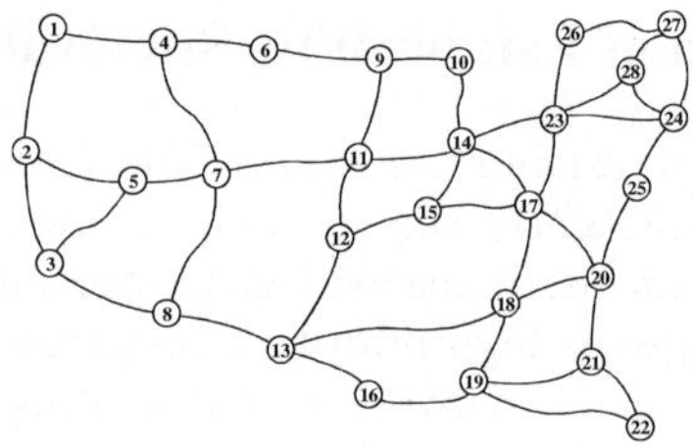

Fig. 3. U.S. Long-Distance Network

For each connection, the following properties were assumed:

- protection against a single link failure (each of the failure states consisted of a failure of one link at a time, the rest being fully operational)
- a demand of resource allocation equal to the capacity of one channel
- provisioning 100% of the requested bandwidth after a failure (preplanned scheme)
- a demand to assure unsplittable flows (both for primary and backup paths)
- the distance metrics and Dijkstra's shortest path algorithm [1] in all path computations
- the three-way handshake protocol of restoring the connections (the exchange of `LINK FAIL`, `SETUP` and `CONFIRM` messages), described in [14].

Repeat 30 times the following steps:

1. Randomly choose 30 pairs of source *s* and destination *d* nodes.
2. Try to establish the connections by applying the *SCMTO* algorithm.
3. Store the level of link capacity utilization and the lengths of the backups.
4. 30 times simulate random failures of single links. For each failure state, restore connections that were broken and note the values of connection restoration time.

Fig. 4. Research plan

[2] Integer Linear Programming formulations of the NP-complete problem of finding optimal active and backup paths for the path protection scheme, can be found in [11].

During a single research, described in Fig. 4, one particular variant of optimization strength as well as one type of protection scope were provided for all the connections. The numbers of low and high-priority demands were assumed to be statistically equal.

5 Modeling Results

5.1 Backup Path Length

Fig. 5 shows the average lengths of backup paths for all variants of optimization strength as well as for various scopes of protection. The results prove that when applying the author's *a posteriori* optimization for the high-priority connections, the average backup path length is not increased and remains at the same level as in case no optimization is performed. Fast restoration of such connections is thus guaranteed. On the contrary, after applying the most capacity-effective *a priori* optimization for the low-priority connections, the length of backup paths is often far from optimal. The obtained backup paths are up to 65% longer, compared to the results of the proposed *a posteriori* optimization.

For the multipriority traffic scheme, an acceptably low average increase of backup path length was observed (up to 29,8%). In general, this increase depends on the number of high-priority connections (here 50%) among all the established connections.

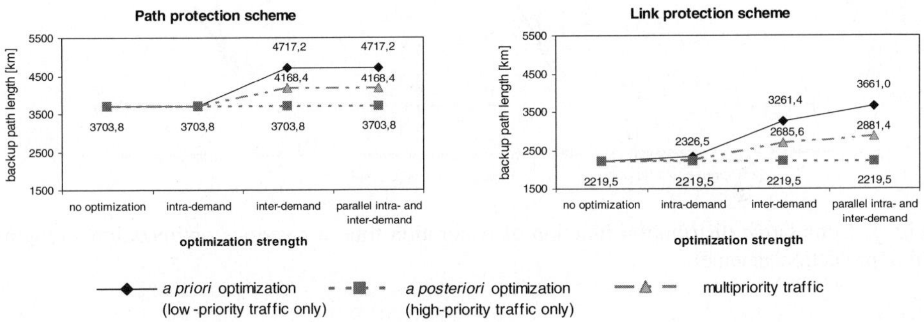

Fig. 5. Average length of backup path as a function of optimization strength

5.2 Restoration Time Values

Figs. 6 ÷ 7 show the cumulative distribution function of restoration time for both all variants of optimization strength and various scopes of protection, while Fig. 8 the respective average values. It is worth mentioning that the mean values of restoration time for the proposed *a posteriori* optimization (high-priority traffic only) were always similar to the shortest ones, achieved when no optimization was performed (about 25 and 50 ms for link and path protection, respectively). When compared to the results of the *a priori* optimization, it takes even about 41% less time on average (25,05 against 41,85 ms for the link protection scheme and the parallel intra- and inter-demand sharing) to restore a high-priority connection, when the *a posteriori* optimization is used. For the multipriority traffic scheme, the use of *SCMTO* algorithm resulted in the average reduction in restoration time values up to 20%, compared to the results of the *a priori* technique only.

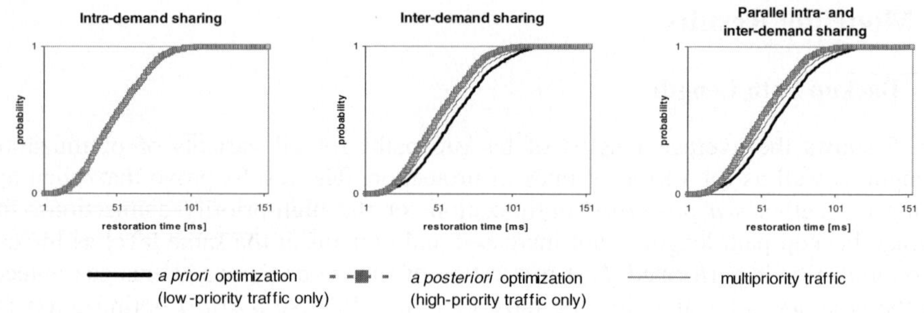

Fig. 6. Cumulative distribution function of restoration time for various optimization strengths (path protection scheme)

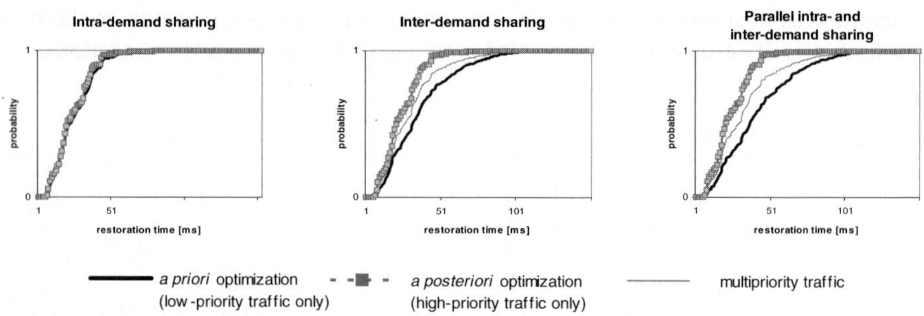

Fig. 7. Cumulative distribution function of restoration time for various optimization strengths (link protection scheme)

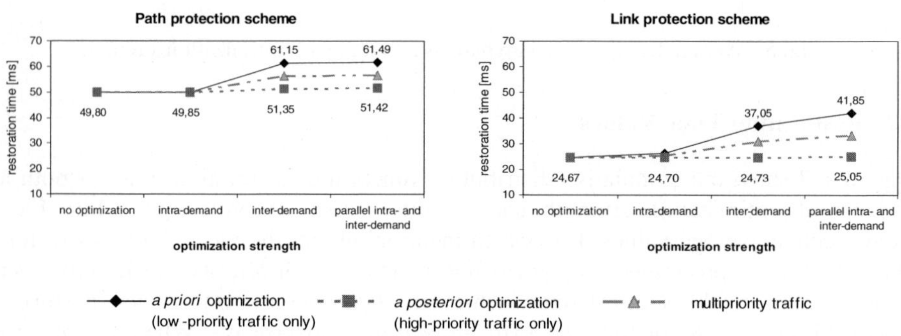

Fig. 8. Average values of restoration time as a function of optimization strength

The results also show that the value of restoration time depends on the protection scope and the length of the backup path to be activated. Additionally, the overall value of restoration time gets increased by reconfigurations of transit nodes of shared backups, that must be performed when restoring each connection [14]. The stronger the optimization is, the more transit nodes of backup paths are to be reconfigured.

Table 1. Confidence intervals of 95 % for the mean values of restoration time [ms]

		intra-demand sharing	inter-demand sharing	parallel intra- and inter-demand sharing
path protection	low-priority traffic only	1,68	2,03	2,01
	high-priority traffic only	1,68	1,75	1,75
	multipriority traffic	1,68	1,89	1,87
link protection	low-priority traffic only	1,12	1,85	2,04
	high-priority traffic only	0,99	1,01	0,99
	multipriority traffic	1,08	1,45	1,53

5.3 Level of Link Capacity Utilization

Fig. 9 shows the average values of link capacity utilization for all variants of optimization strength as well as for various scopes of protection. It proves that the proposed *a posteriori* optimization routine remarkably reduces the level of link capacity utilization

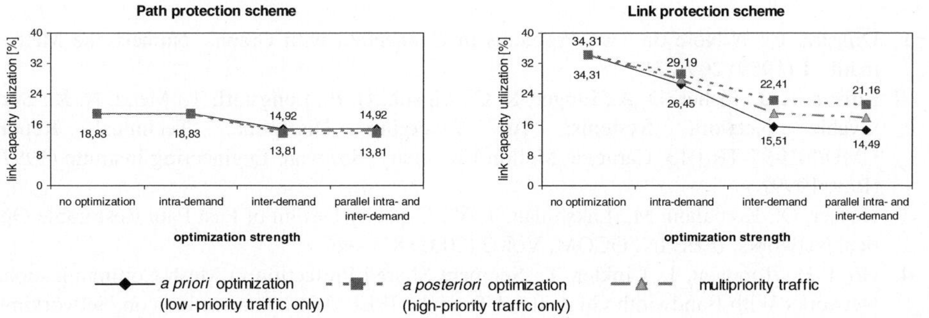

Fig. 9. Average level of link capacity utilization as a function of optimization strength

Table 2. Confidence intervals of 95 % for the mean values of link capacity utilization [%]

		intra-demand sharing	inter-demand sharing	parallel intra- and inter-demand sharing
path protection	low-priority traffic only	0,84	0,61	0,61
	high-priority traffic only	0,84	0,68	0,68
	multipriority traffic	0,84	0,66	0,66
link protection	low-priority traffic only	1,07	0,64	0,64
	high-priority traffic only	1,11	0,99	0,98
	multipriority traffic	1,08	0,78	0,76

for high-priority connections (up to 38 % for the link protection scheme and the parallel intra- and inter-demand sharing), compared to the "no optimization" case. The *a priori* optimization technique, applied to low-priority connections, is, however, always more capacity-effective.

For the multipriority traffic scheme, when using the proposed *SCMTO* algorithm, the level of link capacity utilization was reduced up to 48% (link protection scheme), compared to the "no optimization" case.

6 Conclusions

The common *a priori* optimization of backup path resource utilization, despite being the most capacity-efficient, does not assure fast restoration of connections. However, when using the author's *a posteriori* optimization, fast connection restoration can be achieved for the price of little degradation of capacity utilization.

Finally, for the multipriority traffic, by applying the *a priori* optimization to all the low-priority connections, and the proposed *a posteriori* optimization in all other cases, the *SCMTO* algorithm achieved the significant reduction in resource utilization ratio (up to 48%). Compared to the *a priori* optimization only, it achieved restoration times up to 20% shorter (and even up to 41% for the case of high-priority traffic only).

References

1. Dijkstra, E.: A Note on Two Problems in Connection with Graphs. Numerische Mathematik, 1 (1959) 269-271
2. Ellison, R. J., Fisher, D. A., Linger, R. C., Lipson, H. F., Longstaff, T., Mead, N. R.: Survivable Network Systems: An Emerging Discipline. Technical Report CMU/SEI-97-TR-013. Carnegie Mellon University, Software Engineering Institute (1997) (Rev. 1999)
3. Hauser, O., Kodialam, M., Lakshman, T. V.: Capacity Design of Fast Path Restorable Optical Networks. IEEE INFOCOM, Vol. 2 (2002) 817-826
4. Ho, P-H., Tapolcai, J., Cinkler, T.: Segment Shared Protection in Mesh Communications Networks With Bandwidth Guaranteed Tunnels. IEEE/ACM Transactions on Networking, Vol. 12, No. 6 (2004) 1105-1118
5. Kawamura, R.: Architectures for ATM Network Survivability. IEEE Communications Surveys, Vol. 1, No. 1 (1998) 2-11
6. Kodialam, M., Lakshman, T. V.: Dynamic Routing of Locally Restorable Bandwidth Guaranteed Tunnels Using Aggregated Link Usage Information. IEEE INFOCOM (2001) 376-385
7. Kubale, M. *et al.*: Models and Methods of Graph Coloring. WNT (2002) (in Polish)
8. Mead, N. R., Ellison, R. J., Linger, R. C., Longstaff, T., McHugh, J.: Survivable Network Analysis Method. Technical Report CMU/SEI-2000-TR-013. Carnegie Mellon University, Software Engineering Institute (2000)
9. Mukherjee, B.: WDM Optical Communication Networks: Progress and Challenges, IEEE Journal on Selected Areas in Communications, Vol. 18, No. 10 (2000) 1810-1823
10. Molisz, W.: Survivability Issues in IP-MPLS Networks. Systems Science, Vol. 31, No. 4 (2005) 87-106

11. Molisz, W., Rak, J.: Region Protection/Restoration Scheme in Survivable Networks, 3rd Workshop on Mathematical Methods, Models and Architectures for Computer Network Security (MMM-ACNS'05) St. Petersburg, Russia, Springer-Verlag, LNCS, Vol. 3685 (2005) 442-447
12. Qiao, Ch. *et al.*: Novel Models for Efficient Shared Path Protection. OFC (2002) 545-547
13. Ramamurthy, S., Mukherjee, B.: Survivable WDM Mesh Networks, Part I - Protection. IEEE INFOCOM (1999) 744-751
14. Ramamurthy, S., Mukherjee, B.: Survivable WDM Mesh Networks, Part II - Restoration. Proc. IEEE Integrated Circuits Conference (1999) 2023-2030
15. Ramamurthy, S., Sahasrabuddhe, L., Mukherjee, B.: Survivable WDM Mesh Networks. IEEE Jounral of Lightwave Technology, Vol. 21, No. 4 (2003) 870-883

SHIELD: A Fault-Tolerant MPI for an Infiniband Cluster

Hyuck Han, Hyungsoo Jung, Jai Wug Kim, Jongpil Lee, Youngjin Yu, Shin Gyu Kim, and Heon Y. Yeom

School of Computer Science and Engineering,
Seoul National University,
Seoul, 151-744, South Korea
{hhyuck, jhs, jwkim, jplee, yjyu, sgkim, yeom}@dcslab.snu.ac.kr

Abstract. Today's high performance cluster computing technologies demand extreme robustness against unexpected failures to finish aggressively parallelized work in a given time constraint. Although there has been a steady effort in developing hardware and software tools to increase fault-resilience of cluster environments, a successful solution has yet to be delivered to commercial vendors. This paper presents SHIELD, a practical and easily-deployable fault-tolerant MPI and management system of MPI for an Infiniband cluster. SHIELD provides a novel framework that can be easily used in real cluster systems, and it has different design perspectives than those proposed by other fault-tolerant MPI. We show that SHIELD provides robust fault-resilience to fault-vulnerable cluster systems and that the design features of SHIELD are useful wherever fault-resilience is regarded as the matter of utmost importance.

Keywords: Checkpoint, Consistent recovery, Fault-tolerance, MPI, Infiniband.

1 Introduction

Nearly all parallel applications running on high performance cluster systems are vulnerable to either hardware or software failures. Applications, especially long-running jobs such as a numerical solver in aerodynamics or a CPU-intensive simulator in computational biology, require cluster systems on which they run to be fault-resilient. A single failure that occurs in either the software or hardware can damage entire applications, rendering important computation results useless.

Even though today's cutting-edge technologies in high performance computing provide strong support to make building tera-scale cluster systems possible, the required fault-resilience technology unfortunately has not matured enough to overcome failures. Despite decades of research on fault-tolerant systems, many systems remain susceptible to failures. This is because fault-resilience protocols cannot always achieve the performance predicted by theoretical solutions due to unexpected failures during runtime which were not considered during the initial development stage.

M. Gerndt and D. Kranzlmüller (Eds.): HPCC 2006, LNCS 4208, pp. 874–883, 2006.

It is difficult to build a generic fault-tolerant framework that can be applied to all types of systems, partly due to the increasing complexity of both hardware and software systems. The diversity of hardware and software systems places a heavy burden on a software architect designing and implementing a fault-tolerant system. This complicated work drives us to choose an existing system and to construct a real (or specialized) fault-tolerant system for Infiniband [5].

The primary goal of this paper is to design and implement an easily-deployable and practical fault-tolerant MPI system on InfiniBand. To achieve our goal, certain requirements must be satisfied. First, it must be derived from a well-known MPI standard implementation to ensure wide acceptance and compliance. Second, it must not impede the average user when executing MPI applications. Third, the kernel should not have to be modified. To meet these requirements, (1) the MPI application must run without any modifications, (2) fault-tolerance must be transparent to users, (3) many simultaneous faults must be tolerated, and (4) the system must be scalable.

To this end, we present SHIELD, a fault-tolerant system for MPI based on MVAPICH2 [8]. SHIELD utilizes application-level checkpointing but is nearly user-transparent. Adding fault-tolerance to the MPI library using application-level checkpointing does not lay a burden on MPI application developers because it requires no modifications of application codes. SHIELD inherits the strengths of our previous work M^3 [6], such as multiple fault-tolerance, light-weight job management system, and transparent recovery, while addressing its weaknesses.

The remainder of the paper is organized as follows. Section 2 discusses background information and related works. Section 3 introduces the system architecture. Section 4 concentrates on the implementation of the system. Section 5 presents experimental results. Section 6 concludes the paper.

2 Background and Related Works

2.1 Fault-Tolerance Protocols

Much research on fault-tolerant techniques usually focus on the checkpointing and rollback recovery protocol, which tends to be very theoretical. Checkpointing can be divided into two categories depending on where it takes place: application-level checkpointing and kernel-level checkpointing. Kernel-level checkpointing shows superior performance and requires no modification of application source codes. It, however, lays the burden on system administrators, because either additional kernel modules must be deployed or a checkpointable kernel must be rebuilt and deployed. Application-level checkpointing, although its performance does not compare favorably to that of kernel-level checkpointing, is desirable because it requires no changes to the kernel and is easy to use.

A checkpoint can also be either user-transparent or user-aware. In a user-transparent checkpoint system, the user is oblivious to the checkpoint library, and therefore does not need to modify application codes. On the other hand, in a user-aware checkpoint system, the user must decide where taking a checkpoint will incur the least overhead and modify the code accordingly.

SHIELD performs checkpointing at the MPI library level, and implements application-level and user-transparent checkpointing to keep as much load off the user as possible.

2.2 Related Works

Our previous works, MPICH-GF [11] and M^3, provide fault-tolerance to two widely used standards of communication: Ethernet and Myrinet. MPICH-GF modified MPICH-G2 to add fault-tolerance to TCP/IP, and M^3 modified MPICH-GM to add fault-tolerance to Myrinet.

The management system and MPI library of the MPICH-GF system depend on the Globus Toolkit, which implies that MPICH-GF is not applicable to cluster systems without Globus Toolkit. Moreover, because the management system of MPICH-GF is hierarchical, faults of any managers in any level lead to lose all computation. While M^3 is able to achieve commendable performance, it is vulnerable to faults that can bring the whole system to a halt. SHIELD removes this point of failure while maintaining the performance.

Co-Check MPI [10] was the first attempt to make MPI applications fault-tolerant. Like SHIELD, Co-Check needs applications to checkpoint synchronously. The limitation of Co-Check is that it is dependent on the Condor library and its own version of MPI, known as tuMPI.

MPICH-V [1] is a fault-tolerant MPI implementation designed for very large scale clusters using heterogeneous networks. The design of MPICH-V is based on uncoordinated checkpoint and distributed message logging. One of the shortcomings of MPICH-V is that it reaches only about half the performance of MPICH-P4. MPICH-V2 [2] features a fault tolerant protocol designed for homogeneous network large scale computing (typically large clusters). MPICH-V2 was designed to remove the most limiting factor of the pessimistic message logging approach: reliable logging of in-transit messages. The drawback of MPICH-V2 is that it incurs too much overhead.

Fault Tolerant MPI (FT-MPI) [3] is an implementation of the MPI 1.2 message passing standard that offers process-level fault tolerance. FT-MPI uses optimized data type handling, an efficient point to point communications progress engine, and highly tuned and configurable collective communications. FT-MPI survives the crash of n-1 processes in a n-process job and can restart them, much like SHIELD. However, in FT-MPI, it is up to the application to recover the data structures and the data of the crashed processes.

3 Architecture of SHIELD

Design issues explained in this section should be taken into account when designing fault-tolerant MPI systems for Infiniband cluster systems. The issues include consistent distributed checkpoint protocols, effective methods for reducing checkpoint overhead, and capabilities of management systems.

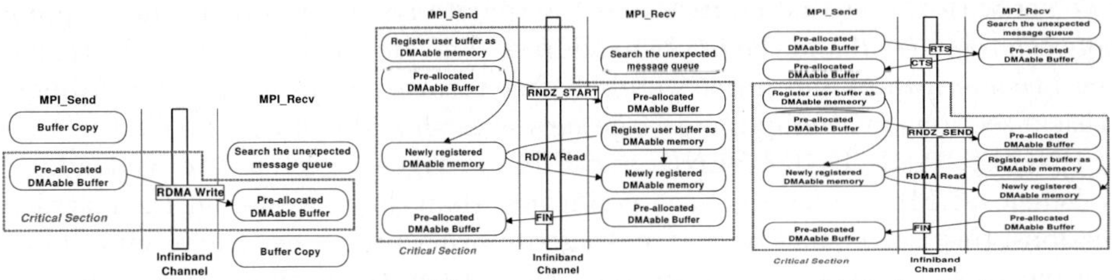

(a) For Small Messages (b) For Medium Messages (c) For Large Messages

Fig. 1. Three Types of Communication Patterns

3.1 Distributed Checkpoint Protocol

To design a robust and consistent distributed checkpoint protocol of parallel processes running on cluster systems, we can choose one of two widely-used protocols - coordinated checkpointing and message logging with checkpointing protocols. Our previous research works [11,12] show that message logging cannot outperform coordinated checkpointing, especially when parallel processes exchange a great deal of messages with each other. We adopt the coordinated checkpointing technique because Infiniband or Myrinet cluster systems are likely to execute parallel applications that frequently exchange large messages with each other.

In cluster environments without a software- or hardware-based oracle, it is important to ensure that no in-transit messages in exist in the communication channels. In order to be guaranteed from in-transit messages, our system uses barrier messages, which is similar to CoCheck's [10] *Ready Message*.

MVAPICH has three communication patterns depending on the size of messages. Figure 1 shows the communication protocol for small messages that are less than 32kb, medium messages between 32kb and 128kb, and large messages greater than 128kb. In all cases, the red dotted lines represent critical sections, in which no checkpoint should be taken.

Small messages are transferred using RDMA Write operations on the preallocted DMAable memory. No checkpoint should be taken during the execution of RDMA Write operations. Where SHIELD differs from M^3 is how it handles the transfer of medium and large messages. For medium and large messages, control messages such as RNDZ_START, FIN, Ready-To-Send(RTS), Clear-To-Send(CTS), and RNDZ_SEND are transferred in the same way small messages are sent, but transmission of the payload is initiated by the receiver using RDMA Read operations. M^3 supports only RDMA Write while SHIELD supports both RDMA Write and Read. Therefore, unlike in M^3, the receiver must know the virtual memory address of the sender and the remote key for resolving the address. This presents a significant distinction from the architecture of M^3.

After the sender registers buffers as DMAable memory, RNDZ_START for medium messages and RNDZ_SEND for large messages are transmitted to the receiver with the virtual address and the remote key for resolving the mapping between the virtual and the physical address. The virtual address and the remote key become invalid and unrecoverable after recovery from failure. Accordingly, critical sections in medium and large messages should contain the transmission of RNDZ_START, RNDZ_SEND, and FIN messages as well as the payload.

When all MPI processes have confirmed that they are not in their critical sections, they send small barrier messages to other processes. The receipt of barrier messages means no in-transit messages exist in the communication channel. Therefore, after receiving barrier messages from all other MPI processes, each MPI process can take a checkpoint locally, and this checkpoint is consistent.

3.2 Job Management System

The job management system manages and monitors parallel processes running on multiple nodes and their running environments. The main function of the management system is to detect failures and to support recovery from failures. To detect and recover from failures, a process separate from the MPI processes must be executed to monitor the MPI processes, conventionally called local job managers. To monitor local job managers, a leader job manager can be executed.

In our previous work, Local Job Manager failures can be recovered from, but Leader Job Manager failures could cause the system to crash. To overcome this problem, we incorporate a leader election algorithm into the recovery procedure. All Local Job Managers periodically send heartbeats to the Leader Job Manager, and the Leader Job Manager responds to each heartbeat. When a Local Job Manager does not receive a response from the Leader Job Manager, it suspects that the Leader Job Manager has failed. If a consensus is reached among Local Job Managers that the Leader Job Manager has failed, the Local Job Manager with the highest rank is chosen to be the temporary leader using the bully algorithm [4]. The temporary leader restarts the Leader Job Manager on a remote, unused node. Because MPI processes continue their computation while the Local Job Managers recover the Leader Job Manager, this does not incur additional overhead to the running time.

The management system of SHIELD can support a job suspend/resume mechanism. Most cluster systems use third party schedulers such as PBS, LSF, and LoadLeveler, which require users to submit a job description file with the number of required CPUs, the name of the executable file, and the execution time to the scheduler. Wrong estimation of the execution time can result in loss of computation or computing resources if the the scheduler aborts the job before it finishes or lets a faulty process execute for an inordinate amount of time. To account for these cases, our system allows jobs to be suspended. When the Leader Job Manager receives a special signal such as `SIGTERM`, the Leader Job Manager orders all participants to take a checkpoint and to terminate themselves. Afterwards, the user or system administrator can resume suspended jobs, if necessary, from the last checkpoint. This mechanism greatly alleviates the drawbacks of job schedulers such as low utilization and high completion time.

4 Implementation

4.1 Fault Tolerant Components of MPI Library

The MPI library of SHIELD is composed of four components: (1) the checkpoint/restart library, which is used to periodically save checkpoint images that are used during recovery, (2) the consistent distributed snapshot module, which ensures that checkpoint images are consistent, (3) ReFS [7], a file system that prevents files from being corrupted by failures, and (4) the recovery module, which restores the system to the last consistent state before failure.

Zandy's checkpoint/restart library [13], which allows the user program's state to be saved without modifications to the user's source code, was modified to fit the standards set by MVAPICH. ReFS, a user-transparent recoverable file system for distributed computing environments, was integrated into SHIELD to protect files from being corrupted by failures. The recovery module was designed specifically to fit the needs of Infiniband. To ensure that the whole system is consistent, all MPI processes, not just the failed process, are restarted. All processes execute the following procedure after fetching the last consistent checkpoint: (1) Open the HCA, create a completion queue, and create queue pairs; (2) Re-register existing buffers as DMAable memory; (3) Send communication information to the Leader Job Manager through Local Job Managers; (4) Receive communication information of other processes from the Leader Job Manager; (5) Establish communication channels with other processes.

5 Experimental Results

5.1 Experimental Environment

The experiments were performed on 8 nodes equipped with dual Intel Xeon 3.0 GHz CPU and 4 GB RAM running Red Hat Linux Enterprise 3 with 2.4.21 kernel. We assess SHIELD's characteristics by running LU and BT of the Numerical Aerodynamics Simulation (NAS) Parallel Benchmark 2 suite. We then gauge the practicality of our system by applying it to a parallel application that are currently being run on commercial clusters.

5.2 Experimental Results

The nature of a fault-tolerant system dictates that there be additional steps taken during the execution of a process. In SHIELD, these extra steps include starting and coordinating Job Managers, saving a checkpoint image, transferring the checkpoint image to a stable storage device, and recovering from failures.

Benchmarks. Table 1 shows the normalized running times of LU and BT. "No Ckpt" refers to the running time of applications using MVAPICH2-0.6.0. "Local Disk Only" signifies that checkpoints were saved on only the local disk. "Local Disk To NFS" refers to saving the checkpoint image on the local storage device

Table 1. Normalized Running Time of Benchmarks

	No Ckpt	Local Disk	Local Disk to NFS	NFS Only
bt.A.4	1	1.27	1.45	1.71
bt.B.4	1	1.29	1.44	1.87
lu.A.4	1	1.09	1.29	1.52
lu.B.4	1	1.14	1.30	1.88
lu.A.8	1	1.16	1.28	1.60
lu.B.8	1	1.04	1.15	2.18
lu.A.16	1	1.25	1.39	1.67
lu.B.16	1	1.26	1.37	2.16

before transferring it to a remote storage. The variable m in Equation 1 was set as 1 for all experiments. Future research will develop a method to determine the optimum value of m in different environments. "NFS Only" is when the checkpoint image is transferred directly to the remote storage device. As can be seen in the table, except for "NFS Only," the performance of SHIELD is comparable to the performance of non-fault-tolerant MVAPICH2.

Table 2. Benchmark Results

	Startup Cost (sec)	Synchronization (sec)	Disk Overhead (sec)
lu.A.4	4.5467	0.0031	1.5966
lu.A.8	7.8027	.0.004	1.36
lu.A.16	9.1353	0.005	1.3

Table 2 shows the specific overheads of SHIELD. The startup cost is the time required for MPI processes to exchange communication information with each other. The disk overhead is the time required to save a checkpoint image. To ensure that the checkpoint image is consistent, the MPI processes must be synchronized before the checkpoint image can be taken. LU and BT were executed with various checkpoint frequencies to result in one checkpoint per experiment.

Real-world Applications. Benchmarks can measure performance but cannot be used to assess whether the system is practical. Therefore, tests were run using a communication- and computation-intensive parallel application used in commercial clusters to determine the applicability of SHIELD to real-world clusters.

mm_par[9] is a molecular dynamics simulatior developed by the Korea Institute of Science and Technology Information (KISTI). mm_par is parallelized to enhance performance. Each processor owns N/P particles, where N is total number of particles and P is total number of processors. This requires a processor to have particle positions updated before calculation, leading it to communicate with all other processors. As a result, the communication cost is significant.

Table 3 shows the running time of each mode and the normalized overhead. mm_par was executed with a checkpoint frequency adjusted to allow just one checkpoint to occur. The performance of SHIELD compares favorably to that of MVAPICH2. Due to the large checkpoint image size (832 MB), writing the checkpoint image to disk took longer than with the benchmarks. However, because the

Table 3. Real World Application

	Running Time (sec)	Normalized Running Time
No Ckpt	3582	1.0
Local Disk Only	3621	1.01089
Local Disk to NFS	3648	1.01843

execution time of the application is much longer, the checkpoint overhead is still barely noticeable. Because the MPI process continues its computation while the Local Job Manager transfers the checkpoint image to the stable storage device, the difference between "Local Disk Only" and "Local Disk to NFS" is minimal.

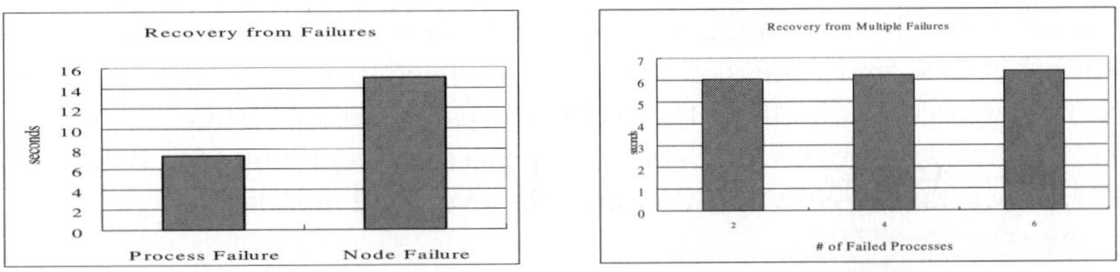

(a) Single Failure (b) Multiple Failures

Fig. 2. Overhead of Recovery from Failures

The cost of recovery is presented in Figure 2. As described in Section 3.2, SHIELD can recover from multiple process and node failures. The cost of recovering from node failures is higher than that of recovering from process failures because a new process is spawned in a new node, and the checkpoint image has to be retrieved from the stable storage device. In the case of multiple simultaneous failures, all processes restart after the final failure. Therefore, the cost to recover from multiple failures is not proportional to the number of failures.

mm_par uses files to log simulation results. For it to recover correctly from failure, the files used must be in the same state they were before failure, i.e., files must also be checkpointed. SHIELD allows the application to recover from failures by supporting file checkpointing with ReFS.

5.3 Discussion

Because the startup and recovery overheads only appear at the beginning of a process and after a failure, its effect on the system is considerably less than that of the checkpoint overhead, which occurs periodically. Disk overhead, or the time required to write the checkpoint image to disk, is proportional to the size of the checkpoint image, which is determined by the memory size of the MPI process, The overheads for synchronization and transfer of checkpoint images are related to the number of processes participating in the parallel computation. Because today's cluster environments are furnished with high-speed networks and high-performance storage servers, the overheads for synchronization and transfer of

Table 4. Cost of Leader Election

# of Processes	Consensus (sec)	Election (sec)
2	0.000852	0.000093
4	0.003936	0.000144
8	0.004776	0.000167
16	0.209537	0.000427
32	3.298679	0.003513

checkpoint images are low enough to be tolerated as shown in Table 2 and our previous work, M^3.

Dummy MPI processes were spawned to test the overhead of the Leader Election procedure. The cost tends to grow as the number of Local Job Managers increases because more messages have to be exchanged to reach a consensus. These messages do not affect the MPI processes because they are exchanged on the Ethernet network while the MPI processes communicate on InfiniBand. Table 4 shows that even with 32 nodes, the overhead is still small.

6 Conclusion

Current fault-resilience techniques for high-performance clusters' reliability are impractical in the context of actual runtime overheads they incur. This paper presents SHIELD, a fault-tolerant system for MPI, to protect parallelized programs from unexpected failure incidents. SHIELD exerts favorable performance results and guarantees a robust fault-resilience property even under various failure conditions. Using experiments with real applications, our results show that SHIELD is very efficient when it is used in failure-free conditions and that the consensus and leader election algorithms that deal with multiple failures have low overhead as well, making SHIELD practical in real cluster systems. Although the experiments were conducted on a relatively small Infiband cluster, we may safely assume that SHIELD is suitable for large applications because scalability is dependent on the job management system and the MPI implementation, and both of these are scalable in SHIELD. SHIELD's job management system is based on M^3's job management system and MVAPICH2, both of which have been shown to be scalable.

We conclude (1) that SHIELD can be an outstanding fault-tolerant solution for high-performance InfiniBand clusters and (2) that design features of the SHIELD framework provide useful advice to software architects when designing or implementing their own fault-tolerant frameworks to be applied to their existing parallel programming systems.

Acknowledgment

The ICT at Seoul National University provides research facilities for this study.

References

1. G. Bosilca, A. Bouteiller, F. Cappello, S. Djilali, G. Fedak, C. Germain, T. Herault, P. Lemarinier, O. Lodygensky, F. Magniette, V. Neri, and A. Selikhov, *MPICH-V: Toward a Scalable Fault Tolerant MPI for Volatile Nodes* Proceedings of the 2002 ACM/IEEE Supercomputing Conference, 2002.
2. B. Bouteiller, F. Cappello, T. Herault, K. Krawezik, P. Lemarinier, and M. Magniette, *MPICH-V2: a Fault Tolerant MPI for Volatile Nodes based on Pessimistic Sender Based Message Logging* Proceedings of the 2003 ACM/IEEE Supercomputing Conference, 2003.
3. G.E. Fagg and J. Dongarra, *FT-MPI: Fault Tolerant MPI, Supporting Dynamic Applications in a Dynamic World* Proceedings of the 7th European PVM/MPI Users' Group Meeting on Recent Advances in Parallel Virtual Machine and Message Passing Interface, 2000.
4. H. Garcia-Molina, *Elections in a Distributed Computing System* IEEE Transactions on Computers. , 1982.
5. InfiniBand Trade Association, *InfiniBand Architecture Specification*, Release 1.2. http://www.infinibandta.org, 2004.
6. H. Jung, D. Shin, H. Han, J.W. Kim, H.Y. Yeom, and J. Lee, *Design and Implementation of Multiple Fault-Tolerant MPI over Myrinet* Proceedings of the 2005 ACM/IEEE Supercomputing Conference, 2005.
7. H.S. Kim and H.Y. Yeom, *A User-Transparent Recoverable File System for Distributed Computing Environment* Challenges of Large Applications in Distributed Environments (CLADE 2005), 2005.
8. J. Liu, J. Wu, S.P. Kini, P. Wyckoff, and D.K. Panda, *High Performance RDMA-based MPI Implementation over InfiniBand* ICS '03: Proceedings of the 17th annual international conference on Supercomputing, 2003.
9. K.J. Oh and M.L. Klein, *A General Purpose Parallel Molecular Dynamics Simulation Program* Computer Physics Communication, 2006.
10. G. Stellner, *CoCheck: Checkpointing and Process Migration for MPI* Proceedings of the International Parallel Processing Symposium, 1996.
11. N. Woo, H. Jung, H.Y. Yeom, T. Park, and H. Park, *MPICH-GF: Transparent Checkpointing and Rollback-Recovery for Grid-Enabled MPI Processes* IEICE Transactions on Information and Systems, 2004.
12. N. Woo, H. Jung, D. Shin, H. Han, H.Y. Yeom, and T. Park, *Performance Evaluation of Consistent Recovery Protocols Using MPICH-GF* Proceedings of the 5th European Dependable Computing Conference, 2005.
13. V. Zandy *ckpt* http://www.cs.wisc.edu/ zandy/ckpt/

Priority-Based Event Message Scheduling in Distributed Virtual Environment

Seokjong Yu

Dept. of Computer Science, Sookmyung Women's University,
Chungpa-dong, Yongsan-gu, Seoul 140-742, Korea
yusjong@sookmyung.ac.kr

Abstract. Event message processing is important in DVE because collaboration among distributed participants is exploited only by sharing event messages over the network. Existing DVE systems process event messages in FCFS algorithm based on the occurrence time of event, and do not reflect the priority of event. It implies how serious an event is to the receivers, and is dependent on the type of event, and the location of event occurrence, and so on. This paper proposes a new event scheduling algorithm in DVE introducing this priority information of event. The advantage of the suggested algorithm is the improvement of system response time to user's interaction, compared with previous FCFS scheduling, especially when high DOI events occur. It considers event occurrence time as well as the priority simultaneously not so as to cause starvation situations of low DOI events. For performance evaluation, this work has measured and compared average waiting time of events in event queue.

1 Introduction

Distributed virtual environment(DVE) is a 3D model-based virtual world that supports collaborative activities among clients distributed over the network, and there are important application fields such as online game, tele-conferencing, and virtual community[1, 2, 3, 4]. MMORPG(Massively Multiplayer Online Role-Playing Game) is one of successful DVE application, which enables more than hundreds of thousands of people to collaborate one another over the network[5, 6]. In DVE, response time until a user receives back a feedback after he/she interacts with a system is a key element to evaluate the performance and satisfaction index of the system. For maintaining the consistency of DVE, all state changes caused by user's behaviors, called as events, should be sent to others in a form of event message packet via a central server[2, 4]. Event message traffic tends to increase proportionally to the number of users participating in DVE. In client-server model, all event messages are sent to others via a central server, and, to reduce the amount of message traffic, less meaningful messages among them are filtered by message filtering algorithm before they are forwarded.

Area of Interest(AOI) models[2, 4, 7, 8] and dead reckoning[3, 4] are popular message filtering methods widely used in DVE. AOI model is able to reduce event message traffic flexibly by controlling the size of AOI based on the situation of system

M. Gerndt and D. Kranzlmüller (Eds.): HPCC 2006, LNCS 4208, pp. 884 – 893, 2006.

load. Traditional DVE systems commonly process message traffic by FCFS(First Come First Served) policy, which handles event messages sequentially according to the order of the time they arrived at event queue[1, 2, 3, 4]. This method could be not fair in some situation in DVE. For example, because FCFS policy does not consider the priority level of each event, some important events related to user interactions should be waiting long time until they are finished if the server is pre-occupied by other casual events arrived earlier. Especially, if a bottleneck situation occurs in the server by heavy message traffic, waiting delay time of messages get much longer, and users might feel system responding slowly unless events for user's interaction are processed immediately. To improve this limitation of existing DVE using FCFS policy, this paper proposes a novel event scheduling policy in DVE, called as the priority event scheduling, introducing the priority concept of event.

In section 2, related works are presented, and in section 3, priority-queue scheduling is described. In section 4, the proposed model is evaluated by experiments comparing with FCFS one.

2 Distributed Virtual Environment

DVE is a cyberspace shared by multiple participants distributed over the network [1, 2, 3, 4]. Virtual worlds and participants of DVE are presented in 3D graphical objects, and avatar, user's agent, plays a role in expressing the behavior and presence of a participant to others [4, 7]. DVE provides multiple participants with environments for collaboration, competition, and communication in the application fields of MMORPG, tele-conference, and virtual city[1, 2, 3, 5, 6].

2.1 Event Sharing Model

Event is an accident that is occurred in DVE directly or indirectly by a participant's behavior. Event message is used to inform the occurrence of the accident to the related participants, and it contains the information on the producer, behavior type, and values of an event. Event sharing is mandatory and important because DVE must be consistent to all participants to perform distributed collaboration. The consistency of DVE is maintained by continuous exchange of event messages over the network. Therefore, A scalable DVE should cope with the possible number of participants so as not to degrade the performance of system [4, 8, 10]. Event message filtering is necessary to keep DVE handling a proper level of message traffic. Spatial partitioning[2, 4, 7] and dead reckoning[3, 4] belong to popular message filtering algorithms.

In client-server model, each client connects to central server, and the central server take charge of relaying and filtering event messages among clients. This model is easy to implement and filter message traffic in DVE, but it has a problem that causes a bottleneck situation because of single central server, and it might cause the crash of whole DVE system. In Peer-to-Peer model[4], event messages are transmitted to other clients directly instead of server message relay. This reduces message latency time, but there is limitation on increasing the number of simultaneous participants. Peer-Server model has been implemented in ATLAS[3], which introduces message multicasting to reduce message traffic. In this model, there is a crucial constraint that multicasting is available in limited domain of the Internet. Distributed server model[11] becomes de facto standard of message relay model, which manages DVE by distributed servers. It is advantageous to extend DVE into large-scale, but, instead, it demands additional communication load among distributed servers when a participants crosses the boundaries between partitioned worlds of DVE[3]. Spatial partitioning is a method that is widely used to improve the scalability of DVE. It divides an entire DVE into several sub-regions to reduce system loads for maintaining DVE to be consistent [1, 2, 4]. One of spatial partitioning methods is AOI, which is an area on which a participants focuses to listen events around him/her, and events occurred outside AOI are discarded. AOI can improve the scalability of DVE by filtering less important message traffic effectively. AOI models is categorized into fixed and movable types by its mobility, and divided into hexagonal, rectangle, circle, and irregular types by its shape[1, 2, 3, 8]. Hexagon cell has been implemented in NPSNET, a military simulation system[4].

2.2 Population Distribution

In DVE using client-server structure, event message is sent to the central server and forwarded sequentially to other participants. If message traffic is produced more than server's capacity, users might suffer from a bottleneck situation in server and event message forwarding are delayed. Message traffic overload is highly related to local population density. Because when an event occurs in a crowded environment. More message traffic is created than normal situation. According to the studies about population distribution in DVE, there are three possible distributions: uniform, skewed, and clustered distributions[8]. Unlike uniform distribution, skewed and clustered ones belong to the cases that are likely to cause a server bottleneck[11]. As researches trying to solve this problem, there are AOI resizing[2] and load balancing among distributed servers[8, 10].

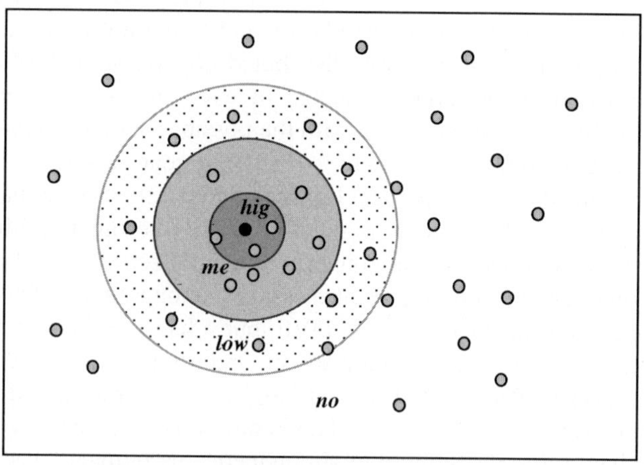

Fig. 1. Degree of Interest in AOI : high, medium, low, and none

3 Priority-Based Event Message Scheduling

In existing AOI models[7], single criteria to process events is occurrence time of event. This work focuses on the priority of event(degree of interest: DOI) as shown in Fig. 1, which has not been considered before in DVE field. This paper classifies events of DVE into two groups: *urgent events for direct interaction*, and *common events for indirect interaction*. If a user operates a menu, gestures, or talks to neighbors, this is where events for direct interaction happen. However, if a user sees someone moving from a distance, it is a kind of indirect interactions. The former is events with high DOI, The latter is ones with low DOI. The problem is that existing models process events based only on single criteria of the occurrence time, and the importance of events are neglected. In particular, if message traffic reaches maximum capacity of server, all event messages are postponed for a bottleneck situation in the server. In this case, effect on direct interaction events is more serious than indirect one. This is why we consider the priority of event as well as the occurrence time. Now, this paper introduces algorithms of two event scheduling models, and compares which one is more efficient when processing event messages.

3.1 Single Queue Event Scheduling

Fig. 2 is a simulation example of event scheduling that events are processed by traditional AOI models with single event queue. All the events in the queue are processed by FCFS(first come first served) policy. Since the server's capacity during

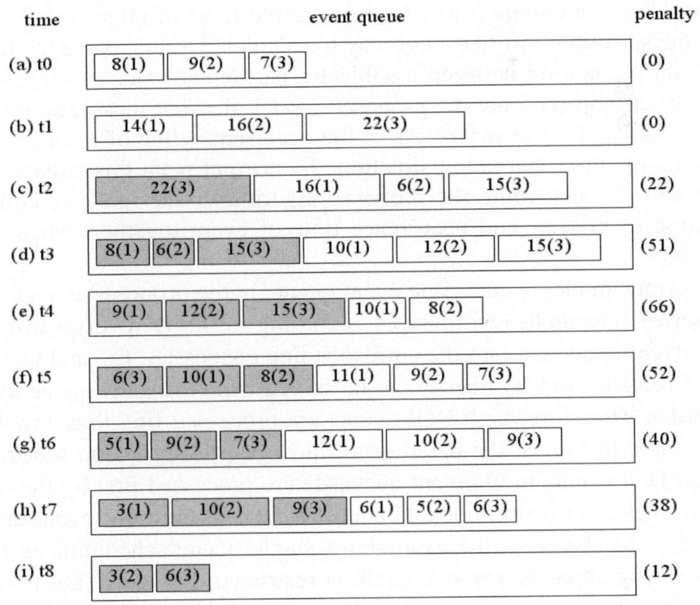

Fig. 2. Single queue scheduling (M_{limit}=30, $QT_{average}$=31.2)

processed by FCFS policy. Since the server's capacity during one cycle is limited, excessive events than this capacity have to be delayed until the next server cycle. This paper defines a new concept, called 'event collision', which is a situation that an event arrives at the server while processing other event messages. The more often event collision happens, the more longer processing delay time is. In Fig. 2a, the notation of '8(1)' means the 'workload (priority level)' of an event message waiting for processing at the server. In Fig. 2a, to process three messages without delay, the server requires at least 24 processing unit per cycle. The total penalty for processing delay is calculated as *all remaining workloads * level penalty*. In this work, level penalty are defined, by the principle of high penalty to high level, as penalties 3, 2, and 1 are imposed on workloads of level 1, 2, and 3, respectively. In Fig. 2b, new workloads arrive at the queue and workload 22 with level 3, 22(3), remains after the cycle t_1. The server processes old workload 22(3) and new ones 16(1), 6(2), 15(3) at cycle t_2. At cycle t_3, workloads 8(1), 6(2), 15(3) remained from cycle t_2, and new workloads 10(1), 12(2), 15(3) are waiting for being processed by the server. In this way, if more event messages are queued than server's capacity, the waiting time of events increases gradually. After all the event messages are processed during cycle t_8 by single queue scheduling(M_{limit}=30), average waiting time of events is 31.2(=281/9).

3.2 Priority Queue Event Scheduling

To improve efficiency of traditional AOI models using single event queue, this paper proposes a new event scheduling algorithm in DVE field based on the priority, which classifies the priority of events into high, medium, and low levels according to DOI of a participant. DOI is a common way to measure the level of interest that a user pays attention to the surroundings. One easy way to calculate DOI is to use the information of distance and orientation between a subject(a participant) and an object(other participant, or virtual object). Like the proposed model, if event messages are processed only by the priority of event regardless of the occurrence time of event, some of low DOI events might suffer starvation situation. To prevent from this situation and give more fairness to the algorithm, the proposed algorithm is designed to consider both the information of priority and occurrence time of event together. when processing events of DVE.

Unlike previous model using single event queue, in the proposed model event message are inserted into multi-level queues according to the DOI, or priority. In other words, each event is queued into the corresponding queues(Q_1, Q_2, and Q_3) according to the DOI. Events in priority queue is to be served from top level queue to low level one by its order. Therefore, high DOI events are processed first than low DOI ones. With the example in Fig. 3, the detail procedures of priority queue scheduling is as follows. High DOI events in Q_1(event queue#1) are processed first by the server, and medium DOI ones in Q_2(event queue#2)and low DOI ones in Q_3 (event queue#3)are served after Q_1. As shown in the example of single queue scheduling in Fig. 2, the server's processing capacity per one cycle is restricted to M_{limit}. That is to say, the sever can process events in the order from Q_1 to Q_2 and Q_3 until the processing capacity per cycle, M_{limit}, is exhausted. Remained workloads after a cycle t_i must wait until the next cycle t_{i+1}. To prevent from a starvation situation for the priority-oriented policy of the algorithm, after each processing cycle, all remained events in each queue move up to its right upper level queue except Q_1 as follows (Q_3->Q_2, Q_2->Q_1).

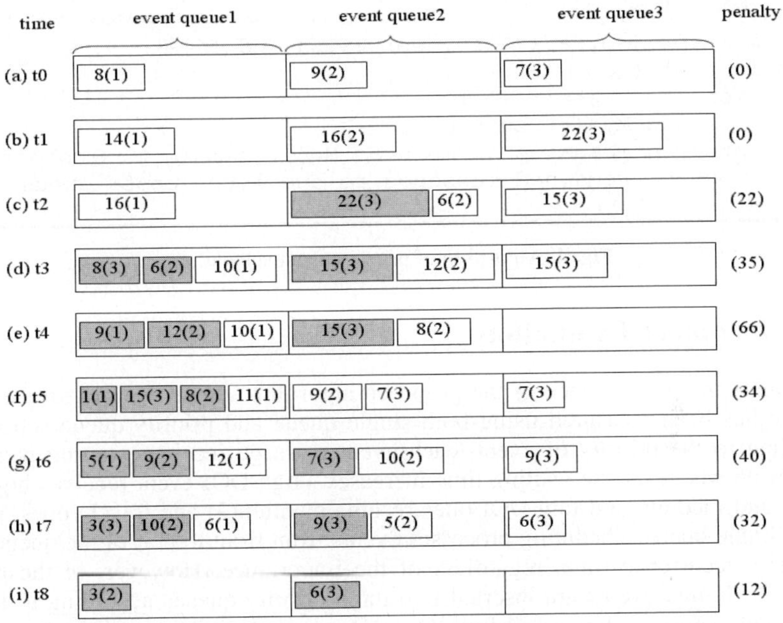

Fig. 3. Priority Queue Scheduling (M_{limit}=30, $QT_{average}$=26.7)

At the next cycle, old events (waited for one cycle) will be added with new comers events, and they are processed by the server in the same way as the previous stage.

By applying this principle to message processing, the algorithm is able to reduce queue waiting time of direct interactive events with high DOI and to improve system responsiveness. As described in Fig. 2 and Fig. 3, average queue waiting times, QT, of single and priority queue models after 8 cycles are 31.2 and 26.7, respectively. It implies that the proposed priority event scheduling is more efficient than single queue method by reducing 14.42% of queue waiting time. Reduction rate of queue waiting time is computed as follows.

$$QT_{reduced} = \frac{QT_{single} - QT_{priority}}{QT_{single}} = \frac{31.2 - 26.7}{31.2} = 14.42\% \quad (1)$$

Fig. 4 lists the algorithm of the priority queue scheduling.

```
Algorithm (Priority queue scheduling)
Mlimit : maximum number of messages that the server is able t
o process during a time interval, v.
AOIi : an i-th AOI group composed of a set of participants t
o be received an event message from a participanti.
Mi,j: number of event messages which occurred in sub-layers
```

```
Li,j of AOIi, for 1≤j≤3
Mtotal(t)= Σ Mi,j(t), total number of event messages occurred
 from all AOIs at time t, for 1≤i≤n and 1≤j≤3.
Ma,j(t) = Σ Mi,j(t), total number of event messages occurred
from layer j of all AOIs, for 1≤i≤n and 1≤j≤3.
while(t++) {
    for (j=1; j<=3; j++)
        add Ma,j(t) into Qj.
    process messages as much as Mlimit from Q1 to Q3 order.
    move up remained messages to one level upper queue as (
Q3->Q2, Q2->Q1)}
```

Fig. 4. Algorithm of priority event scheduling

4 Performance Evaluation

To evaluate the performance of the proposed algorithm, queue waiting time of event message has been measured using both single queue and priority queue scheduling. According to the priority of event, each event message receives waiting penalty at every cycle whenever the waiting time increases. High DOI event receives high penalty(3), and medium and low DOI ones receive medium(2) and low(1) ones, respectively. Single queue scheduling processes events from front to rear of the queue using only their occurrence time, regardless of the importance. However, in the priority queue scheduling, events are inserted into three priority queues according to the priorities, and events are processed from Q_1 to Q_3 as much as the capacity of server per one cycle. Table 1 outlines the characteristics of two experimental algorithms.

Table 1. Characteristics of two comparison models

Event scheduling type	Event processing criteria	# of event queues	Queue waiting penalty to layer 1, 2, and 3
Single queue scheduling	occurrence time	1	3:2:1
Priority queue scheduling	occurrence time + DOI	3	3:2:1

Total queue waiting time of events is calculated as the following formula.

$$QT_{total} = \sum (QT_1(i)w_1 + QT_2(i)w_2 + QT_3(i)w_3) \quad (2)$$

for $0 \le i \le n$, n= number of AOI group, where w_1, w_2, and w_3 are queue time penalty to level 1, 2 and 3, respectively. As a comparison factor to two algorithms, average queue waiting times have been measured, with changing the ratio of high DOI event messages(occurred from layer 1) from 0% to 90%. The measurement is performed, assuming that total number of AOIs is 10, and maximum number of participants is 150. Sample event messages has been supposed to be generated only in uniform distribution, and experiments with skewed and clustered ones has been remained for future researches.

Table 2 and graphs of Fig. 5 created from the experimental data describe the change of queue waiting time over the ratio of high DOI event messages. Fig. 5a, 5b, and 5c show queue waiting times measured from experiments using two comparison algorithms while the ratio of high DOI event messages has been changed from 20%, 50%, to 80%. When high DOI ratio is 20%, difference of queue waiting time between two methods are 539.30, and it implies that 29% of queue waiting time is reduced by priority queue scheduling. Likewise in cases of 50% and 80%, priority queue scheduling has reduced 17% and 5% of queue waiting times, compared with single queue scheduling. Averagely, the proposed algorithm has required 19% less queue waiting time than traditional one. From this experimental results, it is confirmed that priority queue scheduling is more efficient and rational than single queue method.

Table 2. Measurement of queue waiting time

high DOI ratio	Single queue scheduling (S)	Priority queue scheduling (P)	Difference (S-P)	Reduction ratio ((S-P)/S)
0%	1798.60	1204.63	593.98	33%
10%	1903.75	1344.70	559.05	29%
20%	2055.25	1515.95	539.30	26%
30%	2206.43	1685.50	520.93	24%
40%	2427.93	1934.43	493.50	20%
50%	2632.85	2176.90	455.95	17%
60%	2848.13	2486.15	361.98	13%
70%	3114.30	2817.15	297.15	10%
80%	3321.43	3150.33	171.10	5%
90%	3686.58	3578.43	108.15	3%
Average	2599.52	2189.42	410.11	19%

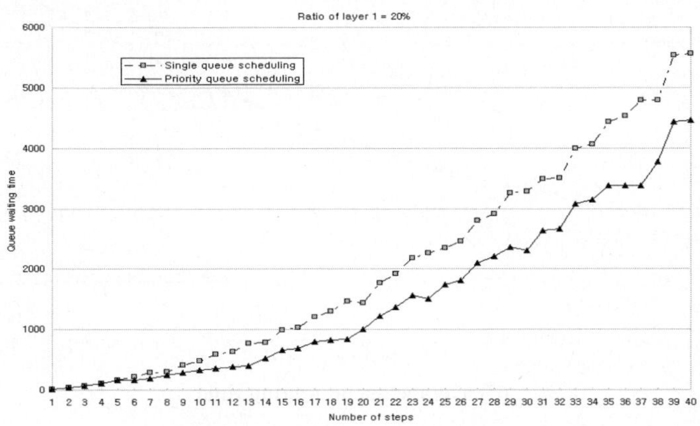

(a) queue waiting time (high DOI ratio = 20%)

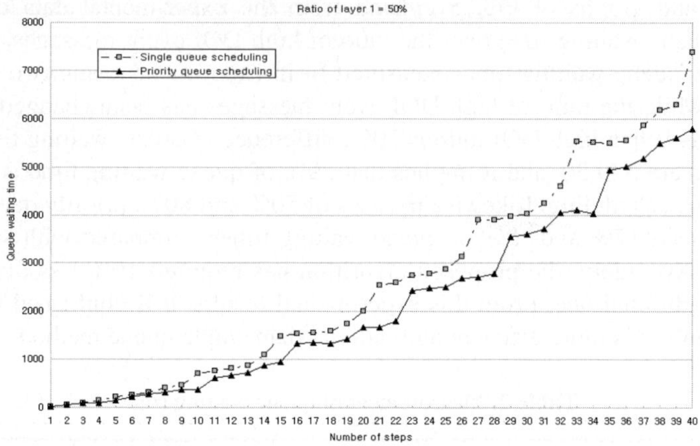

(b) queue waiting time (high DOI ratio = 50%)

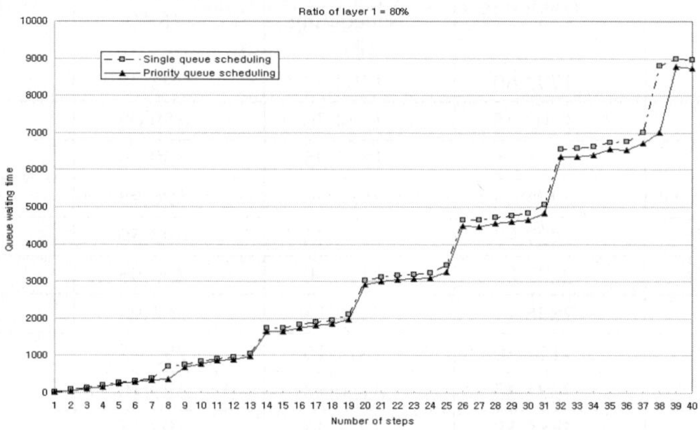

(c) queue waiting time (high DOI ratio = 80%)

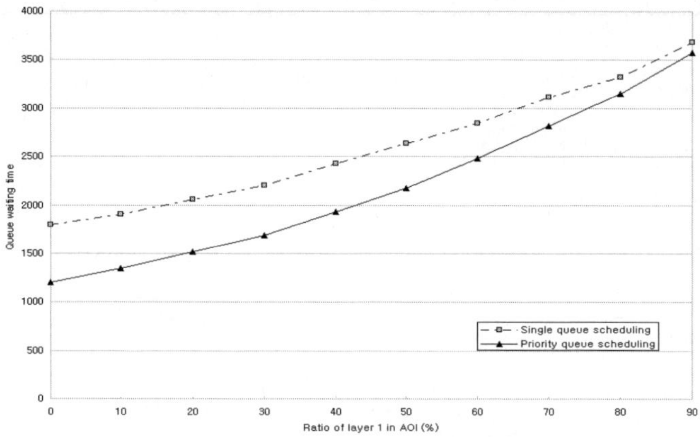

(d) average queue waiting time

Fig. 5. Measurement of queue waiting time

5 Conclusion

The ultimate purpose of DVE is to enable geographically distributed people to collaborate one another by efficient utilization of system resources such as network bandwidth and CPU computation time. To minimize the amount of network bandwidth consumption, many DVEs have developed various message filtering techniques, for example, AOI and dead reckoning. Traditional single queue scheduling has potential problems that urgent events related to direct interactions among close participants are not processed on time because of heavy message traffic of other casual events. The priority event scheduling proposed in this paper is a novel filtering method for the purpose of improving the limitation of existing algorithm. As shown in the performance evaluation section, it has been confirmed that it is useful to shorten system response time by reducing about 19% of queue waiting time of event messages. In addition, the proposed scheduling model supports event migration of events between queues so as not to cause stavation of low priority events in the queue.

Acknowledgements

This Research was supported by the Sookmyung Women's University Research Grants 2006.

References

[1] J.W. Barrus, R.C. Waters, and D.B. Anderson, "Locales and Beacons: Efficient and Precise Support for Large Multi-User Virtual Environments," Proceedings of the IEEE Virtual Reality Annual International Symposium, pp. 204-213, 1996.

[2] C. M. Greenhalgh, and S. D. Benford, "MASSIVE: A Distributed Virtual Reality System Incorporating Spatial Trading," Proceedings of 15th International Conference on Distributed Computing Systems, Los Alamitos CA, ACM Press, pp. 27-34, 1995.

[3] D. Lee, M. Lim, S. Han, "ATLAS A Scalable Network Framework for Distributed Virtual Environments," ACM Collaborative Virtual Environment (CVE2002), Bonn Germany, pp. 47-54, 2002.

[4] S. Singhal, M. Zyda, "Networked Virtual Environments," *ACM Press*, New York, 1999.

[5] Lineage. http://www.lineage.com/.

[6] Ultima online. http://www.uo.com/.

[7] T. K. Capin, I. S. Pandzic, N. Magnenat-Thalmann, and D. Thalmann, "Networking Data for Virtual Humans," *Avatars in Networked Virtual Environments*, Wiley, 1999.

[8] C.S. John, M. Lui, F. Chan, "An Efficient Partitioning Algorithm for Distributed Virtual Environment Systems," *IEEE Transactions on Parallel and Distributed Systems*, Vol. 13, No. 3, pp. 193- 211, 2002.

[9] J. Huang, Y. Du, and C. Wang, "Design of the Server Cluster to Support Avatar Migration," IEEE VR2003, Los Angeles, USA, pp.7-14, 2003.

[10] E. F. Churchill, D. Snowdon, "Collaborative Virtual Environments: An Introductory Review of Issues and Systems," *Virtual Reality*, Vol. 3, No. 1, Springer-Verlag, pp. 3-15, 1998.

[11] B. Ng, A. Si, R. W.H. Lau, F. W.B. Li, "A Multi-Server Architecture for Distributed Virtual Walkthrough," Proceedings of the ACM symposium on Virtual Reality Software and Technology , pp. 163-170, 2002.

inVRs - A Framework for Building Interactive Networked Virtual Reality Systems

Christoph Anthes and Jens Volkert

GUP, Institute of Graphics and Parallel Programming
Johannes Kepler University, Altenbergerstrasse 69, A-4040 Linz, Austria
canthes@gup.uni-linz.ac.at

Abstract. In the recent years a growing interest in Collaborative Virtual Environments (CVEs) can be observed. Users at different locations on the Globe are able to communicate and interact in the same virtual space as if they were in the same physical location. For the implementation of CVEs several approaches exist. General ideas for the design of Virtual Environments (VEs) are analyzed and a novel approach in the form of a highly extensible, flexible, and modular framework - inVRs - is presented.

1 Introduction

Virtual Reality (VR) has become a technology which is widely used in the area of research projects and in industry. Many different application domains exist which can be categorized in visualizations, simulations and VEs. Visualizations are mainly used in the area of industrial design, medicine, chemistry and astrophysics. A variety of simulations like vehicle, plane and ship simulations exist and are most common within the military sector. VEs are used for applications like construction scenarios, safety training, virtual meeting places or phobia treatment. Although the applications of these different domains may use a variety of approaches for navigation and interaction, all of these application areas overlap in many aspects and require similar VR hardware and software. For example visualizations rely on menu driven interaction while VEs normally focus on natural interaction with 3D input devices.

In every application specific functionality has to be tailored which makes it a challenging task to design and develop a general framework that allows for all kinds of VR applications. The approach we present covers the general requirements needed in Networked Virtual Environments (NVEs). It provides clearly structured interaction, navigation and communication mechanisms while keeping up the flexibility needed to develop specific applications within the domain of VEs and NVEs.

This section has given a brief overview of the different types of VR applications. The next section will take a look at various approaches to develop VEs and explain their advantages and disadvantages. Section three will clearly identify the characteristics of a VE as we understand it. And in section four an overview

M. Gerndt and D. Kranzlmüller (Eds.): HPCC 2006, LNCS 4208, pp. 894–904, 2006.

over the inVRs framework is given. The following sections explain the functionality of the inVRs interfaces and modules in more detail and explain their use in NVEs. The final sections conclude the paper and give an outlook into future work.

2 Related Work

Designing VEs is a challenging task. One has to take the areas of Human Computer Interaction (HCI) and hardware- and software design into account. The development of VEs can be approached in different ways, which all carry their advantages and disadvantages.

Customizable GUI intense systems like EON or VIRTools approach the design of the VE via drag and drop mechanisms. This requires predefined communication mechanisms and relatively static structures inside the system, which allow for the easy design of general VEs. It difficult to extend these mechanisms to easily solve the more domain specific problems.

Another approach relies on the usage of scripting languages like TCL or Python and the extension of the VR System by writing plugins in C++ or C. DIVE [6] for example uses TCL and allows to write plugins in C. ALICE [7] makes use of Python as a scripting language. These approaches make it easy for users to develop their own networked VR systems, but restrict the user fully to the mechanisms of the APIs. Through the nature of VR applications, using a variety of HW setups and having different application areas and application types, monolithic systems like the ones described above can not satisfy all of the needs.

Many VR applications are developed fully from scratch using OpenGL and the CAVELib or scene graph APIs like Inventor [14], OpenGL Performer [13], OpenSceneGraph or OpenSG [12] for displaying and structuring the VE. This approach requires an extensive amount of time and cost since most components of the VE have to newly be designed. Other approaches provide abstraction for input and output devices like VRJuggler [4] or they support network communication for a specific scene graph like in CAVERNSoft[9].

Numerous research has been done in the area of NVEs and CVEs. Good overviews on these topics are given in [10] and [11] on a technical level. They describe how NVEs have to be designed to be scalable, responsive and interactive.

3 Characteristics of VEs

An NVE can be defined as a VE in which several users on different VR systems interact in real time. These VR systems can be in the same location or can be geographically dislocated. A VE as we understand it is user centered, the navigation is typically set to a walk mode. Natural interaction is predominant, and parameter changes for application control are possible, but they are not as important as in visualizations for example.

Typical VEs exist in the field of safety training, which can contain a high amount of simulation processing, for example, to visualize the spreading of fire and smoke. Another common application area for VEs would be architecture walkthroughs which contain very little interaction and rely heavily on good and intuitive navigation modes. Virtual meeting rooms focus on communication aspects and application sharing. Arts and entertainment projects are as well very common in the field of VEs. Psychology looks into two different aspects of VEs. The psychologists are interested in analyzing human performance within a VE as a way to gather more information about human behaviour in virtual worlds or they try to incorporate VR technology for a variety of medical treatments such as phobia treatment and burn injury treatment. Figure 1 shows some example VEs from the field of catastrophe simulation, architectural design, and entertainment which use navigation and parts of the core module of the inVRs framework.

Fig. 1. Three example applications using parts of the inVRs framework

4 inVRs Architecture

The inVRs architecture is a highly modular system, which allows the independent usage of its different modules and interfaces as separate libraries. It can act as a full VR framework if the modules are connected together with the central system core module. The architecture provides three independent modules for interaction, navigation, and networking, an additional system core module, and two interfaces for the abstraction of input devices and output devices.

The interaction and navigation modules are based on high level abstractions which consist of expandable and interchangeable components. The high level approach in the networking module allows for a variety of networking topologies which can be implemented by modifying or exchanging the networking module. Abstraction of input and output allows for a variety of devices which can be used with the inVRs architecture. The inVRs libraries are fully written in standard C++ and make use of platform independent Open Source libraries, therefore it is possible to run an inVRs application on Windows, Unix or Irix. The current implementation uses OpenSG as a scene graph and OpenAL for audio output.

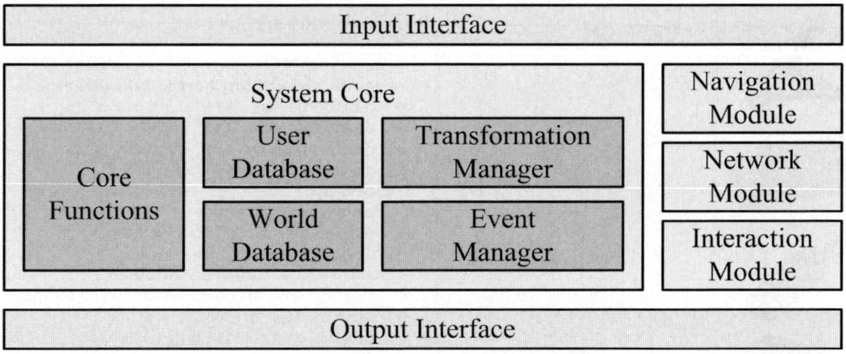

Fig. 2. An overview of the inVRs framework

An overview of the inVRs framework showing the interfaces, modules and components of the system core module is given in Figure 2.

5 Interfaces

Two categories of interfaces exist in the inVRs framework. The input interface takes care of various VR input devices, while the output interface generates output for a scene graph abstraction layer which can be connected to a variety of scene graphs in order to generate stereoscopic images of VR graphics HW. Other output devices like audio systems and motion platforms can also be accessed by using of the output interface.

5.1 Input Interface

The input interface provides an abstraction layer for a variety of input devices. These input devices are split up into their components resulting in three different categories: buttons, axes and sensors. Every input device can be represented by a combination of these components. Axes generate linear values between two thresholds, buttons generate boolean values and can trigger callback functions, e.g. if a button is pressed or released, and sensors generate three dimensional position values with additional rotational information. A collection of the components of different input devices is wrapped into a logical controller object, which can be accessed later by the navigation module and the interaction module. The controller object can be configured via an XML configuration scheme. This XML file describes which component of the physical input device is mapped onto which attribute of the logical controller. A logical controller can for example consist of two buttons and two axes of a mouse as well as five keys from the keyboard. Several controller setups can be defined and exchanged dynamically during runtime. If newly developed devices are to be used as an inVRs input device the input interface library has to be enhanced by following the concept

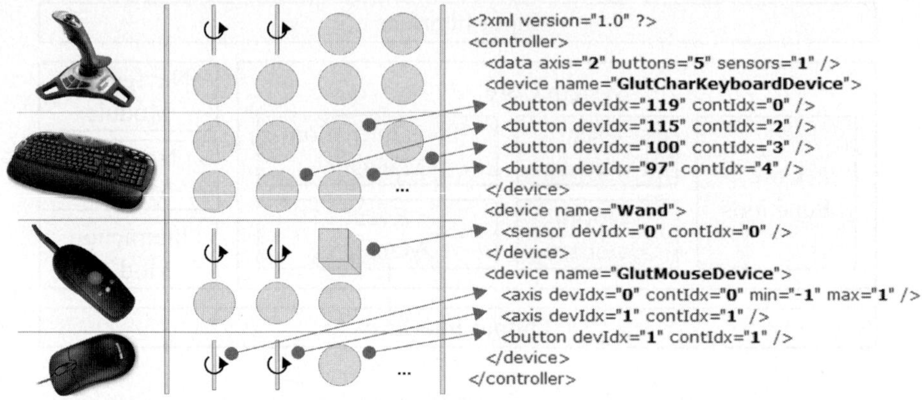

Fig. 3. Mapping of devices and components on the controller

of hardware devices, logical components, and a logical controller. In Figure 3 a mapping of some devices and their components on the XML-specification of the controller is illustrated.

5.2 Output Interface

The output interface is used for graphics and audio rendering. As an audio library currently only OpenAL is supported but other libraries are planned as well. The graphical rendering makes use of OpenSG as a scene graph and the CAVE Scene Manager [8], a tool built upon OpenSG, which is used for the easier access to multiple window displays. Between the scene graph layer and OpenSG an additional interface layer is implemented which abstracts the access to transformations and bounding volumes in the scene graph. This layer is introduced to support other scene graph systems like OpenSceneGraph, Inventor, OpenGL Performer or low level APIs like OpenGL.

6 Modules

The inVRs framework consists of modules for interaction, navigation and network communication, which can be used either independently as libraries or can be individually connected to the system core module. Additional modules like the tools module provide useful functionality, such as graphical effects or collaboration support and visualization. Each of the modules is designed as a separate library which has a clearly defined interfaces to the system core module. An application using only parts of the inVRs framework can implement this interface and use a module without using the core structure. An example would be the connection between the navigation module and the system core. The navigation module provides a transformation matrix as an output which is mapped by the system core on the camera transformation and can be distributed by the network automatically to other participants. By using the navigation module separately it is possible to map the transformation matrix as a steering input for a vehicle in a application without making use of any of the other modules.

6.1 System

The system module is the central module of the inVRs framework. In this module the logical database of the VE - the World Database - is stored, its counterpart the User Database stores information about the local user and the remote users participating in the VE. Discrete events which are used to communicate between the databases and the other modules are handled by the Event Manager. A Transformation Manager takes care of the continuous transformations and manipulations of objects in the VE. To provide a general interface to the other modules commonly used functionality like a platform independent timer, logging mechanisms and debug output are also managed in the system module.

World Database. The data structures needed for the representation and manipulation of the VE are stored in singleton pools inside the World Database component. This database contains objects of the types environment, tile, entity and entity transformation. A VE consists of a set of environments, which are used for structuring regions. A city or a house could be stored as an environment. Each of these environments contain a set of tiles which could be for example elements of a street or rooms inside a house. The tiles and environments have a rectangular shape and are mainly used for helping the developer structure the environment. Having a hierarchical structure helps the scene graph below enormously with frustum culling. Another advantage in structuring the VE in this manor can be found in the area of NVEs. By having a structure like the one described above the VE could be distributed on several servers [1]. To allow for full flexibility, entities which represent the objects used for interaction purposes can be placed arbitrarily inside the VE using entity transformation nodes. The whole setup of the World Database is stored inside XML definition files and can be loaded dynamically during runtime. To design applications it is possible to receive 3D geometries from modelers and arrange them in the XML files. The VE can be altered without changing any application code or any modification of the 3D geometries. With such a mechanism the world setup is fully flexible and configurable. Figure 4 gives an overview of the arrangement of environments, tiles and entities.

User Database. This database contains information about the local user and the other remote participants of the NVE. A user can be identified by their unique ID which is generated by a combination of the IP address, port and process number. The user is represented in the VE by an avatar, which has several animation cycles (e. g. for walking). This data as well as the user position is stored in the User Database. An extension for artificial intelligences which controls characters in the VE would also store the information about these characters in the User Database.

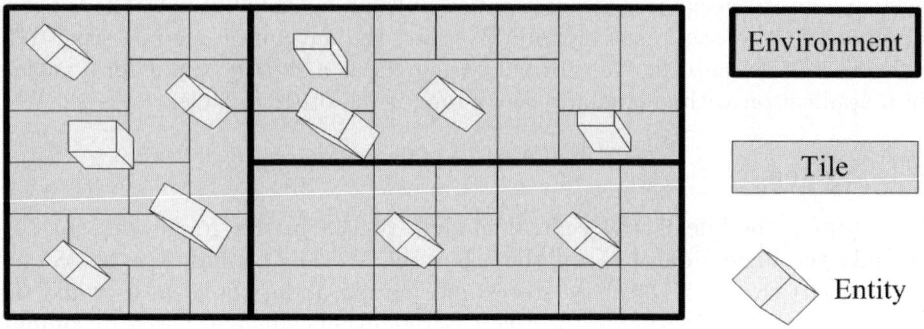

Fig. 4. Environments with tiles and entities

Event Manager. The Event Manager is the communication unit of the inVRs framework which deals with discrete events. Events are generated by the interaction module, they can be received via the network module or they can be generated from another arbitrary user defined component. The components and modules that want to receive the events have to register at the Event Manager using a piping mechanism. In the Event Manager, events are stored in an event queue and can be if necessary reordered or dropped which is useful in the case of late arrival of network messages. The Event Manager distributes the events in its queue to the different modules where they are interpreted and executed. Additionally the locally generated events are sent to the network module if available and are distributed among the other participants of the NVE.

Transformation Manager. The Transformation Manager controls the transformation of objects inside the VE. It takes input from the interaction module, the networking module, the navigation module and other user generated modules which change the layout of the VE. Whereas the Event Manager processes and distributes discrete data, the Transformation Manager processes continuous transformation data and applies these changes in the World Database and the User Database. Using several input streams (e.g. from the interaction module and the network), it is possible to support concurrent object manipulation. Transformation data is stored in several queues which allow for different mechanisms to generate resulting transformations. For example, in the case of concurrent object manipulation the latest transformation, initiated by a remote user could be merged together with a transformation of the local user, which lies a few milliseconds in the past, to calculate the position of the actual object. Another method would be using the latest transformation of the local user and merge it with an extrapolated transformation, which is based on a list of the latest transformations of the remote user. If the navigation module is set to relative transformation mode the actual position is added to the relative transformation to calculate the users position. An additional collision detection module for example, can interfere in the Transformation Manager. In the case of navigation a collision between the users avatar and the environment is detected the collision module can prevent the users representation from moving further in that direction.

6.2 Navigation

Often navigation is seen as a part of interaction. It has been researched with a focus on different navigation methodologies and their effect on human performance by Bowman et. al. [5]. In the inVRs framework navigation or travel is defined as an individual module which is independent from the interaction module. The navigation module provides a variety of travel modes through the VE by mapping the input from the abstract controller of the Input Interface on a combination of models of speed, direction and orientation. These navigation models each generate independent output which is based on the chosen model. Speed for example could be constant if a button is pressed or it could be increased and decreased depending on axial values. The different navigation, orientation and speed models are interchangeable during runtime and are with the help of the controller abstraction fully independent of the input devices. They only take abstract sensors, axes and buttons as input. The navigation module can provide two different types of output. It can generate a full transformation matrix for the camera in absolute coordinates or it can generate a transformation matrix without taking the last position into account. The such a transformation matrix can be used as an input for a physics engine and act as an impulse for a vehicle simulation. More detail on the different navigation models can be found in [2]. Taking into account that the same abstract controller is used for the navigation module and the interaction module and the different navigation models are independent from each other the application designer has to carefully choose his desired interaction and navigation modes to avoid conflicts. For example using the same button for the input device for acceleration and for picking up objects will lead to an undesired application behaviour.

6.3 Interaction

Interaction can be separated into the two tasks of object selection and object manipulation. In the context of the inVRs framework each of the interaction tasks is split into subtasks. Object selection consists of the actual selection process and the positive or negative confirmation of the selection. The manipulation consists of the actual manipulation and the termination of the manipulation. These four interaction processes are implemented by interaction techniques which can be exchanged during runtime.

The interaction is designed as a deterministic finite automaton (DFA) A defined by the 5-tuple $(Q, \Sigma, \delta, q_0, F)$. With $Q = \{q_0, q_1, q_2\}$ as the threes states of interaction, where $q_0 = F$ is the initial and final state of interaction, in which the user simply navigates through the VE or is idle. For every state in Q a transition function $\widehat{\delta}$ exists which is defined as

$$\widehat{\delta} : Q \times \Sigma^* \rightarrow Q$$

with Σ^* as the Kleene Closure of Σ. The transitions $\widehat{\delta}(q,a) = q'$ are the different interaction techniques for selection, unselection, selection confirmation, manipulation or manipulation termination. Some variations of these techniques are initially provided by the framework like raycasting as a selection technique for picking objects. They can easily be enhanced by the user. The input alphabet Σ consists of abstract input from the controller in combination with state information from the World Database and the User Database. Figure 5 shows the interaction DFA.

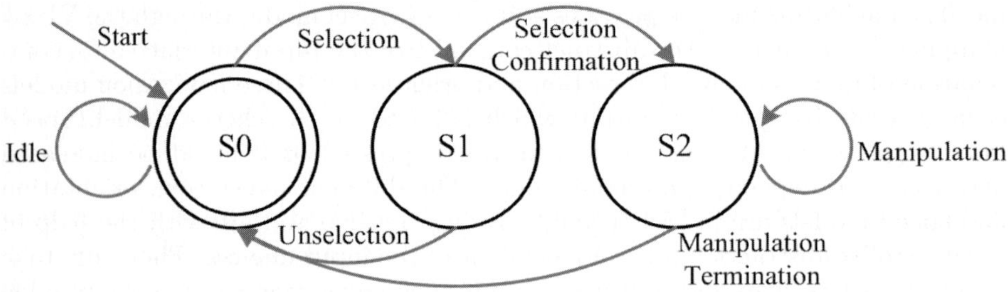

Fig. 5. Interaction state machine

6.4 Network

The networking module is responsible for the distribution of events and continuous information among the other participants of the NVE. Continuous information like position data, transformation data or data generated by the tracking system is streamed via UDP. Events have to be reliable and therefore are sent via TCP. They are less common than the continuous data. The actual implementation of the networking module provides a low level p2p-system, which allows for users to connect to any host in the system. The data of the newly connected peer is then broadcasted to the other peers and and vice versa. Using p2p-architectures guarantees very low latencies but causes high network traffic and is not very scalable. It should be sufficient for smaller link-ups with 5 to 10 users but it is not a large scale approach. Other approaches like the one described in [3] and [1] are harder to implement but guarantee a highly improved scalability. This approach could be integrated in the network module on a higher level on top of the existing p2p architecture.

6.5 Miscellaneous Tools

An additional tools module which does not provide core functionality for the creation of a working NVE has been designed. The tools inside this module include graphical effect packages like swarming algorithms, particle systems, and water simulation. For presentation purposes predefined camera paths are a useful addition. During tasks involving collaboration an additional part of the tools module allows visualization of movements and actions of the remote participants.

Height maps and basic physics in the field of rigid body dynamics simplify the creation of architectural walkthroughs, construction scenarios and safety training applications.

7 Conclusions

This paper has given a brief overview of the inVRs architecture and introduced it's different modules and interfaces. The high flexibility of different input and output devices is generated through the usage of the interfaces. It is focused on the development of VEs and NVEs and does not specifically support the creation of visualizations. The inVRs framework presented in this paper provides a well structured framework for the design of NVEs in general. Basic functionality like navigation, interaction and world structuring is accessible for the application designer by using the inVRs libraries or by altering the XML-configuration files.

The advantages of the inVRs framework lie in the abstraction of input and output, and a clear structure for communication of discrete and continuous data. Centralized replicated databases of the the user and VE data guarantee for a high responsiveness. Clear definition of navigation and interaction methodologies allow high reusability and expandability of navigation models and interaction techniques.

The framework does not support advanced animation, scripting possibilities, or menu driven interaction. It provides the core functionality needed in most VR applications especially in NVEs. Application specific functionality still has to be individually developed.

8 Future Work

In the future the modules have to be enhanced. Different high level network architectures have to be implemented in the network module to guarantee an improved scalability. The framework can be extended with a visualization module which will allow the support of different data formats and provide a variety of visualization techniques. A physics module would be interesting for training environments or vehicle simulation but is very challenging to design with network support. Developing an external editor for configuring the XML definitions and performing the layout of the VE would allow application designers without any VR and programming experience to develop basic VR applications or to design and enhance advanced domain specific applications.

Acknowledgments

Several of our colleagues contributed to this work, most notably Helmut Bressler, Roland Landertshamer and Johannes Zarl at the ZID and GUP of Johannes Kepler University.

References

1. Christoph Anthes, Paul Heinzlreiter, Adrian Haffegee, and Jens Volkert. Message traffic in a distributed virtual environment for close-coupled collaboration. In *PDCS*, pages 484–490, San Francisco, CA, USA, September 2004.
2. Christoph Anthes, Paul Heinzlreiter, Gerhard Kurka, and Jens Volkert. Navigation models for a flexible, multi-mode vr navigation framework. In *VRCAI*, pages 476–479, Singapore, June 2004.
3. Christoph Anthes, Paul Heinzlreiter, and Jens Volkert. An adaptive network architecture for close-coupled collaboration in distributedvirtual environments. In *VRCAI*, pages 382–385, Singapore, June 2004.
4. Allen Douglas Bierbaum. Vr juggler: A virtual platform for virtual reality application development. Master's thesis, Iowa State University, Ames, Iowa, 2000.
5. Douglas A. Bowman, David Koller, and Larry F. Hodges. Travel in immersive virtual environments: An evaluation of viewpoint motioncontrol techniques. In *Virtual Reality Annual International Symposium (VRAIS)*, pages 45–52, 1997.
6. Christer Carlsson and Olof Hagsand. Dive - a platform for multiuser virtual environments. *Computers and Graphics*, 17(6):663–669, 1993.
7. Matthew Conway, Randy Pausch, Rich Gossweiler, and Tommy Burnette. Alice: A rapid prototyping system for building virtual environments. In *ACM CHI '94 Conf. Human Factors in Computing, Conf. Companion*, volume 2, pages 295–296, April 1994.
8. Adrian Haffegee, Ronan Jamieson, Christoph Anthes, and Vassil N. Alexandrov. Tools for collaborative vr application development. In *ICCS*, pages 350–358, May 2005.
9. Jason Leigh, Andrew E. Johnson, and Thomas A. DeFanti. Issues in the design of a flexible distributed architecture for supportingpersistence and interoperability in collaborative virtual environments. In *Supercomputing'97*, 1997.
10. Michael R. Macedonia and Michael J. Zyda. A taxonomy for networked virtual environments. *IEEE MultiMedia*, 4(1):48–56, Jan-Mar 1997.
11. Maja Matijasevic. A review of networked multi-user virtual environments. Technical report tr97-8-1, Center for Advanced Computer Studies, Virtual Reality and Multimedia Laboratory, University of Southwestern Lousiana, USA, 1997.
12. Dirk Reiners. *OpenSG: A Scene Graph System for Flexible and Efficient Realtime Rendering for Virtual and Augmented Reality Applications.* PhD thesis, Technische Universität Darmstadt, Mai 2002.
13. John Rohlf and James Helman. Iris performer: A high performance multiprocessing toolkit for real-time3d graphics. In *SIGGRAPH*, pages 381–394. ACM Press, July 1994.
14. Paul S. Strauss. Iris inventor, a 3d graphics toolkit. In A. Paepcke, editor, *8th Annual Conference on Object-Oriented Programming Systems, Languages,and Applications*, pages 192–200, Washington, D.C., United States, 1993. ACM Press.

JaDiMa: Java Applications Distributed Management on Grid Platforms

Yudith Cardinale, Eduardo Blanco, and Jesús De Oliveira

Universidad Simón Bolívar,
Departamento de Computación y Tecnología de la Información,
Apartado 89000, Caracas 1080-A, Venezuela
{yudith, eduardo}@ldc.usb.ve, jesus@bsc.co.ve

Abstract. This paper describes JaDiMa (*Java Distributed Machine*), a collaborative framework to construct high performance distributed Java applications on grid platforms. JaDiMa is a system that automatically manages remote libraries used in a Java application. It leverages on the advantages of portability, modularity, object oriented model and flexibility of Java, while incorporating well known techniques of communication and security. JaDiMa allows users to compile and execute Java applications which use distributed libraries, without the need of keeping them in the developer and user hosts. The result is a simple and efficient distributed environment upon which applications and data are easily shared and highly portable amongst heterogeneous platforms and multiple users. We describe an implementation of JaDiMa as part of suma/g, a Globus-based grid environment. We also show experiences of executing an application, which uses libraries for managing graph and network data, on several scenarios with suma/g and JaDiMa.

Keywords: Collaborative Environments, High Performance Java Applications, Grid Platforms, Distributed Library Management, Compilation and Execution.

1 Introduction

Grid platforms increase the possibility of environments in which multiple users, geographically distant, may share data, pieces of software, computation resources, and even specialized devices [1]. In this direction, it is very common that programmers use several library components developed by third parties to achieve the global goal required by an application. Following the principle of reusability, it is much more efficient for developers to delegate specific functionalities to already available software, extensively proven and developed exclusively for such function, and concentrate themselves in the resolution of their specific problem.

However, when the pieces of reusable code are distributed (e.g. in a grid platform), developers have to find and obtain the suitable libraries, then reference them in their applications so that the compilation process can be accomplished. In addition, end user must have the same libraries in order to be able to execute

M. Gerndt and D. Kranzlmüller (Eds.): HPCC 2006, LNCS 4208, pp. 905–914, 2006.

the application. In general, these applications are distributed along with libraries on which they depend on. This implies that reused pieces of software must remain local to the compilation and execution environment. In a distributed environment, this approach presents serious disadvantages: i) waste of disk space, when a library is used by several applications, or when only a small portion of this library is used; ii) difficulty in handling and updating the versions of the library, by leaving, in the hands of the developers and end users, the responsibility of updating its local versions with newer ones; and iii) in the case of development of scientific applications for grids, libraries are only required for local compilation, since the application is not going to be executed locally but in some of the remote execution nodes of the grid platform. Downloading the libraries locally only for compilation represents a waste of space and time for developers.

Motivated by these considerations, we have focused on developing a Java-based framework, called JADIMA (*Java Distributed Machine*, sourceforge.net/projects/jadima), a collaborative platform to construct high performance JAVA applications on grid platforms. JADIMA is a system that automatically manages remote libraries used in a JAVA application. It leverages on the advantages of portability, modularity, object oriented model and flexibility of JAVA, while incorporating well known techniques of communication and security (i.e., SOAP protocol and X.509 certificates). The result is a simple and efficient distributed environment upon which applications and data are easily shared and highly portable among heterogeneous platforms and multiple users, therefore motivating the reusability and sharing of libraries by inexperienced and advanced programmers. JADIMA allows users to compile and execute JAVA applications using distributed libraries, without the need of keeping them in the developer hosts.

We describe an implementation of JADIMA as part of SUMA/G [2], a Globus-based middleware. SUMA/G is a grid platform specifically targeted at executing JAVA bytecode on Globus grids. We also show experiences of executing an application, which uses libraries for managing graph and network data, on several scenarios with SUMA/G and JADIMA.

2 JaDiMa Architecture

The design of JADIMA is oriented to satisfy requirements such as: easy to install, configure and use; flexibility, adaptability and modularity; platform independence; transparency; high performance and scalability; and security. Following sections describe how JADIMA satisfies these requirements.

2.1 Components

Our design and implementation can be roughly separated into four components: Repository, Publishing Agent, Compilation Agent and Execution Agent; corresponding respectively to administration of libraries, their publication, the compilation and the execution of applications that require published libraries. Figure 1 presents the general scheme of JADIMA architecture.

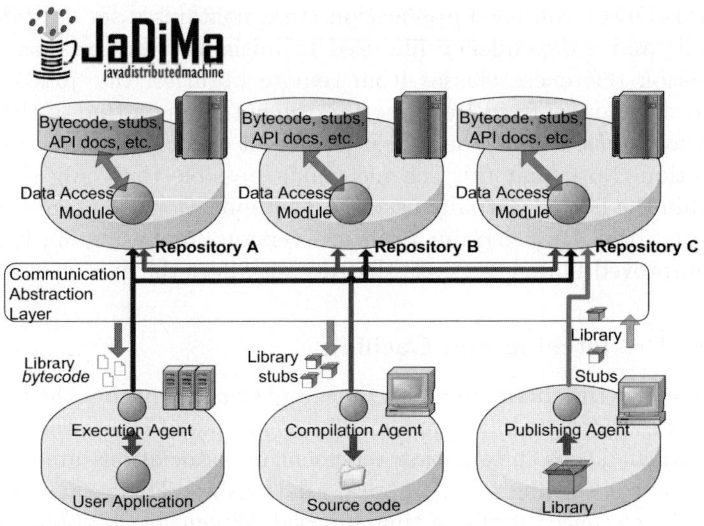

Fig. 1. JaDiMa Architecture

Library Repository
The Library Repository component is in charge of managing the remote libraries that will be used in compilation and execution processes defined in JaDiMa. For each published library, the Repository will maintain three different but related data sets: i) the actual class implementations provided by the *publishing user*, which will be used in the execution phase, ii) the *stubs*, which will be used during the compilation process, and iii) the library documentation (*API*).

Publishing Agent
The library publishing process in JaDiMa consists of two phases: i) Automatic generation of *stubs*. A *stub* is automatically generated for each class of the library and is compiled using JaDiMa Compilation Agent in order to solve dependencies with other previously published libraries. The *stubs* will be used by the developers to compile their applications; and ii) Transmission of *stubs*, documentation and packages of the libraries from the publishing node to the Repository.

Compilation Agent
The Compilation Agent is in charge of obtaining application dependency information provided by the developer, and making requests to specified Repositories in order to obtain library *stubs*. Through a graphical interface, the programmer can query the information about the published libraries on different Repositories and select those he needs; then the graphical interface automatically generates the metadata file that describes the application dependencies.

Execution Agent
The Execution Agent is in charge of setting up the environment in which an application can be transparently executed. When an execution is requested, the

user has a JaDiMa compiled application (that contains a set of *stubs* for each library used) and a dependency file used to initialize `jdmClassLoader`. When the application references classes from remote libraries, the `jdmClassLoader` locates and downloads, from Repositories, the actual definition of these classes, replacing their *stubs*. Hence, the Execution Agent ensures a normal execution of the application. Note that this scheme makes possible that only the subset of classes within the library actually used by the application are downloaded to the execution platform. By using our proposed version numbering [3], it is possible to obtain improved library versions that do not affect the execution.

2.2 Class Pre-fetching and Caching

In order to reduce the communication time, and thus to improve the performance during the execution, we have implemented a scheme of class pre-fetching. The Execution Agent keeps information representing associations amongst the application classes according to a temporal relationship. This relationship defines sets of classes referenced within a time interval. Whenever an application is executed, this information is updated by averaging the elapse times at which a class is referenced, from the beginning of the execution. With this information, the Execution Agent defines *clusters* of classes referenced within a time interval of δ. Hence, when a class is referenced and is not present in the execution node, all the classes that belong to its *cluster* are requested. Information of class associations is returned to the client at the end of the execution, such that it can be submitted in future executions. Class pre-fetching module is defined as a thread, so it can be executed concurrently with the application.

The Execution Agent manages the persistence of the downloaded classes in a local cache. This means that after execution, remotely loaded classes remain in the Execution Agent. In this way, it is possible to reduce communication delay and overhead in future executions. The implementation of cache policies are considered with a *plug-in* scheme that reinforces JaDiMa adaptability.

3 JaDiMa on SUMA/G

In this section we describe a scheme to integrate JaDiMa to a Globus-based grid platform, called SUMA/G. JaDiMa features (such as transparency, security, accessibility, scalability and high performance), in addition to its execution model makes it ideal to leverage execution requirements in grid environments. We identify some benefits of incorporating JaDiMa in grid platforms:

- Grid services could be updated simultaneously in multiple nodes with no intervention of system administrators. These services will be automatically updated with newer and corrected versions of used packages, by using the JaDiMa versioning scheme which guarantees version compatibility.
- Grid users share software easily, supported by centralized administration of libraries. By providing access to Repositories, JaDiMa is virtually distributing every published library to each execution node on the grid.

- Grid users save time, bandwidth, and disk space. They may develop their applications locally to latter execute them on the grid. The use of library *stubs* makes possible for users to relay in JADIMA mechanisms for library administration. Hence programmers may only download the *stubs* needed in the compilation process. When requesting the execution on a grid node, these *stubs* will travel as a part of the application to the selected Execution Agent where they will be substituted for most recent and compatible library versions.

3.1 SUMA/G Overview

SUMA/G (Scientific Ubiquitous Metacomputing Architecture/Globus) [2] is a grid platform that transparently executes JAVA bytecode on remote machines. It extends the JAVA execution model to grid platforms; in particular, classes and data are dynamically loaded.

SUMA/G middleware was originally built on top of commodity software and communication technologies, including JAVA and CORBA [4]. It has been gradually incorporating Globus general services by using the Java CoG Kit [5]. SUMA/G architecture is depicted in Figure 2.

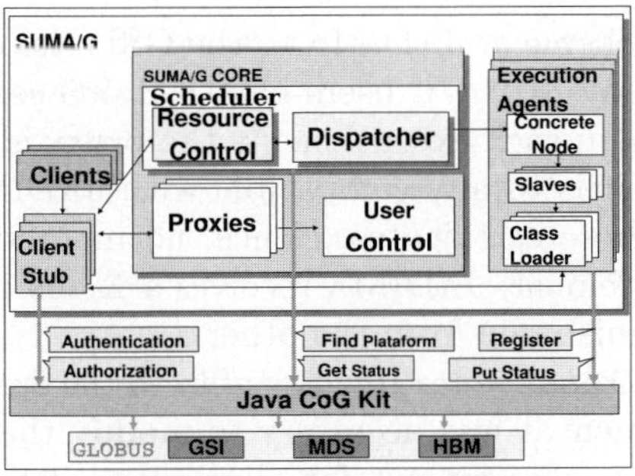

Fig. 2. SUMA/G Architecture

The basics of executing JAVA programs in SUMA/G are simple. Users can start program execution through a shell running on the client machine. They can invoke either `sumag Execute`, corresponding to the on-line execution mode, or `sumag Submit`, which allows off-line execution (batch jobs). At this time a proxy credential is generated (by using GSI) that allows processes created on behalf of the user to acquire resources, without additional user intervention (single sign-on). Once the SUMA/G core receives the request from the client machine, it authenticates the user (through GSI), transparently finds a platform for execution (by querying the MDS), and sends a request message to that platform. An `Execution Agent` at the designated platform receives an object representing the application and starts, in an independent JVM, an `Execution Agent Slave`, who actually executes the application. The `sumagClassLoader` is started in that new JVM, whose function is to load classes and data during the execution. Supported classes and input files sources, and output destinations, include: a) the machine (client) where the application execution command is run and, b) a remote file server on which the user has an account. A pluggable schema allows for implementing several protocols to manage classes and files (e.g., CORBA, sftp, gridFTP).

In case of invoking `sumag Execute` service, user has only to specify the main class. The rest of classes and data files are loaded at run-time, on demand. Standard input and output are handled transparently, as if the user were running the bytecode in the local machine. For the `sumag Submit` service, the client transparently packs all classes together with input files and delivers them to SUMA/G; the output is kept in SUMA/G until the user requests it. A `Proxy` is designated to act on behalf of the client to supply classes and data to the `Execution Agent`.

3.2 Extending the SUMA/G Execution Model with JaDiMa

The integration of JaDiMa on SUMA/G extends its execution model by adding a new class source: JaDiMa Repositories. In order to integrate JaDiMa to SUMA/G, it was necessary to:

- Adapt the security scheme of JaDiMa to the GSI scheme. This requirement could be reached easily. SUMA/G users have a X.509 certificates from which a proxy is generated on each execution. As the proxy is a "reduced version" of X.509 certificate, it is totally compatible with JaDiMa security scheme, which is based on those certificates. Hence, no modification was necessary to achieve this requirement. JaDiMa receives a X.509 certificate or a proxy and does not distinguish one from the other.
- Incorporate the `jdmClassLoader` functionality to the `sumagClassLoader`. To achieve this requirement it was necessary to modify the `sumagClassLoader` in order to add *stubs* management for allowing to detect library *stubs* and request actual libraries from pre-defined Repositories.

Figure 3 shows the SUMA/G extended execution model. According to SUMA/G traditional execution model, `sumagClassLoader` gets classes from the Client (in case of `Execute` service), or from a designated `Proxy` (in case of `Submit` service), or from remote accounts specified by users. With JaDiMa some classes can be got from remote Repositories. The elements denoted with pentagons represent additions to the original mechanism:

a. On the client side there are library *stubs* on which the application depends on. These *stubs* are transferred to the SUMA/G Execution Node as they are referenced, exactly as it happens with application classes.
b. `sumagClassLoader` verifies each class loaded to determine if it is a *stub*.
c. In case a *stub* is loaded, `sumagClassLoader` requests the actual class to the specified Repository (it is specified in the *stub*).

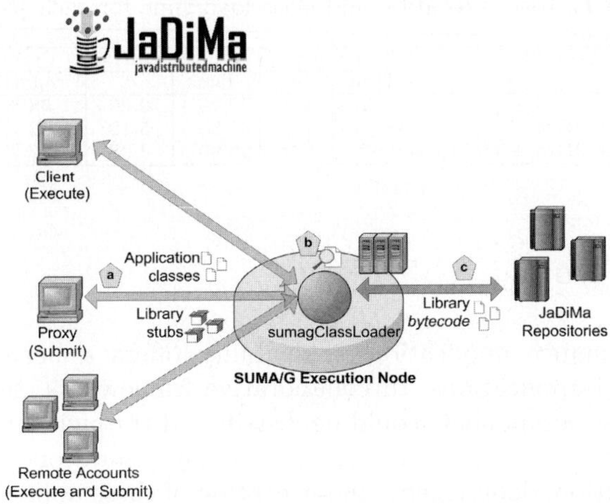

Fig. 3. SUMA/G Execution Model Extended with JADIMA

4 An Experience: Using JUNG Library

JUNG [6] (*Java Universal Network/Graph*) is an open-source software library that provides a language for the modeling, analysis, and visualization of graph and network data. The main packages used by JUNG are `COLT` (http://cern.ch/hoschek/colt/), `commons-collections` (http://jakarta.apache.org/commons/collections/), and `xerces` (http://xml.apache.org/).

We have executed a JUNG-based application in a distributed platform: SUMA/G core components running on a double processor 1.8 GHz, 1GB; an Execution Agent running on a dedicated cluster of PC's, double processor 800 MHz, 512 MB, 100 Mbps Ethernet; a JADIMA Repository on MySQL back-end in a 1.5MHz, 750 MB; a JADIMA Repository on SQLServer back-end in a double processor 600MHz, 1GB, 100 Mbps Ethernet; and a SUMA/G Client running on a 2.2GHz, 1GB, 768Kbps network connection.

We tested 3 scenarios in order to get performance information: i) the execution takes place using SUMA/G original execution model, e.i. with no JADIMA support, all classes are transfered from Client to Execution Agent; ii) the libraries used by the application are published at JADIMA Repositories close to SUMA/G deployment site; and iii) as the second one, but with JADIMA facilities activated, e.i. class pre-fetching and caching. In all of them, the client and the execution node were in different networks. Table 1 shows total execution time, class loading time, and time proportion spent loading classes, for each scenario.

The results show that with the third scenario the total execution time is reduced in 20% with respect the first scenario. The second scenario shows the worst performance because it is necessary to transfer *stubs* from the Client to the Execution Agent, and then actual classes from Repositories.

Table 1. Total execution and class load time for each scenario

Scenario	Class Loading		Total Execution		Proportion	
	Time	σ	Time	σ	Proportion	σ
SUMA/G	12.40 s	6.73	30.70 s	9.26	38 %	0.08
SUMA/G + JDM	18.79 s	3.52	35.80 s	5.18	52 %	0.03
SUMA/G + JDM + facilities	9.97 s	1.58	26.90 s	3.98	37 %	0.03

5 Related Work

For high performance applications, reusability offers a benefit to construct component-based applications. In collaborative frameworks, such as computational grids, those components could be distributed through the platform. However, the current grid enabled tools present many limitations to support compilation, distribution, deployment, and execution of these applications. JADIMA overcomes some of these limitations and provides a framework to build component-based JAVA applications in which the components are distributed.

Apache Maven [7] and Krysalis Centipede [8] extend the functionalities of compilation common tools (i.e. ant and make) to allow distributed management of libraries at compilation time. As well as JADIMA, both projects can automatically download jar files (according to the package dependencies specified at the configuration files) from remote Web repositories. The entire packages are mirrored in a local repository, while JADIMA downloads only *stubs* to the local host. In JADIMA, web repositories are not mandatory because JADIMA supports several communication protocols. Apache Maven and Krysalis Centipede can replace the packages for new versions during the compilation only if this is specified on the dependency file. Based on our proposed versioning specification, `jdmClassLoader` can load new library versions at runtime, which does not affect the execution flow. With respect to security, JADIMA allows fine-grain access control by defining access permissions to packages or libraries instead of repositories. Security in Apache Maven and Krysalis Centipede is managed at the communication protocol level.

DistAnt [9] and GridAnt [10] specifically address the problem of helping users deploy their applications on the grid resources. They extend Ant with grid specific tasks, which users can employ to define their application's deployment workflow (e.g., compiling; authentication; resource location; code and data packaging, transport and unpackaging/installation, etc.). These tools can be combined with Apache Maven to leverage in the compilation activities. They operate at a higher level than JADIMA; however, they are oriented to isolated application deployment, instead of the compilation and execution support for distributed applications provided by JADIMA. As well as JADIMA, GridAnt uses X.509 certificates for user authentication, authorization and delegation.

There are several projects oriented to remote execution of JAVA programs on grid platforms. Some examples are Addistant [11], Unicorn [12], Javelin++ [13], JNLP [14], Bayanihan [15], HORB [16] and SUMA/G [2].

Addistant is a system which enables the distributed execution of software originally developed to be executed on a single JVM, so that some objects of that software are executed on a remote host. That means, it provides functional distribution on multiple JVM. The Addistant execution model is contrary JADIMA model because it allows the execution on a single platform (sequential or parallel) of applications with distributed components. JADIMA does not modify the original execution model of applications. Developers using Addistant have to specify (in a *policy file*) the host where instances of each class are allocated and how remote references are implemented. According to that specification, Addistant automatically transforms the bytecode at load time by generating proxy classes, which take charge of making remote references at run-time. JADIMA uses *stubs* at compilation time and the substitution for actual classes is made transparently at run-time.

Unicorn, Javelin++, Bayanihan, HORB and SUMA/G define architectures using JAVA to harness the vast processing power on the Internet for distributed computing applications. Users can either make use of the suite of applications provided by the systems or upload their own applications to the server together with the data to be processed. These tasks are distributed (may be in parallel) to idle hosts logged on to the corresponding servers. Most of these projects involve the need of using specific programing models and run-time environments for distributing and invoking remote JAVA classes. JADIMA runs unmodified JAVA applications. None of those projects support compilation activities, then they could leverage on JADIMA compilation power.

6 Conclusions and Future Work

JADIMA provides an adequate collaborative framework to construct high performance JAVA applications on distributed platforms. JADIMA is a system that automatically manages remote libraries used in a JAVA application. On the other hand, SUMA/G execution model is very attractive to users, since the grid is used in the same fashion a local JVM is used.

We have shown a feasible integration of JADIMA to SUMA/G platform, keeping the advantages of both systems. We extended the JVM model to provide seamless access to distributed high performance resources and support the reusability of distributed software components. The SUMA/G+JADIMA execution model meets JAVA user expectations, facilitating application porting to the grid, since applications can be run first on local machines, then executed on the grid without any change to classes. This execution model improve application performance by reducing class transfer overhead.

Plans for future work include conducting experiments with other kinds of applications (e.g., different combination of packages distribution), as well as exploring alternatives of *stubs* management on SUMA/G to eliminate *stub* transfer from the client to the execution node. This will reduce the communication overhead during the execution.

References

1. Berman, F., Fox, G., Hey, A., eds.: Grid Computing: Making the Global Infrastructure a Reality. Wiley (2003)
2. Cardinale, Y., Hernández, E.: Parallel Checkpointing on a Grid-enabled Java Platform. Lecture Notes in Computer Science **3470** (2005) 741 – 750
3. Cardinale, Y., Blanco, E., DeOliveira, J.: JaDiMa: Arquitectura de Máquina Virtual para la Construcción de Aplicaciones JAVA en Plataformas Grids. In: XXXI Conferencia Latinoamericana de Informática (CLEI-2005), Colombia (2005)
4. Cardinale, Y., Curiel, M., Figueira, C., García, P., Hernández, E.: Implementation of a CORBA-based metacomputing system. Lecture Notes in Computer Science **2110** (2001) Workshop on Java in High Performance Computing.
5. von Laszewski, G., Foster, I., Gawor, J., Smith, W., Tuecke, S.: CoG Kits: A Bridge between Commodity Distributed Computing and High-Performance Grids. In: ACM Java Grande 2000 Conference, San Francisco, CA (2000) 97–106
6. O'Madadhain, J., Fisher, D., Nelson, T., Krefeldt, J.: JUNG: Java Universal Network/Graph Framework (2003) http://jung.sourceforge.net/index.html.
7. Apache Software Foundation: Apache Maven Project (2005) http://maven.apache.org/.
8. Krysalis Community Project: Krysalis centipede (2004) http://krysalis.org/centipede/.
9. Goscinski, W., Abramson, D.: Distributed Ant: A system to support application deployment in the grid. In: Proceedings of the Fifth IEEE/ACM International Workshop on Grid Computing (GRID'04). (2004)
10. Amin, K., von Laszewski amd Mihael Hategan, G., Zaluzec, N.J., Hampton, S., Rossi, A.: GridAnt- A Client-Controllable Workflow System. In: Proceedings of the 37th Hawaii International Conference on System Sciences. (2004)
11. Tatsubori, M., Sasaki, T., Chiba, S., Itano, K.: A Bytecode Translator for Distributed Execution of "Legacy" Java Software. In: Proceedings of the 15th European Conference on Object Oriented Programming (ECOOP 2001). Volume 2072., Budapest, Hungary, Springer-Verlag (2001)
12. Ong, T.M., Lim, T.M., Lee, B.S., Yeo, C.K.: Unicorn: voluntary computing over Internet. ACM SIGOPS Operating Systems Review **36** (2002) 36–51
13. Neary, M.O., Brydon, S.P., Kmiec, P., Rollins, S., Cappello, P.: Javelin++: Scalability issues in global computing. In: Proceedings of the ACM 1999 conference on Java Grande, San Francisco, California (1999) 171–180
14. Zukowski, J.: Deploying Software with JNLP and Java Web Start (2002) http://java.sun.com/developer/technicalArticles/Programming/jnlp/.
15. Sarmenta, L.F.G., Hirano, S.: Bayanihan: building and studying web-based volunteer computing systems using java. Future Generation Computer Systems **15** (1999) 675–686
16. Satoshi, H.: HORB: Distributed Execution of Java Programs. In: Proceedings of the International Conference on Worldwide Computing and Its Applications. (1997) 29–42

Reducing Data Replication Overhead in DHT Based Peer-to-Peer System

Kyungbaek Kim and Daeyeon Park

Department of Electrical Engineering & Computer Science,
Division of Electrical Engineering,
Korea Advanced Institute of Science and Technology (KAIST),
373-1 Guseong-dong, Yuseong-gu, Daejeon 305-701, Republic of Korea
kbkim@sslab.kaist.ac.kr, daeyeon@ee.kaist.ac.kr

Abstract. DHT based p2p systems are appeared to provide scalable storage services with idle resource from many unreliable clients. If a DHT is used in storage intensive applications where data loss must be minimized, quick replication is especially important to replace lost redundancy on other nodes in reaction to failures. To achieve this easily, the simple replication method directly uses the consistent set such as the leaf set and the successor list. However, this set is tightly coupled to the current state of nodes and the traffic needed to support this replication can be high and bursty under churn.

This paper explores efficient replication methods that only glimpse the consistent set to select a new replica. We propose two types of replication methods : *Quorum based replication* and *Availability based replication*. The replicas are loosely coupled to the consistent set and can eliminate the compulsory replication under churn. Results from a simulation study suggest that our methods can reduce network traffic enormously and achieve high data availability in a DHT based p2p storage system.

Keywords: Peer-to-Peer, Replication, Data availability.

1 Introduction

In these days, peer-to-peer systems have become an extremely popular platform for large-scale content sharing, even the p2p based file systems appear. Unlike client/server model based storage systems, which centralize the management of data in a few highly reliable servers, peer-to-peer storage systems distribute the burden of data storage and communications among tens of thousands of clients. The wide-spread attraction of this model arises from the promise that idle resources may be efficiently harvested to provide scalable storage services. To promise that, a lot of research papers discussed the Distributed Hash Table (DHT) based p2p routing algorithms [1] [2] [3] [4] and we call the p2p system which uses the DHT based p2p routing algorithm the structured p2p system.

These structured p2p systems achieve the efficient and bounded lookup for the requested object. However, they poorly support the acceptable levels of data availability [11] [12]. The main problem is the ad hoc manner in which p2p systems are constructed. In contrast to traditional systems, peer-to-peer systems are composed of components with extremely heterogeneous availabilities - individually administered host

M. Gerndt and D. Kranzlmüller (Eds.): HPCC 2006, LNCS 4208, pp. 915–924, 2006.

PC's may be turned on and off, join and leave the system, have intermittent connectivity, and are constructed from low-cost low reliability components. For example, one recent study of a popular peer-to-peer file sharing system[7] found that the majority of peers had application-level availability rates of under 20% and only 20% nodes have server-like profiles. In such an environment, failure is no longer an exceptional event, but is a pervasive condition. At any point in time the majority of hosts in the system are unavailable and those hosts that are available may soon stop servicing requests.

Many structured p2p systems use the simple replication method to cope with the massively failures of nodes [5] [6] [8]. This simple approach exploits the consistent set of the successive nodeIDs such as successor list of chord [1] and leaf set of pastry [2]. Basically, these sets are used to conserve the routing information to cope with the failures and when any node joins or leaves, the consistent set of every node which detects the change of the membership must update to preserve the current state of the p2p system. Because of this consistent and automated update, many systems use this set to replicate the responsible objects of each node for the simple lookup and the simple data management. However, this simple approach makes high network traffic because of the dynamic membership of the p2p system. For example, if the size of the consistent set is 8 and the p2p system needs 6 replicas for the target data availability, when one new node changes its state, the 8 nodes which are the members of the consistent set of the new node should update their consistent sets. In this case, the simple replication approach is tightly coupled to the consistent set and 6 replicas should be updated without any relation of the current data availability or the node characteristics. According to this behavior, the heavily dynamic membership change of p2p systems causes the compulsory data replication and generates very high network traffic for this replication. Because of this heavy replication overhead, until now, DHT algorithms are not widely used in commercial systems yet.

In our paper, we suggest the efficient replication methods to achieve the highly durable p2p storage system with small maintenance cost. The replicas are loosely coupled to the consistent set and they are interleaved on the consistent set to reduce the compulsory copies which occur under churn. The method with this concept is called the Quorum based replication. In this replication, each node keeps the number of replicas more than the target quorum to achieve target data availability. Moreover, we exploit the node availability and select more reliable nodes as replicas to delay the replication and to reduce the network cost. This Availability based replication calculates the data availability whenever the consistent set changes and guarantees the high data availability by the numerical value. This replication should predict the node availability of each node. To do this, each node manages its availability and advertises it to all members of the consistent set by piggybacking it to the periodic ping message which has been already used to detect node failures on the consistent set.

Our replication methods need additional information for replicas and interleave the replicas. Unlike the simple replication, sometimes, each node does not have the objects which are serviced by the right next neighbors such as successor or predecessor of chord [1]. In this case, each node should contain the information for the replicas of all members on the consistent set to guarantee the correct data routing whenever any node fails or leaves. Because of this complication of replication methods, subtle data management should be needed under churn.

We evaluate the effect of our replication method for the p2p system by using an event driven simulation. We compare the network traffic for various target availability and various node characteristics between the simple replication and the proposed replication. We show that our methods enormously reduce the network traffic to achieve the same target availability.

The rest of this paper is organized as follow. Section 2 briefly presents the DHT based p2p system and problems of the simple replication. Section 3 presents our proposed replication methods for the durable p2p storage system. The performance evaluation is on section 4. We mention other related works in section 5. Finally, we conclude this paper on section 6.

2 Background

There are many DHT based p2p algorithms such as chord, pastry, tapestry and can. Each node in the DHT based p2p system gets a 128-bit node identifier (nodeID). The nodeID is used to indicate a node's position in a circular ID space and it is assumed that nodeIDs are generated such that the resulting set of nodeIDs is uniformly distributed in the 128-bit ID space. Each node is responsible for storing and servicing the objects which are on the range between its node and a neighbor node. This object range of a node changes dynamically under churn. Assuming a network consisting of N nodes, the DHT based p2p system can route to the numerically closest node to a given object key in less than $O(log_2 N)$ steps.

These algorithms can lookup any data efficiently with DHT, but when the massively node failures occurs and spoils the information of DHT without any notification, this efficient lookup can not guarantee the correctness. To cope with the massively node failures, they use the consistent set such as the successor list of chord and the leaf set of pastry. This consistent set of a node is composed of the neighbor nodes which locate numerically near to the node on ID space. This set is tightly coupled to the current state of nodes and when any node joins or leaves, the consistent set of every node which detects the change of the membership must update to preserve the current state of the p2p system. The p2p system guarantees the correctness unless all members of consistent set fail simultaneously.

This consistent set is used for not only the routing correctness but also the data availability on durable p2p storage systems such as p2p file systems [5] [6] and p2p file sharing systems [8]. Data Availability means the total availability when the multiple nodes have the data. This availability is obtained by subtracting the probability of that all nodes which have the data leave from 1. That means if only one node is alive, the data is available. Like the figure 1(a), one node replicates stored objects to the neighbor nodes which are the member of consistent set until the replicas are enough to achieve target data availability. This simple replication guarantees the simple data availability management and the simple lookup under churn easily and automatically. Because the consistent set has the current state of nodes and updates immediately under churn, the p2p system keeps the target data availability automatically. Moreover, because neighbor nodes of a node already have the replicas of its objects, even if this node leaves, the

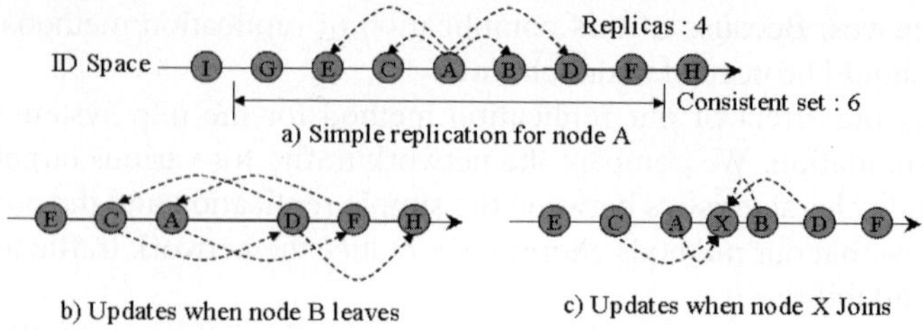

Fig. 1. Simple replication in DHT based p2p

neighbor node automatically replaces it as a servicing node for its object range without additional object copies.

However, the simple replication causes the more maintenance traffic under churn. If the number of replicas is N and a node leaves, the new $N + 1$ replicas are needed for the affected nodes. In figure 1(b), when a node B leaves, the nodes A, C, D, F which already have the replicas on node B should make new replicas and node D which is newly responsible for the object range of node B makes the replica for this range additionally. Moreover, when a node joins, each affected node copies the objects to it as a new replica like figure 1(c). In this case, when node B joins and leaves very frequently, the compulsory data replications occurs and the heavy data traffic wastes even if the dynamic behavior of node B can not affect the data availability. According to this simple behavior and the heavy churn of the p2p participants, the data traffic needed to support the simple replication is very high and bursty.

3 Proposed Idea

3.1 Quorum Based Replication

The simple replication method basically uses the concept of the quorum. The quorum means that the fixed minimum number of members of a set which must be present for its objective to be valid. That is, if the number of the replica for an object is more than the target quorum, the p2p system considers that the object is available under massive failures. However, this simple method directly uses the consistent set such as a successor list or a leaf set which is tightly coupled to the state of the current network. Under churn, to keep the right information of the network, the affected node should update its consistent set and the members of this set change very dynamically. Consequently, the simple replication method is affected by the change of the consistent set and needs too much traffic to keep the availability of an object. Sometimes, this compulsory copy for the replica is meaningless to the availability because the new replica leaves soon.

To prevent this compulsory copy, we modify the replication method which is loosely coupled to the consistent set. Like the figure 2, we add the new information; the replication set that indicates which node replicates the object. The range of this set is same to the consistent set, but the update of this set occurs individually. Like the simple method, the replication only occurs when the number of replicas is fewer than the target quorum. However, if the leaving node is not a member of the replication set, there is no need to

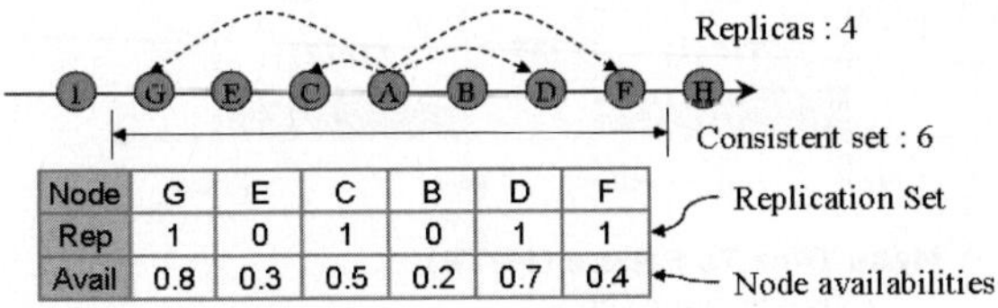

Fig. 2. Metadata for our replication methods

find a new replica and the p2p system can reduce the compulsory copies. When a node needs a new replica, it selects the numerically closest node from it, because the edge of the consistent set may change more easily and more frequently than the inner side.

When a new node joins, it gets not only routing information such as DHT and consistent set but also the replication set. Unlike the simple method, a new node already knows the information of replicas and the data copy only occurs for the object range which is responsible for it. Other nodes whose consistent sets are affected by the new node check whether the replication is needed and if it is, the node makes a new replica. However, because in the general DHT p2p the size of the consistent set is bigger than the number of replicas, the replication does not occur frequently and the replication set can interleave the replicas on the consistent set. This behavior increases the chance to reduce the compulsory copies.

3.2 Availability Based Replication

The quorum based replication considers that each node has the same availability and it tries to keep the number of replicas above the target quorum to achieve the target data availability. However, if a new replica is assigned by the node which has low availability, this node may leave soon and we need another new replica. If we select a new replica with the node availability, we can select the more available node as a replica and can reduce the overhead. To achieve this, the consistent set has the availability information of all members like figure 2.

We assume that the node availability is the prediction value how long a node is alive after it joins the p2p system, because in other research[7] the long lived nodes generally have the large bandwidth and the big computing power. The figure 3 shows this availability prediction mechanism. We use the Mean Time To Failure and the Mean Time To Recover to estimate the node availability. MTTF is the average value how long a node is alive after it joins and MTTR is the average value how long a node is sleep after it leaves. We can get MTTF and MTTR by using the last join time, the last leave time. Unlike MTTR, we periodically update the MTTF by using the current time and the join time because MTTF can change during a node join the p2p system. The average value of MTTF and MTTR is obtained by the sum of the weighted value estimation process. According to these values, we compute the node availability with the equation, MTTF/(MTTF + MTTR).

The availability information is computed by each node and each node advertises this information to all members of the consistent set by using the piggyback method. To detect the node failure, a node sends a ping message to all member of the consistent set.

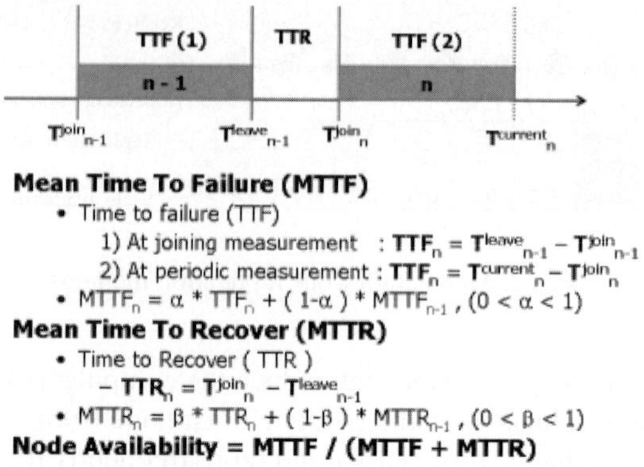

Fig. 3. Node availability prediction

We piggyback the availability information to this ping message and each node manages the consistent set with the node availability.

Like the quorum based replication, the availability based replication use the replication set to make that the replicas are loosely coupled to the consistent set. The main difference of these replications is the selection of a new replica. In this approach, the replication only occurs when the data availability is below the target availability. If the leaving node is not a member of the replication set, there is no need to replicate the data. Otherwise, if it is a member, we select the most available node among non-members of the replication set as a new replica.

When a new node joins, the basic operation is similar to the quorum based replication. The routing table, consistent set and replication set are copied and the other nodes whose consistent sets are affected by the new node decide whether they make new replicas. However, because the availability based replication takes care of selecting new replicas by computing the availability, if all members of a consistent set have averagely low availability, it need more replicas than the quorum based replication. Sometimes this behavior takes more bandwidth, but when nodes leave, this subtle replication can reduce much more bandwidth than the quorum based replication.

3.3 Management of the Replication Set

Unlike the simple replication, our replications interleave the replicas on the consistent set. When a node fails and its neighbor gets the lookup request, this neighbor may not have the replicas for the requested object. In this case, the neighbors must forward the request to the replicas of the failed node for the routing correctness. To do this, each node should have the replication sets of all members of its consistent set by piggybacking this information to the periodic ping message for its consistent set.

Moreover, we should consider that the change of the object range affects the replication set. When a new node joins and a target node gets this join request, the object range of the target node is divided into two object range and the new node is responsible for one of them. In this case, the new node simply copies the replication set and adds the target node as a new replica because it already has the object for

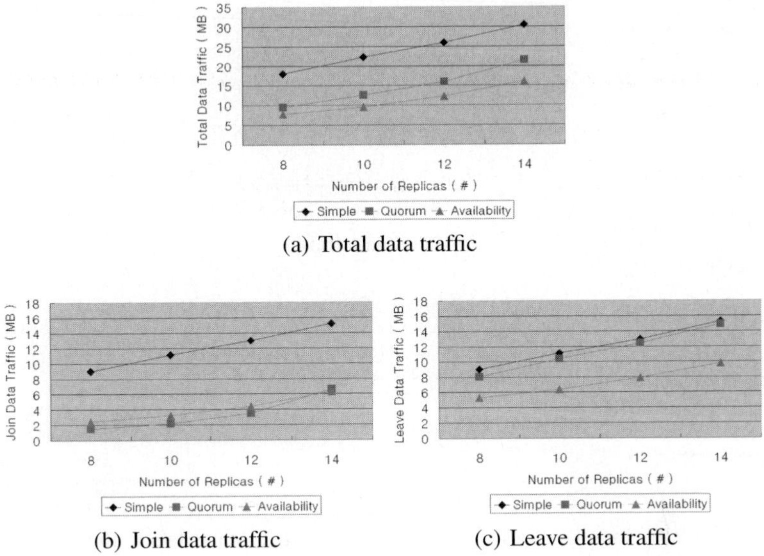

Fig. 4. Data Traffic with various number of replicas

this range. When a node leaves or fails, its neighbor node is responsible for its object range. In this case, both replication sets of the failed node and its neighbor node are merged.

4 Evaluation

4.1 Simulation Setup

We make our p2p simulator which emulates the node behavior on the application layer. We implement the previous DHT based p2p algorithm, Pastry. We apply the simple replication and our new replications to the pastry. We use the 160 bit ID space and use 2000 nodes to organize the p2p system. The size of the consistent set is 16 and the number of replicas is variable from 8 to 14. The target availability for the availability based replication is decided by the number of replicas. We use the Poisson distribution to make the dynamic characteristics of nodes and use the exponential distribution to assign join/leave duration of a node. According to this Poisson distribution, 80% of total nodes have short lifetime and frequently join/leave and only 20% of total nodes have the reliable server-like profile. Recent research [7] measures the life distribution of the p2p nodes and its result is similar to our distribution, and we can tell that our distribution is similar to the real world.

4.2 Reduction of Data Traffic

The figure 4 shows the comparison of average data traffic per a node with various replication methods. As we expect, the simple replication needs much more data traffic to

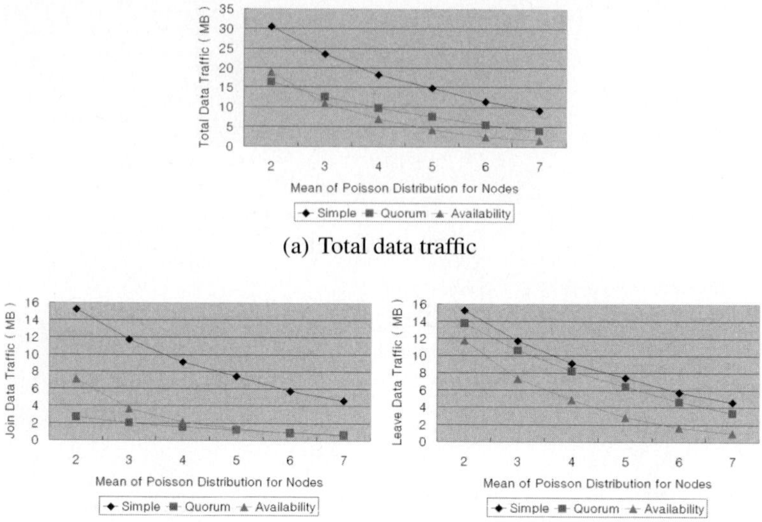

(a) Total data traffic

(b) Join data traffic

(c) Leave data traffic

Fig. 5. Data Traffic with various node characteristics

achieve the same data availability than our replications. The quorum based replication reduces the data traffic by about 40% and the availability based replication reduces the data traffic by about 60%.

To find out the detailed effect of our replications, we separate the join data traffic from the leave data traffic. When a node joins, the affected nodes update their replicas and we call the needed traffic the join data traffic. When a node leaves, the needed data traffic is called the leave data traffic. In the figure 4(b) and figure 4(c), the simple replication uses the similar amount of traffic for join and leave. The main reason is that the replication is tightly coupled to the consistent set and in both type of the changes, the similar amount of compulsory copies is needed to support the simple replication. However, in our two replications, the join data traffic is less than leave data traffic. The replication set is loosely coupled to the consistent set and when a join occurs, a node can decide which it makes a new replica. According to this behavior, the replicas are interleaved on the replication set and we can reduce the number of compulsory copies when a new node joins.

However, like figure 4(c), the quorum based replication needs similar amount of leave data traffic to the simple replication. The quorum based replication does not consider the node characteristics and some new replicas leave the system early after they are chosen by the other nodes. The figure 4(c) shows that the availability based replication solves this problem and needs less leave data traffic than the quorum based replication. On the other hand, the availability based replication computes the data availability whenever the membership changes. According to this, it takes more care of selecting new replicas and it needs more join data traffic than the quorum based replication until the number of replicas is similar to the size of the consistent set.

4.3 Effect of Node Dynamicity

The previous results show that our replications reduce the data traffic needed to achicvc the high data availability. In this result, we try to find out the effect of our replications on the various node characteristics. The figure 5 shows the needed data traffic for each replication method when the mean of the Poisson distribution changes from 2 to 7. When the mean value increases, the average life time of a node increases. In this simulation, the target number of replica is 8. In the figure 5(a), when the mean value increases every replication method takes less data traffic, because there are more reliable nodes and they do not join/leave frequently.

As described on previous results, our replications can save more data traffic for any cases. We pay attention to the difference between the quorum based replication and the availability based replication. Generally, the availability based replication reduces more traffic, however when the mean value is 2, the quorum based replication saves more data traffic. The main reason of this fact is the join data traffic in figure 5(b). As we mentioned, the availability based replication takes more care of selecting the new replicas when a new node joins. When the most of nodes join/leave frequently this subtle care needs more replicas than the quorum and takes much more traffic.

5 Related Work

The commercial p2p file sharing systems leave the data replication up to the popularity of the data. The popular data is replicated on many clients and the data availability of this data is very high. However, the unpopular data are stored on few clients and it is very hard to find this data because of very low data availability. To make the p2p storage system durable, the smart data replication methods is needed and the each inserted data is available for any time and has the similar data availability.

In the paper[9], they stores the replicas on the random nodes on the ID space and periodically checks their availability. This behavior reduces the compulsory copy because the replication has no relation to the consistent set. However, this approach takes too much control traffic to keep the node availability of all replicas for every object on the system. Moreover, they do not use the consistent set and the change of the data availability caused by the node failure is detected slowly. The paper [10] shows that the erasure coding approach reduces the traffic of the replication by using the computing power. This approach is orthogonal to our approaches and we can use this coding with our replication methods.

6 Conclusion

We explore the efficient replication methods to make the DHT based p2p storage system more durable. Our replication methods are loosely coupled to the consistent set such as a successor list of chord and a leaf set of pastry and interleave the replicas on it. Because the consistent set updates the current state of nodes automatically, we can update the data availability immediately under churn. Moreover, we use the node availability to select more reliable replicas and we can reduce more data traffic when a node leaves.

According to these behaviors, the DHT based p2p storage system with our replication methods achieves the high data availability with small data traffic. This can encourage that the DHT based p2p algorithms are applied to the durable storage system.

References

1. I.Stoica, R.Morris, D.Karger, M.F.Kaashoek, and H.Balakrishnan. *Chord: a scalable peer-to-peer lookup service for internet applications*, In Proceedings of ACM SIGCOMM 2001, August 2001.
2. A.Rowstron and P.Druschel. *Pastry: scalable, decentralized object location and routing for large-scale peer-to-peer systems*, In Proceedings of Middleware, November 2001.
3. B.Y.Zhao, J.Kubiatowicz, and A.Joseph. *Tapestry: An infrastructure for fault-tolerant wide-area location and routing*, UCB Technical Report UCB/CSD-01-114, 2001.
4. S.Ratnasamy, P.Francis, M.Handley, R.Karp, and S.Shenker. *A scalable content-addressable network*, In Proceedings of ACM SIGCOMM 2001, 2001.
5. P. Druschel and A. rowstron. *PAST: A large-scale, persistent peer-to-peer storage utility*, In Proceedings of HotOS VIII, May 2001.
6. F. Dabek, M.F. Kaashoek, D. Karger, R. Morris, and I. Stoica. *Wide-area cooperative storage with CFS*, In Proceedings of SOSP 2001, Oct 2001.
7. S. Saroiu et al. *A measurement study of peer-to-peer file sharing systems*, In Proceedings of MMCN 2002, 2002.
8. K.Kim and D.Park. *Efficient and Scalable Client Clustering For Web Proxy Cache*, IEICE Transaction on Information and Systems, E86-D(9), September 2003.
9. R. Bhagwan, K. Tati, Y. Cheng, S. Savage and G. M. Voelker. *Total Recall: System Support for Automated Availability Management*, In Proceedings of NSDI 2004, 2004
10. R. Bhagwan, S. Savage, and G. M. Voelker. *Replication Strategies for Highly Available Peer-to-peer Storage Systems*, In Proceedings of FuDiCo, June 2002.
11. C. Blake and R. Rodrigues. *High Availability, Scalable Storage, Dynamic Peer Networks : Pick Two*, In Proceedings of HotOS-IX, May 2003.
12. R. Bhagwan, S. Savage, and G. M. Voelker. *Understanding Availability*, In Proceedings of IPTPS 03, 2003.

Improving Resiliency Using Capacity-Aware Multicast Tree in P2P-Based Streaming Environments*

Eunseok Kim[1], Jiyong Jang[1], Sungyoung Park[1], Alan Sussman[2], and Jae Soo Yoo[3]

[1] Department of Computer Science and Interdisciplinary Program of Integrated Biotechnology, Sogang University, Seoul, Korea
{wannajump, jiyong97, parksy}@sogang.ac.kr
[2] UMIACS and Department of Computer Science, University of Maryland, USA
[3] School of Electrical and Computer Engineering, Chungbuk National University, Cheongju Chungbuk, Korea

Abstract. This paper presents a capacity-aware multicast tree construction scheme for P2P-based live streaming environments (R-CAT). The R-CAT builds resilient multicast trees by considering the capacity of participating nodes and locating the high capacity nodes toward the upper parts of the trees. The resulting trees become shallow and more resilient since the number of affected nodes is minimized when nodes leave or fail. We have implemented R-CAT on top of Splitstream, a high bandwidth streaming system using Pastry, and evaluated our scheme in terms of packet loss rate and end-to-end delay. The performance results show that R-CAT is more resilient and provides lower end-to-end delay under various group dynamics.

1 Introduction

Video streaming services over the Internet have recently received great attentions and are becoming an important traffic class as the broadband Internet is widely available. As the number of participants is increased and the devices for the streaming become more ubiquitous, a highly scalable and flexible infrastructure is needed to support the large-scale services with unpredictable group dynamics. The inherent nature of peer-to-peer(P2P) infrastructure such as CAN[10], Chord[9], or Pastry[4] allows service developers to use the P2P services for building large-scale streaming services over the Internet[2][3][5] using their scalable and self-organizing features.

The P2P-based streaming systems can be broadly classified into two categories based on how the multicast tree is constructed and maintained: *centralized* or *distributed*. In the *centralized* approach[2], a centralized server collects and maintains the information of all participating peers and distributes the corresponding

* This work was supported by the Brain Korea 21 Project in 2006 and grant No. R01-2003-000-10627-0 from the KOSEF.

M. Gerndt and D. Kranzlmüller (Eds.): HPCC 2006, LNCS 4208, pp. 925–934, 2006.

tree information (i.e., information for parent node and children nodes) to each peer. Since the whole information is maintained at the single server and the tree is constructed by using all relevant information, an efficient and optimal multicast tree can be generated. Despite the fact that this approach provides an efficient and optimal solution to building multicast tree, it can hardly be used in large-scale environments because of the lack of scalability. In the *distributed* approach[3][5], each peer maintains the information of its parent and children nodes only and builds the multicast tree from the local information it collects. Based on the collected information, the multicast tree can be self-organized to reflect current conditions. Some systems such as SplitStream[3] split the media stream into separate stripes and multicast each stripe over a separate tree. This ensures that the forwarding load is distributed across all nodes and also increases the resiliency at the cost of redundancy. Although the generated tree is not guaranteed to be optimal, this approach is more suitable for large-scale, dynamic live streaming environments that we are targeting in this paper.

On the other hand, while conventional P2P-based streaming systems have mainly focused on either reducing end-to-end delay using proximity information or making the services robust by creating multiple multicast trees, they fail to address the problems caused by the heterogeneous and asymmetric nature of Internet resources. It has been reported in previous research efforts[1][6][7] that a large portion of P2P traffic is asymmetric and the node capacities are skewed. Considering that the nodes with good computing power and high bandwidth (i.e., high capacity nodes) are typically stable and provide better performance compared to the low capacity nodes, it is likely that we can build a more resilient tree by locating high capacity nodes at the upper part of a tree.

In this paper, we present a capacity-aware multicast tree construction scheme for P2P-based live streaming environments (R-CAT) that allows us to build resilient multicast trees by considering the capacity of participating nodes and locating the high capacity nodes toward the upper parts of the trees. The resulting trees become shallow and more resilient since the number of affected nodes is minimized when nodes leave or fail. We have implemented R-CAT on top of Splitstream[3], a high bandwidth streaming system using Pastry[4], and evaluated our scheme in terms of packet loss rate and end-to-end delay. The performance results show that R-CAT is more resilient and provides lower end-to-end delay under various group dynamics.

The organization of this paper is as follows. Section 2 outlines the overall architecture of R-CAT in more detail. Section 3 presents the benchmarking results and performance analysis. Section 4 summarizes and concludes the paper.

2 Resilient Capacity-Aware Multicast Tree

In this section, we present an overview of tree construction schemes used in R-CAT design and discuss the R-CAT algorithms in detail.

Table 1. Gnutella-like Capacity Distribution

Capacity Level	Percentage of Nodes
1x	20%
10x	45%
100x	30%
1000x	4.9%
10000x	0.1%

2.1 Assumptions

We assume that all nodes in a peer-to-peer network have the similar capacity distribution to that of Gnutella [7][11] as shown in Table 1. As we can see from Table 1, there are few high-capacity nodes in the network and the capacities of most nodes are within low to medium level. It is also assumed that each node i has a maximum capacity C_i and an available capacity C_{ai}, where C_i is always greater than or equal to C_{ai}. The maximum capacity in this case is a capacity assigned to each node derived from the Gnutella-like capacity distribution. The available capacity of a node is defined as the capacity for uploading streaming data, which is the ability to forward streaming data to other nodes. This is further described as the *degree* of a node, where the *degree* of a node represents the floor of the result by dividing the outgoing link bandwidth by the encoding rate of the stream[1]. For example, if we assume that a node has an outgoing link bandwidth of 120 kbps and the encoding rate of 100 kbs, the *degree* of that node is 1 *degree*, which means that this node can support one child at the full encoding rate. For simplicity, each node is assumed to have enough capacity for downloading streaming data.

The multicast tree in R-CAT is composed of parents, grand-parents, and children. An undirected edge exists between a parent and a child. An end-to-end delay between a parent node i and a child node j is described as d_{ij}. Nodes can form a multicast group and the members of each group periodically exchange routing messages and heartbeat messages.

2.2 Algorithm

The R-CAT algorithm is composed of three main components to improve the resiliency of multicast tree: *information gathering algorithm*, *parent selection algorithm*, and *victim selection algorithm*.

The *information gathering algorithm* is used for each node to collect and maintain the latest information of its children. The *parent selection algorithm* is used to select an appropriate parent using the capacity information of a new node when it joins into a multicast tree. If the prospective parent (i.e., a parent node receiving the join request message) cannot afford to accommodate the new node, it forwards the join request message to its parent so that high capacity node can be located at the upper part of a tree. The *victim selection algorithm* is used when a parent node reaches its maximum out-degree bound. In this case, the parent node has to choose a victim node to improve the resiliency of the multicast tree. In what follows, we present each algorithm in detail.

Information Gathering Algorithm. Gathering precise information of the neighboring nodes is a key to making better policy decisions and designing an efficient algorithm. Considering that more frequent exchanges of messages between nodes can provide more accurate information but with more overhead, it is sometimes difficult to design an efficient information gathering algorithm with little overhead.

The information gathering algorithm in R-CAT is invoked either when a new node joins into a multicast tree or when existing nodes exchange heartbeat messages to check the health of other nodes. For example, when a new node joins, it forwards a join request message to its prospective parent with the node information piggybacked into the message. If the parent node can afford to accommodate the new node, it transmits the ACK message back to the joining node with its information piggybacked. Moreover, when a parent node exchanges heartbeat messages with its child node, the corresponding information is piggybacked with the heartbeat messages.

The piggybacked information includes the capacity of a node, aliveness, goodness value, and the number of sub-children in the tree. Followings are the definitions of each information.

$$C_i \; : \quad \text{maximum capacity of node } i$$

$$h_i \; : \quad \text{direct children set of node } i$$

$$A_i = 1 - \frac{1}{C_i + k} \; , \quad \text{aliveness of node } i (k : \text{scaling factor}) \tag{1}$$

$$S_i = \sum_{j \in h_i} S_j + 1 \; , \quad \text{total number of sub-children of node } i \tag{2}$$

$$G_i = A_i \times S_i \; , \quad \text{goodness value of node } i \tag{3}$$

The capacity of node i, C_i, represents the maximum number of children that a node can accommodate. Since any node cannot afford to accommodate children nodes exceeding its maximum capacity, $|h_i| \leq C_i$ is always guaranteed. In the equations, the direct children set is a set of children nodes directly connected to a parent node i, and the sub-children set is a set of all the children and sub-children nodes rooted at a parent node i. The aliveness factor, A_i, is defined by using C_i. Since it is generally accepted that the nodes with higher capacities are more stable than the nodes with lower capacities, the reciprocal value of C_i can be used to represent the aliveness of a node. The scaling factor k is the constant value to calibrate the C_i value. The goodness value G_i of node i is an indication of fitness to be used in *parent selection algorithm* and *victim selection algorithm*. This value is calculated by multiplying the node's aliveness factor and the total number of sub-children nodes, which means that any node with larger sub-children nodes or higher aliveness factor will have better goodness value.

Parent Selection Algorithm. In a typical P2P-based streaming environment, when a new node joins to a multicast tree, a parent selection algorithm is activated to select an appropriate parent for the joining node.

The parent selection algorithm in R-CAT is based on a simple idea that the shallow tree minimizes the effects of node dynamics and is more stable than a tree with longer depth. When a prospective parent receives a join request message from a new node, it first checks the capacity of the new node and decides whether it can accommodate the new node as its children or not. If the prospective parent doesn't have enough capacity to support the incoming node, it forwards the join request message to its parent and repeats the same steps in order to select an appropriate parent. We call this *climbing.* On the other hand, if the prospective parent has enough capacity to accommodate the new node, it registers the new node as a child node. If the prospective parent has enough capacity but the overall capacity exceeds the maximum capacity, a victim is selected by the *victim selection algorithm* and the selected victim is *pushed down* until the request is accepted.

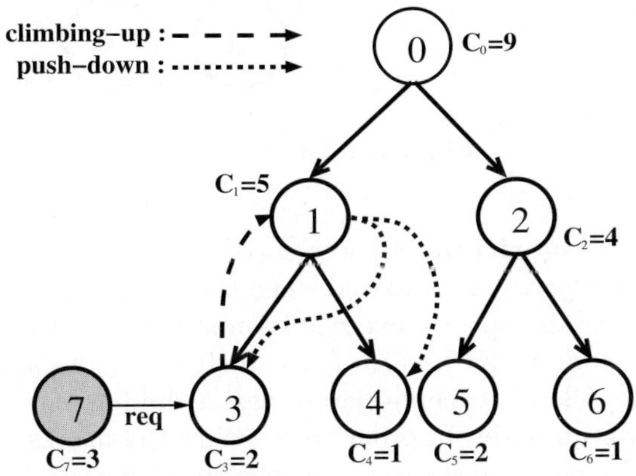

Fig. 1. Climbing and Push Down in Parent Selection

Figure 1 depicts how this algorithm works. Assume that node 7 joins to the tree by sending a join request message to the prospective parent node 3. Once the node 3 receives the message, it compares its capacity (i.e., capacity 2) with that of node 7 (i.e., capacity 3). Since the capacity of the incoming node exceeds the capacity of the prospective parent, the join request message is *climbed up* to the parent of node 3 (i.e., node 1). Now that the capacity of node 1 (i.e., capacity 5) is larger than that of the incoming node (i.e., capacity 3), the node 7 can be one of children nodes rooted at node 1. However, adding node 7 under node 1 exceeds the maximum out-bound capacity, and the *victim selection algorithm* is activated to choose a victim node to be *pushed down* until the capacity can be accommodated.

It should be noted that the the two techniques (i.e., *climbing up* and *push down*) used in this algorithm favor high capacity nodes and put them toward the upper part of the tree, which makes the multicast tree more resilient.

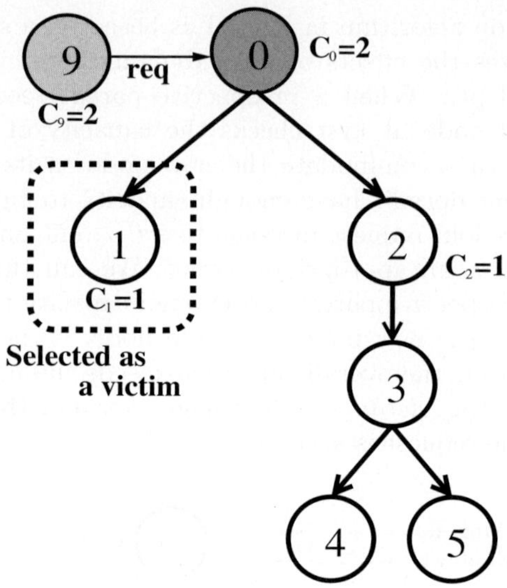

Fig. 2. Victim Selection

Victim Selection Algorithm. As described in the *parent selection algorithm*, the *victim selection algorithm* is invoked when the prospective parent has enough capacity but the overall capacity exceeds its maximum capacity, thereby needs to select a victim node to be *pushed down*. In R-CAT, the *goodness* value of a child node and the delay between a parent and a child are two essential metrics used to select a victim node. In order to measure the fitness of a victim node, we defined a metric called *resiliency* as shown in Equation 4. Since the bigger *goodness* value and smaller delay value can maximize the *resiliency*, this metric indicates the ability of reducing packet loss and end-to-end delay in dynamic multicast group.

$$H_i \ , \quad \text{the set of children that node } i \text{ can choose}$$

$$\forall h_i \in H_i \ , \ |h_i| \le C_i$$

$$R_i = \max_{h_i \in H_i} \left(\sum_{j \in h_i} (G_j - \lambda d_{ij}) \right) \tag{4}$$

$$h_i^* = \arg\max \left(\sum_{j \in h_i, \ h_i \in H_i} (G_j - \lambda d_{ij}) \right) \tag{5}$$

$$victim = \left(h_i' \bigcup j \right) - h_i^* \tag{6}$$

Based on the Equation 4, a parent node can build an optimal children set, h_i^*, that maximizes the *goodness* and minimizes the delay between the parent and the child as shown in Equation 5. When the optimal children set, h_i^*, is chosen by the victim selection algorithm, the victim is selected as shown in Equation 6, where h_i' is the original children set and j is the new child node.

The key idea behind this algorithm is that (1) by choosing a victim node that minimizes the delay between a parent node and a child node, the end-to-end delay from a root node to a leaf node can be reduced; (2) by choosing a victim node that maximizes the *goodness*, we can possibly exclude nodes with larger sub-children group or better stability metric from the victim candidate list. The trade-off parameter λ is introduced to weigh the importance of the delay factor in the equations.

Figure 2 shows an example of this algorithm. Assume that node 9 transmits a request message to node 0 to join. Since the node 0 has enough capacity to accommodate the node 9 but the overall capacity exceeds the maximum capacity, the *victim selection algorithm* is activated and the node 0 has to decide a victim node among node1, node2, and node 9. If we assume that the scaling factor k in Equation 1 is 1, the goodness value of each node is 0.67, 0.5, and 2.0, respectively. Therefore, if we further assume that the delay factor is same for all nodes, the node 0 has to select node 1 as a victim node based on the Equations 4, 5, 6.

3 Performance Evaluation

This section presents the performance of R-CAT and compares it with that of Splitstream. For the experiments, we implement our approach over FreePastry [12], a well-known Pastry simulator, and compare the performance in terms of packet loss and end-to-end delay. The simulation is conducted on Pentium 4 (2.2G CPU) running Linux and is averaged over 10 iterations with the following parameters.

First, for the node capacity distribution, the values in Table 2 are used. These are the modified values from [1] by removing the case for unknown hosts (i.e., hosts with unknown measurements). Table 2 shows that almost half of the nodes are *free riders* (i.e., nodes at the leaf and thereby contributing no resources), and very few nodes have high capacities.

Second, for the arrival, departure, and failure processes of P2P nodes, we use six categories from case 0 to case 5 for the simulation, with the case 0 being the static case (i.e., no arrival, departure, and failure) and the case 5 being the highly dynamic case. In case 5, we assume that almost 1% out of the total

Table 2. Node Capacity Distribution

Node Degree Bound	Percentage (%)
0 (Free Riders)	56
1	21
2	9.5
3-19	5.6
20	7.9

number of nodes leave from or join to the network based on the previous results in [7] and [11]. The Gnutella failure rates reported in [13] are also used for the simulation. The simulation is conducted over 10,000 nodes and all nodes are brought up when the simulation is started and other nodes start to join or leave by a Poisson process.

In what follows, the performance of R-CAT and the comparison with Splitstream are presented.

3.1 Packet Loss

Given that the reliable transfer protocol is used for streaming services, packets can be lost because of the node's dynamics (i.e., leave, join, failure) or the bottlenecks at the inner nodes of the tree. Therefore, measuring the packet loss is a valuable metric to decide whether the multicast tree is resilient or not.

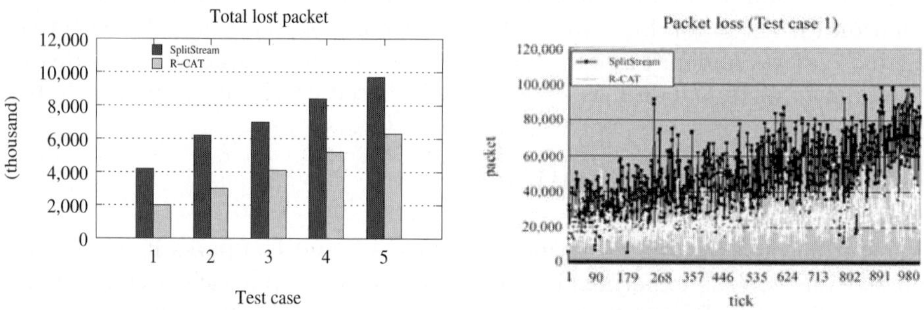

(a) Total Lost Packets (b) Packet Loss Changes

Fig. 3. Simulation Results for Packet Loss

Figures 3(a) and 3(b) illustrate the total number of lost packets of R-CAT and Splitstream for 5 test cases (case 1 to case 5), and their progresses of losing packets. As we can see from Figures 3(a) and 3(b), the number of lost packets in R-CAT is about 50% that of Splitstream. This can be explained by the simple fact that the *parent selection algorithm* and *victim selection algorithm* used in R-CAT favor high-capacity nodes, which results in locating them at the upper part of multicast tree. This in turn creates more shallow and resilient tree, where packets can be transfered more reliably with less packet loss.

3.2 End-to-End Delay

The average end-to-end delay, together with the packet loss rates, is an important factor to measure the resiliency of a multicast tree, and to decide the quality of streaming data. As we can see from Figure 4, the average end-to-end delay of R-CAT is smaller than that of Splitstream for all 6 test cases, which is also the result of shallow tree.

Since the buffering behavior in the client node is not correctly modeled in R-CAT, it is impossible to check the effects of end-to-end delay for the video quality. However, this result can be an implicit indication of good video quality and also an indirect metric of measuring the resiliency of a multicast tree.

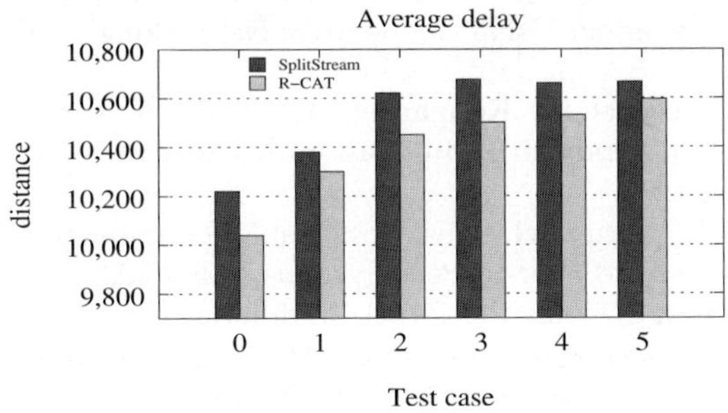

Fig. 4. Average End-to-End Delay

4 Conclusion

Existing P2P-based streaming systems have been developed without focusing on the heterogeneity of node's capacity. Since they are largely focused on reducing the end-to-end delay, the resulting systems provide good video quality in a relatively stable environment. As the P2P computing environments become larger and their traffic workloads move toward more dynamic environments, building a resilient multicast tree for a highly dynamic environment has been a challenging design issue for large-scale P2P applications such as P2P-based live streaming systems.

In this paper, we have presented an overview of R-CAT schemes and its algorithms, and evaluated its performance by comparing it with that of Splitstream. The benchmarking results showed that the R-CAT outperformed the Splitstream in terms of packet loss rate and average end-to-end delay. These results mainly attribute to the outcome of shallow, capacity-aware multicast tree generated by the *parent selection algorithm* and *victim selection algorithm* in R-CAT.

Due to space limitations, the overhead analysis of R-CAT has been omitted from the paper. However, the preliminary results showed that the worst case overhead in R-CAT is almost 30%-40% in highly dynamic environments and this is mainly due to the control message overheads used for improving the resiliency of multicast tree such as *climbing* and *push down*. Although these overheads can be amortized by large improvement in resiliency, continuous efforts need to be made to reduce the control overheads.

Currently, we are working on more detailed analysis using real traffic workloads and parameters. Further research about efficient, scalable membership protocol is needed to better optimize the membership protocol, and thereby reduce the overall overheads.

References

1. K. Sripanidkulchai, A. Ganjam, B. Maggs, and H. Zhang, "The Feasibility of Supporting Large-scale Live Streaming Applications with Dynamic Application Endpoints," ACM Computer Commun. Rev. pp.107–120, Aug. 2004.
2. N. Padmanabhan, J. Wang, A. Chou, and K. Sripanidkulchai, "Distributing Streaming Media Content Using Cooperative Networking," ACM NOSSDAV, May 2002.
3. M. Castro, P. Druschel, A. Kermarrec, A. Nandi, A. Rowstron, and A. Singh, "SplitStream: High-bandwidth Multicast in a Cooperative Environment," SOSP 2003, October 2003.
4. A. Rowstron and P. Druschel, "Pastry: Scalable, Distributed Object Location and Routing for Large-scale Peer-to-Peer Systems," IFIP/ACM International Conference on Distributed Systems Platforms (Middleware), pp.329–350, Nov. 2001.
5. H. Deshpande, M. Bawa, and H. Garcia-Molina, "Streaming Live Media over a Peer-to-Peer Network," Technical Report, Stanford University, Aug. 2001.
6. K. Sripanidkulchai, B. Maggs, and H. Zhang, "An Analysis of Live Streaming Workloads on the Internet," ACM IMC 2004, pp.41–54, Oct. 2004.
7. Y. Chawathe, S. Ratnasamy, L. Breslau, N. Lanham, and S. Shenker, "Making Gnutella-like P2P Systems Scalable," ACM SIGCOMM 2003, pp.407–418, Aug. 2003.
8. T. Nguyen, D. Tran, and S. Cheung, "Efficient P2P Dissemination in a Homogeneous Capaciy Network Using Structured Mesh," International Conference on Multimedia Services Access Networks, June 2005.
9. I. Stoica, R. Morris, D. Karger, M. Kaashoek, and H. Balakrishnan, "Chord: A Scalable Peer-to-Ppeer Lookup Service for Internet Applications," ACM SIGCOMM 2001, pp.149–160, Aug. 2001.
10. S. Ratnasamy, P. Francis, M. Handley, R. Karp, and S. Shenker, "A Scalable Content-Addressable Network," ACM SIGCOMM 2001, Aug. 2001.
11. S. Saroiu, P. Gummadi, and S. Gribble, "A Measurement Study of Peer-to-Peer File Sharing Systems," Multimedia Computing and Networking (MMCN), Jan. 2002.
12. FreePastry, http://freepastry.rice.edu/
13. R. Mahajan, M. Castro, and A. Rowstron, "Controlling the Cost of Reliability in Peer-to-Peer Overlays," IPTPS 2003, 2003.

Author Index

Lecture Notes in Computer Science

For information about Vols. 1–4083

please contact your bookseller or Springer

Vol. 4228: D.E. Lightfoot, C.A. Szyperski (Eds.), Modular Programming Languages. X, 415 pages. 2006.

Vol. 4208: M. Gerndt, D. Kranzlmüller (Eds.), High Performance Computing and Communications. XXII, 938 pages. 2006.

Vol. 4206: P. Dourish, A. Friday (Eds.), UbiComp 2006: Ubiquitous Computing. XIX, 526 pages. 2006.

Vol. 4193: T.P. Runarsson, H.-G. Beyer, E. Burke, J.J. Merelo-Guervós, L. D. Whitley, X. Yao (Eds.), Parallel Problem Solving from Nature - PPSN IX. XIX, 1061 pages. 2006.

Vol. 4192: B. Mohr, J.L. Träff, J. Worringen, J. Dongarra (Eds.), Recent Advances in Parallel Virtual Machine and Message Passing Interface. XVI, 414 pages. 2006.

Vol. 4188: P. Sojka, I. Kopeček, K. Pala (Eds.), Text, Speech and Dialogue. XIV, 721 pages. 2006. (Sublibrary LNAI).

Vol. 4187: J.J. Alferes, J. Bailey, W. May, U. Schwertel (Eds.), Principles and Practice of Semantic Web Reasoning. XI, 277 pages. 2006.

Vol. 4186: C. Jesshope, C. Egan (Eds.), Advances in Computer Systems Architecture. XIV, 605 pages. 2006.

Vol. 4185: R. Mizoguchi, Z. Shi, F. Giunchiglia (Eds.), The Semantic Web – ASWC 2006. XX, 778 pages. 2006.

Vol. 4184: M. Bravetti, M. Núñez, G. Zavattaro (Eds.), Web Services and Formal Methods. X, 289 pages. 2006.

Vol. 4183: J. Euzenat, J. Domingue (Eds.), Artificial Intelligence: Methodology, Systems, and Applications. XIII, 291 pages. 2006. (Sublibrary LNAI).

Vol. 4180: M. Kohlhase, OMDoc – An Open Markup Format for Mathematical Documents [version 1.2]. XIX, 428 pages. 2006. (Sublibrary LNAI).

Vol. 4178: A. Corradini, H. Ehrig, U. Montanari, L. Ribeiro, G. Rozenberg (Eds.), Graph Transformations. XII, 473 pages. 2006.

Vol. 4176: S.K. Katsikas, J. Lopez, M. Backes, S. Gritzalis, B. Preneel (Eds.), Information Security. XIV, 548 pages. 2006.

Vol. 4175: P. Bücher, B.M.E. Moret (Eds.), Algorithms in Bioinformatics. XII, 402 pages. 2006. (Sublibrary LNBI).

Vol. 4174: K. Franke, K.-R. Müller, B. Nickolay, R. Schäfer (Eds.), Pattern Recognition. XX, 773 pages. 2006.

Vol. 4169: H.L. Bodlaender, M.A. Langston (Eds.), Parameterized and Exact Computation. XI, 279 pages. 2006.

Vol. 4168: Y. Azar, T. Erlebach (Eds.), Algorithms – ESA 2006. XVIII, 843 pages. 2006.

Vol. 4165: W. Jonker, M. Petković (Eds.), Secure, Data Management. X, 185 pages. 2006.

Vol. 4163: H. Bersini, J. Carneiro (Eds.), Artificial Immune Systems. XII, 460 pages. 2006.

Vol. 4162: R. Královič, P. Urzyczyn (Eds.), Mathematical Foundations of Computer Science 2006. XV, 814 pages. 2006.

Vol. 4159: J. Ma, H. Jin, L.T. Yang, J.J.-P. Tsai (Eds.), Ubiquitous Intelligence and Computing. XXII, 1190 pages. 2006.

Vol. 4158: L.T. Yang, H. Jin, J. Ma, T. Ungerer (Eds.), Autonomic and Trusted Computing. XIV, 613 pages. 2006.

Vol. 4156: S. Amer-Yahia, Z. Bellahsène, E. Hunt, R. Unland, J.X. Yu (Eds.), Database and XML Technologies. IX, 123 pages. 2006.

Vol. 4155: O. Stock, M. Schaerf (Eds.), Reasoning, Action and Interaction in AI Theories and Systems. XVIII, 343 pages. 2006. (Sublibrary LNAI).

Vol. 4153: N. Zheng, X. Jiang, X. Lan (Eds.), Advances in Machine Vision, Image Processing, and Pattern Analysis. XIII, 506 pages. 2006.

Vol. 4152: Y. Manolopoulos, J. Pokorný, T. Sellis (Eds.), Advances in Databases and Information Systems. XV, 448 pages. 2006.

Vol. 4151: A. Iglesias, N. Takayama (Eds.), Mathematical Software - ICMS 2006. XVII, 452 pages. 2006.

Vol. 4150: M. Dorigo, L.M. Gambardella, M. Birattari, A. Martinoli, R. Poli, T. Stützle (Eds.), Ant Colony Optimization and Swarm Intelligence. XVI, 526 pages. 2006.

Vol. 4149: M. Klusch, M. Rovatsos, T.R. Payne (Eds.), Cooperative Information Agents X. XII, 477 pages. 2006. (Sublibrary LNAI).

Vol. 4148: J. Vounckx, N. Azemard, P. Maurine (Eds.), Integrated Circuit and System Design. XVI, 677 pages. 2006.

Vol. 4146: J.C. Rajapakse, L. Wong, R. Acharya (Eds.), Pattern Recognition in Bioinformatics. XIV, 186 pages. 2006. (Sublibrary LNBI).

Vol. 4144: T. Ball, R.B. Jones (Eds.), Computer Aided Verification. XV, 564 pages. 2006.

Vol. 4139: T. Salakoski, F. Ginter, S. Pyysalo, T. Pahikkala, Advances in Natural Language Processing. XVI, 771 pages. 2006. (Sublibrary LNAI).

Vol. 4138: X. Cheng, W. Li, T. Znati (Eds.), Wireless Algorithms, Systems, and Applications. XVI, 709 pages. 2006.

Vol. 4137: C. Baier, H. Hermanns (Eds.), CONCUR 2006 – Concurrency Theory. XIII, 525 pages. 2006.

Vol. 4136: R.A. Schmidt (Ed.), Relations and Kleene Algebra in Computer Science. XI, 433 pages. 2006.

Vol. 4135: C.S. Calude, M.J. Dinneen, G. Păun, G. Rozenberg, S. Stepney (Eds.), Unconventional Computation. X, 267 pages. 2006.

Vol. 4134: K. Yi (Ed.), Static Analysis. XIII, 443 pages. 2006.

Vol. 4133: J. Gratch, M. Young, R. Aylett, D. Ballin, P. Olivier (Eds.), Intelligent Virtual Agents. XIV, 472 pages. 2006. (Sublibrary LNAI).

Vol. 4132: S. Kollias, A. Stafylopatis, W. Duch, E. Oja (Eds.), Artificial Neural Networks – ICANN 2006, Part II. XXXIV, 1028 pages. 2006.

Vol. 4131: S. Kollias, A. Stafylopatis, W. Duch, E. Oja (Eds.), Artificial Neural Networks – ICANN 2006, Part I. XXXIV, 1008 pages. 2006.

Vol. 4130: U. Furbach, N. Shankar (Eds.), Automated Reasoning. XV, 680 pages. 2006. (Sublibrary LNAI).

Vol. 4129: D. McGookin, S. Brewster (Eds.), Haptic and Audio Interaction Design. XII, 167 pages. 2006.

Vol. 4128: W.E. Nagel, W.V. Walter, W. Lehner (Eds.), Euro-Par 2006 Parallel Processing. XXXIII, 1221 pages. 2006.

Vol. 4127: E. Damiani, P. Liu (Eds.), Data and Applications Security XX. X, 319 pages. 2006.

Vol. 4126: P. Barahona, F. Bry, E. Franconi, N. Henze, U. Sattler, Reasoning Web. X, 269 pages. 2006.

Vol. 4124: H. de Meer, J.P. G. Sterbenz (Eds.), Self-Organizing Systems. XIV, 261 pages. 2006.

Vol. 4121: A. Biere, C.P. Gomes (Eds.), Theory and Applications of Satisfiability Testing - SAT 2006. XII, 438 pages. 2006.

Vol. 4119: C. Dony, J.L. Knudsen, A. Romanovsky, A. Tripathi (Eds.), Advanced Topics in Exception Handling Components. X, 302 pages. 2006.

Vol. 4117: C. Dwork (Ed.), Advances in Cryptology - CRYPTO 2006. XIII, 621 pages. 2006.

Vol. 4116: R. De Prisco, M. Yung (Eds.), Security and Cryptography for Networks. XI, 366 pages. 2006.

Vol. 4115: D.-S. Huang, K. Li, G.W. Irwin (Eds.), Computational Intelligence and Bioinformatics, Part III. XXI, 803 pages. 2006. (Sublibrary LNBI).

Vol. 4114: D.-S. Huang, K. Li, G.W. Irwin (Eds.), Computational Intelligence, Part II. XXVII, 1337 pages. 2006. (Sublibrary LNAI).

Vol. 4113: D.-S. Huang, K. Li, G.W. Irwin (Eds.), Intelligent Computing, Part I. XXVII, 1331 pages. 2006.

Vol. 4112: D.Z. Chen, D. T. Lee (Eds.), Computing and Combinatorics. XIV, 528 pages. 2006.

Vol. 4111: F.S. de Boer, M.M. Bonsangue, S. Graf, W.-P. de Roever (Eds.), Formal Methods for Components and Objects. VIII, 447 pages. 2006.

Vol. 4110: J. Díaz, K. Jansen, J.D.P. Rolim, U. Zwick (Eds.), Approximation, Randomization, and Combinatorial Optimization. XII, 522 pages. 2006.

Vol. 4109: D.-Y. Yeung, J.T. Kwok, A. Fred, F. Roli, D. de Ridder (Eds.), Structural, Syntactic, and Statistical Pattern Recognition. XXI, 939 pages. 2006.

Vol. 4108: J.M. Borwein, W.M. Farmer (Eds.), Mathematical Knowledge Management. VIII, 295 pages. 2006. (Sublibrary LNAI).

Vol. 4106: T.R. Roth-Berghofer, M.H. Göker, H. A. Güvenir (Eds.), Advances in Case-Based Reasoning. XIV, 566 pages. 2006. (Sublibrary LNAI).

Vol. 4105: B. Gunsel, A.K. Jain, A. M. Tekalp, B. Sankur (Eds.), Multimedia, Content Representation, Classification and Security. XIX, 804 pages. 2006.

Vol. 4104: T. Kunz, S.S. Ravi (Eds.), Ad-Hoc, Mobile, and Wireless Networks. XII, 474 pages. 2006.

Vol. 4103: J. Eder, S. Dustdar (Eds.), Business Process Management Workshops. XI, 508 pages. 2006.

Vol. 4102: S. Dustdar, J.L. Fiadeiro, A. Sheth (Eds.), Business Process Management. XV, 486 pages. 2006.

Vol. 4099: Q. Yang, G. Webb (Eds.), PRICAI 2006: Trends in Artificial Intelligence. XXVIII, 1263 pages. 2006. (Sublibrary LNAI).

Vol. 4098: F. Pfenning (Ed.), Term Rewriting and Applications. XIII, 415 pages. 2006.

Vol. 4097: X. Zhou, O. Sokolsky, L. Yan, E.-S. Jung, Z. Shao, Y. Mu, D.C. Lee, D. Kim, Y.-S. Jeong, C.-Z. Xu (Eds.), Emerging Directions in Embedded and Ubiquitous Computing. XXVII, 1034 pages. 2006.

Vol. 4096: E. Sha, S.-K. Han, C.-Z. Xu, M.H. Kim, L.T. Yang, B. Xiao (Eds.), Embedded and Ubiquitous Computing. XXIV, 1170 pages. 2006.

Vol. 4095: S. Nolfi, G. Baldassarre, R. Calabretta, J.C. T. Hallam, D. Marocco, J.-A. Meyer, O. Miglino, D. Parisi (Eds.), From Animals to Animats 9. XV, 869 pages. 2006. (Sublibrary LNAI).

Vol. 4094: O. H. Ibarra, H.-C. Yen (Eds.), Implementation and Application of Automata. XIII, 291 pages. 2006.

Vol. 4093: X. Li, O.R. Zaïane, Z. Li (Eds.), Advanced Data Mining and Applications. XXI, 1110 pages. 2006. (Sublibrary LNAI).

Vol. 4092: J. Lang, F. Lin, J. Wang (Eds.), Knowledge Science, Engineering and Management. XV, 664 pages. 2006. (Sublibrary LNAI).

Vol. 4091: G.-Z. Yang, T. Jiang, D. Shen, L. Gu, J. Yang (Eds.), Medical Imaging and Augmented Reality. XIII, 399 pages. 2006.

Vol. 4090: S. Spaccapietra, K. Aberer, P. Cudré-Mauroux (Eds.), Journal on Data Semantics VI. XI, 211 pages. 2006.

Vol. 4089: W. Löwe, M. Südholt (Eds.), Software Composition. X, 339 pages. 2006.

Vol. 4088: Z.-Z. Shi, R. Sadananda (Eds.), Agent Computing and Multi-Agent Systems. XVII, 827 pages. 2006. (Sublibrary LNAI).

Vol. 4087: F. Schwenker, S. Marinai (Eds.), Artificial Neural Networks in Pattern Recognition. IX, 299 pages. 2006. (Sublibrary LNAI).

Vol. 4085: J. Misra, T. Nipkow, E. Sekerinski (Eds.), FM 2006: Formal Methods. XV, 620 pages. 2006.

Vol. 4084: M.A. Wimmer, H.J. Scholl, Å. Grönlund, K.V. Andersen (Eds.), Electronic Government. XV, 353 pages. 2006.